The Nucleus

SECOND EDITION

A subject collection from *Cold Spring Harbor Perspectives in Biology*

OTHER SUBJECT COLLECTIONS FROM *COLD SPRING HARBOR PERSPECTIVES IN BIOLOGY*

Auxin Signaling: From Synthesis to Systems Biology, Second Edition
T-Cell Memory
Stem Cells: From Biological Principles to Regenerative Medicine
Heart Development and Disease
Cell Survival and Cell Death, Second Edition
Calcium Signaling, Second Edition
Engineering Plants for Agriculture
Protein Homeostasis, Second Edition
Translation Mechanisms and Control
Cytokines: From Basic Mechanisms of Cellular Control to New Therapeutics
Circadian Rhythms
Immune Memory and Vaccines: Great Debates
Cell–Cell Junctions, Second Edition
Prion Biology
The Biology of the TGF-β Family
Synthetic Biology: Tools for Engineering Biological Systems
Cell Polarity
Cilia

SUBJECT COLLECTIONS FROM *COLD SPRING HARBOR PERSPECTIVES IN MEDICINE*

Influenza: The Cutting Edge
Leukemia and Lymphoma: Molecular and Therapeutic Insights
Addiction, Second Edition
Hepatitis C Virus: The Story of a Scientific and Therapeutic Revolution
The PTEN Family
Metastasis: Mechanism to Therapy
Genetic Counseling: Clinical Practice and Ethical Considerations
Bioelectronic Medicine
Function and Dysfunction of the Cochlea: From Mechanisms to Potential Therapies
Next-Generation Sequencing in Medicine
Prostate Cancer
RAS and Cancer in the 21st Century
Enteric Hepatitis Viruses
Bone: A Regulator of Physiology
Multiple Sclerosis
Cancer Evolution
The Biology of Exercise
Prion Diseases

The Nucleus

SECOND EDITION

A subject collection from *Cold Spring Harbor Perspectives in Biology*

EDITED BY

Ana Pombo

Humboldt University of Berlin, Max Delbrück Centre for Molecular Medicine

Martin W. Hetzer

The Salk Institute for Biological Studies

Tom Misteli

National Cancer Institute, National Institutes of Health

COLD SPRING HARBOR LABORATORY PRESS

Cold Spring Harbor, New York • www.cshlpress.org

The Nucleus, Second Edition

A subject collection from *Cold Spring Harbor Perspectives in Biology*
Articles online at www.cshperspectives.org

Printed in the United States of America

Executive Editor	Richard Sever
Managing Editor	Maria Smit
Senior Project Manager	Barbara Acosta
Permissions Administrator	Carol Brown
Production Editor	Diane Schubach
Production Manager/Cover Designer	Denise Weiss
Publisher	John Inglis

Front cover artwork: Three-dimensional model of a diploid genome in the nucleus of a GM12878 cell. Included are schematic volume representations for a selection of nuclear bodies, representing the nucleolus, nuclear speckles, paraspeckles, PML bodies, and Cajal bodies. Territories of different chromosomes are represented by chromatin fibers of different colors. (Cover created and kindly provided by Asli Yildirim and Frank Alber.)

Library of Congress Cataloging-in-Publication Data

Names: Pombo, Ana, editor. | Hetzer, Martin W., editor. | Misteli, Tom, editor.
Title: The nucleus / edited by Ana Pombo, Humboldt University of Berlin, Max Delbrueck Centre for Molecular Medicine, Martin W. Hetzer, The Salk Institute for Biological Studies, and Tom Misteli, National Cancer Institute, National Institutes of Health.
Description: Second edition. | Cold Spring Harbor, NY : Cold Spring Harbor Laboratory Press, [2022] | A subject collection from Cold Spring Harbor Perspectives in Biology. | Includes bibliographical references and index. | Summary: "The nucleus of the cell is the control center that contains its DNA. This volume explores how the nucleus is organized, examining how different subcompartments form and interact and the role these play in gene regulation and processes such as splicing of RNA"-- Provided by publisher.
Identifiers: LCCN 2021030371 (print) | LCCN 2021030372 (ebook) | ISBN 9781621823896 (hardcover) | ISBN 9781621823902 (epub)
Subjects: LCSH: Cell nuclei.
Classification: LCC QH595 .N85 2021 (print) | LCC QH595 (ebook) | DDC 571.6/6--dc23
LC record available at https://lccn.loc.gov/2021030371
LC ebook record available at https://lccn.loc.gov/2021030372

10 9 8 7 6 5 4 3 2 1

All World Wide Web addresses are accurate to the best of our knowledge at the time of printing.

For a complete catalog of all Cold Spring Harbor Laboratory Press publications, visit our website at www.cshlpress.org.

Contents

Preface, vii

NUCLEAR ENVELOPE

The Nuclear Lamina, 1
Xianrong Wong, Ashley J. Melendez-Perez, and Karen L. Reddy

Structure, Maintenance, and Regulation of Nuclear Pore Complexes: The Gatekeepers of the Eukaryotic Genome, 27
Marcela Raices and Maximiliano A. D'Angelo

The Nuclear Pore Complex as a Transcription Regulator, 49
Michael Chas Sumner and Jason Brickner

The Diverse Cellular Functions of Inner Nuclear Membrane Proteins, 59
Sumit Pawar and Ulrike Kutay

PRINCIPLES OF ORGANIZATION

Mechanical Forces in Nuclear Organization, 93
Yekaterina A. Miroshnikova and Sara A. Wickström

Liquid–Liquid Phase Separation in Chromatin, 109
Karsten Rippe

The Stochastic Genome and Its Role in Gene Expression, 133
Christopher H. Bohrer and Daniel R. Larson

Emerging Properties and Functions of Actin and Actin Filaments Inside the Nucleus, 153
Svenja Ulferts, Bina Prajapati, Robert Grosse, and Maria K. Vartiainen

CHROMATIN

Physical Nature of Chromatin in the Nucleus, 169
Kazuhiro Maeshima, Shiori Iida, and Sachiko Tamura

Mechanisms of Chromosome Folding and Nuclear Organization: Their Interplay and Open Questions, 189
Leonid Mirny and Job Dekker

Uncovering the Principles of Genome Folding by 3D Chromatin Modeling, 209
Asli Yildirim, Lorenzo Boninsegna, Yuxiang Zhan, and Frank Alber

Transcription Factor Dynamics, 229
Feiyue Lu and Timothée Lionnet

Essential Roles for RNA in Shaping Nuclear Organization, 251
Sofia A. Quinodoz and Mitchell Guttman

Nuclear Compartments: An Incomplete Primer to Nuclear Compartments, Bodies, and Genome Organization Relative to Nuclear Architecture, 269
Andrew S. Belmont

NUCLEAR PROCESSES

The Impact of Space and Time on the Functional Output of the Genome, 303
Marcelo Nollmann, Isma Bennabi, Markus Götz, and Thomas Gregor

Mammalian DNA Replication Timing, 327
Athanasios E. Vouzas and David M. Gilbert

Imaging Organization of RNA Processing within the Nucleus, 349
Jeetayu Biswas, Weihan Li, Robert H. Singer, and Robert A. Coleman

PHYSIOLOGICAL PROCESSES AND DISEASE

Chromatin Mechanisms Driving Cancer, 369
Berkley Gryder, Peter C. Scacheri, Thomas Ried, and Javed Khan

Viruses in the Nucleus, 389
Bojana Lucic, Ines J. de Castro, and Marina Lusic

3D or Not 3D: Shaping the Genome during Development, 409
Juliane Glaser and Stefan Mundlos

Epigenetic Reprogramming in Early Animal Development, 427
Zhenhai Du, Ke Zhang, and Wei Xie

The Molecular and Nuclear Dynamics of X-Chromosome Inactivation, 457
François Dossin and Edith Heard

Organization of the Pluripotent Genome, 489
Patrick S.L. Lim and Eran Meshorer

Index, 511

Preface

THE NUCLEUS IS ARGUABLY THE MOST eye-catching structure and characteristic feature in most eukaryotic cells. It houses the vast majority of an organism's hereditary material and is the cellular site where critical functions occur such as the execution of precisely regulated gene expression programs, the accurate duplication of the genome during replication, and the maintenance of genome integrity by highly efficient DNA repair mechanisms. Despite its prominence, our cell biological understanding of how the cell nucleus and the genome it contains are organized in space and time and how this organization contributes to faithful genome function has long lagged behind that of other cellular organelles. While early microscopy studies uncovered the presence of the most prominent intra-nuclear compartments, more detailed characterization of the architecture of the nucleus, particularly the 3D organization of chromatin, proved challenging. The past two decades have witnessed the development of remarkable technologies that now allow the high-precision mapping of genome organization and interactions at a global scale, the imaging of intranuclear structure at high resolution and in living cells and tissues, and the three-dimensional reconstruction and modeling of chromatin organization and its fundamental properties.

These dramatic advances in our understanding of how genomes are organized and how the cell nucleus functions are reflected in this second edition of *The Nucleus*, published 10 years after the inaugural publication. The progress captured in the 23 chapters, written by leaders in the field, is impressive. The chapters describe some of the basic principles that organize the nucleus and the genome, including the presence of prominent chromatin domains and the ubiquitous presence of chromatin loops, but also the importance of dynamics, biophysical properties, and the stochastic nature of genome organization and function. In separate sections, the properties of chromatin and its interactions with nuclear landmarks such as the nuclear envelope are described, and their relevance to nuclear processes, development, and disease are discussed. Our goal in the selection of these topics was to paint a broad, but representative, picture of where we are in our understanding of nuclear and genome cell biology, and to outline the challenges that lie ahead.

This book would not have been possible without the dedication and expertise of many. First and foremost, we thank our colleagues who authored their chapters with much care and who responded to our editorial requests with diligence and understanding. We were fortunate to be able to rely, as always, on the outstanding staff at Cold Spring Harbor Press. We thank our Managing Editor Maria Smit, our Production Editor Diane Schubach, and especially our Senior Project Manager Barbara Acosta, who kept us on track and solved all editorial problems large and small. We appreciate the opportunity to take this snapshot of the field given to us by our Publisher John Inglis and our Executive Editor Richard Sever.

The biology of the nucleus and of genomes has come a long way, yet many questions remain for those in the field and for the next generations of genome cell biologists. Our hope is that these chapters will inform and serve as a basis for future work and will also inspire many to continue the fearless exploration of the cell nucleus.

ANA POMBO
MARTIN HETZER
TOM MISTELI

The Nuclear Lamina

Xianrong Wong,[1] Ashley J. Melendez-Perez,[2] and Karen L. Reddy[2,3]

[1]Laboratory of Developmental and Regenerative Biology, Skin Research Institute of Singapore, Agency for Science, Technology and Research (A*STAR), Singapore 138648

[2]Department of Biological Chemistry and Center for Epigenetics, Johns Hopkins University of Medicine, Baltimore, Maryland 21205, USA

[3]Sidney Kimmel Cancer Institute, Johns Hopkins University School of Medicine, Baltimore, Maryland 21231, USA

Correspondence: kreddy4@jhmi.edu

Lamins interact with a host of nuclear membrane proteins, transcription factors, chromatin regulators, signaling molecules, splicing factors, and even chromatin itself to form a nuclear subcompartment, the nuclear lamina, that is involved in a variety of cellular processes such as the governance of nuclear integrity, nuclear positioning, mitosis, DNA repair, DNA replication, splicing, signaling, mechanotransduction and -sensation, transcriptional regulation, and genome organization. Lamins are the primary scaffold for this nuclear subcompartment, but interactions with lamin-associated peptides in the inner nuclear membrane are self-reinforcing and mutually required. Lamins also interact, directly and indirectly, with peripheral heterochromatin domains called lamina-associated domains (LADs) and help to regulate dynamic 3D genome organization and expression of developmentally regulated genes.

The nucleus is structurally organized into functional domains and, as an interpreter and regulator of cellular function, is unique between cell types. In addition to expressing cell-type-specific transcription factors and genome regulators, this diversity is also quite evident at the periphery of the nucleus. The nuclear periphery encompasses the nuclear envelope and the underlying nuclear lamina (Aebi et al. 1986). The nuclear envelope is a dual membrane barrier composed of the inner nuclear membrane (INM) and the outer nuclear membrane (ONM), which is contiguous with the endoplasmic reticulum (ER). The nuclear envelope (NE) is studded with nuclear pore complexes (NPCs) that span the ONM and INM. Underlying the INM is the nuclear lamina, a 10- to 30-nm-thick filamentous meshwork, with its thickness varying between different cell types (Höger et al. 1991). The principal components of the lamina are the type V intermediate filament proteins—the lamins (Fig. 1; Gerace and Huber 2012). In mammals, the lamins are grouped into two classes: A-type (LMNA, LMNAΔ10, and LMNC) and B-type (LMNB1, LMNB2, and LMNB3) (Peter et al. 1989; Vorburger et al. 1989). All metazoans express at least one B-type lamin, with most invertebrates expressing a single lamin gene, although *Drosophila* has one B-type (Dm0 or Dmel/Lam) and one A-type (DmeI/LamC). In mammals, most adult

Cite this article as *Cold Spring Harb Perspect Biol* doi: 10.1101/cshperspect.a040113

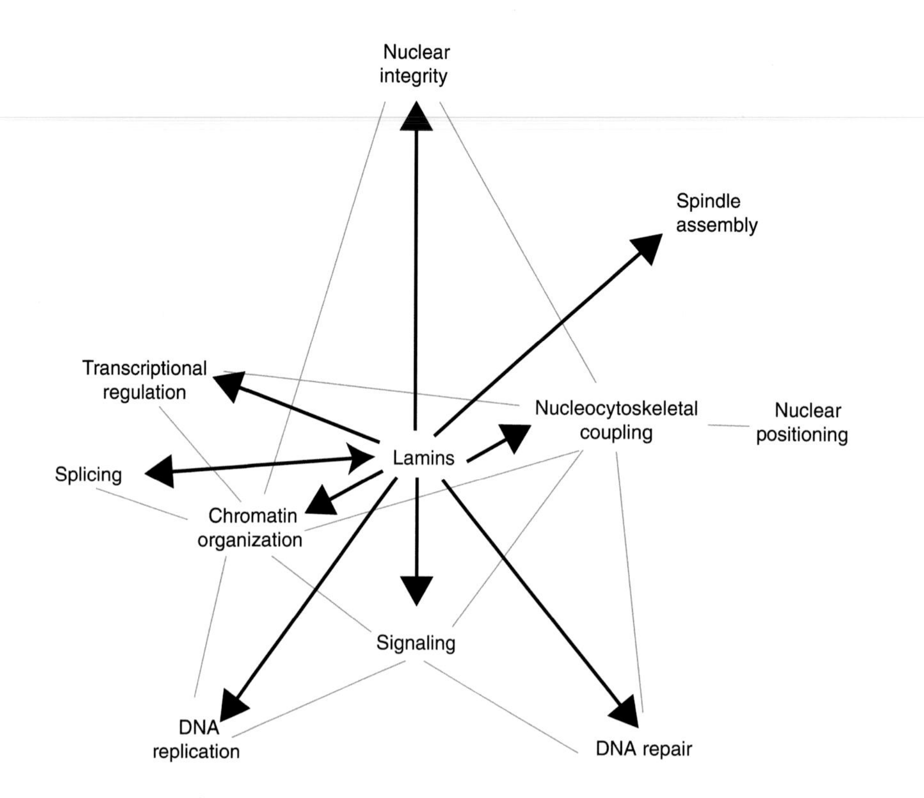

Figure 1. Lamins—a critical nexus in the regulation of cellular processes. Lamins and their associated proteins (lamin-associated proteins [LAPs] and nuclear envelope transmembrane proteins [NETs]) have been implicated in a wide range of interdependent cellular processes placing lamins at the center of an integrative hub at the nuclear lamina. Black arrows indicate lamin influences on specific pathways and processes and the gray lines indicate cross talk between these processes. These processes include maintaining structural integrity of the nucleus (Shin et al. 2013; Swift et al. 2013; Harada et al. 2014), regulation of transcription through direct interactions with transcription factors as well as signaling pathways (Andrés and González 2009; Ho and Lammerding 2012), and three-dimensional chromatin organization (scaffolding of lamina-associated domains [LADs]) (van Steensel and Belmont 2017; Briand and Collas 2020), which in turn impacts nuclear integrity and structure, since chromatin and lamins collaborate to confer mechanical properties to the nucleus (Chalut et al. 2012; Furusawa et al. 2015; Shimamoto et al. 2017; Stephens et al. 2017; Strickfaden et al. 2020). The nuclear lamina and interacting proteins also regulate other DNA-based processes such as DNA replication (Dorner et al. 2006; Shumaker et al. 2008; Pope et al. 2014) and DNA repair (Redwood et al. 2011; Aymard et al. 2014; Gibbs-Seymour et al. 2015; Lottersberger et al. 2015; Li et al. 2018; Marnef et al. 2019). Linker of nucleoskeleton and cytoskeleton complex (LINC) proteins mediate nucleocytoskeletal coupling, which in turn influences other processes such as signaling, chromatin organization, DNA repair, and gene regulation (through mechanosensation and mechanotransduction) as well as nuclear position through direct interaction of LINCs with cytoskeletal motors (Razafsky et al. 2014; Wong et al. 2021a). Other identified but less commonly known roles of the lamins include splicing and spindle pole assembly (Georgatos et al. 1997; Maison et al. 1997; Kumaran et al. 2002; Tsai et al. 2006; Ma et al. 2009).

differentiated somatic cells contain four major lamin proteins (LMNA, LMNB1, LMNB2, and LMNC). A single gene, *LMNA*, encodes the A-type lamins LMNA and LMNC, along with other minor variants including LMNAΔ10, which are all generated by alternative splicing of a common pre-mRNA (Lin and Worman 1993; Machiels et al. 1996; DeBoy et al. 2017). A minor spliced variant, LMNC2, is also produced in the testis (Furukawa et al. 1994). Separate genes encode LMNB1 and LMNB2 with LMNB3 being produced as a minor spliced variant of *LMNB2* and, as with LMNC2, is found in the testis (Furukawa and Hotta 1993; Lin and Worman 1993). All lamin isotypes share a similar overall structure: a head domain, a central α-helical rod domain comprising four coiled-coil domains that enable lamin dimer formation, interspersed by unstructured linkers, followed by a nuclear localization signal (NLS), an Ig-fold domain ending in an unstructured carboxy-terminal tail (detailed in Fig. 2). The basic unit of the lamins is the lamin dimer, which forms protofilaments in a head-to-tail manner and form higher-order assemblies to make the final lamin filament structure (Dechat et al. 2010). In mammals, recent high-resolution light and electron microscopy studies have revealed that the A-, B-, and C-lamin isotypes form their own spatially separate but interacting and overlapping filament networks of 3.5-A tetrameric filaments (Shimi et al. 2008; Turgay et al. 2017), with lamin C showing an apparent preferential association with nuclear pores (Xie et al. 2016). This makes lamin organization in mammalian somatic cells more complex and less regular than previously observed in the frog oocyte lamina (Aebi et al. 1986).

REGULATION OF LAMINS

Lamin functions are regulated by a myriad of posttranslational modifications (PTMs) (see Fig. 1), especially higher-order assembly into filaments and disassembly during nuclear envelope breakdown as cells enter mitosis. Most prominently, lamins are regulated by phosphorylation, but are also subjected to farnesylation, ubiquitination, sumoylation, *O*-linked sugar modification (*O*-GlcNAc), and acetylation (Gerace and Blobel 1980; Simon and Wilson 2013; Kochin et al. 2014; Gruenbaum and Foisner 2015; Torvaldson et al. 2015). While the roles of the vast majority of phosphorylation and other modifications to lamin proteins are not understood, there is a clear role for phosphorylation in lamin disassembly in mitosis. Cyclin-dependent kinase 1 (Cdkl or cdc2) directly phosphorylates both A- and B-type lamins during mitosis. Phosphorylation of amino-terminal Serine 22 (S22P) and carboxy-terminal Serine 392 (S392P) in human lamin A/C or analogous residues in B-type lamins near the coiled-coil domains induce the disassembly of the nuclear lamina. Intriguingly, a subset of A-type lamins remains phosphorylated, soluble, and nucleoplasmic during interphase (Gerace and Blobel 1980; Kochin et al. 2014; Torvaldson et al. 2015; Ikegami et al. 2020). While S22P is necessary for lamin depolymerization, it is not sufficient; thus, CDK1 is likely working in concert with other kinases, such as protein kinase C (PKC), cyclin-dependent kinase 5 (CDK5), and AKT to facilitate complete disassembly and to further regulate lamin functions (Torvaldson et al. 2015). It is important to note that the majority of PTM sites of lamins have been identified by mass spectrometry without accompanying mechanistic and functional studies, so there is still much to be understood about the dynamic regulation of lamins through these different modifications and modification sites.

In addition to being regulated by PTMs (and splicing), expression of lamin isotypes is also developmentally regulated, with expression of A-type lamins (LMNA and LMNC) being restricted to more differentiated cells, and each cell type displaying different relative ratios of the different lamin isotypes. All nucleated mammalian cells express at least one B-type lamin, whereas A-type lamins are absent during the early pre- and postimplantation embryonic stages and in embryonic stem (ES) cells, with these types expressing high levels of LMNB1 and LMNB2 (Stewart and Burke 1987; Wong and Stewart 2020). A-type lamins are expressed as different tissues form in the postimplantation embryo, although some tissues and cell types do not express A-type lamins until after birth (Solovei et al. 2013). Lamins appear to be nonessential for ES cells, since ES cells and their derivative progenitors lacking LMNA, LMNC, and LMNB1 proliferate normally in culture, maintain euploidy, and differentiate normally into multiple cell types (Kim et al. 2013). However, it should be noted that the ES culture system reflects early pre-organ development. The story is not as clear-cut in vivo, although A-type lamins are dispensable for early development. Mice lacking either LMNA, LMNC, or both are indistinguishable from normal siblings at birth, pos-

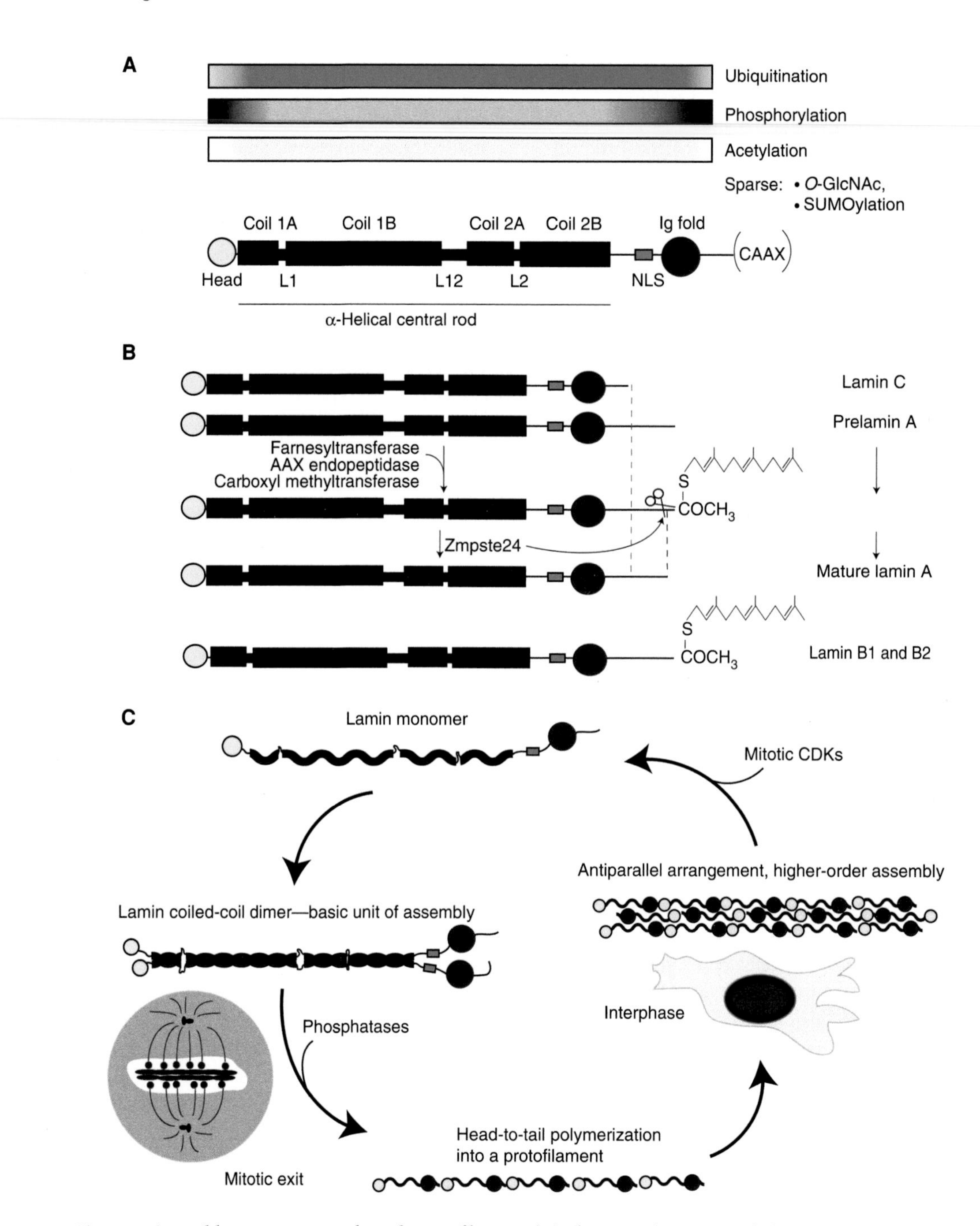

Figure 2. Assembly, processing, and regulation of lamins. (*A*) The general structure of a lamin protein, consisting of a short unstructured head domain (yellow), a central α-helical rod domain comprised of four helical subregions (coils 1A, 1B, 2A, and 2B in purple) interspersed by unstructured linkers (L1, L12, and L2), a tail region that includes a nuclear localization signal (NLS) (green), an immunoglobulin-fold domain (Ig-fold) (red), and a fairly unstructured carboxy-terminal end that in most cases terminates with a carboxy-terminal CAAX motif (Stuurman et al. 1998; Krimm et al. 2002; Herrmann and Aebi 2004). (*Legend continues on following page.*)

sibly because of some functional redundancy with the developmentally regulated INM protein lamin B receptor (LBR) that is expressed (or over- or re-expressed) in cells engineered to lack *Lmna* (Sullivan et al. 1999; Jahn et al. 2012; Solovei et al. 2013). Loss of *Lmna* does, however, result in severe postnatal growth retardation and death within 3 weeks, which is correlated to some extent with the normal silencing of LBR expression in many postnatal tissues (Sullivan et al. 1999; Jahn et al. 2012; Solovei et al. 2013). Intriguingly, the ratio of LMNA/LMNC varies across different cell types suggesting cell-type-specific and developmental regulation of *LMNA* splicing or posttranscriptional regulation. One striking example of differential A/C expression is in neurons where *LMNA* (and not the *LMNC* splice variant) translation is inhibited by the microRNA miR-9, which binds specifically to the longer *LMNA* mRNA 3′ UTR (untranslated region) (Jung et al. 2012). Less is known about the cell and tissue variation of expression of somatic B-type lamins. Both LMNB1 and LMNB2 are required for normal neuronal development (Vergnes et al. 2004; Coffinier et al. 2011). In addition, both human and murine fibroblasts with reduced levels of LMNB1 undergo senescence and display reduced proliferation and loss of LMNB1 correlates with keratinocyte senescence in vivo (Shimi et al. 2011; Dreesen et al. 2013; Shah et al. 2013). In adults, B-type lamin expression appears to be nonessential in some tissues, such as the skin epidermis and liver (Yang et al. 2011). These findings reveal that an absolute dependence on lamin expression in mammals varies among different cell types and that early embryos, including pluripotent cells and some of their differentiated derivatives, may not require any of the lamins for their proliferation and early differentiation.

Figure 2. (*Continued*) While the rod domain is involved in the dimerization of lamins through coiled-coil interactions, the head and tail domains play a significant role in regulating the higher assembly of lamins (Gieffers and Krohne 1991; Heitlinger et al. 1992; Stuurman et al. 1996; Klapper et al. 1997; Sasse et al. 1998; Izumi et al. 2000; Moir et al. 2000a; Ben-Harush et al. 2009). Also shown are potential posttranslational modifications (PTMs) and their documented density occurring throughout the monomer where the color intensity is proportional to the probability of finding the PTMs. The most highly occurring PTMs, as shown, are phosphorylation and ubiquitination, which exhibit an inverse distribution, with phosphorylation more often occurring near and at the head and tail domains while ubiquitination more frequently found within the central rod domain and extending toward the Ig fold. The distribution pattern of acetylation is similar to that of ubiquitination, albeit less frequently occurring. In contrast, *O*-GlcNAcylation and SUMOylation are sparse and have been found only on a few residues (Simon and Wilson 2013). (*B*) A-type lamins (LMNA and LMNC) are the result of alternative splicing from the *LMNA* gene, while the major B-type isoforms (LMNB1 and LMNB2) are transcribed from two different genes. Both LMNA and LMNB1/2 are farnesylated at the cysteine residue of the –CAAX motif by a farnesyltransferase, followed by the removal of the last three amino acids by means of an AAX endopeptidase and finally carboxymethylation via carboxyl methyltransferase (Dechat et al. 2008). Lamin B1 and B2 remain farnesylated, facilitating their anchorage to the nuclear envelope. LMNA undergoes an additional step: removal of the farnesyl group and the 15 most carboxy-terminal residues by the protease Zmpste24, rendering a fully functional and mature, lamin A protein (Dechat et al. 2008). (*C*) Depicts principles of lamin assembly and their regulation through the cell cycle. Lamin monomers associate in a parallel, head-to-head manner, leading to a coiled-coil dimer formation—the basic unit for higher-order assembly. As cells enter mitosis, higher-order lamin filaments localized at the nuclear lamina become phosphorylated by cyclin-dependent kinase 1 (CDK1) and potentially other kinases triggering the disassembly of the nuclear lamina as lamins depolymerize and become cytosolic. A-type lamins are found in the cytosol, while depolymerized B-type remain anchored to lipid membranes (Gerace and Blobel 1980). As the cell approaches interphase, lamins are dephosphorylated by protein phosphatase 1a (PP1a) and are recruited to and reassembled at the reforming NE (Thompson et al. 1997; Dechat et al. 2010). This reincorporation at the NE is temporally regulated, with B-type lamins organizing to the lamin first, followed by lamin A and then lamin C networks (Pugh et al. 1997; Shimi et al. 2015; Xie et al. 2016; Wong et al. 2020). Importantly, and particularly for A-type lamins, a subpopulation of lamins can remain nucleoplasmic, due to the persistence of mitotic PTMs such as S22P (Simon and Wilson 2013; Gruenbaum and Medalia 2015; Ikegami et al. 2020; Wong et al. 2020).

LAMINS AS A SCAFFOLD FOR THE LAMINA INTERFACE

Lamins interact with a host of nuclear membrane proteins, transcription factors, chromatin regulators, signaling molecules, splicing factors, and even chromatin itself to form a nuclear subcompartment that is involved in a variety of cellular processes such as the governance of nuclear integrity, nuclear positioning, mitosis, DNA repair, DNA replication, splicing, mechanotransduction and -sensation, transcriptional regulation, and genome organization. The INM and the nuclear lamina are interdependent structures held together by mutual molecular interactions between a myriad of INM proteins and the underlying lamina meshwork (Fig. 3; Simon and Wilson 2013; Wong et al. 2014, 2021b). The protein composition at the nuclear periphery has been profiled in various cell types using different approaches. Initially, lamin-associated proteins (LAPs) were identified by cofractionation in nuclei extracted with high concentrations of monovalent salts and ionic detergents (Senior and Gerace 1988; Gerace and Foisner 1994). More recently, nuclear envelope transmembrane proteins (NETs) have been identified by mass spectrometry identification proteins extracted from the NE (Korfali et al. 2012; de Las Heras et al. 2013, 2017; Worman and Schirmer 2015). Additional LAPs have been identified by BioID, a technique to biotinylate proximal proteins, followed by mass spectrometry (Kim et al. 2016; Mehus et al. 2016; Xie et al. 2016; Wong et al. 2021b). While the terms may be used interchangeably for the most part, it is important to note that LAPs are presumed to interact with lamins, but not necessarily the NE, since some lamins are nucleoplasmic and, conversely, NETs are present in the nuclear envelope, but may not interact with lamins or be at the INM. Many new tissue-specific NETs have not been fully characterized.

The LEM (Lap2-Emerin-Man1) domain family of proteins are LAPs/NETs containing a LEM domain that binds a conserved metazoan chromatin protein, barrier to autointegration factor (BAF), allowing their indirect interaction with chromatin (Furukawa 1999; Cai et al. 2001, 2007; Lee et al. 2001; Shumaker et al. 2001; Shimi et al. 2004; Margalit et al. 2007; Kind and van Steensel 2014). Mouse and human LEM domain proteins include LAP2 (lamina-associated polypeptide 2), emerin, and MAN1 along with other LEM and LEM-like proteins (Lin et al. 2000; Schirmer et al. 2003; Lee and Wilson 2004; Brachner et al. 2005; Chen et al. 2006; Ulbert et al. 2006; Simon and Wilson 2013). These LEM domain proteins, like lamins, are strongly conserved, reflecting their fundamental roles in the nucleus. LEM proteins primarily reside in the INM, but LAP2 has several isoforms, at least one of which, LAP2α, lacks a transmembrane domain and is nucleoplasmic; thus, LAP2 can interact with both nucleoplasmic and INM-associated lamins (Harris et al. 1994; Berger et al. 1996; Dechat et al. 1998). Nucleoplasmic LAP2α interacts with nucleoplasmic lamin A (likely phosphorylated) and has been implicated in gene activation (Gesson et al. 2016). LEM domain proteins are translated into the ER membrane and diffuse into the ONM and then retained in the INM via interactions with lamin proteins (Holmer and Worman 2001; Wilson and Foisner 2010; Berk et al. 2013); thus, perturbations of the nuclear lamina can lead to delocalization of these proteins.

Additional well-studied lamin-interacting proteins include INM proteins such as LBR and linker of nucleoskeleton and cytoskeleton complex (LINC) proteins (Fig. 2). LBR is an eight-pass transmembrane protein localized at the INM that contains a nucleoplasmic domain that codes for a sterol reductase. LBR is required for nuclear shape changes, as seen in neutrophils, and is developmentally regulated, with higher expression in ES and progenitor cells. LBR has been shown to interact with heterochromatin organizers HP1 (heterochromatin protein 1) and PRR14 (proline-rich protein 14), and to interact with the repressive heterochromatin modification histone H4 lysine 20 dimethylation (H4K20me2) (Ye et al. 1997; Polioudaki et al. 2001; Hirano et al. 2012; Dunlevy et al. 2020). The LINC complex spans the NE and is formed by interactions between SUN domain proteins of the INM (SUN1 and SUN2 in somatic cells) and the KASH domain proteins of the ONM (nes-

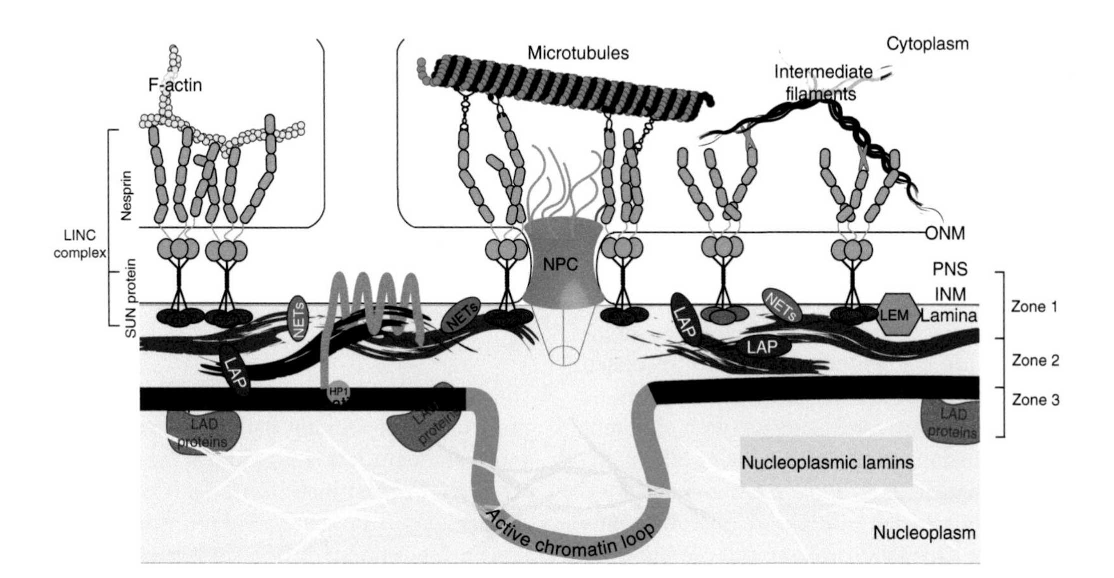

Figure 3. Structure and complexity at the nuclear periphery. The nuclear lamina serves as an integrative hub for various aspects of nuclear and cellular functions (Wong et al. 2021b). The lamin meshwork(s) (gray), serves to maintain nuclear integrity as well as act as a structural scaffold, anchoring a diverse range of proteins and heterochromatin domains (lamina-associated domains [LADs]) to the nuclear envelope (NE). Lamin-associated proteins (LAPs) and some nuclear envelope transmembrane proteins (NETs) interact with lamins and chromatin. Relatively more well studied, NETs/LAPs include the components that form the linker of cytoskeleton and nucleoskeleton complexes (LINCs) (SUN and KASH domain-containing nesprins), LEM domain proteins, and lamin B receptor (LBR) (pink), shown above. Additionally, the NE also associates with interphase heterochromatin (red chromatin—LADs) providing a structural basis for establishing interphase chromosome topology and at the same time providing a means for transcriptional gene regulation through repression of genes sequestered at the periphery. The tethering of chromatin at the nuclear periphery can be achieved via LBR interactions or lamin A/C. LBR is known to interact directly with chromatin binders (such as Hp1α and PRR14) while A-type lamins may mediate chromatin contacts through LAPs and chromatin binders and modifiers (orange circles) interacting with LADs. A recent comparative proteomics study querying the interaction of lamins, LAP2β and LADs, further resolved the nuclear periphery into three functional zones with respect to chromatin regulation. Zone 1 includes proteins at the INM/lamina that do not interact with LADs, Zone 2 is comprised of proteins that bind the INM/lamina network and LADs—the "middlemen," and Zone 3 contains only proteins associating uniquely with LADs and chromatin (Wong et al. 2021b).

prins) (Méjat and Misteli 2010; Rothballer and Kutay 2013; Chang et al. 2015; Kim et al. 2015). KASH domain proteins are able to interact with cytoskeletal elements (actin, myosin, and cytosolic intermediate filaments); thus, LINCs physically link the nucleoplasm and lamina to the cytoskeleton, ultimately connecting to the extracellular matrix (ECM) via integrins. It is through these interactions that the nuclear lamina acts as a critical mechanosensing and transduction node (Lombardi and Lammerding 2011; Osmanagic-Myers et al. 2015; Belaadi et al. 2018; Wang et al. 2018; Maurer and Lammerding 2019). LINCs are also involved in the cytoplasmic motor-driven movement of the nucleus within the cell (Luxton et al. 2010; Wu et al. 2014) and have been implicated in gene regulation as well (Kim and Wirtz 2015; Tajik et al. 2016). Specific isoforms of nesprins are expressed in different cell types (Duong et al. 2014) and some KASH-less nesprin isotypes can be found at the INM and interact with lamins and emerin directly (Rajgor and Shanahan 2013; Kim et al. 2015).

As mentioned above, during mitosis, lamins are phosphorylated and disassembled. Likewise, LAPs and BAFs are also phosphorylated, thus leading to loss of LAP–lamin interactions (Haraguchi et al. 2001; Foisner 2003; Ellenberg 2013). This leads to the vesiculation or movement into the mitotic ER of these INM proteins. At the onset of anaphase/telophase, these proteins are dephosphorylated and ER/vesicle membranes are recruited back onto chromatin to reform the nuclear envelope and subsequently recruit the lamins. Interestingly, the LEM domain proteins emerin and LAP2β appear to interact with condensed chromatin (mediated through BAF, which is also regulated by mitotic phosphorylation) and interact with different chromatin domains than LBR during NE assembly (Haraguchi et al. 2001; Dechat et al. 2004). Thus, while in the interphase, nuclear lamins retain the LAP INM proteins and prevent their redistribution to the ONM and ER, after mitosis membrane-bound LAPs scaffold on chromatin and subsequently recruit the nuclear lamins, highlighting the interdependence of interactions and the lamina interface.

CELL- AND TISSUE-SPECIFIC INM/LAMINA PROTEOMES

Many of the proteins that are resident to the peripheral zone of the nucleus—the lamins, LAPs, and NETs—are differentially expressed (Figs. 3 and 4; Furukawa and Hotta 1993; Furukawa et al. 1994; Alsheimer et al. 1999; Schütz et al. 2005a,b; Chen et al. 2006; Korfali et al. 2010, 2012; Jung et al. 2012; Solovei et al. 2013). This has been highlighted in several proteomic studies using isolated NE from different cell and tissue types that have identified many new tissue- and cell-specific NET proteins. These experiments uncovered an impressive proteomic diversity of this cellular compartment. In particular, a comparative study across three disparate tissues (liver, leukocytes, and muscle) revealed that the majority of the 598 NETs displayed distinct expression profiles between the tissues examined, with only a modest 16% of these identified NETs being shared across all three tissue types (Korfali et al. 2012). Other high throughput studies have documented differential NE proteomes even in less disparate cell types, such as during T-cell activation and in addition, using mRNA expression analyses, during myogenesis. These types of studies will undoubtedly continue to identify additional novel and differential NET proteins and profiles. How such changes to the nuclear peripheral proteome affects cellular process and behavior and how their deregulation would allow the manifestation of disease phenotypes would be of particular interest (Wong et al. 2014).

LAMINA-ASSOCIATED DOMAINS (LADs)

The genome is functionally organized, with regions of late-replicating heterochromatin positioned at the nuclear lamina and at the nucleolar periphery in most metazoan cells (Cremer and Cremer 2001). Conversely, more euchromatic regions are positioned in the nuclear interior or interact with NPC (Buchwalter et al. 2019). Not surprisingly, lamins have long been implicated in regulating the three-dimensional (3D) organization of chromatin, particularly via these heterochromatin domains found at the nuclear lamina. 3D genome organization and chromatin compartments were initially identified and studied by microscopy and cytological tools using DNA stains, DNA hybridization techniques such as fluorescent in situ hybridization (FISH), or electron microscopy. These early studies led to an understanding of nonrandom organization of chromatin, including identification of chromosome territories (CTs) and putative regulatory subdomains within the nucleus, including the lamina. These studies relied heavily on identifying spatial relationships between a known sequence (or sequences) and a protein compartment (such as the lamina) in the nucleus. More recently, deep-sequencing-based approaches have led to genome-wide and molecular-level understanding of genome organization and identification of specific DNA sequences and features that accompany such 3D architecture. Three different molecular methods and their derivatives are most routinely used to measure genome organization: chromatin immunoprecipitation (ChIP) (Ma and Zhang 2020), the proximity-labeling technique DNA adenine methyltransferase identification (DamID) (Orian et al. 2009), and chromatin conformation capture methods such as Hi-C (Lieberman-Aiden et al. 2009; Be-

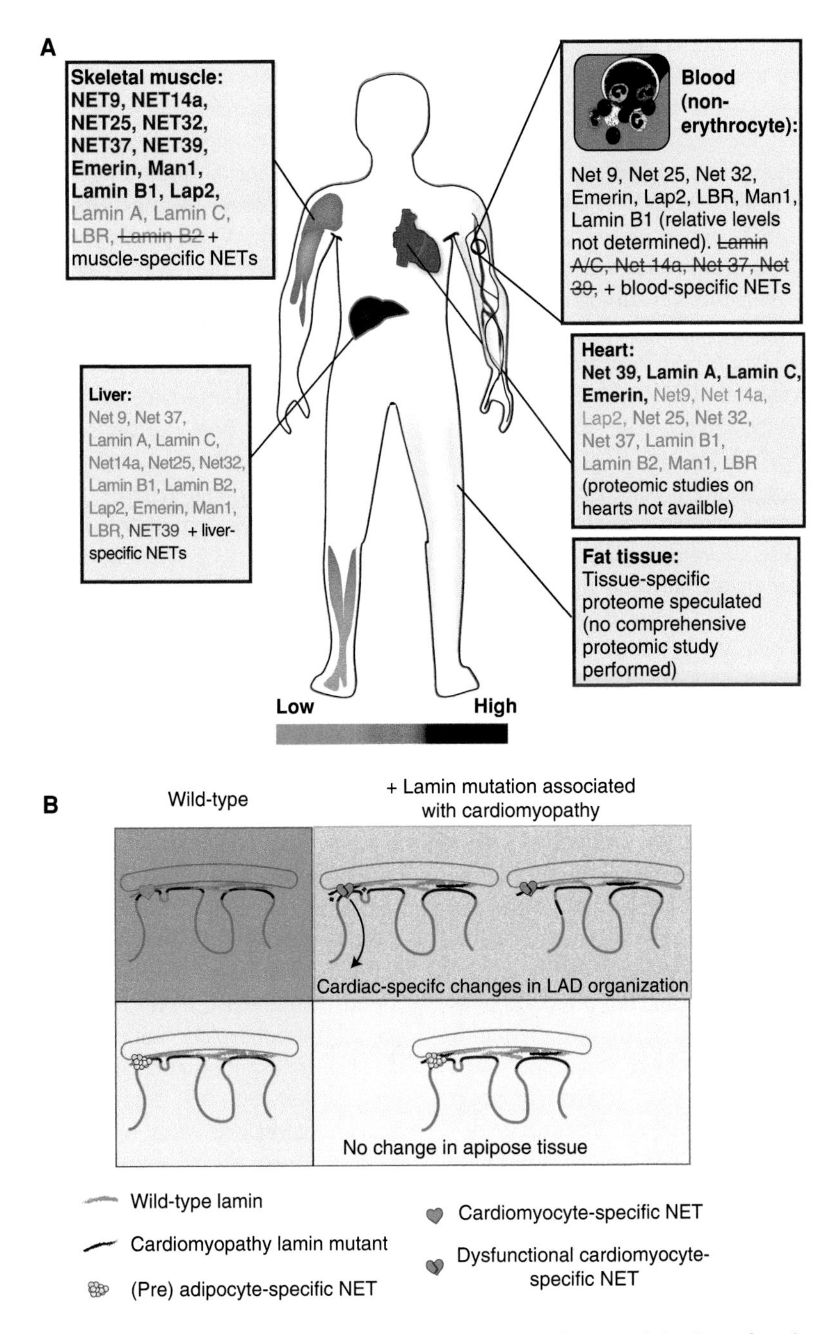

Figure 4. Tissue- and cell-type proteomic diversity at the nuclear lamina. (*A*) The inferred tissue-specific expression of nuclear envelope transmembrane proteins (NETs) determined by a transcriptomic analysis (Chen et al. 2006). For this representation, the proteins were binned into three expression groups for each tissue type examined: low (green), medium (orange), and high (red). It is clear, even given the small number of proteins considered here, how different the NE proteome is in divergent cell types. It has been shown for a few cardiomyopathy lamin mutations that chromatin organization is perturbed in induced pluripotent stem cells derived cardiomyocytes in the form of gain and loss of lamina-associated domains (LADs), thereby affecting transcriptional output and lineage commitment. (*Legend continues on following page.*)

laghzal et al. 2017). ChIP is used to identify chromatin domains and binding sites of chromatin interactors, while DamID and Hi-C directly measure 3D genome organization, with DamID and related techniques measuring proximity to a protein nuclear compartment and Hi-C-related techniques measuring chromatin–chromatin interactions. DamID is most often used to detect peripheral heterochromatin domains, the so-called LADs (Greil et al. 2006; Vogel et al. 2007). Hi-C is used to identify both local and long-range chromatin interactions (Dekker 2002; Lieberman-Aiden et al. 2009; Dixon et al. 2012; Phillips-Cremins et al. 2013; Rao et al. 2014; Belaghzal et al. 2017), in particular, local self-interacting regions called topologically associated domains (TADs) and, in active regions of the genome, promoter–enhancer interactions. Of importance to LAD organization, Hi-C also identifies longer-range chromosome and genome-wide self-interacting domains: the A (active) and B (inactive) compartments, with activity state of the compartments being defined via intersections of other data (such as RNA-seq or chromatin state maps by ChIP) (Dekker 2002; Lieberman-Aiden et al. 2009; Dixon et al. 2012; Rao et al. 2014). It is important to note that special care must be taken in ChIP assays to identify chromatin state and lamin interactions in heterochromatic regions since these regions are resistant to traditional ChIP fragmentation and isolation protocols (Das et al. 2004; Gesson et al. 2016). An alternative method to DamID is TSA-seq (tyramide signal amplification–seq), another proximity-based method to identify geographic chromatin domains, including LADs (Chen et al. 2018b). These molecular methods are often combined with advanced microscopy approaches, including super-resolution FISH technologies and live cell imaging, to complement and extend the molecular findings and to map single-cell chromatin conformations and locus positioning in single cells in situ (Boettiger and Murphy 2020; Kempfer and Pombo 2020). While genome-wide sequencing approaches have the advantage of higher resolution (in base pairs) and numbers of cells processed, imaging approaches are inherently single cell, thus giving an immediate output of cell–cell variability, but also provide additional contextual information, such as cellular state and changes over time. Integration of different data types is key to understanding genome organization generally and this also applies to LADs.

As identified by DamID, LAD regions are large (>100 kb), AT-rich, mostly heterochromatic, and largely correlate with the inactive B compartment as determined by Hi-C (Fig. 5; Dixon et al. 2012; Rao et al. 2014; Fraser et al. 2015; Robson et al. 2017). Although CTCF is found at the boundaries between LADs and non- or inter-LAD (iLAD) regions (Guelen et al. 2008), thus demarcating the transition between B and A compartments, it is depleted within the LADs themselves (Guelen et al. 2008; Harr et al. 2015), suggesting that LADs (and the B compartment) have a fundamentally different organization than the CTCF and cohesin-mediated looping structures found in the active iLAD regions (A compartment). Indeed, depletion of either CTCF or cohesin leads to loss of observable TAD and sub-TAD organization, but A/B compartmentalization is largely maintained with only minor changes, most strikingly the movement of some inactive regions previously constrained in the A compart-

Figure 4. (*Continued*) The same mutations, however, have no impact on LAD architecture in hepatocytes and adipocytes, underscoring the tissue specificity of lamin mutations (Shah et al. 2021). (*B*) How tissue-specific proteomes at the nuclear periphery can allow differential susceptibility to lamin mutations. In the *top* panel, the expression of a lamin mutation associated with cardiomyopathy disrupts the function of a cardiomyocyte-specific NET protein (heart shape to broken heart) that may be involved in LAD anchorage, leading to the loss of LADs at the periphery. In adipocytes, however, a different tissue-specific NET may be involved in LAD tethering (depicted as lipid droplets), the function of which is not affected by the same lamin mutation and hence LAD architecture remains unperturbed. It is important to note that such structural reorganization could result from a mutation in either lamin or the NET protein itself. Crossed-out proteins indicate a documented absence of these NETs from the indicated tissue. (Figure adapted from Wong et al. 2014.)

 Cite this article as *Cold Spring Harb Perspect Biol* doi: 10.1101/cshperspect.a040113

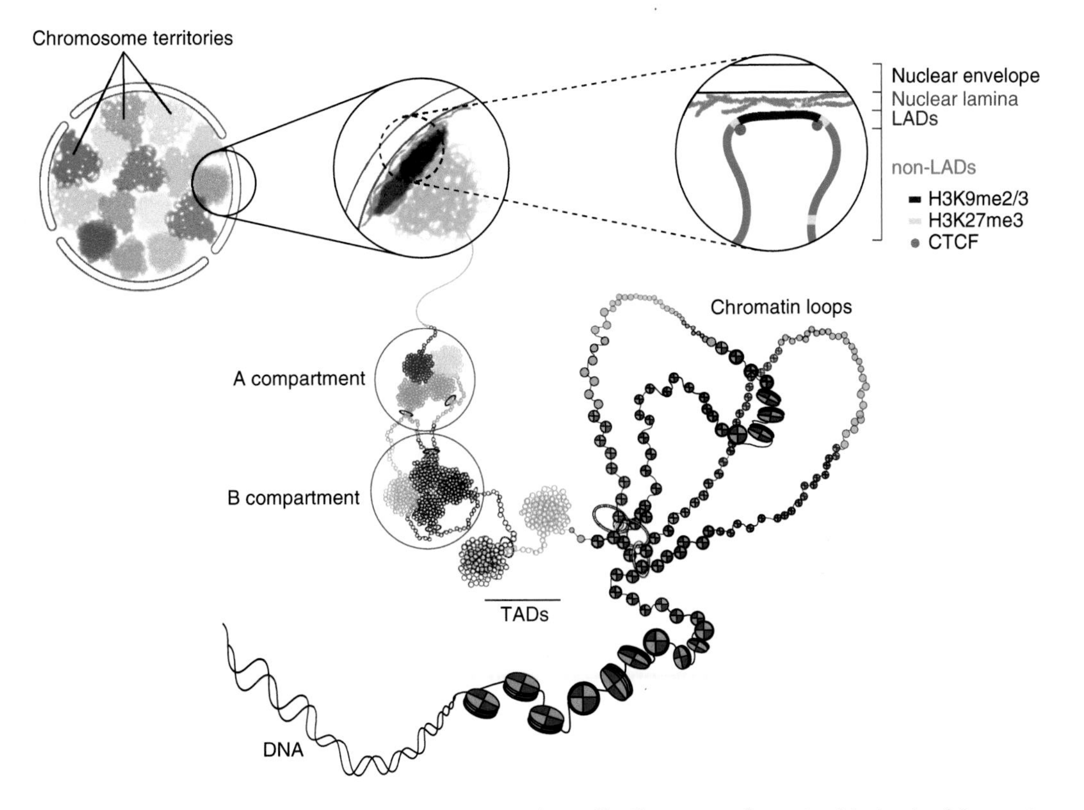

Figure 5. Higher-order genome organization in mammalian cells. Shown is a schematic of the levels of chromatin folding within the higher-order 3D genome organization. The DNA interacts with histone octamers and aggregates forming nucleosome arrays that are more or less compacted, depending on the histone variants present and the posttranslational modifications (PTMs) to their amino-terminal tails. The next level of organization is the formation of topologically associated domains (TADs), which in active chromatin domains are formed by loop extrusion via cohesin and stabilized by CCCTC-binding factor (CTCF) (Dixon et al. 2012; Rao et al. 2014; Sanborn et al. 2015). TADs and sub-TADs range from tens of kilobases to megabase-sized and have delimited boundaries and high rates of interactions inside of these domains. TADs segregate based on their transcriptional status into active A and inactive B compartments, with A compartments mostly occupying the nuclear interior and B compartments associated with transcriptionally repressive nuclear domains enriched in histone H3 lysine 9 di- and trimethylation (H3K9me2/3) and histone H3 lysine 27 trimethylation (H3K7me3) at the NE and the periphery of the nucleoli (Lieberman-Aiden et al. 2009; Harr et al. 2015, 2016; Guelen et al. 2008). It is important to note that A/B compartments exist even if TADs are depleted, suggesting that segregation into A/B compartments is not driven by TAD architecture (Nora et al. 2017; Schwarzer et al. 2017b). The geographic organization of the B compartment to the nuclear envelope (NE) aids in the establishment and/or maintenance of interphase chromosome topology and hence overall genome organization. Moreover, since the nuclear lamina is a transcriptionally repressive compartment, it represents a manipulative model for transcriptional regulation via spatial organizational changes relative to the lamina. The territorial organization of chromosomes in interphase (chromosome territories [CTs]) constitutes a basic feature of nuclear architecture and these other levels of organization occur within the context of CTs (Cremer and Cremer 2010). (Figure based on data in Maeshima et al. 2020.)

ment to the B compartment (Nora et al. 2017; Schwarzer et al. 2017a). Thus, the genome remains partitioned, even if reconfigured (Nora et al. 2017; Schwarzer et al. 2017b; Falk et al. 2019). It remains unclear, however, whether LADs remain geographically positioned at the nuclear lamina in the absence of CTCF or cohesin.

GEOGRAPHIC ORGANIZATION OF LADs IS REGULATED BY CHROMATIN STATE

LADs are enriched in histone H3 lysine 9 di- and trimethylation (H3K9me2/3) and this modification is also required for LAD recruitment to and maintenance at the nuclear lamina (Guelen et al. 2008; Towbin et al. 2012; Harr et al. 2015). In mammals, proline-rich protein 14 (PRR14) has been shown to anchor these regions to the lamina through its interactions with heterochromatin protein 1 (HP1), a chromodomain protein that binds H3K9me3 (Poleshko and Katz 2014; Dunlevy et al. 2020), while in *Caenorhabditis elegans* this function is carried out by the INM-bound chromodomain protein CEC-4 (Towbin et al. 2012; Gonzalez-Sandoval et al. 2015; Gonzalez-Sandoval and Gasser 2016; Harr et al. 2016, 2020; Bian et al. 2020). Intriguingly, heterochromatic regions have been shown to accrete due to phase separation directed by HP1α, and the organization of LADs at the lamina likely depends upon such biophysical forces (Larson et al. 2017; Strom et al. 2017; Falk et al. 2019). In particular, biophysical models predict that, in the absence of an active constraint at the nuclear lamina, LADs would form large clusters in the nucleoplasm (Falk et al. 2019). This is supported by studies showing that in cells lacking lamin A/C and LBR (either naturally or engineered) heterochromatin configuration is inverted (Solovei et al. 2013). Thus, the heterochromatic nature of LADs leads to their separation from euchromatic regions, but other forces and interactions are likely at play in geographically organizing these regions to the lamina. There is a less clear role for the facultative heterochromatin modification histone H3 lysine 27 trimethylation (H3K27me3) in LAD organization, since this modification is found mostly outside of LADs, but is enriched at LAD borders (Guelen et al. 2008; Harr et al. 2015) and necessary to target an inserted test locus to the lamina in coordination with H3K9me2/3 (Harr et al. 2015). In contrast, in *C. elegans*, H3K27me3 modifications seem to be dispensable for driving peripheral localization, but there is a second pathway for lamina localization independent of Cec-4/H3K9me3 anchoring (Towbin et al. 2012). In a clever screen using *cec-4* null worms, this second pathway was discovered to be a sequestration of euchromatin away from the lamina (Cabianca et al. 2019). In this case, the delocalization of CBP/p300 causes aberrant H3K27Ac to replace H3K27me3, suggesting that sequestration of active chromatin modifiers is a driving force for tethering of heterochromatin to the lamina in differentiated cells. Similarly, in a recent study in mouse T-cells in which a LAD border region at the T-cell receptor (TCR) locus was deleted, invading H3K27ac was found to drive an enhancer region away from the nuclear lamina and subsequently led to altered enhancer–promoter interactions (Chen et al. 2018a). The role of histone H4 lysine 20 dimethylation (H4K20me2), a repressive chromatin mark bound by the tudor domain of LBR (Hirano et al. 2012), in LAD regulation and organization remains unclear. H4K20me2 has a wide genome distribution with enrichment on centromeric and telomeric chromatin and does not appear to be particularly enriched on LADs (Mattout et al. 2015). However, this modification does seem to play a role in LAD regulation/organization under certain conditions, particularly senescence and in particular laminopathies (Shumaker et al. 2006; Barski et al. 2007; Bártová et al. 2008; Mattout et al. 2015; Nelson et al. 2016). Taken together, these studies suggest a complex interplay between chromatin state and chromatin-binding proteins in the separation and organization of LADs to the periphery.

GENE REGULATION IN LADs

LAD regions contain relatively few genes, but are, conversely, enriched in developmental and lineage-specific genes, leading to the hypothesis that the epigenetic state of these regions is tied to both organization and developmental control of gene expression (Guelen et al. 2008; Peric-Hupkes et al. 2010; Bian et al. 2013; Harr et al. 2015). Indeed, early studies of regulation of individual genes by the nuclear lamina focused on developmentally regulated gene loci, such as the immunoglobulin heavy chain locus (*Igh*) in B-cell development (Kosak et al. 2002), the β-globin locus in erythroid development

(Bian et al. 2013), and *MyoD* in muscle development (Yao et al. 2011) among others. These early studies pointed to the dynamic nature of loci associated with (and presumably regulated by) the nuclear lamina during development, but it was not until genome-wide mapping through DamID that the shifting LAD organization, chromatin, and expression landscape could be fully appreciated. LADs are dynamic between cell states and in a simplistic view can be divided into constitutive or facultative LADs, cLADs, and fLADs, respectively (Peric-Hupkes et al. 2010; Meuleman et al. 2012; van Steensel and Belmont 2017). cLADs are LADs that remain irrespective of cell type, while fLADs reorganize between cell types or developmental stages. The majority of LADs do not change between individual cell types, although 70% of LADs are fLADs; thus, reorganization is restricted by cell type. fLADs are more gene rich and in cell types where they undergo reorganization away from the lamina, genes in these regions are activated or poised for activation (Peric-Hupkes et al. 2010). Even within a given cell type or population there is some cell–cell heterogeneity in LAD organization, as detected by low-resolution, single-cell DamID, with more gene-dense LADs displaying greater fluctuations in their association with the lamina, implying that these disruptions may be due to differential gene usage between cells (Kind et al. 2015). A level of heterogeneity in reorganizing to the lamina after cell division has also been noted, with some LADs appearing to become nucleolar-proximal in the daughter cells, although there is some expected overlap between nucleolar-associated domains (NADs) and LADs given the invaginations of nuclear lamins at nucleoli noted in many cell types (Cremer and Cremer 2001; Legartová et al. 2014; Padeken and Heun 2014).

In addition to LAD heterogeneity between cells, it has been documented that different promoters respond differently to integration into a LAD and to different environments within a LAD. Early ectopic lamina tethering experiments suggested that recruitment to the nuclear lamina reduces expression of the tethered locus. In one report, an integrated reporter gene in mouse cells showed a two- to threefold reduction in expression after tethering using a truncated emerin to tether a *lac* operator array (*lacO*) (Reddy et al. 2008). A similar study, using Lap2β to target a *lacO* array in a human cell line, found only a minimal reduction in the recruited reporter gene expression, while a subset of endogenous flanking genes on the same chromosome did show reduced expression (Finlan et al. 2008). In *Drosophila*, the tethering of two *lacO* tagged reporter genes by Dme/LamC to the nuclear lamina caused substantial repression, but the level of repression was integration site and reporter gene dependent (Dialynas et al. 2010). These early results suggested that recruitment to the lamina could cause repression, but that this repression was variable depending upon local chromatin context and different promoters. A more recent study sought to investigate exactly this question and found that upon integration into LADs some promoters are more sensitive to the more repressive environment and are almost universally repressed, while other promoters can "escape" silencing within LAD regions (Leemans et al. 2019). These escapers were generally less sensitive repressive domains, particularly H3K27me3. All promoters showed some variation in regulation by LADs and less repressed promoters were found to reside in more weakly (and perhaps transiently) lamina-bound regions within LADs (Leemans et al. 2019). Thus, even though LADs are generally repressive, genes respond differently depending upon promoter type and local LAD context.

ROLE OF TRANSCRIPTION FACTORS IN LAD ORGANIZATION

While both lamins and chromatin are important for organization of LADs to the nuclear lamina, there is some evidence that transcription factors and machinery may be more directly involved in peripheral localization (and regulation) of genes at the lamina. Several transcription factors have been found to interact with lamins and/or lamin-interacting proteins, and some of these have been proposed to be more directly involved in organizing specific loci to the lamina. One study found that sequences from two different

LAD-regulated loci enriched in GA dinucleotides were able to drive an ectopic region to the lamina (Zullo et al. 2012). This is interesting given that GA dinucleotides are not enriched in the relatively AT-rich LADs (Guelen et al. 2008). These motifs were found to bind a cell-type-specific transcription factor (ZBTB7B, also known as THPOK) in complex with a histone deacetylase (HDAC3), and an inner nuclear membrane protein (LAP2β) to mediate de novo interactions with the nuclear lamina. Both ZBTB7B and HDAC3 were also found to be important for myogenesis and in regulating myogenic genes and their organization to the lamina (Poleshko et al. 2017), suggesting that specific transcription factor complexes can regulate lamina interactions, potentially directly. ZBTB7B features a BTB–POZ domain that recruits Polycomb repressive complex 2 (PRC2) to chromatin, thereby initiating trimethylation of histone H3 on lysine 27 (H3K27me3) (Boulay et al. 2012). Interestingly, there are numerous BTB-POZ domain transcription factors that are expressed in a cell-type-specific manner, and these have been found to both activate and repress gene expression and to interact with HDAC3. HDAC3 itself, which interacts with numerous transcription factor complexes, has been shown in independent studies to interact with both LAP2β and emerin (Somech et al. 2005; Demmerle et al. 2012). Whether this complex of ZBTB7B, HDAC3, and LAP2β initiates locus interaction at the lamina or simply maintains a heterochromatin state is unclear. Other studies suggest that altered chromatin state, including both H3K9me2/3 and H3K27me3, drive lamina association (Bian et al. 2013, 2020; Kind et al. 2013; Harr et al. 2015). Taken together, these findings suggest that the epigenetic state may be the ultimate factor in determining organization to the lamina. Lamina-associating sequences may recruit repressive transcriptional complexes to promote a local chromatin signature favorable for lamina association. While at the lamina, these complexes would then maintain or reinforce this repressive state, in agreement with studies showing that integration of promoter sequences into LADs leads to their repression.

LAD ORGANIZATION HELPS ENFORCE CHROMATIN AND GENE EXPRESSION PROGRAMS

If chromatin state drives lamin association, is association with the nuclear lamina simply a by-product of this chromatin state and biophysical properties? It is difficult to tease apart the role that chromatin state itself versus organization to the lamina plays. As discussed previously, recruiting inserted reporter genes to the lamina is able to repress some genes, depending on promoters and integration site (Finlan et al. 2008; Reddy et al. 2008). In addition, biophysical modeling shows that in the absence of nuclear periphery "tethers," heterochromatin will not organize to the lamina, but is instead predicted to collapse into the center of the nucleus and form an "inverted" type of chromatin organization (Falk et al. 2019). This implies that there are active mechanisms to organize LADs to the nuclear lamina. In support of this, several studies have implicated A-type lamins as especially important for the geographical organization of LADs to the lamina in differentiated cells (Solovei et al. 2013; Harr et al. 2015; Zheng et al. 2018). In particular, the predicted inverted chromatin organization is noted in the absence of both A-type lamins and LBRs (Solovei et al. 2013); thus, chromatin organization to the lamina is not the default. But is there evidence that loss of lamin association affects gene regulation—that such localization is itself important? In *C. elegans*, a global reduction of lamin–chromatin interactions through depletion of the chromobox protein CEC-4, which is required for lamina association, resulted in up-regulation of only one single gene, while depletion of H3K9 methylation caused widespread disruption (Gonzalez-Sandoval et al. 2015). From this experiment, one could conclude that organization of chromatin into LADs at the lamina is not important for gene repression (although there is a second non-CEC-4-dependent anchoring pathway that confounds these results (Cabianca et al. 2019). However, loss of CEC-4 did impair induced differentiation into muscle cells, reminiscent of mouse studies, in which the effect of loss of lamins is seen later in muscle differentiation (Sullivan et al. 1999). In addition, several developmentally regulated tissue-restricted NETs have been shown to cause reorganization of lineage-specific genes and altered expression when ectopically expressed, suggesting that these proteins directly influence organization and regulation (de Las Heras et al. 2017). In cells induced to undergo senescence, alterations in LAD organization were noted, including altered chromatin state outside and inside of LADs, particularly an encroachment across boundaries, suggesting a "blurring" of chromatin domains (Shah et al. 2013). This disorganization and dysregulation could be mimicked by acute depletion of lamin B1, suggesting that, unlike ES cells, more differentiated cells may rely on the geographical organization of heterochromatin to the lamina to maintain a cell-type-specific chromatin landscape. This is in agreement with the idea of LAD organization functioning as a mechanism to maintain a cell-type-specific epigenetic state; some cell types and states will be more sensitive to disruption of LAD organization.

DYNAMIC ORGANIZATION OF LADs THROUGH THE CELL CYCLE

Interphase genome organization, including LAD and lamin organization, is ablated during mitosis and re-established after mitosis (Burke and Ellenberg 2002; Salina et al. 2003; Kind et al. 2013; Naumova et al. 2013; Kind and van Steensel 2014; Gibcus et al. 2018; Luperchio et al. 2018). Several studies have suggested that A-type lamins organize to the reforming nuclear envelope with different kinetics, although there is some discrepancy on how lamin A and C might differ in their timing of association (Pugh et al. 1997; Moir et al. 2000b; Vaughan et al. 2001; González-Cruz et al. 2018). Recent work has not only highlighted how chromosomes condense into their mitotic configuration, but also how these regions reorganize into their interphase configuration postmitosis (Gibcus et al. 2018; Zhang et al. 2019). These studies focused on chromatin configuration as measured by Hi-C and it is notable that chromatin A/B compartment organization precedes TAD organization, and that compartmentalization intensifies as the cells proceed further into interphase. Given that LADs are highly enriched in the B compartment, such findings highlight a dynamic process of organization of LADs postmitosis. In a clever study using an antibody variant of DamID technology called pA-DamID on synchronized cells, dynamics of LAD organization after mitotic exit were revealed (van Schaik et al. 2020). LADs organize to the nuclear lamina in early G1, with more telomere-proximal LADs interacting first, followed by more centromere-proximal LADs. Because these studies were done in human cells, this stepwise association of different chromosomal positions suggests that the orientation of the chromosomes postmitosis influences LAD organization to the lamina. These associations generally increase as cells progress further into interphase, in agreement with the increased A/B compartmentalization noted in the earlier Hi-C studies. The relationship between LAD organization and lamin reorganization after mitosis remains to be determined.

LAMINOPATHIES

Disruption of the unique protein environment of the nuclear lamina often leads to disease (Table 1). The diseases that result from mutations in LMNA, or other NET or LAP proteins that heavily interact with lamins, are collectively termed "laminopathies" or "envelopathies." Because of differential expression of lamins during development and in different cell types along with the complex and cell-type-specific interactomes (NETs and LAPs), laminopathies display a wide breadth of phenotypes. Most laminopathies are autosomal dominant and generally cause late-onset degeneration of mesenchymal-derived cells (Wong et al. 2021a). Nearly 500 disease-causing mutations have mapped to the *LMNA* gene, each with its own specific phenotype, and many of these mutations are dominant. A variety of models have been suggested to explain the variety of cell- and tissue-specific phenotypes. Indeed, it appears that disruption of lamins or lamin-binding proteins affect numerous cellular functions, depending upon the particular mutation and cell type, including perturbations in mechanosensation and resilience, DNA repair, signaling pathways, interactions with specific transcription factor, altered interactions with NETs/LAPs, altered chromatin

Table 1. Diseases of the nuclear periphery

Gene	Disease/anomaly	Other notes
Primary laminopathies		
LMNA	Autosomal-dominant form of Emery–Dreifuss muscular dystrophy (AD-EDMD)	Affects striated muscle tissues
	Dilated cardiomyopathy (DCM)	
	Limb-girdle muscular dystrophy 1B (LMG1B).	
	Peripheral neuropathy (R298C), a recessive form of Charcot-Marie-Tooth disease (AR-CMT2A)	
	Dunnigan-type familial partial lipodystrophy (FPLD2)	Affects fat distribution and skeletal development
	Mandibuloacral dysplasia (MAD)	
	Hutchinson–Gilford progeria syndrome (HGPS)	Premature aging
	Atypical Werner's syndrome	
LMNB1	Adult-onset autosomal-dominant leukodystrophy (ADLD)	Neural
	Ataxia telangiectasia	
Gene	Function	Disease/anomaly
Secondary envelopathies		
Emerin (EMD)	NE associated, may regulate β-catenin nuclear entry and MKL1 nuclear localization	X-linked Emery–Dreifuss muscular dystrophy
Man1 (LEMD3)	Regulates TGF-β signaling by modulating Smad phosphorylation	Buschke–Ollendorff syndrome; excessive bone nodule formation, required for development of the vascular system
Lap1/Traf3/TOR1AIP1	Interacting protein with Torsin, LULL1, and emerin	Myopathy exacerbated by EMD loss
LEMD2/Net25	Chromatin organization MAP/AKT signaling	Homozygote nulls embryonic lethal progeroid; symptoms result from missense mutations
Lap2	Chromatin organization, telomere maintenance	Dilated cardiomyopathy, reduced epidermal proliferation, ameliorates LMNA-induced MD/DCM
Torsin	AAA$^+$ATPAse interacts closely with Lap1	DYT dystonia in CNS, steatohepatitis
Lamin B receptor (LBR)	Multifunctional-reduced sterol reductase activity, heterochromatin organization	Pelger–Huët anomaly, Greenberg dysplasia
Nesprin1 (SYNE1)	LINC complex tethers nucleus to cytoskeleton	Limb girdle muscular dystrophy; ARCA1 cerebellar ataxia required for nuclear migration during CNS development
Nesprin2 (SYNE2)	LINC complex tethers nucleus to cytoskeleton	EDMD variants required for nuclear migration during CNS development
Nesprin/Kash4	Interacts with MT motor proteins	Required for nuclear positioning in cochlear hair cells, mutations result in deafness

Continued

Table 1. *Continued*

Gene	Disease/anomaly	Other notes
SUN1	Anchors LINC complex to the INM, regulates miRNA synthesis during muscle regeneration	Ameliorates LMNA-induced DCM/MD, missense mutations associated with MD
BANF1	Postmitotic nuclear reassembly, chromatin organization	Nestor–Guillermo progeria

Data adapted from Wong and Stewart (2020).

state, and genome organization, none of which are mutually exclusive (Mewborn et al. 2010; Osmanagic-Myers et al. 2015; Vadrot et al. 2015; Worman and Schirmer 2015; Perovanovic et al. 2016; Le Dour et al. 2017; Briand and Collas 2018; Bianchi et al. 2018; and reviewed in Wong et al. 2021a).

CONCLUDING REMARKS

The multiple functions of lamins and the nuclear lamina make it very difficult to assess the impact of lamina abnormalities on diseases initiated by LMNA or NET mutations. In laminopathies, both gene dysregulation and the loss of nuclear integrity seem to play a role in observed phenotypes. Since this interface controls both the mechanical "fitness" of the nucleus (through LINCs, lamins, and heterochromatin scaffolding) as well as genome regulation, it will be interesting to determine how mechanical force relates to genome organization. In addition, while it is clear that lamina association frequently represses genes, how LADs are mechanistically maintained at and/or reorganized to the nuclear periphery during interphase and particularly during mitotic exit are only recently getting unraveled. Systematic study of how candidate factors (including lamin isotypes, chromatin interactors, NETs, and mechanical inputs) influence LAD and lamina organization during mitosis to G1 may help provide more definitive answers on the role of these regulators in lamina and LAD organization and regulation, and the consequences of their disruption in disease.

ACKNOWLEDGMENTS

Funding sources were NIHR01GM132427 to K.L.R. and A.J.M-P. and NIH R25GM109441 to A.J.M.-P.

REFERENCES

Aebi U, Cohn J, Buhle L, Gerace L. 1986. The nuclear lamina is a meshwork of intermediate-type filaments. *Nature* **323:** 560–564. doi:10.1038/323560a0

Alsheimer M, von Glasenapp E, Hock R, Benavente R. 1999. Architecture of the nuclear periphery of rat pachytene spermatocytes: distribution of nuclear envelope proteins in relation to synaptonemal complex attachment sites. *Mol Biol Cell* **10:** 1235–1245. doi:10.1091/mbc.10.4.1235

Andrés V, González JM. 2009. Role of A-type lamins in signaling, transcription, and chromatin organization. *J Cell Biol* **187:** 945–957. doi:10.1083/jcb.200904124

Aymard F, Bugler B, Schmidt CK, Guillou E, Caron P, Briois S, Iacovoni JS, Daburon V, Miller KM, Jackson SP, et al. 2014. Transcriptionally active chromatin recruits homologous recombination at DNA double-strand breaks. *Nat Struct Mol Biol* **21:** 366–374. doi:10.1038/nsmb.2796

Barski A, Cuddapah S, Cui K, Roh TY, Schones DE, Wang Z, Wei G, Chepelev I, Zhao K. 2007. High-resolution profiling of histone methylations in the human genome. *Cell* **129:** 823–837. doi:10.1016/j.cell.2007.05.009

Bártová E, Krejcí J, Harnicarová A, Galiová G, Kozubek S. 2008. Histone modifications and nuclear architecture: a review. *J Histochem Cytochem* **56:** 711–721. doi:10.1369/jhc.2008.951251

Belaadi N, Millon-Frémillon A, Aureille J, Guilluy C. 2018. Analyzing mechanotransduction through the LINC complex in isolated nuclei. *The LINC Complex* **1840:** 73–80. doi:10.1007/978-1-4939-8691-0_7

Belaghzal H, Dekker J, Gibcus JH. 2017. Hi-C 2.0: an optimized Hi-C procedure for high-resolution genome-wide mapping of chromosome conformation. *Methods* **123:** 56–65. doi:10.1016/j.ymeth.2017.04.004

Ben-Harush K, Wiesel N, Frenkiel-Krispin D, Moeller D, Soreq E, Aebi U, Herrmann H, Gruenbaum Y, Medalia O. 2009. The supramolecular organization of the *C. elegans* nuclear lamin filament. *J Mol Biol* **386:** 1392–1402. doi:10.1016/j.jmb.2008.12.024

Berger R, Theodor L, Shoham J, Gokkel E, Brok-Simoni F, Avraham KB, Copeland NG, Jenkins NA, Rechavi G, Simon AJ. 1996. The characterization and localization of the mouse thymopoietin/lamina-associated polypeptide 2 gene and its alternatively spliced products. *Genome Res* **6:** 361–370. doi:10.1101/gr.6.5.361

Berk JM, Tifft KE, Wilson KL. 2013. The nuclear envelope LEM-domain protein emerin. *Nucleus* **4:** 298–314. doi:10.4161/nucl.25751

Bian Q, Khanna N, Alvikas J, Belmont AS. 2013. β-Globin *cis*-elements determine differential nuclear targeting through epigenetic modifications. *J Cell Biol* **203:** 767–783. doi:10.1083/jcb.201305027

Bian Q, Anderson EC, Yang Q, Meyer BJ. 2020. Histone H3K9 methylation promotes formation of genome compartments in *Caenorhabditis elegans* via chromosome compaction and perinuclear anchoring. *Proc Natl Acad Sci* **117:** 11459–11470. doi:10.1073/pnas.2002068117

Bianchi A, Manti PG, Lucini F, Lanzuolo C. 2018. Mechanotransduction, nuclear architecture and epigenetics in emery dreifuss muscular dystrophy: tous pour un, un pour tous. *Nucleus* **9:** 321–335. doi:10.1080/19491034.2018.1460044

Boettiger A, Murphy S. 2020. Advances in chromatin imaging at kilobase-scale resolution. *Trends Genet* **36:** 273–287. doi:10.1016/j.tig.2019.12.010

Boulay G, Dubuissez M, Van Rechem C, Forget A, Helin K, Ayrault O, Leprince D. 2012. Hypermethylated in cancer 1 (HIC1) recruits polycomb repressive complex 2 (PRC2) to a subset of its target genes through interaction with human polycomb-like (hPCL) proteins. *J Biol Chem* **287:** 10509–10524. doi:10.1074/jbc.M111.320234

Brachner A, Reipert S, Foisner R, Gotzmann J. 2005. LEM2 is a novel MAN1-related inner nuclear membrane protein associated with A-type lamins. *J Cell Sci* **118:** 5797–5810. doi:10.1242/jcs.02701

Briand N, Collas P. 2018. Laminopathy-causing lamin A mutations reconfigure lamina-associated domains and local spatial chromatin conformation. *Nucleus* **9:** 216–226. doi:10.1080/19491034.2018.1449498

Briand N, Collas P. 2020. Lamina-associated domains: peripheral matters and internal affairs. *Genome Biol* **21:** 85. doi:10.1186/s13059-020-02003-5

Buchwalter A, Kaneshiro JM, Hetzer MW. 2019. Coaching from the sidelines: the nuclear periphery in genome regulation. *Nat Rev Genet* **20:** 39–50. doi:10.1038/s41576-018-0063-5

Burke B, Ellenberg J. 2002. Remodelling the walls of the nucleus. *Nat Rev Mol Cell Biol* **3:** 487–497. doi:10.1038/nrm860

Cabianca DS, Muñoz-Jiménez C, Kalck V, Gaidatzis D, Padeken J, Seeber A, Askjaer P, Gasser SM. 2019. Active chromatin marks drive spatial sequestration of heterochromatin in *C. elegans* nuclei. *Nature* **569:** 734–739. doi:10.1038/s41586-019-1243-y

Cai M, Huang Y, Ghirlando R, Wilson KL, Craigie R, Clore GM. 2001. Solution structure of the constant region of nuclear envelope protein LAP2 reveals two LEM-domain structures: one binds BAF and the other binds DNA. *EMBO J* **20:** 4399–4407. doi:10.1093/emboj/20.16.4399

Cai M, Huang Y, Suh JY, Louis JM, Ghirlando R, Craigie R, Clore GM. 2007. Solution NMR structure of the barrier-to-autointegration factor-Emerin complex. *J Biol Chem* **282:** 14525–14535. doi:10.1074/jbc.M700576200

Chalut KJ, Höpfler M, Lautenschläger F, Boyde L, Chan CJ, Ekpenyong A, Martinez-Arias A, Guck J. 2012. Chromatin decondensation and nuclear softening accompany Nanog downregulation in embryonic stem cells. *Biophys J* **103:** 2060–2070. doi:10.1016/j.bpj.2012.10.015

Chang W, Worman HJ, Gundersen GG. 2015. Accessorizing and anchoring the LINC complex for multifunctionality. *J Cell Biol* **208:** 11–22. doi:10.1083/jcb.201409047

Chen IHB, Huber M, Guan T, Bubeck A, Gerace L. 2006. Nuclear envelope transmembrane proteins (NETs) that are up-regulated during myogenesis. *BMC Cell Biol* **7:** 38. doi:10.1186/1471-2121-7-38

Chen S, Luperchio TR, Wong X, Doan EB, Byrd AT, Roy Choudhury K, Reddy KL, Krangel MS. 2018a. A lamina-associated domain border governs nuclear lamina interactions, transcription, and recombination of the Tcrb locus. *Cell Rep* **25:** 1729–1740.e6. doi:10.1016/j.celrep.2018.10.052

Chen Y, Zhang Y, Wang Y, Zhang L, Brinkman EK, Adam SA, Goldman R, van Steensel B, Ma J, Belmont AS. 2018b. Mapping 3D genome organization relative to nuclear compartments using TSA-Seq as a cytological ruler. *J Cell Biol* **217:** 4025–4048. doi: 10.1083/jcb.201807108.

Coffinier C, Jung HJ, Nobumori C, Chang S, Tu Y, Barnes RH II, Yoshinaga Y, de Jong PJ, Vergnes L, Reue K, et al. 2011. Deficiencies in lamin B1 and lamin B2 cause neurodevelopmental defects and distinct nuclear shape abnormalities in neurons. *Mol Biol Cell* **22:** 4683–4693. doi:10.1091/mbc.e11-06-0504

Cremer T, Cremer C. 2001. Chromosome territories, nuclear architecture and gene regulation in mammalian cells. *Nat Rev Genet* **2:** 292–301. doi:10.1038/35066075

Cremer T, Cremer M. 2010. Chromosome territories. *Cold Spring Harb Perspect Biol* **2:** a003889. doi:10.1101/cshperspect.a003889

Das PM, Ramachandran K, vanWert J, Singal R. 2004. Chromatin immunoprecipitation assay. *BioTechniques* **37:** 961–969. doi:10.2144/04376RV01

DeBoy E, Puttaraju M, Jailwala P, Kasoji M, Cam M, Misteli T. 2017. Identification of novel RNA isoforms of *LMNA*. *Nucleus* **8:** 573–582. doi:10.1080/19491034.2017.1348449

Dechat T, Gotzmann J, Stockinger A, Harris CA, Talle MA, Siekierka JJ, Foisner R. 1998. Detergent-salt resistance of LAP2α in interphase nuclei and phosphorylation-dependent association with chromosomes early in nuclear assembly implies functions in nuclear structure dynamics. *EMBO J* **17:** 4887–4902. doi:10.1093/emboj/17.16.4887

Dechat T, Gajewski A, Korbei B, Gerlich D, Daigle N, Haraguchi T, Furukawa K, Ellenberg J, Foisner R. 2004. LAP2α and BAF transiently localize to telomeres and specific regions on chromatin during nuclear assembly. *J Cell Sci* **117:** 6117–6128. doi:10.1242/jcs.01529

Dechat T, Pfleghaar K, Sengupta K, Shimi T, Shumaker DK, Solimando L, Goldman RD. 2008. Nuclear lamins: major factors in the structural organization and function of the nucleus and chromatin. *Genes Dev* **22:** 832–853. doi:10.1101/gad.1652708

Dechat T, Adam SA, Taimen P, Shimi T, Goldman RD. 2010. Nuclear lamins. *Cold Spring Harb Perspect Biol* **2:** a000547. doi:10.1101/cshperspect.a000547

Dekker J. 2002. Capturing chromosome conformation. *Science* **295:** 1306–1311. doi:10.1126/science.1067799

de Las Heras JI, Meinke P, Batrakou DG, Srsen V, Zuleger N, Kerr AR, Schirmer EC. 2013. Tissue specificity in the nuclear envelope supports its functional complexity. *Nucleus* **4:** 460–477. doi:10.4161/nucl.26872 de Las Heras JI, Zuleger N, Batrakou DG, Czapiewski R, Kerr ARW, Schirmer EC. 2017. Tissue-specific NETs alter genome organization and regulation even in a heterologous system. *Nucleus* **8:** 81–97. doi:10.1080/19491034.2016.1261230

Demmerle J, Koch AJ, Holaska JM. 2012. The nuclear envelope protein emerin binds directly to histone deacetylase 3 (HDAC3) and activates HDAC3 activity. *J Biol Chem* **287:** 22080. doi:10.1074/jbc.M111.325308

Dialynas G, Speese S, Budnik V, Geyer PK, Wallrath LL. 2010. The role of *Drosophila* lamin C in muscle function and gene expression. *Development* **137:** 3067–3077. doi:10.1242/dev.048231

Dixon JR, Selvaraj S, Yue F, Kim A, Li Y, Shen Y, Hu M, Liu JS, Ren B. 2012. Topological domains in mammalian genomes identified by analysis of chromatin interactions. *Nature* **485:** 376–380. doi:10.1038/nature11082

Dorner D, Vlcek S, Foeger N, Gajewski A, Makolm C, Gotzmann J, Hutchison CJ, Foisner R. 2006. Lamina-associated polypeptide 2α regulates cell cycle progression and differentiation via the retinoblastoma-E2F pathway. *J Cell Biol* **173:** 83–93. doi:10.1083/jcb.200511149

Dreesen O, Chojnowski A, Ong PF, Zhao TY, Common JE, Lunny D, Lane EB, Lee SJ, Vardy LA, Stewart CL, et al. 2013. Lamin B1 fluctuations have differential effects on cellular proliferation and senescence. *J Cell Biol* **200:** 605–617. doi:10.1083/jcb.201206121

Dunlevy KL, Medvedeva V, Wilson JE, Hoque M, Pellegrin T, Maynard A, Kremp MM, Wasserman JS, Poleshko A, Katz RA. 2020. The PRR14 heterochromatin tether encodes modular domains that mediate and regulate nuclear lamina targeting. *J Cell Sci* **133:** jcs240416. doi:10.1242/jcs.240416

Duong NT, Morris GE, Lam LT, Zhang Q, Sewry CA, Shanahan CM, Holt I. 2014. Nesprins: tissue-specific expression of epsilon and other short isoforms. *PLoS ONE* **9:** e94380. doi:10.1371/journal.pone.0094380

Ellenberg J. 2013. *Dynamics of nuclear envelope proteins during the cell cycle in mammalian cells.* Landes Bioscience, Austin, TX.

Falk M, Feodorova Y, Naumova N, Imakaev M, Lajoie BR, Leonhardt H, Joffe B, Dekker J, Fudenberg G, Solovei I, et al. 2019. Heterochromatin drives compartmentalization of inverted and conventional nuclei. *Nature* **570:** 395–399. doi:10.1038/s41586-019-1275-3

Finlan LE, Sproul D, Thomson I, Boyle S, Kerr E, Perry P, Ylstra B, Chubb JR, Bickmore WA. 2008. Recruitment to the nuclear periphery can alter expression of genes in human cells. *PLoS Genet* **4:** e1000039. doi:10.1371/journal.pgen.1000039

Foisner R. 2003. Cell cycle dynamics of the nuclear envelope. *ScientificWorldJournal* **3:** 1–20. doi:10.1100/tsw.2003.06

Fraser J, Ferrai C, Chiariello AM, Schueler M, Rito T, Laudanno G, Barbieri M, Moore BL, Kraemer DCA, Aitken S, et al. 2015. Hierarchical folding and reorganization of chromosomes are linked to transcriptional changes in cellular differentiation. *Mol Syst Biol* **11:** 852. doi:10.15252/msb.20156492

Furukawa K. 1999. LAP2 binding protein 1 (L2BP1/BAF) is a candidate mediator of LAP2-chromatin interaction. *J Cell Sci* **112:** 2485–2492. doi:10.1242/jcs.112.15.2485

Furukawa K, Hotta Y. 1993. cDNA cloning of a germ cell specific lamin B3 from mouse spermatocytes and analysis of its function by ectopic expression in somatic cells. *EMBO J* **12:** 97–106. doi:10.1002/j.1460-2075.1993.tb05635.x

Furukawa K, Inagaki H, Hotta Y. 1994. Identification and cloning of an mRNA coding for a germ cell-specific A-type lamin in mice. *Exp Cell Res* **212:** 426–430. doi:10.1006/excr.1994.1164

Furusawa T, Rochman M, Taher L, Dimitriadis EK, Nagashima K, Anderson S, Bustin M. 2015. Chromatin decompaction by the nucleosomal binding protein HMGN5 impairs nuclear sturdiness. *Nat Commun* **6:** 6138. doi:10.1038/ncomms7138

Georgatos SD, Pyrpasopoulou A, Theodoropoulos PA. 1997. Nuclear envelope breakdown in mammalian cells involves stepwise lamina disassembly and microtubule-drive deformation of the nuclear membrane. *J Cell Sci* **110:** 2129–2140. doi:10.1242/jcs.110.17.2129

Gerace L, Blobel G. 1980. The nuclear envelope lamina is reversibly depolymerized during mitosis. *Cell* **19:** 277–287. doi:10.1016/0092-8674(80)90409-2

Gerace L, Foisner R. 1994. Integral membrane proteins and dynamic organization of the nuclear envelope. *Trends Cell Biol* **4:** 127–131. doi:10.1016/0962-8924(94)90067-1

Gerace L, Huber MD. 2012. Nuclear lamina at the crossroads of the cytoplasm and nucleus. *J Struct Biol* **177:** 24–31. doi:10.1016/j.jsb.2011.11.007

Gesson K, Rescheneder P, Skoruppa MP, von Haeseler A, Dechat T, Foisner R. 2016. A-type lamins bind both hetero- and euchromatin, the latter being regulated by lamina-associated polypeptide 2α. *Genome Res* **26:** 462–473. doi:10.1101/gr.196220.115

Gibbs-Seymour I, Markiewicz E, Bekker-Jensen S, Mailand N, Hutchison CJ. 2015. Lamin A/C-dependent interaction with 53BP1 promotes cellular responses to DNA damage. *Aging Cell* **14:** 162–169. doi:10.1111/acel.12258

Gibcus JH, Samejima K, Goloborodko A, Samejima I, Naumova N, Nuebler J, Kanemaki MT, Xie L, Paulson JR, Earnshaw WC, et al. 2018. A pathway for mitotic chromosome formation. *Science* **359:** eaao6135. doi:10.1126/science.aao6135

Gieffers C, Krohne G. 1991. In vitro reconstitution of recombinant lamin A and a lamin A mutant lacking the carboxy-terminal tail. *Eur J Cell Biol* **55:** 191–199.

González-Cruz RD, Dahl KN, Darling EM. 2018. The emerging role of lamin C as an important LMNA isoform in mechanophenotype. *Front Cell Dev Biol* **6:** 151. doi:10.3389/fcell.2018.00151

Gonzalez-Sandoval A, Gasser SM. 2016. Mechanism of chromatin segregation to the nuclear periphery in *C. elegans* embryos. *Worm* **5:** e1190900. doi:10.1080/21624054.2016.1190900

Gonzalez-Sandoval A, Towbin BD, Kalck V, Cabianca DS, Gaidatzis D, Hauer MH, Geng L, Wang L, Yang T, Wang X, et al. 2015. Perinuclear anchoring of H3K9-methylated chromatin stabilizes induced cell fate in *C. elegans* embryos. *Cell* **163:** 1333–1347. doi:10.1016/j.cell.2015.10.066

Greil F, Moorman C, van Steensel B. 2006. DamID: mapping of in vivo protein-genome interactions using tethered DNA adenine methyltransferase. *Methods Enzymol* **410:** 342–359. doi:10.1016/S0076-6879(06)10016-6

Gruenbaum Y, Foisner R. 2015. Lamins: nuclear intermediate filament proteins with fundamental functions in nuclear mechanics and genome regulation. *Annu Rev Biochem* **84:** 131–164. doi:10.1146/annurev-biochem-060614-034115

Gruenbaum Y, Medalia O. 2015. Lamins: the structure and protein complexes. *Curr Opin Cell Biol* **32:** 7–12. doi:10.1016/j.ceb.2014.09.009

Guelen L, Pagie L, Brasset E, Meuleman W, Faza MB, Talhout W, Eussen BH, de Klein A, Wessels L, de Laat W, et al. 2008. Domain organization of human chromosomes revealed by mapping of nuclear lamina interactions. *Nature* **453:** 948–951. doi:10.1038/nature06947

Harada T, Swift J, Irianto J, Shin J-W, Spinler KR, Athirasala A, Diegmiller R, Dingal PCDP, Ivanovska IL, Discher DE. 2014. Nuclear lamin stiffness is a barrier to 3D migration, but softness can limit survival. *J Cell Biol* **204:** 669–682. doi:10.1083/jcb.201308029

Haraguchi T, Koujin T, Segura-Totten M, Lee KK, Matsuoka Y, Yoneda Y, Wilson KL, Hiraoka Y. 2001. BAF is required for emerin assembly into the reforming nuclear envelope. *J Cell Sci* **114:** 4575–4585. doi:10.1242/jcs.114.24.4575

Harr JC, Luperchio TR, Wong X, Cohen E, Wheelan SJ, Reddy KL. 2015. Directed targeting of chromatin to the nuclear lamina is mediated by chromatin state and A-type lamins. *J Cell Biol* **208:** 33–52. doi:10.1083/jcb.201405110

Harr JC, Gonzalez-Sandoval A, Gasser SM. 2016. Histones and histone modifications in perinuclear chromatin anchoring: from yeast to man. *EMBO Rep* **17:** 139–155. doi:10.15252/embr.201541809

Harr JC, Schmid CD, Muñoz-Jiménez C, Romero-Bueno R, Kalck V, Gonzalez-Sandoval A, Hauer MH, Padeken J, Askjaer P, Mattout A, et al. 2020. Loss of an H3K9me anchor rescues laminopathy-linked changes in nuclear organization and muscle function in an Emery–Dreifuss muscular dystrophy model. *Genes Dev* **34:** 560–579. doi:10.1101/gad.332213.119

Harris CA, Andryuk PJ, Cline S, Chan HK, Natarajan A, Siekierka JJ, Goldstein G. 1994. Three distinct human thymopoietins are derived from alternatively spliced mRNAs. *Proc Natl Acad Sci* **91:** 6283–6287. doi:10.1073/pnas.91.14.6283

Heitlinger E, Peter M, Lustig A, Villiger W, Nigg EA, Aebi U. 1992. The role of the head and tail domain in lamin structure and assembly: analysis of bacterially expressed chicken lamin A and truncated B2 lamins. *J Struct Biol* **108:** 74–91. doi:10.1016/1047-8477(92)90009-Y

Herrmann H, Aebi U. 2004. Intermediate filaments: molecular structure, assembly mechanism, and integration into functionally distinct intracellular scaffolds. *Annu Rev Biochem* **73:** 749–789. doi:10.1146/annurev.biochem.73.011303.073823

Hirano Y, Hizume K, Kimura H, Takeyasu K, Haraguchi T, Hiraoka Y. 2012. Lamin B receptor recognizes specific modifications of histone H4 in heterochromatin formation. *J Biol Chem* **287:** 42654–42663. doi:10.1074/jbc.M112.397950

Ho CY, Lammerding J. 2012. Lamins at a glance. *J Cell Sci* **125:** 2087–2093. doi:10.1242/jcs.087288

Höger TH, Grund C, Franke WW, Krohne G. 1991. Immunolocalization of lamins in the thick nuclear lamina of human synovial cells. *Eur J Cell Biol* **54:** 150–156.

Holmer L, Worman HJ. 2001. Inner nuclear membrane proteins: functions and targeting. *Cell Mol Life Sci* **58:** 1741–1747. doi:10.1007/PL00000813

Ikegami K, Secchia S, Almakki O, Lieb JD, Moskowitz IP. 2020. Phosphorylated lamin A/C in the nuclear interior binds active enhancers associated with abnormal transcription in progeria. *Dev Cell* **52:** 699–713.e11. doi:10.1016/j.devcel.2020.02.011

Izumi M, Vaughan OA, Hutchison CJ, Gilbert DM. 2000. Head and/or CaaX domain deletions of lamin proteins disrupt preformed lamin A and C but not lamin B structure in mammalian cells. *Mol Biol Cell* **11:** 4323–4337. doi:10.1091/mbc.11.12.4323

Jahn D, Schramm S, Schnölzer M, Heilmann CJ, de Koster CG, Schütz W, Benavente R, Alsheimer M. 2012. A truncated lamin A in the Lmna$^{-/-}$ mouse line: implications for the understanding of laminopathies. *Nucleus* **3:** 463–474. doi:10.4161/nucl.21676

Jung HJ, Coffinier C, Choe Y, Beigneux AP, Davies BS, Yang SH, Barnes RH II, Hong J, Sun T, Pleasure SJ, et al. 2012. Regulation of prelamin A but not lamin C by miR-9, a brain-specific microRNA. *Proc Natl Acad Sci* **109:** E423–E431. doi:10.1073/pnas.1111780109

Kempfer R, Pombo A. 2020. Methods for mapping 3D chromosome architecture. *Nat Rev Genet* **21:** 207–226. doi:10.1038/s41576-019-0195-2

Kim DH, Wirtz D. 2015. Cytoskeletal tension induces the polarized architecture of the nucleus. *Biomaterials* **48:** 161–172. doi:10.1016/j.biomaterials.2015.01.023

Kim Y, Zheng X, Zheng Y. 2013. Proliferation and differentiation of mouse embryonic stem cells lacking all lamins. *Cell Res* **23:** 1420–1423. doi:10.1038/cr.2013.118

Kim DI, Birendra KC, Roux KJ. 2015. Making the LINC: SUN and KASH protein interactions. *Biol Chem* **396:** 295–310. doi:10.1515/hsz-2014-0267

Kim DI, Jensen SC, Roux KJ. 2016. Identifying protein-protein associations at the nuclear envelope with BioID. *Methods Mol Biol* **1411:** 133–146. doi:10.1007/978-1-4939-3530-7_8

Kind J, van Steensel B. 2014. Stochastic genome-nuclear lamina interactions: modulating roles of lamin A and BAF. *Nucleus* **5:** 124–130. doi:10.4161/nucl.28825

Kind J, Pagie L, Ortabozkoyun H, Boyle S, de Vries SS, Janssen H, Amendola M, Nolen LD, Bickmore WA, van Steensel B. 2013. Single-cell dynamics of genome-nuclear lamina interactions. *Cell* **153:** 178–192. doi:10.1016/j.cell.2013.02.028

Kind J, Pagie L, de Vries SS, Nahidiazar L, Dey SS, Bienko M, Zhan Y, Lajoie B, de Graaf CA, Amendola M, et al. 2015. Genome-wide maps of nuclear lamina interactions in single human cells. *Cell* **163:** 134–147. doi:10.1016/j.cell.2015.08.040

Klapper M, Exner K, Kempf A, Gehrig C, Stuurman N, Fisher PA, Krohne G. 1997. Assembly of A- and B-type lamins studied in vivo with the baculovirus system. *J Cell Sci* **110:** 2519–2532. doi:10.1242/jcs.110.20.2519

Kochin V, Shimi T, Torvaldson E, Adam SA, Goldman A, Pack CG, Melo-Cardenas J, Imanishi SY, Goldman RD, Eriksson JE. 2014. Interphase phosphorylation of lamin A. *J Cell Sci* **127:** 2683–2696. doi:10.1242/jcs.141820

Korfali N, Wilkie GS, Swanson SK, Srsen V, Batrakou DG, Fairley EAL, Malik P, Zuleger N, Goncharevich A, de las Heras J, et al. 2010. The leukocyte nuclear envelope proteome varies with cell activation and contains novel transmembrane proteins that affect genome architecture. *Mol Cell Proteomics* **9:** 2571–2585. doi:10.1074/mcp.M110.002915

Korfali N, Wilkie GS, Swanson SK, Srsen V, de Las Heras J, Batrakou DG, Malik P, Zuleger N, Kerr AR, Florens L, et al. 2012. The nuclear envelope proteome differs notably between tissues. *Nucleus* **3:** 552–564.

Kosak ST, Skok JA, Medina KL, Riblet R, Le Beau MM, Fisher AG, Singh H. 2002. Subnuclear compartmentalization of immunoglobulin loci during lymphocyte development. *Science* **296:** 158–162. doi:10.1126/science.1068768

Krimm I, Östlund C, Gilquin B, Couprie J, Hossenlopp P, Mornon J-P, Bonne G, Courvalin J-C, Worman HJ, Zinn-Justin S. 2002. The Ig-like structure of the C-terminal domain of lamin A/C, mutated in muscular dystrophies, cardiomyopathy, and partial lipodystrophy. *Structure* **10:** 811–823. doi:10.1016/S0969-2126(02)00777-3

Kumaran RI, Muralikrishna B, Parnaik VK. 2002. Lamin A/C speckles mediate spatial organization of splicing factor compartments and RNA polymerase II transcription. *J Cell Biol* **159:** 783–793. doi:10.1083/jcb.200204149

Larson AG, Elnatan D, Keenen MM, Trnka MJ, Johnston JB, Burlingame AL, Agard DA, Redding S, Narlikar GJ. 2017. Liquid droplet formation by HP1α suggests a role for phase separation in heterochromatin. *Nature* **547:** 236–240. doi:10.1038/nature22822

Le Dour C, Wu W, Béréziat V, Capeau J, Vigouroux C, Worman HJ. 2017. Extracellular matrix remodeling and transforming growth factor-β signaling abnormalities induced by lamin A/C variants that cause lipodystrophy. *J Lipid Res* **58:** 151–163. doi:10.1194/jlr.M071381

Lee KK, Wilson KL. 2004. All in the family: evidence for four new LEM-domain proteins Lem2 (NET-25), Lem3, Lem4 and Lem5 in the human genome. *Symp Soc Exp Biol* **56:** 329–339.

Lee KK, Haraguchi T, Lee RS, Koujin T, Hiraoka Y, Wilson KL. 2001. Distinct functional domains in emerin bind lamin A and DNA-bridging protein BAF. *J Cell Sci* **114:** 4567–4573. doi:10.1242/jcs.114.24.4567

Leemans C, van der Zwalm MCH, Brueckner L, Comoglio F, van Schaik T, Pagie L, van Arensbergen J, van Steensel B. 2019. Promoter-intrinsic and local chromatin features determine gene repression in LADs. *Cell* **177:** 852–864.e14. doi:10.1016/j.cell.2019.03.009

Legartová S, Stixová L, Laur O, Kozubek S, Sehnalová P, Bártová E. 2014. Nuclear structures surrounding internal lamin invaginations. *J Cell Biochem* **115:** 476–487. doi:10.1002/jcb.24681

Li BX, Chen J, Chao B, Zheng Y, Xiao X. 2018. A lamin-binding ligand inhibits homologous recombination repair of DNA double-strand breaks. *ACS Cent Sci* **4:** 1201–1210.

Lieberman-Aiden E, van Berkum NL, Williams L, Imakaev M, Ragoczy T, Telling A, Amit I, Lajoie BR, Sabo PJ, Dorschner MO, et al. 2009. Comprehensive mapping of long-range interactions reveals folding principles of the human genome. *Science* **326:** 289–293. doi:10.1126/science.1181369

Lin F, Worman HJ. 1993. Structural organization of the human gene encoding nuclear lamin A and nuclear lamin C. *J Biol Chem* **268:** 16321–16326. doi:10.1016/S0021-9258(19)85424-8

Lin F, Blake DL, Callebaut I, Skerjanc IS, Holmer L, McBurney MW, Paulin-Levasseur M, Worman HJ. 2000. MAN1, an inner nuclear membrane protein that shares the LEM domain with lamina-associated polypeptide 2 and emerin. *J Biol Chem* **275:** 4840–4847. doi:10.1074/jbc.275.7.4840

Lombardi ML, Lammerding J. 2011. Keeping the LINC: the importance of nucleocytoskeletal coupling in intracellular force transmission and cellular function. *Biochem Soc Trans* **39:** 1729–1734. doi:10.1042/BST20110686

Lottersberger F, Karssemeijer RA, Dimitrova N, de Lange T. 2015. 53BP1 and the LINC complex promote microtubule-dependent DSB mobility and DNA repair. *Cell* **163:** 880–893. doi:10.1016/j.cell.2015.09.057

Luperchio TR, Sauria MEG, Hoskins VE, Wong X, DeBoy E, Gaillard M-C, Tsang P, Pekrun K, Ach RA, Yamada NA, et al. 2018. The repressive genome compartment is established early in the cell cycle before forming the lamina associated domains. bioRxiv doi:10.1101/481598v1

Luxton GWG, Gomes ER, Folker ES, Vintinner E, Gundersen GG. 2010. Linear arrays of nuclear envelope proteins harness retrograde actin flow for nuclear movement. *Science* **329:** 956–959. doi:10.1126/science.1189072

Ma S, Zhang Y. 2020. Profiling chromatin regulatory landscape: insights into the development of ChIP-seq and ATAC-seq. *Molecular Biomedicine* **1:** 9. doi:10.1186/s43556-020-00009-w

Ma L, Tsai MY, Wang S, Lu B, Chen R, Yates JR III, Zhu X, Zheng Y. 2009. Requirement for nudel and dynein for assembly of the lamin B spindle matrix. *Nat Cell Biol* **11:** 247–256. doi:10.1038/ncb1832

Machiels BM, Zorenc AH, Endert JM, Kuijpers HJ, van Eys GJ, Ramaekers FC, Broers JL. 1996. An alternative splicing product of the lamin A/C gene lacks exon 10. *J Biol Chem* **271:** 9249–9253. doi:10.1074/jbc.271.16.9249

Maeshima K, Tamura S, Hansen JC, Itoh Y. 2020. Fluid-like chromatin: toward understanding the real chromatin organization present in the cell. *Curr Opin Cell Biol* **64:** 77–89. doi:10.1016/j.ceb.2020.02.016

Maison C, Pyrpasopoulou A, Theodoropoulos PA, Georgatos SD. 1997. The inner nuclear membrane protein LAP1 forms a native complex with B-type lamins and partitions with spindle-associated mitotic vesicles. *EMBO J* **16:** 4839–4850. doi:10.1093/emboj/16.16.4839

Margalit A, Brachner A, Gotzmann J, Foisner R, Gruenbaum Y. 2007. Barrier-to-autointegration factor—a BAFfling little protein. *Trends Cell Biol* **17:** 202–208. doi:10.1016/j.tcb.2007.02.004

Marnef A, Finoux AL, Arnould C, Guillou E, Daburon V, Rocher V, Mangeat T, Mangeot PE, Ricci EP, Legube G. 2019. A cohesin/HUSH- and LINC-dependent pathway controls ribosomal DNA double-strand break repair. *Genes Dev* **33:** 1175–1190. doi:10.1101/gad.324012.119

Mattout A, Cabianca DS, Gasser SM. 2015. Chromatin states and nuclear organization in development—a view from the nuclear lamina. *Genome Biol* **16:** 174. doi:10.1186/s13059-015-0747-5

Maurer M, Lammerding J. 2019. The driving force: nuclear mechanotransduction in cellular function, fate, and disease. *Annu Rev Biomed Eng* **21:** 443–468. doi:10.1146/annurev-bioeng-060418-052139

Mehus AA, Anderson RH, Roux KJ. 2016. BioID identification of lamin-associated proteins. *Methods Enzymol* **569:** 3–22. doi:10.1016/bs.mie.2015.08.008

Méjat A, Misteli T. 2010. LINC complexes in health and disease. *Nucleus* **1:** 40–52. doi:10.4161/nucl.1.1.10530

Meuleman W, Peric-Hupkes D, Kind J, Beaudry JBB, Pagie L, Kellis M, Reinders M, Wessels L, van Steensel B. 2012. Constitutive nuclear lamina–genome interactions are highly conserved and associated with A/T-rich sequence. *Genome Res* **23:** 270–280. doi:10.1101/gr.141028.112

Mewborn SK, Puckelwartz MJ, Abuisneineh F, Fahrenbach JP, Zhang Y, MacLeod H, Dellefave L, Pytel P, Selig S, Labno CM, et al. 2010. Altered chromosomal positioning, compaction, and gene expression with a lamin A/C gene mutation. *PLoS ONE* **5:** e14342. doi:10.1371/journal.pone.0014342

Moir RD, Spann TP, Herrmann H, Goldman RD. 2000a. Disruption of nuclear lamin organization blocks the elongation phase of DNA replication. *J Cell Biol* **149:** 1179–1192. doi:10.1083/jcb.149.6.1179

Moir RD, Yoon M, Khuon S, Goldman RD. 2000b. Nuclear lamins A and B1: different pathways of assembly during nuclear envelope formation in living cells. *J Cell Biol* **151:** 1155–1168. doi:10.1083/jcb.151.6.1155

Naumova N, Imakaev M, Fudenberg G, Zhan Y, Lajoie BR, Mirny LA, Dekker J. 2013. Organization of the mitotic chromosome. *Science* **342:** 948–953. doi:10.1126/science.1236083

Nelson DM, Jaber-Hijazi F, Cole JJ, Robertson NA, Pawlikowski JS, Norris KT, Criscione SW, Pchelintsev NA, Piscitello D, Stong N, et al. 2016. Mapping H4K20me3 onto the chromatin landscape of senescent cells indicates a function in control of cell senescence and tumor suppression through preservation of genetic and epigenetic stability. *Genome Biol* **17:** 158. doi:10.1186/s13059-016-1017-x

Nora EP, Goloborodko A, Valton AL, Gibcus JH, Uebersohn A, Abdennur N, Dekker J, Mirny LA, Bruneau BG. 2017. Targeted degradation of CTCF decouples local insulation of chromosome domains from genomic compartmentalization. *Cell* **169:** 930–944.e22. doi:10.1016/j.cell.2017.05.004

Orian A, Abed M, Kenyagin-Karsenti D, Boico O. 2009. DamID: a methylation-based chromatin profiling approach. *Methods Mol Biol* **567:** 155–169. doi:10.1007/978-1-60327-414-2_11

Osmanagic-Myers S, Dechat T, Foisner R. 2015. Lamins at the crossroads of mechanosignaling. *Genes Dev* **29:** 225–237. doi:10.1101/gad.255968.114

Padeken J, Heun P. 2014. Nucleolus and nuclear periphery: velcro for heterochromatin. *Curr Opin Cell Biol* **28:** 54–60. doi:10.1016/j.ceb.2014.03.001

Peric-Hupkes D, Meuleman W, Pagie L, Bruggeman SWM, Solovei I, Brugman W, Gräf S, Flicek P, Kerkhoven RM, van Lohuizen M, et al. 2010. Molecular maps of the reorganization of genome-nuclear lamina interactions during differentiation. *Mol Cell* **38:** 603–613. doi:10.1016/j.molcel.2010.03.016

Perovanovic J, Dell'Orso S, Gnochi VF, Jaiswal JK, Sartorelli V, Vigouroux C, Mamchaoui K, Mouly V, Bonne G, Hoffman EP. 2016. Laminopathies disrupt epigenomic developmental programs and cell fate. *Sci Transl Med* **8:** 335ra58. doi:10.1126/scitranslmed.aad4991

Peter M, Kitten GT, Lehner CF, Vorburger K, Bailer SM, Maridor G, Nigg EA. 1989. Cloning and sequencing of cDNA clones encoding chicken lamins A and B1 and comparison of the primary structures of vertebrate A- and B-type lamins. *J Mol Biol* **208:** 393–404. doi:10.1016/0022-2836(89)90504-4

Phillips-Cremins JE, Sauria MEG, Sanyal A, Gerasimova TI, Lajoie BR, Bell JSK, Ong CT, Hookway TA, Guo C, Sun Y, et al. 2013. Architectural protein subclasses shape 3D organization of genomes during lineage commitment. *Cell* **153:** 1281–1295. doi:10.1016/j.cell.2013.04.053

Poleshko A, Katz RA. 2014. Specifying peripheral heterochromatin during nuclear lamina reassembly. *Nucleus* **5:** 32–39. doi:10.4161/nucl.28167

Poleshko A, Shah PP, Gupta M, Babu A, Morley MP, Manderfield LJ, Ifkovits JL, Calderon D, Aghajanian H, Sierra-Pagán JE, et al. 2017. Genome-nuclear lamina interactions regulate cardiac stem cell lineage restriction. *Cell* **171:** 573–587.e14. doi:10.1016/j.cell.2017.09.018

Polioudaki H, Kourmouli N, Drosou V, Bakou A, Theodoropoulos PA, Singh PB, Giannakouros T, Georgatos SD. 2001. Histones H3/H4 form a tight complex with the inner nuclear membrane protein LBR and heterochromatin protein 1. *EMBO Rep* **2:** 920–925. doi:10.1093/embo-reports/kve199

Pope BD, Ryba T, Dileep V, Yue F, Wu W, Denas O, Vera DL, Wang Y, Hansen RS, Canfield TK, et al. 2014. Topologically associating domains are stable units of replication-timing regulation. *Nature* **515:** 402–405. doi:10.1038/nature13986

Pugh GE, Coates PJ, Lane EB, Raymond Y, Quinlan RA. 1997. Distinct nuclear assembly pathways for lamins A and C lead to their increase during quiescence in Swiss 3T3 cells. *J Cell Sci* **110:** 2483–2493. doi:10.1242/jcs.110.19.2483

Rajgor D, Shanahan CM. 2013. Nesprins: from the nuclear envelope and beyond. *Expert Rev Mol Med* **15:** e5. doi:10.1017/erm.2013.6

Rao SSP, Huntley MH, Durand NC, Stamenova EK, Bochkov ID, Robinson JT, Sanborn AL, Machol I, Omer AD, Lander ES, et al. 2014. A 3D map of the human genome at kilobase resolution reveals principles of chromatin looping. *Cell* **159:** 1665–1680. doi:10.1016/j.cell.2014.11.021

Razafsky D, Wirtz D, Hodzic D. 2014. Nuclear envelope in nuclear positioning and cell migration. *Adv Exp Med Biol* **773:** 471–490. doi:10.1007/978-1-4899-8032-8_21

Reddy KL, Zullo JM, Bertolino E, Singh H. 2008. Transcriptional repression mediated by repositioning of genes to the nuclear lamina. *Nature* **452:** 243–247. doi:10.1038/nature06727

Redwood AB, Perkins SM, Vanderwaal RP, Feng Z, Biehl KJ, Gonzalez-Suarez I, Morgado-Palacin L, Shi W, Sage J, Roti-Roti JL, et al. 2011. A dual role for A-type lamins in DNA double-strand break repair. *Cell Cycle* **10:** 2549–2560. doi:10.4161/cc.10.15.16531

Robson MI, de Las Heras JI, Czapiewski R, Sivakumar A, Kerr ARW, Schirmer EC. 2017. Constrained release of lamina-associated enhancers and genes from the nuclear envelope during T-cell activation facilitates their association in chromosome compartments. *Genome Res* **27:** 1126–1138. doi:10.1101/gr.212308.116

Rothballer A, Kutay U. 2013. The diverse functional LINCs of the nuclear envelope to the cytoskeleton and chromatin. *Chromosoma* **122:** 415–429. doi:10.1007/s00412-013-0417-x

Salina D, Enarson P, Rattner JB, Burke B. 2003. Nup358 integrates nuclear envelope breakdown with kinetochore assembly. *J Cell Biol* **162:** 991–1001. doi:10.1083/jcb.200304080

Sanborn AL, Rao SSP, Huang S-C, Durand NC, Huntley MH, Jewett AI, Bochkov ID, Chinnappan D, Cutkosky A, Li J, et al. 2015. Chromatin extrusion explains key features of loop and domain formation in wild-type and engineered genomes. *Proc Natl Acad Sci* **112:** E6456–E6465. doi:10.1073/pnas.1518552112

Sasse B, Aebi U, Stuurman N. 1998. A tailless *Drosophila* lamin Dm0 fragment reveals lateral associations of dimers. *J Struct Biol* **123:** 56–66. doi:10.1006/jsbi.1998.4006

Schirmer EC, Florens L, Guan T, Yates JR 3rd, Gerace L. 2003. Nuclear membrane proteins with potential disease links found by subtractive proteomics. *Science* **301:** 1380–1382. doi:10.1126/science.1088176

Schütz W, Alsheimer M, Öllinger R, Benavente R. 2005a. Nuclear envelope remodeling during mouse spermiogenesis: postmeiotic expression and redistribution of germline lamin B3. *Exp Cell Res* **307:** 285–291. doi:10.1016/j.yexcr.2005.03.023

Schütz W, Benavente R, Alsheimer M. 2005b. Dynamic properties of germ line-specific lamin B3: the role of the shortened rod domain. *Eur J Cell Biol* **84:** 649–662. doi:10.1016/j.ejcb.2005.03.001

Schwarzer W, Abdennur N, Goloborodko A, Pekowska A, Fudenberg G, Loe-Mie Y, Fonseca NA, Huber W, Haering CH, Mirny L, et al. 2017a. Two independent modes of chromatin organization revealed by cohesin removal. *Nature* **551:** 51–56. doi:10.1038/nature24281

Schwarzer W, Abdennur N, Goloborodko A, Pekowska A, Fudenberg G, Loe-Mie Y, Fonseca NA, Huber W, Haering CH, Mirny L, et al. 2017b. Two independent modes of chromatin organization revealed by cohesin removal. *Nature* **551:** 51–56. doi:10.1038/nature24281

Senior A, Gerace L. 1988. Integral membrane proteins specific to the inner nuclear membrane and associated with the nuclear lamina. *J Cell Biol* **107:** 2029–2036. doi:10.1083/jcb.107.6.2029

Shah PP, Donahue G, Otte GL, Capell BC, Nelson DM, Cao K, Aggarwala V, Cruickshanks HA, Rai TS, McBryan T, et al. 2013. Lamin B1 depletion in senescent cells triggers large-scale changes in gene expression and the chromatin landscape. *Genes Dev* **27:** 1787–1799. doi:10.1101/gad.223834.113

Shah PP, Lv W, Rhoades JH, Poleshko A, Abbey D, Caporizzo MA, Linares-Saldana R, Heffler JG, Sayed N, Thomas D, et al. 2021. Pathogenic LMNA variants disrupt cardiac lamina-chromatin interactions and de-repress alternative fate genes. *Cell Stem Cell* **28:** 938–954.e9. doi:10.1016/j.stem.2020.12.016

Shimamoto Y, Tamura S, Masumoto H, Maeshima K. 2017. Nucleosome-nucleosome interactions via histone tails and linker DNA regulate nuclear rigidity. *Mol Biol Cell* **28:** 1580–1589. doi:10.1091/mbc.e16-11-0783

Shimi T, Koujin T, Segura-Totten M, Wilson KL, Haraguchi T, Hiraoka Y. 2004. Dynamic interaction between BAF and emerin revealed by FRAP, FLIP, and FRET analyses in living HeLa cells. *J Struct Biol* **147:** 31–41. doi:10.1016/j.jsb.2003.11.013

Shimi T, Pfleghaar K, Kojima S, Pack CG, Solovei I, Goldman AE, Adam SA, Shumaker DK, Kinjo M, Cremer T, et al. 2008. The A- and B-type nuclear lamin networks: microdomains involved in chromatin organization and transcription. *Genes Dev* **22:** 3409–3421. doi:10.1101/gad.1735208

Shimi T, Butin-Israeli V, Adam SA, Hamanaka RB, Goldman AE, Lucas CA, Shumaker DK, Kosak ST, Chandel NS, Goldman RD. 2011. The role of nuclear lamin B1 in cell proliferation and senescence. *Genes Dev* **25:** 2579–2593. doi:10.1101/gad.179515.111

Shimi T, Kittisopikul M, Tran J, Goldman AE, Adam SA, Zheng Y, Jaqaman K, Goldman RD. 2015. Structural organization of nuclear lamins A, C, B1, and B2 revealed by superresolution microscopy. *Mol Biol Cell* **26:** 4075–4086. doi:10.1091/mbc.E15-07-0461

Shin J-W, Spinler KR, Swift J, Chasis JA, Mohandas N, Discher DE. 2013. Lamins regulate cell trafficking and lineage maturation of adult human hematopoietic cells. *Proc Natl Acad Sci* **110:** 18892–18897. doi:10.1073/pnas.1304996110

Shumaker DK, Lee KK, Tanhehco YC, Craigie R, Wilson KL. 2001. LAP2 binds to BAFDNA complexes: requirement for the LEM domain and modulation by variable regions. *EMBO J* **20:** 1754–1764. doi:10.1093/emboj/20.7.1754

Shumaker DK, Dechat T, Kohlmaier A, Adam SA, Bozovsky MR, Erdos MR, Eriksson M, Goldman AE, Khuon S, Collins FS, et al. 2006. Mutant nuclear lamin A leads to progressive alterations of epigenetic control in premature aging. *Proc Natl Acad Sci* **103:** 8703–8708. doi:10.1073/pnas.0602569103

Shumaker DK, Solimando L, Sengupta K, Shimi T, Adam SA, Grunwald A, Strelkov SV, Aebi U, Cardoso MC, Goldman RD. 2008. The highly conserved nuclear lamin Ig-fold binds to PCNA: its role in DNA replication. *J Cell Biol* **181:** 269–280. doi:10.1083/jcb.200708155

Simon DN, Wilson KL. 2013. Partners and post-translational modifications of nuclear lamins. *Chromosoma* **122:** 13–31. doi:10.1007/s00412-013-0399-8

Solovei I, Wang AS, Thanisch K, Schmidt CS, Krebs S, Zwerger M, Cohen TV, Devys D, Foisner R, Peichl L, et al. 2013. LBR and lamin A/C sequentially tether peripheral heterochromatin and inversely regulate differentiation. *Cell* **152:** 584–598. doi:10.1016/j.cell.2013.01.009

Somech R, Shaklai S, Geller O, Amariglio N, Simon AJ, Rechavi G, Gal-Yam EN. 2005. The nuclear-envelope protein and transcriptional repressor LAP2β interacts with HDAC3 at the nuclear periphery, and induces histone H4 deacetylation. *J Cell Sci* **118:** 4017–4025. doi:10.1242/jcs.02521

Stephens AD, Banigan EJ, Adam SA, Goldman RD, Marko JF. 2017. Chromatin and lamin A determine two different mechanical response regimes of the cell nucleus. *Mol Biol Cell* **28:** 1984–1996. doi:10.1091/mbc.e16-09-0653

Stewart C, Burke B. 1987. Teratocarcinoma stem cells and early mouse embryos contain only a single major lamin polypeptide closely resembling lamin B. *Cell* **51:** 383–392. doi:10.1016/0092-8674(87)90634-9

Strickfaden H, Tolsma TO, Sharma A, Underhill DA, Hansen JC, Hendzel MJ. 2020. Condensed chromatin behaves like a solid on the mesoscale in vitro and in living cells. *Cell* **183:** 1772–1784.e13. doi:10.1016/j.cell.2020.11.027

Strom AR, Emelyanov AV, Mir M, Fyodorov DV, Darzacq X, Karpen GH. 2017. Phase separation drives heterochromatin domain formation. *Nature* **547:** 241–245. doi:10.1038/nature22989

Stuurman N, Sasse B, Fisher PA. 1996. Intermediate filament protein polymerization: molecular analysis of *Drosophila* nuclear lamin head-to-tail binding. *J Struct Biol* **117:** 1–15. doi:10.1006/jsbi.1996.0064

Stuurman N, Heins S, Aebi U. 1998. Nuclear lamins: their structure, assembly, and interactions. *J Struct Biol* **122:** 42–66. doi:10.1006/jsbi.1998.3987

Sullivan T, Escalante-Alcalde D, Bhatt H, Anver M, Bhat N, Nagashima K, Stewart CL, Burke B. 1999. Loss of A-type lamin expression compromises nuclear envelope integrity leading to muscular dystrophy. *J Cell Biol* **147:** 913–920. doi:10.1083/jcb.147.5.913

Swift J, Ivanovska IL, Buxboim A, Harada T, Dingal PC, Pinter J, Pajerowski JD, Spinler KR, Shin JW, Tewari M, et al. 2013. Nuclear lamin-A scales with tissue stiffness and enhances matrix-directed differentiation. *Science* **341:** 1240104. doi:10.1126/science.1240104

Tajik A, Zhang Y, Wei F, Sun J, Jia Q, Zhou W, Singh R, Khanna N, Belmont AS, Wang N. 2016. Transcription upregulation via force-induced direct stretching of chromatin. *Nat Mater* **15:** 1287–1296. doi:10.1038/nmat4729

Thompson LJ, Bollen M, Fields AP. 1997. Identification of protein phosphatase 1 as a mitotic lamin phosphatase. *J Biol Chem* **272:** 29693–29697. doi:10.1074/jbc.272.47.29693

Torvaldson E, Kochin V, Eriksson JE. 2015. Phosphorylation of lamins determine their structural properties and signaling functions. *Nucleus* **6:** 166–171. doi:10.1080/19491034.2015.1017167

Towbin BD, González-Aguilera C, Sack R, Gaidatzis D, Kalck V, Meister P, Askjaer P, Gasser SM. 2012. Step-wise methylation of histone H3K9 positions heterochromatin at the nuclear periphery. *Cell* **150:** 934–947. doi:10.1016/j.cell.2012.06.051

Tsai M-Y, Wang S, Heidinger JM, Shumaker DK, Adam SA, Goldman RD, Zheng Y. 2006. A mitotic lamin B matrix induced by RanGTP required for spindle assembly. *Science* **311:** 1887–1893. doi:10.1126/science.1122771

Turgay Y, Eibauer M, Goldman AE, Shimi T, Khayat M, Ben-Harush K, Dubrovsky-Gaupp A, Sapra KT, Goldman RD, Medalia O. 2017. The molecular architecture of lamins in somatic cells. *Nature* **543:** 261–264. doi:10.1038/nature21382

Ulbert S, Antonin W, Platani M, Mattaj IW. 2006. The inner nuclear membrane protein Lem2 is critical for normal nuclear envelope morphology. *FEBS Lett* **580:** 6435–6441. doi:10.1016/j.febslet.2006.10.060

Vadrot N, Duband-Goulet I, Cabet E, Attanda W, Barateau A, Vicart P, Gerbal F, Briand N, Vigouroux C, Oldenburg AR, et al. 2015. The p.R482W substitution in A-type lamins deregulates SREBP1 activity in Dunnigan-type familial partial lipodystrophy. *Hum Mol Genet* **24:** 2096–2109. doi:10.1093/hmg/ddu728

van Schaik T, Vos M, Peric-Hupkes D, Celie PHN, van Steensel B. 2020. Cell cycle dynamics of lamina-associated DNA. *EMBO Rep* **21:** e50636. doi:10.15252/embr.202050636

van Steensel B, Belmont AS. 2017. Lamina-associated domains: links with chromosome architecture, heterochromatin, and gene repression. *Cell* **169:** 780–791. doi:10.1016/j.cell.2017.04.022

Vaughan A, Alvarez-Reyes M, Bridger JM, Broers JL, Ramaekers FC, Wehnert M, Morris GE, Whitfield WGF, Hutchison CJ. 2001. Both emerin and lamin C depend on lamin A for localization at the nuclear envelope. *J Cell Sci* **114:** 2577–2590. doi:10.1242/jcs.114.14.2577

Vergnes L, Peterfy M, Bergo MO, Young SG, Reue K. 2004. Lamin B1 is required for mouse development and nuclear integrity. *Proc Natl Acad Sci* **101:** 10428–10433. doi:10.1073/pnas.0401424101

Vogel MJ, Peric-Hupkes D, van Steensel B. 2007. Detection of in vivo protein–DNA interactions using DamID in mammalian cells. *Nat Protoc* **2:** 1467–1478. doi:10.1038/nprot.2007.148

Vorburger K, Lehner CF, Kitten GT, Eppenberger HM, Nigg EA. 1989. A second higher vertebrate B-type lamin. cDNA sequence determination and in vitro processing of chicken lamin B2. *J Mol Biol* **208:** 405–415. doi:10.1016/0022-2836(89)90505-6

Wang S, Stoops E, Cp U, Markus B, Reuveny A, Ordan E, Volk T. 2018. Mechanotransduction via the LINC complex regulates DNA replication in myonuclei. *J Cell Biol* **217:** 2005–2018. doi:10.1083/jcb.201708137

Wilson KL, Foisner R. 2010. Lamin-binding proteins. *Cold Spring Harb Perspect Biol* **2:** a000554. doi:10.1101/cshperspect.a000554

Wong X, Stewart CL. 2020. The laminopathies and the insights they provide into the structural and functional organization of the nucleus. *Annu Rev Genomics Hum Genet* **21:** 263–288. doi:10.1146/annurev-genom-121219-083616

Wong X, Luperchio TR, Reddy KL. 2014. NET gains and losses: the role of changing nuclear envelope proteomes in genome regulation. *Curr Opin Cell Biol* **28:** 105–120. doi:10.1016/j.ceb.2014.04.005

Wong X, Hoskins VE, Harr JC, Gordon M, Reddy KL. 2020. Lamin C regulates genome organization after mitosis. bioRxiv doi:10.1101/2020.07.28.213884.

Wong X, Loo TH, Stewart CL. 2021a. LINC complex regulation of genome organization and function. *Curr Opin Genet Dev* **67:** 130–141. doi:10.1016/j.gde.2020.12.007

Wong X, Cutler JA, Hoskins VE, Gordon M, Madugundu AK, Pandey A, Reddy KL. 2021b. Mapping the microproteome of the nuclear lamina and lamina-associated domains. *Life Sci Alliance* **4:** e202000774. doi:10.26508/lsa.202000774

Worman HJ, Schirmer EC. 2015. Nuclear membrane diversity: underlying tissue-specific pathologies in disease?

Curr Opin Cell Biol **34:** 101–112. doi:10.1016/j.ceb.2015.06.003

Wu J, Kent IA, Shekhar N, Chancellor TJ, Mendonca A, Dickinson RB, Lele TP. 2014. Actomyosin pulls to advance the nucleus in a migrating tissue cell. *Biophys J* **106:** 7–15. doi:10.1016/j.bpj.2013.11.4489

Xie W, Chojnowski A, Boudier T, Lim JS, Ahmed S, Ser Z, Stewart C, Burke B. 2016. A-type lamins form distinct filamentous networks with differential nuclear pore complex associations. *Curr Biol* **26:** 2651–2658. doi:10.1016/j.cub.2016.07.049

Yang SH, Chang SY, Yin L, Tu Y, Hu Y, Yoshinaga Y, de Jong PJ, Fong LG, Young SG. 2011. An absence of both lamin B1 and lamin B2 in keratinocytes has no effect on cell proliferation or the development of skin and hair. *Hum Mol Genet* **20:** 3537–3544. doi:10.1093/hmg/ddr266

Yao J, Fetter RD, Hu P, Betzig E, Tjian R. 2011. Subnuclear segregation of genes and core promoter factors in myogenesis. *Genes Dev* **25:** 569–580. doi:10.1101/gad.2021411

Ye Q, Callebaut I, Pezhman A, Courvalin JC, Worman HJ. 1997. Domain-specific interactions of human HP1-type chromodomain proteins and inner nuclear membrane protein LBR. *J Biol Chem* **272:** 14983–14989. doi:10.1074/jbc.272.23.14983

Zhang H, Emerson DJ, Gilgenast TG, Titus KR, Lan Y, Huang P, Zhang D, Wang H, Keller CA, Giardine B, et al. 2019. Chromatin structure dynamics during the mitosis-to-G1 phase transition. *Nature* **576:** 158–162. doi:10.1038/s41586-019-1778-y

Zheng X, Hu J, Yue S, Kristiani L, Kim M, Sauria M, Taylor J, Kim Y, Zheng Y. 2018. Lamins organize the global three-dimensional genome from the nuclear periphery. *Mol Cell* **71:** 802–815.e7. doi:10.1016/j.molcel.2018.05.017

Zullo JM, Demarco IA, Piqué-Regi R, Gaffney DJ, Epstein CB, Spooner CJ, Luperchio TR, Bernstein BE, Pritchard JK, Reddy KL, et al. 2012. DNA sequence-dependent compartmentalization and silencing of chromatin at the nuclear lamina. *Cell* **149:** 1474–1487. doi:10.1016/j.cell.2012.04.035

Structure, Maintenance, and Regulation of Nuclear Pore Complexes: The Gatekeepers of the Eukaryotic Genome

Marcela Raices and Maximiliano A. D'Angelo

Cell and Molecular Biology of Cancer Program, NCI-Designated Cancer Center, Sanford Burnham Prebys Medical Discovery Institute, La Jolla, California 92037, USA

Correspondence: mdangelo@sbpdiscovery.org

In eukaryotic cells, the genetic material is segregated inside the nucleus. This compartmentalization of the genome requires a transport system that allows cells to move molecules across the nuclear envelope, the membrane-based barrier that surrounds the chromosomes. Nuclear pore complexes (NPCs) are the central component of the nuclear transport machinery. These large protein channels penetrate the nuclear envelope, creating a passage between the nucleus and the cytoplasm through which nucleocytoplasmic molecule exchange occurs. NPCs are one of the largest protein assemblies of eukaryotic cells and, in addition to their critical function in nuclear transport, these structures also play key roles in many cellular processes in a transport-independent manner. Here we will review the current knowledge of the NPC structure, the cellular mechanisms that regulate their formation and maintenance, and we will provide a brief description of a variety of processes that NPCs regulate.

A signature of eukaryotic cells is the presence of a nucleus, a membrane-based organelle that houses the cellular genome. The membrane structure that encloses the genetic material called the nuclear envelope (NE) is composed of two concentric lipid bilayers. Because the NE provides a physical barrier that completely surrounds the genome, eukaryotic cells have developed a sophisticated transport system that allows them to efficiently and selectively move molecules between the nucleus and the cytoplasm. Nucleocytoplasmic molecule exchange occurs through nuclear pore complexes (NPCs), large aqueous channels that penetrate the two membranes of the NE. As NPCs provide the only conduit into the nucleus, these large multiprotein structures are essential, and their proper formation and function is critical for cellular homeostasis. In addition to their canonical role mediating nucleocytoplasmic transport, a large amount of evidence shows that NPCs and their components play multiple transport-independent functions. Moreover, the expression levels of NPC components can vary significantly among different cell types and tissues, and the residence time at NPCs is widely different between different components. These findings have exposed that NPCs are dynamic in nature, can change their composition or stoichiometry in different cell types, and can regulate a multitude of cellular processes either in a

Cite this article as *Cold Spring Harb Perspect Biol* doi: 10.1101/cshperspect.a040691

transport-dependent or independent fashion. Our goal here is to provide an overview of the latest advances in the understanding of NPC formation, dynamics, maintenance, and functions.

NPC STRUCTURE

In the 1950s, Callan and Tomlin used electron microscopy (EM) of amphibian oocyte nuclei to provide the first evidence for the existence of pores at the NE (Fig. 1; Callan and Tomlin 1950). With these initial images of the nuclear membranes, the authors suggested for the first time that the rims of the pores in the outer nuclear membrane, now known as the outer scaffold ring, were built up above the nuclear membrane. Seventeen years later, Joseph Gall provided the first description of the eightfold rotational symmetry of NPCs (Fig. 1; Gall 1967). A multitude of subsequent studies using more powerful imaging and structural analysis techniques helped to further understand the three-dimensional structure of NPCs over the past half century (Fig. 1; Hinshaw et al. 1992; Akey and Radermacher 1993; Stoffler et al. 2003; Beck et al. 2004, 2007; Maco et al. 2006; Lim et al. 2008; Elad et al. 2009; Frenkiel-Krispin et al. 2010; Maimon et al. 2012; Szymborska et al. 2013; Löschberger et al. 2014; Eibauer et al. 2015; von Appen et al. 2015; Schwartz 2016; von Appen and Beck 2016; Beck and Hurt 2017; Kim et al. 2018; Mosalaganti et al. 2018; Vallotton et al. 2019; Zhang et al. 2020). But because NPCs are such large protein complexes, dissecting the architecture of these channels at the molecular level represented a challenging undertaking, and it was not until more recently that the exact position of nucleoporins within the NPC structure was determined with high resolution. The recent crystallization of many NPC compo-

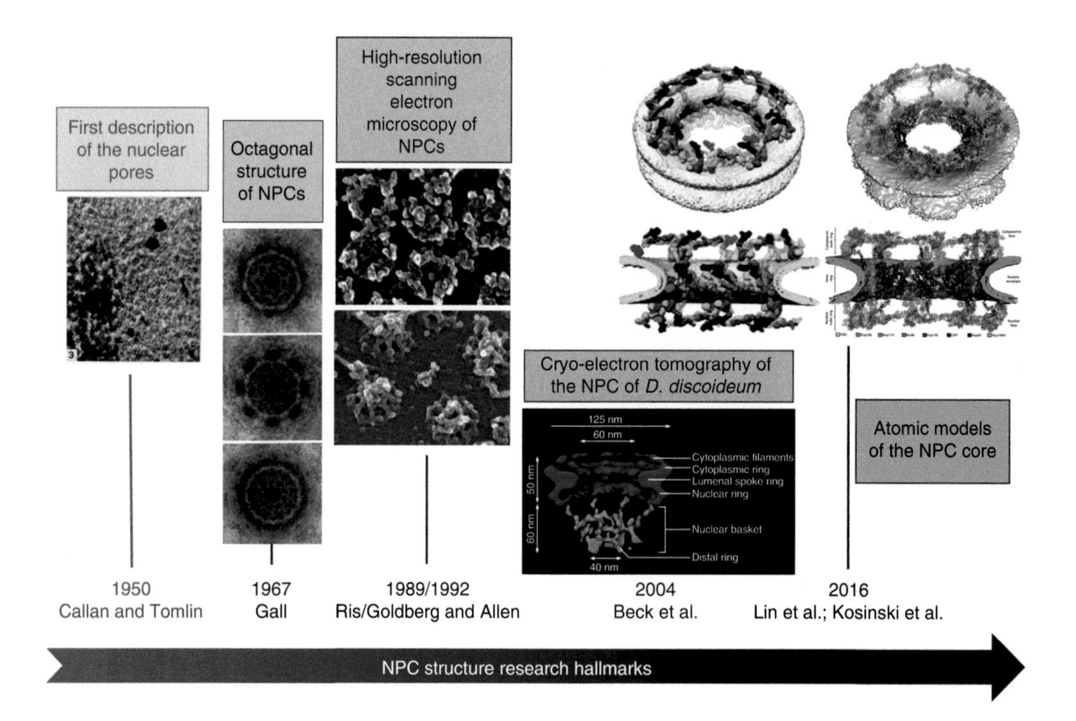

Figure 1. Timeline depicting hallmark structural findings of the nuclear pore complex (NPC). For a better illustration of key findings, images were obtained from original publications or provided by the authors. (Image of first electron microscopy [EM] of NPCs by Callan and Tomlin 1950, reprinted with permission from The Royal Society © 1950. Original high-resolution scanning images kindly provided by Dr. Martin Goldberg [Ris 1989; Goldberg and Allen 1992]. Cryo-electron tomography of NPCs, image reprinted from Beck et al. 2004 with permission from the American Association for the Advancement of Science [AAAS] © 2004. Atomic models of the NPC core, image reprinted from Lin et al. 2016 and Kosinski et al. 2016 with permission from AAAS © 2016.)

nents, the breakthrough generation of high-resolution images of the NPC structure by cryo-electron tomography, and the determination of nucleoporin interactions have allowed several groups to develop highly accurate models of the NPC molecular organization, in particular of the NPC scaffold (Fig. 1; Alber et al. 2007; Bui et al. 2013; Stuwe et al. 2015; von Appen et al. 2015; Kosinski et al. 2016; Lin et al. 2016; Kim et al. 2018; Allegretti et al. 2020). The detailed structural organization of the NPC and its components has been extensively reviewed (Kabachinski and Schwartz 2015; Hoelz et al. 2016; Beck and Hurt 2017; Hampoelz et al. 2019a; Lin and Hoelz 2019) and will be described here concisely.

It is important to note that the massive NPCs are highly modular complexes built by the repetition of ~30 different proteins called nucleoporins (the exact numbers varies slightly in different species) (Table 1; Rout et al. 2000; Cronshaw et al. 2002; Obado et al. 2016; Field and Rout 2019). NPCs are channels of eightfold rotational symmetry that sit on pores generated at the NE by the fusion of the inner and outer nuclear membranes. These channels consist of a central scaffold built by an inner ring embedded in the NE and two outer rings. The scaffold has eight identical spokes and twofold symmetry parallel to the NE plane and surrounds the NPC central transport channel, which is filled with nucleoporins rich in

Table 1. Nuclear pore complex (NPC) composition in budding yeast and vertebrates[a]

NPC domains	Mammalian components	Yeast components
Cytoplasmic filaments	Nup358 Nup214 Nup88 Nup98 Rae1 Aladdin	Nup159 Nup82 Nup42 Nup100 Nup116
Cytoplasmic ring	Nup37 Nup43 Nup75 Nup96 Nup107 Nup133 Nup160 Seh1 Sec13 Elys	Nup84 Nup85 Nup120 Nup133 Nup145 CSeh1 Sec13
Inner ring	Nup35 Nup93 Nup155 Nup188 Nup205	Nup53 Nup59 Nup157 Nup170 Nup188 Nup192 Nic96
Transmembrane nucleoporins	Nup210 Ndc1 Pom121	Pom152 Ndc1 Pom34 Pom33
Central channel	Nup45 Nup54 Nup58 Nup62	Nup49 Nup57 Nsp1
Nuclear ring	Nup37 Nup43 Nup75 Nup96 Nup107 Nup133 Nup160 Seh1 Sec13 Elys	Nup84 Nup85 Nup120 Nup133 Nup145 CSeh1 Sec13
Nuclear basket	Nup50 Nup98 Nup153 Rae1 TPR	Nup1 Nup2 Nup60 NUP145N Mlp1/2

[a]The NPC domains are color coded. Nucleoporins from the different NPC subdomains are shown in groups.

phenylglycine (FG) repeats (GLFG, FxFG, PxFG, or SxFG). Through weak hydrophobic interactions between their FG domains, central channel nucleoporins set the diffusion barrier of the NPC (Ribbeck and Görlich 2002; Frey et al. 2006; Frey and Görlich 2007, 2009; Xu and Powers 2013; Schmidt and Görlich 2015; Zahn et al. 2016; Fisher et al. 2018; Gu et al. 2019; Celetti et al. 2020; Dormann 2020), interact with transport receptors to mediate the nuclear transport (Iovine et al. 1995; Bayliss et al. 2000, 2002; Sträßer et al. 2000; Fribourg et al. 2001; Isgro and Schulten 2007a,b; Terry and Wente 2007; Zahn et al. 2016; Tan et al. 2018; Hayama et al. 2019), and help to stabilize the structure of the NPC (Onischenko et al. 2017). Finally, specific nucleoporins associate on either side of the symmetric core to form the cytoplasmic filaments and the nuclear basket structures.

At the molecular level, and because of the eightfold symmetry of the NPC, nucleoporins are present in eight or multiples of eight copies (Ori et al. 2013; Kim et al. 2018; Rajoo et al. 2018). While the outer rings of the NPC are mainly formed by copies of the Nup107-160 complex, also known as the Y-complex, the inner ring is formed by the repetition of the Nup93-Nup205 complex (recently, for review, see Lin and Hoelz 2019). In humans, the inner and outer rings are sporadically linked through a component of the Nup93-Nup205 complex, nucleoporin Nup155 (Kosinski et al. 2016; Kim et al. 2018; Lin and Hoelz 2019). The main nucleoporins that fill the central channel are the components of the Nup62 complex (Nup62, Nup54, and Nup58) (Finlay et al. 1991; Strawn et al. 2004; Sharma et al. 2015). These three nucleoporins are anchored to the inner ring of the NPC through Nup93 and extend their disordered FG-containing domains into the central channel (Sachdev et al. 2012; Fischer et al. 2015; Stuwe et al. 2015). The Nup62 complex together with Nup98, an FG nucleoporin that associates with the cytoplasmic and nuclear sides of the NPC (Griffis et al. 2003), define the selective permeability of the NPC (Frey and Görlich 2007; Ader et al. 2010). Notably, high concentrations of Nup98 FG repeats have been shown to undergo a phase separation in vitro and to form a hydrogel that recapitulates the permeability and transport properties of the NPC (Frey et al. 2006; Frey and Görlich 2007; Hülsmann et al. 2012). These findings support the idea that phase transitions induced by the interactions of FG repeats might govern the permeability barrier of the NPC (Schmidt and Görlich 2015, 2016; Fisher et al. 2018; Celetti et al. 2020; Dormann 2020). In this model, nuclear transport receptors can move through the molecular condensate barrier by outcompeting the nucleoporin FG–FG interactions (Ribbeck and Görlich 2001; Antonin 2013; Zahn et al. 2016). Even though a molecular condensate environment represents a highly likely scenario for the central channel of the NPC, the intrinsically disordered nature of the FG-containing regions of nucleoporins have prevented their structural definition using conventional techniques, and several other possibilities have also been proposed to explain the permeability barrier of the NPC (D'Angelo and Hetzer 2008; Mincer and Simon 2011; Li et al. 2016; Hayama et al. 2017).

In metazoans, the cytoplasmic side of the NPC is built by the recruitment of Nup358, Nup214, and Nup88, Nup358 being the main component of the cytoplasmic filaments. This protein is not present in budding yeast, *Saccharomyces cerevisiae* (Kim et al. 2018), which might explain why these cells show shorter (or different) filament structures (Kiseleva et al. 2004). The nuclear side of the NPC is mainly composed of the nucleoporin translocated promoter region (TPR), and its associated proteins Nup153 and Nup50, which form a nuclear basket-like structure tethered to the outer nuclear ring. This nuclear basket consists of eight filaments and a distal intranuclear ring.

Although the overall structure of the NPC is evolutionarily conserved, differences between species do exist, size being the most obvious variance. For example, while the human NPC core is 120 nM wide by 80 nM in height, and ~110 MDa (Kosinski et al. 2016), its *S. cerevisiae* counterpart is 95 nM by 60 nM and 52 MDa (Kim et al. 2018). A difference in size is partly explained by the outer scaffold rings. While the human NPC outer rings have 16 copies of the Y complex, the *S. cerevisiae* outer rings only have eight (Bui et al. 2013; Ori et al. 2013; von Appen et al. 2015; Rajoo

 Cite this article as *Cold Spring Harb Perspect Biol* doi: 10.1101/cshperspect.a040691

et al. 2018). This different stoichiometry is a result of humans having double outer rings while yeast has single ones. Interestingly, the *Chlamydomonas reinhardtii* alga has an "in-between" structure with a single cytoplasmic outer ring like yeast, but a double nuclear outer ring like humans (Mosalaganti et al. 2018). A recent study also identified that *S. cerevisiae* NPCs lack the extra molecules Nup157 or Nup170 (homologs of human Nup155) connecting the inner to the outer rings and do not a show Nup188 or Nup192 (homolog of human Nup205) in the outer rings (Kim et al. 2018), as described for the human NPC (Kosinski et al. 2016). This challenges a previous model suggesting that core scaffolds were identical between both species (Lin et al. 2016).

Although NPCs are membrane-embedded channels, only three to four nucleoporins are transmembrane proteins and all the rest are soluble (Table 1; Wozniak and Blobel 1992; Hallberg et al. 1993; Wozniak et al. 1994; Chial et al. 1998; Miao et al. 2006; Stavru et al. 2006; Chadrin et al. 2010). But some soluble nucleoporins have amphipathic lipid-packing sensor (ALPS) domains, which are sequences 20–40 amino acids long containing hydrophobic residues that can bind to curved lipid bilayers, providing additional links to the nuclear membranes (Drin et al. 2007; Leksa et al. 2009; Doucet et al. 2010; Mitchell et al. 2010; Kim et al. 2014; Mészáros et al. 2015; Lin et al. 2016; Lin and Hoelz 2019; Nordeen et al. 2020). Even though the structure and domains of nucleoporins seem to be conserved among many species, these proteins show poor sequence conservation, which has made it difficult to identify them in some organisms. Moreover, some nucleoporins are restricted to specific organisms. For example, the cytoplasmic filament protein Nup358/RanBP2 has so far only been described in metazoans, while the transmembrane nucleoporin Pom121 is restricted to vertebrates (Funakoshi et al. 2007, 2011; Kim et al. 2018). With a few exceptions, nucleoporins have a remarkably limited set of domains mostly limited to structure and protein–protein interactions, such as α-solenoids, β-propellers, coiled-coiled motifs, WD40, and Zn-finger domains (see Lin and Hoelz 2019 for details).

At first sight, the NPC looks like a modular structure of very simple composition built by the repetition of roughly 30 different proteins (Rout et al. 2000; Cronshaw et al. 2002). But increasing evidence points toward a much higher complexity for this structure. First, several nucleoporins have differential expression among cell types and tissues, and NPCs of different composition and stoichiometry of nucleoporins have been reported, indicating that NPCs can vary among different cell types (D'Angelo et al. 2012; Ori et al. 2013, 2016). Moreover, recent evidence exposed that NPCs also change during specific cellular processes and in response to different cellular conditions (Asally et al. 2011; D'Angelo et al. 2012; Raices and D'Angelo 2012; Rodriguez-Bravo et al. 2018; Liu et al. 2019). Adding to this complexity, many nucleoporins show multiple splicing isoforms, most of which have not yet been characterized (Capitanchik et al. 2018; Hampoelz et al. 2019a). Although some of these splice variants could work "off pore," as recently described for Pom121 (Franks et al. 2016), it is possible that some might be used to change the properties or functions of NPCs. A significant number of nucleoporins also have short residence time at NPCs, meaning that they are constantly exchanged, and further highlighting the dynamic nature of these structures (Rabut et al. 2004). Additionally, multiple proteomic studies have shown that most nucleoporins have a significant number of posttranslational modifications, including phosphorylation, ubiquitination, *N*- and *O*-glycosylation, sumoylation, and acetylation (for review, see Hampoelz et al. 2019a). The physiological relevance of most of these modifications is still unknown; some of these modifications might be used to alter the structure, dynamics, and functions of NPCs. Consistent with this idea, phosphorylation of nucleoporins was shown to modulate their interaction with transport receptors (Kosako and Imamoto 2010; Wigington et al. 2020), *O*-glycosylation modifies NPC integrity, permeability, and nuclear transport (Mizuguchi-Hata et al. 2013; Zhu et al. 2016; Eustice et al. 2017), acetylation of NPC components regulates cell-cycle entry and asymmetric cell division (Kumar et al. 2018), and ubiquitylation and sumoylation of nuclear basket nucleoporins modulate their association, transcriptional regulation, nuclear migration, and the cellular response to stress (Hayakawa et al. 2012; Texari and Stutz 2015; Niño et al. 2016; Folz et al. 2019).

NPC LIFE CYCLE IN DIVIDING CELLS

Proliferating cells have a constant need for NPC assembly. When cells divide, they split their contents into two daughter cells that receive only half the NPCs of the mother. Therefore, these cells need to assemble a new set of channels for the next round of cell division to avoid a progressive dilution of NPC numbers. In organisms undergoing closed mitosis, like *S. cerevisiae* and *Schizosaccharomyces pombe*, cell division was long considered to occur with no NE or NPC breakdown, meaning that these cells only assemble new NPCs into an intact NE (Boettcher and Barral 2013). Interestingly, evidence for a local disassembly of NPCs and the breakdown/remodeling of the NE in the bridge that connects the segregating *S. pombe* daughter nuclei were recently provided (Dey et al. 2020; Expósito-Serrano et al. 2020). Although these findings do not indicate a second mechanism of NPC formation in these cells, they provide evidence for similarities in the process of NE remodeling between open and closed mitosis.

In metazoans, mitosis requires a total breakdown of the NE including the full disassembly of NPCs (open mitosis) (Anderson and Hetzer 2008). Consequently, daughter cells need to reassemble their NPCs by recycling maternal components when the NEs reform during M-phase (postmitotic NPC assembly), and they also need to assemble NPCs into an intact NE later during interphase to double their number before the next cell division (interphase NPC assembly). Although mitotic and interphase NPC assembly result in the formation of indistinguishable structures, significant evidence indicates that these processes are fundamentally different (Otsuka and Ellenberg 2018).

In metazoans, NPC disassembly during mitosis is triggered by massive phosphorylation of nucleoporins by mitotic kinases, mostly cyclin-dependent kinase 1 (CDK1), Polo-Like Kinase 1 (PLK1), and the NIMA-related kinase (NEK) family (Macaulay et al. 1995; Favreau et al. 1996; Miller et al. 1999; De Souza et al. 2004; Onischenko et al. 2005; Glavy et al. 2007; Lusk et al. 2007; Rajanala et al. 2014; Linder et al. 2017; de Castro et al. 2018). Mitotic phosphorylation of nucleoporins induces the dissociation of protein complexes and leads to their release from the NE. NPC disassembly occurs synchronously and through a stepwise process that is faster than NPC formation, and does not follow the exact reverse order (Dultz et al. 2008). In mammalian cells, hyperphosphorylation of Nup98 has been found to be an early and rate-limiting step in NPC disassembly (Laurell et al. 2011; Linder et al. 2017). Phosphorylation of this nucleoporin results in its loss from NPCs, followed by the loss of peripheral nucleoporins and disruption of the nuclear permeability barrier (Linder et al. 2017). Phosphorylation of Nup35 (also referred to as Nup53) is another early event in NPC disassembly that leads to its dissociation from Nup155 and NDC1, and to the destabilization of nuclear pores (Linder et al. 2017). Similar stepwise disassembly mechanisms have been observed in other organisms (Kiseleva et al. 2001; Cotter et al. 2007; Katsani et al. 2008).

As mitosis progresses and the NEs of daughter cells begin to reform, so do NPCs. Mitotic NPC assembly is considerably faster than interphase assembly, taking minutes versus approximately an hour; synchronized, meaning that hundreds of NPCs form at the same time; and critical for the establishment of the nuclear permeability barrier (D'Angelo et al. 2006; Dultz et al. 2008; Dultz and Ellenberg 2010; Otsuka et al. 2016, 2018; Onischenko et al. 2020). Mitotic NPC assembly also occurs in a mitotic environment, where the cytoplasmic and nuclear contents are mixed. Reassembly is triggered by dephosphorylation of nucleoporins (Antonin et al. 2008; Huguet et al. 2019), and NPCs are built in a stepwise process with components inherited from the mother cell (Fig. 2A; Maul 1977; Burke and Ellenberg 2002; Dultz et al. 2008). In the mitotic cytoplasm, many nucleoporins are bound to importin β, an association that blocks the interaction between these proteins and prevents NPC reassembly (Zhang et al. 2002; Harel et al. 2003; Walther et al. 2003b). In the vicinity of chromatin, importin

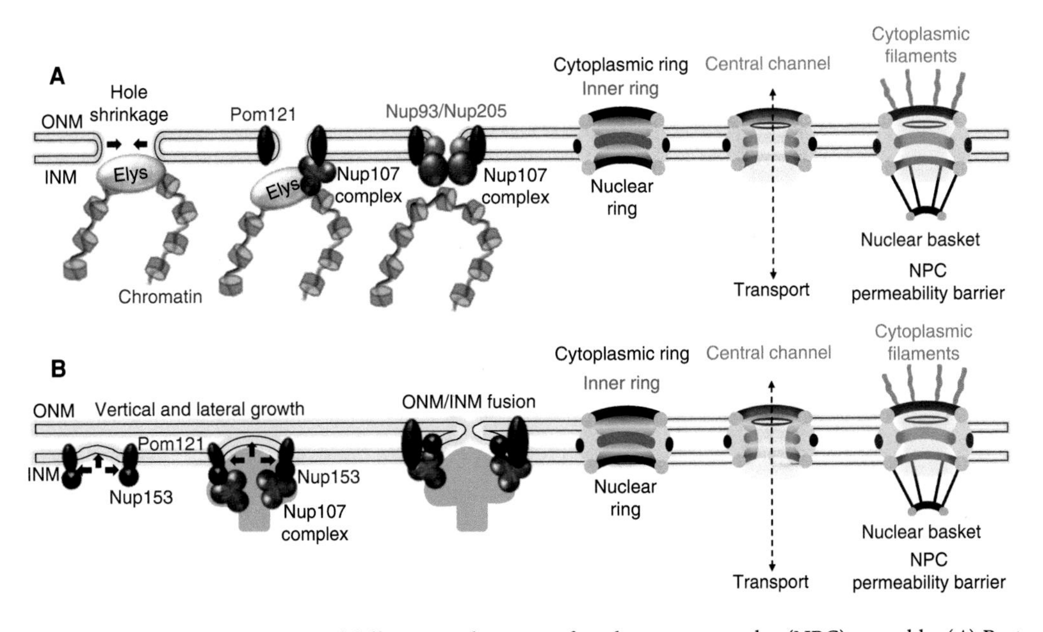

Figure 2. Stepwise representation of different mechanisms of nuclear pore complex (NPC) assembly. (*A*) Postmitotic NPC assembly. (*B*) Interphase NPC assembly. (ONM) Outer nuclear membrane, (INM) inner nuclear membrane.

β-nucleoporin complexes encounter high concentrations of RanGTP, which is generated by Ran's DNA-bound guanine-exchange factor RCC1 (Ohtsubo et al. 1989; Li et al. 2003). RanGTP binding to importin β releases nucleoporins allowing them to initiate NPC formation on chromosomes (Harel et al. 2003; Walther et al. 2003b; Franz et al. 2007; Rotem et al. 2009). A similar role for the export receptors CRM1, exportin-t, and exportin-5 in nucleoporin sequestration and release during postmitotic NPC assembly has recently been reported (Nord et al. 2020). Evidence from several laboratories indicate that the process starts during anaphase with the recruitment of the Y complex nucleoporin Elys to decondensing chromatin (Fig. 2A; Rasala et al. 2006, 2008; Franz et al. 2007; Gillespie et al. 2007; Dultz et al. 2008; Inoue and Zhang 2014). This seeding step is followed by the sequential association of the scaffold Nup107-160 complex, the membrane protein Pom121, components of the Nup93-Nup205 complex, and later, components of the central channel, and other nucleoporins (Fig. 2A; Bodoor et al. 1999; Belgareh et al. 2001; Daigle et al. 2001; Walther et al. 2003a,b; Antonin et al. 2005; Franz et al. 2005, 2007; Rasala et al. 2006, 2008; Gillespie et al. 2007; Dultz et al. 2008; Theisen et al. 2008; Mitchell et al. 2010; Sachdev et al. 2012; Vollmer et al. 2012; Eisenhardt et al. 2014). The stepwise recruitment of nucleoporins to chromatin results in the formation of intermediate structures, known as prepores, that were initially observed by EM in *Xenopus* oocytes and *Drosophila* embryos, and later confirmed in mammalian cells (Sheehan et al. 1988; Macaulay and Forbes 1996; Goldberg et al. 1997; Drummond et al. 2006; Otsuka et al. 2018). Using correlative single-cell live imaging and high-resolution scanning electron tomography, a recent study shed light into the mechanisms of postmitotic NPC formation. This work shows that during NE reformation, nuclear membranes form from highly fenestrated endoplasmic reticulum membrane sheets, and NPCs assemble into small preexisting membrane holes that progressively dilate as the nuclear pore rings and the central channel mature (Fig. 2A; Otsuka et al. 2018). This progressive dilation of membrane openings explains some of the early structural intermediates observed using a *Xenopus* in vitro nuclear assembly system (Goldberg et al. 1997). On the other hand, the formation of NPCs into preexisting membrane holes of the fenestrated ER membranes contradicts previous findings, suggesting that the formation of the NE membranes precedes NPC assembly (Macaulay and Forbes 1996; Fichtman et al. 2010; Lu et al. 2011).

Once the daughter nuclei have formed, cells start assembling new NPCs into an intact NE. Even though this is known as interphase NPC assembly, the process starts during telophase and occurs continuously until G2 (Winey et al. 1997; Maeshima et al. 2006; Otsuka and Ellenberg 2018). Interphase NPC assembly is sporadic and requires the fusion of the inner and outer nuclear membranes. Like mitotic NPC assembly, this is also a stepwise process, although it neither requires Elys nor follows the exact same order of nucleoporin recruitment (Fig. 2B; Doucet et al. 2010; Dultz and Ellenberg 2010; Onischenko et al. 2020). Although interphase NPC assembly requires importin β, RanGTP, and the Nup107-160 complex from the cytoplasmic and nuclear sides of the NE (D'Angelo et al. 2006), a recent study exposed that interphase NPC assembly does not occur symmetrically from both sides of the NE but rather through an inside-out mechanism (Otsuka et al. 2016). Interphase NPC assembly is believed to start with the recruitment of the Nup107-160 complex to the inner nuclear membrane by Nup153 (Vollmer et al. 2015) and requires the Pom121 transmembrane nucleoporin and the Sun1 inner nuclear membrane protein (Talamas and Hetzer 2011). The recruited nucleoporins start to form a dome-shaped intermediate structure that progressively grows in diameter and depth, pushing the inner membrane inward until it contacts the outer nuclear membrane (Fig. 2B). Strikingly, the early mushroom-shaped intermediate structure already shows an eightfold rotational symmetry and the presence of Nup107, suggesting that the outer ring is one of the first structures to assemble. Once the growing structure pushes the nuclear membranes in close proximity, they fuse, and the process continues with the recruitment of additional nucleoporins, such as the Nup358, to form the mature NPC structure (Otsuka et al. 2016). Several factors required for NPC assembly into an intact NE, such as karyopherin β (importin β), RanGTP, and key nucleoporins, were originally identified in *S. cerevisiae* (Vasu and Forbes 2001; Lusk et al. 2002; Ryan et al. 2003, 2007; Madrid et al. 2006), indicating conservation of this assembly mechanism across species.

It is important to note that NPC formation is not restricted to the NE, and stacks of nuclear pores in ER membrane sheets, known as annulate lamellae, can be observed during embryogenesis and in many cancer cells (Maul 1970; Dabauvalle et al. 1991; Cordes et al. 1995, 1996). In *Drosophila* embryos, annulate lamellae NPCs (AL-NPCs) were found to lack several components of the central channel, cytoplasmic filaments, and nuclear basket, suggesting that they mostly consist of the core structure (Fig. 3; Hampoelz et al. 2016). These partially assembled structures were found to be added to the NE during the nuclei expansion that occurs during the fast cell-cycle divisions in early embryogenesis (Fig. 3), suggesting that these structures represent a stock of partially assembled NPCs that are used to support fast proliferation (Hampoelz et al. 2016). Even though the assembly of annulate lamellae NPCs also requires Ran and CRM1, the partially assembled channels were found to form from nucleoporin condensates (Hampoelz et al. 2016, 2019b). Notably, granules of Nup358, a nucleoporin recruited at the late stages of interphase NPC assembly, were identified as the main drivers of NPC formation in the ER. Moreover, AL-NPCs do not contain Nup153, a key player in NPC assembly into the NE membranes (Vollmer et al. 2015; Hampoelz et al. 2016). These findings suggest that NPC assembly in annulate lamellae might employ a mechanism different from the mitotic and interphase assembly processes.

NPC MAINTENANCE AND TURNOVER

The building of the massive NPC machinery is a complicated process that can sporadically fail leading to defective or nonfunctional structures that can affect cellular physiology. Recent studies in yeasts uncovered a novel surveillance mechanism used by cells to identify and eliminate aberrantly assembled NPCs (Webster et al. 2014, 2016; Thaller et al. 2019). In this mechanism, the inner nuclear membrane proteins Heh1 and Heh2 recruit CHM7 and members of the ESCRT

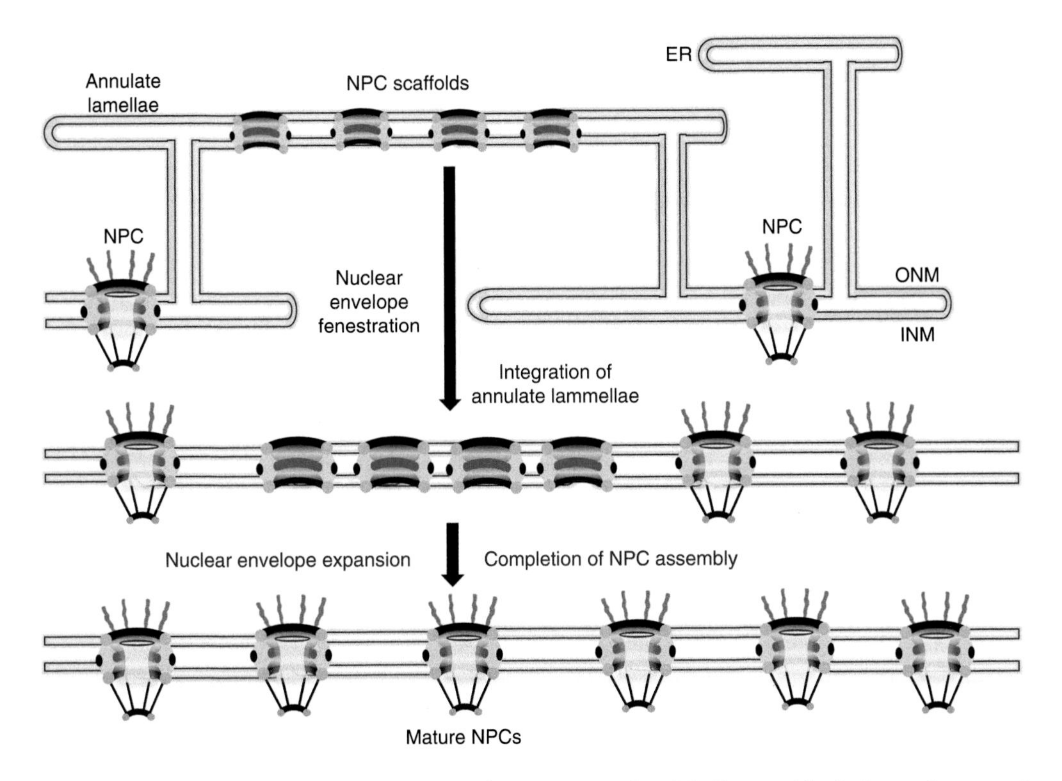

Figure 3. Annulate lamellae contribution to nuclear pore complex (NPC) assembly during embryogenesis. Individual components involved in the processes are shown in the figure. (ER) Endoplasmic reticulum, (ONM) outer nuclear membrane, (INM) inner nuclear membrane.

complex to defective NPC assembly intermediates. These proteins work in concert to seal defective NPCs, as well as destabilize and eliminate the aberrant structures. Disrupting the function of this surveillance machinery leads to the accumulation of defective assembly intermediates, affects nucleocytoplasmic transport, and results in the loss of nuclear compartmentalization (Webster et al. 2014, 2016; Thaller et al. 2019).

Selective autophagy of NPCs, termed NPC-phagy, a mechanism that might contribute to turnover nuclear pores, was recently identified in *S. cerevisiae*. This process requires the autophagy core component ATG11 and is mediated by the interaction of the autophagy marker protein ATG8 with Nup159, the homolog of human Nup214 (Allegretti et al. 2020; Lee et al. 2020; Tomioka et al. 2020). Nucleoporin mutant strains that either interrupt NPC assembly or lead to the aggregation of mature NPCs at the NE suggest that selective autophagy might be required for degradation of fully assembled NPCs rather than elimination of stalled NPC assembly intermediates (Allegretti et al. 2020; Lee et al. 2020; Tomioka et al. 2020). This selectivity is based on the fact that mature NPCs are exposed to the cytoplasm and can contact the autophagy machinery while aberrant assembly intermediates are sealed and hidden from it. These intermediates are covered by membranes, either because their formation stalled before the inner and outer nuclear membranes fused, or because they were sealed by the ESCRT-dependent surveillance mechanism (Webster et al. 2016).

In nondividing cells, the dynamics and maintenance of NPCs is different. First, as cells exit the cell cycle, they shut down the expression of essential nucleoporins and nuclear pore formation is strongly down-regulated (D'Angelo et al. 2009). Whereas in quiescent cells some level of NPC assembly and turnover is maintained (Toyama et al. 2019), in differentiated postmitotic cells the NPC scaffolds are one of the longest-lived protein complexes, and can be maintained for months, maybe even years, without turnover (D'Angelo et al. 2009; Savas et al. 2012; Toyama et al. 2013). In these cells, the NPC scaffolds seem to be renewed by the gradual replacement of individual subunits rather than the replacement of the entire structure (Toyama et al. 2013). Whether autophagy contributes to NPC turnover or to the elimination of malfunctioning transport channels in differentiated cells has not yet been investigated.

NPC FUNCTIONS

Because NPCs are the sole connection between the nucleus and the cytoplasm, their canonical function has traditionally been considered the regulation of nucleocytoplasmic transport. Although this is still recognized as the main function of NPCs, these structures are also one the largest protein assemblies of cells, and a great amount of evidence indicates that they play many other cellular roles. Because of space limitations, we will provide a summarized description of the NPC roles, which by no means represents a comprehensive account of the extensive literature in the field.

As gatekeepers of the nucleus, NPCs play two main functions: (1) they establish the permeability barrier of the NE, and (2) they work in coordination with transport receptors to ferry molecules in and out of the nucleus. As described above, both functions are mediated mostly by the FG nucleoporins of the central channel, although components of the cytoplasmic filaments and nuclear basket also play a role in nucleocytoplasmic transport (Bastos et al. 1996; Ullman et al. 1999; Walther et al. 2001; Forler et al. 2004; Makise et al. 2012). Ions and small molecules diffuse freely through NPCs, but larger molecules need to be actively transported through these channels. There is not an exact diffusion limit for NPCs, but a range between 30 and 60 kDa is largely accepted. Interestingly, several reports show that a permeability limit of NPCs can change in response to different cellular conditions during the cell cycle, as cells age, and in disease (Feldherr and Akin 1990; Roehrig et al. 2003; Belov et al. 2004; Shahin et al. 2005; D'Angelo et al. 2009; Dultz et al. 2009; Eftekharzadeh et al. 2018). In addition, transport receptors were recently reported to contribute to the permeability barrier of these channels (Lowe et al. 2015; Barbato et al. 2020). Although a few proteins have been found to move through NPCs by direct interaction with nucleoporins (Fagotto et al. 1998; Yokoya et al. 1999; Xu et al. 2003; Tsuji et al. 2007; Wagstaff and Jans 2009), the majority of the transported cargoes require transport receptors (Mosammaparast and Pemberton 2004; Kimura and Imamoto 2014). These are molecules that recognize nuclear localization signals (NLS), in the case of importins, or nuclear export signals (NES), in the case of exportins, in the cargo molecules and ferry them through NPCs. The family of nuclear transport receptors (collectively known as karyopherins or importin β family members) encompasses >20 different importins and exportins (Mosammaparast and Pemberton 2004; Kimura and Imamoto 2014). Specific exportins can also bind and export RNAs including microRNAs, ribosomal RNAs, tRNAs, and small nuclear RNAs, but the nuclear export of the majority of mRNAs is carried out through a completely different mechanism. The processes of protein and RNA nuclear transport have been the subject of several reviews (Köhler and Hurt 2007; Carmody and Wente 2009; Wente and Rout 2010; Cautain et al. 2015; Okamura et al. 2015; Scott et al. 2019) and will not be described in detail here.

But NPCs are far more than just the doors of the nucleus. Evidence accumulated over the past few decades, and particularly in the last one, have exposed that NPCs act as scaffolds for the regulation of a myriad of cellular process in a transport-independent manner. For example, a large amount of data shows that NPCs play a direct role in the regulation of gene expression and chromatin organization. They do this by recruiting transcription factors and chromatin modulators to the nuclear periphery to regulate local enhancers, gene activity, and chromatin state by promoting gene looping, and by modulating gene positioning (for reviews, see Randise-Hinchliff and Brickner 2016; Buchwalter et al. 2019; Pascual-Garcia and Capelson 2019; Sun et al. 2019). Nucleopor-

 Cite this article as *Cold Spring Harb Perspect Biol* doi: 10.1101/cshperspect.a040691

ins have also been shown to regulate gene expression "off pore" (in the nuclear interior) by interacting and modulating the activity of chromatin and transcriptional regulators (Capelson et al. 2010b; Kalverda et al. 2010; Liang et al. 2013; Pascual-Garcia et al. 2014; Jacinto et al. 2015; Franks et al. 2016). There is clear evidence that NPCs also play a key role in the response to DNA damage and replication stress (Bukata et al. 2013; Freudenreich and Su 2016; Duheron et al. 2017; Mackay et al. 2017; Rodriguez-Berriguete et al. 2018; Gaillard et al. 2019; Horigome et al. 2019; Aguilera et al. 2020; Pinzaru et al. 2020), and contribute to RNA processing and surveillance (Lewis et al. 2007; Ikegami and Lieb 2013; Bonnet and Palancade 2015).

Several protein-modifying enzymes, including histone deacetylases (Kehat et al. 2011), SUMO proteases (Panse et al. 2003; Chow et al. 2012, 2014), poly(ADP-ribose) polymerases (PARPs) (Meyer-Ficca et al. 2015; Carter-O'Connell et al. 2016; Kirby et al. 2018), phosphatases (Sales Gil et al. 2018; Wigington et al. 2020), and kinases (Faustino et al. 2011; Martino et al. 2017) have been found associated with NPCs, and nucleoporin Nup358 has been shown to be an E3 SUMO ligase important for the sumoylation of many cellular factors (Pichler et al. 2004). Even proteosomes have been recently found to be associated with NPCs (Albert et al. 2017). These findings indicate that NPCs also act as hubs for signal transduction, posttranslational modifications, and potentially localized protein degradation. Cell-cycle regulatory proteins Mad1 and Mad2 associate with NPCs during interphase (Campbell et al. 2001; Iouk et al. 2002), and the presence of Mad1 at the pores has been shown to be important for proper cell-cycle timing (Rodriguez-Bravo et al. 2014; Jackman et al. 2020). In addition, several nucleoporins have been shown to localize to kinetochores and/or the mitotic spindle during mitosis where they contribute to regulate their functions (Belgareh et al. 2001; Stukenberg and Macara 2003; Joseph et al. 2004; Loïodice et al. 2004; Blower et al. 2005; Galy et al. 2006; Rasala et al. 2006; Schetter et al. 2006; Zuccolo et al. 2007; Lussi et al. 2010; Mishra et al. 2010; Schweizer et al. 2013). These findings directly link NPCs and nucleoporins to cell-cycle regulation. Examples of additional functions that nucleoporins have been shown to play off-pore include a role for Nup188 in centriole duplication (Vishnoi et al. 2020), Nup358 function in cytoskeletal organization (Joseph and Dasso 2008), the regulation of mTORC1 activity by Seh1 and Sec13 being part of the GATOR2 complex (Shaw 2013), and of course the regulation of gene expression and chromatin organization as described above (Capelson et al. 2010b; Kalverda et al. 2010; Liang et al. 2013; Pascual-Garcia et al. 2014; Jacinto et al. 2015; Franks et al. 2016).

CONCLUDING REMARKS

In the past two decades there has been an explosion in our understanding of NPCs. We now know where almost every nucleoporin sits within the NPC structure, how specific nucleoporins build the permeability barrier and regulate nuclear transport, and how these complex structures are formed and maintained in dividing and differentiated cells. We also have learned that NPCs play many different functions, independent of their role in nucleocytoplasmic transport. Because these functions span several different fields—some examples are the recently identified roles of NPCs in the development and function of immune, muscle, and brain cells (for reviews, see Guglielmi et al. 2020); their contribution to cancer, heart disease, and neurodegeneration (Simon and Rout 2014; Beck et al. 2017; Jühlen and Fahrenkrog 2018; Hutten and Dormann 2019; Burdine et al. 2020); and their key roles in gene expression regulation, and the maintenance of genome integrity (Capelson et al. 2010a; Bukata et al. 2013; Raices and D'Angelo 2017; Sun et al. 2019)—investigators with a variety of backgrounds have become interested in the biology of NPCs, and it can be expected that our knowledge of these structures will continue to expand significantly in the coming decade.

ACKNOWLEDGMENTS

We thank Dr. Martin Goldberg for providing NPC EM images for Figure 1. We apologize to all colleagues whose work could not be cited directly owing to space limitation. M.A.D. is supported by a Research Scholar Grant RSG-17-148-01-CCG from the American Cancer Society. This work was also supported by the National Institutes of Health (Awards R01 AI148668-01 and R21 CA244028) and Department of Defense (Award PR191142). The content is solely the responsibility of the authors and does not necessarily represent the official views of the National Institutes of Health.

REFERENCES

Ader C, Frey S, Maas W, Schmidt HB, Gorlich D, Baldus M. 2010. Amyloid-like interactions within nucleoporin FG hydrogels. *Proc Natl Acad Sci* **107:** 6281–6285. doi:10.1073/pnas.0910163107

Aguilera P, Whalen J, Minguet C, Churikov D, Freudenreich C, Simon MN, Géli V. 2020. The nuclear pore complex prevents sister chromatid recombination during replicative senescence. *Nat Commun* **11:** 160. doi:10.1038/s41467-019-13979-5

Akey CW, Radermacher M. 1993. Architecture of the *Xenopus* nuclear pore complex revealed by three-dimensional cryo-electron microscopy. *J Cell Biol* **122:** 1–19. doi:10.1083/jcb.122.1.1

Alber F, Dokudovskaya S, Veenhoff LM, Zhang W, Kipper J, Devos D, Suprapto A, Karni-Schmidt O, Williams R, Chait BT, et al. 2007. The molecular architecture of the nuclear pore complex. *Nature* **450:** 695–701. doi:10.1038/nature06405

Albert S, Schaffer M, Beck F, Mosalaganti S, Asano S, Thomas HF, Plitzko JM, Beck M, Baumeister W, Engel BD. 2017. Proteasomes tether to two distinct sites at the nuclear pore complex. *Proc Natl Acad Sci* **114:** 13726–13731. doi:10.1073/pnas.1716305114

Allegretti M, Zimmerli CE, Rantos V, Wilfling F, Ronchi P, Fung HKH, Lee CW, Hagen W, Turonova B, Karius K, et al. 2020. In-cell architecture of the nuclear pore and snapshots of its turnover. *Nature* **586:** 796–800. doi:10.1038/s41586-020-2670-5

Anderson DJ, Hetzer MW. 2008. Reshaping of the endoplasmic reticulum limits the rate for nuclear envelope formation. *J Cell Biol* **182:** 911–924. doi:10.1083/jcb.200805140

Antonin W. 2013. Don't get stuck in the pore. *EMBO J* **32:** 173–175. doi:10.1038/emboj.2012.338

Antonin W, Franz C, Haselmann U, Antony C, Mattaj IW. 2005. The integral membrane nucleoporin pom121 functionally links nuclear pore complex assembly and nuclear envelope formation. *Mol Cell* **17:** 83–92. doi:10.1016/j.molcel.2004.12.010

Antonin W, Ellenberg J, Dultz E. 2008. Nuclear pore complex assembly through the cell cycle: regulation and membrane organization. *FEBS Lett* **582:** 2004–2016. doi:10.1016/j.febslet.2008.02.067

Asally M, Yasuda Y, Oka M, Otsuka S, Yoshimura SH, Takeyasu K, Yoneda Y. 2011. Nup358, a nucleoporin, functions as a key determinant of the nuclear pore complex structure remodeling during skeletal myogenesis. *FEBS J* **278:** 610–621. doi:10.1111/j.1742-4658.2010.07982.x

Barbato S, Kapinos LE, Rencurel C, Lim RYH. 2020. Karyopherin enrichment at the nuclear pore complex attenuates Ran permeability. *J Cell Sci* **133:** jcs238121. doi:10.1242/jcs.238121

Bastos R, Lin A, Enarson M, Burke B. 1996. Targeting and function in mRNA export of nuclear pore complex protein Nup153. *J Cell Biol* **134:** 1141–1156. doi:10.1083/jcb.134.5.1141

Bayliss R, Littlewood T, Stewart M. 2000. Structural basis for the interaction between FxFG nucleoporin repeats and importin-β in nuclear trafficking. *Cell* **102:** 99–108. doi:10.1016/S0092-8674(00)00014-3

Bayliss R, Littlewood T, Strawn LA, Wente SR, Stewart M. 2002. GLFG and FxFG nucleoporins bind to overlapping sites on importin-β. *J Biol Chem* **277:** 50597–50606. doi:10.1074/jbc.M209037200

Beck M, Hurt E. 2017. The nuclear pore complex: understanding its function through structural insight. *Nat Rev Mol Cell Biol* **18:** 73–89. doi:10.1038/nrm.2016.147

Beck M, Forster F, Ecke M, Plitzko JM, Melchior F, Gerisch G, Baumeister W, Medalia O. 2004. Nuclear pore complex structure and dynamics revealed by cryoelectron tomography. *Science* **306:** 1387–1390. doi:10.1126/science.1104808

Beck M, Lucic V, Förster F, Baumeister W, Medalia O. 2007. Snapshots of nuclear pore complexes in action captured by cryo-electron tomography. *Nature* **449:** 611–615. doi:10.1038/nature06170

Beck M, Schirmacher P, Singer S. 2017. Alterations of the nuclear transport system in hepatocellular carcinoma—new basis for therapeutic strategies. *J Hepatol* **67:** 1051–1061. doi:10.1016/j.jhep.2017.06.021

Belgareh N, Rabut G, Baï SW, van Overbeek M, Beaudouin J, Daigle N, Zatsepina OV, Pasteau F, Labas V, Fromont-Racine M, et al. 2001. An evolutionarily conserved NPC subcomplex, which redistributes in part to kinetochores in mammalian cells. *J Cell Biol* **154:** 1147–1160. doi:10.1083/jcb.200101081

Belov GA, Lidsky PV, Mikitas OV, Egger D, Lukyanov KA, Bienz K, Agol VI. 2004. Bidirectional increase in permeability of nuclear envelope upon poliovirus infection and accompanying alterations of nuclear pores. *J Virol* **78:** 10166–10177. doi:10.1128/JVI.78.18.10166-10177.2004

Blower MD, Nachury M, Heald R, Weis K. 2005. A Rae1-containing ribonucleoprotein complex is required for mitotic spindle assembly. *Cell* **121:** 223–234. doi:10.1016/j.cell.2005.02.016

Bodoor K, Shaikh S, Salina D, Raharjo WH, Bastos R, Lohka M, Burke B. 1999. Sequential recruitment of NPC proteins to the nuclear periphery at the end of mitosis. *J Cell Sci* **112:** 2253–2264. doi:10.1242/jcs.112.13.2253

Boettcher B, Barral Y. 2013. The cell biology of open and closed mitosis. *Nucleus* **4:** 160–165. doi:10.4161/nucl.24676

Bonnet A, Palancade B. 2015. Intron or no intron: a matter for nuclear pore complexes. *Nucleus* **6:** 455–461. doi:10.1080/19491034.2015.1116660

Buchwalter A, Kaneshiro JM, Hetzer MW. 2019. Coaching from the sidelines: the nuclear periphery in genome regulation. *Nat Rev Genet* **20:** 39–50. doi:10.1038/s41576-018-0063-5

Bui KH, von Appen A, DiGuilio AL, Ori A, Sparks L, Mackmull MT, Bock T, Hagen W, Andres-Pons A, Glavy JS, et al. 2013. Integrated structural analysis of the human nuclear pore complex scaffold. *Cell* **155:** 1233–1243. doi:10.1016/j.cell.2013.10.055

Bukata L, Parker SL, D'Angelo MA. 2013. Nuclear pore complexes in the maintenance of genome integrity. *Curr Opin Cell Biol* **25:** 378–386. doi:10.1016/j.ceb.2013.03.002

Burdine RD, Preston CC, Leonard RJ, Bradley TA, Faustino RS. 2020. Nucleoporins in cardiovascular disease. *J Mol Cell Cardiol* **141:** 43–52. doi:10.1016/j.yjmcc.2020.02.010

Burke B, Ellenberg J. 2002. Remodelling the walls of the nucleus. *Nat Rev Mol Cell Biol* **3:** 487–497. doi:10.1038/nrm860

Callan HG, Tomlin SG. 1950. Experimental studies on amphibian oocyte nuclei. I: Investigation of the structure of the nuclear membrane by means of the electron microscope. *Proc R Soc Lond B Biol Sci* **137:** 367–378. doi:10.1098/rspb.1950.0047

Campbell MS, Chan GK, Yen TJ. 2001. Mitotic checkpoint proteins HsMAD1 and HsMAD2 are associated with nuclear pore complexes in interphase. *J Cell Sci* **114:** 953–963. doi:10.1242/jcs.114.5.953

Capelson M, Doucet C, Hetzer MW. 2010a. Nuclear pore complexes: guardians of the nuclear genome. *Cold Spring Harb Symp Quant Biol* **75:** 585–597. doi:10.1101/sqb.2010.75.059

Capelson M, Liang Y, Schulte R, Mair W, Wagner U, Hetzer MW. 2010b. Chromatin-bound nuclear pore components regulate gene expression in higher eukaryotes. *Cell* **140:** 372–383. doi:10.1016/j.cell.2009.12.054

Capitanchik C, Dixon CR, Swanson SK, Florens L, Kerr ARW, Schirmer EC. 2018. Analysis of RNA-Seq datasets reveals enrichment of tissue-specific splice variants for nuclear envelope proteins. *Nucleus* **9:** 410–430. doi:10.1080/19491034.2018.1469351

Carmody SR, Wente SR. 2009. mRNA nuclear export at a glance. *J Cell Sci* **122:** 1933–1937. doi:10.1242/jcs.041236

Carter-O'Connell I, Jin H, Morgan RK, Zaja R, David LL, Ahel I, Cohen MS. 2016. Identifying family-member-specific targets of mono-ARTDs by using a chemical genetics approach. *Cell Rep* **14:** 621–631. doi:10.1016/j.celrep.2015.12.045

Cautain B, Hill R, de Pedro N, Link W. 2015. Components and regulation of nuclear transport processes. *FEBS J* **282:** 445–462. doi:10.1111/febs.13163

Celetti G, Paci G, Caria J, VanDelinder V, Bachand G, Lemke EA. 2020. The liquid state of FG-nucleoporins mimics permeability barrier properties of nuclear pore complexes. *J Cell Biol* **219:** e201907157. doi:10.1083/jcb.201907157

Chadrin A, Hess B, San Roman M, Gatti X, Lombard B, Loew D, Barral Y, Palancade B, Doye V. 2010. Pom33, a novel transmembrane nucleoporin required for proper nuclear pore complex distribution. *J Cell Biol* **189:** 795–811. doi:10.1083/jcb.200910043

Chial HJ, Rout MP, Giddings TH, Winey M. 1998. *Saccharomyces cerevisiae* Ndc1p is a shared component of nuclear pore complexes and spindle pole bodies. *J Cell Biol* **143:** 1789–1800. doi:10.1083/jcb.143.7.1789

Chow KH, Elgort S, Dasso M, Ullman KS. 2012. Two distinct sites in Nup153 mediate interaction with the SUMO proteases SENP1 and SENP2. *Nucleus* **3:** 349–358. doi:10.4161/nucl.20822

Chow KH, Elgort S, Dasso M, Powers MA, Ullman KS. 2014. The SUMO proteases SENP1 and SENP2 play a critical role in nucleoporin homeostasis and nuclear pore complex function. *Mol Biol Cell* **25:** 160–168. doi:10.1091/mbc.e13-05-0256

Cordes VC, Gajewski A, Stumpp S, Krohne G. 1995. Immunocytochemistry of annulate lamellae: potential cell biological markers for studies of cell differentiation and pathology. *Differentiation* **58:** 307–312. doi:10.1046/j.1432-0436.1995.5840307.x

Cordes VC, Reidenbach S, Franke WW. 1996. Cytoplasmic annulate lamellae in cultured cells: composition, distribution, and mitotic behavior. *Cell Tissue Res* **284:** 177–191. doi:10.1007/s004410050578

Cotter L, Allen TD, Kiseleva E, Goldberg MW. 2007. Nuclear membrane disassembly and rupture. *J Mol Biol* **369:** 683–695. doi:10.1016/j.jmb.2007.03.051

Cronshaw JM, Krutchinsky AN, Zhang W, Chait BT, Matunis MJ. 2002. Proteomic analysis of the mammalian nuclear pore complex. *J Cell Biol* **158:** 915–927. doi:10.1083/jcb.200206106

Dabauvalle MC, Loos K, Merkert H, Scheer U. 1991. Spontaneous assembly of pore complex-containing membranes ("annulate lamellae") in *Xenopus* egg extract in the absence of chromatin. *J Cell Biol* **112:** 1073–1082. doi:10.1083/jcb.112.6.1073

Daigle N, Beaudouin J, Hartnell L, Imreh G, Hallberg E, Lippincott-Schwartz J, Ellenberg J. 2001. Nuclear pore complexes form immobile networks and have a very low turnover in live mammalian cells. *J Cell Biol* **154:** 71–84. doi:10.1083/jcb.200101089

D'Angelo MA, Hetzer MW. 2008. Structure, dynamics and function of nuclear pore complexes. *Trends Cell Biol* **18:** 456–466. doi:10.1016/j.tcb.2008.07.009

D'Angelo MA, Anderson DJ, Richard E, Hetzer MW. 2006. Nuclear pores form de novo from both sides of the nuclear envelope. *Science* **312:** 440–443. doi:10.1126/science.1124196

D'Angelo MA, Raices M, Panowski SH, Hetzer MW. 2009. Age-dependent deterioration of nuclear pore complexes causes a loss of nuclear integrity in postmitotic cells. *Cell* **136:** 284–295. doi:10.1016/j.cell.2008.11.037

D'Angelo MA, Gomez-Cavazos JS, Mei A, Lackner DH, Hetzer MW. 2012. A change in nuclear pore complex composition regulates cell differentiation. *Dev Cell* **22:** 446–458. doi:10.1016/j.devcel.2011.11.021

de Castro IJ, Gil RS, Ligammari L, Di Giacinto ML, Vagnarelli P. 2018. CDK1 and PLK1 coordinate the disassembly and reassembly of the nuclear envelope in vertebrate mitosis. *Oncotarget* **9:** 7763–7773. doi:10.18632/oncotarget.23666

De Souza CP, Osmani AH, Hashmi SB, Osmani SA. 2004. Partial nuclear pore complex disassembly during closed mitosis in *Aspergillus nidulans*. *Curr Biol* **14:** 1973–1984. doi:10.1016/j.cub.2004.10.050

Dey G, Culley S, Curran S, Schmidt U, Henriques R, Kukulski W, Baum B. 2020. Closed mitosis requires local disassembly of the nuclear envelope. *Nature* **585:** 119–123. doi:10.1038/s41586-020-2648-3

Dormann D. 2020. FG-nucleoporins caught in the act of liquid-liquid phase separation. *J Cell Biol* **219:** e201910211. doi:10.1083/jcb.201910211

Doucet CM, Talamas JA, Hetzer MW. 2010. Cell cycle-dependent differences in nuclear pore complex assembly in metazoa. *Cell* **141:** 1030–1041. doi:10.1016/j.cell.2010.04.036

Drin G, Casella JF, Gautier R, Boehmer T, Schwartz TU, Antonny B. 2007. A general amphipathic α-helical motif for sensing membrane curvature. *Nat Struct Mol Biol* **14:** 138–146. doi:10.1038/nsmb1194

Drummond SP, Rutherford SA, Sanderson HS, Allen TD. 2006. High resolution analysis of mammalian nuclear structure throughout the cell cycle: implications for nuclear pore complex assembly during interphase and mitosis. *Can J Physiol Pharmacol* **84:** 423–430. doi:10.1139/y05-148

Duheron V, Nilles N, Pecenko S, Martinelli V, Fahrenkrog B. 2017. Localisation of Nup153 and SENP1 to nuclear pore complexes is required for 53BP1-mediated DNA double-strand break repair. *J Cell Sci* **130:** 2306–2316. doi:10.1242/jcs.198390

Dultz E, Ellenberg J. 2010. Live imaging of single nuclear pores reveals unique assembly kinetics and mechanism in interphase. *J Cell Biol* **191:** 15–22. doi:10.1083/jcb.201007076

Dultz E, Zanin E, Wurzenberger C, Braun M, Rabut G, Sironi L, Ellenberg J. 2008. Systematic kinetic analysis of mitotic dis- and reassembly of the nuclear pore in living cells. *J Cell Biol* **180:** 857–865. doi:10.1083/jcb.200707026

Dultz E, Huet S, Ellenberg J. 2009. Formation of the nuclear envelope permeability barrier studied by sequential photoswitching and flux analysis. *Biophys J* **97:** 1891–1897. doi:10.1016/j.bpj.2009.07.024

Eftekharzadeh B, Daigle JG, Kapinos LE, Coyne A, Schiantarelli J, Carlomagno Y, Cook C, Miller SJ, Dujardin S, Amaral AS, et al. 2018. Tau protein disrupts nucleocytoplasmic transport in Alzheimer's disease. *Neuron* **99:** 925–940.e7. doi:10.1016/j.neuron.2018.07.039

Eibauer M, Pellanda M, Turgay Y, Dubrovsky A, Wild A, Medalia O. 2015. Structure and gating of the nuclear pore complex. *Nat Commun* **6:** 7532. doi:10.1038/ncomms8532

Eisenhardt N, Redolfi J, Antonin W. 2014. Interaction of Nup53 with Ndc1 and Nup155 is required for nuclear pore complex assembly. *J Cell Sci* **127:** 908–921. doi:10.1242/jcs.141739

Elad N, Maimon T, Frenkiel-Krispin D, Lim RY, Medalia O. 2009. Structural analysis of the nuclear pore complex by integrated approaches. *Curr Opin Struct Biol* **19:** 226–232. doi:10.1016/j.sbi.2009.02.009

Eustice M, Bond MR, Hanover JA. 2017. O-GlcNAc cycling and the regulation of nucleocytoplasmic dynamics. *Biochem Soc Trans* **45:** 427–436. doi:10.1042/BST20160171

Expósito-Serrano M, Sánchez-Molina A, Gallardo P, Salas-Pino S, Daga RR. 2020. Selective nuclear pore complex removal drives nuclear envelope division in fission yeast. *Curr Biol* **30:** 3212–3222.e2. doi:10.1016/j.cub.2020.05.066

Fagotto F, Glück U, Gumbiner BM. 1998. Nuclear localization signal-independent and importin/karyopherin-independent nuclear import of β-catenin. *Curr Biol* **8:** 181–190. doi:10.1016/S0960-9822(98)70082-X

Faustino RS, Maddaford TG, Pierce GN. 2011. Mitogen activated protein kinase at the nuclear pore complex. *J Cell Mol Med* **15:** 928–937. doi:10.1111/j.1582-4934.2010.01093.x

Favreau C, Worman HJ, Wozniak RW, Frappier T, Courvalin JC. 1996. Cell cycle-dependent phosphorylation of nucleoporins and nuclear pore membrane protein Gp210. *Biochemistry* **35:** 8035–8044. doi:10.1021/bi9600660

Feldherr CM, Akin D. 1990. The permeability of the nuclear envelope in dividing and nondividing cell cultures. *J Cell Biol* **111:** 1–8. doi:10.1083/jcb.111.1.1

Fichtman B, Ramos C, Rasala B, Harel A, Forbes DJ. 2010. Inner/outer nuclear membrane fusion in nuclear pore assembly: biochemical demonstration and molecular analysis. *Mol Biol Cell* **21:** 4197–4211. doi:10.1091/mbc.e10-04-0309

Field MC, Rout MP. 2019. Pore timing: the evolutionary origins of the nucleus and nuclear pore complex. *F1000Res* **8:** 369. doi:10.12688/f1000research.16402.1

Finlay DR, Meier E, Bradley P, Horecka J, Forbes DJ. 1991. A complex of nuclear pore proteins required for pore function. *J Cell Biol* **114:** 169–183. doi:10.1083/jcb.114.1.169

Fischer J, Teimer R, Amlacher S, Kunze R, Hurt E. 2015. Linker nups connect the nuclear pore complex inner ring with the outer ring and transport channel. *Nat Struct Mol Biol* **22:** 774–781. doi:10.1038/nsmb.3084

Fisher PDE, Shen Q, Akpinar B, Davis LK, Chung KKH, Baddeley D, Šarić A, Melia TJ, Hoogenboom BW, Lin C, et al. 2018. A programmable DNA origami platform for organizing intrinsically disordered nucleoporins within nanopore confinement. *ACS Nano* **12:** 1508–1518. doi:10.1021/acsnano.7b08044

Folz H, Niño CA, Taranum S, Caesar S, Latta L, Waharte F, Salamero J, Schlenstedt G, Dargemont C. 2019. SUMOylation of the nuclear pore complex basket is involved in sensing cellular stresses. *J Cell Sci* **132:** jcs224279. doi:10.1242/jcs.224279

Forler D, Rabut G, Ciccarelli FD, Herold A, Köcher T, Niggeweg R, Bork P, Ellenberg J, Izaurralde E. 2004. RanBP2/Nup358 provides a major binding site for NXF1-p15 dimers at the nuclear pore complex and functions in nuclear mRNA export. *Mol Cell Biol* **24:** 1155–1167. doi:10.1128/MCB.24.3.1155-1167.2004

Franks TM, Benner C, Narvaiza I, Marchetto MC, Young JM, Malik HS, Gage FH, Hetzer MW. 2016. Evolution of a transcriptional regulator from a transmembrane nucleoporin. *Genes Dev* **30:** 1155–1171.

Franz C, Askjaer P, Antonin W, Iglesias CL, Haselmann U, Schelder M, de Marco A, Wilm M, Antony C, Mattaj IW. 2005. Nup155 regulates nuclear envelope and nuclear pore complex formation in nematodes and vertebrates. *EMBO J* **24:** 3519–3531. doi:10.1038/sj.emboj.7600825

Franz C, Walczak R, Yavuz S, Santarella R, Gentzel M, Askjaer P, Galy V, Hetzer M, Mattaj IW, Antonin W. 2007. MEL-28/ELYS is required for the recruitment of nucleoporins to chromatin and postmitotic nuclear pore complex assembly. *EMBO Rep* **8:** 165–172. doi:10.1038/sj.embor.7400889

Frenkiel-Krispin D, Maco B, Aebi U, Medalia O. 2010. Structural analysis of a metazoan nuclear pore complex reveals a fused concentric ring architecture. *J Mol Biol* **395:** 578–586. doi:10.1016/j.jmb.2009.11.010

Freudenreich CH, Su XA. 2016. Relocalization of DNA lesions to the nuclear pore complex. *FEMS Yeast Res* **16:** fow095. doi:10.1093/femsyr/fow095

Frey S, Görlich D. 2007. A saturated FG-repeat hydrogel can reproduce the permeability properties of nuclear pore complexes. *Cell* **130:** 512–523. doi:10.1016/j.cell.2007.06.024

Frey S, Görlich D. 2009. FG/FxFG as well as GLFG repeats form a selective permeability barrier with self-healing properties. *EMBO J* **28:** 2554–2567. doi:10.1038/emboj.2009.199

Frey S, Richter RP, Gorlich D. 2006. FG-rich repeats of nuclear pore proteins form a three-dimensional meshwork with hydrogel-like properties. *Science* **314:** 815–817. doi:10.1126/science.1132516

Fribourg S, Braun IC, Izaurralde E, Conti E. 2001. Structural basis for the recognition of a nucleoporin FG repeat by the NTF2-like domain of the TAP/p15 mRNA nuclear export factor. *Mol Cell* **8:** 645–656. doi:10.1016/S1097-2765(01)00348-3

Funakoshi T, Maeshima K, Yahata K, Sugano S, Imamoto F, Imamoto N. 2007. Two distinct human POM121 genes: requirement for the formation of nuclear pore complexes. *FEBS Lett* **581:** 4910–4916. doi:10.1016/j.febslet.2007.09.021

Funakoshi T, Clever M, Watanabe A, Imamoto N. 2011. Localization of Pom121 to the inner nuclear membrane is required for an early step of interphase nuclear pore complex assembly. *Mol Biol Cell* **22:** 1058–1069. doi:10.1091/mbc.e10-07-0641

Gaillard H, Santos-Pereira JM, Aguilera A. 2019. The Nup84 complex coordinates the DNA damage response to warrant genome integrity. *Nucleic Acids Res* **47:** 4054–4067. doi:10.1093/nar/gkz066

Gall JG. 1967. Octagonal nuclear pores. *J Cell Biol* **32:** 391–399. doi:10.1083/jcb.32.2.391

Galy V, Askjaer P, Franz C, López-Iglesias C, Mattaj IW. 2006. MEL-28, a novel nuclear-envelope and kinetochore protein essential for zygotic nuclear-envelope assembly in *C. elegans*. *Curr Biol* **16:** 1748–1756. doi:10.1016/j.cub.2006.06.067

Gillespie PJ, Khoudoli GA, Stewart G, Swedlow JR, Blow JJ. 2007. ELYS/MEL-28 chromatin association coordinates nuclear pore complex assembly and replication licensing. *Curr Biol* **17:** 1657–1662. doi:10.1016/j.cub.2007.08.041

Glavy JS, Krutchinsky AN, Cristea IM, Berke IC, Boehmer T, Blobel G, Chait BT. 2007. Cell-cycle-dependent phosphorylation of the nuclear pore Nup107-160 subcomplex. *Proc Natl Acad Sci* **104:** 3811–3816. doi:10.1073/pnas.0700058104

Goldberg MW, Allen TD. 1992. High resolution scanning electron microscopy of the nuclear envelope: demonstration of a new, regular, fibrous lattice attached to the baskets of the nucleoplasmic face of the nuclear pores. *J Cell Biol* **119:** 1429–1440. doi:10.1083/jcb.119.6.1429

Goldberg MW, Wiese C, Allen TD, Wilson KL. 1997. Dimples, pores, star-rings, and thin rings on growing nuclear envelopes: evidence for structural intermediates in nuclear pore complex assembly. *J Cell Sci* **110:** 409–420. doi:10.1242/jcs.110.4.409

Griffis ER, Xu S, Powers MA. 2003. Nup98 localizes to both nuclear and cytoplasmic sides of the nuclear pore and binds to two distinct nucleoporin subcomplexes. *Mol Biol Cell* **14:** 600–610. doi:10.1091/mbc.e02-09-0582

Gu C, Vovk A, Zheng T, Coalson RD, Zilman A. 2019. The role of cohesiveness in the permeability of the spatial assemblies of FG nucleoporins. *Biophys J* **116:** 1204–1215. doi:10.1016/j.bpj.2019.02.028

Guglielmi V, Sakuma S, D'Angelo MA. 2020. Nuclear pore complexes in development and tissue homeostasis. *Development* **147:** dev183442. doi:10.1242/dev.183442

Hallberg E, Wozniak RW, Blobel G. 1993. An integral membrane protein of the pore membrane domain of the nuclear envelope contains a nucleoporin-like region. *J Cell Biol* **122:** 513–521. doi:10.1083/jcb.122.3.513

Hampoelz B, Mackmull MT, Machado P, Ronchi P, Bui KH, Schieber N, Santarella-Mellwig R, Necakov A, Andres-Pons A, Philippe JM, et al. 2016. Pre-assembled nuclear pores insert into the nuclear envelope during early development. *Cell* **166:** 664–678. doi:10.1016/j.cell.2016.06.015

Hampoelz B, Andres-Pons A, Kastritis P, Beck M. 2019a. Structure and assembly of the nuclear pore complex. *Annu Rev Biophys* **48:** 515–536. doi:10.1146/annurev-biophys-052118-115308

Hampoelz B, Schwarz A, Ronchi P, Bragulat-Teixidor H, Tischer C, Gaspar I, Ephrussi A, Schwab Y, Beck M. 2019b. Nuclear pores assemble from nucleoporin condensates during oogenesis. *Cell* **179:** 671–686.e17. doi:10.1016/j.cell.2019.09.022

Harel A, Chan RC, Lachish-Zalait A, Zimmerman E, Elbaum M, Forbes DJ. 2003. Importin β negatively regulates nuclear membrane fusion and nuclear pore complex assembly. *Mol Biol Cell* **14:** 4387–4396. doi:10.1091/mbc.e03-05-0275

Hayakawa A, Babour A, Sengmanivong L, Dargemont C. 2012. Ubiquitylation of the nuclear pore complex controls nuclear migration during mitosis in *S. cerevisiae*. *J Cell Biol* **196:** 19–27. doi:10.1083/jcb.201108124

Hayama R, Rout MP, Fernandez-Martinez J. 2017. The nuclear pore complex core scaffold and permeability barrier: variations of a common theme. *Curr Opin Cell Biol* **46:** 110–118. doi:10.1016/j.ceb.2017.05.003

Hayama R, Sorci M, Keating Iv JJ, Hecht LM, Plawsky JL, Belfort G, Chait BT, Rout MP. 2019. Interactions of nuclear transport factors and surface-conjugated FG nucleoporins: insights and limitations. *PLoS ONE* **14:** e0217897. doi:10.1371/journal.pone.0217897

Hinshaw JE, Carragher BO, Milligan RA. 1992. Architecture and design of the nuclear pore complex. *Cell* **69:** 1133–1141. doi:10.1016/0092-8674(92)90635-P

Hoelz A, Glavy JS, Beck M. 2016. Toward the atomic structure of the nuclear pore complex: when top down meets bottom up. *Nat Struct Mol Biol* **23:** 624–630. doi:10.1038/nsmb.3244

Horigome C, Unozawa E, Ooki T, Kobayashi T. 2019. Ribosomal RNA gene repeats associate with the nuclear pore complex for maintenance after DNA damage. *PLoS Genet* **15:** e1008103. doi:10.1371/journal.pgen.1008103

Huguet F, Flynn S, Vagnarelli P. 2019. The role of phosphatases in nuclear envelope disassembly and reassembly and their relevance to pathologies. *Cells* **8:** 687. doi:10.3390/cells8070687

Hülsmann BB, Labokha AA, Görlich D. 2012. The permeability of reconstituted nuclear pores provides direct evidence for the selective phase model. *Cell* **150:** 738–751. doi:10.1016/j.cell.2012.07.019

Hutten S, Dormann D. 2019. Nucleocytoplasmic transport defects in neurodegeneration—cause or consequence? *Semin Cell Dev Biol* **99:** 151–162. doi:10.1016/j.semcdb.2019.05.020

Ikegami K, Lieb JD. 2013. Integral nuclear pore proteins bind to Pol III-transcribed genes and are required for Pol III transcript processing in *C. elegans*. *Mol Cell* **51:** 840–849. doi:10.1016/j.molcel.2013.08.001

Inoue A, Zhang Y. 2014. Nucleosome assembly is required for nuclear pore complex assembly in mouse zygotes. *Nat Struct Mol Biol* **21:** 609–616. doi:10.1038/nsmb.2839

Iouk T, Kerscher O, Scott RJ, Basrai MA, Wozniak RW. 2002. The yeast nuclear pore complex functionally interacts with components of the spindle assembly checkpoint. *J Cell Biol* **159:** 807–819. doi:10.1083/jcb.200205068

Iovine MK, Watkins JL, Wente SR. 1995. The GLFG repetitive region of the nucleoporin Nup116p interacts with Kap95p, an essential yeast nuclear import factor. *J Cell Biol* **131:** 1699–1713. doi:10.1083/jcb.131.6.1699

Isgro TA, Schulten K. 2007a. Association of nuclear pore FG-repeat domains to NTF2 import and export complexes. *J Mol Biol* **366:** 330–345. doi:10.1016/j.jmb.2006.11.048

Isgro TA, Schulten K. 2007b. Cse1p-binding dynamics reveal a binding pattern for FG-repeat nucleoporins on transport receptors. *Structure* **15:** 977–991. doi:10.1016/j.str.2007.06.011

Jacinto FV, Benner C, Hetzer MW. 2015. The nucleoporin Nup153 regulates embryonic stem cell pluripotency through gene silencing. *Genes Dev* **29:** 1224–1238. doi:10.1101/gad.260919.115

Jackman M, Marcozzi C, Barbiero M, Pardo M, Yu L, Tyson AL, Choudhary JS, Pines J. 2020. Cyclin B1-Cdk1 facilitates MAD1 release from the nuclear pore to ensure a robust spindle checkpoint. *J Cell Biol* **219:** e201907082. doi:10.1083/jcb.201907082

Joseph J, Dasso M. 2008. The nucleoporin Nup358 associates with and regulates interphase microtubules. *FEBS Lett* **582:** 190–196. doi:10.1016/j.febslet.2007.11.087

Joseph J, Liu ST, Jablonski SA, Yen TJ, Dasso M. 2004. The RanGAP1-RanBP2 complex is essential for microtubule-kinetochore interactions in vivo. *Curr Biol* **14:** 611–617. doi:10.1016/j.cub.2004.03.031

Jühlen R, Fahrenkrog B. 2018. Moonlighting nuclear pore proteins: tissue-specific nucleoporin function in health and disease. *Histochem Cell Biol* **150:** 593–605. doi:10.1007/s00418-018-1748-8

Kabachinski G, Schwartz TU. 2015. The nuclear pore complex—structure and function at a glance. *J Cell Sci* **128:** 423–429. doi:10.1242/jcs.083246

Kalverda B, Pickersgill H, Shloma VV, Fornerod M. 2010. Nucleoporins directly stimulate expression of developmental and cell-cycle genes inside the nucleoplasm. *Cell* **140:** 360–371. doi:10.1016/j.cell.2010.01.011

Katsani KR, Karess RE, Dostatni N, Doye V. 2008. In vivo dynamics of *Drosophila* nuclear envelope components. *Mol Biol Cell* **19:** 3652–3666. doi:10.1091/mbc.e07-11-1162

Kehat I, Accornero F, Aronow BJ, Molkentin JD. 2011. Modulation of chromatin position and gene expression by HDAC4 interaction with nucleoporins. *J Cell Biol* **193:** 21–29. doi:10.1083/jcb.201101046

Kim SJ, Fernandez-Martinez J, Sampathkumar P, Martel A, Matsui T, Tsuruta H, Weiss TM, Shi Y, Markina-Inarrairaegui A, Bonanno JB, et al. 2014. Integrative structure–function mapping of the nucleoporin Nup133 suggests a conserved mechanism for membrane anchoring of the nuclear pore complex. *Mol Cell Proteomics* **13:** 2911–2926. doi:10.1074/mcp.M114.040915

Kim SJ, Fernandez-Martinez J, Nudelman I, Shi Y, Zhang W, Raveh B, Herricks T, Slaughter BD, Hogan JA, Upla P, et al. 2018. Integrative structure and functional anatomy of a nuclear pore complex. *Nature* **555:** 475–482. doi:10.1038/nature26003

Kimura M, Imamoto N. 2014. Biological significance of the importin-β family-dependent nucleocytoplasmic transport pathways. *Traffic* **15:** 727–748. doi:10.1111/tra.12174

Kirby IT, Kojic A, Arnold MR, Thorsell AG, Karlberg T, Vermehren-Schmaedick A, Sreenivasan R, Schultz C, Schüler H, Cohen MS. 2018. A potent and selective PARP11 inhibitor suggests coupling between cellular localization and catalytic activity. *Cell Chem Biol* **25:** 1547–1553.e12. doi:10.1016/j.chembiol.2018.09.011

Kiseleva E, Rutherford S, Cotter LM, Allen TD, Goldberg MW. 2001. Steps of nuclear pore complex disassembly and reassembly during mitosis in early *Drosophila* embryos. *J Cell Sci* **114:** 3607–3618. doi:10.1242/jcs.114.20.3607

Kiseleva E, Allen TD, Rutherford S, Bucci M, Wente SR, Goldberg MW. 2004. Yeast nuclear pore complexes have a cytoplasmic ring and internal filaments. *J Struct Biol* **145:** 272–288. doi:10.1016/j.jsb.2003.11.010

Köhler A, Hurt E. 2007. Exporting RNA from the nucleus to the cytoplasm. *Nat Rev Mol Cell Biol* **8:** 761–773. doi:10.1038/nrm2255

Kosako H, Imamoto N. 2010. Phosphorylation of nucleoporins: signal transduction-mediated regulation of their interaction with nuclear transport receptors. *Nucleus* **1:** 309–313. doi:10.4161/nucl.1.4.11744

Kosinski J, Mosalaganti S, von Appen A, Teimer R, DiGuilio AL, Wan W, Bui KH, Hagen WJ, Briggs JA, Glavy JS, et al. 2016. Molecular architecture of the inner ring scaffold of the human nuclear pore complex. *Science* **352:** 363–365. doi:10.1126/science.aaf0643

Kumar A, Sharma P, Gomar-Alba M, Shcheprova Z, Daulny A, Sanmartin T, Matucci I, Funaya C, Beato M, Mendoza M. 2018. Daughter-cell-specific modulation of nuclear pore complexes controls cell cycle entry during asymmetric division. *Nat Cell Biol* **20:** 432–442. doi:10.1038/s41556-018-0056-9

Laurell E, Beck K, Krupina K, Theerthagiri G, Bodenmiller B, Horvath P, Aebersold R, Antonin W, Kutay U. 2011. Phosphorylation of Nup98 by multiple kinases is crucial for NPC disassembly during mitotic entry. *Cell* **144:** 539–550. doi:10.1016/j.cell.2011.01.012

Lee CW, Wilfling F, Ronchi P, Allegretti M, Mosalaganti S, Jentsch S, Beck M, Pfander B. 2020. Selective autophagy degrades nuclear pore complexes. *Nat Cell Biol* **22:** 159–166. doi:10.1038/s41556-019-0459-2

Leksa NC, Brohawn SG, Schwartz TU. 2009. The structure of the scaffold nucleoporin Nup120 reveals a new and unexpected domain architecture. *Structure* **17:** 1082–1091. doi:10.1016/j.str.2009.06.003

Lewis A, Felberbaum R, Hochstrasser M. 2007. A nuclear envelope protein linking nuclear pore basket assembly, SUMO protease regulation, and mRNA surveillance. *J Cell Biol* **178:** 813–827. doi:10.1083/jcb.200702154

Li HY, Wirtz D, Zheng Y. 2003. A mechanism of coupling RCC1 mobility to RanGTP production on the chromatin in vivo. *J Cell Biol* **160:** 635–644. doi:10.1083/jcb.200211004

Li C, Goryaynov A, Yang W. 2016. The selective permeability barrier in the nuclear pore complex. *Nucleus* **7:** 430–446. doi:10.1080/19491034.2016.1238997

Liang Y, Franks TM, Marchetto MC, Gage FH, Hetzer MW. 2013. Dynamic association of NUP98 with the human genome. *PLoS Genet* **9:** e1003308. doi:10.1371/journal.pgen.1003308

Lim RY, Aebi U, Fahrenkrog B. 2008. Towards reconciling structure and function in the nuclear pore complex. *Histochem Cell Biol* **129:** 105–116. doi:10.1007/s00418-007-0371-x

Lin DH, Hoelz A. 2019. The structure of the nuclear pore complex (an update). *Annu Rev Biochem* **88:** 725–783. doi:10.1146/annurev-biochem-062917-011901

Lin DH, Stuwe T, Schilbach S, Rundlet EJ, Perriches T, Mobbs G, Fan Y, Thierbach K, Huber FM, Collins LN, et al. 2016. Architecture of the symmetric core of the nuclear pore. *Science* **352:** aaf1015. doi:10.1126/science.aaf1015

Linder MI, Köhler M, Boersema P, Weberruss M, Wandke C, Marino J, Ashiono C, Picotti P, Antonin W, Kutay U. 2017. Mitotic disassembly of nuclear pore complexes involves CDK1- and PLK1-mediated phosphorylation of key interconnecting nucleoporins. *Dev Cell* **43:** 141–156.e7. doi:10.1016/j.devcel.2017.08.020

Liu Z, Yan M, Liang Y, Liu M, Zhang K, Shao D, Jiang R, Li L, Wang C, Nussenzveig DR, et al. 2019. Nucleoporin Seh1 interacts with Olig2/Brd7 to promote oligodendrocyte differentiation and myelination. *Neuron* **102:** 587–601.e7. doi:10.1016/j.neuron.2019.02.018

Loïodice I, Alves A, Rabut G, Van Overbeek M, Ellenberg J, Sibarita JB, Doye V. 2004. The entire Nup107-160 complex, including three new members, is targeted as one entity to kinetochores in mitosis. *Mol Biol Cell* **15:** 3333–3344. doi:10.1091/mbc.e03-12-0878

Löschberger A, Franke C, Krohne G, van de Linde S, Sauer M. 2014. Correlative super-resolution fluorescence and electron microscopy of the nuclear pore complex with molecular resolution. *J Cell Sci* **127:** 4351–4355. doi:10.1242/jcs.156620

Lowe AR, Tang JH, Yassif J, Graf M, Huang WY, Groves JT, Weis K, Liphardt JT. 2015. Importin-β modulates the permeability of the nuclear pore complex in a Ran-dependent manner. *eLife* **4:** e04052. doi:10.7554/eLife.04052

Lu L, Ladinsky MS, Kirchhausen T. 2011. Formation of the postmitotic nuclear envelope from extended ER cisternae precedes nuclear pore assembly. *J Cell Biol* **194:** 425–440. doi:10.1083/jcb.201012063

Lusk CP, Makhnevych T, Marelli M, Aitchison JD, Wozniak RW. 2002. Karyopherins in nuclear pore biogenesis: a role for Kap121p in the assembly of Nup53p into nuclear pore complexes. *J Cell Biol* **159:** 267–278. doi:10.1083/jcb.200203079

Lusk CP, Waller DD, Makhnevych T, Dienemann A, Whiteway M, Thomas DY, Wozniak RW. 2007. Nup53p is a target of two mitotic kinases, Cdk1p and Hrr25p. *Traffic* **8:** 647–660. doi:10.1111/j.1600-0854.2007.00559.x

Lussi YC, Shumaker DK, Shimi T, Fahrenkrog B. 2010. The nucleoporin Nup153 affects spindle checkpoint activity due to an association with Mad1. *Nucleus* **1:** 71–84. doi:10.4161/nucl.1.1.10244

Macaulay C, Forbes DJ. 1996. Assembly of the nuclear pore: biochemically distinct steps revealed with NEM, GTP γ-S, and BAPTA. *J Cell Biol* **132:** 5–20. doi:10.1083/jcb.132.1.5

Macaulay C, Meier E, Forbes DJ. 1995. Differential mitotic phosphorylation of proteins of the nuclear pore complex. *J Biol Chem* **270:** 254–262. doi:10.1074/jbc.270.1.254

Mackay DR, Howa AC, Werner TL, Ullman KS. 2017. Nup153 and Nup50 promote recruitment of 53BP1 to DNA repair foci by antagonizing BRCA1-dependent events. *J Cell Sci* **130:** 3347–3359. doi:10.1242/jcs.203513

Maco B, Fahrenkrog B, Huang NP, Aebi U. 2006. Nuclear pore complex structure and plasticity revealed by electron and atomic force microscopy. *Methods Mol Biol* **322:** 273–288. doi:10.1007/978-1-59745-000-3_19

Madrid AS, Mancuso J, Cande WZ, Weis K. 2006. The role of the integral membrane nucleoporins Ndc1p and Pom152p in nuclear pore complex assembly and function. *J Cell Biol* **173:** 361–371. doi:10.1083/jcb.200506199

Maeshima K, Yahata K, Sasaki Y, Nakatomi R, Tachibana T, Hashikawa T, Imamoto F, Imamoto N. 2006. Cell-cycle-dependent dynamics of nuclear pores: pore-free islands and lamins. *J Cell Sci* **119:** 4442–4451. doi:10.1242/jcs.03207

Maimon T, Elad N, Dahan I, Medalia O. 2012. The human nuclear pore complex as revealed by cryo-electron tomography. *Structure* **20:** 998–1006. doi:10.1016/j.str.2012.03.025

Makise M, Mackay DR, Elgort S, Shankaran SS, Adam SA, Ullman KS. 2012. The Nup153-Nup50 protein interface and its role in nuclear import. *J Biol Chem* **287:** 38515–38522. doi:10.1074/jbc.M112.378893

Martino L, Morchoisne-Bolhy S, Cheerambathur DK, Van Hove L, Dumont J, Joly N, Desai A, Doye V, Pintard L. 2017. Channel nucleoporins recruit PLK-1 to nuclear pore complexes to direct nuclear envelope breakdown in *C. elegans*. *Dev Cell* **43:** 157–171.e7. doi:10.1016/j.devcel.2017.09.019

Maul GG. 1970. Ultrastructure of pore complexes of annulate lamellae. *J Cell Biol* **46:** 604–610. doi:10.1083/jcb.46.3.604

Maul GG. 1977. Nuclear pore complexes. elimination and reconstruction during mitosis. *J Cell Biol* **74:** 492–500. doi:10.1083/jcb.74.2.492

Mészáros N, Cibulka J, Mendiburo MJ, Romanauska A, Schneider M, Köhler A. 2015. Nuclear pore basket proteins are tethered to the nuclear envelope and can regulate membrane curvature. *Dev Cell* **33:** 285–298. doi:10.1016/j.devcel.2015.02.017

Meyer-Ficca ML, Ihara M, Bader JJ, Leu NA, Beneke S, Meyer RG. 2015. Spermatid head elongation with normal nuclear shaping requires ADP-ribosyltransferase PARP11 (ARTD11) in mice. *Biol Reprod* **92:** 80.

Miao M, Ryan KJ, Wente SR. 2006. The integral membrane protein Pom34p functionally links nucleoporin subcomplexes. *Genetics* **172:** 1441–1457. doi:10.1534/genetics.105.052068

Miller MW, Caracciolo MR, Berlin WK, Hanover JA. 1999. Phosphorylation and glycosylation of nucleoporins. *Arch Biochem Biophys* **367:** 51–60. doi:10.1006/abbi.1999.1237

Mincer JS, Simon SM. 2011. Simulations of nuclear pore transport yield mechanistic insights and quantitative predictions. *Proc Natl Acad Sci* **108:** E351–E358. doi:10.1073/pnas.1104521108

Mishra RK, Chakraborty P, Arnaoutov A, Fontoura BM, Dasso M. 2010. The Nup107-160 complex and γ-TuRC regulate microtubule polymerization at kinetochores. *Nat Cell Biol* **12:** 164–169. doi:10.1038/ncb2016

Mitchell JM, Mansfeld J, Capitanio J, Kutay U, Wozniak RW. 2010. Pom121 links two essential subcomplexes of the nuclear pore complex core to the membrane. *J Cell Biol* **191:** 505–521. doi:10.1083/jcb.201007098

Mizuguchi-Hata C, Ogawa Y, Oka M, Yoneda Y. 2013. Quantitative regulation of nuclear pore complex proteins by O-GlcNAcylation. *Biochim Biophys Acta* **1833:** 2682–2689. doi:10.1016/j.bbamcr.2013.06.008

Mosalaganti S, Kosinski J, Albert S, Schaffer M, Strenkert D, Salome PA, Merchant SS, Plitzko JM, Baumeister W, Engel BD, et al. 2018. In situ architecture of the algal nuclear pore complex. *Nat Commun* **9:** 2361. doi:10.1038/s41467-018-04739-y

Mosammaparast N, Pemberton LF. 2004. Karyopherins: from nuclear-transport mediators to nuclear-function regulators. *Trends Cell Biol* **14:** 547–556. doi:10.1016/j.tcb.2004.09.004

Niño CA, Guet D, Gay A, Brutus S, Jourquin F, Mendiratta S, Salamero J, Geli V, Dargemont C. 2016. Posttranslational marks control architectural and functional plasticity of the nuclear pore complex basket. *J Cell Biol* **212:** 167–180. doi:10.1083/jcb.201506130

Nord MS, Bernis C, Carmona S, Garland DC, Travesa A, Forbes DJ. 2020. Exportins can inhibit major mitotic assembly events in vitro: membrane fusion, nuclear pore formation, and spindle assembly. *Nucleus* **11:** 178–193. doi:10.1080/19491034.2020.1798093

Nordeen SA, Turman DL, Schwartz TU. 2020. Yeast Nup84-Nup133 complex structure details flexibility and reveals conservation of the membrane anchoring ALPS motif. *Nat Commun* **11:** 6060. doi:10.1038/s41467-020-19885-5

Obado SO, Brillantes M, Uryu K, Zhang W, Ketaren NE, Chait BT, Field MC, Rout MP. 2016. Interactome mapping reveals the evolutionary history of the nuclear pore complex. *PLoS Biol* **14:** e1002365. doi:10.1371/journal.pbio.1002365

Ohtsubo M, Okazaki H, Nishimoto T. 1989. The RCC1 protein, a regulator for the onset of chromosome condensation locates in the nucleus and binds to DNA. *J Cell Biol* **109:** 1389–1397. doi:10.1083/jcb.109.4.1389

Okamura M, Inose H, Masuda S. 2015. RNA export through the NPC in eukaryotes. *Genes (Basel)* **6:** 124–149. doi:10.3390/genes6010124

Onischenko EA, Gubanova NV, Kiseleva EV, Hallberg E. 2005. Cdk1 and okadaic acid-sensitive phosphatases control assembly of nuclear pore complexes in *Drosophila* embryos. *Mol Biol Cell* **16:** 5152–5162. doi:10.1091/mbc.e05-07-0642

Onischenko E, Tang JH, Andersen KR, Knockenhauer KE, Vallotton P, Derrer CP, Kralt A, Mugler CF, Chan LY, Schwartz TU, et al. 2017. Natively unfolded FG repeats stabilize the structure of the nuclear pore complex. *Cell* **171:** 904–917.e19. doi:10.1016/j.cell.2017.09.033

Onischenko E, Noor E, Fischer JS, Gillet L, Wojtynek M, Vallotton P, Weis K. 2020. Maturation kinetics of a multiprotein complex revealed by metabolic labeling. *Cell* **183:** 1785–1800.e26. doi:10.1016/j.cell.2020.11.001

Ori A, Banterle N, Iskar M, Andrés-Pons A, Escher C, Khanh Bui H, Sparks L, Solis-Mezarino V, Rinner O, Bork P, et al. 2013. Cell type-specific nuclear pores: a case in point for context-dependent stoichiometry of molecular machines. *Mol Syst Biol* **9:** 648. doi:10.1038/msb.2013.4

Ori A, Iskar M, Buczak K, Kastritis P, Parca L, Andrés-Pons A, Singer S, Bork P, Beck M. 2016. Spatiotemporal variation of mammalian protein complex stoichiometries. *Genome Biol* **17:** 47. doi:10.1186/s13059-016-0912-5

Otsuka S, Ellenberg J. 2018. Mechanisms of nuclear pore complex assembly—two different ways of building one molecular machine. *FEBS Lett* **592:** 475–488. doi:10.1002/1873-3468.12905

Otsuka S, Bui KH, Schorb M, Hossain MJ, Politi AZ, Koch B, Eltsov M, Beck M, Ellenberg J. 2016. Nuclear pore assembly proceeds by an inside-out extrusion of the nuclear envelope. *eLife* **5:** e19071. doi:10.7554/eLife.19071

Otsuka S, Steyer AM, Schorb M, Hériché JK, Hossain MJ, Sethi S, Kueblbeck M, Schwab Y, Beck M, Ellenberg J. 2018. Postmitotic nuclear pore assembly proceeds by radial dilation of small membrane openings. *Nat Struct Mol Biol* **25:** 21–28. doi:10.1038/s41594-017-0001-9

Panse VG, Küster B, Gerstberger T, Hurt E. 2003. Unconventional tethering of Ulp1 to the transport channel of the nuclear pore complex by karyopherins. *Nat Cell Biol* **5:** 21–27. doi:10.1038/ncb893

Pascual-Garcia P, Capelson M. 2019. Nuclear pores in genome architecture and enhancer function. *Curr Opin Cell Biol* **58:** 126–133. doi:10.1016/j.ceb.2019.04.001

Pascual-Garcia P, Jeong J, Capelson M. 2014. Nucleoporin Nup98 associates with Trx/MLL and NSL histone-modifying complexes and regulates Hox gene expression. *Cell Rep* **9:** 433–442. doi:10.1016/j.celrep.2014.09.002

Pichler A, Knipscheer P, Saitoh H, Sixma TK, Melchior F. 2004. The RanBP2 SUMO E3 ligase is neither HECT- nor RING-type. *Nat Struct Mol Biol* **11:** 984–991. doi:10.1038/nsmb834

Pinzaru AM, Kareh M, Lamm N, Lazzerini-Denchi E, Cesare AJ, Sfeir A. 2020. Replication stress conferred by POT1 dysfunction promotes telomere relocalization to the nuclear pore. *Genes Dev* **34:** 1619–1636. doi:10.1101/gad.337287.120

Rabut G, Doye V, Ellenberg J. 2004. Mapping the dynamic organization of the nuclear pore complex inside single living cells. *Nat Cell Biol* **6:** 1114–1121. doi:10.1038/ncb1184

Raices M, D'Angelo MA. 2012. Nuclear pore complex composition: a new regulator of tissue-specific and developmental functions. *Nat Rev Mol Cell Biol* **13:** 687–699. doi:10.1038/nrm3461

Raices M, D'Angelo MA. 2017. Nuclear pore complexes and regulation of gene expression. *Curr Opin Cell Biol* **46:** 26–32. doi:10.1016/j.ceb.2016.12.006

Rajanala K, Sarkar A, Jhingan GD, Priyadarshini R, Jalan M, Sengupta S, Nandicoori VK. 2014. Phosphorylation of nucleoporin Tpr governs its differential localization and is required for its mitotic function. *J Cell Sci* **127:** 3505–3520. doi:10.1242/jcs.149112

Rajoo S, Vallotton P, Onischenko E, Weis K. 2018. Stoichiometry and compositional plasticity of the yeast nuclear pore complex revealed by quantitative fluorescence microscopy. *Proc Natl Acad Sci* **115:** E3969–E3977. doi:10.1073/pnas.1719398115

Randise-Hinchliff C, Brickner JH. 2016. Transcription factors dynamically control the spatial organization of the yeast genome. *Nucleus* **7:** 369–374. doi:10.1080/19491034.2016.1212797

Rasala BA, Orjalo AV, Shen Z, Briggs S, Forbes DJ. 2006. ELYS is a dual nucleoporin/kinetochore protein required for nuclear pore assembly and proper cell division. *Proc Natl Acad Sci* **103:** 17801–17806. doi:10.1073/pnas.0608484103

Rasala BA, Ramos C, Harel A, Forbes DJ. 2008. Capture of AT-rich chromatin by ELYS recruits POM121 and NDC1 to initiate nuclear pore assembly. *Mol Biol Cell* **19:** 3982–3996. doi:10.1091/mbc.e08-01-0012

Ribbeck K, Görlich D. 2001. Kinetic analysis of translocation through nuclear pore complexes. *EMBO J* **20:** 1320–1330. doi:10.1093/emboj/20.6.1320

Ribbeck K, Görlich D. 2002. The permeability barrier of nuclear pore complexes appears to operate via hydrophobic exclusion. *EMBO J* **21:** 2664–2671. doi:10.1093/emboj/21.11.2664

Ris H. 1989. Three-dimensional imaging of cell ultrastructure with high resolution low voltage SEM. *Inst Phys, Conf Ser* **98:** 657–662.

Rodriguez-Berriguete G, Granata G, Puliyadi R, Tiwana G, Prevo R, Wilson RS, Yu S, Buffa F, Humphrey TC, McKenna WG, et al. 2018. Nucleoporin 54 contributes to homologous recombination repair and post-replicative DNA integrity. *Nucleic Acids Res* **46:** 7731–7746. doi:10.1093/nar/gky569

Rodriguez-Bravo V, Maciejowski J, Corona J, Buch HK, Collin P, Kanemaki MT, Shah JV, Jallepalli PV. 2014. Nuclear pores protect genome integrity by assembling a premitotic and Mad1-dependent anaphase inhibitor. *Cell* **156:** 1017–1031. doi:10.1016/j.cell.2014.01.010

Rodriguez-Bravo V, Pippa R, Song WM, Carceles-Cordon M, Dominguez-Andres A, Fujiwara N, Woo J, Koh AP, Ertel A, Lokareddy RK, et al. 2018. Nuclear pores promote lethal prostate cancer by increasing POM121-driven E2F1, MYC, and AR nuclear import. *Cell* **174:** 1200–1215.e20. doi:10.1016/j.cell.2018.07.015

Roehrig S, Tabbert A, Ferrando-May E. 2003. In vitro measurement of nuclear permeability changes in apoptosis. *Anal Biochem* **318:** 244–253. doi:10.1016/S0003-2697(03)00242-2

Rotem A, Gruber R, Shorer H, Shaulov L, Klein E, Harel A. 2009. Importin β regulates the seeding of chromatin with initiation sites for nuclear pore assembly. *Mol Biol Cell* **20:** 4031–4042. doi:10.1091/mbc.e09-02-0150

Rout MP, Aitchison JD, Suprapto A, Hjertaas K, Zhao Y, Chait BT. 2000. The yeast nuclear pore complex: composition, architecture, and transport mechanism. *J Cell Biol* **148:** 635–652. doi:10.1083/jcb.148.4.635

Ryan KJ, McCaffery JM, Wente SR. 2003. The Ran GTPase cycle is required for yeast nuclear pore complex assembly. *J Cell Biol* **160:** 1041–1053. doi:10.1083/jcb.200209116

Ryan KJ, Zhou Y, Wente SR. 2007. The karyopherin Kap95 regulates nuclear pore complex assembly into intact nuclear envelopes in vivo. *Mol Biol Cell* **18:** 886–898. doi:10.1091/mbc.e06-06-0525

Sachdev R, Sieverding C, Flotenmeyer M, Antonin W. 2012. The C-terminal domain of Nup93 is essential for assembly of the structural backbone of nuclear pore complexes. *Mol Biol Cell* **23:** 740–749. doi:10.1091/mbc.e11-09-0761

Sales Gil R, de Castro IJ, Berihun J, Vagnarelli P. 2018. Protein phosphatases at the nuclear envelope. *Biochem Soc Trans* **46:** 173–182. doi:10.1042/BST20170139

Savas JN, Toyama BH, Xu T, Yates JR III, Hetzer MW. 2012. Extremely long-lived nuclear pore proteins in the rat brain. *Science* **335:** 942. doi:10.1126/science.1217421

Schetter A, Askjaer P, Piano F, Mattaj I, Kemphues K. 2006. Nucleoporins NPP-1, NPP-3, NPP-4, NPP-11 and NPP-13 are required for proper spindle orientation in *C. elegans*. *Dev Biol* **289:** 360–371. doi:10.1016/j.ydbio.2005.10.038

Schmidt HB, Görlich D. 2015. Nup98 FG domains from diverse species spontaneously phase-separate into particles with nuclear pore-like permselectivity. *eLlife* **4:** e04251. doi:10.7554/eLife.04251

Schmidt HB, Görlich D. 2016. Transport selectivity of nuclear pores, phase separation, and membraneless organelles. *Trends Biochem Sci* **41:** 46–61. doi:10.1016/j.tibs.2015.11.001

Schwartz TU. 2016. The structure inventory of the nuclear pore complex. *J Mol Biol* **428:** 1986–2000. doi:10.1016/j.jmb.2016.03.015

Schweizer N, Ferrás C, Kern DM, Logarinho E, Cheeseman IM, Maiato H. 2013. Spindle assembly checkpoint robustness requires Tpr-mediated regulation of Mad1/Mad2 proteostasis. *J Cell Biol* **203:** 883–893. doi:10.1083/jcb.201309076

Scott DD, Aguilar LC, Kramar M, Oeffinger M. 2019. It's not the destination, it's the journey: heterogeneity in mRNA export mechanisms. *Adv Exp Med Biol* **1203:** 33–81. doi:10.1007/978-3-030-31434-7_2

Shahin V, Ludwig Y, Schafer C, Nikova D, Oberleithner H. 2005. Glucocorticoids remodel nuclear envelope structure and permeability. *J Cell Sci* **118:** 2881–2889. doi:10.1242/jcs.02429

Sharma A, Solmaz SR, Blobel G, Melčák I. 2015. Ordered regions of channel nucleoporins Nup62, Nup54, and

Nup58 form dynamic complexes in solution. *J Biol Chem* **290:** 18370–18378. doi:10.1074/jbc.M115.663500

Shaw RJ. 2013. Cell biology. GATORs take a bite out of mTOR. *Science* **340:** 1056–1057. doi:10.1126/science.1240315

Sheehan MA, Mills AD, Sleeman AM, Laskey RA, Blow JJ. 1988. Steps in the assembly of replication-competent nuclei in a cell-free system from *Xenopus* eggs. *J Cell Biol* **106:** 1–12. doi:10.1083/jcb.106.1.1

Simon DN, Rout MP. 2014. Cancer and the nuclear pore complex. *Adv Exp Med Biol* **773:** 285–307. doi:10.1007/978-1-4899-8032-8_13

Stavru F, Hulsmann BB, Spang A, Hartmann E, Cordes VC, Gorlich D. 2006. NDC1: a crucial membrane-integral nucleoporin of metazoan nuclear pore complexes. *J Cell Biol* **173:** 509–519. doi:10.1083/jcb.200601001

Stoffler D, Feja B, Fahrenkrog B, Walz J, Typke D, Aebi U. 2003. Cryo-electron tomography provides novel insights into nuclear pore architecture: implications for nucleocytoplasmic transport. *J Mol Biol* **328:** 119–130. doi:10.1016/S0022-2836(03)00266-3

Sträßer K, Baßler J, Hurt E. 2000. Binding of the Mex67p/Mtr2p heterodimer to FXFG, GLFG, and FG repeat nucleoporins is essential for nuclear mRNA export. *J Cell Biol* **150:** 695–706. doi:10.1083/jcb.150.4.695

Strawn LA, Shen T, Shulga N, Goldfarb DS, Wente SR. 2004. Minimal nuclear pore complexes define FG repeat domains essential for transport. *Nat Cell Biol* **6:** 197–206. doi:10.1038/ncb1097

Stukenberg PT, Macara IG. 2003. The kinetochore NUPtials. *Nat Cell Biol* **5:** 945–947. doi:10.1038/ncb1103-945

Stuwe T, Correia AR, Lin DH, Paduch M, Lu VT, Kossiakoff AA, Hoelz A. 2015. Nuclear pores. Architecture of the nuclear pore complex coat. *Science* **347:** 1148–1152. doi:10.1126/science.aaa4136

Sun J, Shi Y, Yildirim E. 2019. The nuclear pore complex in cell type-specific chromatin structure and gene regulation. *Trends Genet* **35:** 579–588. doi:10.1016/j.tig.2019.05.006

Szymborska A, de Marco A, Daigle N, Cordes VC, Briggs JA, Ellenberg J. 2013. Nuclear pore scaffold structure analyzed by super-resolution microscopy and particle averaging. *Science* **341:** 655–658. doi:10.1126/science.1240672

Talamas JA, Hetzer MW. 2011. POM121 and Sun1 play a role in early steps of interphase NPC assembly. *J Cell Biol* **194:** 27–37. doi:10.1083/jcb.201012154

Tan PS, Aramburu IV, Mercadante D, Tyagi S, Chowdhury A, Spitz D, Shammas SL, Gräter F, Lemke EA. 2018. Two differential binding mechanisms of FG-nucleoporins and nuclear transport receptors. *Cell Rep* **22:** 3660–3671. doi:10.1016/j.celrep.2018.03.022

Terry LJ, Wente SR. 2007. Nuclear mRNA export requires specific FG nucleoporins for translocation through the nuclear pore complex. *J Cell Biol* **178:** 1121–1132. doi:10.1083/jcb.200704174

Texari L, Stutz F. 2015. Sumoylation and transcription regulation at nuclear pores. *Chromosoma* **124:** 45–56. doi:10.1007/s00412-014-0481-x

Thaller DJ, Allegretti M, Borah S, Ronchi P, Beck M, Lusk CP. 2019. An ESCRT-LEM protein surveillance system is poised to directly monitor the nuclear envelope and nuclear transport system. *eLife* **8:** e45284. doi:10.7554/eLife.45284

Theisen U, Straube A, Steinberg G. 2008. Dynamic rearrangement of nucleoporins during fungal "open" mitosis. *Mol Biol Cell* **19:** 1230–1240. doi:10.1091/mbc.e07-02-0130

Tomioka Y, Kotani T, Kirisako H, Oikawa Y, Kimura Y, Hirano H, Ohsumi Y, Nakatogawa H. 2020. TORC1 inactivation stimulates autophagy of nucleoporin and nuclear pore complexes. *J Cell Biol* **219:** e201910063. doi:10.1083/jcb.201910063

Toyama BH, Savas JN, Park SK, Harris MS, Ingolia NT, Yates JR, Hetzer MW. 2013. Identification of long-lived proteins reveals exceptional stability of essential cellular structures. *Cell* **154:** 971–982. doi:10.1016/j.cell.2013.07.037

Toyama BH, Arrojo EDR, Lev-Ram V, Ramachandra R, Deerinck TJ, Lechene C, Ellisman MH, Hetzer MW. 2019. Visualization of long-lived proteins reveals age mosaicism within nuclei of postmitotic cells. *J Cell Biol* **218:** 433–444. doi:10.1083/jcb.201809123

Tsuji T, Sheehy N, Gautier VW, Hayakawa H, Sawa H, Hall WW. 2007. The nuclear import of the human T lymphotropic virus type I (HTLV-1) tax protein is carrier- and energy-independent. *J Biol Chem* **282:** 13875–13883. doi:10.1074/jbc.M611629200

Ullman KS, Shah S, Powers MA, Forbes DJ. 1999. The nucleoporin nup153 plays a critical role in multiple types of nuclear export. *Mol Biol Cell* **10:** 649–664. doi:10.1091/mbc.10.3.649

Vallotton P, Rajoo S, Wojtynek M, Onischenko E, Kralt A, Derrer CP, Weis K. 2019. Mapping the native organization of the yeast nuclear pore complex using nuclear radial intensity measurements. *Proc Natl Acad Sci* **116:** 14606–14613. doi:10.1073/pnas.1903764116

Vasu SK, Forbes DJ. 2001. Nuclear pores and nuclear assembly. *Curr Opin Cell Biol* **13:** 363–375. doi:10.1016/S0955-0674(00)00221-0

Vishnoi N, Dhanasekeran K, Chalfant M, Surovstev I, Khokha MK, Lusk CP. 2020. Differential turnover of Nup188 controls its levels at centrosomes and role in centriole duplication. *J Cell Biol* **219:** e201906031. doi:10.1083/jcb.201906031

Vollmer B, Schooley A, Sachdev R, Eisenhardt N, Schneider AM, Sieverding C, Madlung J, Gerken U, Macek B, Antonin W. 2012. Dimerization and direct membrane interaction of Nup53 contribute to nuclear pore complex assembly. *EMBO J* **31:** 4072–4084. doi:10.1038/emboj.2012.256

Vollmer B, Lorenz M, Moreno-Andrés D, Bodenhöfer M, De Magistris P, Astrinidis SA, Schooley A, Flötenmeyer M, Leptihn S, Antonin W. 2015. Nup153 recruits the Nup107-160 complex to the inner nuclear membrane for interphasic nuclear pore complex assembly. *Dev Cell* **33:** 717–728. doi:10.1016/j.devcel.2015.04.027

von Appen A, Beck M. 2016. Structure determination of the nuclear pore complex with three-dimensional cryo electron microscopy. *J Mol Biol* **428:** 2001–2010. doi:10.1016/j.jmb.2016.01.004

von Appen A, Kosinski J, Sparks L, Ori A, DiGuilio AL, Vollmer B, Mackmull MT, Banterle N, Parca L, Kastritis

P, et al. 2015. In situ structural analysis of the human nuclear pore complex. *Nature* **526:** 140–143. doi:10.1038/nature15381

Wagstaff KM, Jans DA. 2009. Importins and beyond: non-conventional nuclear transport mechanisms. *Traffic* **10:** 1188–1198. doi:10.1111/j.1600-0854.2009.00937.x

Walther TC, Fornerod M, Pickersgill H, Goldberg M, Allen TD, Mattaj IW. 2001. The nucleoporin Nup153 is required for nuclear pore basket formation, nuclear pore complex anchoring and import of a subset of nuclear proteins. *EMBO J* **20:** 5703–5714. doi:10.1093/emboj/20.20.5703

Walther TC, Alves A, Pickersgill H, Loı¨odice I, Hetzer M, Galy V, Hülsmann BB, Köcher T, Wilm M, Allen T, et al. 2003a. The conserved Nup107-160 complex is critical for nuclear pore complex assembly. *Cell* **113:** 195–206. doi:10.1016/S0092-8674(03)00235-6

Walther TC, Askjaer P, Gentzel M, Habermann A, Griffiths G, Wilm M, Mattaj IW, Hetzer M. 2003b. RanGTP mediates nuclear pore complex assembly. *Nature* **424:** 689–694. doi:10.1038/nature01898

Webster BM, Colombi P, Jäger J, Lusk CP. 2014. Surveillance of nuclear pore complex assembly by ESCRT-III/Vps4. *Cell* **159:** 388–401. doi:10.1016/j.cell.2014.09.012

Webster BM, Thaller DJ, Jäger J, Ochmann SE, Borah S, Lusk CP. 2016. Chm7 and Heh1 collaborate to link nuclear pore complex quality control with nuclear envelope sealing. *EMBO J* **35:** 2447–2467. doi:10.15252/embj.201694574

Wente SR, Rout MP. 2010. The nuclear pore complex and nuclear transport. *Cold Spring Harb Perspect Biol* **2:** a000562. doi:10.1101/cshperspect.a000562

Wigington CP, Roy J, Damle NP, Yadav VK, Blikstad C, Resch E, Wong CJ, Mackay DR, Wang JT, Krystkowiak I, et al. 2020. Systematic discovery of short linear motifs decodes calcineurin phosphatase signaling. *Mol Cell* **79:** 342–358.e12. doi:10.1016/j.molcel.2020.06.029

Winey M, Yarar D, Giddings TH Jr, Mastronarde DN. 1997. Nuclear pore complex number and distribution throughout the *Saccharomyces cerevisiae* cell cycle by three-dimensional reconstruction from electron micrographs of nuclear envelopes. *Mol Biol Cell* **8:** 2119–2132. doi:10.1091/mbc.8.11.2119

Wozniak RW, Blobel G. 1992. The single transmembrane segment of gp210 is sufficient for sorting to the pore membrane domain of the nuclear envelope. *J Cell Biol* **119:** 1441–1449. doi:10.1083/jcb.119.6.1441

Wozniak RW, Blobel G, Rout MP. 1994. POM152 is an integral protein of the pore membrane domain of the yeast nuclear envelope. *J Cell Biol* **125:** 31–42. doi:10.1083/jcb.125.1.31

Xu S, Powers MA. 2013. In vivo analysis of human nucleoporin repeat domain interactions. *Mol Biol Cell* **24:** 1222–1231. doi:10.1091/mbc.e12-08-0585

Xu L, Alarcon C, Çöl S, Massaguè J. 2003. Distinct domain utilization by Smad3 and Smad4 for nucleoporin interaction and nuclear import. *J Biol Chem* **278:** 42569–42577. doi:10.1074/jbc.M307601200

Yokoya F, Imamoto N, Tachibana T, Yoneda Y. 1999. β-Catenin can be transported into the nucleus in a Ran-unassisted manner. *Mol Biol Cell* **10:** 1119–1131. doi:10.1091/mbc.10.4.1119

Zahn R, Osmanović D, Ehret S, Araya Callis C, Frey S, Stewart M, You C, Görlich D, Hoogenboom BW, Richter RP. 2016. A physical model describing the interaction of nuclear transport receptors with FG nucleoporin domain assemblies. *eLife* **5:** e14119. doi:10.7554/eLife.14119

Zhang C, Hutchins JR, Mühlhäusser P, Kutay U, Clarke PR. 2002. Role of importin-β in the control of nuclear envelope assembly by Ran. *Curr Biol* **12:** 498–502. doi:10.1016/S0960-9822(02)00714-5

Zhang Y, Li S, Zeng C, Huang G, Zhu X, Wang Q, Wang K, Zhou Q, Yan C, Zhang W, et al. 2020. Molecular architecture of the luminal ring of the xenopus laevis nuclear pore complex. *Cell Res* **30:** 532–540. doi:10.1038/s41422-020-0320-y

Zhu Y, Liu TW, Madden Z, Yuzwa SA, Murray K, Cecioni S, Zachara N, Vocadlo DJ. 2016. Post-translational O-GlcNAcylation is essential for nuclear pore integrity and maintenance of the pore selectivity filter. *J Mol Cell Biol* **8:** 2–16. doi:10.1093/jmcb/mjv033

Zuccolo M, Alves A, Galy V, Bolhy S, Formstecher E, Racine V, Sibarita JB, Fukagawa T, Shiekhattar R, Yen T, et al. 2007. The human Nup107-160 nuclear pore subcomplex contributes to proper kinetochore functions. *EMBO J* **26:** 1853–1864. doi:10.1038/sj.emboj.7601642

The Nuclear Pore Complex as a Transcription Regulator

Michael Chas Sumner and Jason Brickner

Department of Molecular Biosciences, Northwestern University, Evanston, Illinois 60208, USA

Correspondence: j-brickner@northwestern.edu

The nuclear pore complex (NPC) is a highly conserved channel in the nuclear envelope that mediates mRNA export to the cytosol and bidirectional protein transport. Many chromosomal loci physically interact with nuclear pore proteins (Nups), and interactions with Nups can promote transcriptional repression, transcriptional activation, and transcriptional poising. Interaction with the NPC also affects the spatial arrangement of genes, interchromosomal clustering, and folding of topologically associated domains. Thus, the NPC is a spatial organizer of the genome and regulator of genome function.

Eukaryotic cells spatially organize their genomes in a nonrandom fashion that both reflects and facilitates transcription regulation (Misteli 2020). Electron micrographs of metazoan nuclei show that heterochromatin associates with the nuclear lamina at the nuclear periphery in many cell types (Fig. 1; Jost et al. 2012). While it is true that heterochromatin is primarily positioned near the nuclear lamina or chromocenters, this can vary with cell type and organism (Zykova et al. 2018; Falk et al. 2019). While some organisms lack lamins, all eukaryotic nuclei are punctuated by hundreds to thousands of nuclear pore complexes (NPCs) that facilitate exchange between the nucleoplasm and cytoplasm. These structures also physically interact with specific sites in the genome, impacting their positioning and their expression.

NPCs act as a gate between the cell's two major compartments: the cytoplasm and the nucleus. Ubiquitous and essential, pores facilitating the transfer of material across the nuclear envelope were discovered at a time when there was still an ongoing debate about the existence of an organized nuclear membrane (Callan and Tomlin 1950; Watson 1955). Work over the intervening decades has revealed NPC function, organization, mechanism, and structure (Gall 1967; Goldberg and Allen 1992; von Appen and Beck 2016). Approximately 30 unique proteins make up the core eightfold radially symmetrical channel, with subcomplexes that extend from its cytoplasmic and nucleoplasmic faces, comprising additional proteins. NPC subunits with access to the nucleoplasm physically interact with chromatin and can impact their transcriptional regulation.

Based on electron micrographs showing that chromatin near NPCs was less condensed than adjacent, lamin-associated heterochromatin, Günter Blobel hypothesized that positioning of active genes near NPCs would enhance mRNA

Cite this article as *Cold Spring Harb Perspect Biol* doi: 10.1101/cshperspect.a039438

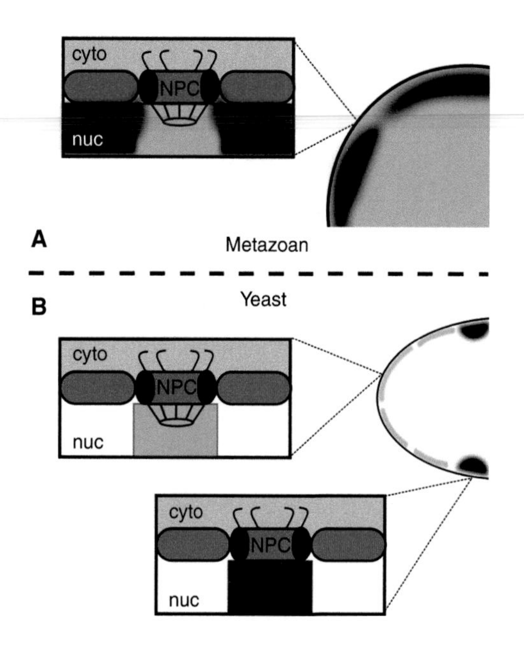

Figure 1. Spatial organization of nuclear pore complex (NPC) component interactions with chromatin. (*A*) (*Right*) Portion of metazoan nucleus where green represents euchromatin and red denotes heterochromatin. (*Left*) Peripheral chromatin adjacent to the nuclear lamina (red) and the NPC (green). (*B*) (*Right*) Portion of a *Saccharomyces cerevisiae* nucleus. Color indicates the space where NPC–chromatin interactions can occur (telomere silencing interactions in red, euchromatin transcription initiation in green). (*Left*) *Insets* show that the effects on transcription only occur at the nucleoplasmic face of the nuclear pore.

export (Blobel 1985). Years later, work from several systems confirmed that the NPC physically associates with active genes, impacting the spatial organization of genes and gene expression (Brickner and Walter 2004; Casolari et al. 2004; Guglielmi et al. 2020). To date, a convincing role for this interaction in promoting efficient mRNA export has not been established. However, data from yeast, flies, and mammals suggests that nuclear pore protein (Nup) interaction can alter gene expression to improve viability during stress conditions, maintain epigenetic memory of previous expression states, and promote tissue differentiation (Brickner and Walter 2004; Brickner et al. 2007; Liang et al. 2013a).

Here we review multiple, evolutionarily conserved ways the NPC regulates transcription. Widening our collective understanding of how the NPC regulates gene expression will help guide future research in epigenetics and transcriptional mechanisms.

MOLECULAR MECHANISM OF INTERACTION WITH THE NPC

Repositioning and interaction of active genes with the yeast NPC requires both nucleoplasmic nuclear pore proteins such as Nup1, Nup2, Nup60, Mlp1, and Mlp2 (Cabal et al. 2006; Luthra et al. 2007; Ahmed et al. 2010; Light et al. 2010). Furthermore, interaction with the NPC requires binding of sequence-specific DNA-binding transcription factors (TFs) to *cis*-acting DNA sequences near promoters (Schmid et al. 2006; Ahmed et al. 2010; Randise-Hinchliff et al. 2016). The binding sites of such transcription factors function as DNA zip codes; they are both necessary and sufficient to mediate repositioning to the periphery and interaction with Nups (Ahmed et al. 2010; Brickner et al. 2012). Of course, the NPC interacts with many complexes involved in transcription initiation, elongation, and RNA maturation. TFs, Nups, and complexes involved in mRNA export (i.e., TREX2 and Mex67) and transcription (Mediator, SAGA histone acetyltransferase) are required for targeting the nuclear periphery (Dieppois et al. 2006; Ahmed et al. 2010; Jani et al. 2014). However, several experiments argue that TFs and Nups play a direct role, while the others do not. Conditional inactivation of TFs and Nups, but not TREX-2, Mex67, Mediator, or SAGA, leads to rapid loss of peripheral localization (Brickner et al. 2019). Furthermore, tethering a TF at an ectopic site leads to peripheral localization and physical interaction with Nups but does not lead to chromatin binding with Mediator, TREX-2, or Mex67. Finally, for Gcn4-targeted genes, TF overexpression bypassed the requirement for SAGA and Mediator component null mutants for targeting chromatin to the nuclear periphery but still required Nup2 (Brickner et al. 2019). These results suggest that TFs and Nups play a

 Cite this article as *Cold Spring Harb Perspect Biol* doi: 10.1101/cshperspect.a039438

direct role in targeting genes to the nuclear periphery.

This is a common function of TFs: when tethered to a chromosomal locus, most yeast TFs can induce Nup-dependent peripheral localization (Brickner et al. 2019). Importantly, these TFs include activators, repressors, and chromatin factors, suggesting that the interaction with the NPC may impinge upon transcription and chromatin structure in more than one way.

THE NPC AND GENE SILENCING

The chromosomal position of a gene can impact its expression; genes near centromeres or telomeres show reduced recombination and transcription, a phenomenon known as "position effect" (Weiler and Wakimoto 1995). Position effects often reflect the spreading of silencing factors from sites of recruitment (Gottschling et al. 1990). For example, in budding yeast, telomeres and adjacent sequences are transcriptionally silenced by the recruitment of Rap1, and the silencing factors Sir2, Sir3, and Sir4 (Fig. 2A; Gotta et al. 1996). Rap1 is a sequence-specific DNA-binding protein that binds to telomeres and recruits Sir2, Sir3, and Sir4. The Sir proteins deacetylate histones and spread down the chromosome arms into subtelomeric regions, establishing and maintaining silenced chromatin. Telomeres localize at the nuclear periphery, and nuclear envelope membrane proteins such as Esc1 and Mps4, as well as Nups like those that make up the inner and outer ring subcomplexes (i.e., Nup170, Nup145, and Nup60) as well as nucleoplasmic TPR homologs (Mlp1 and Mlp2; Galy et al. 2000), act as a physical anchor to recruit and stabilize the Rap1/Sir3/Sir4 complex with chromatin and maintain telomere positioning at the nuclear periphery (Van de Vosse et al. 2013; Lapetina et al. 2017). Loss of these nuclear pore proteins leads to a defect in the silencing of telomeres and the silent mating-type loci (Feuerbach et al. 2002). In this way, the nuclear envelope and Nups contribute to both the spatial organization and transcriptional silencing of yeast telomeres.

In metazoan cells, genes interact with both NPC-associated Nups at the nuclear periphery and soluble Nups localized throughout the nucleoplasm (Griffis et al. 2002; Capelson et al. 2010; Kalverda et al. 2010; Liang et al. 2013b). At the fly NPC, active and silent genes interact with distinct Nups (Nup107 and Nup93, respectively; Gozalo et al. 2020). Polycomb repressive complexes (PRCs) catalyze methylation of H3K27 to establish and maintain facultative heterochromatin (Fig. 2A). More than a third of Polycomb-associated domains also physically interact with Nup93. Compared to other Polycomb domains, those that interacted with Nup93 had increased PRC presence and were more likely to be positioned at the nuclear periphery (Gozalo et al. 2020). Finally, Nup93 contributes to transcriptional silencing of these regions. Thus, interaction of Nups can also promote transcriptional silencing and heterochromatin formation in animals.

Interaction of Nups with the genome can also impact chromosome folding. Boundary elements and the chromatin architectural proteins that localize to them (CTCF and cohesin) interact with Nup153 to stabilize their organization in physical space and enhance insulation between TADs (Fig. 2B; Kadota et al. 2020). Knockdown of Nup153 leads to improper TAD formation and ectopic enhancer function across boundary domains. Embryonic stem cells are insensitive or slow to respond to epidermal growth factor following Nup153 knockdown. Likewise, looping of promoters with certain enhancers is stimulated by Nup98 in *Drosophila* (Pascual-Garcia et al. 2017). Finally, Nups are implicated in the formation of senescence-associated heterochromatin foci (SAHF) during oncogene-induced senescence (Boumendil et al. 2019). During this process, the heterochromatin reorganizes from the periphery to coalesce in the nucleoplasm, followed by cell cycle arrest and secretion of inflammatory cytokines. Knockdown of TPR blocks SAHF formation and reorganization of heterochromatin and cytokine secretion, suggesting that TPR influences the position and expression of heterochromatin.

The NPC and Transcriptional Activation

The NPC can also positively affect transcription. In budding yeast, genome-wide chromatin im-

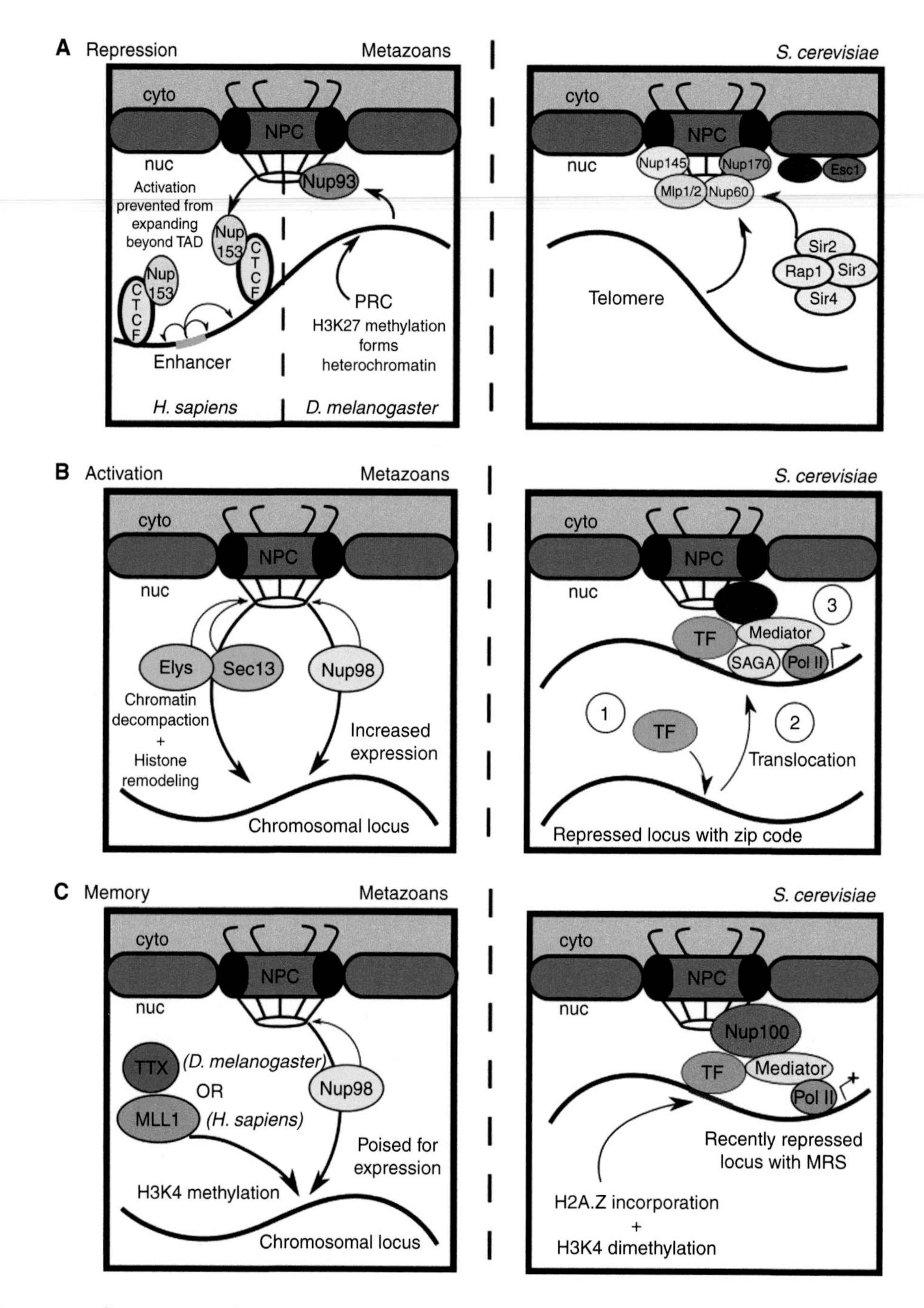

Figure 2. Nuclear pore complex (NPC) proteins affect transcription. (*A–C*) Schematic of metazoan nucleus (*left*) or yeast nucleus (*right*). (*A*) Nup-mediated transcriptional repression. (*Left*) Soluble Nup153 interacts with CTCF at topologically associated domain (TAD) boundaries in *Homo sapiens*. Heterochromatin at the nuclear periphery interacts with Nup93 and maintains Polycomb repressive complex (PRC) histone modification in *Drosophila melanogaster*. (*Right*) Telomere recruitment to the *Saccharomyces cerevisiae* NPC facilitates binding of chromatin silencing Sir factors. (*B*) Nup-mediated transcriptional activation. (*Left*) Chromatin decompaction by recruitment of Nups and transcriptional activation by recruitment of Nup98. (*Right*) Transcription factor (TF)- and Nup-dependent stimulation of transcription in *S. cerevisiae*. (*C*) Nup-dependent transcriptional poising. (*Left*) Nup98 recruitment can both enhance promoter–enhancer looping and, potentially, through recruitment of specific histone methyltransferases. (*Right*) Schematic of a Nu100-dependent chromatin changes leading to transcriptional poising during memory in *S. cerevisiae*. (MRS) Memory recruitment sequence.

munoprecipitation studies showed that hundreds of transcriptionally active loci interact with Nups (Casolari et al. 2004, 2005) and inducible genes reposition from the nucleoplasm to the nuclear periphery upon activation (Brickner and Walter 2004; Casolari et al. 2004). Likewise, in flies and mammals, thousands of chromosomal sites, including many euchromatic, transcriptionally active regions, interact with nuclear pore proteins (Brown et al. 2008; Capelson et al. 2010; Kalverda et al. 2010; Liang et al. 2013a; Jacinto et al. 2015; Pascual-Garcia et al. 2017). These interactions enhance transcription and increase the rate of expression; disrupting the interaction with nuclear pore proteins reduces the rate and extent of transcriptional induction of many genes in budding yeast and animals (Brickner et al. 2007, 2012, 2019; Ahmed et al. 2010; Capelson et al. 2010; Light et al. 2010, 2013; Liang et al. 2013a; Jacinto et al. 2015; Pascual-Garcia et al. 2017). Moving a DNA zip code from upstream of the promoter to downstream of the coding sequence had an interesting effect. Such a locus was targeted to the periphery but was still defective for transcription (Ahmed et al. 2010), suggesting that the physical interaction with the promoter is critical for promoting transcription. However, tethering of inducible genes to the nuclear envelope or the NPC is not sufficient to cause transcriptional activation (Brickner and Walter 2004; Green et al. 2012; Texari et al. 2013).

How do Nups promote transcription? Single molecule RNA FISH experiments suggest that disrupting the interaction with the yeast NPC reduces the fraction of cells expressing the *GAL1* gene (Brickner et al. 2016). For the subset of cells that express *GAL1*, the level of expression is normal. Because transcription occurs through stochastic bursts, mRNA output is the product of burst frequency, burst duration, and burst amplitude (Rodriguez and Larson 2020). This observation suggests that interaction with Nup2 quantitatively increases transcription by increasing the burst frequency or burst duration, without affecting burst amplitude. Because enhancers have been implicated in burst frequency, while core promoter strength has been implicated in burst amplitude (Tunnacliffe et al. 2018; Larsson et al. 2019), this suggests that Nups stimulate enhancer function.

An additional hint into the role of yeast Nups in transcriptional activation comes from structure–function analysis of the Gcn4 TF. Gcn4 is essential for both transcriptional activation and NPC interaction of many target genes (Hope and Struhl 1985; Rawal et al. 2018; Brickner et al. 2019). Tethering Gcn4 to a nucleoplasmic locus is sufficient to target that locus to the nuclear periphery, allowing identification of a minimal portion of Gcn4 that is necessary and sufficient to promote interaction with Nups. This strategy identified a *positioning domain* within the Gcn4 TF (PD_{GCN4}), a 27 amino acid peptide, separable from the activation domains. Mutation of three amino acids within this sequence disrupts interaction of Gcn4 target genes with the NPC and results in a global defect in transcriptional activation of Gcn4 target genes (Brickner et al. 2019). Tethering of the PD_{GCN4} to an ectopic locus led to interaction with Nup2, but not with coactivators like SAGA or Mediator. These results argue that Nup interaction can enhance activator domain-dependent transcription, but cannot activate transcription in the absence of an activator domain.

Interaction with metazoan Nups can also positively regulate transcription (Fig. 2B). Knockdown of Nups in flies leads to widespread decrease in transcription (Capelson et al. 2010). Likewise, loss of Nup98 in embryonic stem cells impacts transcription and developmental potential (Liang et al. 2013a) and loss of the tissue-specific Nup210 inhibits proper muscle cell gene expression and differentiation (D'Angelo et al. 2012; Raices et al. 2017). Although Nups interact throughout the genome, they bind strongly at super-enhancers (Ibarra et al. 2016). These effects correlate with impacts on chromatin folding and structure. Interaction with Nups facilitates recruitment of cohesin and formation of topologically associated domains (TADs) as well as promoter–enhancer looping (Pascual-Garcia et al. 2017; Kadota et al. 2020). Tethering of Nups such as Sec13 or Elys to polytene chromosomes in *Drosophila* leads to chromatin decondensation (Fig. 2B; Kuhn et al. 2019). The tethered Nups recruit PBAP/Brm and GAGA, which are required for decondensation (Kuhn et al. 2019). Thus, the role of Nups in animals in promoting transcription may relate to their effects on chromatin folding and condensation.

The NPC's role in interacting with chromatin to modulate transcription has also been explored in plants. In *Arabidopsis thaliana*, tethering of the Nup Seh1 to a reporter transgene caused positioning at the nuclear periphery and stimulated transcription (Smith et al. 2015). Likewise, *Arabidopsis* Nup1 is necessary for pollen and ovule development, and loss of Nup1 leads to a significant decrease in the expression of gametogenesis genes, although it is unknown whether these genes physically interact with the NPC (Bao et al. 2019). Also, the chlorophyll a/b gene locus undergoes light-dependent repositioning from the nuclear interior to the nuclear periphery upon transcriptional activation (Feng et al. 2014). However, a connection to the NPC has not been explored and this repositioning requires a set of proteins not implicated in yeast or metazoan gene positioning, raising the possibility that other mechanisms impact localization to the nuclear periphery in plants.

THE NPC AS A REGULATOR OF EPIGENETIC MEMORY

Nups interact with both active and silent loci and can promote both transcription and repression. But Nups also play a conserved role in a type of epigenetic poising following specific expression states. In yeast, the *INO1* locus is targeted to the NPC both when active and when recently repressed (Brickner et al. 2007). This latter state is called "transcriptional memory" and is inherited for several generations. The molecular mechanism for targeting (i.e., the TFs, the *cis*-acting zip codes, and the Nups required) of active *INO1* and recently repressed *INO1* are different (Fig. 2C; Light et al. 2010). Whereas active *INO1* is targeted to the NPC by the gene recruitment sequences GRS1 and GRS2 and the TFs Put3 and Cbf1, recently repressed *INO1* is targeted to the NPC by the memory recruitment sequence (MRS), which binds the Sfl1 TF.

Furthermore, memory requires chromatin modifications and Nup100, both of which are not required for recruiting active *INO1* to the periphery (Brickner et al. 2007; Light et al. 2010; D'Urso et al. 2016). This suggests that there are at least two mechanisms by which genes interact with the NPC in yeast, one that requires Nup100 and one that does not. This conclusion is bolstered by the finding that, while tethering of 121 yeast TFs to a chromosomal site is sufficient to cause Nup2-dependent targeting to the nuclear periphery, only 76 of those TFs also require Nup100 (Brickner et al. 2019).

Transcriptional memory leads to changes in the chromatin state of the promoter, allowing the recruitment of a poised form of RNA polymerase II preinitiation complex (RNAPII PIC). This poised state of the promoter is activated more rapidly than it would be otherwise, presumably providing an adaptive fitness advantage. In the case of the *INO1* gene, loss of Sfl1, the MRS, or Nup100 blocks memory, leading to slower reactivation. Thus, the same locus can be targeted to the NPC by two distinct mechanisms, producing two distinct outcomes, depending on the state and history of the cell.

The phenomenon of yeast transcriptional memory is widespread. In yeast, many genes exhibit an enhanced activation rate if previously expressed, which can enhance adaptive fitness (Sood and Brickner 2017). Furthermore, memory is generally associated with changes in chromatin modifications (H2A.Z incorporation, H3 lysine 4 dimethylation [H4K4me2]) and recruitment of a novel, poised RNA polymerase II preinitation complex (Fig. 2C; D'Urso et al. 2016; Sood et al. 2017). However, it does not always require interaction with the NPC, suggesting that the interaction with the NPC regulates a core memory mechanism involving chromatin changes and promoter poising.

Nup-dependent memory has also been observed in metazoan cells. In HeLa cells, genes induced by interferon γ (IFN-γ) interact with Nup98 (the Nup100 homolog) upon removal of IFN-γ, and this interaction persists for >4 days. These genes show faster/more robust expression if cells are exposed to IFN-γ again. Promoters of such poised genes are marked with H3K4me2 and bind RNA polymerase II. Transient knockdown of Nup98 during memory led to a loss of H3K4me2 and RNA polymerase II from promoters and disrupted the faster rate of reactivation (Light et al. 2013). Thus, Nups have an ancient, conserved role in controlling chromatin states that facilitate epigenetic transcriptional regulation.

In flies, ecdysone-induced genes also interact with Nup98 and exhibit transcriptional memory. Brief exposure of S2 cells to ecdysone leads to Nup98 binding and poises target genes for induction (Pascual-Garcia et al. 2017). The knockdown of Nup98 specifically disrupts this effect, leading to no memory. The effect of Nup98 in this system (and perhaps others) is to stabilize a promoter–enhancer loop. This loop is strengthened by previous treatment with ecdysone and by binding to Nup98, suggesting that Nup98-dependent chromatin folding can facilitate the establishment and inheritance of epigenetic states.

The impact of Nups on chromatin and transcription also has important effects on human health. Chromosomal translocations that lead to translational fusions of Nup98 with several proteins such as HOXA9, HOXD13, Top1, and Nsd1 lead to acute myeloid leukemia (AML) (Franks et al. 2017). Why? Nup98 is associated with H3K4 methylation in flies and humans and interacts with the H3K4 methyltransferases Trithorax and MLL1, respectively (Fig. 2C; Kaltenbach et al. 2010; Gough et al. 2011; Pascual-Garcia et al. 2014). This led to the hypothesis that AML is due to excessive H3K4 methylation of target genes induced by ectopic recruitment of Nup98 at those loci (Franks et al. 2017). Indeed, the Nup98-Nsd1 fusion protein expressed in myeloid cells from AML patients binds to Wdr82, a component of the H3K4 methyltransferase Set1A/B-COMPASS. Forming this aberrant complex results in transcription-associated histone modifications at Nsd1 target genes, such as the HOXA locus, and leads to an increase in expression (Michmerhuizen et al. 2020). These findings show the critical role of Nup98/Nup100 in epigenetic regulation in regulating normal and pathogenic transcription.

CLOSING REMARKS

The NPC is an ancient structural component of eukaryotic cells. In addition to mediating nucleocytoplasmic trafficking, a role for nuclear pore proteins in regulating transcription and chromatin structure is now appreciated. Additionally, the interaction of the NPC with the genome impacts its spatial arrangement. Specific interactions between genes and Nups are associated with transcriptional silencing, transcriptional activation, transcriptional poising, changes in chromatin modifications and changes in chromatin folding. While the field is still exploring the precise molecular nature of these roles, it is clear that the NPC impacts genome function and gene expression and multiple levels.

ACKNOWLEDGMENTS

We would like to acknowledge our colleagues cited herein for their contributions to the field and the editors of this edition for their patience. M.C.S. was supported by National Institutes of Health (NIH) Grant T32 GM008061. M.C.S. and J.B. were supported by NIH Grants R01 GM118712 (to J.B.) and R35 GM136419 (to J.B.).

REFERENCES

Ahmed S, Brickner DG, Light WH, Cajigas I, McDonough M, Froyshteter AB, Volpe T, Brickner JH. 2010. DNA zip codes control an ancient mechanism for gene targeting to the nuclear periphery. *Nat Cell Biol* **12:** 111–118. doi:10.1038/ncb2011

Bao S, Shen G, Li G, Liu Z, Arif M, Wei Q, Men S. 2019. The *Arabidopsis* nucleoporin NUP1 is essential for megasporogenesis and early stages of pollen development. *Plant Cell Rep* **38:** 59–74. doi:10.1007/s00299-018-2349-7

Blobel G. 1985. Gene gating: a hypothesis. *Proc Natl Acad Sci* **82:** 8527–8529. doi:10.1073/pnas.82.24.8527

Boumendil C, Hari P, Olsen KCF, Acosta JC, Bickmore WA. 2019. Nuclear pore density controls heterochromatin reorganization during senescence. *Gene Dev* **33:** 144–149. doi:10.1101/gad.321117.118

Brickner JH, Walter P. 2004. Gene recruitment of the activated INO1 locus to the nuclear membrane. *Plos Biol* **2:** e342. doi:10.1371/journal.pbio.0020342

Brickner DG, Cajigas I, Fondufe-Mittendorf Y, Ahmed S, Lee PC, Widom J, Brickner JH. 2007. H2a.Z-mediated localization of genes at the nuclear periphery confers epigenetic memory of previous transcriptional state. *Plos Biol* **5:** e81. doi:10.1371/journal.pbio.0050081

Brickner DG, Ahmed S, Meldi L, Thompson A, Light W, Young M, Hickman TL, Chu F, Fabre E, Brickner JH. 2012. Transcription factor binding to a DNA zip code controls interchromosomal clustering at the nuclear periphery. *Dev Cell* **22:** 1234–1246. doi:10.1016/j.devcel.2012.03.012

Brickner DG, Sood V, Tutucci E, Coukos R, Viets K, Singer RH, Brickner JH. 2016. Subnuclear positioning and interchromosomal clustering of the GAL1-10 locus are controlled by separable, interdependent mechanisms. *Mol Biol Cell* **27:** 2980–2993. doi:10.1091/mbc.E16-03-0174

Brickner DG, Randise-Hinchliff C, Corbin ML, Liang JM, Kim S, Sump B, D'Urso A, Kim SH, Satomura A, Schmit H, et al. 2019. The role of transcription factors and nuclear pore proteins in controlling the spatial organization of the yeast genome. *Dev Cell* **49:** 936–947.e4. doi:10.1016/j.devcel.2019.05.023

Brown JM, Green J, das Neves RP, Wallace HA, Smith AJ, Hughes J, Gray N, Taylor S, Wood WG, Higgs DR, et al. 2008. Association between active genes occurs at nuclear speckles and is modulated by chromatin environment. *J Cell Biol* **182:** 1083–1097. doi:10.1083/jcb.200803174

Cabal GG, Genovesio A, Rodriguez-Navarro S, Zimmer C, Gadal O, Lesne A, Buc H, Feuerbach-Fournier F, Olivo-Marin JC, Hurt EC, et al. 2006. SAGA interacting factors confine sub-diffusion of transcribed genes to the nuclear envelope. *Nature* **441:** 770–773. doi:10.1038/nature04752

Callan HG, Tomlin SG. 1950. Experimental studies on amphibian oocyte nuclei. I: Investigation of the structure of the nuclear membrane by means of the electron microscope. *Proc R Soc Lond B Biol Sci* **137:** 367–378.

Capelson M, Liang Y, Schulte R, Mair W, Wagner U, Hetzer MW. 2010. Chromatin-bound nuclear pore components regulate gene expression in higher eukaryotes. *Cell* **140:** 372–383. doi:10.1016/j.cell.2009.12.054

Casolari JM, Brown CR, Komili S, West J, Hieronymus H, Silver PA. 2004. Genome-wide localization of the nuclear transport machinery couples transcriptional status and nuclear organization. *Cell* **117:** 427–439. doi:10.1016/S0092-8674(04)00448-9

Casolari JM, Brown CR, Drubin DA, Rando OJ, Silver PA. 2005. Developmentally induced changes in transcriptional program alter spatial organization across chromosomes. *Genes Dev* **19:** 1188–1198. doi:10.1101/gad.1307205

D'Angelo MA, Gomez-Cavazos JS, Mei A, Lackner DH, Hetzer MW. 2012. A change in nuclear pore complex composition regulates cell differentiation. *Dev Cell* **22:** 446–458. doi:10.1016/j.devcel.2011.11.021

Dieppois G, Iglesias N, Stutz F. 2006. Cotranscriptional recruitment to the mRNA export receptor Mex67p contributes to nuclear pore anchoring of activated genes. *Mol Cell Biol* **26:** 7858–7870. doi:10.1128/MCB.00870-06

D'Urso A, Takahashi Y, Xiong B, Marone J, Coukos R, Randise-Hinchliff C, Wang J-P, Shilatifard A, Brickner JH. 2016. Set1/COMPASS and Mediator are repurposed to promote epigenetic transcriptional memory. *eLife* **5:** e16691. doi:10.7554/eLife.16691

Falk M, Feodorova Y, Naumova N, Imakaev M, Lajoie BR, Leonhardt H, Joffe B, Dekker J, Fudenberg G, Solovei I, et al. 2019. Heterochromatin drives compartmentalization of inverted and conventional nuclei. *Nature* **570:** 395–399. doi:10.1038/s41586-019-1275-3

Feng CM, Qiu Y, Buskirk EKV, Yang EJ, Chen M. 2014. Light-regulated gene repositioning in *Arabidopsis*. *Nat Commun* **5:** 3027. doi:10.1038/ncomms4027

Feuerbach F, Galy V, Trelles-Sticken E, Fromont-Racine M, Jacquier A, Gilson E, Olivo-Marin JC, Scherthan H, Nehrbass U. 2002. Nuclear architecture and spatial positioning help establish transcriptional states of telomeres in yeast. *Nat Cell Biol* **4:** 214–221. doi:10.1038/ncb756

Franks TM, McCloskey A, Shokirev MN, Benner C, Rathore A, Hetzer MW. 2017. Nup98 recruits the Wdr82-Set1 A/COMPASS complex to promoters to regulate H3K4 trimethylation in hematopoietic progenitor cells. *Gene Dev* **31:** 2222–2234. doi:10.1101/gad.306753.117

Gall JG. 1967. Octagonal nuclear pores. *J Cell Biol* **32:** 391–399. doi:10.1083/jcb.32.2.391

Galy V, Olivo-Marin JC, Scherthan H, Doye V, Rascalou N, Nehrbass U. 2000. Nuclear pore complexes in the organization of silent telomeric chromatin. *Nature* **403:** 108–112. doi:10.1038/47528

Goldberg MW, Allen TD. 1992. High resolution scanning electron microscopy of the nuclear envelope: demonstration of a new, regular, fibrous lattice attached to the baskets of the nucleoplasmic face of the nuclear pores. *J Cell Biol* **119:** 1429–1440. doi:10.1083/jcb.119.6.1429

Gotta M, Laroche T, Formenton A, Maillet L, Scherthan H, Gasser SM. 1996. The clustering of telomeres and colocalization with Rap1, Sir3, and Sir4 proteins in wild-type *Saccharomyces cerevisiae*. *J Cell Biol* **134:** 1349–1363. doi:10.1083/jcb.134.6.1349

Gottschling DE, Aparicio OM, Billington BL, Zakian VA. 1990. Position effect at *S. cerevisiae* telomeres: reversible repression of Pol II transcription. *Cell* **63:** 751–762. doi:10.1016/0092-8674(90)90141-Z

Gough SM, Slape CI, Aplan PD. 2011. NUP98 gene fusions and hematopoietic malignancies: common themes and new biologic insights. *Blood* **118:** 6247–6257. doi:10.1182/blood-2011-07-328880

Gozalo A, Duke A, Lan Y, Pascual-Garcia P, Talamas JA, Nguyen SC, Shah PP, Jain R, Joyce EF, Capelson M. 2020. Core components of the nuclear pore bind distinct states of chromatin and contribute to Polycomb repression. *Mol Cell* **77:** 67–81.e7. doi:10.1016/j.molcel.2019.10.017

Green EM, Jiang Y, Joyner R, Weis K. 2012. A negative feedback loop at the nuclear periphery regulates *GAL* gene expression. *Mol Biol Cell* **23:** 1367–1375. doi:10.1091/mbc.e11-06-0547

Griffis ER, Altan N, Lippincott-Schwartz J, Powers MA. 2002. Nup98 is a mobile nucleoporin with transcription-dependent dynamics. *Mol Biol Cell* **13:** 1282–1297. doi:10.1091/mbc.01-11-0538

Guglielmi V, Sakuma S, D'Angelo MA. 2020. Nuclear pore complexes in development and tissue homeostasis. *Development* **147:** dev183442. doi:10.1242/dev.183442

Hope IA, Struhl K. 1985. GCN4 protein, synthesized in vitro, binds HIS3 regulatory sequences: implications for general control of amino acid biosynthetic genes in yeast. *Cell* **43:** 177–188. doi:10.1016/0092-8674(85)90022-4

Ibarra A, Benner C, Tyagi S, Cool J, Hetzer MW. 2016. Nucleoporin-mediated regulation of cell identity genes. *Gene Dev* **30:** 2253–2258. doi:10.1101/gad.287417.116

Jacinto FV, Benner C, Hetzer MW. 2015. The nucleoporin Nup153 regulates embryonic stem cell pluripotency through gene silencing. *Gene Dev* **29:** 1224–1238. doi:10.1101/gad.260919.115

Jani D, Valkov E, Stewart M. 2014. Structural basis for binding the TREX2 complex to nuclear pores, *GAL1* localisation and mRNA export. *Nucleic Acids Res* **42:** 6686–6697. doi:10.1093/nar/gku252

Jost KL, Bertulat B, Cardoso MC. 2012. Heterochromatin and gene positioning: inside, outside, any side? *Chromosoma* **121:** 555–563. doi:10.1007/s00412-012-0389-2

Kadota S, Ou J, Shi Y, Lee JT, Sun J, Yildirim E. 2020. Nucleoporin 153 links nuclear pore complex to chromatin architecture by mediating CTCF and cohesin binding. *Nat Commun* **11:** 2606. doi:10.1038/s41467-020-16394-3

Kaltenbach S, Soler G, Barin C, Gervais C, Bernard OA, Penard-Lacronique V, Romana SP. 2010. NUP98-MLL fusion in human acute myeloblastic leukemia. *Blood* **116:** 2332–2335. doi:10.1182/blood-2010-04-277806

Kalverda B, Pickersgill H, Shloma VV, Fornerod M. 2010. Nucleoporins directly stimulate expression of developmental and cell-cycle genes inside the nucleoplasm. *Cell* **140:** 360–371. doi:10.1016/j.cell.2010.01.011

Kuhn TM, Pascual-Garcia P, Gozalo A, Little SC, Capelson M. 2019. Chromatin targeting of nuclear pore proteins induces chromatin decondensation. *J Cell Biol* **218:** 2945–2961. doi:10.1083/jcb.201807139

Lapetina DL, Ptak C, Roesner UK, Wozniak RW. 2017. Yeast silencing factor Sir4 and a subset of nucleoporins form a complex distinct from nuclear pore complexes. *J Cell Biol* **216:** 3145–3159. doi:10.1083/jcb.201609049

Larsson AJM, Johnsson P, Hagemann-Jensen M, Hartmanis L, Faridani OR, Reinius B, Segerstolpe Å, Rivera CM, Ren B, Sandberg R. 2019. Genomic encoding of transcriptional burst kinetics. *Nature* **565:** 251–254. doi:10.1038/s41586-018-0836-1

Liang Y, Franks TM, Marchetto MC, Gage FH, Hetzer MW. 2013a. Dynamic association of NUP98 with the human genome. *PLoS Genet* **9:** e1003308. doi:10.1371/journal.pgen.1003308

Liang Y, Franks TM, Marchetto MC, Gage FH, Hetzer MW. 2013b. Dynamic association of NUP98 with the human genome. *Plos Genet* **9:** e1003308. doi:10.1371/journal.pgen.1003308

Light WH, Brickner DG, Brand VR, Brickner JH. 2010. Interaction of a DNA zip code with the nuclear pore complex promotes H2A.Z incorporation and INO1 transcriptional memory. *Mol Cell* **40:** 112–125. doi:10.1016/j.molcel.2010.09.007

Light WH, Freaney J, Sood V, Thompson A, D'Urso A, Horvath CM, Brickner JH. 2013. A conserved role for human Nup98 in altering chromatin structure and promoting epigenetic transcriptional memory. *Plos Biol* **11:** e1001524. doi:10.1371/journal.pbio.1001524

Luthra R, Kerr SC, Harreman MT, Apponi LH, Fasken MB, Ramineni S, Chaurasia S, Valentini SR, Corbett AH. 2007. Actively transcribed *GAL* genes can be physically linked to the nuclear pore by the SAGA chromatin modifying complex. *J Biol Chem* **282:** 3042–3049. doi:10.1074/jbc.M608741200

Michmerhuizen NL, Klco JM, Mullighan CG. 2020. Mechanistic insights and potential therapeutic approaches for NUP98-rearranged hematologic malignancies. *Blood* **136:** 2275–2289. doi:10.1182/blood.2020007093

Misteli T. 2020. The self-organizing genome: principles of genome architecture and function. *Cell* **183:** 28–45. doi:10.1016/j.cell.2020.09.014

Pascual-Garcia P, Jeong J, Capelson M. 2014. Nucleoporin Nup98 associates with Trx/MLL and NSL histone-modifying complexes and regulates Hox gene expression. *Cell Rep* **9:** 433–442. doi:10.1016/j.celrep.2014.09.002

Pascual-Garcia P, Debo B, Aleman JR, Talamas JA, Lan Y, Nguyen NH, Won KJ, Capelson M. 2017. Metazoan nuclear pores provide a scaffold for poised genes and mediate induced enhancer–promoter contacts. *Mol Cell* **66:** 63–76.e6. doi:10.1016/j.molcel.2017.02.020

Raices M, Bukata L, Sakuma S, Borlido J, Hernandez LS, Hart DO, D'Angelo MA. 2017. Nuclear pores regulate muscle development and maintenance by assembling a localized Mef2C complex. *Dev Cell* **41:** 540–554.e7. doi:10.1016/j.devcel.2017.05.007

Randise-Hinchliff C, Coukos R, Sood V, Sumner MC, Zdraljevic S, Sholl LM, Brickner DG, Ahmed S, Watchmaker L, Brickner JH. 2016. Strategies to regulate transcription factor–mediated gene positioning and interchromosomal clustering at the nuclear periphery. *J Cell Biol* **212:** 633–646. doi:10.1083/jcb.201508068

Rawal Y, Chereji RV, Valabhoju V, Qiu H, Ocampo J, Clark DJ, Hinnebusch AG. 2018. Gcn4 binding in coding regions can activate internal and canonical 5′ promoters in yeast. *Mol Cell* **70:** 297–311.e4. doi:10.1016/j.molcel.2018.03.007

Rodriguez J, Larson DR. 2020. Transcription in living cells: molecular mechanisms of bursting. *Annu Rev Biochem* **89:** 189–212. doi:10.1146/annurev-biochem-011520-105250

Schmid M, Arib G, Laemmli C, Nishikawa J, Durussel T, Laemmli UK. 2006. Nup-PI: the nucleopore–promoter interaction of genes in yeast. *Mol Cell* **21:** 379–391. doi:10.1016/j.molcel.2005.12.012

Smith S, Galinha C, Desset S, Tolmie F, Evans D, Tatout C, Graumann K. 2015. Marker gene tethering by nucleoporins affects gene expression in plants. *Nucleus* **6:** 471–478. doi:10.1080/19491034.2015.1126028

Sood V, Brickner JH. 2017. Genetic and epigenetic strategies potentiate Gal4 activation to enhance fitness in recently diverged yeast species. *Curr Biol* **27:** 3591–3602.e3. doi:10.1016/j.cub.2017.10.035

Sood V, Cajigas I, D'Urso A, Light WH, Brickner JH. 2017. Epigenetic transcriptional memory of *GAL* genes depends on growth in glucose and the Tup1 transcription factor in *Saccharomyces cerevisiae*. *Genetics* **206:** 1895–1907. doi:10.1534/genetics.117.201632

Texari L, Dieppois G, Vinciguerra P, Contreras MP, Groner A, Letourneau A, Stutz F. 2013. The nuclear pore regulates *GAL1* gene transcription by controlling the localization of the SUMO protease Ulp1. *Mol Cell* **51:** 807–818. doi:10.1016/j.molcel.2013.08.047

Tunnacliffe E, Corrigan AM, Chubb JR. 2018. Promoter-mediated diversification of transcriptional bursting dynamics following gene duplication. *Proc National Acad Sci* **115:** 8364–8369. doi:10.1073/pnas.1800943115

Van de Vosse DW, Wan Y, Lapetina DL, Chen WM, Chiang JH, Aitchison JD, Wozniak RW. 2013. A role for the nucleoporin Nup170p in chromatin structure and gene silencing. *Cell* **152:** 969–983. doi:10.1016/j.cell.2013.01.049

von Appen A, Beck M. 2016. Structure determination of the nuclear pore complex with three-dimensional cryo electron microscopy. *J Mol Biol* **428:** 2001–2010. doi:10.1016/j.jmb.2016.01.004

Watson ML. 1955. The nuclear envelope its structure and relation to cytoplasmic membranes. *J Biophys Biochem Cytol* **1:** 257–270. doi:10.1083/jcb.1.3.257

Weiler KS, Wakimoto BT. 1995. Heterochromatin and gene expression in *Drosophila. Annu Rev Genet* **29:** 577–605. doi:10.1146/annurev.ge.29.120195.003045

Zykova TY, Levitsky VG, Belyaeva ES, Zhimulev IF. 2018. Polytene chromosomes—a portrait of functional organization of the *Drosophila* genome. *Curr Genomics* **19:** 179–191. doi:10.2174/1389202918666171016123830

The Diverse Cellular Functions of Inner Nuclear Membrane Proteins

Sumit Pawar[1] and Ulrike Kutay

Institute of Biochemistry, Department of Biology, ETH Zurich, 8093 Zurich, Switzerland

Correspondence: ulrike.kutay@bc.biol.ethz.ch

The nuclear compartment is delimited by a specialized expanded sheet of the endoplasmic reticulum (ER) known as the nuclear envelope (NE). Compared to the outer nuclear membrane and the contiguous peripheral ER, the inner nuclear membrane (INM) houses a unique set of transmembrane proteins that serve a staggering range of functions. Many of these functions reflect the exceptional position of INM proteins at the membrane–chromatin interface. Recent research revealed that numerous INM proteins perform crucial roles in chromatin organization, regulation of gene expression, genome stability, and mediation of signaling pathways into the nucleus. Other INM proteins establish mechanical links between chromatin and the cytoskeleton, help NE remodeling, or contribute to the surveillance of NE integrity and homeostasis. As INM proteins continue to gain prominence, we review these advancements and give an overview on the functional versatility of the INM proteome.

During evolution, the nuclear envelope (NE) arose as a separating membrane sheath between the genetic material and the cytoplasm. This physical barrier became one of the hallmarks of eukaryotic cells and defined nuclear compartmentalization. The emergence of the NE did not only generate a protective cover of the genome but also spatially separated transcription and translation, allowing for new forms of their regulation. Over the course of a billion years, the NE got equipped with a set of membrane-embedded proteins that endowed it with diverse tasks, yielding a functionally versatile compartment interface.

One of the most prominent features of the NE is its asymmetry. The NE is built by two double lipid bilayers, an outer nuclear membrane (ONM) and an inner nuclear membrane (INM), which are compositionally distinct with respect to their membrane protein assortment although both membranes are interconnected at numerous sites. Compared to the ONM and the connected endoplasmic reticulum (ER), the INM is enriched in a unique collection of membrane proteins that maintain close associations with chromatin and a network of intermediate filament proteins called nuclear lamins.

Clearly, the molecular identity of the INM is reflective of its functional specialization. Early work had identified only a handful of INM proteins (Senior and Gerace 1988; Worman et al. 1988; Foisner and Gerace 1993; Manilal et al. 1996; Lin et al. 2000). Later, the mass spectrometric characterization of the NE proteome

[1]Present address: X4 Pharmaceuticals GmbH, Vienna A-1030, Austria.

Cite this article as *Cold Spring Harb Perspect Biol* doi: 10.1101/cshperspect.a040477

from different sources, including rat liver and muscle (Schirmer et al. 2003; Wilkie et al. 2011; Korfali et al. 2012), human leukocytes (Korfali et al. 2010), and mouse neuroblastoma cells (Dreger et al. 2001), identified nearly 1000 putative nuclear envelope transmembrane proteins (NETs). However, as the NE is interconnected with ER, a fraction of the identified putative NETs may not be NE-specific and represent membrane proteins of the peripheral ER. Therefore, protein localization studies are a decisive element of NET definition. Remarkably, many of these NETs showed a high level of tissue specificity, suggesting that the NE proteome is adapted to cell functionality. Only a small fraction of NETs has so far been confirmed to be enriched at the NE, and most of those are also residents of the INM. To date, we know roughly 35 integral membrane proteins that were found to localize at the INM of mammalian cells (Table 1). These include prominent, well-studied INM proteins like the lamin B receptor (LBR), the LAP2-Emerin-Man1 (LEM)-domain family members lamina-associated polypeptide 2 (LAP2), emerin, MAN1, and LEM2, the Sad1p, Unc-84 (SUN)-domain proteins SUN1 and SUN2, as well as LAP1 and NET5 (Table 1; Fig. 1). Various others are less well characterized (Table 1) and the biological significance of their INM localization remains unexplored.

Most transmembrane proteins of the INM are cotranslationally inserted into the ER membrane and assume either a type II or a polytopic topology, with their extended amino termini facing the cytosol (Fig. 1). From the ER, these membrane proteins distribute to the ONM and INM by diffusion, passing through narrow membrane-proximal openings of nuclear pore complexes (NPCs) (Soullam and Worman 1995; Ohba et al. 2004; Theerthagiri et al. 2010; Zuleger et al. 2011; Boni et al. 2015; Ungricht et al. 2015; Pawar et al. 2017). These peripheral channels prevent the passage of membrane proteins with extraluminal domains larger than 60 kDa. The accumulation of INM-destined proteins at the nuclear face of the NE is determined by their retention on nuclear lamins and/or chromatin (Soullam and Worman 1995; Zuleger et al. 2011; Boni et al. 2015; Ungricht et al. 2015; Pawar et al. 2017). Thus, in principle, ER proteins with small extraluminal domains can also partition to the INM (Soullam and Worman 1995; Zuleger et al. 2011; Boni et al. 2015; Ungricht et al. 2015; Pawar et al. 2017), but in contrast to bona fide INM proteins, they are not enriched at the NE. Notably, in dividing cells undergoing open mitosis, both INM and ER proteins disperse in the mitotic ER network. During mitotic exit, a nearly synchronous enrichment of INM proteins on chromatin drives the process of NE reformation during late anaphase and telophase, rapidly reestablishing the INM membrane territory, representing a second, postmitotic mode of targeting proteins to the INM (Mattaj 2004).

The INM can be viewed as a functionally multifaceted membrane domain (Fig. 2). INM proteins engage in an array of genome-regulatory functions ranging from spatial genome organization (Solovei et al. 2013), epigenetic silencing (Somech et al. 2005), DNA replication (Martins et al. 2003), DNA damage repair (Lei et al. 2012), and transcriptional regulation. They also play important roles in diverse signaling pathways (Berk et al. 2013), in lipid synthesis (Tsai et al. 2016), and mediate mechanobiological tasks, including nuclear anchorage, migration, and mechanotransduction (Lammerding et al. 2005; Rothballer and Kutay 2013; Cho et al. 2017; Lee and Burke 2018). Interestingly, many of these functions converge and are essential for differentiation and development. They are realized through an intricate spectrum of biochemical interactions of INM proteins with chromatin, the nuclear lamina, and regulatory factors (Fig. 2).

An appreciation of the functional complexity of INM proteome is imperative to achieve an integrated view of INM biology. This is of particular importance in light of the numerous links that exist between INM protein dysfunction and human diseases. The idea that INM proteins are physiologically relevant was first realized by the discovery that emerin, mutations of which cause X-linked Emery–Dreifuss muscular dystrophy, is located at the INM (Bione et al. 1994; Manilal et al. 1996). This placed an INM protein at the causative epicenter of a disease, which paved ways for the intensification of research directed toward the cellular functions of INM proteins

 Cite this article as *Cold Spring Harb Perspect Biol* doi: 10.1101/cshperspect.a040477

Table 1. Well-characterized (in bold) and other putative inner nuclear membrane (INM) proteins, listing their predicted molecular weight, the number of transmembrane segments, membrane orientation, and references pertaining to their localization at the INM

Gene name/NET designation	Protein name	Predicted molecular weight (kD)	Predicted transmembrane (TM) segments	Orientation	References for localization at the INM
LBR	Lamin B receptor (LBR)	58 (observed), 70.7 (predicted)	8	Multipass	Worman et al. 1988
SUN1	SUN domain-containing protein 1 (SUN1)	90	1	Single-pass, type II	Dreger et al. 2001
SUN2	SUN domain-containing protein 2 (SUN2)	80.3	1	Single-pass, type II	Hodzic et al. 2004
TMPO	Lamina-associated polypeptide 2 (LAP2)	50.6 (isoform β)	1	Single-pass, type II	Foisner and Gerace 1993
EMD	Emerin (EDMD)/LEMD5	28.9	1	Single-pass, type II	Manilal et al. 1996
LEMD3/MAN1	INM protein MAN1 (MAN1)/LEMD3	99.9	2	Multipass	Lin et al. 2000
LEMD2	LEM-domain-containing protein 2 (LEMD2)	56.9	2	Multipass	Brachner et al. 2005
TMEM201/NET5	TM protein 201/SAMP1	43.3	4	Multipass	Buch et al. 2009
TOR1AIP1/LAP1	Torsin-1A-interacting protein 1 (TOR1AIP1)/ Lamina-associated polypeptide 1 (LAP1)	66.2 (isoform B)	1	Single-pass, type II	Senior and Gerace 1988
NRM	Nurim	29.3	6	Multipass	Rolls et al. 1999
LUMA/TMEM43	TM protein 43 (TMEM43)/LUMA	44.8	4	Multipass	Dreger et al. 2001
TMEM120A/ NET29	TM protein 120A (TMEM120A)/TMPIT	42.8	5	Multipass	Malik et al. 2010
MOSPD3/NET30	Motile sperm domain containing 3 (MOSPD3)	25.5	2	Multipass	Malik et al. 2010
SCARA5/NET33[a]	Scavenger receptor class A, member 5 (SCARA5)	43.2	1	Single-pass, type II	Malik et al. 2010

Continued

Table 1. *Continued*

Gene name/NET designation	Protein name	Predicted molecular weight (kD)	Predicted transmembrane (TM) segments	Orientation	References for localization at the INM
SLC39A14/ NET34[a]	Metal cation symporter ZIP14 (ZIP14)/SLC39A14	52.8	7	Multipass	Malik et al. 2010
MYORG/NET37	Myogenesis-regulating glycosidase (MYORG)/ KIAA1161	81	1	Single-pass, type II	Malik et al. 2010
PLPP7/NET39	Inactive phospholipid phosphatase 7 (PLPP7)/ PPAPDC3/C9orf67	21.9	4	Multipass	Liu et al. 2009; Malik et al. 2010
TM7SF2/NET47[b]	Delta (14)-sterol reductase TM7SF2/ANG1/ DHCR14A	46.4	7	Multipass	Malik et al. 2010
DHRS7/NET50	Dehydrogenase/reductase (SDR family) member 7 (DHRS7)	38.2	1	N.D.	Malik et al. 2010
ERG28/NET51[b]	Ergosterol biosynthetic protein 28 (ERG28)/C14orf1	15.8	4	Multipass	Malik et al. 2010
APH1B/NET55[b]	Anterior pharynx defective 1B (APH1B)	28.4	6	Multipass	Malik et al. 2010
NCLN/NET59[b]	Nicalin (NCLN)	62.9	2	Multipass	Malik et al. 2010
NCEH1[b]	Neutral cholesterol ester hydrolase 1 (NCEH1)/ AADACL1/KIAA1363	45.8	1	Single-pass, type II	Korfali et al. 2010
MAGT1[a]	Magnesium transporter protein 1 (MAGT1)/IAG2	38	4	Multipass	Korfali et al. 2010
METTL7A[b]	Methyltransferase-like protein 7A (METTL7A)	28.3	1	N.D.	Korfali et al. 2010
TMEM41A	TM protein 41A (TMEM41A)	29.7	5	Multipass	Korfali et al. 2010
STT3A[b]	Dolichyl-diphosphooligosaccharide—protein glycosyltransferase subunit STT3A(STT3-A)/ ITM1/TMC	80.5	12	Multipass	Korfali et al. 2010

Cite this article as *Cold Spring Harb Perspect Biol* doi: 10.1101/cshperspect.a040477

TMEM38A[b]	Trimeric intracellular cation channel type A (TRIC-A)/TMEM38A	33.6	7	Multipass	Wilkie et al. 2011
LPCAT3[b]	Lysophospholipid acyltransferase 5 (LPCAT3)/ MBOAT5/OACT5	56	9	Multipass	Wilkie et al. 2011
NEMP1	Nuclear envelope integral membrane protein 1 (NEMP1)/TMEM194/KIAA0286	50.6	5	Multipass	Wilkie et al. 2011
TMEM214[c]	TM protein 214 (TMEM214)	77	2	Multipass	Wilkie et al. 2011
LRRC59[b]	Leucine-rich repeat-containing protein 59 (LRRC59)/p34	35	1	Single-pass, type II	Blenski and Kehlenbach 2019

All the putative INM proteins listed here fulfilled two criteria: (1) resistance to prefixation extraction with triton X-100, and (2) localization of a fraction of the total population at the INM observed through superresolution microscopy. Please note that possibly all membrane proteins that are inserted in the endoplasmic reticulum (ER) and possess extraluminal domains <60 kDa can access the INM. However, this does not ensure their enrichment and functional relevance at the INM. The other predominant cellular localizations for the putative INM proteins are annotated in the footnotes. Please note that SUN3, SUN4 (Spag4), and SUN5 (Frohnert et al. 2011; Calvi et al. 2015; Pasch et al. 2015) are specifically expressed in testis where they enrich at the NE. However, their localization to the INM remains to be formally proven.

(N.D.) No data available.

[a]Proteins also localize at the plasma membrane as observed for SCARA5 (Huang et al. 2010), SLC39A14 (Taylor et al. 2004; Tuschl et al. 2016), and MAGT1 (Zhou and Clapham 2009).

[b]Proteins primarily localize at the ER as observed for SLC39A14 (Malik et al. 2010), TM7SF2 (Holmer et al. 1998), ERG28 (Malik et al. 2010), APH1B (Malik et al. 2010), NCLN (Dettmer et al. 2010), NCEH1 (Igarashi et al. 2010), MAGT1 (Cherepanova et al. 2014), METTL7A (McKinnon and Mellor 2017), STT3A (Braunger et al. 2018; Zhu et al. 2019), TMEM38A (Yazawa et al. 2007; Shrestha et al. 2020), LPCAT3 (Zhao et al. 2008), and LRRC 59 (Zhen et al. 2012).

[c]Predominantly localizes at the ER and the Golgi complex (Li et al. 2013).

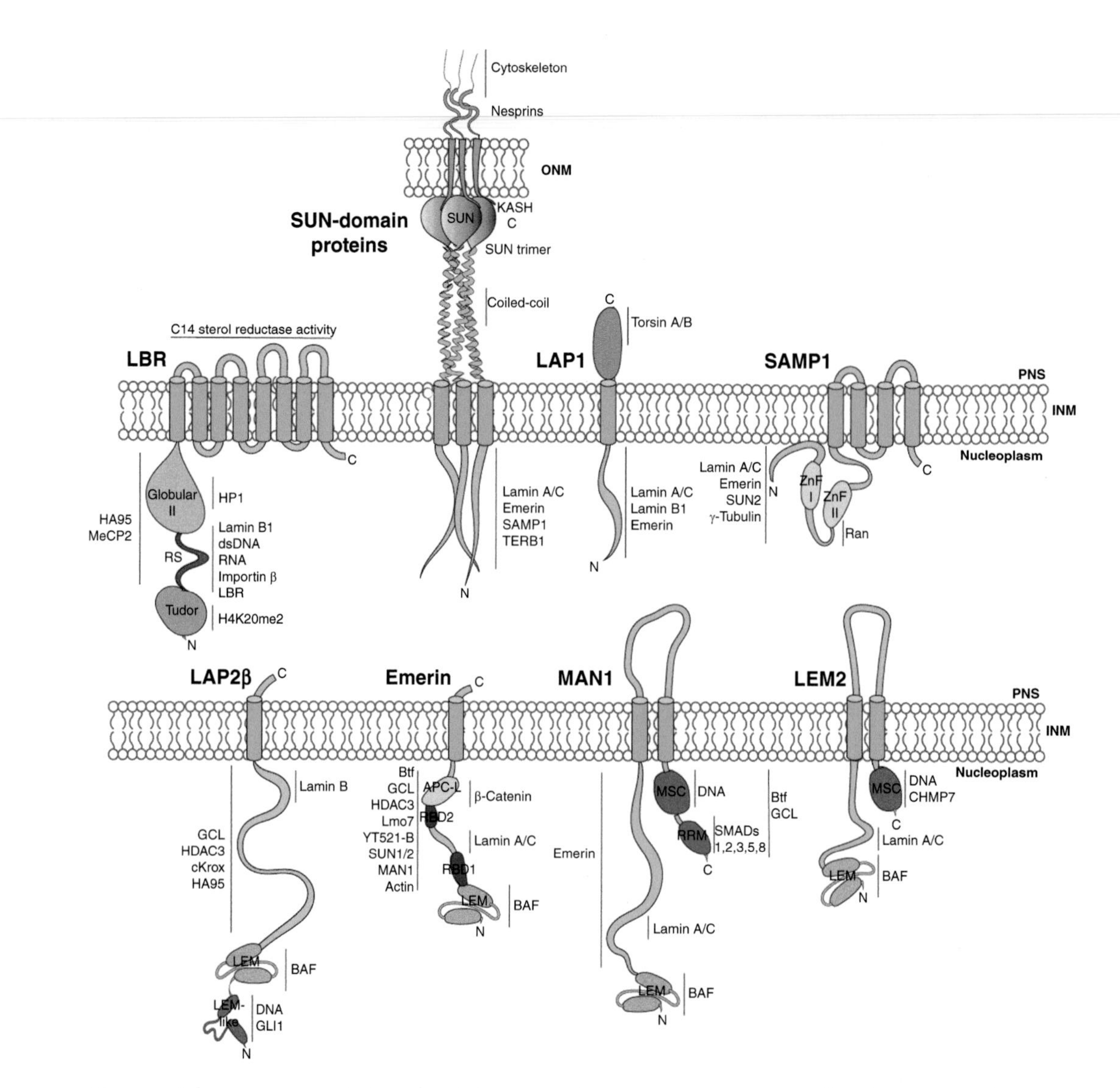

Figure 1. Schematic representation of well-characterized inner nuclear membrane (INM) proteins indicating their membrane topology, domain organization, and known interaction partners. INM proteins bind chromatin, chromatin-associated factors, and nuclear lamins as indicated. Certain INM proteins like lamin B receptor (LBR) and LAP2-Emerin-Man1 (LEM)-domain proteins bind transcription factors and thereby modulate gene expression. Additionally, some INM proteins bind to each other while others, like LBR, self-interact, altogether creating a complex network of interactions. At the INM, Sad1p, Unc-84 (SUN) domain proteins form trimers via their coiled-coil domains and interact with three KASH peptides of outer nuclear membrane (ONM)-resident Nesprins to form the LINC (linker of nucleoskeleton and cytoskeleton) complex.

and their implications in human health. Presently, over 15 syndromes have been linked to the dysfunction of INM proteins (Table 2), and the list of diseases continues to expand. Every new disease link reiterates the significance of the INM proteome and stimulates novel ways to perceive how human physiology is influenced by the cellular functions of INM proteins. In this review, we use the examples of well-characterized, widely expressed INM proteins to portray their broad functional repertoire, giving priority to mammalian representatives.

 Cite this article as *Cold Spring Harb Perspect Biol* doi: 10.1101/cshperspect.a040477

LAMIN B RECEPTOR (LBR)

When radioactive lamins were incubated with erythrocyte NE membranes, B-type lamins bound a 58 kDa protein with high specificity (Worman et al. 1988). This protein, designated as the "lamin B receptor," was the first integral INM protein to be identified and remains one of the most well-characterized proteins of the INM to date. Structurally, LBR is a polytopic membrane protein with a hydrophilic amino-terminal domain that protrudes into the nucleoplasm and is composed of a Tudor domain, a serine/arginine-rich (RS) hinge region, and a second globular domain. This is followed by a hydrophobic region containing eight membrane-spanning helices, and a short nucleoplasmic carboxy-terminal tail (Fig. 1). Functionally, LBR is a versatile protein. Whereas its extraluminal domain contributes to chromatin organization in the nuclear periphery, the transmembrane segments exhibit sterol reductase activity.

Early biochemical studies had revealed a striking enrichment of heterochromatin marks on chromatin pulled down by LBR (Makatsori et al. 2004). Later, a combination of histone tail peptide arrays, chromatin immunoprecipitation, and direct binding experiments indicated that LBR can directly recognize the heterochromatin mark H4K20me2 via its amino-terminal Tudor domain (Hirano et al. 2012). Besides nucleosomes, LBR engages in a multitude of other interactions with nuclear partners, including lamin B1 (Ye and Worman 1994), dsDNA (Ye and Worman 1994; Duband-Goulet and Courvalin 2000), and RNA (Chen et al. 2016), which all bind to the RS domain of LBR; HA95 (Martins et al. 2000) and the methylated DNA-binding protein MeCP2 (Guarda et al. 2009), as well as the heterochromatin organizer HP1 that associates with LBR's membrane-proximal globular domain (Ye et al. 1997; Lechner et al. 2005). The nucleoplasmic domain of LBR thereby constitutes a multifaceted molecular backbone for heterochromatin tethering to the NE. In addition, homo-oligomerization of LBR may even contribute to the compaction of heterochromatin at the nuclear periphery (Hirano et al. 2012).

The functional importance of LBR's interaction with chromatin is highlighted by studies on chromatin organization during mammalian development and differentiation. The spatial segregation of transcriptionally repressed heterochromatin to the nuclear periphery of differentiated mouse cells was shown to rely on two molecular assemblies, referred to as the A-type and B-type tethers (Solovei et al. 2013). LBR constitutes the molecular pillar of the B-type tether, whereas the A-type tether consists of lamins A/C and (an) unknown protein(s) of the INM. One of the two tethers is sufficient to maintain heterochromatin at the nuclear periphery. While developing tissues primarily rely on LBR for peripheral heterochromatin tethering, differentiated tissues seem to be more dependent on lamin A/C (Solovei et al. 2013). Remarkably, loss of both LBR and lamin A/C can lead to "nuclear inversion," an atypical collapse of heterochromatin into the nuclear interior, which is physiologically relevant for light perception in the photoreceptor rod cells of nocturnal animals (Solovei et al. 2013).

Beyond tethering and compacting chromatin at the nuclear periphery, LBR also facilitates transcriptional repression. Myoblast transcriptome analyses, for example, revealed that LBR is involved in suppression of muscle-specific gene expression during early stages of differentiation (Solovei et al. 2013). Furthermore, LBR has been suggested to promote X-chromosome inactivation (XCI) during development in mammals. The association of LBR with the long noncoding RNA *Xist* contributes to the recruitment of the inactivated X chromosome to the nuclear periphery and was also suggested to ensure its transcriptional silencing (McHugh et al. 2015; Chen et al. 2016). However, in a recent X-chromosome-wide analysis of *Xist*-mediated silencing, loss of LBR had only a minor effect on silencing (Nesterova et al. 2019). Discrepancies in these observations may be linked to the different silencing assays, different times of *Xist* induction, as well as allelic versus nonallelic analyses of XCI used in these studies. Thus, although the LBR–*Xist* interaction might be necessary for positioning of the inactive X chromosome and/or to stabilize gene expression, the

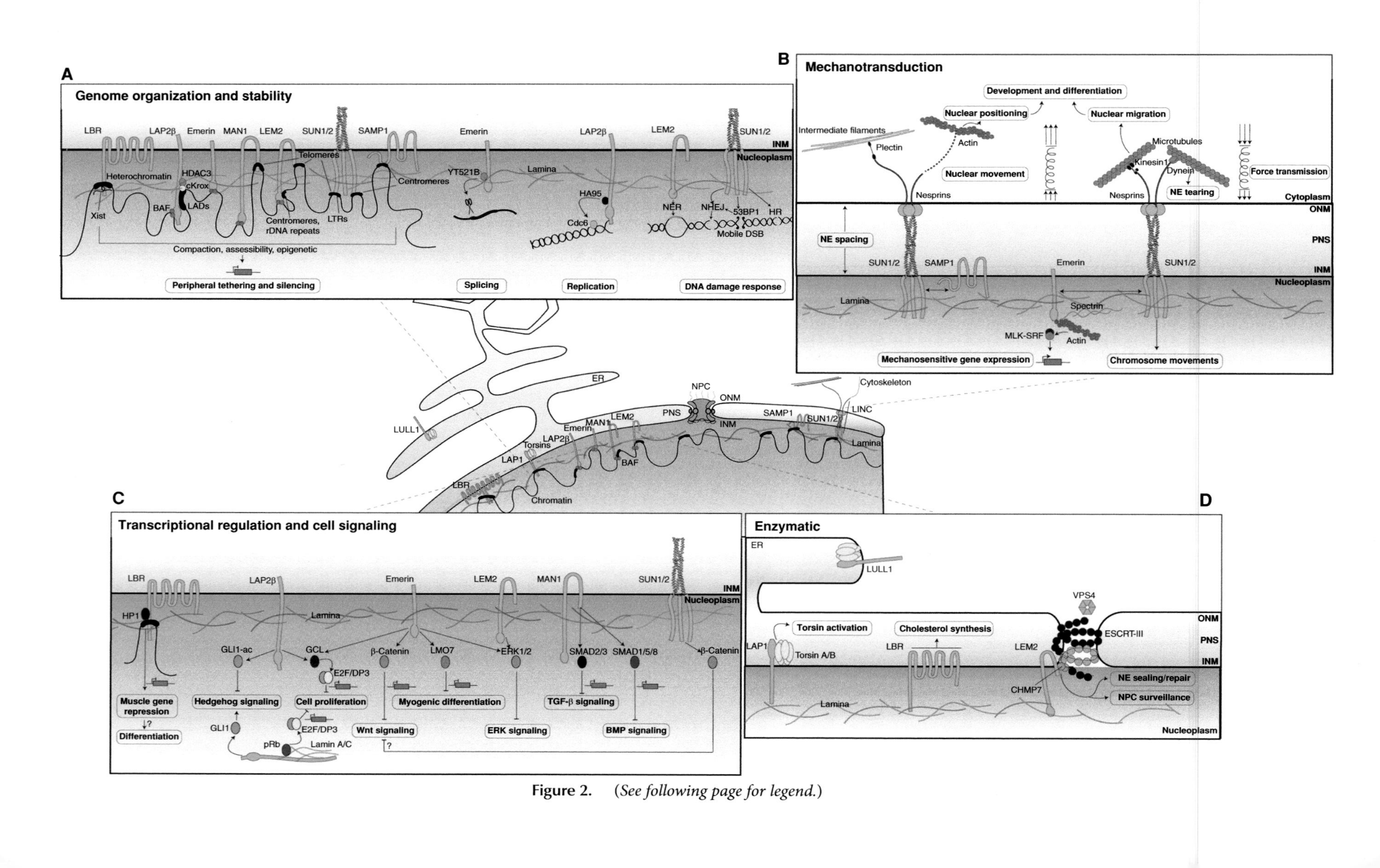

Figure 2. (*See following page for legend.*)

exact impact of LBR on XCI still remains to be determined.

LBR–chromatin association may also be relevant for cellular senescence. Initial studies established a link between a reduction of lamin B1 and cellular senescence, suggesting that the nuclear lamina undergoes profound changes as cells enter senescence (Shimi et al. 2011; Freund et al. 2012; Dreesen et al. 2013; Shah et al. 2013; Dou et al. 2015). More recent studies have revealed that LBR is also down-regulated during transition into senescence (Ivanov et al. 2013; Lenain et al. 2015; Lukášová et al. 2017; Arai et al. 2019; En et al. 2020). The cellular transition to senescence is associated with an extensive reorganization of chromatin and changes in gene expression. Induction of senescence through γ-irradiation of cancer cells significantly decreased the levels of LBR and caused relocation of centromeric heterochromatin from the nuclear periphery to the nuclear interior (Lukášová et al. 2017). Upon induction of senescence via replicative or oncogene-induced stress, heterochromatin-rich fragments bud off from nuclei associated with a down-regulation of lamin B1 and LBR (Ivanov et al. 2013; Dou et al. 2015). The significant decrease in the levels of both LBR and lamin B1 in all these studies may imply that the regulation of their levels during senescence is interrelated. Importantly, however, down-regulation of LBR and lamin B1 levels alone is insufficient to induce senescence phenotypes (Dreesen et al. 2013; Lukášová et al. 2017), indicating that reduction of lamin B1 and LBR is a hallmark and the consequence of senescence rather than its cause.

Last, LBR–chromatin interactions have also been implicated in postmitotic NE assembly. This is supported by early observations showing that immunodepletion of LBR reduces chromatin recruitment of the NE vesicles in vitro (Pyrpasopoulou et al. 1996) and that deletion of LBR slightly delays NE assembly in cultured mammalian cells (Anderson et al. 2009).

Compared to the plethora of studies on the chromatin-associated functions of the nucleoplasmic domain of LBR, its carboxy-terminal membrane domain, which shares extensive homology to the human C14 sterol reductase TM7SF2 (Holmer et al. 1998) (also called DHCR14 or SR-1), has remained neglected for a long time. Because most enzymes involved in the cholesterol biosynthesis pathway localize to the ER, such a function seemed unexpected for an INM protein (Schuler et al. 1994). However, human LBR complemented yeast C14 reductase mutants, which provided first evidence for the enzymatic activity of LBR (Silve et al. 1998; Prakash et al. 1999). Recent findings have now unraveled that the C14-sterol reductase activity of LBR is even essential for the viability of mammalian cells, which can explain its relevance to human physiology (Tsai et al. 2016). Two congenital diseases are associated with mutations in LBR: Pelger–Huët anomaly, an autosomal dominant disorder resulting in abnormal hypolobulation of granulocyte nuclei, and Greenberg skeletal dysplasia, an autosomal-recessive condition resulting in abnormal bone development, fetal hydrops, and the ultimate nonviability of the fetus (Turner and Schlieker 2016). Most of these mutations map to the enzymatic mem-

Figure 2. The diverse functions of inner nuclear membrane (INM) proteins. (*A*) Many INM proteins tether chromatin to the nuclear envelope (NE) and facilitate repression of genomic elements. Some INM proteins assist processes such as splicing, DNA replication, or the DNA damage response. (*B*) Members of the LINC (linker of nucleoskeleton and cytoskeleton) complex have mechanical and structural roles, interlinking the nucleoskeleton and cytoskeleton and thereby acting as force transmission devices across the NE. LINC complexes perform a wide range of functions, including nuclear anchorage and migration, chromosome movements, and the maintenance of NE membrane spacing. Other INM proteins (e.g., emerin and SAMP1) may functionally collaborate with LINC complexes in mechanotransduction. (*C*) INM proteins interact with transcription factors to regulate gene expression, thereby contributing to diverse signaling pathways. (*D*) Certain INM proteins possess enzymatic functions. Lamin B receptors (LBRs) exhibit C14 sterol reductase activity that is used in cholesterol synthesis. Others, like LAP1 and LEM2, directly or indirectly regulate the activities of partner proteins such as Torsins and the ESCRT-III complex, respectively.

Table 2. Human diseases (in bold) related to either mutations in genes encoding for inner nuclear membrane (INM) proteins or a misregulation of INM protein levels

Disease	Pathology/clinical features	Implicated INM protein	Associated mutations/defects	References
Pelger-Huët anomaly	Hyposegmentation and abnormal chromatin organization in granulocyte nuclei	**Lamin B receptor (LBR)**	P119L, P569R, R377X*, W436X*, L534X*, V11EfsX24, G382DfsX39, S167TfsX176, loss of splicing of exon 3, 11, 14, altered splicing of exon 13	Hoffmann et al. 2002; Best et al. 2003; Clayton et al. 2010; Tsai et al. 2016
Greenberg skeletal dysplasia	Abnormal sterol metabolism, hydrops, ectopic calcification, and moth-eaten (HEM) skeletal dysplasia	**LBR**	N547D, R583Q, L534X, Y468TfsX475, V11EfsX24	Waterham et al. 2003; Clayton et al. 2010; Tsai et al. 2016; Turner and Schlieker 2016
Pelger-Huët anomaly with mild skeletal anomaly	Bilobed neutrophil nuclei with mild skeletal dysplasia	**LBR**	I218DfsX19, R586H, R76X, N547S	Borovik et al. 2013; Sobreira et al. 2015
Reynolds syndrome	Primary biliary cirrhosis, cutaneous systemic sclerosis	**LBR**	R372C	Gaudy-Marqueste et al. 2010
Emery–Dreifuss muscular dystrophy (EDMD) or EDMD-related myopathies	Joint contracture, muscle wasting and weakness, cardiac diseases with conduction defects, and arrhythmia	**Emerin**	~100 mutations in emerin have been associated with X-linked EDMD, including M1V, S54F, Q133H, P183H, P183T, S143X, W200X, Q44X, Q86X, S171X, Q219X, Q228X, Y41X, S66X, S120X, W226X, Δ95–99, R207GfsX30, L84fsX7, D9GfsX24, F39SfsX17, R45KfsX16, F190YfsX19, R203AfsX34, R204PfsX7, S52QfsX9, L84PfsX7, Y85LfsX8, E11SfsX2, S49LfsX11, and others	Bione et al. 1994; Mora et al. 1997; Ellis et al. 1999; Morris and Manilal 1999; Astejada et al. 2007; Ben Yaou et al. 2007; Brown et al. 2011; Dai et al. 2019; Zhou et al. 2019
		LUMA	E85K, I91V	Liang et al. 2011
		MAN1	G88V, R230T occur along with mutations in dysferlin, plectin, ryanodine receptor 3, ankyrin2, nesprin1, integrator complex subunit 1, or titin	Meinke et al. 2020

		NET39	M92K, R252P occur along with mutations in plectin, collagen, dysferlin, or integrator complex subunit 1	Meinke et al. 2020
		SUN1	A230V, G76A, and W377C occur along with mutations in either emerin or lamin A/C: G68D, G338S, W377C	Meinke et al. 2014
		SUN2	A56P and V378I occur along with mutations in lamin A/C: R620C, E438D	Meinke et al. 2014
		TMEM201/ Samp1	G15A, G18S, and G597S occur along with mutations in lamin A/C, nesprin1, or SUN1	Meinke et al. 2020
		TMEM38A	V247M and D260N occur along with mutations in plectin, collagen, dysferlin, or integrator complex subunit 1	Meinke et al. 2020
		TMEM214	G236S and R179H occur along with mutations in nesprin3 or collagen	Meinke et al. 2020
Buschke–Ollendorff syndrome	Connective tissue nevi combined with osteopoikilosis	**MAN1**	W621X, Y441X, R625X, L611X, S619X, W855X, R655X, and L638fs mutation at exon 12/intron 12 boundary causing defective splicing	Hellemans et al. 2006; Kobayashi et al. 2007; Zhang et al. 2009; Burger et al. 2010; Korekawa et al. 2012; Brodbeck et al. 2016
Isolated osteopoikilosis	Small and round spots of increased bone density that are mainly located in the epiphyseal regions of the tubular bones	**MAN1**	E601X, Q153X, C1323A, R537X, L478X, R678X, and R735X	Hellemans et al. 2004, 2006; Couto et al. 2007; Mumm et al. 2007; Baasanjav et al. 2010
Isolated melorheostosis	"Dripping wax" appearance in the cortex of long bones, accompanied by abnormalities of adjacent soft tissues	**MAN1**	L638X	Hellemans et al. 2006

Continued

Table 2. *Continued*

Disease	Pathology/clinical features	Implicated INM protein	Associated mutations/defects	References
Hutterite-type cataract (with or without arrhythmic cardiomyopathy)	Clouding of the lens of the eye; some patients have suffered sudden death presumably of arrhythmogenic origin	**LEM2**	L13R	Boone et al. 2016
Severe dystonia with cerebellar atrophy	Progressive dystonia, severe contractures of the Achilles tendons, and feet deformations	**LAP1**	E482A	Dorboz et al. 2014
Limb–girdle muscular dystrophy type 2	Proximal and distal weakness and atrophy, rigid spine and contractures of the proximal and distal interphalangeal hand joints, cardiomyopathy, and respiratory involvement	**LAP1**	L394P, E62DfsX25, P43fsX15	Kayman-Kurekci et al. 2014; Ghaoui et al. 2016
Unnamed multisystemic disease	Psychomotor retardation, cataract, heart malformation, sensorineural deafness	**LAP1**	R321X	Fichtman et al. 2019
Arrhythmogenic right ventricular cardiomyopathy/ dysplasia (ARVC5)	Cardiomyocyte apoptosis, lethal ventricular tachycardia, and fibro-fatty infiltration mainly in the right ventricle	**LUMA**	S358L	Stroud et al. 2018
Dilated cardiomyopathy	Increase in left ventricular systolic and diastolic diameter and decrease in ejection fraction	**LAP2α**	R690C	Taylor et al. 2005

Hutchinson–Gilford progeria syndrome	Premature aging disease: individuals exhibit low body weight, decreased joint mobility, scleroderma, and die around the age of 13	**SUN1**	Up-regulated levels of SUN1 in patients expressing mutant lamin A/C	Chen et al. 2012
Hypertrophic cardiomyopathy	Cardiac arrythmia, supraventricular extrasystoles	**SUN2**	M50T, V378I occur along with a mutation in myosin binding protein C	Meinke et al. 2014
Mandibuloacral dysplasia type A	Growth retardation, craniofacial anomalies, mottled cutaneous pigmentation, skin rigidity, partial lipodystrophy, insulin resistance	**SUN2**	Mislocalization of SUN2 due to mutation in LMNA	Camozzi et al. 2012
Cancers	Subtype: cervical, colon, colorectal, gastric, pancreatic cancers, myeloma, lymphoma, medulloblastoma, and others	**LAP2α**	Up-regulated levels of LAP2α	Reviewed in Brachner and Foisner (2014)
	Subtype: digestive tract cancers	**LAP2β**	Up-regulated levels of LAP2β	Kim et al. 2012
	Subtype: ovarian cancer	**Emerin**	Reduced expression of emerin	Capo-chichi et al. 2009
	Subtype: hepatocellular carcinoma and neoplasms	**NET33**	Down-regulation of NET33	Liu et al. 2018; Ulker et al. 2018
	Subtype: aggressive breast, ovarian and prostate cancers	**AADACL1**	Up-regulated levels/activity of AADACL1	Chiang et al. 2006; Chang et al. 2011

The table does not cover diseases associated with nuclear lamins and barrier to autointegration factor (BAF).
(X) Termination codon, (fs) frameshift, (*) the amino acids at which premature termination occurs were allocated using Uniprot.

brane domain of LBR and strongly perturb LBR's ability to engage in cholesterol synthesis, albeit through two different mechanisms. Whereas some mutations interfere with LBR's ability to bind the enzymatic cofactor NADPH, others lead to LBR degradation. Importantly, all investigated disease-causing mutants fail to complement the cholesterol auxotrophy imposed by LBR deficiency in cultured mammalian cells, thus establishing LBR as a major sterol reductase required for cholesterol synthesis (Tsai et al. 2016). Peculiarly, the sterol reductase domain of LBR also resembles a conserved domain of isoprenylcysteine carboxyl methyltransferases, which methylate the carboxyl group of prenylated cysteine as a last step of CaaX modification. Such a function for LBR has been speculated to aid the final modification steps of substrates such as prelamin A and/or lamin B (Li et al. 2015), but still awaits supporting evidence.

SUN DOMAIN PROTEINS

Members of the conserved family of SUN (Sad1p, Unc-84)-domain proteins localize at the INM and interact with KASH (Klarsicht/ANC-1/Syne-1 homology)-domain proteins of the ONM in the perinuclear space, forming NE-spanning LINC (linker of nucleoskeleton and cytoskeleton) complexes (Crisp et al. 2006). SUN domain proteins oligomerize into homotrimers with the help of their long intraluminal coiled-coil regions that also force the carboxy-terminal SUN domains into a trimeric arrangement (Sosa et al. 2012; Wang et al. 2012). The formation of LINC complexes relies on the tight intercalation of structurally extended KASH peptides at the SUN domain interfaces, building force-resistant coupling devices within the NE. The extraluminal domains of different SUN and KASH proteins, in turn, engage in a variety of interactions with nuclear components and the cytoskeleton, respectively, conferring the LINC complex with a diverse range of functions. These include nuclear anchorage, nuclear migration, insertion of NPCs or spindle pole bodies into the NE, the coupling of centrosomes to the nucleus, NE membrane spacing, and NE remodeling at the onset of mitosis (Malone et al. 1999; Starr et al. 2001; Crisp et al. 2006; Liu et al. 2007; Hiraoka and Dernburg 2009; Friederichs et al. 2011; Talamas and Hetzer 2011; Gundersen and Worman 2013; Cain et al. 2014; Guilluy et al. 2014; Turgay et al. 2014). As these processes have been covered in elaborate reviews (Starr and Fridolfsson 2010; Gundersen and Worman 2013; Rothballer and Kutay 2013; Rothballer et al. 2013; Burke and Stewart 2014; Chang et al. 2015; Burke 2018; Lee and Burke 2018), we will limit ourselves here to some prominent intranuclear functions of LINC complexes.

From yeast to humans, SUN domain proteins establish specialized links between the INM and chromatin. The best-defined role is the anchorage of telomeres. In *Saccharomyces cerevisiae*, the SUN domain protein Mps3 mediates telomere tethering to the NE by chromatin silencing factors like Sir4 (Bupp et al. 2007; Horigome et al. 2011) or telomerase subunits like Est1 (Antoniacci et al. 2007; Schober et al. 2009). The association between telomeres and the NE is thought to suppress telomeric transcription and protect telomeres from harmful recombination (Gartenberg 2009; Schober et al. 2009; Mekhail and Moazed 2010).

In addition, the association between telomeres and SUN proteins is of particular importance for meiosis. During meiotic prophase I, rapid prophase movements (RPMs) led by telomeres facilitate the alignment and pairing of homologous chromosomes to ensure their faithful partitioning. Clustering of telomeres requires the concerted effort of LINC complexes, the cytoskeleton, and telomere-associated proteins. In mammalian cells, a specialized LINC complex consisting of SUN1 and KASH5 is required to couple telomeres via cytoplasmic dynein to the microtubule system (Morimoto et al. 2012; Horn et al. 2013; Lee et al. 2015). Telomere association of SUN1 relies on a complex of the membrane-associated junction protein (MAJIN1) and the telomere repeat-binding proteins 1 and 2 (TERB1 and 2) (Shibuya et al. 2014, 2015), and is regulated by cyclin-dependent kinase 2 (CDK2) (Viera et al. 2015). Mice deficient in either SUN1 or KASH5 are infertile due to meiotic arrest, highlighting the physiological relevance of RPMs and subsequent meiotic bouquet formation (Ding et al. 2007; Horn et al. 2013).

Similarly, in *Schizosaccharomyces pombe*, the SUN and KASH domain proteins Sad1 and Kms1 promote the formation of the meiotic chromosome bouquet. They mediate RPMs through an association of Kms1 with cytoplasmic dynein and of Sad1 with the telomere bouquet proteins Bqt1-4 and Rap1 and Taz1, components of the shelterin complex (Chikashige et al. 2006, 2009; Hiraoka and Dernburg 2009). The function of LINC complexes in bouquet formation is also conserved in the budding yeast *S. cerevisiae* (Trelles-Sticken et al. 2000; Conrad et al. 2008), in which the telomere-led RPMs are actin dependent (Trelles-Sticken et al. 2005; Koszul et al. 2008), and in *Caenorhabditis elegans* (Penkner et al. 2007, 2009; Sato et al. 2009; Woglar and Jantsch 2014), with the difference that RPMs rely on the association of specialized chromosome pairing centers with SUN1 (MacQueen et al. 2005). The evolutionary conserved meiotic chromosome movements nicely illustrate how cytoskeletal forces are transmitted via LINC complexes across the NE, here to impart changes in intranuclear organization.

SUN domain proteins have also been prominently implicated in NE-associated DNA repair. The yeast SUN domain protein Mps3, for instance, assists the tethering of persistent DNA double-strand breaks (DSBs) to the NE, potentially providing an environment supportive for alternative repair pathways (Oza et al. 2009; Horigome et al. 2014). Also in higher organisms, several lines of evidence suggest a role for LINC complexes in the DNA damage response (DDR) (Aymard et al. 2017; Marnef et al. 2019). DDR is accomplished either through error-prone nonhomologous end joining (NHEJ) or by a more precise repair pathway involving homologous recombination (HR). Initial studies revealed that mouse SUN1 and SUN2 interact with the DNA-dependent protein kinase (DNAPK), which plays a role in NHEJ repair and thereby assists the process (Lei et al. 2012). A later study then demonstrated that human cells use the DNA repair factor 53BP1, SUN1/2, and dynamic microtubules to promote the mobility of both dysfunctional telomeres and DSBs, potentially to facilitate NHEJ of DSBs (Lottersberger et al. 2015). Whether the mobility of DSBs is facilitated via direct and specific interactions between 53BP1-bound DSBs and the LINC complex or is the result of LINC complex-mediated transduction of forces onto chromatin in an untargeted manner remain interesting possibilities to be explored. On the other hand, the *C. elegans* SUN domain protein, UNC-84 sequesters components of the NHEJ pathway at the NE while promoting the recruitment of Fanconi anemia nuclease (FAN-1) to the sites of DNA interstrand cross-links. Thereby, UNC-84 has been proposed to guide repair pathway choice through inactivation of NHEJ and promotion of FAN-1-mediated HR at chromosomal breaks (Lawrence et al. 2016).

A surge of recent work reaffirmed the significance of LINC complexes in coupling nuclear mechanics and chromatin-associated processes. First, shear stress applied to the mammalian cell surface was found to stretch chromatin regions and led to transcriptional up-regulation of a transgene inserted in a stretched region, perhaps due to stretch-enabled chromatin opening and binding of RNA polymerase (RNAP) II (Tajik et al. 2016). Importantly, both chromatin displacement and transgene expression were affected upon depletion of SUN proteins, supporting the requirement of SUN proteins in coupling mechanical signals to transcription. Second, LINC complexes were suggested to influence a mechanically induced blockade of adipogenesis in mesenchymal stem cells (MSCs) (Uzer et al. 2018). Here, depletion of SUN1 and SUN2 diminished the nuclear localization of β-catenin, which normally counteracts adipocytic commitment by down-regulation of adipogenic transcription factors (Sen et al. 2008). Finally, extreme mechanical stimulation imposed by a high frequency tensile strain in human MSCs causes a rapid phosphorylation and turnover of SUN2. This contributes to a decoupling of the cytoskeleton and the nucleus, thereby imparting protection from strain-induced DNA damage (Gilbert et al. 2019).

Interestingly, through its association with lamin A/C, human SUN2 is also known to contribute to the maintenance of the latent HIV provirus in a transcriptionally repressed chromatin domain at the NE, thereby promoting the latency of HIV infection (Sun et al. 2018). Yet, SUN proteins may not only modulate viral latency but also affect early steps of HIV infection. The mechanistic details are, however, not well understood (Donahue et al. 2016; Lahaye et al. 2016; Schaller et al. 2017; Luo et al. 2018). Likely, even to date, many other exciting roles of LINC complexes remain to be discovered.

LEM-DOMAIN PROTEINS

LEM-domain proteins, named after the founding family members LAP2, emerin, and MAN1, are defined by the presence of a bihelical motif called the LEM domain (Lin et al. 2000). This domain confers interaction with the widespread metazoan DNA-binding protein, the barrier to autointegration factor (BAF), which has been implicated in important processes such as gene regulation, chromatin condensation, and nuclear assembly (Margalit et al. 2007; Samwer et al. 2017). Whereas BAF and LEM domains are restricted to metazoans, some INM proteins in both lower and higher eukaryotes possess a structurally related helix-extension-helix fold (Heh) that confers direct DNA interaction.

Metazoan LEM-domain proteins strongly associate with the nuclear lamina (Sullivan et al. 1999; Östlund et al. 2006; Ulbert et al. 2006) and are organizers of a complex interaction network at the NE–lamina–chromatin interface. In this network, LEM-domain proteins functionally overlap in recruiting chromatin-modifying proteins such as the histone deacetylase HDAC3 (Somech et al. 2005; Demmerle et al. 2012), and transcriptional regulators such as germ cell-less (GCL) and Btf (Holaska et al. 2003; Haraguchi et al. 2004; Mansharamani and Wilson 2005) (see also Fig. 1). Thereby, they coordinate the peripheral immobilization and repression of genomic elements, and the inhibition of transcription (Nili et al. 2001; Brachner and Foisner 2011; Ho et al. 2013; Guilluy et al. 2014; Lee et al. 2017). However, individually, they also perform some specialized functions, which will be discussed in the following sections.

LAP2

LAP2 proteins comprise at least six distinct protein isoforms (α, β, γ, δ, ε, ζ) that are generated by alternative splicing in mammals (Dechat et al. 2000). The most abundant among them are LAP2α and LAP2β, both of which interact with lamins and chromatin. LAP2α diverges significantly from LAP2β and the other LAP2 isoforms, as it is a nucleoplasmic protein, whereas the others are membrane embedded. LAP2β, the longest membrane-bound isoform, possesses an extended nucleoplasmic domain composed of an amino-terminal LEM-like domain through which it interacts with DNA, followed by the LEM domain that interacts with BAF, a low complexity region encompassing a lamina-binding domain, a single transmembrane segment, and a short luminal domain (Fig. 1). Early studies demonstrated a role for LAP2β in postmitotic targeting of membranes to chromatin and postmitotic nuclear growth (Foisner and Gerace 1993; Yang et al. 1997; Gant et al. 1999; Anderson et al. 2009).

In interphase cells, Lap2β plays an important role in fundamental cellular processes such as transcription and DNA replication. First, LAP2β contributes to the formation of a repressive chromatin environment at the INM (Cutler et al. 2019). Both on its own and in complex with the transcriptional repressor GCL, it represses the transcriptional activity of the E2F-DP3 heterodimer (Nili et al. 2001). Transcriptional repression by LAP2β is further regulated by its interaction with epigenetic modifiers like the histone deacetylase 3 (HDAC3) (Somech et al. 2005). Importantly, the LAP2β-HDAC3 complex also binds the transcriptional repressor cKrox that recognizes GAGA-type DNA elements within lamina-associated chromatin domains, for instance those spanning the developmentally regulated gene loci IgH and Cyp3a in human cells, leading to their peripheral localization and transcriptional silencing (Zullo et al. 2012). Second, LAP2β has also been suggested to control DNA replication by direct association with the chromatin factor HA95 that in turn interacts with Cdc6, an essential component of the prereplication complex (preRC). Abolishing the HA95-LAP2β interaction induces proteasome-mediated degradation of Cdc6 and thereby inhibits the initiation of DNA replication (Martins et al. 2003).

In LAP2α, the transmembrane domain is replaced by a unique coiled-coil domain that confers A-type lamin-binding activity (Dechat et al. 1998; Vlcek et al. 1999). Notably, both LAP2α and the nucleoplasmic pool of lamin A/C associate with euchromatin (Gesson et al. 2016). Furthermore, LAP2α is known to stabilize the tumor suppressor retinoblastoma protein pRb by anchoring it to nucleoplasmic lamin A/C (Markiewicz et al. 2002; Dorner et al. 2006). Thereby LAP2α regulates several functions of pRb such as cell cycle control and terminal differentiation of adipose and muscle tissues (Hansen et al. 2004; Huh et al. 2004). Overexpression of LAP2α in pre-adipocytes promotes cell cycle exit and initiation of differentiation to adipocytes in vitro (Dorner et al. 2006). On the other hand, LAP2α loss impairs pRb function causing inefficient cell cycle arrest in mouse fibroblast cultures and hyperproliferation of epidermal and erythroid progenitor cells in vivo (Naetar et al. 2008).

But do the soluble and membrane-bound LAP2 isoforms have completely independent functions? A recent study has demonstrated that LAP2α and LAP2β use their common LEM-like domain to modulate the activity of GLI1 (Mirza et al. 2019), a zinc finger transcription factor that controls the hedgehog pathway during tumorigenesis (Oro et al. 1997). Acetylated GLI1 is anchored at the INM by LAP2β, creating an inactive but dynamic nuclear reserve of GLI1 at the NE. Nucleoplasmic LAP2α does not only compete for binding of GLI1 but also drives GLI1 activation together with its binding partner HDAC1, which converts GLI1 into the active, deacetylated form on chromatin. Thus, LAP2 proteins form an isoform-based nuclear chaperoning system that controls the balance between inactive GLI1 in the nuclear periphery and active GLI1 in the nucleoplasm, promoted by the release of GLI1 from LAP2β by aPKC to ensure maximal GLI activation. As LAP2 proteins were found to interact quite broadly with zinc-finger proteins via their LEM-like domain, this may suggest a novel and general nuclear scaffolding function toward zinc-finger transcription factors (Mirza et al. 2019).

EMERIN

Emerin is one of the best-studied INM proteins, as mutations in the *EMD* gene causes X-linked Emery–Dreifuss muscular dystrophy (EDMD) (Bione et al. 1994). It is a tail-anchored membrane protein possessing an amino-terminal LEM domain followed by a so-called regulator-binding domain (RBD), a lamin-binding region, and a second RBD overlapping with an adenomatous polyposis coli-like (APC-L) domain (Berk et al. 2013; Koch and Holaska 2014).

Emerin perfectly epitomizes the degree of biochemical complexity at the NE as it can interact with a plethora of partners and is involved in diverse cellular processes. First, it associates with factors involved in genome organization and regulation. These include epigenetic regulators, proteins involved in signaling, transcription, mRNA splicing, and of course BAF (Holaska and Wilson 2007; Wilson and Foisner 2010; Koch and Holaska 2014). Following the paradigm of formation of repressive chromatin at the NE, emerin epigenetically modulates gene expression by interacting with and regulating the enzymatic activity of HDAC3, which forms the catalytic subunit of nuclear corepressor complex (NCoR). Emerin-null fibroblasts exhibit epigenetic changes that strikingly resemble those of HDAC3 knockout (KO) cells. These include a global increase in H4K5 acetylation and decreased H3K27 and H3K9 trimethylation (Demmerle et al. 2012). Besides epigenetic modulation, emerin influences gene expression by scaffolding a variety of gene-regulatory partners at the INM through its RBDs and APC-L domain. Prominent among them are GCL (Holaska et al. 2003), Btf (Haraguchi et al. 2004), Lim domain only (Lmo7) (Holaska et al. 2006), β-catenin (Markiewicz et al. 2006), and the splicing factor YT521-B (Wilkinson et al. 2003). Through these interactions, emerin regulates fundamental processes such as cell cycle progression, apoptosis, myogenic differentiation, and mRNA splicing. Not surprisingly, due to its involvement in several gene regulatory networks, emerin also has an established role in signaling. A large number of genes regulated by the transforming growth factor β (TGF-β), Notch, JNK, MAPK, NF-κB, integrin, and IGF pathways, are misexpressed upon loss of emerin (Muchir et al. 2007; Koch and Holaska 2012; Berk et al. 2013). Interestingly, most of these signaling pathways regulate myogenic differentiation—an intriguing aspect given the involvement of emerin in EDMD (Massague et al. 1986; Bione et al. 1994; Polesskaya et al. 2003).

Second, emerin associates with a variety of structural components such as lamin A/C (Clements et al. 2000), the LINC complex (Mislow et al. 2002; Haque et al. 2010), actin (Holaska et al. 2004), myosin (Holaska and Wilson 2007), and spectrin (Holaska and Wilson 2007). Owing to these interactions, emerin is perfectly positioned to integrate mechanical impetuses at the INM. Emerin-null mouse embryonic fibroblasts display alterations in nuclear morphology, NE plasticity, and response to mechanical stimulation as well as an impaired viability under mechanical strain (Lammerding et al. 2005; Rowat et al. 2006). This suggests that emerin may help cells to adapt to mechanical load. Indeed, upon mechanical stimulation, emerin gets phosphorylated at tyrosine residues 74 and 95. This strengthens the interaction between the LINC complex and lamin A/C, and initiates actin bundle formation at the nuclear periphery, which may provide structural rigidity to the nucleus (Guilluy et al. 2014). Additionally, in response to the mechanically induced increase in actin polymerization, emerin regulates nuclear accumulation and activity of mechanosensitive transcription factors such as MLK1-SRF, and thus stimulates mechanosensitive gene expression (Olson and Nordheim 2010; Ho et al. 2013; Willer and Carroll 2017). Interestingly, the function of emerin as a mechanosensor also impinges on the positioning of chromosome territories. Growing cells on softer matrices leads to phosphorylation of emerin at tyrosine residue 99, a spatial repositioning of chromatin domains to the nuclear interior and transcriptional deregulation of associated genes, albeit to different extents (Pradhan et al. 2018).

Further investigations into how perturbation of signaling pathways and mechanosensing influence muscle regeneration and cardiac conduction may prove beneficial for the development of EDMD treatments.

MAN1

MAN1/LEMD3 is the longest of all LEM family members. It is anchored at the INM by two transmembrane segments encompassing a 130-residues-long protein segment that is located in the perinuclear space. The LEM domain of MAN1 resides at the amino terminus of its first, long nucleoplasmic domain, whereas the shorter carboxy-terminal nucleoplasmic domain comprises a DNA-binding winged-helix domain, also known as the Man1-Src1p carboxy-terminal (MSC) domain, and a carboxy-terminal RNA recognition motif (RRM) (Lin et al. 2000; Caputo et al. 2006).

MAN1 has been shown to interact with lamin A/C, BAF, R-SMADs, and PPM1A (Smad2/3 phosphatase). Much of the cellular function of MAN1 is attributed to the interaction of its RRM with R-SMADs (Osada et al. 2003; Raju et al. 2003; Hellemans et al. 2004; Lin et al. 2005; Pan et al. 2005; Cohen et al. 2007). R-SMAD proteins are key regulators of multiple signaling pathway. Two types of R-SMADs exist in mammals: TGF-β-responsive (SMAD2 and SMAD3) and bone morphogenic protein (BMP)-responsive (SMAD1, SMAD5, and SMAD8) Smads. Interaction of the carboxyl terminus of MAN1 with SMAD2/3 sequesters them at the NE and competes for their binding to transcriptional activator complexes, thereby regulating expression of TGF-β target genes. Indeed, MAN1 has been shown to antagonize TGF-β signaling in mammalian cells (Lin et al. 2005; Pan et al. 2005; Chambers et al. 2018) and mice embryos (Ishimura et al. 2006; Cohen et al. 2007). Inactivation of MAN1 in mouse embryos is associated with increased transcriptional activity of SMAD2/3 and an increased expression of TGF-β, causing perturbations in vascular remodeling and angiogenesis, which leads to embryonic lethality (Ishimura et al. 2006; Cohen et al. 2007). Similarly, MAN1 also negatively regulates the BMP signaling pathway via interactions with SMADs 1, 5, and 8. In *Drosophila*, MAN1-mediated inhibition of the BMP pathway has been shown to ensure proper synaptic growth and the integrity of neuromuscular junctions, which is required for proper locomotor activity (Wagner et al. 2010; Laugks et al. 2017).

Notably, MAN1 has also been found to activate the promoter of *BMAL1*, which is one of the small numbers of "clock genes" that are responsible for generating the internal circadian rhythm. Consistently, depletion of MAN1 resulted in a prolonged circadian period, whereas its overexpression led to a reduced period length in mammalian cells and *Drosophila* (Lin et al. 2014). This study brings to light two interesting findings: first, it presents the possibility that INM proteins can also affect gene transcription in an activating manner. Second, it reveals an unexpected function of an INM protein in determining the circadian rhythm.

Beyond transcriptional control, MAN1 also influences the dynamic organization of the NE during the cell cycle in various organisms. In the fission yeast *Schizosaccharomyces japonicus*, the NE partially ruptures during late anaphase and reseals following mitotic exit. During anaphase, NPCs redistribute toward the spindle poles where they cosegregate with chromatin prior to NE breakage. Here, Man1 has been implicated in connecting segregating chromatin to NPCs to ensure equal partitioning of the nucleus and NPCs between daughters. Cells lacking Man1 exhibit a failure in anaphase NPC distribution, evident by an irregular clustering of NPCs between segregated chromosomes in daughter nuclei (Yam et al. 2013). In *S. pombe*, Man1 collaborates with Lem2 to perform some key functions of the missing nuclear lamina, including the maintenance of NE structure and stability as well as the anchorage of telomeres at the nuclear periphery (Gonzalez et al. 2012). Finally, MAN1 functionally cooperates with other LEM-domain proteins also in *C. elegans*, where Ce-MAN1 and Ce-emerin ensure proper chromosome segregation and cell division (Liu et al. 2003).

LEM2

Similar to MAN1, LEM2 is anchored in the INM by two transmembrane segments and possesses two nucleoplasmic domains (Fig. 1), which comprise an amino-terminal LEM domain and a winged-helix MSC domain, respectively. Both domains have been proposed to mediate interactions with DNA and/or chromatin. The LEM2 homolog of fission yeast is part of a network of silencing factors at the nuclear periphery that collectively ensure perinuclear heterochromatin repression (Banday et al. 2016; Barrales et al. 2016). *S. pombe* Lem2 mediates centromere tethering via its LEM domain, whereas heterochromatin silencing and the anchorage of telomeres requires its MSC domain. For transcriptional silencing, Lem2 cooperates with other factors, such as the RNAi machinery and the telomere-associated protein Taz1 (Barrales et al. 2016). Additionally, Lem2, via its MSC domain, promotes the binding of the Snf2-like/HDAC repressor complex (SHREC) to chromatin, which in turn suppresses the recruitment of the anti-silencing factor Epe1 (Banday et al. 2016; Barrales et al. 2016). This function seems key to the role of Lem2 in silencing of telomeres, as deletion of *epe1*$^{+}$ completely suppresses the respective defect of a *lem2*Δ mutant. Lem2 has also been implicated in the maintenance of genome stability in *S. cerevisiae*, where the Lem2 homolog Heh1 along with its interacting partner Nur1 forms the "chromosome linkage INM protein" (CLIP) complex that physically links rDNA repeats to the nuclear periphery (Mekhail et al. 2008). Deletion of either *heh1* or *nur1* causes release of rDNA repeats from the NE and leads to chromosome instability by promoting aberrant recombination events in the rDNA repeats.

A suite of recent studies in both yeast and mammalian cells have placed LEM2 into a new spotlight for its role in the recruitment of the ESCRT-III complex to the NE. Here, ESCRT-III supports the constriction of tubular membrane structures culminating in membrane fission at different occasions, in line with the general role of ESCRT-III in membrane scission events (Henne et al. 2013; McCullough et al. 2018). NE-associated functions of ESCRT-III include the surveillance of NPC assembly (Webster et al. 2014), postmitotic NE closure (Olmos et al. 2015; Vietri et al. 2015), repair of NE ruptures (Denais et al. 2016; Raab et al. 2016), and the remodeling of heterochromatin-INM contacts (Pieper et al. 2020). Recruitment of the ESCRT-III complex to its site of action depends on specific adaptor proteins, and LEM2 has emerged as one player that promotes the recruitment of ESCRT-III to a specific site at the NE.

The role of LEM2/Lem2 in directing the ESCRT-III complex to the NE initially emerged from studies on NPC assembly in yeast (Webster et al. 2014). The inhibition of NPC assembly in *S. cerevisiae* leads to formation of an NE subdomain named "storage of improperly assembled NPCs" (SINC) compartment, which is enriched in NPCs that may represent improper assembly intermediates. Heh1, the yeast homolog of LEM2, recruits Chm7 (the CHMP7 homolog) to these sites and promotes SINC formation (Webster et al. 2016). Recent work has now even revealed that the Heh1-Chm7 axis does not only support the surveillance of NPC formation but also the sealing of NE ruptures. Heh1 and Chm7 are usually physically separated on opposite sides of the nuclear border. However, damage of the NE leads to the local encounter of Heh1 and Chm7 at the site of perturbation. Heh1 then activates the membrane shaping function of Chm7, leading to rapid repair of the nuclear border (Thaller et al. 2019). However, there seems to be an additional backup mechanism for the repair of NE ruptures in cultured mammalian cells where repair of laser-induced NE ruptures can occur independently of Chmp7 but requires recruitment of LEM-domain proteins including LEM2 to chromatin through BAF (Halfmann et al. 2019).

LEM2 does not only cooperate with ESCRT-III in interphase cells, but also during mitosis. In mammalian cells, postmitotic NE sealing depends on ESCRT-III, which is recruited to sites where spindle microtubules penetrate the reforming NE during anaphase (Olmos et al. 2015; Vietri et al. 2015). ESCRT-III recruitment is mediated by its component CHMP7, which interacts with the MSC domain of LEM2 (Gu et al. 2017). Initially, LEM2 tethers membranes to postmitotic chromatin disks by binding of its LEM domain to BAF. At the same time, a low-complexity domain within LEM2 undergoes liquid–liquid phase separation to coat bundles of spindle microtubules that need to be cleared. Finally, the MSC domain of LEM2 activates CHMP7. These two proteins copolymerize around microtubule bundles to form a molecular O-ring that initiates recruitment of spastin and ESCRT-III to severe the microtubules and promote nuclear compartmentalization, respectively (von Appen et al. 2020).

The contribution of LEM2 in establishing nuclear compartmentalization is also evident in *S. japonicus* that undergoes semi-open mitosis during which the NE is broken and then resealed at a single site. For NE sealing, the intersecting spindle becomes tightly enwrapped by the nuclear membrane, and both Lem2 and its interacting partner Nur1 are enriched at these sites called NE "tails." However, the accumulation of LEM2 at these "tails" demands the release of a pool of LEM2 from their interaction with pericentromeric and telomeric heterochromatin. This is mediated by the ESCRT-III-associated AAA$^+$-ATPase Vps4, which ensures that the association between LEM2 and peripheral heterochromatin is transient and subject to continuous remodeling. In the absence of Vps4, the interactions between Lem2 and heterochromatin are locked, causing a defect in bulk release of chromosomes from the NE at mitotic entry and a failure in the reestablishment of nuclear compartmentalization during mitotic exit (Yam et al. 2011; Pieper et al. 2020). Taken together, LEM2/Lem2 safeguards the integrity of the cell nucleus in response to numerous insults that are associated with a lack of proper nuclear compartmentalization.

LEM2 might also protect cells from DNA damage by interacting with several components of the nucleotide excision repair (NER) machinery. Indeed, cells depleted of LEM2 show increased sensitivity to UV-induced DNA damage and increased phosphorylated γ-H2AX protein levels, indicating an impaired or delayed DNA damage response (Moser et al. 2020). Furthermore, LEM2 plays a role in regulating MAP kinase and AKT signaling pathway during embryonic development in mice (Tapia et al. 2015) and acts together with emerin to regulate the ERK signaling during myoblast differentiation

(Huber et al. 2009). And, last but not least, *S. pombe* Lem2 was suggested to restrict changes in nuclear size imposed by alterations in membrane synthesis and nucleocytoplasmic transport, thereby contributing to the maintenance of a constant nucleus to cell volume ratio, perhaps by acting as a barrier to membrane flow into and out of the NE (Kume et al. 2019).

NET5 ALIAS SAMP1

SAMP1 is a polytopic membrane protein with four central transmembrane segments that is well conserved from yeast to human. The amino-terminal nucleoplasmic domain of SAMP1 contains two conserved zinc-finger domains that are required for NE localization (Gudise et al. 2011). From a functional perspective, the zinc-fingers might be used for chromatin interaction. In line with this, Ima1, the SAMP1 homolog in *S. pombe*, attaches centromeric heterochromatin to the nuclear periphery (King et al. 2008). Likewise, human SAMP1 has been suggested to contribute to the organization of peripheral heterochromatin in U2OS cells (Bergqvist et al. 2019) and its overexpression was shown to enhance the peripheral localization of chromosome 5 (Zuleger et al. 2013).

Originally, mammalian SAMP1 had been linked to mitotic functions (Buch et al. 2009; Larsson et al. 2018). During mitosis, it localizes to membranes in the vicinity of the spindle poles. The depletion of SAMP1 was associated with some spindle defects, prolonged mitosis, and chromosome mis-segregation. Because SAMP1 interacts with γ-tubulin and HAUS6, a subunit of the augmin complex involved in microtubule nucleation, it has been suggested that SAMP1 may promote the recruitment of the augmin complex and γ-tubulin to the mitotic spindle (Larsson et al. 2018).

Other interaction partners of SAMP1 include the LINC complex, emerin, and lamin A/C (Gudise et al. 2011). The association of SAMP1 with the LINC complex plays an important role in nuclear movement during fibroblast polarization and centrosome positioning (Gudise et al. 2011; Borrego-Pinto et al. 2012). Similar to the defects observed upon depletion of lamin A and emerin, knockdown of SAMP1 perturbs the differentiation of myoblasts (Jafferali et al. 2017; Le Thanh et al. 2017). Supporting the role of SAMP1 in early differentiation, recent studies have revealed that SAMP1 levels increase in differentiating induced pluripotent stem cells (iPSCs) concomitantly with those of lamin A/C and ectopic expression of SAMP1 induces a rapid differentiation of iPSCs even under pluripotent culturing conditions. Whereas the idea that an INM protein can drive differentiation of iPSCs is interesting, mechanistic insights into how Samp1 may regulate this process are lacking and require further investigation (Bergqvist et al. 2017).

LAP1

LAP1/TOR1AIP1 is a lamina-binding type II membrane protein that is restricted to metazoans. It exists in three isoforms (A, B, C) in rodents (Foisner and Gerace 1993), of which only two, LAP1B and LAP1C, are expressed in human cells (Santos et al. 2014). Whereas LAP1 is a bona fide component of the INM, its paralog LULL1 is distributed in the peripheral ER. Both possess a well-conserved luminal domain that adopts a RecA-like fold and is used for the activation of Torsin AAA^+ ATPase family members, first and foremost Torsin-1A and Torsin-1B (Goodchild and Dauer 2005; Jungwirth et al. 2010; Kim et al. 2010; Zhao et al. 2013; Brown et al. 2014; Sosa et al. 2014). Torsins reside in the perinuclear space and the ER lumen; however, their precise molecular function remains elusive to date. Torsin activation by LAP1 and LULL1 relies on an active site complementation mechanism in which a critical arginine finger complements the active site of Torsins (Brown et al. 2014; Sosa et al. 2014; Rose et al. 2015). Interestingly, Torsin-1A, the predominant Torsin isoform in the brain, is prominently associated with early onset torsion dystonia (DYT1), a severe movement disorder (Ozelius et al. 1997). Similarly, mutations in the *TOR1AIP1* gene have been linked to primary dystonia (Table 2; Dorboz et al. 2014; Rebelo et al. 2015), and the ablation of *Lap1* phenocopies both the nuclear blebbing and perinatal lethality observed in *Tor1A* KO mice (Kim et al. 2010).

In comparison, the nucleoplasmic domain of LAP1, which interacts with lamins A/C and B1 (Foisner and Gerace 1993; Senior and Gerace 1988) and emerin (Shin et al. 2013), has remained less well studied. Functional investigations have suggested that LAP1 collaborates with emerin in the maintenance of skeletal muscle cells, supported by the observation that combined loss of LAP1 and emerin causes a significantly more pronounced myopathy than that observed upon loss of the individual proteins (Shin et al. 2013). Some recent work uncovered that the nucleoplasmic domain of LAP1 also directly interacts with chromatin, similar to other INM proteins. Surprisingly, however, it was observed that LAP1 cannot be released from chromatin during mitosis if Torsin functionality is compromised or LAP1 is overexpressed, resulting in chromosome segregation defects and binucleation (Luithle et al. 2020), revealing an unexpected function of Torsins in the ER lumen in modulating chromatin association of LAP1 in the nucleoplasm.

Future work aimed at characterizing the molecular function of LAP1, LULL1, and the enigmatic Torsin family of proteins will hopefully allow rationalizing the cellular and organismal phenotypes associated with their dysfunction.

CONCLUDING REMARKS

In the past years, a wealth of research has pushed our perception of the NE from a mere shield for the genome to a dynamic biochemical factory that supports a multitude of essential cellular processes. Several converging themes emerge from the functional characterization of INM proteins. Principally, most of the INM proteins can be appreciated for their association with chromatin and for their role in the establishment of a repressive environment for gene expression. Yet, we still miss the full picture of the mechanisms governing the formation, organization, and functionality of the peripheral heterochromatin vis-à-vis INM proteins. That being said, there have been many fresh insights into important roles of INM proteins like lipid synthesis, mechanosensing, and functional collaborations with the ESCRT-III and Torsin machineries. Understanding the molecular principles governing these functions remains an exciting challenge for the future. Finally, a large number of INM proteins, especially those expressed in a tissue-specific manner, remain uncharacterized. This demands attention in light of nuclear envelopathies, many of which are characterized by tissue-specific defects. We emphasize that convergence of fundamental research driven toward unraveling the mechanistic basis of INM protein functions with translational research directed toward understanding disease etiology will provide a great promise toward developing future therapeutic strategies.

ACKNOWLEDGMENTS

We thank Dr. Madhav Jagannathan, Jelmi uit de Bos, and Renard Lewis for comments on the manuscript, and the Swiss National Science Foundation (SNSF, Grant 310030_184801) for financial support. We apologize for not being able to cite all original publications.

REFERENCES

Anderson DJ, Vargas JD, Hsiao JP, Hetzer MW. 2009. Recruitment of functionally distinct membrane proteins to chromatin mediates nuclear envelope formation in vivo. *J Cell Biol* **186:** 183–191. doi:10.1083/jcb.200901106

Antoniacci LM, Kenna MA, Skibbens RV. 2007. The nuclear envelope and spindle pole body-associated Mps3 protein bind telomere regulators and function in telomere clustering. *Cell Cycle* **6:** 75–79. doi:10.4161/cc.6.1.3647

Arai R, En A, Takauji Y, Maki K, Miki K, Fujii M, Ayusawa D. 2019. Lamin B receptor (LBR) is involved in the induction of cellular senescence in human cells. *Mech Ageing Dev* **178:** 25–32. doi:10.1016/j.mad.2019.01.001

Astejada MN, Goto K, Nagano A, Ura S, Noguchi S, Nonaka I, Nishino I, Hayashi YK. 2007. Emerinopathy and laminopathy clinical, pathological and molecular features of muscular dystrophy with nuclear envelopathy in Japan. *Acta Myol* **26:** 159–164.

Aymard F, Aguirrebengoa M, Guillou E, Javierre BM, Bugler B, Arnould C, Rocher V, Iacovoni JS, Biernacka A, Skrzypczak M, et al. 2017. Genome-wide mapping of long-range contacts unveils clustering of DNA double-strand breaks at damaged active genes. *Nat Struct Mol Biol* **24:** 353–361. doi:10.1038/nsmb.3387

Baasanjav S, Jamsheer A, Kolanczyk M, Horn D, Latos T, Hoffmann K, Latos-Bielenska A, Mundlos S. 2010. Osteopoikilosis and multiple exostoses caused by novel mutations in LEMD3 and EXT1 genes respectively—coincidence within one family. *BMC Med Genet* **11:** 110. doi:10.1186/1471-2350-11-110

Banday S, Farooq Z, Rashid R, Abdullah E, Altaf M. 2016. Role of inner nuclear membrane protein complex Lem2-Nur1 in heterochromatic gene silencing. *J Biol Chem* **291:** 20021–20029. doi:10.1074/jbc.M116.743211

Barrales RR, Forn M, Georgescu PR, Sarkadi Z, Braun S. 2016. Control of heterochromatin localization and silencing by the nuclear membrane protein Lem2. *Genes Dev* **30:** 133–148.

Ben Yaou R, Toutain A, Arimura T, Demay L, Massart C, Peccate C, Muchir A, Llense S, Deburgrave N, Leturcq F, et al. 2007. Multitissular involvement in a family with LMNA and EMD mutations: role of digenic mechanism? *Neurology* **68:** 1883–1894. doi:10.1212/01.wnl.0000263138.57257.6a

Bergqvist C, Jafferali MH, Gudise S, Markus R, Hallberg E. 2017. An inner nuclear membrane protein induces rapid differentiation of human induced pluripotent stem cells. *Stem Cell Res* **23:** 33–38. doi:10.1016/j.scr.2017.06.008

Bergqvist C, Niss F, Figueroa RA, Beckman M, Maksel D, Jafferali MH, Kulyté A, Ström AL, Hallberg E. 2019. Monitoring of chromatin organization in live cells by FRIC. Effects of the inner nuclear membrane protein Samp1. *Nucleic Acids Res* **47:** e49. doi:10.1093/nar/gkz123

Berk JM, Tifft KE, Wilson KL. 2013. The nuclear envelope LEM-domain protein emerin. *Nucleus* **4:** 298–314. doi:10.4161/nucl.25751

Best S, Salvati F, Kallo J, Garner C, Height S, Thein SL, Rees DC. 2003. Lamin B-receptor mutations in Pelger-Huët anomaly. *Br J Haematol* **123:** 542–544. doi:10.1046/j.1365-2141.2003.04621.x

Bione S, Maestrini E, Rivella S, Mancini M, Regis S, Romeo G, Toniolo D. 1994. Identification of a novel X-linked gene responsible for Emery–Dreifuss muscular dystrophy. *Nat Genet* **8:** 323–327. doi:10.1038/ng1294-323

Blenski M, Kehlenbach RH. 2019. Targeting of LRRC59 to the endoplasmic reticulum and the inner nuclear membrane. *Int J Mol Sci* **20:** 334. doi:10.3390/ijms20020334

Boni A, Politi AZ, Strnad P, Xiang W, Hossain MJ, Ellenberg J. 2015. Live imaging and modeling of inner nuclear membrane targeting reveals its molecular requirements in mammalian cells. *J Cell Biol* **209:** 705–720. doi:10.1083/jcb.201409133

Boone PM, Yuan B, Gu S, Ma Z, Gambin T, Gonzaga-Jauregui C, Jain M, Murdock TJ, White JJ, Jhangiani SN, et al. 2016. Hutterite-type cataract maps to chromosome 6p21.32-p21.31, cosegregates with a homozygous mutation in *LEMD2*, and is associated with sudden cardiac death. *Mol Genet Genomic Med* **4:** 77–94. doi:10.1002/mgg3.181

Borovik L, Modaff P, Waterham HR, Krentz AD, Pauli RM. 2013. Pelger–Huet anomaly and a mild skeletal phenotype secondary to mutations in *LBR*. *Am J Med Genet A* **161A:** 2066–2073. doi:10.1002/ajmg.a.36019

Borrego-Pinto J, Jegou T, Osorio DS, Aurade F, Gorjanacz M, Koch B, Mattaj IW, Gomes ER. 2012. Samp1 is a component of TAN lines and is required for nuclear movement. *J Cell Sci* **125:** 1099–1105. doi:10.1242/jcs.087049

Brachner A, Foisner R. 2011. Evolvement of LEM proteins as chromatin tethers at the nuclear periphery. *Biochem Soc Trans* **39:** 1735–1741. doi:10.1042/BST20110724

Brachner A, Foisner R. 2014. Lamina-associated polypeptide (LAP)2α and other LEM proteins in cancer biology. *Adv Exp Med Biol* **773:** 143–163. doi:10.1007/978-1-4899-8032-8_7

Brachner A, Reipert S, Foisner R, Gotzmann J. 2005. LEM2 is a novel MAN1-related inner nuclear membrane protein associated with A-type lamins. *J Cell Sci* **118:** 5797–5810. doi:10.1242/jcs.02701

Braunger K, Pfeffer S, Shrimal S, Gilmore R, Berninghausen O, Mandon EC, Becker T, Förster F, Beckmann R. 2018. Structural basis for coupling protein transport and *N*-glycosylation at the mammalian endoplasmic reticulum. *Science* **360:** 215–219. doi:10.1126/science.aar7899

Brodbeck M, Yousif Q, Diener PA, Zweier M, Gruenert J. 2016. The Buschke–Ollendorff syndrome: a case report of simultaneous osteo-cutaneous malformations in the hand. *BMC Res Notes* **9:** 294. doi:10.1186/s13104-016-2095-2

Brown CA, Scharner J, Felice K, Meriggioli MN, Tarnopolsky M, Bower M, Zammit PS, Mendell JR, Ellis JA. 2011. Novel and recurrent EMD mutations in patients with Emery–Dreifuss muscular dystrophy, identify exon 2 as a mutation hot spot. *J Hum Genet* **56:** 589–594. doi:10.1038/jhg.2011.65

Brown RS, Zhao C, Chase AR, Wang J, Schlieker C. 2014. The mechanism of Torsin ATPase activation. *Proc Natl Acad Sci* **111:** E4822–E4831. doi:10.1073/pnas.1415271111

Buch C, Lindberg R, Figueroa R, Gudise S, Onischenko E, Hallberg E. 2009. An integral protein of the inner nuclear membrane localizes to the mitotic spindle in mammalian cells. *J Cell Sci* **122:** 2100–2107. doi:10.1242/jcs.047373

Bupp JM, Martin AE, Stensrud ES, Jaspersen SL. 2007. Telomere anchoring at the nuclear periphery requires the budding yeast Sad1-UNC-84 domain protein Mps3. *J Cell Biol* **179:** 845–854. doi:10.1083/jcb.200706040

Burger B, Hershkovitz D, Indelman M, Kovac M, Galambos J, Haeusermann P, Sprecher E, Itin PH. 2010. Buschke–Ollendorff syndrome in a three-generation family: influence of a novel LEMD3 mutation to tropoelastin expression. *Eur J Dermatol* **20:** 693–697.

Burke B. 2018. LINC complexes as regulators of meiosis. *Curr Opin Cell Biol* **52:** 22–29. doi:10.1016/j.ceb.2018.01.005

Burke B, Stewart CL. 2014. Functional architecture of the cell's nucleus in development, aging, and disease. *Curr Top Dev Biol* **109:** 1–52. doi:10.1016/B978-0-12-397920-9.00006-8

Cain NE, Tapley EC, McDonald KL, Cain BM, Starr DA. 2014. The SUN protein UNC-84 is required only in force-bearing cells to maintain nuclear envelope architecture. *J Cell Biol* **206:** 163–172. doi:10.1083/jcb.201405081

Calvi A, Wong AS, Wright G, Wong ES, Loo TH, Stewart CL, Burke B. 2015. SUN4 is essential for nuclear remodeling during mammalian spermiogenesis. *Dev Biol* **407:** 321–330. doi:10.1016/j.ydbio.2015.09.010

Camozzi D, D'Apice MR, Schena E, Cenni V, Columbaro M, Capanni C, Maraldi NM, Squarzoni S, Ortolani M, Novelli G, et al. 2012. Altered chromatin organization and

SUN2 localization in mandibuloacral dysplasia are rescued by drug treatment. *Histochem Cell Biol* **138:** 643–651. doi:10.1007/s00418-012-0977-5

Capo-chichi CD, Cai KQ, Testa JR, Godwin AK, Xu XX. 2009. Loss of GATA6 leads to nuclear deformation and aneuploidy in ovarian cancer. *Mol Cell Biol* **29:** 4766–4777. doi:10.1128/MCB.00087-09

Caputo S, Couprie J, Duband-Goulet I, Kondé E, Lin F, Braud S, Gondry M, Gilquin B, Worman HJ, Zinn-Justin S. 2006. The carboxyl-terminal nucleoplasmic region of MAN1 exhibits a DNA binding winged helix domain. *J Biol Chem* **281:** 18208–18215. doi:10.1074/jbc.M601980200

Chambers DM, Moretti L, Zhang JJ, Cooper SW, Chambers DM, Santangelo PJ, Barker TH. 2018. LEM domain-containing protein 3 antagonizes TGFβ-SMAD2/3 signaling in a stiffness-dependent manner in both the nucleus and cytosol. *J Biol Chem* **293:** 15867–15886. doi:10.1074/jbc.RA118.003658

Chang JW, Nomura DK, Cravatt BF. 2011. A potent and selective inhibitor of KIAA1363/AADACL1 that impairs prostate cancer pathogenesis. *Chem Biol* **18:** 476–484. doi:10.1016/j.chembiol.2011.02.008

Chang W, Worman HJ, Gundersen GG. 2015. Accessorizing and anchoring the LINC complex for multifunctionality. *J Cell Biol* **208:** 11–22. doi:10.1083/jcb.201409047

Chen CY, Chi YH, Mutalif RA, Starost MF, Myers TG, Anderson SA, Stewart CL, Jeang KT. 2012. Accumulation of the inner nuclear envelope protein Sun1 is pathogenic in progeric and dystrophic laminopathies. *Cell* **149:** 565–577. doi:10.1016/j.cell.2012.01.059

Chen CK, Blanco M, Jackson C, Aznauryan E, Ollikainen N, Surka C, Chow A, Cerase A, McDonel P, Guttman M. 2016. Xist recruits the X chromosome to the nuclear lamina to enable chromosome-wide silencing. *Science* **354:** 468–472. doi:10.1126/science.aae0047

Cherepanova NA, Shrimal S, Gilmore R. 2014. Oxidoreductase activity is necessary for N-glycosylation of cysteine-proximal acceptor sites in glycoproteins. *J Cell Biol* **206:** 525–539. doi:10.1083/jcb.201404083

Chiang KP, Niessen S, Saghatelian A, Cravatt BF. 2006. An enzyme that regulates ether lipid signaling pathways in cancer annotated by multidimensional profiling. *Chem Biol* **13:** 1041–1050. doi:10.1016/j.chembiol.2006.08.008

Chikashige Y, Tsutsumi C, Yamane M, Okamasa K, Haraguchi T, Hiraoka Y. 2006. Meiotic proteins bqt1 and bqt2 tether telomeres to form the bouquet arrangement of chromosomes. *Cell* **125:** 59–69. doi:10.1016/j.cell.2006.01.048

Chikashige Y, Yamane M, Okamasa K, Tsutsumi C, Kojidani T, Sato M, Haraguchi T, Hiraoka Y. 2009. Membrane proteins Bqt3 and -4 anchor telomeres to the nuclear envelope to ensure chromosomal bouquet formation. *J Cell Biol* **187:** 413–427. doi:10.1083/jcb.200902122

Cho S, Irianto J, Discher DE. 2017. Mechanosensing by the nucleus: from pathways to scaling relationships. *J Cell Biol* **216:** 305–315. doi:10.1083/jcb.201610042

Clayton P, Fischer B, Mann A, Mansour S, Rossier E, Veen M, Lang C, Baasanjav S, Kieslich M, Brossuleit K, et al. 2010. Mutations causing Greenberg dysplasia but not Pelger anomaly uncouple enzymatic from structural functions of a nuclear membrane protein. *Nucleus* **1:** 354–366. doi:10.4161/nucl.1.4.12435

Clements L, Manilal S, Love DR, Morris GE. 2000. Direct interaction between emerin and lamin A. *Biochem Biophys Res Commun* **267:** 709–714. doi:10.1006/bbrc.1999.2023

Cohen TV, Kosti O, Stewart CL. 2007. The nuclear envelope protein MAN1 regulates TGFβ signaling and vasculogenesis in the embryonic yolk sac. *Development* **134:** 1385–1395. doi:10.1242/dev.02816

Conrad MN, Lee CY, Chao G, Shinohara M, Kosaka H, Shinohara A, Conchello JA, Dresser ME. 2008. Rapid telomere movement in meiotic prophase is promoted by NDJ1, MPS3, and CSM4 and is modulated by recombination. *Cell* **133:** 1175–1187. doi:10.1016/j.cell.2008.04.047

Couto AR, Bruges-Armas J, Peach CA, Chapman K, Brown MA, Wordsworth BP, Zhang Y. 2007. A novel LEMD3 mutation common to patients with osteopoikilosis with and without melorheostosis. *Calcif Tissue Int* **81:** 81–84. doi:10.1007/s00223-007-9043-z

Crisp M, Liu Q, Roux K, Rattner JB, Shanahan C, Burke B, Stahl PD, Hodzic D. 2006. Coupling of the nucleus and cytoplasm: role of the LINC complex. *J Cell Biol* **172:** 41–53. doi:10.1083/jcb.200509124

Cutler JA, Wong X, Hoskins VE, Gordon M, Madugundu AK, Pandey A, Reddy KL. 2019. Mapping the micro-proteome of the nuclear lamina and lamin associated domains. bioRxiv doi:10.1101/828210

Dai X, Zheng C, Chen X, Tang Y, Zhang H, Yan C, Ma H, Li X. 2019. Targeted next-generation sequencing identified a known EMD mutation in a Chinese patient with Emery–Dreifuss muscular dystrophy. *Hum Genome Var* **6:** 42. doi:10.1038/s41439-019-0072-8

Dechat T, Gotzmann J, Stockinger A, Harris CA, Talle MA, Siekierka JJ, Foisner R. 1998. Detergent-salt resistance of LAP2α in interphase nuclei and phosphorylation-dependent association with chromosomes early in nuclear assembly implies functions in nuclear structure dynamics. *EMBO J* **17:** 4887–4902. doi:10.1093/emboj/17.16.4887

Dechat T, Vlcek S, Foisner R. 2000. Review: lamina-associated polypeptide 2 isoforms and related proteins in cell cycle-dependent nuclear structure dynamics. *J Struct Biol* **129:** 335–345. doi:10.1006/jsbi.2000.4212

Demmerle J, Koch AJ, Holaska JM. 2012. The nuclear envelope protein emerin binds directly to histone deacetylase 3 (HDAC3) and activates HDAC3 activity. *J Biol Chem* **287:** 22080–22088. doi:10.1074/jbc.M111.325308

Denais CM, Gilbert RM, Isermann P, McGregor AL, te Lindert M, Weigelin B, Davidson PM, Friedl P, Wolf K, Lammerding J. 2016. Nuclear envelope rupture and repair during cancer cell migration. *Science* **352:** 353–358. doi:10.1126/science.aad7297

Dettmer U, Kuhn PH, Abou-Ajram C, Lichtenthaler SF, Krüger M, Kremmer E, Haass C, Haffner C. 2010. Transmembrane protein 147 (TMEM147) is a novel component of the Nicalin-NOMO protein complex. *J Biol Chem* **285:** 26174–26181. doi:10.1074/jbc.M110.132548

Ding X, Xu R, Yu J, Xu T, Zhuang Y, Han M. 2007. SUN1 is required for telomere attachment to nuclear envelope and gametogenesis in mice. *Dev Cell* **12:** 863–872. doi:10.1016/j.devcel.2007.03.018

Donahue DA, Amraoui S, di Nunzio F, Kieffer C, Porrot F, Opp S, Diaz-Griffero F, Casartelli N, Schwartz O. 2016. SUN2 overexpression deforms nuclear shape and inhibits HIV. *J Virol* **90:** 4199–4214. doi:10.1128/JVI.03202-15

Dorboz I, Coutelier M, Bertrand AT, Caberg JH, Elmaleh-Bergès M, Lainé J, Stevanin G, Bonne G, Boespflug-Tanguy O, Servais L. 2014. Severe dystonia, cerebellar atrophy, and cardiomyopathy likely caused by a missense mutation in TOR1AIP1. *Orphanet J Rare Dis* **9:** 174. doi:10.1186/s13023-014-0174-9

Dorner D, Vlcek S, Foeger N, Gajewski A, Makolm C, Gotzmann J, Hutchison CJ, Foisner R. 2006. Lamina-associated polypeptide 2α regulates cell cycle progression and differentiation via the retinoblastoma-E2F pathway. *J Cell Biol* **173:** 83–93. doi:10.1083/jcb.200511149

Dou Z, Xu C, Donahue G, Shimi T, Pan JA, Zhu J, Ivanov A, Capell BC, Drake AM, Shah PP, et al. 2015. Autophagy mediates degradation of nuclear lamina. *Nature* **527:** 105–109. doi:10.1038/nature15548

Dreesen O, Chojnowski A, Ong PF, Zhao TY, Common JE, Lunny D, Lane EB, Lee SJ, Vardy LA, Stewart CL, et al. 2013. Lamin B1 fluctuations have differential effects on cellular proliferation and senescence. *J Cell Biol* **200:** 605–617. doi:10.1083/jcb.201206121

Dreger M, Bengtsson L, Schoneberg T, Otto H, Hucho F. 2001. Nuclear envelope proteomics: novel integral membrane proteins of the inner nuclear membrane. *Proc Natl Acad Sci* **98:** 11943–11948. doi:10.1073/pnas.211201898

Duband-Goulet I, Courvalin JC. 2000. Inner nuclear membrane protein LBR preferentially interacts with DNA secondary structures and nucleosomal linker. *Biochemistry* **39:** 6483–6488. doi:10.1021/bi992908b

Ellis JA, Yates JR, Kendrick-Jones J, Brown CA. 1999. Changes at P183 of emerin weaken its protein–protein interactions resulting in X-linked Emery–Dreifuss muscular dystrophy. *Hum Genet* **104:** 262–268. doi:10.1007/s004390050946

En A, Takauji Y, Ayusawa D, Fujii M. 2020. The role of lamin B receptor in the regulation of senescence-associated secretory phenotype (SASP). *Exp Cell Res* **390:** 111927. doi:10.1016/j.yexcr.2020.111927

Fichtman B, Zagairy F, Biran N, Barsheshet Y, Chervinsky E, Ben Neriah Z, Shaag A, Assa M, Elpeleg O, Harel A, et al. 2019. Combined loss of LAP1B and LAP1C results in an early onset multisystemic nuclear envelopathy. *Nat Commun* **10:** 605. doi:10.1038/s41467-019-08493-7

Foisner R, Gerace L. 1993. Integral membrane proteins of the nuclear envelope interact with lamins and chromosomes, and binding is modulated by mitotic phosphorylation. *Cell* **73:** 1267–1279. doi:10.1016/0092-8674(93)90355-T

Freund A, Laberge RM, Demaria M, Campisi J. 2012. Lamin B1 loss is a senescence-associated biomarker. *Mol Biol Cell* **23:** 2066–2075. doi:10.1091/mbc.e11-10-0884

Friederichs JM, Ghosh S, Smoyer CJ, McCroskey S, Miller BD, Weaver KJ, Delventhal KM, Unruh J, Slaughter BD, Jaspersen SL. 2011. The SUN protein Mps3 is required for spindle pole body insertion into the nuclear membrane and nuclear envelope homeostasis. *PLoS Genet* **7:** e1002365. doi:10.1371/journal.pgen.1002365

Frohnert C, Schweizer S, Hoyer-Fender S. 2011. SPAG4L/SPAG4L-2 are testis-specific SUN domain proteins restricted to the apical nuclear envelope of round spermatids facing the acrosome. *Mol Hum Reprod* **17:** 207–218. doi:10.1093/molehr/gaq099

Gant TM, Harris CA, Wilson KL. 1999. Roles of LAP2 proteins in nuclear assembly and DNA replication: truncated LAP2β proteins alter lamina assembly, envelope formation, nuclear size, and DNA replication efficiency in *Xenopus laevis* extracts. *J Cell Biol* **144:** 1083–1096. doi:10.1083/jcb.144.6.1083

Gartenberg MR. 2009. Life on the edge: telomeres and persistent DNA breaks converge at the nuclear periphery. *Genes Dev* **23:** 1027–1031. doi:10.1101/gad.1805309

Gaudy-Marqueste C, Roll P, Esteves-Vieira V, Weiller PJ, Grob JJ, Cau P, Levy N, De Sandre-Giovannoli A. 2010. LBR mutation and nuclear envelope defects in a patient affected with Reynolds syndrome. *J Med Genet* **47:** 361–370. doi:10.1136/jmg.2009.071696

Gesson K, Rescheneder P, Skoruppa MP, von Haeseler A, Dechat T, Foisner R. 2016. A-type lamins bind both hetero- and euchromatin, the latter being regulated by lamina-associated polypeptide 2α. *Genome Res* **26:** 462–473. doi:10.1101/gr.196220.115

Ghaoui R, Benavides T, Lek M, Waddell LB, Kaur S, North KN, MacArthur DG, Clarke NF, Cooper ST. 2016. TOR1AIP1 as a cause of cardiac failure and recessive limb-girdle muscular dystrophy. *Neuromuscul Disord* **26:** 500–503. doi:10.1016/j.nmd.2016.05.013

Gilbert HTJ, Mallikarjun V, Dobre O, Jackson MR, Pedley R, Gilmore AP, Richardson SM, Swift J. 2019. Nuclear decoupling is part of a rapid protein-level cellular response to high-intensity mechanical loading. *Nat Commun* **10:** 4149. doi:10.1038/s41467-019-11923-1

Gonzalez Y, Saito A, Sazer S. 2012. Fission yeast Lem2 and Man1 perform fundamental functions of the animal cell nuclear lamina. *Nucleus* **3:** 60–76. doi:10.4161/nucl.18824

Goodchild RE, Dauer WT. 2005. The AAA$^+$ protein torsinA interacts with a conserved domain present in LAP1 and a novel ER protein. *J Cell Biol* **168:** 855–862. doi:10.1083/jcb.200411026

Gu M, LaJoie D, Chen OS, von Appen A, Ladinsky MS, Redd MJ, Nikolova L, Bjorkman PJ, Sundquist WI, Ullman KS, et al. 2017. LEM2 recruits CHMP7 for ESCRT-mediated nuclear envelope closure in fission yeast and human cells. *Proc Natl Acad Sci* **114:** E2166–E2175. doi:10.1073/pnas.1613916114

Guarda A, Bolognese F, Bonapace IM, Badaracco G. 2009. Interaction between the inner nuclear membrane lamin B receptor and the heterochromatic methyl binding protein, MeCP2. *Exp Cell Res* **315:** 1895–1903. doi:10.1016/j.yexcr.2009.01.019

Gudise S, Figueroa RA, Lindberg R, Larsson V, Hallberg E. 2011. Samp1 is functionally associated with the LINC complex and A-type lamina networks. *J Cell Sci* **124:** 2077–2085. doi:10.1242/jcs.078923

Guilluy C, Osborne LD, Van Landeghem L, Sharek L, Superfine R, Garcia-Mata R, Burridge K. 2014. Isolated nuclei adapt to force and reveal a mechanotransduction pathway in the nucleus. *Nat Cell Biol* **16:** 376–381. doi:10.1038/ncb2927

Gundersen GG, Worman HJ. 2013. Nuclear positioning. *Cell* **152:** 1376–1389. doi:10.1016/j.cell.2013.02.031

Halfmann CT, Sears RM, Katiyar A, Busselman BW, Aman LK, Zhang Q, O'Bryan CS, Angelini TE, Lele TP, Roux KJ. 2019. Repair of nuclear ruptures requires barrier-to-autointegration factor. *J Cell Biol* **218:** 2136–2149. doi:10.1083/jcb.201901116

Hansen JB, Jorgensen C, Petersen RK, Hallenborg P, De Matteis R, Boye HA, Petrovic N, Enerback S, Nedergaard J, Cinti S, et al. 2004. Retinoblastoma protein functions as a molecular switch determining white versus brown adipocyte differentiation. *Proc Natl Acad Sci* **101:** 4112–4117. doi:10.1073/pnas.0301964101

Haque F, Mazzeo D, Patel JT, Smallwood DT, Ellis JA, Shanahan CM, Shackleton S. 2010. Mammalian SUN protein interaction networks at the inner nuclear membrane and their role in laminopathy disease processes. *J Biol Chem* **285:** 3487–3498. doi:10.1074/jbc.M109.071910

Haraguchi T, Holaska JM, Yamane M, Koujin T, Hashiguchi N, Mori C, Wilson KL, Hiraoka Y. 2004. Emerin binding to Btf, a death-promoting transcriptional repressor, is disrupted by a missense mutation that causes Emery–Dreifuss muscular dystrophy. *Eur J Biochem* **271:** 1035–1045. doi:10.1111/j.1432-1033.2004.04007.x

Hellemans J, Preobrazhenska O, Willaert A, Debeer P, Verdonk PC, Costa T, Janssens K, Menten B, Van Roy N, Vermeulen SJ, et al. 2004. Loss-of-function mutations in LEMD3 result in osteopoikilosis, Buschke–Ollendorff syndrome and melorheostosis. *Nat Genet* **36:** 1213–1218. doi:10.1038/ng1453

Hellemans J, Debeer P, Wright M, Janecke A, Kjaer KW, Verdonk PC, Savarirayan R, Basel L, Moss C, Roth J, et al. 2006. Germline LEMD3 mutations are rare in sporadic patients with isolated melorheostosis. *Hum Mutat* **27:** 290. doi:10.1002/humu.9403

Henne WM, Stenmark H, Emr SD. 2013. Molecular mechanisms of the membrane sculpting ESCRT pathway. *Cold Spring Harb Perspect Biol* **5:** a016766. doi:10.1101/cshperspect.a016766

Hirano Y, Hizume K, Kimura H, Takeyasu K, Haraguchi T, Hiraoka Y. 2012. Lamin B receptor recognizes specific modifications of histone H4 in heterochromatin formation. *J Biol Chem* **287:** 42654–42663. doi:10.1074/jbc.M112.397950

Hiraoka Y, Dernburg AF. 2009. The SUN rises on meiotic chromosome dynamics. *Dev Cell* **17:** 598–605. doi:10.1016/j.devcel.2009.10.014

Ho CY, Jaalouk DE, Vartiainen MK, Lammerding J. 2013. Lamin A/C and emerin regulate MKL1-SRF activity by modulating actin dynamics. *Nature* **497:** 507–511. doi:10.1038/nature12105

Hodzic DM, Yeater DB, Bengtsson L, Otto H, Stahl PD. 2004. Sun2 is a novel mammalian inner nuclear membrane protein. *J Biol Chem* **279:** 25805–25812. doi:10.1074/jbc.M313157200

Hoffmann K, Dreger CK, Olins AL, Olins DE, Shultz LD, Lucke B, Karl H, Kaps R, Müller D, Vayá A, et al. 2002. Mutations in the gene encoding the lamin B receptor produce an altered nuclear morphology in granulocytes (Pelger–Huët anomaly). *Nat Genet* **31:** 410–414. doi:10.1038/ng925

Holaska JM, Wilson KL. 2007. An emerin "proteome": purification of distinct emerin-containing complexes from HeLa cells suggests molecular basis for diverse roles including gene regulation, mRNA splicing, signaling, mechanosensing, and nuclear architecture. *Biochemistry* **46:** 8897–8908. doi:10.1021/bi602636m

Holaska JM, Lee KK, Kowalski AK, Wilson KL. 2003. Transcriptional repressor germ cell-less (GCL) and barrier to autointegration factor (BAF) compete for binding to emerin in vitro. *J Biol Chem* **278:** 6969–6975. doi:10.1074/jbc.M208811200

Holaska JM, Kowalski AK, Wilson KL. 2004. Emerin caps the pointed end of actin filaments: evidence for an actin cortical network at the nuclear inner membrane. *PLoS Biol* **2:** E231. doi:10.1371/journal.pbio.0020231

Holaska JM, Rais-Bahrami S, Wilson KL. 2006. Lmo7 is an emerin-binding protein that regulates the transcription of emerin and many other muscle-relevant genes. *Hum Mol Genet* **15:** 3459–3472. doi:10.1093/hmg/ddl423

Holmer L, Pezhman A, Worman HJ. 1998. The human lamin B receptor/sterol reductase multigene family. *Genomics* **54:** 469–476. doi:10.1006/geno.1998.5615

Horigome C, Okada T, Shimazu K, Gasser SM, Mizuta K. 2011. Ribosome biogenesis factors bind a nuclear envelope SUN domain protein to cluster yeast telomeres. *EMBO J* **30:** 3799–3811. doi:10.1038/emboj.2011.267

Horigome C, Oma Y, Konishi T, Schmid R, Marcomini I, Hauer MH, Dion V, Harata M, Gasser SM. 2014. SWR1 and INO80 chromatin remodelers contribute to DNA double-strand break perinuclear anchorage site choice. *Mol Cell* **55:** 626–639. doi:10.1016/j.molcel.2014.06.027

Horn HF, Kim DI, Wright GD, Wong ES, Stewart CL, Burke B, Roux KJ. 2013. A mammalian KASH domain protein coupling meiotic chromosomes to the cytoskeleton. *J Cell Biol* **202:** 1023–1039. doi:10.1083/jcb.201304004

Huang J, Zheng DL, Qin FS, Cheng N, Chen H, Wan BB, Wang YP, Xiao HS, Han ZG. 2010. Genetic and epigenetic silencing of SCARA5 may contribute to human hepatocellular carcinoma by activating FAK signaling. *J Clin Invest* **120:** 223–241. doi:10.1172/JCI38012

Huber MD, Guan T, Gerace L. 2009. Overlapping functions of nuclear envelope proteins NET25 (Lem2) and emerin in regulation of extracellular signal-regulated kinase signaling in myoblast differentiation. *Mol Cell Biol* **29:** 5718–5728. doi:10.1128/MCB.00270-09

Huh MS, Parker MH, Scimè A, Parks R, Rudnicki MA. 2004. Rb is required for progression through myogenic differentiation but not maintenance of terminal differentiation. *J Cell Biol* **166:** 865–876. doi:10.1083/jcb.200403004

Igarashi M, Osuga J, Isshiki M, Sekiya M, Okazaki H, Takase S, Takanashi M, Ohta K, Kumagai M, Nishi M, et al. 2010. Targeting of neutral cholesterol ester hydrolase to the endoplasmic reticulum via its N-terminal sequence. *J Lipid Res* **51:** 274–285. doi:10.1194/jlr.M900201-JLR200

Ishimura A, Ng JK, Taira M, Young SG, Osada S. 2006. Man1, an inner nuclear membrane protein, regulates vascular remodeling by modulating transforming growth factor β signaling. *Development* **133:** 3919–3928. doi:10.1242/dev.02538

Ivanov A, Pawlikowski J, Manoharan I, van Tuyn J, Nelson DM, Rai TS, Shah PP, Hewitt G, Korolchuk VI, Passos JF, et al. 2013. Lysosome-mediated processing of chromatin in senescence. *J Cell Biol* **202:** 129–143. doi:10.1083/jcb.201212110

Jafferali MH, Figueroa RA, Hasan M, Hallberg E. 2017. Spindle associated membrane protein 1 (Samp1) is required for the differentiation of muscle cells. *Sci Rep* **7:** 16655. doi:10.1038/s41598-017-16746-y

Jungwirth M, Dear ML, Brown P, Holbrook K, Goodchild R. 2010. Relative tissue expression of homologous torsinB correlates with the neuronal specific importance of DYT1 dystonia-associated torsinA. *Hum Mol Genet* **19:** 888–900. doi:10.1093/hmg/ddp557

Kayman-Kurekci G, Talim B, Korkusuz P, Sayar N, Sarioglu T, Oncel I, Sharafi P, Gundesli H, Balci-Hayta B, Purali N, et al. 2014. Mutation in TOR1AIP1 encoding LAP1B in a form of muscular dystrophy: a novel gene related to nuclear envelopathies. *Neuromuscul Disord* **24:** 624–633. doi:10.1016/j.nmd.2014.04.007

Kim CE, Perez A, Perkins G, Ellisman MH, Dauer WT. 2010. A molecular mechanism underlying the neural-specific defect in torsinA mutant mice. *Proc Natl Acad Sci* **107:** 9861–9866. doi:10.1073/pnas.0912877107

Kim HJ, Hwang SH, Han ME, Baek S, Sim HE, Yoon S, Baek SY, Kim BS, Kim JH, Kim SY, et al. 2012. LAP2 is widely overexpressed in diverse digestive tract cancers and regulates motility of cancer cells. *PLoS ONE* **7:** e39482. doi:10.1371/journal.pone.0039482

King MC, Drivas TG, Blobel G. 2008. A network of nuclear envelope membrane proteins linking centromeres to microtubules. *Cell* **134:** 427–438. doi:10.1016/j.cell.2008.06.022

Kobayashi H, Kasahara M, Hino M, Takahara S, Ikeda K, Son C, Iwakura T, Matsuoka N, Yoshimoto A, Ohgo N, et al. 2007. A novel heterozygous splice-site mutation of LEM domain-containing 3 in a Japanese kindred with Buschke–Ollendorff syndrome. *J Endocrinol Invest* **30:** 263–265. doi:10.1007/BF03347437

Koch AJ, Holaska JM. 2012. Loss of emerin alters myogenic signaling and miRNA expression in mouse myogenic progenitors. *PLoS ONE* **7:** e37262. doi:10.1371/journal.pone.0037262

Koch AJ, Holaska JM. 2014. Emerin in health and disease. *Semin Cell Dev Biol* **29:** 95–106. doi:10.1016/j.semcdb.2013.12.008

Korekawa A, Nakano H, Toyomaki Y, Takiyoshi N, Rokunohe D, Akasaka E, Nakajima K, Sawamura D. 2012. Buschke–Ollendorff syndrome associated with hypertrophic scar formation: a possible role for LEMD3 mutation. *Br J Dermatol* **166:** 900–903. doi:10.1111/j.1365-2133.2011.10691.x

Korfali N, Wilkie GS, Swanson SK, Srsen V, Batrakou DG, Fairley EA, Malik P, Zuleger N, Goncharevich A, de Las Heras J, et al. 2010. The leukocyte nuclear envelope proteome varies with cell activation and contains novel transmembrane proteins that affect genome architecture. *Mol Cell Proteomics* **9:** 2571–2585. doi:10.1074/mcp.M110.002915

Korfali N, Wilkie GS, Swanson SK, Srsen V, de Las Heras J, Batrakou DG, Malik P, Zuleger N, Kerr AR, Florens L, et al. 2012. The nuclear envelope proteome differs notably between tissues. *Nucleus* **3:** 552–564. doi:10.4161/nucl.22257

Koszul R, Kim KP, Prentiss M, Kleckner N, Kameoka S. 2008. Meiotic chromosomes move by linkage to dynamic actin cables with transduction of force through the nuclear envelope. *Cell* **133:** 1188–1201. doi:10.1016/j.cell.2008.04.050

Kume K, Cantwell H, Burrell A, Nurse P. 2019. Nuclear membrane protein Lem2 regulates nuclear size through membrane flow. *Nat Commun* **10:** 1871. doi:10.1038/s41467-019-09623-x

Lahaye X, Satoh T, Gentili M, Cerboni S, Silvin A, Conrad C, Ahmed-Belkacem A, Rodriguez EC, Guichou JF, Bosquet N, et al. 2016. Nuclear envelope protein SUN2 promotes cyclophilin-A-dependent steps of HIV replication. *Cell Rep* **15:** 879–892. doi:10.1016/j.celrep.2016.03.074

Lammerding J, Hsiao J, Schulze PC, Kozlov S, Stewart CL, Lee RT. 2005. Abnormal nuclear shape and impaired mechanotransduction in emerin-deficient cells. *J Cell Biol* **170:** 781–791. doi:10.1083/jcb.200502148

Larsson VJ, Jafferali MH, Vijayaraghavan B, Figueroa RA, Hallberg E. 2018. Mitotic spindle assembly and γ-tubulin localisation depend on the integral nuclear membrane protein Samp1. *J Cell Sci* **131:** jcs211664. doi:10.1242/jcs.211664

Laugks U, Hieke M, Wagner N. 2017. MAN1 restricts BMP signaling during synaptic growth in *Drosophila*. *Cell Mol Neurobiol* **37:** 1077–1093. doi:10.1007/s10571-016-0442-4

Lawrence KS, Tapley EC, Cruz VE, Li Q, Aung K, Hart KC, Schwartz TU, Starr DA, Engebrecht J. 2016. LINC complexes promote homologous recombination in part through inhibition of nonhomologous end joining. *J Cell Biol* **215:** 801–821. doi:10.1083/jcb.201604112

Lechner MS, Schultz DC, Negorev D, Maul GG, Rauscher FJ III. 2005. The mammalian heterochromatin protein 1 binds diverse nuclear proteins through a common motif that targets the chromoshadow domain. *Biochem Biophys Res Commun* **331:** 929–937. doi:10.1016/j.bbrc.2005.04.016

Lee YL, Burke B. 2018. LINC complexes and nuclear positioning. *Semin Cell Dev Biol* **82:** 67–76. doi:10.1016/j.semcdb.2017.11.008

Lee CY, Horn HF, Stewart CL, Burke B, Bolcun-Filas E, Schimenti JC, Dresser ME, Pezza RJ. 2015. Mechanism and regulation of rapid telomere prophase movements in mouse meiotic chromosomes. *Cell Rep* **11:** 551–563. doi:10.1016/j.celrep.2015.03.045

Lee B, Lee TH, Shim J. 2017. Emerin suppresses Notch signaling by restricting the Notch intracellular domain to the nuclear membrane. *Biochim Biophys Acta Mol Cell Res* **1864:** 303–313. doi:10.1016/j.bbamcr.2016.11.013

Lei K, Zhu X, Xu R, Shao C, Xu T, Zhuang Y, Han M. 2012. Inner nuclear envelope proteins SUN1 and SUN2 play a prominent role in the DNA damage response. *Curr Biol* **22:** 1609–1615. doi:10.1016/j.cub.2012.06.043

Lenain C, Gusyatiner O, Douma S, van den Broek B, Peeper DS. 2015. Autophagy-mediated degradation of nuclear envelope proteins during oncogene-induced senescence. *Carcinogenesis* **36:** 1263–1274. doi:10.1093/carcin/bgv124

Le Thanh P, Meinke P, Korfali N, Srsen V, Robson MI, Wehnert M, Schoser B, Sewry CA, Schirmer EC. 2017. Immunohistochemistry on a panel of Emery–Dreifuss muscular dystrophy samples reveals nuclear envelope proteins as inconsistent markers for pathology. *Neuro-*

muscul Disord **27:** 338–351. doi:10.1016/j.nmd.2016.12.003

Li C, Wei J, Li Y, He X, Zhou Q, Yan J, Zhang J, Liu Y, Liu Y, Shu HB. 2013. Transmembrane protein 214 (TMEM214) mediates endoplasmic reticulum stress-induced caspase 4 enzyme activation and apoptosis. *J Biol Chem* **288:** 17908–17917. doi:10.1074/jbc.M113.458836

Li X, Roberti R, Blobel G. 2015. Structure of an integral membrane sterol reductase from *Methylomicrobium alcaliphilum*. *Nature* **517:** 104–107. doi:10.1038/nature13797

Liang WC, Mitsuhashi H, Keduka E, Nonaka I, Noguchi S, Nishino I, Hayashi YK. 2011. TMEM43 mutations in Emery–Dreifuss muscular dystrophy-related myopathy. *Ann Neurol* **69:** 1005–1013. doi:10.1002/ana.22338

Lin F, Blake DL, Callebaut I, Skerjanc IS, Holmer L, McBurney MW, Paulin-Levasseur M, Worman HJ. 2000. MAN1, an inner nuclear membrane protein that shares the LEM domain with lamina-associated polypeptide 2 and emerin. *J Biol Chem* **275:** 4840–4847. doi:10.1074/jbc.275.7.4840

Lin F, Morrison JM, Wu W, Worman HJ. 2005. MAN1, an integral protein of the inner nuclear membrane, binds Smad2 and Smad3 and antagonizes transforming growth factor-β signaling. *Hum Mol Genet* **14:** 437–445. doi:10.1093/hmg/ddi040

Lin ST, Zhang L, Lin X, Zhang LC, Garcia VE, Tsai CW, Ptáček L, Fu YH. 2014. Nuclear envelope protein MAN1 regulates clock through BMAL1. *eLife* **3:** e02981. doi:10.7554/eLife.02981

Liu J, Lee KK, Segura-Totten M, Neufeld E, Wilson KL, Gruenbaum Y. 2003. MAN1 and emerin have overlapping function(s) essential for chromosome segregation and cell division in *Caenorhabditis elegans*. *Proc Natl Acad Sci* **100:** 4598–4603. doi:10.1073/pnas.0730821100

Liu Q, Pante N, Misteli T, Elsagga M, Crisp M, Hodzic D, Burke B, Roux KJ. 2007. Functional association of Sun1 with nuclear pore complexes. *J Cell Biol* **178:** 785–798. doi:10.1083/jcb.200704108

Liu GH, Guan T, Datta K, Coppinger J, Yates J III, Gerace L. 2009. Regulation of myoblast differentiation by the nuclear envelope protein NET39. *Mol Cell Biol* **29:** 5800–5812. doi:10.1128/MCB.00684-09

Liu H, Hu J, Wei R, Zhou L, Pan H, Zhu H, Huang M, Luo J, Xu W. 2018. SPAG5 promotes hepatocellular carcinoma progression by downregulating SCARA5 through modifying β-catenin degradation. *J Exp Clin Cancer Res* **37:** 229. doi:10.1186/s13046-018-0891-3

Lottersberger F, Karssemeijer RA, Dimitrova N, de Lange T. 2015. 53BP1 and the LINC complex promote microtubule-dependent DSB mobility and DNA repair. *Cell* **163:** 880–893. doi:10.1016/j.cell.2015.09.057

Luithle N, de Bos JU, Hovius R, Maslennikova D, Lewis RT, Ungricht R, Fierz B, Kutay U. 2020. Torsin ATPases influence chromatin interaction of the torsin regulator LAP1. *eLife* **9:** e63614. doi:10.7554/eLife.63614

Lukášová E, Kovařík A, Bačíková A, Falk M, Kozubek S. 2017. Loss of lamin B receptor is necessary to induce cellular senescence. *Biochem J* **474:** 281–300. doi:10.1042/BCJ20160459

Luo X, Yang W, Gao G. 2018. SUN1 regulates HIV-1 nuclear import in a manner dependent on the interaction between the viral capsid and cellular cyclophilin A. *J Virol* **92:** JVI.00229-18.

MacQueen AJ, Phillips CM, Bhalla N, Weiser P, Villeneuve AM, Dernburg AF. 2005. Chromosome sites play dual roles to establish homologous synapsis during meiosis in *C. elegans*. *Cell* **123:** 1037–1050. doi:10.1016/j.cell.2005.09.034

Makatsori D, Kourmouli N, Polioudaki H, Shultz LD, McLean K, Theodoropoulos PA, Singh PB, Georgatos SD. 2004. The inner nuclear membrane protein lamin B receptor forms distinct microdomains and links epigenetically marked chromatin to the nuclear envelope. *J Biol Chem* **279:** 25567–25573. doi:10.1074/jbc.M313606200

Malik P, Korfali N, Srsen V, Lazou V, Batrakou DG, Zuleger N, Kavanagh DM, Wilkie GS, Goldberg MW, Schirmer EC. 2010. Cell-specific and lamin-dependent targeting of novel transmembrane proteins in the nuclear envelope. *Cell Mol Life Sci* **67:** 1353–1369. doi:10.1007/s00018-010-0257-2

Malone CJ, Fixsen WD, Horvitz HR, Han M. 1999. UNC-84 localizes to the nuclear envelope and is required for nuclear migration and anchoring during *C. elegans* development. *Development* **126:** 3171–3181.

Manilal S, Nguyen TM, Sewry CA, Morris GE. 1996. The Emery–Dreifuss muscular dystrophy protein, emerin, is a nuclear membrane protein. *Hum Mol Genet* **5:** 801–808. doi:10.1093/hmg/5.6.801

Mansharamani M, Wilson KL. 2005. Direct binding of nuclear membrane protein MAN1 to emerin in vitro and two modes of binding to barrier-to-autointegration factor. *J Biol Chem* **280:** 13863–13870. doi:10.1074/jbc.M413020200

Margalit A, Brachner A, Gotzmann J, Foisner R, Gruenbaum Y. 2007. Barrier-to-autointegration factor—a BAFfling little protein. *Trends Cell Biol* **17:** 202–208. doi:10.1016/j.tcb.2007.02.004

Markiewicz E, Dechat T, Foisner R, Quinlan RA, Hutchison CJ. 2002. Lamin A/C binding protein LAP2α is required for nuclear anchorage of retinoblastoma protein. *Mol Biol Cell* **13:** 4401–4413. doi:10.1091/mbc.e02-07-0450

Markiewicz E, Tilgner K, Barker N, van de Wetering M, Clevers H, Dorobek M, Hausmanowa-Petrusewicz I, Ramaekers FC, Broers JL, Blankesteijn WM, et al. 2006. The inner nuclear membrane protein emerin regulates β-catenin activity by restricting its accumulation in the nucleus. *EMBO J* **25:** 3275–3285. doi:10.1038/sj.emboj.7601230

Marnef A, Finoux AL, Arnould C, Guillou E, Daburon V, Rocher V, Mangeat T, Mangeot PE, Ricci EP, Legube G. 2019. A cohesin/HUSH- and LINC-dependent pathway controls ribosomal DNA double-strand break repair. *Genes Dev* **33:** 1175–1190. doi:10.1101/gad.324012.119

Martins SB, Eide T, Steen RL, Jahnsen T, Skalhegg BS, Collas P. 2000. HA95 is a protein of the chromatin and nuclear matrix regulating nuclear envelope dynamics. *J Cell Sci* **113** (Pt 21): 3703–3713.

Martins S, Eikvar S, Furukawa K, Collas P. 2003. HA95 and LAP2β mediate a novel chromatin-nuclear envelope interaction implicated in initiation of DNA replication. *J Cell Biol* **160:** 177–188. doi:10.1083/jcb.200210026

Massague J, Cheifetz S, Endo T, Nadal-Ginard B. 1986. Type β transforming growth factor is an inhibitor of myogenic differentiation. *Proc Natl Acad Sci* **83:** 8206–8210. doi:10.1073/pnas.83.21.8206

Mattaj IW. 2004. Sorting out the nuclear envelope from the endoplasmic reticulum. *Nat Rev Mol Cell Biol* **5:** 65–69. doi:10.1038/nrm1263

McCullough J, Frost A, Sundquist WI. 2018. Structures, functions, and dynamics of ESCRT-III/Vps4 membrane remodeling and fission complexes. *Annu Rev Cell Dev Biol* **34:** 85–109. doi:10.1146/annurev-cellbio-100616-060600

McHugh CA, Chen CK, Chow A, Surka CF, Tran C, McDonel P, Pandya-Jones A, Blanco M, Burghard C, Moradian A, et al. 2015. The *Xist* lncRNA interacts directly with SHARP to silence transcription through HDAC3. *Nature* **521:** 232–236. doi:10.1038/nature14443

McKinnon CM, Mellor H. 2017. The tumor suppressor RhoBTB1 controls Golgi integrity and breast cancer cell invasion through METTL7B. *BMC Cancer* **17:** 145. doi:10.1186/s12885-017-3138-3

Meinke P, Mattioli E, Haque F, Antoku S, Columbaro M, Straatman KR, Worman HJ, Gundersen GG, Lattanzi G, Wehnert M, et al. 2014. Muscular dystrophy-associated SUN1 and SUN2 variants disrupt nuclear-cytoskeletal connections and myonuclear organization. *PLoS Genet* **10:** e1004605. doi:10.1371/journal.pgen.1004605

Meinke P, Kerr ARW, Czapiewski R, de Las Heras JI, Dixon CR, Harris E, Kölbel H, Muntoni F, Schara U, Straub V, et al. 2020. A multistage sequencing strategy pinpoints novel candidate alleles for Emery–Dreifuss muscular dystrophy and supports gene misregulation as its pathomechanism. *EBioMedicine* **51:** 102587. doi:10.1016/j.ebiom.2019.11.048

Mekhail K, Moazed D. 2010. The nuclear envelope in genome organization, expression and stability. *Nat Rev Mol Cell Biol* **11:** 317–328. doi:10.1038/nrm2894

Mekhail K, Seebacher J, Gygi SP, Moazed D. 2008. Role for perinuclear chromosome tethering in maintenance of genome stability. *Nature* **456:** 667–670. doi:10.1038/nature07460

Mirza AN, McKellar SA, Urman NM, Brown AS, Hollmig T, Aasi SZ, Oro AE. 2019. LAP2 proteins chaperone GLI1 movement between the lamina and chromatin to regulate transcription. *Cell* **176:** 198–212.e15. doi:10.1016/j.cell.2018.10.054

Mislow JM, Holaska JM, Kim MS, Lee KK, Segura-Totten M, Wilson KL, McNally EM. 2002. Nesprin-1α self-associates and binds directly to emerin and lamin A in vitro. *FEBS Lett* **525:** 135–140. doi:10.1016/S0014-5793(02)03105-8

Mora M, Cartegni L, Di Blasi C, Barresi R, Bione S, Raffaele di Barletta M, Morandi L, Merlini L, Nigro V, Politano L, et al. 1997. X-linked Emery–Dreifuss muscular dystrophy can be diagnosed from skin biopsy or blood sample. *Ann Neurol* **42:** 249–253. doi:10.1002/ana.410420218

Morimoto A, Shibuya H, Zhu X, Kim J, Ishiguro K, Han M, Watanabe Y. 2012. A conserved KASH domain protein associates with telomeres, SUN1, and dynactin during mammalian meiosis. *J Cell Biol* **198:** 165–172. doi:10.1083/jcb.201204085

Morris GE, Manilal S. 1999. Heart to heart: from nuclear proteins to Emery–Dreifuss muscular dystrophy. *Hum Mol Genet* **8:** 1847–1851. doi:10.1093/hmg/8.10.1847

Moser B, Basílio J, Gotzmann J, Brachner A, Foisner R. 2020. Comparative interactome analysis of emerin, MAN1 and LEM2 reveals a unique role for LEM2 in nucleotide excision repair. *Cells* **9:** 463. doi:10.3390/cells9020463

Muchir A, Pavlidis P, Bonne G, Hayashi YK, Worman HJ. 2007. Activation of MAPK in hearts of EMD null mice: similarities between mouse models of X-linked and autosomal dominant Emery Dreifuss muscular dystrophy. *Hum Mol Genet* **16:** 1884–1895. doi:10.1093/hmg/ddm137

Mumm S, Wenkert D, Zhang X, McAlister WH, Mier RJ, Whyte MP. 2007. Deactivating germline mutations in LEMD3 cause osteopoikilosis and Buschke–Ollendorff syndrome, but not sporadic melorheostosis. *J Bone Miner Res* **22:** 243–250. doi:10.1359/jbmr.061102

Naetar N, Korbei B, Kozlov S, Kerenyi MA, Dorner D, Kral R, Gotic I, Fuchs P, Cohen TV, Bittner R, et al. 2008. Loss of nucleoplasmic LAP2α-lamin A complexes causes erythroid and epidermal progenitor hyperproliferation. *Nat Cell Biol* **10:** 1341–1348. doi:10.1038/ncb1793

Nesterova TB, Wei G, Coker H, Pintacuda G, Bowness JS, Zhang T, Almeida M, Bloechl B, Moindrot B, Carter EJ, et al. 2019. Systematic allelic analysis defines the interplay of key pathways in X chromosome inactivation. *Nat Commun* **10:** 3129. doi:10.1038/s41467-019-11171-3

Nili E, Cojocaru GS, Kalma Y, Ginsberg D, Copeland NG, Gilbert DJ, Jenkins NA, Berger R, Shaklai S, Amariglio N, et al. 2001. Nuclear membrane protein LAP2β mediates transcriptional repression alone and together with its binding partner GCL (germ-cell-less). *J Cell Sci* **114:** 3297–3307.

Ohba T, Schirmer EC, Nishimoto T, Gerace L. 2004. Energy- and temperature-dependent transport of integral proteins to the inner nuclear membrane via the nuclear pore. *J Cell Biol* **167:** 1051–1062. doi:10.1083/jcb.200409149

Olmos Y, Hodgson L, Mantell J, Verkade P, Carlton JG. 2015. ESCRT-III controls nuclear envelope reformation. *Nature* **522:** 236–239. doi:10.1038/nature14503

Olson EN, Nordheim A. 2010. Linking actin dynamics and gene transcription to drive cellular motile functions. *Nat Rev Mol Cell Biol* **11:** 353–365. doi:10.1038/nrm2890

Oro AE, Higgins KM, Hu Z, Bonifas JM, Epstein EH Jr, Scott MP. 1997. Basal cell carcinomas in mice overexpressing sonic hedgehog. *Science* **276:** 817–821. doi:10.1126/science.276.5313.817

Osada S, Ohmori SY, Taira M. 2003. XMAN1, an inner nuclear membrane protein, antagonizes BMP signaling by interacting with Smad1 in *Xenopus* embryos. *Development* **130:** 1783–1794. doi:10.1242/dev.00401

Östlund C, Sullivan T, Stewart CL, Worman HJ. 2006. Dependence of diffusional mobility of integral inner nuclear membrane proteins on A-type lamins. *Biochemistry* **45:** 1374–1382. doi:10.1021/bi052156n

Oza P, Jaspersen SL, Miele A, Dekker J, Peterson CL. 2009. Mechanisms that regulate localization of a DNA double-strand break to the nuclear periphery. *Genes Dev* **23:** 912–927. doi:10.1101/gad.1782209

Ozelius LJ, Hewett JW, Page CE, Bressman SB, Kramer PL, Shalish C, de Leon D, Brin MF, Raymond D, Corey DP, et al. 1997. The early-onset torsion dystonia gene (DYT1) encodes an ATP-binding protein. *Nat Genet* **17:** 40–48. doi:10.1038/ng0997-40

Pan D, Estévez-Salmerón LD, Stroschein SL, Zhu X, He J, Zhou S, Luo K. 2005. The integral inner nuclear membrane protein MAN1 physically interacts with the R-Smad proteins to repress signaling by the transforming growth factor-β superfamily of cytokines. *J Biol Chem* **280:** 15992–16001. doi:10.1074/jbc.M411234200

Pasch E, Link J, Beck C, Scheuerle S, Alsheimer M. 2015. The LINC complex component Sun4 plays a crucial role in sperm head formation and fertility. *Biol Open* **4:** 1792–1802. doi:10.1242/bio.015768

Pawar S, Ungricht R, Tiefenboeck P, Leroux JC, Kutay U. 2017. Efficient protein targeting to the inner nuclear membrane requires Atlastin-dependent maintenance of ER topology. *eLife* **6:** e28202.

Penkner A, Tang L, Novatchkova M, Ladurner M, Fridkin A, Gruenbaum Y, Schweizer D, Loidl J, Jantsch V. 2007. The nuclear envelope protein Matefin/SUN-1 is required for homologous pairing in *C. elegans* meiosis. *Dev Cell* **12:** 873–885. doi:10.1016/j.devcel.2007.05.004

Penkner AM, Fridkin A, Gloggnitzer J, Baudrimont A, Machacek T, Woglar A, Csaszar E, Pasierbek P, Ammerer G, Gruenbaum Y, et al. 2009. Meiotic chromosome homology search involves modifications of the nuclear envelope protein Matefin/SUN-1. *Cell* **139:** 920–933. doi:10.1016/j.cell.2009.10.045

Pieper GH, Sprenger S, Teis D, Oliferenko S. 2020. ESCRT-III/Vps4 controls heterochromatin-nuclear envelope attachments. *Dev Cell* **53:** 27–41.e6. doi:10.1016/j.devcel.2020.01.028

Polesskaya A, Seale P, Rudnicki MA. 2003. Wnt signaling induces the myogenic specification of resident $CD45^+$ adult stem cells during muscle regeneration. *Cell* **113:** 841–852. doi:10.1016/S0092-8674(03)00437-9

Pradhan R, Ranade D, Sengupta K. 2018. Emerin modulates spatial organization of chromosome territories in cells on softer matrices. *Nucleic Acids Res* **46:** 5561–5586. doi:10.1093/nar/gky288

Prakash A, Sengupta S, Aparna K, Kasbekar DP. 1999. The erg-3 (sterol $\Delta^{14,15}$-reductase) gene of *Neurospora crassa*: generation of null mutants by repeat-induced point mutation and complementation by proteins chimeric for human lamin B receptor sequences. *Microbiology* **145:** 1443–1451. doi:10.1099/13500872-145-6-1443

Pyrpasopoulou A, Meier J, Maison C, Simos G, Georgatos SD. 1996. The lamin B receptor (LBR) provides essential chromatin docking sites at the nuclear envelope. *EMBO J* **15:** 7108–7119. doi:10.1002/j.1460-2075.1996.tb01102.x

Raab M, Gentili M, de Belly H, Thiam HR, Vargas P, Jimenez AJ, Lautenschlaeger F, Voituriez R, Lennon-Dumenil AM, Manel N, et al. 2016. ESCRT III repairs nuclear envelope ruptures during cell migration to limit DNA damage and cell death. *Science* **352:** 359–362. doi:10.1126/science.aad7611

Raju GP, Dimova N, Klein PS, Huang HC. 2003. SANE, a novel LEM domain protein, regulates bone morphogenetic protein signaling through interaction with Smad1. *J Biol Chem* **278:** 428–437. doi:10.1074/jbc.M210505200

Rebelo S, da Cruz ESEF, da Cruz ESOA. 2015. Genetic mutations strengthen functional association of LAP1 with DYT1 dystonia and muscular dystrophy. *Mutat Res Rev Mutat Res* **766:** 42–47. doi:10.1016/j.mrrev.2015.07.004

Rolls MM, Stein PA, Taylor SS, Ha E, McKeon F, Rapoport TA. 1999. A visual screen of a GFP-fusion library identifies a new type of nuclear envelope membrane protein. *J Cell Biol* **146:** 29–44. doi:10.1083/jcb.146.1.29

Rose AE, Brown RS, Schlieker C. 2015. Torsins: not your typical AAA^+ ATPases. *Crit Rev Biochem Mol Biol* **50:** 532–549. doi:10.3109/10409238.2015.1091804

Rothballer A, Kutay U. 2013. The diverse functional LINCs of the nuclear envelope to the cytoskeleton and chromatin. *Chromosoma* **122:** 415–429. doi:10.1007/s00412-013-0417-x

Rothballer A, Schwartz TU, Kutay U. 2013. LINCing complex functions at the nuclear envelope: what the molecular architecture of the LINC complex can reveal about its function. *Nucleus* **4:** 29–36. doi:10.4161/nucl.23387

Rowat AC, Lammerding J, Ipsen JH. 2006. Mechanical properties of the cell nucleus and the effect of emerin deficiency. *Biophys J* **91:** 4649–4664. doi:10.1529/biophysj.106.086454

Samwer M, Schneider MWG, Hoefler R, Schmalhorst PS, Jude JG, Zuber J, Gerlich DW. 2017. DNA cross-bridging shapes a single nucleus from a set of mitotic chromosomes. *Cell* **170:** 956–972.e23. doi:10.1016/j.cell.2017.07.038

Santos M, Domingues SC, Costa P, Muller T, Galozzi S, Marcus K, da Cruz e Silva EF, da Cruz e Silva OA, Rebelo S. 2014. Identification of a novel human LAP1 isoform that is regulated by protein phosphorylation. *PLoS ONE* **9:** e113732. doi:10.1371/journal.pone.0113732

Sato A, Isaac B, Phillips CM, Rillo R, Carlton PM, Wynne DJ, Kasad RA, Dernburg AF. 2009. Cytoskeletal forces span the nuclear envelope to coordinate meiotic chromosome pairing and synapsis. *Cell* **139:** 907–919. doi:10.1016/j.cell.2009.10.039

Schaller T, Bulli L, Pollpeter D, Betancor G, Kutzner J, Apolonia L, Herold N, Burk R, Malim MH. 2017. Effects of inner nuclear membrane proteins SUN1/UNC-84A and SUN2/UNC-84B on the early steps of HIV-1 infection. *J Virol* **91:** e00463-17. doi:10.1128/JVI.00463-17

Schirmer EC, Florens L, Guan T, Yates JR III, Gerace L. 2003. Nuclear membrane proteins with potential disease links found by subtractive proteomics. *Science* **301:** 1380–1382. doi:10.1126/science.1088176

Schober H, Ferreira H, Kalck V, Gehlen LR, Gasser SM. 2009. Yeast telomerase and the SUN domain protein Mps3 anchor telomeres and repress subtelomeric recombination. *Genes Dev* **23:** 928–938. doi:10.1101/gad.1787509

Schuler E, Lin F, Worman HJ. 1994. Characterization of the human gene encoding LBR, an integral protein of the nuclear envelope inner membrane. *J Biol Chem* **269:** 11312–11317. doi:10.1016/S0021-9258(19)78127-7

Sen B, Xie Z, Case N, Ma M, Rubin C, Rubin J. 2008. Mechanical strain inhibits adipogenesis in mesenchymal stem cells by stimulating a durable β-catenin signal. *Endocrinology* **149:** 6065–6075. doi:10.1210/en.2008-0687

Senior A, Gerace L. 1988. Integral membrane proteins specific to the inner nuclear membrane and associated with the nuclear lamina. *J Cell Biol* **107:** 2029–2036. doi:10.1083/jcb.107.6.2029

Shah PP, Donahue G, Otte GL, Capell BC, Nelson DM, Cao K, Aggarwala V, Cruickshanks HA, Rai TS, McBryan T, et al. 2013. Lamin B1 depletion in senescent cells triggers large-scale changes in gene expression and the chromatin landscape. *Genes Dev* **27:** 1787–1799. doi:10.1101/gad.223834.113

Shibuya H, Ishiguro K, Watanabe Y. 2014. The TRF1-binding protein TERB1 promotes chromosome movement and telomere rigidity in meiosis. *Nat Cell Biol* **16:** 145–156. doi:10.1038/ncb2896

Shibuya H, Hernández-Hernández A, Morimoto A, Negishi L, Höög C, Watanabe Y. 2015. MAJIN links telomeric DNA to the nuclear membrane by exchanging telomere cap. *Cell* **163:** 1252–1266. doi:10.1016/j.cell.2015.10.030

Shimi T, Butin-Israeli V, Adam SA, Hamanaka RB, Goldman AE, Lucas CA, Shumaker DK, Kosak ST, Chandel NS, Goldman RD. 2011. The role of nuclear lamin B1 in cell proliferation and senescence. *Genes Dev* **25:** 2579–2593. doi:10.1101/gad.179515.111

Shin JY, Méndez-López I, Wang Y, Hays AP, Tanji K, Lefkowitch JH, Schulze PC, Worman HJ, Dauer WT. 2013. Lamina-associated polypeptide-1 interacts with the muscular dystrophy protein emerin and is essential for skeletal muscle maintenance. *Dev Cell* **26:** 591–603. doi:10.1016/j.devcel.2013.08.012

Shrestha N, Bacsa B, Ong HL, Scheruebel S, Bischof H, Malli R, Ambudkar IS, Groschner K. 2020. TRIC-A shapes oscillatory Ca^{2+} signals by interaction with STIM1/Orai1 complexes. *PLoS Biol* **18:** e3000700. doi:10.1371/journal.pbio.3000700

Silve S, Dupuy PH, Ferrara P, Loison G. 1998. Human lamin B receptor exhibits sterol C14-reductase activity in *Saccharomyces cerevisiae. Biochim Biophys Acta* **1392:** 233–244. doi:10.1016/S0005-2760(98)00041-1

Sobreira N, Modaff P, Steel G, You J, Nanda S, Hoover-Fong J, Valle D, Pauli RM. 2015. An anadysplasia-like, spontaneously remitting spondylometaphyseal dysplasia secondary to lamin B receptor (*LBR*) gene mutations: further definition of the phenotypic heterogeneity of *LBR*-bone dysplasias. *Am J Med Genet A* **167:** 159–163. doi:10.1002/ajmg.a.36808

Solovei I, Wang AS, Thanisch K, Schmidt CS, Krebs S, Zwerger M, Cohen TV, Devys D, Foisner R, Peichl L, et al. 2013. LBR and lamin A/C sequentially tether peripheral heterochromatin and inversely regulate differentiation. *Cell* **152:** 584–598. doi:10.1016/j.cell.2013.01.009

Somech R, Shaklai S, Geller O, Amariglio N, Simon AJ, Rechavi G, Gal-Yam EN. 2005. The nuclear-envelope protein and transcriptional repressor LAP2β interacts with HDAC3 at the nuclear periphery, and induces histone H4 deacetylation. *J Cell Sci* **118:** 4017–4025. doi:10.1242/jcs.02521

Sosa BA, Rothballer A, Kutay U, Schwartz TU. 2012. LINC complexes form by binding of three KASH peptides to domain interfaces of trimeric SUN proteins. *Cell* **149:** 1035–1047. doi:10.1016/j.cell.2012.03.046

Sosa BA, Demircioglu FE, Chen JZ, Ingram J, Ploegh HL, Schwartz TU. 2014. How lamina-associated polypeptide 1 (LAP1) activates Torsin. *eLife* **3:** e03239. doi:10.7554/eLife.03239

Soullam B, Worman HJ. 1995. Signals and structural features involved in integral membrane protein targeting to the inner nuclear membrane. *J Cell Biol* **130:** 15–27. doi:10.1083/jcb.130.1.15

Starr DA, Fridolfsson HN. 2010. Interactions between nuclei and the cytoskeleton are mediated by SUN-KASH nuclear-envelope bridges. *Annu Rev Cell Dev Biol* **26:** 421–444. doi:10.1146/annurev-cellbio-100109-104037

Starr DA, Hermann GJ, Malone CJ, Fixsen W, Priess JR, Horvitz HR, Han M. 2001. unc-83 encodes a novel component of the nuclear envelope and is essential for proper nuclear migration. *Development* **128:** 5039–5050.

Stroud MJ, Fang X, Zhang J, Guimarães-Camboa N, Veevers J, Dalton ND, Gu Y, Bradford WH, Peterson KL, Evans SM, et al. 2018. Luma is not essential for murine cardiac development and function. *Cardiovasc Res* **114:** 378–388. doi:10.1093/cvr/cvx205

Sullivan T, Escalante-Alcalde D, Bhatt H, Anver M, Bhat N, Nagashima K, Stewart CL, Burke B. 1999. Loss of A-type lamin expression compromises nuclear envelope integrity leading to muscular dystrophy. *J Cell Biol* **147:** 913–920. doi:10.1083/jcb.147.5.913

Sun WW, Jiao S, Sun L, Zhou Z, Jin X, Wang JH. 2018. SUN2 modulates HIV-1 infection and latency through association with lamin A/C to maintain the repressive chromatin. *MBio* **9:** e02408-17. doi:10.1128/mBio.02408-17

Tajik A, Zhang Y, Wei F, Sun J, Jia Q, Zhou W, Singh R, Khanna N, Belmont AS, Wang N. 2016. Transcription upregulation via force-induced direct stretching of chromatin. *Nat Mater* **15:** 1287–1296. doi:10.1038/nmat4729

Talamas JA, Hetzer MW. 2011. POM121 and Sun1 play a role in early steps of interphase NPC assembly. *J Cell Biol* **194:** 27–37. doi:10.1083/jcb.201012154

Tapia O, Fong LG, Huber MD, Young SG, Gerace L. 2015. Nuclear envelope protein Lem2 is required for mouse development and regulates MAP and AKT kinases. *PLoS ONE* **10:** e0116196. doi:10.1371/journal.pone.0116196

Taylor KM, Morgan HE, Johnson A, Nicholson RI. 2004. Structure-function analysis of HKE4, a member of the new LIV-1 subfamily of zinc transporters. *Biochem J* **377:** 131–139. doi:10.1042/bj20031183

Taylor MR, Slavov D, Gajewski A, Vlcek S, Ku L, Fain PR, Carniel E, Di Lenarda A, Sinagra G, Boucek MM, et al. 2005. Thymopoietin (lamina-associated polypeptide 2) gene mutation associated with dilated cardiomyopathy. *Hum Mutat* **26:** 566–574. doi:10.1002/humu.20250

Thaller DJ, Allegretti M, Borah S, Ronchi P, Beck M, Lusk CP. 2019. An ESCRT-LEM protein surveillance system is poised to directly monitor the nuclear envelope and nuclear transport system. *eLife* **8:** e45284. doi:10.7554/eLife.45284

Theerthagiri G, Eisenhardt N, Schwarz H, Antonin W. 2010. The nucleoporin Nup188 controls passage of membrane proteins across the nuclear pore complex. *J Cell Biol* **189:** 1129–1142. doi:10.1083/jcb.200912045

Trelles-Sticken E, Dresser ME, Scherthan H. 2000. Meiotic telomere protein Ndj1p is required for meiosis-specific telomere distribution, bouquet formation and efficient homologue pairing. *J Cell Biol* **151:** 95–106. doi:10.1083/jcb.151.1.95

Trelles-Sticken E, Adelfalk C, Loidl J, Scherthan H. 2005. Meiotic telomere clustering requires actin for its formation and cohesin for its resolution. *J Cell Biol* **170:** 213–223. doi:10.1083/jcb.200501042

Tsai PL, Zhao C, Turner E, Schlieker C. 2016. The lamin B receptor is essential for cholesterol synthesis and perturbed by disease-causing mutations. *eLife* **5:** e16011. doi:10.7554/eLife.16011

Turgay Y, Champion L, Balazs C, Held M, Toso A, Gerlich DW, Meraldi P, Kutay U. 2014. SUN proteins facilitate the removal of membranes from chromatin during nuclear envelope breakdown. *J Cell Biol* **204:** 1099–1109. doi:10.1083/jcb.201310116

Turner EM, Schlieker C. 2016. Pelger-Huët anomaly and Greenberg skeletal dysplasia: LBR-associated diseases of cholesterol metabolism. *Rare Dis* **4:** e1241363. doi:10.1080/21675511.2016.1241363

Tuschl K, Meyer E, Valdivia LE, Zhao N, Dadswell C, Abdul-Sada A, Hung CY, Simpson MA, Chong WK, Jacques TS, et al. 2016. Mutations in SLC39A14 disrupt manganese homeostasis and cause childhood-onset parkinsonism-dystonia. *Nat Commun* **7:** 11601. doi:10.1038/ncomms11601

Ulbert S, Antonin W, Platani M, Mattaj IW. 2006. The inner nuclear membrane protein Lem2 is critical for normal nuclear envelope morphology. *FEBS Lett* **580:** 6435–6441. doi:10.1016/j.febslet.2006.10.060

Ulker D, Ersoy YE, Gucin Z, Muslumanoglu M, Buyru N. 2018. Downregulation of SCARA5 may contribute to breast cancer via promoter hypermethylation. *Gene* **673:** 102–106. doi:10.1016/j.gene.2018.06.036

Ungricht R, Klann M, Horvath P, Kutay U. 2015. Diffusion and retention are major determinants of protein targeting to the inner nuclear membrane. *J Cell Biol* **209:** 687–704. doi:10.1083/jcb.201409127

Uzer G, Bas G, Sen B, Xie Z, Birks S, Olcum M, McGrath C, Styner M, Rubin J. 2018. Sun-mediated mechanical LINC between nucleus and cytoskeleton regulates β catenin nuclear access. *J Biomech* **74:** 32–40. doi:10.1016/j.jbiomech.2018.04.013

Viera A, Alsheimer M, Gomez R, Berenguer I, Ortega S, Symonds CE, Santamaria D, Benavente R, Suja JA. 2015. CDK2 regulates nuclear envelope protein dynamics and telomere attachment in mouse meiotic prophase. *J Cell Sci* **128:** 88–99. doi:10.1242/jcs.154922

Vietri M, Schink KO, Campsteijn C, Wegner CS, Schultz SW, Christ L, Thoresen SB, Brech A, Raiborg C, Stenmark H. 2015. Spastin and ESCRT-III coordinate mitotic spindle disassembly and nuclear envelope sealing. *Nature* **522:** 231–235. doi:10.1038/nature14408

Vlcek S, Just H, Dechat T, Foisner R. 1999. Functional diversity of LAP2α and LAP2β in postmitotic chromosome association is caused by an α-specific nuclear targeting domain. *EMBO J* **18:** 6370–6384. doi:10.1093/emboj/18.22.6370

von Appen A, LaJoie D, Johnson IE, Trnka MJ, Pick SM, Burlingame AL, Ullman KS, Frost A. 2020. LEM2 phase separation promotes ESCRT-mediated nuclear envelope reformation. *Nature* **582:** 115–118. doi:10.1038/s41586-020-2232-x

Wagner N, Weyhersmüller A, Blauth A, Schuhmann T, Heckmann M, Krohne G, Samakovlis C. 2010. The *Drosophila* LEM-domain protein MAN1 antagonizes BMP signaling at the neuromuscular junction and the wing crossveins. *Dev Biol* **339:** 1–13. doi:10.1016/j.ydbio.2009.11.036

Wang W, Shi Z, Jiao S, Chen C, Wang H, Liu G, Wang Q, Zhao Y, Greene MI, Zhou Z. 2012. Structural insights into SUN-KASH complexes across the nuclear envelope. *Cell Res* **22:** 1440–1452. doi:10.1038/cr.2012.126

Waterham HR, Koster J, Mooyer P, Noort Gv G, Kelley RI, Wilcox WR, Wanders RJ, Hennekam RC, Oosterwijk JC. 2003. Autosomal recessive HEM/Greenberg skeletal dysplasia is caused by 3β-hydroxysterol Δ^{14}-reductase deficiency due to mutations in the lamin B receptor gene. *Am J Hum Genet* **72:** 1013–1017. doi:10.1086/373938

Webster BM, Colombi P, Jäger J, Lusk CP. 2014. Surveillance of nuclear pore complex assembly by ESCRT-III/Vps4. *Cell* **159:** 388–401. doi:10.1016/j.cell.2014.09.012

Webster BM, Thaller DJ, Jäger J, Ochmann SE, Borah S, Lusk CP. 2016. Chm7 and Heh1 collaborate to link nuclear pore complex quality control with nuclear envelope sealing. *EMBO J* **35:** 2447–2467. doi:10.15252/embj.201694574

Wilkie GS, Korfali N, Swanson SK, Malik P, Srsen V, Batrakou DG, de las Heras J, Zuleger N, Kerr AR, Florens L, et al. 2011. Several novel nuclear envelope transmembrane proteins identified in skeletal muscle have cytoskeletal associations. *Mol Cell Proteomics* **10:** M110.003129. doi:10.1074/mcp.M110.003129

Wilkinson FL, Holaska JM, Zhang Z, Sharma A, Manilal S, Holt I, Stamm S, Wilson KL, Morris GE. 2003. Emerin interacts in vitro with the splicing-associated factor, YT521-B. *Eur J Biochem* **270:** 2459–2466. doi:10.1046/j.1432-1033.2003.03617.x

Willer MK, Carroll CW. 2017. Substrate stiffness-dependent regulation of the SRF-Mkl1 co-activator complex requires the inner nuclear membrane protein Emerin. *J Cell Sci* **130:** 2111–2118. doi:10.1242/jcs.197517

Wilson KL, Foisner R. 2010. Lamin-binding proteins. *Cold Spring Harb Perspect Biol* **2:** a000554. doi:10.1101/cshperspect.a000554

Woglar A, Jantsch V. 2014. Chromosome movement in meiosis I prophase of *Caenorhabditis elegans*. *Chromosoma* **123:** 15–24. doi:10.1007/s00412-013-0436-7

Worman HJ, Yuan J, Blobel G, Georgatos SD. 1988. A lamin B receptor in the nuclear envelope. *Proc Natl Acad Sci* **85:** 8531–8534. doi:10.1073/pnas.85.22.8531

Yam C, He Y, Zhang D, Chiam KH, Oliferenko S. 2011. Divergent strategies for controlling the nuclear membrane satisfy geometric constraints during nuclear division. *Curr Biol* **21:** 1314–1319. doi:10.1016/j.cub.2011.06.052

Yam C, Gu Y, Oliferenko S. 2013. Partitioning and remodeling of the *Schizosaccharomyces japonicus* mitotic nucleus require chromosome tethers. *Curr Biol* **23:** 2303–2310. doi:10.1016/j.cub.2013.09.057

Yang L, Guan T, Gerace L. 1997. Lamin-binding fragment of LAP2 inhibits increase in nuclear volume during the cell cycle and progression into S phase. *J Cell Biol* **139:** 1077–1087. doi:10.1083/jcb.139.5.1077

Yazawa M, Ferrante C, Feng J, Mio K, Ogura T, Zhang M, Lin PH, Pan Z, Komazaki S, Kato K, et al. 2007. TRIC channels are essential for $Ca2^{+}$ handling in intracellular stores. *Nature* **448:** 78–82. doi:10.1038/nature05928

Ye Q, Worman HJ. 1994. Primary structure analysis and lamin B and DNA binding of human LBR, an integral protein of the nuclear envelope inner membrane. *J Biol Chem* **269:** 11306–11311. doi:10.1016/S0021-9258(19)78126-5

Ye Q, Callebaut I, Pezhman A, Courvalin JC, Worman HJ. 1997. Domain-specific interactions of human HP1-type chromodomain proteins and inner nuclear membrane protein LBR. *J Biol Chem* **272:** 14983–14989. doi:10.1074/jbc.272.23.14983

Zhang Y, Castori M, Ferranti G, Paradisi M, Wordsworth BP. 2009. Novel and recurrent germline *LEMD3* mutations causing Buschke–Ollendorff syndrome and osteopoikilosis but not isolated melorheostosis. *Clin Genet* **75:** 556–561. doi:10.1111/j.1399-0004.2009.01177.x

Zhao Y, Chen YQ, Bonacci TM, Bredt DS, Li S, Bensch WR, Moller DE, Kowala M, Konrad RJ, Cao G. 2008. Identification and characterization of a major liver lysophosphatidylcholine acyltransferase. *J Biol Chem* **283:** 8258–8265. doi:10.1074/jbc.M710422200

Zhao C, Brown RS, Chase AR, Eisele MR, Schlieker C. 2013. Regulation of Torsin ATPases by LAP1 and LULL1. *Proc Natl Acad Sci* **110:** E1545–E1554. doi:10.1073/pnas.1300676110

Zhen Y, Sørensen V, Skjerpen CS, Haugsten EM, Jin Y, Wälchli S, Olsnes S, Wiedlocha A. 2012. Nuclear import of exogenous FGF1 requires the ER-protein LRRC59 and the importins Kpnα1 and Kpnβ1. *Traffic* **13:** 650–664. doi:10.1111/j.1600-0854.2012.01341.x

Zhou H, Clapham DE. 2009. Mammalian *MagT1* and *TUSC3* are required for cellular magnesium uptake and vertebrate embryonic development. *Proc Natl Acad Sci* **106:** 15750–15755. doi:10.1073/pnas.0908332106

Zhou J, Li H, Li X, Li Y, Yang M, Shi G, Xu D, Shi X. 2019. A novel *EMD* mutation in a Chinese family with initial diagnosis of conduction cardiomyopathy. *Brain Behav* **9:** e01167. doi:10.1002/brb3.1167

Zhu S, Wan W, Zhang Y, Shang W, Pan X, Zhang LK, Xiao G. 2019. Comprehensive interactome analysis reveals that STT3B is required for N-glycosylation of Lassa virus glycoprotein. *J Virol* **93:** e01443-19. doi:10.1128/JVI.01443-19

Zuleger N, Kelly DA, Richardson AC, Kerr AR, Goldberg MW, Goryachev AB, Schirmer EC. 2011. System analysis shows distinct mechanisms and common principles of nuclear envelope protein dynamics. *J Cell Biol* **193:** 109–123. doi:10.1083/jcb.201009068

Zuleger N, Boyle S, Kelly DA, de las Heras JI, Lazou V, Korfali N, Batrakou DG, Randles KN, Morris GE, Harrison DJ, et al. 2013. Specific nuclear envelope transmembrane proteins can promote the location of chromosomes to and from the nuclear periphery. *Genome Biol* **14:** R14. doi:10.1186/gb-2013-14-2-r14

Zullo JM, Demarco IA, Piqué-Regi R, Gaffney DJ, Epstein CB, Spooner CJ, Luperchio TR, Bernstein BE, Pritchard JK, Reddy KL, et al. 2012. DNA sequence-dependent compartmentalization and silencing of chromatin at the nuclear lamina. *Cell* **149:** 1474–1487. doi:10.1016/j.cell.2012.04.035

Mechanical Forces in Nuclear Organization

Yekaterina A. Miroshnikova[1,2,3,4,5] **and Sara A. Wickström**[1,2,3,4,6]

[1]Helsinki Institute of Life Science, Biomedicum Helsinki, University of Helsinki, Helsinki 00014, Finland

[2]Wihuri Research Institute, Biomedicum Helsinki, University of Helsinki, Helsinki 00290, Finland

[3]Stem Cells and Metabolism Research Program, Faculty of Medicine, University of Helsinki, Helsinki 00014, Finland

[4]Max Planck Institute for Biology of Ageing, Cologne 50931, Germany

[5]Laboratory of Molecular Biology, National Institute of Diabetes and Digestive and Kidney Diseases, National Institutes of Health, Bethesda, Maryland 20892, USA

[6]Cluster of Excellence Cellular Stress Responses in Aging-Associated Diseases (CECAD), University of Cologne, Cologne 50931, Germany

Correspondence: yekaterina.miroshnikova@helsinki.fi; sara.wickstrom@helsinki.fi

Cells generate and sense mechanical forces that trigger biochemical signals to elicit cellular responses that control cell fate changes. Mechanical forces also physically distort neighboring cells and the surrounding connective tissue, which propagate mechanochemical signals over long distances to guide tissue patterning, organogenesis, and adult tissue homeostasis. As the largest and stiffest organelle, the nucleus is particularly sensitive to mechanical force and deformation. Nuclear responses to mechanical force include adaptations in chromatin architecture and transcriptional activity that trigger changes in cell state. These force-driven changes also influence the mechanical properties of chromatin and nuclei themselves to prevent aberrant alterations in nuclear shape and help maintain genome integrity. This review will discuss principles of nuclear mechanotransduction and chromatin mechanics and their role in DNA damage and cell fate regulation.

MECHANICAL FORCES IN CELLS AND TISSUES

Tissue and cell dynamics and movement generate physical forces, including compression or stretching, which are transferred to the nucleus and chromatin, resulting in their deformation. Substantial deformation occurs not only in mechanically active tissues, such as the muscle, where cells and nuclei are directly exposed to constant contractile forces, but also in tissues such as the lung, skin, and vasculature whose surface area changes during respiration, body motion, and blood flow (Fig. 1). In addition to these high-amplitude changes that affect the tissue as a whole, most cells experience deformation at smaller scales. For instance, cell migration, especially within confined spaces of the connective tissue, causes substantial cell deformation (Friedl et al. 2011). Cell motility, death, division, and extrusion also inflict dynamic changes in cell shape,

Cite this article as *Cold Spring Harb Perspect Biol* doi: 10.1101/cshperspect.a039685

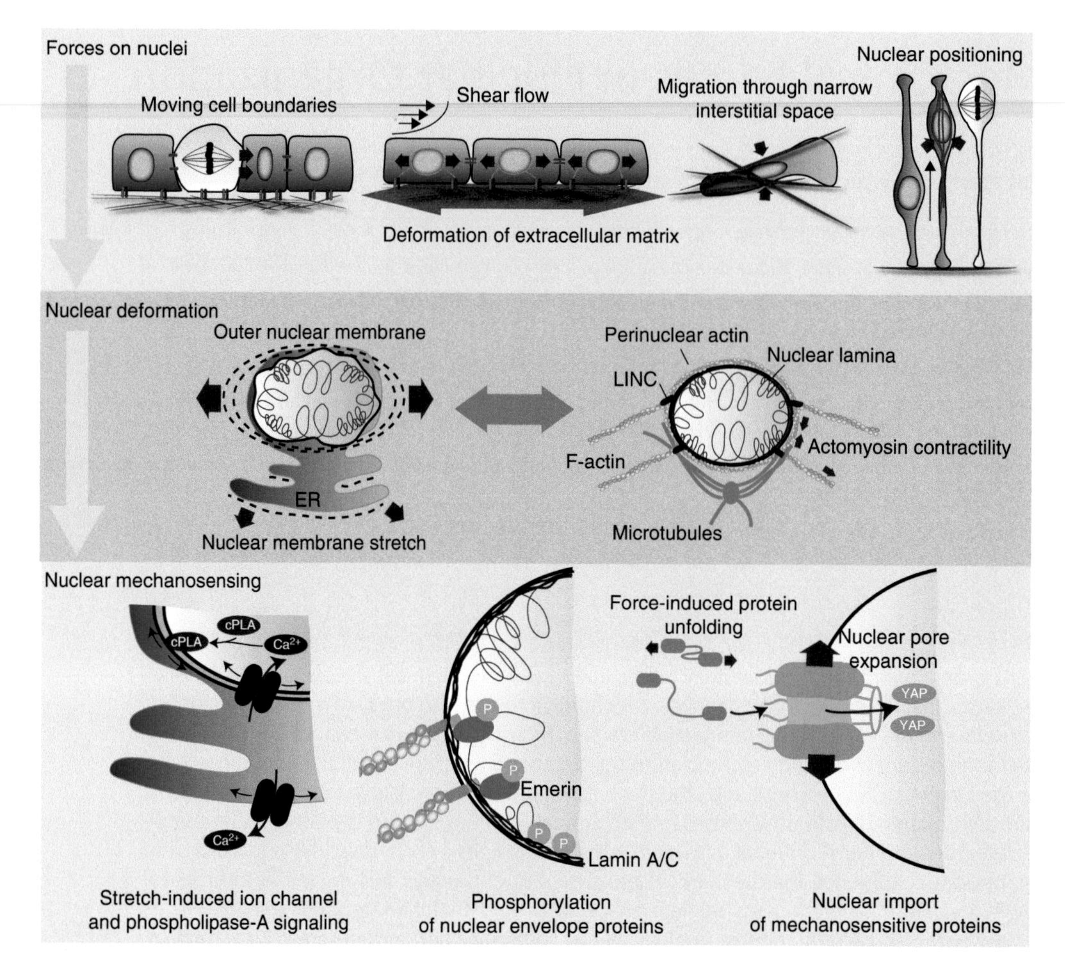

Figure 1. The nucleus as a mechanosensor. The nucleus is subject to deformation forces (red block arrows) when cells are compressed, stretched, or actively contracting during various tissue processes (*upper* panel). Nuclear deformation causes stretching of the nuclear envelope and the associated endoplasmic reticulum (ER). The amount of force transmitted, and the degree of deformation is modulated by the nuclear-cytoplasmic linkage and the stiffness of the nuclear lamina and chromatin. Perinuclear actin and microtubule networks exert contractile and compressive forces on the nucleus to counteract deformation and regulate nuclear shape (*middle* panel). Nuclear deformation and stretching of nuclear membranes activate stretch-induced ion channels causing elevated intracellular Ca^{2+} levels and incorporation of phospholipase A2 (cPLA) into the nuclear membrane (*bottom left*). Mechanical force transmitted by the linker of nucleoskeleton and cytoskeleton (LINC) complex induces phosphorylation of emerin and lamin A/C through unknown mechanisms (*bottom middle*). Nuclear deformation can lead to expansion of the nuclear pore to facilitate nuclear import. Import of mechanically unstable and mechanosensitive proteins such as YAP are induced by force (*bottom right*).

which can simultaneously trigger deformation of neighboring cells (Chen et al. 2018; Wickström and Niessen 2018). Importantly, local compression and stretching events have profound effects on cell state. Cell compression can trigger cell differentiation and extrusion, whereas stretching can prevent differentiation and promote cell division (Ruprecht et al. 2015; Le et al. 2016; Gudipaty et al. 2017; Miroshnikova et al. 2018; Lam et al. 2020). Thus, mechanical forces relay information on tissue dynamics across length scales and provide efficient means to communicate tissue needs to individual cells.

PRINCIPLES OF MECHANOTRANSDUCTION

The dynamic deformation of cells is transferred to various intracellular, nuclear, and nucleus-associated macromolecular complexes to trigger mechanochemical signaling cascades that regulate cell morphology, metabolism, and gene expression. These processes are collectively termed mechanotransduction. Nuclear mechanotransduction refers to specific events that are triggered through structures associated with or located within the nucleus and that specifically impact nuclear components, most importantly chromatin itself. Upon particularly large-scale deformations, such as immune cell infiltration and cancer cell invasion through the narrow space of the interstitium, mechanical stresses have the potential to physically damage the nucleus and the genetic material (Denais et al. 2016; Raab et al. 2016; Irianto et al. 2017). Recent studies have also identified mechanisms by which cells are able to counteract mechanical stress to prevent damage by altering nuclear mechanical properties (Stephens et al. 2019; Nava et al. 2020). This review will address both of these aspects of nuclear mechanotransduction.

Until recently, mechanosensing has been thought to occur mainly at the plasma membrane through transmembrane receptor complexes, such as integrin-based cell–matrix adhesions and cadherin-based cell–cell adhesions (Iskratsch et al. 2014). These adhesive complexes are connected with the contractile actomyosin cytoskeleton, through which cells can both exert forces on their surroundings, and sense mechanical properties or dynamic deformation of neighboring cells or the connective tissue substrate (Charras and Yap 2018; Gauthier and Roca-Cusachs 2018). Mechanical forces applied on these multiprotein adhesion complexes can be converted into biochemical signals, for example through their mechanical unfolding and subsequent activation of signaling molecules (Hu et al. 2017). Stretch-induced ion channels, such as the Piezo channels, can also be activated at the plasma membrane to trigger Ca^{2+}-dependent intracellular mechanosignaling (Murthy et al. 2017). Further downstream, remodeling of the actomyosin cytoskeleton that occurs in response to force also controls the activity of a number of signaling molecules and transcriptional regulators, of which the MRTF/SRF and YAP signaling pathways are best understood. For more detailed insights on these mechanochemical signaling pathways, which are potent regulators of cell behavior, we refer to other recent reviews (Totaro et al. 2018; Sidorenko and Vartiainen 2019).

Mechanical forces and deformation of the plasma membrane are directly transmitted to the nucleus through the contractile actomyosin cytoskeleton, which connects adhesion complexes to the nucleus and thus can cause nuclear deformation. This connection occurs through specialized receptors collectively termed the linker of nucleoskeleton and cytoskeleton (LINC) complex (Starr and Fridolfsson 2010; Luxton and Starr 2014). The LINC complex is composed of cytoplasmic components of the nesprin family members. These proteins link the actomyosin and microtubule cytoskeletons to the nuclear membrane. On the inner side of the nuclear membrane, nesprins bind SUN domain proteins, which in turn provide a direct mechanochemical link to chromatin through their association with components of the nuclear lamina and various inner nuclear membrane proteins, such as emerin, torsinA, lamina-associated polypeptide 1, and spectrin-repeat-containing proteins (Rothballer and Kutay 2013; Hao and Starr 2019).

Lamins are intermediate filament proteins that form the nuclear lamina, and have been directly implicated in establishing the mechanical properties and stability of the nucleus. Together with B-type lamins, the A-type lamins, lamin A and C, form a filamentous network underlying the inner nuclear membrane. The nuclear lamina, and specifically lamin A, are critical for determining nuclear shape, stiffness, and deformability (Liu et al. 2000). Lamin A/C levels positively correlate with nuclear stiffness, whereas reduced lamin A levels result in softer, more deformable, and fragile nuclei (Lammerding et al. 2004). Lamins also interact directly and indirectly, through chromatin-binding proteins, with specific genomic regions to generate lamina-associated domains (LADs) (Guelen et al.

2008), which are rich in silenced heterochromatin and are involved in regulating chromatin organization and gene expression (Briand and Collas 2020). Reflecting their central function in nuclear mechanics and chromatin organization, mutations in the LINC complex and lamins are linked to developmental disorders such as Emery–Dreifuss muscular dystrophy, dilated cardiomyopathy, and Hutchinson–Gilford progeria syndrome. On the cellular level, these diseases manifest with nuclear shape abnormalities and fragility, gene regulation defects, and DNA damage (Davidson and Lammerding 2014). Intriguingly, despite ubiquitous expression of the mutant proteins, the diseases specifically affect mechanically active or mechanically loaded tissues such as skeletal muscle, heart, and skin (Davidson and Lammerding 2014; Miroshnikova et al. 2019). DNA damage has also been shown to occur as a response to mechanically induced nuclear rupture, at least in some muscular dystrophies (Earle et al. 2020), highlighting the role of nuclear deformation and nuclear mechanics in tissue development and function, and the important but incompletely understood relationship between chromatin and mechanical stress.

THE NUCLEUS AS A MECHANOSENSOR

Through its unique elastic properties and compressibility, the nucleus functions as a mechanosensor by detecting dynamic changes in cell volume and by acting as a mechanical shock absorber (Dahl et al. 2004, 2005; Lomakin et al. 2020). While the plasma membrane, cytoplasm, and the associated cytoskeleton are all highly deformable, the nucleus is up to tenfold stiffer than the rest of the cell, making its deformation a potentially rate-limiting process and thus ideal for mechanosensing (Harada et al. 2014; Davidson et al. 2015; Renkawitz et al. 2019).

Sensing Nuclear Deformation

Migrating cells use the nucleus as a "ruler" to measure both the dimensions of the surrounding microenvironment and acute changes in their own volume due to extrinsic forces, such as physical confinement (Fig. 1; Renkawitz et al. 2019; Lomakin et al. 2020). For example, migrating leukocytes apply cytoskeletal forces to insert their nuclei into multiple adjacent connective tissue pores to measure pore sizes, after which the largest pore will be selected as the path for migration (Renkawitz et al. 2019). The proposed ruler function is beautifully simple: the nucleus can deform until the membrane reservoirs of the outer nuclear membrane are fully unfolded. Any further deformation will stretch and tense the outer nuclear membrane and most likely its continuous endoplasmic reticulum (ER), activating stretch-induced calcium channels and subsequent downstream effects such as actomyosin contractility (Lomakin et al. 2020). A similar nuclear membrane deformation pathway operates in epithelial monolayers in response to substrate stretch (Nava et al. 2020). Here, stretch-activated calcium signaling, activated by nuclear membrane stretching via Piezo-1 ion channel, reduces the chromatin occupancy of trimethylation of lysine 9 on histone 3 (H3K9me3) genome wide resulting in nucleus softening, which protects it from mechanical damage. Nuclear deformation in response to cell spreading or osmotic swelling of the nucleus has also been shown to trigger perinuclear calcium release, which results in elevated nuclear calcium levels that change gene expression (Itano et al. 2003). Osmotic swelling of the nucleus also triggers activation of phospholipase A_2 by promoting its hydrophobic membrane insertion, which together with raised calcium signals enhance inflammatory signaling and cell contractility (Fig. 1; Enyedi and Niethammer 2017; Lomakin et al. 2020).

Regulation of Nuclear Transport and Signaling by Mechanical Forces

Nuclear pore complexes, which are large protein complexes that traverse the nuclear envelope and regulate nuclear transport of RNA and proteins, are also sensitive to mechanical force (Donnaloja et al. 2019; Infante et al. 2019). These structures have been shown to dilate and assume a more open conformation in response to mechanical force, which enhances the import of transcriptional regulators such as YAP (Elosegui-Artola et al. 2017). Nuclear transport can further be modulated by mechanical unfolding of proteins to enhance their translocation rate into the nucleus (Infante et al. 2019), leading to changes in gene expression. Finally, phosphorylation of components of the nucleus, such as lamin A/C or emerin, is also altered by mechanical forces acting on nuclei, through molecular mechanisms that remain unclear. Lamin A/C phosphorylation is triggered by low cytoskeletal tension on soft adhesive substrates, resulting in increased lamin A/C mobility and turnover (Swift et al. 2013; Buxboim et al. 2014; Kochin et al. 2014). Emerin phosphorylation, on the other hand, is triggered by high tension, resulting in strengthening interactions of the lamina with the cytoskeleton and in mechanosensitive signaling thorough YAP (Fig. 1; Guilluy et al. 2014).

Cytoskeletal Modulation of Nuclear Mechanosensing

Direct deformation of perinuclear membranes and/or the nuclear pore by extrinsic forces can trigger downstream signaling, and thus nuclear mechanosensing may occur independent of the cytoskeleton and other mechanosensitive signaling pathways. However, the lamina and the LINC complex are likely involved in modulating "nuclear mechanosensitivity" also in cases where signaling is triggered through nuclear deformation. The LINC complex is critical for force transmission from the cytoskeleton to the nucleus and can control the amplitude of nuclear deformation in response to cell contractility (Fig. 1; Lombardi et al. 2011). Further, as lamin A is central to determining nuclear elasticity, its levels influence the degree of force-induced nuclear deformation, and help adjust nuclear membrane tension and membrane reservoir (Enyedi and Niethammer 2017; Lomakin et al. 2020; Nava et al. 2020). Importantly, cell-type-specific effects of nuclear mechanotransduction can be modulated through the lamina, especially considering marked differences in lamin A levels between cell types and differentiation states, by defining the nuclear deformation threshold that triggers downstream signaling (Nava et al. 2020).

The actin cytoskeleton, in particular perinuclear actin, is also involved in nuclear mechanosensing. Perinuclear actin is arranged in a cell-type- and cell-state-specific manner and can consist of (1) an actin cap formed by dorsal actin stress fibers that are attached to adhesions in both ends and to the nuclear envelope through the LINC complex, or (2) a perinuclear actin ring that associates with perinuclear membranes and the ER. Both structures are dynamic, and form rapidly in response to force application through Ca^{2+} or small GTPase signaling (Woroniuk et al. 2018; Wang et al. 2019), thus most likely directly responding to the stretching of nuclear membranes. The central functions of perinuclear actin are to stabilize nuclear shape and volume in the presence of mechanical stress, and to regulate mechanosensitive gene expression through controlling nuclear and cytoplasmic pools of free G-actin and MAL/SRF and YAP pathways (Fig. 1; Le et al. 2016; Wales et al. 2016; Kim et al. 2017; Shiu et al. 2018; Nava et al. 2020). Importantly, both G- and F-forms of actin also exist in the nucleus and are critical mediators of chromatin architecture, as will be discussed in the following section.

Direct Chromatin Deformation by Force

Mechanical forces transmitted via the cytoskeleton to the nucleus may directly stretch chromatin and activate gene expression. Application of ~20 kPa shear stress at the cell surface using a magnetic bead leads to chromatin displacement and rapid transcriptional activation of a set of mechanosensitive genes (Tajik et al. 2016; Sun et al. 2020). While direct force-mediated regulation of chromatin is an exciting concept, the mechanism of specificity that render certain genes sensitive to force without activating others remains a key open question.

FORCE-MEDIATED REGULATION OF TRANSCRIPTION AND CELL STATE

The nucleus lies in the direct path of cellular force sensing and transduction. Recent data show that mechanical forces transmitted to the nucleus regulate chromatin architecture and transcription to guide cell fate changes. Dynamic changes in cell and nuclear geometry are powerful regulators of cell states. This connection is highly relevant, as changes in shape/geometry occur downstream of most mechanical forces and are core to many developmental and homeostatic processes that involve fate transitions. Perhaps the clearest illustration of this phenomenon can be observed during early embryonic development where mechanical forces are critical for the establishment of the anterior–posterior axis, as well as for the sorting of the germ layers (Vining and Mooney 2017).

Simplest experimental demonstrations of the direct effects of cell geometry on chromatin and cell state come from in vitro experiments using adhesive micropatterns that can be used to force cultured cells to assume specific shapes. Among the first studies, restricting adhesive area of mammary epithelial cells to induce "rounded" shapes, which are reminiscent of the cellular shape observed in 3D versus flat 2D culture conditions, resulted in global histone deacetylation, chromatin condensation, and reduction in gene expression (Le Beyec et al. 2007). Similarly, culturing mesenchymal stem cells on anisotropic, elongated micropatterns led to increased histone deacetylase activity and decreased histone acetylation (Li et al. 2011), whereas increasing cell spread area of fibroblasts triggered increased histone acetylation and modifies gene expression patterns related to actin cytoskeleton (Jain et al. 2013). Further, restriction of adhesive area, or even a change from an isotropic circular geometry to an anisotropic elongated geometry in epidermal stem cells induced transcription of epidermal differentiation genes (Connelly et al. 2010; Miroshnikova et al. 2018).

The mechanisms by which cell geometry changes drive epigenetic and transcriptional adaptation are beginning to unravel. For instance, stretch-induced ion channels are triggered by changes in cell and nuclear shape, implicating the role of intracellular Ca^{2+} signaling in heterochromatin regulation to occur downstream of mechanical deformation (Stephens et al. 2017; Nava et al. 2020). Calcium signaling can trigger chromatin and gene expression changes through various mechanisms, where the source and the duration of the calcium signals are of critical importance. Imaging-based studies have shown that artificially elevating nuclear calcium levels by ionophore treatment induces chromatin hypercompaction (Phengchat et al. 2016). However, genome-wide chromatin accessibility studies find changes in only relatively few sites in response to artificial calcium influx, compared to the effects of biochemical signaling events that trigger elevation of intracellular Ca^{2+} (Brignall et al. 2017), indicating a more complex involvement of Ca^{2+} in chromatin regulation. While the precise mechanisms are yet to be elucidated, the nucleus is thought to have autonomy in regulating its own calcium levels (Leite et al. 2003; Bootman et al. 2009; Rodrigues et al. 2009). Going forward, it will thus be important to define the specific Ca^{2+} oscillation patterns within the cytoplasmic and nuclear compartments that are directly triggered by nuclear deformation, and to dissect the precise intranuclear effects of Ca^{2+}, considering the myriad mechanisms by which Ca^{2+} signaling impacts the core transcriptional machinery (Vilborg et al. 2016).

The actin cytoskeleton also plays an important role in communicating changes in cell geometry to the nucleus. Actin dynamically senses changes in geometry through regulation of the balance between filamentous F-actin and free, nuclear-import compatible G-actin, which communicates mechanoregulation between the cytoplasm and the nucleus. In particular, nuclear actin has important functions in chromatin organization and cellular transcriptional activity (Grosse and Vartiainen 2013). Nuclear actin is required for nuclear reprogramming of oocytes via its ability to transcriptionally reactivate the otherwise silenced Oct4 pluripotency gene (Miyamoto et al. 2011). Downstream of mechanical cues, nuclear actin regulates the mammary epithelial cell fate as these cells fluctuate between periods of quiescence and active growth in response to environmental and hormonal cues (Spencer et al. 2011). Specifically, signals from the basement membrane, which are absent in regions of active growth in the developing mammary end buds, decrease levels of nuclear G-actin, which destabilizes RNA polymerases II/III, in turn leading to reduced transcription and thereby triggering cell quiescence (Spencer et al. 2011). Further increased perinuclear F-actin polymerization, and the resulting decreased nuclear G-actin, attenuates RNA polymerase II–driven transcription in stretched epidermal stem cells, resulting in increased deposition of H3K27me3 on lineage commitment genes, thereby preventing differentiation (Le et al. 2016). Collectively, mechanical regulation of nuclear actin is implicated in determining stem cell quiescence and commitment to differentiation through its effects on global transcription levels. In general, low nuclear actin levels correlate with reduced transcription and promote quiescence/stemness, whereas high levels increase transcription and promote activated/differentiated states. Interestingly, the formation of perinuclear actin is Ca^{2+} dependent, providing an intriguing link between Ca^{2+} signaling induced by nuclear deformation and mechanotransduction pathways dependent on actin dynamics (Le et al. 2016; Wales et al. 2016; Nava et al. 2020).

Actin is not the only relevant force-responsive and force-transducing cytoskeletal element that plays a role in transcription. Compressive forces from the microtubules have also been recently shown to deform the nucleus, inducing lobulated nuclear shapes and local loss of H3K9me3-marked heterochromatin from within nuclear envelope invaginations. These local heterochromatin changes drive specific gene expression changes in human hematopoietic stem cells during their early differentiation (Biedzinski et al. 2020), although the precise mechanisms are unclear.

Thus, cellular cytoskeleton both efficiently relays cell-extrinsic mechanical forces into the nucleus through its dynamic assembly cycles, as well as applies direct mechanical forces to the nucleus to reorganize chromatin and regulate gene expression. In addition to these mechanisms, there is a potentially highly relevant, but currently understudied link between mechanotransduction, metabolism, and epigenetic regulation of gene expression. For instance, shear stress in endothelial cells activates AMP-activated protein kinase (AMPK), the master regulator of energy homeostasis (Bays et al. 2017), whereas matrix rigidity can control lipid metabolism and glycolysis (Romani et al. 2019; Park et al. 2020). Given the direct link between metabolites feeding into epigenetic reactions of methylation and acetylation (Su et al. 2016), it seems plausible that mechanical regulation of metabolism is involved in facilitating epigenetic changes caused by cell and nuclear deformation, although the precise mechanisms require further investigation. While the mechanisms by which cell and nuclear shape changes regulate chromatin and transcription are being unraveled, the question of specificity still remains. How do force-induced global changes in transcription or histone modifications lead to specific state transitions or cell-type-specific responses? We postulate that the specificity could arise from cross talk between mechanical and biochemical pathways, where mechanics, through its effect on transcription or chromatin state, plays a role in thresholding, amplifying, or attenuating the specific signals that are propagated by growth factors, hormones, and their downstream transcription factors (Fig. 2).

MECHANICAL PROPERTIES OF CHROMATIN AND THEIR IMPLICATIONS IN GENOME ORGANIZATION AND FUNCTION

To understand how chromatin is impacted by mechanical forces, it is also important to consider their influence on core chromatin processes, such as transcription factor binding, enhancer engagement and DNA damage repair, all of which are tightly linked to 3D genome structure and influenced by the physical properties of chromatin. Chromatin is a disordered, variably compacted polymer chain that can interact with itself on multiple scales (compartments, topologically associating domains, loops) as well as with the lamina or nucleolar periphery (Pombo and Dillon 2015; Szabo et al. 2019). Initial chromatin rheological studies used particle nanotracking of injected beads, whereas recent work mainly use fluorescently labeled histones or specifically tagged genomic loci, such as telomeres, coupled to mechanical manipulation by micropipette aspiration or atomic force microscopy to map chromatin movement in intact cells (Tseng et al. 2004; Lammerding 2011; Spagnol and Dahl 2016; Stephens et al. 2017, 2018; Hobson et al. 2020). Collectively, these studies have revealed several in-

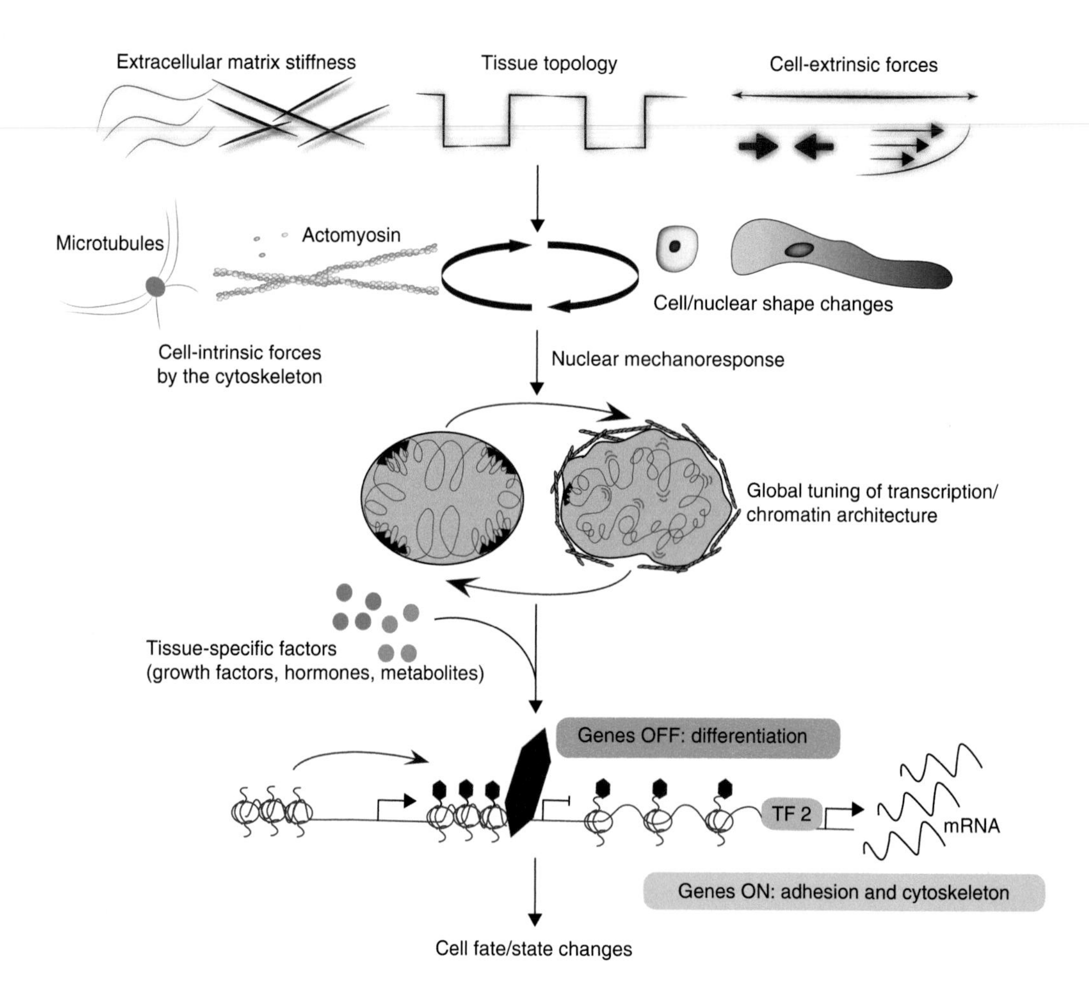

Figure 2. Force-mediated regulation of transcription and cell state. Cells sense a spectrum of cell-extrinsic forces and biophysical cues from their microenvironment, such as tissue stiffness, topology, shear stress, compression, and stretch. These cues induce cell and organelle shape changes, triggering cellular mechanosensing and mechanosignaling, facilitated by cell-intrinsic force generation and force distribution through the microtubule and actomyosin cytoskeletons. Nuclear mechanosignaling involves changes in global chromatin architecture and transcriptional activity. We propose that specificity in nuclear mechanoresponses is achieved via collaborative, bidirectional cross talk with tissue-specific factors to facilitate specific transcriptional changes to alter cell fate/state. As examples, extrinsic stretch on epidermal stem cells reduces transcription of differentiation genes but promotes transcription of cytoskeleton and adhesion genes.

triguing features of chromatin. First, instead of being a purely viscous "melt" of liquid-like polymers, the nuclear interior is actually significantly stiffer than the cytoplasm, where chromatin displays elastic and solid-like behaviors. While isolated chromosomes respond elastically in response to force (Cui and Bustamante 2000; Marko 2008; Strickfaden et al. 2020), micropipette aspiration-based measurements from intact, locally deformed nuclei show that chromatin can also flow and locally condense as would be expected of a spongy polymer from which solvent is locally squeezed out (Pajerowski et al. 2007). It is further evident that chromatin motion is spatially correlated over length scales ranging up to several micrometers (Tseng et al. 2004; Zidovska et al. 2013). Thus, based on current data, chromatin can be considered viscoelastic (i.e., displaying mechanical properties of both an elastic solid and a viscous liquid).

When mechanically deformed, both nuclei and chromatin show an elastic deformation response (i.e., they have the ability to return to their original shape after the applied force is removed) (Maeshima et al. 2019). Interestingly, the mechanical resistance of chromatin governs elastic deformations of the nucleus under small (<3 μm) extensions, while the mechanical properties of lamins govern the elastic deformation under larger extensions (Stephens et al. 2018). Thus, it seems that the nucleus consists of at least two spring-like mechanical elements, the chromatin and the nuclear lamina, which operate on distinct deformation scales. This interesting phenomenon might be related to the fact that the nuclear lamina is wrinkled, and requires a large deformation to be fully stretched, together with the inherent mechanical properties of intermediate filaments, which are easy to bend but hard to stretch (Li et al. 2015; Turgay et al. 2017; Lomakin et al. 2020).

Importantly, mechanical properties of chromatin and thus nuclei are dictated by the chromatin compaction state. Nuclei with abundant heterochromatin are rigid, whereas the nuclei with decondensed chromatin are soft (Stephens et al. 2018; Maeshima et al. 2019; Nava et al. 2020). Further, less compacted chromatin is more mobile and deformable (Booth-Gauthier et al. 2012; Spagnol and Dahl 2016; Whitefield et al. 2018; Ghosh et al. 2019; Nava et al. 2020). The association of chromatin to the nuclear lamina also modulates nuclear mechanical properties, and untethering of chromatin from the nuclear lamina increases chromatin flow, enhances nuclear deformability and alters nuclear shape (Schreiner et al. 2015; Stephens et al. 2019; Hobson et al. 2020; Nava et al. 2020).

Collectively, these mechanical properties of chromatin have profound functional implications. The spring-like behavior of chromatin points to chromatin mechanics as a central stabilizer of chromatin architecture and nuclear shape in response to extrinsic mechanical forces that deform the nucleus. The specific mechanical properties of H3K9me3-containing and lamina-associated heterochromatin, and their significant contribution to bulk nuclear stiffness, indicate that chromatin and nuclear mechanical properties can be dynamically regulated to match the force environment of a given cell/tissue. Conversely, the fact that local deformation can be transmitted over micrometers implies that a sustained localized force acting on the nucleus and/or chromatin could, in principle, specifically modulate chromatin interactions and thus gene activity. Both of these aspects will be discussed in the following sections.

MECHANICAL FORCE AND DNA DAMAGE

As discussed above, nuclei undergo substantial deformation during physiological processes such as during migration of leukocytes or cancer cells through small pores of connective tissue. Importantly, nuclear deformation is also associated with DNA damage. The first observations of deformation-induced damage come from the field of cancer cell migration, where migration of tumor cells in confined environments triggers pressurization, which in turn induces formation of local inflations of the nuclear envelope as "blebs." Blebs can eventually burst to cause leakage of nuclear factors and protrusion of chromatin into the cytoplasm (Denais et al. 2016; Raab et al. 2016; Deviri et al. 2017). The exposure of genomic DNA to cytoplasmic components such as nucleases leads to DNA damage, which is particularly enriched at these herniated sites. Consequently, depletion of cytoplasmic exonuclease TREX1 is sufficient to abolish DNA damage in disrupted nuclei in human breast cancer cells (Fig. 3; Nader et al. 2020).

Depletion of lamins increases the likelihood of nuclear envelope ruptures, consistent with their importance in stabilizing the nuclear envelope (Schreiber and Kennedy 2013). On the other hand, increasing heterochromatin to increase nuclear stiffness can decrease blebbing, further underlining the important role of chromatin as a mechanical regulator of the nucleus (Stephens et al. 2019). Cells are able to rapidly repair the ruptured nuclear envelope using the ESCRT pathway (Denais et al. 2016; Raab et al. 2016), which is also used for sealing plasma membrane ruptures and the postmitotic nuclear envelope (Jimenez et al. 2014; Olmos et al. 2015; Vietri et al. 2015). However, in cases where deforma-

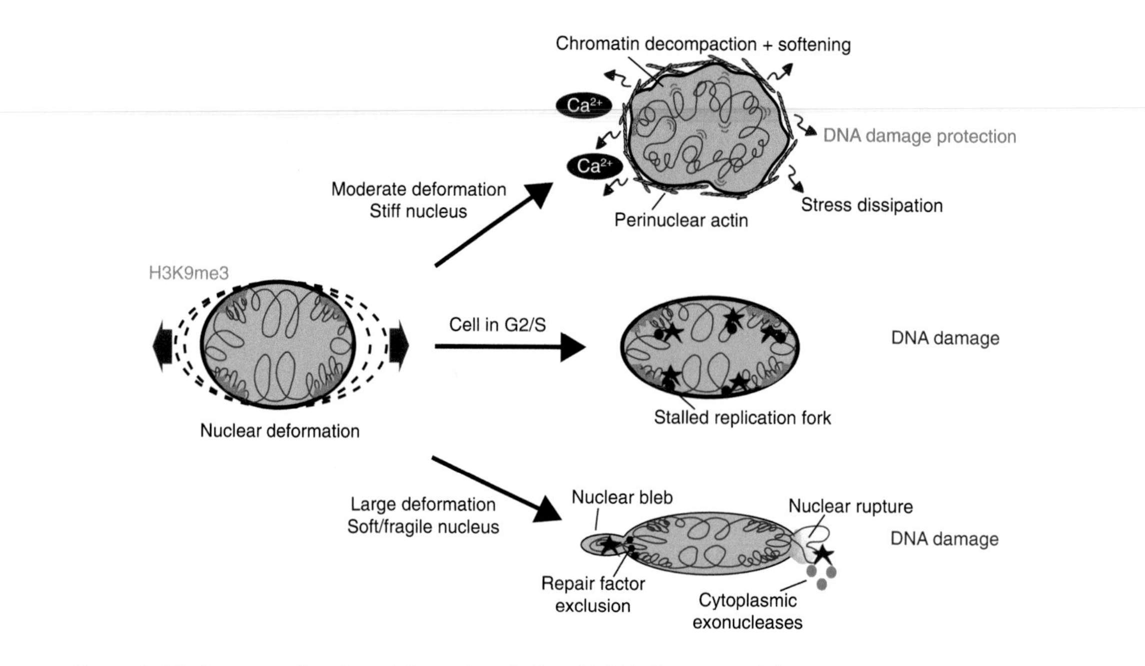

Figure 3. Mechanisms of nuclear deformation–induced DNA damage and damage protection. The impact of nuclear deformation on genome integrity depends on the mechanical properties of the nuclear lamina and on the cell cycle state of the cell. Cells with stiff nuclei respond to deformation by reducing lamina-associated H3K9me3 heterochromatin in a Ca^{2+}-dependent manner, which increases chromatin mobility to facilitate dissipation of the mechanical energy to prevent DNA damage. Simultaneous formation of a perinuclear actin ring prevents further nuclear shape and volume changes. Nuclear deformation in cells in G2/S phase cause replication fork stalling and subsequent DNA damage. Particularly large deformations in cells with soft/fragile nuclei can cause blebbing and bleb rupture. Resulting exposure of herniating chromatin to cytoplasmic exonucleases triggers DNA damage. Exclusion of repair factors from blebs may hamper damage repair.

tion occurs repeatedly and the mechanical DNA damage becomes chronic, the long-term consequences include induction of senescence in nontransformed cells and induced invasive behavior in cancer cells (Nader et al. 2020). Cells that have repeatedly migrated through narrow pores have also been shown to display cell cycle arrest, chromosome copy number alteration, and loss of heterozygosity (Irianto et al. 2017; Pfeifer et al. 2018), although the direct evidence that this is caused by the physical damage, and not selection for cells particularly capable of migration through narrow pores, is missing.

DNA damage can also occur in deformed nuclei in the absence of nuclear envelope rupture (Denais et al. 2016; Nava et al. 2020; Shah et al. 2021), and even in the case of rupture, DNA damage can occasionally be observed distant from the damage site (Irianto et al. 2017). Two main mechanisms by which this rupture-independent damage occurs have been proposed: One posits that since chromatin is elastic and thus behaves like a solid, nuclear constriction may function to expel the liquid component that contains soluble repair factors. Thus, the mechanism of damage would be the limitation of repair factors (Irianto et al. 2017). This type of factor exclusion most likely operates only upon extreme deformation. Another proposed mechanism, which could also explain observed damage in moderately deformed nuclei, is replicative stress-induced DNA damage. Both confined migration and experimental compression of the nucleus have been shown to increase replication fork stalling, triggering DNA damage at these sites (Fig. 3; Shah et al. 2021). Going forward, it will be important to dissect the mechanisms by which deformation induces replication fork stalling and to understand whether certain genomic regions are

 Cite this article as *Cold Spring Harb Perspect Biol* doi: 10.1101/cshperspect.a039685

more susceptible than others to force-induced DNA damage.

Interestingly, despite in vitro studies showing that nuclear deformation can also induce DNA damage in nontransformed cells, tissues that undergo large-scale deformation during normal physiology, such as muscle, heart, lung, and skin, do not display frequent DNA damage. This indicates that mechanisms exist to counteract nuclear deformation as well as deformation-induced DNA damage. Recent studies show that cells are able to dissipate mechanical stress both on the tissue scale and at the chromatin scale to prevent nuclear deformation and damage. Skin epithelial cell monolayers exposed to uniaxial stretch display moderate (1–3 µm scale) nuclear deformation in the direction of stretch. Despite this deformation, no DNA damage is induced. Instead, cells respond by decreasing levels of H3K9me3, in particular in the proximity of the nuclear lamina. This reduction in H3K9me3 has no immediate transcriptional consequences but renders the nucleus and chromatin more elastic, allowing dissipation of mechanical energy (Nava et al. 2020). Only if this chromatin remodeling is prevented, DNA becomes damaged, indicating that decreased H3K9me3 and chromatin softening is a mechanism to protect the genome from damage (Fig. 3). Interestingly, when the stretch persists over hours, the entire cell monolayer or even an intact skin tissue explant, aligns according to the direction of stretch, redistributing mechanical energy within the tissue to prevent nuclear deformation. This rearrangement allows chromatin to resume its steady-state architecture for more sustainable long-term mechanoprotection (Nava et al. 2020). The precise mechanism by which DNA is damaged in this scenario is unclear, but it is likely related to torsional stress and replication fork stalling. This notion is supported by studies on long-term (>6 h) biaxial stretch, where cells cannot reorient themselves due to the absence of a single direction of strain. This results not only in reduced H3K9me3 and repositioning of chromatin away from the lamina but also in global transcriptional repression (Le et al. 2016), which could collectively reduce torsional stress on chromatin (Nelson 1999).

CONCLUDING REMARKS

Mechanical forces reorganize chromatin and impact nuclear architecture and mechanics by engaging signaling cascades capable of modifying the levels and distribution of charged entities such as Ca^{2+} ions or ATP, and by modifying chromatin accessibility, compaction, and association with nuclear landmarks such as the lamina. With major advances in our understanding of the mechanisms by which mechanical forces mediate transcriptional regulation, it becomes increasingly important to study how forces organize chromatin and how specificity in chromatin rearrangements is achieved. Further technological developments will be needed, such as integrated systems capable of exerting calibrated forces on specific genomic loci or nuclear structures, while measuring both chromatin state/mechanics and transcriptional output in real time with high spatiotemporal resolution.

Major advances have been made in understanding of mechanotransduction into the cell nucleus to directly act on gene regulation and chromatin functions. A large body of in vitro experiments in multiple systems have demonstrated that dynamic, bidirectional cross talk between cell and nuclear mechanics and chromatin state modulate stem cell differentiation. The next step will be to construct genetic or otherwise manipulatable animal and organ models to challenge and refine these mechanisms in complex, multicellular tissues. Besides development and homeostasis, nucleomechanical regulation of cell states is likely relevant more broadly during processes such as aging, which leads to modified tissue mechanics, as well as many diseases, such as cancers, where abnormal nuclear mechanics has already been shown to be beneficial for aggression within the primary site and during invasion across a spectrum of tissues (Levental et al. 2009; Denais and Lammerding 2014; Bell and Lammerding 2016; Miroshnikova et al. 2016; Emon et al. 2018; Pfeifer et al. 2019). A better understanding of the dynamic relationship between cell-extrinsic forces and their effect on nuclear shape, mechanics, chromatin organization, and cell state during disease onset and evolution, is potentially of high clinical relevance in diagnosis and even treatment. To that effect, a number of high throughout imaging and screening efforts have begun to unravel the mechanisms regulating nuclear shape and mechanics in homeostasis and disease progression (Kume et al. 2017; Hwang et al. 2019; Tamashunas et al. 2020), but their clinical relevance remains to be fully explored.

ACKNOWLEDGMENTS

We apologize to all investigators whose work could not be cited due to space constraints, and thank Clementine Villeneuve and Matthias Rübsam for feedback on the manuscript. Work on nuclear mechanotransduction in the Wickström laboratory is supported by the Helsinki Institute of Life Science, Wihuri Research Institute, Academy of Finland, Juselius Foundation, and European Research Council (ERC) under the European Union's Horizon 2020 research and innovation programme (Grant agreement 770877 - STEMpop). Y.A.M. is the recipient of the EMBO Long-Term fellowship ALTF 728-2017 and Human Frontier Science Program fellowship LT000861/2018.

REFERENCES

Bays JL, Campbell HK, Heidema C, Sebbagh M, Demali KA. 2017. Linking E-cadherin mechanotransduction to cell metabolism through force-mediated activation of AMPK. *Nat Cell Biol* **19:** 724–731. doi:10.1038/ncb3537

Bell ES, Lammerding J. 2016. Causes and consequences of nuclear envelope alterations in tumour progression. *Eur J Cell Biol* **95:** 449–464. doi:10.1016/j.ejcb.2016.06.007

Biedzinski S, Agsu C, Vianay B, Delord M, Blanchoin L, Larghero J, Faivre LThéry M, Brunet S. 2020. Microtubules control nuclear shape and gene expression during stages of hematopoietic differentiation. *EMBO J* **39:** e103957. doi:10.15252/embj.2019103957

Booth-Gauthier EA, Alcoser TA, Yang G, Dahl KN. 2012. Force-induced changes in subnuclear movement and rheology. *Biophys J* **103:** 2423–2431. doi:10.1016/j.bpj.2012.10.039

Bootman MD, Fearnley C, Smyrnias I, MacDonald F, Roderick HL. 2009. An update on nuclear calcium signalling. *J Cell Sci* **122:** 2337–2350. doi:10.1242/jcs.028100

Briand N, Collas P. 2020. Lamina-associated domains: peripheral matters and internal affairs. *Genome Biol* **21:** 85. doi:10.1186/s13059-020-02003-5

Brignall R, Cauchy P, Bevington SL, Gorman B, Pisco AO, Bagnall J, Boddington C, Rowe W, England H, Rich K, et al. 2017. Integration of kinase and calcium signaling at the level of chromatin underlies inducible gene activation in T cells. *J Immunol* **199:** 2652–2667. doi:10.4049/jimmunol.1602033

Buxboim A, Swift J, Irianto J, Spinler KR, Dingal PCDP, Athirasala A, Kao YRC, Cho S, Harada T, Shin JW, et al. 2014. Matrix elasticity regulates lamin-A,C phosphorylation and turnover with feedback to actomyosin. *Curr Biol* **24:** 1909–1917. doi:10.1016/j.cub.2014.07.001

Charras G, Yap AS. 2018. Tensile forces and mechanotransduction at cell–cell junctions. *Curr Biol* **28:** R445–R457. doi:10.1016/j.cub.2018.02.003

Chen T, Saw TB, Mège RM, Ladoux B. 2018. Mechanical forces in cell monolayers. *J Cell Sci* **131:** jcs218156. doi:10.1242/jcs.218156

Connelly JT, Gautrot JE, Trappmann B, Tan DWM, Donati G, Huck WTS, Watt FM. 2010. Actin and serum response factor transduce physical cues from the microenvironment to regulate epidermal stem cell fate decisions. *Nat Cell Biol* **12:** 711–718. doi:10.1038/ncb2074

Cui Y, Bustamante C. 2000. Pulling a single chromatin fiber reveals the forces that maintain its higher-order structure. *Proc Natl Acad Sci* **97:** 127–132. doi:10.1073/pnas.97.1.127

Dahl KN, Kahn SM, Wilson KL, Discher DE. 2004. The nuclear envelope lamina network has elasticity and a compressibility limit suggestive of a molecular shock absorber. *J Cell Sci* **117:** 4779–4786. doi:10.1242/jcs.01357

Dahl KN, Engler AJ, Pajerowski JD, Discher DE. 2005. Power-law rheology of isolated nuclei with deformation mapping of nuclear substructures. *Biophys J* **89:** 2855–2864. doi:10.1529/biophysj.105.062554

Davidson PM, Lammerding J. 2014. Broken nuclei—lamins, nuclear mechanics, and disease. *Trends Cell Biol* **24:** 247–256. doi:10.1016/j.tcb.2013.11.004

Davidson PM, Sliz J, Isermann P, Denais C, Lammerding J. 2015. Design of a microfluidic device to quantify dynamic intra-nuclear deformation during cell migration through confining environments. *Integr Biol (Camb)* **7:** 1534–1546. doi:10.1039/C5IB00200A

Denais C, Lammerding J. 2014. Nuclear mechanics in cancer. *Adv Exp Med Biol* **773:** 435–470. doi:10.1007/978-1-4899-8032-8_20

Denais CM, Gilbert RM, Isermann P. McGregor AL, te Lindert M, Weigelin B, Davidson PM, Friedl P, Wolf K, Lammerding J. 2016. Nuclear envelope rupture and repair during cancer cell migration. *Science* **352:** 353–358. doi:10.1126/science.aad7297

Deviri D, Discher DE, Safran SA. 2017. Rupture dynamics and chromatin herniation in deformed nuclei. *Biophys J* **113:** 1060–1071. doi:10.1016/j.bpj.2017.07.014

Donnaloja F, Jacchetti E, Soncini M, Raimondi MT. 2019. Mechanosensing at the nuclear envelope by nuclear pore complex stretch activation and its effect in physiology and pathology. *Front Physiol* **10:** 896. doi:10.3389/fphys.2019.00896

Earle AJ, Kirby TJ, Fedorchak GR, Isermann P, Patel J, Iruvanti S, Moore SA, Bonne G, Wallrath LL, Lammerding J. 2020. Mutant lamins cause nuclear envelope rupture and DNA damage in skeletal muscle cells. *Nat Mater* **19:** 464–473. doi:10.1038/s41563-019-0563-5

Elosegui-Artola A, Andreu I, Beedle AEM, Lezamiz A, Uroz M, Kosmalska AJ, Oria R, Kechagia JZ, Rico-Lastres P, Le Roux AL, et al. 2017. Force triggers YAP nuclear entry by regulating transport across nuclear pores. *Cell* **171:** 1397–1410.e14. doi:10.1016/j.cell.2017.10.008

Emon B, Bauer J, Jain Y, Jung B, Saif T. 2018. Biophysics of tumor microenvironment and cancer metastasis—a mini review. *Comput Struct Biotechnol J* **16:** 279–287. doi:10.1016/j.csbj.2018.07.003

Enyedi B, Niethammer P. 2017. Nuclear membrane stretch and its role in mechanotransduction. *Nucleus* **8:** 156–161. doi:10.1080/19491034.2016.1263411

Friedl P, Wolf K, Lammerding J. 2011. Nuclear mechanics during cell migration. *Curr Opin Cell Biol* **23:** 55–64. doi:10.1016/j.ceb.2010.10.015

Gauthier NC, Roca-Cusachs P. 2018. Mechanosensing at integrin-mediated cell–matrix adhesions: from molecular to integrated mechanisms. *Curr Opin Cell Biol* **50:** 20–26. doi:10.1016/j.ceb.2017.12.014

Ghosh S, Seelbinder B, Henderson JT, Watts RD, Scott AK, Veress AI, Neu CP. 2019. Deformation microscopy for dynamic intracellular and intranuclear mapping of mechanics with high spatiotemporal resolution. *Cell Rep* **27:** 1607–1620.e4. doi:10.1016/j.celrep.2019.04.009

Grosse R, Vartiainen MK. 2013. To be or not to be assembled: progressing into nuclear actin filaments. *Nat Rev Mol Cell Biol* **14:** 693–697. doi:10.1038/nrm3681

Gudipaty SA, Lindblom J, Loftus PD, Redd MJ, Edes K, Davey CF, Krishnegowda V, Rosenblatt J. 2017. Mechanical stretch triggers rapid epithelial cell division through Piezo1. *Nature* **543:** 118–121. doi:10.1038/nature21407

Guelen L, Pagie L, Brasset E, Meuleman W, Faza MB, Talhout W, Eussen BH, De Klein A, Wessels L, De Laat W, et al. 2008. Domain organization of human chromosomes revealed by mapping of nuclear lamina interactions. *Nature* **453:** 948–951. doi:10.1038/nature06947

Guilluy C, Osborne LD, Van Landeghem L, Sharek L, Superfine R, Garcia-Mata R, Burridge K. 2014. Isolated nuclei adapt to force and reveal a mechanotransduction pathway in the nucleus. *Nat Cell Biol* **16:** 376–381. doi:10.1038/ncb2927

Hao H, Starr DA. 2019. SUN/KASH interactions facilitate force transmission across the nuclear envelope. *Nucleus* **10:** 73–80. doi:10.1080/19491034.2019.1595313

Harada T, Swift J, Irianto J, Shin JW, Spinler KR, Athirasala A, Diegmiller R, Dingal PCDP, Ivanovska IL, Discher DE. 2014. Nuclear lamin stiffness is a barrier to 3D migration, but softness can limit survival. *J Cell Biol* **204:** 669–682. doi:10.1083/jcb.201308029

Hobson CM, Kern M, O'Brien ET, Stephens AD, Falvo MR, Superfine R. 2020. Correlating nuclear morphology and external force with combined atomic force microscopy and light sheet imaging separates roles of chromatin and lamin A/C in nuclear mechanics. *Mol Biol Cell* **31:** 1788–1801. doi:10.1091/mbc.E20-01-0073

Hu X, Margadant FM, Yao M, Sheetz MP. 2017. Molecular stretching modulates mechanosensing pathways. *Protein Sci* **26:** 1337–1351. doi:10.1002/pro.3188

Hwang S, Williams JF, Kneissig M, Lioudyno M, Rivera I, Helguera P, Busciglio J, Storchova Z, King MC, Torres EM. 2019. Suppressing aneuploidy-associated phenotypes improves the fitness of trisomy 21 cells. *Cell Rep* **29:** 2473–2488.e5. doi:10.1016/j.celrep.2019.10.059

Infante E, Stannard A, Board SJ, Rico-Lastres P, Rostkova E, Beedle AEM, Lezamiz A, Wang YJ, Gulaidi Breen S, Panagaki F, et al. 2019. The mechanical stability of proteins regulates their translocation rate into the cell nucleus. *Nat Phys* **15:** 973–981. doi:10.1038/s41567-019-0551-3

Irianto J, Xia Y, Pfeifer CR, Athirasala A, Ji J, Alvey C, Tewari M, Bennett RR, Harding SM, Liu AJ, et al. 2017. DNA damage follows repair factor depletion and portends genome variation in cancer cells after pore migration. *Curr Biol* **27:** 210–223. doi:10.1016/j.cub.2016.11.049

Iskratsch T, Wolfenson H, Sheetz MP. 2014. Appreciating force and shape—the rise of mechanotransduction in cell biology. *Nat Rev Mol Cell Biol* **15:** 825–833. doi:10.1038/nrm3903

Itano N, Okamoto Si, Zhang D, Lipton SA, Ruoslahti E. 2003. Cell spreading controls endoplasmic and nuclear calcium: a physical gene regulation pathway from the cell surface to the nucleus. *Proc Natl Acad Sci* **100:** 5181–5186. doi:10.1073/pnas.0531397100

Jain N, Iyer KV, Kumar A, Shivashankar GV. 2013. Cell geometric constraints induce modular gene-expression patterns via redistribution of HDAC3 regulated by actomyosin contractility. *Proc Natl Acad Sci* **110:** 11349–11354. doi:10.1073/pnas.1300801110

Jimenez AJ, Maiuri P, Lafaurie-Janvore J, Divoux S, Piel M, Perez F. 2014. ESCRT machinery is required for plasma membrane repair. *Science* **343:** 1247136. doi:10.1126/science.1247136

Kim KD, Bae S, Capece T, Nedelkovska H, De Rubio RG, Smrcka AV, Jun CD, Jung W, Park B, Kim TI, et al. 2017. Targeted calcium influx boosts cytotoxic T lymphocyte function in the tumour microenvironment. *Nat Commun* **8:** 1–10. doi:10.1038/ncomms15365

Kochin V, Shimi T, Torvaldson E, Adam SA, Goldman A, Pack CG, Melo-Cardenas J, Imanishi SY, Goldman RD, Eriksson JE. 2014. Interphase phosphorylation of lamin A. *J Cell Sci* **127:** 2683–2696. doi:10.1242/jcs.141820

Kume K, Cantwell H, Neumann FR, Jones AW, Snijders AP, Nurse P. 2017. A systematic genomic screen implicates nucleocytoplasmic transport and membrane growth in nuclear size control. *PLoS Genet* **13:** e1006767. doi:10.1371/journal.pgen.1006767

Lam MSY, Lisica A, Ramkumar N, Hunter G, Mao Y, Charras G, Baum B. 2020. Isotropic myosin-generated tissue tension is required for the dynamic orientation of the mitotic spindle. *Mol Biol Cell* **31:** 1370–1379. doi:10.1091/mbc.E19-09-0545

Lammerding J. 2011. Mechanics of the nucleus. *Compr Physiol* **1:** 783–807.

Lammerding J, Schulze PC, Takahashi T, Kozlov S, Sullivan T, Kamm RD, Stewart CL, Lee RT. 2004. Lamin A/C deficiency causes defective nuclear mechanics and mechanotransduction. *J Clin Invest* **113:** 370–378. doi:10.1172/JCI200419670

Le HQ, Ghatak S, Yeung CYC, Tellkamp F, Günschmann C, Dieterich C, Yeroslaviz A, Habermann B, Pombo A, Niessen CM, et al. 2016. Mechanical regulation of transcription controls Polycomb-mediated gene silencing during lineage commitment. *Nat Cell Biol* **18:** 864–875. doi:10.1038/ncb3387

Le Beyec J, Xu R, Lee SY, Nelson CM, Rizki A, Alcaraz J, Bissell MJ. 2007. Cell shape regulates global histone acetylation in human mammary epithelial cells. *Exp Cell Res* **313:** 3066–3075. doi:10.1016/j.yexcr.2007.04.022

Leite MF, Thrower EC, Echevarria W, Koulen P, Hirata K, Bennett AM, Ehrlich BE, Nathanson MH. 2003. Nuclear and cytosolic calcium are regulated independently. *Proc Natl Acad Sci* **100:** 2975–2980. doi:10.1073/pnas.0536590100

Levental KR, Yu H, Kass L, Lakins JN, Egeblad M, Erler JT, Fong SFT, Csiszar K, Giaccia A, Weninger W, et al. 2009. Matrix crosslinking forces tumor progression by enhancing integrin signaling. *Cell* **139:** 891–906. doi:10.1016/j.cell.2009.10.027

Li Y, Chu JS, Kurpinski K, Li X, Bautista DM, Yang L, Paul Sung KL, Li S. 2011. Biophysical regulation of histone acetylation in mesenchymal stem cells. *Biophys J* **100:** 1902–1909. doi:10.1016/j.bpj.2011.03.008

Li Y, Lovett D, Zhang Q, Neelam S, Kuchibhotla RA, Zhu R, Gundersen GG, Lele TP, Dickinson RB. 2015. Moving cell boundaries drive nuclear shaping during cell spreading. *Biophys J* **109:** 670–686. doi:10.1016/j.bpj.2015.07.006

Liu J, Ben-Shahar TR, Riemer D, Treinin M, Spann P, Weber K, Fire A, Gruenbaum Y. 2000. Essential roles for *Caenorhabditis elegans* lamin gene in nuclear organization, cell cycle progression, and spatial organization of nuclear pore complexes. *Mol Biol Cell* **11:** 3937–3947. doi:10.1091/mbc.11.11.3937

Lomakin AJ, Cattin CJ, Cuvelier D, Alraies Z, Molina M, Nader GPF, Srivastava N, Sáez PJ, Garcia-Arcos JM, Zhitnyak IY, et al. 2020. The nucleus acts as a ruler tailoring cell responses to spatial constraints. *Science* **370:** eaba2894. doi:10.1126/science.aba2894

Lombardi ML, Jaalouk DE, Shanahan CM, Burke B, Roux KJ, Lammerding J. 2011. The interaction between nesprins and sun proteins at the nuclear envelope is critical for force transmission between the nucleus and cytoskeleton. *J Biol Chem* **286:** 26743–26753. doi:10.1074/jbc.M111.233700

Luxton GG, Starr DA. 2014. KASHing up with the nucleus: novel functional roles of KASH proteins at the cytoplasmic surface of the nucleus. *Curr Opin Cell Biol* **28:** 69–75. doi:10.1016/j.ceb.2014.03.002

Maeshima K, Ide S, Babokhov M. 2019. Dynamic chromatin organization without the 30-nm fiber. *Curr Opin Cell Biol* **58:** 95–104. doi:10.1016/j.ceb.2019.02.003

Marko JF. 2008. Micromechanical studies of mitotic chromosomes. *Chromosom Res* **16:** 469–497. doi:10.1007/s10577-008-1233-7

Miroshnikova YA, Mouw JK, Barnes JM, Pickup MW, Lakins JN, Kim Y, Lobo K, Persson AI, Reis GF, McKnight TR, et al. 2016. Tissue mechanics promote IDH1-dependent HIF1α–tenascin C feedback to regulate glioblastoma aggression. *Nat Cell Biol* **18:** 1336–1345. doi:10.1038/ncb3429

Miroshnikova YA, Le HQ, Schneider D, Thalheim T, Rübsam M, Bremicker N, Polleux J, Kamprad N, Tarantola M, Wang I, et al. 2018. Adhesion forces and cortical tension couple cell proliferation and differentiation to drive epidermal stratification. *Nat Cell Biol* **20:** 69–80. doi:10.1038/s41556-017-0005-z

Miroshnikova YA, Hammesfahr T, Wickström SA. 2019. Cell biology and mechanopathology of laminopathic cardiomyopathies. *J. Cell Biol* **218:** 393–394. doi:10.1083/jcb.201805079

Miyamoto K, Pasque V, Jullien J, Gurdon JB. 2011. Nuclear actin polymerization is required for transcriptional reprogramming of Oct4 by oocytes. *Genes Dev* **25:** 946–958. doi:10.1101/gad.615211

Murthy SE, Dubin AE, Patapoutian A. 2017. Piezos thrive under pressure: mechanically activated ion channels in health and disease. *Nat Rev Mol Cell Biol* **18:** 771–783. doi:10.1038/nrm.2017.92

Nader GP, Agüera-Gonzalez S, Routet F, Gratia M, Maurin M, Cancila V, Cadart C, Gentili M, Yamada A, Lodillinsky C, et al. 2020. Compromised nuclear envelope integrity drives tumor cell invasion. bioRxiv doi:10.1101/2020.05.22.110122

Nava MM, Miroshnikova YA, Biggs LC, Whitefield DB, Metge F, Boucas J, Vihinen H, Jokitalo E, Li X, García Arcos JM, et al. 2020. Heterochromatin-driven nuclear softening protects the genome against mechanical stress-induced damage. *Cell* **181:** 800–817.e22. doi:10.1016/j.cell.2020.03.052

Nelson P. 1999. Transport of torsional stress in DNA. *Proc Natl Acad Sci* **96:** 14342–14347. doi:10.1073/pnas.96.25.14342

Olmos Y, Hodgson L, Mantell J, Verkade P, Carlton JG. 2015. ESCRT-III controls nuclear envelope reformation. *Nature* **522:** 236–239. doi:10.1038/nature14503

Pajerowski JD, Dahl KN, Zhong FL, Sammak PJ, Discher DE. 2007. Physical plasticity of the nucleus in stem cell differentiation. *Proc Natl Acad Sci* **104:** 15619–15624. doi:10.1073/pnas.0702576104

Park JS, Burckhardt CJ, Lazcano R, Solis LM, Isogai T, Li L, Chen CS, Gao B, Minna JD, Bachoo R, et al. 2020. Mechanical regulation of glycolysis via cytoskeleton architecture. *Nature* **578:** 621–626. doi:10.1038/s41586-020-1998-1

Pfeifer CR, Xia Y, Zhu K, Liu D, Irianto J, Morales García VM, Santiago Millán LM, Niese B, Harding S, Deviri D, et al. 2018. Constricted migration increases DNA damage and independently represses cell cycle. *Mol Biol Cell* **29:** 1948–1962. doi:10.1091/mbc.E18-02-0079

Pfeifer CR, Irianto J, Discher DE. 2019. Nuclear mechanics and cancer cell migration. *Adv Exp Med Biol* **1146:** 117–130. doi:10.1007/978-3-030-17593-1_8

Phengchat R, Takata H, Morii K, Inada N, Murakoshi H, Uchiyama S, Fukui K. 2016. Calcium ions function as a booster of chromosome condensation. *Sci Rep* **6:** 1–10. doi:10.1038/srep38281

Pombo A, Dillon N. 2015. Three-dimensional genome architecture: players and mechanisms. *Nat Rev Mol Cell Biol* **16:** 245–257. doi:10.1038/nrm3965

Raab M, Gentili M, de Belly H, Thiam HR, Vargas P, Jimenez AJ, Lautenschlaeger F, Voituriez R, Lennon-Duménil AM, Manel N, et al. 2016. ESCRT III repairs nuclear envelope ruptures during cell migration to limit DNA damage and cell death. *Science* **352:** 359–362. doi:10.1126/science.aad7611

Renkawitz J, Kopf A, Stopp J, de Vries I, Driscoll MK, Merrin J, Hauschild R, Welf ES, Danuser G, Fiolka R, et al. 2019. Nuclear positioning facilitates amoeboid migration along the path of least resistance. *Nature* **568:** 546–550. doi:10.1038/s41586-019-1087-5

Rodrigues MA, Gomes DA, Nathanson MH, Leite MF. 2009. Nuclear calcium signaling: a cell within a cell. *Braz J Med Biol Res* **42:** 17–20. doi:10.1590/S0100-879X2008005000050

Romani P, Brian I, Santinon G, Pocaterra A, Audano M, Pedretti S, Mathieu S, Forcato M, Bicciato S, Manneville JB, et al. 2019. Extracellular matrix mechanical cues regulate lipid metabolism through Lipin-1 and SREBP. *Nat Cell Biol* **21:** 338–347. doi:10.1038/s41556-018-0270-5

Rothballer A, Kutay U. 2013. The diverse functional LINCs of the nuclear envelope to the cytoskeleton and chromatin. *Chromosoma* **122:** 415–429. doi:10.1007/s00412-013-0417-x

Ruprecht V, Wieser S, Callan-Jones A, Smutny M, Morita H, Sako K, Barone V, Ritsch-Marte M, Sixt M, Voituriez R, et al. 2015. Cortical contractility triggers a stochastic switch to fast amoeboid cell motility. *Cell* **160:** 673–685. doi:10.1016/j.cell.2015.01.008

Schreiber KH, Kennedy BK. 2013. When lamins go bad: nuclear structure and disease. *Cell* **152:** 1365–1375. doi:10.1016/j.cell.2013.02.015

Schreiner SM, Koo PK, Zhao Y, Mochrie SGJ, King MC. 2015. The tethering of chromatin to the nuclear envelope supports nuclear mechanics. *Nat Commun* **6:** 7159. doi:10.1038/ncomms8159

Shah P, Hobson CM, Cheng S, Colville MJ, Paszek MJ, Superfine R, Lammerding J. 2021. Nuclear deformation causes DNA damage by increasing replication stress. *Curr Biol* **31:** 753–765.e6. doi:10.1016/j.cub.2020.11.037

Shiu JY, Aires L, Lin Z, Vogel V. 2018. Nanopillar force measurements reveal actin-cap-mediated YAP mechanotransduction. *Nat Cell Biol* **20:** 262–271. doi:10.1038/s41556-017-0030-y

Sidorenko E, Vartiainen MK. 2019. Nucleoskeletal regulation of transcription: actin on MRTF. *Exp Biol Med* **244:** 1372–1381. doi:10.1177/1535370219854669

Spagnol ST, Dahl KN. 2016. Spatially resolved quantification of chromatin condensation through differential local rheology in cell nuclei fluorescence lifetime imaging. *PLoS ONE* **11:** e0146244. doi:10.1371/journal.pone.0146244

Spencer VA, Costes S, Inman JL, Xu R, Chen J, Hendzel MJ, Bissell MJ. 2011. Depletion of nuclear actin is a key mediator of quiescence in epithelial cells. *J Cell Sci* **124:** 123–132.

Starr DA, Fridolfsson HN. 2010. Interactions between nuclei and the cytoskeleton are mediated by SUN-KASH nuclear-envelope bridges. *Annu Rev Cell Dev Biol* **26:** 421–444. doi:10.1146/annurev-cellbio-100109-104037

Stephens AD, Banigan EJ, Adam SA, Goldman RD, Marko JF. 2017. Chromatin and lamin A determine two different mechanical response regimes of the cell nucleus. *Mol Biol Cell* **28:** 1984–1996. doi:10.1091/mbc.e16-09-0653

Stephens AD, Liu PZ, Banigan EJ, Almassalha LM, Backman Vadim, Adam SA, Goldman RD, Marko JF. 2018. Chromatin histone modifications and rigidity affect nuclear morphology independent of lamins. *Mol Biol Cell* **29:** 220–233. doi:10.1091/mbc.E17-06-0410

Stephens AD, Liu PZ, Kandula V, Chen H, Almassalha LM, Herman C, Backman V, O'Halloran T, Adam SA, Goldman RD, et al. 2019. Physicochemical mechanotransduction alters nuclear shape and mechanics via heterochromatin formation. *Mol Biol Cell* **30:** 2320–2330. doi:10.1091/mbc.E19-05-0286

Strickfaden H, Tolsma TO, Sharma A, Underhill DA, Hansen JC, Hendzel MJ. 2020. Condensed chromatin behaves like a solid on the mesoscale in vitro and in living cells. *Cell* **183:** 1772–1784.e13. doi:10.1016/j.cell.2020.11.027

Su X, Wellen KE, Rabinowitz JD. 2016. Metabolic control of methylation and acetylation. *Curr Opin Chem Biol* **30:** 52–60. doi:10.1016/j.cbpa.2015.10.030

Sun J, Chen J, Mohagheghian E, Wang N. 2020. Force-induced gene up-regulation does not follow the weak power law but depends on H3K9 demethylation. *Sci Adv* **6:** eaay9095. doi:10.1126/sciadv.aay9095

Swift J, Ivanovska IL, Buxboim A, Harada T, Dingal PCDP, Pinter J, Pajerowski JD, Spinler KR, Shin JW, Tewari M, et al. 2013. Nuclear lamin-A scales with tissue stiffness and enhances matrix-directed differentiation. *Science* **341:** 1240104. doi:10.1126/science.1240104

Szabo Q, Bantignies F, Cavalli G. 2019. Principles of genome folding into topologically associating domains. *Sci Adv* **5:** eaaw1668. doi:10.1126/sciadv.aaw1668

Tajik A, Zhang Y, Wei F, Sun J, Jia Q, Zhou W, Singh R, Khanna N, Belmont AS, Wang N. 2016. Transcription upregulation via force-induced direct stretching of chromatin. *Nat Mater* **15:** 1287–1296. doi:10.1038/nmat4729

Tamashunas AC, Tocco VJ, Matthews J, Zhang Q, Atanasova KR, Paschall L, Pathak S, Ratnayake R, Stephens AD, Luesch H, et al. 2020. High-throughput gene screen reveals modulators of nuclear shape. *Mol Biol Cell* **31:** 1392–1402. doi:10.1091/mbc.E19-09-0520

Totaro A, Panciera T, Piccolo S. 2018. YAP/TAZ upstream signals and downstream responses. *Nat Cell Biol* **20:** 888–899. doi:10.1038/s41556-018-0142-z

Tseng Y, Lee JSH, Kole TP, Jiang I, Wirtz D. 2004. Micro-organization and visco-elasticity of the interphase nucleus revealed by particle nanotracking. *J Cell Sci* **117:** 2159–2167. doi:10.1242/jcs.01073

Turgay Y, Eibauer M, Goldman AE, Shimi T, Khayat M, Ben-Harush K, Dubrovsky-Gaupp A, Sapra KT, Goldman RD, Medalia O. 2017. The molecular architecture of lamins in somatic cells. *Nature* **543:** 261–264. doi:10.1038/nature21382

Vietri M, Schink KO, Campsteijn C, Wegner CS, Schultz SW, Christ L, Thoresen SB, Brech A, Raiborg C, Stenmark H. 2015. Spastin and ESCRT-III coordinate mitotic spindle disassembly and nuclear envelope sealing. *Nature* **522:** 231–235. doi:10.1038/nature14408

Vilborg A, Passarelli M, Steitz J. 2016. Calcium signaling and transcription: elongation, DoGs, and eRNAs. *Recept Clin Investig* **3:** e1169.

Vining KH, Mooney DJ. 2017. Mechanical forces direct stem cell behaviour in development and regeneration. *Nat Rev Mol Cell Biol* **18:** 728–742. doi:10.1038/nrm.2017.108

Wales P, Schuberth CE, Aufschnaiter R, Fels J, García-Aguilar I, Janning A, Dlugos CP, Schäfer-Herte M, Klingner C, Wälte M, et al. 2016. Calcium-mediated actin reset (CaAR) mediates acute cell adaptations. *eLife* **5:** e19850. doi:10.7554/eLife.19850

Wang Y, Sherrard A, Zhao B, Melak M, Trautwein J, Kleinschnitz EM, Tsopoulidis N, Fackler OT, Schwan C, Grosse R. 2019. GPCR-induced calcium transients trigger nuclear actin assembly for chromatin dynamics. *Nat Commun* **10:** 5271. doi:10.1038/s41467-019-13322-y

Whitefield DB, Spagnol ST, Armiger TJ, Lan L, Dahl KN. 2018. Quantifying site-specific chromatin mechanics and DNA damage response. *Sci Rep* **8:** 18084. doi:10.1038/s41598-018-36343-x

Wickström SA, Niessen CM. 2018. Cell adhesion and mechanics as drivers of tissue organization and differentiation: local cues for large scale organization. *Curr Opin Cell Biol* **54:** 89–97. doi:10.1016/j.ceb.2018.05.003

Woroniuk A, Porter A, White G, Newman DT, Diamantopoulou Z, Waring T, Rooney C, Strathdee D, Marston DJ, Hahn KM, et al. 2018. STEF/TIAM2-mediated Rac1 activity at the nuclear envelope regulates the perinuclear actin cap. *Nat Commun* **9:** 2124. doi:10.1038/s41467-018-04404-4

Zidovska A, Weitz DA, Mitchison TJ. 2013. Micron-scale coherence in interphase chromatin dynamics. *Proc Natl Acad Sci* **110:** 15555–15560. doi:10.1073/pnas.1220313110

Liquid–Liquid Phase Separation in Chromatin

Karsten Rippe

Division of Chromatin Networks, German Cancer Research Center (DKFZ) and Bioquant, 69120 Heidelberg, Germany

Correspondence: Karsten.Rippe@dkfz.de

In eukaryotic cells, protein and RNA factors involved in genome activities like transcription, RNA processing, DNA replication, and repair accumulate in self-organizing membraneless chromatin subcompartments. These structures contribute to efficiently conduct chromatin-mediated reactions and to establish specific cellular programs. However, the underlying mechanisms for their formation are only partly understood. Recent studies invoke liquid–liquid phase separation (LLPS) of proteins and RNAs in the establishment of chromatin activity patterns. At the same time, the folding of chromatin in the nucleus can drive genome partitioning into spatially distinct domains. Here, the interplay between chromatin organization, chromatin binding, and LLPS is discussed by comparing and contrasting three prototypical chromatin subcompartments: the nucleolus, clusters of active RNA polymerase II, and pericentric heterochromatin domains. It is discussed how the different ways of chromatin compartmentalization are linked to transcription regulation, the targeting of soluble factors to certain parts of the genome, and to disease-causing genetic aberrations.

In a simplified and coarse-grained view, the interior of the eukaryotic cell nucleus can be separated into two main compartments. One is chromatin, consisting of the large supramolecular complex of genomic DNA wrapped around histone proteins and bound by a large number of chromosomal proteins as well as chromatin-associated RNAs. The other compartment is the soluble, liquid nucleoplasmic fraction, which is referred to here simply as the nucleoplasm. It is a highly viscous fluid, rich in dissolved proteins and RNAs, which surrounds the chromatin compartment. Inert proteins diffuse in a few seconds across the complete nucleus with the accessible space being dependent on their size (Baum et al. 2014). Thus, one would expect that proteins and RNA are homogeneously distributed in the nucleus unless locally excluded due to their size or bound to chromatin. Remarkably, the genome naturally self-organizes on the mesoscale by enriching protein and RNA factors into chromatin subcompartments (CSCs) that are ~0.1–1 µm in size (Misteli 2001, 2007, 2020; Cook 2002; Spector 2003; Wachsmuth et al. 2008; Caudron-Herger and Rippe 2012; Cremer et al. 2015; Cook and Marenduzzo 2018; Belmont 2021). CSCs are associated with a variety of activities and direct genome functions like transcription, DNA replication, recombination, and repair. The exchange of marker proteins between a CSC and the nucleoplasm is surprisingly fast and frequently on the second scale, pointing to

Cite this article as *Cold Spring Harb Perspect Biol* doi: 10.1101/cshperspect.a040683

highly dynamic structures. This process can be observed in fluorescence recovery after photobleaching (FRAP) experiments as demonstrated in pioneering studies for nucleolar factors like fibrillarin (Phair and Misteli 2000) and RNA polymerase I (Pol I) (Dundr et al. 2002) in the nucleolus, the RNA polymerase II (Pol II) preinitiation complex (Kimura et al. 2002), linker histone H1 (Lever et al. 2000; Misteli et al. 2000), and heterochromatin protein 1 (HP1) at transcriptionally silenced pericentromeric heterochromatin (Cheutin et al. 2003; Festenstein et al. 2003). It is noted that these studies also identified more immobile protein fractions that were bound to chromatin on a minute timescale. Thus, there appears to be a complex interplay of transient and more long-lived interactions that targets proteins to certain parts of the genome to assemble CSCs in a self-organizing manner reliably across the cell cycle as discussed previously (Wachsmuth et al. 2008).

To describe the process of CSC formation a definition of the relevant terms in the context of this review appears to be warranted. The general description of membraneless cellular subcompartments as "biomolecular condensates" has been used rather broadly for the local accumulation of biological macromolecules independent of the formation mechanism (Banani et al. 2017; Sabari et al. 2020). On the other hand, in physics, the term "condensation" and "condensate" is mostly used for a phase transition. Thus, we here suggest applying "condensate" specifically for the assembly of subcompartments that are the product of a phase separation process. In contrast, the CSC designation makes no assumptions on the formation mechanism and only refers to the local enrichment of protein and/or RNA into a distinct chromatin domain on the mesoscopic scale of 0.1–1 µm. The term "liquid" is used here for a state in which biological macromolecules can independently change their location randomly in all dimensions like molecules in a fluid. Accordingly, the nucleosomes themselves by definition cannot be liquid as they are linked via the DNA into a polymeric chain, which constrains their individual translocations. This definition differs from other studies that refer to nucleosomes or chromatin as "liquid" or "fluid" if they are in a dynamic and disordered state where they retain some configurational flexibility relative to each other (Maeshima et al. 2016a, 2020; Sanulli et al. 2019). Here, this type of dynamic organization is referred to as "transient interactions" and the fast exchange of factors between the free and bound state in CSC as "transient binding" but not as "liquid."

MECHANISMS OF CHROMATIN SUBCOMPARTMENT FORMATION

Soluble protein and RNA factors are mostly homogeneously distributed in the nucleoplasm (Fig. 1A). Their local enrichment by binding to chromatin can be mapped along the linear DNA sequence. This sequencing-based analysis has been conducted for chromosomal proteins (Filion et al. 2010), histone modifications (Barski et al. 2007; Ernst et al. 2011), or associated RNAs (Li and Fu 2019). Thus, protein or RNA binding at clustered sites leads to the local enrichment of these factors (Fig. 1B). Furthermore, it is well established that the nucleosome chain folds into distinct 3D conformations via interactions between protein and RNA factors bound at distant parts of the nucleosome chain (Fig. 1C). This type of interaction drives the dynamic folding of the genome on multiple scales, which could additionally also involve associations via liquid droplets (Misteli 2020; Mirny and Dekker 2021).

One well-established structure on the scale of 1 Mb are topologically associating domains (TADs) (Beagan and Phillips-Cremins 2020; Cavalheiro et al. 2021) and their substructures (Krietenstein et al. 2020; Szabo et al. 2020; Mirny and Dekker 2021). The dynamic features of TADs observed in living cells are compatible with different polymer-folding models (Wachsmuth et al. 2016). Transcriptionally active or inactive TADs segregate into distinct A-/B-compartments as inferred from chromosome conformation capture analysis, which measures the in situ cross-linking efficiency of genomic loci (Lieberman-Aiden et al. 2009). If the protein/RNA-mediated bridging between parts of the chain exceeds a certain threshold a sharp transition from an open random coil conformation into a collapsed chromatin globule can occur. This polymer–polymer

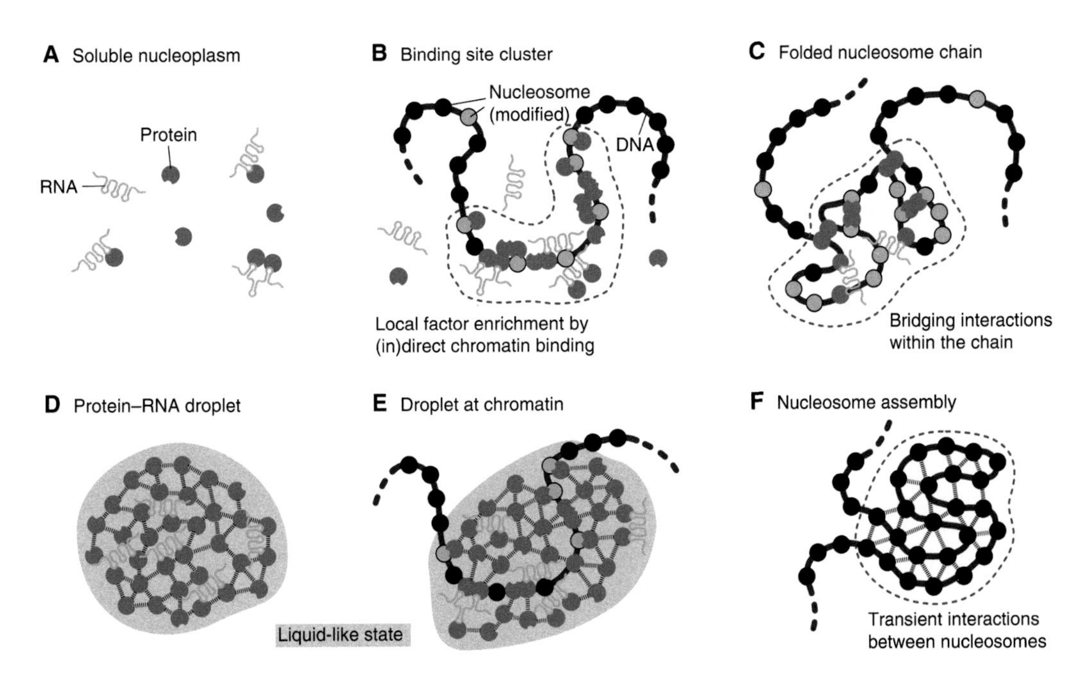

Figure 1. Multiple mechanisms for formation of chromatin subcompartments (CSCs). (*A*) Macromolecules in the soluble nucleoplasm are homogeneously distributed as diffusion quickly equilibrates concentration gradients. (*B*) Direct or indirect binding to clustered sites on the nucleosome chain can locally enrich protein/RNA into a CSC indicated by the dashed line. (*C*) Bridging interactions induced by proteins and/or RNA fold the nucleosome chain into a spatially distinct domain. If a sufficient number of these attractive interactions between chain segments are present, they can induce a polymer–polymer phase transition into a condensed chromatin globule. (*D*) Protein and RNA can separate in the nucleoplasm or cytoplasm by undergoing liquid–liquid phase separation (LLPS) into a liquid-like droplet that is mediated by multivalent interactions. (*E*) Chromatin-bound proteins and RNA could nucleate an LLPS to accumulate additional protein and RNA factors into a liquid droplet. (*F*) Nucleosomes themselves assemble locally into a disordered state where they transiently interact with each other to form an irregular structure that excludes other macromolecules based on their size. It is noted that this state would not be called "liquid" here as the DNA connection between nucleosomes constrains their translocations relative to each other.

phase-separation process is driven by attractive interaction between segments of the chain that induce the transition into a more densely folded chromatin domain (Fig. 1C; Leibler 1980; Williams et al. 1981; Bates 1991; Nicodemi and Pombo 2014; Michieletto et al. 2016; Jost et al. 2017; MacPherson et al. 2018).

The CSCs depicted in Figure 1B and C arise predominantly from the direct chromatin binding of protein and RNA factors. Thus, the "null hypothesis" for forming a CSC against which a potential phase-separation mechanism should be tested is the enrichment of protein and RNA factors by (cooperative) binding to a cluster of sites on the nucleosome chain (Fig. 1B). This process may also include additional indirect binding of proteins and RNA and can be described by well-established ligand-binding models (Teif and Rippe 2010; Gutierrez et al. 2012; Phillips 2015). For example, the DNA sequence-dependent formation of heterochromatin nanodomains marked by the histone modification H3K9me2/me3 can be rationalized by this type of approach (Thorn et al. 2020). To explain how mesoscale proteins and RNA assemblies form with sharp boundaries against the surrounding regions, the mechanism of liquid–liquid phase separation (LLPS) has been applied (Hyman et al. 2014; Banani et al. 2017; Shin and Brangwynne 2017; Boeynaems et al. 2018). It describes the reversible demixing of an originally homogeneous solution of proteins and RNA into two distinct fluid-like phases. This process can drive the formation of cellular subcompartments by sequestering certain proteins and RNAs into a liquid droplet-like state that segregates them from the surrounding solution like oil drops in water. A molecular description of this process in the cell is given by the "stickers-and-spacers" model (Choi et al. 2020). It represents protein and RNA as flexible polymers where sequence motifs of one or more residues, the "stickers," mediate attractive interactions between different molecules while other parts of the chain act as mostly inert "spacers" between them. Above a critical concentration threshold, the stickers on the protein/RNA chain can induce a separation into a dense phase that coexists with a dilute phase in which the interacting macromolecules are depleted. If interactions in the dense phase are weak and transient it has liquid-like properties. However, the same framework can be used to also describe gel- or solid-like states with reduced protein/RNA mobility as their interaction strength increases (Choi et al. 2020). This type of LLPS description rationalizes the formation of cytoplasmic P granules, membraneless organelles formed by RNA, and proteins that are involved in RNA processing (Fig. 1D; Brangwynne et al. 2009). LLPS arises via transient multivalent interactions that frequently involve intrinsically disordered protein regions (IDRs) and RNA, creating an exclusionary local protein–RNA environment with distinct physicochemical properties (Weber and Brangwynne 2012; Uversky et al. 2015; Banani et al. 2017; Drino and Schaefer 2018). It has also been suggested to be a crucial driver of genome organization (Erdel and Rippe 2018; McSwiggen et al. 2019b; Strom and Brangwynne 2019; Frank and Rippe 2020; Hildebrand and Dekker 2020; Narlikar 2020; Sabari et al. 2020). LLPS at chromatin directly involves chromatin-bound protein and RNA factors as nucleation sites so that a liquid droplet assembles at a specific chromatin locus (Fig. 1E). Macromolecules not directly bound to chromatin can constantly rearrange and mix within the droplet, and access to this type of CSC depends on the chemical nature of the CSC components. In contrast, access to a CSC formed by bridging interactions of the nucleosome chain (Fig. 1C) is controlled by particle size. Other properties like the response to concentration changes of constituting components in terms of size change or buffering also differ. Finally, reconstituted mono- and oligonucleosome particles have been shown to undergo LLPS in vitro and it has been proposed that this state exists also in the cell (Fig. 1F; Gibson et al. 2019; Sanulli et al. 2019; Wang et al. 2019). However, within a chromosome, the DNA linkage between nucleosomes imposes a number of constraints with respect to their mobility relative to each other. Confined random translocations of the nucleosome chain can occur on the scale of 10–100 nm but on the mesoscopic CSC scale chromatin displays solid-like properties (Kimura and Cook 2001; Chubb et al. 2002; Gerlich et al. 2003; Walter et al. 2003; Levi et al. 2005; Jegou et al. 2009; Strickfaden et al. 2010; Chen et al. 2013; Wachsmuth et al. 2016; Maeshima et al. 2020, 2021; Strickfaden et al. 2020). Thus, liquid-like protein and RNA droplets could nucleate at certain points of a mostly immobile chromatin scaffold with confined motions of nucleosomes or parts of the chain within this droplet (Fig. 1E). It is noted that the mechanisms depicted in Figure 1 are not mutually exclusive. For example, the binding to clustered sites (Fig. 1B) would be part of both the chain folding (Fig. 1C) and LLPS (Fig. 1E) mechanism. In addition, liquid droplets as well as nucleosome–nucleosome interactions (Fig. 1F) could also act as bridging factors to promote folding of the chain into a compacted state.

FORMATION OF TRANSCRIPTIONALLY ACTIVE OR SILENCED CSCs

In the following, we will not consider phase separation into mostly irreversible gel or aggregated states as it is a crucial feature of functional CSCs that they are dynamic and can form reversibly in a self-organizing manner across the cell cycle. Rather, the focus is on three prototypical CSCs: the nucleolus, clusters of Pol II referred to as transcription factories or transcriptional condensates, as well as chromocenters. LLPS has been suggested to be operative for all three of

Table 1. Features of exemplary chromatin subcompartments (CSCs) for which formation by a phase-separation mechanism has been proposed in relation to the surrounding nucleoplasm

CSC	Nucleolus[a]	Pol II transcription factories[b]	Chromocenters[c]
Organism	Human	Human, mouse	Mouse, *Drosophila*
Marker proteins	Pol I, NPM1, NCL, FBL, UBF	Pol II, TAF15, BRD4, MED1/19, specific transcription factors (TFs)	HP1α, MeCP2, H1
Structure	Tripartite	Diverse	Granular (HP1α, DNA)
Exchange with nucleoplasm	Seconds-minutes	Seconds-minutes	Seconds-minutes
Internal mixing	Yes	?	No
Fusion	Yes	?	Yes
Accessibility	Chemical properties	?	Size
Protein/DNA ratio	High	High	Average
RNA/DNA ratio	Very high	High	Average
Local viscosity	Increased	?	Average
Architectural RNA component	rRNA, aluRNA	Nascent RNAs, enhancer RNAs, LINE1, aluRNA	Major satellite RNA

[a]Andersen et al. 2005; Nemeth et al. 2010; Brangwynne et al. 2011; Caudron-Herger et al. 2015b; Martin et al. 2015; Feric et al. 2016; Nemeth and Grummt 2018; Caragine et al. 2019; Frottin et al. 2019; Yao et al. 2019; Ide et al. 2020; Lafontaine et al. 2021; Lawrimore et al. 2021.

[b]Melnik et al. 2011; Ghamari et al. 2013; Papantonis and Cook 2013; Caudron-Herger et al. 2015a; Hnisz et al. 2017; Cho et al. 2018; Chong et al. 2018; Sabari et al. 2018; Guo et al. 2019; Nair et al. 2019; Quintero-Cadena et al. 2020; Sabari et al. 2020; Wei et al. 2020; Garcia et al. 2021b; Hilbert et al. 2021; Ma et al. 2021.

[c]Peters et al. 2001; Brero et al. 2005; Lu et al. 2009; Cao et al. 2013; Muller-Ott et al. 2014; Saksouk et al. 2014; Bosch-Presegué et al. 2017; Strom et al. 2017; Ostromyshenskii et al. 2018; Jagannathan et al. 2019; Erdel et al. 2020; Kochanova et al. 2020.

them (Table 1), and several of their purified constituting marker proteins can undergo LLPS in vitro (Table 2). The review will use them as exemplary cases to discuss how their dynamic structure, material properties, and biological activities are related to an LLPS process for their formation in comparison to alternative mechanisms. More general discussions of phase-separated processes that involve chromatin can be found elsewhere (Erdel and Rippe 2018; McSwiggen et al. 2019b; Strom and Brangwynne 2019; Frank and Rippe 2020; Hildebrand and Dekker 2020; Narlikar 2020; Sabari et al. 2020).

Nucleolus

The nucleolus is a prototypic CSC for an LLPS-driven formation mechanism (Brangwynne et al. 2011; Feric et al. 2016; Caragine et al. 2019; Lafontaine et al. 2021). Its structure is characterized by the association of hundreds of nucleolar proteins around the nucleolar organizer regions containing the ribosomal DNA (rDNA) gene repeats from different chromosomes from which large amounts of ribosomal RNA (rRNA) are transcribed (Mangan et al. 2017; Németh and Grummt 2018; Lafontaine et al. 2021). In the nucleolus, key marker proteins like Pol I, fibrillarin (FBL), nucleolin (NCL), and nucleophosmin (NPM1) are highly enriched together with the rRNA and form a sharp concentration boundary to the surrounding nucleoplasm.

Pol II Transcription Factories

Transcriptionally active CSCs enriched with Pol II have been characterized as transcription factories (Jackson et al. 1993; Iborra et al. 1996; Osborne et al. 2004). They accumulate transcription factors (TFs), RNA, and both promoter/enhancer DNA loci (Jackson et al. 1993; Iborra et al. 1996; Osborne et al. 2004). A number of previous studies have studied their features as well as their function as self-assembling organizers of the genome (Cook 2002; Chakalova et al. 2005; Papantonis and Cook 2013; Buckley and Lis 2014; Cook

Table 2. Chromatin subcompartment (CSC) marker proteins that can undergo liquid–liquid phase separation (LLPS) in vitro

Protein	Abbreviation	CSC	References
Nucleophosmin	NPM1	Nucleolus	Feric et al. 2016; Mitrea et al. 2016, 2018
Fibrillarin	FBL/FIB		Berry et al. 2015; Feric et al. 2016
Carboxy-terminal domain (CTD) of Pol II	CTD	Pol II transcription factories/transcriptional condensates	Kwon et al. 2013; Boehning et al. 2018; Lu et al. 2018
TATA-Box-binding protein-associated factor 15	TAF15		Chong et al. 2018
p300/CREB-binding protein	p300/CBP		Ma et al. 2021
Bromodomain-containing protein 4	BRD4		Sabari et al. 2018
Mediator subunits 1/19	MED1, MED19		Cho et al. 2018; Sabari et al. 2018; Guo et al. 2019; Zamudio et al. 2019
Heterochromatin protein 1	HP1/α/β/γ, HP1a	Chromocenter (pericentric heterochromatin)	Larson et al. 2017; Strom et al. 2017; Wang et al. 2019; Erdel et al. 2020; Qin et al. 2021
Methyl CpG-binding protein 2	MeCP2		Fan et al. 2020; Li et al. 2020a; Wang et al. 2020
Linker histone H1	H1		Gibson et al. 2019; Shakya et al. 2020; Muzzopappa et al. 2021

and Marenduzzo 2018). In recent studies, the IDR-mediated assembly of specific TFs like SP1, OCT4, β-catenin, STAT3, estrogen receptor (ER), and SMAD3, the TBP-associated general TF TAF15, as well as transcriptional coactivators like MED1/19, GCN4, and BRD4 and the unstructured carboxy-terminal domain (CTD) of Pol II into so-called transcriptional condensates has been described as a phase separation process (Hnisz et al. 2017; Frank and Rippe 2020; Peng et al. 2020; Sabari et al. 2020).

Chromocenters

Pericentric repeat sequences assemble into compact heterochromatin domains in mouse and *Drosophila* cells called chromocenters due to their strong fluorescence after DAPI staining (Probst and Almouzni 2008; Fodor et al. 2010). They contain mostly major satellite repeat sequences but also other types of repeats (Ostromyshenskii et al. 2018; Jagannathan et al. 2019). Recent work concluded that this type of CSC arises from HP1-driven LLPS that condenses chromatin (Larson et al. 2017; Strom et al. 2017; Fan et al. 2020; Li et al. 2020a; Wang et al. 2020) according to the scheme shown in Figure 1E. However, another study reported that chromocenters form independently of HP1 by polymer–polymer phase separation into a chromatin globule (Fig. 1C; Erdel et al. 2020).

HIGH-RESOLUTION STRUCTURE

A CSC formed by LLPS would be expected to show a homogeneous distribution of a given marker protein within the droplet (Fig. 1E). However, other types of local protein enrichment (Fig. 1B,C) could also appear like a dense spherical structure at the limited resolution of light microscopy. This is shown by labeling an endogenous intronic repeat sequence in the

MUC4 gene with dCas9-GFP, which results in punctate structures with an apparent size of 0.5–0.8 µm (Chen et al. 2013). Thus, high-resolution CSC structures obtained with electron microscopy or fluorescence superresolution microscopy methods are more informative to distinguish between protein-/RNA-filled droplets as opposed to chromatin-bound factors.

Nucleolus

In mammals, the nucleolus is structured into three domains that are clearly distinguishable on electron microscopy images (Thiry et al. 2011). Pol I is enriched in the fibrillar centers (FCs), and the actively transcribed rRNA genes (rDNA) are located at the interface between FCs and dense fibrillar components (DFCs). The upstream binding factor (UBF), a key regulatory factor of rDNA transcription, is associated with both active and poised repeats at the FC/DFC border (Maiser et al. 2020). The resulting pre-rRNA is processed and assembled with ribosomal proteins in the DFC and in the granular component (GC), which is enriched in NPM1 and NCL. This internal compartmentalization can be rationalized as three coexisting, immiscible liquid-like phases (Feric et al. 2016; Lafontaine et al. 2021). Fluorescence microscopy superresolution images are in line with this model as the distribution of marker proteins such as Pol I, FBL, NPM1, and NCL is quite homogeneous in the respective nucleolar subcompartments (Yao et al. 2019; Maiser et al. 2020; Lafontaine et al. 2021). However, it is also apparent that further fine structures exist for the organization of the actively described rDNA. These loci adopt a ring-shaped conformation of ~170 nm and ~240 nm in diameter in human and mouse fibroblasts, respectively (Maiser et al. 2020). Another study shows that FBL forms small clusters in the DFC of 50 nm in size spaced 100–200 nm apart (Yao et al. 2019).

Pol II Transcription Factories

Clusters of Pol II have been described as comprising 4–30 active polymerases that assemble around a protein-rich core with two or more transcription units with diameters of 50–180 nm in diploid human cells (Rieder et al. 2012; Papantonis and Cook 2013). The initial characterization of Pol II factories was conducted in fixed cells. Subsequent fluorescence microscopy analysis in living cells yielded similarly sized Pol II clusters of 220 nm (Cisse et al. 2013) as well as foci of CDK9, a kinase associated with active Pol II (Ghamari et al. 2013). Furthermore, active Pol II constrains chromatin movements, supporting the view that transcription factories link chromatin loci (Nagashima et al. 2019). Recent studies investigated the structure of active Pol II compartments in the context of a phase separation mechanism (Cho et al. 2018; Hilbert et al. 2021). The analysis of endogenously tagged MED1 and Pol II in mouse embryonic stem cells points to the existence of two different types of supramolecular complexes (Cho et al. 2018). One is relatively small (~100 nm) and instable with average lifetimes on the 10-sec scale. The other population of larger clusters (>300 nm) with ~200 to 400 molecules persists for several minutes. Another study characterized Pol II transcription compartments in zebrafish cells (Hilbert et al. 2021). Clusters of active Pol II were present in micrometer-sized regions enriched in RNA but depleted of chromatin with the active transcription sites of 100–200 nm in size being located at the RNA–chromatin interface.

Chromocenters

The current high-resolution structural data on chromocenters comprise electron and superresolution fluorescence microscopy (Fussner et al. 2012; Erdel et al. 2020; Kochanova et al. 2020; Miron et al. 2020; Strickfaden et al. 2020; Xu et al. 2020). The results point to irregularly shaped domains with condensed chromatin in a granular structure in mouse cells with HP1 and H3K9me3 enrichment following the chromatin density (Erdel et al. 2020). Methyl-CpG-binding protein 2 (MeCP2) and linker histone H1 are also enriched in chromocenters but their fine structure is difficult to assess in the analysis conducted so far (Misteli et al. 2000; Cao et al. 2013; Muller-Ott et al. 2014; Linhoff et al. 2015). In

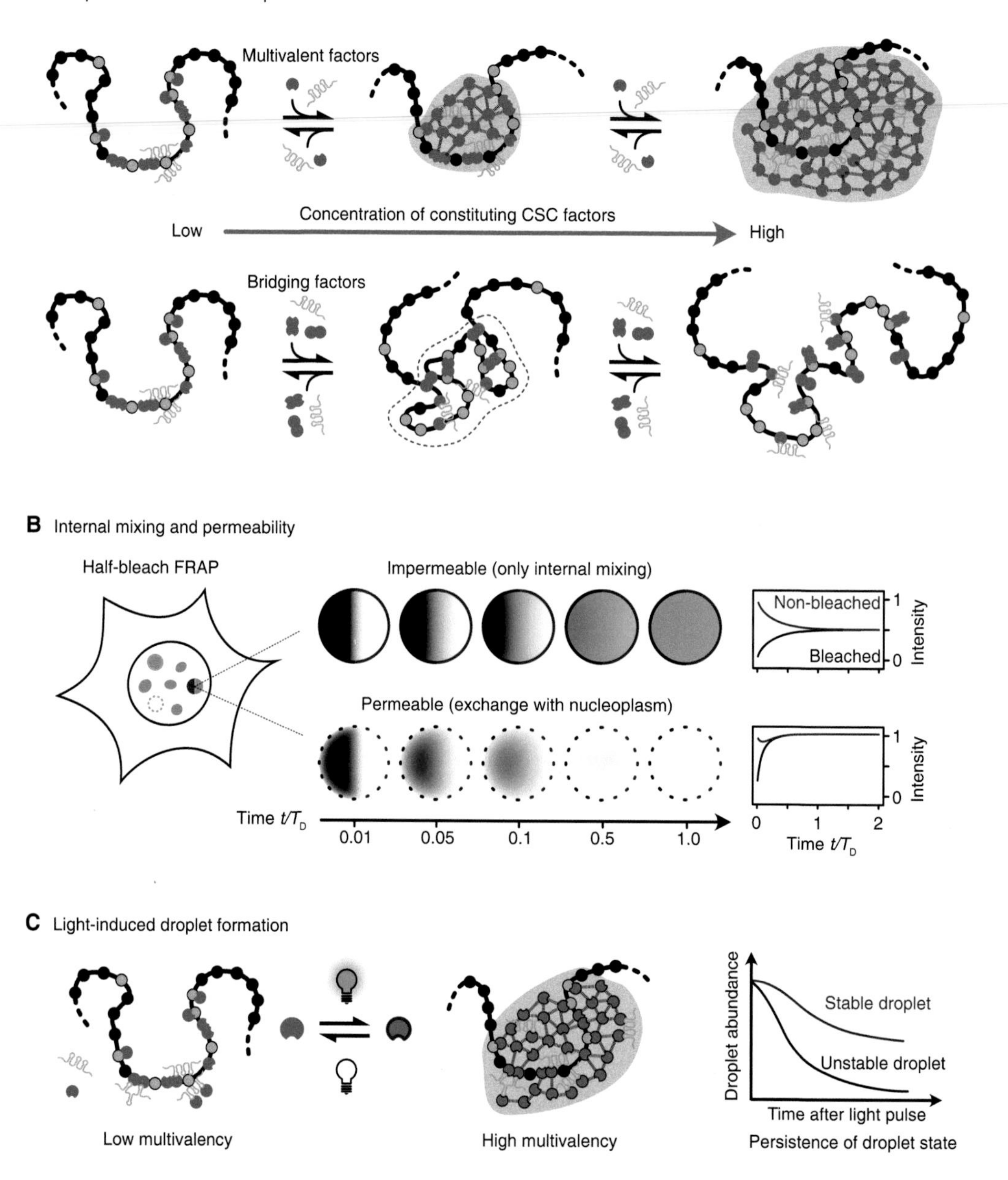

Figure 2. Experimental approaches to analyze chromatin subcompartment (CSC) assembly in the cell nucleus. (*A*) Response of CSCs to concentration changes. (*Top*) Increasing the concentration of constituting proteins/RNAs is expected to expand liquid droplets while maintaining their internal composition (Banani et al. 2017). (*Bottom*) Bivalent chromatin cross-linking could be disrupted at high concentration of bridging factors (Malhotra et al. 2021). (*B*) Half-bleach fluorescence recovery after photobleaching (FRAP) evaluates internal mixing and permeability of the boundary (Erdel et al. 2020). Simulated temporal intensity traces for low, intermediate, and high permeability are depicted for a time axis normalized for differences in the diffusion coefficient by division to the diffusion time τ_D. (*C*) Light-induced formation of liquid droplets (Shin et al. 2017). In this assay, the protein of interest is fused to the photolyase homologous region (PHR) domain, which promotes multivalent interactions and droplet formation upon illumination with blue light. This allows it to evaluate the effect of an artificially induced liquid–liquid phase separation (LLPS) on the activity of a chromatin locus of interest (e.g., to study transcription activation). Furthermore, the stability of the resulting droplets can be assessed from their persistence in the absence of the light trigger.

Drosophila, the chromocenter organization appears to be less granular with a multilayer organization of marker proteins (Jagannathan et al. 2019; Kochanova et al. 2020).

INTERNAL MIXING OF MARKER PROTEINS IN CSCs AND EXCHANGE WITH THE NUCLEOPLASM

The fast exchange of a large fraction of CSC marker proteins points to highly dynamic structures that nevertheless stably direct genome-associated activities to specific loci. LLPS could confine the translocations of protein and factors to the interior of the resulting liquid droplets so that they become segregated from the surrounding nucleoplasm (Fig. 1E). In this environment, they are concentration-buffered and maintain a steady concentration of molecules against external fluctuations that would only affect the droplet size (Fig. 2A; Banani et al. 2017). The CSC types depicted in Figure 1B and C on the other hand are permeated by soluble factors from the surrounding nucleoplasm. Access to the CSC is determined by the size of the macromolecule. For this type of CSC, factors can quickly exchange with the surrounding nucleoplasm and the domain size should be mostly unaffected by concentration changes (Erdel and Rippe 2018; Frank and Rippe 2020). However, at sufficiently high concentrations the bivalent attractive bridging interactions between chromatin segments could be competed out by monovalent chromatin interactions of the linking factors (Fig. 2A; Malhotra et al. 2021). A fast exchange of bound proteins with the surrounding nucleoplasm that is measured in conventional FRAP can be explained simply by a short residence time in the chromatin-bound state and does not represent evidence for the formation of a liquid droplet (McSwiggen et al. 2019b). The hallmark feature of LLPS that molecules can mix within the compartment like in a fluid can be evaluated by bleaching only half of the subcompartment and analyzing exchange of molecules with the unbleached half (Fig. 2B; Brangwynne et al. 2009; Patel et al. 2015; Erdel et al. 2020). The resulting part of fluorescence recovery is then compared to the exchange with molecules from the surrounding nucleoplasm, which provides information on the permeability of the compartment boundary (Erdel et al. 2020).

An alternative approach to FRAP is the tracking of single fluorescently labeled particles as has been done for TFs (Chen et al. 2014; Kent et al. 2020; Garcia et al. 2021a,b). It provides direct information on the confinement of particle mobility but is typically limited to observation periods in the ~20 sec range due to loss of the fluorescence signal over time.

Nucleolus

Pol I and UBF have residence times on the 10-sec to minute scale in the nucleolus, with prolonged retention at rDNA promoters upon activation (Chen and Huang 2001; Dundr et al. 2002; Gorski et al. 2008). Likewise, FBL, NPM1, and NCL show complete recovery in FRAP experiments on the 10–20 sec timescale (Phair and Misteli 2000; Chen and Huang 2001; Dundr et al. 2002; Gorski et al. 2008; Frottin et al. 2019; Erdel et al. 2020). Interestingly, NPM1 displays preferred internal mixing within the nucleolus, a feature indicative of liquid droplet formation, which was less pronounced for NCL (Fig. 2B; Erdel et al. 2020). Nucleolar access or exclusion is dependent on the chemical nature of a protein and less so on its size as expected for an LLPS compartment. In particular, certain peptides can carry a nucleolar localization signal that lacks defined sequence motifs and the exclusion of wild-type GFP was reverted by fusion of a small arginine-rich and positively charged peptide (Martin et al. 2015). Finally, it is noted that nucleoli are remarkably stable during their purification, which includes dilution/washing through multiple steps and allows for characterization of their protein, DNA, and RNA content (Andersen et al. 2005; Németh et al. 2010; Caudron-Herger et al. 2015b). This property is difficult to reconcile with a reversible liquid droplet state, which would disassemble upon removing its constituting components from the surrounding solution.

RNA Polymerase II Transcription Factories

Several FRAP studies evaluated the dynamic properties of Pol II complexes at chromatin

(Becker et al. 2002; Kimura et al. 2002; Hieda et al. 2005; Darzacq et al. 2007). In these experiments, the Pol II fraction recovering over 10–20 min was assigned to the elongating state. In contrast the putative preinitiation complex was very dynamic and recovered within seconds after bleaching. These findings are in line with studies that report Pol II residence times in clusters of 5–10 sec (Cisse et al. 2013; Cho et al. 2018) and that foci of the CDK9 kinase, which associates with active Pol II, exchange within seconds (Ghamari et al. 2013). In addition, these studies also report the existence of more long-lived complexes stable on the minute timescale or even for hours. In general, TFs show highly dynamic and stochastic binding with typical residence times of seconds (Mueller et al. 2013; Lionnet and Wu 2021; Lu and Lionnet 2021). In many instances, the residence times at their target promoter sites appear to be in the range of less than a minute although longer times have also been reported. Likewise, estrogen receptor α (ERα) (Nair et al. 2019) as well as SOX2 (Chen et al. 2014) were specifically bound for 10–20 sec at their enhancers together with other TFs. However, the view that the distribution of TF residence times is bimodal and reflects essentially either specifically or nonspecifically bound complexes might be too simplistic. A recent study concluded that several TFs (including ERα, FOXA1, and CTCF) follow a power-law distribution of residence times and may involve longer binding events in the right-skewed tail of the distribution than previously derived from bi-exponential models (Garcia et al. 2021a).

Chromocenters

HP1, a marker protein enriched at transcriptionally silenced pericentromeric heterochromatin domains, exchanges within seconds with the nucleoplasm (Cheutin et al. 2003; Festenstein et al. 2003). However, this exchange arises in mouse fibroblasts predominantly by diffusion of factors from the nucleoplasm surrounding the chromocenters. In these cells, neither HP1 nor MeCP2 displayed preferential internal mixing within chromocenters in half-bleach FRAP experiments (Fig. 2A) as expected for a liquid-like droplet (Erdel et al. 2020). MeCP2 is relatively stably bound in chromocenters with 65% of the protein displaying a residence time of 25 sec and ~20% of protein binding for more than 4 min (Ghosh et al. 2010; Agarwal et al. 2011; Muller-Ott et al. 2014). In *Drosophila*, the mobility of HP1a in chromocenters as measured by FRAP was highest at the early embryo stage (Strom et al. 2017). Subsequently, the fraction of immobile HP1a increased from 0% (nuclear division cycle 10) to 30% (cycle 14), pointing to a change of chromocenter organization during differentiation. Another interesting observation with respect to protein mobility is that KMT5C (SUV4-20H2), which trimethylates histone H4 at lysine 20, shows preferential mixing within mouse chromocenters (Strickfaden et al. 2020). Furthermore, its FRAP dynamics are dependent on the three different HP1 isoforms (Bosch-Presegué et al. 2017). The formation of a liquid droplet state of SUV4-20H2, however, is difficult to reconcile with its very tight binding (immobile fraction >90% on the minute timescale) and low abundance of 200 nM concentration in chromocenters (Muller-Ott et al. 2014). It will be therefore important to further characterize the origin of the confined mobility of SUV4-20H2. Another important factor for the dynamic structure of chromocenters is linker histone H1 that displays complex isoform-specific interactions with chromatin and is involved in its compaction (Prendergast and Reinberg 2021). In the initial characterization of H1 binding by FRAP, the immobile fraction at chromocenters was increased by 10%–25% (Misteli et al. 2000). Subsequent FRAP studies provided evidence for at least two different H1 chromatin-bound states established by simultaneous interactions of the H1 globular and CTD to different DNA regions (Brown et al. 2006; Stasevich et al. 2010; Wachsmuth et al. 2016). The longer-lived fraction shows a residence time of ~100 sec and is likely to drive the linker histone-mediated packaging of nucleosomes (Maeshima et al. 2016b).

DNA, RNA, AND PROTEIN CONTENT AND LOCAL VISCOSITY

CSC formed by an LLPS mechanism (Fig. 1E) are expected to have a particularly high protein/DNA or RNA/DNA ratio as compared to the nuclear average. These parameters are compared for the nucleolus, Pol II factories, and chromocenters in Table 1. The DNA (Németh et al. 2010), protein (Andersen et al. 2005), and RNA (Caudron-Herger et al. 2015b) content of the nucleolus have been mapped and it is estimated that the DNA concentration in the nucleolus is about 20-fold lower while its protein content is twofold higher than in the surrounding parts of the nucleus. At the same time, the nucleolus is filled with ribosomal and other RNAs leading to a ~2000-fold higher RNA/DNA ratio and ~40-fold higher protein/DNA ratio (Frank and Rippe 2020). Despite its low relative concentration, however, the rDNA sequences play an important role in nucleating the RNA-dependent assembly of the nucleolus (Grob et al. 2014; Berry et al. 2015; Falahati et al. 2016; Németh and Grummt 2018; Lafontaine et al. 2021). Thus, the composition of the nucleolus is quite similar to that of a cytoplasmic protein–RNA body (Fig. 1D) and fits well to a chromatin-nucleated LLPS mechanism (Fig. 1E; Lafontaine et al. 2021). Analysis of the protein (Melnik et al. 2011) and RNA content (Caudron-Herger et al. 2015a) of Pol II transcription show that for a relatively small factory size of 50–180 nm diameter the RNA/DNA ratio could be almost as high as that in the nucleolus (Jackson et al. 1998) and a high protein/DNA ratio is also estimated (Melnik et al. 2011). For mouse chromocenters, their DNA content has been determined after purification with major satellite repeats being the dominating component but also contain a 2 kb LINE element (Zatsepina et al. 2008; Ostromyshenskii et al. 2018). The total DNA concentration in chromocenters is about twofold higher than the nuclear average (Muller-Ott et al. 2014). The proteins associated with the major satellite repeats have been mapped (Saksouk et al. 2014) and their chromocenter concentration is in general <5% of the nucleosome concentration (Muller-Ott et al. 2014). Thus, compared to the nuclear average, chromocenters display an average protein/DNA and low/average RNA/DNA ratio as their transcriptional activity is silenced under normal conditions.

In summary, the high protein/DNA and RNA/DNA ratios of the nucleolus and Pol II transcription factories distinguish these CSCs from the surrounding nucleoplasm. A protein and RNA enrichment by LLPS is expected to lead to an increased viscosity of the dense phase as shown previously for NPM1 and the nucleolus (Hyman et al. 2014; Feric et al. 2016). In contrast, the local intracellular viscosities in chromocenters as measured by polarization-dependent FCS are similar to that of the surrounding euchromatic regions (Erdel et al. 2020). Thus, it appears that high-protein/DNA and RNA/DNA ratios will correlate with liquid-like CSC features and an increased local viscosity. Vice versa, CSCs like mouse chromocenters that display average RNA/DNA and protein/DNA ratios and no significant viscosity differences may be less likely to be formed by LLPS.

STRUCTURE–FUNCTION RELATIONSHIPS

The different mechanisms that confine genome-associated activities by establishing CSCs (Fig. 1) lead to distinct structure–function relationships. In general, two main functional aspects are apparent. One is to target macromolecules to certain parts of the genome, while the other is the formation of a specific local environment that enhances chromatin-mediated reactions.

Nucleolus

A number of findings show that the intact nucleolus structure and LLPS properties are directly linked to efficient ribosome biogenesis (Lafontaine et al. 2021). The tripartite nucleolar architecture of FC, DFC, and GC is disrupted if Pol I or Pol II transcription is inhibited (Caudron-Herger et al. 2015b, 2016). At the same time, dispersed pre-nucleolar bodies containing NCL, NPM, and FBL that assemble postmitotically at the nucleolar organizer regions to reform the nucleolus only have a low rRNA content (Carron et al. 2012; Németh and Grummt 2018). Highly proliferating tumor cells, on the other hand, harbor larger and more active nucleoli for high rRNA and ribosome production (Derenzini et al. 2000; Montanaro et al. 2008; Weeks et al. 2019). In addition, cells from patients suffering from neurodegenerative diseases often present with less active nucleoli with structural aberrations (Parlato and Kreiner 2013). Such a correlation of size and activity would be expected for an LLPS-driven mechanism in which a concentration increase of rRNA could increase the droplet size (Fig. 2A). Furthermore, it is well established that LLPS can create an environment with an increased local concentration of protein and RNA factors and enhance enzymatic activity (O'Flynn and Mittag 2021). Within the fully assembled nucleolus, a multiphase LLPS event could serve to compartmentalize rDNA transcription, rRNA processing, and rRNA–ribosomal protein assembly (Feric et al. 2016). Interestingly, repression of Pol I in the nucleolar cap has also been reported by formation of a phase-separated subcompartment (Ide et al. 2020). The liquid-like properties of these distinct subcompartments within the nucleolus could also be important for quality control of misfolded proteins (Frottin et al. 2019). According to the latter study, the GC of the nucleolus with its liquid-like state prevents the irreversible aggregation of misfolding of proteins during heat shock.

Pol II Transcription Factories

For Pol II, transcription factories providing specificity of gene regulation as well as promoting efficient transcription are important functional aspects (Papantonis and Cook 2013). It is, however, currently not clear what the driving mechanism of formation for this type of CSC is and how the formation mechanism would affect transcription. For example, the enrichment of Pol II and TFs in replication compartments of the herpes simplex virus appears to be mostly driven by locally enhanced chromatin binding (Fig. 1B) due to creating nucleosome-free regions (McSwiggen et al. 2019a). Furthermore, modeling studies show that bridging interactions of TFs as depicted in Figure 1C would suffice for the formation of Pol II transcription factories with a 3D organization similar to that found in the cell (Brackley et al. 2013).

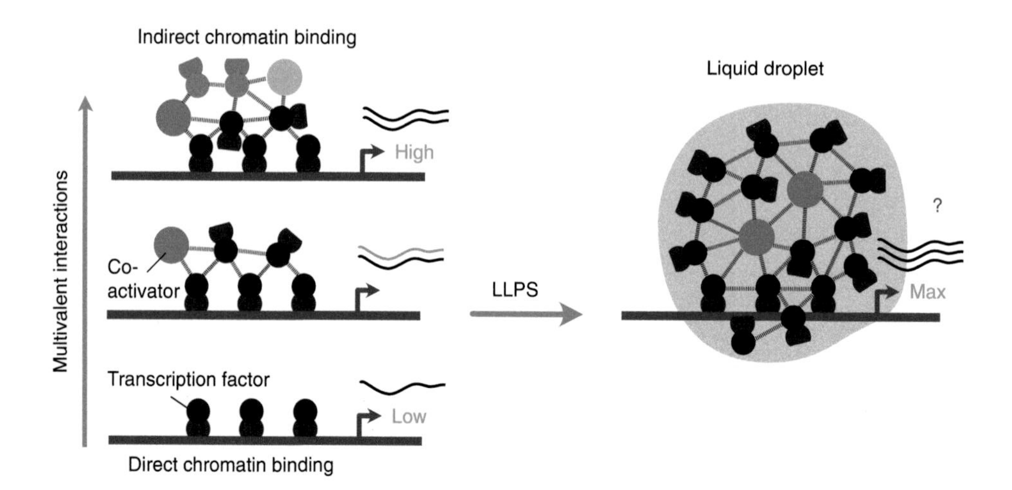

Figure 3. Multivalent interactions, chromatin binding, and liquid–liquid phase separation (LLPS). Direct chromatin binding of a transcription factor (TF) is accompanied with indirect interactions of coactivators like histone acetylases, BRD4, or components of the mediator complex that enhance transcription. LLPS would largely increase the amount of indirectly bound factors. It would also lead to a sharp concentration boundary between the droplet and the nucleoplasm while indirectly chromatin-bound factors would otherwise be expected to show a concentration decrease as the distance from the directly chromatin-bound TFs becomes larger. Furthermore, it is currently not clear whether the formation of a liquid droplet around a given promoter would indeed increase transcription as proposed in a number of studies as compared to the indirect binding of coactivators depicted on the *left* side of the scheme.

The functional consequences of TF liquid droplets were studied with synthetic activator constructs using the approach of light-induced droplet formation shown in Figure 2C (Wei et al. 2020; Schneider et al. 2021). In these studies, it was concluded that droplets formed by TF fusion constructs increase gene expression or transcription activation, supporting the view that LLPS of TFs and coactivators induces high transcription activity (Hnisz et al. 2017; Sabari et al. 2018, 2020). However, corroborating these conclusions would require a comparison of TF activation capacity of the same factor in the presence/absence of LLPS under identical conditions (Fig. 3). It is noted that the propensity of a given TF or coactivator to undergo LLPS in vitro might simply reflect its ability to engage in multivalent interactions. These multivalent interactions could also promote interactions and enhance transcription activation in the absence of phase separation (Cho et al. 2018; Trojanowski et al. 2021). One alternative function would be that IDRs increase the kinetic rate for the formation of a specific complex between proteins and/or nucleic acids (Pontius 1993). In such a mechanism, IDRs stabilize an intermediate state that allows the interacting factors to sample different orientations to each other, which increases the probability of specific complex formation during a diffusive encounter. Accordingly, it will be important to further dissect how IDRs modulate the interplay of interactions that differ in strength and specificity between TFs, coactivators, and parts of the general transcription machinery in relation to the transcriptional output.

On the mesoscale, liquid droplet formation itself could also accelerate the binding reaction of TFs and/or coactivators to their target sites (Brodsky et al. 2020; Kent et al. 2020; Garcia et al. 2021b). Confining a random search process to a chromatin-associated droplet and increasing the local concentration of a given factor could greatly increase its kinetic binding rate. Finally, several studies link the IDR-mediated formation of liquid droplets to the phenomenon of "transcriptional bursting" where the promoter enters a refractory state after being in a period of active transcription for several minutes (Rodriguez and Larson 2020). The propensity of TF activation domains to form liquid droplets with the TF p300 as well as the length of the Pol II CTD correlates with an increased frequency and longer duration of transcriptional bursts (Quintero-Cadena et al. 2020; Ma et al. 2021). Remarkably, Quintero-Cadena et al. also show in their study that the loss of Pol II activity due to shortening the CTD can be partially rescued by fusion with an IDR from FUS or TAF15. The stability of the putative LLPS-driven condensates formed via these IDR interactions could be dependent on their RNA content as shown for MED1-IDR droplets in vitro (Henninger et al. 2021). These findings raise the possibility that transcriptional bursting arises from the periodic formation and disruption of an activating liquid droplet state formed between IDRs of Pol II, TFs, and coactivators and nascent RNA. However, the switching of a given gene between an active and silent state can also be explained by the promoter proximal and distal binding and dissociation of regulators and their chromatin-mediated interactions with the transcription machinery (Rodriguez and Larson 2020).

Chromocenters

The assembly of intact chromocenters is linked to chromosome segregation and silencing of repeat transcription (Probst and Almouzni 2008; Fodor et al. 2010; Janssen et al. 2018). How these functions might be affected by proposed LLPS events of relevant chromocenter proteins like HP1, MeCP2, or H1 is currently not clear. It is noted that a number of studies show that global compaction, accessibility, and size of mouse chromocenters is largely independent on HP1 (Peters et al. 2001; Schotta et al. 2004; Mateos-Langerak et al. 2007; Bosch-Presegué et al. 2017; Erdel et al. 2020). Notably, the knockout of HP1α, which has been proposed to be crucial for LLPS in mammalian heterochromatin (Larson et al. 2017; Wang et al. 2019), has no apparent phenotype in mice (Aucott et al. 2008; Singh 2010; Mattout et al. 2015). The chromocenter structure in HP1$\alpha^{-/-}$, HP1$\beta^{-/-}$, and HP1$\gamma^{-/-}$ knockouts in mouse embryonic fibroblasts was mostly unaffected on the mesoscale in terms of DNA compaction as compared to wild-type cells

(Bosch-Presegué et al. 2017). MNase digestion experiments in the latter study point to a decrease in accessibility at nucleosome resolution of chromocenters if HP1α is lost. Likewise, structural phenotypes of chromocenters in differentiated *Drosophila* cells at the mesoscale are not associated with HP1 but rather with two sequence-specific satellite DNA-binding proteins, D1 and Prod (Jagannathan et al. 2019). In *Drosophila* embryonic cells, however, HP1a binding is required to establish the clustering of pericentromeric regions and the overall chromosome folding, while it is dispensable in differentiated cells for these functions (Zenk et al. 2021). Thus, multiple studies arrive at the conclusion that HP1 does not induce chromatin compaction in differentiated cells. Rather, chromocenter-specific interactions of HP1, which can act as a transcription repressor (Hathaway et al. 2012), might prevent spurious induction of satellite repeat transcription (Erdel et al. 2020). In this manner, HP1 would stabilize the transcriptional silencing of a collapsed chromatin globule (Fig. 1C) rather than forming liquid droplets (Fig. 1E). On the other hand, MeCP2, linker histones, and KMT5C are important for the structural integrity of chromocenters. MeCP2 induces clustering of pericentric heterochromatin upon overexpression in mouse myoblasts (Brero et al. 2005) and it could thus be involved in changes of chromocenter structure. In fact, mutations of MeCP2 that cause Rett syndrome, a severe neurological disorder, have recently been proposed to be detrimental because they prevent the formation of liquid droplets in vitro (Fan et al. 2020; Li et al. 2020a; Wang et al. 2020). It is noted, however, that the structural phenotype of these MeCP2 mutations has been rationalized previously as being the result of perturbed chromatin interactions that decrease the ability of MeCP2 to compact heterochromatin (Agarwal et al. 2011). In addition, in neurons, loss of MeCP2 is accompanied by redistribution of the H3K20me3 modification at chromocenters (Linhoff et al. 2015). Linker histones are highly abundant in the nucleus at a stoichiometry of about 0.7 H1 per nucleosome (Fan et al. 2003) and enriched at mouse chromocenters (Cao et al. 2013). Their depletion leads to chromocenter clustering and de-repression of the major satellite repeat sequences in them (Cao et al. 2013; Healton et al. 2020). In *Drosophila*, H1 is also required for the structural integrity of chromocenters (Lu et al. 2009). Since linker histones have been shown to undergo LLPS in vitro (Gibson et al. 2019; Shakya et al. 2020; Muzzopappa et al. 2021), it will be important to investigate whether H1 at chromocenters in the cell displays material properties indicative of its accumulation via LLPS. It is noted, however, that the ability of H1 to form liquid droplets is lost with increasing DNA length, which promotes the formation of more solid-like aggregates (Muzzopappa et al. 2021). Finally, KMT5C is enriched at chromocenters and mediates changes of pericentric repeat organization and chromatin accessibility (Hahn et al. 2013). As discussed above, its mobility appears to be confined to chromocenters (Strickfaden et al. 2020), which makes it an interesting protein for further investigation of LLPS at chromocenters. Apart from dissecting the contributions of factors beyond HP1 to the dynamic chromocenter organization, it will be important to further investigate embryonic cells. In these cells in *Drosophila*, the chromocenter mobility of HP1a is increased (Strom et al. 2017) and the protein is required for the 3D organization of pericentromeric heterochromatin (Zenk et al. 2021).

ASSESSING THE CONTRIBUTION OF LLPS TO THE STRUCTURE OF THE NUCLEOLUS, POL II TRANSCRIPTION FACTORIES, AND CHROMOCENTERS

With respect to the three CSCs compared here, the following tentative assignment is made: Evidence for a CSC formed in the cell by LLPS is currently strongest for the nucleolus, which has a number of features in support of this mechanism. These comprise liquid-like properties of constituting factors, transitions between coalescent and dispersed states, and an increased local viscosity as discussed above. These features are likely to be related to its unusual composition with respect to the high enrichment of RNA and proteins and very low DNA content. Thus, the overall properties of the nucleolus are dominated by multivalent interactions of protein and

RNA. The direct association of these factors with the DNA of the nucleolar organizer regions makes a relatively small contribution albeit being important for nucleating and targeting the assembly. It is noted that the transcribed rDNA locus adopts a folded conformation (Maiser et al. 2020) and a recent study in budding yeast reports that it forms distinct condensates by a polymer–polymer phase separation (Fig. 1C) within an LLPS subcompartment of ribonucleoproteins (Lawrimore et al. 2021).

For Pol II transcription factories or transcriptional condensates, it is difficult to conclude at this stage by which mechanism they form. Several lines of evidence indicate that multivalent interactions mediated by IDRs are important to form the active transcription machinery. Many of these IDR-containing factors have a high propensity to undergo LLPS as demonstrated with purified proteins in vitro. However, evidence that such a phase separation indeed occurs under endogenous conditions in the cell is scarce. Rather, multivalent interactions of IDRs might simply mediate protein–protein interactions between specific and general TFs as well as coactivators (Fig. 3; Chong et al. 2018; Trojanowski et al. 2021). Furthermore, it is currently an open question whether the formation of a liquid droplet state induced by sufficiently high endogenous cellular protein concentration would indeed amplify gene expression or increase transcription activation.

For chromocenters, a number of criteria and corresponding experimental tests to dissect how this type of CSC is formed in mouse fibroblasts have been presented (Erdel et al. 2020). The results argue against HP1-driven LLPS as a major driver of chromocenter formation. A similar type of analysis appears to be warranted to make conclusions about LLPS at chromocenters in other organisms or cell types such as embryonic stem cells in *Drosophila* where HP1a can affect chromatin organization (Zenk et al. 2021). In differentiated cells, HP1 appears to be irrelevant for chromocenter structure as corresponding phenotypes are lacking as discussed above (Peters et al. 2001; Mateos-Langerak et al. 2007; Aucott et al. 2008; Singh 2010; Mattout et al. 2015; Bosch-Presegué et al. 2017; Erdel et al. 2020; Zenk et al. 2021). These observations lead to the model that HP1 binds and bridges H3K9me3-modified nucleosomes without inducing chromatin compaction (Fig. 1B,C). In mouse cells, the latter process is likely to be driven by linker histone H1 that mediates the interchromosomal packing of the nucleosome chain (Hansen 2020) and counteracts clustering of chromocenters from different chromosomes (Cao et al. 2013). This clustering could be mediated by DNA methylation-dependent chromatin binding of MeCP2 (Brero et al. 2005; Agarwal et al. 2011), which competes with H1 for binding sites (Ghosh et al. 2010). In *Drosophila*, which lacks DNA methylation, chromocenter clustering is dependent on D1 and Prod (Jagannathan et al. 2019). The resulting chromocenter conformation in mouse fibroblasts would be that of a collapsed chromatin globule induced by H1- and MeCP2-mediated interactions between the nucleosome chain (Fig. 1C) and HP1 binding providing an additional safeguard against spurious transcription activation (Erdel et al. 2020).

CONCLUSIONS

The concept of LLPS-driven assembly of chromatin compartments provides a novel and inspiring perspective on how the cell organizes genome-associated activities. Such a mechanism could have far-reaching implications and has been associated with a variety of human pathologies like Rett syndrome (Fan et al. 2020; Li et al. 2020a; Wang et al. 2020), oncogenic RNA splicing (Li et al. 2020b), and various neurodegenerative diseases (Zbinden et al. 2020). The latter, together with developmental disorders, could involve deregulated LLPS due to the expansion of repeat sequences within TFs (Basu et al. 2020). Another study linked the formation of nuclear droplets to drug targeting and metabolism via the preferential enrichment of anticancer drugs in CSCs (Klein et al. 2020). However, as discussed here, a number of considerations and findings challenge the general application of the LLPS mechanism to chromatin: (1) The formation of a CSC is clearly different from the assembly of a complex that only comprises protein and RNA, such as a cytoplasmic P body, which is devoid of chromatin. The binding of proteins and RNA to clustered sites on a mostly immobile chromatin scaffold could be fully sufficient to target genome-associated activities to specific loci in the nucleus. Thus, invoking LLPS to rationalize local chromatin enrichment might be a solution to a problem that does not exist for chromatin patterning in many instances. (2) CSCs have very heterogeneous properties as shown here for three exemplary cases. Thus, a "one-size-fits-all" approach does not seem appropriate to rationalize how CSCs are formed. Accordingly, a more systematic comparison of different mechanisms and cell types against each other is needed that considers the scenarios depicted in Figure 1. (3) Informative material properties like high-resolution structure, mixing within the CSC versus the exchange with the surrounding nucleoplasm, RNA/DNA/protein content and local viscosity need to be determined in a consistent and well-defined manner in living cells. In some instances, the currently available results argue in favor of an LLPS while in others against it. (4) A general challenge in the field of chromatin organization is to derive structure–function relationships for a given CSC. This is exemplified by the well-established organization of the genome into TADs. Despite their ubiquitous presence across organisms, defining the specific functions of TADs has proven to be difficult (Beagan and Phillips-Cremins 2020; Cavalheiro et al. 2021). Likewise, for LLPS, even for artificial systems with ectopic expression of factors, evidence is often lacking that the transition from direct and indirect chromatin binding to a phase-separated droplet state is associated with functional changes. (5) Perturbation experiments of proteins and RNA factors as well as chromatin states are highly informative to reveal underlying organization principles and could be integrated with structural features in high-content screening approaches (Berchtold et al. 2018). In combination with appropriate readouts, structure–function relationships can be revealed. Thus, perturbation analyses should be integrated more frequently into studies of phase separation in chromatin. In summary, an integrative approach that considers different mechanisms across a variety of CSCs is needed to elucidate the role of phase separation as a self-organizing principle of chromatin domains. Toward this goal, the "infusion" of the field by biophysical experimental methods and quantitative mechanistic models in the context of phase separation studies creates a unique opportunity to take our understanding of chromatin patterning and its functional consequences to the next level.

ACKNOWLEDGMENTS

I am grateful to Maïwen Caudron-Herger, Akis Papantonis, Jorge Trojanowski, Lukas Frank, and Robin Weinmann for discussion. Work from my laboratory is supported by DFG Priority Program 2191 "Molecular Mechanisms of Functional Phase Separation" via Grant RI1283/16-1.

REFERENCES

*Reference is also in this collection.

Agarwal N, Becker A, Jost KL, Haase S, Thakur BK, Brero A, Hardt T, Kudo S, Leonhardt H, Cardoso MC. 2011. MeCP2 Rett mutations affect large scale chromatin organization. *Hum Mol Genet* **20:** 4187–4195. doi:10.1093/hmg/ddr346

Andersen JS, Lam YW, Leung AK, Ong SE, Lyon CE, Lamond AI, Mann M. 2005. Nucleolar proteome dynamics. *Nature* **433:** 77–83. doi:10.1038/nature03207

Aucott R, Bullwinkel J, Yu Y, Shi W, Billur M, Brown JP, Menzel U, Kioussis D, Wang G, Reisert I, et al. 2008. HP1-β is required for development of the cerebral neocortex and neuromuscular junctions. *J Cell Biol* **183:** 597–606. doi:10.1083/jcb.200804041

Banani SF, Lee HO, Hyman AA, Rosen MK. 2017. Biomolecular condensates: organizers of cellular biochemistry. *Nat Rev Mol Cell Biol* **18:** 285–298. doi:10.1038/nrm.2017.7

Barski A, Cuddapah S, Cui K, Roh TY, Schones DE, Wang Z, Wei G, Chepelev I, Zhao K. 2007. High-resolution profiling of histone methylations in the human genome. *Cell* **129:** 823–837. doi:10.1016/j.cell.2007.05.009

Basu S, Mackowiak SD, Niskanen H, Knezevic D, Asimi V, Grosswendt S, Geertsema H, Ali S, Jerkovic I, Ewers H, et al. 2020. Unblending of transcriptional condensates in human repeat expansion disease. *Cell* **181:** 1062–1079.e30. doi:10.1016/j.cell.2020.04.018

Bates FS. 1991. Polymer-polymer phase behavior. *Science* **251:** 898–905. doi:10.1126/science.251.4996.898

Baum M, Erdel F, Wachsmuth M, Rippe K. 2014. Retrieving the intracellular topology from multi-scale protein mobility mapping in living cells. *Nat Commun* **5:** 4494. doi:10.1038/ncomms5494

Beagan JA, Phillips-Cremins JE. 2020. On the existence and functionality of topologically associating domains. *Nat Genet* **52:** 8–16. doi:10.1038/s41588-019-0561-1

Becker M, Baumann C, John S, Walker DA, Vigneron M, McNally JG, Hager GL. 2002. Dynamic behavior of transcription factors on a natural promoter in living cells. *EMBO Rep* **3:** 1188–1194. doi:10.1093/embo-reports/kvf244

* Belmont A. 2021. Nuclear compartments: an incomplete primer to nuclear compartments, bodies, and genome organization relative to nuclear architecture. *Cold Spring Harb Perspect Biol* doi: 10.1101/cshperspect.a040154

Berchtold D, Battich N, Pelkmans L. 2018. A systems-level study reveals regulators of membrane-less organelles in human cells. *Mol Cell* **72:** 1035–1049.e5. doi:10.1016/j.molcel.2018.10.036

Berry J, Weber SC, Vaidya N, Haataja M, Brangwynne CP. 2015. RNA transcription modulates phase transition-driven nuclear body assembly. *Proc Natl Acad Sci* **112:** E5237–E5245. doi:10.1073/pnas.1509317112

Boehning M, Dugast-Darzacq C, Rankovic M, Hansen AS, Yu T, Marie-Nelly H, McSwiggen DT, Kokic G, Dailey GM, Cramer P, et al. 2018. RNA polymerase II clustering through carboxy-terminal domain phase separation. *Nat Struct Mol Biol* **25:** 833–840. doi:10.1038/s41594-018-0112-y

Boeynaems S, Alberti S, Fawzi NL, Mittag T, Polymenidou M, Rousseau F, Schymkowitz J, Shorter J, Wolozin B, Van Den Bosch L, et al. 2018. Protein phase separation: a new phase in cell biology. *Trends Cell Biol* **28:** 420–435. doi:10.1016/j.tcb.2018.02.004

Bosch-Presegué L, Raurell-Vila H, Thackray JK, Gonzalez J, Casal C, Kane-Goldsmith N, Vizoso M, Brown JP, Gómez A, Ausio J, et al. 2017. Mammalian HP1 isoforms have specific roles in heterochromatin structure and organization. *Cell Rep* **21:** 2048–2057. doi:10.1016/j.celrep.2017.10.092

Brackley CA, Taylor S, Papantonis A, Cook PR, Marenduzzo D. 2013. Nonspecific bridging-induced attraction drives clustering of DNA-binding proteins and genome organization. *Proc Natl Acad Sci* **110:** E3605–E3611. doi:10.1073/pnas.1302950110

Brangwynne CP, Eckmann CR, Courson DS, Rybarska A, Hoege C, Gharakhani J, Julicher F, Hyman AA. 2009. Germline P granules are liquid droplets that localize by controlled dissolution/condensation. *Science* **324:** 1729–1732. doi:10.1126/science.1172046

Brangwynne CP, Mitchison TJ, Hyman AA. 2011. Active liquid-like behavior of nucleoli determines their size and shape in *Xenopus laevis* oocytes. *Proc Natl Acad Sci* **108:** 4334–4339. doi:10.1073/pnas.1017150108

Brero A, Easwaran HP, Nowak D, Grunewald I, Cremer T, Leonhardt H, Cardoso MC. 2005. Methyl CpG-binding proteins induce large-scale chromatin reorganization during terminal differentiation. *J Cell Biol* **169:** 733–743. doi:10.1083/jcb.200502062

Brodsky S, Jana T, Mittelman K, Chapal M, Kumar DK, Carmi M, Barkai N. 2020. Intrinsically disordered regions direct transcription factor in vivo binding specificity. *Mol Cell* **79:** 459–471.e4. doi:10.1016/j.molcel.2020.05.032

Brown DT, Izard T, Misteli T. 2006. Mapping the interaction surface of linker histone $H1^0$ with the nucleosome of native chromatin in vivo. *Nat Struct Mol Biol* **13:** 250–255. doi:10.1038/nsmb1050

Buckley MS, Lis JT. 2014. Imaging RNA polymerase II transcription sites in living cells. *Curr Opin Genet Dev* **25:** 126–130. doi:10.1016/j.gde.2014.01.002

Cao K, Lailler N, Zhang Y, Kumar A, Uppal K, Liu Z, Lee EK, Wu H, Medrzycki M, Pan C, et al. 2013. High-resolution mapping of h1 linker histone variants in embryonic stem cells. *PLoS Genet* **9:** e1003417. doi:10.1371/journal.pgen.1003417

Caragine CM, Haley SC, Zidovska A. 2019. Nucleolar dynamics and interactions with nucleoplasm in living cells. *eLife* **8:** e47533. doi:10.7554/eLife.47533

Carron C, Balor S, Delavoie F, Plisson-Chastang C, Faubladier M, Gleizes PE, O'Donohue MF. 2012. Post-mitotic dynamics of pre-nucleolar bodies is driven by pre-rRNA processing. *J Cell Sci* **125:** 4532–4542. doi:10.1242/jcs.106419

Caudron-Herger M, Rippe K. 2012. Nuclear architecture by RNA. *Curr Opin Genet Dev* **22:** 179–187. doi:10.1016/j.gde.2011.12.005

Caudron-Herger M, Cook PR, Rippe K, Papantonis A. 2015a. Dissecting the nascent human transcriptome by analysing the RNA content of transcription factories. *Nucleic Acids Res* **43:** e95. doi:10.1093/nar/gkv390

Caudron-Herger M, Pankert T, Seiler J, Nemeth A, Voit R, Grummt I, Rippe K. 2015b. Alu element-containing RNAs maintain nucleolar structure and function. *EMBO J* **34:** 2758–2774. doi:10.15252/embj.201591458

Caudron-Herger M, Pankert T, Rippe K. 2016. Regulation of nucleolus assembly by non-coding RNA polymerase II transcripts. *Nucleus* **7:** 308–318. doi:10.1080/19491034.2016.1190890

Cavalheiro GR, Pollex T, Furlong EE. 2021. To loop or not to loop: what is the role of TADs in enhancer function and gene regulation? *Curr Opin Genet Dev* **67:** 119–129. doi:10.1016/j.gde.2020.12.015

Chakalova L, Debrand E, Mitchell JA, Osborne CS, Fraser P. 2005. Replication and transcription: shaping the landscape of the genome. *Nat Rev Genet* **6:** 669–677. doi:10.1038/nrg1673

Chen D, Huang S. 2001. Nucleolar components involved in ribosome biogenesis cycle between the nucleolus and nucleoplasm in interphase cells. *J Cell Biol* **153:** 169–176. doi:10.1083/jcb.153.1.169

Chen B, Gilbert LA, Cimini BA, Schnitzbauer J, Zhang W, Li GW, Park J, Blackburn EH, Weissman JS, Qi LS, et al. 2013. Dynamic imaging of genomic loci in living human cells by an optimized CRISPR/Cas system. *Cell* **155:** 1479–1491. doi:10.1016/j.cell.2013.12.001

Chen J, Zhang Z, Li L, Chen BC, Revyakin A, Hajj B, Legant W, Dahan M, Lionnet T, Betzig E, et al. 2014. Single-molecule dynamics of enhanceosome assembly in embryonic stem cells. *Cell* **156:** 1274–1285. doi:10.1016/j.cell.2014.01.062

Cheutin T, McNairn AJ, Jenuwein T, Gilbert DM, Singh PB, Misteli T. 2003. Maintenance of stable heterochromatin domains by dynamic HP1 binding. *Science* **299:** 721–725. doi:10.1126/science.1078572

Cho WK, Spille JH, Hecht M, Lee C, Li C, Grube V, Cisse II. 2018. Mediator and RNA polymerase II clusters associate in transcription-dependent condensates. *Science* **361:** 412–415. doi:10.1126/science.aar4199

Choi JM, Holehouse AS, Pappu RV. 2020. Physical principles underlying the complex biology of intracellular phase transitions. *Annu Rev Biophys* **49:** 107–133. doi:10.1146/annurev-biophys-121219-081629

Chong S, Dugast-Darzacq C, Liu Z, Dong P, Dailey GM, Cattoglio C, Heckert A, Banala S, Lavis L, Darzacq X, et al. 2018. Imaging dynamic and selective low-complexity domain interactions that control gene transcription. *Science* **361:** eaar2555. doi:10.1126/science.aar2555

Chubb JR, Boyle S, Perry P, Bickmore WA. 2002. Chromatin motion is constrained by association with nuclear compartments in human cells. *Curr Biol* **12:** 439–445. doi:10.1016/S0960-9822(02)00695-4

Cisse II, Izeddin I, Causse SZ, Boudarene L, Senecal A, Muresan L, Dugast-Darzacq C, Hajj B, Dahan M, Darzacq X. 2013. Real-time dynamics of RNA polymerase II clustering in live human cells. *Science* **341:** 664–667. doi:10.1126/science.1239053

Cook PR. 2002. Predicting three-dimensional genome structure from transcriptional activity. *Nat Genet* **32:** 347–352. doi:10.1038/ng1102-347

Cook PR, Marenduzzo D. 2018. Transcription-driven genome organization: a model for chromosome structure and the regulation of gene expression tested through simulations. *Nucleic Acids Res* **46:** 9895–9906. doi:10.1093/nar/gky763

Cremer T, Cremer M, Hübner B, Strickfaden H, Smeets D, Popken J, Sterr M, Markaki Y, Rippe K, Cremer C. 2015. The 4D nucleome: evidence for a dynamic nuclear landscape based on co-aligned active and inactive nuclear compartments. *FEBS Lett* **589:** 2931–2943. doi:10.1016/j.febslet.2015.05.037

Darzacq X, Shav-Tal Y, de Turris V, Brody Y, Shenoy SM, Phair RD, Singer RH. 2007. In vivo dynamics of RNA polymerase II transcription. *Nat Struct Mol Biol* **14:** 796–806. doi:10.1038/nsmb1280

Derenzini M, Trere D, Pession A, Govoni M, Sirri V, Chieco P. 2000. Nucleolar size indicates the rapidity of cell proliferation in cancer tissues. *J Pathol* **191:** 181–186.

Drino A, Schaefer MR. 2018. RNAs, phase separation, and membrane-less organelles: are post-transcriptional modifications modulating organelle dynamics? *Bioessays* **40:** 1800085. doi:10.1002/bies.201800085

Dundr M, Hoffmann-Rohrer U, Hu Q, Grummt I, Rothblum LI, Phair RD, Misteli T. 2002. A kinetic framework for a mammalian RNA polymerase in vivo. *Science* **298:** 1623–1626. doi:10.1126/science.1076164

Erdel F, Rippe K. 2018. Formation of chromatin subcompartments by phase separation. *Biophys J* **114:** 2262–2270. doi:10.1016/j.bpj.2018.03.011

Erdel F, Rademacher A, Vlijm R, Tünnermann J, Frank L, Weinmann R, Schweigert E, Yserentant K, Hummert J, Bauer C, et al. 2020. Mouse heterochromatin adopts digital compaction states without showing hallmarks of HP1-driven liquid-liquid phase separation. *Mol Cell* **78:** 236–249.e7. doi:10.1016/j.molcel.2020.02.005

Ernst J, Kheradpour P, Mikkelsen TS, Shoresh N, Ward LD, Epstein CB, Zhang X, Wang L, Issner R, Coyne M, et al. 2011. Mapping and analysis of chromatin state dynamics in nine human cell types. *Nature* **473:** 43–49. doi:10.1038/nature09906

Falahati H, Pelham-Webb B, Blythe S, Wieschaus E. 2016. Nucleation by rRNA dictates the precision of nucleolus assembly. *Curr Biol* **26:** 277–285. doi:10.1016/j.cub.2015.11.065

Fan Y, Nikitina T, Morin-Kensicki EM, Zhao J, Magnuson TR, Woodcock CL, Skoultchi AI. 2003. H1 linker histones are essential for mouse development and affect nucleosome spacing in vivo. *Mol Cell Biol* **23:** 4559–4572. doi:10.1128/MCB.23.13.4559-4572.2003

Fan C, Zhang H, Fu L, Li Y, Du Y, Qiu Z, Lu F. 2020. Rett mutations attenuate phase separation of MeCP2. *Cell Discov* **6:** 38. doi:10.1038/s41421-020-0172-0

Feric M, Vaidya N, Harmon TS, Mitrea DM, Zhu L, Richardson TM, Kriwacki RW, Pappu RV, Brangwynne CP. 2016. Coexisting liquid phases underlie nucleolar subcompartments. *Cell* **165:** 1686–1697. doi:10.1016/j.cell.2016.04.047

Festenstein R, Pagakis SN, Hiragami K, Lyon D, Verreault A, Sekkali B, Kioussis D. 2003. Modulation of heterochromatin protein 1 dynamics in primary mammalian cells. *Science* **299:** 719–721. doi:10.1126/science.1078694

Filion GJ, van Bemmel JG, Braunschweig U, Talhout W, Kind J, Ward LD, Brugman W, de Castro IJ, Kerkhoven RM, Bussemaker HJ, et al. 2010. Systematic protein location mapping reveals five principal chromatin types in *Drosophila* cells. *Cell* **143:** 212–224. doi:10.1016/j.cell.2010.09.009

Fodor BD, Shukeir N, Reuter G, Jenuwein T. 2010. Mammalian *Su(var)* genes in chromatin control. *Annu Rev Cell Dev Biol* **26:** 471–501. doi:10.1146/annurev.cellbio.042308.113225

Frank L, Rippe K. 2020. Repetitive RNAs as regulators of chromatin-associated subcompartment formation by phase separation. *J Mol Biol* **432:** 4270–4286. doi:10.1016/j.jmb.2020.04.015

Frottin F, Schueder F, Tiwary S, Gupta R, Körner R, Schlichthaerle T, Cox J, Jungmann R, Hartl FU, Hipp MS. 2019. The nucleolus functions as a phase-separated protein quality control compartment. *Science* **365:** 342–347. doi:10.1126/science.aaw9157

Fussner E, Strauss M, Djuric U, Li R, Ahmed K, Hart M, Ellis J, Bazett-Jones DP. 2012. Open and closed domains in the mouse genome are configured as 10-nm chromatin fibres. *EMBO Rep* **13:** 992–996. doi:10.1038/embor.2012.139

Garcia DA, Fettweis G, Presman DM, Paakinaho V, Jarzynski C, Upadhyaya A, Hager GL. 2021a. Power-law behavior of transcription factor dynamics at the single-molecule level implies a continuum affinity model. *Nucleic Acids Res* doi:10.1093/nar/gkab072

Garcia DA, Johnson TA, Presman DM, Fettweis G, Wagh K, Rinaldi L, Stavreva DA, Paakinaho V, Jensen RAM, Mandrup S, et al. 2021b. An intrinsically disordered region-mediated confinement state contributes to the dynamics and function of transcription factors. *Mol Cell* **81:** 1484–1498.e6. doi:10.1016/j.molcel.2021.01.013

Gerlich D, Beaudouin J, Kalbfuss B, Daigle N, Eils R, Ellenberg J. 2003. Global chromosome positions are transmitted through mitosis in mammalian cells. *Cell* **112:** 751–764. doi:10.1016/S0092-8674(03)00189-2

Ghamari A, van de Corput MP, Thongjuea S, van Cappellen WA, van Ijcken W, van Haren J, Soler E, Eick D, Lenhard B, Grosveld FG. 2013. In vivo live imaging of RNA polymerase II transcription factories in primary cells. *Genes Dev* **27:** 767–777. doi:10.1101/gad.216200.113

Ghosh RP, Horowitz-Scherer RA, Nikitina T, Shlyakhtenko LS, Woodcock CL. 2010. MeCP2 binds cooperatively to its substrate and competes with histone H1 for chromatin binding sites. *Mol Cell Biol* **30:** 4656–4670. doi:10.1128/MCB.00379-10

Gibson BA, Doolittle LK, Schneider MWG, Jensen LE, Gamarra N, Henry L, Gerlich DW, Redding S, Rosen MK. 2019. Organization of chromatin by intrinsic and regulated phase separation. *Cell* **179:** 470–484.e21. doi:10.1016/j.cell.2019.08.037

Gorski SA, Snyder SK, John S, Grummt I, Misteli T. 2008. Modulation of RNA polymerase assembly dynamics in transcriptional regulation. *Mol Cell* **30:** 486–497. doi:10.1016/j.molcel.2008.04.021

Grob A, Colleran C, McStay B. 2014. Construction of synthetic nucleoli in human cells reveals how a major functional nuclear domain is formed and propagated through cell division. *Genes Dev* **28:** 220–230. doi:10.1101/gad.234591.113

Guo YE, Manteiga JC, Henninger JE, Sabari BR, Dall'Agnese A, Hannett NM, Spille JH, Afeyan LK, Zamudio AV, Shrinivas K, et al. 2019. Pol II phosphorylation regulates a switch between transcriptional and splicing condensates. *Nature* **572:** 543–548. doi:10.1038/s41586-019-1464-0

Gutierrez PS, Monteoliva D, Diambra L. 2012. Cooperative binding of transcription factors promotes bimodal gene expression response. *PLoS ONE* **7:** e44812. doi:10.1371/journal.pone.0044812

Hahn M, Dambacher S, Dulev S, Kuznetsova AY, Eck S, Worz S, Sadic D, Schulte M, Mallm JP, Maiser A, et al. 2013. Suv4-20h2 mediates chromatin compaction and is important for cohesin recruitment to heterochromatin. *Genes Dev* **27:** 859–872. doi:10.1101/gad.210377.112

Hansen JC. 2020. Silencing the genome with linker histones. *Proc Natl Acad Sci* **117:** 15388–15390. doi:10.1073/pnas.2009513117

Hathaway NA, Bell O, Hodges C, Miller EL, Neel DS, Crabtree GR. 2012. Dynamics and memory of heterochromatin in living cells. *Cell* **149:** 1447–1460. doi:10.1016/j.cell.2012.03.052

Healton SE, Pinto HD, Mishra LN, Hamilton GA, Wheat JC, Swist-Rosowska K, Shukeir N, Dou Y, Steidl U, Jenuwein T, et al. 2020. H1 linker histones silence repetitive elements by promoting both histone H3K9 methylation and chromatin compaction. *Proc Natl Acad Sci* **117:** 14251–14258.

Henninger JE, Oksuz O, Shrinivas K, Sagi I, LeRoy G, Zheng MM, Andrews JO, Zamudio AV, Lazaris C, Hannett NM, et al. 2021. RNA-mediated feedback control of transcriptional condensates. *Cell* **184:** 207–225.e24. doi:10.1016/j.cell.2020.11.030

Hieda M, Winstanley H, Maini P, Iborra FJ, Cook PR. 2005. Different populations of RNA polymerase II in living mammalian cells. *Chromosome Res* **13:** 135–144. doi:10.1007/s10577-005-7720-1

Hilbert L, Sato Y, Kuznetsova K, Bianucci T, Kimura H, Jülicher F, Honigmann A, Zaburdaev V, Vastenhouw NL. 2021. Transcription organizes euchromatin via microphase separation. *Nat Commun* **12:** 1360. doi:10.1038/s41467-021-21589-3

Hildebrand EM, Dekker J. 2020. Mechanisms and functions of chromosome compartmentalization. *Trends Biochem Sci* **45:** 385–396. doi:10.1016/j.tibs.2020.01.002

Hnisz D, Shrinivas K, Young RA, Chakraborty AK, Sharp PA. 2017. A phase separation model for transcriptional control. *Cell* **169:** 13–23. doi:10.1016/j.cell.2017.02.007

Hyman AA, Weber CA, Jülicher F. 2014. Liquid–liquid phase separation in biology. *Annu Rev Cell Dev Biol* **30:** 39–58. doi:10.1146/annurev-cellbio-100913-013325

Iborra FJ, Pombo A, Jackson DA, Cook PR. 1996. Active RNA polymerases are localized within discrete transcription "factories" in human nuclei. *J Cell Sci* **109:** 1427–1436.

Ide S, Imai R, Ochi H, Maeshima K. 2020. Transcriptional suppression of ribosomal DNA with phase separation. *Sci Adv* **6:** eabb5963. doi:10.1126/sciadv.abb5953

Jackson DA, Hassan AB, Errington RJ, Cook PR. 1993. Visualization of focal sites of transcription within human nuclei. *EMBO J* **12:** 1059–1065. doi:10.1002/j.1460-2075.1993.tb05747.x

Jackson DA, Iborra FJ, Manders EM, Cook PR. 1998. Numbers and organization of RNA polymerases, nascent transcripts, and transcription units in HeLa nuclei. *Mol Biol Cell* **9:** 1523–1536. doi:10.1091/mbc.9.6.1523

Jagannathan M, Cummings R, Yamashita YM. 2019. The modular mechanism of chromocenter formation in *Drosophila*. *eLife* **8:** e43938. doi:10.7554/eLife.43938

Janssen A, Colmenares SU, Karpen GH. 2018. Heterochromatin: guardian of the genome. *Annu Rev Cell Dev Biol* **34:** 265–288. doi:10.1146/annurev-cellbio-100617-062653

Jegou T, Chung I, Heuvelman G, Wachsmuth M, Görisch SM, Greulich-Bode KM, Boukamp P, Lichter P, Rippe K. 2009. Dynamics of telomeres and promyelocytic leukemia nuclear bodies in a telomerase-negative human cell line. *Mol Biol Cell* **20:** 2070–2082. doi:10.1091/mbc.e08-02-0108

Jost D, Vaillant C, Meister P. 2017. Coupling 1D modifications and 3D nuclear organization: data, models and function. *Curr Opin Cell Biol* **44:** 20–27. doi:10.1016/j.ceb.2016.12.001

Kent S, Brown K, Yang CH, Alsaihati N, Tian C, Wang H, Ren X. 2020. Phase-separated transcriptional condensates accelerate target-search process revealed by live-cell single-molecule imaging. *Cell Rep* **33:** 108248. doi:10.1016/j.celrep.2020.108248

Kimura H, Cook PR. 2001. Kinetics of core histones in living human cells: little exchange of H3 and H4 and some rapid exchange of H2B. *J Cell Biol* **153:** 1341–1354. doi:10.1083/jcb.153.7.1341

Kimura H, Sugaya K, Cook PR. 2002. The transcription cycle of RNA polymerase II in living cells. *J Cell Biol* **159:** 777–782. doi:10.1083/jcb.200206019

Klein IA, Boija A, Afeyan LK, Hawken SW, Fan M, Dall' Agnese A, Oksuz O, Henninger JE, Shrinivas K, Sabari BR, et al. 2020. Partitioning of cancer therapeutics in nuclear condensates. *Science* **368:** 1386–1392. doi:10.1126/science.aaz4427

Kochanova NY, Schauer T, Mathias GP, Lukacs A, Schmidt A, Flatley A, Schepers A, Thomae AW, Imhof A. 2020. A multi-layered structure of the interphase chromocenter revealed by proximity-based biotinylation. *Nucleic Acids Res* **48:** 4161–4178. doi:10.1093/nar/gkaa145

Krietenstein N, Abraham S, Venev SV, Abdennur N, Gibcus J, Hsieh TS, Parsi KM, Yang L, Maehr R, Mirny LA, et al. 2020. Ultrastructural details of mammalian chromosome architecture. *Mol Cell* **78:** 554–565.e7. doi:10.1016/j.molcel.2020.03.003

Kwon I, Kato M, Xiang S, Wu L, Theodoropoulos P, Mirzaei H, Han T, Xie S, Corden JL, McKnight SL. 2013. Phosphorylation-regulated binding of RNA polymerase II to fibrous polymers of low-complexity domains. *Cell* **155:** 1049–1060. doi:10.1016/j.cell.2013.10.033

Lafontaine DLJ, Riback JA, Bascetin R, Brangwynne CP. 2021. The nucleolus as a multiphase liquid condensate. *Nat Rev Mol Cell Biol* **22:** 165–182. doi:10.1038/s41580-020-0272-6

Larson AG, Elnatan D, Keenen MM, Trnka MJ, Johnston JB, Burlingame AL, Agard DA, Redding S, Narlikar GJ. 2017. Liquid droplet formation by HP1α suggests a role for phase separation in heterochromatin. *Nature* **547:** 236–240. doi:10.1038/nature22822

Lawrimore J, Kolbin D, Stanton J, Khan M, de Larminat SC, Lawrimore C, Yeh E, Bloom K. 2021. The rDNA is biomolecular condensate formed by polymer-polymer phase separation and is sequestered in the nucleolus by transcription and R-loops. *Nucleic Acids Res* **49:** 4586–4598. doi:10.1093/nar/gkab229

Leibler L. 1980. Theory of microphase separation in block co-polymers. *Macromolecules* **13:** 1602–1617. doi:10.1021/ma60078a047

Lever MA, Th'ng JP, Sun X, Hendzel MJ. 2000. Rapid exchange of histone H1.1 on chromatin in living human cells. *Nature* **408:** 873–876. doi:10.1038/35048603

Levi V, Ruan Q, Plutz M, Belmont AS, Gratton E. 2005. Chromatin dynamics in interphase cells revealed by tracking in a two-photon excitation microscope. *Biophys J* **89:** 4275–4285. doi:10.1529/biophysj.105.066670

Li X, Fu XD. 2019. Chromatin-associated RNAs as facilitators of functional genomic interactions. *Nat Rev Genet* **20:** 503–519. doi:10.1038/s41576-019-0135-1

Li CH, Coffey EL, Dall'Agnese A, Hannett NM, Tang X, Henninger JE, Platt JM, Oksuz O, Zamudio AV, Afeyan LK, et al. 2020a. MeCP2 links heterochromatin condensates and neurodevelopmental disease. *Nature* **586:** 440–444. doi:10.1038/s41586-020-2574-4

Li W, Hu J, Shi B, Palomba F, Digman MA, Gratton E, Jiang H. 2020b. Biophysical properties of AKAP95 protein condensates regulate splicing and tumorigenesis. *Nat Cell Biol* **22:** 960–972. doi:10.1038/s41556-020-0550-8

Lieberman-Aiden E, van Berkum NL, Williams L, Imakaev M, Ragoczy T, Telling A, Amit I, Lajoie BR, Sabo PJ, Dorschner MO, et al. 2009. Comprehensive mapping of long-range interactions reveals folding principles of the human genome. *Science* **326:** 289–293. doi:10.1126/science.1181369

Linhoff MW, Garg SK, Mandel G. 2015. A high-resolution imaging approach to investigate chromatin architecture in complex tissues. *Cell* **163:** 246–255. doi:10.1016/j.cell.2015.09.002

Lionnet T, Wu C. 2021. Single-molecule tracking of transcription protein dynamics in living cells: seeing is believing, but what are we seeing? *Curr Opin Genet Dev* **67:** 94–102. doi:10.1016/j.gde.2020.12.001

* Lu F, Lionnet T. 2021. Transcription factor dynamics. *Cold Spring Harb Perspect Biol* doi: 10.1101/cshperspect.a040949

Lu X, Wontakal SN, Emelyanov AV, Morcillo P, Konev AY, Fyodorov DV, Skoultchi AI. 2009. Linker histone H1 is essential for *Drosophila* development, the establishment of pericentric heterochromatin, and a normal polytene chromosome structure. *Genes Dev* **23:** 452–465. doi:10.1101/gad.1749309

Lu H, Yu D, Hansen AS, Ganguly S, Liu R, Heckert A, Darzacq X, Zhou Q. 2018. Phase-separation mechanism for C-terminal hyperphosphorylation of RNA polymerase II. *Nature* **558:** 318–323. doi:10.1038/s41586-018-0174-3

Ma L, Gao Z, Wu J, Zhong B, Xie Y, Huang W, Lin Y. 2021. Co-condensation between transcription factor and coactivator p300 modulates transcriptional bursting kinetics. *Mol Cell* **81:** 1682–1697.e7. doi:10.1016/j.molcel.2021.01.031

MacPherson Q, Beltran B, Spakowitz AJ. 2018. Bottom-up modeling of chromatin segregation due to epigenetic modifications. *Proc Natl Acad Sci* **115:** 12739–12744. doi:10.1073/pnas.1812268115

Maeshima K, Ide S, Hibino K, Sasai M. 2016a. Liquid-like behavior of chromatin. *Curr Opin Genet Dev* **37:** 36–45. doi:10.1016/j.gde.2015.11.006

Maeshima K, Rogge R, Tamura S, Joti Y, Hikima T, Szerlong H, Krause C, Herman J, Seidel E, DeLuca J, et al. 2016b. Nucleosomal arrays self-assemble into supramolecular globular structures lacking 30-nm fibers. *EMBO J* **35:** 1115–1132. doi:10.15252/embj.201592660

Maeshima K, Tamura S, Hansen JC, Itoh Y. 2020. Fluid-like chromatin: toward understanding the real chromatin organization present in the cell. *Curr Opin Cell Biol* **64:** 77–89. doi:10.1016/j.ceb.2020.02.016

* Maeshima K, Iida S, Tamura S. 2021. Physical nature of chromatin in the nucleus. *Cold Spring Harb Perspect Biol* **13:** a040675. doi:10.1101/cshperspect.a040675

Maiser A, Dillinger S, Langst G, Schermelleh L, Leonhardt H, Nemeth A. 2020. Super-resolution in situ analysis of active ribosomal DNA chromatin organization in the nucleolus. *Sci Rep* **10:** 7462. doi:10.1038/s41598-020-64589-x

Malhotra I, Oyarzún B, Mognetti BM. 2021. Unfolding of the chromatin fiber driven by overexpression of noninteracting bridging factors. *Biophys J* **120:** 1247–1256. doi:10.1016/j.bpj.2020.12.027

Mangan H, Gailín MO, McStay B. 2017. Integrating the genomic architecture of human nucleolar organizer regions with the biophysical properties of nucleoli. *FEBS J* **284:** 3977–3985. doi:10.1111/febs.14108

Martin RM, Ter-Avetisyan G, Herce HD, Ludwig AK, Lättig-Tünnemann G, Cardoso MC. 2015. Principles of protein targeting to the nucleolus. *Nucleus* **6:** 314–325. doi:10.1080/19491034.2015.1079680

Mateos-Langerak J, Brink MC, Luijsterburg MS, van der Kraan I, van Driel R, Verschure PJ. 2007. Pericentromeric heterochromatin domains are maintained without accumulation of HP1. *Mol Biol Cell* **18:** 1464–1471. doi:10.1091/mbc.e06-01-0025

Mattout A, Aaronson Y, Sailaja BS, Raghu Ram EV, Harikumar A, Mallm JP, Sim KH, Nissim-Rafinia M, Supper E, Singh PB, et al. 2015. Heterochromatin protein 1β (HP1β) has distinct functions and distinct nuclear distribution in pluripotent versus differentiated cells. *Genome Biol* **16:** 213. doi:10.1186/s13059-015-0760-8

McSwiggen DT, Hansen AS, Teves SS, Marie-Nelly H, Hao Y, Heckert AB, Umemoto KK, Dugast-Darzacq C, Tjian R, Darzacq X. 2019a. Evidence for DNA-mediated nuclear compartmentalization distinct from phase separation. *eLife* **8:** e47098. doi:10.7554/eLife.47098

McSwiggen DT, Mir M, Darzacq X, Tjian R. 2019b. Evaluating phase separation in live cells: diagnosis, caveats, and functional consequences. *Genes Dev* **33:** 1619–1634. doi:10.1101/gad.331520.119

Melnik S, Deng B, Papantonis A, Baboo S, Carr IM, Cook PR. 2011. The proteomes of transcription factories containing RNA polymerases I, II or III. *Nat Methods* **8:** 963–968. doi:10.1038/nmeth.1705

Michieletto D, Orlandini E, Marenduzzo D. 2016. Polymer model with epigenetic recoloring reveals a pathway for the de novo establishment and 3D organization of chromatin domains. *Phys Rev X* **6:** 041047.

* Mirny L, Dekker J. 2021. Mechanisms of chromosome folding and nuclear organization: their interplay and open questions. *Cold Spring Harb Perspect Biol* doi:10.1101/cshperspect.a040147

Miron E, Oldenkamp R, Brown JM, Pinto DMS, Xu CS, Faria AR, Shaban HA, Rhodes JDP, Innocent C, de Ornellas S, et al. 2020. Chromatin arranges in chains of mesoscale domains with nanoscale functional topography independent of cohesin. *Sci Adv* **6:** eaba8811. doi:10.1126/sciadv.aba8811

Misteli T. 2001. The concept of self-organization in cellular architecture. *J Cell Biol* **155:** 181–186. doi:10.1083/jcb.200108110

Misteli T. 2007. Beyond the sequence: cellular organization of genome function. *Cell* **128:** 787–800. doi:10.1016/j.cell.2007.01.028

Misteli T. 2020. The self-organizing genome: principles of genome architecture and function. *Cell* **183:** 28–45. doi:10.1016/j.cell.2020.09.014

Misteli T, Gunjan A, Hock R, Bustin M, Brown DT. 2000. Dynamic binding of histone H1 to chromatin in living cells. *Nature* **408:** 877–881. doi:10.1038/35048610

Mitrea DM, Cika JA, Guy CS, Ban D, Banerjee PR, Stanley CB, Nourse A, Deniz AA, Kriwacki RW. 2016. Nucleophosmin integrates within the nucleolus via multi-modal interactions with proteins displaying R-rich linear motifs and rRNA. *eLife* **5:** e13571. doi:10.7554/eLife.13571

Mitrea DM, Cika JA, Stanley CB, Nourse A, Onuchic PL, Banerjee PR, Phillips AH, Park CG, Deniz AA, Kriwacki RW. 2018. Self-interaction of NPM1 modulates multiple mechanisms of liquid–liquid phase separation. *Nat Commun* **9:** 842. doi:10.1038/s41467-018-03255-3

Montanaro L, Treré D, Derenzini M. 2008. Nucleolus, ribosomes, and cancer. *Am J Pathol* **173:** 301–310. doi:10.2353/ajpath.2008.070752

Mueller F, Stasevich TJ, Mazza D, McNally JG. 2013. Quantifying transcription factor kinetics: at work or at play? *Crit Rev Biochem Mol Biol* **48:** 492–514. doi:10.3109/10409238.2013.833891

Muller-Ott K, Erdel F, Matveeva A, Mallm JP, Rademacher A, Hahn M, Bauer C, Zhang Q, Kaltofen S, Schotta G, et al. 2014. Specificity, propagation, and memory of pericentric heterochromatin. *Mol Syst Biol* **10:** 746. doi:10.15252/msb.20145377

Muzzopappa F, Hertzog M, Erdel F. 2021. DNA length tunes the fluidity of DNA-based condensates. *Biophys J* **120:** 1288–1300. doi:10.1016/j.bpj.2021.02.027

Nagashima R, Hibino K, Ashwin SS, Babokhov M, Fujishiro S, Imai R, Nozaki T, Tamura S, Tani T, Kimura H, et al. 2019. Single nucleosome imaging reveals loose genome chromatin networks via active RNA polymerase II. *J Cell Biol* **218:** 1511–1530. doi:10.1083/jcb.201811090

Nair SJ, Yang L, Meluzzi D, Oh S, Yang F, Friedman MJ, Wang S, Suter T, Alshareedah I, Gamliel A, et al. 2019. Phase separation of ligand-activated enhancers licenses cooperative chromosomal enhancer assembly. *Nat Struct Mol Biol* **26:** 193–203. doi:10.1038/s41594-019-0190-5

Narlikar GJ. 2020. Phase-separation in chromatin organization. *J Biosci* **45:** 5. doi:10.1007/s12038-019-9978-z

Németh A, Grummt I. 2018. Dynamic regulation of nucleolar architecture. *Curr Opin Cell Biol* **52:** 105–111. doi:10.1016/j.ceb.2018.02.013

Németh A, Conesa A, Santoyo-Lopez J, Medina I, Montaner D, Peterfia B, Solovei I, Cremer T, Dopazo J, Langst G. 2010. Initial genomics of the human nucleolus. *PLoS Genet* **6:** e1000889. doi:10.1371/journal.pgen.1000889

Nicodemi M, Pombo A. 2014. Models of chromosome structure. *Curr Opin Cell Biol* **28:** 90–95. doi:10.1016/j.ceb.2014.04.004

O'Flynn BG, Mittag T. 2021. The role of liquid–liquid phase separation in regulating enzyme activity. *Curr Opin Cell Biol* **69:** 70–79. doi:10.1016/j.ceb.2020.12.012

Osborne CS, Chakalova L, Brown KE, Carter D, Horton A, Debrand E, Goyenechea B, Mitchell JA, Lopes S, Reik W, et al. 2004. Active genes dynamically colocalize to shared sites of ongoing transcription. *Nat Genet* **36:** 1065–1071. doi:10.1038/ng1423

Ostromyshenskii DI, Chernyaeva EN, Kuznetsova IS, Podgornaya OI. 2018. Mouse chromocenters DNA content: sequencing and in silico analysis. *BMC Genomics* **19:** 151. doi:10.1186/s12864-018-4534-z

Papantonis A, Cook PR. 2013. Transcription factories: genome organization and gene regulation. *Chem Rev* **113:** 8683–8705. doi:10.1021/cr300513p

Parlato R, Kreiner G. 2013. Nucleolar activity in neurodegenerative diseases: a missing piece of the puzzle? *J Mol Med (Berl)* **91:** 541–547. doi:10.1007/s00109-012-0981-1

Patel A, Lee HO, Jawerth L, Maharana S, Jahnel M, Hein MY, Stoynov S, Mahamid J, Saha S, Franzmann TM, et al. 2015. A liquid-to-solid phase transition of the ALS protein FUS accelerated by disease mutation. *Cell* **162:** 1066–1077. doi:10.1016/j.cell.2015.07.047

Peng L, Li EM, Xu LY. 2020. From start to end: phase separation and transcriptional regulation. *Biochim Biophys Acta Gene Regul Mech* **1863:** 194641. doi:10.1016/j.bbagrm.2020.194641

Peters AH, O'Carroll D, Scherthan H, Mechtler K, Sauer S, Schofer C, Weipoltshammer K, Pagani M, Lachner M, Kohlmaier A, et al. 2001. Loss of the Suv39h histone methyltransferases impairs mammalian heterochromatin and genome stability. *Cell* **107:** 323–337. doi:10.1016/S0092-8674(01)00542-6

Phair RD, Misteli T. 2000. High mobility of proteins in the mammalian cell nucleus. *Nature* **404:** 604–609. doi:10.1038/35007077

Phillips R. 2015. Napoleon is in equilibrium. *Annu Rev Condens Matter Phys* **6:** 85–111. doi:10.1146/annurev-conmatphys-031214-014558

Pontius BW. 1993. Close encounters: why unstructured, polymeric domains can increase rates of specific macromolecular association. *Trends Biochem Sci* **18:** 181–186. doi:10.1016/0968-0004(93)90111-Y

Prendergast L, Reinberg D. 2021. The missing linker: emerging trends for H1 variant-specific functions. *Genes Dev* **35:** 40–58. doi:10.1101/gad.344531.120

Probst AV, Almouzni G. 2008. Pericentric heterochromatin: dynamic organization during early development in mammals. *Differentiation* **76:** 15–23. doi:10.1111/j.1432-0436.2007.00220.x

Qin W, Stengl A, Ugur E, Leidescher S, Ryan J, Cardoso MC, Leonhardt H. 2021. HP1β carries an acidic linker domain and requires H3K9me3 for phase separation. *Nucleus* **12:** 44–57. doi:10.1080/19491034.2021.1889858

Quintero-Cadena P, Lenstra TL, Sternberg PW. 2020. RNA pol II length and disorder enable cooperative scaling of transcriptional bursting. *Mol Cell* **79:** 207–220.e8. doi:10.1016/j.molcel.2020.05.030

Rieder D, Trajanoski Z, McNally JG. 2012. Transcription factories. *Front Genet* **3:** 221. doi:10.3389/fgene.2012.00221

Rodriguez J, Larson DR. 2020. Transcription in living cells: molecular mechanisms of bursting. *Annu Rev Biochem* **89:** 189–212. doi:10.1146/annurev-biochem-011520-105250

Sabari BR, Dall'Agnese A, Boija A, Klein IA, Coffey EL, Shrinivas K, Abraham BJ, Hannett NM, Zamudio AV, Manteiga JC, et al. 2018. Coactivator condensation at super-enhancers links phase separation and gene control. *Science* **361:** eaar3958. doi:10.1126/science.aar3958

Sabari BR, Dall'Agnese A, Young RA. 2020. Biomolecular condensates in the nucleus. *Trends Biochem Sci* **45:** 961–977. doi:10.1016/j.tibs.2020.06.007

Saksouk N, Barth TK, Ziegler-Birling C, Olova N, Nowak A, Rey E, Mateos-Langerak J, Urbach S, Reik W, Torres-Padilla ME, et al. 2014. Redundant mechanisms to form silent chromatin at pericentromeric regions rely on BEND3 and DNA methylation. *Mol Cell* **56:** 580–594. doi:10.1016/j.molcel.2014.10.001

Sanulli S, Trnka MJ, Dharmarajan V, Tibble RW, Pascal BD, Burlingame AL, Griffin PR, Gross JD, Narlikar GJ. 2019. HP1 reshapes nucleosome core to promote phase separation of heterochromatin. *Nature* **575:** 390–394. doi:10.1038/s41586-019-1669-2

Schneider N, Wieland FG, Kong D, Fischer AAM, Hörner M, Timmer J, Ye H, Weber W. 2021. Liquid–liquid phase separation of light-inducible transcription factors increases transcription activation in mammalian cells and mice. *Sci Adv* **7:** eabd3568. doi:10.1126/sciadv.abd3568

Schotta G, Lachner M, Sarma K, Ebert A, Sengupta R, Reuter G, Reinberg D, Jenuwein T. 2004. A silencing pathway to induce H3-K9 and H4-K20 trimethylation at constitutive heterochromatin. *Genes Dev* **18:** 1251–1262. doi:10.1101/gad.300704

Shakya A, Park S, Rana N, King JT. 2020. Liquid–liquid phase separation of histone proteins in cells: role in chromatin organization. *Biophys J* **118:** 753–764. doi:10.1016/j.bpj.2019.12.022

Shin Y, Brangwynne CP. 2017. Liquid phase condensation in cell physiology and disease. *Science* **357:** eaaf4382. doi:10.1126/science.aaf4382

Shin Y, Berry J, Pannucci N, Haataja MP, Toettcher JE, Brangwynne CP. 2017. Spatiotemporal control of intracellular phase transitions using light-activated optoDroplets. *Cell* **168:** 159–171.e14. doi:10.1016/j.cell.2016.11.054

Singh PB. 2010. HP1 proteins—what is the essential interaction? *Genetika* **46:** 1257–1262.

Spector DL. 2003. The dynamics of chromosome organization and gene regulation. *Annu Rev Biochem* **72:** 573–608. doi:10.1146/annurev.biochem.72.121801.161724

Stasevich TJ, Mueller F, Brown DT, Mcnally JG. 2010. Dissecting the binding mechanism of the linker histone in live cells: an integrated FRAP analysis. *EMBO J* **29:** 1225–1234. doi:10.1038/emboj.2010.24

Strickfaden H, Zunhammer A, van Koningsbruggen S, Kohler D, Cremer T. 2010. 4D chromatin dynamics in cycling cells: Theodor Boveri's hypotheses revisited. *Nucleus* **1:** 284–297.

Strickfaden H, Tolsma TO, Sharma A, Underhill DA, Hansen JC, Hendzel MJ. 2020. Condensed chromatin behaves like a solid on the mesoscale in vitro and in living cells. *Cell* **183:** 1772–1784.e13. doi:10.1016/j.cell.2020.11.027

Strom AR, Brangwynne CP. 2019. The liquid nucleome—phase transitions in the nucleus at a glance. *J Cell Sci* **132:** jcs235093. doi:10.1242/jcs.235093

Strom AR, Emelyanov AV, Mir M, Fyodorov DV, Darzacq X, Karpen GH. 2017. Phase separation drives heterochromatin domain formation. *Nature* **547:** 241–245. doi:10.1038/nature22989

Szabo Q, Donjon A, Jerković I, Papadopoulos GL, Cheutin T, Bonev B, Nora EP, Bruneau BG, Bantignies F, Cavalli G. 2020. Regulation of single-cell genome organization into TADs and chromatin nanodomains. *Nat Genet* **52:** 1151–1157. doi:10.1038/s41588-020-00716-8

Teif VB, Rippe K. 2010. Statistical-mechanical lattice models for protein-DNA binding in chromatin. *J Phys Condens Matter* **22:** 414105. doi:10.1088/0953-8984/22/41/414105

Thiry M, Lamaye F, Lafontaine DL. 2011. The nucleolus: when 2 became 3. *Nucleus* **2:** 289–293. doi:10.4161/nucl.2.4.16806

Thorn GJ, Clarkson CT, Rademacher A, Mamayusupova H, Schotta G, Rippe K, Teif VB. 2020. DNA sequence-dependent formation of heterochromatin nanodomains. bioRxiv doi:10.1101/2020.12.20.423673

Trojanowski J, Frank L, Rademacher A, Grigaitis P, Rippe K. 2021. Transcription activation is enhanced by multivalent interactions independent of phase separation. bioRxiv doi:10.1101/2021.01.27.428421

Uversky VN, Kuznetsova IM, Turoverov KK, Zaslavsky B. 2015. Intrinsically disordered proteins as crucial constituents of cellular aqueous two phase systems and coacervates. *FEBS Lett* **589:** 15–22. doi:10.1016/j.febslet.2014.11.028

Wachsmuth M, Caudron-Herger M, Rippe K. 2008. Genome organization: balancing stability and plasticity. *Biochim Biophys Acta* **1783:** 2061–2079. doi:10.1016/j.bbamcr.2008.07.022

Wachsmuth M, Knoch TA, Rippe K. 2016. Dynamic properties of independent chromatin domains measured by correlation spectroscopy in living cells. *Epigenetics Chromatin* **9:** 57. doi:10.1186/s13072-016-0093-1

Walter J, Schermelleh L, Cremer M, Tashiro S, Cremer T. 2003. Chromosome order in HeLa cells changes during mitosis and early G1, but is stably maintained during subsequent interphase stages. *J Cell Biol* **160:** 685–697. doi:10.1083/jcb.200211103

Wang L, Gao Y, Zheng X, Liu C, Dong S, Li R, Zhang G, Wei Y, Qu H, Li Y, et al. 2019. Histone modifications regulate chromatin compartmentalization by contributing to a phase separation mechanism. *Mol Cell* **76:** 646–659 e646. doi:10.1016/j.molcel.2019.08.019

Wang L, Hu M, Zuo MQ, Zhao J, Wu D, Huang L, Wen Y, Li Y, Chen P, Bao X, et al. 2020. Rett syndrome-causing mutations compromise MeCP2-mediated liquid–liquid phase separation of chromatin. *Cell Res* **30:** 393–407. doi:10.1038/s41422-020-0288-7

Weber SC, Brangwynne CP. 2012. Getting RNA and protein in phase. *Cell* **149:** 1188–1191. doi:10.1016/j.cell.2012.05.022

Weeks SE, Metge BJ, Samant RS. 2019. The nucleolus: a central response hub for the stressors that drive cancer progression. *Cell Mol Life Sci* **76:** 4511–4524. doi:10.1007/s00018-019-03231-0

Wei MT, Chang YC, Shimobayashi SF, Shin Y, Strom AR, Brangwynne CP. 2020. Nucleated transcriptional condensates amplify gene expression. *Nat Cell Biol* **22:** 1187–1196. doi:10.1038/s41556-020-00578-6

Williams C, Brochard F, Frisch HL. 1981. Polymer collapse. *Ann Rev Phys Chem* **32:** 433–451. doi:10.1146/annurev.pc.32.100181.002245

Xu J, Ma H, Ma H, Jiang W, Mela CA, Duan M, Zhao S, Gao C, Hahm ER, Lardo SM, et al. 2020. Super-resolution imaging reveals the evolution of higher-order chromatin folding in early carcinogenesis. *Nat Commun* **11:** 1899. doi:10.1038/s41467-020-15718-7

Yao RW, Xu G, Wang Y, Shan L, Luan PF, Wang Y, Wu M, Yang LZ, Xing YH, Yang L, et al. 2019. Nascent pre-rRNA sorting via phase separation drives the assembly of dense fibrillar components in the human nucleolus. *Mol Cell* **76:** 767–783.e11. doi:10.1016/j.molcel.2019.08.014

Zamudio AV, Dall'Agnese A, Henninger JE, Manteiga JC, Afeyan LK, Hannett NM, Coffey EL, Li CH, Oksuz O, Sabari BR, et al. 2019. Mediator condensates localize signaling factors to key cell identity genes. *Mol Cell* **76:** 753–766.e6. doi:10.1016/j.molcel.2019.08.016

Zatsepina OV, Zharskaya OO, Prusov AN. 2008. Isolation of the constitutive heterochromatin from mouse liver nuclei. *Methods Mol Biol* **463:** 169–180. doi:10.1007/978-1-59745-406-3_12

Zbinden A, Pérez-Berlanga M, De Rossi P, Polymenidou M. 2020. Phase separation and neurodegenerative diseases: a disturbance in the force. *Dev Cell* **55:** 45–68. doi:10.1016/j.devcel.2020.09.014

Zenk F, Zhan Y, Kos P, Loser E, Atinbayeva N, Schachtle M, Tiana G, Giorgetti L, Iovino N. 2021. HP1 drives de novo 3D genome reorganization in early *Drosophila* embryos. *Nature* **593:** 289–293. doi:10.1038/s41586-021-03460-z

The Stochastic Genome and Its Role in Gene Expression

Christopher H. Bohrer and Daniel R. Larson

Laboratory of Receptor Biology and Gene Expression, Center for Cancer Research, National Cancer Institute, National Institutes of Health, Bethesda, Maryland 20892, USA

Correspondence: dan.larson@nih.gov

Mammalian genomes have distinct levels of spatial organization and structure that have been hypothesized to play important roles in transcription regulation. Although much has been learned about these architectural features with ensemble techniques, single-cell studies are showing a new universal trend: Genomes are stochastic and dynamic at every level of organization. Stochastic gene expression, on the other hand, has been studied for years. In this review, we probe whether there is a causative link between the two phenomena. We specifically discuss the functionality of chromatin state, topologically associating domains (TADs), and enhancer biology in light of their stochastic nature and their specific roles in stochastic gene expression. We highlight persistent fundamental questions in this area of research.

Gene expression varies from cell to cell through programmed and stochastic processes (Elowitz et al. 2002; Swain et al. 2002; Raj and van Oudenaarden 2008). This phenomenon has been shown extensively based on observations of RNA and protein copy number per cell. Live-cell imaging of RNA production from single genes reveals the dynamic nature of this process in human cells (Rodriguez et al. 2019), and single-cell RNA sequencing shows the pervasiveness of copy number variations (Chen et al. 2018b; Larsson et al. 2019). At the molecular level, many genes toggle between active and inactive states, and the resulting "transcriptional bursting" is indeed often described by a two-state model, although more complicated models are sometimes needed to quantitatively describe gene expression at the single-cell level (Fig. 1; Shahrezaei and Swain 2008; Tantale et al. 2016; Rodriguez and Larson 2020; Tunnacliffe and Chubb 2020). Heterogeneous gene expression has been linked to changes in cell phenotype, cancer, and aging (Raj and van Oudenaarden 2008). Additionally, transcription variation can serve as a readout of the underlying biochemistry: The characterization of specific perturbations on transcription with mathematical modeling can lead to a deeper understanding of the mechanism. Therefore, understanding the mechanisms that control the properties of a transcriptional burst is vital. For example, laboratories have used the bursting paradigm to dissect the role of *trans*-acting factors, noncoding RNA (ncRNA), and *cis*-acting motifs in modulating transcription (Suter et al. 2011; Hornung et al. 2012; Donovan et al. 2019; Stavreva et al. 2019). Recent studies have also begun to probe the connection between enhancers and tran-

Cite this article as *Cold Spring Harb Perspect Biol* doi: 10.1101/cshperspect.a040386

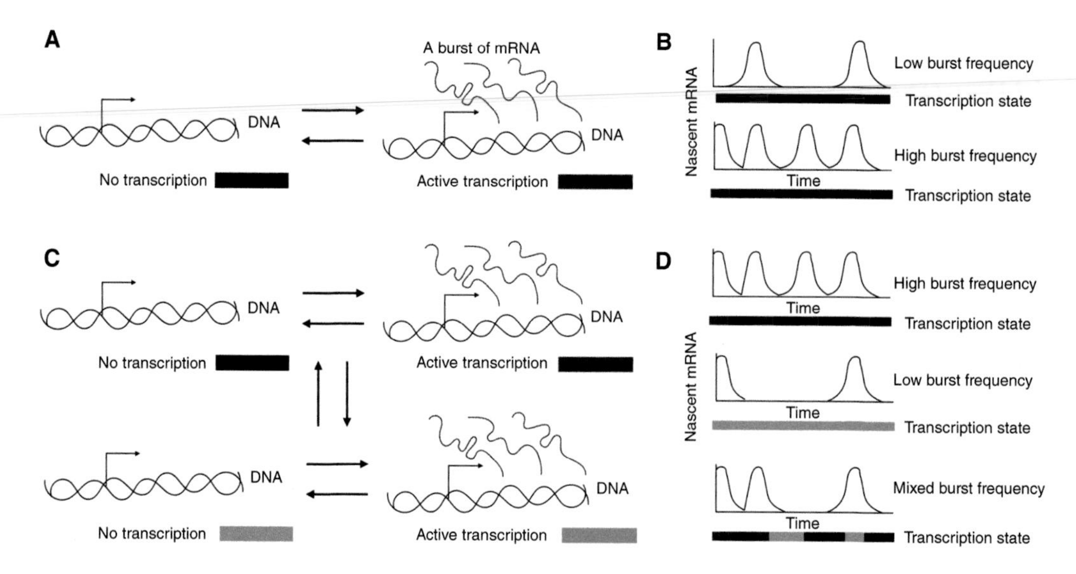

Figure 1. Transcription varies in time. (*A*) The two-state model for transcription in which a gene can switch between an inactive state and active state leading to bursts of RNA. (*B*) Illustration of a gene with a low burst frequency and a high burst frequency. The colors underneath each time trace indicate the underlying state of the gene through time. (*C*) A multistate model of transcription in which a gene can switch between states with different burst frequencies. The gene can occupy one of four different states illustrated with the different colors. (*D*) The diverse time traces produced by a multistate model of transcription dynamics. The colors underneath the time traces show the actual state of the gene.

scriptional bursting in cell lines and even whole organisms (Fukaya et al. 2016; Chen et al. 2018a; Lim et al. 2018; Alexander et al. 2019). More broadly, a study in single cells observed a distinct multimodal distribution of burst frequencies that was directly correlated with chromatin "state" (heterochromatin or euchromatin) (Su et al. 2020). Thus, an emerging theme in the study of stochastic gene regulation is the potential role of nuclear architecture for modulating transcription dynamics. This link between genomic structure and transcriptional regulation as revealed through single-cell studies is the subject of this review.

Although much has been learned about genomic organization and functionality at the ensemble level using methods such as Hi-C, ChIA-PET, ATAC-seq, etc., single-cell Hi-C and imaging studies are revolutionizing our understanding of these structures and how they vary between cells. Specific properties of the genome have been shown to play a pivotal role in directing transcription. From 10 kb bacterial DNA loops causing the temporal buildup and release of supercoiling leading to transcriptional bursts (Chong et al. 2014) to the various ways chromatin is able to regulate the binding of transcription factors (TFs) to direct transcription (Bulger and Groudine 2010)—the importance of understanding chromatin structure and its regulatory mechanisms has been clearly shown with our enhanced comprehension of disease (Krijger and De Laat 2016). A new trend is that genomic structures are probabilistic at almost every level of organization, and this stochasticity is suggestively linked to gene expression (Finn and Misteli 2019). Here we discuss some of the methodologies for deciphering single-cell genomic structure while highlighting possible links to stochastic gene expression. Specifically, we review both the evidence for genomic structure in single cells at the level of A/B "compartments" and "topologically associating domains (TADs)" and also the potential for such structures to regulate transcription. Similarly, we highlight what our knowledge of stochastic gene expression might reveal about the functionality of genomic structures and highlight critical questions for the field.

 Cite this article as *Cold Spring Harb Perspect Biol* doi: 10.1101/cshperspect.a040386

ENSEMBLE STUDIES OF GENOME STRUCTURE

Nonrandom nuclear organization is visible at multiple length scales. At the highest level of organization, individual chromosomes occupy chromosomal territories within the nucleus (Fig. 2A; Bolzer et al. 2005; Stevens et al. 2017; Tan et al. 2018), and chromatin is separated into two major conformations or states: euchromatin and heterochromatin. Euchromatin is predominantly made up of "open" gene rich fibers (Stevens et al. 2017) and is often enriched in the center of the nucleus (Gilbert et al. 2004). Heterochromatin fibers have a condensed structure (Ricci et al. 2015), a low density of genes with little expression, and often make direct contact with the nuclear lamina (Guelen et al. 2008; Kind and van Steensel 2010). These chromatin states are enriched in particular histone modifications, DNA modifications, and proteins (Jenuwein and Allis 2001; Janssen et al. 2018), all of which play a role in the formation and propagation of the compartments; the driving forces behind these compartments are still an active area of investigation (Ganai et al. 2014; Strom et al. 2017; van Steensel and Belmont 2017; Abramo et al. 2019; Falk et al. 2019; Wang et al. 2019).

With ensemble Hi-C contact maps, chromosomal regions can also be classified into two compartments (A and B), defined by their propensity to have higher contact frequencies between DNA segments within the same compartment than with loci in the other compartment (Lieberman-Aiden et al. 2009). These A and B compartments were shown to directly correspond to the euchromatin (A) and heterochromatin (B) states (Bickmore and Van Steensel 2013), allowing one to correlate a particular ensemble chromatin state to segments of DNA. However, as we discuss below, these compartments have a different interpretation in single cells.

On a smaller scale, another prominent feature of ensemble Hi-C contact maps are ~1 Mb regions that show an enrichment in pairwise contacts termed topologically associating domains (TADs) (Fig. 2C,D). For clarity, here we refer to these structures as eTADs (ensemble TADs) when derived from ensemble techniques. eTADs have an average size of 1 Mb with a range of 100 kb to 5 Mb (Dixon et al. 2012; Nora et al. 2012; Sexton et al. 2012; Rao et al. 2014), with a twofold enrichment in pairwise contacts within an eTAD compared to interactions with regions outside of eTADs (Dixon et al. 2012; Hou et al. 2012; Nora et al. 2012; Sexton et al. 2012; Nagano et al. 2013). The difference in contact frequency is also reflected in the distances between loci, with smaller distances between loci that share an ensemble domain (Bintu et al. 2018; Szabo et al. 2018; Finn et al. 2019; Mateo et al. 2019). Whereas some eTADs are invariant throughout various cells and tissues, others have been shown to form in a tissue-specific manner (Rao et al. 2014; Bintu et al. 2018; Mateo et al. 2019), and can also be correlated with cell fate (Bonev et al. 2017).

The central mechanism behind the formation of eTADs is "loop extrusion," an active process in the cell that occurs through the interplay between the architecture proteins CTCF and cohesin (Mizuguchi et al. 2014; Nora et al. 2017; Rao et al. 2017). Originally believed to be specific to vertebrates, this mechanism has now been shown to also be at work in *Drosophila* (Mateo et al. 2019). Note fly embryos lacking CTCF are still able to develop (Gambetta and Furlong 2018), suggesting that CTCF-mediated eTADs may not be necessary for function, at least in these organisms (Rowley et al. 2017). The simplest form of the loop extrusion model posits the following: The Nipbl-Mau4 complex loads cohesin rings onto specific Nipbl sites and extrudes the chromatin (using energy from ATP) until coming into contact with a pair of CTCF-binding sites in a convergent orientation, forming a "stable" complex (Rao et al. 2014; Sanborn et al. 2015; Fudenberg et al. 2016; Gassler et al. 2017; Vian et al. 2018). The cohesin release factor, WAPL, can aid in the dissociation of cohesin at all stages (Haarhuis et al. 2017), and certain TFs likely influence this process, especially for cell-type-specific eTADs (Phanstiel et al. 2017). Support for this model comes from many different studies, which describe the fusion of neighboring eTADs with the deletion of a CTCF-binding

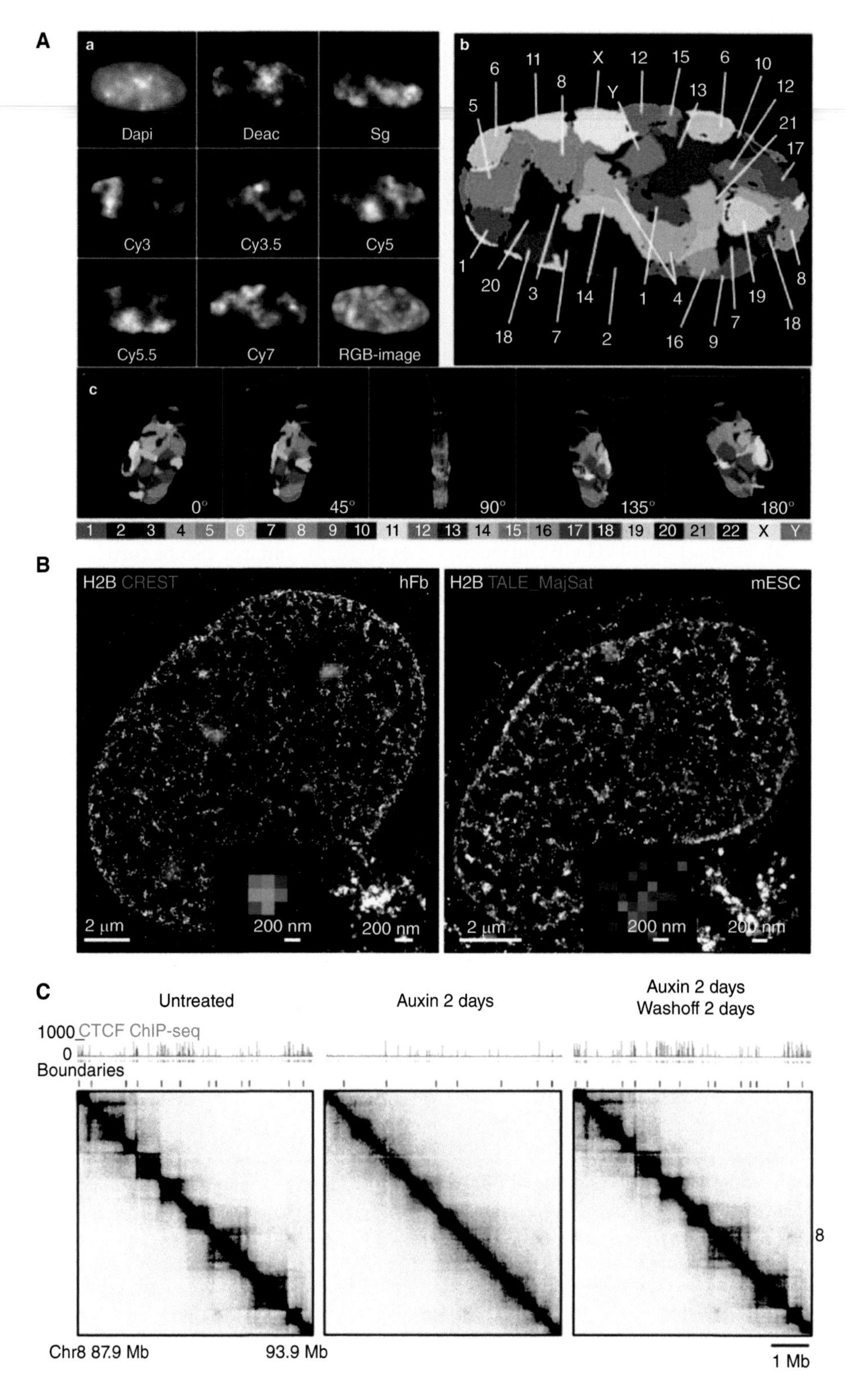

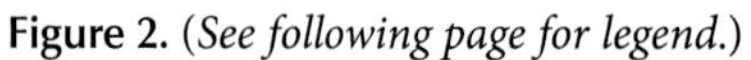
Figure 2. (*See following page for legend.*)

site (Lupiáñez et al. 2015; Sanborn et al. 2015; Mateo et al. 2019) and the elimination of eTADs upon the depletion of cohesin or CTCF (Fig. 2C, D; Nora et al. 2017; Rao et al. 2017; Bintu et al. 2018). Inversions of CTCF sites likewise "flip" the topological structure in regard to the favorable convergent CTCF orientation (Guo et al. 2015). There are also interdependencies between CTCF sites, evidenced by the observation that deletion of one site will influence the binding occupancy of neighboring CTCF sites (Narendra et al. 2015). Last, the extrusion process has now been directly visualized in vitro (Fig. 2E; Davidson et al. 2019; Kim et al. 2019), solidifying a central concept of the model.

Variations on these general architectural motifs have also been reported. Most studies have focused on pairwise interactions between segments of DNA, but even at the ensemble level a more complicated picture is emerging. Cooperative three-way interactions between three CTCF-binding sites have been shown to take place, leading to the formation of complex topological structures (Narendra et al. 2015). At an even finer scale (as small as that of a single locus), small compartments (sub-eTADs) have been shown to form within the eTADs (Rowley et al. 2017; Hsieh et al. 2020).

Taken together, nuclear organization is evident at almost any length scale that can be experimentally interrogated, but the function of these architectures remains an enduring question in cell biology.

SINGLE-CELL STUDIES OF GENOME STRUCTURE

Possible clues to the function of nuclear architecture might come from observing these same structures in single cells. Two main methodologies allow one to quantify heterogeneous chromosomal structure in individual cells: single-cell Hi-C and DNA fluorescence-based in situ hybridization (DNA-FISH). The number of labeled DNA segments for traditional DNA-FISH is limited by the number of colors one can image, greatly limiting the technique. This limitation can be partially overcome with multiple rounds of hy-

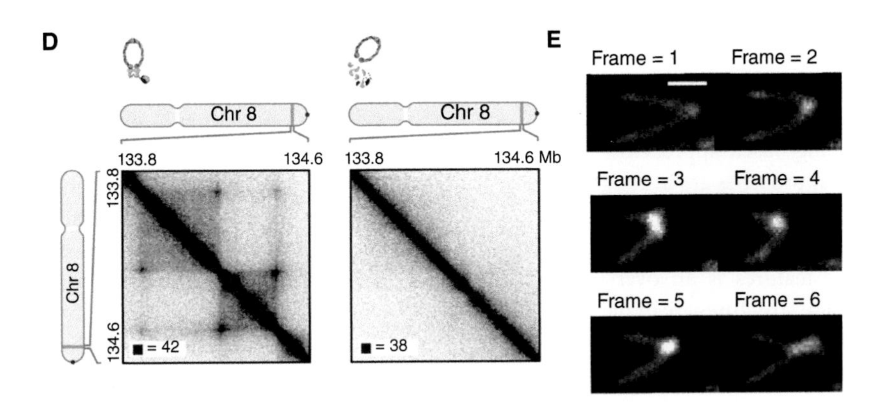

Figure 2. (*Continued*) Hierarchical nuclear organization. (*A*) Individual chromosomes occupy distinct territories. (Panel *A* reprinted from Bolzer et al. 2005 under the terms of a Creative Commons Attribution License.) (*B*) Superresolution imaging of chromatin (labeled H2B) showing a wide distribution of densities, with Crest and TALE_ MajSat (heterochromatin markers), showing that dense clusters of chromatin correspond to heterochromatin. (Panel *B* from Ricci et al. 2015; reprinted, with permission, from Elsevier © 2015.) (*C*) CTCF is needed for ensemble topologically associated domain (eTAD) formation in ensemble Hi-C maps. Here, CTCF is degraded in the presence of auxin illustrated with the ChIP-seq data. (Panel *C* from Nora et al. 2017; reprinted, with permission, from Elsevier © 2017.) (*D*) Cohesin is needed for eTAD formation in ensemble Hi-C maps. Here, auxin eliminates functional cohesion. (Panel *D* from Rao et al. 2017; reprinted, with permission, from Elsevier © 2017.) (*E*) Direct visualization of loop extrusion with cohesion. (Panel *E* from Davidson et al. 2019; reprinted, with permission, from the American Association for the Advancement of Science © 2019.)

bridization, leading to chromatin-tracing techniques that rely on determining the positions of many different chromosomal positions inside each cell after each iterative hybridization. Additionally, DNA-FISH-based methodologies can be augmented with RNA-FISH, allowing one to simultaneously quantify how transcriptional state relates to chromosomal structure.

Both single-cell Hi-C and DNA-FISH have revealed that the higher-order organization of the chromatin varies from cell to cell in the overall folding of the chromosome and the specific interfaces between chromosomes (Nagano et al. 2013; Stevens et al. 2017; Tan et al. 2018). A seminal study in this area came from Boettiger et al., which was the first to show that compartment-folding patterns are fundamentally different from each other and that the volumes of specific chromosomal regions can vary among cells (Boettiger et al. 2016). More recently, Su et al. showed that global chromosomal structure takes on a variety of different conformations (Fig. 3A; Su et al. 2020). Moreover, they extended the technique by looking at nascent RNA in parallel, imaging more than 1000 different genomic loci and more than 1000 genes in individual cells (Su et al. 2020). Similar approaches (but on a smaller scale) have also been performed in *Drosophila* (Cardozo Gizzi et al. 2019; Larson 2019; Mateo et al. 2019). In general, sequencing-based approaches and microscopy-based approaches are in rough agreement. However, it is now clear that the static picture of chromosomal structures as distinct, well-separated, stable features is an oversimplification. In reality, these structures are dynamic and these structures relate to transcriptional activity (Cardozo Gizzi et al. 2019).

First, truly quantifying whether a locus is in a heterochromatin or euchromatin state in individual cells is still a central goal of the field. Specifically, in terms of chromatin state, assigning a locus to an A or B compartment in single cells can only be performed with additional assumptions. For example, the DNA sequences of euchromatin are CpG-rich (Xie et al. 2017), and, consequently, the local density of CpG has been used as a proxy of chromatin state in single-cell Hi-C studies (Tan et al. 2018). Nonetheless, in individual cells, Wang et al. (2016) showed that the loci assigned to the ensemble-A compartments and ensemble-B compartments, on average per cell, do generally separate from each other. This separation was also apparent in Su et al. where they were able to trace an entire chromosome. However, large variations were observed in the arrangement of ensemble-A and -B loci in individual cells (Fig. 3A; Su et al. 2020). Furthermore, even when averaged over entire chromosomes, the separation between the two types of loci (assigned at the ensemble level) showed overlap with the randomized control, suggesting that the level of intermixing at the single locus level is substantial (Fig. 3B). With additional assumptions, the single-cell Hi-C work of Tan et al. was able to show that chromosomal regions could occupy either A and B compartments (Tan et al. 2018). This result is also supported with the superresolution work of Szabo et al. (2018) where heterogeneity was seen in the higher-order structure: "ranging from a compact conformation to rarer unfolded chromosomes." Interestingly, autosomal alleles varied in their compartments to the same extent as that between cells, suggesting the noise is intrinsic, that is, not attributable to the varying amounts of specific proteins between cells (Tan et al. 2018).

Second, a major question that the first single-cell Hi-C experiments sought to investigate was whether TADs in individual cells were consistently formed in the locations of the eTADs (Nagano et al. 2013). Specifically, the enrichment of contacts at the eTAD locations in individual cells was quantified and compared with that of the ensemble. Interestingly, they found no difference, suggesting that the TADs that form eTADs were consistently formed in individual cells. Here we should note that the limited number of contacts per cell in single-cell Hi-C data often lead to the use of additional assumptions to reach conclusions. The logic of this particular analysis was dependent on the formation (or absence) of complete TADs (at the locations of the eTADS) in individual cells. Interestingly, with further single-cell Hi-C experiments (Flyamer et al. 2017; Stevens et al. 2017; Tan et al. 2018), various superresolution DNA-FISH methodologies (Bintu et al. 2018; Szabo et al. 2018; Cardozo Gizzi et al. 2019; Finn et al. 2019; Mateo et al. 2019; Su et al. 2020), and now with high-resolution electron microscopy technologies (Peddie and Collinson 2014; Trzaskoma et al. 2020), the formation of TADs and their barriers were found to be extremely stochastic in individual cells. For example, a 1.7 Mb genomic region observed at the single-cell level clearly showed one domain in 25% of cells, two domains in 39% of cells, three domains in 31% of cells, and four or five domains in 5% of cells. Furthermore, even for the individual chromosome regions that had the same number of domains there were clearly different folding patterns (Fig. 3C; Trzaskoma et al. 2020), suggesting again a very dynamic and stochastic structure. Additionally, TAD formation at different alleles in individual cells was shown to be independent (Finn et al. 2019), indicating that the noise in TAD formation is also intrinsic.

These probabilistic structures do still have structural properties that emerge with different analyses. Of particular note, the eTADs were found to result from cohesin introducing biases in which boundaries could form (Bintu et al. 2018). Surprisingly, even in the absence of cohesin, TADs were still formed, but with no bias in boundary formation (Fig. 3D). This result could be interpreted as suggesting that the forces directing the chromatin to A and B compartments do not play a significant role in the formation of TADs. Furthermore, TADs can have both ensemble-A loci and ensemble-B loci, again suggesting the ensemble compartmentalization may not play a dominating role in TAD formation (Fig. 3E; Su et al. 2020). Individual TADs appear to show globular structures (Bintu et al. 2018; Szabo et al. 2018) and *Drosophila* Polycomb TADs have been shown to be extremely compact and organized like that of a random coil (Mateo et al. 2019). Notably, Polycomb proteins are important for propagating and maintaining chromatin modifications, and these modifications were shown to play a role in "strong long-range" chromosomal contacts (Bonev et al. 2017) and form "loops" with "patches" of the Polycomb chromatin forming the contact (Hsieh et al. 2020; see Aranda et al. 2015 for information on Polycomb proteins and their influence on TAD formation). Last, the fact that TADs are probabilistic structures causes their DNA to frequently contact their neighboring eTADs; stochastic boundary formation can cause DNA segments within neighboring eTADs to be located within the same TAD (Bintu et al. 2018; Finn et al. 2019).

In summary, it appears that there is a great deal of heterogeneity in the overall conformation of chromosomes, chromatin state, and TAD structure. Understanding how this heterogeneous structural genomic landscape influences transcription and cellular decisions thus becomes a central question of future studies (Misteli 2020).

TRANSCRIPTION IN THE LIGHT OF STOCHASTIC GENOMIC STRUCTURES

Does the stochastic nature of the genome have any influence on transcription or vice versa? This is still a very much open question, and the answer ultimately lies in the functionality of genomic structure. Here we focus primarily on compartment- and TAD-level organization and what new methodologies reveal about this question.

The role of the local surroundings of a gene on transcriptional regulation has a long history, starting first with the suggestion that location of a gene in a physically compact region of chromatin might be indicative of low activity (Schultz 1947; Raser and O'Shea 2004; Raj et al. 2006). Studies hoping to investigate how the local chromatin state influences transcription sought to investigate perturbations to nuclear position. Multiple experiments have redirected/tethered specific chromosomal regions to the different regions of the nucleus and have quantified its impact on transcription. Interestingly, whether or not transcription was influenced was context specific; repositioning an interior gene to the periphery of the nucleus did not always lead to its inactivation (Williams et al. 2006; Finlan et al. 2008; Kumaran and Spector 2008; Reddy et al. 2008; Therizols et al. 2014; Wijchers et al. 2016). In terms of chromosomal territories, regions of active transcription are enriched in *trans*-chromosomal contacts (Lieberman-Aiden et al. 2009; Yaffe and Tanay 2011), owing to active transcription at the interfaces of

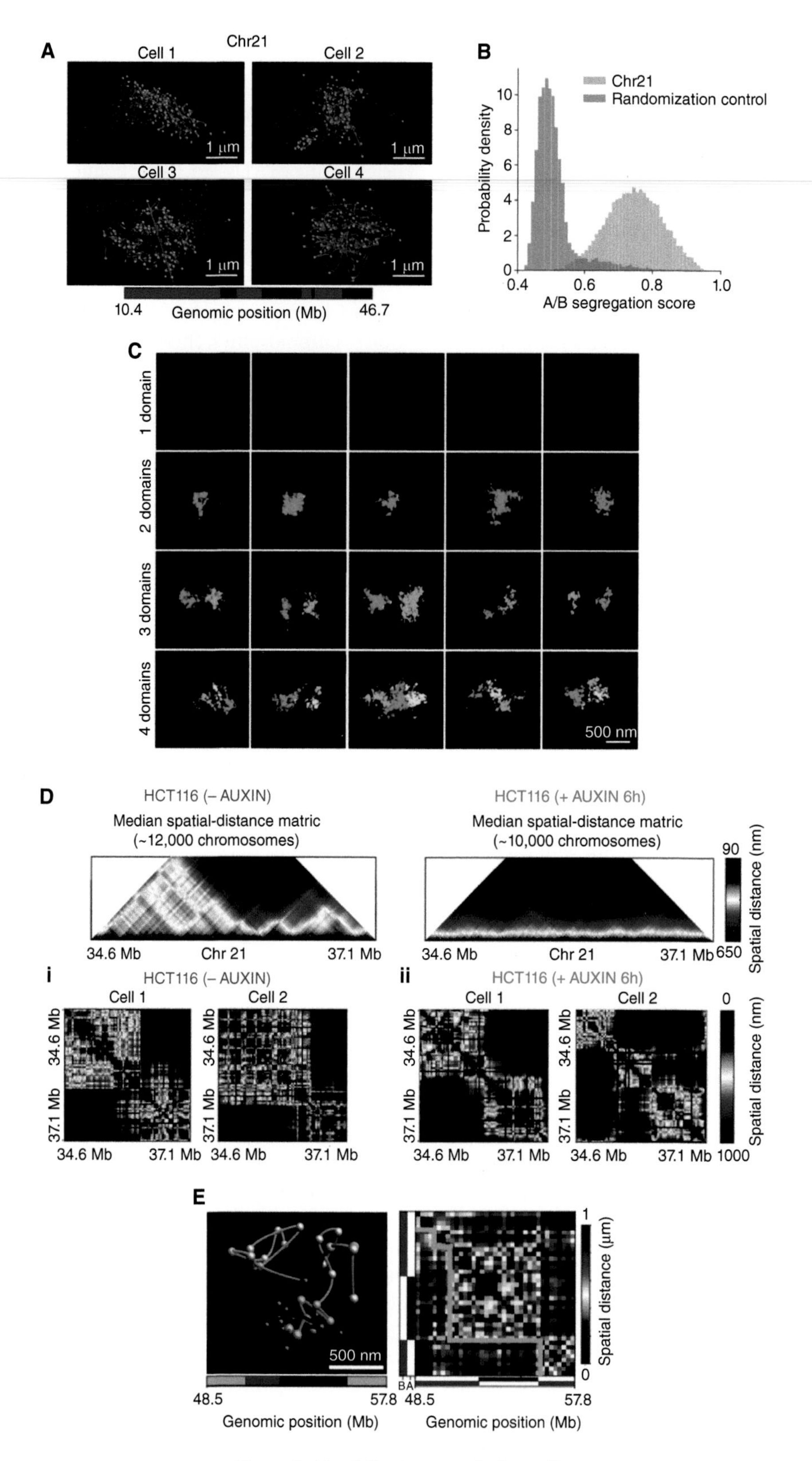

Figure 3. (*See following page for legend.*)

chromosomes (Nagano et al. 2013; Stevens et al. 2017). To date, to the best of our knowledge, it is unclear whether general rules can be inferred about the role of chromosome interfaces or the nuclear periphery in influencing transcriptional state or bursting behavior.

Moreover, even within the context dependence and nondeterministic role of nuclear position on transcription, the higher-order structure could have a long timescale impact on the variation of transcription owing to the constraints on chromatin mobility. The diffusion of DNA-bound fluorescently tagged TFs (Robinett et al. 1996) has shown that the majority of chromatin exhibit confined diffusion (Chubb et al. 2002). Memory of past positions and specific dynamics are evident well beyond 20 min with the dynamic behavior specific to individual cells (Alexander et al. 2019), suggesting it would take a long time for specifically bound TFs to reach a new location through chromatin diffusion. Here we should also note that with transcriptional activation, chromatin was able to move directionally from the periphery to the center of the nucleus (Chuang et al. 2006). Last, the repositioning of chromatin to different nuclear positions was shown to largely depend on mitosis (Kumaran and Spector 2008). It is therefore tempting to state that if changing nuclear position does change the chromatin state, the characteristic timescales of switching between different transcriptional states would be on the order of the cell cycle.

At the compartment/euchromatin/heterochromatin level, global methodologies found that the influence of ensemble heterochromatin on bursting was noisy expression resulting from a low burst frequency (Dey et al. 2015). This result was reemphasized by the finding that genes that occupied the two ensemble chromatin states had a bimodal distribution of burst frequencies (Fig. 4A). The power of this recent work is the ability to directly quantify how the stochastic nature of the chromatin state at the single-cell level influences transcription. Specifically, instead of assigning loci to A and B compartments in individual cells, they quantified the density of *trans*-ensemble-A loci relative to *trans*-ensemble-B loci for each gene, excluding the ensemble chromatin state of the gene in question, and investigated whether there was a link to its transcriptional state. This "*trans* A/B density" can be thought of as an approximation for whether or not a gene is in an A or B state in individual cells. Importantly, there was a strong correlation between the *trans* A/B density and transcription, where 86% of genes showed an enrichment in *trans* A/B density when they were transcribing and 89% showed a higher burst frequency with a higher *trans* A/B ratio (Fig. 4B,C; Su et al. 2020).

These results suggest the modulation of burst frequency through chromatin state could be the mechanism leading to correlations in the steady-state RNA levels between genomically co-located genes (Raj et al. 2006; Sun and Zhang 2019; Ibragimov et al. 2020), and should be further quantified in terms of specific mechanism. Nonetheless, these results suggest a transcriptional model with multiple burst frequencies for genes that stochastically switch between chromatin states (Fig. 4D). A potential mecha-

Figure 3. Chromosome organization varies in single cells. (*A*) Diverse structures of whole individual chromosomes with the loci that occupy the ensemble-A and -B compartments shown in red and blue. (Panel *A* from Su et al. 2020; reprinted, with permission, from Elsevier © 2020.) (*B*) The degree of separation of chromatin for individual chromosomes in the two ensemble-A and -B states compared with a random control. (Panel *B* from Su et al. 2020; reprinted, with permission, from Elsevier © 2020.) (*C*) Direct visualization of chromatin with electron microscopy showing the large amount of stochasticity of individual topologically associating domains (TADs)—each assigned domain is color-coded. (Panel *C* is reprinted from Trzaskoma et al. 2020 under a Creative Commons Attribution 4.0 International License.) (*D*) Individual TADs still form in the absence of cohesin. The first row shows the ensemble median distances with (no auxin) and without cohesin (auxin). (Panel *D* from Bintu et al. 2018; reprinted, with permission, from the American Association for the Advancement of Science © 2018.) (*i*) TADs in individual cells with cohesion, and (*ii*) without cohesin. (*E*) TADs in individual cells can contain loci assigned to the ensemble-A and -B compartments. (Panel *E* from Su et al. 2020; reprinted, with permission, from Elsevier © 2020.)

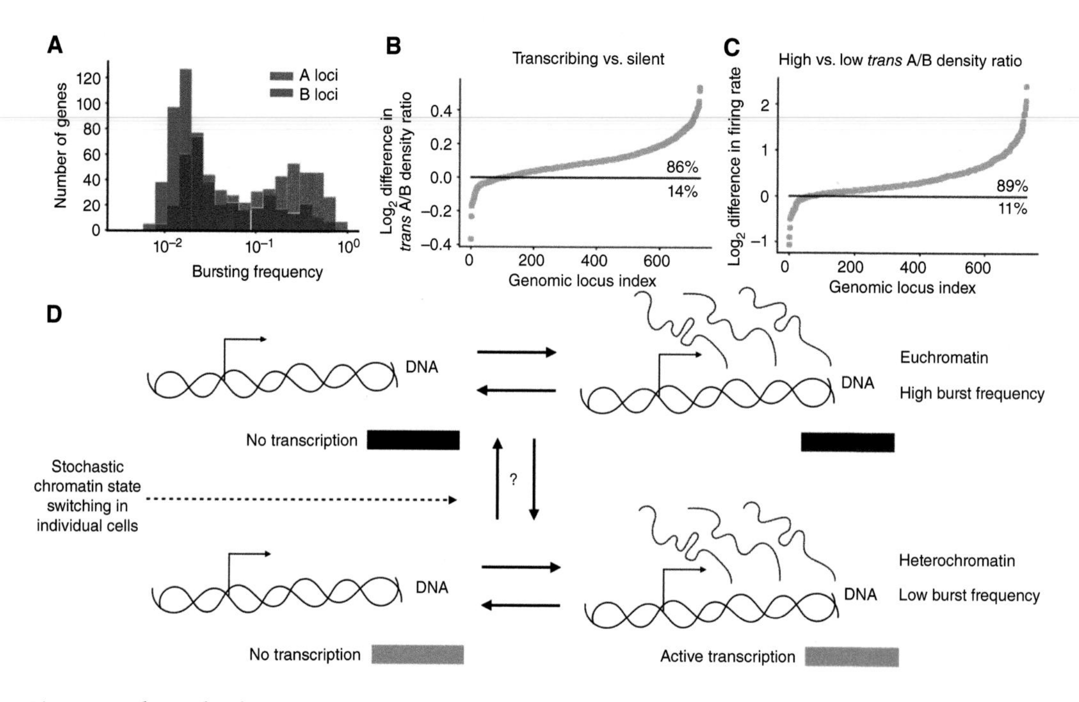

Figure 4. Relationship between transcriptional bursting and compartmentalization. (*A*) The bimodal bursting frequency distribution of genes naturally found in the ensemble-A and -B chromatin states. (*B*) The difference in the *trans* A/B ratio (the local density of *trans*-ensemble-A loci to *trans*-ensemble-B loci in individual cells around each gene) when a gene was transcribing versus when it was not, showing there is a strong correlation between chromatin state and transcription state in individual cells. (*C*) Similar to *B* but with the difference in bursting rate on the *y*-axis between high and low *trans* A/B density, showing a strong correlation of chromatin state in individual cells and burst frequency. (Figures and data in *A–C* from Su et al. 2020; reprinted, with permission, from Elsevier © 2020.) (*D*) Hypothetical model of transcription for a gene with two different burst frequencies owing to variations in chromatin state within individual cells (the same as Fig. 1A).

nism where the stochasticity of the chromatin state could be propagated through to transcription could be enhancer activation and deactivation (see below), possibly through specific histone modifications/chromatin conformations that are concomitant with activation of enhancers (Krijger and De Laat 2016). Last, if the chromatin state is what drives burst frequency, could one use transcriptional bursts as a monitor of single-cell chromatin state?

FUNCTIONALITY OF TOPOLOGICALLY ASSOCIATING DOMAINS (TADs) AND ENSEMBLE TADs (eTADs)

Topological domains arise through the interplay between cohesin and *cis*- and *trans*-acting boundary elements. CTCF is perhaps the best characterized of these *trans*-acting factors (Ghirlando and Felsenfeld 2016), consistent with its role as an insulator (Bell and Felsenfeld 2000). Thus, in relation to the previously discussed higher-order structure of the genome, the transcriptional states of individual eTADs are often believed to be partitioned into either heterochromatin or euchromatin (Le Dily et al. 2014; Wang et al. 2016; Wijchers et al. 2016) and correlated with chromatin state (Sexton et al. 2012; Rao et al. 2014; Ulianov et al. 2016). The functionality of eTADs is believed to be in (1) promoting specific intracontacts such as specific enhancer–promoter interactions, and (2) limiting the contacts between adjacent eTADs, avoiding enhancer cross talk. Indeed, the expression of tissue-specific expression is correlated with eTAD structure (Symmons et al. 2014) and perturbations of individual eTADs lead to the dysregulation of expression (Northcott et al. 2014;

Lupiáñez et al. 2015; Flavahan et al. 2016; Franke et al. 2016; Hnisz et al. 2016; Krijger and De Laat 2016; Weischenfeldt et al. 2017; Mateo et al. 2019) providing strong support for this view.

Yet, experiments that have globally eliminated eTAD structure through cohesin (Rao et al. 2017; Vian et al. 2018) or CTCF (Nora et al. 2017) depletion have shown minimal ensemble transcription changes. Similarly, although eTADs might be expected to constrain the spread of chromatin state, the compartmentalization of chromatin into the ensemble-A and -B states was maintained and even strengthened with the depletion of eTADs (Nora et al. 2017; Rao et al. 2017; Vian et al. 2018). Also, the concept of eTADs directing specific DNA contacts and preventing others is a difficult model to grasp, considering TADs have been shown to be stochastic structures with ill-defined boundaries in individual cells. Furthermore, as pointed out in the work of Ghavi-Helm et al., most studies that found a strong correlation of perturbations to eTAD structure and transcription generally started with a phenotype (like cancer) and "worked backward to explain the misexpression" (Northcott et al. 2014; Lupiáñez et al. 2015; Flavahan et al. 2016; Franke et al. 2016; Hnisz et al. 2016; Weischenfeldt et al. 2017). Therefore, the general influence of eTAD structure on transcription appears minimal. Indeed, in their work, they were able to quantify the effects of a large number of unbiased chromosomal perturbations using balancer chromosomes in *Drosophila* and found that only a small fraction of genes was sensitive to disruptions in chromosome topology (Ghavi-Helm et al. 2019).

Indeed, recent data suggest a more nuanced view. Within TADs, one observes smaller "sub-TADs." The boundaries of these sub-eTADs are enriched in TFs, coactivators, and RNA polymerase II (RNAP II), and the boundary strengths of sub-TADs are directly proportional to transcription activity (Hsieh et al. 2020). Importantly, RNAP II inhibition (elongation and initiation) did not perturb these structures, suggesting it is the other factors involved in transcription directing the formation of these sub-eTAD structures (Hsieh et al. 2020). Here, we should note that there is a similar relationship between eTADs and active transcription (Du et al. 2017; Hug et al. 2017; Ke et al. 2017; Hsieh et al. 2020) and some TFs have been shown to form eTAD boundaries (Hug et al. 2017; Weintraub et al. 2017). Additionally, recent investigations showed that active transcription within an eTAD seemed to unfold the region (evident at the ensemble level) (Cardozo Gizzi et al. 2019), directly linking eTAD structure to transcription, likely attributable to the unfolding of chromosomal domains with supercoiling (Naughton et al. 2013).

Distinguishing the role of compartments versus TADs as the primary determinant of transcriptional regulation is emerging as a central concept in the field. The interesting recent work of Luppino et al. (2020) was able to show that cohesin is responsible for the intermixing of neighboring eTADs and through this intermixing, cohesin modulates the burst frequency of genes near the borders of eTADs, suggesting an interesting direction of research. Also, the strengthening of A and B compartmentalization seen with cohesin depletion (Rao et al. 2017; Vian et al. 2018) could indicate that eTADs play a role in the stochasticity of A and B compartmentalization and the higher organization of the genome. However, modeling studies of Nuebler et al. (2018) suggest that it is the activity of loop extrusion that leads to the active mixing of the chromatin and the weakening of A and B compartmentalization, potentially leading to this stochasticity and bringing into question the specific role of eTADs. Still, this stochasticity in A and B compartmentalization would not necessarily lead to changes in mean expression levels but could in turn influence the noise and (potentially) the correlations between certain genes, but this remains to be seen. Overall, much needs to be probed in terms of TAD and eTAD structure as a general process controlling gene expression and whether their stochastic nature influences enhancer biology.

ENHANCER-DEPENDENT TRANSCRIPTION ACTIVATION

An appealing connection between genomic structure and transcription centers on the concept of the enhancer. Enhancers were initially discovered in 1981 on plasmids with the observation that DNA elements "far" from the promoter greatly influenced transcription (Banerji et al. 1981; Benoist and Chambon 1981; Schaffner 2015). Enhancers were then shown in metazoans (mouse) in 1983 (Banerji et al. 1983) and are now believed to be a central mechanism of transcription regulation. Mammalian genomes are predicted to have hundreds of thousands of enhancers (The 1000 Genomes Project Consortium et al. 2012; Shen et al. 2012), and the majority of genes have more than one enhancer (Fishilevich et al. 2017) presenting an enormous challenge for dissecting their mechanism of action. Enhancers can act synergistically, thus raising expression to a higher extent than their sum, repress the action of neighboring enhancers, and can show a degree of "hierarchical logic" (Long et al. 2016); for example, the activation of one enhancer is necessary for the activation of others. Additionally, some enhancers seem to act on any gene, while some show specificity (Furlong and Levine 2018). Still, given their dominant role in regulation, the variations within the steps of enhancer-mediated transcriptional activation are a clear knob for the cell to control the timescales of transcriptional bursting and the extent of noise within expression (Larsson et al. 2019).

At the molecular level, the "parts list" of enhancers has been well characterized. DNA sequences of enhancers are enriched in accessible DNA (Buenrostro et al. 2013) and the H3K4me1 and H3K27ac histone modifications (Rada-Iglesias et al. 2011). A majority of enhancers are cell/tissue specific, in which the expression of specific TFs and chromatin modifications are believed to lead to the specific activation of the enhancer. Coactivators (Mediator, BRG1, and p300), TFs (Spitz and Furlong 2012), enhancer RNA, and RNAP II binding are all characteristics of active enhancers (Long et al. 2016). The confluence of these factors has engendered a view that for proper enhancer activation, multiple different TFs bind cooperatively and displace the nucleosomes. An alternative first step is that pioneer factors can bind and modify specific chromatin sites paving the way for the future factors to bind and work (Zaret and Carroll 2011), although it is unclear whether pioneering activity is a unique activity reserved for certain proteins (Voss et al. 2011). Nevertheless, the majority of experiments have indeed shown that coactivators trigger transcription through guided recruitment (Hilton et al. 2015; Stampfel et al. 2015).

ENHANCER–PROMOTER PROXIMITY: THE QUESTION OF "RANGE OF ACTION"

If there is a general mechanism of enhancer transcription activation, it has yet to be clearly shown. A variety of different models have been proposed, but given a lack of direct experimental evidence, we do not exhaustively discuss them here (Bulger and Groudine 2010; Furlong and Levine 2018). However, enhancer-activated transcription does seem to be dependent on the physical proximity of a promoter and its enhancer; whether or not direct contact between a promoter and an enhancer is needed for transcription activation is still a matter of debate. Proximity is believed to lead to an increase in coactivators and TFs around the promoter, increasing the likelihood of a transcriptional burst (Rodriguez and Larson 2020). Indeed, the dominating mechanistic role of enhancers in stochastic gene expression has been shown in living cells, where the major effect of enhancer function is modulating burst frequency (Bartman et al. 2016; Fukaya et al. 2016; Larsson et al. 2019; Rodriguez et al. 2019). Different mechanisms have different degrees of proximity needed for activation; regardless of the exact mechanism, the degree of proximity (or "range of action" for the enhancer) needed for transcription must be relatively small (hundreds of nm) to ensure precision as argued in Furlong and Levine (2018).

This simple model has been quite useful, as enhancer–promoter proximity has been extremely important for our understanding of oncogene dysregulation. Specific examples include, but are not limited to (Fang et al. 2020) (1) A single chromosomal rearrangement that moves the GATA2 enhancer proximal to *EVI1* (a stem-cell regulator) directly results in leukemia (Gröschel et al. 2014); (2) The up-regulation of *MYC* in some forms of Burkitt lymphoma is attributable to the repositioning of immunoglobin heavy chain enhancer (Dalla-Favera et al. 1982); (3) Mutations within the superenhancer of the *LMO1* oncogene was shown to modulate its TF-binding sites leading to neuroblastoma (Oldridge et al. 2015); and (4) Copy number variations of superenhancers around various oncogenes (*KLF5, USP12, PARD6B*, and *MYC*) lead to their dysregulation in 12 different tumor types, providing strong evidence that perturbations to enhancer biology is a common driving force behind cancer.

The genomic distances between enhancers and their target genes can be very large, such as the Shh limb enhancer (ZRS), which is >1 Mb away from the target gene (Lettice et al. 2003). Therefore, understanding how chromosomal rearrangements direct/limit enhancers to a particular promoter and which conformations actually trigger transcription have become central questions. eTADs are believed to play a role in directing enhancers as perturbations in individual eTAD structures can lead to large changes in gene expression and disease (Northcott et al. 2014; Lupiáñez et al. 2015; Flavahan et al. 2016; Franke et al. 2016; Hnisz et al. 2016; Krijger and De Laat 2016; Weischenfeldt et al. 2017; Mateo et al. 2019). Additional support for the previous statement is clearly shown in the pivotal work of Symmons et al. where they monitored the expression level of a minimal promoter with the *lacZ* gene randomly inserted into hundreds of different positions within the mouse genome (Symmons et al. 2014; developed in Ruf et al. 2011). Notably, a large majority of these insertions showed tissue-specific expression similar to that of the neighboring genes' tissue-specific expression. Further investigation found the regions over which the reporter gene showed similar expression was directly correlated with the eTADs, suggesting that the "region of action" for the enhancers directing the tissue-specific expression was confined to eTADs (Symmons et al. 2014).

Directed looping is an attractive model for modulating proximity through chromosomal conformation. For example, in the work of Guo et al. (2015), ~50% of their identified enhancers had a CTCF-binding site nearby, suggesting that upon TAD formation, the CTCF-adjacent enhancer could be repositioned for transcription activation. Looping between a promoter and its enhancer has been shown to directly increase mean expression levels (Deng et al. 2012, 2014; Williamson et al. 2016; Morgan et al. 2017), attributable to an increase in burst frequency (Bartman et al. 2016). Also, the oscillatory expression of circadian genes was shown to correlate with promoter–enhancer contacts (Mermet et al. 2018). Similar to a looping mechanism, cohesin could promote a direct enhancer–promoter contact through loop extrusion, which has been proposed as a mechanism to constrain enhancer–promoter interactions to the same eTAD (Dixon et al. 2012). Last, in ensemble Hi-C experiments, enriched contacts of enhancer–promoters have been observed, suggesting looping could promote direct contact (Rao et al. 2014). However, the majority of active enhancer–promoters did not show an enrichment in contact frequency (Rao et al. 2014; Long et al. 2016), suggesting that for the most general form of enhancer regulation, direct contact is not needed. The work of Benabdallah et al. 2019. clearly demonstrated this phenomenon by showing the exact opposite. Expression of *Shh* was correlated with its promoter–enhancer distances increasing, bringing into question this general model. Thus, obstacles to a unified view of eTAD functionality may be the gaps in our understanding of enhancer biology.

CONNECTIONS BETWEEN ENHANCER–PROMOTER PROXIMITY AND BURSTING REVEALED IN SINGLE-CELL STUDIES

Ideally, an experimentalist would want to monitor DNA conformation and nascent RNA production in living cells with millisecond time resolution and nanometer spatial precision. Current microscopy technology does not yet allow these experiments. However, the progress in recent years in simultaneously imaging both DNA and RNA has been substantial. Pivotal live-cell experiments have provided invaluable information about the dynamics of chromosomal enhancer transcription regulation (Larson et al. 2011). Two studies have now shown that a single enhancer is able to activate two adjacent genes at the same time, arguing against direct enhancer–promoter contact for activation (Fukaya et al. 2016; Lim et al. 2018). In the work of Chen et al. (2018a) (*Drosophila*), it was clearly shown that loop formation was needed to allow for enhancer–promoter proximity activation. Upon loop formation, the enhancer–promoter distances showed a Gaussian distribution with a mean of ~375 nm, whereas the unlooped conformation showed a Gaussian distribution with a mean of ~720 nm. Interestingly, for transcription activation, the promoter–enhancer distances showed a Gaussian distribution with a slightly smaller mean of ~330 nm, suggesting once a promoter is within this distance the enhancer could trigger transcription. Note, the investigators attributed this observation to DNA compaction with transcriptional activation, which is opposite to the finding of Cardozo Gizzi et al. (2019). Similarly, high-resolution work within embryonic stem cells partially supports this result, where they investigated the placement of an enhancer close to a promoter. This experiment resulted in an enhancer–promoter distance distribution with a mean of 300 nm and an extended tail up to 750 nm. Importantly, they found that there was no dependence between transcription and enhancer–promoter distance (Fig. 5A), suggesting that for this system the promoter is always within the range of action of the enhancer (Alexander et al. 2019). Yet, the bursting behavior of the gene was still extremely variable, in which the majority (>65%) of the cells were nontranscribing for a period of >60 min (Fig. 5A). If the bursts within individual cells were independent events, it is clear that the transcribing cells and nontranscribing cells did not show the same bursting frequency. Taken together, these results suggest that a large proportion of the variation observed within transcription is not dependent on the modulation of enhancer–promoter proximity and is likely governed by other mechanisms, such as enhancer activation, local variations in TF concentrations, etc.

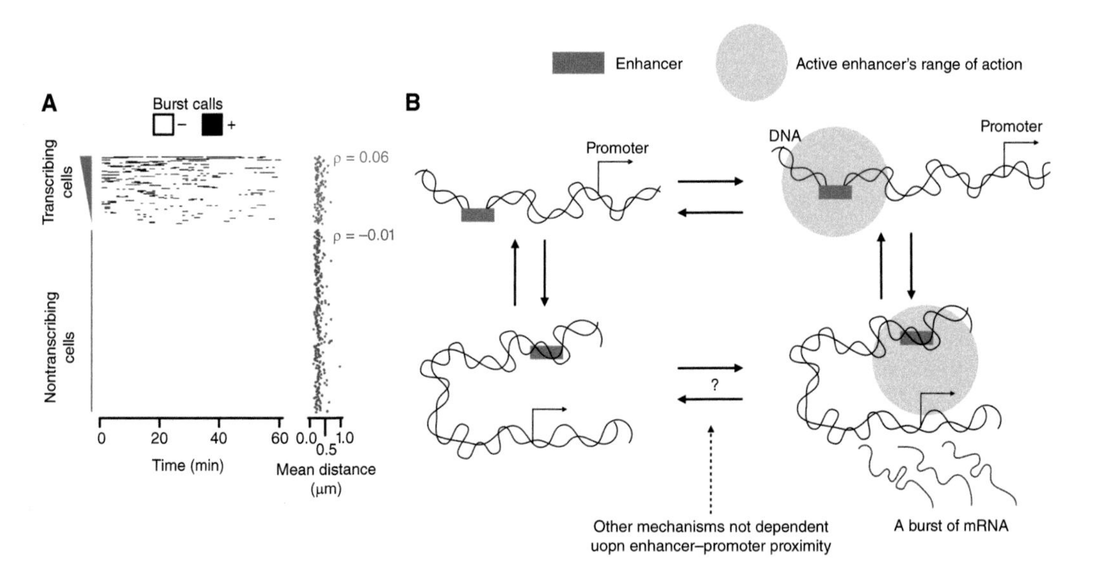

Figure 5. Enhancer-driven transcription. (*A*) The transcription state over time for individual cells and the distance between the enhancer and promoter to the right showing little correlation with transcription. (Panel *A* reprinted from Alexander et al. 2019 under the terms of a Creative Commons Attribution License.) (*B*) Model of transcription with the range of action of the enhancer taken into consideration, showing that if enhancer–promoter proximity is within the range of action of the enhancer, other mechanisms must be responsible for the stochastic nature of transcription.

In addition to the live-cell studies, the single-cell imaging work of Mateo et al. (2019) was able to directly visualize the 3D structure of a local region of the chromosome as well as the nascent RNA from the different genes within the region, providing the experimental means of investigating enhancer–promoter proximity and chromosomal structure. The enhancer–promoter distances between the two were predictive of nascent transcription, but only weakly. Importantly, the DNA segments that were predictive of transcription were not specific to the enhancer location but seemed to be confined to certain domains of the chromosome. Additionally, genes were active when separated from their enhancers and inactive when in proximity to their enhancers (Mateo et al. 2019). Again, we offer the interpretation that a large amount of the stochasticity is not directly dictated by enhancer–promoter proximity as long as the promoter is within the range of action of the enhancer.

OUTLOOK

In summary, the genome is stochastic at every level of organization. Stochastic A and B compartmentalization in individual cells is propagated through to transcription, but the specifics of this process are unclear. Enhancers seem to have a range of action around 300 nm potentially related to local ensemble chromosomal domains. Even so, enhancer-regulated transcription has a large amount of variation that is not dependent on fluctuations in enhancer–promoter proximity. In Figure 5B, we provide a hypothetical model of transcription to show these findings.

In terms of the higher-order structure of the chromosome, the lack of studies investigating A and B compartmentalization in individual cells and transcription clearly indicates the need for further work. A primary experimental impediment is that there is still no clear way to define the chromatin state in individual cells. Additionally, the conflicting results surrounding the importance of TADs and eTADs indicate that understanding what they do and how they work is still a central question within the field. For instance, how do these probabilistic structures convey the range of action of enhancers to the ensemble domains? There have been proposals that enhancers are able to nucleate "phase-separated" droplets, suggesting that maybe some of the probabilistic structures could create or limit this phenomenon. Although, even this recent idea has to be questioned as the expression of Shh was shown to be extremely resistant to perturbations of the eTAD structure (Williamson et al. 2019).

Overall, it should be noted that our understanding of stochastic genomic organization is dependent on very new technologies, and only time will tell whether our current model of the stochastic genome is attributable to real biological variability or experimental noise. Still, we have shown here that there are plausible links between the stochastic chromosome and stochastic transcription. Presently, single-cell studies are much better at visualizing proteins such as TFs and RNA than the high-resolution structure of the chromatin fiber. Future efforts should be aimed at dissecting the contributions of the myriad factors involved in transcriptional regulation as they work in the dynamic milieu of the nucleus.

REFERENCES

Abramo K, Valton AL, Venev SV, Ozadam H, Fox AN, Dekker J. 2019. A chromosome folding intermediate at the condensin-to-cohesin transition during telophase. *Nat Cell Biol* **21:** 1393–1402. doi:10.1038/s41556-019-0406-2

Alexander JM, Guan J, Li B, Maliskova L, Song M, Shen Y, Huang B, Lomvardas S, Weiner OD. 2019. Live-cell imaging reveals enhancer-dependent *Sox2* transcription in the absence of enhancer proximity. *eLife* **8:** e41769. doi:10.7554/eLife.41769

Aranda S, Mas G, Di Croce L. 2015. Regulation of gene transcription by Polycomb proteins. *Sci Adv* **1:** e1500737. doi:10.1126/sciadv.1500737

Banerji J, Rusconi S, Schaffner W. 1981. Expression of a β-globin gene is enhanced by remote SV40 DNA sequences. *Cell* **27:** 299–308. doi:10.1016/0092-8674(81)90413-X

Banerji J, Olson L, Schaffner W. 1983. A lymphocyte-specific cellular enhancer is located downstream of the joining region in immunoglobulin heavy chain genes. *Cell* **33:** 729–740. doi:10.1016/0092-8674(83)90015-6

Bartman CR, Hsu SC, Hsiung CCS, Raj A, Blobel GA. 2016. Enhancer regulation of transcriptional bursting parameters revealed by forced chromatin looping. *Mol Cell* **62:** 237–247. doi:10.1016/j.molcel.2016.03.007

Bell AC, Felsenfeld G. 2000. Methylation of a CTCF-dependent boundary controls imprinted expression of the *Igf2* gene. *Nature* **405:** 482–485. doi:10.1038/35013100

Benabdallah NS, Williamson I, Illingworth RS, Kane L, Boyle S, Sengupta D, Grimes GR, Therizols P, Bickmore WA. 2019. Decreased enhancer-promoter proximity accompanying enhancer activation. *Mol Cell* **76:** 473–484.

Benoist C, Chambon P. 1981. In vivo sequence requirements of the SV40 early promoter region. *Nature* **290:** 304–310. doi:10.1038/290304a0

Bickmore WA, Van Steensel B. 2013. Genome architecture: domain organization of interphase chromosomes. *Cell* **152:** 1270–1284. doi:10.1016/j.cell.2013.02.001

Bintu B, Mateo LJ, Su JH, Sinnott-Armstrong NA, Parker M, Kinrot S, Yamaya K, Boettiger AN, Zhuang X. 2018. Super-resolution chromatin tracing reveals domains and cooperative interactions in single cells. *Science* **362:** eaau1783. doi:10.1126/science.aau1783

Boettiger AN, Bintu B, Moffitt JR, Wang S, Beliveau BJ, Fudenberg G, Imakaev M, Mirny LA, Wu CT, Zhuang X. 2016. Super-resolution imaging reveals distinct chromatin folding for different epigenetic states. *Nature* **529:** 418–422. doi:10.1038/nature16496

Bolzer A, Kreth G, Solovei I, Koehler D, Saracoglu K, Fauth C, Müller S, Eils R, Cremer C, Speicher MR, et al. 2005. Three-dimensional maps of all chromosomes in human male fibroblast nuclei and prometaphase rosettes. *PLoS Biol* **3:** e157. doi:10.1371/journal.pbio.0030157

Bonev B, Mendelson Cohen N, Szabo Q, Fritsch L, Papadopoulos GL, Lubling Y, Xu X, Lv X, Hugnot JP, Tanay A, et al. 2017. Multiscale 3D genome rewiring during mouse neural development. *Cell* **171:** 557–572.e24. doi:10.1016/j.cell.2017.09.043

Buenrostro JD, Giresi PG, Zaba LC, Chang HY, Greenleaf WJ. 2013. Transposition of native chromatin for fast and sensitive epigenomic profiling of open chromatin, DNA-binding proteins and nucleosome position. *Nat Methods* **10:** 1213–1218. doi:10.1038/nmeth.2688

Bulger M, Groudine M. 2010. Enhancers: the abundance and function of regulatory sequences beyond promoters. *Dev Biol* **339:** 250–257. doi:10.1016/j.ydbio.2009.11.035

Cardozo Gizzi AM, Cattoni DI, Fiche JB, Espinola SM, Gurgo J, Messina O, Houbron C, Ogiyama Y, Papadopoulos GL, Cavalli G, et al. 2019. Microscopy-based chromosome conformation capture enables simultaneous visualization of genome organization and transcription in intact organisms. *Mol Cell* **74:** 212–222.e5. doi:10.1016/j.molcel.2019.01.011

Chen H, Levo M, Barinov L, Fujioka M, Jaynes JB, Gregor T. 2018a. Dynamic interplay between enhancer–promoter topology and gene activity. *Nat Genet* **50:** 1296–1303. doi:10.1038/s41588-018-0175-z

Chen X, Teichmann SA, Meyer KB. 2018b. From tissues to cell types and back: single-cell gene expression analysis of tissue architecture. *Ann Rev Biomed Data Sci* **1:** 29–51. doi:10.1146/annurev-biodatasci-080917-013452

Chong S, Chen C, Ge H, Xie XS. 2014. Mechanism of transcriptional bursting in bacteria. *Cell* **158:** 314–326. doi:10.1016/j.cell.2014.05.038

Chuang CH, Carpenter AE, Fuchsova B, Johnson T, de Lanerolle P, Belmont AS. 2006. Long-range directional movement of an interphase chromosome site. *Curr Biol* **16:** 825–831. doi:10.1016/j.cub.2006.03.059

Chubb JR, Boyle S, Perry P, Bickmore WA. 2002. Chromatin motion is constrained by association with nuclear compartments in human cells. *Curr Biol* **12:** 439–445. doi:10.1016/S0960-9822(02)00695-4

Dalla-Favera R, Bregni M, Erikson J, Patterson D, Gallo RC, Croce CM. 1982. Human c-myc *onc* gene is located on the region of chromosome 8 that is translocated in Burkitt lymphoma cells. *Proc Natl Acad Sci* **79:** 7824–7827. doi:10.1073/pnas.79.24.7824

Davidson IF, Bauer B, Goetz D, Tang W, Wutz G, Peters JM. 2019. DNA loop extrusion by human cohesin. *Science* **366:** 1338–1345. doi:10.1126/science.aaz3418

Deng W, Lee J, Wang H, Miller J, Reik A, Gregory PD, Dean A, Blobel GA. 2012. Controlling long-range genomic interactions at a native locus by targeted tethering of a looping factor. *Cell* **149:** 1233–1244. doi:10.1016/j.cell.2012.03.051

Deng W, Rupon JW, Krivega I, Breda L, Motta I, Jahn KS, Reik A, Gregory PD, Rivella S, Dean A, et al. 2014. Reactivation of developmentally silenced globin genes by forced chromatin looping. *Cell* **158:** 849–860. doi:10.1016/j.cell.2014.05.050

Dey SS, Foley JE, Limsirichai P, Schaffer DV, Arkin AP. 2015. Orthogonal control of expression mean and variance by epigenetic features at different genomic loci. *Mol Syst Biol* **11:** 806. doi:10.15252/msb.20145704

Dixon JR, Selvaraj S, Yue F, Kim A, Li Y, Shen Y, Hu M, Liu JS, Ren B. 2012. Topological domains in mammalian genomes identified by analysis of chromatin interactions. *Nature* **485:** 376–380. doi:10.1038/nature11082

Donovan BT, Huynh A, Ball DA, Patel HP, Poirier MG, Larson DR, Ferguson ML, Lenstra TL. 2019. Live-cell imaging reveals the interplay between transcription factors, nucleosomes, and bursting. *EMBO J* **38:** e100809. doi:10.15252/embj.2018100809

Du Z, Zheng H, Huang B, Ma R, Wu J, Zhang X, He J, Xiang Y, Wang Q, Li Y, et al. 2017. Allelic reprogramming of 3D chromatin architecture during early mammalian development. *Nature* **547:** 232–235. doi:10.1038/nature23263

Elowitz MB, Levine AJ, Siggia ED, Swain PS. 2002. Stochastic gene expression in a single cell. *Science* **297:** 1183–1186. doi:10.1126/science.1070919

Falk M, Feodorova Y, Naumova N, Imakaev M, Lajoie BR, Leonhardt H, Joffe B, Dekker J, Fudenberg G, Solovei I, et al. 2019. Heterochromatin drives compartmentalization of inverted and conventional nuclei. *Nature* **570:** 395–399. doi:10.1038/s41586-019-1275-3

Fang C, Rao S, Crispino JD, Ntziachristos P. 2020. Determinants and role of chromatin organization in acute leukemia. *Leukemia* **34:** 2561–2575. doi: 10.1038/s41375-020-0981-z

Finlan LE, Sproul D, Thomson I, Boyle S, Kerr E, Perry P, Ylstra B, Chubb JR, Bickmore WA. 2008. Recruitment to the nuclear periphery can alter expression of genes in human cells. *PLoS Genet* **4:** e1000039. doi:10.1371/journal.pgen.1000039

Finn EH, Misteli T. 2019. Molecular basis and biological function of variability in spatial genome organization. *Science* **365:** eaaw9498. doi:10.1126/science.aaw9498

Finn EH, Pegoraro G, Brandão HB, Valton AL, Oomen ME, Dekker J, Mirny L, Misteli T. 2019. Extensive heterogeneity and intrinsic variation in spatial genome organization. *Cell* **176:** 1502–1515.e10. doi:10.1016/j.cell.2019.01.020

Fishilevich S, Nudel R, Rappaport N, Hadar R, Plaschkes I, Stein TI, Rosen N, Kohn A, Twik M, Safran M, et al. 2017. GeneHancer: genome-wide integration of enhancers and target genes in GeneCards. *Database (Oxford)* **2017:** bax028. doi:10.1093/database/bax028

Flavahan WA, Drier Y, Liau BB, Gillespie SM, Venteicher AS, Stemmer-Rachamimov AO, Suvà ML, Bernstein BE. 2016. Insulator dysfunction and oncogene activation in *IDH* mutant gliomas. *Nature* **529:** 110–114. doi:10.1038/nature16490

Flyamer IM, Gassler J, Imakaev M, Brandão HB, Ulianov SV, Abdennur N, Razin SV, Mirny LA, Tachibana-Konwalski K. 2017. Single-nucleus Hi-C reveals unique chromatin reorganization at oocyte-to-zygote transition. *Nature* **544:** 110–114. doi:10.1038/nature21711

Franke M, Ibrahim DM, Andrey G, Schwarzer W, Heinrich V, Schöpflin R, Kraft K, Kempfer R, Jerković I, Chan WL, et al. 2016. Formation of new chromatin domains determines pathogenicity of genomic duplications. *Nature* **538:** 265–269. doi:10.1038/nature19800

Fudenberg G, Imakaev M, Lu C, Goloborodko A, Abdennur N, Mirny LA. 2016. Formation of chromosomal domains by loop extrusion. *Cell Rep* **15:** 2038–2049. doi:10.1016/j.celrep.2016.04.085

Fukaya T, Lim B, Levine M. 2016. Enhancer control of transcriptional bursting. *Cell* **166:** 358–368. doi:10.1016/j.cell.2016.05.025

Furlong EEM, Levine M. 2018. Developmental enhancers and chromosome topology. *Science* **361:** 1341–1345. doi:10.1126/science.aau0320

Gambetta MC, Furlong EEM. 2018. The insulator protein CTCF is required for correct *Hox* gene expression, but not for embryonic development in *Drosophila*. *Genetics* **210:** 129–136. doi:10.1534/genetics.118.301350

Ganai N, Sengupta S, Menon GI. 2014. Chromosome positioning from activity-based segregation. *Nucleic Acids Res* **42:** 4145–4159. doi:10.1093/nar/gkt1417

Gassler J, Brandão HB, Imakaev M, Flyamer IM, Ladstätter S, Bickmore WA, Peters JM, Mirny LA, Tachibana K. 2017. A mechanism of cohesin-dependent loop extrusion organizes zygotic genome architecture. *EMBO J* **36:** 3600–3618. doi:10.15252/embj.201798083

Ghavi-Helm Y, Jankowski A, Meiers S, Viales RR, Korbel JO, Furlong EEM. 2019. Highly rearranged chromosomes reveal uncoupling between genome topology and gene expression. *Nat Genet* **51:** 1272–1282. doi:10.1038/s41588-019-0462-3

Ghirlando R, Felsenfeld G. 2016. CTCF: making the right connections. *Genes Dev* **30:** 881–891. doi:10.1101/gad.277863.116

Gilbert N, Boyle S, Fiegler H, Woodfine K, Carter NP, Bickmore WA. 2004. Chromatin architecture of the human genome: gene-rich domains are enriched in open chromatin fibers. *Cell* **118:** 555–566. doi:10.1016/j.cell.2004.08.011

Gröschel S, Sanders MA, Hoogenboezem R, de Wit E, Bouwman BAM, Erpelinck C, van der Velden VHJ, Havermans M, Avellino R, van Lom K, et al. 2014. A single oncogenic enhancer rearrangement causes concomitant *EVI1* and *GATA2* deregulation in leukemia. *Cell* **157:** 369–381. doi:10.1016/j.cell.2014.02.019

Guelen L, Pagie L, Brasset E, Meuleman W, Faza MB, Talhout W, Eussen BH, de Klein A, Wessels L, de Laat W, et al. 2008. Domain organization of human chromosomes revealed by mapping of nuclear lamina interactions. *Nature* **453:** 948–951. doi:10.1038/nature06947

Guo Y, Xu Q, Canzio D, Shou J, Li J, Gorkin DU, Jung I, Wu H, Zhai Y, Tang Y, et al. 2015. CRISPR inversion of CTCF sites alters genome topology and enhancer/promoter function. *Cell* **162:** 900–910. doi:10.1016/j.cell.2015.07.038

Haarhuis JHI, van der Weide RH, Blomen VA, Yáñez-Cuna JO, Amendola M, van Ruiten MS, Krijger PHL, Teunissen H, Medema RH, van Steensel B, et al. 2017. The cohesin release factor WAPL restricts chromatin loop extension. *Cell* **169:** 693–707.e14. doi:10.1016/j.cell.2017.04.013

Hilton IB, D'Ippolito AM, Vockley CM, Thakore PI, Crawford GE, Reddy TE, Gersbach CA. 2015. Epigenome editing by a CRISPR-Cas9-based acetyltransferase activates genes from promoters and enhancers. *Nat Biotechnol* **33:** 510–517. doi:10.1038/nbt.3199

Hnisz D, Weintraub AS, Day DS, Valton AL, Bak RO, Li CH, Goldmann J, Lajoie BR, Fan ZP, Sigova AA, et al. 2016. Activation of proto-oncogenes by disruption of chromosome neighborhoods. *Science* **351:** 1454–1458. doi:10.1126/science.aad9024

Hornung G, Bar-Ziv R, Rosin D, Tokuriki N, Tawfik DS, Oren M, Barkai N. 2012. Noise–mean relationship in mutated promoters. *Genome Res* **22:** 2409–2417. doi:10.1101/gr.139378.112

Hou C, Li L, Qin ZS, Corces VG. 2012. Gene density, transcription, and insulators contribute to the partition of the *Drosophila* genome into physical domains. *Mol Cell* **48:** 471–484. doi:10.1016/j.molcel.2012.08.031

Hsieh THS, Cattoglio C, Slobodyanyuk E, Hansen AS, Rando OJ, Tjian R, Darzacq X. 2020. Resolving the 3D landscape of transcription-linked mammalian chromatin folding. *Mol Cell* **78:** 539–553.e8. doi:10.1016/j.molcel.2020.03.002

Hug CB, Grimaldi AG, Kruse K, Vaquerizas JM. 2017. Chromatin architecture emerges during zygotic genome activation independent of transcription. *Cell* **169:** 216–228.e19. doi:10.1016/j.cell.2017.03.024

Ibragimov AN, Bylino OV, Shidlovskii YV. 2020. Molecular basis of the function of transcriptional enhancers. *Cells* **9:** 1620. doi:10.3390/cells9071620

Janssen A, Colmenares SU, Karpen GH. 2018. Heterochromatin: guardian of the genome. *Ann Rev Cell Dev Biol* **34:** 265–288. doi:10.1146/annurev-cellbio-100617-062653

Jenuwein T, Allis CD. 2001. Translating the histone code. *Science* **293:** 1074–1080. doi:10.1126/science.1063127

Ke Y, Xu Y, Chen X, Feng S, Liu Z, Sun Y, Yao X, Li F, Zhu W, Gao L, et al. 2017. 3D chromatin structures of mature gametes and structural reprogramming during mammalian embryogenesis. *Cell* **170:** 367–381.e20. doi:10.1016/j.cell.2017.06.029

Kim Y, Shi Z, Zhang H, Finkelstein IJ, Yu H. 2019. Human cohesin compacts DNA by loop extrusion. *Science* **366:** 1345–1349. doi:10.1126/science.aaz4475

Kind J, van Steensel B. 2010. Genome–nuclear lamina interactions and gene regulation. *Curr Opin Cell Biol* **22:** 320–325. doi:10.1016/j.ceb.2010.04.002

Krijger PHL, De Laat W. 2016. Regulation of disease-associated gene expression in the 3D genome. *Nat Rev Mol Cell Biol* **17:** 771–782. doi:10.1038/nrm.2016.138

Kumaran RI, Spector DL. 2008. A genetic locus targeted to the nuclear periphery in living cells maintains its transcriptional competence. *J Cell Biol* **180:** 51–65. doi:10.1083/jcb.200706060

Larson DR. 2019. Structure and function in *Drosophila* chromosomes: visualizing topological domains. *Mol Cell* **74:** 3–4. doi:10.1016/j.molcel.2019.03.017

Larson DR, Zenklusen D, Wu B, Chao JA, Singer RH. 2011. Real-time observation of transcription initiation and elongation on an endogenous yeast gene. *Science* **332:** 475–478. doi:10.1126/science.1202142

Larsson AJM, Johnsson P, Hagemann-Jensen M, Hartmanis L, Faridani OR, Reinius B, Segerstolpe Å, Rivera CM, Ren B, Sandberg R. 2019. Genomic encoding of transcriptional burst kinetics. *Nature* **565:** 251–254. doi:10.1038/s41586-018-0836-1

Le Dily F, Baù D, Pohl A, Vicent GP, Serra F, Soronellas D, Castellano G, Wright RHG, Ballare C, Filion G, et al. 2014. Distinct structural transitions of chromatin topological domains correlate with coordinated hormone-induced gene regulation. *Genes Dev* **28:** 2151–2162. doi:10.1101/gad.241422.114

Lettice LA, Heaney SJH, Purdie LA, Li L, de Beer P, Oostra BA, Goode D, Elgar G, Hill RE, de Graaff E. 2003. A long-range *Shh* enhancer regulates expression in the developing limb and fin and is associated with preaxial polydactyly. *Hum Mol Genet* **12:** 1725–1735. doi:10.1093/hmg/ddg180

Lieberman-Aiden E, van Berkum NL, Williams L, Imakaev M, Ragoczy T, Telling A, Amit I, Lajoie BR, Sabo PJ, Dorschner MO, et al. 2009. Comprehensive mapping of long-range interactions reveals folding principles of the human genome. *Science* **326:** 289–293. doi:10.1126/science.1181369

Lim B, Heist T, Levine M, Fukaya T. 2018. Visualization of transvection in living *Drosophila* embryos. *Mol Cell* **70:** 287–296.e6. doi:10.1016/j.molcel.2018.02.029

Long HK, Prescott SL, Wysocka J. 2016. Ever-changing landscapes: transcriptional enhancers in development and evolution. *Cell* **167:** 1170–1187. doi:10.1016/j.cell.2016.09.018

Lupiáñez DG, Kraft K, Heinrich V, Krawitz P, Brancati F, Klopocki E, Horn D, Kayserili H, Opitz JM, Laxova R, et al. 2015. Disruptions of topological chromatin domains cause pathogenic rewiring of gene-enhancer interactions. *Cell* **161:** 1012–1025. doi:10.1016/j.cell.2015.04.004

Luppino JM, Park DS, Nguyen SC, Lan Y, Xu Z, Yunker R, Joyce EF. 2020. Cohesin promotes stochastic domain intermingling to ensure proper regulation of boundary-proximal genes. *Nat Genet* **52:** 840–848. doi:10.1038/s41588-020-0647-9

Mateo LJ, Murphy SE, Hafner A, Cinquini IS, Walker CA, Boettiger AN. 2019. Visualizing DNA folding and RNA in embryos at single-cell resolution. *Nature* **568:** 49–54. doi:10.1038/s41586-019-1035-4

Mermet J, Yeung J, Hurni C, Mauvoisin D, Gustafson K, Jouffe C, Nicolas D, Emmenegger Y, Gobet C, Franken P, et al. 2018. Clock-dependent chromatin topology modulates circadian transcription and behavior. *Genes Dev* **32:** 347–358. doi:10.1101/gad.312397.118

Misteli T. 2020. The self-organizing genome: principles of genome architecture and function. *Cell* **183:** 28–45. doi:10.1016/j.cell.2020.09.014

Mizuguchi T, Fudenberg G, Mehta S, Belton JM, Taneja N, Folco HD, FitzGerald P, Dekker J, Mirny L, Barrowman J, et al. 2014. Cohesin-dependent globules and heterochromatin shape 3D genome architecture in *S. pombe*. *Nature* **516:** 432–435. doi:10.1038/nature13833

Morgan SL, Mariano NC, Bermudez A, Arruda NL, Wu F, Luo Y, Shankar G, Jia L, Chen H, Hu JF, et al. 2017. Manipulation of nuclear architecture through CRISPR-mediated chromosomal looping. *Nat Commun* **8:** 15993. doi:10.1038/ncomms15993

Nagano T, Lubling Y, Stevens TJ, Schoenfelder S, Yaffe E, Dean W, Laue ED, Tanay A, Fraser P. 2013. Single-cell Hi-C reveals cell-to-cell variability in chromosome structure. *Nature* **502:** 59–64. doi:10.1038/nature12593

Narendra V, Rocha PP, An D, Raviram R, Skok JA, Mazzoni EO, Reinberg D. 2015. CTCF establishes discrete functional chromatin domains at the *Hox* clusters during differentiation. *Science* **347:** 1017–1021. doi:10.1126/science.1262088

Naughton C, Avlonitis N, Corless S, Prendergast JG, Mati IK, Eijk PP, Cockroft SL, Bradley M, Ylstra B, Gilbert N. 2013. Transcription forms and remodels supercoiling domains unfolding large-scale chromatin structures. *Nat Struct Mol Biol* **20:** 387–395. doi:10.1038/nsmb.2509

Nora EP, Lajoie BR, Schulz EG, Giorgetti L, Okamoto I, Servant N, Piolot T, van Berkum NL, Meisig J, Sedat J, et al. 2012. Spatial partitioning of the regulatory landscape of the X-inactivation centre. *Nature* **485:** 381–385. doi:10.1038/nature11049

Nora EP, Goloborodko A, Valton AL, Gibcus JH, Uebersohn A, Abdennur N, Dekker J, Mirny LA, Bruneau BG. 2017. Targeted degradation of CTCF decouples local insulation of chromosome domains from genomic compartmentalization. *Cell* **169:** 930–944.e22. doi:10.1016/j.cell.2017.05.004

Northcott PA, Lee C, Zichner T, Stütz AM, Erkek S, Kawauchi D, Shih DJH, Hovestadt V, Zapatka M, Sturm D, et al. 2014. Enhancer hijacking activates GFI1 family oncogenes in medulloblastoma. *Nature* **511:** 428–434. doi:10.1038/nature13379

Nuebler J, Fudenberg G, Imakaev M, Abdennur N, Mirny LA. 2018. Chromatin organization by an interplay of loop extrusion and compartmental segregation. *Proc Natl Acad Sci* **115:** E6697–E6706.

Oldridge DA, Wood AC, Weichert-Leahey N, Crimmins I, Sussman R, Winter C, McDaniel LD, Diamond M, Hart LS, Zhu S, et al. 2015. Genetic predisposition to neuroblastoma mediated by a *LMO1* super-enhancer polymorphism. *Nature* **528:** 418–421. doi:10.1038/nature15540

Peddie CJ, Collinson LM. 2014. Exploring the third dimension: volume electron microscopy comes of age. *Micron* **61:** 9–19. doi:10.1016/j.micron.2014.01.009

Phanstiel DH, Van Bortle K, Spacek D, Hess GT, Saad Shamim M, Machol I, Love MI, Lieberman Aiden E, Bassik MC, Snyder MP. 2017. Static and dynamic DNA loops form AP-1-bound activation hubs during macrophage development. *Mol Cell* **67:** 1037–1048.e6. doi:10.1016/j.molcel.2017.08.006

Rada-Iglesias A, Bajpai R, Swigut T, Brugmann SA, Flynn RA, Wysocka J. 2011. A unique chromatin signature uncovers early developmental enhancers in humans. *Nature* **470:** 279–283. doi:10.1038/nature09692

Raj A, van Oudenaarden A. 2008. Nature, nurture, or chance: stochastic gene expression and its consequences. *Cell* **135:** 216–226. doi:10.1016/j.cell.2008.09.050

Raj A, Peskin CS, Tranchina D, Vargas DY, Tyagi S. 2006. Stochastic mRNA synthesis in mammalian cells. *PLoS Biol* **4:** e309.

Rao SSP, Huntley MH, Durand NC, Stamenova EK, Bochkov ID, Robinson JT, Sanborn AL, Machol I, Omer AD, Lander ES, et al. 2014. A 3D map of the human genome at kilobase resolution reveals principles of chromatin looping. *Cell* **159:** 1665–1680. doi:10.1016/j.cell.2014.11.021

Rao SSP, Huang SC, St Hilaire BG, Engreitz JM, Perez EM, Kieffer-Kwon KR, Sanborn AL, Johnstone SE, Bascom GD, Bochkov ID, et al. 2017. Cohesin loss eliminates all loop domains. *Cell* **171:** 305–320.e24. doi:10.1016/j.cell.2017.09.026

Raser JM, O'Shea EK. 2004. Control of stochasticity in eukaryotic gene expression. *Science* **304:** 1811–1814. doi:10.1126/science.1098641

Reddy KL, Zullo JM, Bertolino E, Singh H. 2008. Transcriptional repression mediated by repositioning of genes to the nuclear lamina. *Nature* **452:** 243–247.

Ricci MA, Manzo C, García-Parajo MF, Lakadamyali M, Cosma MP. 2015. Chromatin fibers are formed by heterogeneous groups of nucleosomes in vivo. *Cell* **160:** 1145–1158. doi:10.1016/j.cell.2015.01.054

Robinett CC, Straight A, Li G, Willhelm C, Sudlow G, Murray A, Belmont AS. 1996. In vivo localization of DNA sequences and visualization of large-scale chromatin organization using lac operator/repressor recognition. *J Cell Biol* **135:** 1685–1700. doi:10.1083/jcb.135.6.1685

Rodriguez J, Larson DR. 2020. Transcription in living cells: molecular mechanisms of bursting. *Annu Rev Biochem* **89:** 189–212. doi:10.1146/annurev-biochem-011520-105250

Rodriguez J, Ren G, Day CR, Zhao K, Chow CC, Larson DR. 2019. Intrinsic dynamics of a human gene reveal the basis of expression heterogeneity. *Cell* **176:** 213–226.e18. doi:10.1016/j.cell.2018.11.026

Rowley MJ, Nichols MH, Lyu X, Ando-Kuri M, Rivera ISM, Hermetz K, Wang P, Ruan Y, Corces VG. 2017. Evolutionarily conserved principles predict 3D chromatin organization. *Mol Cell* **67:** 837–852.e7. doi:10.1016/j.molcel.2017.07.022

Ruf S, Symmons O, Uslu VV, Dolle D, Hot C, Ettwiller L, Spitz F. 2011. Large-scale analysis of the regulatory architecture of the mouse genome with a transposon-associated sensor. *Nat Genet* **43:** 379–386. doi:10.1038/ng.790

Sanborn AL, Rao SSP, Huang SC, Durand NC, Huntley MH, Jewett AI, Bochkov ID, Chinnappan D, Cutkosky A, Li J, et al. 2015. Chromatin extrusion explains key features of loop and domain formation in wild-type and engineered genomes. *Proc Natl Acad Sci* **112:** E6456–E6465. doi:10.1073/pnas.1518552112

Schaffner W. 2015. Enhancers, enhancers—from their discovery to today's universe of transcription enhancers. *Biol Chem* **396:** 311–327. doi:10.1515/hsz-2014-0303

Schultz J. 1947. The nature of heterochromatin. *Cold Spring Harb Symp Quant Biol* **12:** 179–191. doi:10.1101/SQB.1947.012.01.021

Sexton T, Yaffe E, Kenigsberg E, Bantignies F, Leblanc B, Hoichman M, Parrinello H, Tanay A, Cavalli G. 2012. Three-dimensional folding and functional organization principles of the *Drosophila* genome. *Cell* **148:** 458–472. doi:10.1016/j.cell.2012.01.010

Shahrezaei V, Swain PS. 2008. Analytical distributions for stochastic gene expression. *Proc Natl Acad Sci* **105:** 17256–17261. doi:10.1073/pnas.0803850105

Shen Y, Yue F, McCleary DF, Ye Z, Edsall L, Kuan S, Wagner U, Dixon J, Lee L, Lobanenkov VV, et al. 2012. A map of the *cis*-regulatory sequences in the mouse genome. *Nature* **488:** 116–120. doi:10.1038/nature11243

Spitz F, Furlong EEM. 2012. Transcription factors: from enhancer binding to developmental control. *Nat Rev Genet* **13:** 613–626. doi:10.1038/nrg3207

Stampfel G, Kazmar T, Frank O, Wienerroither S, Reiter F, Stark A. 2015. Transcriptional regulators form diverse groups with context-dependent regulatory functions. *Nature* **528:** 147–151. doi:10.1038/nature15545

Stavreva DA, Garcia DA, Fettweis G, Gudla PR, Zaki GF, Soni V, McGowan A, Williams G, Huynh A, Palangat M, et al. 2019. Transcriptional bursting and co-bursting regulation by steroid hormone release pattern and transcription factor mobility. *Mol Cell* **75:** 1161–1177.e11. doi:10.1016/j.molcel.2019.06.042

Stevens TJ, Lando D, Basu S, Atkinson LP, Cao Y, Lee SF, Leeb M, Wohlfahrt KJ, Boucher W, O'Shaughnessy-Kirwan A, et al. 2017. 3D structures of individual mammalian genomes studied by single-cell Hi-C. *Nature* **544:** 59–64. doi:10.1038/nature21429

Strom AR, Emelyanov AV, Mir M, Fyodorov DV, Darzacq X, Karpen GH. 2017. Phase separation drives heterochromatin domain formation. *Nature* **547:** 241–245. doi:10.1038/nature22989

Su JH, Zheng P, Kinrot SS, Bintu B, Zhuang X. 2020. Genome-scale imaging of the 3D organization and transcriptional activity of chromatin. *Cell* **182:** 1641–1659.e26. doi: 10.1016/j.cell.2020.07.032

Sun M, Zhang J. 2019. Chromosome-wide co-fluctuation of stochastic gene expression in mammalian cells. *PLoS Genet* **15:** e1008389.

Suter DM, Molina N, Gatfield D, Schneider K, Schibler U, Naef F. 2011. Mammalian genes are transcribed with widely different bursting kinetics. *Science* **332:** 472–474. doi:10.1126/science.1198817

Swain PS, Elowitz MB, Siggia ED. 2002. Intrinsic and extrinsic contributions to stochasticity in gene expression. *Proc Natl Acad Sci* **99:** 12795–12800. doi:10.1073/pnas.162041399

Symmons O, Uslu VV, Tsujimura T, Ruf S, Nassari S, Schwarzer W, Ettwiller L, Spitz F. 2014. Functional and topological characteristics of mammalian regulatory domains. *Genome Res* **24:** 390–400. doi:10.1101/gr.163519.113

Szabo Q, Jost D, Chang JM, Cattoni DI, Papadopoulos GL, Bonev B, Sexton T, Gurgo J, Jacquier C, Nollmann M, et al. 2018. TADs are 3D structural units of higher-order chromosome organization in *Drosophila*. *Science Adv* **4:** eaar8082. doi:10.1126/sciadv.aar8082

Tan L, Xing D, Chang CH, Li H, Xie XS. 2018. Three-dimensional genome structures of single diploid human cells. *Science* **361:** 924–928. doi:10.1126/science.aat5641

Tantale K, Mueller F, Kozulic-Pirher A, Lesne A, Victor JM, Robert MC, Capozi S, Chouaib R, Bäcker V, Mateos-Langerak J, et al. 2016. A single-molecule view of transcription reveals convoys of RNA polymerases and multiscale bursting. *Nat Commun* **7:** 12248. doi:10.1038/ncomms12248

The 1000 Genomes Project Consortium; Abecasis GR, Auton A, Brooks LD, DePristo MA, Durbin RM, Handsaker RE, Kang HM, Marth GT, McVean GA. 2012. An integrated map of genetic variation from 1,092 human genomes. *Nature* **491:** 56–65. doi:10.1038/nature11632

Therizols P, Illingworth RS, Courilleau C, Boyle S, Wood AJ, Bickmore WA. 2014. Chromatin decondensation is sufficient to alter nuclear organization in embryonic stem cells. *Science* **346:** 1238–1242. doi:10.1126/science.1259587

Trzaskoma P, Ruszczycki B, Lee B, Pels KK, Krawczyk K, Bokota G, Szczepankiewicz AA, Aaron J, Walczak A, Śliwińska MA, et al. 2020. Ultrastructural visualization of 3D chromatin folding using volume electron microscopy and DNA in situ hybridization. *Nat Commun* **11:** 2120. doi:10.1038/s41467-020-15987-2

Tunnacliffe E, Chubb JR. 2020. What is a transcriptional burst? *Trends Genet* **36:** 288–297. doi:10.1016/j.tig.2020.01.003

Ulianov SV, Khrameeva EE, Gavrilov AA, Flyamer IM, Kos P, Mikhaleva EA, Penin AA, Logacheva MD, Imakaev MV, Chertovich A, et al. 2016. Active chromatin and transcription play a key role in chromosome partitioning into topologically associating domains. *Genome Res* **26:** 70–84. doi:10.1101/gr.196006.115

van Steensel B, Belmont AS. 2017. Lamina-associated domains: links with chromosome architecture, heterochromatin, and gene repression. *Cell* **169:** 780–791. doi:10.1016/j.cell.2017.04.022

Vian L, Pękowska A, Rao SSP, Kieffer-Kwon KR, Jung S, Baranello L, Huang SC, El Khattabi L, Dose M, Pruett N, et al. 2018. The energetics and physiological impact of cohesin extrusion. *Cell* **173:** 1165–1178.e20. doi:10.1016/j.cell.2018.03.072

Voss TC, Schiltz RL, Sung MH, Yen PM, Stamatoyannopoulos JA, Biddie SC, Johnson TA, Miranda TB, John S, Hager GL. 2011. Dynamic exchange at regulatory elements during chromatin remodeling underlies assisted loading mechanism. *Cell* **146:** 544–554. doi:10.1016/j.cell.2011.07.006

Wang S, Su JH, Beliveau BJ, Bintu B, Moffitt JR, Wu CT, Zhuang X. 2016. Spatial organization of chromatin domains and compartments in single chromosomes. *Science* **353:** 598–602. doi:10.1126/science.aaf8084

Wang L, Gao Y, Zheng X, Liu C, Dong S, Li R, Zhang G, Wei Y, Qu H, Li Y, et al. 2019. Histone modifications regulate chromatin compartmentalization by contributing to a phase separation mechanism. *Mol Cell* **76:** 646–659.e6. doi:10.1016/j.molcel.2019.08.019

Weintraub AS, Li CH, Zamudio AV, Sigova AA, Hannett NM, Day DS, Abraham BJ, Cohen MA, Nabet B, Buckley DL, et al. 2017. YY1 is a structural regulator of enhancer-promoter loops. *Cell* **171:** 1573–1588.e28. doi:10.1016/j.cell.2017.11.008

Weischenfeldt J, Dubash T, Drainas AP, Mardin BR, Chen Y, Stütz AM, Waszak SM, Bosco G, Halvorsen AR, Raeder B, et al. 2017. Pan-cancer analysis of somatic copy-number alterations implicates *IRS4* and *IGF2* in enhancer hijacking. *Nat Genet* **49:** 65–74. doi:10.1038/ng.3722

Wijchers PJ, Krijger PHL, Geeven G, Zhu Y, Denker A, Verstegen MJAM, Valdes-Quezada C, Vermeulen C, Janssen M, Teunissen H, et al. 2016. Cause and consequence of tethering a subTAD to different nuclear compartments. *Mol Cell* **61:** 461–473. doi:10.1016/j.molcel.2016.01.001

Williams RRE, Azuara V, Perry P, Sauer S, Dvorkina M, Jørgensen H, Roix J, McQueen P, Misteli T, Merkenschlager M, et al. 2006. Neural induction promotes large-scale chromatin reorganisation of the *Mash1* locus. *J Cell Sci* **119:** 132–140. doi:10.1242/jcs.02727

Williamson I, Lettice LA, Hill RE, Bickmore WA. 2016. *Shh* and ZRS enhancer colocalisation is specific to the zone of polarising activity. *Development* **143:** 2994–3001. doi:10.1242/dev.139188

Williamson I, Kane L, Devenney PS, Flyamer IM, Anderson E, Kilanowski F, Hill RE, Bickmore WA, Lettice LA. 2019. Developmentally regulated *Shh* expression is robust to TAD perturbations. *Development* **146:** dev179523. doi:10.1242/dev.179523

Xie WJ, Meng L, Liu S, Zhang L, Cai X, Gao YQ. 2017. Structural modeling of chromatin integrates genome features and reveals chromosome folding principle. *Sci Rep* **7:** 2818. doi:10.1038/s41598-017-02923-6

Yaffe E, Tanay A. 2011. Probabilistic modeling of Hi-C contact maps eliminates systematic biases to characterize global chromosomal architecture. *Nat Genet* **43:** 1059–1065. doi:10.1038/ng.947

Zaret KS, Carroll JS. 2011. Pioneer transcription factors: establishing competence for gene expression. *Genes Dev* **25:** 2227–2241. doi:10.1101/gad.176826.111

Emerging Properties and Functions of Actin and Actin Filaments Inside the Nucleus

Svenja Ulferts,[1] Bina Prajapati,[2] Robert Grosse,[1,3] and Maria K. Vartiainen[2]

[1]Institute for Clinical and Experimental Pharmacology and Toxicology I, University of Freiburg, 79104 Freiburg, Germany

[2]Institute of Biotechnology, Helsinki Institute for Life Science, University of Helsinki, 00014 Helsinki, Finland

[3]Centre for Integrative Biological Signalling Studies (CIBSS), 79104 Freiburg, Germany

Correspondence: robert.grosse@pharmakol.uni-freiburg.de; maria.vartiainen@helsinki.fi

Recent years have provided considerable insights into the dynamic nature of the cell nucleus, which is constantly reorganizing its genome, controlling its size and shape, as well as spatiotemporally orchestrating chromatin remodeling and transcription. Remarkably, it has become clear that the ancient and highly conserved cytoskeletal protein actin plays a crucial part in these processes. However, the underlying mechanisms, regulations, and properties of actin functions inside the nucleus are still not well understood. Here we summarize the diverse and distinct roles of monomeric and filamentous actin as well as the emerging roles for actin dynamics inside the nuclear compartment for genome organization and nuclear architecture.

Actin is a highly conserved and abundant protein present in two conformations in all eukaryotic cells. Its monomeric form, globular G-actin, reversibly assembles into long helical filaments (F-actin) controlled via the interaction with a myriad of actin-binding proteins (ABPs) (Pollard 2016). As microfilaments, they constitute one of the three major components of the cytoskeleton and play a fundamental role in cell shape and motility, intracellular transport, muscle contraction, and organelle dynamics, among others (Pollard 2016; Titus 2018). It is now known that both actin and actin regulatory factors constantly shuttle between the cytosolic and nuclear compartments (Grosse and Vartiainen 2013; Hyrskyluoto and Vartiainen 2020). Even though a plethora of cytosolic functions of actin have been extensively studied in the past, the diverse and distinct roles for actin dynamics and the actin cytoskeleton within the nuclear compartment are just beginning to emerge (Plessner and Grosse 2019).

In the past decade, the concept of the cell nucleus as a mere repository for genomic information has been replaced by the perception of the nucleus as a highly dynamic organelle, where spatially separated processes strongly affect nuclear and chromosome architecture and organization to guard and control genome functions (Rout and Karpen 2014). However, a potential role for actin structures in these dynamic nuclear events, including transport of molecules, nucle-

Cite this article as *Cold Spring Harb Perspect Biol* doi: 10.1101/cshperspect.a040121

ar shape, and maintenance of nuclear integrity, is surfacing only now. Better microscopy techniques and the use of modified actin-binding probes targeted to the nucleus have allowed for the first direct visualizations of nuclear dynamic actin assembly in live somatic cells and within the interchromatin region (Baarlink et al. 2013, 2017; Belin et al. 2013; Melak et al. 2017). Since then, evidence is accumulating for the involvement of actin dynamics in fundamental nuclear processes like chromatin reorganization (Baarlink et al. 2017), DNA damage repair (DDR) (Caridi et al. 2018; Schrank et al. 2018), transcription regulation and initiation (Wei et al. 2020), or functional control of chromatin remodeling complexes (Jungblut et al. 2020). Here we summarize recent progress in our understanding of the roles and functions of the ancient molecule actin for nuclear dynamics, genome organization, and architecture.

MONOMERIC ACTIN IN THE NUCLEUS

Dynamically Connected Actin Pools in the Cytoplasm and the Nucleus

Because actin is an important protein in both the cytoplasm and the nucleus, there must be mechanisms to ensure appropriate balance and regulation between the cellular actin pools. Indeed, actin uses an active transport mechanism to constantly and rapidly shuttle in and out of the nucleus (Dopie et al. 2012), and this process is subject to regulation at multiple levels to fine-tune the cellular distribution of actin (Fig. 1). First, an actin monomer binds to the small ABP cofilin, and this cargo is imported into the nucleus by the import factor importin-9 (Fig. 1; Dopie et al. 2012). Many proteins that regulate phosphorylation of cofilin, and thus its ability to interact with actin, can influence nuclear import of actin (Dopie et al. 2015). Actin is exported out of the nucleus as a monomer in complex with profilin by RanGTP-bound export factor exportin-6 (Fig. 1; Stuven et al. 2003), which is subject to regulation by several mechanisms that in turn influence nuclear actin levels and play important roles in both development and disease. Suppression of exportin-6 expression leads to massive accumulation of actin in the nucleus of *Xenopus* oocytes (Bohnsack et al. 2006), and the resulting nuclear F-actin meshwork is required to protect the ribonucleoprotein droplets against gravity within these huge nuclei (Feric and Brangwynne 2013). Laminin-111 is an essential component of the basement membrane that regulates cell death and quiescence in many tissues. It enhances exportin-6 activity via attenuation of the PI3K pathway, thereby increasing nuclear export of actin (Fig. 1), which leads to reduction in nuclear actin levels and quiescence in mammary epithelial cells (Spencer et al. 2011; Fiore et al. 2017). Interestingly, this pathway is disrupted in human breast cancer cells resulting in continuous proliferation (Fiore et al. 2017). Moreover, recent results show that the tumor suppressor RASSF1A (Ras association domain family 1 isoform A), which is frequently silenced in many different types of cancers, is required for the efficient interaction between exportin-6 and the RanGTPase (Fig. 1) and consequently for the maintenance of nuclear actin levels (Chatzifrangkeskou et al. 2019).

Because actin is transported across the nuclear envelope as a monomer, and actin monomer levels limit the transport rate of actin in both directions (Dopie et al. 2012), processes that regulate the actin monomer pool, including actin polymerization, influence nuclear actin levels. For example, MICAL-2 is an actin-regulatory protein that is enriched in the nucleus, where it oxidizes actin at methionine 44, leading to depolymerization of nuclear actin and subsequent reduction in nuclear actin levels (Lundquist et al. 2014). On the cytoplasmic side of the nuclear envelope, mechanical strain induces polymerization of actin at the outer nuclear membrane via emerin and nonmuscle myosin. This leads to decreased nuclear actin levels, which plays a role in Polycomb-mediated gene silencing required for lineage commitment of epidermal stem cells (Le et al. 2016). On the other hand, multiple stimuli that induce cyclic AMP (cAMP) regulate nuclear actin levels by inhibiting Rho GTPase-dependent actin polymerization in the cytoplasm. Increased availability of actin monomers for nuclear import thus results in elevated nuclear actin levels (McNeill et al. 2020). Taken

Cite this article as *Cold Spring Harb Perspect Biol* doi: 10.1101/cshperspect.a040121

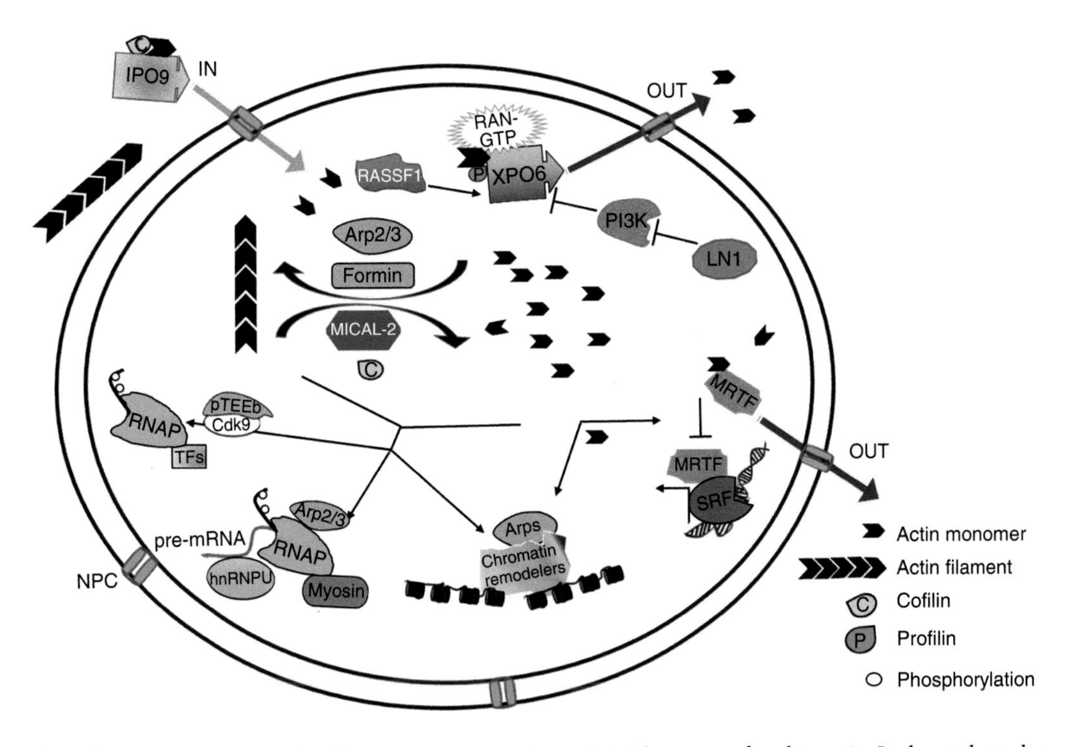

Figure 1. Nucleocytoplasmic shuttling and gene expression–related functions of nuclear actin. In the nucleus, the balance between monomeric and filamentous actin pools is regulated by different actin regulatory proteins, such as Arp2/3 complex and formins, which promote actin polymerization (see also Table 1) and MICAL-2 and cofilin (C), which promote actin filament depolymerization. Actin constantly and rapidly shuttles in and out of the nucleus through nuclear pore complexes (NPCs). Nuclear import is mediated by importin-9 (IPO9) bound to actin monomer and an unphosphorylated cofilin, while monomeric actin is transported out of the nucleus in complex with RanGTP, profilin (P), and exportin-6 (XPO6). RASSF1 enhances the binding of RanGTP to exportin-6, and is thereby required for nuclear export of actin. Laminin-111 (LN1) promotes exportin-6 activity by inhibiting PI3K, thus enhancing nuclear export of actin. Both actin monomers and filaments play functional roles in regulation of gene expression. Actin monomers are integral subunits of several chromatin-remodeling and -modifying complexes together with actin-related proteins (Arps), and actin monomers regulate the subcellular localization and nuclear activity of MRTF-A, the transcription cofactor of SRF. During transcription, actin may influence elongation via positive transcription elongation factor (pTEFb-Cdk9). In addition to actin, actin-binding proteins, such as Arp2/3 complex and nuclear myosins, also interact with the transcription machinery, but the functional form of actin required for RNA polymerase function remains unclear. Polymerized actin may also influence chromatin remodeling.

together, the nuclear pool of actin is in constant and active communication with the cytoplasmic actin networks, and this dynamic process can be used to transmit information from the cytoplasmic cytoskeleton to the nucleus.

Actin as a Signaling Molecule in MRTF/SRF-Mediated Transcription

One of the best-described mediators between the cytoskeleton and nucleus is the myocardin-related transcription factor A (MRTF-A, also known as MKL1 or MAL). MRTF-A is a transcription cofactor of serum response factor (SRF), which regulates many immediate-early, muscle-specific, and cytoskeletal genes (Posern and Treisman 2006). MRTF-A operates as an actin monomer sensor (Vartiainen et al. 2007). In unstimulated conditions, when actin monomer levels are high, MRTF-A is predominantly cytoplasmic, because actin binding occludes the bipartite nuclear localization signal (NLS) em-

Table 1. Signal-regulated F-actin functions in mammalian cell nuclei

Signal input	Actin regulator	Nuclear F-actin properties	Function	References
Serum stimulation	mDia1 and mDia2, MICAL-2	F-actin network	MRTF-A/SRF transcriptional activity	Baarlink et al. 2013; Lundquist et al. 2014
Serum stimulation	N-WASP/Arp2/3	F-actin clusters	Pol II clustering for transcriptional modulation	Wei et al. 2020
Cell spreading	mDia1/2	Thick and short filaments, long-lived (up to 3 h)	MRTF-A/SRF transcriptional activity	Plessner et al. 2015
Mitosis	mDia2	Short dynamic actin filaments	Centromere movement during CENP-A loading	Liu et al. 2018
Mitotic exit	mDia formins	F-actin network	Regulate PCNA loading onto chromatin and initiation of DNA replication	Parisis et al. 2017
Mitotic exit	Cofilin-1, ACTN4	Dynamic and long-lived; bundled short and thick filaments	Actin bundling for nuclear expansion, chromatin decondensation	Baarlink et al. 2017; Krippner et al. 2020
UV/genotoxic agent–induced DNA damage	Spire-1/2, Formin-2	Long, nucleoplasmic filaments; short, nucleolus-associated filaments; dense, nucleoplasmic clusters	DNA double-strand break (DSB) clearance	Belin et al. 2015
Irradiation-/drug-induced DNA damage	Arp2/3, WASP	Long filaments spanning the nucleus	Relocalization of heterochromatic DSB to repair sites at nuclear periphery for homology-directed repair	Caridi et al. 2018; Schrank et al. 2018
Replication stress	Not determined	Long filaments spanning the nucleus	Promote replication fork repair	Lamm et al. 2018
GPCR agonists	INF2, mDia formins	Rapidly assembled and short-lived (60–120 sec)	Rapid Ca^{2+} transient, dynamic chromatin organization	Wang et al. 2019
Ca^{2+} elevation	Not determined	Long actin filaments	Long-range movements of myosin VI along F-actin	Grosse-Berkenbusch et al. 2020
T-cell receptor (TCR) activation/immunological synapse formation	Arp2/3	Rapidly assembled and short-lived actin network (60 sec to 8 min)	Ca^{2+} elevation, immune response/effector cytokine expression	Tsopoulidis et al. 2019
Mouse embryo fertilization	Cofilin-1	F-actin network/nucleoskeleton	Regulation of pronuclei formation and function, efficient DNA damage repair	Okuno et al. 2020

Cite this article as *Cold Spring Harb Perspect Biol* doi: 10.1101/cshperspect.a040121

bedded within the actin-binding RPEL domain (Vartiainen et al. 2007; Pawłowski et al. 2010; Mouilleron et al. 2011). Moreover, actin binding also promotes MRTF-A nuclear export by Crm1 via an unknown mechanism (Fig. 1; Vartiainen et al. 2007). Upon mechanical or mitogenic signaling that activates RhoA-dependent actin polymerization, actin monomer levels decrease, thus liberating MRTF-A from actin and exposing the NLS for nuclear import and preventing nuclear export. This leads to the accumulation of MRTF-A in the nucleus, and activation of MRTF/SRF target genes (Vartiainen et al. 2007; Pawłowski et al. 2010; Mouilleron et al. 2011). Because actin also regulates MRTF-A in the nucleus, factors affecting nuclear actin levels, such as MICAL-2 (Fig. 1; Lundquist et al. 2014), RASSF1A (Chatzifrangkeskou et al. 2019), and cAMP signaling (McNeill et al. 2020), influence MRTF/SRF-mediated transcription. Finally, signal-induced polymerization of actin in the nucleus (discussed in detail below) has emerged as an important regulator of MRTF/SRF activity (Baarlink et al. 2013; Plessner et al. 2015; Wang et al. 2019). Many cytoskeletal genes, including actin itself, are MRTF/SRF targets (Posern and Treisman 2006), creating a feedback loop where cytoskeletal dynamics regulate the expression of its own constituents.

Monomeric Actin in Allosteric Regulation of Chromatin Remodeling Complexes

In addition to the regulation of MRTF/SRF activity, monomeric actin has a well-established role as an integral component of many chromatin-remodeling (including Ino80 and SWI/SNF families) and -modifying complexes (NuA4 and possibly hATAC) (Fig. 1). These complexes regulate chromatin accessibility for replication, transcription, and repair of DNA damage, and thus play critical roles in most chromatin-related processes. Structural studies by both X-ray crystallography, and more recently by CryoEM, have helped to resolve the role of actin in these multisubunit complexes (Jungblut et al. 2020). Here, actin operates together with actin-related proteins (Arps), and most complexes contain an actin–Arp4 pair bound by the Helicase–SANT-associated (HSA) domain (Szerlong et al. 2008). In addition, the Ino80 HSA domain interacts also with Arp8, while Arp5 forms a distinct module with Ies6. In yeast SWI/SNF and RSC complexes, the actin–Arp4 pair is replaced by the Arp7–Arp9 pair (Jungblut et al. 2020). Binding to the HSA domain prevents actin from polymerizing. For instance, in the yeast SWR1, which is the ATPase of a chromatin remodeling complex belonging to the Ino80-family, the barbed end of actin is sequestered by Arp4 and the HSA domain, and the actin molecule adopts a twisted orientation that prevents its contact with another actin molecule and frees it from ATP binding and regulation (Cao et al. 2016). Also, in the Ino80 complex, the HSA domain makes contact along the barbed ends of actin and the Arps and actin is sandwiched between Arp4 and Arp8, which interact with actin using distinct binding modes (Knoll et al. 2018). Functionally, the actin-containing modules play a role in establishing the allosteric control of the motor subunit in these complexes (Jungblut et al. 2020). In the Ino80 complex, the actin–Arp-bound HSA module binds and senses the length of the extranucleosomal/linker DNA to orient the complex, and especially the ATPase, on the nucleosome for remodeling. This is achieved by functional interplay with the Arp5–Ies6 module, which interacts with the acidic patch of the H2A–H2B dimer, and is required for gripping the remodeler on the nucleosome (Brahma et al. 2018; Eustermann et al. 2018; Knoll et al. 2018; Zhang et al. 2019). Also, in the human BAF (SWI/SNF family) complex, the actin–Arp4-containing module bridges the ATPase and the base module, and thus likely couples their motions during chromatin remodeling (He et al. 2020). The functional significance of actin for appropriate SWI/SNF function is further highlighted by studies using β-actin, null-mouse embryonic fibroblasts, which display reduced chromatin association of Brg1, the ATPase subunit of the BAF complex, consequently leading to defects in gene expression (Xie et al. 2018a,b).

Chromatin-modifying complexes controlling histone acetylation have also been reported to interact with actin. NuA4 histone acetyltransferase (HAT) plays important roles in DNA repair and transcription, and similarly to the chromatin-remodeling complexes, interacts with the actin–Arp4 pair via the HSA domain of Eaf1, which operates as an assembly platform for the complex. CryoEM studies of the yeast NuA4 complex have revealed that the actin–Arp4 pair is located at the peripheral head region of the complex (Wang et al. 2018). Mass spectrometry studies on nuclear actin revealed a potential interaction between actin and the human Ada-Two-A-containing (hATAC) HAT complex, and further experiments demonstrated that actin binds directly, and modulates, the HAT activity of KAT14, which is one of the HAT enzymes in the hATAC complex (Viita et al. 2019). Further structural studies are needed to elucidate the molecular and functional details of this interaction. Interestingly, actin also interacts with histone deacetylases (HDACs) 1 and 2 to modulate its activity. Although this association appears not to be direct (Serebryannyy et al. 2016a), it highlights the multifunctional nature of actin in regulating various proteins and complexes that influence chromatin accessibility for essential nuclear processes.

Although structural data clearly pinpoint a role for monomeric actin in chromatin-remodeling and -modifying complexes, there are some indications that filamentous actin may also play a role. Early studies suggested that Brg1-containing BAF chromatin-remodeling complexes interact with actin filaments in a phosphatidylinositol 4,5-biphosphate (PIP_2)-dependent manner (Zhao et al. 1998; Rando et al. 2002). Similarly, yeast INO80 and SWR1 complexes associate with actin filaments. This interaction depends on the Arp subunits in these complexes and is regulated by the Hsp90 and p23 chaperones (Wang et al. 2020). Moreover, in mammalian cells, ARID1A-containing SWI/SNF complexes associate with nuclear F-actin and this interaction serves as a mechanosensitive switch, which regulates the activity of YAP/TAZ transcriptional coactivators that control cell proliferation (Chang et al. 2018). Binding of ARID1A-containing complexes to actin prevents their association with YAP/TAZ, which are then free to bind the transcription factor TEAD and activate transcription. This pathway may also play a role in tumorigenesis, because ARID1A, and other SWI/SNF subunits, are often mutated in various cancers (Chang et al. 2018). Further biochemical and functional studies are required to elucidate how actin filaments interact with SWI/SNF, and whether the putative roles for both monomeric and filamentous actin on chromatin remodelers are functionally connected.

Actin in RNA Polymerase Functions

Actin has also been linked directly to transcription, mainly through its propensity to copurify with all three RNA polymerases (Fig. 1; Egly et al. 1984; Hofmann et al. 2004; Hu et al. 2004; Philimonenko et al. 2004). Decreased availability of monomeric actin, for example, upon inhibition of its nuclear import (Dopie et al. 2012; Le et al. 2016) by increasing its nuclear export (Spencer et al. 2011) or by persistently forcing nuclear actin into stable filaments (Serebryannyy et al. 2016b) results in impaired transcription. Furthermore, genome-wide binding studies demonstrate that actin is bound to promoter regions of essentially all transcribed genes in *Drosophila* ovaries (Sokolova et al. 2018), suggesting a general role for actin in Pol II–mediated transcription.

The molecular mechanisms by which actin promotes RNA polymerase function still remain somewhat unclear, but roles in several steps of the transcription process have been proposed (Hofmann et al. 2004; Obrdlik et al. 2008). In support, mass spectrometry-based identification of nuclear actin-binding partners revealed association of actin with proteins implicated in preinitiation complex (PIC) assembly, transcription elongation, and pre-mRNA splicing and processing (Viita et al. 2019). Interference with nucleocytoplasmic shuttling of actin results in defects in alternative splicing of reporter gene constructs, potentially implying a role in pre-mRNA splicing, be it direct or indirect. One possible hypothesis is that actin regulates the transcription elongation rate (Viita et al. 2019), which has been shown to affect splice site selection. A role in elongation is also sup-

 Cite this article as *Cold Spring Harb Perspect Biol* doi: 10.1101/cshperspect.a040121

ported by the observation that on highly transcribed genes, actin binds, together with Pol II, along the gene body (Sokolova et al. 2018). An interesting candidate for connecting actin to the transcription elongation machinery is the positive transcription elongation factor b (P-TEFb), which interacts with actin (Qi et al. 2011), although the biochemical details are still unknown.

To date, it remains unclear whether the functional form of actin, which impinges on RNA polymerase function, is its monomeric or polymeric form. In laminin-111-treated cells that display reduced nuclear actin, the transcription defect can be rescued not only with NLS-tagged wild-type actin, but also with the NLS-tagged actin-R62D mutant, which cannot polymerize (Spencer et al. 2011). This is in contrast to studies done on β-actin knockout MEFs in the context of Pol I–mediated transcription, where actin-R62D does not improve rRNA synthesis unlike wild-type actin (Almuzzaini et al. 2016). Interestingly, actin regulatory proteins, which can control actin polymerization, such as the Arp2/3 complex (Yoo et al. 2007) and its activators N-WASP (Wu et al. 2006), WAVE1 (Miyamoto et al. 2013), and WASH (Xia et al. 2014), as well as motor protein myosins, have also been linked to transcription or transcription-related processes. Myosin-1C and its isoform nuclear myosin 1 (NM1) have been implicated in both Pol I– and Pol II–mediated transcription, possibly via the B-WICH chromatin remodeling complex (Philimonenko et al. 2004; Hofmann et al. 2006; Sarshad et al. 2013; Almuzzaini et al. 2015). Moreover, NM1 is required for activation of *p21* transcription upon DNA damage, which is required for cell-cycle arrest. Here NM1 seems to partner with p53 to control the *p21* promoter through epigenetic mechanisms via the HAT PCAF and histone methyltransferase Set1 (Venit et al. 2020). Curiously, myosin VI, the only myosin traveling toward the minus end of the actin filament, has also been implicated in transcription. Early studies have already indicated the existence of myosin VI in the Pol II transcription complex, as well as the presence of myosin VI on gene promoters and intragenic regions (Vreugde et al. 2006). More recent studies have demonstrated that the transcription coactivator NDP52 regulates myosin VI activity by relieving its autoinhibition and allowing the interaction with DNA through the cargo-binding domain of myosin VI (Fili et al. 2017). The NDP52–myosin VI complex associates with Pol II and is required, for example, for the expression of nuclear receptor target genes (Fili et al. 2017).

Actin, and its binding partners, have therefore a multitude of connections with the RNA polymerase machineries, and could thereby influence the transcription process both as a monomer and as a filament. Whether actin operates here as a cofactor similarly as in chromatin remodeling complexes remains to be elucidated. Further studies are needed to establish the biochemical basis of these interactions, and how nuclear actin dynamics are precisely regulated to facilitate transcription.

FILAMENTOUS ACTIN IN THE NUCLEUS

Multiple Functions of Dynamic Intranuclear Actin Assembly

Although the presence of actin within the nuclear compartment has long been recognized, its dynamic assembly into filamentous structures is still only poorly understood (Percipalle and Vartiainen 2019). Recent methodological and technical advances in imaging have allowed for the first convincing visualization of nuclear F-actin in living mammalian cells; however, its regulation, roles, and functions remain largely unexplored.

Earlier studies exploited *Xenopus laevis* oocytes as a model system to study nuclear F-actin as they are relatively large in size (Dumont 1972) and, as mentioned above, contain high nuclear actin levels because of a lack of expression of exportin-6, thereby drastically increasing the concentration of actin for polymerization (Bohnsack et al. 2006). Wave1 (Wiskott–Aldrich syndrome protein family member 1) belongs to the family of nucleation-promoting factors (NPFs) that facilitate Arp2/3-mediated actin polymerization. Several studies revealed an essential role for Wave1-dependent dynamics and prolonged nuclear F-actin assembly in transcriptional reprogramming by inducing pluripotency genes such as *Oct4* (Miyamoto et al. 2011, 2013) as well as a role in chromatin tethering to the nuclear envelope (Oda et al. 2017).

The high abundance of actin in the cytoplasm makes it challenging to visualize F-actin structures within the nuclear compartment when using traditional actin probes like phalloidin, LifeAct, or utrophin. Progress was made when fluorescently labeled actin-binding domains fused to an NLS were used (Baarlink et al. 2013; Belin et al. 2013; Melak et al. 2017), which helped overcome the predominance of cytosolic over the nuclear actin signal as exemplified in Figure 2. Since then, a multitude of functions have been described for dynamic actin polymerization in the nuclei of living cells in response to several stimuli including serum stimulation (Baarlink et al. 2013; Wei et al. 2020), DNA damage (Belin et al. 2015; Caridi et al. 2018; Schrank et al. 2018), T-cell receptor (TCR) activation (Tsopoulidis et al. 2019), integrin signaling (Plessner et al. 2015), or viral infection (see Table 1; Ohkawa and Welch 2018).

First evidence and visualization of dynamic and transient assembly of F-actin structures in somatic cell nuclei came from a study showing that serum stimulation leads to the rapid formation of an endogenous nuclear actin network (Baarlink et al. 2013). This signal-dependent F-actin polymerization could be spatiotemporally controlled using a light-activatable optogenetic system to release autoinhibition of diaphanous formins (mDia1 and mDia2) localized to the nuclear compartment, thereby demonstrating the dynamic and reversible character of F-actin assembly in somatic cell nuclei.

In another study, nuclear actin filament assembly was observed during cell spreading (Plessner et al. 2015), using a nuclear-targeted nanobody recognizing endogenous actin. Integrin signaling during cell adhesion or spreading or via addition of soluble fibronectin induced the formation of nuclear F-actin depending on

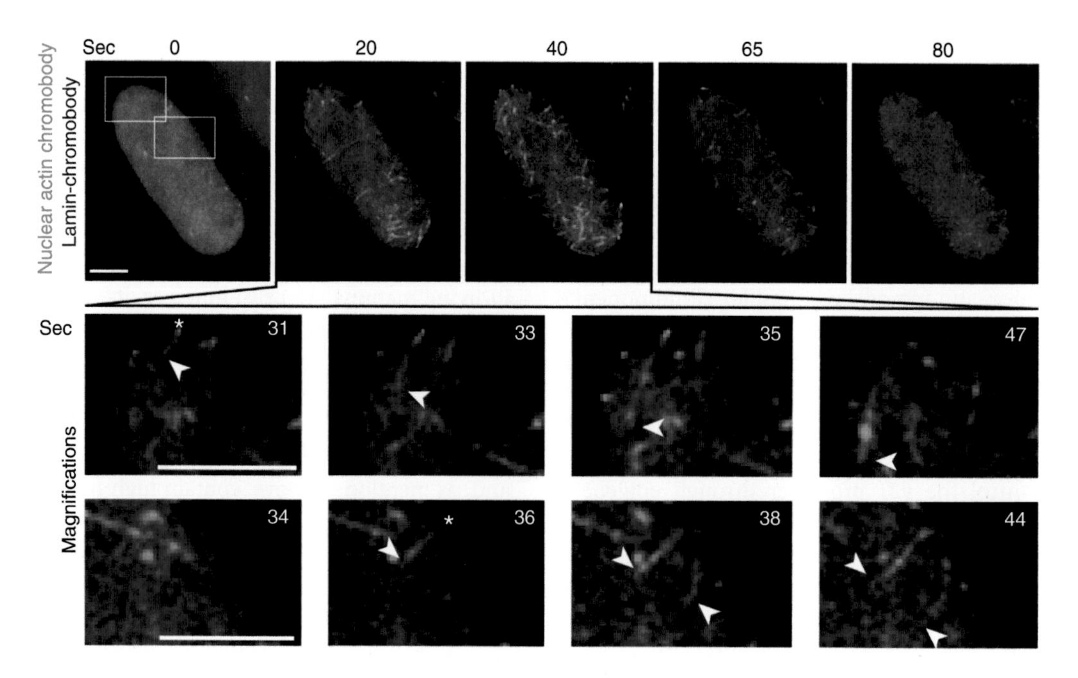

Figure 2. Calcium elevation transiently induces dynamic intranuclear actin polymerization. NIH3T3 cells stably expressing a nuclear actin chromobody (green) were transfected with lamin-chromobody-mCherry (red). Upon stimulation with A23187, actin polymerizes originating from the nuclear envelope (depicted by asterisks). Arrowheads show the tips of growing actin filaments. Scale bar, 5 µm.

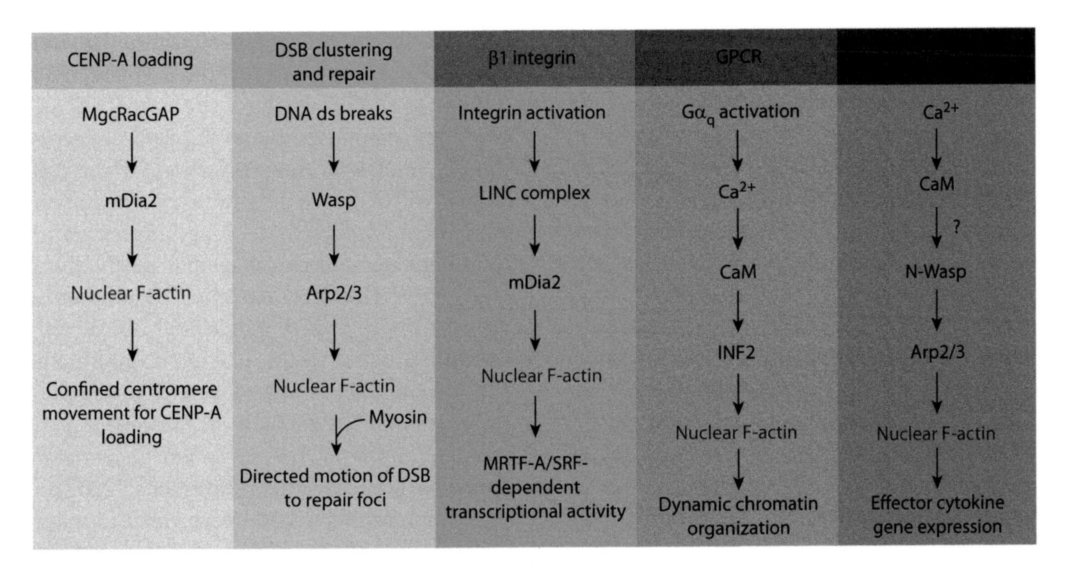

Figure 3. Signaling mechanisms for nuclear F-actin polymerization. Scheme of different signaling events controlling distinct actin assembly factors for nuclear F-actin formation.

the linker of nucleoskeleton and cytoskeleton (LINC) complex and mDia formins (Fig. 3), thereby promoting nuclear retention of MRTF-A (Plessner et al. 2015). Thus, different extracellular stimuli that induce nuclear actin filament formation appear to converge on nuclear formin function (Baarlink et al. 2013; Belin et al. 2015; Plessner et al. 2015). Recently, F-actin assembly upon serum stimulation has also been implicated in transcription regulation by enhancing RNA polymerase II (Pol II) clustering (Wei et al. 2020). Using next-generation transcriptome sequencing and superresolution microscopy, it was demonstrated that the formation of discrete RNA Pol II foci is facilitated through dynamic N-WASP/Arp2/3-dependent de novo polymerization of short actin filaments. Additionally, stimulating cells with IFN-γ, a crucial cytokine in innate and adaptive immunity, led to a similar response where nuclear F-actin promoted enhanced Pol II clustering. Interestingly, these F-actin structures did not have an influence on Pol II clustering and transcription initiation under normal growth conditions, suggesting a general role for dynamic nuclear F-actin assembly and disassembly in providing a dynamic scaffold and thereby regulating Pol II clustering in a signal-dependent manner in response to external stimuli (Wei et al. 2020).

Nuclear Actin Filaments: From Cell Cycle to Chromatin Reorganization

Cellular Signaling

It has become clear in the past few years that dynamic nuclear F-actin polymerization can be triggered and regulated by distinct signaling mechanisms and during different cellular processes. This has revealed not only different temporal characteristics but also differing actin assembly factors and F-actin structures inside the nucleus.

One critical role for nuclear actin assembly has been described after TCR activation, which leads to cytokine expression to drive T-cell proliferation and antibody production (Tsopoulidis et al. 2019). After TCR engagement, a rapid Ca^{2+} increase induces the formation of a dynamic nuclear actin network possibly through nuclear N-Wasp/Arp2/3, and Nck-interacting kinase (NIK) activity (Fig. 3). Importantly, this study links dynamic F-actin structures in the nucleus to critical immune defense mechanisms, allowing for the rapid response of $CD4^+$ T cells to TCR signaling to facilitate T-cell helper functions (Tsopoulidis et al. 2019). However, how these nuclear filaments are mechanistically connected to immediate early transcriptional responses requires further investigation. Another receptor mechanism of how extracellular signals are transmitted from the plasma membrane to the nuclear compartment for functional F-actin assembly has recently been identified (Wang et al. 2019). Stimulation of cells with physiological ligands for G protein–coupled receptors (GPCRs) and downstream activation of $G\alpha_q$ led to transient nuclear Ca^{2+} elevations and Inverted-Formin-2 (INF2)-dependent nuclear F-actin assembly, thereby promoting rapid changes in chromatin dynamics (Fig. 3; Wang et al. 2019). Polymerization of these Ca^{2+}-triggered linear actin filaments appeared to originate from the inner nuclear membrane (INM) (Fig. 2). These findings provide evidence for the INM as a potential signaling hub for intranuclear F-actin assembly (Wang et al. 2019). Even though an interaction between INF2 and the calcium sensor protein calmodulin could be observed in IP experiments (Wang et al. 2019), it remains elusive how precisely INF2 is activated, which receptors of the nuclear envelope might be involved in signal transduction, and what functions the actin filaments might have apart from altering chromatin dynamics.

Long-Range Chromatin Motion

Actin polymerization could also influence transcription by controlling the subnuclear localization of target genes. In budding yeast, the *INO1* locus moves from the center of the nucleus toward the periphery by long-range directed motion upon its transcriptional activation. This movement is dependent on chromatin remodelers INO80 and SWR, as well as on actin-polymerizing proteins, such as the Diaphanous formin homologue Bnr1, which are likely required to create a dynamic pool of short actin filaments in the nucleus (Wang et al. 2020). Interestingly, the movement of the *INO1* locus requires motor activity in the form of myosin 3, which is tethered to the locus via interactions with the transcription factor Put3. The interaction between the chromatin remodelers and actin filaments is thought to further reinforce the connection between the moving locus and the actin filaments, thus fostering long-range transport (Wang et al. 2020). Notably, actin polymerization had previously been implicated in long-range intranuclear movement of chromosomal loci (Chuang et al. 2006; Dundr et al. 2007; Khanna et al. 2014). Although these studies did not directly visualize nuclear actin filaments, long-range chromosomal motion was sensitive to actin depolymerizing drugs such as Latrunculin or could be inhibited by expression of nonpolymerizable actin mutants (Posern et al. 2002).

Only recently, a study performed in mammalian cells presented compelling evidence for ATP-dependent, directed long-range movements of myosin VI along F-actin within the nucleus, possibly enhancing transcription by supporting long-range chromatin rearrangements (Grosse-Berkenbusch et al. 2020). Hence, it is worth speculating whether the long linear actin filaments observed in the nucleus after Ca^{2+} increase also play a role in myosin-mediated transport along F-actin within the nucleus.

Cell Cycle

The histone H3 variant centromere protein A (CENP-A) epigenetically defines centromere regions of the chromosome and needs to be replenished in every cell cycle. Nuclear localization and activity of the formin mDia2 downstream of the MgcRacGAP-dependent GTPase pathway has been identified as a crucial factor for CENP-A loading and maintenance at centrosomes (Liu and Mao 2016, 2017). In a follow-up study, using a utrophin-based probe, mDia2-dependent formation of short dynamic nuclear actin filaments was found to be required for constrained centromere movement during CENP-A loading in G_1 nuclei (Fig. 3; Liu et al. 2018). However, the exact mechanism and potentially spatial regulation of F-actin formation remains to be identified. Interestingly, transient nuclear actin filament formation has been observed during a similar time window in early G_1 phase of the cell cycle in another context (Baarlink et al.

 Cite this article as *Cold Spring Harb Perspect Biol* doi: 10.1101/cshperspect.a040121

2017). Using PALM and STORM superresolution imaging these filaments were found to consist of both bundled and longer single F-actin structures that assembled dynamically in daughter cell nuclei after mitotic exit to promote nuclear protrusions during nuclear volume expansion and thereby enabling chromatin decondensation. Interfering with the polymerization-competent nuclear actin pool by overexpression of the nonpolymerizable actin-R62D-NLS mutant or the actin exporter exportin-6 resulted in a decreased volume of daughter cell nuclei as well as an increase in chromatin compaction at mitotic exit (Baarlink et al. 2017). Timing and turnover of the observed nuclear actin filaments, which reorganize the postmitotic mammalian nucleus, appear to be tightly controlled by the actin-depolymerizing factor cofilin-1 (Baarlink et al. 2017). Its nuclear activity was observed to be cell-cycle-dependent with an increase of phosphorylated and thus inactivated cofilin-1 during mitotic exit by as-yet-unknown mechanisms. A parallel study implicated a critical role for formin-dependent nuclear F-actin assembly in promoting cyclin-dependent kinase (CDK) and proliferating cell nuclear antigen (PCNA) loading onto chromatin and initiation of DNA replication by as-yet-unknown mechanisms (Parisis et al. 2017). Recently, an involvement of the actin bundling factor α-actinin 4 (ACTN4) in regulating postmitotic nuclear volume expansion and actin assembly has been proposed (Krippner et al. 2020). Performing superresolution STORM imaging, cells expressing an ACTN-4 mutant lacking its actin-binding domain were found to have fewer and thinner actin filaments as opposed to ACTN4-wt expressing cells, that displayed abundant thick F-actin structures (Krippner et al. 2020). Together, these studies emphasize the transient and highly dynamic character of nuclear actin assembly controlled by complex mechanisms as a response to various stimuli or cellular processes.

Viral Replication

Well-described functions of F-actin in the cytosol, including scaffolding and transport along the filaments, have long not been observed within the nuclear compartment. Interestingly, several viruses use both cytosolic and nuclear actin-based motility in their life cycle (Wilkie et al. 2016; Ohkawa and Welch 2018). For instance, the baculovirus-type species *Autographa californica* M nucleopolyhedrovirus (AcMNPV) promotes actin polymerization to facilitate its nuclear egress (Ohkawa and Welch 2018). It uses viral Wasp-like proteins to hijack the host system and induce Arp2/3-dependent nuclear actin comet-tails that push the virus against the nuclear envelope (NE) creating nuclear protrusions and finally leading to NE rupture and viral egress into the cytosol where it continues to exploit nuclear-based motility to reach the cell periphery. The human cytomegalovirus (CMV), on the other hand, uses transiently assembled extensive nuclear F-actin networks to control intranuclear polymerization by spatially segregating viral DNA from inactive histones and host DNA to facilitate viral replication (Procter et al. 2020).

Nuclear F-Actin in DNA Damage Repair Mechanisms

Earlier studies had indicated a potential role for nuclear actin in long-range movements and repositioning of chromosomal loci (Chuang et al. 2006; Dundr et al. 2007). More recent work now elucidated a novel regulatory role for actin polymerization in directed chromatin dynamics upon DNA damage and double-strand break (DSB) repair progression (Fig. 3; Andrin et al. 2012; Belin et al. 2015; Caridi et al. 2018; Schrank et al. 2018). As a consequence of DNA damage induced by various endogenous or exogenous triggers, the tightly regulated DDR machinery is activated and recruits and assembles various DDR factors at the site of the lesion (Chatterjee and Walker 2017). Faithfully restoring genome integrity is of particular importance as unresolved DSBs are implicated in a variety of human disorders and cancers (Jackson and Bartek 2009; Chatterjee and Walker 2017). While single-strand breaks are often cleared by different excision repair mechanisms, cells usually employ two main mechanisms to repair DSB: nonhomologous end joining (NHEJ), which is more error-prone and executed throughout the cell cycle, and homologous recombination (HR), which is thought to be restricted to late G_2/S phases, when a sister chromatid is present to function as a template for DNA repair (Phillips and McKinnon 2007; Ceccaldi et al. 2016; Chatterjee and Walker 2017). First evidence for a potential involvement of F-actin in DSB repair came from an in vitro study performing pull-down experiments using nuclear cell extracts incubated with purified polymeric actin showing binding of DNA repair proteins including Ku80, Mre11, and Rad51 (Andrin et al. 2012). Later, DNA damage induced formation of both long actin filaments and dense actin clusters in the nucleoplasm as well as short, nucleolus-associated filaments were identified in living cells using a variety of actin probes (Belin et al. 2015). Simultaneous global knockdown of actin nucleators Spire-1 and Spire-2 or single knockdown of formin-2 completely abolished clearance of chemically induced DNA DSBs in HeLa cells uncovering a new signaling pathway for formin-induced nuclear actin filament formation (Belin et al. 2015). Whether actin nucleation actually takes place inside the nucleus and which function F-actin fibers have in the DNA damage response, however, remains to be elucidated. Only a few years later, using different model organisms, two publications independently uncovered a role for Arp2/3-dependent, actin-based mobility in DSB repair (Caridi et al. 2018; Schrank et al. 2018). It was shown that the actin-NPF WASP was recruited to DSBs independently of the repair mechanism, while Arp2/3 was only enriched and activated at chromatin lesions in G_2 phase, which are cleared by homology-directed repair (Schrank et al. 2018). Arp2/3-mediated actin filaments locally assemble these DSBs and are required for their migration into discrete subnuclear clusters for repair (Schrank et al. 2018). On the other hand, actin filaments observed in *Drosophila melanogaster* cells in response to irradiation-induced DNA damage were longer in size spanning the nucleus (Caridi et al. 2018). Actin filament assembly could be visualized at DSBs in heterochromatic regions through the actin nucleator Arp2/3 that allowed for myosin-dependent, directed movement of repair sites toward the nuclear periphery (Caridi et al. 2018).

CONCLUDING REMARKS

During the last decade, actin has emerged as an essential protein in the nucleus with functional roles in fundamental nuclear processes from gene expression to DNA repair. At the same time, spatially and temporally controlled nuclear organization has emerged as a key concept in nuclear biology. Actin has the inherent capacity to produce force that can be used to drive motility or to create scaffolds for higher-order assemblies, and is therefore an ideal candidate for organizing the dynamic nuclear landscape. With the aid of advanced microscopy and chromosome conformation capture techniques, novel concepts have been established for a dynamic nuclear organization where chromosome loci cluster into large compartments and subcompartments called topologically associating domains (TADs) (Dekker et al. 2013; Schrank and Gautier 2019). Keeping these and the recent findings of F-actin-mediated chromosome repositioning in mind, it is tempting to speculate that future research might uncover a link between spatiotemporally controlled, dynamic nuclear F-actin assembly and the formation of nuclear domains or promoter-enhancer loops, among others. Another interesting avenue for future research are the posttranslational modifications on actin, and how they impinge on nuclear functions of actin (Kumar et al. 2020; Mu et al. 2020). Thus, the actin monomer is not a mere building block for the filament but has itself an important role as a signaling molecule or as a cofactor (e.g., in allosteric control of chromatin-remodeling complexes). How the functions of actin monomers and filaments are connected and how the polymerization process impinges on these activities are important questions for the future.

ACKNOWLEDGMENTS

We thank Carsten Schwan for providing Figure 2. Work in the laboratory of R.G. is funded by the DFG, under Germany's Excellence Strategy

(EXC-2189, project ID: 390939984) and the HFSP program (grant ID: RGP0021/2016). The work in the laboratory of M.K.V. is funded by the Academy of Finland, Helsinki Institute for Life Science, as well as the Jane and Aatos Erkko and Sigrid Juselius Cancer foundations of Finland.

REFERENCES

Almuzzaini B, Sarshad AA, Farrants AK, Percipalle P. 2015. Nuclear myosin 1 contributes to a chromatin landscape compatible with RNA polymerase II transcription activation. *BMC Biol* **13:** 35. doi:10.1186/s12915-015-0147-z

Almuzzaini B, Sarshad AA, Rahmanto AS, Hansson ML, Von Euler A, Sangfelt O, Visa N, Farrants AK, Percipalle P. 2016. In β-actin knockouts, epigenetic reprogramming and rDNA transcription inactivation lead to growth and proliferation defects. *FASEB J* **30:** 2860–2873. doi:10.1096/fj.201600280R

Andrin C, McDonald D, Attwood KM, Rodrigue A, Ghosh S, Mirzayans R, Masson JY, Dellaire G, Hendzel MJ. 2012. A requirement for polymerized actin in DNA double-strand break repair. *Nucl (United States)* **3:** 384–395.

Baarlink C, Wang H, Grosse R. 2013. Nuclear actin network assembly by formins regulates the SRF coactivator MAL. *Science* **340:** 864–867. doi:10.1126/science.1235038

Baarlink C, Plessner M, Sherrard A, Morita K, Misu S, Virant D, Kleinschnitz EM, Harniman R, Alibhai D, Baumeister S, et al. 2017. A transient pool of nuclear F-actin at mitotic exit controls chromatin organization. *Nat Cell Biol* **19:** 1389–1399. doi:10.1038/ncb3641

Belin BJ, Cimini BA, Blackburn EH, Mullins RD. 2013. Visualization of actin filaments and monomers in somatic cell nuclei. *Mol Biol Cell* **24:** 982–994. doi:10.1091/mbc.e12-09-0685

Belin BJ, Lee T, Mullins RD. 2015. DNA damage induces nuclear actin filament assembly by Formin-2 and Spire-1/2 that promotes efficient DNA repair. *eLife* **4:** e07735.

Bohnsack MT, Stüven T, Kuhn C, Cordes VC, Görlich D. 2006. A selective block of nuclear actin export stabilizes the giant nuclei of *Xenopus* oocytes. *Nat Cell Biol* **8:** 257–263. doi:10.1038/ncb1357

Brahma S, Ngubo M, Paul S, Udugama M, Bartholomew B. 2018. The Arp8 and Arp4 module acts as a DNA sensor controlling INO80 chromatin remodeling. *Nat Commun* **9:** 3309. doi:10.1038/s41467-018-05710-7

Cao T, Sun L, Jiang Y, Huang S, Wang J, Chen Z. 2016. Crystal structure of a nuclear actin ternary complex. *Proc Natl Acad Sci* **113:** 8985–8990. doi:10.1073/pnas.1602818113

Caridi CP, D'agostino C, Ryu T, Zapotoczny G, Delabaere L, Li X, Khodaverdian VY, Amaral N, Lin E, Rau AR, et al. 2018. Nuclear F-actin and myosins drive relocalization of heterochromatic breaks. *Nature* **559:** 54–60. doi:10.1038/s41586-018-0242-8

Ceccaldi R, Rondinelli B, D'Andrea AD. 2016. Repair pathway choices and consequences at the double-strand break. *Trends Cell Biol* **26:** 52–64. doi:10.1016/j.tcb.2015.07.009

Chang L, Azzolin L, Di Biagio D, Zanconato F, Battilana G, Lucon Xiccato R, Aragona M, Giulitti S, Panciera T, Gandin A, et al. 2018. The SWI/SNF complex is a mechano-regulated inhibitor of YAP and TAZ. *Nature* **563:** 265–269. doi:10.1038/s41586-018-0658-1

Chatterjee N, Walker GC. 2017. Mechanisms of DNA damage, repair, and mutagenesis. *Environ Mol Mutagen* **58:** 235–263. doi:10.1002/em.22087

Chatzifrangkeskou M, Pefani DE, Eyres M, Vendrell I, Fischer R, Pankova D, O'Neill E. 2019. RASSF1A is required for the maintenance of nuclear actin levels. *EMBO J* **38:** e101168. doi:10.15252/embj.2018101168

Chuang CH, Carpenter AE, Fuchsova B, Johnson T, de Lanerolle P, Belmont AS. 2006. Long-range directional movement of an interphase chromosome site. *Curr Biol* **16:** 825–831. doi:10.1016/j.cub.2006.03.059

Dekker J, Marti-Renom MA, Mirny LA. 2013. Exploring the three-dimensional organization of genomes: interpreting chromatin interaction data. *Nat Rev Genet* **14:** 390–403. doi:10.1038/nrg3454

Dopie J, Skarp K-P, Kaisa Rajakylä E, Tanhuanpää K, Vartiainen MK. 2012. Active maintenance of nuclear actin by importin 9 supports transcription. *Proc Natl Acad Sci* **109:** E544–E552. doi:10.1073/pnas.1118880109

Dopie J, Rajakylä EK, Joensuu MS, Huet G, Ferrantelli E, Xie T, Jäälinoja H, Jokitalo E, Vartiainen MK. 2015. Genome-wide RNAi screen for nuclear actin reveals a network of cofilin regulators. *J Cell Sci* **128:** 2388–2400. doi:10.1242/jcs.169441

Dumont JN. 1972. Oogenesis in *Xenopus laevis* (Daudin). I: Stages of oocyte development in laboratory maintained animals. *J Morphol* **136:** 153–179. doi:10.1002/jmor.1051360203

Dundr M, Ospina JK, Sung MH, John S, Upender M, Ried T, Hager GL, Matera AG. 2007. Actin-dependent intranuclear repositioning of an active gene locus in vivo. *J Cell Biol* **179:** 1095–1103. doi:10.1083/jcb.200710058

Egly JM, Miyamoto NG, Moncollin V, Chambon P. 1984. Is actin a transcription initiation factor for RNA polymerase B? *EMBO J* **3:** 2363–2371. doi:10.1002/j.1460-2075.1984.tb02141.x

Eustermann S, Schall K, Kostrewa D, Lakomek K, Strauss M, Moldt M, Hopfner K. 2018. Structural basis for ATP-dependent chromatin remodelling by the INO80 complex. *Nature* **556:** 386–390. doi:10.1038/s41586-018-0029-y

Feric M, Brangwynne CP. 2013. A nuclear F-actin scaffold stabilizes ribonucleoprotein droplets against gravity in large cells. *Nat Cell Biol* **15:** 1253–1259. doi:10.1038/ncb2830

Fili N, Hari-Gupta Y, Dos Santos A, Cook A, Poland S, Ameer-Beg SM, Parsons M, Toseland CP. 2017. NDP52 activates nuclear myosin VI to enhance RNA polymerase II transcription. *Nat Commun* **8:** 1871. doi:10.1038/s41467-017-02050-w

Fiore A, Spencer VA, Mori H, Carvalho HF, Bissell MJ, Bruni-Cardoso A. 2017. Laminin-111 and the level of nuclear actin regulate epithelial quiescence via exportin-6. *Cell Rep* **19:** 2102–2115. doi:10.1016/j.celrep.2017.05.050

Grosse R, Vartiainen MK. 2013. To be or not to be assembled: progressing into nuclear actin filaments. *Nat Rev Mol Cell Biol* **14:** 693–697. doi:10.1038/nrm3681

Grosse-Berkenbusch A, Hettich J, Kuhn T, Fili N, Cook AW, Hari-Gupta Y, Palmer A, Streit L, Ellis PJI, Toseland C, et al. 2020. Myosin VI moves on nuclear actin filaments and supports long-range chromatin rearrangements. bioRxiv doi:10.1101/2020.04.03.023614

He S, Wu Z, Tian Y, Yu Z, Yu J, Wang X, Li J, Liu B, Xu Y. 2020. Structure of nucleosome-bound human BAF complex. *Science* **367:** 875–881. doi:10.1126/science.aaz9761

Hofmann WA, Stojiljkovic L, Fuchsova B, Vargas GM, Mavrommatis E, Philimonenko V, Kysela K, Goodrich JA, Lessard JL, Hope TJ, et al. 2004. Actin is part of preinitiation complexes and is necessary for transcription by RNA polymerase II. *Nat Cell Biol* **6:** 1094–1101. doi:10.1038/ncb1182

Hofmann WA, Vargas GM, Ramchandran R, Stojiljkovic L, Goodrich JA, de Lanerolle P. 2006. Nuclear myosin I is necessary for the formation of the first phosphodiester bond during transcription initiation by RNA polymerase II. *J Cell Biochem* **99:** 1001–1009. doi:10.1002/jcb.21035

Hu P, Wu S, Hernandez N. 2004. A role for β-actin in RNA polymerase III transcription. *Genes Dev* **18:** 3010–3015. doi:10.1101/gad.1250804

Hyrskyluoto A, Vartiainen MK. 2020. Regulation of nuclear actin dynamics in development and disease. *Curr Opin Cell Biol* **64:** 18–24. doi:10.1016/j.ceb.2020.01.012

Jackson SP, Bartek J. 2009. The DNA-damage response in human biology and disease. *Nature* **461:** 1071–1078. doi:10.1038/nature08467

Jungblut A, Hopfner KP, Eustermann S. 2020. Megadalton chromatin remodelers: common principles for versatile functions. *Curr Opin Struct Biol* **64:** 134–144. doi:10.1016/j.sbi.2020.06.024

Khanna N, Hu Y, Belmont AS. 2014. HSP70 transgene directed motion to nuclear speckles facilitates heat shock activation. *Curr Biol* **24:** 1138–1144. doi:10.1016/j.cub.2014.03.053

Knoll KR, Eustermann S, Niebauer V, Oberbeckmann E, Stoehr G, Schall K, Tosi A, Schwarz M, Buchfellner A, Korber P, et al. 2018. The nuclear actin-containing Arp8 module is a linker DNA sensor driving INO80 chromatin remodeling. *Nat Struct Mol Biol* **25:** 823–832. doi:10.1038/s41594-018-0115-8

Krippner S, Winkelmeier J, Knerr J, Brandt DT, Virant D, Schwan C, Endesfelder U, Grosse R. 2020. Postmitotic expansion of cell nuclei requires nuclear actin filament bundling by α-actinin 4. *EMBO Rep* e50758.

Kumar A, Zhong Y, Albrecht A, Sang PB, Maples A, Liu Z, Vinayachandran V, Reja R, Lee C-F, Kumar A, et al. 2020. Actin R256 mono-methylation is a conserved post-translational modification involved in transcription. *Cell Rep* **32:** 108172. doi:10.1016/j.celrep.2020.108172

Lamm N, Masamsetti VP, Read MN, Biro M, Cesare AJ. 2018. ATR and mTOR regulate F-actin to alter nuclear architecture and repair replication stress. bioRxiv doi:10.1101/451708

Le HQ, Ghatak S, Yeung CY, Tellkamp F, Günschmann C, Dieterich C, Yeroslaviz A, Habermann B, Pombo A, Niessen CM, et al. 2016. Mechanical regulation of transcription controls Polycomb-mediated gene silencing during lineage commitment. *Nat Cell Biol* **18:** 864–875. doi:10.1038/ncb3387

Liu C, Mao Y. 2016. Diaphanous formin mDia2 regulates CENP-A levels at centromeres. *J Cell Biol* **213:** 415–424. doi:10.1083/jcb.201512034

Liu C, Mao Y. 2017. Formin-mediated epigenetic maintenance of centromere identity. *Small GTPases* **8:** 245–250. doi:10.1080/21541248.2016.1215658

Liu C, Zhu R, Mao Y. 2018. Nuclear actin polymerized by mDia2 confines centromere movement during CENP-A loading. *iScience* **9:** 314–327. doi:10.1016/j.isci.2018.10.031

Lundquist MR, Storaska AJ, Liu TC, Larsen SD, Evans T, Neubig RR, Jaffrey SR. 2014. Redox modification of nuclear actin by MICAL-2 regulates SRF signaling. *Cell* **156:** 563–576. doi:10.1016/j.cell.2013.12.035

McNeill MC, Wray J, Sala-Newby GB, Hindmarch CCT, Smith SA, Ebrahimighaei R, Newby AC, Bond M. 2020. Nuclear actin regulates cell proliferation and migration via inhibition of SRF and TEAD. *Biochim Biophys Acta Mol Cell Res* **1867:** 118691. doi:10.1016/j.bbamcr.2020.118691

Melak M, Plessner M, Grosse R. 2017. Correction: actin visualization at a glance. *J Cell Sci* **130:** 1688–1688. doi:10.1242/jcs.204487

Miyamoto K, Pasque V, Jullien J, Gurdon JB. 2011. Nuclear actin polymerization is required for transcriptional reprogramming of Oct4 by oocytes. *Genes Dev* **25:** 946–958. doi:10.1101/gad.615211

Miyamoto K, Teperek M, Yusa K, Allen GE, Bradshaw CR, Gurdon JB. 2013. Nuclear Wave1 is required for reprogramming transcription in oocytes and for normal development. *Science* **341:** 1002–1005. doi:10.1126/science.1240376

Mouilleron S, Langer CA, Guettler S, McDonald NQ, Treisman R. 2011. Structure of a pentavalent G-actin*MRTF-A complex reveals how G-actin controls nucleocytoplasmic shuttling of a transcriptional coactivator. *Sci Signal* **4:** ra40. doi:10.1126/scisignal.2001750

Mu A, Fung TS, Francomacaro LM, Huynh T, Kotila T, Svindrych Z, Higgs HN. 2020. Regulation of INF2-mediated actin polymerization through site-specific lysine acetylation of actin itself. *Proc Natl Acad Sci* **117:** 439–447. doi:10.1073/pnas.1914072117

Obrdlik A, Kukalev A, Louvet E, Östlund Farrants AK, Caputo L, Percipalle P. 2008. The histone acetyltransferase PCAF associates with actin and hnRNP U for RNA polymerase II transcription. *Mol Cell Biol* **28:** 6342–6357. doi:10.1128/MCB.00766-08

Oda H, Shirai N, Ura N, Ohsumi K, Iwabuchi M. 2017. Chromatin tethering to the nuclear envelope by nuclear actin filaments: A novel role of the actin cytoskeleton in the *Xenopus* blastula. *Genes Cells* **22:** 376–391. doi:10.1111/gtc.12483

Ohkawa T, Welch MD. 2018. Baculovirus actin-based motility drives nuclear envelope disruption and nuclear egress. *Curr Biol* **28:** 2153–2159.e4. doi:10.1016/j.cub.2018.05.027

Okuno T, Li WY, Hatano Y, Takasu A, Sakamoto Y, Yamamoto M, Ikeda Z, Shindo T, Plessner M, Morita K, et al. 2020. Zygotic nuclear F-actin safeguards embryonic development. *Cell Rep* **31:** 107824. doi:10.1016/j.celrep.2020.107824

Parisis N, Krasinska L, Harker B, Urbach S, Rossignol M, Camasses A, Dewar J, Morin N, Fisher D. 2017. Initiation of DNA replication requires actin dynamics and formin activity. *EMBO J* **36:** 3212–3231. doi:10.15252/embj.201796585

Pawłowski R, Rajakylä EK, Vartiainen MK, Treisman R. 2010. An actin-regulated importin α/β-dependent extended bipartite NLS directs nuclear import of MRTF-A. *EMBO J* **29:** 3448–3458. doi:10.1038/emboj.2010.216

Percipalle P, Vartiainen M. 2019. Cytoskeletal proteins in the cell nucleus: a special nuclear actin perspective. *Mol Biol Cell* **30:** 1781–1785. doi:10.1091/mbc.E18-10-0645

Philimonenko VV, Zhao J, Iben S, Dingová H, Kyselá K, Kahle M, Zentgraf H, Hofmann WA, de Lanerolle P, Hozák P, et al. 2004. Nuclear actin and myosin I are required for RNA polymerase I transcription. *Nat Cell Biol* **6:** 1165–1172. doi:10.1038/ncb1190

Phillips ER, McKinnon PJ. 2007. DNA double-strand break repair and development. *Oncogene* **26:** 7799–7808. doi:10.1038/sj.onc.1210877

Plessner M, Grosse R. 2019. Dynamizing nuclear actin filaments. *Curr Opin Cell Biol* **56:** 1–6. doi:10.1016/j.ceb.2018.08.005

Plessner M, Melak M, Chinchilla P, Baarlink C, Grosse R. 2015. Nuclear F-actin formation and reorganization upon cell spreading. *J Biol Chem* **290:** 11209–11216. doi:10.1074/jbc.M114.627166

Pollard TD. 2016. Actin and actin-binding proteins. *Cold Spring Harb Perspect Biol* **8:** a018226. doi:10.1101/cshperspect.a018226

Posern G, Treisman R. 2006. Actin' together: serum response factor, its cofactors and the link to signal transduction. *Trends Cell Biol* **16:** 588–596. doi:10.1016/j.tcb.2006.09.008

Posern G, Sotiropoulos A, Treisman R. 2002. Mutant actins demonstrate a role for unpolymerized actin in control of transcription by serum response factor. *Mol Biol Cell* **13:** 4167–4178. doi:10.1091/mbc.02-05-0068

Procter DJ, Furey C, Garza-gongora AG, Kosak ST, Walsh D. 2020. Cytoplasmic control of intranuclear polarity by human cytomegalovirus. *Nature* doi:10.1038/s41586-020-2714-x

Qi T, Tang W, Wang L, Zhai L, Guo L, Zeng X. 2011. G-actin participates in RNA polymerase II-dependent transcription elongation by recruiting positive transcription elongation factor b (P-TEFb). *J Biol Chem* **286:** 15171–15181. doi:10.1074/jbc.M110.184374

Rando OJ, Zhao K, Janmey P, Crabtree GR. 2002. Phosphatidylinositol-dependent actin filament binding by the SWI/SNF-like BAF chromatin remodeling complex. *Proc Natl Acad Sci* **99:** 2824–2829. doi:10.1073/pnas.032662899

Rout MP, Karpen GH. 2014. Editorial overview: cell nucleus: the nucleus: a dynamic organelle. *Curr Opin Cell Biol* **28:** iv–vii. doi:10.1016/j.ceb.2014.05.005

Sarshad A, Sadeghifar F, Louvet E, Mori R, Böhm S, Al-Muzzaini B, Vintermist A, Fomproix N, Östlund AK, Percipalle P. 2013. Nuclear myosin 1c facilitates the chromatin modifications required to activate rRNA gene transcription and cell cycle progression. *PLoS Genet* **9:** e1003397. doi:10.1371/journal.pgen.1003397

Schrank B, Gautier J. 2019. Assembling nuclear domains: lessons from DNA repair. *J Cell Biol* **218:** 2444–2455. doi:10.1083/jcb.201904202

Schrank BR, Aparicio T, Li Y, Chang W, Chait BT, Gundersen GG, Gottesman ME, Gautier J. 2018. Nuclear ARP2/3 drives DNA break clustering for homology-directed repair. *Nature* **559:** 61–66. doi:10.1038/s41586-018-0237-5

Serebryannyy LA, Cruz CM, de Lanerolle P. 2016a. A role for nuclear actin in HDAC 1 and 2 regulation. *Sci Rep* **6:** 28460. doi:10.1038/srep28460

Serebryannyy LA, Parilla M, Annibale P, Cruz CM, Laster K, Gratton E, Kudryashov D, Kosak ST, Gottardi CJ, de Lanerolle P. 2016b. Persistent nuclear actin filaments inhibit transcription by RNA polymerase II. *J Cell Sci* **129:** 3412–3425. doi:10.1242/jcs.195867

Sokolova M, Moore HM, Prajapati B, Dopie J, Meriläinen L, Honkanen M, Matos RC, Poukkula M, Hietakangas V, Vartiainen MK. 2018. Nuclear actin is required for transcription during *Drosophila* oogenesis. *iScience* **9:** 63–70. doi:10.1016/j.isci.2018.10.010

Spencer VA, Costes S, Inman JL, Xu R, Chen J, Hendzel MJ, Bissell MJ. 2011. Depletion of nuclear actin is a key mediator of quiescence in epithelial cells. *J Cell Sci* **124:** 123–132. doi:10.1242/jcs.073197

Stuven T, Hartmann E, Gorlich D. 2003. Exportin 6: a novel nuclear export receptor that is specific for profilin·actin complexes. *EMBO J* **22:** 5928–5940. doi:10.1093/emboj/cdg565

Szerlong H, Hinata K, Viswanathan R, Erdjument-Bromage H, Tempst P, Cairns BR. 2008. The HSA domain binds nuclear actin-related proteins to regulate chromatin-remodeling ATPases. *Nat Struct Mol Biol* **15:** 469–476. doi:10.1038/nsmb.1403

Titus MA. 2018. Myosin-driven intracellular transport. *Cold Spring Harb Perspect Biol* **10:** a021972. doi:10.1101/cshperspect.a021972

Tsopoulidis N, Kaw S, Laketa V, Kutscheidt S, Baarlink C, Stolp B, Grosse R, Fackler OT. 2019. T cell receptor–triggered nuclear actin network formation drives $CD4^+$ T cell effector functions. *Sci Immunol* **4:** eaav1987. doi:10.1126/sciimmunol.aav1987

Vartiainen MK, Guettler S, Larijani B, Treisman R. 2007. Nuclear actin regulates dynamic subcellular localization and activity of the SRF cofactor MAL. *Science* **316:** 1749–1752. doi:10.1126/science.1141084

Venit T, Semesta K, Farrukh S, Endara-Coll M, Havalda R, Hozak P, Percipalle P. 2020. Nuclear myosin 1 activates p21 gene transcription in response to DNA damage through a chromatin-based mechanism. *Commun Biol* **3:** 115. doi:10.1038/s42003-020-0836-1

Viita T, Kyheröinen S, Prajapati B, Virtanen J, Frilander MJ, Varjosalo M, Vartiainen MK. 2019. Nuclear actin interactome analysis links actin to KAT14 histone acetyl transferase and mRNA splicing. *J Cell Sci* **132:** jcs226852. doi:10.1242/jcs.226852

Vreugde S, Ferrai C, Miluzio A, Hauben E, Marchisio PC, Crippa MP, Bussi M, Biffo S. 2006. Nuclear myosin VI enhances RNA polymerase II-dependent transcription. *Mol Cell* **23:** 749–755. doi:10.1016/j.molcel.2006.07.005

Wang X, Ahmad S, Zhang Z, Côté J, Cai G. 2018. Architecture of the *Saccharomyces cerevisiae* NuA4/TIP60 complex. *Nat Commun* **9:** 1147. doi:10.1038/s41467-018-03504-5

Wang Y, Sherrard A, Zhao B, Melak M, Trautwein J, Kleinschnitz EM, Tsopoulidis N, Fackler OT, Schwan C, Grosse R. 2019. GPCR-induced calcium transients trigger nuclear actin assembly for chromatin dynamics. *Nat Commun* **10:** 5271. doi:10.1038/s41467-019-13322-y

Wang A, Kolhe JA, Gioacchini N, Baade I, Brieher WM, Peterson CL, Freeman BC. 2020. Mechanism of long-range chromosome motion triggered by gene activation. *Dev Cell* **52:** 309–320.e5. doi:10.1016/j.devcel.2019.12.007

Wei M, Fan X, Ding M, Li R, Shao S, Hou Y, Meng S, Tang F, Li C, Sun Y. 2020. Nuclear actin regulates inducible transcription by enhancing RNA polymerase II clustering. *Sci Adv* **6:** eaay6515.

Wilkie AR, Lawler JL, Coen DM. 2016. A role for nuclear F-actin induction in human cytomegalovirus nuclear egress. *MBio* **7:** e01254. doi:10.1128/mBio.01254-16

Wu X, Yoo Y, Okuhama NN, Tucker PW, Liu G, Guan JL. 2006. Regulation of RNA-polymerase-II-dependent transcription by N-WASP and its nuclear-binding partners. *Nat Cell Biol* **8:** 756–763. doi:10.1038/ncb1433

Xia P, Wang S, Huang G, Zhu P, Li M, Ye B, Du Y, Fan Z. 2014. WASH is required for the differentiation commitment of hematopoietic stem cells in a c-Myc-dependent manner. *J Exp Med* **211:** 2119–2134. doi:10.1084/jem.20140169

Xie X, Almuzzaini B, Drou N, Kremb S, Yousif A, Farrants AO, Gunsalus K, Percipalle P. 2018a. β-Actin-dependent global chromatin organization and gene expression programs control cellular identity. *FASEB J* **32:** 1296–1314. doi:10.1096/fj.201700753R

Xie X, Jankauskas R, Mazari AMA, Drou N, Percipalle P. 2018b. β-Actin regulates a heterochromatin landscape essential for optimal induction of neuronal programs during direct reprograming. *PLoS Genet* **14:** e1007846. doi:10.1371/journal.pgen.1007846

Yoo Y, Wu X, Guan JL. 2007. A novel role of the actin-nucleating Arp2/3 complex in the regulation of RNA polymerase II-dependent transcription. *J Biol Chem* **282:** 7616–7623. doi:10.1074/jbc.M607596200

Zhang X, Wang X, Zhang Z, Cai G. 2019. Structure and functional interactions of INO80 actin/Arp module. *J Mol Cell Biol* **11:** 345–355. doi:10.1093/jmcb/mjy062

Zhao K, Wang W, Rando OJ, Xue Y, Swiderek K, Kuo A, Crabtree GR. 1998. Rapid and phosphoinositol-dependent binding of the SWI/SNF-like BAF complex to chromatin after T lymphocyte receptor signaling. *Cell* **95:** 625–636. doi:10.1016/S0092-8674(00)81633-5

Physical Nature of Chromatin in the Nucleus

Kazuhiro Maeshima,[1,2] Shiori Iida,[1,2] and Sachiko Tamura[1]

[1]Genome Dynamics Laboratory, National Institute of Genetics, Mishima, Shizuoka 411-8540, Japan

[2]Department of Genetics, School of Life Science, Sokendai (Graduate University for Advanced Studies), Mishima, Shizuoka 411-8540, Japan

Correspondence: kmaeshim@nig.ac.jp

Genomic information is encoded on long strands of DNA, which are folded into chromatin and stored in a tiny nucleus. Nuclear chromatin is a negatively charged polymer composed of DNA, histones, and various nonhistone proteins. Because of its highly charged nature, chromatin structure varies greatly depending on the surrounding environment (e.g., cations, molecular crowding, etc.). New technologies to capture chromatin in living cells have been developed over the past 10 years. Our view on chromatin organization has drastically shifted from a regular and static one to a more variable and dynamic one. Chromatin forms numerous compact dynamic domains that act as functional units of the genome in higher eukaryotic cells and locally appear liquid-like. By changing DNA accessibility, these domains can govern various functions. Based on new evidences from versatile genomics and advanced imaging studies, we discuss the physical nature of chromatin in the crowded nuclear environment and how it is regulated.

DNA, GENOME, AND INFORMATION STORAGE IN THE NUCLEUS

How much information can be stored in the nucleus of a cell? Biological (or genetic) information is inscribed in deoxyribonucleic acid. DNA is a negatively charged double-helix polymer, composed of four bases (guanine, G; adenine, A; thymine, T; cytosine, C), which are linked by a negatively charged phosphate backbone (Fig. 1A). A single human cell has ~2 m of DNA (0.33 nm $\times 3 \times 10^9$ bp $\times$ 2 sets of chromosomes), which is $\sim 2 \times 10^5$-fold longer than the typical diameter of a nucleus (~10 µm) (Fig. 1B). Given that the human genome has 3×10^9 bp composed of four bases, there are $4^{(3\times10^\wedge9)}$ combinations of information in it, which is equivalent to $2^{(6\times10^\wedge9)}$, or 750 M (0.75 G) bytes. This roughly corresponds to only a single compact disc (CD) (~700 M bytes) and is much smaller than our current smartphones' memory storage.

However, the memory density of the human genome in the nucleus can be calculated as $\sim 7.5 \times 10^{14}$ bytes/mm^3, considering that the nuclear volume is roughly 1000 µm^3. This memory density is several orders of magnitude higher than a CD (5.2×10^4 bytes/mm^3), blu-ray disc (QL) (9.4×10^6 bytes/mm^3), or even a flash memory (16 Gb, 6.3×10^8 bytes/mm^3) (Church et al. 2012). Including possible chemical modifications on DNA and histones, such as methyl-

Cite this article as *Cold Spring Harb Perspect Biol* doi: 10.1101/cshperspect.a040675

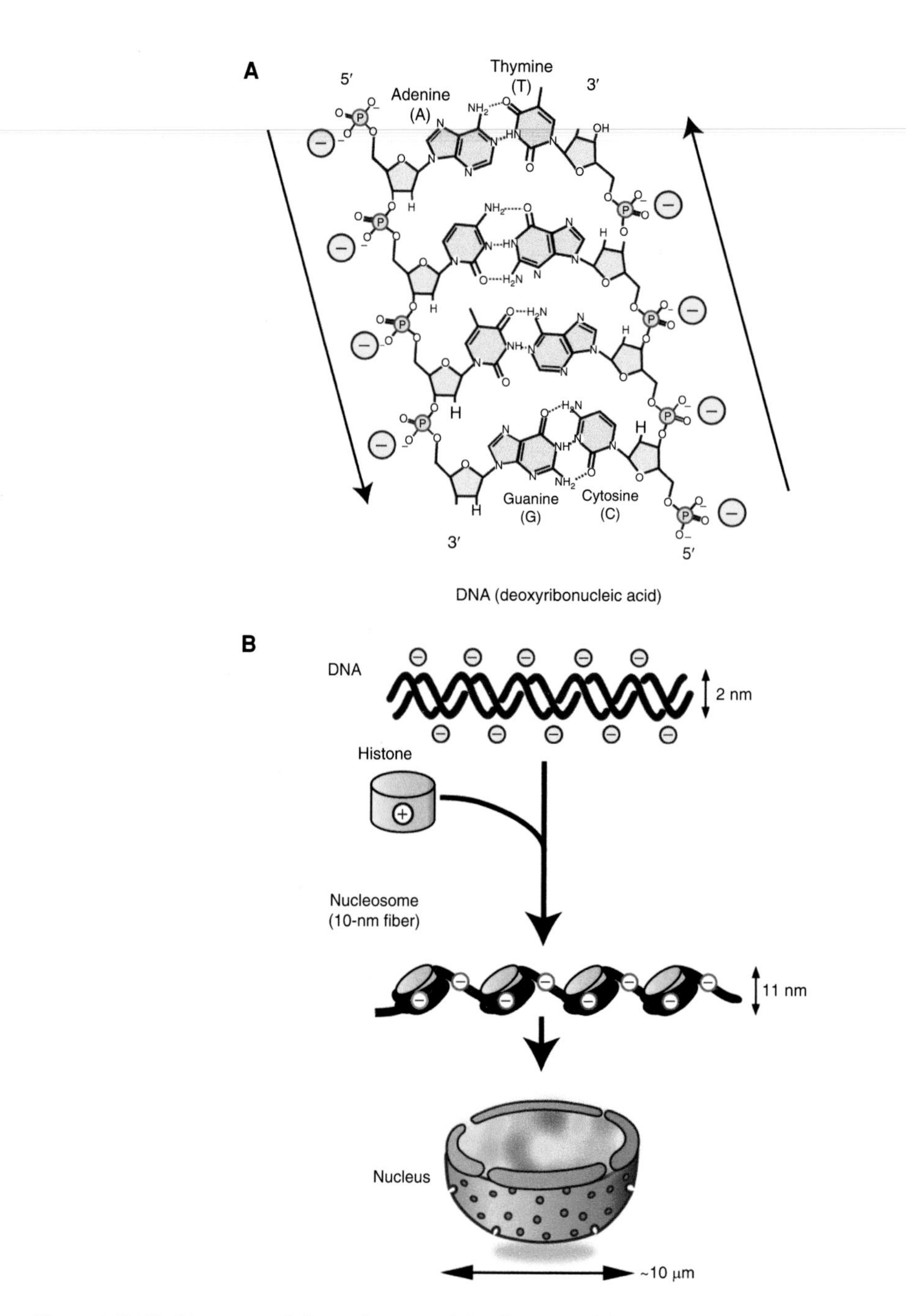

Figure 1. DNA, histones, and the nucleosome. (*A*) Schematic of the structure of deoxyribonucleic acid (DNA). DNA is a long double-helix polymer made from repeating units called nucleotides, that consists of four bases: adenine (A), cytosine (C), guanine (G), and thymine (T), which are attached to the sugar/phosphate. Adenine pairs with thymine and guanine pairs with cytosine by hydrogen bonds. The two strands of DNA run in opposite directions to each other in an antiparallel manner (shown by arrows). Phosphate backbones are negatively charged (marked with "-"). (Illustration and legend from Nozaki et al. 2018; reproduced with modifications, with permission, from the authors.) (*B*) Negatively charged DNA (*top* row) is wrapped around basic (or positive) core histone octamers (yellow on *second* row) to form 10-nm fiber, and it is further organized in a cell nucleus. Note that the fiber is still negatively charged.

ation and acetylation, this calculation increased dramatically to encompass an enormous memory size and density. Reading of information requires energy and if one assumes that a 1 Gb flash memory needs roughly ~1 J (1 W per 1 sec) of electric power and a similar level of energy was required in each cell of our body (~40×10^{12} cells; Bianconi et al. 2013) to read its genetic information, the total energy required would be ~4×10^{13} J, which is far greater than the total electric energy produced by all the nuclear power plants on earth (one plant can produce ~1×10^{9} J). In other words, the nuclear genome works with amazing energy efficiency.

It is thus appropriate to call the nucleus a truly "amazing memory device" in terms of both information storage and readout. Although our above example used the human cell nucleus, this remarkable property is shared widely across all eukaryotic nuclei. To function correctly, genome DNA needs to be carefully and properly organized in the nucleus, where differences in DNA folding can bring about dramatic changes in information readout. In this article, we review and discuss how genomic DNA is organized and behaves in the nucleus.

THE NUCLEOSOME

The electric charge of its components is an essential part of understanding the organization of the genomic information in the nucleus. Negatively charged DNA is wrapped around a core histone octamer (Fig. 1B), which consists of four positively charged histone proteins (H2A, H2B, H3, and H4), and forms a nucleosome (Figs. 1B and 2; Kornberg 1974; Olins and Olins 1974; Woodcock et al. 1976). The nucleosome structure has been resolved to 1.9 Å (Fig. 2B; Davey et al. 2002), in which 147 base pairs of DNA are wrapped in 1.7 left-handed superhelical turns around the histone octamer to generate a 225 kDa nucleosome core structure. The core histone octamer contains two copies of each of the four core histones, which are highly conserved in both length and amino acid sequences among eukaryotic species (Fig. 2A; Luger et al. 1997). H2A/H2B and H3/H4 form stable dimers at physiological conditions. Two H3/H4 dimers also form a stable tetramer under such conditions. Each core histone contains an ~60 amino acid residue histone-fold domain, which accounts for ~70%–75% of the mass of each protein (Fig. 2A) and provides a highly structured and extensive dimerization interface between the H3/H4 and H2A/H2B pairs (Fig. 2B). Each histone also has an amino-terminal tail domain, which is an intrinsically disordered region (IDR) and extends to the exterior of the nucleosome, while H2A contains an additional carboxy-terminal tail domain. Thus, a single nucleosome has 10 tails (Fig. 2C; Luger et al. 1997; Davey et al. 2002), which are highly positively charged due to a preponderance of lysines and arginines. For example, the ~35-residue H3 tail and the ~20-residue H4 tail contain 13 and 9 positively charged residues, respectively. Because of this positive charge abundance, the long, disordered tail domains contribute to the thermal stability of the nucleosome core (Ausio et al. 1989), and play important roles in the interactions between nucleosomes to contribute higher-order organization of chromatin. Chromatin is composed of nucleosomes and associated nonhistone proteins (Figs. 1B and 3; discussed later; Olins and Olins 2003).

In most eukaryotic species, an additional family of histone proteins bind to the nucleosome. Linker histones (also referred to as H1s) are highly basic proteins but share no structural homology with the core histones (Fig. 2D; Woodcock et al. 2006; Bednar et al. 2017). Interestingly, two IDRs in the carboxy- and amino-terminal regions of H1s connected by an ~80-residue "globular" domain are important for chromatin condensation (Turner et al. 2018). Whereas it is assumed that H1s bind to the exterior of nucleosomes and stabilize the wrapping of DNA within nucleosomes (Fig. 2D), H1s seem to be highly mobile in the cell (Misteli et al. 2000).

Each nucleosome particle is connected by linker DNA segments (~20–80 bp; ~6.6–27 nm) that generate repetitive motifs of ~200 bp and are described as "beads on a string" or as a "10-nm fiber" (Figs. 1B and 3A; Olins and Olins 2003). This 10-nm fiber produces electrostatic repulsion between adjacent regions because only about half of the DNA negative charges derived from the phosphate backbone are neutralized by the basic

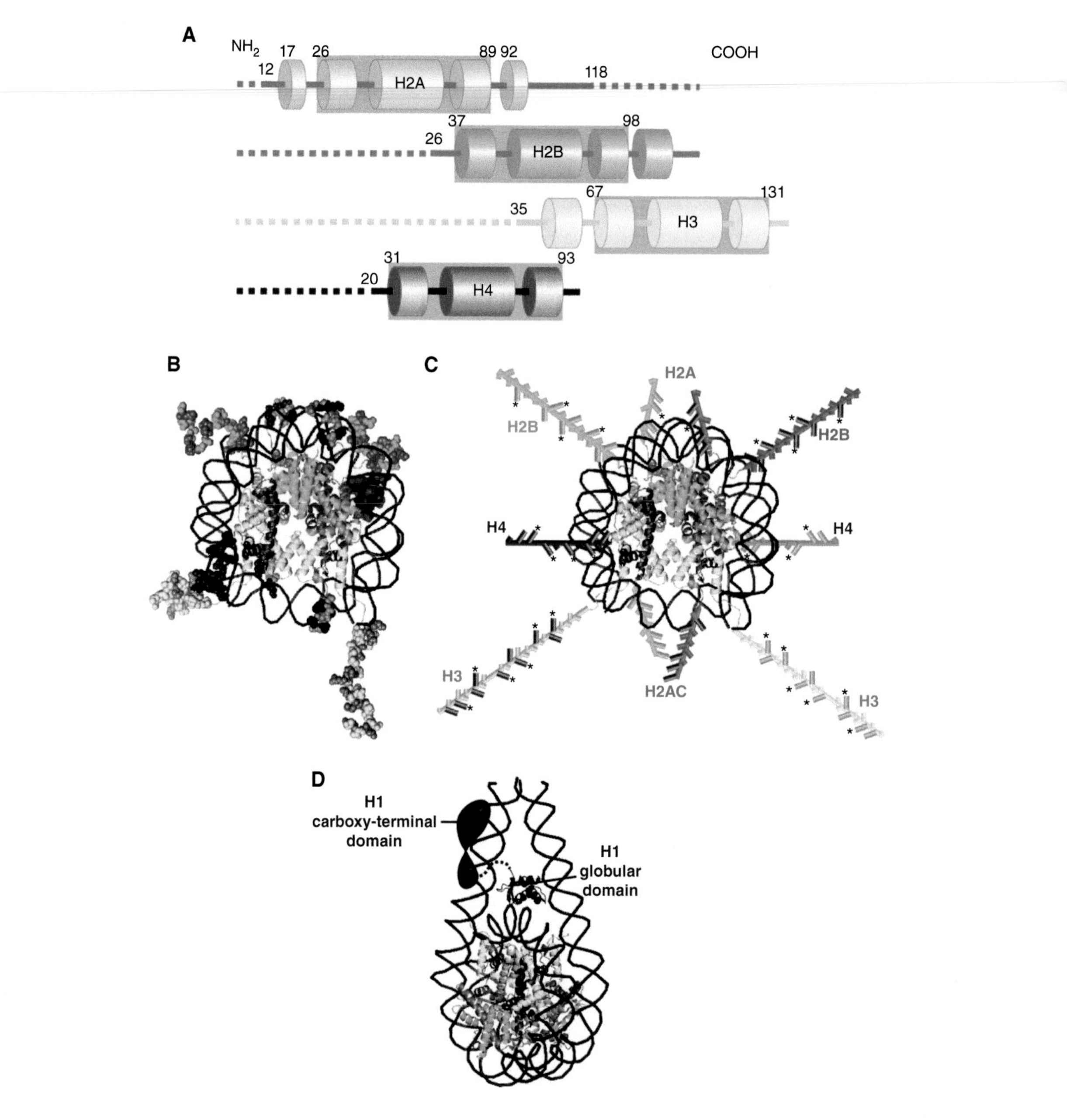

Figure 2. Structures of core histones and nucleosome. (*A*) Schematic of the core histone structures; histone-fold domains are enclosed by gray boxes, tail domains are denoted by dashed lines, and α-helices are shown by columns. The positions of residues flanking the histone fold in each protein are shown. The approximate residue at the end of the tail domain nearest the histone fold is also shown. (*B*) The nucleosome structure at 1.9 Å resolution (Davey et al. 2002). Note that histone colors correspond to *A*. Amino acid residues of histone tails (intrinsically disordered regions [IDRs]) are shown as ball models. Basic residues (lysines and arginines) in the tail domains are highlighted with a darker color. (*C*) Amino-terminal tail domains in *B* are extended away from the nucleosome core. H2A also has a carboxy-terminal tail. Basic residues (lysines and arginines) in the tail domains are colored in orange. An asterisk indicates sites of lysine acetylation. (Illustrations are based on data in Wolffe and Hayes 1999 and Pepenella et al. 2014a.) (*D*) Structural model of a nucleosome with a linker histone H1. (The model in *D* was created from PDB data in Bednar et al. 2017.)

 Cite this article as *Cold Spring Harb Perspect Biol* doi: 10.1101/cshperspect.a040675

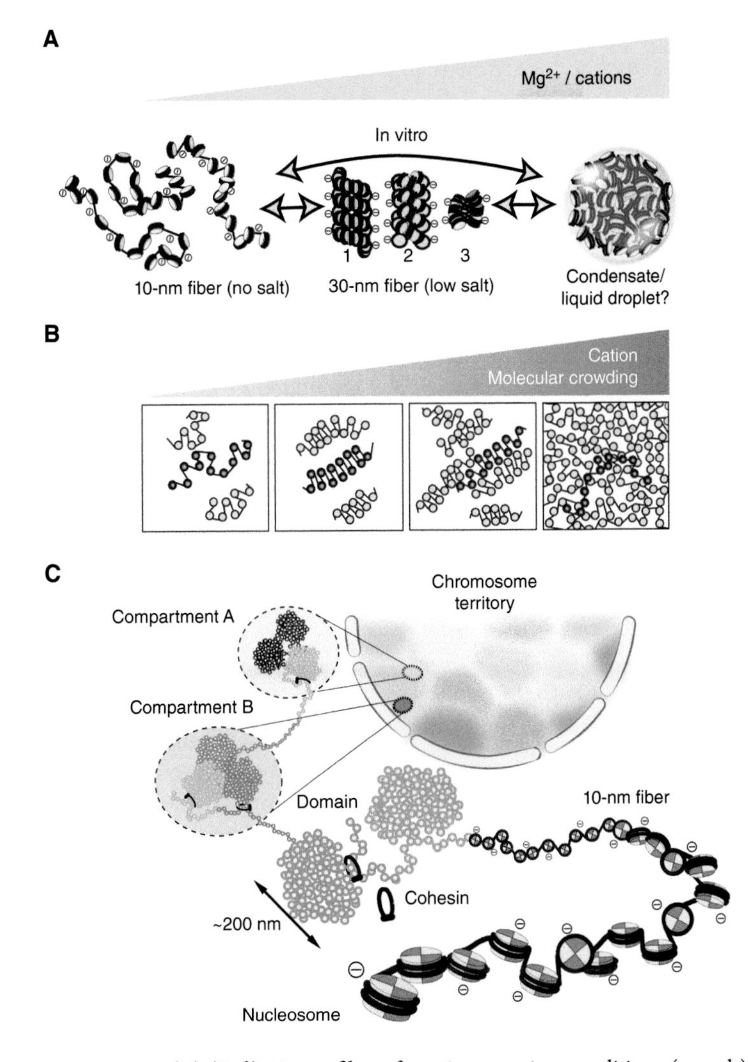

Figure 3. Chromatin structure. (*A*) (*Left*) 10-nm fibers form in no cation conditions (no salt). (*Center*) Three types of 30-nm fibers form with a low concentration of cations/Mg^{2+}: (1) a solenoid (one-start) model, (2) a two-start zigzag, and (3) a zigzag tetranucleosomal model (Song et al. 2014). (*Right*) Large chromatin condensates with interdigitated 10-nm fibers that form in more physiological cations or higher cations. Note, this condensate lacks the 30-nm structures (Maeshima et al. 2016b). Recently, under a certain specific condition, the condensates were shown to be liquid droplets formed by liquid–liquid phase separation (Gibson et al. 2019). (Panels from Maeshima et al. 2020; reproduced with modifications, with permission, from the authors.) (*B*) Polymer melt model. Under low cation and/or molecular crowding conditions, 10-nm fibers could form 30-nm chromatin fibers via intrafiber nucleosome associations. An increase in cation concentration and/or molecular crowding results in interfiber nucleosomal contacts that interfere with intrafiber nucleosomal associations, leading to a melted polymer-like or "sea of nucleosomes" state (Dubochet et al. 1986; Eltsov et al. 2008; Maeshima et al. 2010). In this state, tetranucleosome zigzag motifs are occasionally seen (highlighted in red). Note that we show a highly simplified two-dimensional nucleosome model in these illustrations. (*C*) A simplified scheme for hierarchical chromatin organization in the nucleus. The 10-nm fiber is compacted into chromatin domains (e.g., topologically associating domain [TAD]/contact domain/loop domain) (Dixon et al. 2012; Nora et al. 2012; Sexton et al. 2012; Dekker and Heard 2015), which interact over long distances to form chromatin compartments (Lieberman-Aiden et al. 2009). Compartments generally represent a transcriptionally active or open chromatin state (compartment A) and an inactive or closed chromatin state (compartment B). A single interphase chromosome is occupied in a chromosome territory (highlighted as different colors) (Cremer and Cremer 2001).

core histones (Fig. 3A; Maeshima et al. 2014). For further folding of the 10-nm fiber, the remaining negative charges have to be screened by other factors, such as linker histones, nonhistone proteins, cations, and other positively charged molecules. Chromatin, which consists of 10-nm fibers and various nonhistone proteins, is overall still a negatively charged polymer because the negative charge of DNA is not completely neutralized by these positively charged molecules. Therefore, chromatin can dynamically change its local structure depending on the electrostatic state of its environment (Fig. 3A; Cole 1967; Earnshaw and Laemmli 1983; Hansen 2002; Hudson et al. 2003; Takata et al. 2013). Such changes in local chromatin structure are critical for gene expression because they directly govern access to the DNA and therefore impact how the genome is scanned and read in the nucleus.

STRUCTURAL VARIATIONS OF CHROMATIN IN VITRO: 10-NM FIBERS, 30-NM FIBERS, AND CHROMATIN CONDENSATES

A crucial factor that influences local chromatin structure is the ionic condition of the surrounding environment. For instance, in the presence of linker histones and a low concentration of cations (e.g., $<\sim1$ mM Mg^{2+} or $<\sim50$ mM Na^{+}), the purified 10-nm fiber (Fig. 3A, left) is folded into a zigzag or solenoidal fiber structure with a diameter of 30 nm, which is called the 30-nm chromatin fiber (30-nm fiber) (Fig. 3A, center; Finch and Klug 1976; Woodcock et al. 1984; Schalch et al. 2005; Robinson et al. 2006; Song et al. 2014). Such low ionic conditions partially screen the fiber's electrostatic charge, leading to local contacts between nucleosomes in the fiber and the formation of the folded 30-nm structure (Fig. 3A, center). Note, the concentration of cations required for this folding depends on the size and concentration of the fiber.

More cations are present at physiological concentrations that result in further charge screening. At these levels, electrostatic repulsion between neighboring nucleosomes almost disappears and large chromatin condensates form (Hansen 2002; Maeshima et al. 2016b; Gibson et al. 2019; Strickfaden et al. 2020). Such condensates are organized like a melted polymer or "sea of nucleosomes" and do not contain the 30-nm structure (Fig. 3A,B, right; McDowall et al. 1986; Eltsov et al. 2008; Maeshima et al. 2016a). Nucleosomes are free to contact long distal nucleosome partners, leading to interdigitation of their fibers (Fig. 3A,B, right; Zheng et al. 2005; Kan et al. 2009; Maeshima et al. 2010).

This structural variation scheme of chromatin (Fig. 3A,B) was demonstrated in vitro (Maeshima et al. 2016b) using a well-defined chromatin model, the 12-mer nucleosome array (Hansen 2002). Previous biochemical studies using the nucleosome arrays can explain why such large condensates lack the 30-nm fiber (Dorigo et al. 2003; Kan et al. 2009; Sinha and Shogren-Knaak 2010). The long tail domain of histone H4 (Fig. 2B,C) mediates the formation of both the 30-nm fiber and large condensates. Consequently, organization into chromatin condensates can prevent the formation of 30-nm fibers by sequestering the H4 tail. It is also important to note that this interdigitated chromatin folding might contribute to forming large chromatin structures such as mitotic chromosomes (Hansen et al. 2018).

The physical state of chromatin condensates is intriguing and a subject of ongoing research. Gibson et al. showed that nucleosome arrays with cations developed into large chromatin condensates ~10 µm in size. The resultant large condensates behaved like liquid droplets (Fig. 3A, right; Gibson et al. 2019), in which the nucleosomes self-organized through a process termed liquid–liquid phase separation (LLPS) (Hyman et al. 2014; Mitrea and Kriwacki 2016; Rippe 2021). The droplets rapidly fused to each other and the fluorescence recovery after photobleaching (FRAP) showed that fluorescently labeled nucleosome arrays freely diffused in the droplets (Gibson et al. 2019). However, Strickfaden et al. (2020) revealed that the condensates formed from similar nucleosome arrays show a rather solid-like property and that the liquid chromatin condensates, as shown by Gibson et al. (2019), were only observed under a highly specific set of conditions (Strickfaden et al. 2020). The condensates thus seem to have solid-like features at a micron scale. Nonetheless, we infer that the condensates might have weak intrinsic droplet properties at a nanoscale (100–200 nm) range, given that each nucleosome possesses 10 IDRs (Fig. 2B, C; Luger et al. 1997; Pepenella et al. 2014b). IDRs often mediate multivalent contacts for droplet formation (Lin et al. 2015). Emerging evidence suggests that functional compartmental droplets are formed by LLPS in the cell (Hyman et al. 2014; Mitrea and Kriwacki 2016). Further investigation will be required to examine the intrinsic droplet property in the condensates.

CHROMATIN ORGANIZATION IN THE NUCLEUS

In 1976, the 30-nm fiber was discovered in vitro and has since been believed to be the basic chromatin structure in eukaryotic cells (Fig. 3A, center; Finch and Klug 1976; Woodcock et al. 1984; Schalch et al. 2005; Robinson et al. 2006; Song et al. 2014). However, during the past 10 years, extensive data from various techniques, including cryo-electron microscopy (EM) (McDowall et al. 1986; Eltsov et al. 2008; Chen et al. 2016; Cai et al. 2018), X-ray scattering (Joti et al. 2012; Nishino et al. 2012), electron spectroscopic imaging (ESI) (Fussner et al. 2012; Visvanathan et al. 2013), EM tomography (Ou et al. 2017), stochastic optical reconstruction microscopy (STORM) (Ricci et al. 2015), chromosome conformation capture (3C)/Hi-C (Dekker 2008; Hsieh et al. 2015; Sanborn et al. 2015; Ohno et al. 2019), and computational modeling (Collepardo-Guevara and Schlick 2014; Bajpai et al. 2017), demonstrated that chromatin consists of rather irregular and variable nucleosome arrangements like "sea of nucleosomes" (Fig. 3A,B, right). In addition to the cation issue discussed above, many factors in native chromatin, including irregular nucleosome spacing, histone modifications and variants, and nonhistone protein binding, would preclude the formation of regular 30-nm fibers, especially in actively transcribing cells. Thus, the 30-nm fiber is unlikely to be the basic structure or unit of chromatin in vivo. Nonetheless, the formation of tetranucleosome motifs is highly possible (Baldi et al. 2020; Krietenstein and Rando 2020), as reported by new genomics studies that found occasional zigzag nucleosome configurations in budding yeast cells (Hsieh et al. 2015; Ohno et al. 2019) and in heterochromatin regions of human cells (Risca et al. 2017).

At a higher level of organization, a large-scale chromatin structure with a diameter ~200 nm has long been observed by light and electron microscopic imaging in more complex eukaryotes (Fig. 3C; Hu et al. 2009; Nozaki et al. 2017; Olins and Olins 2018; Xu et al. 2018), while yeast cells seem to have predominately open chromatin arrangements with clusters of only a few nucleosomes (Hsieh et al. 2015; Chen et al. 2016; Ohno et al. 2019). Recently, a combination of 3D super-resolution and scanning EM revealed "chromatin domains," which were observed as irregularly sized clusters of nucleosomes (~200 nm) (Hoffman et al. 2020; Miron et al. 2020). Interestingly, the chromatin domains in euchromatic regions maintained compact chromatin organizations with active epigenetic marks, but their sizes were smaller than those of heterochromatin, which is consistent with the previous reports (Maeshima et al. 2015; Nozaki et al. 2017). ESI, which maps phosphorus and nitrogen atoms with contrast and resolution sufficient to visualize 10-nm fibers (Hendzel et al. 1999), also exhibited such chromatin domain structures (Strickfaden et al. 2020).

3C/Hi-C methods can be used to generate a fine contact probability map of genome DNA (Dekker and Heard 2015; Dekker 2021), and this technology has revealed that the higher eukaryotic genome is partitioned into chromatin domains (Fig. 3C): topologically associating domains (TADs) at the scale of several hundreds of kilobytes (Dixon et al. 2012; Nora et al. 2012; Sexton et al. 2012) and smaller contact/loop domains (mean size of ~185 kb) (Rao et al. 2014). These structures were confirmed visually by superresolution imaging (Bintu et al. 2018). The loop domains seemed to be held together by cohesin (Figs. 3C and 5C,E), in cooperation with CCCTC-binding factor (CTCF) (Zuin et al. 2014; Rao et al. 2017; Wutz et al. 2017; Bintu et al. 2018; for more details, see Dekker 2021). These domains are often clustered as two distinct, megabase scale compartments A and B (Lieberman-Aiden et al. 2009), which likely represent transcriptionally active chromatin and inactive chromatin, respectively (Fig. 3C). Collectively, chromatin organization in the nucleus appears to have a hierarchy (Fig. 3C): from 10-nm fibers to domains, domains to compartments A and B, and from these compartments to chromosomes to form the "chromosome territory" of the nucleus (Cremer and Cremer 2001).

The large-scale arrangements/domains described above seem to be fundamental chromatin features in the nucleus (Misteli 2020). Notably, the domains have a compact organization (Fig. 3C), irrespective of euchromatin or heterochromatin (Nozaki et al. 2017; Miron et al. 2020; Strickfaden et al. 2020). To form these compact domains, local nucleosome–nucleosome interactions seem to be critical (Nozaki et al. 2017; Strickfaden et al. 2020). Cations like magnesium ion (Mg^{2+}) can facilitate these interactions, as shown in the formation of cation (Mg^{2+})-dependent chromatin condensates in vitro, which requires nucleosome–nucleosome interactions by histone tails (Fig. 2C; Maeshima et al. 2016b; Gibson et al. 2019; Sanulli et al. 2019; Strickfaden et al. 2020).

Furthermore, it is very likely that non-nucleosomal proteins and RNAs, as well as cations, contribute to the formation of compact chromatin domains through macromolecular crowding effects (Figs. 3B and 5E; Asakura and Oosawa 1954; Marenduzzo et al. 2006; Hancock 2007). Imai et al. (2017) found that the total density of heterochromatin (208 mg/mL) was only 1.53-fold higher than that of the euchromatic regions (136 mg/mL), while the DNA density of heterochromatin was 5.5–7.5-fold higher (Imai et al. 2017). This finding suggests that non-nucleosomal materials (proteins, RNAs, etc.), which are dominant in both heterochromatin and euchromatin (~120 mg/mL), play an important role in compact organization of the domains in the nuclear crowded environments.

The organization of chromatin may serve to modulate the accessibility of larger protein complexes to target sites (Maeshima et al. 2015; Miron et al. 2020). The compact domains might hinder the access of large protein complexes such as transcription factors and replication initiation complexes to the inner core of chromatin domains (Maeshima et al. 2015; Miron et al. 2020). Decompaction of such domains with histone modifications or the action of other proteins can increase accessibility to the complexes to turn on gene transcription (Toth et al. 2004; Bian and Belmont 2012; Kieffer-Kwon et al. 2017; Nozaki et al. 2017; Miron et al. 2020; Strickfaden et al. 2020). Interestingly, it has been suggested that various RNAs may be involved in the regulation of these compact chromatin domains (Nozawa and Gilbert 2019). Nozawa et al. (2017) showed that scaffold attachment factor A (SAF-A)/heterogeneous nuclear ribonucleoprotein U (hnRNP U) decompacts transcriptionally active chromatin domains by forming a filament-shaped oligomer with nuclear RNA (Nozawa et al. 2017). How RNAs act on the domain formation remains unclear and is an area to be investigated further.

Whereas the compact organization of the genome appears to be primarily designed for storage efficiency, it seems to have additional functional roles to protect genomic information. The compact organization can generate a spring-like restoring force that resists nuclear deformation by mechanical stress and plays an important role in maintaining genomic integrity. Nuclei with condensed chromatin possess significant elastic rigidity, while those with decondensed chromatin are considerably softer (Shimamoto et al. 2017; Stephens et al. 2017; Maeshima et al. 2018b; Agbleke et al. 2020; also see Miroshnikova and Wickström 2021). Interestingly, compact chromatin seems more resistant to radiation and chemical damage than the decondensed form, probably because condensed chromatin has lower reactive radical generation and is less prone to chemical attack (Takata et al. 2013). The maintenance of genomic DNA integrity by greater compaction may have contributed to the evolution of eukaryotic organisms.

LOCAL MOTION OF CHROMATIN IN THE NUCLEUS

The view of chromatin organization discussed so far suggests that chromatin is less physically constrained and more dynamic than expected in the regular static structure model with helical coiling (Maeshima et al. 2010). In addition, given that each nucleosome possesses 10 IDRs (Fig.

2C; Luger et al. 1997; Pepenella et al. 2014b), it is tempting to consider that chromatin locally behaves like a liquid (Fig. 3A,B, right).

Live-cell imaging studies using a LacO/LacI-GFP system (Fig. 4A; Robinett et al. 1996) or related systems (Germier et al. 2017; Tasan et al. 2018; Eykelenboom et al. 2019) have long revealed dynamic movements of chromatin in numerous cells such as yeast, nematodes, flies, and mammals (Marshall et al. 1997; Heun et al. 2001; Chubb et al. 2002; Levi et al. 2005; Meister et al. 2010; Hajjoul et al. 2013; Arai et al. 2017). This system can track the motion of bacterial operator (LacO) repeats (<256 repeats) inserted into the genome of living cells to which fluorescent LacI proteins bind (Fig. 4A; Robinett et al. 1996). Analyses of FRAP have also demonstrated turnover rates of histones or chromatin-associated proteins at bleaching sites in living cells and provided indirect information on chromatin motion (Misteli et al. 2000; Kimura and Cook 2001; Meshorer et al. 2006). Furthermore, CRISPR-based genome editing technology has allowed a targeted region of interest in genome chromatin to be specifically labeled (e.g., Tasan et al. 2018) to visualize motion of that region (Fig. 4B; Chen et al. 2013; Gu et al. 2018; Guo et al. 2019; Ma et al. 2019).

Moreover, genome-wide local chromatin motion has been investigated using single-nucleosome imaging and tracking in living cells through fluorescently tagged histone proteins (Fig. 4C; Hihara et al. 2012; Nozaki et al. 2017; Nagashima et al. 2019; Lerner et al. 2020; Gómez-García et al. 2021). This powerful imaging modality can trace the movement of nucleosomes throughout the nucleus at the single-molecule level with high resolution. Similar global chromatin movements have been described using fluorescently labeled bulk histones, or DNA-binding dye, to analyze their displacement correlation (Fig. 4D; Zidovska et al. 2013; Shaban et al. 2020). Development of imaging techniques that record local chromatin motion at a genome-wide scale enabled visualization of heterogeneity of local chromatin motion in a whole nucleus as a chromatin heat map (Fig. 4E,F; Nozaki et al. 2017; Nagashima et al. 2019) or chromatin dynamics maps (Zidovska et al. 2013; Lerner et al. 2020; Shaban et al. 2020). These maps show overall distributions of heterochromatin and euchromatin. As expected, the local chromatin motion at the nuclear periphery (heterochromatin) is less mobile (Fig. 4F). Consistently, statistical analysis of genome-wide single-nucleosome tracking data revealed heterogeneity of movements, which roughly categorized chromatin into two types: slow chromatin that moves under structurally constrained environments and fast chromatin that moves with fewer constraints (Fig. 4G; Ashwin et al. 2019, 2020). Lerner et al. (2020) suggested a correlation between chromatin mobility and transcription factor binding. Furthermore, statistical analysis of single-nucleosome tracking data has unveiled that the organized nucleosomes within chromatin are locally mobile and behave like a liquid at the scale of less than hundreds of nanometers (Ashwin et al. 2019, 2020). In other words, when considered at the scale of the chromatin domain, chromatin is very dynamic and exhibits a liquid-like diffusion rather than a crystal-like structure, which has a long-range order. A physical basis of the liquid-like behavior of chromatin is provided by polymorphism of the 10-nm fiber with the long histone tails (Fig. 2C), which can turn into various structures, including extended, folded, interdigitated, bent, and looped structures (Collepardo-Guevara and Schlick 2014).

FACTORS THAT CONSTRAIN CHROMATIN MOTION IN THE NUCLEUS

Whereas local chromatin behaves like a liquid (Ashwin et al. 2019, 2020), each chromosome occupied in the territory appears to be relatively stable, without mixing of chromosomes (Fig. 3C; Strickfaden et al. 2020). Certainly, the mean square displacement (MSD) analysis of local nucleosome motion in living human cells revealed that the plots almost reached a plateau value after 3 sec (Fig. 5A,B), which is proportional to the square of the radius of constraint (Rc; $P = 6/5 \times Rc^2$) (Dion and Gasser 2013). In addition, the estimated radius of the nucleosome motion constraint in living human RPE-1 cells is 141 ± 19.2 nm (mean ± SD), which roughly corresponds to the size of the chromatin domain mentioned above (i.e., local chromatin is constrained

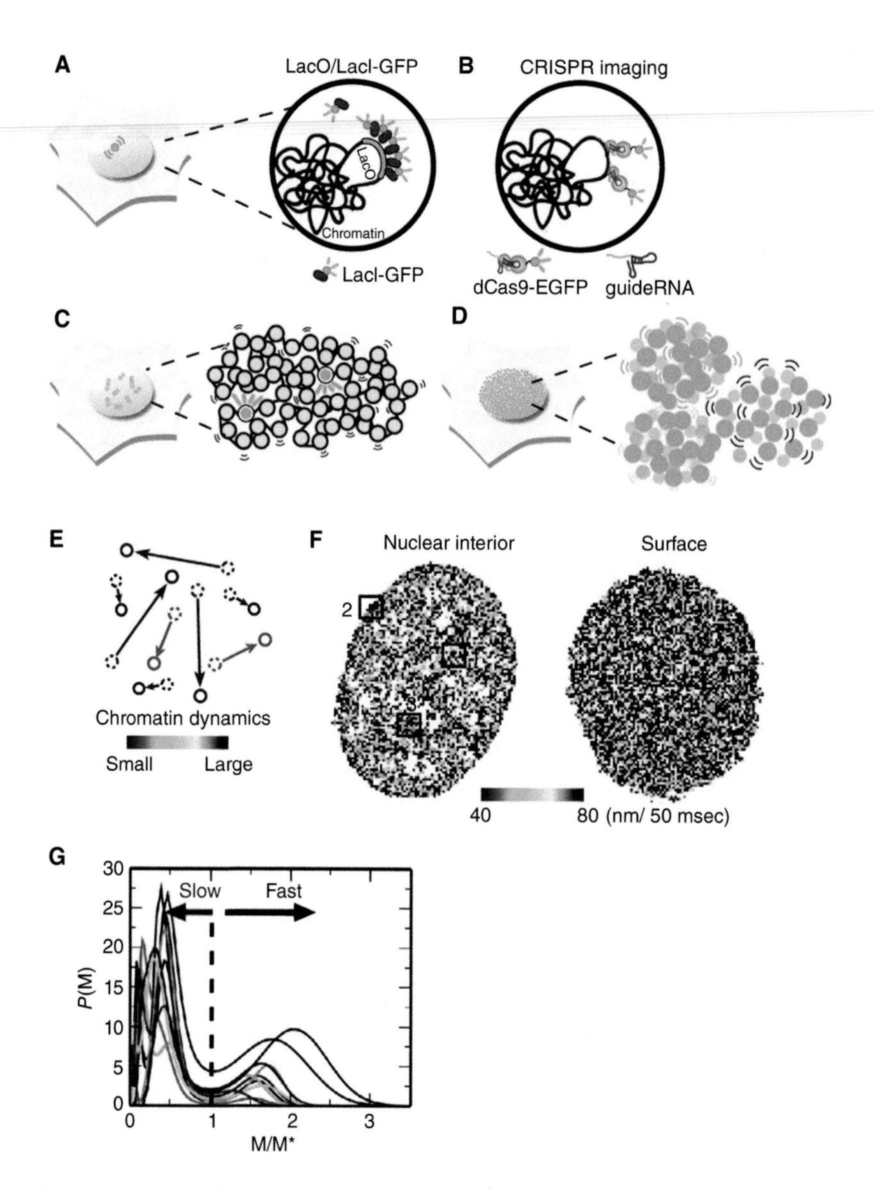

Figure 4. Visualization of local chromatin motion. Schematics for (*A*) LacO/LacI-GFP, (*B*) CRISPR-based chromatin labeling, (*C*) single nucleosome imaging and tracking, and (*D*) bulk chromatin labeling for displacement correlation analyses. A small number of nucleosomes are labeled with photoactivatable GFP or other fluorescent tags for single nucleosome imaging in *C* while much more are labeled in *D*. (*E*) Example of chromatin dynamics used to generate a heat map. Movements are measured over 50 msec. Small movements are shown in blue and large movements are shown in red. Heat maps allow the visualization of the heterogeneity of local chromatin motion in the nucleus. (*F*) Diagrams of a chromatin heat map where the nuclear interior (*left*) and nuclear periphery (*right*) of a living HeLa cell are shown. The boxed regions 1–3 show nucleoplasm, nuclear periphery, and around nucleolus periphery, respectively. Note that regions 2 and 3 are presumably heterochromatin regions shown in dark blue. The map of the nuclear periphery (surface) appears more bluish, showing less mobile chromatin in the heterochromatin-rich regions. (*G*) Mean square displacement (MSD) distribution of local chromatin motion. The MSD at the first minimum between the peaks is denoted by M^*. P (M, 0.5 sec) from 10 cell samples were scaled by M^* and shown, indicating fast and slow peaks. For details, see Ashwin et al. (2019). (Illustrations and legend from Maeshima et al. 2020; reproduced with modifications, with permission, from the authors.)

 Cite this article as *Cold Spring Harb Perspect Biol* doi: 10.1101/cshperspect.a040675

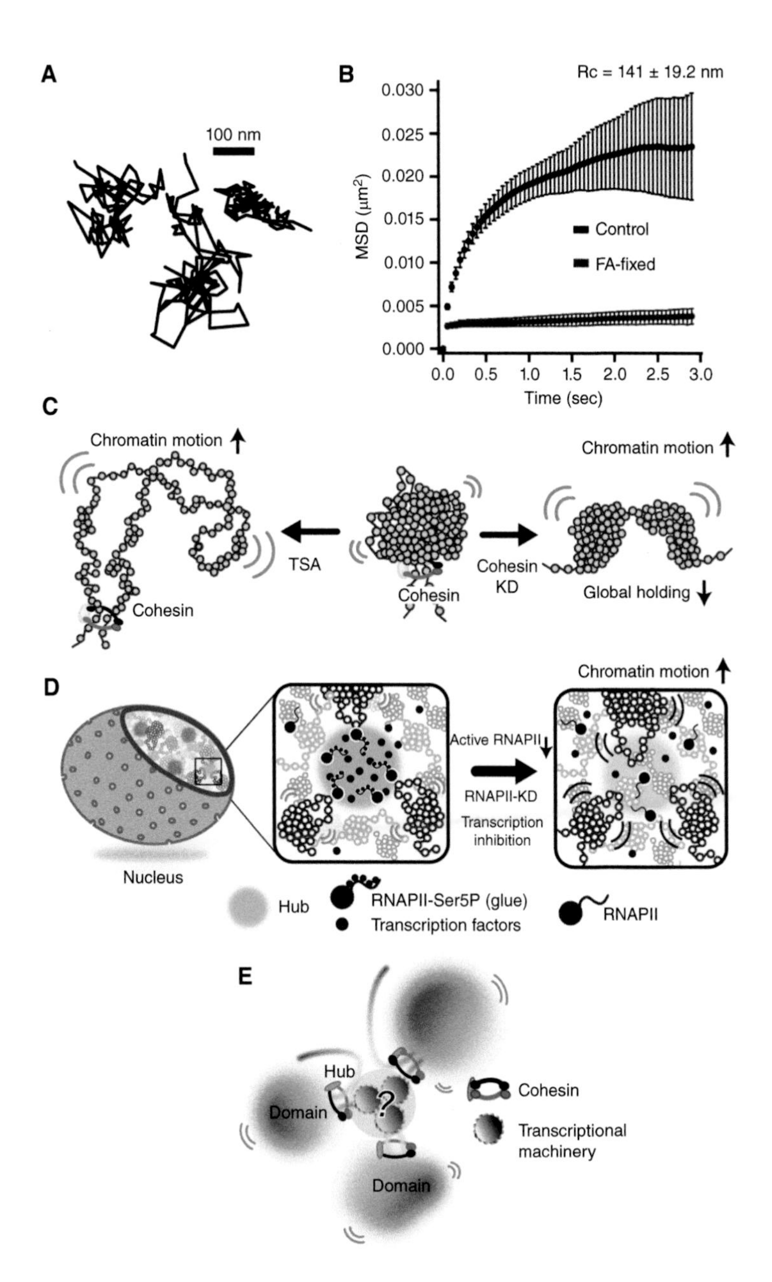

Figure 5. Local chromatin motion and chromatin domains in the nucleus. (*A*) Three representative trajectories of single nucleosomes (Nagashima et al. 2019). (*B*) Mean square displacement (MSD) plots (±SD among cells) of single nucleosomes in living (black) and formaldehyde-fixed (red) RPE-1 cells over time (0.05 to 3 sec). For each sample, $n = 20$ cells. (Data from Nagashima et al. 2019.) (*C*) (*Left*) Decondensed chromatin in cells treated with histone deacetylase (HDAC) inhibitor trichostatin A (TSA) shows increased chromatin movements resulting from weakened nucleosome–nucleosome interactions and subsequently less local chromatin constraining. (*Center*) Usual state of chromatin with reduced movements. The chromatin domain is organized by local nucleosome–nucleosome interactions and global holding by cohesin. (*Right*) Loss of cohesin leads to less constraining and a resultant increase in chromatin motion. (*D*) (*Left* and *center*), Cluster/condensate of active RNA polymerase II (RNAPII) and transcription factors (blue spheres) can work as a transient hub (cyan sphere) to weakly connect multiple chromatin domains and to globally constrain chromatin motion. (*Right*) RNAPII inhibition, or its rapid depletion, releases chromatin constraints and increases chromatin motion. (*E*) A current model of chromatin domains. Chromatin is locally very dynamic, but the local chromatin motion is constrained by various physical or geometrical factors to keep each chromosome in the chromosome territories. (Diagrams in panels *C* and *D* from Babokhov et al. 2020; reproduced with modifications, with permission, from the authors.)

to a range of the chromatin domain size and does not appear to travel long distances) (Fig. 3C; Nagashima et al. 2019). Furthermore, the DNA replication foci observed via pulse fluorescent labeling revealed constrained euchromatin and heterochromatin behavior in living mouse cells (Strickfaden et al. 2020). These properties are not consistent with liquid behavior and suggest highly constrained motion. If chromatin behaves like a liquid locally but does not move over long distances, what constrains the chromatin in a nuclear aqueous environment? How are chromosome territories maintained? How can chromatin motion be adjusted depending on the physiological state(s) of the cell?

In contrast to ATP-dependent long-range movements of chromatin (see reviews by Soutoglou and Misteli 2007; Seeber et al. 2018), local chromatin motion is likely to be isotropic and primarily driven by thermal fluctuation (Fig. 5A; Marshall et al. 1997; Chubb et al. 2002; Levi et al. 2005; Nozaki et al. 2017; Ashwin et al. 2019). Variability in this motion can be created by constraints from several physical or geometrical factors (Fig. 4G; Ashwin et al. 2019).

Nucleosome Interactions

Nucleosome–nucleosome interactions are one of the factors to constrain chromatin motion. Chromatin becomes more mobile when treated with the histone deacetylase (HDAC) inhibitor trichostatin A (TSA) (Fig. 5C, left). This drug increases histone H3 and H4 tail acetylation (Fig. 2C) and decondenses chromatin (Fig. 5C, left), presumably by decreasing the number of nucleosome–nucleosome interactions (Gorisch et al. 2005; Ricci et al. 2015; Nozaki et al. 2017; Strickfaden et al. 2020). Such artificial decondensation of chromatin (i.e., extended fibers or fibers with nucleosome-free regions) reduces constraints that restrict thermal-fluctuating motion and results in elevated local chromatin movement (Fig. 5C, left; Amitai et al. 2017; Nozaki et al. 2017; Ashwin et al. 2019).

Nucleoplasmic Milieu

Another factor is the quantity of free cations in the nucleus. An increase in cations, such as Mg^{2+}, can condense chromatin and lower its local motion, since nucleosomes have residual negative charges and tend to repulse one another as discussed above (Fig. 3A,B). Consistently, intracellular ATP reduction, which releases Mg^{2+} from ATP-Mg (Maeshima et al. 2018a), hypercondenses chromatin and decreases its motion (Visvanathan et al. 2013; Nozaki et al. 2017; Maeshima et al. 2018a). These findings offer a regulatory mechanism for the higher-order chromatin organization by the intracellular Mg^{2+}-ATP balance. Since intracellular ATP levels vary depending on cell types (Pecqueur et al. 2013; Qian et al. 2014; Kieffer-Kwon et al. 2017), they may regulate cellular functions such as differentiation via changing higher-order chromatin organization. Moreover, hypertonic treatment (~570 mOsm) had similar effects (Albiez et al. 2006; Nozaki et al. 2017; Strickfaden et al. 2020). This may be due to an increase in intracellular cations and molecular crowding following hypertonic treatment (Albiez et al. 2006).

Chromatin Proteins

Various chromatin proteins constrain the local chromatin motion driven by thermal fluctuation, contributing to the retention of each chromosome in its "chromosome territory" (Fig. 3C; Cremer and Cremer 2001) as well as genome functions. The cohesin complex is an important protein complex that can capture chromatin fibers with its ring structure to form loops and for sister chromatid cohesion (Nasmyth and Haering 2005; Morales and Losada 2018; Nishiyama 2019). Loss of cohesin leads to fewer chromatin constraints and thereby a drastic increase in local chromatin motion (Fig. 5C, right; Dion et al. 2013; Nozaki et al. 2017; Ashwin et al. 2019; Itoh et al. 2021).

Transcription

Interestingly, the RNA transcriptional machinery has a constraining role for local chromatin motion in the nucleus (Fig. 5D,E; Nagashima et al. 2019; Babokhov et al. 2020; Itoh et al. 2021). Knockdown of active RNA polymerase II (RNAPII) or other transcription regulatory factors, as well as RNAPII inhibitor treatments, releases RNAPII from chromatin and raises local chromatin movements (Nagashima et al. 2019). Consistent with this finding, some specific genomic loci in human breast cancer and fly embryos became less dynamic when actively transcribed (Germier et al. 2017; Chen et al. 2018). Furthermore, it was recently shown that active RNA polymerase I (RNAPI) also forms clusters (Andrews et al. 2018) and constrains ribosomal DNA chromatin in the nucleoli (Ide et al. 2020). In this case, transcription inhibition dissociated RNAPI from rDNA chromatin and moved like a liquid within the newly formed droplet in the nucleolus (Ide et al. 2020). The concept that the transcriptional machinery constrains chromatin (Fig. 5D,E) is consistent with the classic transcription factory hypothesis (Cook 1999; Edelman and Fraser 2012; Feuerborn and Cook 2015), where clusters of RNAPII and other transcription factors immobilize genome chromatin to facilitate transcription. Recent studies also suggest that RNAPII, Mediator, and other factors form condensate/clusters upon transcription (Cho et al. 2018; Chong et al. 2018; Lu et al. 2018; Sabari et al. 2018).

Association with Chromatin Domains

Other prominent chromatin constraining factors are the proteins associated with heterochromatin, where the chromatin is less mobile (Fig. 4F; Chubb et al. 2002; Shinkai et al. 2016; Nozaki et al. 2017; Nagashima et al. 2019; Lerner et al. 2020). The nuclear periphery lamina-associated chromatin domains (LADs) enriched regions (van Steensel and Belmont 2017) have more heterochromatin marker di- and trimethylation of histone H3 lysine 9 (H3K9me2/3) and HP1 protein (Machida et al. 2018). Inner nuclear membrane proteins (Pawar and Kutay 2021) such as lamins (Reddy 2021), together with HP1, may constrain chromatin movement (Fig. 4F; Chubb et al. 2002; Shinkai et al. 2016; Nozaki et al. 2017; Nagashima et al. 2019; Lerner et al. 2020). The pericentric heterochromatin regions in mouse cells (Maison et al. 2010) are also enriched with the H3K9me2/3 and HP1 protein, which can cross-link nucleosomes (Machida et al. 2018) and restrict chromatin motion (Fig. 4F; Nozaki et al. 2017; Lerner et al. 2020). Whereas HP1 was suggested to form liquid droplets by LLPS (Larson et al. 2017; Strom et al. 2017), Erdel et al. (2020) recently reported that LLPS may not be a prerequisite for the chromatin cross-link function of HP1 (for details, see Rippe 2021). Whether condensates/clusters/liquid droplets of other nuclear bodies can stabilize the genome chromatin for their functions remains unclear and an area to be investigated.

PERSPECTIVES

The physical nature of chromatin in the nucleus, discussed in this review, is functionally important for gene regulation and control of gene expression. Therefore, defects in the physical nature may be relevant for physiological and pathological processes (Misteli 2010; see Khan 2021). It is known that mutations and transcriptional misregulation of several "global genome organizers," such as the cohesin complex (Misteli 2010; Kline et al. 2018), are linked to human diseases. Interestingly, some mutations in core histones act as oncogenic drivers, creating the term "oncohistones" (Bennett et al. 2019; Nacev et al. 2019). For example, several mutations in the histone variant H3.3 have been reported in pediatric gliomas (Schwartzentruber et al. 2012; Wu et al. 2012). Among these, the dominant-negative H3.3 K27M mutant, where the 27th Lys is replaced by Met, inhibits the activity of the Polycomb repressor complex 2 (PRC2) to methylate H3K27, resulting in impaired transcriptional silencing (Chan et al. 2013; Lewis et al. 2013). H2B Glu76 and H3.1 Glu97 mutations are also frequently found in cancer cells, and replacement of these residues with Lys leads to nucleosome instability (Arimura et al. 2018). Furthermore, mutations in the genes that encode several linker histone H1 isoforms are highly recurrent in B-cell lymphomas (Yusufova et al. 2021). Whereas how these mutations enhance tumorigenesis remains unclear, their resultant altered behaviors may be involved in the process. Such alterations might affect transcription, DNA repair efficiency, or plasticity of chromatin in the nucleus (Shimamoto et al.

2017; Stephens et al. 2017; Nagashima et al. 2019; also see Miroshnikova and Wickström 2021). Discovering how altering the physical properties of chromatin contributes to creating cell abnormalities and subsequent human disorders, including tumorigenesis, is an exciting area for future research.

ACKNOWLEDGMENTS

We are grateful to Dr. Y. Hiromi, Dr. Y. Itoh, and Dr. K.M. Marshall for critical reading and editing of this manuscript. We thank Dr. M. Sasai, Dr. Belmont, Dr. Rando, and Maeshima laboratory members for helpful discussions and support. We apologize that space limitations rendered us unable to mention many important works and papers on chromatin structure and dynamics. This work was supported by the Japan Society for the Promotion of Science (JSPS) and MEXT KAKENHI Grants (19H05273 and 20H05936), a Japan Science and Technology Agency CREST Grant (JPMJCR15G2), the Takeda Science Foundation, and the Uehara Memorial Foundation.

REFERENCES

*Reference is also in this collection.

Agbleke AA, Amitai A, Buenrostro JD, Chakrabarti A, Chu L, Hansen AS, Koenig KM, Labade AS, Liu S, Nozaki T, et al. 2020. Advances in chromatin and chromosome research: perspectives from multiple fields. *Mol Cell* **79:** 881–901. doi:10.1016/j.molcel.2020.07.003

Albiez H, Cremer M, Tiberi C, Vecchio L, Schermelleh L, Dittrich S, Küpper K, Joffe B, Thormeyer T, von Hase J, et al. 2006. Chromatin domains and the interchromatin compartment form structurally defined and functionally interacting nuclear networks. *Chromosome Res* **14:** 707–733. doi:10.1007/s10577-006-1086-x

Amitai A, Seeber A, Gasser SM, Holcman D. 2017. Visualization of chromatin decompaction and break site extrusion as predicted by statistical polymer modeling of single-locus trajectories. *Cell Rep* **18:** 1200–1214. doi:10.1016/j.celrep.2017.01.018

Andrews JO, Conway W, Cho WK, Narayanan A, Spille JH, Jayanth N, Inoue T, Mullen S, Thaler J, Cissé II. 2018. qSR: a quantitative super-resolution analysis tool reveals the cell-cycle dependent organization of RNA polymerase I in live human cells. *Sci Rep* **8:** 7424. doi:10.1038/s41598-018-25454-0

Arai R, Sugawara T, Sato Y, Minakuchi Y, Toyoda A, Nabeshima K, Kimura H, Kimura A. 2017. Reduction in chromosome mobility accompanies nuclear organization during early embryogenesis in *Caenorhabditis elegans*. *Sci Rep* **7:** 3631. doi:10.1038/s41598-017-03483-5

Arimura Y, Ikura M, Fujita R, Noda M, Kobayashi W, Horikoshi N, Sun J, Shi L, Kusakabe M, Harata M, et al. 2018. Cancer-associated mutations of histones H2B, H3.1 and H2A.Z.1 affect the structure and stability of the nucleosome. *Nucleic Acids Res* **46:** 10007–10018.

Asakura S, Oosawa F. 1954. On interaction between two bodies immersed in a solution of macromolecules. *J Chem Phys* **22:** 1255–1256. doi:10.1063/1.1740347

Ashwin SS, Nozaki T, Maeshima K, Sasai M. 2019. Organization of fast and slow chromatin revealed by single-nucleosome dynamics. *Proc Natl Acad Sci* **116:** 19939–19944. doi:10.1073/pnas.1907342116

Ashwin SS, Maeshima K, Sasai M. 2020. Heterogeneous fluid-like movements of chromatin and their implications to transcription. *Biophys Rev* **12:** 461–468. doi:10.1007/s12551-020-00675-8

Ausio J, Dong F, van Holde KE. 1989. Use of selectively trypsinized nucleosome core particles to analyze the role of the histone "tails" in the stabilization of the nucleosome. *J Mol Biol* **206:** 451–463. doi:10.1016/0022-2836(89)90493-2

Babokhov M, Hibino K, Itoh Y, Maeshima K. 2020. Local chromatin motion and transcription. *J Mol Biol* **432:** 694–700. doi:10.1016/j.jmb.2019.10.018

Bajpai G, Jain I, Inamdar MM, Das D, Padinhateeri R. 2017. Binding of DNA-bending non-histone proteins destabilizes regular 30-nm chromatin structure. *PLoS Comput Biol* **13:** e1005365. doi:10.1371/journal.pcbi.1005365

Baldi S, Korber P, Becker PB. 2020. Beads on a string-nucleosome array arrangements and folding of the chromatin fiber. *Nat Struct Mol Biol* **27:** 109–118. doi:10.1038/s41594-019-0368-x

Bednar J, Garcia-Saez I, Boopathi R, Cutter AR, Papai G, Reymer A, Syed SH, Lone IN, Tonchev O, Crucifix C, et al. 2017. Structure and dynamics of a 197 bp nucleosome in complex with linker histone H1. *Mol Cell* **66:** 729. doi:10.1016/j.molcel.2017.05.018

Bennett RL, Bele A, Small EC, Will CM, Nabet B, Oyer JA, Huang X, Ghosh RP, Grzybowski AT, Yu T, et al. 2019. A mutation in histone H2B represents a new class of oncogenic driver. *Cancer Discov* **9:** 1438–1451. doi:10.1158/2159-8290.CD-19-0393

Bian Q, Belmont AS. 2012. Revisiting higher-order and large-scale chromatin organization. *Curr Opin Cell Biol* **24:** 359–366. doi:10.1016/j.ceb.2012.03.003

Bianconi E, Piovesan A, Facchin F, Beraudi A, Casadei R, Frabetti F, Vitale L, Pelleri MC, Tassani S, Piva F, et al. 2013. An estimation of the number of cells in the human body. *Annals Hum Biol* **40:** 463–471. doi:10.3109/03014460.2013.807878

Bintu B, Mateo LJ, Su JH, Sinnott-Armstrong NA, Parker M, Kinrot S, Yamaya K, Boettiger AN, Zhuang X. 2018. Super-resolution chromatin tracing reveals domains and cooperative interactions in single cells. *Science* **362:** eaau1783. doi:10.1126/science.aau1783

Cai S, Chen C, Tan ZY, Huang Y, Shi J, Gan L. 2018. Cryo-ET reveals the macromolecular reorganization of *S. pombe* mitotic chromosomes in vivo. *Proc Natl Acad Sci* **115:** 10977–10982. doi:10.1073/pnas.1720476115

Chan KM, Fang D, Gan H, Hashizume R, Yu C, Schroeder M, Gupta N, Mueller S, James CD, Jenkins R, et al. 2013. The histone H3.3K27M mutation in pediatric glioma reprograms H3K27 methylation and gene expression. *Genes Dev* **27:** 985–990. doi:10.1101/gad.217778.113

Chen B, Gilbert LA, Cimini BA, Schnitzbauer J, Zhang W, Li GW, Park J, Blackburn EH, Weissman JS, Qi LS, et al. 2013. Dynamic imaging of genomic loci in living human cells by an optimized CRISPR/Cas system. *Cell* **155:** 1479–1491. doi:10.1016/j.cell.2013.12.001

Chen C, Lim HH, Shi J, Tamura S, Maeshima K, Surana U, Gan L. 2016. Budding yeast chromatin is dispersed in a crowded nucleoplasm in vivo. *Mol Biol Cell* **27:** 3357–3368. doi:10.1091/mbc.E16-07-0506

Chen H, Levo M, Barinov L, Fujioka M, Jaynes JB, Gregor T. 2018. Dynamic interplay between enhancer-promoter topology and gene activity. *Nat Genet* **50:** 1296–1303. doi:10.1038/s41588-018-0175-z

Cho WK, Spille JH, Hecht M, Lee C, Li C, Grube V, Cisse II. 2018. Mediator and RNA polymerase II clusters associate in transcription-dependent condensates. *Science* **361:** 412–415. doi:10.1126/science.aar4199

Chong S, Dugast-Darzacq C, Liu Z, Dong P, Dailey GM, Cattoglio C, Heckert A, Banala S, Lavis L, Darzacq X, et al. 2018. Imaging dynamic and selective low-complexity domain interactions that control gene transcription. *Science* **361:** eaar2555. doi:10.1126/science.aar2555

Chubb JR, Boyle S, Perry P, Bickmore WA. 2002. Chromatin motion is constrained by association with nuclear compartments in human cells. *Curr Biol* **12:** 439–445. doi:10.1016/S0960-9822(02)00695-4

Church GM, Gao Y, Kosuri S. 2012. Next-generation digital information storage in DNA. *Science* **337:** 1628. doi:10.1126/science.1226355

Cole A. 1967. Chromosome structure. *Theoret Biophys* **1:** 305–375.

Collepardo-Guevara R, Schlick T. 2014. Chromatin fiber polymorphism triggered by variations of DNA linker lengths. *Proc Natl Acad Sci* **111:** 8061–8066. doi:10.1073/pnas.1315872111

Cook PR. 1999. The organization of replication and transcription. *Science* **284:** 1790–1795. doi:10.1126/science.284.5421.1790

Cremer T, Cremer C. 2001. Chromosome territories, nuclear architecture and gene regulation in mammalian cells. *Nat Rev Genet* **2:** 292–301. doi:10.1038/35066075

Davey CA, Sargent DF, Luger K, Maeder AW, Richmond TJ. 2002. Solvent mediated interactions in the structure of the nucleosome core particle at 1.9Å resolution. *J Mol Biol* **319:** 1097–1113. doi:10.1016/S0022-2836(02)00386-8

Dekker J. 2008. Mapping in vivo chromatin interactions in yeast suggests an extended chromatin fiber with regional variation in compaction. *J Biol Chem* **283:** 34532–34540. doi:10.1074/jbc.M806479200

* Dekker J. 2021. 3D chromatin organization. *Cold Spring Harb Perspect Biol* doi:10.1101/cshperspect.a040147

Dekker J, Heard E. 2015. Structural and functional diversity of topologically associating domains. *FEBS Lett* **589:** 2877–2884. doi:10.1016/j.febslet.2015.08.044

Dion V, Gasser SM. 2013. Chromatin movement in the maintenance of genome stability. *Cell* **152:** 1355–1364. doi:10.1016/j.cell.2013.02.010

Dion V, Kalck V, Seeber A, Schleker T, Gasser SM. 2013. Cohesin and the nucleolus constrain the mobility of spontaneous repair foci. *EMBO Rep* **14:** 984–991. doi:10.1038/embor.2013.142

Dixon JR, Selvaraj S, Yue F, Kim A, Li Y, Shen Y, Hu M, Liu JS, Ren B. 2012. Topological domains in mammalian genomes identified by analysis of chromatin interactions. *Nature* **485:** 376–380. doi:10.1038/nature11082

Dorigo B, Schalch T, Bystricky K, Richmond TJ. 2003. Chromatin fiber folding: requirement for the histone H4 N-terminal tail. *J Mol Biol* **327:** 85–96. doi:10.1016/S0022-2836(03)00025-1

Dubochet J, Adrian M, Schultz P, Oudet P. 1986. Cryo-electron microscopy of vitrified SV40 minichromosomes: the liquid drop model. *EMBO J* **5:** 519–528. doi:10.1002/j.1460-2075.1986.tb04241.x

Earnshaw WC, Laemmli UK. 1983. Architecture of metaphase chromosomes and chromosome scaffolds. *J Cell Biol* **96:** 84–93. doi:10.1083/jcb.96.1.84

Edelman LB, Fraser P. 2012. Transcription factories: genetic programming in three dimensions. *Curr Opin Genet Dev* **22:** 110–114. doi:10.1016/j.gde.2012.01.010

Eltsov M, Maclellan KM, Maeshima K, Frangakis AS, Dubochet J. 2008. Analysis of cryo-electron microscopy images does not support the existence of 30-nm chromatin fibers in mitotic chromosomes in situ. *Proc Natl Acad Sci* **105:** 19732–19737. doi:10.1073/pnas.0810057105

Erdel F, Rademacher A, Vlijm R, Tunnermann J, Frank L, Weinmann R, Schweigert E, Yserentant K, Hummert J, Bauer C, et al. 2020. Mouse heterochromatin adopts digital compaction states without showing hallmarks of HP1-driven liquid-liquid phase separation. *Mol Cell* **78:** 1236–249.e7.

Feuerborn A, Cook PR. 2015. Why the activity of a gene depends on its neighbors. *Trends Genet* **31:** 483–490. doi:10.1016/j.tig.2015.07.001

Finch JT, Klug A. 1976. Solenoidal model for superstructure in chromatin. *Proc Natl Acad Sci* **73:** 1897–1901. doi:10.1073/pnas.73.6.1897

Fussner E, Strauss M, Djuric U, Li R, Ahmed K, Hart M, Ellis J, Bazett-Jones DP. 2012. Open and closed domains in the mouse genome are configured as 10-nm chromatin fibres. *EMBO Rep* **13:** 992–996. doi:10.1038/embor.2012.139

Germier T, Kocanova S, Walther N, Bancaud A, Shaban HA, Sellou H, Politi AZ, Ellenberg J, Gallardo F, Bystricky K. 2017. Real-time imaging of a single gene reveals transcription-initiated local confinement. *Biophys J* **113:** 1383–1394. doi:10.1016/j.bpj.2017.08.014

Gibson BA, Doolittle LK, Schneider MWG, Jensen LE, Gamarra N, Henry L, Gerlich DW, Redding S, Rosen MK. 2019. Organization of chromatin by intrinsic and regulated phase separation. *Cell* **179:** 470–484.e21. doi:10.1016/j.cell.2019.08.037

Gómez-García PA, Portillo-Ledesma S, Neguembor MV, Pesaresi M, Oweis W, Rohrlich T, Wieser S, Meshorer E, Schlick T, Cosma MP, et al. 2021. Mesoscale modeling and single-nucleosome tracking reveal remodeling of clutch folding and dynamics in stem cell differentiation. *Cell Rep* **34:** 108614. doi:10.1016/j.celrep.2020.108614

Gorisch SM, Wachsmuth M, Toth KF, Lichter P, Rippe K. 2005. Histone acetylation increases chromatin accessibility. *J Cell Sci* **118:** 5825–5834. doi:10.1242/jcs.02689

Gu B, Swigut T, Spencley A, Bauer MR, Chung M, Meyer T, Wysocka J. 2018. Transcription-coupled changes in nuclear mobility of mammalian *cis*-regulatory elements. *Science* **359:** 1050–1055. doi:10.1126/science.aao3136

Guo YE, Manteiga JC, Henninger JE, Sabari BR, Dall'Agnese A, Hannett NM, Spille JH, Afeyan LK, Zamudio AV, Shrinivas K, et al. 2019. Pol II phosphorylation regulates a switch between transcriptional and splicing condensates. *Nature* **572:** 543–548. doi:10.1038/s41586-019-1464-0

Hajjoul H, Mathon J, Ranchon H, Goiffon I, Mozziconacci J, Albert B, Carrivain P, Victor JM, Gadal O, Bystricky K, et al. 2013. High-throughput chromatin motion tracking in living yeast reveals the flexibility of the fiber throughout the genome. *Genome Res* **23:** 1829–1838. doi:10.1101/gr.157008.113

Hancock R. 2007. Packing of the polynucleosome chain in interphase chromosomes: evidence for a contribution of crowding and entropic forces. *Semin Cell Dev Biol* **18:** 668–675. doi:10.1016/j.semcdb.2007.08.006

Hansen JC. 2002. Conformational dynamics of the chromatin fiber in solution: determinants, mechanisms, and functions. *Annu Rev Biophys Biomol Struct* **31:** 361–392. doi:10.1146/annurev.biophys.31.101101.140858

Hansen JC, Connolly M, McDonald CJ, Pan A, Pryamkova A, Ray K, Seidel E, Tamura S, Rogge R, Maeshima K. 2018. The 10-nm chromatin fiber and its relationship to interphase chromosome organization. *Biochem Soc Trans* **46:** 67–76. doi:10.1042/BST20170101

Hendzel MJ, Boisvert F, Bazett-Jones DP. 1999. Direct visualization of a protein nuclear architecture. *Mol Biol Cell* **10:** 2051–2062. doi:10.1091/mbc.10.6.2051

Heun P, Laroche T, Shimada K, Furrer P, Gasser SM. 2001. Chromosome dynamics in the yeast interphase nucleus. *Science* **294:** 2181–2186. doi:10.1126/science.1065366

Hihara S, Pack CG, Kaizu K, Tani T, Hanafusa T, Nozaki T, Takemoto S, Yoshimi T, Yokota H, Imamoto N, et al. 2012. Local nucleosome dynamics facilitate chromatin accessibility in living mammalian cells. *Cell Rep* **2:** 1645–1656. doi:10.1016/j.celrep.2012.11.008

Hoffman DP, Shtengel G, Xu CS, Campbell KR, Freeman M, Wang L, Milkie DE, Pasolli HA, Iyer N, Bogovic JA, et al. 2020. Correlative three-dimensional super-resolution and block-face electron microscopy of whole vitreously frozen cells. *Science* **367:** eaaz5357. doi:10.1126/science.aaz5357

Hsieh TH, Weiner A, Lajoie B, Dekker J, Friedman N, Rando OJ. 2015. Mapping nucleosome resolution chromosome folding in yeast by micro-C. *Cell* **162:** 108–119. doi:10.1016/j.cell.2015.05.048

Hu Y, Kireev I, Plutz M, Ashourian N, Belmont AS. 2009. Large-scale chromatin structure of inducible genes: transcription on a condensed, linear template. *J Cell Biol* **185:** 87–100. doi:10.1083/jcb.200809196

Hudson DF, Vagnarelli P, Gassmann R, Earnshaw WC. 2003. Condensin is required for nonhistone protein assembly and structural integrity of vertebrate mitotic chromosomes. *Dev Cell* **5:** 323–336. doi:10.1016/S1534-5807(03)00199-0

Hyman AA, Weber CA, Jülicher F. 2014. Liquid–liquid phase separation in biology. *Ann Rev Cell Dev Biol* **30:** 39–58. doi:10.1146/annurev-cellbio-100913-013325

Ide S, Imai R, Ochi H, Maeshima K. 2020. Transcriptional suppression of ribosomal DNA with phase separation. *Sci Adv* **6:** eabb5953. doi:10.1126/sciadv.abb5953

Imai R, Nozaki T, Tani T, Kaizu K, Hibino K, Ide S, Tamura S, Takahashi K, Shribak M, Maeshima K. 2017. Density imaging of heterochromatin in live cells using orientation-independent-DIC microscopy. *Mol Biol Cell* **28:** 3349–3359. doi:10.1091/mbc.e17-06-0359

Itoh Y, Iida S, Tamura S, Nagashima R, Shiraki K, Goto T, Hibino K, Ide S, Maeshima K. 2021. 1,6-hexanediol rapidly immobilizes and condenses chromatin in living human cells. *Life Sci Alliance* **4:** e202001005. doi:10.26508/lsa.202001005

Joti Y, Hikima T, Nishino Y, Kamada F, Hihara S, Takata H, Ishikawa T, Maeshima K. 2012. Chromosomes without a 30-nm chromatin fiber. *Nucleus* **3:** 404–410. doi:10.4161/nucl.21222

Kan PY, Caterino TL, Hayes JJ. 2009. The H4 tail domain participates in intra- and internucleosome interactions with protein and DNA during folding and oligomerization of nucleosome arrays. *Mol Cell Biol* **29:** 538–546. doi:10.1128/MCB.01343-08

* Khan J. 2021. Chromatin in cancer. *Cold Spring Harb Perspect Biol* doi:10.1101/cshperspect.a040956

Kieffer-Kwon KR, Nimura K, Rao SSP, Xu J, Jung S, Pekowska A, Dose M, Stevens E, Mathe E, Dong P, et al. 2017. Myc regulates chromatin decompaction and nuclear architecture during B cell activation. *Mol Cell* **67:** 566–578.e10. doi:10.1016/j.molcel.2017.07.013

Kimura H, Cook PR. 2001. Kinetics of core histones in living human cells: little exchange of H3 and H4 and some rapid exchange of H2B. *J Cell Biol* **153:** 1341–1354. doi:10.1083/jcb.153.7.1341

Kline AD, Moss JF, Selicorni A, Bisgaard AM, Deardorff MA, Gillett PM, Ishman SL, Kerr LM, Levin AV, Mulder PA, et al. 2018. Diagnosis and management of Cornelia de Lange syndrome: first international consensus statement. *Nat Rev Genet* **19:** 649–666. doi:10.1038/s41576-018-0031-0

Kornberg RD. 1974. Chromatin structure: a repeating unit of histones and DNA. *Science* **184:** 868–871. doi:10.1126/science.184.4139.868

Krietenstein N, Rando OJ. 2020. Mesoscale organization of the chromatin fiber. *Curr Opin Genet Dev* **61:** 32–36. doi:10.1016/j.gde.2020.02.022

Larson AG, Elnatan D, Keenen MM, Trnka MJ, Johnston JB, Burlingame AL, Agard DA, Redding S, Narlikar GJ. 2017. Liquid droplet formation by HP1α suggests a role for phase separation in heterochromatin. *Nature* **547:** 236–240. doi:10.1038/nature22822

Lerner J, Gomez-Garcia PA, McCarthy RL, Liu Z, Lakadamyali M, Zaret KS. 2020. Two-parameter mobility assessments discriminate diverse regulatory factor behaviors in chromatin. *Mol Cell* **79:** 677–688.e6. doi:10.1016/j.molcel.2020.05.036

Levi V, Ruan Q, Plutz M, Belmont AS, Gratton E. 2005. Chromatin dynamics in interphase cells revealed by tracking in a two-photon excitation microscope. *Biophys J* **89:** 4275–4285. doi:10.1529/biophysj.105.066670

Lewis PW, Müller MM, Koletsky MS, Cordero F, Lin S, Banaszynski LA, Garcia BA, Muir TW, Becher OJ, Allis CD. 2013. Inhibition of PRC2 activity by a gain-of-function H3 mutation found in pediatric glioblastoma. *Science* **340:** 857–861. doi:10.1126/science.1232245

Lieberman-Aiden E, van Berkum NL, Williams L, Imakaev M, Ragoczy T, Telling A, Amit I, Lajoie BR, Sabo PJ, Dorschner MO, et al. 2009. Comprehensive mapping of long-range interactions reveals folding principles of the human genome. *Science* **326:** 289–293. doi:10.1126/science.1181369

Lin Y, Protter DS, Rosen MK, Parker R. 2015. Formation and maturation of phase-separated liquid droplets by RNA-binding proteins. *Mol Cell* **60:** 208–219. doi:10.1016/j.molcel.2015.08.018

Lu H, Yu D, Hansen AS, Ganguly S, Liu R, Heckert A, Darzacq X, Zhou Q. 2018. Phase-separation mechanism for C-terminal hyperphosphorylation of RNA polymerase II. *Nature* **558:** 318–323. doi:10.1038/s41586-018-0174-3

Luger K, Mäder AW, Richmond RK, Sargent DF, Richmond TJ. 1997. Crystal structure of the nucleosome core particle at 2.8 Å resolution. *Nature* **389:** 251–260. doi:10.1038/38444

Ma H, Tu LC, Chung YC, Naseri A, Grunwald D, Zhang S, Pederson T. 2019. Cell cycle- and genomic distance-dependent dynamics of a discrete chromosomal region. *J Cell Biol* **218:** 1467–1477. doi:10.1083/jcb.201807162

Machida S, Takizawa Y, Ishimaru M, Sugita Y, Sekine S, Nakayama JI, Wolf M, Kurumizaka H. 2018. Structural basis of heterochromatin formation by human HP1. *Mol Cell* **69:** 385–397.e8. doi:10.1016/j.molcel.2017.12.011

Maeshima K, Hihara S, Eltsov M. 2010. Chromatin structure: does the 30-nm fibre exist in vivo? *Curr Opin Cell Biol* **22:** 291–297. doi:10.1016/j.ceb.2010.03.001

Maeshima K, Imai R, Tamura S, Nozaki T. 2014. Chromatin as dynamic 10-nm fibers. *Chromosoma* **123:** 225–237. doi:10.1007/s00412-014-0460-2

Maeshima K, Kaizu K, Tamura S, Nozaki T, Kokubo T, Takahashi K. 2015. The physical size of transcription factors is key to transcriptional regulation in the chromatin domains. *J Phys* **27:** 064116.

Maeshima K, Ide S, Hibino K, Sasai M. 2016a. Liquid-like behavior of chromatin. *Curr Opin Genet Dev* **37:** 36–45. doi:10.1016/j.gde.2015.11.006

Maeshima K, Rogge R, Tamura S, Joti Y, Hikima T, Szerlong H, Krause C, Herman J, Seidel E, DeLuca J, et al. 2016b. Nucleosomal arrays self-assemble into supramolecular globular structures lacking 30-nm fibers. *EMBO J* **35:** 1115–1132. doi:10.15252/embj.201592660

Maeshima K, Matsuda T, Shindo Y, Imamura H, Tamura S, Imai R, Kawakami S, Nagashima R, Soga T, Noji H, et al. 2018a. A transient rise in free Mg^{2+} ions released from ATP-Mg hydrolysis contributes to mitotic chromosome condensation. *Curr Biol* **28:** 444–451.e6. doi:10.1016/j.cub.2017.12.035

Maeshima K, Tamura S, Shimamoto Y. 2018b. Chromatin as a nuclear spring. *Biophys Physicobiol* **15:** 189–195. doi:10.2142/biophysico.15.0_189

Maeshima K, Tamura S, Hansen JC, Itoh Y. 2020. Fluid-like chromatin: toward understanding the real chromatin organization present in the cell. *Curr Opin Cell Biol* **64:** 77–89. doi:10.1016/j.ceb.2020.02.016

Maison C, Quivy JP, Probst AV, Almouzni G. 2010. Heterochromatin at mouse pericentromeres: a model for de novo heterochromatin formation and duplication during replication. *Cold Spring Harb Symp Quant Biol* **75:** 155–165. doi:10.1101/sqb.2010.75.013

Marenduzzo D, Finan K, Cook PR. 2006. The depletion attraction: an underappreciated force driving cellular organization. *J Cell Biol* **175:** 681–686. doi:10.1083/jcb.200609066

Marshall WF, Straight A, Marko JF, Swedlow J, Dernburg A, Belmont A, Murray AW, Agard DA, Sedat JW. 1997. Interphase chromosomes undergo constrained diffusional motion in living cells. *Curr Biol* **7:** 930–939. doi:10.1016/S0960-9822(06)00412-X

McDowall AW, Smith JM, Dubochet J. 1986. Cryo-electron microscopy of vitrified chromosomes in situ. *EMBO J* **5:** 1395–1402. doi:10.1002/j.1460-2075.1986.tb04373.x

Meister P, Towbin BD, Pike BL, Ponti A, Gasser SM. 2010. The spatial dynamics of tissue-specific promoters during *C. elegans* development. *Genes Dev* **24:** 766–782. doi:10.1101/gad.559610

Meshorer E, Yellajoshula D, George E, Scambler PJ, Brown DT, Misteli T. 2006. Hyperdynamic plasticity of chromatin proteins in pluripotent embryonic stem cells. *Dev Cell* **10:** 105–116. doi:10.1016/j.devcel.2005.10.017

Miron E, Oldenkamp R, Brown JM, Pinto DMS, Xu CS, Faria AR, Shaban HA, Rhodes JDP, Innocent C, de Ornellas S, et al. 2020. Chromatin arranges in chains of mesoscale domains with nanoscale functional topography independent of cohesin. *Sci Adv* **6:** eaba8811. doi:10.1126/sciadv.aba8811

* Miroshnikova YA, Wickström SA. 2021. Mechanical forces in nuclear organization. *Cold Spring Harb Perspect Biol* doi:10.1101/cshperspect.a039685

Misteli T. 2010. Higher-order genome organization in human disease. *Cold Spring Harb Perspect Biol* **2:** a000794. doi:10.1101/cshperspect.a000794

Misteli T. 2020. The self-organizing genome: principles of genome architecture and function. *Cell* **183:** 28–45. doi:10.1016/j.cell.2020.09.014

Misteli T, Gunjan A, Hock R, Bustin M, Brown DT. 2000. Dynamic binding of histone H1 to chromatin in living cells. *Nature* **408:** 877–881. doi:10.1038/35048610

Mitrea DM, Kriwacki RW. 2016. Phase separation in biology; functional organization of a higher order. *Cell Commun Signal* **14:** 1. doi:10.1186/s12964-015-0125-7

Morales C, Losada A. 2018. Establishing and dissolving cohesion during the vertebrate cell cycle. *Curr Opin Cell Biol* **52:** 51–57. doi:10.1016/j.ceb.2018.01.010

Nacev BA, Feng L, Bagert JD, Lemiesz AE, Gao J, Soshnev AA, Kundra R, Schultz N, Muir TW, Allis CD. 2019. The expanding landscape of "oncohistone" mutations in human cancers. *Nature* **567:** 473–478. doi:10.1038/s41586-019-1038-1

Nagashima R, Hibino K, Ashwin SS, Babokhov M, Fujishiro S, Imai R, Nozaki T, Tamura S, Tani T, Kimura H, et al. 2019. Single nucleosome imaging reveals loose genome chromatin networks via active RNA polymerase II. *J Cell Biol* **218:** 1511–1530. doi:10.1083/jcb.201811090

Nasmyth K, Haering CH. 2005. The structure and function of SMC and kleisin complexes. *Annu Rev Biochem* **74:** 595–648. doi:10.1146/annurev.biochem.74.082803.133219

Nishino Y, Eltsov M, Joti Y, Ito K, Takata H, Takahashi Y, Hihara S, Frangakis AS, Imamoto N, Ishikawa T, et al. 2012. Human mitotic chromosomes consist predominantly of irregularly folded nucleosome fibres without a 30-nm chromatin structure. *EMBO J* **31:** 1644–1653. doi:10.1038/emboj.2012.35

Nishiyama T. 2019. Cohesion and cohesin-dependent chromatin organization. *Curr Opin Cell Biol* **58:** 8–14. doi:10.1016/j.ceb.2018.11.006

Nora EP, Lajoie BR, Schulz EG, Giorgetti L, Okamoto I, Servant N, Piolot T, van Berlum NL, Meisig J, Sedat JW, et al. 2012. Spatial partitioning of the regulatory landscape of the X-inactivation centre. *Nature* **485:** 381–385. doi:10.1038/nature11049

Nozaki T, Imai R, Tanbo M, Nagashima R, Tamura S, Tani T, Joti Y, Tomita M, Hibino K, Kanemaki MT, et al. 2017. Dynamic organization of chromatin domains revealed by super-resolution live-cell imaging. *Mol Cell* **67:** 282–293.e7. doi:10.1016/j.molcel.2017.06.018

Nozaki T, Hudson DF, Tamura S, Maeshima K. 2018. Dynamic chromatin folding in the cell. In *Nuclear architecture and dynamics* (ed. C Lavelle, JM Victor), pp. 101–122. Academic, New York.

Nozawa RS, Gilbert N. 2019. RNA: nuclear glue for folding the genome. *Trends Cell Biol* **29:** 201–211. doi:10.1016/j.tcb.2018.12.003

Nozawa RS, Boteva L, Soares DC, Naughton C, Dun AR, Buckle A, Ramsahoye B, Bruton PC, Saleeb RS, Arnedo M, et al. 2017. SAF-A regulates interphase chromosome structure through oligomerization with chromatin-associated RNAs. *Cell* **169:** 1214–1227.e18. doi:10.1016/j.cell.2017.05.029

Ohno M, Ando T, Priest DG, Kumar V, Yoshida Y, Taniguchi Y. 2019. Sub-nucleosomal genome structure reveals distinct nucleosome folding motifs. *Cell* **176:** 520–534.e25. doi:10.1016/j.cell.2018.12.014

Olins AL, Olins DE. 1974. Spheroid chromatin units (v bodies). *Science* **183:** 330–332. doi:10.1126/science.183.4122.330

Olins DE, Olins AL. 2003. Chromatin history: our view from the bridge. *Nat Rev Mol Cell Biol* **4:** 809–814. doi:10.1038/nrm1225

Olins DE, Olins AL. 2018. Epichromatin and chromomeres: a "fuzzy" perspective. *Open Biol* **8:** 180058. doi:10.1098/rsob.180058

Ou HD, Phan S, Deerinck TJ, Thor A, Ellisman MH, O'Shea CC. 2017. ChromEMT: visualizing 3D chromatin structure and compaction in interphase and mitotic cells. *Science* **357:** eaag0025. doi:10.1126/science.aag0025

* Pawar S, Kutay U. 2021. The diverse cellular functions of inner nuclear membrane proteins. *Cold Spring Harb Perspect Biol* doi:10.1101/cshperspect.a040477.

Pecqueur C, Oliver L, Oizel K, Lalier L, Vallette FM. 2013. Targeting metabolism to induce cell death in cancer cells and cancer stem cells. *Int J Cell Biol* **2013:** 805975. doi:10.1155/2013/805975

Pepenella S, Murphy KJ, Hayes JJ. 2014a. Intra- and inter-nucleosome interactions of the core histone tail domains in higher-order chromatin structure. *Chromosoma* **123:** 3–13. doi:10.1007/s00412-013-0435-8

Pepenella S, Murphy KJ, Hayes JJ. 2014b. A distinct switch in interactions of the histone H4 tail domain upon salt-dependent folding of nucleosome arrays. *J Biol Chem* **289:** 27342–27351. doi:10.1074/jbc.M114.595140

Qian Y, Wang X, Liu Y, Li Y, Colvin RA, Tong L, Wu S, Chen X. 2014. Extracellular ATP is internalized by macropinocytosis and induces intracellular ATP increase and drug resistance in cancer cells. *Cancer Lett* **351:** 242–251. doi:10.1016/j.canlet.2014.06.008

Rao SS, Huntley MH, Durand NC, Stamenova EK, Bochkov ID, Robinson JT, Sanborn AL, Machol I, Omer AD, Lander ES, et al. 2014. A 3D map of the human genome at kilobase resolution reveals principles of chromatin looping. *Cell* **159:** 1665–1680. doi:10.1016/j.cell.2014.11.021

Rao SSP, Huang SC, Glenn St Hilaire B, Engreitz JM, Perez EM, Kieffer-Kwon KR, Sanborn AL, Johnstone SE, Bascom GD, Bochkov ID, et al. 2017. Cohesin loss eliminates all loop domains. *Cell* **171:** 305–320.e24. doi:10.1016/j.cell.2017.09.026

* Reddy K. 2021. Nuclear lamina. *Cold Spring Harb Perspect Biol* doi:10.1101/cshperspect.a040113.

Ricci MA, Manzo C, García-Parajo MF, Lakadamyali M, Cosma MP. 2015. Chromatin fibers are formed by heterogeneous groups of nucleosomes in vivo. *Cell* **160:** 1145–1158. doi:10.1016/j.cell.2015.01.054

* Rippe K. 2021. Phase separation in chromatin. *Cold Spring Harb Perspect Biol* doi:10.1101/cshperspect.a040683

Risca VI, Denny SK, Straight AF, Greenleaf WJ. 2017. Variable chromatin structure revealed by in situ spatially correlated DNA cleavage mapping. *Nature* **541:** 237–241. doi:10.1038/nature20781

Robinett CC, Straight A, Li G, Willhelm C, Sudlow G, Murray A, Belmont AS. 1996. In vivo localization of DNA sequences and visualization of large-scale chromatin organization using lac operator/repressor recognition. *J Cell Biol* **135:** 1685–1700. doi:10.1083/jcb.135.6.1685

Robinson PJ, Fairall L, Huynh VA, Rhodes D. 2006. EM measurements define the dimensions of the "30-nm" chromatin fiber: evidence for a compact, interdigitated structure. *Proc Natl Acad Sci* **103:** 6506–6511. doi:10.1073/pnas.0601212103

Sabari BR, Dall'Agnese A, Boija A, Klein IA, Coffey EL, Shrinivas K, Abraham BJ, Hannett NM, Zamudio AV, Manteiga JC, et al. 2018. Coactivator condensation at super-enhancers links phase separation and gene control. *Science* **361:** eaar3958. doi:10.1126/science.aar3958

Sanborn AL, Rao SS, Huang SC, Durand NC, Huntley MH, Jewett AI, Bochkov ID, Chinnappan D, Cutkosky A, Li J, et al. 2015. Chromatin extrusion explains key features of loop and domain formation in wild-type and engineered genomes. *Proc Natl Acad Sci* **112:** E6456–E6465. doi:10.1073/pnas.1518552112

Sanulli S, Trnka MJ, Dharmarajan V, Tibble RW, Pascal BD, Burlingame AL, Griffin PR, Gross JD, Narlikar GJ. 2019. HP1 reshapes nucleosome core to promote phase separation of heterochromatin. *Nature* **575:** 390–394. doi:10.1038/s41586-019-1669-2

Schalch T, Duda S, Sargent DF, Richmond TJ. 2005. X-ray structure of a tetranucleosome and its implications for the chromatin fibre. *Nature* **436:** 138–141. doi:10.1038/nature03686

Schwartzentruber J, Korshunov A, Liu XY, Jones DT, Pfaff E, Jacob K, Sturm D, Fontebasso AM, Quang DA, Tönjes M, et al. 2012. Driver mutations in histone H3.3 and chromatin remodelling genes in paediatric glioblastoma. *Nature* **482:** 226–231. doi:10.1038/nature10833

Seeber A, Hauer MH, Gasser SM. 2018. Chromosome dynamics in response to DNA damage. *Annu Rev Genet* **52:** 295–319. doi:10.1146/annurev-genet-120417-031334

Sexton T, Yaffe E, Kenigsberg E, Bantignies F, Leblanc B, Hoichman M, Parrinello H, Tanay A, Cavalli G. 2012. Three-dimensional folding and functional organization principles of the *Drosophila* genome. *Cell* **148:** 458–472. doi:10.1016/j.cell.2012.01.010

Shaban HA, Barth R, Recoules L, Bystricky K. 2020. Hi-D: nanoscale mapping of nuclear dynamics in single living cells. *Genome Biol* **21:** 95. doi:10.1186/s13059-020-02002-6

Shimamoto Y, Tamura S, Masumoto H, Maeshima K. 2017. Nucleosome–nucleosome interactions via histone tails and linker DNA regulate nuclear rigidity. *Mol Biol Cell* **28:** 1580–1589. doi:10.1091/mbc.e16-11-0783

Shinkai S, Nozaki T, Maeshima K, Togashi Y. 2016. Dynamic nucleosome movement provides structural information of topological chromatin domains in living human cells. *PLoS Comput Biol* **12:** e1005136. doi:10.1371/journal.pcbi.1005136

Sinha D, Shogren-Knaak MA. 2010. Role of direct interactions between the histone H4 tail and the H2A core in long range nucleosome contacts. *J Biol Chem* **285:** 16572–16581. doi:10.1074/jbc.M109.091298

Song F, Chen P, Sun D, Wang M, Dong L, Liang D, Xu RM, Zhu P, Li G. 2014. Cryo-EM study of the chromatin fiber reveals a double helix twisted by tetranucleosomal units. *Science* **344:** 376–380. doi:10.1126/science.1251413

Soutoglou E, Misteli T. 2007. Mobility and immobility of chromatin in transcription and genome stability. *Curr Opin Genet Dev* **17:** 435–442. doi:10.1016/j.gde.2007.08.004

Stephens AD, Banigan EJ, Adam SA, Goldman RD, Marko JF. 2017. Chromatin and lamin A determine two different mechanical response regimes of the cell nucleus. *Mol Biol Cell.*

Strickfaden H, Tolsma TO, Sharma A, Underhill DA, Hansen JC, Hendzel MJ. 2020. Condensed chromatin behaves like a solid on the mesoscale in vitro and in living cells. *Cell* **183:** 1772–1784.e13. doi:10.1016/j.cell.2020.11.027

Strom AR, Emelyanov AV, Mir M, Fyodorov DV, Darzacq X, Karpen GH. 2017. Phase separation drives heterochromatin domain formation. *Nature* **547:** 241–245. doi:10.1038/nature22989

Takata H, Hanafusa T, Mori T, Shimura M, Iida Y, Ishikawa K, Yoshikawa K, Yoshikawa Y, Maeshima K. 2013. Chromatin compaction protects genomic DNA from radiation damage. *PLoS ONE* **8:** e75622. doi:10.1371/journal.pone.0075622

Tasan I, Sustackova G, Zhang L, Kim J, Sivaguru M, HamediRad M, Wang Y, Genova J, Ma J, Belmont AS, et al. 2018. CRISPR/Cas9-mediated knock-in of an optimized TetO repeat for live cell imaging of endogenous loci. *Nucleic Acids Res* **46:** e100. doi:10.1093/nar/gky501

Toth KF, Knoch TA, Wachsmuth M, Frank-Stohr M, Stohr M, Bacher CP, Muller G, Rippe K. 2004. Trichostatin A-induced histone acetylation causes decondensation of interphase chromatin. *J Cell Sci* **117:** 4277–4287. doi:10.1242/jcs.01293

Turner AL, Watson M, Wilkins OG, Cato L, Travers A, Thomas JO, Stott K. 2018. Highly disordered histone H1-DNA model complexes and their condensates. *Proc Natl Acad Sci* **115:** 11964–11969. doi:10.1073/pnas.1805943115

van Steensel B, Belmont AS. 2017. Lamina-associated domains: links with chromosome architecture, heterochromatin, and gene repression. *Cell* **169:** 780–791. doi:10.1016/j.cell.2017.04.022

Visvanathan A, Ahmed K, Even-Faitelson L, Lleres D, Bazett-Jones DP, Lamond AI. 2013. Modulation of higher order chromatin conformation in mammalian cell nuclei can be mediated by polyamines and divalent cations. *PLoS ONE* **8:** e67689. doi:10.1371/journal.pone.0067689

Wolffe AP, Hayes JJ. 1999. Chromatin disruption and modification. *Nucleic Acids Res* **27:** 711–720. doi:10.1093/nar/27.3.711

Woodcock CL, Safer JP, Stanchfield JE. 1976. Structural repeating units in chromatin. I: Evidence for their general occurrence. *Exp Cell Res* **97:** 101–110. doi:10.1016/0014-4827(76)90659-5

Woodcock CL, Frado LL, Rattner JB. 1984. The higher-order structure of chromatin: evidence for a helical ribbon arrangement. *J Cell Biol* **99:** 42–52. doi:10.1083/jcb.99.1.42

Woodcock CL, Skoultchi AI, Fan Y. 2006. Role of linker histone in chromatin structure and function: H1 stoichiometry and nucleosome repeat length. *Chromosome Res* **14:** 17–25. doi:10.1007/s10577-005-1024-3

Wu G, Broniscer A, McEachron TA, Lu C, Paugh BS, Becksfort J, Qu C, Ding L, Huether R, Parker M, et al. 2012. Somatic histone H3 alterations in pediatric diffuse intrinsic pontine gliomas and non-brainstem glioblastomas. *Nat Genet* **44:** 251–253. doi:10.1038/ng.1102

Wutz G, Várnai C, Nagasaka K, Cisneros DA, Stocsits RR, Tang W, Schoenfelder S, Jessberger G, Muhar M, Hossain MJ, et al. 2017. Topologically associating domains and chromatin loops depend on cohesin and are regulated by CTCF, WAPL, and PDS5 proteins. *EMBO J* **36:** 3573–3599. doi:10.15252/embj.201798004

Xu J, Ma H, Jin J, Uttam S, Fu R, Huang Y, Liu Y. 2018. Super-resolution imaging of higher-order chromatin structures at different epigenomic states in single mammalian cells. *Cell Rep* **24:** 873–882. doi:10.1016/j.celrep.2018.06.085

Yusufova N, Kloetgen A, Teater M, Osunsade A, Camarillo JM, Chin CR, Doane AS, Venters BJ, Portillo-Ledesma S, Conway J, et al. 2021. Histone H1 loss drives lymphoma by disrupting 3D chromatin architecture. *Nature* **589:** 299–305. doi:10.1038/s41586-020-3017-y

Zheng C, Lu X, Hansen JC, Hayes JJ. 2005. Salt-dependent intra- and internucleosomal interactions of the H3 tail domain in a model oligonucleosomal array. *J Biol Chem* **280:** 33552–33557. doi:10.1074/jbc.M507241200

Zidovska A, Weitz DA, Mitchison TJ. 2013. Micron-scale coherence in interphase chromatin dynamics. *Proc Natl Acad Sci* **110:** 15555–15560. doi:10.1073/pnas.1220313110

Zuin J, Dixon JR, van der Reijden MIJA, Ye Z, Kolovos P, Brouwer RWW, van de Corput MPC, van de Werken HJG, Knoch TA, van IJcken WFJ, et al. 2014. Cohesin and CTCF differentially affect chromatin architecture and gene expression in human cells. *Proc Natl Acad Sci* **111:** 996–1001. doi:10.1073/pnas.1317788111

Mechanisms of Chromosome Folding and Nuclear Organization: Their Interplay and Open Questions

Leonid Mirny[1] and Job Dekker[2]

[1]Institute for Medical Engineering and Science, and Department of Physics, MIT, Cambridge, Massachusetts 02139, USA

[2]Howard Hughes Medical Institute, and Program in Systems Biology, Department of Biochemistry and Molecular Pharmacology, University of Massachusetts Medical School, Worcester, Massachusetts 01605, USA

Correspondence: leonid@mit.edu; Job.Dekker@umassmed.edu

Microscopy and genomic approaches provide detailed descriptions of the three-dimensional folding of chromosomes and nuclear organization. The fundamental question is how activity of molecules at the nanometer scale can lead to complex and orchestrated spatial organization at the scale of chromosomes and the whole nucleus. At least three key mechanisms can bridge across scales: (1) tethering of specific loci to nuclear landmarks leads to massive reorganization of the nucleus; (2) spatial compartmentalization of chromatin, which is driven by molecular affinities, results in spatial isolation of active and inactive chromatin; and (3) loop extrusion activity of SMC (structural maintenance of chromosome) complexes can explain many features of interphase chromatin folding and underlies key phenomena during mitosis. Interestingly, many features of chromosome organization ultimately result from collective action and the interplay between these mechanisms, and are further modulated by transcription and topological constraints. Finally, we highlight some outstanding questions that are critical for our understanding of nuclear organization and function. We believe many of these questions can be answered in the coming years.

THREE MECHANISMS OF CHROMOSOME FOLDING AND THEIR INTERPLAY IN NUCLEAR ORGANIZATION

The organization of the cell nucleus is directly related to the folding of chromosomes and their associations with subnuclear structures such as the nuclear lamina. Microscopy and genomic approaches now allow detailed descriptions of the three-dimensional (3D) organization of chromosomes, the presence of nuclear bodies such as the nucleolus, speckles, and Cajal bodies around specific loci, and the formation of subnuclear compartments enriched in sets of loci and specific *trans* factors. Principles of genome organization, the dynamics in this organization during the cell cycle (Dileep et al. 2015; Gibcus et al. 2018; Abramo et al. 2019; Zhang et al. 2019; Kang et al. 2020) and development (Hug et al. 2017; Wike et al. 2021), and

Cite this article as *Cold Spring Harb Perspect Biol* doi: 10.1101/cshperspect.a040147

variation in folding between single cells (Nagano et al. 2013; Ramani et al. 2017) are now being described at increasing resolution.

In recent years, at least three mechanisms that contribute to chromosome folding and nuclear organization have come into focus, allowing going beyond descriptive studies (Fig. 1). First, loci can be tethered to specific nuclear features such as the nuclear periphery (Guelen et al. 2008). For instance, genetic perturbation experiments have identified specific factors (e.g., lamin B receptor and lamin A/C) required for the tethering of heterochromatic loci to the nuclear lamina (Solovei et al. 2013). Second, the nucleus is compartmentalized so that active and inactive chromatin domains are spatially

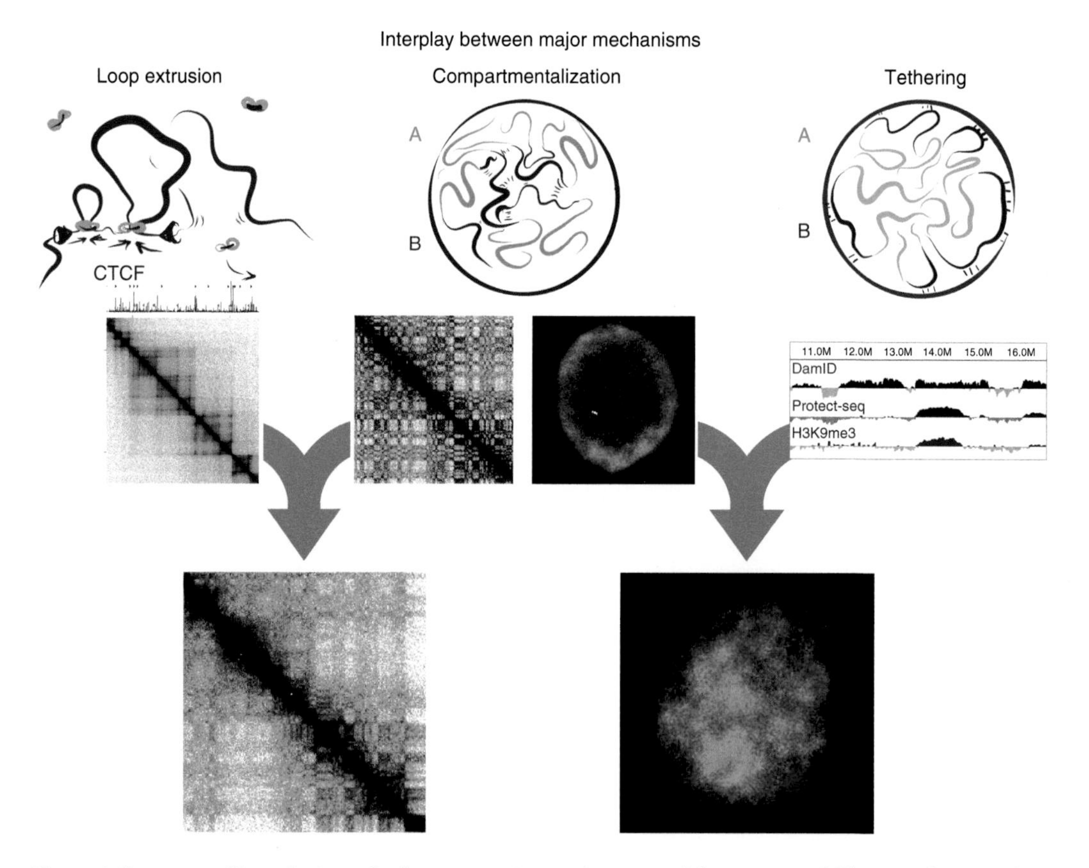

Figure 1. Summary of how the interplay between major mechanisms of chromosome folding results in nuclear organization. Loop extrusion (*left* column) occluded by CTCF results in topologically associating domains (TADs), stripes, and dots, but also influences compartmentalization of euchromatin and heterochromatin. (*Middle* column) In the absence of extrusion (owing to cohesin depletion), compartmentalization gets stronger and finer, and better follows patterns of histone modifications. Compartmentalization observed in wild-type cells (*bottom* row) is a result of the interplay between such modification-dependent compartmentalization and loop extrusion. In the absence of tethering, attractions between heterochromatic (red) regions result in a phase-separated but inverted nucleus (*middle* column) in which constitutive heterochromatin (blue) is located in the center of the nucleus, surrounded by the facultative heterochromatin (B compartment, red), with euchromatin (A compartment) at the nuclear periphery. Lamina tethering, in turn, leads to peripheral location of heterochromatin as evident from DamID and Protect-seq (Spracklin and Pradhan 2020). The interplay of attraction between regions of heterochromatin and its tethering to the nuclear lamina results in conventional nuclear organization. (Hi-C data from Falk et al. 2019; microscopy images courtesy of Irina Solovei.)

segregated (Simonis et al. 2006; Lieberman-Aiden et al. 2009). Interestingly, the spatial segregation of active and inactive chromatin is not dependent on tethering of heterochromatin and occurs also in inverted nuclei in which such tethering is absent and heterochromatin is clustered in the center of the nucleus (Solovei et al. 2009; Falk et al. 2019). The spatial separation of active and inactive chromatin (compartmentalization) and the formation of subnuclear bodies both can be understood as the result of phase separation (Falk et al. 2019; Hildebrand and Dekker 2020). This biophysical process can drive clustering of specific types of chromatin and aggregation of sets of proteins through weak multivalent interactions. Theoretical considerations that such a process could explain aspects of nuclear organization are now leading to testable predictions of how experimental perturbations would alter compartmentalization. Third, chromatin is folded into dynamically growing and disappearing loops. Theoretical considerations predicted that specific molecular machines form chromatin loops through a process of loop extrusion (Fudenberg et al. 2016, 2017). This mechanism is increasingly understood in molecular detail. Loop extrusion activity has been directly observed for the SMC (structural maintenance of chromosome) complexes cohesin and condensin that can extrude chromatin loops in an ATP-dependent manner (Ganji et al. 2018; Davidson et al. 2019; Kim et al. 2019; Golfier et al. 2020; Kong et al. 2020). Loop extrusion can explain many features of chromatin folding in interphase and in mitosis.

Ultimately, the folded state of the genome, and the organization of the nucleus in general, is the result of interplay between different biophysical and molecular mechanisms that fold chromosomes. For instance, tethering and phase separation together result in the spatial organization of heterochromatin and euchromatin in the nucleus (Falk et al. 2019). Another example discussed below is the interference of loop extrusion with compartmentalization (Schwarzer et al. 2017; Nuebler et al. 2018). Moreover, different loop extruding complexes can interfere with each other (e.g., condensin and cohesin), and the relative contributions of the two condensin complexes determine whether mitotic chromosomes are long and thin, or short and wide (Shintomi and Hirano 2011).

In this review, we outline the different mechanisms that fold chromosomes, and focus on their interplay that results in the folding patterns observed in the cell. We also highlight outstanding questions. Given the enormous progress over the past years in development of both experimental methods and powerful predictive theoretical models, we believe many of these questions can be answered in the coming years.

COMPARTMENTALIZATION

One of the most prominent patterns in Hi-C data of higher eukaryotes is the checkerboard pattern that consists of ~0.1–2 Mb size rectangular "checkers" visible in both intrachromosomal (*cis*) and interchromosomal (*trans*) sections of Hi-C maps (Lieberman-Aiden et al. 2009). Prominent during interphase, the checkerboard patterns rapidly (<15 min) disappear when cells go into mitosis (Gibcus et al. 2018), and gradually reemerge (~2–4 h) after exit from mitosis (Abramo et al. 2019; Zhang et al. 2019). Although weak during early stages of embryonic development (Du et al. 2017; Ke et al. 2017), the checkerboard pattern appears as early as in mouse zygotes (Flyamer et al. 2017) and increases in intensity through early development. Interestingly, the pattern is present in the paternal, but not in the maternal zygotic pronuclei. Compartments are absent in yeast and in bacteria. Compartmentalization in archaea has been reported by some groups (Takemata et al. 2019), whereas others (Cockram et al. 2021) suggested that domains in archaea are of a different nature. Positions and intensity of the checkers vary between metazoan cell types suggesting their connection to transcriptional and/or chromatin states of the genome.

This checkerboard pattern reflects the presence of two (or more) types of genomic regions that alternate along the chromosomes and have an enrichment of interactions between regions of the same type and depletion of interactions between regions of different types. Such alternating regions are referred to as compartments, compartmental domains, or compartmental regions, and the pattern of preferential interactions between them is referred to as compartmentalization.

Genomic locations of compartmental regions can be directly inferred from Hi-C maps. One approach is to infer the "compartment genomic track" is by eigenvector decomposition (Lieberman-Aiden et al. 2009; Imakaev et al. 2012). Identified this way, compartments show a striking correlation with local transcriptional activity and specific histone marks. Regions of one type (compartment A) correspond to transcriptionally active and open regions of the genome, and the other type (compartment B) corresponds to silent, gene-poor, or repressed regions (Lieberman-Aiden et al. 2009). Regions of type A are rich in marks of active chromatin: H3K27ac, H3K4me1/3, H3K36me3, etc., whereas B regions are enriched in inactive marks (e.g., H3K9me2/3). As such, A compartments closely match euchromatin and B matches heterochromatin. Note that repeat-rich parts of heterochromatin, typically pericentromeric and peri-telomeric, are not visible in Hi-C, and hence cannot be classified as B compartment. Compartmentalization is also highly correlated with features of genomic sequence that are known to be associated with hetero- and euchromatin (e.g., GC composition in warm-blooded vertebrates [high in A, lower in B], short interspersed nuclear elements [SINEs] associated with A compartment, and long interspersed nuclear elements [LINEs] with B compartment) (Solovei et al. 2016). Nevertheless, the transcriptional and epigenetic states of a locus determine its compartmental type; hence, A/B compartments can be different in different cell types.

Enrichment of contacts between compartmental regions of the same type as seen in Hi-C, reflect spatial segregation of euchromatin (A compartment) and heterochromatin (B compartment). In the vast majority of nuclei, euchromatin occupies the center of the nucleus, and heterochromatin is located at the nuclear periphery and surrounds nucleoli (Solovei et al. 2016; van Steensel and Belmont 2017).

What mechanisms can guide spatial segregation of A and B regions? Such spatial segregation is reminiscent of phase separation in polymers made of A- and B-type monomers with A-A and/or B-B attraction (Rubinstein and Colby 2003). Consistently, several studies showed that the spatial segregation of compartments and the checkerboard patterns observed in Hi-C can be reproduced in polymer simulations in which regions of the same type attract each other (some studies erroneously referred to such regions as topologically associating domains [TADs], not compartments) and in which B regions were additionally tethered to the lamina (Barbieri et al. 2012; Jerabek and Heermann 2012; Jost et al. 2014; Di Pierro et al. 2016; MacPherson et al. 2018). A recent study (Falk et al. 2019) allowed to disentangle contributions of these factors indicating that compartmentalization is mostly driven by strong attractions between B regions, with weaker attractions between A regions, and is independent from tethered to the lamina (see below).

Not all regions, however, fall under the simple A/B classification. Regions with an intermediate value of the compartment signal (the first eigenvector) seem to avoid interactions with either A or B, likely forming a separate compartment (referred as I compartment) (Schwarzer et al. 2017; Johnstone et al. 2020). Another interesting exception are Polycomb-repressed H3K27me3 regions that tend to interact with each other (Vieux-Rochas et al. 2015; Boyle et al. 2020), but that can show little enrichment in interactions with either A or B compartments, and are spatially located within the central (euchromatic/A compartment) part of the nuclear volume (Girelli et al. 2020). Another pattern that stands out from the A/B compartmentalization is the enrichment of interactions between exon-rich genes, likely mediated by interactions with splicing factories (Bonev et al. 2017; Kerpedjiev et al. 2018; Zhang et al. 2021). This observation implies that the process of transcription and splicing can contribute to at least some aspects of nuclear compartmentalization. Likely more exceptions from the A/B compartmentalization will be discovered, suggesting that more compartment types will be necessary to describe them.

Several questions about mechanisms and functional role of compartmentalization remain open:

1. What is the functional role of spatial segregation between different types of genomic regions? Is spatial compartmentalization necessary for their function: gene silencing in B and gene activation in A, important for the maintenance of histone marks, or necessary for some other genome maintenance functions (e.g., silencing of transposable elements)?
2. How do repressive compartments repress? Which factors are critical for this: high volume-density of heterochromatin, its peripheral location, or presence of specific histone marks, DNA modification, or recruitment of repressors?
3. What molecular players are key for mediating attractions between different types of chromatin? Can affinities be driven by direct interactions between methylated histones or do they require "bridging" by proteins (e.g., PRC and HP1 family) or RNAs?
4. How many types of compartmental regions, beyond eu-/heterochromatin are present in the cell? Can some compartment types emerge or become widespread only in specific cell types? What are their mutual interactions? What functional roles do multitudes of compartment types play?
5. Is compartmentalization a result of transcription and histone modifications or is it a driver for establishment of transcription and histone modification patterns? Are there feedback mechanisms between compartmentalization and processes on the DNA? Does compartmentalization help to "memorize" the transcriptional state of a cell or does it drive changes in cellular states?
6. What is the evolutionary origin of compartmental organization? Is it essential for establishment and maintenance of the many different cell types in multicellular organisms?

LOOP EXTRUSION

During interphase, Hi-C maps of higher eukaryotes show rich patterns of contact enrichments near the diagonal (i.e., between regions separated by less than a few megabases) (Hsieh et al. 2020; Krietenstein et al. 2020). When first characterized in 2012, such patterns were limited by the resolution of the data (~100 kb) and largely revealed segments of local contact enrichment referred to as TADs (Dixon et al. 2012; Nora et al. 2012). Neighboring TADs show approximately two- to threefold enrichment of contacts within a domain as compared with interactions between neighboring domains. In contrast to compartments, they do not "checkerboard" with each other, and do not show specific enrichment of interactions when far away on the same or on different chromosomes (Mirny et al. 2019).

As the resolution and the depth of Hi-C data increased (Rao et al. 2014), new patterns such as focal enrichments of interactions between CTCF sites at TAD borders started to emerge. Commonly referred to as "loops," such enrichments may be better called "dots" as they represent transient (Cattoni et al. 2017; Finn et al. 2019; Luppino et al. 2020; Su et al. 2020), rather than stable loops as was originally believed. Another frequent, near-diagonal pattern is a "stripe" or a "line" that emanates from a CTCF site (Fudenberg et al. 2016, 2017; Vian et al. 2018; Barrington et al. 2019), and reflects an enrichment of contacts between the CTCF and its neighborhood, extending sometimes for up to ~1–3 Mb. Most recent and highest resolution Micro-C and Hi-C data revealed the abundance of such "dot" and "stripe" patterns, and that many stripes appear to include a series of dots (Hsieh et al. 2020; Krietenstein et al. 2020; Oksuz et al. 2020).

High-resolution microscopy extended our understanding of these patterns. First, microscopy has shown that maps of the median distance between loci closely resemble Hi-C maps (Bintu et al. 2018), with intra-TAD distances being smaller than inter-TAD ones (Finn et al. 2019). Second, and consistent with single-cell Hi-C (Nagano et al. 2013; Flyamer et al. 2017;

Stevens et al. 2017), microscopy has observed a great deal of cell-to-cell variation in distances for pairs of loci within the same or in different TADs (Bintu et al. 2018; Finn et al. 2019; Su et al. 2020), further supporting the notion that TADs reflect enrichments of contacts rather than solid or even spatially discernable structures. Third, microscopy has clearly shown that focal enrichments of contacts between CTCF sites ("dots" or "loops") do not represent stable loops, with their ends being in detectable proximity (<200–300 nm) in merely 10% of cells in the population at any given moment in time (Cattoni et al. 2017; Finn et al. 2019).

A surprising result of the last few years is that all these intricate patterns of contact enrichments are produced by a single process: an active (ATP-dependent) process of loop extrusion by cohesins that is occluded by extrusion barriers that are created largely by CTCF proteins (for review, see Fudenberg et al. 2017), and possibly modulated by interplay with other processes such as transcription.

THE MECHANICS OF LOOP EXTRUSION

Loop extrusion is a process whereby a molecular motor binds the chromatin fiber or DNA (e.g., in bacteria) and reels it in from one or both sides forming a progressively larger loop (Alipour and Marko 2012). Hypotheses about such a mechanism and its role in somatic recombination (Wood and Tonegawa 1983), enhancer–promoter interactions and chromosome organization during interphase (Riggs 1990), and DNA compaction and segregation during mitosis (Nasmyth 2001) were appearing in early literature, about once per decade, but remained largely unexplored until recently. The first computational models of loop extrusion suggested that it can generate arrays of nested or consecutive loops (Alipour and Marko 2012). Polymer models further showed that such arrays of loops can reproduce Hi-C data for mitotic chromosomes (Naumova et al. 2013). Polymer models further showed that extrusion can compact chromosomes and segregate sister chromatids (Goloborodko et al. 2016a,b), and help them disentangle topologically (Brahmachari and Marko 2019; Orlandini et al. 2019), thus suggesting a critical role of the loop extrusion process in mitosis.

Critically, simulations have shown that when extrusion is combined with barriers that can stop or pause it at specific genomic positions (Sanborn et al. 2015; Fudenberg et al. 2016), extrusion generates a broad range of patterns of local contact enrichment, such as TADs, dots, and stripes. Computational studies (Fudenberg et al. 2016) further hypothesized that SMC complexes (cohesins, condensins, etc.), believed at the time to be passive rings (Nasmyth and Haering 2009), are actually loop-extruding motors, and DNA-bound CTCF proteins are extrusion barriers. A broad range of experimental evidence from in vivo depletion of cohesin (Gassler et al. 2017; Haarhuis et al. 2017; Rao et al. 2017; Schwarzer et al. 2017; Wutz et al. 2017) and CTCF (Nora et al. 2017) to direct single-molecule visualizations (Lazar-Stefanita et al. 2017; Ganji et al. 2018; Davidson et al. 2019; Kim et al. 2019; Golfier et al. 2020; Kong et al. 2020; Cockram et al. 2021) support the loop extrusion mechanisms in eukaryotes.

Studies in bacteria, in turn, provided extensive evidence of loop extrusion by bacterial SMC complexes that juxtapose chromosomal arms (Wang et al. 2017; Gruber 2018; Böhm et al. 2020) and facilitate lengthwise compaction of chromosomes (Mäkelä and Sherratt 2020a,b). The presence of a specific loading (ParS) site for bacterial condensins (bSMC) in several bacteria lead to emergence of a distinct pattern on a Hi-C map, a secondary diagonal orthogonal to the main one (Gruber 2014; Wang et al. 2017; Böhm et al. 2020). Studies in bacteria further allowed (1) direction of extrusion in vivo and direct measurement of the extrusion speed (~25 kb/min in *Caulobacter* [Tran et al. 2017] and ~50 kb/min in *Bacillus* [Wang et al. 2017]); (2) characterizing specific loading and randomly loading of different extruding SMCs (Lioy et al. 2020; for review, see Gruber 2018); (3) measuring slow-down caused by interactions with transcriptional machinery (Brandão et al. 2019); and (4) genomic engineering experiments to create a "collision track" for examining interactions between SMCs (Brandão et al. 2021), suggesting that SMCs can bypass each other. Bypassing condensins has been observed in single-molecule experiments with yeast condensin on naked DNA (Kim et al. 2020).

Together Hi-C, microscopy, and single-molecule experiments draw the following picture of chromosome organization by loop extrusion in many eukaryotes including mammals:

1. By forming ∼100–200 kb loops, extrusion reduces contour length between every pair of loci, and hence increases the contact frequency between all regions on the same chromosome. This increase in contact frequency is most prominent for loci separated by <3 Mb as seen in Hi-C, and accompanied by reduced spatial distance as seen in superresolution microscopy (Bintu et al. 2018). Because loops are unlikely to get formed across TAD borders (depending on the permeability of CTCF barriers to cohesins), the contact frequency within TADs is increased more than between TADs.
2. Individual extruded loops have not been directly observed in vivo neither in microscopy nor in Hi-C because they are transient and dynamic (cohesin exchanges every 5–20 min [Gerlich et al. 2006a; Hansen et al. 2017; Wutz et al. 2017]); interactions they create can be practically indiscernible from myriads of other random interactions. Extruded loops, however, modify Hi-C maps in a predictable manner, so loops sizes (∼100–200 kb) can be inferred from the scaling $P(s)$ curves of Hi-C data (Gassler et al. 2017).
3. Elevated contact frequency within a TAD and its insulation from neighboring TADs is caused by different positions of individual extruded loops in different cells and at different times, rather than caused by a fixed loop between two CTCF borders. Simulations show that a fixed loop does not increase contact frequency within a loop interior, nor insulates its interior from the flanks or neighboring loops (Fudenberg et al. 2016).
4. Similar to TADs, other local patterns such as dots and stripes do not reflect stable loops and constitute enrichments of contact frequency. They emerge as a result of pausing of extruding cohesin at specific genomic locations.

Functional Roles of Loop Extrusion

What possible functions can this versatile mechanism and intricate pattern of contact enrichments that it generates have? It is possible that during interphase the dynamic process of extrusion, with 5–20 min turnover time (Gerlich et al. 2006a; Hansen et al. 2017; Wutz et al. 2017), can be more important than the loops it produces (Liu et al. 2021). Extrusion can bring together distant genomic elements, such as enhancers and promoters, and do so more reliably than 3D search alone because it acts strictly in *cis*. Importantly, such extrusion-mediated contacts can be controlled by placing CTCF barriers between elements that should not interact. CTCF barriers can also facilitate contacts: When cohesin stalls on CTCF, it can continue reeling DNA in on the other side, thus scanning long genomic regions (Fudenberg et al. 2017). Such scanning is implicated in stochastic promoter choice in the protocadherin gene cluster (Guo et al. 2012; Canzio et al. 2019), and stochastic choice of partner sites for somatic recombination in the V(D)J locus (Ba et al. 2020). One possibility is that cohesin/CTCF-mediated scanning (Kraft et al. 2019) can also allow a locus (e.g., a promoter) to scan a long genomic region in search for its enhancer and/or integrating information from multiple scanned enhancers. Because CTCF bound sites constitute directional barriers (Rao et al. 2014; Vietri Rudan et al. 2015) (i.e., halt extrusion only if cohesin approaches CTCF from one side, but not the other) CTCF can direct scanning in a particular direction. Other mechanisms such as 3D spatial contacts and phase-separation/affinity-mediated contacts can be hardly controlled by insulators or established in a specific direction along the genome. Broadly, such a system of extruders and barriers opens many possibilities for controlled and targeted long-range communication along the genome.

Open questions:

1. Molecular mechanism of loop extrusion by SMCs remains mysterious. Recent cryo-electron microscopy (EM) structures (Hassler et al. 2019; Higashi et al. 2020; Lee et al. 2020; Li et al. 2020b; Muir et al. 2020; Shi et al. 2020) and single-molecule measurements (Ryu et al. 2020) inspired different molecular models (Marko et al. 2019; Higashi et al. 2021), but more detailed physical and structural characterization of SMC complexes at different steps of the extrusion cycle is essential for understanding this vital process.

2. Loop extrusion dynamics in live cells remains elusive. Despite the wealth of evidence from static perturbation experiments, direct visualization of loop extrusion in single-molecule experiments in vitro, and observation of extrusion in bacteria, dynamics of loop extrusion in live eukaryotic cells have not been captured. Although challenging because of fluctuations of live chromosomes, such experiments can provide direct evidence of extrusion in vivo. Moreover, in vivo characterization of extrusion dynamics (Brandão et al. 2021) could allow measuring the speed and the processivity of different SMC complexes, their interactions with CTCF and other factors, their activity through the cell cycle, and interactions with each other and with other processes on crowded chromatin templates and DNA.

3. Functional roles of the loop extension and patterns of contacts that it generates during interphase remain a subject of debate. On the one hand, loop extrusion is an attractive mechanism for mediating enhancer–promoter interactions because of its ability to facilitate contacts, and owing to control over direction and extent of such contacts exerted by CTCF and other extrusion barriers. On the other hand, depletion of cohesin and CTCF appear to be essential for transcriptional response to stimuli (Cuartero et al. 2018; Stik et al. 2020; Weiss et al. 2021), but dispensable for the maintenance of transcription of the majority of constitutive genes (Rao et al. 2017; Schwarzer et al. 2017).

4. What are the rules and regulation of SMC traffic along the genome? Understanding these rules is essential for understanding their possibly broad-range functional roles. These include understanding interactions between extruders when they meet each other, effects of positioned and random barriers (Dequeker et al. 2020) on extruders and vice versa. Moreover, it is crucial to understand how extrusion operates on crowded chromatin and interferes with and/or is modulated by other processes such as transcription of genic and nongenic regions, replication, condensate formation, and tethering of loci.

5. Are there other roles for loop extrusion? Other possible function of extrusion may be roles in facilitation of double-strand break (DSB) repair (Piazza et al. 2020; Arnould et al. 2021; Mirny 2021) and homology search on DNA damage and in meiosis (Patel et al. 2019; Schalbetter et al. 2019; Grey and de Massy 2021; Jin et al. 2021), its potential role in spreading of histone marks (Collins et al. 2020; Li et al. 2020a), etc.

TETHERING TO SUBNUCLEAR STRUCTURES

Besides loop extrusion and compartmentalization, a third mechanism of tethering can act on chromosomes to determine where loci are positioned with respect to other nuclear structures. Most prominently, loci can become tethered or anchored near the nuclear periphery. Early microscopic observations already established that densely staining heterochromatin is located at the nuclear periphery, and also around nucleoli. Genomic assays have been instrumental in identifying and characterizing chromosomal regions that are positioned at the nuclear periphery. The DamID method (van Steensel et al. 2001) has been used to label and then identify genomic loci (lamina-associated domains [LADs]) that are spatially proximal to the nuclear lamina. These studies found that loci near the lamina are repressed or expressed at very low levels (Pickersgill et al. 2006; Guelen et al. 2008). Chromatin in these domains is enriched for histone modifications such as H3K9me2 and H3K9me3, is late replicating, and is enriched in LINE elements (van Steensel and Belmont 2017). These are all features that are typical for inactive heterochromatin.

Following single cells through mitosis showed that loci that are located at the periphery in one cell cycle, can become repositioned in daughter cells away from the periphery, suggesting that nuclear positioning for these loci is not strictly heritable but more stochastic (Kind et al. 2013). Some LADs can be also localized around the nucleolus. This is confirmed by other experiments in which loci at or around the nucleolus were mapped by isolating nucleoli and analyzing the associated DNA (van Koningsbruggen et al. 2010). These studies identified a set of loci, nucleolus-associated domains (NADs) that have chromatin features that are similar to LADs (van Koningsbruggen et al. 2010; Vertii et al. 2019). More recent studies in mouse embryonic stem cells identified two types of NADs: one that resembles LADs (type I) and a second type (type II) that is characterized by higher levels of gene expression, the facultative heterochromatin mark H3K27me3, and early replication (Vertii et al. 2019). Thus, a subset of heterochromatic domains can be either tethered to the lamina or is associated with the nucleolus (and probably alternates between these locations in different cells in the population), whereas a second set of facultative heterochromatin is specifically localized at the nucleolus.

The mechanisms by which loci are tethered to either the nuclear periphery of the nucleolus are not fully understood in detail. Little is known about the factors involved in localizing chromatin domains near the nucleolus. Some factors have been identified to regulate or directly mediate the association of LADs with the nuclear periphery. Although LADs have been identified by their proximity to lamin B, lamins appear not directly required for association of LADs with the nuclear periphery at least in mouse embryonic stem cells (Amendola and van Steensel 2015). There are likely other factors that may contribute to the tethering of these domains. One candidate is the lamin B receptor (LBR). Studies in mouse cells showed that LBR acts sequentially with lamin A/C to tether heterochromatin (Solovei et al. 2013). When both factors are absent, heterochromatin localizes at the center of the nucleus with euchromatin at the periphery (see above). This unusual organization has been referred to as an inverted nucleus given that the positions of heterochromatin and euchromatin are reversed compared with their canonical organization in most cells. In natural settings, such inverted organization is observed in rod cells of nocturnal animals (Solovei et al. 2009). Other potential factors contributing to peripheral localization of heterochromatin include histone methyltransferases (e.g., MET-2 and SET-25 in *Caenorhabditis elegans*) (Towbin et al. 2012). Similarly, in mammalian cells, the histone methyltransferase G9a has been shown to regulate association of loci to the nuclear lamina (Kind et al. 2013).

Although tethering of heterochromatin is well-characterized, tethering of euchromatin is less understood. For instance, highly active loci can be localized at nuclear speckles (Zhang et al. 2021). Similarly, genes with many exons are enriched in contacts with other similar genes in *cis* and in *trans*, suggesting a role of splicing machinery in mediating such interactions. Whether this localization is caused by tethering of loci to these splicing factor-enriched nuclear bodies or whether speckles form around clusters of active loci is currently not known.

Open questions:

1. Why is heterochromatin tethered to the periphery? What are functional advantages of the conventional nuclear architecture with heterochromatin tethered to the lamina at the nuclear periphery? The inverted organization appears to be the default and to be compatible with gene expression.

2. Does tethering loci to the nuclear periphery help maintain transcriptional repression and repressive histone marks of heterochromatic regions? Does tethering of active regions to nuclear speckles play similar roles in maintaining transcription, its histone marks, and cotranscriptional splicing?

3. To what extent is chromatin tethered to intranuclear lamins? Lamins are found in the nucleoplasm but their roles are not well understood.

4. Does tethering of heterochromatin change mechanical properties to the nucleus? Mechanical properties of chromosomes and nuclei are increasingly understood to be important for cells within tissues and when cells are mobile. Does tethering reduce general mobility of chromatin and is this important for control of genomic processes such as transcription, replication, and repair?

TOPOLOGICAL EFFECTS

Topological effects do not constitute a mechanism of folding by themselves but they constrain and modulate what other mechanisms can achieve. Topological constraints prevent chromatin fibers from passing each other, unless topoisomerase II facilitates such passing. Activity of topoisomerase II during interphase is believed to be modest (Canela et al. 2019), arguing that topological constraints can play an important role in the shaping of interphase chromosomes. As such, topological effects can constrain and modulate chromosome folding driven by the three major mechanisms described above.

Topological constraints are known to have two major effects on polymer systems: (1) dramatic slowdown of equilibration, thus producing long-lived nonequilibrium states, and (2) generating of a different equilibrium state in a topologically constrained system (e.g., an unknotted ring that remains unknotted can behave differently than a ring in which strands can pass by each other) (Halverson et al. 2014).

Topological effects have been implicated in a range of chromosome phenomena. Formation of chromosomal territories have been attributed to formation of a nonequilibrium state after exit from mitosis (Abramo et al. 2019) when chromosomes do not have time to mix with each other (Rosa and Everaers 2008; Rosa et al. 2010). Formation of the largely unknotted fractal globule, as evident from Hi-C data (Lieberman-Aiden et al. 2009), is also attributed to a nonequilibrium state in which a polymer chain is crumpled attributed to topological interactions (Grosberg et al. 1988). Theoretically, such nonequilibrium states in which chains form territories and crumpled globules may nevertheless gradually, yet very slowly, equilibrate. Alternatively, nonconcatenated ring polymers form an equilibrium state with highly territorial rings, each forming a crumpled (fractal) globule (Halverson et al. 2014). Whether topological interactions lead to formation of long-lived nonequilibrium states or to equilibrium states remains to be understood. Recent studies that used modeling of Hi-C (Goundaroulis et al. 2020) and a complementary approach of multicontact 3C came to the conclusion that interphase chromatin is indeed largely unknotted (Tavares-Cadete et al. 2020). The interplay between topological constraints and major mechanisms discussed above are yet to be understood.

THE INTERPLAY OF MAJOR FOLDING MECHANISMS

Several aspects of interphase organization result from complex interplay of the three major mechanisms described above.

Extrusion versus Compartmentalization

One of the first successful cohesin and Wapl depletion studies (Haarhuis et al. 2017; Schwarzer et al. 2017) observed that, surprisingly, on depletion or enrichment of chromatin associated cohesin, not only cohesin-dependent features, but also compartmentalization was affected. Specifically, cohesin depletion resulted not only in the disappearance of extrusion-mediated features (TADs, dots, and stripes), but also in the strengthening of compartments (Schwarzer et al. 2017). An increase in cohesin residence time upon Wapl depletion, in turn, results in the weakening of compartmentalization and less-defined LADs (Haarhuis et al. 2017). These and the following studies suggested extensive interplay between loop extrusion and compartmentalization (Nuebler et al. 2018).

Cohesin-depleted chromosomes showed stronger and finer patterns of compartmentalization: Longer compartment regions got split into shorter regions. Interestingly, such finer compartments better reflect patterns of histone modification with H3K9me3- and H3K27me3-containing regions becoming shorter B compartment regions. Similarly, shorter H3K27ac islands become A compartmental regions. In wild-type, B and A regions are more stretched and less correlated with histone marks of repression and activity. These results suggested that cohesin depletion revealed "innate" compartmentalization preferences that were partially washed off by loop extrusion in the wild-type (Schwarzer et al. 2017). Simulations further showed that such masking of fine compartmentalization results from loop extrusion interfering with compartmental phase separation (Nuebler et al. 2018). Furthermore, simulations showed that increased cohesin residence time, as observed after Wapl depletion, leads to further weakening of compartmentalization.

The underlying physics of this interference is being actively explored. Polymer simulations in which chromosomes experience both compartmentalizing interactions (attraction between B regions) and active loop extrusion show that extrusion indeed weakens compartmentalization. Simulations further show that short compartmental regions are more sensitive to extrusion than long ones. Consistently, cohesin depletion experiments observed a similar phenomenon: small compartments are washed off in the wild-type while visible after cohesin depletion (Nuebler et al. 2018). Simulations of an increased extrusion activity reproduce phenomena observed in Wapl-depleted cells (Tedeschi et al. 2013; Haarhuis et al. 2017) (i.e., formation of overcompacted interphase ["vermicelli"] chromosomes accompanied by weakening of compartmentalization). Importantly, simulations show that reduction of compartmentalization by extrusion is not owing to loops per se, but rather owing to the active process of extrusion that "stirs" and mixes chromosomal regions, making it difficult for them to phase-separate (Nuebler et al. 2018).

Together, these results indicate that patterns of compartmentalization observed in wild-type interphase cells result from the interplay of phase separation of compartments and active loop extrusion.

Compartmentalization versus Tethering

Loss of tethering results in "nuclear inversion" in both natural systems such as the rods of nocturnal animals and in mutants that lack both LBR and lamin A/C (Solovei et al. 2013). In the inverted nucleus, heterochromatic regions occupy the center of the nucleus, while the euchromatic compartment is located at the nuclear periphery, thus inverting the conventional organization (Solovei et al. 2016). Hi-C and microscopy showed that despite inversion, A and B compartmental regions remain spatially segregated (Falk et al. 2019). Thus, compartmentalization by itself is independent of tethering of loci to the periphery. Polymer models showed that compartmentalization, when disentangled from tethering in inverted nuclei, requires strong attractive interactions between heterochromatic regions and weak, if any, attractions between euchromatic regions. Polymer simulations that combine mechanisms for compartmentalization and tethering of heterochromatin to the nuclear periphery are sufficient to explain the observed nuclear organization in typical mammalian cells (Falk et al. 2019b). Furthermore, restoration of lamin A/C activity in rod cells leads to partial deinversion. Together, these results indicate that conventional nuclear organization results from the joined activity of compartmentalization and tethering.

Interplay between Transcription, Compartmentalization, and Loop Extruders

Transcription can sometimes play a dominant role in chromosome folding. For instance, domain formation in bacteria is reduced or absent when transcription is blocked (Le et al. 2013). Transcription and RNA processing machinery can also influence chromosome folding in more subtle ways by interfering with folding mechanisms described above (Hnisz et al. 2017; Hilbert et al. 2018; Trojanowski et al. 2021). This is an area of active research, and insights into this interplay can further our still limited understanding of how transcription affects chromosome folding and vice versa. For example, in mammalian cells, transcription can have effects on nuclear positioning and compartment association of loci. As mentioned above, highly expressed genes can cluster together especially when extensively spliced. This may be related to the observed positioning of active genes around nuclear speckles (Zhang et al. 2021). Large and highly expressed genes can become covered with nascent RNA–protein complexes and this results in stiffening of the chromatin fiber and the relocalization of the gene away from its chromosomal territory and into the nuclear center (Leidescher et al. 2020). It is important to point out that the overall effect of transcription on genome-wide chromosome organization can be quite subtle: Blocking transcription, or entirely removing RNA polymerases using inducible degron approaches has only minor effects on compartmentalization and other features per se (Vian et al. 2018; Barutcu et al. 2019; Jiang et al. 2020; Olan et al. 2020).

The RNA transcription machinery and the loop extrusion machinery also directly interact along chromosomes. For instance, early studies in budding yeast have shown that RNA polymerase can push cohesin rings toward sites of convergent transcription (Lengronne et al. 2004). Given that in *Saccharomyces cerevisiae* cohesin is located on chromosomes predominantly in the S phase and G2/M, this could be cohesive cohesin. In mammalian cells, RNA polymerase can modulate the position of cohesin, either by acting as a passive extrusion barrier or by actively pushing cohesin complexes toward 3′ ends of the active gene (Busslinger et al. 2017; Olan et al. 2020). In bacteria, polymerase can act as a moving barrier to extruding SMC complexes resulting in the slower extrusion against the direction of transcription (Brandão et al. 2019).

Open questions:

1. What other mechanisms shape chromosome organization and how do they interfere with these mechanisms? For example, can random and transient interactions (e.g., mediated by HP1-family proteins) lead to weak gelation of chromatin?
2. What is the impact of heterochromatic interactions and tethering on loop extrusion by cohesin and binding/activity of CTCF? Conversely, can extrusion interfere with tethering or maintenance of heterochromatin and associated histone marks?
3. What kinds of interplay between the various folding mechanisms and transcription can enhance or weaken chromosome territoriality? Does territoriality benefit from extrusion, tethering or compartmentalization? Conversely, can territoriality interfere or modulate compartmentalization?

DIFFERENTIAL IMPLEMENTATION OF THE SAME FOLDING MECHANISMS DRIVES CELL-CYCLE CHANGES IN CHROMOSOME SHAPE

The three mechanisms of chromosome folding described above can explain folding properties observed for chromosomes during interphase. However, chromosomes change their shape dramatically through the cell cycle. In mitosis, chromosomes form compacted rods with sister chromatids largely separated but running side-by-side. The large morphological differences between interphase and mitotic chromosomes suggest that different mechanisms may be involved in folding chromosomes during these different cell-cycle stages. However, it now appears that in many eukaryotes the same mechanisms operate, but performed by different SMC complexes (Dekker and Mirny 2016; Goloborodko et al. 2016a; Fudenberg et al. 2017).

As cells enter prophase, condensin II complexes take over the role of cohesin to form mitotic chromatin loops. In contrast to cohesin, the residence time of condensin II on chromatin is much longer (Gerlich et al. 2006b; Hansen et al. 2017), leading to more stable extruded loops. Tightly packed arrays of stable loops lead to rod-shaped mitotic chromosomes (Goloborodko et al. 2016a,b). Although during prophase, chromatin remains tethered to the periphery, as the nuclear envelope breaks down chromosomes are mostly untethered during prometaphase.

As cells exit mitosis, condensins become inactivated after the metaphase–anaphase transition, and during cytokinesis, cohesin takes over again as the main loop extrusion factor (Abramo et al. 2019; Zhang et al. 2019; Kang et al. 2020). The more transient nature of cohesin-mediated loops, and their lower density along the chromosome, leads to a more decondensed chromosome that may be sufficient to allow long-range compartmental interactions to reform. Chromosomes also become tethered again at the nuclear envelope during the subsequent G1 (van Schaik et al. 2020; Wong et al. 2020).

In the above picture, it is the alternation between cohesin-driven loop extrusion and condensin-driven loop extrusion that leads to chromosome morphologies that appear very different. Interestingly, in mutants that stabilize cohesin-mediated loops (e.g., by removal of the cohesin-unloading factor WAPL), interphase chromosomes form so-called vermicelli chromosomes that resemble prophase threads (Tedeschi et al. 2013). This supports the proposal that similar mechanisms are at play in interphase and mitosis but that they are implemented in quantitatively different ways.

Compartmentalization of chromosomes rapidly disappears as cells enter prophase. Whether the biophysical process of compartmentalization is still active in mitosis, but simply overridden by other processes is not known at this moment. For instance, the formation of relatively stiff rods could be sufficient to prevent long-range compartmental interactions. However, the observation that meiotic prophase chromosomes are elongated rods but still show compartmentalization by Hi-C would argue rod formation itself is not sufficient to erase compartments. Similarly, in vermicelli interphase chromosomes (in cells depleted for WAPL), the chromosomes form long threads but compartmentalization is still detectable (although reduced). Alternatively, the process of compartmentalization may be actively turned off during mitosis, and then turned on again after cells exit mitosis. One way compartmentalization could be turned off is by removing or inactivating key factors that facilitate phase separation by acting as bridging factors linking active or inactive chromatin domains. Such factors can include the histones themselves, or nonhistone factors that bind either active or inactive chromatin domains. During mitosis, several histone residues become phosphorylated and perhaps this leads to loss of attractions between chromatin domains. Alternatively, or in addition, it has been shown that phosphorylation of histone tail residues including H310 and H3T3 prevents binding of other factors such as heterochromatin protein 1 (HP1) (Fischle et al. 2005; Hirota et al. 2005) or TFIID (Varier et al. 2010), respectively, which may in turn lead to loss of compartmentalization.

Open questions:

1. How is compartmentalization modulated during the cell cycle, differentiation, and aging? Compartmentalization appears highly dynamic during the cell cycle, cell-state transitions, and as cells age. Is this a regulated process? If so, how is it regulated?
2. What is the interplay between cohesin and condensin during prophase and prometaphase? During prophase, both cohesins and condensins are acting along chromosomes. Is there interference or collaboration between the complexes and how does that affect folding and segregating chromosomes during mitosis?
3. Is there an interplay between condensins and cohesins during mitotic exit? During mitotic exit, condensin inactivation and cohesin reloading appear to be temporally separated (Abramo et al. 2019). Is this a regulated process to avoid interference between these folding machines? If so, how are these processes coordinated?

CELL-TYPE-SPECIFIC CHROMOSOME FOLDING AND NUCLEAR ORGANIZATION

In different cell types, different parts of the genome are active or repressed. For instance, in each cell type specific genes, regulatory elements, and other functional elements such as CTCF sites and enhancers are active. The same folding mechanisms described above will act along chromosomes leading to cell-type-specific patterns of chromosome structures. For instance, in most cell types, compartmentalization of active and inactive chromatin domains is observed, but given that different loci are active and inactive dependent on the cell type, the composition of the A and B compartments is different as well. Similarly, cohesin-mediated loop extrusion will occur in most cell types, but the positions in which cohesin is loaded, and the genomic position of blocks to extrusion, most notably CTCF-bound sites, can differ between cell types. Finally, promoter–enhancer interactions that could be mediated by both loop extrusion or phase separation will differ between cell types given that different promoters and enhancers are active.

Interestingly, there are examples of cell types that display unique global nuclear organizations because specific mechanisms of chromosome folding themselves have been altered. One example is that of cells with inverted nuclei (see above). In rod photoreceptor cells of nocturnal mammals such as mice, the nucleus is inverted so that all heterochromatic loci are located in the center with active chromatin located peripherally (Solovei et al. 2009). Such inverted organization changes the optical properties of the nucleus and this is beneficial for photodetection. As mentioned above, this organization arises when the tethering of heterochromatin to the periphery is turned off.

Pro-B and pre-B cells represent another example of cells in which global chromosome organization is altered by modulating a specific mechanism of nuclear organization, in this case by changing cohesin-mediated loop extrusion. In these cells, Pax5 represses the expression of the cohesin-unloading factor WAPL. As a result, cohesin is more stably associated with chromatin and generates larger and more stable loops genome-wide (Hill et al. 2020). In pro-B and pre-B cells, this is thought to be important to facilitate the very long-range interactions required for contraction of the 2.8-Mb-long immunoglobulin heavy chain (Igh) locus during V(D)J recombination.

Open questions:

1. What other mechanisms drive chromosome folding and nuclear organization? Some cell types display unique chromosome conformations that could be formed by additional mechanisms. For instance, in olfactory neurons especially dense heterochromatic clusters are formed in the central part of the nucleus.
2. How malleable and/or reversible is genome and nuclear organization (e.g., during reprogramming, senescence, aging, and disease)?

ACKNOWLEDGMENTS

We are grateful to members of Mirny and Dekker laboratories for many discussions of these mechanisms and chromosome phenomena, to Irina Solovei for deep knowledge of the field she shared with us and for microscopy images of conventional and inverted nuclei, and to George Spracklin for producing DamID/Protect-seq/H3K9me3 panel of Figure 1. L.M. and J.D. are supported by a grant from the National Institutes of Health Common Fund 4D Nucleome Program (U54-DK107980, UM1-HG011536) and a grant from the National Human Genome Research Institute (NHGRI) to J.D. (HG003143). J.D. is an investigator of the Howard Hughes Medical Institute. L.M. is Blaize Pascal Chair of Ile-de-France, visiting Institut Curie.

REFERENCES

Abramo K, Valton AL, Venev SV, Ozadam H, Fox AN, Dekker J. 2019. A chromosome folding intermediate at the condensin-to-cohesin transition during telophase. *Nat Cell Biol* **21:** 1393–1402. doi:10.1038/s41556-019-0406-2

Alipour E, Marko JF. 2012. Self-organization of domain structures by DNA-loop-extruding enzymes. *Nucleic Acids Res* **40:** 11202–11212. doi:10.1093/nar/gks925

Amendola M, van Steensel B. 2015. Nuclear lamins are not required for lamina-associated domain organization in mouse embryonic stem cells. *EMBO Rep* **16:** 610–617. doi:10.15252/embr.201439789

Arnould C, Rocher V, Finoux AL, Clouaire T, Li K, Zhou F, Caron P, Mangeot PE, Ricci EP, Mourad R, et al. 2021. Loop extrusion as a mechanism for formation of DNA damage repair foci. *Nature* **590:** 660–665. doi:10.1038/s41586-021-03193-z

Ba Z, Lou J, Ye AY, Dai HQ, Dring EW, Lin SG, Jain S, Kyritsis N, Kieffer-Kwon KR, Casellas R, et al. 2020. CTCF orchestrates long-range cohesin-driven V(D)J recombinational scanning. *Nature* **586:** 305–310. doi:10.1038/s41586-020-2578-0

Barbieri M, Chotalia M, Fraser J, Lavitas LM, Dostie J, Pombo A, Nicodemi M. 2012. Complexity of chromatin folding is captured by the strings and binders switch model. *Proc Natl Acad Sci* **109:** 16173–16178. doi:10.1073/pnas.1204799109

Barrington C, Georgopoulou D, Pezic D, Varsally W, Herrero J, Hadjur S. 2019. Enhancer accessibility and CTCF occupancy underlie asymmetric TAD architecture and cell type specific genome topology. *Nat Commun* **10:** 2908. doi:10.1038/s41467-019-10725-9

Barutcu AR, Blencowe BJ, Rinn JL. 2019. Differential contribution of steady-state RNA and active transcription in chromatin organization. *EMBO Rep* **20:** e48068. doi:10.15252/embr.201948068

Bintu B, Mateo LJ, Su JH, Sinnott-Armstrong NA, Parker M, Kinrot S, Yamaya K, Boettiger AN, Zhuang X. 2018. Super-resolution chromatin tracing reveals domains and cooperative interactions in single cells. *Science* **362:** eaau1783. doi:10.1126/science.aau1783

Böhm K, Giacomelli G, Schmidt A, Imhof A, Koszul R, Marbouty M, Bramkamp M. 2020. Chromosome organization by a conserved condensin-ParB system in the actinobacterium *Corynebacterium glutamicum*. *Nat Commun* **11:** 1485. doi:10.1038/s41467-020-15238-4

Bonev B, Mendelson Cohen N, Szabo Q, Fritsch L, Papadopoulos GL, Lubling Y, Xu X, Lv X, Hugnot JP, Tanay A, et al. 2017. Multiscale 3D genome rewiring during mouse neural development. *Cell* **171:** 557–572.e24. doi:10.1016/j.cell.2017.09.043

Boyle S, Flyamer IM, Williamson I, Sengupta D, Bickmore WA, Illingworth RS. 2020. A central role for canonical PRC1 in shaping the 3D nuclear landscape. *Genes Dev* **34:** 931–949. doi:10.1101/gad.336487.120

Brahmachari S, Marko JF. 2019. Chromosome disentanglement driven via optimal compaction of loop-extruded brush structures. *Proc Natl Acad Sci* **116:** 24956–24965. doi:10.1073/pnas.1906355116

Brandão HB, Paul P, van den Berg AA, Rudner DZ, Wang X, Mirny LA. 2019. RNA polymerases as moving barriers to condensin loop extrusion. *Proc Natl Acad Sci* **116:** 20489–20499. doi:10.1073/pnas.1907009116

Brandão HB, Ren Z, Karaboja X, Mirny LA, Wang X. 2021. DNA-loop-extruding SMC complexes can traverse one another in vivo. *Nat Struct Mol Biol* doi:10.1038/s41594-021-00626-1

Brandão HB, Gabriele M, Hansen AS. 2021. Tracking and interpreting long-range chromatin interactions with super-resolution live-cell imaging. *Curr Opin Cell Biol* **70:** 18–26. doi:10.1016/j.ceb.2020.11.002

Busslinger GA, Stocsits RR, van der Lelij P, Axelsson E, Tedeschi A, Galjart N, Peters JM. 2017. Cohesin is positioned in mammalian genomes by transcription, CTCF and Wapl. *Nature* **544:** 503–507. doi:10.1038/nature22063

Canela A, Maman Y, Huang SYN, Wutz G, Tang W, Zagnoli-Vieira G, Callen E, Wong N, Day A, Peters JM, et al. 2019. Topoisomerase II-induced chromosome breakage and translocation is determined by chromosome architecture and transcriptional activity. *Mol Cell* **75:** 252–266.e8. doi:10.1016/j.molcel.2019.04.030

Canzio D, Nwakeze CL, Horta A, Rajkumar SM, Coffey EL, Duffy EE, Duffié R, Monahan K, O'Keeffe S, Simon MD, et al. 2019. Antisense lncRNA transcription mediates DNA demethylation to drive stochastic protocadherin α promoter choice. *Cell* **177:** 639–653.e15. doi:10.1016/j.cell.2019.03.008

Cattoni DI, Cardozo Gizzi AM, Georgieva M, Di Stefano M, Valeri A, Chamousset D, Houbron C, Déjardin S, Fiche JB, González I, et al. 2017. Single-cell absolute contact probability detection reveals chromosomes are organized by multiple low-frequency yet specific interactions. *Nat Commun* **8:** 1753. doi:10.1038/s41467-017-01962-x

Cockram C, Thierry A, Gorlas A, Lestini R, Koszul R. 2021. Euryarchaeal genomes are folded into SMC-dependent loops and domains, but lack transcription-mediated compartmentalization. *Mol Cell* **81:** 459–472.e10. doi:10.1016/j.molcel.2020.12.013

Collins PL, Purman C, Porter SI, Nganga V, Saini A, Hayer KE, Gurewitz GL, Sleckman BP, Bednarski JJ, Bassing CH, et al. 2020. DNA double-strand breaks induce H2Ax phosphorylation domains in a contact-dependent manner. *Nat Commun* **11:** 3158. doi:10.1038/s41467-020-16926-x

Cuartero S, Weiss FD, Dharmalingam G, Guo Y, Ing-Simmons E, Masella S, Robles-Rebollo I, Xiao X, Wang YF, Barozzi I, et al. 2018. Control of inducible gene expression links cohesin to hematopoietic progenitor self-renewal and differentiation. *Nat Immunol* **19:** 932–941. doi:10.1038/s41590-018-0184-1

Davidson IF, Bauer B, Goetz D, Tang W, Wutz G, Peters JM. 2019. DNA loop extrusion by human cohesin. *Science* **366:** 1338–1345. doi:10.1126/science.aaz3418

Dekker J, Mirny L. 2016. The 3D genome as moderator of chromosomal communication. *Cell* **164:** 1110–1121. doi:10.1016/j.cell.2016.02.007

Dequeker BJH, Brandão HB, Scherr MJ, Gassler J, Powell S, Gaspar I, Flyamer IM, Tang W, Stocsits R, Davidson IF, et al. 2020. MCM complexes are barriers that restrict cohesin-mediated loop extrusion. bioRxiv doi:10.1101/2020.10.15.340356

Dileep V, Ay F, Sima J, Vera DL, Noble WS, Gilbert DM. 2015. Topologically associating domains and their long-range contacts are established during early G1 coincident with the establishment of the replication-timing program. *Genome Res* **25:** 1104–1113. doi:10.1101/gr.183699.114

Di Pierro M, Zhang B, Aiden EL, Wolynes PG, Onuchic JN. 2016. Transferable model for chromosome architecture. *Proc Natl Acad Sci* **113:** 12168–12173. doi:10.1073/pnas.1613607113

Dixon JR, Selvaraj S, Yue F, Kim A, Li Y, Shen Y, Hu M, Liu JS, Ren B. 2012. Topological domains in mammalian genomes identified by analysis of chromatin interactions. *Nature* **485:** 376–380. doi:10.1038/nature11082

Du Z, Zheng H, Huang B, Ma R, Wu J, Zhang X, He J, Xiang Y, Wang Q, Li Y, et al. 2017. Allelic reprogramming of 3D chromatin architecture during early mammalian development. *Nature* **547:** 232–235. doi:10.1038/nature23263

Falk M, Feodorova Y, Naumova N, Imakaev M, Lajoie BR, Leonhardt H, Joffe B, Dekker J, Fudenberg G, Solovei I, et al. 2019. Heterochromatin drives compartmentalization of inverted and conventional nuclei. *Nature* **570:** 395–399. doi:10.1038/s41586-019-1275-3

Finn EH, Pegoraro G, Brandão HB, Valton AL, Oomen ME, Dekker J, Mirny L, Misteli T. 2019. Extensive heterogeneity and intrinsic variation in spatial genome organization. *Cell* **176:** 1502–1515.e10. doi:10.1016/j.cell.2019.01.020

Fischle W, Tseng BS, Dormann HL, Ueberheide BM, Garcia BA, Shabanowitz J, Hunt DF, Funabiki H, Allis CD. 2005. Regulation of HP1-chromatin binding by histone H3 methylation and phosphorylation. *Nature* **438:** 1116–1122. doi:10.1038/nature04219

Flyamer IM, Gassler J, Imakaev M, Brandão HB, Ulianov SV, Abdennur N, Razin SV, Mirny LA, Tachibana-Konwalski K. 2017. Single-nucleus Hi-C reveals unique chromatin reorganization at oocyte-to-zygote transition. *Nature* **544:** 110–114. doi:10.1038/nature21711

Fudenberg G, Imakaev M, Lu C, Goloborodko A, Abdennur N, Mirny LA. 2016. Formation of chromosomal domains by loop extrusion. *Cell Rep* **15:** 2038–2049. doi:10.1016/j.celrep.2016.04.085

Fudenberg G, Abdennur N, Imakaev M, Goloborodko A, Mirny LA. 2017. Emerging evidence of chromosome folding by loop extrusion. *Cold Spring Harb Symp Quant Biol* **82:** 45–55. doi:10.1101/sqb.2017.82.034710

Ganji M, Shaltiel IA, Bisht S, Kim E, Kalichava A, Haering CH, Dekker C. 2018. Real-time imaging of DNA loop extrusion by condensin. *Science* **360:** 102–105. doi:10.1126/science.aar7831

Gassler J, Brandão HB, Imakaev M, Flyamer IM, Ladstätter S, Bickmore WA, Peters JM, Mirny LA, Tachibana K. 2017. A mechanism of cohesin-dependent loop extrusion organizes zygotic genome architecture. *EMBO J* **36:** 3600–3618. doi:10.15252/embj.201798083

Gerlich D, Koch B, Dupeux F, Peters JM, Ellenberg J. 2006a. Live-cell imaging reveals a stable cohesin-chromatin interaction after but not before DNA replication. *Curr Biol* **16:** 1571–1578. doi:10.1016/j.cub.2006.06.068

Gerlich D, Hirota T, Koch B, Peters JM, Ellenberg J. 2006b. Condensin I stabilizes chromosomes mechanically through a dynamic interaction in live cells. *Curr Biol* **16:** 333–344. doi:10.1016/j.cub.2005.12.040

Gibcus JH, Samejima K, Goloborodko A, Samejima I, Naumova N, Nuebler J, Kanemaki MT, Xie L, Paulson JR, Earnshaw WC, et al. 2018. A pathway for mitotic chromosome formation. *Science* **359:** eaao6135. doi:10.1126/science.aao6135

Girelli G, Custodio J, Kallas T, Agostini F, Wernersson E, Spanjaard B, Mota A, Kolbeinsdottir S, Gelali E, Crosetto N, et al. 2020. GPSeq reveals the radial organization of chromatin in the cell nucleus. *Nat Biotechnol* **38:** 1184–1193. doi:10.1038/s41587-020-0519-y

Golfier S, Quail T, Kimura H, Brugués J. 2020. Cohesin and condensin extrude DNA loops in a cell cycle-dependent manner. *eLife* **9:** e53885. doi:10.7554/eLife.53885

Goloborodko A, Imakaev MV, Marko JF, Mirny L. 2016a. Compaction and segregation of sister chromatids via active loop extrusion. *eLife* **5:** e14864. doi:10.7554/eLife.14864

Goloborodko A, Marko JF, Mirny LA. 2016b. Chromosome compaction by active loop extrusion. *Biophys J* **110:** 2162–2168. doi:10.1016/j.bpj.2016.02.041

Goundaroulis D, Lieberman Aiden E, Stasiak A. 2020. Chromatin is frequently unknotted at the megabase scale. *Biophys J* **118:** 2268–2279. doi:10.1016/j.bpj.2019.11.002

Grey C, de Massy B. 2021. Chromosome organization in early meiotic prophase. *Front Cell Dev Biol* **9:** 688878. doi:10.3389/fcell.2021.688878

Grosberg AY, Nechaev SK, Shakhnovich EI. 1988. The role of topological constraints in the kinetics of collapse of macromolecules. *J Phys (France)* **49:** 2095–2100. doi:10.1051/jphys:0198800490120209500

Gruber S. 2014. Multilayer chromosome organization through DNA bending, bridging and extrusion. *Curr Opin Microbiol* **22:** 102–110. doi:10.1016/j.mib.2014.09.018

Gruber S. 2018. SMC complexes sweeping through the chromosome: going with the flow and against the tide. *Curr Opin Microbiol* **42:** 96–103. doi:10.1016/j.mib.2017.10.004

Guelen L, Pagie L, Brasset E, Meuleman W, Faza MB, Talhout W, Eussen BH, de Klein A, Wessels L, de Laat W, et al. 2008. Domain organization of human chromosomes revealed by mapping of nuclear lamina interactions. *Nature* **453:** 948–951. doi:10.1038/nature06947

Guo Y, Monahan K, Wu H, Gertz J, Varley KE, Li W, Myers RM, Maniatis T, Wu Q. 2012. CTCF/cohesin-mediated DNA looping is required for protocadherin α promoter choice. *Proc Natl Acad Sci* **109:** 21081–21086. doi:10.1073/pnas.1219280110

Haarhuis JHI, van der Weide RH, Blomen VA, Yáñez-Cuna JO, Amendola M, van Ruiten MS, Krijger PHL, Teunissen H, Medema RH, van Steensel B, et al. 2017. The cohesin release factor WAPL restricts chromatin loop extension. *Cell* **169:** 693–707.e14. doi:10.1016/j.cell.2017.04.013

Halverson JD, Smrek J, Kremer K, Grosberg AY. 2014. From a melt of rings to chromosome territories: the role of topological constraints in genome folding. *Rep Prog Phys* **77:** 022601. doi:10.1088/0034-4885/77/2/022601

Hansen AS, Pustova I, Cattoglio C, Tjian R, Darzacq X. 2017. CTCF and cohesin regulate chromatin loop stability with distinct dynamics. *eLife* **6:** e25776. doi:10.7554/eLife.25776

Hassler M, Shaltiel IA, Kschonsak M, Simon B, Merkel F, Thärichen L, Bailey HJ, Macošek J, Bravo S, Metz J, et al. 2019. Structural basis of an asymmetric condensin ATPase cycle. *Mol Cell* **74:** 1175–1188.e9. doi:10.1016/j.molcel.2019.03.037

Higashi TL, Eickhoff P, Simoes JS, Locke J, Nans A, Flynn HR, Snijders AP, Papageorgiou G, O'Reilly N, Chen ZA, et al. 2020. A structure-based mechanism for DNA entry into the cohesin ring. *Mol Cell* **79:** 917–933.e9. doi:10.1016/j.molcel.2020.07.013

Higashi TL, Pobegalov G, Tang M, Molodtsov MI, Uhlmann F. 2021. A Brownian ratchet model for DNA loop extrusion by the cohesin complex. *eLife* **10:** e67530. doi:10.7554/eLife.67530

Hilbert L, Sato Y, Kimura H, Jülicher F, Honigmann A. 2021. Transcription organizes euchromatin similar to an active microemulsion. *Nat Commun* **12:** 1360. doi:10.1038/s41467-021-21589-3

Hildebrand EM, Dekker J. 2020. Mechanisms and functions of chromosome compartmentalization. *Trends Biochem Sci* **45:** 385–396. doi:10.1016/j.tibs.2020.01.002

Hill L, Ebert A, Jaritz M, Wutz G, Nagasaka K, Tagoh H, Kostanova-Poliakova D, Schindler K, Sun Q, Bönelt P, et al. 2020. *Wapl* repression by Pax5 promotes *V* gene recombination by *Igh* loop extrusion. *Nature* **584:** 142–147. doi:10.1038/s41586-020-2454-y

Hirota T, Lipp JJ, Toh BH, Peters JM. 2005. Histone H3 serine 10 phosphorylation by Aurora B causes HP1 dissociation from heterochromatin. *Nature* **438:** 1176–1180. doi:10.1038/nature04254

Hnisz D, Shrinivas K, Young RA, Chakraborty AK, Sharp PA. 2017. A phase separation model for transcriptional control. *Cell* **169:** 13–23. doi:10.1016/j.cell.2017.02.007

Hsieh THS, Cattoglio C, Slobodyanyuk E, Hansen AS, Rando OJ, Tjian R, Darzacq X. 2020. Resolving the 3D landscape of transcription-linked mammalian chromatin folding. *Mol Cell* **78:** 539–553.e8. doi:10.1016/j.molcel.2020.03.002

Hug CB, Grimaldi AG, Kruse K, Vaquerizas JM. 2017. Chromatin architecture emerges during zygotic genome activation independent of transcription. *Cell* **169:** 216–228.e19. doi:10.1016/j.cell.2017.03.024

Imakaev M, Fudenberg G, McCord RP, Naumova N, Goloborodko A, Lajoie BR, Dekker J, Mirny LA. 2012. Iterative correction of Hi-C data reveals hallmarks of chromosome organization. *Nat Methods* **9:** 999–1003. doi:10.1038/nmeth.2148

Jerabek H, Heermann DW. 2012. Expression-dependent folding of interphase chromatin. *PLoS ONE* **7:** e37525. doi:10.1371/journal.pone.0037525

Jiang Y, Huang J, Lun K, Li B, Zheng H, Li Y, Zhou R, Duan W, Wang C, Feng Y, et al. 2020. Genome-wide analyses of chromatin interactions after the loss of Pol I, Pol II, and Pol III. *Genome Biol* **21:** 158. doi:10.1186/s13059-020-02067-3

Jin X, Fudenberg G, Pollard KS. 2021. Genome-wide variability in recombination activity is associated with meiotic chromatin organization. *Genome Res* doi:10.1101/gr.275358.121

Johnstone SE, Reyes A, Qi Y, Adriaens C, Hegazi E, Pelka K, Chen JH, Zou LS, Drier Y, Hecht V, et al. 2020. Large-scale topological changes restrain malignant progression in colorectal cancer. *Cell* **182:** 1474–1489.e23. doi:10.1016/j.cell.2020.07.030

Jost D, Carrivain P, Cavalli G, Vaillant C. 2014. Modeling epigenome folding: formation and dynamics of topologically associated chromatin domains. *Nucleic Acids Res* **42:** 9553–9561. doi:10.1093/nar/gku698

Kang H, Shokhirev MN, Xu Z, Chandran S, Dixon JR, Hetzer MW. 2020. Dynamic regulation of histone modifications and long-range chromosomal interactions during postmitotic transcriptional reactivation. *Genes Dev* **34:** 913–930. doi:10.1101/gad.335794.119

Ke Y, Xu Y, Chen X, Feng S, Liu Z, Sun Y, Yao X, Li F, Zhu W, Gao L, et al. 2017. 3D chromatin structures of mature gametes and structural reprogramming during mammalian embryogenesis. *Cell* **170:** 367–381.e20. doi:10.1016/j.cell.2017.06.029

Kerpedjiev P, Abdennur N, Lekschas F, McCallum C, Dinkla K, Strobelt H, Luber JM, Ouellette SB, Azhir A, Kumar N, et al. 2018. HiGlass: web-based visual exploration and analysis of genome interaction maps. *Genome Biol* **19:** 125. doi:10.1186/s13059-018-1486-1

Kim Y, Shi Z, Zhang H, Finkelstein IJ, Yu H. 2019. Human cohesin compacts DNA by loop extrusion. *Science* **366:** 1345–1349. doi:10.1126/science.aaz4475

Kim E, Kerssemakers J, Shaltiel IA, Haering CH, Dekker C. 2020. DNA-loop extruding condensin complexes can traverse one another. *Nature* **579:** 438–442. doi:10.1038/s41586-020-2067-5

Kind J, Pagie L, Ortabozkoyun H, Boyle S, de Vries SS, Janssen H, Amendola M, Nolen LD, Bickmore WA, van Steensel B. 2013. Single-cell dynamics of genome-nuclear lamina interactions. *Cell* **153:** 178–192. doi:10.1016/j.cell.2013.02.028

Kong M, Cutts EE, Pan D, Beuron F, Kaliyappan T, Xue C, Morris EP, Musacchio A, Vannini A, Greene EC. 2020. Human condensin I and II drive extensive ATP-dependent compaction of nucleosome-bound DNA. *Mol Cell* **79:** 99–111.e9. doi:10.1016/j.molcel.2020.04.026

Kraft K, Magg A, Heinrich V, Riemenschneider C, Schöpflin R, Markowski J, Ibrahim DM, Acuna-Hidalgo R, Despang A, Andrey G, et al. 2019. Serial genomic inversions induce tissue-specific architectural stripes, gene misexpression and congenital malformations. *Nat Cell Biol* **21:** 305–310. doi:10.1038/s41556-019-0273-x

Krietenstein N, Abraham S, Venev SV, Abdennur N, Gibcus J, Hsieh THS, Parsi KM, Yang L, Maehr R, Mirny LA, et al. 2020. Ultrastructural details of mammalian chromosome architecture. *Mol Cell* **78:** 554–565.e7. doi:10.1016/j.molcel.2020.03.003

Lazar-Stefanita L, Scolari VF, Mercy G, Muller H, Guérin TM, Thierry A, Mozziconacci J, Koszul R. 2017. Cohesins and condensins orchestrate the 4D dynamics of yeast chromosomes during the cell cycle. *EMBO J* **36:** 2684–2697. doi:10.15252/embj.201797342

Le TBK, Imakaev MV, Mirny LA, Laub MT. 2013. High-resolution mapping of the spatial organization of a bacterial chromosome. *Science* **342:** 731–734. doi:10.1126/science.1242059

Lee BG, Merkel F, Allegretti M, Hassler M, Cawood C, Lecomte L, O'Reilly FJ, Sinn LR, Gutierrez-Escribano P, Kschonsak M, et al. 2020. Cryo-EM structures of holo condensin reveal a subunit flip-flop mechanism. *Nat Struct Mol Biol* **27:** 743–751. doi:10.1038/s41594-020-0457-x

Leidescher S, Ribisel J, Ullrich S, Feodorova Y, Hildebrand E, Bultmann S, Link S, Thanisch K, Mulholland C, Dekker J, et al. 2020. Spatial organization of transcribed eukaryotic genes. bioRxiv doi:10.1101/2020.05.20.106591

Lengronne A, Katou Y, Mori S, Yokobayashi S, Kelly GP, Itoh T, Watanabe Y, Shirahige K, Uhlmann F. 2004. Cohesin relocation from sites of chromosomal loading to places of convergent transcription. *Nature* **430:** 573–578. doi:10.1038/nature02742

Li K, Bronk G, Kondev J, Haber JE. 2020a. Yeast ATM and ATR kinases use different mechanisms to spread histone H2A phosphorylation around a DNA double-strand break. *Proc Natl Acad Sci* **117:** 21354–21363. doi:10.1073/pnas.2002126117

Li Y, Haarhuis JHI, Cacciatore ÁS, Oldenkamp R, van Ruiten MS, Willems L, Teunissen H, Muir KW, de Wit E, Rowland BD, et al. 2020b. The structural basis for cohesin-CTCF-anchored loops. *Nature* **578:** 472–476. doi:10.1038/s41586-019-1910-z

Lieberman-Aiden E, van Berkum NL, Williams L, Imakaev M, Ragoczy T, Telling A, Amit I, Lajoie BR, Sabo PJ, Dorschner MO, et al. 2009. Comprehensive mapping of long-range interactions reveals folding principles of the human genome. *Science* **326:** 289–293. doi:10.1126/science.1181369

Lioy VS, Junier I, Lagage V, Vallet I, Boccard F. 2020. Distinct activities of bacterial condensins for chromosome management in *Pseudomonas aeruginosa*. *Cell Rep* **33:** 108344. doi:10.1016/j.celrep.2020.108344

Liu NQ, Maresca M, van den Brand T, Braccioli L, Schijns MMGA, Teunissen H, Bruneau BG, Nora EP, de Wit E. 2021. WAPL maintains a cohesin loading cycle to preserve cell-type-specific distal gene regulation. *Nat Genet* **53:** 100–109. doi:10.1038/s41588-020-00744-4

Luppino JM, Park DS, Nguyen SC, Lan Y, Xu Z, Yunker R, Joyce EF. 2020. Cohesin promotes stochastic domain intermingling to ensure proper regulation of boundary-proximal genes. *Nat Genet* **52:** 840–848. doi:10.1038/s41588-020-0647-9

MacPherson Q, Beltran B, Spakowitz AJ. 2018. Bottom-up modeling of chromatin segregation due to epigenetic modifications. *Proc Natl Acad Sci* **115:** 12739–12744. doi:10.1073/pnas.1812268115

Mäkelä J, Sherratt DJ. 2020a. Organization of the *Escherichia coli* chromosome by a MukBEF axial core. *Mol Cell* **78:** 250–260.e5. doi:10.1016/j.molcel.2020.02.003

Mäkelä J, Sherratt D. 2020b. SMC complexes organize the bacterial chromosome by lengthwise compaction. *Curr Genet* **66:** 895–899. doi:10.1007/s00294-020-01076-w

Marko JF, De Los Rios P, Barducci A, Gruber S. 2019. DNA-segment-capture model for loop extrusion by structural maintenance of chromosome (SMC) protein complexes. *Nucleic Acids Res* **47:** 6956–6972. doi:10.1093/nar/gkz497

Mirny LA. 2021. Cells use loop extrusion to weave and tie the genome. *Nature* **590:** 554–555. doi:10.1038/d41586-021-00351-1

Mirny LA, Imakaev M, Abdennur N. 2019. Two major mechanisms of chromosome organization. *Curr Opin Cell Biol* **58:** 142–152. doi:10.1016/j.ceb.2019.05.001

Muir KW, Li Y, Weis F, Panne D. 2020. The structure of the cohesin ATPase elucidates the mechanism of SMC-kleisin ring opening. *Nat Struct Mol Biol* **27:** 233–239. doi:10.1038/s41594-020-0379-7

Nagano T, Lubling Y, Stevens TJ, Schoenfelder S, Yaffe E, Dean W, Laue ED, Tanay A, Fraser P. 2013. Single-cell Hi-C reveals cell-to-cell variability in chromosome structure. *Nature* **502:** 59–64. doi:10.1038/nature12593

Nasmyth K. 2001. Disseminating the genome: joining, resolving, and separating sister chromatids during mitosis and meiosis. *Annu Rev Genet* **35:** 673–745. doi:10.1146/annurev.genet.35.102401.091334

Nasmyth K, Haering CH. 2009. Cohesin: its roles and mechanisms. *Annu Rev Genet* **43:** 525–558. doi:10.1146/annurev-genet-102108-134233

Naumova N, Imakaev M, Fudenberg G, Zhan Y, Lajoie BR, Mirny LA, Dekker J, 2013. Organization of the mitotic chromosome. *Science* **342:** 948–953.

Nora EP, Lajoie BR, Schulz EG, Giorgetti L, Okamoto I, Servant N, Piolot T, van Berkum NL, Meisig J, Sedat J, et al. 2012. Spatial partitioning of the regulatory landscape of the X-inactivation centre. *Nature* **485:** 381–385. doi:10.1038/nature11049

Nora EP, Goloborodko A, Valton AL, Gibcus JH, Uebersohn A, Abdennur N, Dekker J, Mirny LA, Bruneau BG. 2017. Targeted degradation of CTCF decouples local insulation of chromosome domains from genomic compartmentalization. *Cell* **169:** 930–944.e22. doi:10.1016/j.cell.2017.05.004

Nuebler J, Fudenberg G, Imakaev M, Abdennur N, Mirny LA. 2018. Chromatin organization by an interplay of loop extrusion and compartmental segregation. *Proc Natl Acad Sci* **115:** E6697–E6706. doi:10.1073/pnas.1717730115

Oksuz BA, Yang L, Abraham S, Venev SV, Krietenstein N, Parsi KM, Ozadam H, Oomen ME, Nand A, Mao H, et al. 2020. Systematic evaluation of chromosome conformation capture assays. bioRxiv doi:10.1101/2020.12.26.424448

Olan I, Parry AJ, Schoenfelder S, Narita M, Ito Y, Chan ASL, Slater GSC, Bihary D, Bando M, Shirahige K, et al. 2020. Transcription-dependent cohesin repositioning rewires chromatin loops in cellular senescence. *Nat Commun* **11:** 6049. doi:10.1038/s41467-020-19878-4

Orlandini E, Marenduzzo D, Michieletto D. 2019. Synergy of topoisomerase and structural-maintenance-of-chromosomes proteins creates a universal pathway to simplify genome topology. *Proc Natl Acad Sci* **116:** 8149–8154. doi:10.1073/pnas.1815394116

Patel L, Kang R, Rosenberg SC, Qiu Y, Raviram R, Chee S, Hu R, Ren B, Cole F, Corbett KD. 2019. Dynamic reorganization of the genome shapes the recombination landscape in meiotic prophase. *Nat Struct Mol Biol* **26:** 164–174. doi:10.1038/s41594-019-0187-0

Piazza A, Bordelet H, Dumont A, Thierry A, Savocco J, Girard F, Koszul R. 2020. Cohesin regulates homology search during recombinational DNA repair. bioRxiv doi:10.1101/2020.12.17.423195

Pickersgill H, Kalverda B, de Wit E, Talhout W, Fornerod M, van Steensel B. 2006. Characterization of the *Drosophila melanogaster* genome at the nuclear lamina. *Nat Genet* **38:** 1005–1014. doi:10.1038/ng1852

Ramani V, Deng X, Qiu R, Gunderson KL, Steemers FJ, Disteche CM, Noble WS, Duan Z, Shendure J. 2017. Massively multiplex single-cell Hi-C. *Nat Methods* **14:** 263–266. doi:10.1038/nmeth.4155

Rao SSP, Huntley MH, Durand NC, Stamenova EK, Bochkov ID, Robinson JT, Sanborn AL, Machol I, Omer AD, Lander ES, et al. 2014. A 3D map of the human genome at kilobase resolution reveals principles of chromatin looping. *Cell* **159:** 1665–1680. doi:10.1016/j.cell.2014.11.021

Rao SSP, Huang SC, Glenn St Hilaire B, Engreitz JM, Perez EM, Kieffer-Kwon KR, Sanborn AL, Johnstone SE, Bascom GD, Bochkov ID, et al. 2017. Cohesin loss eliminates all loop domains. *Cell* **171:** 305–320.e24. doi:10.1016/j.cell.2017.09.026

Riggs AD. 1990. DNA methylation and late replication probably aid cell memory, and type I DNA reeling could aid chromosome folding and enhancer function. *Philos Trans R Soc Lond B Biol Sci* **326:** 285–297. doi:10.1098/rstb.1990.0012

Rosa A, Everaers R. 2008. Structure and dynamics of interphase chromosomes. *PLoS Comput Biol* **4:** e1000153. doi:10.1371/journal.pcbi.1000153

Rosa A, Becker NB, Everaers R. 2010. Looping probabilities in model interphase chromosomes. *Biophys J* **98:** 2410–2419. doi:10.1016/j.bpj.2010.01.054

Rubinstein M, Colby R. 2003. *Polymer physics.* Oxford University Press, Oxford.

Ryu JK, Rah SH, Janissen R, Kerssemakers JWJ, Dekker C. 2020. Resolving the step size in condensin-driven DNA loop extrusion identifies ATP binding as the step-generating process. bioRxiv doi:10.1101/2020.11.04.368506

Sanborn AL, Rao SSP, Huang SC, Durand NC, Huntley MH, Jewett AI, Bochkov ID, Chinnappan D, Cutkosky A, Li J, et al. 2015. Chromatin extrusion explains key features of loop and domain formation in wild-type and engineered genomes. *Proc Natl Acad Sci* **112:** E6456–E6465. doi:10.1073/pnas.1518552112

Schalbetter SA, Fudenberg G, Baxter J, Pollard KS, Neale MJ. 2019. Principles of meiotic chromosome assembly revealed in *S. cerevisiae*. *Nat Commun* **10:** 4795. doi:10.1038/s41467-019-12629-0

Schwarzer W, Abdennur N, Goloborodko A, Pekowska A, Fudenberg G, Loe-Mie Y, Fonseca NA, Huber W, Haering CH, Mirny L, et al. 2017. Two independent modes of chromatin organization revealed by cohesin removal. *Nature* **551:** 51–56. doi:10.1038/nature24281

Shi Z, Gao H, Bai XC, Yu H. 2020. Cryo-EM structure of the human cohesin-NIPBL-DNA complex. *Science* **368:** 1454–1459. doi:10.1126/science.abb0981

Shintomi K, Hirano T. 2011. The relative ratio of condensin I to II determines chromosome shapes. *Genes Dev* **25:** 1464–1469. doi:10.1101/gad.2060311

Simonis M, Klous P, Splinter E, Moshkin Y, Willemsen R, de Wit E, van Steensel B, de Laat W. 2006. Nuclear organization of active and inactive chromatin domains uncovered by chromosome conformation capture-on-chip (4C). *Nat Genet* **38:** 1348–1354. doi:10.1038/ng1896

Solovei I, Kreysing M, Lanctôt C, Kösem S, Peichl L, Cremer T, Guck J, Joffe B. 2009. Nuclear architecture of rod photoreceptor cells adapts to vision in mammalian evolution. *Cell* **137:** 356–368. doi:10.1016/j.cell.2009.01.052

Solovei I, Wang AS, Thanisch K, Schmidt CS, Krebs S, Zwerger M, Cohen TV, Devys D, Foisner R, Peichl L, et al. 2013. LBR and lamin A/C sequentially tether peripheral heterochromatin and inversely regulate differentiation. *Cell* **152:** 584–598. doi:10.1016/j.cell.2013.01.009

Solovei I, Thanisch K, Feodorova Y. 2016. How to rule the nucleus: *divide et impera*. *Curr Opin Cell Biol* **40:** 47–59. doi:10.1016/j.ceb.2016.02.014

Spracklin G, Pradhan S. 2020. Protect-seq: genome-wide profiling of nuclease inaccessible domains reveals physical properties of chromatin. *Nucleic Acids Res* **48:** e16. doi:10.1093/nar/gkz1150

Stevens TJ, Lando D, Basu S, Atkinson LP, Cao Y, Lee SF, Leeb M, Wohlfahrt KJ, Boucher W, O'Shaughnessy-Kirwan A, et al. 2017. 3D structures of individual mammalian genomes studied by single-cell Hi-C. *Nature* **544:** 59–64. doi:10.1038/nature21429

Stik G, Vidal E, Barrero M, Cuartero S, Vila-Casadesús M, Mendieta-Esteban J, Tian TV, Choi J, Berenguer C, Abad A, et al. 2020. CTCF is dispensable for immune cell transdifferentiation but facilitates an acute inflammatory response. *Nat Genet* **52:** 655–661. doi:10.1038/s41588-020-0643-0

Su JH, Zheng P, Kinrot SS, Bintu B, Zhuang X. 2020. Genome-scale imaging of the 3D organization and transcriptional activity of chromatin. *Cell* **182:** 1641–1659.e26. doi:10.1016/j.cell.2020.07.032

Takemata N, Samson RY, Bell SD. 2019. Physical and functional compartmentalization of archaeal chromosomes. *Cell* **179:** 165–179.e18. doi:10.1016/j.cell.2019.08.036

Tavares-Cadete F, Norouzi D, Dekker B, Liu Y, Dekker J. 2020. Multi-contact 3C reveals that the human genome during interphase is largely not entangled. *Nat Struct Mol Biol* **27:** 1105–1114. doi:10.1038/s41594-020-0506-5

Tedeschi A, Wutz G, Huet S, Jaritz M, Wuensche A, Schirghuber E, Davidson IF, Tang W, Cisneros DA, Bhaskara V, et al. 2013. Wapl is an essential regulator of chromatin structure and chromosome segregation. *Nature* **501:** 564–568. doi:10.1038/nature12471

Towbin BD, González-Aguilera C, Sack R, Gaidatzis D, Kalck V, Meister P, Askjaer P, Gasser SM. 2012. Step-wise methylation of histone H3K9 positions heterochromatin at the nuclear periphery. *Cell* **150:** 934–947. doi:10.1016/j.cell.2012.06.051

Tran NT, Laub MT, Le TBK. 2017. SMC progressively aligns chromosomal arms in *Caulobacter crescentus* but is antagonized by convergent transcription. *Cell Rep* **20:** 2057–2071. doi:10.1016/j.celrep.2017.08.026

Trojanowski J, Frank L, Rademacher A, Grigaitis P, Rippe K. 2021. Transcription activation is enhanced by multivalent interactions independent of phase separation. bioRxiv doi:10.1101/2021.01.27.428421

van Koningsbruggen S, Gierlinski M, Schofield P, Martin D, Barton GJ, Ariyurek Y, den Dunnen JT, Lamond AI. 2010. High-resolution whole-genome sequencing reveals that specific chromatin domains from most human chromosomes associate with nucleoli. *Mol Biol Cell* **21:** 3735–3748. doi:10.1091/mbc.e10-06-0508

van Schaik T, Vos M, Peric-Hupkes D, Hn Celie P, van Steensel B. 2020. Cell cycle dynamics of lamina-associated DNA. *EMBO Rep* **21:** e50636. doi:10.15252/embr.202050636

van Steensel B, Belmont AS. 2017. Lamina-associated domains: links with chromosome architecture, heterochromatin, and gene repression. *Cell* **169:** 780–791. doi:10.1016/j.cell.2017.04.022

van Steensel B, Delrow J, Henikoff S. 2001. Chromatin profiling using targeted DNA adenine methyltransferase. *Nat Genet* **27:** 304–308. doi:10.1038/85871

Varier RA, Outchkourov NS, de Graaf P, van Schaik FMA, Ensing HJL, Wang F, Higgins JMG, Kops GJPL, Timmers HTM. 2010. A phospho/methyl switch at histone H3 regulates TFIID association with mitotic chromosomes. *EMBO J* **29:** 3967–3978. doi:10.1038/emboj.2010.261

Vertii A, Ou J, Yu J, Yan A, Pagès H, Liu H, Zhu LJ, Kaufman PD. 2019. Two contrasting classes of nucleolus-associated domains in mouse fibroblast heterochromatin. *Genome Res* **29:** 1235–1249. doi:10.1101/gr.247072.118

Vian L, Pękowska A, Rao SSP, Kieffer-Kwon KR, Jung S, Baranello L, Huang SC, El Khattabi L, Dose M, Pruett N, et al. 2018. The energetics and physiological impact of cohesin extrusion. *Cell* **173:** 1165–1178.e20. doi:10.1016/j.cell.2018.03.072

Vietri Rudan M, Barrington C, Henderson S, Ernst C, Odom DT, Tanay A, Hadjur S. 2015. Comparative Hi-C reveals that CTCF underlies evolution of chromosomal domain architecture. *Cell Rep* **10:** 1297–1309. doi:10.1016/j.celrep.2015.02.004

Vieux-Rochas M, Fabre PJ, Leleu M, Duboule D, Noordermeer D. 2015. Clustering of mammalian *Hox* genes with other H3K27me3 targets within an active nuclear domain. *Proc Natl Acad Sci* **112:** 4672–4677. doi:10.1073/pnas.1504783112

Wang X, Brandão HB, Le TBK, Laub MT, Rudner DZ. 2017. Bacillus subtilis SMC complexes juxtapose chromosome arms as they travel from origin to terminus. *Science* **355:** 524–527. doi:10.1126/science.aai8982

Weiss FD, Calderon L, Wang YF, Georgieva R, Guo Y, Cvetesic N, Kaur M, Dharmalingam G, Krantz ID, Lenhard B, et al. 2021. Neuronal genes deregulated in Cornelia de Lange syndrome respond to removal and re-expression of cohesin. *Nat Commun* **12:** 2919. doi:10.1038/s41467-021-23141-9

Wike CL, Guo Y, Tan M, Nakamura R, Shaw DK, Díaz N, Whittaker-Tademy AF, Durand NC, Aiden EL, Vaquerizas JM, et al. 2021. Chromatin architecture transitions from zebrafish sperm through early embryogenesis. *Genome Res* **31:** 981–994. doi:10.1101/gr.269860.120

Wong X, Hoskins VE, Harr JC, Gordon M, Reddy KL. 2020. Lamin C regulates genome organization after mitosis. bioRxiv doi:10.1101/2020.07.28.213884

Wood C, Tonegawa S. 1983. Diversity and joining segments of mouse immunoglobulin heavy chain genes are closely linked and in the same orientation: implications for the joining mechanism. *Proc Natl Acad Sci* **80:** 3030–3034. doi:10.1073/pnas.80.10.3030

Wutz G, Várnai C, Nagasaka K, Cisneros DA, Stocsits RR, Tang W, Schoenfelder S, Jessberger G, Muhar M, Hossain MJ, et al. 2017. Topologically associating domains and chromatin loops depend on cohesin and are regulated by CTCF, WAPL, and PDS5 proteins. *EMBO J* **36:** 3573–3599. doi:10.15252/embj.201798004

Zhang H, Emerson DJ, Gilgenast TG, Titus KR, Lan Y, Huang P, Zhang D, Wang H, Keller CA, Giardine B, et al. 2019. Chromatin structure dynamics during the mitosis-to-G1 phase transition. *Nature* **576:** 158–162. doi:10.1038/s41586-019-1778-y

Zhang L, Zhang Y, Chen Y, Gholamalamdari O, Wang Y, Ma J, Belmont AS. 2021. TSA-seq reveals a largely conserved genome organization relative to nuclear speckles with small position changes tightly correlated with gene expression changes. *Genome Res* **31:** 251–264. doi:10.1101/gr.266239.120

Uncovering the Principles of Genome Folding by 3D Chromatin Modeling

Asli Yildirim,[1] Lorenzo Boninsegna,[1] Yuxiang Zhan,[1,2] and Frank Alber[1,2]

[1]Institute for Quantitative and Computational Biosciences, Department of Microbiology, Immunology and Molecular Genetics, University of California Los Angeles, Los Angeles, California 90095, USA

[2]Quantitative and Computational Biology, Department of Biological Sciences, University of Southern California, Los Angeles, California 90089, USA

Correspondence: falber@g.ucla.edu

Our understanding of how genomic DNA is tightly packed inside the nucleus, yet is still accessible for vital cellular processes, has grown dramatically over recent years with advances in microscopy and genomics technologies. Computational methods have played a pivotal role in the structural interpretation of experimental data, which helped unravel some organizational principles of genome folding. Here, we give an overview of current computational efforts in mechanistic and data-driven 3D chromatin structure modeling. We discuss strengths and limitations of different methods and evaluate the added value and benefits of computational approaches to infer the 3D structural and dynamic properties of the genome and its underlying mechanisms at different scales and resolution, ranging from the dynamic formation of chromatin loops and topological associated domains to nuclear compartmentalization of chromatin and nuclear bodies.

Over the last decade, a proliferation of improved sequencing-based genomics and microscopy technologies has led to new opportunities to unravel the intertwined relationship between the three-dimensional (3D) organization of genomes and key biological processes (Dekker et al. 2017; Kempfer and Pombo 2020). Experimental evidence points to a complex and polymorphic chromatin fiber at the intermediate folding level, which is in agreement with an irregular dynamic folding state for interphase chromatin (Ou et al. 2017; Bintu et al. 2018). A major advance in our understanding came from the realization of the ubiquitous nature of chromatin loops and topologically associating domains (TADs) from genome-wide mapping methods such as Hi-C (Dixon et al. 2012; Nora et al. 2012; Sexton et al. 2012; Rao et al. 2014), genome architecture mapping (GAM) (Beagrie et al. 2017), chromatin interaction analysis by paired-end tag sequencing (Chia-PET) (Fullwood et al. 2009), and by superresolution microscopy (Kempfer and Pombo 2020; McCord et al. 2020) (Fig. 1A,B). Chromatin loops are inherently dynamic structures and, together with TAD borders, emerge in ensemble data at binding site locations of insulator proteins from weak preferences of loop positions summed over many cells (Rao et al. 2014). Insulator proteins act as enhancer-block-

Cite this article as *Cold Spring Harb Perspect Biol* doi: 10.1101/cshperspect.a039693

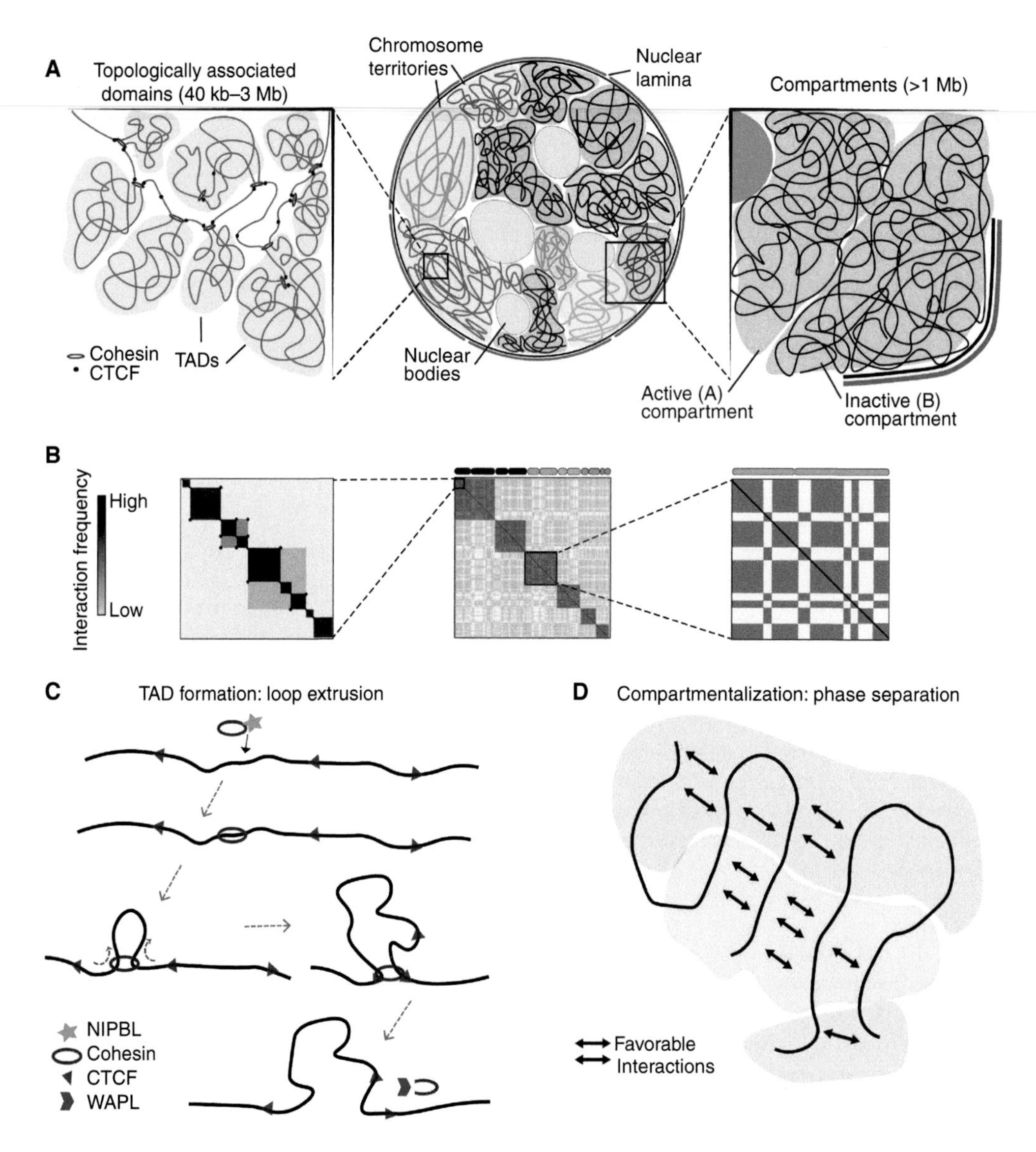

Figure 1. Hallmarks of spatial genome organization. (*A*) Chromosomes form distinct territories in the nucleus (*middle* panel). Active and inactive chromatin are further segregated into nuclear compartments (*right*) and organized around nuclear bodies such as nuclear speckles, nucleoli, and nuclear lamina. At local and intermediate folding levels (40 kb–3 Mb), chromatin forms loops and topologically associating domains (TADs), the boundaries of which are enriched with CTCF and cohesin (*left*). (*B*) Schematic view of Hi-C contact map patterns characteristic of chromosome territories (*middle*), nuclear compartmentalization (*right*), and TADs and loops (*left*). TADs appear as blocks with enriched contacts along the diagonal; loops appear as dots of locally enriched contact frequencies, often at corners of chromatin domains. Domains can be nested with patterns of "sub-TADs" and nested loops. Nuclear compartmentalization is characterized by a checkerboard pattern for long-range and interchromosomal interactions. (*C*) Formation of TADs and loops by loop extrusion. Cohesin-loader protein NIPBL facilitates the initial association of cohesin with chromatin. Once cohesin is loaded, it starts extruding chromatin and stops when it encounters a bound CTCF protein with binding motif in the proper directional orientation. Other proteins and cohesin complexes may also terminate the extrusion process. Cohesin is released from chromatin by the cohesin release factor, WAPL. (*D*) Nuclear compartmentalization arises from phase separation of chromatin via favorable interactions between similar chromatin types.

Cite this article as *Cold Spring Harb Perspect Biol* doi: 10.1101/cshperspect.a039693

ing elements to prevent communication between enhancer and promoters and/or as barriers to prevent the spreading of heterochromatin. At the global level, mammalian genome folding is governed by affinity-driven spatial compartmentalization (Fig. 1A,B). Chromatin in different functional classes are spatially segregated by microphase separation (Hildebrand and Dekker 2020) and chromatin associations to nuclear bodies, including nuclear speckles and lamina domains (Steensel and Henikoff 2000; Guelen et al. 2008; Chen et al. 2018).

Nuclear genome organization is therefore the result of a multitude of competing processes at various organizational levels (Misteli 2020). For instance, compartment and loop formation processes act antagonistically and are driven by distinct mechanisms (Fig. 1C, D), which prevents a common structural hierarchy for the description of chromatin folding across scales (Nuebler et al. 2018). Moreover, critical components and underlying physical mechanisms are often not yet fully understood. It is evident that the complexities of nuclear organization cannot be explained by a single experimental method and require a combination of complementary data and methods (Dekker et al. 2017). Computational simulations offer the unique opportunity to address some of these challenges: they provide a robust platform to integrate different data types, can assist in the interpretation of experimental data, and are capable to test hypotheses of physical processes that may be responsible for shaping genome organization. Over recent years, a multitude of different computational methods have been published, each with individual strengths and limitations. These complementary methods provide a renewed opportunity to gain better quantitative understanding of the nuclear organization. Here we provide a detailed overview of computational methods available for various applications in genome modeling.

MODELING APPROACHES

Computational simulations are uniquely suited to validate proposed mechanisms of genome folding if their outcome agrees with experimental observations. Simulating nuclear processes with all their complexities is a challenging task, and choosing the appropriate structural representation is a crucial component. An effective model requires the identification of descriptors, which are structural units that are most informative at the chosen level of resolution and whose interactions can be faithfully described. Such an approach is known as coarse-graining. For coarse-grained genome-scale models, the chromatin fiber is typically represented as a polymer (Fig. 2A), defined as a chain of connected beads, each representative of a specific chromatin region typically with hundreds to millions of base pairs depending on the chosen model of granularity (i.e., base-pair resolution) (Lin et al. 2019; Parmar et al. 2019; Brackey et al. 2020). Bead monomers can be assigned to different types (often referred to as colors), to account for epigenetic or biochemical identities.

The chromatin bead chain dynamics is defined by a set of mathematical functions (also referred to as energy terms), which describe the energy of the model as a function of its coordinates. The complete set of energy functions is often referred to as the system Hamiltonian. A polymer Hamiltonian involves at least harmonic bonded terms (i.e., "springs" connecting the beads), which ensure chain connectivity, and excluded volume terms, which prevent spatial overlap between beads.

Additional terms are added to the Hamiltonian to account for model-specific interactions, which are either derived from physical principles or experimental data. For instance, a short-range attractive force between beads of the same type can reproduce phase separation of chromatin types and spatial constraints can account for chromatin tethering to the nuclear landmarks.

Genome structures show a high degree of cell-to-cell variation (Finn and Misteli 2019). Therefore, polymer models must sample large numbers of chromosome conformations, which are typically explored by standard computational techniques such as molecular dynamics (MD) (Allen and Tildesley 1987; Schlick 1996), simulated annealing (Kirkpatrick et al. 1983), or

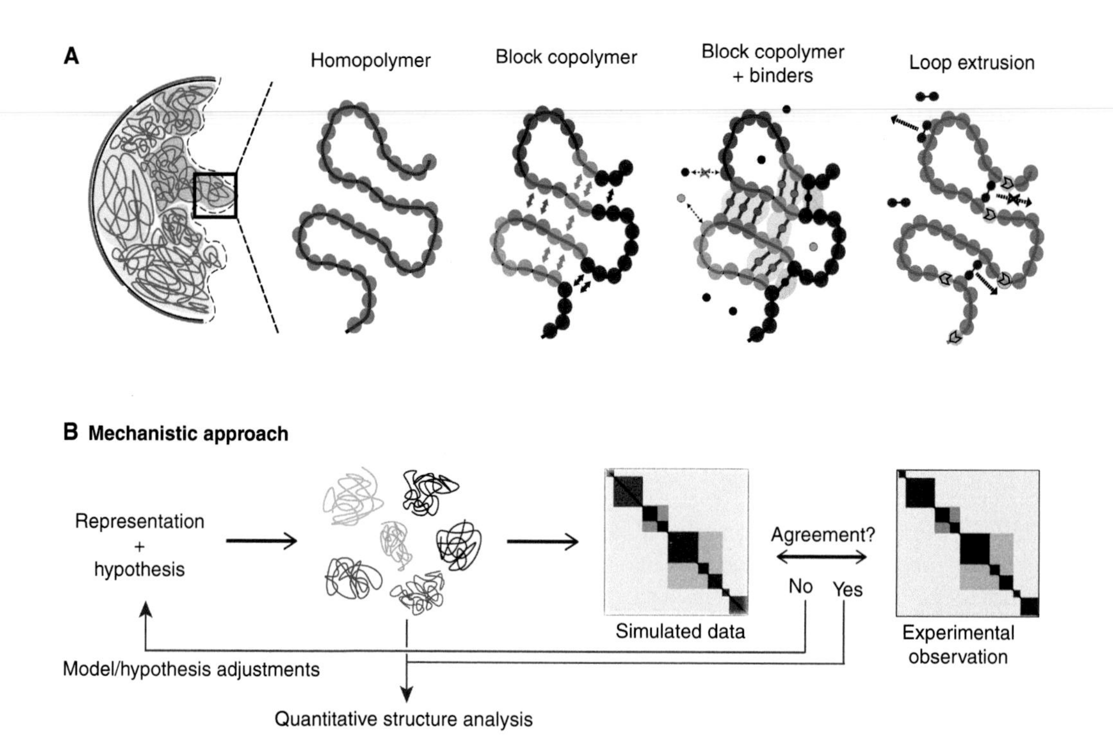

Figure 2. 3D chromatin modeling with a mechanistic approach. (*A*) Different polymer models for simulating chromatin structures. A homopolymer chain models the chromatin fiber as a chain of beads connected by "springs" (i.e., harmonic potentials). Excluded volume potentials prevent spatial overlap of beads. Block copolymer models classify bead monomers into distinct types, which share similar physical properties. Beads of the same type experience favorable attractive interactions to each other. Some approaches include binder particles, which can mediate interactions between chromatin beads of the same type only if they are bound to the chain. The loop extrusion model imitates the loop extrusion mechanism. Extrusion factors are diffusive particles that can bind to regions on the fiber (orange links). The extrusion factor is represented by an additional bond between beads that form a loop base. This bond is iteratively shifted along the chain to gradually extrude the chromatin loop. The extrusion process is terminated when the extrusion factor encounters a boundary element with proper orientation (yellow beads) or another extrusion factor. (*B*) Schematic flowchart of a mechanistic modeling approach. The approach relies on prior knowledge about a folding hypothesis that can explain the experimental observations. The polymer model is defined to simulate the hypothesis. If the collected structures do not agree with experimental observations, the initial hypothesis and/or the model parameters are modified and new simulations are performed until the resulting structures are able to reproduce the experimental data. Finally, the collected models are used for quantitative structure analysis.

Monte Carlo simulations (Metropolis et al. 1953). Quantitative analysis of structural features is then performed on the ensemble of all sampled conformations.

Because of their versatility, polymer models have been highly successful in modeling structure and dynamics of biomolecules. Here, we distinguish between two classes of modeling approaches, namely, the mechanistic (bottom-up) and the data-driven (top-down) approach. In the mechanistic approach (Fig. 2B), a physical process is postulated based on empirical evidence or physical intuition, and is explicitly included in the Hamiltonian. This approach enables testing of folding mechanisms by estimating observables to be compared with experiments and predicting the response upon perturbations of the process. Parametrization has to be carefully calibrated for simulations to faithfully reproduce experimental evidence, which is far from trivial. Because ge-

 Cite this article as *Cold Spring Harb Perspect Biol* doi: 10.1101/cshperspect.a039693

nome structure is the result of a collective interplay between various processes across hierarchical scales, these models are not easily scalable and not all the molecular players nor potential mechanisms are known yet.

The second class of models are data-driven (Fig. 3B), for which the Hamiltonian calibration can be performed in an almost fully unsupervised fashion. Here, no prior knowledge about chromatin folding processes is required; instead, experimental data, such as Hi-C, are fed to a protocol that generates structures recapitulating the data. The Hamiltonian will include energy terms to model the effect of each data point. Special care has to be devoted to a faithful interpretation of experimental data and its representation in the Hamiltonian. Key components are adequate model assessment strategies, a description of model uncertainties, and detection of false data points, which could affect the outcome. Integration of data modalities from independent experiments can overcome some of the limitations to increase accuracy and structural coverage. Data-driven approaches resemble those in structural biology, in which 3D structures are examined to infer mechanistic insights and make quantitative structure–function predictions for specific genomic regions.

In the following sections, we will first review a number of mechanistic modeling approaches, which explain the formation of chromatin loops, TADs, and phase-separated compartments. Then, we focus on data-driven approaches, which use experimental data to generate genome structures to gain functional insights.

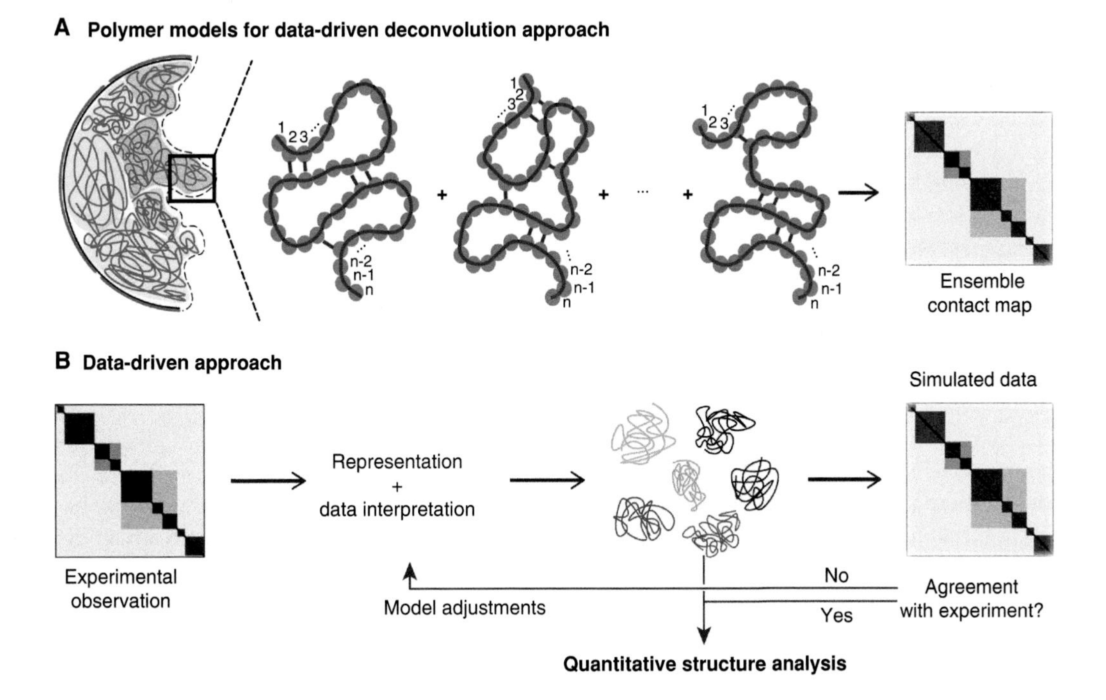

Figure 3. 3D chromatin modeling with a data-driven approach. (*A*) In the data-driven deconvolution approach, chromatin is modeled as a chain of beads with additional contacts derived from the Hi-C data. Each structure in the population contains a subset of chromatin contacts, the summation of which recapitulates the ensemble Hi-C data. (*B*) Data-driven methods use the experimental data explicitly to build the 3D models. They generate structures that are able to fully reproduce the experimental data. Adjustments of the initial models and/or data interpretation can be performed to ensure better agreement with experimental data. The collected models are finally used for quantitative structure analysis.

Mechanistic (Bottom-Up) Modeling Methods

Studying the Origins of Chromatin Loops and TADs

A breakthrough in our understanding of chromatin folding came from the genome-wide detection of chromatin loops, contact domains (Rao et al. 2014), and TADs from Hi-C experiments (Dixon et al. 2012; Nora et al. 2012; Sexton et al. 2012), and subsequently other genomics (e.g., SPRITE [Quinodoz et al. 2018], GAM [Beagrie et al. 2017], and imaging data [Bintu et al. 2018; Nir et al. 2018]). Chromatin loops are identified as locally enriched peaks in the Hi-C maps (Rao et al. 2014), often located at domain borders. TADs are revealed in Hi-C maps as squared blocks of enriched frequencies along the diagonal (Fig. 1B), indicating preferred interactions within and depleted interactions between TAD regions (Dixon et al. 2012; Nora et al. 2012; Sexton et al. 2012). Larger TADs are often nested with smaller "sub-TADs" (i.e., contact domains) and shared loops at their boundaries (Shen et al. 2012; Phillips-Cremins et al. 2013; Rao et al. 2014). Loop and most TAD boundaries are enriched with CTCF (CCCTC-binding factor protein) with binding motifs in convergent orientation at the two loop anchors as well as cohesin-binding sites (Rao et al. 2014).

Loop formation was initially explained by random polymer collisions, which could form small loops and condense chromosomes efficiently (Marko and Siggia 1997; Sankararaman and Marko 2005), but would be ineffective for the reproducible formation of larger loops and could not explain the enrichment of convergent CTCF-binding motifs at TAD boundaries.

The Loop Extrusion Model. Over the last decade, loop extrusion has emerged as a potential mechanism for chromosome condensation and chromatin looping, initially proposed as a statistical mechanical model (Nasmyth 2001; Alipour and Marko 2012). Only recently, a series of landmark papers have introduced polymer simulations to manifest this process as a primary mechanism for TAD formation by recapitulating a wealth of experimental observations from Hi-C data and superresolution microscopy (Sanborn et al. 2015; Fudenberg et al. 2016, 2017; Goloborodko et al. 2016a,b). Loop extrusion is initiated when a loop-extruding factor selectively attaches to the chromatin fiber and actively reels in the chromatin fiber in both directions, thus leading to a progressively growing loop until the loop-extruder falls off or encounters either another extrusion factor or an extrusion barrier at specific boundary elements (Fig. 1C). Several loop extruder complexes have been identified, including structural maintenance of chromosomes (SMCs) complexes, condensin for compaction of mitotic chromosomes (Hirano 2002), and cohesin (Wood et al. 2010; Uhlmann 2016) for loop extrusion in interphase (Kagey et al. 2010). The insulator protein CTCF acts as an extrusion barrier only if the CTCF-binding motifs are oriented in a convergent sequence orientation at the loop anchors (Rao et al. 2014). Other proteins may also act as extrusion barriers, including Znf143 (Bailey et al. 2015) and YY1 (Weintraub et al. 2017).

To simulate loop extrusion, the chromatin polymer can be represented as a self-avoiding chain of beads (Fig. 2A), which are connected by harmonic potentials, and typically have a base-pair resolution of ~0.1 kb (Gibcus et al. 2018) to ~1 kb (Sanborn et al. 2015). The interaction between the extrusion factor and the polymer chain is modeled by an additional bond between two beads, which forms the loop base (Fudenberg et al. 2016); active extrusion is then implemented by iteratively shifting the bond at a given rate to increasingly separated pairs of beads, one bead at a time. The process stops when the extruder runs into either an extrusion barrier or another extruder, or stochastically dissociates from the fiber, in which case the bond is removed. Size and position of loops are controlled, among others, by the separation and permeability of the boundary elements, and by the lifetime and processivity of the extruding factors. These simulation parameters play a critical role in recapitulating experimental observations, and are either inferred from Chip-seq, Chia-PET, or Hi-C data or selected so that models best match the experiments (Fudenberg et al. 2017; Gassler et al. 2017). Polymer simulations of loop extrusion have been very successful at recapitulating a wide variety of chromosomal phenomena, including TAD and loop formation in interphase (Sanborn et al. 2015; Fudenberg et al. 2016), structure changes upon perturbations (Fudenberg et al. 2017; Nuebler et al. 2018), and compaction of mitotic chromatids (Alipour and Marko 2012; Goloborodko et al. 2016b; Gibcus et al. 2018). Simulations naturally explain CTCF directionality and predict almost all characteristic Hi-C patterns of nested sub-TADs, peaks at TAD corners, and stripes from unidirectional chromatin extrusion when cohesin lands near a CTCF site (Fudenberg et al. 2016). Models correctly predicted changes in Hi-C maps upon degradation of protein factors (Fudenberg et al. 2017; Nuebler et al. 2018), including the loss of peaks and TADs upon induced degradation of cohesin (Rao et al. 2017; Wutz et al. 2017) and cohesin-loading complexes (Schwarzer et al. 2017), together with an overall structure decompaction (Nozaki et al. 2017). Simulating CTCF depletion (Fudenberg et al. 2017; Nuebler et al. 2018) blurred TAD boundaries but maintained chromatin compaction from ongoing loop extrusion (Nora et al. 2017; Nozaki et al. 2017; Wutz et al. 2017), while depletion of the cohesin release factor WAPL correctly predicted a proliferation in the number of peaks (Haarhuis et al. 2017; Wutz et al. 2017), resulting in prophase-like elongated nuclear chromatin structures, also known as vermicelli chromatids ("vermicelli" is Italian for "small worms") (Tedeschi et al. 2013).

A debate has been ongoing whether loop extrusion is an equilibrium or energy-driven process (Banigan and Mirny 2020). Whereas most works support an active, ATP-driven process, others (Brackley et al. 2017, 2018) proposed a process without motor activity, where cohesin complexes slide directionally biased by osmotic pressure from subsequently loaded cohesin complexes (also known as the slip-link model). Recently, single-molecule experiments imaged active DNA loop extrusion by condensin and cohesin in vitro, therefore showing evidence of a nonequilibrium process (Ganji et al. 2018; Davidson et al. 2019; Golfier et al. 2020; Kong et al. 2020).

Block Copolymer Models to Study Global Chromatin Compartmentalization

Loop extrusion is typically limited to local interactions of up to tens of Mb sequence distance, and cannot account for the characteristic checkerboard patterns of long-range and interchromosomal interactions in Hi-C maps (Fig. 1B). Those are a result of spatial segregation of euchromatin and heterochromatin compartments, driven by preferential affinity for chromatin in the same functional state (Lieberman-Aiden et al. 2009; Rao et al. 2014; Solovei et al. 2016; van Steensel and Belmont 2017; Falk et al. 2019; Hildebrand and Dekker 2020).

Phase separation has been extensively studied with a variety of block copolymer models (Leibler 1980), which describe the chromatin fiber as a self-avoiding chain of beads of distinct types (i.e., active and inactive chromatin) and short-range attractive forces acting only between beads of the same type (Jost et al. 2014; Chiariello et al. 2016; Di Pierro et al. 2016; Michieletto et al. 2016; Haddad et al. 2017; MacPherson et al. 2018). Because the chromatin fiber consists of alternating blocks of active and inactive chromatin, the covalent linkages of the chain prevent the complete separation into only two macrophases (Hildebrand and Dekker 2020). Instead, the two compartments spatially segregate (microphase separation) into a larger number of smaller clusters (Fig. 1D). Microphase separation is possibly driven by phase separation (also referred as condensate formation), which creates distinct phases from a single homogeneous mixture through multivalent weak interactions between proteins or nucleic acids that preferentially associate with a given chromatin type and form condensates once a critical concentration is reached (Erdel and Rippe 2018).

Several studies explored phase separation of chromatin compartments as a cause for the spatial segregation of transcriptionally active euchromatin and inactive heterochromatin, driven by preferential affinity of chromatin to their own kind. Jost et al. (2014) modeled chromatin regions in *Drosophila melanogaster* at 10 kb resolution with a block copolymer of four chromatin types derived from epigenetic profiles. MD simulations with weak homotypic interactions produce random coil structures, while strong interactions reproduce the checkerboard-like interaction patterns of phase-separated chromatin (Jost et al. 2014). At specific interaction strengths, metastable configurations emerge in which TAD-like domains exist and transiently interact with each other, while global chromatin compartments undergo microphase separation.

Phase separation can be driven by binding factors, which bridge chromatin through multivalent interactions. Heterochromatin protein 1 (HP1) binds methylated H3K9me3 histone tails and oligomerizes with HP1 at other chromatin regions, which then condenses into a heterochromatic phase (Larson et al. 2017; Strom et al. 2017). Some simulations imitate such a process. MacPherson et al. (2018) modeled human chromosome 16 as a worm-like chain of nucleosomes classified into eu- or heterochromatin based on H3K9me3 occupancy. Heterochromatin interaction strengths varied with the number of HP1 entities bound to nucleosomes. At specific HP1 concentrations, simulations led to phase separation of heterochromatin with chromatin densities similar to those in experiments. Besides appropriate interaction strengths and HP1 concentrations, sufficiently large copolymer block sizes (here ∼20 kb for given modeling parameters) were also crucial requirements to faithfully reproduce compartmental phase separation. Overall, these models reproduced some but not all aspects of the experiment, such as the scaling behavior of contacts in Hi-C maps, which emphasizes that additional physical processes must be considered.

Recently, a pivotal study investigated what type of homotypic interactions are the dominant factor in reproducing compartmental phase separation of rod cells from nocturnal mammalians (Falk et al. 2019). Falk et al. used a copolymer model with three chromatin types (euchromatin, heterochromatin, and pericentromeric heterochromatin) to simulate an artificial nucleus with eight chromosomes at a base-pair resolution of 40 kb per bead (Fig. 4A). Exhaustive screens across all combinations of interaction strengths indicated that heterochromatin attractive interactions played the primary role in phase separation in both inverted (i.e., interior heterochromatin) and conventional nuclei (i.e., peripheral heterochromatin), whereas euchromatic attractive interactions were dispensable. However, the nuclear position of heterochromatin in conventional nuclei cannot be reproduced by phase separation alone, and requires additional attractive forces to the nuclear periphery.

Combining Loop Extrusion with Block Copolymer Models

Phase separation describes an equilibrium state and generally does not recapitulate loops from energy-driven loop extrusions. Nuebler et al. (2018) jointly simulated loop extrusion and copolymer phase separation and demonstrated that active loop extrusion shifts local chromosome structures to a nonequilibrium state and overrides locally the fine-scale compartment patterns that would otherwise be visible at longer-range interactions. Genome compartments and loops are driven by distinct mechanisms and a common hierarchical organization with TADs as building blocks of compartments does not likely exist. For instance, simulating depletion of loop extrusion factor cohesin reduced TADs and revealed finer compartments, while increased processivity of cohesin strengthened large TADs and reduced compartmentalization (Fudenberg et al. 2017; Nuebler et al. 2018). Depletion of extrusion barrier protein CTCF weakened TADs, while leaving compartments and chromatin compaction unaffected. There is growing experimental evidence that TADs can exist without compartments and vice versa (Nuebler et al. 2018; Mirny et al. 2019). A comprehensive discussion of phase separation and its role in genome organization is provided in a recent review (Hildebrand and Dekker 2020).

The minimal chromatin model (MiChroM) (Di Pierro et al. 2016) introduces a transferable force field based on predefined chromatin types. The method combines loops with block copolymer modeling and does not rely on a specific mechanism for loop formation. The chromosome is modeled as a self-avoiding polymer chain at 50 kb resolution, while loop locations

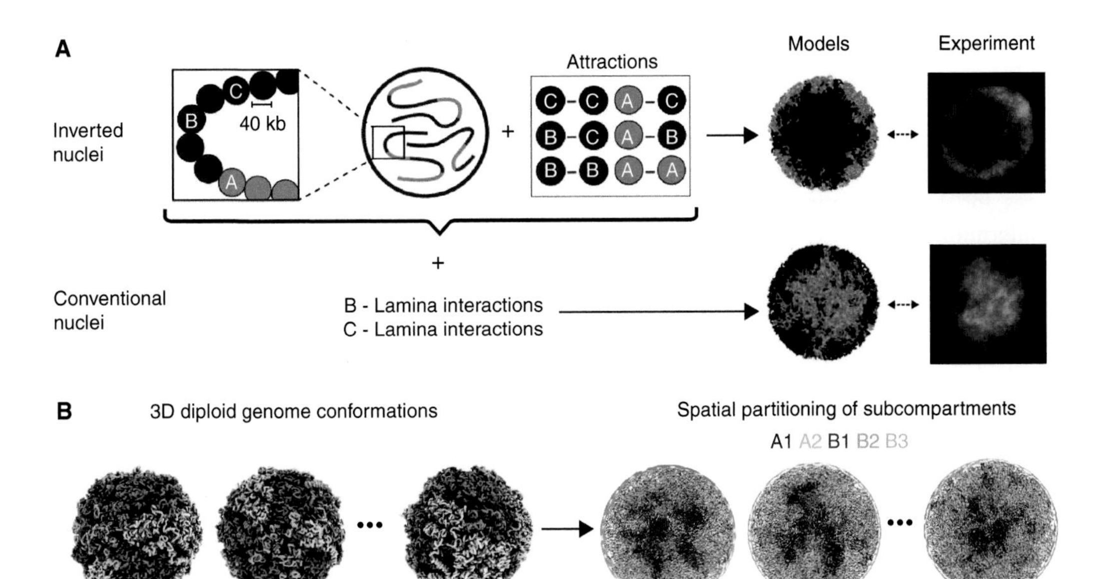

Figure 4. Selected examples of modeling studies investigating chromatin organization. (*A*) Falk et al. (2019) used a mechanistic approach with a block copolymer chromatin model containing three chromatin types. The model predicted the inverted chromatin organization observed in microscopy. Predicting the chromatin organization in conventional nuclei required additional interactions of heterochromatin regions (B and C types) with the nuclear lamina, which highlights the crucial role of lamina interactions in maintaining the conventional nuclei organization. (Panel *A* is from Falk et al. 2019; adapted, with permission, from Springer Nature © 2019.) (*B*) This example showcases our data-driven deconvolution approach, where we simulated a population of diploid genome structures for GM12878 cells from Hi-C data (Yildirim et al. 2021). The models were generated from Hi-C data without explicit subcompartment notations or block copolymer chromatin classes. The resulting structures were able to show the segregation of different chromatin subcompartments and provided additional insights into structural features that distinguish different subcompartments.

and chromatin types are derived from Hi-C data. The energy function contains 27 adjustable parameters, which are determined by maximizing the agreement of contact frequencies from simulated models and Hi-C data by means of Lagrange multipliers. Lagrange multipliers are used to find the extrema of a function that is subject to constraints. They are coefficients in the potential energy function to maximize the agreement between the contact probabilities from models and Hi-C experiments. Parameters trained on one chromosome can be used to model other chromosomes, in good agreement with Hi-C contact frequencies. Chromosome compaction is driven by a generic energy function, mimicking the behavior of a liquid crystal (Di Pierro et al. 2016).

An ensemble of chromosome structures is collected by MD simulations, which recapitulates the characteristic features of chromatin organization, including chromosome territories, phase separation of chromatin types, loops between specific anchors, and knot-free chromatin conformations (Di Pierro et al. 2016). MiChroM was later extended by a machine learning algorithm, MEGABASE, to infer chromatin types from Chip-seq data (Di Pierro et al. 2017). A recent application on six human cell types studied the structural heterogeneity of chromosomes across cell types (Cheng et al. 2020). When MiChroM is combined with Langevin dynamics (i.e., standard MD with additional stochastic and viscous energy terms allowing a thermal equilibration with the environment), models reproduce some dynamic behavior of chromosomes, including chromatin subdiffusion, viscoelasticity, and spatial coherence of chromatin as well as dynamically associated domains (DADs) from phase separation at longer time intervals (Di Pierro et al. 2018).

Qi and Zhang (2019) expanded MiChroM with an energy function for chromatin at 5 kb resolution. The Hamiltonian includes attractive forces for 15 chromatin states from histone modifications, and interaction potentials for intra-TAD chromatin, for a total of ~1800 parameters, which are adjusted to maximize agreement with Hi-C data. Parameters are transferable and, when trained on specific chromosome segments in one cell, produce structures of about 20 Mb in length for other chromosomes with good agreement to Hi-C data. A minimum of six chromatin types were sufficient to reproduce chromatin compartmentalization, while short-range chromatin interactions within TADs required a more detailed energy function. Recently, a generalized energy function was introduced to simulate diploid genome structures at 1 Mb resolution (Qi et al. 2020) and considered intra- and interchromosomal interaction potentials, centromere clustering, A/B compartmentalization, and inactive X-chromosome condensation. Overall, the structures captured global structural features such as chromosome locations and territories, and phase separation of A/B compartments. Their work suggested that correct chromosome positioning requires specific interchromosomal interactions and centromere clustering, and is not driven by phase separation alone.

Shi et al. (2018) developed a transferable chromosome copolymer model (CCM) with one free parameter for chromatin type and loop anchor interactions to reproduce TAD and compartment organization. Brownian dynamics simulations of relatively small chromosomal regions at 1.2 kb resolution revealed a hierarchical folding where chromosome droplets (CDs) similar in size to TADs formed first, which then coalesced to a more compact state.

The strings and binder switch (SBS) model (Nicodemi and Prisco 2009; Barbieri et al. 2012) is an approach for jointly modeling loop formation and chromatin compartmentalization. The model is inspired by transcription factors (TFs) that bridge *cis*-regulatory elements with distal promoters upon chromatin binding (Barbieri et al. 2017; Kundu et al. 2017). The diffusive-bridge model (Brackley et al. 2013, 2016; Buckle et al. 2018) also uses a similar strategy: the model introduces diffusive particles representing protein complexes that form polymer-protein-polymer bridges, which subsequently aggregate and induce local polymer compaction. In the SBS model, chromatin is modeled as a self-avoiding chain of beads surrounded by a cloud of diffusing "binder" particles, which can bind to chromatin and only then mediate interactions between distal chromatin beads (Fig. 2A). Beads and binders are classified into types (i.e., colors), and binders experience an attraction only to cognate beads of the same color. Multivalent binder interactions can generate a variety of loop patterns and TAD boundaries, which depend on the number of binder types, the sequence locations of binding sites, and the concentrations of binder particles. To express binder-mediated chromatin interactions, a truncated finite-range Lennard-Jones potential is added to a traditional polymer chain Hamiltonian. The Lennard-Jones potential describes an interaction between particles and consists of both a repulsive term at very small particle distances (to prevent particle overlap) and a weak attractive term at larger distances (which eventually converges to zero at very large distances). The interaction strength and equilibrium distance can be adjusted by choosing adequate parameter settings. MD simulations will then induce a classic coil-to-globule phase-separation of chromatin at specific binder concentrations, binding affinities, and diffusive properties of binders and chromatin (Annunziatella et al. 2016, 2018; Chiariello et al. 2016; Conte et al. 2020). Binder concentrations and affinities are unknown beforehand, and simulations over a range of values allow selection of those that maximize the agreement between models and experiment (Chiariello et al. 2016; Bianco et al. 2017, 2018).

SBS models are typically applied to chromosomal regions of a few ~Mb in size (Brackley et al. 2013) at a base-pair resolution between ~0.1 kb (Barbieri et al. 2012; Brackley et al. 2016; Bianco et al. 2018) to tens of kb per bead (Chiariello et al. 2016). The optimal number of binder types and binding sites and their sequence locations can be inferred from knowledge about the regulatory landscape (i.e., locations of genes, enhancers, CTCFs, and TFs) (Brackley et al. 2016; Barbieri et al. 2017) or from Hi-C data by a recently developed machine learning approach (PRISMS) (Bianco et al. 2018; Conte et al. 2020).

The strings and binders approach has been used in a broad range of applications, in which chromatin structures recapitulated contact probabilities and scaling from Hi-C (Brackley et al. 2016; Chiariello et al. 2016) and GAM data (Beagrie et al. 2017), and predicted spatial distances from FISH experiments (Barbieri et al. 2012; Nicodemi and Pombo 2014; Fraser et al. 2015; Chiariello et al. 2016; Bianco et al. 2017, 2018). SBS simulations reproduced loops and TADs as well as A/B compartments with a two-color copolymer model (Barbieri et al. 2012). However, more chromatin types are required to reproduce specific details of the Hi-C map at higher resolution (Chiariello et al. 2016; Bianco et al. 2017; Conte et al. 2020). SBS models studied the folding of the Sox9 and HoxB loci in mESC (Chiariello et al. 2016; Barbieri et al. 2017) and characterized the effects of pathogenic variants on the folding of EPHA4 locus in human fibroblasts (Bianco et al. 2017). The structure of the healthy locus was predicted with a model trained on Hi-C data. Then, models of the pathogenic mutants correctly predicted most of the ectopic interactions. SBS models also explored the tissue-specific architecture of the mouse Pitx1 gene, a regulator of hindlimb development (Kragesteen et al. 2018). Observed chromatin refolding in forelimb and hindlimb provided a rationale to explain expression data. Recently, models of human HTC116 and IMB90 loci in wild-type and cohesin depleted cells were validated against superresolution imaging (Bintu et al. 2018; Conte et al. 2020).

Data-Driven Approaches

Mechanistic models simulate the time evolution of chromatin-folding processes. Often, the underlying Hamiltonian uses predefined chromatin classes to generalize chromatin interactions, assuming the same chromatin types share identical physical properties. These generalized energy terms are parameterized so that models best agree with experiments (Fig. 2B).

Data-driven approaches use a different strategy. They do not generalize chromatin into predefined classes and do not require prior knowledge of folding mechanisms. Instead, they use all data points explicitly, assume an appropriate representation of experimental errors and uncertainties, and relate all data to an ensemble of 3D genome structures that are statistically consistent with it (Fig. 3B). These 3D structures are then examined to derive structure–function correlations and make quantitative predictions of structural features for specific genomic regions and study cell-to-cell variabilities of chromosome conformations. There are several data-driven modeling strategies, which differ in the functional interpretation of experimental data and sampling strategies to generate genome structures. We will first focus on data deconvolution methods, which attempt to de-multiplex ensemble data, and then discuss resampling methods.

Data Deconvolution Methods

The population-based genome structure modeling approach (PGS) is a probabilistic framework to model fully diploid genomes from Hi-C data at base-pair resolutions of tens to hundreds of kb and is available as a software package (Kalhor et al. 2012; Tjong et al. 2016; Hua et al. 2018). The approach performs a structure-based deconvolution of ensemble Hi-C data into a population of individual structures, in which the cumulated physical contacts across all structures recapitulate the Hi-C data. Chromosomes are modeled as polymer chains subject to chain connectivity, chromatin contacts, excluded volume and nuclear volume restraints (Fig. 3A). The key step is to infer those chromatin contacts likely to co-occur in the same structure. This problem is formulated as a maximum likelihood estimation problem, which is solved iteratively with a variant of the expectation-maximization algorithm and optimization strategies for efficient and scalable model estimation (Tjong et al. 2016; Li et al. 2017). Each iteration involves two steps: first, finding the optimal allocations of chromatin contacts across all structures by maximizing the log-likelihood over all contact assignments, given the optimized structures from the previous iteration, and second, generating genome structures by imposing physical contact restraints for all contact allocations using a combination of MD simulated annealing (Kirkpatrick et al. 1983) and conjugate gradient minimizations (Hestenes and Stiefel 1952). Each individual structure is described by a unique Hamiltonian, which expresses only a subset of contact restraints according to the optimized contact allocations. At each iteration, contact allocations are reevaluated and the process is repeated until convergence is reached. In addition, Hi-C contacts are gradually added to the iterative process, starting with the most frequent interactions. Gradually fitting an increasing number of contacts can effectively guide the search for the best solution and facilitates the detection of cooperative chromatin interactions. Population-based modeling produces structures of entire diploid genomes with high predictive value. For instance, genome models for GM12878 cells at 200 kb resolution allowed a detailed analysis of the spatial partitioning of chromatin subcompartments (Yildirim et al. 2021), as defined by Rao et al. (2014). Because subcompartment notations were not included as input information, the models allowed an independent characterization of structural features that distinguish chromatin in each subcompartment. Chromatin in the two active subcompartments (A1 and A2) differ in the variability of their nuclear locations: while A1 chromatin is localized in the nuclear interior in most cells, A2 chromatin shows large cell-to-cell variability (Fig. 4B). Models also revealed a relationship between micro-partitions of subcompartment chromatin and nuclear bodies, which made it possible to predict the locations of nuclear speckles in individual models, and predict with good accuracy data from SON tyramide signal amplification sequencing (TSA-seq) experiments (Chen et al. 2018). SON-TSA-seq estimates mean cytological distances of chromatin to nuclear speckles. The SON protein, an mRNA splicing cofactor, is a highly specific marker for nuclear speckles. SON-TSA produces a gradient of diffusible tyramide free radicals, instigated at regions of highest SON concentrations (i.e., the speckle locations), for distance-dependent biotin labeling of DNA. The models also predicted, with good accuracy, other omics data (e.g., laminB1 TSA-seq [Chen et al. 2018] and laminB1 DNA adenine methyltransferase identification [DamID], which maps genome-wide nuclear lamina interactions [Leemans et al. 2019]). The models also indicated a connection between a gene's association frequency to nuclear speckles and its transcript count in single-cell RNA-seq experiments (Osorio et al. 2019) and provided insights into the cell-to-cell variability of TADs (Yildirim et al. 2021).

Population-based modeling has been successfully applied to a variety of cell types and organisms, including human GM12878 lymphocytes (Dai et al. 2016; Tjong et al. 2016; Hua et al. 2018; Yildirim et al. 2021), mouse neutrophils (Zhu et al. 2017), cardiac myocytes, liver tissue (Chapski et al. 2019), and *D. melanogaster* (Li et al. 2017) at base-pair resolutions from 200 kb to ~3 Mb. Genome models in lymphoblastoid cells predicted chromosome-specific centromere clusters toward the nuclear interior, which were confirmed by cryo-soft X-ray tomography and play a pivotal role in chromosome positioning and stabilizing interchromosomal interactions (Tjong et al. 2016). The models also detected hundreds of frequently occurring multivalent chromatin clusters, which were enriched for the same regulatory factors (Dai et al. 2016). Another study characterized structural reorganizations during mouse neutrophil differentiation, leading to the discovery of chromosome supercontraction, driven by long-range heterochromatic interactions, combined with repositioning of centromeres and nucleoli (Zhu et al. 2017).

The scope of population-based modeling has recently been expanded to accommodate additional data modalities that can be cast into a polymer energy term. The resulting implementation is referred to as the Integrated Genome Modeling (IGM) platform (Li et al. 2017; Polles et al. 2019). The method was recently employed to generate diploid genome structures of human HFFc6 cells at 200 kb resolution from Hi-C, laminB1 DamID, SPRITE, and 3D HIPMAp FISH data which demonstrated that heterogeneous data sources can uncover structural features that may not be accessible to Hi-C alone (L. Boninsegna, unpubl.).

By integrating Hi-C and lamina DamID data, models of the *D. melanogaster* genome predicted location preferences for heterochromatin regions of each chromosome in the heterochromatic phase along preferred locations of the nucleolus, which were confirmed by FISH experiments (Li et al. 2017). Even though unphased Hi-C data cannot reveal interactions between chromosome copies, the models correctly show an anticorrelation between predicted pairing frequencies for homolog chromatin regions and the enrichment of binding sites for the mortality factor 4–like protein 1 (Mrg15), which is known to cause homolog unpairing.

Another approach by Giorgetti et al. (2014) models chromatin as a self-avoiding chain of beads, which interact via spherical well potentials. For each contact pair, the strength of the interaction potential is optimized to reproduce 5C contact frequencies. A study of the Xic locus on the inactive X chromosome in mouse embryonic stem (ES) cells at 3 kb resolution showed high structural variability, highlighted the role of cohesin/CTCF in shaping TAD conformations, and revealed insights into the relationship between conformational fluctuations and transcriptional variability. Further studies indicated that structural variations within TADs occur on timescales shorter than the cell cycle (Tiana et al. 2016). A study on ~2500 TADs in mouse ES cells provided further information on the correlations between TAD structures and gene activity (Zhan et al. 2017).

In another approach, Zhang and Wolynes (2015) developed an iterative algorithm to approximate an energy landscape for an ensemble of chromosome conformations that is consistent with the maximum entropy principle and reproduces Hi-C contact frequencies. Chromosome 12 of human ES cells and fibroblasts were modeled as self-avoiding polymer at 40 kb resolution with a potential energy function involving terms for chain connectivity, chromosome confinement, hard/soft core repulsive interactions, and chromatin interaction terms with Lagrangian multipliers. The Lagrangian multipliers were determined through an iterative optimization scheme. Chromosome structures were obtained via a series of independent MD simulations from which tens of thousands of conformations are collected. The resulting structures agreed with Hi-C data and revealed highly variable chromosome configurations in which TADs play a key role to locally rigidify the chain. A subset of TADs showed two-state transitions, possibly to modulate transcriptional activity.

Resampling Methods

Resampling approaches differ from deconvolution methods in the interpretation of the data. They generally express the agreement between data and structures by a solitary scoring function, in which Hi-C contact frequencies are typically expressed as distance restraints, representing either chromatin contacts or mean distances. However, because Hi-C data are accumulated over millions of cells with considerable variability in chromosome structures, they may contain conflicting information when imposed in a single scoring function. This may cause unresolved violations of restraints during optimization and could limit realistic descriptions of structural cell-to-cell variability. This problem is addressed in some approaches by considering only the most significant subsets of interactions, likely to be present in a dominant structural state (Baù et al. 2011; Umbarger et al. 2011; Gehlen et al. 2012; Trieu and Cheng 2014, 2016; Di Stefano et al. 2016; Paulsen et al. 2017; Serra et al. 2017; Yildirim and Feig 2018). To generate representative structures, resampling methods repeat independent optimizations of the solitary scoring function by Monte Carlo sampling, simulated annealing, MD simulations, expectation maximization or Bayesian optimization.

There is a variety of tools available to model small chromatin regions (Baù et al. 2011; Rousseau et al. 2011; Junier et al. 2012; Meluzzi and Arya 2013; Serra et al. 2017), chromosomes (Trieu and Cheng 2014, 2016; Wang et al. 2015; Zhu et al. 2018), whole-genomes (Gehlen et al. 2012; Di Stefano et al. 2016; Paulsen et al. 2017, 2018; Trieu and Cheng 2017), or bacterial chromosomes (Umbarger et al. 2011; Yildirim and Feig 2018). Comprehensive lists of data-driven modeling tools can be found elsewhere (Oluwadare et al. 2019; MacKay and Kusalik 2020). One of the most commonly applied methods, TADbit, uses significant contacts/noncontacts to study the α-globin locus in human K562 and GM12878 hematopoietic cells (Baù et al. 2011), 1 Mb genomic regions in *D. melanogaster* (Serra et al. 2017), and the *Caulobacter crescentus* genome (Umbarger et al. 2011). Resampling methods have also been expanded to incorporate Hi-C with other data sources, including lamin contacts from lamin Chip-seq data (Chrom3D) (Paulsen et al. 2017, 2018) and 3D FISH distances (Zhu et al. 2018; Abbas et al. 2019). Nir et al. (2018) combined Hi-C data with superresolution microscopy by generating 3D chromosome structures with TADbit, which were subsequently fitted to superresolution images from OligoSTORM and Oligo-DNA-PAINT microscopy. It was possible to generate structures for an 8 Mb region of Chromosome 19 in PGP1f cells at 10 kb resolution. These structures shed light on enhancer-promoter clusters, and 3D localization patterns of active and inactive regions.

Resampling methods are well suited to build genome structures from single-cell Hi-C data. The NucDynamics package (Stevens et al. 2017) and most other methods use simulated annealing optimizations of chromatin interaction restraints, polymer chain connectivity, and a noninteracting particle repulsion (Nagano et al. 2013). Most studies comprise 8–15 cells at 50–500 kb resolution and showed considerable structure variations, while chromosome territories and A/B compartment segregation is conserved (Nagano et al. 2013; Stevens et al. 2017; Tan et al. 2018). A study of mouse ES cells at 100 kb resolution revealed changes in chromosome conformations during cell cycle together with the timing of chromatin compartment and TAD formation (Nagano et al. 2017). There are conflicting reports about the existence of TADs in single cells. Nagano et al. (2013) observed them, whereas other studies reached the opposite conclusion (Flyamer et al. 2017; Stevens et al. 2017; Tan et al. 2018). Chromatin loops are dynamic structures and TADs emerge in ensemble Hi-C from weak preferences of loop positions summed over many cells. It is therefore not surprising that in any given cell loops vary. However, recent superresolution microscopy confirmed the existence of chromatin domains in single cells and supported their distinct structural features in individual cells (Bintu et al. 2018; Su et al. 2020).

Other single-cell modeling methods exist, including a manifold-based optimization (Paulsen et al. 2015), Bayesian estimation combined with gradient-based optimization (SIMBA3D) (Rosenthal et al. 2019), and Bayesian inferential structure determination combined with Markov chain Monte Carlo sampling (Carstens et al. 2016). Others use a cubic lattice representation of structures and 2D Gaussian imputation of contact matrices for 3D structure reconstructions (Zhu et al. 2019).

Several challenges remain related to sparsity of the single cell data, presence of false-positive contacts, and distinction of homologous chromosomes along with the unknown nuclear morphology, which could influence the outcome of single-cell modeling.

CONCLUDING REMARKS

Recent developments in experimental technologies provide new opportunities to probe the structure and dynamics of the nuclear genome and its structure–function relationships. However, inferring 3D structures, their dynamic attributes, and the physical processes that establish them from experimental observations alone is not a straightforward task. Consequently, computational methodologies have become a fundamental component in bridging the gap between experiments and their 3D structural interpretation. In this review, we summarized a growing number of state-of-the-art computational strategies for 3D chromatin structure modeling; these approaches are applied at various scopes and structural scales, ranging from simulations of individual gene loci, to chromosomes, to entire genomes. Both mechanistic and

 Cite this article as *Cold Spring Harb Perspect Biol* doi: 10.1101/cshperspect.a039693

data-driven modeling tools have provided critical insights into 3D chromatin organization and underlying mechanisms, which inspired new hypotheses and triggered further experiments.

Despite the successes, major challenges still need to be addressed. Improving accuracy and coverage of structural models requires integration of complementary data modalities: in particular, further efforts are required to integrate microscopy with genomics technologies to overcome limitations of individual techniques. Moreover, it is crucial to develop standardized strategies for model assessment and define an accurate description of model uncertainties. It is also important to advance data sharing efforts, to allow a standardized access to structural models for the community. This would likely require a community effort to develop databanks, common file formats with standardized information about input data and parameters settings to ensure maximal reproducibility.

ACKNOWLEDGMENTS

This work was supported by the National Institutes of Health (Grant U54DK107981 and UM1HG011593 to F.A.), and an NSF CAREER Grant (1150287 to F.A.).

REFERENCES

Abbas A, He X, Niu J, Zhou B, Zhu G, Ma T, Song J, Gao J, Zhang MQ, Zeng J. 2019. Integrating Hi-C and FISH data for modeling of the 3D organization of chromosomes. *Nat Commun* **10:** 2049. doi:10.1038/s41467-019-10005-6

Alipour E, Marko JF. 2012. Self-organization of domain structures by DNA-loop-extruding enzymes. *Nucleic Acids Res* **40:** 11202–11212. doi:10.1093/nar/gks925

Allen MP, Tildesley DJ. 1987. *Computer simulation of liquids*. Clarendon, Oxford.

Annunziatella C, Chiariello AM, Bianco S, Nicodemi M. 2016. Polymer models of the hierarchical folding of the Hox-B chromosomal locus. *Phys Rev E* **94:** 042402–042402. doi:10.1103/PhysRevE.94.042402

Annunziatella C, Chiariello AM, Esposito A, Bianco S, Fiorillo L, Nicodemi M. 2018. Molecular dynamics simulations of the strings and binders switch model of chromatin. *Methods* **142:** 81–88. doi:10.1016/j.ymeth.2018.02.024

Bailey SD, Zhang X, Desai K, Aid M, Corradin O, Cowper-Sal·lari R, Akhtar-Zaidi B, Scacheri PC, Haibe-Kains B, Lupien M. 2015. ZNF143 provides sequence specificity to secure chromatin interactions at gene promoters. *Nat Commun* **6:** 6186. doi:10.1038/ncomms7186

Banigan EJ, Mirny LA. 2020. Loop extrusion: theory meets single-molecule experiments. *Curr Opin Cell Biol* **64:** 124–138. doi:10.1016/j.ceb.2020.04.011

Barbieri M, Chotalia M, Fraser J, Lavitas LM, Dostie J, Pombo A, Nicodemi M. 2012. Complexity of chromatin folding is captured by the strings and binders switch model. *Proc Natl Acad Sci* **109:** 16173–16178. doi:10.1073/pnas.1204799109

Barbieri M, Xie SQ, Torlai Triglia E, Chiariello AM, Bianco S, De Santiago I, Branco MR, Rueda D, Nicodemi M, Pombo A. 2017. Active and poised promoter states drive folding of the extended HoxB locus in mouse embryonic stem cells. *Nat Struct Mol Biol* **24:** 515–524. doi:10.1038/nsmb.3402

Baù D, Sanyal A, Lajoie BR, Capriotti E, Byron M, Lawrence JB, Dekker J, Marti-Renom MA. 2011. The three-dimensional folding of the α-globin gene domain reveals formation of chromatin globules. *Nat Struct Mol Biol* **18:** 107–114. doi:10.1038/nsmb.1936

Beagrie RA, Scialdone A, Schueler M, Kraemer DCA, Chotalia M, Xie SQ, Barbieri M, de Santiago I, Lavitas LM, Branco MR, et al. 2017. Complex multi-enhancer contacts captured by genome architecture mapping. *Nature* **543:** 519–524. doi:10.1038/nature21411

Bianco S, Chiariello AM, Annunziatella C, Esposito A, Nicodemi M. 2017. Predicting chromatin architecture from models of polymer physics. *Chromosome Res* **25:** 25–34. doi:10.1007/s10577-016-9545-5

Bianco S, Lupiáñez DG, Chiariello AM, Annunziatella C, Kraft K, Schöpflin R, Wittler L, Andrey G, Vingron M, Pombo A, et al. 2018. Polymer physics predicts the effects of structural variants on chromatin architecture. *Nat Genet* **50:** 662–667. doi:10.1038/s41588-018-0098-8

Bintu B, Mateo LJ, Su JH, Sinnott-Armstrong NA, Parker M, Kinrot S, Yamaya K, Boettiger AN, Zhuang X. 2018. Super-resolution chromatin tracing reveals domains and cooperative interactions in single cells. *Science* **362:** eaau1783. doi:10.1126/science.aau1783

Brackey CA, Marenduzzo D, Gilbert N. 2020. Mechanistic modeling of chromatin folding to understand function. *Nat Methods* **17:** 767–775. doi:10.1038/s41592-020-0852-6

Brackley CA, Taylor S, Papantonis A, Cook PR, Marenduzzo D. 2013. Nonspecific bridging-induced attraction drives clustering of DNA-binding proteins and genome organization. *Proc Natl Acad Sci* **110:** E3605–E3611. doi:10.1073/pnas.1302950110

Brackley CA, Brown JM, Waithe D, Babbs C, Davies J, Hughes JR, Buckle VJ, Marenduzzo D. 2016. Predicting the three-dimensional folding of *cis*-regulatory regions in mammalian genomes using bioinformatic data and polymer models. *Genome Biol* **17:** 59–59. doi:10.1186/s13059-016-0909-0

Brackley CA, Johnson J, Michieletto D, Morozov AN, Nicodemi M, Cook PR, Marenduzzo D. 2017. Nonequilibrium chromosome looping via molecular slip links. *Phys Rev Lett* **119:** 138101. doi:10.1103/PhysRevLett.119.138101

Brackley CA, Johnson J, Michieletto D, Morozov AN, Nicodemi M, Cook PR, Marenduzzo D. 2018. Extrusion without a motor: a new take on the loop extrusion model of genome organization. *Nucleus* **9:** 95–103. doi:10.1080/19491034.2017.1421825

Buckle A, Brackley CA, Boyle S, Marenduzzo D, Gilbert N. 2018. Polymer simulations of heteromorphic chromatin predict the 3D folding of complex genomic loci. *Mol Cell* **72:** 786–797. doi:10.1016/j.molcel.2018.09.016

Carstens S, Nilges M, Habeck M. 2016. Inferential structure determination of chromosomes from single-cell Hi-C data. *PLoS Comput Biol* **12:** e1005292. doi:10.1371/journal.pcbi.1005292

Chapski DJ, Rosa-Garrido M, Hua N, Alber F, Vondriska TM. 2019. Spatial principles of chromatin architecture associated with organ-specific gene regulation. *Front Cardiovasc Med* **5:** 186. doi:10.3389/fcvm.2018.00186

Chen Y, Zhang Y, Wang Y, Zhang L, Brinkman EK, Adam SA, Goldman R, van Steensel B, Ma J, Belmont AS. 2018. Mapping 3D genome organization relative to nuclear compartments using TSA-Seq as a cytological ruler. *J Cell Biol* **217:** 4025–4048. doi:10.1083/jcb.201807108

Cheng RR, Contessoto VG, Lieberman Aiden E, Wolynes PG, Di Pierro M, Onuchic JN. 2020. Exploring chromosomal structural heterogeneity across multiple cell lines. *eLife* **9:** e60312. doi:10.7554/eLife.60312

Chiariello AM, Annunziatella C, Bianco S, Esposito A, Nicodemi M. 2016. Polymer physics of chromosome large-scale 3D organisation. *Sci Rep* **6:** 29775. doi:10.1038/srep29775

Conte M, Fiorillo L, Bianco S, Chiariello AM, Esposito A, Nicodemi M. 2020. Polymer physics indicates chromatin folding variability across single-cells results from state degeneracy in phase separation. *Nat Commun* **11:** 3289. doi:10.1038/s41467-020-17141-4

Dai C, Li W, Tjong H, Hao S, Zhou Y, Li Q, Chen L, Zhu B, Alber F, Jasmine Zhou X. 2016. Mining 3D genome structure populations identifies major factors governing the stability of regulatory communities. *Nat Commun* **7:** 11549. doi:10.1038/ncomms11549

Davidson IF, Bauer B, Goetz D, Tang W, Wutz G, Peters JM. 2019. DNA loop extrusion by human cohesin. *Science* **366:** 1338–1345. doi:10.1126/science.aaz3418

Dekker J, Belmont AS, Guttman M, Leshyk VO, Lis JT, Lomvardas S, Mirny LA, O'Shea CC, Park PJ, Ren B, et al. 2017. The 4D nucleome project. *Nature* **549:** 219–226. doi:10.1038/nature23884

Di Pierro M, Zhang B, Aiden EL, Wolynes PG, Onuchic JN. 2016. Transferable model for chromosome architecture. *Proc Natl Acad Sci* **113:** 12168–12173. doi:10.1073/pnas.1613607113

Di Pierro M, Cheng RR, Aiden EL, Wolynes PG, Onuchic JN. 2017. De novo prediction of human chromosome structures: epigenetic marking patterns encode genome architecture. *Proc Natl Acad Sci* **114:** 12126–12131. doi:10.1073/pnas.1714980114

Di Pierro M, Potoyan DA, Wolynes PG, Onuchic JN. 2018. Anomalous diffusion, spatial coherence, and viscoelasticity from the energy landscape of human chromosomes. *Proc Natl Acad Sci* **115:** 7753–7758. doi:10.1073/pnas.1806297115

Di Stefano M, Paulsen J, Lien TG, Hovig E, Micheletti C. 2016. Hi-C-constrained physical models of human chromosomes recover functionally-related properties of genome organization. *Sci Rep* **6:** 35985. doi:10.1038/srep35985

Dixon JR, Selvaraj S, Yue F, Kim A, Li Y, Shen Y, Hu M, Liu JS, Ren B. 2012. Topological domains in mammalian genomes identified by analysis of chromatin interactions. *Nature* **485:** 376–380. doi:10.1038/nature11082

Erdel F, Rippe K. 2018. Formation of chromatin subcompartments by phase separation. *Biophys J* **114:** 2262–2270. doi:10.1016/j.bpj.2018.03.011

Falk M, Feodorova Y, Naumova N, Imakaev M, Lajoie BR, Leonhardt H, Joffe B, Dekker J, Fudenberg G, Solovei I, et al. 2019. Heterochromatin drives compartmentalization of inverted and conventional nuclei. *Nature* **570:** 395–399. doi:10.1038/s41586-019-1275-3

Finn EH, Misteli T. 2019. Molecular basis and biological function of variability in spatial genome organization. *Science* **365:** eaaw9498. doi:10.1126/science.aaw9498

Flyamer IM, Gassler J, Imakaev M, Brandão HB, Ulianov SV, Abdennur N, Razin SV, Mirny LA, Tachibana-Konwalski K. 2017. Single-nucleus Hi-C reveals unique chromatin reorganization at oocyte-to-zygote transition. *Nature* **544:** 110–114. doi:10.1038/nature21711

Fraser J, Ferrai C, Chiariello AM, Schueler M, Rito T, Laudanno G, Barbieri M, Moore BL, Kraemer DC, Aitken S, et al. 2015. Hierarchical folding and reorganization of chromosomes are linked to transcriptional changes in cellular differentiation. *Mol Syst Biol* **11:** 852. doi:10.15252/msb.20156492

Fudenberg G, Imakaev M, Lu C, Goloborodko A, Abdennur N, Mirny LA. 2016. Formation of chromosomal domains by loop extrusion. *Cell Rep* **15:** 2038–2049. doi:10.1016/j.celrep.2016.04.085

Fudenberg G, Abdennur N, Imakaev M, Goloborodko A, Mirny LA. 2017. Emerging evidence of chromosome folding by loop extrusion. *Cold Spring Harb Symp Quant Biol* **82:** 45–55. doi:10.1101/sqb.2017.82.034710

Fullwood MJ, Liu MH, Pan YF, Liu J, Xu H, Mohamed YB, Orlov YL, Velkov S, Ho A, Mei PH, et al. 2009. An oestrogen-receptor-α-bound human chromatin interactome. *Nature* **462:** 58–64. doi:10.1038/nature08497

Ganji M, Shaltiel IA, Bisht S, Kim E, Kalichava A, Haering CH, Dekker C. 2018. Real-time imaging of DNA loop extrusion by condensin. *Science* **360:** 102–105. doi:10.1126/science.aar7831

Gassler J, Brandão HB, Imakaev M, Flyamer IM, Ladstätter S, Bickmore WA, Peters J-M, Mirny LA, Tachibana K. 2017. A mechanism of cohesin-dependent loop extrusion organizes zygotic genome architecture. *EMBO J* **36:** 3600–3618. doi:10.15252/embj.201798083

Gehlen LR, Gruenert G, Jones MB, Rodley CD, Langowski J, O'Sullivan JM. 2012. Chromosome positioning and the clustering of functionally related loci in yeast is driven by chromosomal interactions. *Nucleus* **3:** 370–383. doi:10.4161/nucl.20971

Gibcus JH, Samejima K, Goloborodko A, Samejima I, Naumova N, Nuebler J, Kanemaki MT, Xie L, Paulson JR, Earnshaw WC, et al. 2018. A pathway for mitotic chromosome formation. *Science* **359:** eaao6135. doi:10.1126/science.aao6135

Giorgetti L, Galupa R, Nora EP, Piolot T, Lam F, Dekker J, Tiana G, Heard E. 2014. Predictive polymer modeling reveals coupled fluctuations in chromosome conformation and transcription. *Cell* **157:** 950–963. doi:10.1016/j.cell.2014.03.025

Golfier S, Quail T, Kimura H, Brugués J. 2020. Cohesin and condensin extrude DNA loops in a cell-cycle dependent manner. *eLife* **9:** e53885. doi:10.7554/eLife.53885

Goloborodko A, Imakaev MV, Marko JF, Mirny L. 2016a. Compaction and segregation of sister chromatids via active loop extrusion. *eLife* **5:** e14864. doi:10.7554/eLife.14864

Goloborodko A, Marko JF, Mirny LA. 2016b. Chromosome compaction by active loop extrusion. *Biophys J* **110:** 2162–2168. doi:10.1016/j.bpj.2016.02.041

Guelen L, Pagie L, Brasset E, Meuleman W, Faza MB, Talhout W, Eussen BH, de Klein A, Wessels L, de Laat W, et al. 2008. Domain organization of human chromosomes revealed by mapping of nuclear lamina interactions. *Nature* **453:** 948–951. doi:10.1038/nature06947

Haarhuis JHI, Weide Rvd, Blomen VA, Yáñez-Cuna JO, Amendola M, Ruiten Mv, Krijger PHL, Teunissen H, Medema RH, Steensel Bv, et al. 2017. The cohesin release factor WAPL restricts chromatin loop extension. *Cell* **169:** 693–707.e14. doi:10.1016/j.cell.2017.04.013

Haddad N, Jost D, Vaillant C. 2017. Perspectives: using polymer modeling to understand the formation and function of nuclear compartments. *Chromosome Res* **25:** 35–50. doi:10.1007/s10577-016-9548-2

Hestenes MR, Stiefel E. 1952. Methods of conjugate gradients for solving linear systems. *J Res Natl Bur Stand* **49:** 409–436. doi:10.6028/jres.049.044

Hildebrand EM, Dekker J. 2020. Mechanisms and functions of chromosome compartmentalization. *Trends Biochem Sci* **45:** 385–396. doi:10.1016/j.tibs.2020.01.002

Hirano T. 2002. The ABCs of SMC proteins: two-armed ATPases for chromosome condensation, cohesion, and repair. *Genes Dev* **16:** 399–414. doi:10.1101/gad.955102

Hua N, Tjong H, Shin H, Gong K, Zhou XJ, Alber F. 2018. Producing genome structure populations with the dynamic and automated PGS software. *Nat Protoc* **13:** 915–926. doi:10.1038/nprot.2018.008

Jost D, Carrivain P, Cavalli G, Vaillant C. 2014. Modeling epigenome folding: formation and dynamics of topologically associated chromatin domains. *Nucleic Acids Res* **42:** 9553–9561. doi:10.1093/nar/gku698

Junier I, Dale RK, Hou C, Képès F, Dean A. 2012. CTCF-mediated transcriptional regulation through cell type-specific chromosome organization in the β-globin locus. *Nucleic Acids Res* **40:** 7718–7727. doi:10.1093/nar/gks536

Kagey MH, Newman JJ, Bilodeau S, Zhan Y, Orlando DA, van Berkum NL, Ebmeier CC, Goossens J, Rahl PB, Levine SS, et al. 2010. Mediator and cohesin connect gene expression and chromatin architecture. *Nature* **467:** 430–435. doi:10.1038/nature09380

Kalhor R, Tjong H, Jayathilaka N, Alber F, Chen L. 2012. Genome architectures revealed by tethered chromosome conformation capture and population-based modeling. *Nat Biotechnol* **30:** 90–98. doi:10.1038/nbt.2057

Kempfer R, Pombo A. 2020. Methods for mapping 3D chromosome architecture. *Nat Rev Genet* **21:** 207–226. doi:10.1038/s41576-019-0195-2

Kirkpatrick S, Gelatt CD, Vecchi MP. 1983. Optimization by simulated annealing. *Science* **220:** 671–680. doi:10.1126/science.220.4598.671

Kong M, Cutts EE, Pan D, Beuron F, Kaliyappan T, Xue C, Morris EP, Musacchio A, Vannini A, Greene EC. 2020. Human condensin I and II drive extensive ATP-dependent compaction of nucleosome-bound DNA. *Mol Cell* **79:** 99–114.e9. doi:10.1016/j.molcel.2020.04.026

Kragesteen BK, Spielmann M, Paliou C, Heinrich V, Schöpflin R, Esposito A, Annunziatella C, Bianco S, Chiariello AM, Jerković I, et al. 2018. Dynamic 3D chromatin architecture contributes to enhancer specificity and limb morphogenesis. *Nat Genet* **50:** 1463–1473. doi:10.1038/s41588-018-0221-x

Kundu S, Ji F, Sunwoo H, Jain G, Lee JT, Sadreyev RI, Dekker J, Kingston RE. 2017. Polycomb repressive complex 1 generates discrete compacted domains that change during differentiation. *Mol Cell* **65:** 432–446.e5. doi:10.1016/j.molcel.2017.01.009

Larson AG, Elnatan D, Keenen MM, Trnka MJ, Johnston JB, Burlingame AL, Agard DA, Redding S, Narlikar GJ. 2017. Liquid droplet formation by HP1α suggests a role for phase separation in heterochromatin. *Nature* **547:** 236–240. doi:10.1038/nature22822

Leemans C, van der Zwalm MCH, Brueckner L, Comoglio F, van Schaik T, Pagie L, van Arensbergen J, van Steensel B. 2019. Promoter-intrinsic and local chromatin features determine gene repression in LADs. *Cell* **177:** 852–864.e14. doi:10.1016/j.cell.2019.03.009

Leibler L. 1980. Theory of microphase separation in block copolymers. *Macromolecules* **13:** 1602–1617. doi:10.1021/ma60078a047

Li Q, Tjong H, Li X, Gong K, Zhou XJ, Chiolo I, Alber F. 2017. The three-dimensional genome organization of *Drosophila melanogaster* through data integration. *Genome Biol* **18:** 145. doi:10.1186/s13059-017-1264-5

Lieberman-Aiden E, Van Berkum NL, Williams L, Imakaev M, Ragoczy T, Telling A, Amit I, Lajoie BR, Sabo PJ, Dorschner MO, et al. 2009. Comprehensive mapping of long-range interactions reveals folding principles of the human genome. *Science* **326:** 289–293. doi:10.1126/science.1181369

Lin D, Bonora G, Yardımcı GG, Noble WS. 2019. Computational methods for analyzing and modeling genome structure and organization. *Wiley Interdiscip Rev Syst Biol Med* **11:** e1435. doi:10.1002/wsbm.1435

MacKay K, Kusalik A. 2020. Computational methods for predicting 3D genomic organization from high-resolution chromosome conformation capture data. *Brief Funct Genomics* **19:** 292–308. doi:10.1093/bfgp/elaa004

MacPherson Q, Beltran B, Spakowitz AJ. 2018. Bottom–up modeling of chromatin segregation due to epigenetic modifications. *Proc Natl Acad Sci* **115:** 12739–12744. doi:10.1073/pnas.1812268115

Marko JF, Siggia ED. 1997. Polymer models of meiotic and mitotic chromosomes. *Mol Biol Cell* **8:** 2217–2231. doi:10.1091/mbc.8.11.2217

McCord RP, Kaplan N, Giorgetti L. 2020. Chromosome conformation capture and beyond: toward an integrative view of chromosome structure and function. *Mol Cell* **77:** 688–708. doi:10.1016/j.molcel.2019.12.021

Meluzzi D, Arya G. 2013. Recovering ensembles of chromatin conformations from contact probabilities. *Nucleic Acids Res* **41:** 63–75. doi:10.1093/nar/gks1029

Metropolis N, Rosenbluth AW, Rosenbluth MN, Teller AH, Teller E. 1953. Equation of state calculations by fast computing machines. *J Chem Phys* **21:** 1087–1092. doi:10.1063/1.1699114

Michieletto D, Orlandini E, Marenduzzo D. 2016. Polymer model with epigenetic recoloring reveals a pathway for the de novo establishment and 3D organization of chromatin domains. *Phys Rev X* **6:** 041047.

Mirny LA, Imakaev M, Abdennur N. 2019. Two major mechanisms of chromosome organization. *Curr Opin Cell Biol* **58:** 142–152. doi:10.1016/j.ceb.2019.05.001

Misteli T. 2020. The self-organizing genome: principles of genome architecture and function. *Cell* **183:** 28–45. doi:10.1016/j.cell.2020.09.014

Nagano T, Lubling Y, Stevens TJ, Schoenfelder S, Yaffe E, Dean W, Laue ED, Tanay A, Fraser P. 2013. Single-cell Hi-C reveals cell-to-cell variability in chromosome structure. *Nature* **502:** 59–64. doi:10.1038/nature12593

Nagano T, Lubling Y, Várnai C, Dudley C, Leung W, Baran Y, Mendelson Cohen N, Wingett S, Fraser P, Tanay A. 2017. Cell-cycle dynamics of chromosomal organization at single-cell resolution. *Nature* **547:** 61–67. doi:10.1038/nature23001

Nasmyth K. 2001. Disseminating the genome: joining, resolving, and separating sister chromatids during mitosis and meiosis. *Annu Rev Genet* **35:** 673–745. doi:10.1146/annurev.genet.35.102401.091334

Nicodemi M, Pombo A. 2014. Models of chromosome structure. *Curr Opin Cell Biol* **28:** 90–95. doi:10.1016/j.ceb.2014.04.004

Nicodemi M, Prisco A. 2009. Thermodynamic pathways to genome spatial organization in the cell nucleus. *Biophys J* **96:** 2168–2177. doi:10.1016/j.bpj.2008.12.3919

Nir G, Farabella I, Pérez Estrada C, Ebeling CG, Beliveau BJ, Sasaki HM, Lee SH, Nguyen SC, McCole RB, Chattoraj S, et al. 2018. Walking along chromosomes with super-resolution imaging, contact maps, and integrative modeling. *PLoS Genet* **14:** e1007872. doi:10.1371/journal.pgen.1007872

Nora EP, Lajoie BR, Schulz EG, Giorgetti L, Okamoto I, Servant N, Piolot T, van Berkum NL, Meisig J, Sedat J, et al. 2012. Spatial partitioning of the regulatory landscape of the X-inactivation centre. *Nature* **485:** 381–385. doi:10.1038/nature11049

Nora EP, Goloborodko A, Valton AL, Gibcus JH, Uebersohn A, Abdennur N, Dekker J, Mirny LA, Bruneau BG. 2017. Targeted degradation of CTCF decouples local insulation of chromosome domains from genomic compartmentalization. *Cell* **169:** 930–944.e22. doi:10.1016/j.cell.2017.05.004

Nozaki T, Imai R, Tanbo M, Nagashima R, Tamura S, Tani T, Joti Y, Tomita M, Hibino K, Kanemaki MT, et al. 2017. Dynamic organization of chromatin domains revealed by super-resolution live-cell imaging. *Mol Cell* **67:** 282–293.e7. doi:10.1016/j.molcel.2017.06.018

Nuebler J, Fudenberg G, Imakaev M, Abdennur N, Mirny LA. 2018. Chromatin organization by an interplay of loop extrusion and compartmental segregation. *Proc Natl Acad Sci* **115:** E6697–E6706. doi:10.1073/pnas.1717730115

Oluwadare O, Highsmith M, Cheng J. 2019. An overview of methods for reconstructing 3-D chromosome and genome structures from Hi-C data. *Biol Proced Online* **21:** 7. doi:10.1186/s12575-019-0094-0

Osorio D, Yu X, Yu P, Serpedin E, Cai JJ. 2019. Single-cell RNA sequencing of a European and an African lymphoblastoid cell line. *Sci Data* **6:** 112. doi:10.1038/s41597-019-0116-4

Ou HD, Phan S, Deerinck TJ, Thor A, Ellisman MH, O'Shea CC. 2017. ChromEMT: visualizing 3D chromatin structure and compaction in interphase and mitotic cells. *Science* **357:** eaag0025. doi: 10.1126/science.aag0025

Parmar JJ, Woringer M, Zimmer C. 2019. How the genome folds: the biophysics of four-dimensional chromatin organization. *Annu Rev Biophys* **48:** 231–253. doi:10.1146/annurev-biophys-052118-115638

Paulsen J, Gramstad O, Collas P. 2015. Manifold based optimization for single-cell 3D genome reconstruction. *PLoS Comput Biol* **11:** e1004396. doi:10.1371/journal.pcbi.1004396

Paulsen J, Sekelja M, Oldenburg AR, Barateau A, Briand N, Delbarre E, Shah A, Sørensen AL, Vigouroux C, Buendia B, et al. 2017. Chrom3D: three-dimensional genome modeling from Hi-C and nuclear lamin-genome contacts. *Genome Biol* **18:** 21. doi:10.1186/s13059-016-1146-2

Paulsen J, Liyakat Ali TM, Collas P. 2018. Computational 3D genome modeling using Chrom3D. *Nat Protoc* **13:** 1137–1152. doi:10.1038/nprot.2018.009

Phillips-Cremins JE, Sauria MEG, Sanyal A, Gerasimova TI, Lajoie BR, Bell JSK, Ong C-T, Hookway TA, Guo C, Sun Y, et al. 2013. Architectural protein subclasses shape 3D organization of genomes during lineage commitment. *Cell* **153:** 1281–1295. doi:10.1016/j.cell.2013.04.053

Polles G, Hua N, Yildirim A, Alber F. 2019. Genome structure calculation through comprehensive data integration. In *Modeling the 3D conformation of genomes*, p. 253. CRC, Boca Raton, FL.

Qi Y, Zhang B. 2019. Predicting three-dimensional genome organization with chromatin states. *PLoS Comput Biol* **15:** e1007024–e1007024. doi:10.1371/journal.pcbi.1007024

Qi Y, Reyes A, Johnstone SE, Aryee MJ, Bernstein BE, Zhang B. 2020. Data-driven polymer model for mechanistic exploration of diploid genome organization. *Biophys J* **119:** 1905–1916. doi:10.1016/j.bpj.2020.09.009

Quinodoz SA, Ollikainen N, Tabak B, Palla A, Schmidt JM, Detmar E, Lai MM, Shishkin AA, Bhat P, Takei Y, et al. 2018. Higher-order inter-chromosomal hubs shape 3D genome organization in the nucleus. *Cell* **174:** 744–757.e24. doi:10.1016/j.cell.2018.05.024

Rao SSP, Huntley MH, Durand NC, Stamenova EK, Bochkov ID, Robinson JT, Sanborn AL, Machol I, Omer AD, Lander ES, et al. 2014. A 3D map of the human genome at kilobase resolution reveals principles of chromatin looping. *Cell* **159:** 1665–1680. doi:10.1016/j.cell.2014.11.021

Rao SSP, Huang SC, Glenn St Hilaire B, Engreitz JM, Perez EM, Kieffer-Kwon KR, Sanborn AL, Johnstone SE, Bascom GD, Bochkov ID, et al. 2017. Cohesin loss eliminates all loop domains. *Cell* **171:** 305–320.e24. doi:10.1016/j.cell.2017.09.026

Rosenthal M, Bryner D, Huffer F, Evans S, Srivastava A, Neretti N. 2019. Bayesian estimation of three-dimensional chromosomal structure from single-cell Hi-C data. *J Comput Biol* **26:** 1191–1202. doi:10.1089/cmb.2019.0100

Rousseau M, Fraser J, Ferraiuolo MA, Dostie J, Blanchette M. 2011. Three-dimensional modeling of chromatin structure from interaction frequency data using Markov chain Monte Carlo sampling. *BMC Bioinformatics* **12:** 414. doi:10.1186/1471-2105-12-414

Sanborn AL, Rao SSP, Huang SC, Durand NC, Huntley MH, Jewett AI, Bochkov ID, Chinnappan D, Cutkosky A, Li J, et al. 2015. Chromatin extrusion explains key features of loop and domain formation in wild-type and engineered genomes. *Proc Natl Acad Sci* **112:** E6456–E6465. doi:10.1073/pnas.1518552112

Sankararaman S, Marko JF. 2005. Formation of loops in DNA under tension. *Phys Rev E* **71:** 021911. doi:10.1103/PhysRevE.71.021911

Schlick T. 1996. Pursuing Laplace's vision on modern computers. In *Mathematical approaches to biomolecular structure and dynamics, the IMA volumes in mathematics and its applications* (ed. Mesirov JP, Schulten K, Sumners DW), pp. 219–247. Springer, New York.

Schwarzer W, Abdennur N, Goloborodko A, Pekowska A, Fudenberg G, Loe-Mie Y, Fonseca NA, Huber W, Haering CH, Mirny L, et al. 2017. Two independent modes of chromatin organization revealed by cohesin removal. *Nature* **551:** 51–56. doi:10.1038/nature24281

Serra F, Baù D, Goodstadt M, Castillo D, Filion G, Marti-Renom MA. 2017. Automatic analysis and 3D-modelling of Hi-C data using TADbit reveals structural features of the fly chromatin colors. *PLoS Comput Biol* **13:** e1005665. doi:10.1371/journal.pcbi.1005665

Sexton T, Yaffe E, Kenigsberg E, Bantignies F, Leblanc B, Hoichman M, Parrinello H, Tanay A, Cavalli G. 2012. Three-dimensional folding and functional organization principles of the *Drosophila* genome. *Cell* **148:** 458–472. doi:10.1016/j.cell.2012.01.010

Shen Y, Yue F, McCleary DF, Ye Z, Edsall L, Kuan S, Wagner U, Dixon J, Lee L, Lobanenkov VV, et al. 2012. A map of the *cis*-regulatory sequences in the mouse genome. *Nature* **488:** 116–120. doi:10.1038/nature11243

Shi G, Liu L, Hyeon C, Thirumalai D. 2018. Interphase human chromosome exhibits out of equilibrium glassy dynamics. *Nat Commun* **9:** 3161. doi:10.1038/s41467-018-05606-6

Solovei I, Thanisch K, Feodorova Y. 2016. How to rule the nucleus: divide et impera. *Curr Opin Cell Biol* **40:** 47–59. doi:10.1016/j.ceb.2016.02.014

Steensel Bv, Henikoff S. 2000. Identification of in vivo DNA targets of chromatin proteins using tethered Dam methyltransferase. *Nat Biotechnol* **18:** 424–428. doi:10.1038/74487

Stevens TJ, Lando D, Basu S, Atkinson LP, Cao Y, Lee SF, Leeb M, Wohlfahrt KJ, Boucher W, O'Shaughnessy-Kirwan A, et al. 2017. 3D structures of individual mammalian genomes studied by single-cell Hi-C. *Nature* **544:** 59–64. doi:10.1038/nature21429

Strom AR, Emelyanov AV, Mir M, Fyodorov DV, Darzacq X, Karpen GH. 2017. Phase separation drives heterochromatin domain formation. *Nature* **547:** 241–245. doi:10.1038/nature22989

Su JH, Zheng P, Kinrot SS, Bintu B, Zhuang X. 2020. Genome-scale imaging of the 3D organization and transcriptional activity of chromatin. *Cell* **182:** 1641–1659.e26. doi:10.1016/j.cell.2020.07.032

Tan L, Xing D, Chang CH, Li H, Xie XS. 2018. Three-dimensional genome structures of single diploid human cells. *Science* **361:** 924–928. doi:10.1126/science.aat5641

Tedeschi A, Wutz G, Huet S, Jaritz M, Wuensche A, Schirghuber E, Davidson IF, Tang W, Cisneros DA, Bhaskara V, et al. 2013. Wapl is an essential regulator of chromatin structure and chromosome segregation. *Nature* **501:** 564–568. doi:10.1038/nature12471

Tiana G, Amitai A, Pollex T, Piolot T, Holcman D, Heard E, Giorgetti L. 2016. Structural fluctuations of the chromatin fiber within topologically associating domains. *Biophys J* **110:** 1234–1245. doi:10.1016/j.bpj.2016.02.003

Tjong H, Li W, Kalhor R, Dai C, Hao S, Gong K, Zhou Y, Li H, Zhou XJ, Le Gros MA, et al. 2016. Population-based 3D genome structure analysis reveals driving forces in spatial genome organization. *Proc Natl Acad Sci* **113:** E1663–E1672. doi:10.1073/pnas.1512577113

Trieu T, Cheng J. 2014. Large-scale reconstruction of 3D structures of human chromosomes from chromosomal contact data. *Nucleic Acids Res* **42:** e52. doi:10.1093/nar/gkt1411

Trieu T, Cheng J. 2016. MOGEN: a tool for reconstructing 3D models of genomes from chromosomal conformation capturing data. *Bioinformatics* **32:** 1286–1292. doi:10.1093/bioinformatics/btv754

Trieu T, Cheng J. 2017. 3D genome structure modeling by Lorentzian objective function. *Nucleic Acids Res* **45:** 1049–1058. doi:10.1093/nar/gkw1155

Uhlmann F. 2016. SMC complexes: from DNA to chromosomes. *Nat Rev Mol Cell Biol* **17:** 399–412. doi:10.1038/nrm.2016.30

Umbarger MA, Toro E, Wright MA, Porreca GJ, Baù D, Hong SH, Fero MJ, Zhu LJ, Marti-Renom MA, McAdams HH, et al. 2011. The three-dimensional architecture of a bacterial genome and its alteration by genetic perturbation. *Mol Cell* **44:** 252–264. doi:10.1016/j.molcel.2011.09.010

van Steensel B, Belmont AS. 2017. Lamina-associated domains: links with chromosome architecture, heterochromatin, and gene repression. *Cell* **169:** 780–791. doi:10.1016/j.cell.2017.04.022

Wang S, Xu J, Zeng J. 2015. Inferential modeling of 3D chromatin structure. *Nucleic Acids Res* **43:** e54. doi:10.1093/nar/gkv100

Weintraub AS, Li CH, Zamudio AV, Sigova AA, Hannett NM, Day DS, Abraham BJ, Cohen MA, Nabet B, Buckley DL, et al. 2017. YY1 is a structural regulator of enhancer-promoter loops. *Cell* **171:** 1573–1588.e28. doi:10.1016/j.cell.2017.11.008

Wood AJ, Severson AF, Meyer BJ. 2010. Condensin and cohesin complexity: the expanding repertoire of functions. *Nat Rev Genet* **11:** 391–404. doi:10.1038/nrg2794

Wutz G, Várnai C, Nagasaka K, Cisneros DA, Stocsits RR, Tang W, Schoenfelder S, Jessberger G, Muhar M, Hossain

MJ, et al. 2017. Topologically associating domains and chromatin loops depend on cohesin and are regulated by CTCF, WAPL, and PDS5 proteins. *EMBO J* **36:** 3573–3599. doi:10.15252/embj.201798004

Yildirim A, Feig M. 2018. High-resolution 3D models of *Caulobacter crescentus* chromosome reveal genome structural variability and organization. *Nucleic Acids Res* **46:** 3937–3952. doi:10.1093/nar/gky141

Yildirim A, Hua N, Boninsegna L, Polles G, Gong K, Hao S, Li W, Zhou XJ, Alber F. 2021. Mapping the nuclear microenvironment of genes at a genome-wide scale. bioRxiv doi:10.1101/2021.07.11.451976

Zhan Y, Giorgetti L, Tiana G. 2017. Modelling genome-wide topological associating domains in mouse embryonic stem cells. *Chromosome Res* **25:** 5–14. doi:10.1007/s10577-016-9544-6

Zhang B, Wolynes PG. 2015. Topology, structures, and energy landscapes of human chromosomes. *Proc Natl Acad Sci* **112:** 6062–6067. doi:10.1073/pnas.1506257112

Zhu Y, Gong K, Denholtz M, Chandra V, Kamps MP, Alber F, Murre C. 2017. Comprehensive characterization of neutrophil genome topology. *Genes Dev* **31:** 141–153. doi:10.1101/gad.293910.116

Zhu G, Deng W, Hu H, Ma R, Zhang S, Yang J, Peng J, Kaplan T, Zeng J. 2018. Reconstructing spatial organizations of chromosomes through manifold learning. *Nucleic Acids Res* **46:** e50. doi:10.1093/nar/gky065

Zhu H, Wang Z, Valencia A. 2019. SCL: a lattice-based approach to infer 3D chromosome structures from single-cell Hi-C data. *Bioinformatics* **35:** 3981–3988. doi:10.1093/bioinformatics/btz181

Transcription Factor Dynamics

Feiyue Lu and Timothée Lionnet

Institute for Systems Genetics and Cell Biology Department, NYU School of Medicine, New York, New York 10016, USA

Correspondence: Timothee.lionnet@nyulangone.org

To predict transcription, one needs a mechanistic understanding of how the numerous required transcription factors (TFs) explore the nuclear space to find their target genes, assemble, cooperate, and compete with one another. Advances in fluorescence microscopy have made it possible to visualize real-time TF dynamics in living cells, leading to two intriguing observations: first, most TFs contact chromatin only transiently; and second, TFs can assemble into clusters through their intrinsically disordered regions. These findings suggest that highly dynamic events and spatially structured nuclear microenvironments might play key roles in transcription regulation that are not yet fully understood. The emerging model is that while some promoters directly convert TF-binding events into on/off cycles of transcription, many others apply complex regulatory layers that ultimately lead to diverse phenotypic outputs. Cracking this kinetic code is an ongoing and challenging task that is made possible by combining innovative imaging approaches with biophysical models.

Cell-fate control rests with a series of proteins generally termed transcription factors (TFs): cell types are the product of the sequential expression of distinct TFs during development and forced expression of the right TF(s) can reprogram cells into different cell fates (Takahashi and Yamanaka 2006). TFs recognize specific sequences within the promoter or enhancer(s), which regulates a given gene (Suter 2020). Once bound, they recruit coactivators and chromatin remodelers, culminating in the assembly at the gene promoter of the preinitiation complex that loads a functional RNA polymerase II (Pol II). Pol II is then licensed for elongation by regulatory complexes and proceeds to synthesize the nascent transcript (Cramer 2019). For the purpose of this review, we adopt an umbrella definition of TFs that, in addition to sequence-specific TFs, also includes chromatin remodelers, general transcription factors (gTFs), Pol II, coactivators, and repressors.

In vitro biochemical assays and in vivo footprinting assays have provided important insights into the DNA sequences targeted by TFs (Lambert et al. 2018). In higher eukaryotes, sequence-specific TFs typically bind thousands of targets to regulate hundreds of genes, although some TFs display increased specialization (Zolotarev et al. 2017), in extreme cases controlling a single gene such as the TF ZNF410 that uniquely controls γ-globin transcription in erythroid cells (Yang et al. 2017; Lan et al. 2020). Because transcription programs are often long-lived (days), TFs have generally been assumed to bind their targets for long periods (Perlmann et al. 1990), consistent with the complexity of

Cite this article as *Cold Spring Harb Perspect Biol* doi: 10.1101/cshperspect.a040949

the transcription machinery, deemed incompatible with rapid assembly. However, over the last two decades, live imaging studies have demonstrated that most TFs are highly dynamic with residence time of seconds (Hager et al. 2009), that different TF subpopulations exhibit specific mobility dynamics, and that TFs often form nonstoichiometric complexes consisting of many molecules (Liu and Tjian 2018). These complexes have been referred to as clusters, condensates, or hubs across the literature. As condensates are often implied to form through phase separation and hubs hint at a functional role, we restrict ourselves to the term cluster here when we refer to those complexes without assumption of their assembly mechanism or function. In parallel, live-cell mRNA imaging experiments have similarly uncovered complex transcription dynamics. Active genes do not synthesize mRNAs steadily over time; rather, transcription occurs stochastically in bursts that alternate with off periods (Rodriguez and Larson 2020). As a result, expression levels, and thus phenotypes, are probabilistic rather than deterministic (Symmons and Raj 2016). Despite recent progress, we still lack mechanistic models linking stochastic transcription kinetics with upstream TF biophysics. In this article, we review our current understanding of TF mobility and discuss mechanisms linking TF dynamics with stochastic transcription outputs.

TECHNIQUES TO CAPTURE TF DYNAMICS

Understanding TF dynamics requires tools to detect where TFs bind in the genome, at what amount, what percentage of TFs are bound to DNA (% bound), and what their association (k_{on}) and dissociation rates with the DNA (k_{off}, the inverse of the TF residence time) are.

In Vitro Binding Specificity

Systematic evolution of ligands by exponential enrichment (SELEX) probes DNA motifs preferentially bound by TFs in vitro (Jolma et al. 2010). A purified TF is incubated with a large library of DNA molecules, from which TF-bound sequences are identified. SELEX may not accurately reflect in vivo binding, and where TFs compete or cooperate with one another, DNA is folded into chromatin and exposed to a very different ionic milieu than in vitro. Nevertheless, SELEX has been used to determine the binding specificity of hundreds of TFs on free and nucleosome-containing DNA, confirming notable differences (Zhu et al. 2018).

In Vivo Binding Specificity

Chromatin immunoprecipitation (ChIP)-seq is widely used to determine in vivo genome-wide binding profiles (Johnson et al. 2007). Chemical cross-linking of a TF of interest to DNA in cells is followed by ChIP and sequencing of TF-bound DNA fragments. Recent variations of the procedure achieve better spatiotemporal resolution than the original methods; for example, digesting fragmented DNA prior to ChIP enables near base-pair resolution (Rhee and Pugh 2011; He et al. 2015). While ChIP traditionally measures average TF occupancy (Fig. 1), recent modifications of the technique enable measurements of kinetic rates (k_{on}, k_{off}):k_{on} can be determined by varying the cross-linking duration (Poorey et al. 2013), whereas measuring bound TFs at various time points after acute depletion of nuclear TFs provides access to genome-wide k_{off} values (Jonge et al. 2020). ChIP, however, is subject to two technical caveats: first, TFs can artificially dissociate from chromosomes upon cross-linking (Teves et al. 2016; Festuccia et al. 2019), which can be overcome by alternatives bypassing cross-linking (Skene and Henikoff 2017), and second, ChIP accuracy is limited by antibody quality (Shah et al. 2018). ChIP-seq and its derivatives have two further limitations: first, they require tens to millions of cells to produce a robust signal, and therefore only provide binding profiles averaged over many cells, and second, free TFs are lost as only the chromatin-bound fraction is captured. Despite these limitations, ChIP-seq remains the sole approach capable of capturing genome-wide target sites of TFs in cells.

Live-Cell Kinetics

Because transcription bursts are heterogeneous among cell populations (Chubb et al. 2006; Raj

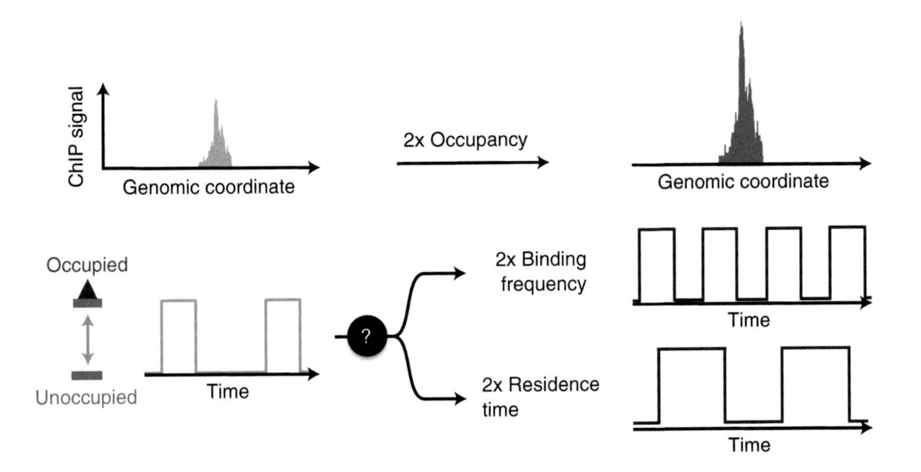

Figure 1. Occupancy versus kinetics. ChIP-seq (*top*) measures the average occupancy of a given transcription factor (TF) at its binding site. Increased occupancy may result from higher TF-binding frequency (higher k_{on}), or longer residence times ($1/k_{off}$, *bottom*). Kinetic profiles can be decoded into distinct transcriptional outputs by promoters.

et al. 2006; Suter et al. 2011), resolving transcription kinetics requires single-cell sensitivity. Given that TFs turn over within seconds, temporal resolution is key for any method to measure in vivo TF dynamics. Furthermore, in situ studies are necessary to recapitulate physiological chromatin states and cofactors. Live imaging satisfies all these conditions, and various methods discussed below have emerged as tools of choice to probe TF dynamics (Mueller et al. 2013).

Fluorescence recovery after photobleaching (FRAP) involves photobleaching fluorescently tagged TFs in a nuclear region of interest, and subsequently measuring fluorescence recovery, typically over seconds to hours, as photobleached TFs exchange with fluorescent TFs freely diffusing into the focal volume from the rest of the nucleus (Phair and Misteli 2000). Fitting recovery curves to reaction-diffusion models provides estimates of average diffusion coefficients of free TFs, residence times of bound TFs, and the relative proportions of the two states (Darzacq et al. 2007; Maiuri et al. 2011).

Fluorescence correlation spectroscopy (FCS) measures the passage of individual molecules through a focused beam (Elf et al. 2007). The number of molecules crossing the beam per unit time provides access to the absolute TF concentration in the chosen region, while the duration each molecule dwells in the focal volume provides similar observables as FRAP, albeit in a different time regime (ms-sec).

Single-molecule tracking (SMT) captures the dynamics of TFs over an entire nuclear plane (Elf and Barkefors 2019). TFs labeled with a photoactivatable fluor are initially dark but upon a brief pulse of blue light, a few TF molecules turn on and are tracked until they either photobleach or diffuse out of the focal volume. Many cycles of activation followed by tracking generate hundreds to thousands of individual trajectories per cell. SMT is usually performed in one of two imaging modes: the fast-tracking mode (exposure times of 1–50 msec per frame) captures fast diffusing TFs in the nucleoplasm, from which one can determine the relative amounts of free versus chromatin-bound TFs and their diffusion coefficients. The slow-tracking mode (exposure times of 100–500 msec per frame) motion-blurs free molecules to calculate the residence times ($1/k_{off}$) of TFs bound to chromatin.

A SHORT STAY AFTER A LONG SEARCH

TF Search Dynamics

Live imaging studies have helped shape a clearer picture of TF behavior. TFs reside on DNA only for short intervals (1–100 sec) and at any given time, less than half the TF population is bound to chromatin (Figs. 2 and 3; Tables 1 and 2). According to the facilitated diffusion model (von Hippel and Berg 1989), while searching for its target(s), a TF molecule diffuses through the nuclear space in 3D, occasionally binding and briefly sliding along accessible DNA regions (<1 sec) (Chen et al. 2014; Marklund et al. 2020). After several attempts, the TF molecule eventually lands on a cognate binding site, where it resides longer (~10 sec). The average time it takes for a TF to travel between two specific binding sites is defined as its search time,

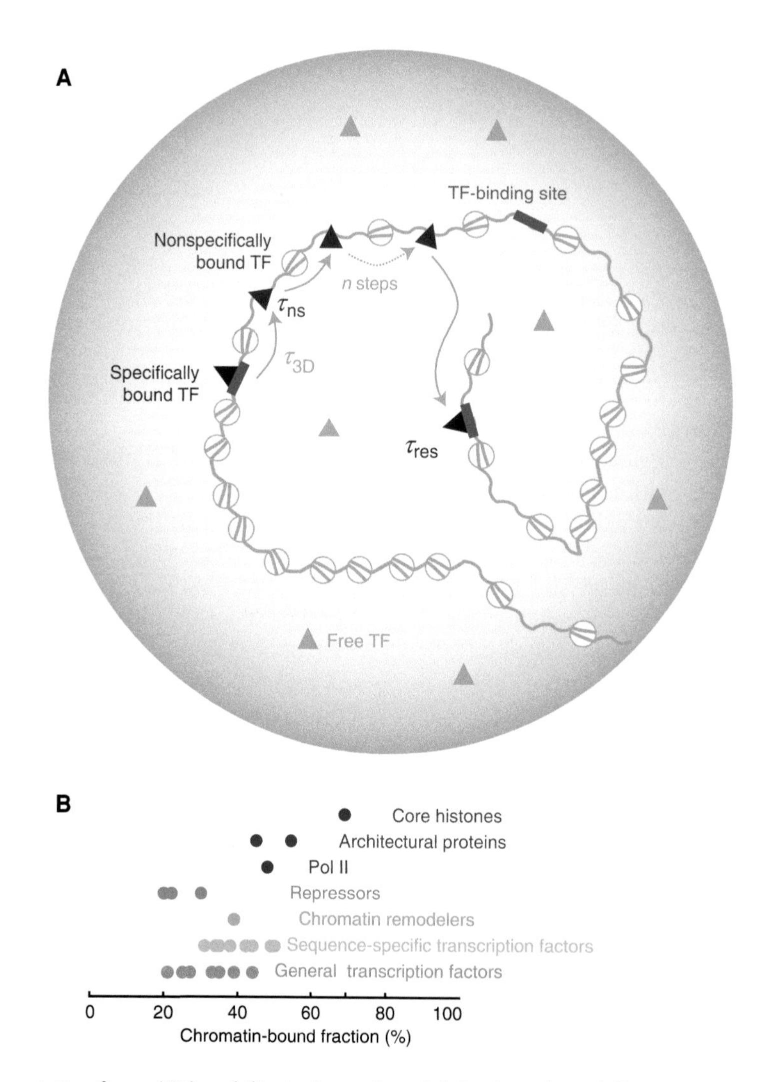

Figure 2. Transcription factor (TF) mobility in the nucleus. (*A*) Facilitated 3D diffusion model. While searching for its target sites (blue), a TF makes multiple, brief, and nonspecific contacts ($\tau_{ns} < 1$ sec; orange) with open chromatin before landing on its cognate site where it dwells longer ($\tau_{res} \sim 1$–100 sec; magenta). The average time between two specific binding events is defined as the search time $\tau_{search} = (n-1)^*(\tau_{3D} + \tau_{ns}) + \tau_{3D}$, where $n \sim 10$–100 is the number of trials and τ_{3D} is the averaged diffusion time between two trials. (Panel *A* is based on data in Chen et al. 2014.) (*B*) Experimentally measured chromatin-bound fraction (circles) for various TFs compiled from the community resource developed by Mir and colleagues (www.mir-lab.com/dynamics-database) and the recent literature. Only factors expressed as knockins or rescuing a knockout background are featured here.

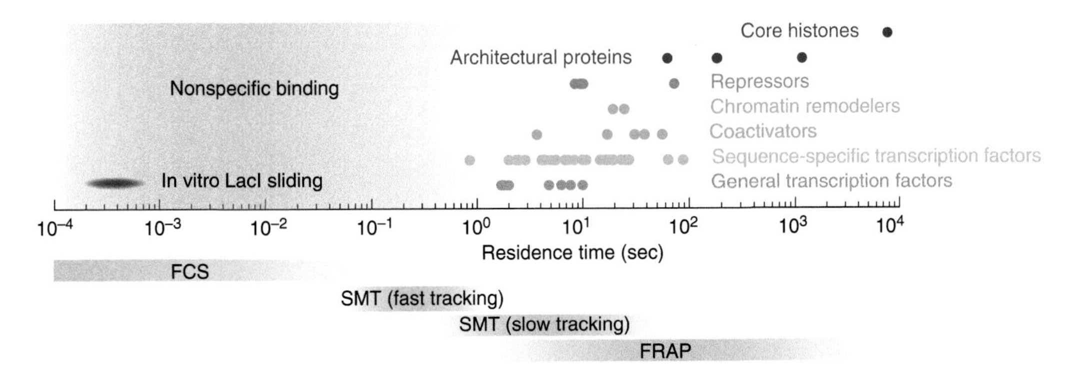

Figure 3. Chromatin association timescales in vivo. Experimentally measured residence times (circles) compiled from the community resource developed by Mir and colleagues (www.mir-lab.com/dynamics-database) and the recent literature, relative to the temporal resolution of different imaging techniques (*bottom*, gray). (FCS) Fluorescence correlation spectroscopy, (SMT) single-molecule tracking, (FRAP) fluorescence recovery after photobleaching.

which is inversely proportional to k_{on}, and sets a limit to how fast a gene can be activated (Elf and Barkefors 2019). As a result of the dozens of nonproductive interactions with chromatin, the search time is often orders of magnitude longer than the TF residence time (Fig. 2). The chromatin-bound fraction consists of both specifically and nonspecifically bound TF molecules and the search time is impacted by the concentration of target sites (Reisser et al. 2018), DNA folding (Cortini and Filion 2018), the chromatin states surrounding the target sites (Mehta et al. 2018), the uneven spatial distribution of protein/DNA barriers in the nucleoplasm (Izeddin et al. 2014; Li et al. 2016), and the presence of coregulators at the target site (Mir et al. 2017). The effective k_{on} is also directly proportional to the nuclear concentration of TFs, enabling dynamical regulation of target occupancy across a large dynamic range (Di Ventura and Kuhlman 2016).

Repressive chromatin complexes are generally smaller than activating ones (Miron et al. 2020), but TF mobility and accessibility to their target sites are largely independent of molecular weights, ruling out steric hindrance as a main partition mechanism (Grünwald et al. 2008; Bancaud et al. 2009; Liu et al. 2014). Lower TF mobility and higher trapping frequencies in heterochromatin appear more likely to explain the observed enrichment of repressive complexes in chromatin-dense regions. Additionally, while 3D diffusing, some TFs move isotropically as expected for free diffusion, while others exhibit anisotropy, making more U-turns than expected by chance, which affects their search times (Izeddin et al. 2014). Anisotropic exploration results in the TF oversampling its nuclear neighborhood. Conversely, isotropic diffusion enables global exploration where any target is equally likely to be reached, regardless of distance. Nuclear exploration is therefore likely more complex than the three discrete states envisioned in the facilitated diffusion model (3D diffusion, 1D DNA sliding, stable binding to a target): interactions with nucleosomes (Lerner et al. 2020) or association into local clusters (Hansen et al. 2020) impact TF mobility and could introduce apparent bound states guiding search in preferred compartments. Single-molecule trajectories constitute rich data sets, and mobility metrics extending beyond the diffusion coefficient bring important insights (Shukron et al. 2019).

TFs Interact Briefly with Chromatin

Unlike core histones that stably bind chromatin for tens of minutes to hours (Kimura and Cook 2001; Dion et al. 2007; Deal et al. 2010), the residence times of sequence-specific TFs, gTFs, and repressive complexes tend to be short, in the range of seconds (Fig. 3; Choi et al. 2017;

Table 1. Experimentally measured residence times for various transcription factors (TFs) related to Figure 3

Protein name	Protein type	Residence time (seconds)	References
Histone H1	Architectural protein	183	Phair et al. 2004
CTCF	Architectural protein	63	Hansen et al. 2017
CTCF	Architectural protein	184.3	Hansen et al. 2017
RAD21	Architectural protein	1170.6	Hansen et al. 2017
HMGN1	Chromatin remodeler	24.8	Phair et al. 2004
BRG1	Chromatin remodeler	19.4	Phair et al. 2004
PCAF	Coactivator	17.1	Phair et al. 2004
CYCT1	Coactivator	56	Lu et al. 2018
BRD4	Coactivator	38.1	Phair et al. 2004
ARNT	Coactivator	30.9	Phair et al. 2004
Mediator	Coactivator	3.7	Nguyen et al. 2020
H2B	Core histone	7800	Kimura and Cook 2001
TFIIA	General TF	6.3	Nguyen et al. 2020
TFIIB	General TF	1.8	Nguyen et al. 2020
TFIID	General TF	4.8	Nguyen et al. 2020
TFIIE	General TF	2	Nguyen et al. 2020
TFIIF	General TF	1.7	Nguyen et al. 2020
TFIIH	General TF	10	Nguyen et al. 2020
TFIIK	General TF	7.7	Nguyen et al. 2020
Cbx7	Repressor	8.5	Tatavosian et al. 2018
HT-Cbx7/F-H3.3	Repressor	9.7	Tatavosian et al. 2018
Eed	Repressor	9.5	Tatavosian et al. 2018
EZH2	Repressor	10	Tatavosian et al. 2018
HT-EZH2/F-H3.3	Repressor	10.2	Tatavosian et al. 2018
HP1β	Repressor	73	Phair et al. 2004
ESRRB	Sequence-specific TF	10	Xie et al. 2017
STAT3	Sequence-specific TF	8.3	Xie et al. 2017
TBP	Sequence-specific TF	88	Teves et al. 2018
Sox2	Sequence-specific TF	14.6	Teves et al. 2016
Estrogen receptor	Sequence-specific TF	4.36	Swinstead et al. 2016
FoxA1	Sequence-specific TF	10.8	Swinstead et al. 2016
CREB1	Sequence-specific TF	2.86	Sugo et al. 2015
Glucocorticoid receptor	Sequence-specific TF	0.85	Stasevich et al. 2010
Sox19b	Sequence-specific TF	2	Reisser et al. 2018
TBP	Sequence-specific TF	6.8	Reisser et al. 2018
Glucocorticoid receptor	Sequence-specific TF	7.25	Presman et al. 2016
AhR	Sequence-specific TF	25.6	Phair et al. 2004
C/EBP	Sequence-specific TF	18.8	Phair et al. 2004
FBP	Sequence-specific TF	63.6	Phair et al. 2004
Fos	Sequence-specific TF	14.6	Phair et al. 2004
Jun	Sequence-specific TF	27.3	Phair et al. 2004
Mad	Sequence-specific TF	19.5	Phair et al. 2004
Myc	Sequence-specific TF	16.3	Phair et al. 2004
NF1	Sequence-specific TF	16.2	Phair et al. 2004
XBP	Sequence-specific TF	23.1	Phair et al. 2004
TetR	Sequence-specific TF	5	Normanno et al. 2015
Bicoid	Sequence-specific TF	2.33	Mir et al. 2018
Zelda	Sequence-specific TF	5.56	Mir et al. 2018
p53	Sequence-specific TF	2.5	Hinow et al. 2006

Continued

Cite this article as *Cold Spring Harb Perspect Biol* doi: 10.1101/cshperspect.a040949

Table 1. *Continued*

Protein name	Protein type	Residence time (seconds)	References
Gal4	Sequence-specific TF	17	Donovan et al. 2019
OCT-4	Sequence-specific TF	14.6	Chen et al. 2014
P65 (NF-κB)	Sequence-specific TF	4.1	Callegari et al. 2019

Youmans et al. 2018; Suter 2020). The linker histone H1 exchanges from chromatin with similarly fast kinetics (~20 sec–3 min) (Lever et al. 2000; Misteli et al. 2000), while the architectural proteins cohesin and CTCF fall in between core histones and TFs (min) (Hansen et al. 2017). The residence time depends on the affinity between a TF and its target (Clauß et al. 2017; Callegari et al. 2019; Donovan et al. 2019; Popp et al. 2020), but can be modulated across loci, cell types, and cell states. For example, the residence time of the gTF TBP (TATA-binding protein) ranges from seconds in the developing zebrafish embryo (Reisser et al. 2018), interphase U2OS cells, and in vitro assays (Zhang et al. 2016), to hours at the histone locus in *Drosophila* cells (Guglielmi et al. 2013). TBP resides longer at active genes on mitotic chromosomes than during interphase in mouse embryonic stem (ES) cells, enabling rapid transcriptional reactivation upon mitotic exit (Teves et al. 2018). SRF (serum response factor) (Hipp et al. 2019), the glucocorticoid receptor (Stavreva et al. 2019), and GAL4 (Donovan et al. 2019) all exhibit increased residence times upon activation of their upstream pathways. In the case of GAL4, the promoter nucleosome is a key modulator of TF residence time (Donovan et al. 2019) but it is not clear whether this is a general mechanism. In most cases studied so far, the changes in TF residence times are small compared to the changes in the transcription output of their downstream genes, suggesting additional regulation of k_{on} and/or downstream amplifying mechanisms (Fig. 1). Short TF residence times mirror the observations that target sites favor low-affinity TF motifs, a feature that confers extended sensitivity to TF concentration (Kribelbauer et al. 2019).

The biological interpretation of residence times measured by SMT faces two challenges: first, photobleaching limits the direct observation of very long events, even though photobleaching contributions can be corrected from measurements (Gebhardt et al. 2013; Chen et al. 2014; Hansen et al. 2017; Reisser et al. 2020; Garcia et al. 2021). Second, in each experiment, binding events are measured across the nucleus without knowledge of the locus bound by each TF. The original separation of binding events into two discrete populations, assumed to represent nonspecific versus specific events, based on the observation of short (<1 sec) and long (~10 sec) subpopulations in residence time distributions, and supported by DNA-binding domain (DBD) deletion experiments (Chen et al. 2014), might not be valid in all cases. Some TFs exhibit broad residence time distributions, consistent with a continuum of affinities across diverse genomic targets (Normanno et al. 2015; Stavreva et al. 2019; Garcia et al. 2021). A novel kinetic analysis suggests on the other hand the existence of 5–6 discrete dissociation rates ranging from subseconds to minutes (Popp et al. 2020; Reisser et al. 2020). Despite their differences, the different SMT analyses converge on the fact that only a minority of TF-binding events extend beyond the seconds regime, consistent with dozens of earlier studies by FRAP (Hemmerich et al. 2011). In contrast, an indirect approach suggests exceptionally long-lived TF binding in the *Xenopus* oocyte (hours to days) (Gurdon et al. 2020) but it remains unclear whether the chromatin environment of the oocyte fosters this unusual behavior. Albeit chromatin motion and microscopy constrain the live-cell-imaging resolution to ~10 kbp (Li et al. 2019), new microscopes able to measure TF binding at a specific locus have validated that seconds-long TF interactions do occur at relevant targets, and that they correlate with productive transcription (Donovan et al. 2019; Li et al. 2019; Stavreva et al. 2019). Combined with biological perturbations, these tools hold great potential to decipher the TF kinetic code.

Table 2. Experimentally measured chromatin-bound fraction for various transcription factor (TFs), related to Figure 2

Protein name	Protein type	Bound (%)	References
RAD21	Architectural protein	45	Hansen et al. 2017
CTCF	Architectural protein	54.5	Hansen et al. 2017
Mediator	Coactivator	39	Nguyen et al. 2020
H2B	Core histone	69	Nguyen et al. 2020
TFIID	General TF	39	Nguyen et al. 2020
TFIIA	General TF	35	Nguyen et al. 2020
TFIIB	General TF	21	Nguyen et al. 2020
TFIIF	General TF	25	Nguyen et al. 2020
TFIIE	General TF	33	Nguyen et al. 2020
TFIIH	General TF	27	Nguyen et al. 2020
TFIIK	General TF	44	Nguyen et al. 2020
TBP	General TF	34	Nguyen et al. 2020
TBP	General TF	31	Teves et al. 2018
Eed	Repressor	22	Tatavosian et al. 2018
Ring1B	Repressor	20	Huseyin and Klose 2021
Cbx7	Repressor	30	Tatavosian et al. 2018
EZH2	Repressor	22	Tatavosian et al. 2018
Zelda	Sequence-specific TF	49	Mir et al. 2018
Bicoid	Sequence-specific TF	50	Mir et al. 2018
STAT3	Sequence-specific TF	35	Xie et al. 2017
Sox2	Sequence-specific TF	38	Liu et al. 2014
OCT-4	Sequence-specific TF	42.2	Chen et al. 2014
ESRRB	Sequence-specific TF	44	Xie et al. 2017
RPB1	Subunit of Pol II	48	Nguyen et al. 2020

From TF Binding to Transcripts

The short TF residence times echo the short bursts that constitute the basic unit of transcription (Rodriguez and Larson 2020). It is thus tempting to propose that TF-binding events coincide with bursts (Fig. 4A). This simple "one-to-one" model predicts that the burst frequency should equal the product of k_{on} and TF concentration and the burst duration should equal the TF residence time. Several observations are consistent with the one-to-one model: (1) the transcription machinery assembles within seconds, well within typical TF residence times (Zhang et al. 2016; Nguyen et al. 2020); (2) increasing TF concentration increases burst frequency (Senecal et al. 2014; Stavreva et al. 2019); (3) search time estimates are inversely proportional to burst frequencies observed in yeast (Larson et al. 2011); (4) decreasing k_{off} increases the number of transcripts per burst (Senecal et al. 2014); (5) enhancers regulate burst frequency (Walters et al. 1995; Bartman et al. 2016; Fukaya et al. 2016; Chen et al. 2018); and (6) elegant experiments directly observe that single TF-binding events correlate temporally with burst firing (Donovan et al. 2019; Stavreva et al. 2019). While the simple two-state model might be valid in simple systems, it often breaks down in higher eukaryotes (Bartman et al. 2016; Corrigan et al. 2016; Rodriguez et al. 2019; Lammers et al. 2020; Popp et al. 2020). In a striking example, pluripotency genes cease to transcribe early in differentiation, long before their enhancers lose TF occupancy (Hamilton et al. 2019). gTFs constitute obvious kinetic intermediates that could mediate this decoupling: TBP binding regulates permissive periods over 5–20 min timescales, while Mediator ensures rapid back-to-back Pol II initiations (seconds) (Tantale et al. 2016), and pausing regulates the number of transcripts per burst (Bartman et al. 2016). Besides gTFs, supercoiling and chromatin also shape burst timing (Muramoto et al. 2010; Chong et al. 2014; Teves and Henikoff 2014). Thus, dis-

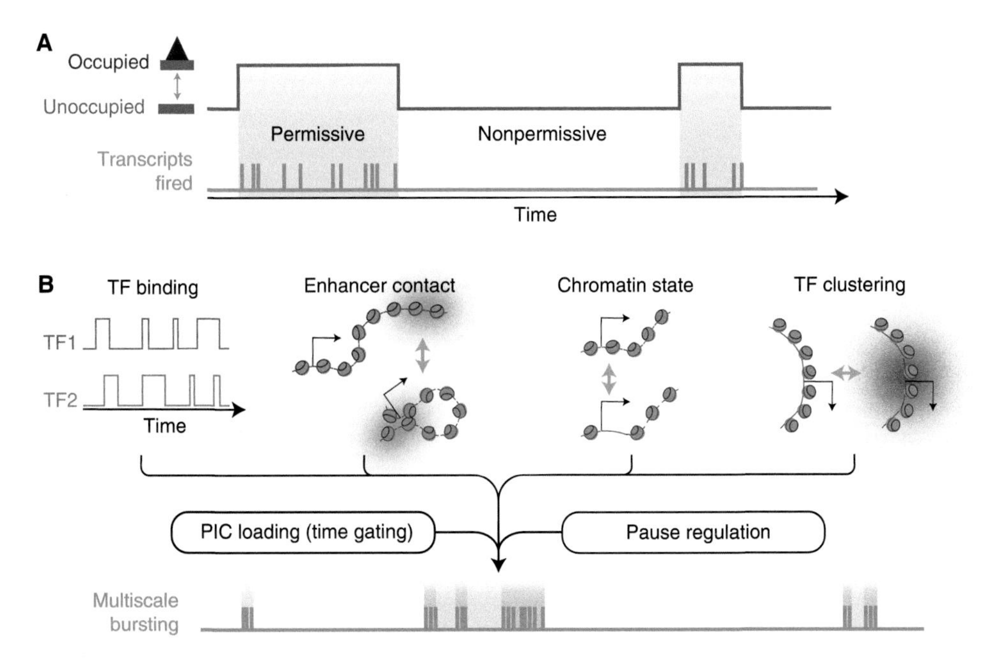

Figure 4. Decoding of transcription factor (TF) kinetics by promoters. (*A*) In simple systems, TF binding directly leads to permissive periods (gray) during which many Pol II are rapidly fired. In this one-to-one model, the TF residence time equals the burst duration. (*B*) Promoters often integrate complex regulation from multiple TFs and enhancers, as well as the state of chromatin at the promoter, and the kinetics of cluster formation. These interdependent inputs are processed by the transcription machinery, which applies further control layers, leading to multistate bursting dynamics. (PIC) Preinitiation complex.

tinct steps of the transcription cycle are controlled by separate TFs (Stasevich et al. 2014), leading to multiscale bursting kinetics (Corrigan et al. 2016) and complex regulatory logic (Fig. 4B; Scholes et al. 2017). These features likely explain why predicting enhancer combinations remains challenging (Vincent et al. 2016). As the enhancer–promoter looping paradigm has recently been called into question (Alexander et al. 2019; Benabdallah et al. 2019), biophysical models of enhancer function remain sorely needed (Bothma et al. 2015). Ideally such descriptions will integrate binding of the various players, DNA organization, and local TF clustering (see last section) to predict bursting kinetics.

Intrinsically Disordered Regions Provide a Flexible Platform for TF Dynamics

Regulatory sequences favor low-affinity TF-binding sites that ensure specificity and sensitivity to TF concentrations over a wide range (Kribelbauer et al. 2019). Could the same principles hold for interactions between TFs and their protein partners? Indeed, it has long been appreciated that the activating domains of TFs contain intrinsically disordered regions (IDRs), which are unstructured peptides that are essential sites for TF–TF interactions and retain function upon extensive mutations (Sigler 1988; van der Lee et al. 2014; Wright and Dyson 2015). Over 80% of all eukaryotic TFs contain one or more IDRs (Liu et al. 2006), also called low complexity regions because of their limited repertoire of amino acids. While protein–protein interactions are classically thought of as stoichiometric complexes forming via a lock–key mechanism relying on complementary structures, IDRs by definition cannot fit this model. In addition, IDR sequences are poorly conserved, challenging our understanding of how TFs associate with specific partner(s) to regulate transcription. The possible role of IDRs in TF clustering (see below) has fueled renewed interest in these elusive domains, with the hope to better detect and interpret pathological mutations, many of which fall within IDRs (Uyar et al. 2014).

Stoichiometric IDR Complexes

While IDRs explore a large ensemble of conformations in solution, they can exhibit a more ordered conformation when bound to a cofactor, enabling the appearance of a traditional protein–protein interface (Fig. 5; Schuler et al. 2020). This process, termed binding-coupled folding, can occur in two ways: (1) the ordered bound state pre-exists in the conformational ensemble of the free IDR and is required for binding to the cofactor ("conformational selection"); or (2) the TF can recognize its partner in its disordered state and fold while binding ("induced fit") (Staby et al. 2017). Binding-coupled folding offers flexibility; for instance, the IDR of the coactivator CBP adopts different structures when bound to different TFs (Demarest et al. 2002; Qin et al. 2005; Waters et al. 2006). IDRs can in some cases form "fuzzy complexes" that do not exhibit a fixed conformation but gain stability through multiple weak and dynamic contacts between the IDR and its partner (Tompa and Fuxreiter 2008; Henley et al. 2020). Thanks to their dynamic nature, fuzzy complexes can be easily remodeled via competitive substitution, enabling rapid gear switching of the transcription machinery (Schuler et al. 2020).

IDR-Driven Phase Separation

Beyond fuzzy complexes, IDRs can mediate the formation of nonstoichiometric clusters (Fig. 6). Pioneering work showing that IDRs can form or associate with hydrogels in vitro suggested that IDRs of TFs may drive clustering via phase separation (Frey et al. 2006; Kwon et al. 2013).

IDRs are often repetitive in sequence, and thus intrinsically multivalent, a prerequisite of phase separation (Choi et al. 2020). For instance, the carboxy-terminal domain (CTD) of the largest Pol II subunit, Rpb1, is a well-studied IDR that consists of repeats of the tyrosine-serine-proline-threonine-serine-proline-serine (YSPTSPS) motif (Corden 2013; Eick and Geyer 2013; Zaborowska et al. 2016; Gibbs et al. 2017; Portz et al. 2017). Pol II and many purified TFs or their IDRs indeed self-organize into phase-separated droplets in vitro (Larson et al. 2017; Boehning et al. 2018; Boija et al. 2018; Lu et al. 2018; Sabari et al. 2018; Guo et al. 2019; Plys et al. 2019; Zamudio et al. 2019; Daneshvar et al. 2020; Li et al. 2020a). In vivo, some features of TF clusters are consistent with liquid–liquid phase separation. TFs exchange dynamically within clusters as revealed by FRAP, and TF clusters undergo fusion and fission and are sensitive to treatment with 1,6-hexanediol, which disrupts some of the weak interactions that can contribute to phase separation (Cho et al. 2018; Chong et al. 2018; Sabari et al. 2018). However, these features alone do not rule out nonphase separation

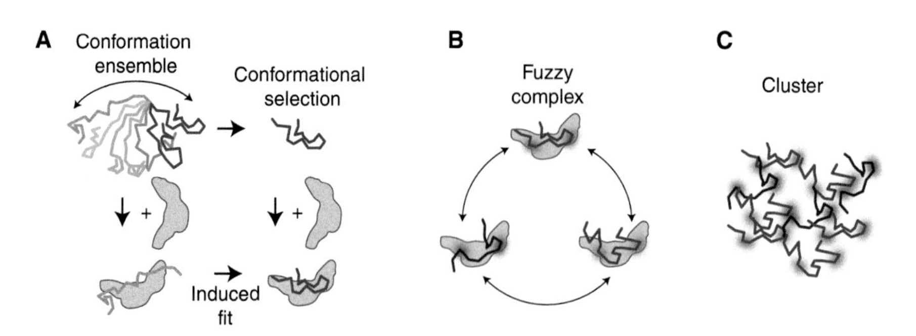

Figure 5. Intrinsically disordered regions (IDRs) mediate different types of complexes. (*A*) Binding-coupled folding: free IDRs explore a vast conformation space (colors), but some IDRs adopt a fixed conformation when bound to a partner (gray). (*B*) In a fuzzy complex, the IDR is dynamic yet remains bound to its partner. (*C*) Nonstoichiometric complexes (clusters) can form via networked interactions between multivalent IDRs.

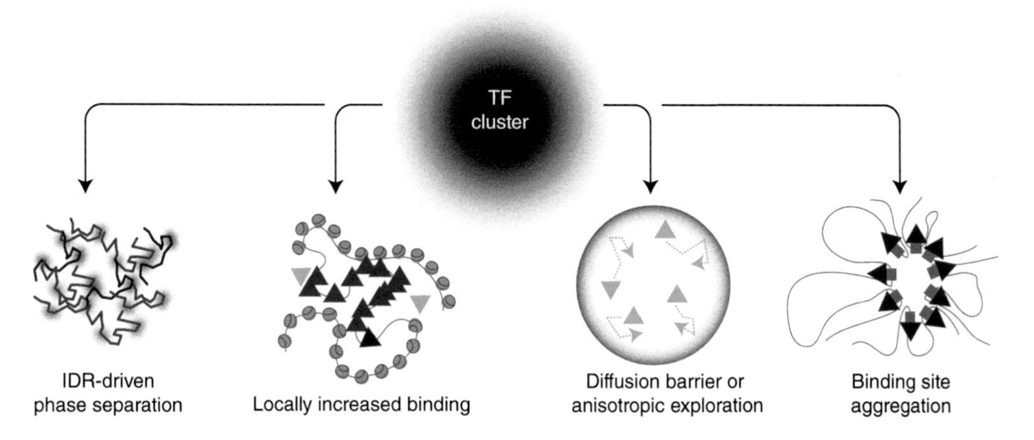

Figure 6. Clustering mechanisms. Transcription factor (TF) clustering can occur via phase separation, thanks to multivalent interactions between intrinsically disordered regions (IDRs) (*left*). Other mechanisms also exist; locally enhanced TF binding on highly accessible chromatin, locally anisotropic diffusion, and collapse of a TF-binding site. Triangles represent TFs; green, orange, and magenta denote, respectively, free, nonspecifically bound, and specifically bound species.

mechanisms (McSwiggen et al. 2019b) (see last section). A substantial caveat is that in vitro assays often use crowding agents and are done at TF concentrations much higher than physiological conditions and thus favor phase separation. For this reason, while these assays provide useful insights into the relative role of different parameters in clustering, they likely offer only a limited representation of in vivo clusters, and do for example not take into account the role of RNA and other TFs in clustering in vivo (Wei et al. 2020; Henninger et al. 2021).

Beyond Complexes: IDRs Facilitate Target Search

IDRs might not just mediate the affinity of TFs to each other, but may also influence search kinetics. The IDRs of sequence-specific TFs Msn2 and Yap1 are both necessary and sufficient for promoter specificity, but their DBDs are not (Brodsky et al. 2020), suggesting a two-step target search. TFs first localize to an open promoter by promiscuously scanning it with their IDR, and later stably bind to target DNA motifs via their DBDs. The separation of tasks between DBD and activation domains might therefore not be binary (Liu et al. 2008). IDRs could boost search efficiency in several ways including (1) the large interaction surface of IDRs may increase TF affinity for DNA, enhancing its sliding propensity; (2) IDRs may facilitate TF translocation to a neighboring DNA segment (Vuzman and Levy 2012); or (3) IDRs could guide TFs by facilitating clustering or anisotropic diffusion in specialized compartments (Izeddin et al. 2014; Hansen et al. 2020; Nguyen et al. 2020).

Sequence Determinants of IDR Interactions

In contrast to lock–key interactions driven by unique motifs, fuzzy complexes and phase separation involve individually weak but multivalent interactions between protein partners and/or nucleic acids, which could explain the low sequence conservation of IDRs. Indeed, mutation scans suggest that IDRs interact through a sum of weak interactions distributed along their entire domain (Wang et al. 2018; Brodsky et al. 2020). Furthermore, the *Drosophila* Pol II CTD, which consists of heptapeptide repeats diverged from the canonical YSPTSPS sequence, can be replaced with a shorter canonical CTD, but not with a canonical CTD at wild-type length (Lu et al. 2019). These findings suggest a model in which the sum interaction strength required for proper biological function can be achieved either through many low affinity sites (noncanonical repeats), or fewer high affinity ones (canonical repeats). Consistent with this picture, truncated CTDs decrease the frequency and size of transcription bursts in yeast (Quintero-Cadena et al. 2020), likely through reduced Pol II clustering (Boehning et al. 2018), while longer CTDs exhibit aberrant clustering (Lu et al. 2019). *Drosophila* Pol II mutants containing exceedingly long or short CTDs are inviable, confirming that sum interaction strength constitutes a key selection pressure. The fact that a specific sum interaction strength can be achieved in many ways (e.g., using distinct residues with similar physicochemical properties, and/or distributing these residues differently along the IDR sequence, accounts for the low sequence conservation of IDRs.

The prevalence of IDRs in TFs presents a conundrum. TFs need to selectively associate with specific sequence(s) or cofactors, yet IDRs mediate promiscuous interactions. How is specificity encoded into the underlying sequence of IDRs? It appears that small motifs or even a single residue within the IDR can convey specificity for a partner for instance by biasing a fuzzy complex toward a subset of its binding modes (Sims et al. 2011; Warfield et al. 2014; Desai et al. 2015; Zhao et al. 2016; Henley et al. 2020). Nonstoichiometric IDR clustering also exhibits specificity (Chong et al. 2018), although it remains unclear what mechanisms apply here. The numbers and positions of acidic, hydrophobic, and aromatic residues all modulate the affinity of IDR interactions (Staller et al. 2018; Erijman et al. 2020), as do posttranslational modifications (Guo et al. 2019). These observations suggest a departure from the dogma of sequence-defining function via a deterministic 3D protein structure. Rather, IDR sequences specify the physicochemical properties of residues such as hydrophobicity, charge distribution, and flexibility, in turn constraining both the conformation space explored by the protein chain as well as its potential for interactions through the number and strength of sticky residues, thus encoding specific TF functions including clustering propensity, exploration mode, and target selectivity (Vernon et al. 2018; Wang et al. 2018; Martin and Holehouse 2020).

Evolution of IDRs

Eukaryotic proteomes contain more disordered segments than those of simpler organisms (Ward et al. 2004; Tompa et al. 2006; Peng et al. 2015) and individual transcription regulators, such as the Pol II CTD (Quintero-Cadena et al. 2020) or Mediator subunits (Tóth-Petróczy et al. 2008) exhibit increasing disordered content over evolutionary timescales. An interesting idea is that regulatory innovation is often gained by adding new components to existing complexes. It is likely faster in evolutionary terms to achieve binding to an existing complex through a flexible domain, rather than creating a de novo specialized 3D structure matching the complex interface. Consistent with this idea, proteins that participate in large complexes are more disordered (Hegyi et al. 2007). Altogether, IDR size selection likely results from a series of tradeoffs between regulatory potential (longer IDRs enabling binding to more targets), nuclear exploration mode (longer positive tails increase affinity for DNA, enhancing a TF sliding propensity at the expense of its ability to hop to a locus in *trans* [Vuzman and Levy 2012]), and clustering potential (IDRs with higher valency or interaction strength generate static aggregates unable to respond to dynamic signals).

TFs ASSEMBLE IN CLUSTERS

Transcription has long been proposed to occur in stable, self-assembled "hubs" that could outlive the binding of individual components (Cook 1999; Edelman and Fraser 2012), similar to larger subnuclear structures such as the nucleolus (Phair and Misteli 2000). Consistent with the hub model, a variety of factors form clusters in cells: sequence-specific TFs (Liu et al. 2014; Mir et al. 2017, 2018; Chong et al. 2018; Basu et al. 2020; Li et al. 2020b), coactivators (Cho et al. 2018; Sabari et al. 2018; Guo et al. 2019; Li et al. 2019, 2020b; Zamudio et al. 2019), Pol II (see below), splicing factors (Guo et al. 2019), corepressors (Treen et al. 2020), repressive complexes (Wollman et al. 2017; Plys et al. 2019; Ruault et al. 2020), chromatin modifiers (Tatavosian et al. 2019), and HP1 (heterochromatin protein 1) (Strom et al. 2017; Erdel et al. 2020; Li et al. 2020a). Contrasting with the model of a stable factory, cluster lifetimes are generally short (Cisse et al. 2013).

Pol II Clustering

Single-molecule imaging has revealed that RNA polymerase II (Pol II) forms clusters in various mammalian cultured cell models (Cisse et al. 2013; Cho et al. 2016, 2016; Boehning et al. 2018; Li et al. 2019). Pol II clusters are short-lived, generally on the order of seconds, but some last minutes, and vary in size (Cisse et al. 2013; Cho et al. 2016, 2018; Boehning et al. 2018), from diffraction-limited foci all the way to micron-sized accumulations at the histone locus body (Guglielmi et al. 2013) and viral replication compartments (McSwiggen et al. 2019a). The number, size, and lifetime of clusters change upon induction (Cisse et al. 2013; Cho et al. 2016; Li et al. 2019) or inhibition of transcription (Cho et al. 2018; Li et al. 2019) and during differentiation (Cho et al. 2018), suggesting that clustering may contribute to transcription control.

The CTD is a key regulator of Pol II clustering. On its own, it forms liquid condensates in vitro and its length regulates Pol II clustering in vivo (Boehning et al. 2018). Phosphorylation of the CTD during early stages of transcription could control CTD clustering via charge modulation (Harlen and Churchman 2017). Indeed, inhibitors against P-TEFb, which triggers Pol II pause release via CTD phosphorylation, stabilize Pol II clusters (Cisse et al. 2013; Cho et al. 2016) while CTD phosphorylation disperses clusters in vitro (Boehning et al. 2018; Lu et al. 2018). These observations place Pol II clustering at the transcription preinitiation or initiation stage. CTD phosphorylation also biases Pol II association with splicing factor condensates versus those containing Mediator (Guo et al. 2019), while electrostatic repulsion by charged nascent RNAs dismantles Pol II clusters (Henninger et al. 2021). Together, these observations suggest a model wherein RNA accumulation and/or CTD phosphorylation force cluster turnover once a Pol II convoy has initiated on a transcribed gene (Quintero-Cadena et al. 2020).

Clustering without Phase Separation

Besides IDR-driven phase separation, other clustering mechanisms exist (Fig. 6). In cells infected with herpes simplex virus (HSV), viral replication compartments form micron-sized Pol II clusters due to locally enhanced Pol II binding to nucleosome-free DNA (McSwiggen et al. 2019a). Strikingly, these Pol II clusters are insensitive to CTD length in the HSV context, confirming their distinctive mechanism. TF clustering can also emerge from locally hindered diffusion as indicated by the observation that CTCF molecules are partially retained in specific nuclear zones, likely via interactions with RNA (Hansen et al. 2020). Similarly, upon heat shock, various factors are retained at induced loci, in a poly(ADP-ribose) polymerase (PARP) activity-dependent manner. This led to the speculation that PAR polymerization could create a diffusion barrier around the locus, favoring local recycling of TF molecules once they finish a round of transcription (Yao et al. 2007; Zobeck et al. 2010). Alternatively, PARP-induced chromatin decondensation could enhance TF binding, and/or PARylation could increase TF mutual affinity (Benabdallah et al. 2019). Finally, TFs may cluster due to the collapsing of their DNA targets, as suggested for HP1 in mouse embryonic fibroblasts, where the formation of chromocenters occurs independently of HP1 (Erdel et al. 2020). Interestingly, HP1 clusters reminiscent of phase separation are observed during *Drosophila* embryogenesis (Strom et al. 2017). The Pol II and HP1 examples suggest that a given factor can evolve distinct biophysical mechanisms to cluster in different contexts.

Functions of Clustering in Transcription

Clusters generate high local TF concentrations, which could ensure robust TF recruitment via mass action law even if TFs exchange from the cluster faster than the cluster lifetime (Dufourt et al. 2018). Indeed, clustering ensures high target site occupancy at low TF concentrations (Mir et al. 2017, 2018), and artificially induced clustering of TFs or IDRs in vivo is sufficient to recruit higher levels of transcription components and increase levels of transcription locally (Wei et al. 2020; Schneider et al. 2021). Increased local concentration of TFs is also expected to favor efficient transcription reinitiation. This prediction is consistent with the observation that increased Pol II clustering leads to increased burst size (Cho et al. 2016; Quintero-Cadena et al. 2020), and that rapid reinitiation is regulated by Mediator, a factor prone to clustering (Cho et al. 2018; Nguyen et al. 2020). TF cluster lifetimes are in the range of seconds, and thus are unlikely to constitute the molecular substrate of long-term transcription memory (Cisse et al. 2013; Cho et al. 2016, 2018; Mir et al. 2018). Clustering may also form a molecular bridge between distant loci (Tsai et al. 2019), which could explain why enhancers do not always directly contact promoters upon activation (Alexander et al. 2019; Benabdallah et al. 2019). Instead TF clusters may generate a regulatory environment shared by *cis*-regulatory elements without the need for molecular contact. Disruption of clustering does not abolish existing enhancer–promoter contacts, suggesting that clusters are not needed to maintain long-range interactions (Crump et al. 2021). Clustered enhancers confer robustness (Tsai et al. 2019) and constitute a flexible platform able to encode a variety of regulatory responses (Ezer et al. 2014). Similar to activators, repressors may also bring together distant loci (Ruault et al. 2020). Finally, clusters could facilitate nuclear exploration by guiding TFs to specialized compartments (Hansen et al. 2020; Nguyen et al. 2020).

CONCLUDING REMARKS

Advances in live imaging have uncovered novel modes of TF exploration and transient assemblies whose function and regulation are just beginning to be understood. Since TF dynamics parameters such as search time, residence time, concentration, fraction bound, etc. all impact transcription levels, experimental separation of individual factors is a challenge that will need to be overcome to build mechanistic models (Popp et al. 2020).

Clustering is emerging as a ubiquitous feature of transcription regulation that locally boosts transcription via mass action, while ensuring a nimble architecture able to rapidly respond to changing cues. One could envision other functions, for instance that TF clustering away from active sites could also titrate out TFs when transcription needs to be globally turned down. The next challenge is to better understand the mechanisms of cluster formation, and how clustering dynamics are decoded by promoters into transcription outputs. Biophysical regulators of clustering identified so far include phase separation, locally enhanced DNA binding, and local diffusion barriers. One difficulty is that these mechanisms are likely intertwined; for instance, transcription-coupled clustering could enhance DNA binding and/or generate local diffusion barriers. Since the respective roles of these mechanisms likely depend on the biological context, a key question is how biochemical pathways interface with TF biophysics. So far, posttranslational modifications, particularly of the Pol II CTD, have been demonstrated to control clustering by rapidly modulating IDR affinities. The role of other pathways or regulators remains to be fully explored.

Overall, the dominating feature of TF dynamics is that they follow a distributed interaction principle, apparent at many scales. First, in stoichiometric fuzzy complexes, multiple weak interaction sites between two partners rapidly exchange without complex dissociation. Second, multivalent interactions distributed across IDRs ensure the formation of nonstoichiometric clusters. Finally, enhancers favor multiple weak TF-binding motifs over high affinity ones to ensure expression specificity (Frankel et al. 2010; Crocker et al. 2015, 2016; Farley et al. 2015). This unifying principle offers many advantages needed for regulatory function, particularly robustness, tunability, and responsiveness.

ACKNOWLEDGMENTS

T.L. is supported by NIH Grant R01 GM127538. F.L. is a recipient of NYSTEM Institutional Training Grant #C032560GG. We thank Mustafa Mir and members of the Lionnet laboratory for critical reading of the manuscript.

REFERENCES

Alexander JM, Guan J, Li B, Maliskova L, Song M, Shen Y, Huang B, Lomvardas S, Weiner OD. 2019. Live-cell imaging reveals enhancer-dependent Sox2 transcription in the absence of enhancer proximity. *eLife* **8:** e41769. doi:10.7554/eLife.41769

Bancaud A, Huet S, Daigle N, Mozziconacci J, Beaudouin J, Ellenberg J. 2009. Molecular crowding affects diffusion and binding of nuclear proteins in heterochromatin and reveals the fractal organization of chromatin. *EMBO J* **28:** 3785–3798. doi:10.1038/emboj.2009.340

Bartman CR, Hsu SC, Hsiung CCS, Raj A, Blobel GA. 2016. Enhancer regulation of transcriptional bursting parameters revealed by forced chromatin looping. *Mol Cell* **62:** 237–247. doi:10.1016/j.molcel.2016.03.007

Basu S, Mackowiak SD, Niskanen H, Knezevic D, Asimi V, Grosswendt S, Geertsema H, Ali S, Jerković I, Ewers H, et al. 2020. Unblending of transcriptional condensates in human repeat expansion disease. *Cell* **181:** 1062–1079.e30. doi:10.1016/j.cell.2020.04.018

Benabdallah NS, Williamson I, Illingworth RS, Kane L, Boyle S, Sengupta D, Grimes GR, Therizols P, Bickmore WA. 2019. Decreased enhancer–promoter proximity accompanying enhancer activation. *Mol Cell* **76:** 473–484.e7. doi:10.1016/j.molcel.2019.07.038

Boehning M, Dugast-Darzacq C, Rankovic M, Hansen AS, Yu T, Marie-Nelly H, McSwiggen DT, Kokic G, Dailey GM, Cramer P, et al. 2018. RNA polymerase II clustering through carboxy-terminal domain phase separation. *Nat Struct Mol Biol* **25:** 833–840. doi:10.1038/s41594-018-0112-y

Boija A, Klein IA, Sabari BR, Dall'Agnese A, Coffey EL, Zamudio AV, Li CH, Shrinivas K, Manteiga JC, Hannett NM, et al. 2018. Transcription factors activate genes through the phase-separation capacity of their activation domains. *Cell* **175:** 1842–1855.e16. doi:10.1016/j.cell.2018.10.042

Bothma JP, Garcia HG, Ng S, Perry MW, Gregor T, Levine M. 2015. Enhancer additivity and non-additivity are determined by enhancer strength in the *Drosophila* embryo. *eLife* **4:** e07956. doi:10.7554/eLife.07956

Brodsky S, Jana T, Mittelman K, Chapal M, Kumar DK, Carmi M, Barkai N. 2020. Intrinsically disordered regions direct transcription factor in vivo binding specificity. *Mol Cell* **79:** 459–471.e4. doi:10.1016/j.molcel.2020.05.032

Callegari A, Sieben C, Benke A, Suter DM, Fierz B, Mazza D, Manley S. 2019. Single-molecule dynamics and genome-wide transcriptomics reveal that NF-κB (p65)-DNA binding times can be decoupled from transcriptional activation. *PLoS Genet* **15:** e1007891. doi:10.1371/journal.pgen.1007891

Chen J, Zhang Z, Li L, Chen BC, Revyakin A, Hajj B, Legant W, Dahan M, Lionnet T, Betzig E, et al. 2014. Single-molecule dynamics of enhanceosome assembly in embryonic stem cells. *Cell* **156:** 1274–1285. doi:10.1016/j.cell.2014.01.062

Chen H, Levo M, Barinov L, Fujioka M, Jaynes JB, Gregor T. 2018. Dynamic interplay between enhancer–promoter topology and gene activity. *Nat Genet* **50:** 1296–1303. doi:10.1038/s41588-018-0175-z

Cho WK, Jayanth N, English BP, Inoue T, Andrews JO, Conway W, Grimm JB, Spille JH, Lavis LD, Lionnet T, et al. 2016. RNA polymerase II cluster dynamics predict mRNA output in living cells. *eLife* **5:** e13617. doi:10.7554/eLife.13617

Cho WK, Spille JH, Hecht M, Lee C, Li C, Grube V, Cisse II. 2018. Mediator and RNA polymerase II clusters associate in transcription-dependent condensates. *Science* **361:** 412–415. doi:10.1126/science.aar4199

Choi J, Bachmann AL, Tauscher K, Benda C, Fierz B, Müller J. 2017. DNA binding by PHF1 prolongs PRC2 residence time on chromatin and thereby promotes H3K27 methylation. *Nat Struct Mol Biol* **24:** 1039–1047. doi:10.1038/nsmb.3488

Choi JM, Holehouse AS, Pappu RV. 2020. Physical principles underlying the complex biology of intracellular phase transitions. *Annu Rev Biophys* **49:** 107–133. doi:10.1146/annurev-biophys-121219-081629

Chong S, Chen C, Ge H, Xie XS. 2014. Mechanism of transcriptional bursting in bacteria. *Cell* **158:** 314–326. doi:10.1016/j.cell.2014.05.038

Chong S, Dugast-Darzacq C, Liu Z, Dong P, Dailey GM, Cattoglio C, Heckert A, Banala S, Lavis L, Darzacq X, et al. 2018. Imaging dynamic and selective low-complexity domain interactions that control gene transcription. *Science* **361:** eaar2555. doi:10.1126/science.aar2555

Chubb JR, Trcek T, Shenoy SM, Singer RH. 2006. Transcriptional pulsing of a developmental gene. *Curr Biol* **16:** 1018–1025. doi:10.1016/j.cub.2006.03.092

Cisse II, Izeddin I, Causse SZ, Boudarene L, Senecal A, Muresan L, Dugast-Darzacq C, Hajj B, Dahan M, Darzacq X. 2013. Real-time dynamics of RNA polymerase II clustering in live human cells. *Science* **341:** 664–667. doi:10.1126/science.1239053

Clauß K, Popp AP, Schulze L, Hettich J, Reisser M, Escoter Torres L, Uhlenhaut NH, Gebhardt JCM. 2017. DNA residence time is a regulatory factor of transcription repression. *Nucleic Acids Res* **45:** 11121–11130. doi:10.1093/nar/gkx728

Cook PR. 1999. The organization of replication and transcription. *Science* **284:** 1790–1795. doi:10.1126/science.284.5421.1790

Corden JL. 2013. RNA polymerase II C-terminal domain: tethering transcription to transcript and template. *Chem Rev* **113:** 8423–8455. doi:10.1021/cr400158h

Corrigan AM, Tunnacliffe E, Cannon D, Chubb JR. 2016. A continuum model of transcriptional bursting. *eLife* **5:** e13051. doi:10.7554/eLife.13051

Cortini R, Filion GJ. 2018. Theoretical principles of transcription factor traffic on folded chromatin. *Nat Commun* **9:** 1740. doi:10.1038/s41467-018-04130-x

Cramer P. 2019. Organization and regulation of gene transcription. *Nature* **573:** 45–54. doi:10.1038/s41586-019-1517-4

Crocker J, Abe N, Rinaldi L, McGregor AP, Frankel N, Wang S, Alsawadi A, Valenti P, Plaza S, Payre F, et al. 2015. Low affinity binding site clusters confer hox specificity and regulatory robustness. *Cell* **160:** 191–203. doi:10.1016/j.cell.2014.11.041

Crocker J, Noon EPB, Stern DL. 2016. The soft touch: low-affinity transcription factor binding sites in development and evolution. *Curr Top Dev Biol* **117:** 455–469. doi:10.1016/bs.ctdb.2015.11.018

Crump NT, Ballabio E, Godfrey L, Thorne R, Repapi E, Kerry J, Tapia M, Hua P, Lagerholm C, Filippakopoulos P, et al. 2021. BET inhibition disrupts transcription but retains enhancer–promoter contact. *Nat Commun* **12:** 223.

Daneshvar K, Behfar Ardehali M, Klein IA, Kratkiewicz AJ, Zhou C, Mahpour A, Cook BM, Li W, Pondick JV, Moran SP, et al. 2020. lncRNA DIGIT and BRD3 protein form phase-separated condensates to regulate endoderm differentiation. *Nat Cell Biol* **22:** 1211–1222. doi:10.1038/s41556-020-0572-2

Darzacq X, Shav-Tal Y, de Turris V, Brody Y, Shenoy SM, Phair RD, Singer RH. 2007. In vivo dynamics of RNA polymerase II transcription. *Nat Struct Mol Biol* **14:** 796–806. doi:10.1038/nsmb1280

Deal RB, Henikoff JG, Henikoff S. 2010. Genome-wide kinetics of nucleosome turnover determined by metabolic labeling of histones. *Science* **328:** 1161–1164. doi:10.1126/science.1186777

de Jonge WJ, Brok M, Lijnzaad P, Kemmeren P, Holstege FCP. 2020. Genome-wide off-rates reveal how DNA binding dynamics shape transcription factor function. *Mol Syst Biol* **16:** e9885. doi:10.15252/msb.20209885

Demarest SJ, Martinez-Yamout M, Chung J, Chen H, Xu W, Dyson HJ, Evans RM, Wright PE. 2002. Mutual synergistic folding in recruitment of CBP/p300 by p160 nuclear receptor coactivators. *Nature* **415:** 549–553. doi:10.1038/415549a

Desai MA, Webb HD, Sinanan LM, Scarsdale JN, Walavalkar NM, Ginder GD, Williams DC Jr. 2015. An intrinsically disordered region of methyl-CpG binding domain protein 2 (MBD2) recruits the histone deacetylase core of the NuRD complex. *Nucleic Acids Res* **43:** 3100–3113. doi:10.1093/nar/gkv168

Dion MF, Kaplan T, Kim M, Buratowski S, Friedman N, Rando OJ. 2007. Dynamics of replication-independent histone turnover in budding yeast. *Science* **315:** 1405–1408. doi:10.1126/science.1134053

Di Ventura B, Kuhlman B. 2016. Go in! Go out! inducible control of nuclear localization. *Curr Opin Chem Biol* **34:** 62–71. doi:10.1016/j.cbpa.2016.06.009

Donovan BT, Huynh A, Ball DA, Patel HP, Poirier MG, Larson DR, Ferguson ML, Lenstra TL. 2019. Live-cell imaging reveals the interplay between transcription factors, nucleosomes, and bursting. *EMBO J* **38:** e100809. doi:10.15252/embj.2018100809

Dufourt J, Trullo A, Hunter J, Fernandez C, Lazaro J, Dejean M, Morales L, Nait-Amer S, Schulz KN, Harrison MM, et al. 2018. Temporal control of gene expression by the pioneer factor Zelda through transient interactions in hubs. *Nat Commun* **9:** 5194. doi:10.1038/s41467-018-07613-z

Edelman LB, Fraser P. 2012. Transcription factories: genetic programming in three dimensions. *Curr Opin Genet Dev* **22:** 110–114. doi:10.1016/j.gde.2012.01.010

Eick D, Geyer M. 2013. The RNA polymerase II carboxy-terminal domain (CTD) code. *Chem Rev* **113:** 8456–8490. doi:10.1021/cr400071f

Elf J, Barkefors I. 2019. Single-molecule kinetics in living cells. *Annu Rev Biochem* **88:** 635–659. doi:10.1146/annurev-biochem-013118-110801

Elf J, Li GW, Xie XS. 2007. Probing transcription factor dynamics at the single-molecule level in a living cell. *Science* **316:** 1191–1194. doi:10.1126/science.1141967

Erdel F, Rademacher A, Vlijm R, Tünnermann J, Frank L, Weinmann R, Schweigert E, Yserentant K, Hummert J, Bauer C, et al. 2020. Mouse heterochromatin adopts digital compaction states without showing hallmarks of HP1-driven liquid–liquid phase separation. *Mol Cell* **78:** 236–249.e7. doi:10.1016/j.molcel.2020.02.005

Erijman A, Kozlowski L, Sohrabi-Jahromi S, Fishburn J, Warfield L, Schreiber J, Noble WS, Söding J, Hahn S. 2020. A high-throughput screen for transcription activation domains reveals their sequence features and permits prediction by deep learning. *Mol Cell* **79:** *1066*. doi:10.1016/j.molcel.2020.08.013

Ezer D, Zabet NR, Adryan B. 2014. Homotypic clusters of transcription factor binding sites: a model system for understanding the physical mechanics of gene expression. *Comput Struct Biotechnol J* **10:** 63–69. doi:10.1016/j.csbj.2014.07.005

Farley EK, Olson KM, Zhang W, Brandt AJ, Rokhsar DS, Levine MS. 2015. Suboptimization of developmental enhancers. *Science* **350:** 325–328. doi:10.1126/science.aac6948

Festuccia N, Owens N, Papadopoulou T, Gonzalez I, Tachtsidi A, Vandoermel-Pournin S, Gallego E, Gutierrez N, Dubois A, Cohen-Tannoudji M, et al. 2019. Transcription factor activity and nucleosome organization in mitosis. *Genome Res* **29:** 250–260. doi:10.1101/gr.243048.118

Frankel N, Davis GK, Vargas D, Wang S, Payre F, Stern DL. 2010. Phenotypic robustness conferred by apparently redundant transcriptional enhancers. *Nature* **466:** 490–493. doi:10.1038/nature09158

Frey S, Richter RP, Görlich D. 2006. FG-rich repeats of nuclear pore proteins form a three-dimensional meshwork with hydrogel-like properties. *Science* **314:** 815–817. doi:10.1126/science.1132516

Fukaya T, Lim B, Levine M. 2016. Enhancer control of transcriptional bursting. *Cell* **166:** 358–368. doi:10.1016/j.cell.2016.05.025

Garcia DA, Fettweis G, Presman DM, Paakinaho V, Jarzynski C, Upadhyaya A, Hager GL. 2021. Power-law behavior of transcription factor dynamics at the single-molecule level implies a continuum affinity model. *Nucleic Acids Res* doi:10.1093/nar/gkab072

Gebhardt JC, Suter DM, Roy R, Zhao ZW, Chapman AR, Basu S, Maniatis T, Xie XS. 2013. Single-molecule imaging of transcription factor binding to DNA in live mammalian cells. *Nat Methods* **10:** 421–426. doi:10.1038/nmeth.2411

Gibbs EB, Lu F, Portz B, Fisher MJ, Medellin BP, Laremore TN, Zhang YJ, Gilmour DS, Showalter SA. 2017. Phosphorylation induces sequence-specific conformational switches in the RNA polymerase II C-terminal domain. *Nat Commun* **8:** 15233. doi:10.1038/ncomms15233

Grünwald D, Martin RM, Buschmann V, Bazett-Jones DP, Leonhardt H, Kubitscheck U, Cardoso MC. 2008. Probing intranuclear environments at the single-molecule level. *Biophys J* **94:** 2847–2858. doi:10.1529/biophysj.107.115014

Guglielmi B, La Rochelle N, Tjian R. 2013. Gene-specific transcriptional mechanisms at the histone gene cluster revealed by single-cell imaging. *Mol Cell* **51:** 480–492. doi:10.1016/j.molcel.2013.08.009

Guo YE, Manteiga JC, Henninger JE, Sabari BR, Dall'Agnese A, Hannett NM, Spille J-H, Afeyan LK, Zamudio AV, Shrinivas K, et al. 2019. Pol II phosphorylation regulates a switch between transcriptional and splicing condensates. *Nature* **572:** 543–548. doi:10.1038/s41586-019-1464-0

Gurdon JB, Javed K, Vodnala M, Garrett N. 2020. Long-term association of a transcription factor with its chromatin binding site can stabilize gene expression and cell fate commitment. *Proc Natl Acad Sci* **117:** 15075–15084. doi:10.1073/pnas.2000467117

Hager GL, McNally JG, Misteli T. 2009. Transcription dynamics. *Mol Cell* **35:** 741–753. doi:10.1016/j.molcel.2009.09.005

Hamilton WB, Mosesson Y, Monteiro RS, Emdal KB, Knudsen TE, Francavilla C, Barkai N, Olsen JV, Brickman JM. 2019. Dynamic lineage priming is driven via direct enhancer regulation by ERK. *Nature* **575:** 355–360. doi:10.1038/s41586-019-1732-z

Hansen AS, Pustova I, Cattoglio C, Tjian R, Darzacq X. 2017. CTCF and cohesin regulate chromatin loop stability with distinct dynamics. *eLife* **6:** e25776. doi:10.7554/eLife.25776

Hansen AS, Amitai A, Cattoglio C, Tjian R, Darzacq X. 2020. Guided nuclear exploration increases CTCF target search efficiency. *Nat Chem Biol* **16:** 257–266. doi:10.1038/s41589-019-0422-3

Harlen KM, Churchman LS. 2017. The code and beyond: transcription regulation by the RNA polymerase II carboxy-terminal domain. *Nat Rev Mol Cell Biol* **18:** 263–273. doi:10.1038/nrm.2017.10

He Q, Johnston J, Zeitlinger J. 2015. ChIP-nexus enables improved detection of in vivo transcription factor binding footprints. *Nat Biotechnol* **33:** 395–401. doi:10.1038/nbt.3121

Hegyi H, Schad E, Tompa P. 2007. Structural disorder promotes assembly of protein complexes. *BMC Struct Biol* **7:** 65. doi:10.1186/1472-6807-7-65

Hemmerich P, Schmiedeberg L, Diekmann S. 2011. Dynamic as well as stable protein interactions contribute to genome function and maintenance. *Chromosome Res* **19:** 131–151. doi:10.1007/s10577-010-9161-8

Henley MJ, Linhares BM, Morgan BS, Cierpicki T, Fierke CA, Mapp AK. 2020. Unexpected specificity within dynamic transcriptional protein–protein complexes. *Proc Natl Acad Sci* **117:** 27346–27353. doi:10.1073/pnas.2013244117

Henninger JE, Oksuz O, Shrinivas K, Sagi I, LeRoy G, Zheng MM, Owen Andrews J, Zamudio AV, Lazaris C, Hannett NM, et al. 2021. RNA-mediated feedback control of transcriptional condensates. *Cell* **184:** 207–225.e24. doi:10.1016/j.cell.2020.11.030

Hinow P, Rogers CE, Barbieri CE, Pietenpol JA, Kenworthy AK, DiBenedetto E. 2006. The DNA binding activity of p53 displays reaction–diffusion kinetics. *Biophys J* **91:** 330–342. doi:10.1529/biophysj.105.078303

Hipp L, Beer J, Kuchler O, Reisser M, Sinske D, Michaelis J, Gebhardt JCM, Knöll B. 2019. Single-molecule imaging of the transcription factor SRF reveals prolonged chromatin-binding kinetics upon cell stimulation. *Proc Natl Acad Sci* **116:** 880–889. doi:10.1073/pnas.1812734116

Huseyin MK, Klose RJ. 2021. Live-cell single particle tracking of PRC1 reveals a highly dynamic system with low target site occupancy. *Nat Commun* **12:** 887.

Izeddin I, Récamier V, Bosanac L, Cissé II, Boudarene L, Dugast-Darzacq C, Proux F, Bénichou O, Voituriez R, Bensaude O, et al. 2014. Single-molecule tracking in live cells reveals distinct target-search strategies of transcription factors in the nucleus. *eLife* **3:** e02230. doi:10.7554/eLife.02230

Johnson DS, Mortazavi A, Myers RM, Wold B. 2007. Genome-wide mapping of in vivo protein–DNA interactions. *Science* **316:** 1497–1502. doi:10.1126/science.1141319

Jolma A, Kivioja T, Toivonen J, Cheng L, Wei G, Enge M, Taipale M, Vaquerizas JM, Yan J, Sillanpää MJ, et al. 2010. Multiplexed massively parallel SELEX for characterization of human transcription factor binding specificities. *Genome Res* **20:** 861–873. doi:10.1101/gr.100552.109

Kimura H, Cook PR. 2001. Kinetics of core histones in living human cells: little exchange of H3 and H4 and some rapid exchange of H2B. *J Cell Biol* **153:** 1341–1354. doi:10.1083/jcb.153.7.1341

Kribelbauer JF, Rastogi C, Bussemaker HJ, Mann RS. 2019. Low-affinity binding sites and the transcription factor specificity paradox in eukaryotes. *Annu Rev Cell Dev Biol* **35:** 357–379. doi:10.1146/annurev-cellbio-100617-062719

Kwon I, Kato M, Xiang S, Wu L, Theodoropoulos P, Mirzaei H, Han T, Xie S, Corden JL, McKnight SL. 2013. Phosphorylation-regulated binding of RNA polymerase II to fibrous polymers of low-complexity domains. *Cell* **155:** 1049–1060. doi:10.1016/j.cell.2013.10.033

Lambert SA, Jolma A, Campitelli LF, Das PK, Yin Y, Albu M, Chen X, Taipale J, Hughes TR, Weirauch MT. 2018. The human transcription factors. *Cell* **175:** 598–599. doi:10.1016/j.cell.2018.09.045

Lammers NC, Galstyan V, Reimer A, Medin SA, Wiggins CH, Garcia HG. 2020. Multimodal transcriptional control of pattern formation in embryonic development. *Proc Natl Acad Sci* **117:** 836–847. doi:10.1073/pnas.1912500117

Lan X, Ren R, Feng R, Ly LC, Lan Y, Zhang Z, Aboreden N, Qin K, Horton JR, Grevet JD, et al. 2020. ZNF410 uniquely activates the NuRD component CHD4 to silence fetal hemoglobin expression. *Blood* **136:** 54–54. doi:10.1182/blood-2020-137564

Larson DR, Zenklusen D, Wu B, Chao JA, Singer RH. 2011. Real-time observation of transcription initiation and elongation on an endogenous yeast gene. *Science* **332:** 475–478. doi:10.1126/science.1202142

Larson AG, Elnatan D, Keenen MM, Trnka MJ, Johnston JB, Burlingame AL, Agard DA, Redding S, Narlikar GJ. 2017. Liquid droplet formation by HP1α suggests a role for phase separation in heterochromatin. *Nature* **547:** 236–240. doi:10.1038/nature22822

Lerner J, Gomez-Garcia PA, McCarthy RL, Liu Z, Lakadamyali M, Zaret KS. 2020. Two-parameter mobility assessments discriminate diverse regulatory factor behaviors in chromatin. *Mol Cell* **79:** 677–688.e6. doi:10.1016/j.molcel.2020.05.036

Lever MA, Th'ng JP, Sun X, Hendzel MJ. 2000. Rapid exchange of histone H1.1 on chromatin in living human cells. *Nature* **408:** 873–876. doi:10.1038/35048603

Li L, Liu H, Dong P, Li D, Legant WR, Grimm JB, Lavis LD, Betzig E, Tjian R, Liu Z. 2016. Real-time imaging of Huntingtin aggregates diverting target search and gene transcription. *eLife* **5:** e17056. doi:10.7554/eLife.17056

Li J, Dong A, Saydaminova K, Chang H, Wang G, Ochiai H, Yamamoto T, Pertsinidis A. 2019. Single-molecule nanoscopy elucidates RNA polymerase II transcription at single genes in live cells. *Cell* **178:** 491–506.e28. doi:10.1016/j.cell.2019.05.029

Li CH, Coffey EL, Dall'Agnese A, Hannett NM, Tang X, Henninger JE, Platt JM, Oksuz O, Zamudio AV, Afeyan LK, et al. 2020a. MeCP2 links heterochromatin condensates and neurodevelopmental disease. *Nature* **586:** 440–444. doi:10.1038/s41586-020-2574-4

Li J, Hsu A, Hua Y, Wang G, Cheng L, Ochiai H, Yamamoto T, Pertsinidis A. 2020b. Single-gene imaging links genome topology, promoter–enhancer communication and transcription control. *Nat Struct Mol Biol* **27:** 1032–1040. doi:10.1038/s41594-020-0493-6

Liu Z, Tjian R. 2018. Visualizing transcription factor dynamics in living cells. *J Cell Biol* **217:** 1181–1191. doi:10.1083/jcb.201710038

Liu J, Perumal NB, Oldfield CJ, Su EW, Uversky VN, Keith Dunker A. 2006. Intrinsic disorder in transcription factors. *Biochemistry* **45:** 6873–6888. doi:10.1021/bi0602718

Liu Y, Matthews KS, Bondos SE. 2008. Multiple intrinsically disordered sequences alter DNA binding by the homeodomain of the *Drosophila* hox protein ultrabithorax. *J Biol Chem* **283:** 20874–20887. doi:10.1074/jbc.M800375200

Liu Z, Legant WR, Chen BC, Li L, Grimm JB, Lavis LD, Betzig E, Tjian R. 2014. 3D imaging of Sox2 enhancer clusters in embryonic stem cells. *eLife* **3:** e04236. doi:10.7554/eLife.04236

Lu H, Yu D, Hansen AS, Ganguly S, Liu R, Heckert A, Darzacq X, Zhou Q. 2018. Phase-separation mechanism for C-terminal hyperphosphorylation of RNA polymerase II. *Nature* **558:** 318–323. doi:10.1038/s41586-018-0174-3

Lu F, Portz B, Gilmour DS. 2019. The C-terminal domain of RNA polymerase II is a multivalent targeting sequence that supports *Drosophila* development with only consensus heptads. *Mol Cell* **73:** 1232–1242.e4. doi:10.1016/j.molcel.2019.01.008

Maiuri P, Knezevich A, Bertrand E, Marcello A. 2011. Real-time imaging of the HIV-1 transcription cycle in single living cells. *Methods* **53:** 62–67. doi:10.1016/j.ymeth.2010.06.015

Marklund E, van Oosten B, Mao G, Amselem E, Kipper K, Sabantsev A, Emmerich A, Globisch D, Zheng X, Lehmann LC, et al. 2020. DNA surface exploration and operator bypassing during target search. *Nature* **583:** 858–861. doi:10.1038/s41586-020-2413-7

Martin EW, Holehouse AS. 2020. Intrinsically disordered protein regions and phase separation: sequence determinants of assembly or lack thereof. *Emerg Top Life Sci* **4:** 307–329. doi:10.1042/ETLS20190164

McSwiggen DT, Hansen AS, Teves SS, Marie-Nelly H, Hao Y, Heckert AB, Umemoto KK, Dugast-Darzacq C, Tjian R, Darzacq X. 2019a. Evidence for DNA-mediated nuclear compartmentalization distinct from phase separation. *eLife* **8:** e47098. doi:10.7554/eLife.47098

McSwiggen DT, Mir M, Darzacq X, Tjian R. 2019b. Evaluating phase separation in live cells: diagnosis, caveats, and functional consequences. *Genes Dev* **33:** 1619–1634. doi:10.1101/gad.331520.119

Mehta GD, Ball DA, Eriksson PR, Chereji RV, Clark DJ, McNally JG, Karpova TS. 2018. Single-molecule analysis reveals linked cycles of RSC chromatin remodeling and Ace1p transcription factor binding in yeast. *Mol Cell* **72:** 875–887.e9. doi:10.1016/j.molcel.2018.09.009

Mir M, Reimer A, Haines JE, Li XY, Stadler M, Garcia H, Eisen MB, Darzacq X. 2017. Dense bicoid hubs accentuate binding along the morphogen gradient. *Genes Dev* **31:** 1784–1794. doi:10.1101/gad.305078.117

Mir M, Stadler MR, Ortiz SA, Hannon CE, Harrison MM, Darzacq X, Eisen MB. 2018. Dynamic multifactor hubs interact transiently with sites of active transcription in *Drosophila* embryos. *eLife* **7:** e40497. doi:10.7554/eLife.40497

Miron E, Oldenkamp R, Brown JM, Pinto DMS, Shan Xu C, Faria AR, Shaban HA, Rhodes JDP, Innocent C, de Ornellas S, et al. 2020. Chromatin arranges in chains of mesoscale domains with nanoscale functional topography independent of cohesin. *Sci Adv* **6:** eaba8811. doi:10.1126/sciadv.aba8811

Misteli T, Gunjan A, Hock R, Bustin M, Brown DT. 2000. Dynamic binding of histone H1 to chromatin in living cells. *Nature* **408:** 877–881. doi:10.1038/35048610

Mueller F, Stasevich TJ, Mazza D, McNally JG. 2013. Quantifying transcription factor kinetics: at work or at play? *Crit Rev Biochem Mol Biol* **48:** 492–514. doi:10.3109/10409238.2013.833891

Muramoto T, Müller I, Thomas G, Melvin A, Chubb JR. 2010. Methylation of H3K4 Is required for inheritance of active transcriptional states. *Curr Biol* **20:** 397–406. doi:10.1016/j.cub.2010.01.017

Nguyen VQ, Ranjan A, Liu S, Tang X, Ling YH, Wisniewski J, Mizuguchi G, Li KY, Jou V, Zheng Q, et al. 2020. Spatio-temporal coordination of transcription preinitiation complex assembly in live cells. bioRxiv doi:10.1101/2020.12.30.424853

Normanno D, Boudarene L, Dugast-Darzacq C, Chen J, Richter C, Proux F, Benichou O, Voituriez R, Darzacq X, Dahan M. 2015. Probing the target search of DNA-binding proteins in mammalian cells using TetR as model searcher. *Nat Commun* **6:** 7357. doi:10.1038/ncomms8357

Peng Z, Yan J, Fan X, Mizianty MJ, Xue B, Wang K, Hu G, Uversky VN, Kurgan L. 2015. Exceptionally abundant exceptions: comprehensive characterization of intrinsic disorder in all domains of life. *Cell Mol Life Sci* **72:** 137–151. doi:10.1007/s00018-014-1661-9

Perlmann T, Eriksson P, Wrange O. 1990. Quantitative analysis of the glucocorticoid receptor–DNA interaction at the mouse mammary tumor virus glucocorticoid response element. *J Biol Chem* **265:** 17222–17229. doi:10.1016/S0021-9258(17)44892-7

Phair RD, Misteli T. 2000. High mobility of proteins in the mammalian cell nucleus. *Nature* **404:** 604–609. doi:10.1038/35007077

Phair RD, Scaffidi P, Elbi C, Vecerová J, Dey A, Ozato K, Brown DT, Hager G, Bustin M, Misteli T. 2004. Global nature of dynamic protein–chromatin interactions in vivo: three-dimensional genome scanning and dynamic interaction networks of chromatin proteins. *Mol Cell Biol* **24:** 6393–6402. doi:10.1128/MCB.24.14.6393-6402.2004

Plys AJ, Davis CP, Kim J, Rizki G, Keenen MM, Marr SK, Kingston RE. 2019. Phase separation of Polycomb-repressive complex 1 is governed by a charged disordered region of CBX2. *Genes Dev* **33:** 799–813. doi:10.1101/gad.326488.119

Poorey K, Viswanathan R, Carver MN, Karpova TS, Cirimotich SM, McNally JG, Bekiranov S, Auble DT. 2013. Measuring chromatin interaction dynamics on the second time scale at single-copy genes. *Science* **342:** 369–372. doi:10.1126/science.1242369

Popp AP, Hettich J, Gebhardt JCM. 2020. Transcription factor residence time dominates over concentration in transcription activation. bioRxiv doi:10.1101/2020.11.26.400069

Portz B, Lu F, Gibbs EB, Mayfield JE, Rachel Mehaffey M, Zhang YJ, Brodbelt JS, Showalter SA, Gilmour DS. 2017. Structural heterogeneity in the intrinsically disordered RNA polymerase II C-terminal domain. *Nat Commun* **8:** 15231. doi:10.1038/ncomms15231

Presman DM, Ganguly S, Schiltz RL, Johnson TA, Karpova TS, Hager GL. 2016. DNA binding triggers tetramerization of the glucocorticoid receptor in live cells. *Proc Natl Acad Sci* **113:** 8236–8241. doi:10.1073/pnas.1606774113

Qin BY, Liu C, Srinath H, Lam SS, Correia JJ, Derynck R, Lin K. 2005. Crystal structure of IRF-3 in complex with CBP. *Structure* **13:** 1269–1277. doi:10.1016/j.str.2005.06.011

Quintero-Cadena P, Lenstra TL, Sternberg PW. 2020. RNA Pol II length and disorder enable cooperative scaling of transcriptional bursting. *Mol Cell* **79:** 207–220.e8. doi:10.1016/j.molcel.2020.05.030

Raj A, Peskin CS, Tranchina D, Vargas DY, Tyagi S. 2006. Stochastic mRNA synthesis in mammalian cells. *PLoS Biol* **4:** e309. doi:10.1371/journal.pbio.0040309

Reisser M, Palmer A, Popp AP, Jahn C, Weidinger G, Gebhardt JCM. 2018. Single-molecule imaging correlates decreasing nuclear volume with increasing TF-chromatin associations during zebrafish development. *Nat Commun* **9:** 5218. doi:10.1038/s41467-018-07731-8

Reisser M, Hettich J, Kuhn T, Popp AP, Große-Berkenbusch A, Gebhardt JCM. 2020. Inferring quantity and qualities of superimposed reaction rates from single molecule survival time distributions. *Sci Rep* **10:** 1758. doi:10.1038/s41598-020-58634-y

Rhee HS, Pugh BF. 2011. Comprehensive genome-wide protein–DNA interactions detected at single-nucleotide resolution. *Cell* **147:** 1408–1419. doi:10.1016/j.cell.2011.11.013

Rodriguez J, Larson DR. 2020. Transcription in living cells: molecular mechanisms of bursting. *Annu Rev Biochem* **89:** 189–212. doi:10.1146/annurev-biochem-011520-105250

Rodriguez J, Ren G, Day CR, Zhao K, Chow CC, Larson DR. 2019. Intrinsic dynamics of a human gene reveal the basis of expression heterogeneity. *Cell* **176:** 213–226.e18. doi:10.1016/j.cell.2018.11.026

Ruault M, Scolari VF, Lazar-Stefanita L, Hocher A. 2020. The silencing factor Sir3 is a molecular bridge that sticks together distant loci. bioRxiv doi:10.1101/2020.06.29.178368

Sabari BR, Dall'Agnese A, Boija A, Klein IA, Coffey EL, Shrinivas K, Abraham BJ, Hannett NM, Zamudio AV, Manteiga JC, et al. 2018. Coactivator condensation at super-enhancers links phase separation and gene control. *Science* **361:** eaar3958. doi:10.1126/science.aar3958

Schneider N, Wieland FG, Kong D, Fischer AAM, Hörner M, Timmer J, Ye H, Weber W. 2021. Liquid–liquid phase separation of light-inducible transcription factors increases transcription activation in mammalian cells and mice. *Sci Adv* **7:** eabd3568. doi:10.1126/sciadv.abd3568

Scholes C, DePace AH, Sánchez Á. 2017. Combinatorial gene regulation through kinetic control of the transcription cycle. *Cell Syst* **4:** 97–108.e9. doi:10.1016/j.cels.2016.11.012

Schuler B, Borgia A, Borgia MB, Heidarsson PO, Holmstrom ED, Nettels D, Sottini A. 2020. Binding without folding—the biomolecular function of disordered polyelectrolyte complexes. *Curr Opin Struct Biol* **60:** 66–76. doi:10.1016/j.sbi.2019.12.006

Senecal A, Munsky B, Proux F, Ly N, Braye FE, Zimmer C, Mueller F, Darzacq X. 2014. Transcription factors modulate c-Fos transcriptional bursts. *Cell Rep* **8:** 75–83. doi:10.1016/j.celrep.2014.05.053

Shah RN, Grzybowski AT, Cornett EM, Johnstone AL, Dickson BM, Boone BA, Cheek MA, Cowles MW, Maryanski D, Meiners MJ, et al. 2018. Examining the roles of H3K4 methylation states with systematically characterized antibodies. *Mol Cell* **72:** 162–177.e7. doi:10.1016/j.molcel.2018.08.015

Shukron O, Seeber A, Amitai A, Holcman D. 2019. Advances using single-particle trajectories to reconstruct chromatin organization and dynamics. *Trends Genet* **35:** 685–705. doi:10.1016/j.tig.2019.06.007

Sigler PB. 1988. Acid blobs and negative noodles. *Nature* **333:** 210–212. doi:10.1038/333210a0

Sims RJ III, Rojas LA, Beck DB, Bonasio R, Schüller R, Drury WJ III, Eick D, Reinberg D. 2011. The C-terminal domain of RNA polymerase II is modified by site-specific methylation. *Science* **332:** 99–103. doi:10.1126/science.1202663

Skene PJ, Henikoff S. 2017. An efficient targeted nuclease strategy for high-resolution mapping of DNA binding sites. *eLife* **6:** e21856. doi:10.7554/eLife.21856

Staby L, O'Shea C, Willemoës M, Theisen F, Kragelund BB, Skriver K. 2017. Eukaryotic transcription factors: paradigms of protein intrinsic disorder. *Biochem J* **474:** 2509–2532. doi:10.1042/BCJ20160631

Staller MV, Holehouse AS, Swain-Lenz D, Das RK, Pappu RV, Cohen BA. 2018. A high-throughput mutational scan of an intrinsically disordered acidic transcriptional activation domain. *Cell Syst* **6:** 444–455.e6. doi:10.1016/j.cels.2018.01.015

Stasevich TJ, Mueller F, Michelman-Ribeiro A, Rosales T, Knutson JR, McNally JG. 2010. Cross-validating FRAP and FCS to quantify the impact of photobleaching on in vivo binding estimates. *Biophys J* **99:** 3093–3101. doi:10.1016/j.bpj.2010.08.059

Stasevich TJ, Hayashi-Takanaka Y, Sato Y, Maehara K, Ohkawa Y, Sakata-Sogawa K, Tokunaga M, Nagase T, Nozaki N, McNally JG, et al. 2014. Regulation of RNA polymerase II activation by histone acetylation in single living cells. *Nature* **516:** 272–275. doi:10.1038/nature13714

Stavreva DA, Garcia DA, Fettweis G, Gudla PR, Zaki GF, Soni V, McGowan A, Williams G, Huynh A, Palangat M, et al. 2019. Transcriptional bursting and co-bursting regulation by steroid hormone release pattern and transcription factor mobility. *Mol Cell* **75:** 1161–1177.e11. doi:10.1016/j.molcel.2019.06.042

Strom AR, Emelyanov AV, Mir M, Fyodorov DV, Darzacq X, Karpen GH. 2017. Phase separation drives heterochromatin domain formation. *Nature* **547:** 241–245. doi:10.1038/nature22989

Sugo N, Morimatsu M, Arai Y, Kousoku Y, Ohkuni A, Nomura T, Yanagida T, Yamamoto N. 2015. Single-molecule imaging reveals dynamics of CREB transcription factor bound to its target sequence. *Sci Rep* **5:** 10662. doi:10.1038/srep10662

Suter DM. 2020. Transcription factors and DNA play hide and seek. *Trends Cell Biol* **30:** 491–500. doi:10.1016/j.tcb.2020.03.003

Suter DM, Molina N, Gatfield D, Schneider K, Schibler U, Naef F. 2011. Mammalian genes are transcribed with widely different bursting kinetics. *Science* **332:** 472–474. doi:10.1126/science.1198817

Swinstead EE, Miranda TB, Paakinaho V, Baek S, Goldstein I, Hawkins M, Karpova TS, Ball D, Mazza D, Lavis LD, et al. 2016. Steroid receptors reprogram FoxA1 occupancy through dynamic chromatin transitions. *Cell* **165:** 593–605. doi:10.1016/j.cell.2016.02.067

Symmons O, Raj A. 2016. What's luck got to do with it: single cells, multiple fates, and biological nondeterminism. *Mol Cell* **62:** 788–802. doi:10.1016/j.molcel.2016.05.023

Takahashi K, Yamanaka S. 2006. Induction of pluripotent stem cells from mouse embryonic and adult fibroblast cultures by defined factors. *Cell* **126:** 663–676. doi:10.1016/j.cell.2006.07.024

Tantale K, Mueller F, Kozulic-Pirher A, Lesne A, Victor JM, Robert MC, Capozi S, Chouaib R, Bäcker V, Mateos-Langerak J, et al. 2016. A single-molecule view of transcription reveals convoys of RNA polymerases and multi-scale bursting. *Nat Commun* **7:** 12248. doi:10.1038/ncomms12248

Tatavosian R, Duc HN, Huynh TN, Fang D, Schmitt B, Shi X, Deng Y, Phiel C, Yao T, Zhang Z, et al. 2018. Live-cell single-molecule dynamics of PcG proteins imposed by the DIPG H3.3K27M mutation. *Nat Commun* **9:** 2080. doi:10.1038/s41467-018-04455-7

Tatavosian R, Kent S, Brown K, Yao T, Duc HN, Huynh TN, Zhen CY, Ma B, Wang H, Ren X. 2019. Nuclear condensates of the Polycomb protein chromobox 2 (CBX2) assemble through phase separation. *J Biol Chem* **294:** 1451–1463. doi:10.1074/jbc.RA118.006620

Teves SS, Henikoff S. 2014. Transcription-generated torsional stress destabilizes nucleosomes. *Nat Struct Mol Biol* **21:** 88–94. doi:10.1038/nsmb.2723

Teves SS, An L, Hansen AS, Xie L, Darzacq X, Tjian R. 2016. A dynamic mode of mitotic bookmarking by transcription factors. *eLife* **5:** e22280. doi:10.7554/eLife.22280

Teves SS, An L, Bhargava-Shah A, Xie L, Darzacq X, Tjian R. 2018. A stable mode of bookmarking by TBP recruits RNA polymerase II to mitotic chromosomes. *eLife* **7:** 35621. doi:10.7554/eLife.35621

Tompa P, Fuxreiter M. 2008. Fuzzy complexes: polymorphism and structural disorder in protein–protein interactions. *Trends Biochem Sci* **33:** 2–8. doi:10.1016/j.tibs.2007.10.003

Tompa P, Dosztanyi Z, Simon I. 2006. Prevalent structural disorder in *E. coli* and *S. cerevisiae* proteomes. *J Proteome Res* **5:** 1996–2000. doi:10.1021/pr0600881

Tóth-Petróczy A, Oldfield CJ, Simon I, Takagi Y, Dunker AK, Uversky VN, Fuxreiter M. 2008. Malleable machines in transcription regulation: the mediator complex. *PLoS Comput Biol* **4:** e1000243. doi:10.1371/journal.pcbi.1000243

Treen N, Shimobayashi SF, Eeftens J, Brangwynne CP, Levine MS. 2020. Regulation of gene expression by repression condensates during development. bioRxiv doi:10.1101/2020.03.03.975680

Tsai A, Alves MR, Crocker J. 2019. Multi-enhancer transcriptional hubs confer phenotypic robustness. *eLife* **8:** e45325. doi:10.7554/eLife.45325

Uyar B, Weatheritt RJ, Dinkel H, Davey NE, Gibson TJ. 2014. Proteome-wide analysis of human disease mutations in short linear motifs: neglected players in cancer? *Mol Biosyst* **10:** 2626–2642. doi:10.1039/C4MB00290C

van der Lee R, Buljan M, Lang B, Weatheritt RJ, Daughdrill GW, Dunker AK, Fuxreiter M, Gough J, Gsponer J, Jones DT, et al. 2014. Classification of intrinsically disordered regions and proteins. *Chem Rev* **114:** 6589–6631. doi:10.1021/cr400525m

Vernon RM, Chong PA, Tsang B, Kim TH, Bah A, Farber P, Lin H, Forman-Kay JD. 2018. Pi–Pi contacts are an overlooked protein feature relevant to phase separation. *eLife* **7:** e31486. doi:10.7554/eLife.31486

Vincent BJ, Estrada J, DePace AH. 2016. The appeasement of Doug: a synthetic approach to enhancer biology. *Integr Biol* **8:** 475–484. doi:10.1039/c5ib00321k

von Hippel PH, Berg OG. 1989. Facilitated target location in biological systems. *J Biol Chem* **264:** 675–678. doi:10.1016/S0021-9258(19)84994-3

Vuzman D, Levy Y. 2012. Intrinsically disordered regions as affinity tuners in protein–DNA interactions. *Mol Biosyst* **8:** 47–57. doi:10.1039/C1MB05273J

Walters MC, Fiering S, Eidemiller J, Magis W, Groudine M, Martin DI. 1995. Enhancers increase the probability but not the level of gene expression. *Proc Natl Acad Sci* **92:** 7125–7129. doi:10.1073/pnas.92.15.7125

Wang J, Choi JM, Holehouse AS, Lee HO, Zhang X, Jahnel M, Maharana S, Lemaitre R, Pozniakovsky A, Drechsel D, et al. 2018. A molecular grammar governing the driving forces for phase separation of prion-like RNA binding proteins. *Cell* **174:** 688–699.e16. doi:10.1016/j.cell.2018.06.006

Ward JJ, Sodhi JS, McGuffin LJ, Buxton BF, Jones DT. 2004. Prediction and functional analysis of native disorder in proteins from the three kingdoms of life. *J Mol Biol* **337:** 635–645. doi:10.1016/j.jmb.2004.02.002

Warfield L, Tuttle LM, Pacheco D, Klevit RE, Hahn S. 2014. A sequence-specific transcription activator motif and powerful synthetic variants that bind Mediator using a fuzzy protein interface. *Proc Natl Acad Sci* **111:** E3506–E3513. doi:10.1073/pnas.1412088111

Waters L, Yue B, Veverka V, Renshaw P, Bramham J, Matsuda S, Frenkiel T, Kelly G, Muskett F, Carr M, et al. 2006. Structural diversity in p160/CREB-binding protein coactivator complexes. *J Biol Chem* **281:** 14787–14795. doi:10.1074/jbc.M600237200

Wei MT, Chang YC, Shimobayashi SF, Shin Y, Strom AR, Brangwynne CP. 2020. Nucleated transcriptional condensates amplify gene expression. *Nat Cell Biol* **22:** 1187–1196. doi:10.1038/s41556-020-00578-6

Wollman AJ, Shashkova S, Hedlund EG, Friemann R, Hohmann S, Leake MC. 2017. Transcription factor clusters regulate genes in eukaryotic cells. *eLife* **6:** e27451. doi:10.7554/eLife.27451

Wright PE, Dyson HJ. 2015. Intrinsically disordered proteins in cellular signalling and regulation. *Nat Rev Mol Cell Biol* **16:** 18–29. doi:10.1038/nrm3920

Xie L, Torigoe SE, Xiao J, Mai DH, Li L, Davis FP, Dong P, Marie-Nelly H, Grimm J, Lavis L, et al. 2017. A dynamic interplay of enhancer elements regulates *Klf4* expression in naïve pluripotency. *Genes Dev* **31:** 1795–1808. doi:10.1101/gad.303321.117

Yang P, Wang Y, Hoang D, Tinkham M, Patel A, Sun M-A, Wolf G, Baker M, Chien HC, Lai KYN, et al. 2017. A placental growth factor is silenced in mouse embryos by the zinc finger protein ZFP568. *Science* **356:** 757–759. doi:10.1126/science.aah6895

Yao J, Ardehali MB, Fecko CJ, Webb WW, Lis JT. 2007. Intranuclear distribution and local dynamics of RNA polymerase II during transcription activation. *Mol Cell* **28:** 978–990. doi:10.1016/j.molcel.2007.10.017

Youmans DT, Schmidt JC, Cech TR. 2018. Live-cell imaging reveals the dynamics of PRC2 and recruitment to chromatin by SUZ12-associated subunits. *Genes Dev* **32:** 794–805. doi:10.1101/gad.311936.118

Zaborowska J, Egloff S, Murphy S. 2016. The pol II CTD: new twists in the tail. *Nat Struct Mol Biol* **23:** 771–777. doi:10.1038/nsmb.3285

Zamudio AV, Dall'Agnese A, Henninger JE, Manteiga JC, Afeyan LK, Hannett NM, Coffey EL, Li CH, Oksuz O, Sabari BR, et al. 2019. Mediator condensates localize signaling factors to key cell identity genes. *Mol Cell* **76:** 753–766.e6. doi:10.1016/j.molcel.2019.08.016

Zhang Z, English BP, Grimm JB, Kazane SA, Hu W, Tsai A, Inouye C, You C, Piehler J, Schultz PG, et al. 2016. Rapid dynamics of general transcription factor TFIIB binding during preinitiation complex assembly revealed by single-molecule analysis. *Genes Dev* **30:** 2106–2118. doi:10.1101/gad.285395.116

Zhao DY, Gish G, Braunschweig U, Li Y, Ni Z, Schmitges FW, Zhong G, Liu K, Li W, Moffat J, et al. 2016. SMN and symmetric arginine dimethylation of RNA polymerase II C-terminal domain control termination. *Nature* **529:** 48–53. doi:10.1038/nature16469

Zhu F, Farnung L, Kaasinen E, Sahu B, Yin Y, Wei B, Dodonova SO, Nitta KR, Morgunova E, Taipale M, et al. 2018. The interaction landscape between transcription factors and the nucleosome. *Nature* **562:** 76–81. doi:10.1038/s41586-018-0549-5

Zobeck KL, Buckley MS, Zipfel WR, Lis JT. 2010. Recruitment timing and dynamics of transcription factors at the Hsp70 loci in living cells. *Mol Cell* **40:** 965–975. doi:10.1016/j.molcel.2010.11.022

Zolotarev N, Maksimenko O, Kyrchanova O, Sokolinskaya E, Osadchiy I, Girardot C, Bonchuk A, Ciglar L, Furlong EEM, Georgiev P. 2017. Opbp is a new architectural/insulator protein required for ribosomal gene expression. *Nucleic Acids Res* **45:** 12285–12300. doi:10.1093/nar/gkx840

Essential Roles for RNA in Shaping Nuclear Organization

Sofia A. Quinodoz[1,2] and Mitchell Guttman[1]

[1]Division of Biology and Biological Engineering, California Institute of Technology, Pasadena, California 91125, USA

Correspondence: quinodoz@princeton.edu; mguttman@caltech.edu

It has long been proposed that nuclear RNAs might play an important role in organizing the structure of the nucleus. Initial experiments performed more than 30 years ago found that global disruption of RNA led to visible rearrangements of nuclear organization. Yet, this idea remained controversial for many years, in large part because it was unclear what specific RNAs might be involved, and which specific nuclear structures might be dependent on RNA. Over the past few years, the contributions of RNA to organizing nuclear structures have become clearer with the discovery that many nuclear bodies are enriched for specific noncoding RNAs (ncRNAs); in specific cases, ncRNAs have been shown to be essential for establishment and maintenance of these nuclear structures. More recently, many different ncRNAs have been shown to play critical roles in initiating the three-dimensional (3D) spatial organization of DNA, RNA, and protein molecules in the nucleus. These examples, combined with global imaging and genomic experiments, have begun to paint a picture of a broader role for RNA in nuclear organization and to uncover a unifying mechanism that may explain why RNA is a uniquely suited molecule for this role. In this review, we provide an overview of the history of RNA and nuclear structure and discuss key examples of RNA-mediated bodies, the global roles of ncRNAs in shaping nuclear structure, and emerging insights into mechanisms of RNA-mediated nuclear organization.

A HISTORICAL OVERVIEW OF RNA AND NUCLEAR STRUCTURE

The history of RNA and nuclear organization is a story of two seemingly parallel lines of research that converged over time. The development of microscopy allowed researchers to observe cells, nuclei, and the distinct structures contained within them (Lafarga et al. 2009, Nizami et al. 2010; Pederson 2011a). This led to the discovery of multiple structures within the nucleus that were initially described primarily based on their morphological features (Pederson 2011b). One of the earliest, largest, and most easily discernible subnuclear structures observed was the nucleolus, which was first described in the 1830s (Wagner 1835; Valentin 1836, 1839). Several decades later, in the early 1900s, Santiago Ramón y Cajal was exploring the cytological features of neurons using histological labeling techniques that made it possible to identify distinct intracellular structures and

[2]Present address: Department of Chemical and Biological Engineering, Princeton University, Princeton, New Jersey 08544, USA.

Cite this article as *Cold Spring Harb Perspect Biol* doi: 10.1101/cshperspect.a039719

observed several nuclear structures, which are now known as Cajal bodies and nuclear speckles, named for their "speckled" pattern throughout the nucleus (Ramón y Cajal 1903, 1910; Lafarga et al. 2009). In 1949, Murray Barr observed the presence of a condensed aggregate of DNA in some, but not all, cat neuronal cells (Barr and Bertram 1949). The key determinant of cells containing this structure was the sex of the cat; only the cells of female cats contain the condensed structure, which is now known to correspond to the inactive X chromosome (Xi) and is referred to as the Barr body.

In the 1930s, Heitz and McClintock observed that nucleoli form at specific chromosomal loci and that the number of nucleoli correlates with the chromosome ploidy within a cell (Heitz 1931; McClintock 1934). This discovery led to a new appreciation that nuclear bodies may represent functionally organized structures rather than simply cytological features.

ncRNAs Play Critical Roles in Various Nuclear Processes

In parallel to these morphological characterizations of the nucleus, the development of molecular and cell biology techniques made it possible to begin exploring the RNA contents of the nucleus. In the 1970s, Darnell, Penman, and others used metabolic labeling to study the complexity of RNA species and track their localization and life cycle in the cell (Darnell 1968, 2011; Weinberg and Penman 1968, 1969). These studies uncovered that the vast majority of RNAs are not translated into proteins and that many of them are retained within the nucleus; these were termed heteronuclear RNAs (hnRNAs) (Warner et al. 1966; Salditt-Georgieff et al. 1981). Although much of the initially described hnRNA was subsequently found to be intronic sequences excised during pre-mRNA splicing, many additional stable, nuclear noncoding RNAs (ncRNAs) were also detected.

In 1968, Sheldon Penman and colleagues discovered a population of "small" ncRNAs ranging in size from 90 to 300 nucleotides (nt) that were abundantly expressed in the nuclei of mammalian cells (Weinberg and Penman 1968). These became known as small nuclear RNAs (snRNAs) and small nucleolar RNAs (snoRNAs) (Zieve and Penman 1976; Weinberg and Penman 1968). Because identifying each RNA within a complex population was technically challenging at the time, the identity of most ncRNA species, with the exception of these highly abundant species within this hnRNA population, remained uncharacterized for decades.

The first clues into the possible function of snRNAs came in 1980 when Joan Steitz and colleagues identified sequence complementary between various snRNAs and sequences on pre-mRNAs (e.g., U1 and 5′ splice site) (Lerner et al. 1980). This discovery led to the purification and characterization of the spliceosome, which consists of snRNAs that directly hybridize to pre-mRNA and the various proteins they recruit to facilitate the splicing reaction (Sharp 2005). Altogether, several of the initially identified snRNAs—including U1, U2, U4, U5, and U6—were shown to play critical roles in mediating mRNA splicing.

The ability to undergo hybridization also provided the first insights into the functional roles of snoRNAs. Specifically, the discovery that several of the snoRNAs can base pair with the 45S pre-ribosomal RNA suggested that they might play a role in ribosomal RNA processing (Calvet and Pederson 1981; Kass 1990; Filipowicz and Kiss 1993; Fournier and Stuart Maxwell 1993; Maxwell and Fournier 1995). Indeed, subsequent experiments depleting snoRNAs in mouse extracts resulted in impaired cleavage of the ribosomal RNA (Maxwell and Fournier 1995). The observation that snoRNAs purified with the nucleolar fraction provided the first indication that the nucleolus might serve as a specialized body associated with ribosome biogenesis (Zieve and Penman 1976).

In the 1990s, long ncRNAs (lncRNAs) were identified. The very first example was the discovery of the H19 gene located within the imprinted IGF2 cluster (Pachnis et al. 1988). Initially, H19 was thought to encode a protein, but analysis of its sequence revealed the absence of any reasonably sized open reading frame and the 2500-nt-long RNA was not detectable on polyribosomes

(Brannan et al. 1990). Combined, these observations suggested that it was unlikely to be translated. A few years later, another lncRNA was identified as an essential regulator of X chromosome inactivation (XCI), the process by which one of the two X chromosomes in female mammals is silenced to achieve dosage balance in X-linked gene expression between males and females (Plath et al. 2002). The process of XCI was first proposed by Mary Lyon in the 1960s, but the molecular basis of this process remained unknown (Lyon 1961, 1992). Efforts by Willard, Brown, Brockdorff, and others focused on identifying the gene(s) involved in regulating this process (Brockdorff et al. 1991; Brown et al. 1991; Penny et al. 1996). To do this, they searched for genes that were expressed from the inactive X chromosome (Xi) and not the active X chromosome, and ultimately discovered a single gene they called *Xist* (Xi specific transcript). It was subsequently shown that the *Xist* gene is essential for XCI and encodes a >17,000 nt lncRNA that coats the inactive X chromosome to silence transcription (Brown et al. 1992; Clemson et al. 1996).

Global Disruption of RNA Leads to Large-Scale Morphological Changes in the Nucleus

The first indication that RNA itself might play a role in shaping nuclear organization came from initial experiments in the 1980s performed by Sheldon Penman and colleagues (Nickerson et al. 1989). Specifically, they showed that a large amount of RNA was associated with the "nuclear matrix"—a term used to describe the insoluble components of the nucleus after detergent, salt, and DNAse extraction and digestion. To explore whether RNA might be important for shaping the structure of the nuclear matrix, they performed biochemical extraction and showed that removing RNA from the isolated matrix led to collapse or aggregation of matrix-associated proteins. To explore this process within intact cells, they crudely disrupted global RNA levels by either treating nuclei with enzymes that degrade RNA or with drugs that inhibit global RNA production. In both cases, they observed large-scale morphological changes in the structure of the nucleus (Fig. 1; Nickerson et al. 1989; Quinodoz and Guttman 2014; Rinn and Guttman 2014; Hall and Lawrence 2016; Melé and Rinn 2016; Nozawa and Gilbert 2019). In contrast, drugs that blocked protein translation did not show observable structural changes (Nickerson et al. 1989). Although these initial experiments suggested that RNA was likely to play a critical role in nuclear organization, the idea remained controversial for some time because it was unclear which specific RNAs might play these roles and which nuclear structures were dependent on RNA.

RNA ORGANIZES NUCLEAR STRUCTURES ASSOCIATED WITH DIVERSE NUCLEAR FUNCTIONS

As biochemical and microscopy methods matured, the parallel discoveries regarding nuclear structures and functional ncRNAs started to converge (Fig. 2). Advances in microscopy methods such as the detection of specific protein localization using antibodies, fluorescent proteins, and immunoelectron microscopy, and DNA and RNA localization using in situ hybridization made it possible to visualize the spatial localization of specific molecules in the nucleus (Gall 2016). This led to the discovery that many ncRNAs are enriched within specific nuclear bodies and, subsequently, that several ncRNAs can play central roles in their organization. For example, specific RNAs are sufficient to seed the formation of nuclear bodies, including the nucleolus, histone locus body (HLB), and Barr body. These RNA-mediated nuclear bodies are associated with different nuclear functions, including RNA processing, heterochromatin regulation, and gene regulation (Fig. 3).

RNA Processing

Several of the most classical nuclear bodies contain DNA, RNA, and protein components associated with specialized processing of different classes of nascent RNA molecules, including ribosomal RNA, mRNA, snRNAs, and histone pre-mRNAs.

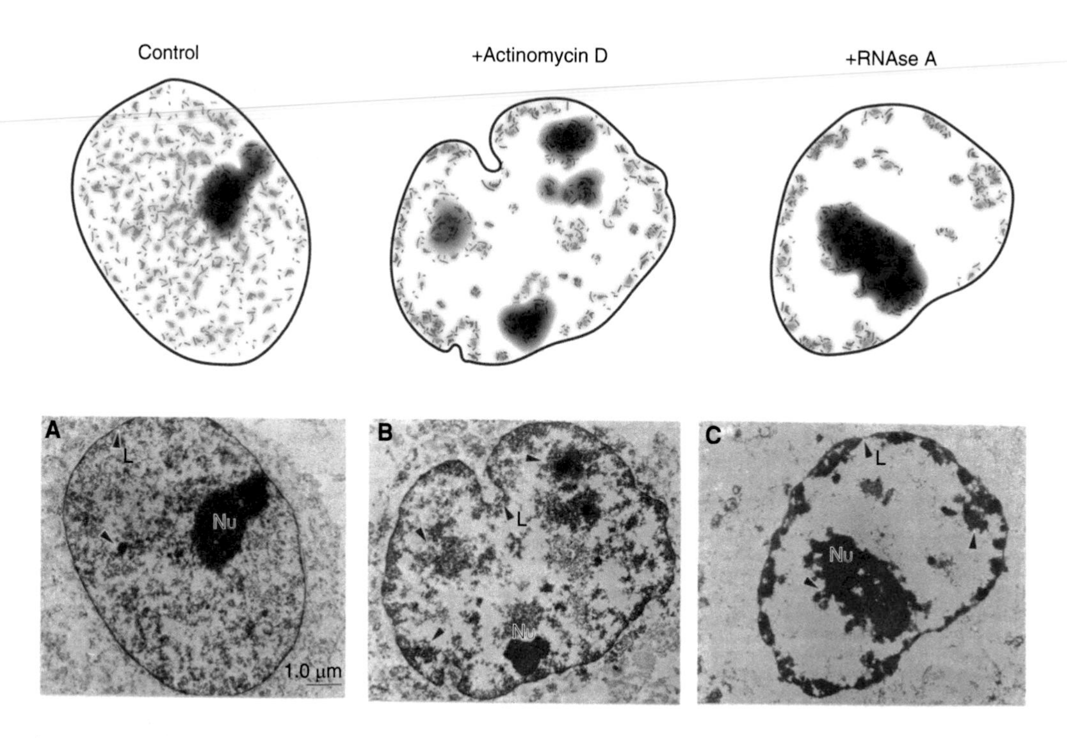

Figure 1. Morphological changes in nuclear organization upon global disruption of RNA. (*A*) Classical experiments explored the role of RNA in maintaining the structural core of the nucleus. To do this, Nickerson and colleagues (*B*) inhibited transcription using actinomycin D or (*C*) degraded RNA using RNAse A. In both cases, they observed large-scale changes in the overall morphology of the nucleus, including collapse or aggregation of chromatin matrix-associated proteins. (*Bottom* row of images are reprinted from Nickerson et al. 1989 courtesy of the CC BY-NC-ND license and The National Academy of Sciences.)

Nucleolus

Key experiments demonstrated that DNA encoding ribosomal RNAs (rDNA) are present in multiple copies on distinct chromosomes throughout the genome (Pederson 2011a). These genes are transcribed by RNA polymerase I as single 45S ribosomal RNA (rRNA) precursors prior to being cleaved by RNase MRP and modified by various snoRNAs to generate the mature 5.8S, 18S, and 28S rRNAs (Maxwell and Fournier 1995; Jády and Kiss 2001; Goldfarb and Cech 2017). Imaging experiments exploring the localization of these components in the nucleus showed that 45S pre-rRNA, snoRNAs, and other ncRNAs involved in ribosome biogenesis are enriched within the nucleolus. Additionally, the multiple rDNA-containing regions from distinct chromosomes come together in 3D space around the nucleolus, resulting in the formation of an interchromosomal RNA-chromatin compartment (Quinodoz et al. 2018). Together, these microscopy and biochemical observations implicated the nucleolus as the site of ribosome biogenesis containing rDNA genes, nascent rRNAs, and several ncRNAs and proteins involved in rRNA processing (Pederson 2011a).

Later work showed that maintenance of the nucleolus is dependent on the ongoing transcription of the 45S pre-rRNA, suggesting that RNA plays a key role in formation of the nucleolus. Specifically, inhibition of Pol I transcription using small molecule inhibitors or global degradation of RNA using RNAse A alters nucleolar morphology by forming "nucleolar caps" and subsequently leads to diffusive localization of the DNA, RNA, and protein components that are normally localized within the nucleolus (Reynolds et al. 1964). These data indicated

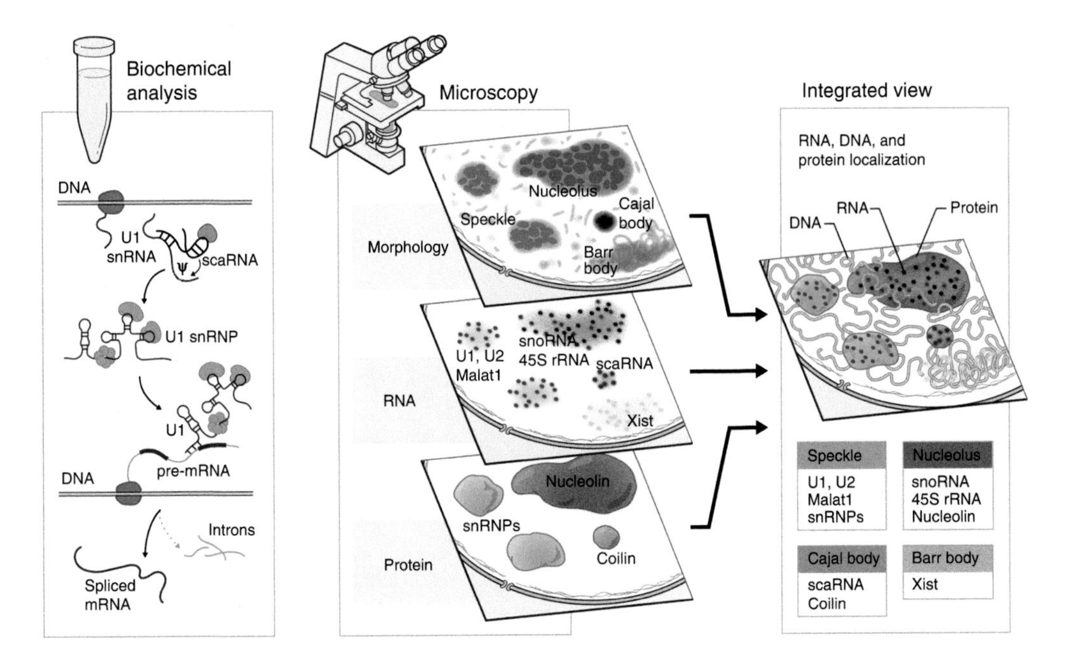

Figure 2. Two parallel approaches converged to uncover a central role for RNA in nuclear organization. As biochemical and microscopy methods matured, the parallel discoveries regarding functional noncoding RNAs (ncRNAs) and nuclear structures started to converge. In the 1970s, biochemical approaches uncovered that many RNAs are not translated into proteins and are retained within the nucleus. Several of these ncRNAs play key roles in nuclear function, including small nuclear RNA (snRNA) biogenesis (e.g., small Cajal body-associated RNAs [scaRNAs]) and pre-mRNA splicing (e.g., snRNAs). In parallel, nuclear bodies were first observed as cytological features in the nucleus more than 100 years ago. Later, advances in microscopy methods made it possible to visualize the spatial localization of specific RNA (via in situ hybridization) and protein molecules (via immunofluorescence or fluorescent proteins) in the nucleus. This led to the discovery that many functional ncRNAs are enriched within specific nuclear bodies and, subsequently, that several ncRNAs can play central roles in their organization. Integrating these biochemical and localization discoveries provided indications of the functional role of RNA in the organization of various nuclear bodies.

that the nascent 45S pre-rRNA plays a critical role in establishing and maintaining the structure of the nucleolus.

Nuclear Speckles

Imaging of several snRNAs (U1, U2, U4, U5, U6) showed them to be concentrated in discrete structures referred to as nuclear speckles based on their appearance (Huang and Spector 1992; Matera and Ward 1993). Using immunofluorescence, many distinct splicing protein components were similarly found to colocalize within nuclear speckles (Perraud et al. 1979; Lerner et al. 1981; Spector et al. 1984). Interestingly, gene-dense Pol II-transcribed regions and their associated nascent pre-mRNAs are positioned close to nuclear speckles where they can form interchromosomal contacts around individual speckles (Chen et al. 2018; Quinodoz et al. 2018). Using live cell imaging, Misteli and Spector showed that splicing proteins within the speckles are primarily inactive and upon activation diffuse from the speckle to nascent pre-mRNAs to catalyze the splicing reaction (Misteli et al. 1997). In addition to snRNAs, there are additional ncRNAs that are specifically enriched within this nuclear compartment, including Malat1 and 7SK (Matera and Ward 1993; Hutchinson et al. 2007).

RNA has been shown to play a structural role in maintaining the morphology of the

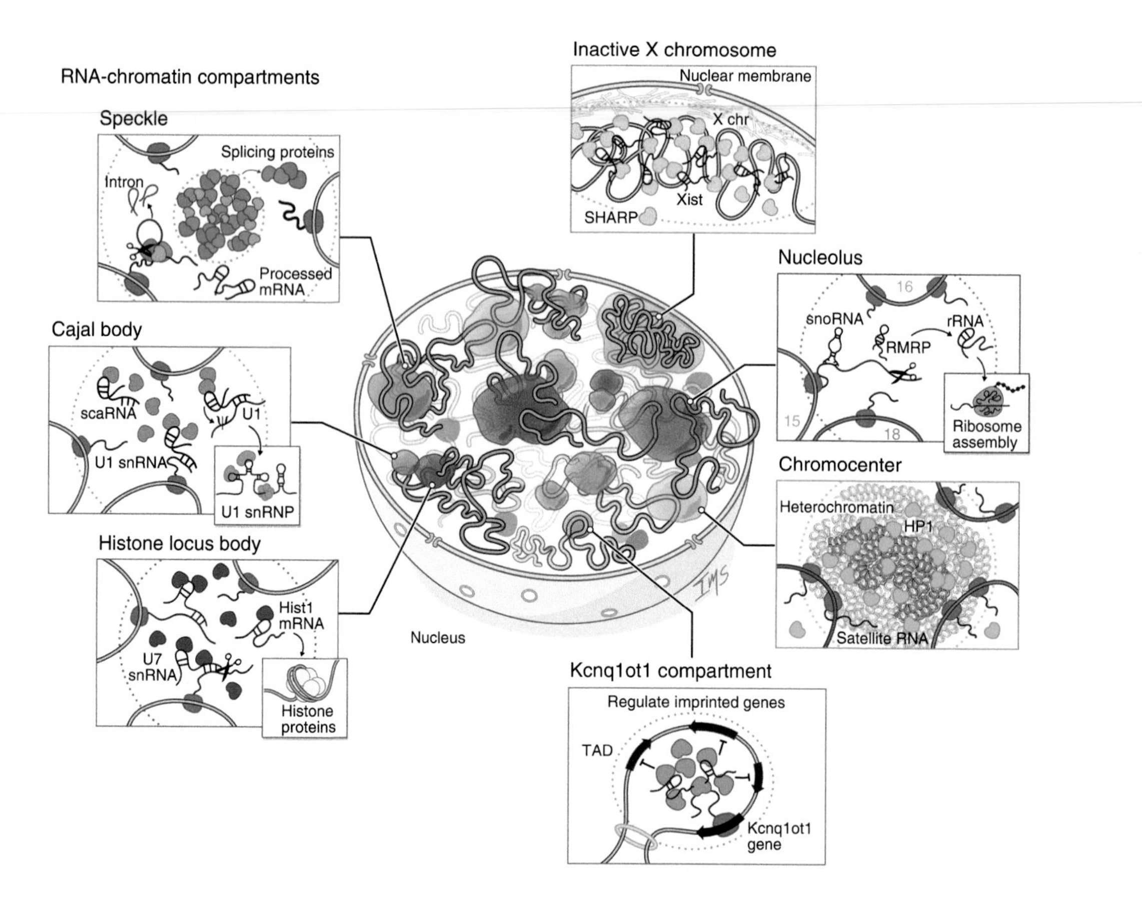

Figure 3. RNA-chromatin compartments regulate various processes throughout the nucleus. RNA has been shown to organize nuclear bodies involved in RNA processing, chromatin regulation, and gene expression. These include RNA processing bodies such as the nucleolus: The nucleolus is the site of ribosome biogenesis within the nucleus. Ribosomal DNA (rDNA) genes transcribed from multicopy genes on distinct chromosomes are positioned within nucleoli where they are transcribed by RNA Pol I. These nascent 45S pre-ribosomal RNAs (rRNAs) are processed by small nucleolar RNAs (snoRNAs) and cleaved by RNAse P (RMRP) into mature 18S, 5.8S, and 28S rRNA and assembled into ribosomes. Histone locus bodies (HLBs): HLBs contain histone gene loci and are sites of histone pre-mRNA transcription and processing. The U7 small nuclear ribonucleoprotein (snRNP) cleaves the 3′ ends of histone pre-mRNAs to produce mature histone mRNAs. Cajal bodies: Cajal bodies are sites of snRNP maturation. Splicing snRNAs are transcribed from genes located at multiple sites across the genome and are modified by small Cajal body-associated RNAs (scaRNAs). Speckles: Nuclear speckles are nuclear bodies with high concentrations of pre-mRNA splicing proteins and snRNAs. Certain transcriptionally active genomic loci are positioned proximal to speckles. Other nuclear structures are associated with chromatin regulation and gene expression, such as the inactive X chromosome: The Xist long noncoding RNA (lncRNA) is responsible for chromosome-wide silencing, compaction, and localization of the inactive X chromosome to the nuclear lamina in female cells. It orchestrates silencing through direct binding and recruitment of the SHARP repressive protein to the X chromosome. Chromocenter: Satellite DNA repeats on multiple chromosomes are compacted into heterochromatin-dense foci. These foci are enriched for Heterochromatin protein 1 (HP1) and typically repressed, but satellite RNAs are also expressed at low levels and transcribed from these compartments. These RNAs are required for HP1 localization to chromocenters. Genomic imprinting: The Kcnq1ot1 lncRNA silences gene expression of imprinted target genes located in a 3D compartment next to its transcriptional locus. It does this by directly binding and recruiting the SHARP repressive protein to these target loci within a topologically associated domain (TAD), which results in silencing of gene expression.

 Cite this article as *Cold Spring Harb Perspect Biol* doi: 10.1101/cshperspect.a039719

nuclear speckle. Specifically, Spector and colleagues showed that global disruption of RNA or inhibition of transcription leads to speckles that display more spherical morphology (Spector et al. 1991; Huang et al. 1994; Misteli et al. 1997). In contrast, degradation of genomic DNA does not lead to changes in observed speckle morphology (Spector et al. 1991). Whereas the core of the speckle is retained in the absence of RNA, disruption of RNA leads to diffusive localization of splicing proteins (small nuclear ribonucleoproteins [snRNPs]) throughout the nucleus (Spector et al. 1991). Together, these results suggested that localization of snRNPs within nuclear speckles is dependent on active transcription of nascent RNA.

Cajal Body

In addition to their localization in the speckle, snRNAs also localize within Cajal bodies (Carmo-Fonseca et al. 1992; Matera and Ward 1993). However, the function of these bodies remained elusive until the discovery of small Cajal body-specific RNAs (scaRNAs) (Darzacq et al. 2002). scaRNAs are similar in sequence and function to snoRNAs, but instead of modifying rRNA, they directly hybridize to and guide modification of snRNA transcripts, rendering them functional within the spliceosome (Jády et al. 2003; Nizami et al. 2010). At the DNA level, snRNA genes that are encoded at multiple genomic locations come together within the Cajal body (Smith et al. 1995; Machyna et al. 2014; Wang et al. 2016; Quinodoz et al. 2020). Together, these observations established the Cajal body as a spatially defined compartment containing the genomic DNA, RNA, and protein components involved in snRNP biogenesis (Gall 2000; Machyna et al. 2013).

Several studies have dissected the role of snRNAs in seeding formation of Cajal bodies (Carmo-Fonseca et al. 1992). For example, Dundr and colleagues found that active transcription of a U2 snRNA array is required for recruitment of various Cajal body components to this locus (Dundr et al. 2007) and that tethering of snRNPs to genomic DNA is sufficient for their formation (Kaiser et al. 2008). These results indicated that transcription of snRNAs is critical for Cajal body organization on chromatin.

Histone Locus Body

Unlike most pre-mRNAs, histone pre-mRNAs are not polyadenylated; instead, their 3′ ends are bound and cleaved by the U7 snRNP complex to produce mature histone mRNAs. The DNA that encodes histone genes, nascent histone pre-mRNAs, U7, and various proteins involved in histone mRNA biogenesis are enriched within HLBs (Nizami et al. 2010). In fact, multiple histone loci that are not linearly close in genomic sequence can organize together in 3D space within HLBs (Quinodoz et al. 2018, 2020).

Histone pre-mRNAs are critical for establishing HLBs. Specifically, Dundr and colleagues demonstrated that synthetically tethering histone pre-mRNAs to chromatin was sufficient to recruit proteins that form the HLB to this genomic locus (Shevtsov and Dundr 2011). Interestingly, recruitment of a histone pre-mRNA containing a deletion in the region that hybridizes to U7 was unable to form the HLB (Shevtsov and Dundr 2011). These observations demonstrated that nascent histone pre-mRNAs, through their ability to interact with U7 snRNA, are important for seeding and maintaining the HLB.

Possible Role of Nuclear Bodies in Promoting Cotranscriptional RNA Processing

Although the functional advantages of organizing RNA, DNA, and proteins of shared functions within nuclear bodies is still unknown, spatial organization of components involved in cotranscriptional RNA processing appears to be a shared feature of many distinct RNA processing pathways. One possible role of such organization is that by spatially organizing these components, nuclear bodies can act to increase the local concentration of critical regulatory RNA and protein molecules near their nascent RNA targets immediately upon transcription. It has been proposed, but not directly demonstrated, that such compartmentalization might act to increase the rate at which cotranscriptional processing can occur and ensure the robustness of processing of RNA targets that are present at dramatically higher concentrations than their regulators (e.g., snRNAs and scaRNAs).

Chromatin Regulation and Gene Expression

In addition to RNA processing, specific nuclear structures are also organized around DNA regions of shared transcriptional regulation and chromatin modifications. These include nuclear structures associated with X chromosome inactivation (XCI), parent-of-origin genomic imprinting, centromeric heterochromatin organization, among others.

Xist Orchestrates X Chromosome Inactivation: A Paradigm of RNA-Mediated Nuclear Organization and Function

The Xi, or Barr body, is a nuclear compartment with several unique features: it is compacted, depleted of RNA Pol II, enriched for various repressive chromatin regulators, and forms a unique 3D structure that is positioned at the nuclear lamina (Wutz 2011; Galupa and Heard 2018; Strehle and Guttman 2020; Żylicz and Heard 2020). Importantly, these structural characteristics of the Xi are driven by expression of the Xist lncRNA. In fact, expression of Xist on autosomes or the X chromosome of male cells is sufficient to trigger chromosome-wide silencing (Wutz and Jaenisch 2000; Wutz et al. 2002). In addition, expression of the Xist lncRNA in male cells can lead to chromosome compaction, remodeling, and repositioning to the lamina (Wutz et al. 2002; Chen et al. 2016; Giorgetti et al. 2016). Because Xist can both silence gene expression and drive structural changes on the Xi, it has propelled our understanding of how an ncRNA can shape nuclear organization and how these changes impact gene regulation.

Upon induction of XCI, Xist spreads to sites on the X chromosome by diffusing in 3D from its sites of transcription to other DNA regions that are in close spatial proximity (Engreitz et al. 2013). At each of these sites, Xist localizes to DNA by binding to the DNA- and RNA-binding protein SAF-A (also known as hnRNP U) that is present across the genome (Fig. 4A; Hasegawa et al. 2010; Chu et al. 2015; McHugh et al. 2015). Xist can spread from these initial sites to more distal sites on the X chromosome by repositioning DNA regions to the nuclear lamina (Chen et al. 2016). This results in DNA sites that are already bound by Xist being positioned away from the actively transcribed Xist locus, and new DNA regions on the chromosome being brought into closer spatial proximity. Xist can then spread to these newly accessible sites and continue spreading and repositioning until it coats the entire X chromosome. At each of these bound sites, Xist binds directly to SHARP (also known as SPEN) and recruits this protein and its associated SMRT and HDAC3 repressors to evict RNA Pol II and silence transcription (Fig. 4A; Chu et al. 2015; Moindrot et al. 2015; Monfort et al. 2015; Żylicz et al. 2019; Dossin et al. 2020). In this way, Xist drives recruitment of these repressive proteins to its spatially enriched RNA compartment to silence target genes.

Many Nuclear Compartments Are Organized by RNA

Beyond Xist, many ncRNAs can diffuse in 3D space to create regulatory compartments within the nucleus. For example, the Kncq1ot1 lncRNA is responsible for silencing imprinted genes near its transcriptional locus but avoids repressing other proximal nonimprinted targets (Nagano and Fraser 2009; Kanduri 2011). Specifically, Kcnq1ot1 localizes within a topologically associating domain (TAD) containing its paternally imprinted gene targets (Cdkn1c, Slc22a18, Phlda2), but excludes neighboring nonimprinted genes (Cars, Nap1l4) (Quinodoz et al. 2020). The Kcnq1ot1 RNA binds directly to the SHARP/SPEN-repressive protein and recruits it and its associated histone deacetylase to silence gene expression within this specific compartment. In fact, loss of the SHARP/SPEN-binding site on the Kcnq1ot1 lncRNA disrupts this ability to silence genes within the TAD (Mohammad et al. 2008; Quinodoz et al. 2020).

 Cite this article as *Cold Spring Harb Perspect Biol* doi: 10.1101/cshperspect.a039719

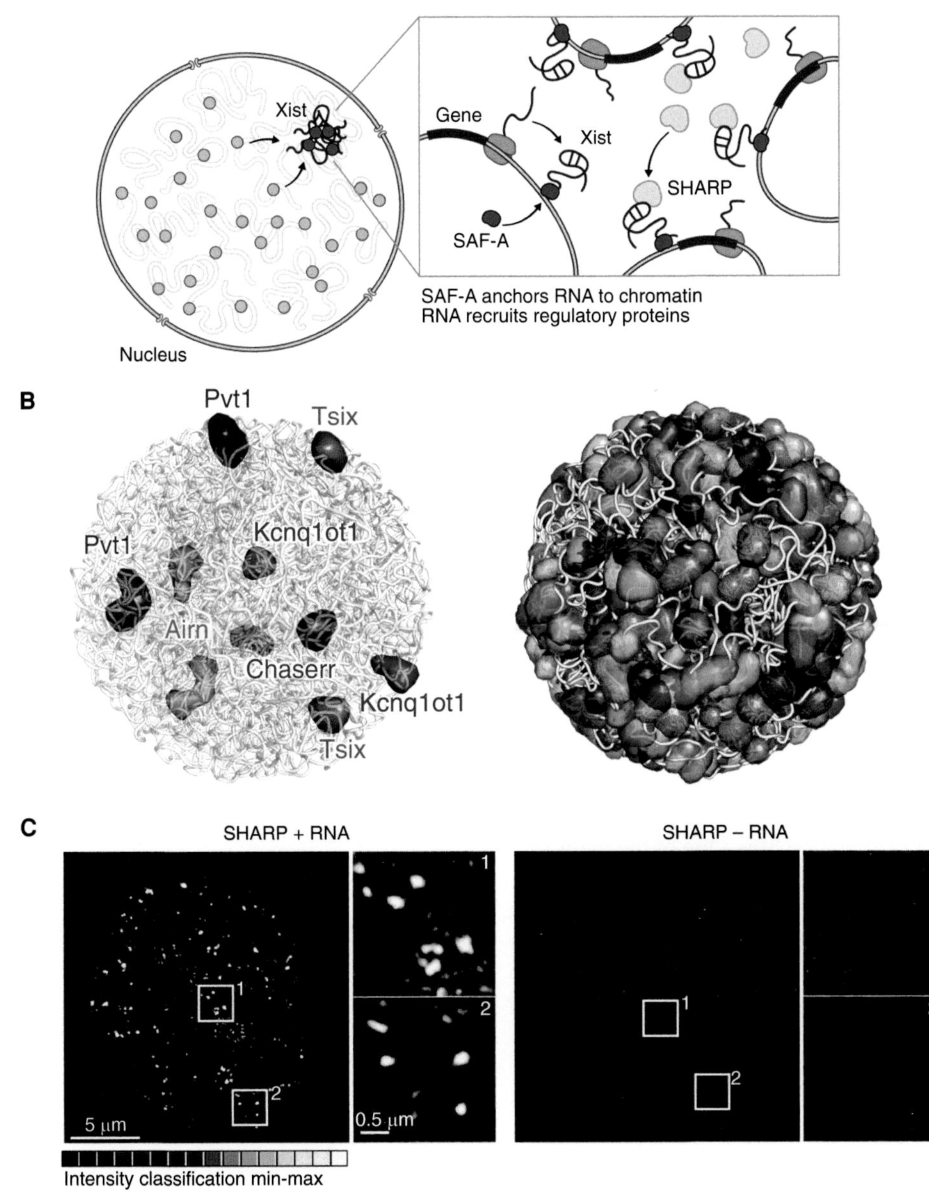

Figure 4. RNA localization and nuclear structure. (*A*) The Xist long noncoding RNA (lncRNA) acts in *cis* to silence transcription of the genes located proximal to its transcriptional locus. It is anchored to chromatin through interactions with the scaffold attachment factor A (SAF-A) DNA/RNA binding protein and, at these sites, directly binds to and recruits SHARP protein to silence gene expression across the X chromosome. (*B*) ~95% of all lncRNAs stably associate in compartments proximal to their transcriptional loci. These include the Kcnq1ot1, Airn, Pvt1, Tsix, and Chaserr lncRNAs, which silence expression of genes located proximal to their transcriptional loci within these compartments (*left*). Overall, these lncRNA compartments are present across the vast majority of DNA regions within the nucleus (*right*). (Panel *B* adapted from Quinodoz et al. 2020 with permission from the authors.) (*C*) SHARP forms dozens of discrete foci in the nucleus and these foci are disrupted upon deletion of its RNA-binding domain, resulting in diffusive localization of SHARP throughout the nucleus. (Panel *C* adapted from Quinodoz et al. 2020 with permission from the authors.)

More recently, genomic methods to map all ncRNAs and their spatial organization uncovered hundreds of ncRNAs that demarcate distinct nuclear territories and showed that the majority of DNA regions within the nucleus are contained within discrete ncRNA-demarcated compartments (Fig. 4B; Quinodoz et al. 2020). In addition, several chromatin regulatory proteins have been shown to form dozens of discrete foci within the nucleus and, in many cases, the focal localization of these chromatin proteins is dependent on their ability to bind to RNA (Maison et al. 2002; Bernstein and Allis 2005; Bernstein et al. 2006). For example, SHARP forms dozens of discrete foci in the nucleus and these foci are disrupted upon deletion of its RNA binding domain, resulting in diffusive localization of SHARP throughout the nucleus (Fig. 4C; Quinodoz et al. 2020).

Similarly, global perturbation of RNA using RNase has been shown to lead to disruption of the compartmentalized localization of HP1 and components of the polycomb repressive complex 1 (PRC1) (Maison et al. 2002; Bernstein and Allis 2005; Bernstein et al. 2006; Caudron-Herger et al. 2011), both associated with repressed heterochromatin. One of the main sites of HP1 localization in the nucleus is over centromeric and pericentromeric DNA regions (Maison and Almouzni 2004). These DNA regions organize across chromosomes into 3D foci referred to as chromocenters (Maison and Almouzni 2004; Jagannathan et al. 2019). Although chromocenters are heterochromatic compartments, the centromeric DNA within this compartment is transcribed by RNA Pol II to produce major and minor satellite-derived ncRNAs (Probst et al. 2010; Casanova et al. 2013). These satellite RNAs are highly enriched around centromeric heterochromatin sites (Quinodoz et al. 2020). Similar to global RNA perturbations, specific knockdown of major or minor satellite-derived ncRNAs leads to loss of HP1 localization over these chromocenter structures (Quinodoz et al. 2020). These observations demonstrated that satellite RNAs are required for recruitment of HP1 to centromere-proximal nuclear compartments.

Possible Functional Roles of RNA-Mediated Nuclear Compartments in Chromatin and Gene Regulation

These examples highlight several possible functional roles for ncRNA-mediated compartments in gene regulation. Specifically, through their ability to diffuse to sites near their transcriptional loci, ncRNAs can enable highly specific regulation by localizing exclusively to target genes contained within a specific 3D compartment while precluding neighboring genes. Moreover, by concentrating regulatory RNAs and proteins within specific nuclear compartments, they may act to increase the effective concentration at these target genes to ensure the robustness of regulation at its target sites. Additionally, because ncRNAs can be spatially enriched and diffuse from their site of transcription, they can spread beyond their immediate transcriptional locus to amplify regulatory signals beyond the limited topological range of DNA elements to regulate multiple genes. This mechanism enables regulation of multiple imprinted genes (e.g., Kcnq1ot1) or entire chromosomes (e.g., Xist) through expression of a single RNA. Finally, such RNA-mediated compartments enable allele-specific gene regulation that cannot be achieved by proteins. This is because proteins are exported to the cytoplasm to be translated and, therefore, lose the positional information encoded at their transcription loci.

SHARED MECHANISMS OF RNA-MEDIATED NUCLEAR ORGANIZATION

These examples highlight common principles by which RNA can seed the formation of spatially anchored DNA-, RNA-, and protein compartments in the nucleus (Fig. 5A). Specifically, as part of the nucleation event, the process of transcription produces a high local concentration of RNA in spatial proximity to its transcription locus. Because these spatially enriched ncRNAs can contain sequence and secondary structure motifs that can bind to diffusible RNA and protein molecules for which they have affinity (e.g., Xist binds SHARP/SPEN, 45S pre-rRNA binds snoRNAs), these high-

 Cite this article as *Cold Spring Harb Perspect Biol* doi: 10.1101/cshperspect.a039719

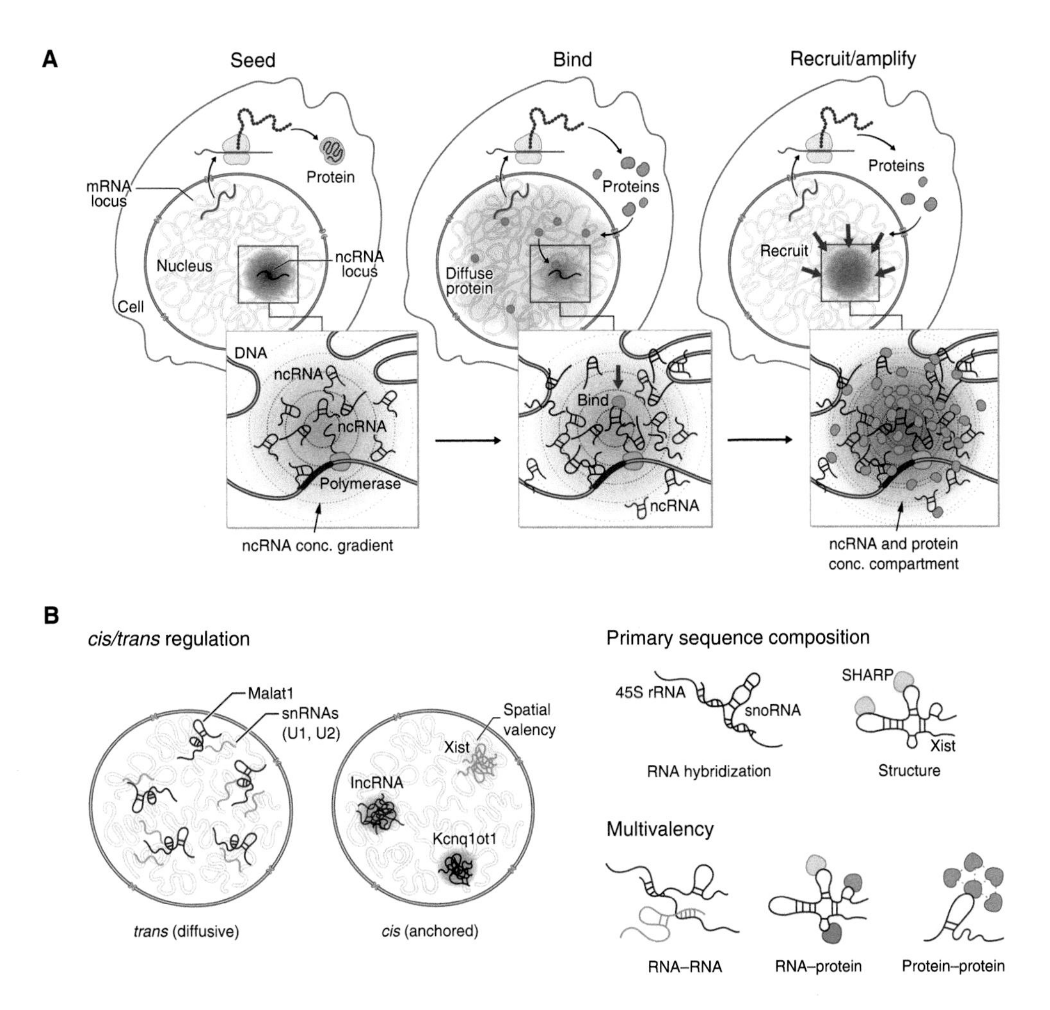

Figure 5. A general model by which noncoding RNAs (ncRNAs) may facilitate nuclear organization. (*A*) Seed: Because transcription creates multiple copies of an RNA species, it can achieve a high local concentration around its transcription locus to "seed" formation of an RNA-chromatin compartment. Bind: RNAs can form high-affinity binding interactions with other diffusive RNAs and proteins and recruit them into these transcriptional loci. Recruit/amplify: Bound proteins can recruit others through protein–protein interactions to form an RNA-chromatin compartment. (*B*) (*Left*) ncRNAs have distinct localization patterns in the nucleus—either diffusive (in *trans*) away from their transcription sites or anchored (in *cis*) at their transcriptional loci. (*Right*) RNAs can act as unique multivalent scaffolds for RNA and proteins. RNAs can bind to and recruit other RNAs or proteins via RNA hybridization or RNA–protein interactions, where RNAs fold into secondary structures and bind proteins. Because of its length, a single RNA may hybridize to more than one additional RNA and bind to multiple proteins, enabling the molecules to assemble into a higher-order RNA–protein assemblies. In addition, some RNA-binding proteins have intrinsically disordered regions that allow them to interact with multiple proteins.

affinity interactions act to recruit diffusible molecules to form spatial compartments (Fig. 5B).

ncRNAs are uniquely suited for this role because, to form a nuclear compartment, molecules need to achieve high concentrations within a spatially enriched subvolume of the nucleus. Unlike proteins that are translated in the cytoplasm and need to diffuse through the nucleus to engage their target, ncRNAs can accumulate at high concentration near their site of transcription. Conversely, DNA is intrinsically spatially localized in the nucleus, but it is present at a single copy and therefore does not achieve sufficiently high local concentration required to seed a nuclear compartment. In this way, the multiple copies of an RNA produced at a locus by transcription can achieve higher concentrations than DNA. Notably, most lncRNAs (~95% mapped) have been shown to be stably associated in spatial compartments around their transcriptional loci (Cabili et al. 2015; Quinodoz et al. 2020). This ability of ncRNAs to remain near their loci may occur through transient tethering by RNA polymerase during transcription or through more stable association by binding to specific DNA-binding proteins. For example, the nucleolus is organized around nascent transcription of 45S rRNA, likely while it is transcribed by RNA Pol I. Conversely, the mature Xist lncRNA is stably associated with the Xi through a high-affinity interaction with the SAF-A RNA/DNA-binding proteins that is present across genomic DNA (Hasegawa et al. 2010; Kolpa et al. 2016).

In addition, unlike DNA, because RNA can diffuse from its transcription locus, it can spread across longer genomic and topological distances and can associate with multiple target sites simultaneously. For example, Xist can spread from its transcription locus on the X chromosome across the entire chromosome (Engreitz et al. 2013; Simon et al. 2013). In fact, several ncRNAs can function primary through diffusion to localize at target sites that are far away from their transcriptional locus (Fig. 5B). For example, the abundant Malat1 lncRNA localizes broadly across the nucleus at Pol II-transcribed genes (Engreitz et al. 2014; Cabili et al. 2015). Accordingly, these two features of RNA—the ability to form sites of high local concentration in the nucleus and the ability to diffuse from their transcription locus to form compartments of different sizes—make it a highly versatile molecule for seeding nuclear compartments. Indeed, many nuclear structures have now been shown to form via interactions between diffusible ncRNAs and proteins and spatially enriched ncRNAs (Engreitz et al. 2016).

We note that this role for ncRNAs in mediating compartment formation is not strictly limited to ncRNAs, but instead is a property of any RNA molecule that can function as an RNA regardless of whether it may also encode a protein product. For example, the histone pre-mRNAs (which code for histone proteins) have been shown to seed the formation of the HLB by achieving a high concentration of the nascent RNA in proximity to its transcribed DNA locus and binding to and recruiting diffusible ncRNAs and protein complexes (e.g., NPAT, FLASH) into this compartment (Nizami et al. 2010; Shevtsov and Dundr 2011).

RNA and Phase Separation

Recently, many nuclear bodies have been described to form liquid-like, phase-separated condensates or droplets, which can locally concentrate molecules and coalesce with neighboring structures (Banani et al. 2017; Strom and Brangwynne 2019). The central premise of this model is that high concentrations of nucleic acids and proteins—especially those containing multivalent low complexity domains—can undergo concentration-dependent phase transitions to "demix" from the surrounding nuclear environment.

ncRNAs play critical roles in seeding phase-separated compartments in the nucleus because they can act as unique multivalent scaffolds for RNA and proteins (Fig. 5B; Roden and Gladfelter 2021). Specifically, RNAs have both primary sequence specificity and secondary structure to bind to and recruit other proteins or RNAs via a variety of mechanisms: (1) RNAs participate in RNA–RNA interactions where they hybridize to each together. In many cases, because of length, a single RNA can hybridize to more than one additional RNA, enabling the molecules to assemble into higher-order RNA–RNA assemblies (Jain and Vale 2017; Van Treeck and Parker 2018). An example of this type of interaction is the hybridization of 45S rRNA to many snoRNAs (Maxwell and Fournier 1995). (2) RNAs participate in RNA–protein interactions where RNAs fold into secondary structures, bind proteins, and recruit them to a given transcriptional locus (Engreitz et al. 2016). An example of this behavior is the Xist lncRNA, which binds and recruits SHARP/SPEN to the Xi (Chu et al.

 Cite this article as *Cold Spring Harb Perspect Biol* doi: 10.1101/cshperspect.a039719

2015; McHugh et al. 2015). (3) Many RNA-binding proteins have large intrinsically disordered regions (IDRs) that can undergo concentration-dependent associations to form higher-order assemblies (e.g., PTBP1, FUS, and SHARP/SPEN) (Lin et al. 2015; Banani et al. 2016; Shin and Brangwynne 2017). (4) Many ncRNAs contain multivalent sites for protein binding that enable high avidity interactions with specific RBPs. For example, the A-repeat of Xist consists of a tandem repeat that enables Xist to bind to multiple copies of SHARP simultaneously and FIRRE contains a tandem repeat that bind to multiple copies of SAF-A (Hacisuleyman et al. 2014, 2016; Lu et al. 2016; Brockdorff 2018; Bansal et al. 2020). (5) Because ncRNAs can achieve high local concentrations upon transcription, the ncRNA can achieve high "spatial valency" within a specific territory in the nucleus. These various features make RNA an ideal molecule for seeding the formation of concentration-dependent nuclear structures.

It has also been found that condensates can act as hubs for DNA interactions, yet how these DNA interactions actively come together is only beginning to be understood. One model for DNA interactions at these bodies involves the coalescence of proteins (Fig. 6). Studies using optogenetic tools revealed that condensates tethered to DNA can fuse, or coalesce, and subsequently bring their respective DNA components together (Shin et al. 2018). For example, multiple nucleoli can coalesce such that nucleoli with multiple rDNA-containing chromosomes come together around the same nuclear body (Quinodoz et al. 2018; Caragine et al. 2019; Lafontaine et al. 2021). This is thought to occur through the process of rRNA transcription on each individual rDNA-containing chromosome, which then forms RNA–protein condensates at each locus; these condensates can then coalesce to bring together several DNA loci from multiple chromosomes into a larger nucleolus (Feric et al. 2016; Quinodoz et al. 2018; Lafontaine et al. 2021). This mechanism likely underlies the formation of many of the DNA contacts around classical nuclear bodies (Misteli 2020).

We note that many nuclear bodies (e.g., nucleoli, HLB) form around DNA regions that are highly repetitive or multicopy. This may enable the DNA itself to achieve high "spatial valency" around a linear genomic locus. For example, the nucleolus is arranged around transcription from arrays of rDNA genes that are clustered in linear space on the chromosome. At a smaller scale, the histone locus body forms around linearly clustered histone genes (~20–30 genes per gene cluster) in a genomic region. At these multicopy DNA regions, multiple nascent RNAs are transcribed within each genomic region,

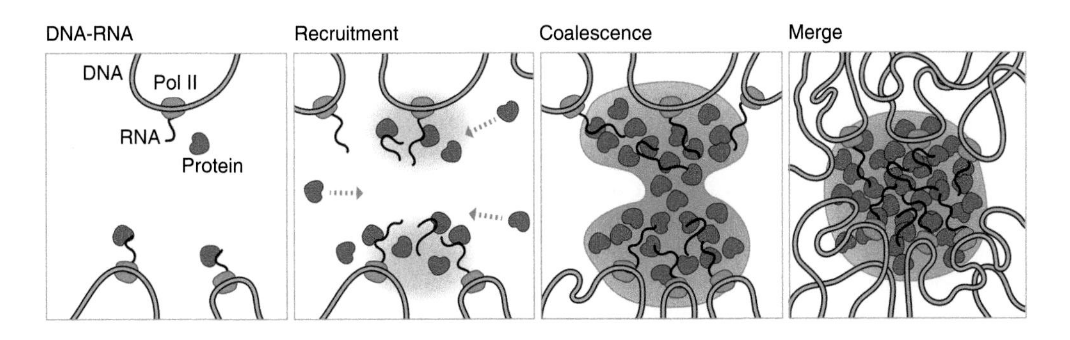

Figure 6. A model for 3D organization of DNA loci within RNA-chromatin compartments. Nuclear bodies can act as hubs for DNA interactions. One model for establishing such DNA interactions involves the coalescence of proteins at a nuclear body. Specifically, this may occur by RNA transcription occurring at multiple DNA loci, followed by recruitment of proteins to these sites, growth and coalescence of droplets, and merging of droplets into a single nuclear body. This series of events would create a hub of DNA interactions around a large nuclear body.

which likely results in a high local concentration of RNAs and protein recruitment (through RNA–protein interactions) to form these compartments.

CONCLUDING REMARKS

In the nearly two centuries since the initial description of nuclear bodies, significant advances have been made in our understanding of the molecular components, organization principles, and functional characteristics of many of these nuclear structures. Whereas specific RNAs have been demonstrated to be essential for organizing specific nuclear structures, these remain limited to a handful of well-defined examples and the vast majority of nuclear-retained ncRNAs remain to be fully explored. The development and application of high-throughput genomic and microscopy methods for mapping the 3D organization of RNA and DNA throughout the nucleus have most recently uncovered hundreds of ncRNAs that localize in precise spatial territories throughout the nucleus (Quinodoz et al. 2020), representing prime candidates to explore the extent of RNA-mediated structural organization.

Additionally, despite the significant progress made to uncover the molecular components of these RNA-chromatin compartments, it remains unclear how spatial organization drives regulatory processing. The challenge is that it is often difficult to genetically distinguish between the role of a given RNA in organizing nuclear structure and its role in a specific function because they are often interconnected in the cell. For example, whether the nucleolus is critical for ribosome biogenesis, or simply the location where it occurs, has been difficult to test because methods that disrupt nucleolar organization, such as transcriptional inhibition, also impact ribosome biogenesis. Newly developed technologies to alter nuclear structure such as optogenetic tools (Bracha et al. 2018; Shin et al. 2018) that can drive the local concentration of molecules through phase separation may enable new ways of perturbing nuclear structure to measure its impact on function. As a result of these observations, RNA has emerged as a key molecule in regulating diverse nuclear structures and functions. We anticipate that new technological advances in high-throughput measurement and perturbation techniques will enable a more complete characterization of the roles of RNA in nuclear organization.

ACKNOWLEDGMENTS

We thank Shawna Hiley for editing, Inna-Marie Strazhnik for illustrations, and members of the Guttman Laboratory for helpful discussions.

REFERENCES

Banani SF, Rice AM, Peeples WB, Lin Y, Jain S, Parker R, Rosen MK. 2016. Compositional control of phase-separated cellular bodies. *Cell* **166:** 651–663. doi:10.1016/j.cell.2016.06.010

Banani SF, Lee HO, Hyman AA, Rosen MK. 2017. Biomolecular condensates: organizers of cellular biochemistry. *Nat Rev Mol Cell Biol* **18:** 285–298. doi:10.1038/nrm.2017.7

Bansal P, Kondaveeti Y, Pinter SF. 2020. Forged by DXZ4, FIRRE, and ICCE: how tandem repeats shape the active and inactive X chromosome. *Front Cell Dev Biol* **7:** 328. doi:10.3389/fcell.2019.00328

Barr ML, Bertram EG. 1949. A morphological distinction between neurones of the male and female, and the behaviour of the nucleolar satellite during accelerated nucleoprotein synthesis. *Nature* **163:** 676–677. doi:10.1038/163676a0

Bernstein E, Allis CD. 2005. RNA meets chromatin. *Genes Dev* **19:** 1635–1655. doi:10.1101/gad.1324305

Bernstein E, Duncan EM, Masui O, Gil J, Heard E, Allis CD. 2006. Mouse polycomb proteins bind differentially to methylated histone H3 and RNA and are enriched in facultative heterochromatin. *Mol Cell Biol* **26:** 2560–2569. doi:10.1128/MCB.26.7.2560-2569.2006

Bracha D, Walls MT, Wei MT, Zhu L, Kurian M, Avalos JL, Toettcher JE, Brangwynne CP. 2018. Mapping local and global liquid phase behavior in living cells using photo-oligomerizable seeds. *Cell* **175:** 1467–1480. doi:10.1016/j.cell.2018.10.048

Brannan CI, Dees EC, Ingram RS, Tilghman SM. 1990. The product of the H19 gene may function as an RNA. *Mol Cell Biol* **10:** 28–36.

Brockdorff N. 2018. Local tandem repeat expansion in Xist RNA as a model for the functionalisation of ncRNA. *Noncoding RNA* **4:** 28. doi:10.3390/ncrna4040028

Brockdorff N, Ashworth A, Kay G, Cooper P, Smith S, McCabe VM, Norris DP, Penny GD, Patel D, Rastan S. 1991. Conservation of position and exclusive expression of mouse *Xist* from the inactive X chromosome. *Nature* **351:** 329–331. doi:10.1038/351329a0

Brown CJ, Ballabio A, Rupert JL, Lafreniere RG, Grompe M, Tonlorenzi R, Willard HF. 1991. A gene from the region of the human X inactivation centre is expressed exclusively from the inactive X chromosome. *Nature* **349:** 38–44. doi:10.1038/349038a0

Brown CJ, Hendrich BD, Rupert JL, Lafrenière RG, Xing Y, Lawrence J, Willard HF. 1992. The human *XIST* gene: analysis of a 17 kb inactive X-specific RNA that contains conserved repeats and is highly localized within the nucleus. *Cell* **71:** 527–542. doi:10.1016/0092-8674(92)90520-M

Cabili MN, Dunagin MC, McClanahan PD, Biaesch A, Padovan-Merhar O, Regev A, Rinn JL, Raj A. 2015. Localization and abundance analysis of human lncRNAs at single-cell and single-molecule resolution. *Genome Biol* **16:** 20. doi:10.1186/s13059-015-0586-4

Calvet JP, Pederson T. 1981. Base-pairing interactions between small nuclear RNAs and nuclear RNA precursors as revealed by psoralen cross-linking in vivo. *Cell* **26:** 363–370. doi:10.1016/0092-8674(81)90205-1

Caragine CM, Haley SC, Zidovska A. 2019. Nucleolar dynamics and interactions with nucleoplasm in living cells. *eLife* **8:** e47533. doi:10.7554/eLife.47533

Carmo-Fonseca M, Pepperkok R, Carvalho M, Lamond A. 1992. Transcription-dependent colocalization of the U1, U2, U4/U6, and U5 snRNPs in coiled bodies. *J Cell Biol* **117:** 1–14. doi:10.1083/jcb.117.1.1

Casanova M, Pasternak M, El Marjou F, Le Baccon P, Probst AV, Almouzni G. 2013. Heterochromatin reorganization during early mouse development requires a single-stranded noncoding transcript. *Cell Rep* **4:** 1156–1167. doi:10.1016/j.celrep.2013.08.015

Caudron-Herger M, Müller-Ott K, Mallm JP, Marth C, Schmidt U, Fejes-Tóth K, Rippe K. 2011. Coding RNAs with a non-coding function: maintenance of open chromatin structure. *Nucleus* **2:** 410–424. doi:10.4161/nucl.2.5.17736

Chen CK, Blanco M, Jackson C, Aznauryan E, Ollikainen N, Surka C, Chow A, Cerase A, McDonel P, Guttman M. 2016. Xist recruits the X chromosome to the nuclear lamina to enable chromosome-wide silencing. *Science* **354:** 468–472. doi:10.1126/science.aae0047

Chen Y, Zhang Y, Wang Y, Zhang L, Brinkman EK, Adams SA, Goldman R, van Steensel B, Ma J, Belmont AS. 2018. Mapping 3D genome organization relative to nuclear compartments using TSA-Seq as a cytological ruler. *J Cell Biol* **217:** 4025–4048. doi:10.1083/jcb.201807108

Chu C, Zhang QC, da Rocha ST, Flynn RA, Bharadwaj M, Calabrese JM, Magnuson T, Heard E, Chang HY. 2015. Systematic discovery of Xist RNA binding proteins. *Cell* **161:** 404–416. doi:10.1016/j.cell.2015.03.025

Clemson CM, McNeil JA, Willard HF, Lawrence JB. 1996. XIST RNA paints the inactive X chromosome at interphase: evidence for a novel RNA involved in nuclear/chromosome structure. *J Cell Biol* **132:** 259–275. doi:10.1083/jcb.132.3.259

Darnell JE. 1968. Ribonucleic acids from animal cells. *Bacteriol Rev* **32:** 262–290. doi:10.1128/br.32.3.262-290.1968

Darnell J. 2011. *RNA: Life's indispensable molecule.* Cold Spring Harbor Laboratory Press, Cold Spring, NY.

Darzacq X, Jády BE, Verheggen C, Kiss AM, Bertrand E, Kiss T. 2002. Cajal body-specific small nuclear RNAs: a novel class of 2′-*O*-methylation and pseudouridylation guide RNAs. *EMBO J* **21:** 2746–2756. doi:10.1093/emboj/21.11.2746

Dossin F, Pinheiro I, Żylicz JJ, Roensch J, Collombet S, Le Saux A, Chelmicki T, Attia M, Kapoor V, Zhan Y, et al. 2020. SPEN integrates transcriptional and epigenetic control of X-inactivation. *Nature* **578:** 455–460. doi:10.1038/s41586-020-1974-9

Dundr M, Ospina JK, Sung MH, John S, Upender M, Ried T, Hager GL, Matera AG. 2007. Actin-dependent intranuclear repositioning of an active gene locus in vivo. *J Cell Biol* **179:** 1095–1103. doi:10.1083/jcb.200710058

Engreitz JM, Pandya-Jones A, McDonel P, Shishkin A, Sirokman K, Surka C, Kadri S, Xing J, Goren A, Lander ES, et al. 2013. The Xist lncRNA exploits three-dimensional genome architecture to spread across the X chromosome. *Science* **341:** 1237973. doi:10.1126/science.1237973

Engreitz JM, Sirokman K, McDonal P, Shishkin AA, Surka C, Russell P, Grossman SR, Chow AY, Guttman M, Lander ES. 2014. RNA–RNA interactions enable specific targeting of noncoding RNAs to nascent pre-mRNAs and chromatin sites. *Cell* **159:** 188–199. doi:10.1016/j.cell.2014.08.018

Engreitz JM, Ollikainen N, Guttman M. 2016. Long noncoding RNAs: spatial amplifiers that control nuclear structure and gene expression. *Nat Rev Mol Cell Biol* **17:** 756–770. doi:10.1038/nrm.2016.126

Feric M, Vaidya N, Harmon TS, Mitrea DM, Zhu L, Richardson TM, Kriwacki RW, Pappu RH, Brangwynne CP. 2016. Coexisting liquid phases underlie nucleolar subcompartments. *Cell* **165:** 1686–1697. doi:10.1016/j.cell.2016.04.047

Filipowicz W, Kiss T. 1993. Structure and function of nucleolar snRNPs. *Mol Biol Rep* **18:** 149–156. doi:10.1007/BF00986770

Fournier MJ, Stuart Maxwell E. 1993. The nucleolar snRNAs: catching up with the spliceosomal snRNAs. *Trends Biochem Sci* **18:** 131–135. doi:10.1016/0968-0004(93)90020-N

Gall JG. 2000. Cajal bodies: the first 100 years. *Annu Rev Cell Dev Biol* **16:** 273–300. doi:10.1146/annurev.cellbio.16.1.273

Gall JG. 2016. The origin of in situ hybridization—A personal history. *Methods* **98:** 4–9. doi:10.1016/j.ymeth.2015.11.026

Galupa R, Heard E. 2018. X-chromosome inactivation: a crossroads between chromosome architecture and gene regulation. *Annu Rev Genet* **52:** 535–566. doi:10.1146/annurev-genet-120116-024611

Giorgetti L, Lajoie BR, Carter AC, Attia M, Zhan Y, Xu J, Chen CJ, Kaplan N, Chang HY, Heard E, et al. 2016. Structural organization of the inactive X chromosome in the mouse. *Nature* **535:** 575–579. doi:10.1038/nature18589

Goldfarb KC, Cech TR. 2017. Targeted CRISPR disruption reveals a role for RNase MRP RNA in human preribosomal RNA processing. *Genes Dev* **31:** 59–71. doi:10.1101/gad.286963.116

Hacisuleyman E, Goff LA, Trapnell C, Williams A, Henao-Mejia J, Sun L, McClanahan P, Hendrickson DG, Sauvageau M, Kelley DR, et al. 2014. Topological organization of multichromosomal regions by the long intergenic noncoding RNA Firre. *Nat Struct Mol Biol* **21:** 198–206. doi:10.1038/nsmb.2764

Hacisuleyman E, Shukla CJ, Weiner CL, Rinn JL. 2016. Function and evolution of local repeats in the Firre locus. *Nat Commun* **7:** 11021. doi:10.1038/ncomms11021

Hall LL, Lawrence JB. 2016. RNA as a fundamental component of interphase chromosomes: could repeats prove key? *Curr Opin Genet Dev* **37:** 137–147. doi:10.1016/j.gde.2016.04.005

Hasegawa Y, Brockdorff N, Kawano S, Tsutui K, Tsutui K, Nakagawa S. 2010. The matrix protein hnRNP U is required for chromosomal localization of Xist RNA. *Dev Cell* **19:** 469–476. doi:10.1016/j.devcel.2010.08.006

Heitz E. 1931. Nukleolen und chromosomen in der gattung vicia. [Nucleoli and chromosomes in the genus *Vicia*.] *Planta* **15:** 495–505. doi:10.1007/BF01909065

Huang S, Spector DL. 1992. U1 and U2 small nuclear RNAs are present in nuclear speckles. *Proc Natl Acad Sci* **89:** 305–308. doi:10.1073/pnas.89.1.305

Huang S, Deerinck TJ, Ellisman MH, Spector DL. 1994. In vivo analysis of the stability and transport of nuclear poly(A)+ RNA. *J Cell Biol* **126:** 877–899. doi:10.1083/jcb.126.4.877

Hutchinson JN, Ensminger AW, Clemson CM, Lynch CR, Lawrence JB, Chess A. 2007. A screen for nuclear transcripts identifies two linked noncoding RNAs associated with SC35 splicing domains. *BMC Genomics* **8:** 39. doi:10.1186/1471-2164-8-39

Jády BE, Kiss T. 2001. A small nucleolar guide RNA functions both in 2′-*O*-ribose methylation and pseudouridylation of the U5 spliceosomal RNA. *EMBO J* **20:** 541–551. doi:10.1093/emboj/20.3.541

Jády BE, Darzacq X, Tucker KE, Matera AG, Bertrand E, Kiss T. 2003. Modification of Sm small nuclear RNAs occurs in the nucleoplasmic Cajal body following import from the cytoplasm. *EMBO J* **22:** 1878–1888. doi:10.1093/emboj/cdg187

Jagannathan M, Cummings R, Yamashita YM. 2019. The modular mechanism of chromocenter formation in *Drosophila*. *eLife* **8:** e43938. doi:10.7554/eLife.43938

Jain A, Vale RD. 2017. RNA phase transitions in repeat expansion disorders. *Nature* **546:** 243–247. doi:10.1038/nature22386

Kaiser TE, Intine RV, Dundr M. 2008. De novo formation of a subnuclear body. *Science* **322:** 1713–1717. doi:10.1126/science.1165216

Kanduri C. 2011. Kcnq1ot1: a chromatin regulatory RNA. *Semin Cell Dev Biol* **22:** 343–350. doi:10.1016/j.semcdb.2011.02.020

Kass S. 1990. The U3 small nucleolar ribonucleoprotein functions in the first step of preribosomal RNA processing. *Cell* **60:** 897–908. doi:10.1016/0092-8674(90)90338-F

Kolpa HJ, Fackelmayer FO, Lawrence JB. 2016. SAF-A requirement in anchoring XIST RNA to chromatin varies in transformed and primary cells. *Dev Cell* **22:** 343–350.

Lafarga M, Casafont I, Bengoechea R, Tapia O, Berciano MT. 2009. Cajal's contribution to the knowledge of the neuronal cell nucleus. *Chromosoma* **118:** 437–443. doi:10.1007/s00412-009-0212-x

Lafontaine DLJ, Riback JA, Bascetin R, Brangwynne CP. 2021. The nucleolus as a multiphase liquid condensate. *Nat Rev Mol Cell Biol* **22:** 165–182. doi:10.1038/s41580-020-0272-6

Lerner MR, Boyle JA, Mount SM, Wolin SL, Steitz JA. 1980. Are snRNPs involved in splicing? *Nature* **283:** 220–224. doi:10.1038/283220a0

Lerner EA, Lerner MR, Janeway CA, Steitz JA. 1981. Monoclonal antibodies to nucleic acid-containing cellular constituents: probes for molecular biology and autoimmune disease. *Proc Natl Acad Sci* **78:** 2737–2741. doi:10.1073/pnas.78.5.2737

Lin Y, Protter DSW, Rosen MK, Parker R. 2015. Formation and maturation of phase-separated liquid droplets by RNA-binding proteins. *Mol Cell* **60:** 208–219. doi:10.1016/j.molcel.2015.08.018

Lu Z, Zhang QC, Lee B, Flynn RA, Smith MA, Robinson JT, Davidovich C, Gooding AR, Goodrich KJ, Mattick JS, et al. 2016. RNA duplex map in living cells reveals higher-order transcriptome structure. *Cell* **165:** 1267–1279. doi:10.1016/j.cell.2016.04.028

Lyon MF. 1961. Gene action in the X-chromosome of the mouse (*Mus musculus* L.). *Nature* **190:** 372–373. doi:10.1038/190372a0

Lyon MF. 1992. Some milestones in the history of X-chromosome inactivation. *Annu Rev Genet* **26:** 17–29. doi:10.1146/annurev.ge.26.120192.000313

Machyna M, Heyn P, Neugebauer KM. 2013. Cajal bodies: where form meets function. *Wiley Interdiscip Rev RNA* **4:** 17–34. doi:10.1002/wrna.1139

Machyna M, Kehr S, Straube K, Kappei D, Buchholz F, Butter F, Ule J, Hertel J, Stadler PF, Neugebauer KM. 2014. The coilin interactome identifies hundreds of small noncoding RNAs that traffic through Cajal bodies. *Mol Cell* **56:** 389–399. doi:10.1016/j.molcel.2014.10.004

Maison C, Almouzni G. 2004. HP1 and the dynamics of heterochromatin maintenance. *Nat Rev Mol Cell Biol* **5:** 296–305. doi:10.1038/nrm1355

Maison C, Bailly D, Peters AHFM, Quivy JP, Roche D, Taddei A, Lachner M, Jenuwein T, Almouzni G. 2002. Higher-order structure in pericentric heterochromatin involves a distinct pattern of histone modification and an RNA component. *Nat Genet* **30:** 329–334. doi:10.1038/ng843

Matera AG, Ward DC. 1993. Nucleoplasmic organization of small nuclear ribonucleoproteins in cultured human cells. *J Cell Biol* **121:** 715–727. doi:10.1083/jcb.121.4.715

Maxwell ES, Fournier MJ. 1995. The small nucleolar RNAs. *Annu Rev Biochem* **64:** 897–934. doi:10.1146/annurev.bi.64.070195.004341

McClintock B. 1934. The relation of a particular chromosomal element to the development of the nucleoli in *Zea mays*. *Z Zellforsch Mikrosk Anat* **21:** 294–326. doi:10.1007/BF00374060

McHugh CA, Chen CK, Chow A, Surka CF, Tran C, McDonel P, Pandya-Jones A, Blanco M, Burghard C, Moradian A, et al. 2015. The Xist lncRNA interacts directly with SHARP to silence transcription through HDAC3. *Nature* **521:** 232–236. doi:10.1038/nature14443

Melé M, Rinn JL. 2016. "Cat's cradling" the 3D genome by the act of LncRNA transcription. *Mol Cell* **62:** 657–664. doi:10.1016/j.molcel.2016.05.011

Misteli T. 2020. The self-organizing genome: principles of genome architecture and function. *Cell* **183:** 28–45. doi:10.1016/j.cell.2020.09.014

Misteli T, Cáceres JF, Spector DL. 1997. The dynamics of a pre-mRNA splicing factor in living cells. *Nature* **387:** 523–527. doi:10.1038/387523a0

Mohammad F, Pandey RR, Nagano T, Chakalova L, Mondal T, Fraser P, Kanduri C. 2008. Kcnq1ot1/Lit1 noncoding RNA mediates transcriptional silencing by targeting to the perinucleolar region. *Mol Cell Biol* **28:** 3713–3728. doi:10.1128/MCB.02263-07

Moindrot B, Cerase A, Coker H, Masui O, Grijzenhout A, Pintacuda G, Schermelleh L, Nesterova TB, Brockdorff N. 2015. A pooled shRNA screen identifies Rbm15, Spen, and Wtap as factors required for Xist RNA-mediated silencing. *Cell Rep* **12:** 562–572. doi:10.1016/j.celrep.2015.06.053

Monfort A, Di Minin G, Postlmayr A, Freimann R, Arieti F, Thore S, Wutz A. 2015. Identification of Spen as a crucial factor for Xist function through forward genetic screening in haploid embryonic stem cells. *Cell Rep* **12:** 554–561. doi:10.1016/j.celrep.2015.06.067

Nagano T, Fraser P. 2009. Emerging similarities in epigenetic gene silencing by long noncoding RNAs. *Mamm Genome* **20:** 557–562. doi:10.1007/s00335-009-9218-1

Nickerson JA, Krochmalnic G, Wan KM, Penman S. 1989. Chromatin architecture and nuclear RNA. *Proc Natl Acad Sci* **86:** 177–181. doi:10.1073/pnas.86.1.177

Nizami Z, Deryusheva S, Gall JG. 2010. The Cajal body and histone locus body. *Cold Spring Harb Perspect Biol* **2:** a000653. doi:10.1101/cshperspect.a000653

Nozawa RS, Gilbert N. 2019. RNA: nuclear glue for folding the genome. *Trends Cell Biol* **29:** 201–211. doi:10.1016/j.tcb.2018.12.003

Pachnis V, Brannan CI, Tilghman SM. 1988. The structure and expression of a novel gene activated in early mouse embryogenesis. *EMBO J* **7:** 673–681. doi:10.1002/j.1460-2075.1988.tb02862.x

Pederson T. 2011a. The nucleolus. *Cold Spring Harb Perspect Biol* **3:** 1–15.

Pederson T. 2011b. The nucleus introduced. *Cold Spring Harb Perspect Biol* **3:** a000521.

Penny GD, Kay GF, Sheardown SA, Rastan S, Brockdorff N. 1996. Requirement for Xist in X chromosome inactivation. *Nature* **379:** 131–137. doi:10.1038/379131a0

Perraud M, Gioud M, Monier JC. 1979. Intranuclear structures of monkey kidney cells recognised by immunofluorescence and immuno-electron microscopy using anti-ribonucleoprotein antibodies (author's transl). *Ann Immunol (Paris)* **130C:** 635–647.

Plath K, Mlynarczyk-Evans S, Nusinow D, Panning B. 2002. Xist RNA and the mechanism of X chromosome inactivation. *Annu Rev Genet* **36:** 233–278. doi:10.1146/annurev.genet.36.042902.092433

Probst AV, Okamoto I, Casanova M, El Marjou F, Le Baccon P, Almouzni G. 2010. A strand-specific burst in transcription of pericentric satellites is required for chromocenter formation and early mouse development. *Dev Cell* **19:** 625–638. doi:10.1016/j.devcel.2010.09.002

Quinodoz S, Guttman M. 2014. Long noncoding RNAs: an emerging link between gene regulation and nuclear organization. *Trends Cell Biol* **24:** 651–663. doi:10.1016/j.tcb.2014.08.009

Quinodoz SA, Ollikainen N, Tabak B, Palla A, Marten Schmidt J, Detmar E, Lai MM, Shishkin AA, Bhat P, Takei Y, et al. 2018. Higher-order inter-chromosomal hubs shape 3D genome organization in the nucleus. *Cell* **174:** 744–757.e24. doi:10.1016/j.cell.2018.05.024

Quinodoz SA, Bhat P, Ollikainen N, Jachowicz JW, Banerjee AK, Chovanec P, Blanco MR, Chow A, Markaki Y, Plath K, et al. 2020. RNA promotes the formation of spatial compartments in the nucleus. bioRxiv doi:10.1101/2020.08.25.267435

Ramón y Cajal S. 1903. Un sencillo metodo de coloracion seletiva del reticulo protoplasmatico y sus efectos en los diversos organos nerviosos de vertebrados e invertebrados. [A simple method for selective staining of the protoplasmic reticulum and its effects on the diverse nervous system organs of vertebrates and invertebrates.] *Trab Lab Invest Biol Madrid* **2:** 129–221.

Ramón y Cajal S. 1910. El núcleo de las células piramidales del cerebro humano y de algunos mamíferos. [The nucleus of the pyramidal cells of the human brain and of certain mammals.] *Trab Lab Invest Biol* **8:** 27–62.

Rinn JL, Guttman M. 2014. RNA and dynamic nuclear organization. *Science* **345:** 1240–1241. doi:10.1126/science.1252966

Reynolds RC, Montomergy PO, Hughes B. 1964. Nucleolar "caps" produced by actinomycin D. *Cancer Res* **24:** 1269–1277.

Roden C, Gladfelter AS. 2021. RNA contributions to the form and function of biomolecular condensates. *Nat Rev Mol Cell Biol* **22:** 183–195. doi:10.1038/s41580-020-0264-6

Salditt-Georgieff M, Harpold MM, Wilson MC, Darnell JE. 1981. Large heterogeneous nuclear ribonucleic acid has three times as many 5′ caps as polyadenylic acid segments, and most caps do not enter polyribosomes. *Mol Cell Biol* **1:** 179–187.

Sharp PA. 2005. The discovery of split genes and RNA splicing. *Trends Biochem Sci* **30:** 279–281. doi:10.1016/j.tibs.2005.04.002

Shevtsov SP, Dundr M. 2011. Nucleation of nuclear bodies by RNA. *Nat Cell Biol* **13:** 167–173. doi:10.1038/ncb2157

Shin Y, Brangwynne CP. 2017. Liquid phase condensation in cell physiology and disease. *Science* **357:** eaaf4382. doi:10.1126/science.aaf4382

Shin Y, Chang YC, Lee DSW, Berry J, Sanders DW, Ronceray P, Wingreen NS, Haataja M, Brangwynne CP. 2018. Liquid nuclear condensates mechanically sense and restructure the genome. *Cell* **175:** 1481–1491.e13. doi:10.1016/j.cell.2018.10.057

Simon MD, Pinter SF, Fang R, Sarma K, Rutenberg-Schoenberg M, Bowman SK, Kesner BA, Maier VK, Kingston RE, Lee JT. 2013. High-resolution Xist binding maps reveal two-step spreading during X-chromosome inactivation. *Nature* **504:** 465–469. doi:10.1038/nature12719

Smith KP, Carter KC, Johnson CV, Lawrence JB. 1995. U2 and U1 snRNA gene loci associate with coiled bodies. *J Cell Biochem* **59:** 473–485. doi:10.1002/jcb.240590408

Spector DL, Schrier WH, Busch H. 1984. Immunoelectron microscopic localization of snRNPs. *Biol Cell* **49:** 1–10. doi:10.1111/j.1768-322X.1984.tb00215.x

Spector DL, Fu XD, Maniatis T. 1991. Associations between distinct pre-mRNA splicing components and the cell nucleus. *EMBO J* **10:** 3467–3481. doi:10.1002/j.1460-2075.1991.tb04911.x

Strehle M, Guttman M. 2020. Xist drives spatial compartmentalization of DNA and protein to orchestrate initiation and maintenance of X inactivation. *Curr Opin Cell Biol* **64:** 139–147. doi:10.1016/j.ceb.2020.04.009

Strom AR, Brangwynne CP. 2019. The liquid nucleome—phase transitions in the nucleus at a glance. *J Cell Sci* **132:** jcs235093. doi:10.1242/jcs.235093

Valentin G. 1836. *Repertorium für Anatomie und Physiologie*, Vol. 1, pp. 1–293. Verlag von Veit und Comp, Berlin.

Valentin G. 1839. *Repertorium für Anatomie und Physiologie*, Vol. 4, pp. 1–275. Verlag von Veit und Comp, Berlin.

Van Treeck B, Parker R. 2018. Emerging roles for intermolecular RNA–RNA interactions in RNP assemblies. *Cell* **174:** 791–802. doi:10.1016/j.cell.2018.07.023

Wagner R. 1835. Einige bemerkungen und fragen über das keimbläschen (vesicular germinativa). [Some remarks and questions about the germinal vesicle (vesicula germinativa).] *Müller's Arch Anat Physiol Wiss Med* **268:** 373–377.

Wang Q, Sawyer IA, Sung MH, Sturgill D, Shevtsov SP, Pegoraro G, Hakim O, Baek S, Hager GL, Dundr M. 2016. Cajal bodies are linked to genome conformation. *Nat Commun* **7:** 10966. doi:10.1038/ncomms10966

Warner JR, Soeiro R, Birnboim HC, Girard M, Darnell JE. 1966. Rapidly labeled HeLa cell nuclear RNA. *J Mol Biol* **19:** 349–361. doi:10.1016/S0022-2836(66)80009-8

Weinberg R, Penman S. 1968. Small molecular weight monodisperse nuclear RNA. *J Mol Biol* **38:** 289–304. doi:10.1016/0022-2836(68)90387-2

Weinberg R, Penman S. 1969. Metabolism of small molecular weight monodisperse nuclear RNA. *Biochim Biophys Acta* **190:** 10–29. doi:10.1016/0005-2787(69)90150-6

Wutz A. 2011. Gene silencing in X-chromosome inactivation: advances in understanding facultative heterochromatin formation. *Nat Rev Genet* **12:** 542–553. doi:10.1038/nrg3035

Wutz A, Jaenisch R. 2000. A shift from reversible to irreversible X inactivation is triggered during ES cell differentiation. *Mol Cell* **5:** 695–705. doi:10.1016/S1097-2765(00)80248-8

Wutz A, Rasmussen TP, Jaenisch R. 2002. Chromosomal silencing and localization are mediated by different domains of Xist RNA. *Nat Genet* **30:** 167–174. doi:10.1038/ng820

Zieve G, Penman S. 1976. Small RNA species of the HeLa cell: metabolism and subcellular localization. *Cell* **8:** 19–31. doi:10.1016/0092-8674(76)90181-1

Żylicz JJ, Heard E. 2020. Molecular mechanisms of facultative heterochromatin formation: an X-chromosome perspective. *Annu Rev Biochem* **89:** 255–282. doi:10.1146/annurev-biochem-062917-012655

Żylicz JJ, Bousard A, Žumer K, Dossin F, Mohammad E, da Rocha S T, Schwalb B, Syx L, Dingli F, Loew D, et al. 2019. The implication of early chromatin changes in X chromosome inactivation. *Cell* **176:** 182–197.e23. doi:10.1016/j.cell.2018.11.041

Nuclear Compartments: An Incomplete Primer to Nuclear Compartments, Bodies, and Genome Organization Relative to Nuclear Architecture

Andrew S. Belmont

Department of Cell and Developmental Biology, University of Illinois, Urbana-Champaign, Urbana, Illinois 61801, USA

Correspondence: asbel@illinois.edu

This work reviews nuclear compartments, defined broadly to include distinct nuclear structures, bodies, and chromosome domains. It first summarizes original cytological observations before comparing concepts of nuclear compartments emerging from microscopy versus genomic approaches and then introducing new multiplexed imaging approaches that promise in the future to meld both approaches. I discuss how previous models of radial distribution of chromosomes or the binary division of the genome into A and B compartments are now being refined by the recognition of more complex nuclear compartmentalization. The poorly understood question of how these nuclear compartments are established and maintained is then discussed, including through the modern perspective of phase separation, before moving on to address possible functions of nuclear compartments, using the possible role of nuclear speckles in modulating gene expression as an example. Finally, the review concludes with a discussion of future questions for this field.

The Oxford Language definition of *compartment* is "a separate section of a structure in which certain items can be kept separate from others." In addition to their diffuse localization throughout the nucleoplasm, many proteins and RNAs concentrate within distinct nuclear bodies, or within less distinct, but still spatially concentrated, condensates. The cumulative volume of all these nuclear bodies and condensates, still unknown, likely occupies a significant fraction of the total interchromosomal nuclear space, suggesting that a large portion of the genome lies within small distances from multiple nuclear compartments with distinct functional properties. Meanwhile, different types of chromatin domains position differentially but specifically near these different nuclear nonchromosomal compartments, while also compacting to form discrete chromosome structures that may themselves function as a distinct kind of nuclear compartment. A classic example would be the inactive mammalian X chromosome, which is positioned preferentially adjacent to either the nuclear or nucleolar periphery while also compacting into a condensed "Barr body" (Barr and Bertram 1951; Barr and Carr 1962; Belmont et al. 1986; Zhang et al. 2007; Rego et al. 2008). Recently, the Barr body has been proposed to function as a phase-separated condensate that would exclude specific proteins and macro-

Cite this article as *Cold Spring Harb Perspect Biol* doi: 10.1101/cshperspect.a041268

molecular complexes based on additional molecular properties beyond simply size (Cerase et al. 2019; Pandya-Jones et al. 2020).

Here, I first survey over 100 years of cytology, describing this multitude of nuclear bodies and structures. I then discuss previous imaging approaches to studying chromosome nuclear compartmentalization and compare this with modern genomic methods for describing the same. Briefly, I review recent new imaging approaches that promise to meld genomic and imaging approaches. This is followed by bringing in the modern perspective of phase separation to the discussion of nuclear compartmentalization. I then address the currently poorly understood question of how these nuclear compartments are established and maintained, before discussing how to approach the functions of nuclear compartments, using recent experiments examining the possible contribution of nuclear speckles to gene regulation as an example. Finally, I comment on challenges in addressing the role of nuclear compartments in nuclear genome organization and function before concluding with a discussion of outstanding questions in the field.

AN INCOMPLETE SURVEY OF NUCLEAR BODIES AND NUCLEAR COMPARTMENTS

The Usual Suspects

One definition of nuclear bodies describes them as "nonmembrane-bound structures that can be visualized as independent domains by transmission electron microscopy without antibody labeling" (Spector 2006). But decades before the invention of transmission electron microscopy (TEM), four major nuclear bodies/compartments—the nuclear periphery/lamina, nucleoli, nuclear speckles, and Cajal bodies—were either inferred from (nuclear envelope and lamina) or specifically stained and visualized by light microscopy (nucleoli, nuclear speckles, and Cajal bodies) (Lafarga et al. 2009).

Nucleolus

The nucleolus is the largest nuclear body and the "factory" for ribosome transcription, representing ~60% of total transcription within the nucleus, and assembly (Schöfer and Weipoltshammer 2018). In tumors, larger nucleoli correlate with higher tumor growth and ribosomal RNA (rRNA) synthesis rates (Derenzini et al. 2000). Numbers and intranuclear positioning of nucleoli vary from a centrally located, single nucleolus to multiple nucleoli located at noncentral locations in different cell types. Although appearing interiorly located in many typical light microscopy images (Fig. 1A), nucleoli typically are attached either directly to the nuclear lamina or indirectly through association with peripheral heterochromatin; frequently, these attachments are associated with invaginations of the nuclear envelope (Bourgeois et al. 1979; Bourgeois and Hubert 1988). TEM revealed the nucleolus's characteristic structure (Fig. 1E,F): a large granular component (GC) within which are embedded dense fibrillar components (DFCs) surrounding fibrillar centers (FCs) (Bernhard et al. 1952; Yasuzumi et al. 1958; Smetana and Busch 1964). With cell stress, FC/DFCs locate more toward the nucleolar exterior (Frottin et al. 2019; Latonen 2019). A consensus has emerged that transcription occurs at the interface between the DFC and FC, and possibly also within the DFCs, with initial rRNA processing occurring in the DFC followed by posttranscriptional processing in the GC (Raška et al. 2006; Boisvert et al. 2007; Schöfer and Weipoltshammer 2018). In many cell types, condensed chromatin coats some of the nucleolar periphery.

On reformation of the nucleus in G1 phase of the cell cycle, many components of the GC colocalize within a number of "prenucleolar bodies" (PNBs) (Hernandez-Verdun 2011), which sometimes localize nearer to the nuclear periphery, and also show chromatin juxtaposed to the PNB periphery (Ochs et al. 1985; Zatsepina et al. 1997). Originally, PBNs were conceptualized as being precursors to nucleolar assembly, with PNBs imagined as migrating and fusing with chromosome nucleolar organizing regions (NORs) early in G1 phase to form the nucleolus. The advent of live-cell imaging using green fluorescent protein (GFP) instead revealed GC components diffusing at different rates out of PNBs and accumulating within reforming nucleoli,

 Cite this article as *Cold Spring Harb Perspect Biol* doi: 10.1101/cshperspect.a041268

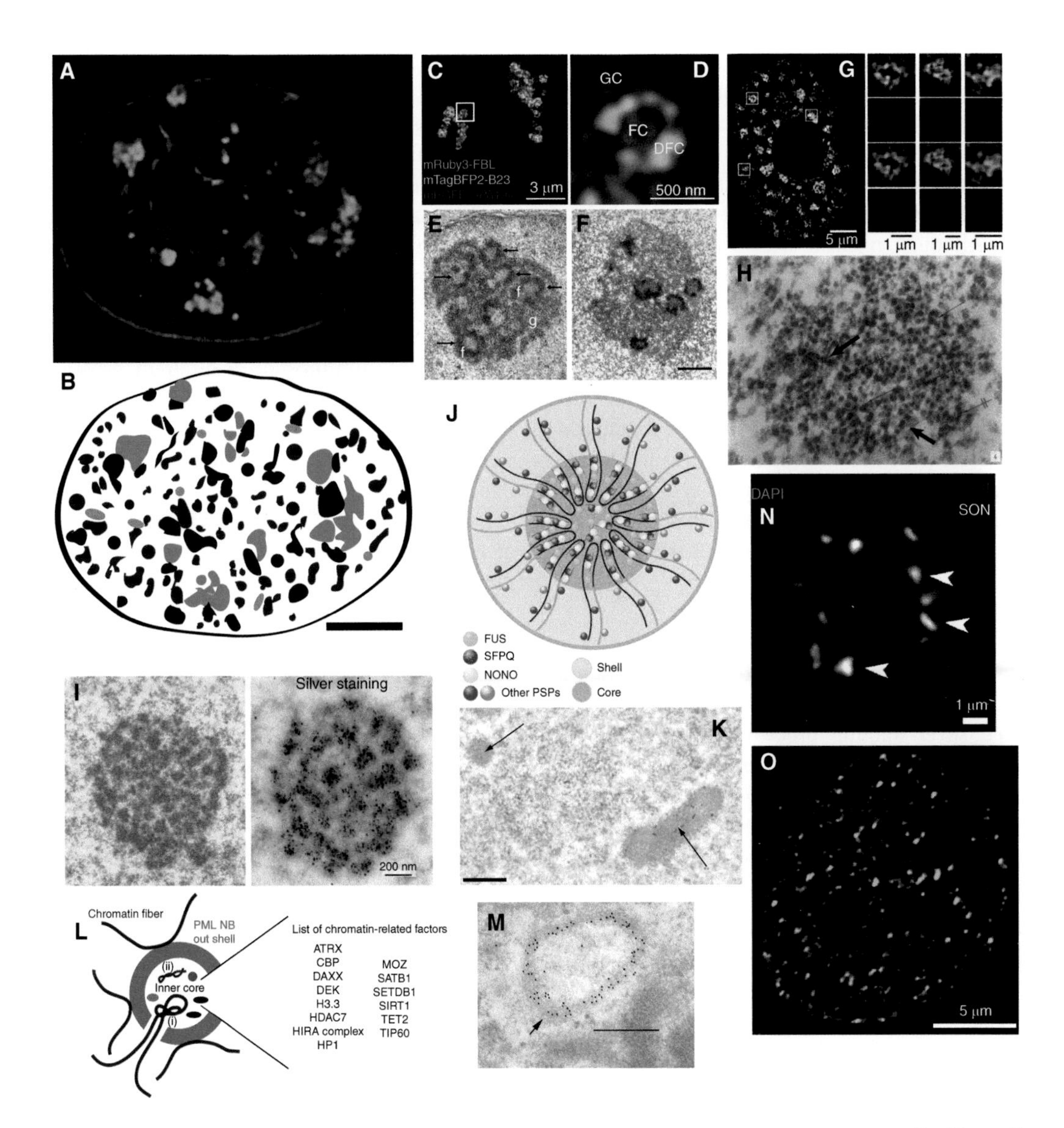

Figure 1. Varied nuclear bodies occupy significant fraction of cell nucleus. (*A*) Mouse NIH 3T3 fibroblast cell stained with DAPI (blue) to highlight DNA-dense bodies, including chromocenters, and expressing fluorescently labeled proteins to identify nuclear lamina (green, lamin B1), nucleoli (green, fibrillarin), and nuclear speckles (Magoh, red) (Zhao et al. 2020). (Image courtesy of Dr. Pankaj Chaturvedi.) (*B*) Same nucleus as *A* was used to outline nuclear lamina (black), nucleoli (green), and nuclear speckles (red) and then superimpose typical numbers for NIH 3T3 cells of additional nuclear bodies: Cajal bodies (brown), paraspeckles (orange), and promyelocytic leukemia (PML) bodies (purple). Scale bar, 5 µm (*A,B*). (*C,D*) Nucleoli stained for granular compartment (GC) (blue, B23), fibrillar center (FC) (purple, RPA194), and dense fibrillar component (DFC) (fibrillarin, FBL). Scale bars, 3 µm (*C*); 500 nm (*D*). (Panels *C* and *D* from Figure 1C in Yao et al. 2019; reprinted, with permission, from Elsevier © 2019.) (*E,F*) Electron microscopy (EM) visualization of nucleoli showing GC (g), DFC (arrows), FC (f). Conventional uranyl and lead staining were used in *E*, and silver staining for NOC proteins used in *F*. (Panels *E* and *F* from Penzo et al. 2019; reprinted under the Creative Commons Attribution License CC BY 4.0.) (*G*) 3D SIM microscopy showing nuclear speckles with partial spatial separation between anti-"SC35" (actually anti-SRRM2, see text) staining (blue) and *MALAT1* (red) and *U2* RNA (green). (*Legend continues on following page.*)

with different GC components shifting from PNBs to nucleoli with different kinetics during G1 phase (Dundr et al. 2000; Hernandez-Verdun 2011). PNBs also appear to be sites of resumed processing of precursor rRNAs present during mitosis (Carron et al. 2012).

Although nucleoli are sites of rRNA transcription, they also have been implicated in a multitude of additional processes ranging from gene silencing to assembly and regulation of other non-ribosomal nucleoproteins (RNPs), including telomerase, regulation of protein activity, including p53, and more generally in biological processes, such as development, organismal aging, and stress responses (Boulon et al. 2010; Pederson 2011; Tiku and Antebi 2018; Iarovaia et al. 2019; Latonen 2019; Weeks et al. 2019).

Nuclear Speckles

Ramon y Cajal in 1910 described "hyaline grumes" as a distinct nuclear body using a modified silver-staining procedure (Lafarga et al. 2009). These nuclear speckles were rediscovered in the mid-twentieth century using TEM as interchromatin granule clusters (IGCs) consisting of ~20–25-nm-diameter RNP particles (granules) lying between and adjacent to chromatin domains (Swift 1959; Bernhard and Granboulan 1963; Monneron and Bernhard 1969). These granules frequently align forming linear chains or apparent rods (Fig. 1H). They were rediscovered yet again as bodies enriched in splicing factors, snRNAs, polyA-RNAs, and immunostaining by an antibody to a phosphorylated epitope tied to several splicing factors, including SC35 (SRSF2) (Fig. 1A,G), but recently revealed to primarily immunostain a different protein, SRRM2 (Ilik et al. 2020). In fibroblasts, nuclear speckles are largely excluded from near the nuclear lamina and concentrated within the equatorial plane of the nuclear interior (Fig. 1A; Carter et al. 1993); exclusion from the nuclear periphery and concentration toward the nuclear interior appears common in many cell types. A recent tyramide signal amplification (TSA)-proximity-labeling proteomics study revealed both SRRM2 and SON, with similar levels of enrichment, as the most highly enriched proteins in nuclear speckles (Dopie et al. 2020). Double SRMM2 and SON RNAi knockdown caused significant dissolution of nuclear speckle staining using multiple speckle markers, suggesting that either protein is sufficient for nucle-

Figure 1. (*Continued*) (Panel *G* from Fei et al. 2017; reprinted, with permission, from Company of Biologists © 2017.) (*H*) EM visualization of interchromatin granule cluster (IGC)/nuclear speckle in rat adrenal cortex cell; granules sometimes appear in linear chains (arrows). (Panel *H* from Monneron and Bernhard 1969; reprinted, with permission, from Elsevier © 1969.) (*I*) EM visualization of large Cajal (accessory) bodies in neuronal cells showing "coiled thread" internal structure using conventional (*left*) or silver (*right*) staining. Scale bar, 200 nm. (Panel *I* from Lafarga et al. 2017; reprinted, with permission, from Taylor & Francis © 2017.) (*J*) Model of paraspeckles showing proposed scaffolding role of NEAT1_2 long noncoding RNA (lncRNA) (red and yellow lines) recruiting/binding different paraspeckle proteins over 5′ and 3′ ends (outer shell regions) versus internal sequences (core region). (Panel *J* from McCluggage and Fox 2021; reprinted, with permission, from John Wiley and Sons © 2017.) (*K*) EM visualization of round or elliptical paraspeckles (arrows). Scale bar, 0.5 µm. (Panel *K* is from Fox and Lamond 2010; image courtesy of Sylvie Souquere and Gerard Pierron, Villejuif, France.) (*L*) Model for PML body structure showing PML protein outer shell and an inner cavity containing many PML proteins, including chromatin factors and, in some cases, chromatin. (Panel *L* from Corpet et al. 2020; reprinted, with permission, from Oxford University Press © 2020.) (*M*) EM visualization of PML body showing immunogold-labeled PML protein outer shell. Scale bar, 0.5 µm. (Panel *M* from Lallemand-Breitenbach and de Thè 2018; reprinted, with permission, from Elsevier © 2018.) (*N*) Immunostained nucleus from human K562 cell showing RNA Pol II CTD Ser5p foci (red) clustered around nuclear speckles (SON, white) as well as other nuclear interior regions but depleted from periphery of nucleus counterstained for DNA with DAPI (blue). (Panel *N* from Chen et al. 2018b; reprinted under the Creative Commons License CC BY-NC-SA 4.0.) (*O*) Live-cell imaging of Mediator (MED1) condensates (green) visualized in mouse embryonic stem cell (mESC) nucleus stained with Hoechst for DNA (blue). (Panel *O* from Sabari et al. 2018; reprinted, with permission, from The American Association for the Advancement of Science © 2018.)

ar speckle formation (Ilik et al. 2020). Immunoelectron microscopy staining tied these variably named structures (nuclear speckles, polyA-islands, SC-35 islands) to the previously described IGCs, and, in turn, these IGCs to the hyaline grumes identified by Cajal through specific silver-staining protocols (Lafarga et al. 2009; Spector and Lamond 2011).

Local concentrations of splicing factors, visualized by light microscopy, are still conflated with "nuclear speckles", a name now reserved for the light microscopy analog of the IGCs visualized by electron microscopy (EM). Whereas all IGCs show accumulation of various splicing factors, not all splicing factor local accumulations represent IGCs, as clearly revealed by analysis of transgene arrays (Hochberg-Laufer et al. 2019).

Nuclear speckles form rapidly after mitosis, beginning in late telophase (Ferreira et al. 1994; Thiry 1995; Tripathi and Parnaik 2008). Analogous perhaps to PNBs, discrete bodies containing the nuclear speckle protein MFAB1 appear in the reforming telophase nucleus even while the SON and SRRM2 proteins are still cytoplasmic in mitotic interchromatin granules (MIGs) (Dopie et al. 2020). Whether these MFAB1 bodies nucleate nuclear speckles or instead accumulate nuclear speckle proteins that then diffuse and concentrate into nuclear speckles remains unknown. Even after SC35 staining (SRRM2), nuclear speckles first appear in late telophase/ early G1 nuclei; SR proteins accumulate first near NORs before then accumulating through an apparent RNA Pol II transcription-dependent mechanism in nuclear speckles (Bubulya et al. 2004).

Nuclear speckles have alternatively been proposed to serve as storage sites for factors involved in RNA Pol II transcription and RNA processing or as gene expression "hubs" for a subset of highly active genes (Hall et al. 2006; Spector and Lamond 2011). These two models are not mutually exclusive. RNA processing factors were proposed to transit from nuclear speckles to active genes and then recycle back to nuclear speckles to be "recharged" for another cycle of RNA processing; this cycling between nuclear speckle and adjacent transcription sites was proposed to be linked to cycles of posttranscriptional modifications, particularly phosphorylation and dephosphorylation, providing "recharging" of these factors for another RNA processing cycle (Misteli and Spector 1997). A unified model then was proposed in which the positioning of a subset of active genes adjacent to nuclear speckles would facilitate this recycling of RNA processing factors from nuclear speckles to transcription, thus supporting high rates of gene expression for these nuclear-speckle-adjacent genes (Hall et al. 2006).

Cajal Bodies

Similar to nuclear speckles, Cajal bodies were first identified as a distinct nuclear body in neurons by Cajal using combinations of histochemical stains or silver-staining procedures (Lafarga et al. 2009). They were rediscovered and named by electron microscopists as "coiled bodies," ~0.2–2 µm in diameter (Cioce and Lamond 2005), owing to their appearance after heavy-metal staining (Fig. 1I; Monneron and Bernhard 1969). They were then identified again through their immunostaining against the marker protein, coilin (Andrade et al. 1991; Raška et al. 1991; Lafarga et al. 2009). Cajal bodies were previously known as nucleolar accessory bodies because of their location close to nucleoli in some cell types. Several Cajal bodies (up to ~10) per nucleus are present in cells associated with high transcriptional activity and/or growth rates, including rapidly dividing embryonic cells, cancer cells, and neurons, but present in fewer numbers or even less than one Cajal body per cell in nontransformed cell types (Ogg and Lamond 2002; Cioce and Lamond 2005; Strzelecka et al. 2010; Machyna et al. 2013).

Cajal body formation is dependent both on the presence of coilin and SMN; Cajal bodies are enriched in RNPs and factors involved in RNP maturation, including spliceosomal snRNPs, scaRNPs, snoRNPs, and the telomerase RNP. Spliceosomal snRNPs are both assembled and recycled within Cajal bodies, and thus Cajal bodies have been proposed as sites of accelerated assembly and modification of multiple small RNA-containing RNPs (Machyna et al. 2013; Meier 2017). Cajal bodies associate with active gene loci, most notably the tandem U2 snRNA gene locus (Machyna et al. 2013; Sawyer et al. 2016).

Nuclear Pores and Nuclear Lamina

Both nuclear pores and the nuclear lamina have been reviewed extensively elsewhere (de Leeuw et al. 2018; Lin and Hoelz 2019; Briand and Collas 2020; Cho and Hetzer 2020; also see Miroshnikova and Wickström 2021; Pawar and Kutay 2021). TEM visualized the nuclear lamina as a fibrous layer lying between the inner nuclear envelope and the peripheral, condensed chromatin (Fawcett 1966). The intermediate filament lamin proteins—lamins A and C, both encoded by the LMNA gene, lamin B1, and lamin B2 in mammalian cells—comprise the major constituents of the nuclear lamina. Recent superresolution light microscopy reveals that these various lamins are concentrated differentially within spatially distinct meshworks, separated by several hundred nanometers, within the lamina (Shimi et al. 2015; Xie et al. 2016). Previous TEM tomography had shown local attachments of chromatin to the nuclear periphery underlying regions of high lamin B concentration within this meshwork (Belmont et al. 1993). A small fraction of lamin A is nucleoplasmic, where it may play diverse roles separate from its function at the nuclear lamina (Briand and Collas 2020). Mutations in lamin A are associated with ~15 diseases collectively termed laminopathies, including one type of premature aging disease (Hutchinson-Gilford progeria syndrome), Emery-Dreifuss and other muscle dystrophies, lipodystrophies, and peripheral neuropathies (Kang et al. 2018; Osmanagic-Myers and Foisner 2019; Briand and Collas 2020).

The nuclear lamina contains hundreds of additional proteins, many of which interact directly or indirectly with lamins (Wilson and Foisner 2010; Mehus et al. 2016; Wong et al. 2021). This includes inner nuclear membrane (INM) proteins as well as proteins concentrated near the lamina and peripheral chromatin. INM proteins notably include LEM-domain proteins (Wilson and Foisner 2010), which in mammals include LAP2 (α, β, and other isoforms), emerin, MAN1, and LEM2/NET25. The INM also includes lamin B receptor (LBR), which combines chromatin-binding and sterol reductase domains, SUN domain proteins that link to KASH domain proteins in the outer nuclear membrane that interact directly with cytoskeleton, centrosome, and organelle proteins (de Leeuw et al. 2018), as well as many other transmembrane proteins that may provide cell-specific links of specific chromosome regions to the nuclear periphery (Robson et al. 2016). These proteins together with lamins interact with chromatin as well as various transcription factors, chromatin modifying, and signaling proteins. At least in some postmitotic cell types, lamin A/C and LBR together anchor peripheral chromatin to the nuclear lamina; their knockout results in "inverted nuclei" with peripheral chromatin now located in the nuclear interior (Solovei et al. 2013).

Proposed functions of the nuclear lamina include imparting mechanical stability to the nucleus, the anchoring of specific chromatin domains (lamin-associated domains [LADs]) to the nuclear periphery (discussed later in this article), repression of gene activity of these LADs and possibly maintenance of epigenetic gene silencing, and cell signaling (Wilson and Foisner 2010; van Steensel and Belmont 2017; de Leeuw et al. 2018; Briand and Collas 2020).

The nuclear envelope is perforated by thousands of nuclear pores. These large protein complexes were first visualized by early EM as ~30 nm holes in the nuclear membrane (Callan and Tomlin 1950). Early observations on sectioned nuclei visualized ~150 nm diameter annuli and already made observations of chromatin-free channels extending from these nuclear pores well into the nuclear interior (Watson 1959). The physiological relevance of nuclear pores to transport in and out of the nucleus was realized immediately and thus they have been a focus of extensive biochemical, molecular, structural, and biophysical research ever since. Cryo-EM reconstructions now provide the highest resolution imaging of intact nuclear pore structure, which includes a central inner pore ring between outer and inner rings, all with eightfold symmetry, plus filaments extending into the cytoplasm on one side and a "fish-trap"-shaped nuclear basket extending into the nucleoplasm (Lin and Hoelz 2019).

Molecular cloning of nuclear pore proteins, or nucleoporins, revealed a subset of NUPs with intrinsically disordered, FG repeats that together create the permeability barrier of the nuclear pore and interact with exportin and importin transport factors (Beck and Hurt 2017; Lin and Hoelz 2019). Biophysical studies suggest the local concentration of these FG repeat NUPs in the nuclear pore channel induce either a liquid–liquid phase separation or hydrogel accounting for this barrier. This type of barrier, involving a high frequency of low-affinity interactions, may account for the combined high selectivity plus high rates of nuclear pore traffic (Schmidt and Görlich 2016; Frey et al. 2018; Celetti et al. 2020). The nuclear basket, comprised largely of the protein TPR, has been implicated in helping to maintain a chromatin-free channel facing the nucleoplasmic side of the nuclear pore. The loss of peripheral heterochromatin during oncogene-induced cell senescence was correlated with increased nuclear pore density and reversed by knockdown of TPR (Boumendil et al. 2019).

Beyond nuclear import and export of proteins and RNPs, nuclear pores and NUPs likely have additional functions. Both gene activation and transcriptional memory and gene silencing have been linked to direct contacts of genes with nuclear pores across a wide range of species from yeast to human (Randise-Hinchliff and Brickner 2018; Cho and Hetzer 2020). Additionally, a subset of NUPs shuttle between nuclear pores and the nuclear interior and contribute to gene regulation (Cho and Hetzer 2020). DNA break repair has been linked as well to chromosome movement to and contact with nuclear pores (Seeber and Gasser 2017). Conversely, certain nuclear compartmentalization, including association with the nuclear lamina, may restrain chromosome movement and the available molecular pathways for DNA repair (Lemaître et al. 2014; Schep et al. 2021).

Given these diverse functions, nuclear pores and/or NUPs have been implicated in a wide range of biological processes ranging from control of differentiation and cell identity, viral infection, cancer, cell senescence and organismal premature and pathological aging, and neurodegenerative diseases (Boumendil et al. 2019; Cho and Hetzer 2020).

Other Well-Known Suspects

Many additional nuclear bodies have been described through imaging of the distribution of specific proteins and/or RNAs. In some cases, TEM has recognized these as distinct domains that can be recognized subsequently without antibody staining. Like Cajal bodies, some of these bodies may be present in only a small fraction of cells and a subset of cell types—for example, cancer cells, rapidly growing normal cell types, and/or metabolically active cells such as neurons.

Paraspeckles

Paraspeckles were first recognized through immunostaining against a protein, PSPC1 (Paraspeckle Protein 1), identified in a nucleolus proteomics screen (Fox et al. 2002; Fox and Lamond 2010). Unexpectedly, PSPC1 localized in distinct nuclear bodies away from nucleoli but near or adjacent to nuclear speckles; however, PSPC1 did localize in perinucleolar "caps" in early G1 nuclei before the onset of significant transcription (Fox et al. 2002, 2005). Core paraspeckle proteins include the three members—PSF/SFPQ, NONO/P54NRB, and PSPC1—of the DBHS family and RNA-binding motif protein 14 (RMB14) (Fig. 1J; Nakagawa et al. 2018). In most mammalian cultured cells, ∼5–20 paraspeckles are present, appearing as ellipsoidal, ∼0.5–1 µm diameter bodies (Fig. 1K; Clemson et al. 2009; Fox and Lamond 2010). Paraspeckles are now known to be nucleated by the NEAT1_2 long noncoding RNA (lncRNA) (Fig. 1J; Hutchinson et al. 2007; Clemson et al. 2009) and to concentrate certain nuclear-retained mRNAs containing long 3′UTRs with A-I edited stretches of inverted repeats (Chen and Carmichael 2009). Regulated cleavage of the 3′UTR of such mRNAs, as described first for *Ctn* mRNA, can lead to rapid nuclear export and a rapid increase in protein expression (Prasanth et al. 2005). Whereas normal paraspeckles are not required for *Ctn* nuclear retention, they do regulate the nuclear compartmentalization of Ctn, may modulate its A-I editing, and have been reported to regulate the export of other structured RNAs (Anantharaman et al. 2016). In mice, only a few tissues normally contain paraspeckles, but they can be induced under special conditions, including after various types of cell stresses (McCluggage and Fox 2021). Paraspeckles have been proposed to act in gene regulation through protein and RNA sequestration, thereby possibly playing a role in modulating various stress responses, including the hypoxic response, the circadian rhythm, and cell proliferation, among other pathways, and may play a role in miRNA processing (Pisani and Baron 2019).

PML Bodies

PML (promyelocytic leukemia) protein bodies are enigmatic nuclear compartments (Fig. 1L) implicated in a wide range of cell responses and processes, with PML body number and size regulated by various cellular stresses, including viral infections, and implicated in chromatin remodeling, DNA repair, apoptosis, cell senescence, stem cell renewal, antiviral activity, and inhibition of neurodegenerative diseases (Lallemand-Breitenbach and de Thé 2018; Corpet et al. 2020). They were discovered originally through immunostaining of the PML tumor suppressor gene product, and then connected to heterogenous-type spherical objects visualized previously by EM (Fig. 1M; Lallemand-Breitenbach and de Thé 2010). PML bodies are round, ~100–1000 nm in diameter, and are present in most mammalian cells at copy numbers of ~1–30 (5–15 in typical cell lines). They are nucleated through a spherical shell assembly of PML protein subunits and appear to transiently recruit a wide range of seemingly unrelated proteins, including the histone chaperone DAXX, the HIRA H3.3-specific histone chaperone complex, SETDB1, the transcriptional coactivator CBP, and PTEN, perhaps at least in part through SUMO conjugation of PML and recruited proteins. Protein sequestration within PML bodies and enhanced protein modifications and/or degradation within PML bodies may all be related to PML function. PML bodies have been observed to interact with particular gene loci, including the *MHC* gene cluster (Shiels et al. 2001; Gialitakis et al. 2010), the TP53 locus (Sun et al. 2003), active histone genes in S-phase, and transcriptionally active genes in general (Wang et al. 2004; Corpet et al. 2020). In at least one case, this gene association was correlated with transcriptional memory, in which repeated gene induction is associated with a faster response and a higher level of gene induction (Gialitakis et al. 2010). Specialized PML bodies are associated with telomeres undergoing alternative lengthening of telomeres (ALTs) (Lallemand-Breitenbach and de Thé 2018; Corpet et al. 2020).

Perinucleolar Compartment (PNC)

These appear as ~250–4000 nm diameter caps, containing 80–180 nm electron-dense threads as visualized by EM, on the nucleolar surface in many cancer cells but not nontransformed cells or tissues (Pollock et al. 2011). Their prevalence and number per nucleus correlates with metastatic potential of primary tumor cells and inversely correlates with patient survival. PNCs contain a subset of RNA Pol III transcribed small, noncoding RNAs and RNA-binding proteins associated with RNA processing of RNA Pol II transcribed transcripts plus nucleolin, a nucleolar-localized protein involved in rRNA processing. In mice, metarrestin, a drug selected for its ability to disassemble PNCs inhibited metastatic development and extended survival in several cancer models (Frankowski et al. 2018).

Histone Locus Body (HLB)

These bodies appear similar to Cajal bodies, sharing some components, but appear specifically adjacent to histone genes and contain many additional components related to histone gene transcription and processing including coactivator NPAT and FLASH, involved in 3′ end processing of histone transcripts (Yang et al. 2009; Machyna et al. 2013).

Cleavage Bodies

Numbering one to several per nucleus, these are bodies ranging from ~0.3 to 1 µm in diameter and enriched in factors involved in 3′ cleavage

 Cite this article as *Cold Spring Harb Perspect Biol* doi: 10.1101/cshperspect.a041268

and polyadenylation of nascent transcripts; they often are found in spatial proximity or overlap with Cajal bodies, HLBs, and GEMs, which are bodies containing the SMN protein (Li et al. 2006). Note that GEMs, HLBs, and cleavage bodies frequently lie adjacent to each other or overlap; they have been proposed to represent "sub-Cajal" bodies that are precursors to Cajal bodies (Machyna et al. 2013), perhaps as multilayered condensates that then fuse into one merged condensate, analogous to what has been proposed for the different nucleolar compartments (Lafontaine et al. 2021).

Less Conventional Suspects

A number of nuclear structures do not meet the traditional definition of nuclear bodies that can be recognized by EM without specific staining but fit the definition of nuclear compartments that concentrate factors. One example would be "transcription factories," defined by immunostaining against RNA Pol II carboxy-terminal domain (CTD) amino acid repeats phosphorylated at specific sites (Ser2, Ser5) in both fixed and living cells (Fig. 1N; Xie and Pombo 2006; Uchino et al. 2021). Several hundred to thousands of punctate foci, ~80–130 nm in diameter, are present per nucleus (Jackson et al. 1998; Cook 1999; Eskiw and Fraser 2011), localizing adjacent to many active gene loci (Osborne et al. 2004, 2007; Ferrai et al. 2010). They were speculated to represent small clusters of RNA Pol II through which DNA is "reeled" during transcriptional elongation (Iborra et al. 1996; Jackson et al. 1998; Cook 1999). Their functional significance—as storage sites for initiating or elongating RNA Pol II polymerases or actual active polymerases engaged on DNA—remains unknown, but they are found clustered around nuclear speckles and other active nuclear regions (Chen and Belmont 2019), show increased numbers adjacent to more active Hsp70 genes (Kim et al. 2020), and show high contact frequencies with many highly active gene loci (Takei et al. 2021).

Condensates of subunits of the transcriptional Mediator complex are a second example of these unconventional nuclear compartments that do not form a nuclear body or structure identifiable by TEM without immunostaining. Mediator aggregates form both in vitro and in vivo (Fig. 1O; Cho et al. 2018; Guo et al. 2019) and super-enhancers in mouse embryonic stem cell (mESC) nuclei frequently associate adjacent to these condensates (Sabari et al. 2018). In vitro, Mediator subunit condensates excluded RNA Pol II with phosphorylated CTDs but were miscible with condensates formed from splicing factors (Guo et al. 2019). Thus, RNA Pol II CTD phosphorylation was proposed to transfer RNA Pol II engaged genes from Mediator condensates involved in transcriptional initiation to condensates enriched in splicing factors involved in transcriptional elongation (Guo et al. 2019).

Additional examples would include specialized chromatin domains—for example, chromocenters and polycomb bodies formed in certain species and/or cell types by coalescence of pericentric, constitutive heterochromatin or polycomb-silenced regions, respectively. Both form long-distance contacts in both *cis* and *trans* with other chromatin regions correlating with their gene silencing (Csink and Henikoff 1996; Dernburg et al. 1996; Brown et al. 1999; Bantignies et al. 2011; Pirrotta and Li 2012).

More broadly, chromatin domains in general, and entire chromosomes such as the mammalian inactive X chromosome, are increasingly being thought of as discrete compartments that may interact in *cis* and *trans* with other similar chromatin compartments as discussed below.

CHROMATIN COMPARTMENTS—INSIGHTS FROM IMAGING

Early Cytology

Key features of metazoan nuclear chromosome were first recognized roughly a century ago. Folding of chromatin into largely discrete, localized interphase chromosomes territories, as reviewed elsewhere (Cremer and Cremer 2010), was first inferred by early cytologists. Decades later, the concept of chromosome territories was resurrected by the Cremer laboratory's observation that microirradiation of local nuclear regions caused DNA damage and repair in only a small number of interphase chromosomes (Cremer and Cremer 2010). With the continued evolution of fluorescence in situ hybridization and light microscopy, chromosome territories were then directly visualized by whole chromosome "paints." Similarly, the existence in many species and cell types of a variant "Rabl" interphase chromosome configuration, in which centromeres and telomeres localize to opposite poles of the nuclear periphery, was observed by early cytologists examining chromosomes exiting and reentering mitosis (Rabl 1885). Recent Hi-C and molecular analysis has revealed that this Rabl configuration likely has appeared in multiple species through convergent evolution driven by condensin II reduced activity through mutations in condensin II subunits (Hoencamp et al. 2021).

The original definition of heterochromatin as chromosome regions that remain condensed after mitosis during interphase was made by Emil Heitz in the 1920s (Heitz 1928; Passarge 1979). Heitz defined heterochromatin as chromosome regions that remained condensed throughout most of the cell cycle, loosening only briefly before mitosis, and condensing again before the mitotic condensation of euchromatin, the chromatin that did decondense during interphase (Heitz 1928; Brown 1966; Passarge 1979). Heitz later made the association between heterochromatin and gene-poor chromosome regions. Subsequently, heterochromatin was divided into constitutive heterochromatin, which remains heterochromatic in all developmental stages and all tissues, and facultative heterochromatin, which does not. Constitutive heterochromatin is found in many species flanking centromeres, near telomeres, adjacent to the NOR, in sex chromosomes, and scattered in blocks throughout the euchromatin chromosome arms.

In the 1960s, TEM revealed the tight apposition of a chromatin layer adjacent to the nuclear lamina as well as adjacent at the nucleolar periphery and in the nuclear interior (e.g., Fawcett 1966). This early TEM also led to the textbook labeling of heterochromatin and euchromatin in which the darkly stained chromatin after heavy metal staining is considered heterochromatin, whereas the lightly stained regions with apparent finely fibrillar and granular texture is considered euchromatin.

This textbook model is almost assuredly incorrect. Most "heterochromatin" regions visualized by EM would fall under the original euchromatin definition of chromosome regions that decondense during interphase, forming "chromomeres," or granular-type structures, and "chromonema," or fiber-type structures, viewed by histological staining and light microscopy and later by EM (Zatsepina et al. 1983; Belmont et al. 1989). Notably, Sklar and Whitock used light microscopy of living cells containing polytene chromosomes to show the fixation-induced appearance of structure in the nuclear "sap" during fixation and then showed that a similar "euchromatin" TEM-staining pattern filled the chromosome-free nucleoplasm in *Drosophila* salivary gland nuclei between the clearly distinguished polytene chromosomes and also in mammalian liver nuclei after centrifuging chromatin to the opposing half of the nucleus (Skaer and Whytock 1976; Skaer and Whytock 1977). Meanwhile, multiple methods suggest that chromatin in many somatic cell types exists in large-scale domains that are likely comparable to the chromomeres visualized by early cytologists; these methods include a variety of alternative DNA staining, sample preparation, and light and electron microscopy imaging approaches (Belmont et al. 1989; Olins et al. 1989; Testillano et al. 1991; Derenzini et al. 1993; Bohrmann and Kellenberger 1994; Biggiogera et al. 1996; Bazett-Jones and Hendzel 1999; Nozaki et al. 2017; Hoffman et al. 2020; Miron et al. 2020). Thus, the textbook "heterochromatin" in heavy-metal stained EM images likely represents the bulk of genomic DNA, with smaller differences between active and inactive genomic regions than suggested by the typically used labeling of "heterochromatin" and "euchromatin." Further confirmation of the existence of stable folding of early DNA replicating, euchromatin into chromonema fibers extending over micron distances comes from combined live-cell visualization and immunogold staining of engineered chromosome regions (Kireev et al. 2008; Hu et al. 2009; Deng et al. 2016).

Emergence of Radial and Binary Models of Nuclear Compartmentalization by Imaging

Distinct, large blocks of heterochromatin, as defined by Heitz, were seen to preferentially associ-

 Cite this article as *Cold Spring Harb Perspect Biol* doi: 10.1101/cshperspect.a041268

ate with the nuclear and nucleolar periphery. This included the Barr body, the mammalian inactive X chromosome that was visualized associated with the nucleolar periphery in neurons and the nuclear periphery, and/or nucleolus in other cell types (Barr and Bertram 1951; Belmont et al. 1986; Zhang et al. 2007). The development of immunostaining, labeled nucleotides, and in situ hybridization (ISH) methods rapidly advanced our appreciation for distinct spatial nuclear compartmentalization to the entire genome.

Early fluorescence in situ hybridization (FISH) experiments exploiting repetitive DNA probes showed preferential association of telomeres and centromeres with the nuclear periphery and nucleoli as a function of cell type and cell-cycle stage and proliferation (Vourc'h et al. 1993; Solovei et al. 2004). The later development of whole chromosome FISH paints revealed the gene-poor human chromosome 18 frequently juxtaposed to the nuclear periphery and more peripheral than the more centrally located, and nucleolar-associated, gene-rich chromosome 19 (Croft et al. 1999). Analysis of all human chromosomes revealed varying distributions relative to the periphery versus nuclear center, with a general dependence both on chromosome size and on gene density, with increasing gene density/activity associated with a more central location (Cremer et al. 2003; Bolzer et al. 2005). This was observed in multiple cell types, although some cell types, especially with flatter nuclei, also showed a dependence on chromosome size, with smaller chromosomes more central and larger chromosomes more peripheral (Bolzer et al. 2005). Analysis of individual gene locations by many laboratories also revealed a correlation between gene activity and radial positioning of gene loci (Takizawa et al. 2008; Bickmore 2013); this relationship between gene activity and radial positioning more recently has been attributed to the gene density and activity of ∼Mbp chromosome regions (Kölbl et al. 2012).

These results led to a radial positioning model of genome organization with more active chromosomal loci located more interiorly and more silent chromosomal loci located more peripherally in the nucleus. Notably, this is a statistical, correlative model and there is a large variability in radial positioning for any particular chromosome locus, as well as a large variability in radial positioning among different genes with similar expression levels (Takizawa et al. 2008). Bias in radial positioning could simply be the indirect result of association of chromosome loci with different nuclear compartments, which themselves are distributed with a radial bias (Chen et al. 2018b; Misteli 2020).

Meanwhile, other imaging observations led instead to an approximately binary model of genome nuclear organization. LINE-1 repeats, enriched in gene-poor, mitotic chromosome G-bands, concentrate in a thin rim adjacent to the nuclear and nucleolar periphery, whereas Alu repeats, enriched in gene-rich, mitotic chromosome R-bands, distribute through much of the nuclear interior (Fig. 2A; Korenberg and Rykowski 1988; Bolzer et al. 2005; Solovei et al. 2009; Lu et al. 2021). Similarly, labeling of DNA replication revealed two main replication patterns—early replicating DNA distributed over most of the nuclear interior but excluded from the nuclear and nucleolar periphery versus middle to late replicating distributed similarly to the LINE-1 repeats in a rim adjacent to the nuclear and nucleolar periphery (Fig. 2B)—plus a minor, late-replicating pattern in a small number of large domains distributed in the nuclear interior (O'Keefe et al. 1992). Whereas the first two patterns were each estimated to occupy several hours of S-phase, the late replicating stage, thought to correspond to constitutive heterochromatin, was estimated to occur in a shorter time window (Dimitrova and Gilbert 1999). Later, visualization of the redistribution of LADs after mitosis revealed their concentration at both the nuclear and nucleolar periphery (Fig. 2C; Kind et al. 2013).

Both the radial genome positioning and binary models of genome organization clearly represent approximations. Indeed, beyond the binary division of early and late DNA replication patterns, a third, very late DNA replication pattern shows large, condensed, and likely heterochromatic regions in the nuclear interior. Chromocenters and the Y chromosome, representing constitutive heterochromatin, as well as the facultative heterochromatic Barr body can be found in the nuclear interior as well as the nuclear and

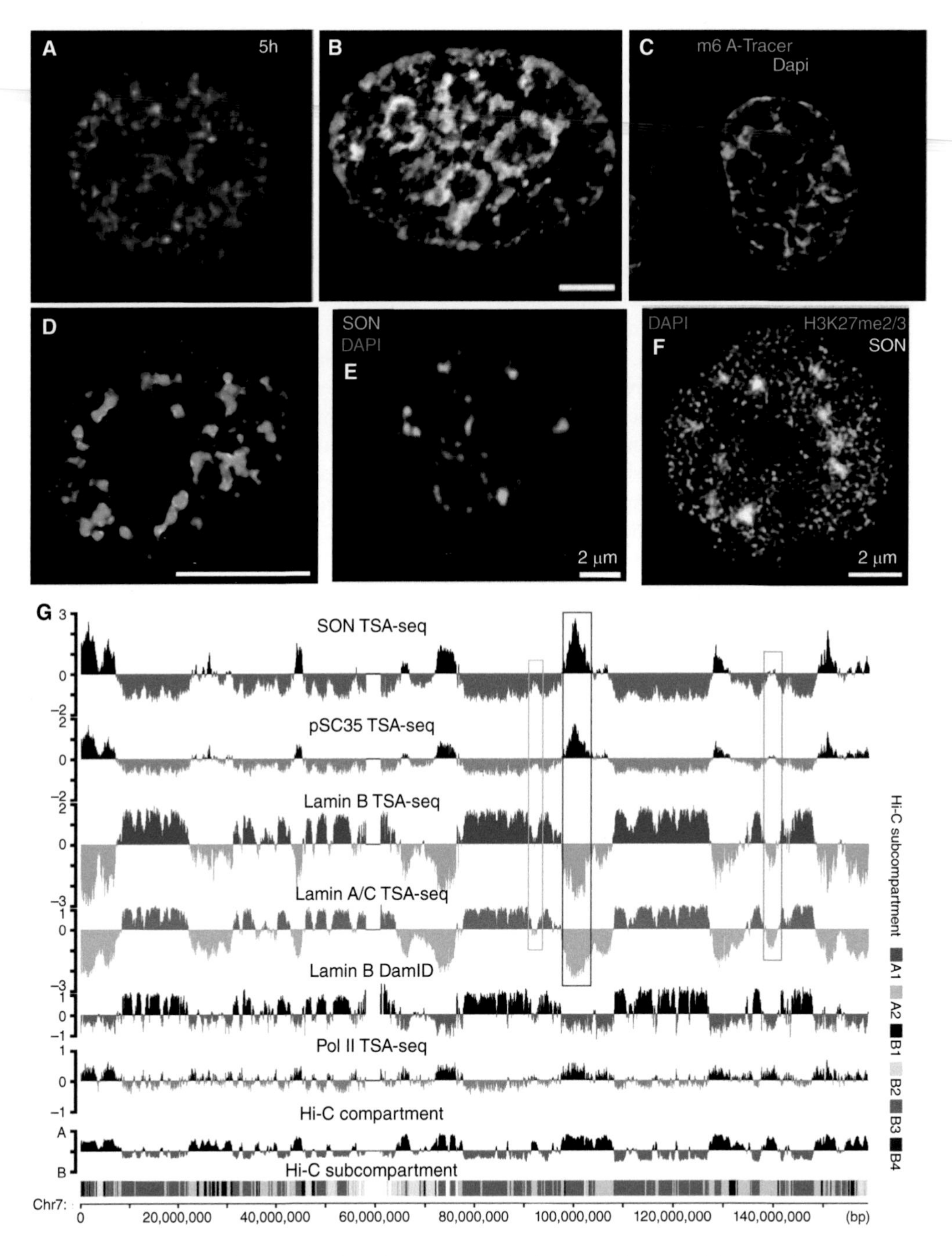

Figure 2. Both imaging and sequencing-based genomics methods suggest binary model for nuclear genome organization as a first approximation to a more complex organization. (*A–C*) An approximately binary nuclear genome organization revealed by imaging L1 repeat enriched chromatin, late-replicating DNA, and lamina-associated domains (LADs) largely at the nuclear and nucleolar peripheries with B1/Alu repeat enriched chromatin, early-replicating DNA, and intervening LADs (iLADs) in the nuclear interior. (*A*) Mouse embryonic fibroblast, early (green) versus late (red) DNA replication pulse labeling (5 h chase between early and late labeling). (Panel *A* from Wu et al. 2006; reprinted, with permission, from The Rockefeller Press © 2006.) (*Legend continues on following page.*)

Cite this article as *Cold Spring Harb Perspect Biol* doi: 10.1101/cshperspect.a041268

nucleolar periphery. More prevalent, however, are the hundreds to thousands of foci located in the nuclear interior with inactive chromatin marks, for example, H3K27me3 (Fig. 2F) or H3K9me3, interspersed with foci located in the nuclear interior with active chromatin marks (Fig. 2D) or even foci of nascent RNA (Fig. 2E).

In an early, unusually insightful study, Shopland and colleagues FISH labeled a 4.3 Mbp chromosome region, painting several, 400–1000 kbp gene-poor domains green and interspersed, 280–890 kbp gene-rich domains red (Shopland et al. 2006). As expected, a "barber-striped" pattern along a linear interphase chromosome trajectory was observed, but only in ~20% of G1 chromosomes. Instead, alternate conformations emerged with red domains and green domains self-segregated with like domains. This included clusters of red adjacent to the nuclear periphery with green domains located more interior, clusters of green surrounded by red domains, and zigzag patterns in which linear arrays of red domains abutted a parallel array of green domains. Thus, gene-poor versus gene-rich chromatin domains, segregated at this light microscopy resolution through apparent preferential interactions of like-domains with each other, independent of treatment with the transcriptional inhibitor, DRB. In contrast, a chromosome paint of a uniformly active, ~4 Mbp, gene-rich region showed a higher percentage of striped patterns and decreased percentages of alternative clustered or zigzag patterns (Shopland et al. 2006).

An analogous self-organization of active versus inactive DNA sequence was revealed in the folding of engineered, large BAC (bacterial artificial chromosome) transgene arrays 10s to 100s of Mbp consisting of multiple copies of single, ~200 kbp BACs (Sinclair et al. 2010). The plasmid backbone and a 10 kb 256mer lac operator repeat inserted within the human DNA BAC inserts from multiple BACs came together into separate heterochromatin foci enriched in the histome modification H3K9me3 and the architectural heterochromatin protein HP1—the vector backbone in one set of clusters and the lac operator repeat in another set of clusters. In contrast, active gene sequences and polycomb-repressed regions within α-globin BAC transgene arrays arrange toward the periphery of the BAC transgene array "territory" but in separate

Figure 2. (*Continued*) (*B*) Mouse C2C12 cell. L1 (green) or B repeat (red) FISH with nucleoli stained by nucleolin (purple). Scale bar, 5 µm. (Panel *B* from Lu et al. 2021; reprinted under the terms of the Creative Commons CC BY license.) (*C*) LADs, whose DNA was methylated by contact with lamin B1 fused to Dam methylase in the preceding interphase, stochastically redistribute early in the next interphase to the nuclear lamina, periphery of the nucleoli (red, NPM1), and nuclear interior (blue, DAPI), as visualized by the binding of the m6A-Tracer protein (green) that binds methylated DNA. (Panel *C* from Kind et al. 2013; reprinted, with permission, from Elsevier © 2013.) (*D–F*) Signs of a more complex nuclear genome organization emerge after staining for nuclear speckles and various marks of active versus repressive chromatin. (*D*) Hyperacetylated histones (red) are distributed nonuniformly within nuclear interior (DNA, blue), including concentrations adjacent to nuclear speckles (green). Scale bar, 10 µm. (Panel *D* from Hendzel et al. 1998; reprinted, with permission, from the American Society for Cell Biology © 1998.) (*E,F*) Local concentrations of EU pulse-labeling of nascent transcripts (red) revealing transcriptionally active chromosome regions dispersed nonuniformly through nuclear interior (DAPI, blue), including surrounding nuclear speckles (SON, green) (*E*); in contrast, repressive H3K27me3 mark (green) for facultative heterochromatin also is present in foci distributed throughout most of the nucleus from the nuclear periphery to the edge of nuclear speckles (*F*). (Panels *E* and *F* from Chen et al. 2018b; reprinted under the Creative Commons License CC BY-NC-SA 4.0.) (*G*) Genome browser view showing how largely binary division of nuclear genome organization based on lamin B1 DamID, Hi-C compartment (EV1) score, or RNA Pol II CTD Ser5p TSA-seq is further subdivided into chromosomal regions with varying distances to nuclear speckles or from nuclear lamina, as seen by varying location and heights/depths of SON/SC35 TSA-seq peaks/valleys, varying depths of lamin A/C and B TSA-seq valleys, as well as Hi-C subcompartments (note correlation of A1 subcompartment with SON/SC35 TSA-seq peaks and varied localization of B1 [enriched in H3K27me3 mark] along chromosome). (Panel *G* from Chen et al. 2018b; reprinted under the Creative Commons License CC BY-NC-SA 4.0.)

apparent clusters. Nascent transgene RNA and RNA Pol II concentrated in an outer rim surrounding the BAC transgene "territory." The segregation of H3K27me3-modified, polycomb-repressed α-globin genes from the GFP-lac repressor stained lac operator repeats was maintained even in mitotic chromosomes. This tendency of self-sorting and self-association of like sequences was particularly enhanced in undifferentiated mESCs (Sinclair et al. 2010).

This observed clustering of ectopic lac operator and vector backbone repeats, described above, may be revealing mechanisms acting on endogenous genomic repeats to shape chromosome folding. Indeed, homotypic clustering of L1 and Alu repeats driven by repeat RNA transcription was proposed recently to contribute to nuclear compartmentalization (Lu et al. 2021). More generally, computer polymer-folding simulations have suggested that this self-association of active with active and inactive with inactive chromatin regions, together with their affinity for different nuclear compartments, is likely a major driver of nuclear genome folding, as reviewed recently (Misteli 2020).

A different question is how compartmentalization of DNA into large-scale chromatin domains might be stable during interphase progression and particularly during DNA replication. Similar to the transcription "factory" model, discussed previously, combined live-cell imaging plus light and electron microscopy imaging of pulse-chased replicated DNA suggested large-scale chromatin domains remain condensed during DNA replication, with DNA instead pulled out of these chromatin domains and into adjacent PCNA-enriched "replication foci" and then "snapping back" into the chromatin domain after replication (Deng et al. 2016).

GENOMIC ANALYSIS OF GENOME NUCLEAR ORGANIZATION

Binary Compartment Model Based on Genome-Wide Mapping of Genomes

More recently developed genome-wide mapping methods are complementary to the imaging methods that generated these initial conceptual frameworks for genome organization. The first genomic method to suggest an approximately binary division of the genome was the measurement of DNA replication timing (Schübeler et al. 2002; White et al. 2004; Hiratani et al. 2008). A two-fraction assay using early versus late pulse-labeling suggested a largely binary division of the genome into early versus late replicating domains with transition zones, possibly corresponding to single, elongating replication forks, connecting early with late regions. In *Drosophila*, constitutive heterochromatin regions were, as in imaging studies, detected as replicating even later (Schübeler et al. 2002).

The subsequent development of molecular proximity mapping methods revealed a striking division of the genome into discrete domains. One early genome-wide proximity-mapping method was DamID, which relies on methylation of DNA regions that interact with a protein of interest (van Steensel and Henikoff 2000; van Steensel et al. 2001). DamID showed differential molecular interaction of genome regions with nuclear lamina proteins such as lamin B1 (Fig. 2G) or emerin (Guelen et al. 2008; Peric-Hupkes et al. 2010). Thus, DamID mapping provided an approximately binary division of the genome into LADs and intervening domains (iLADs), with small genomic regions connecting the LADs and iLADs. Constitutive (cLADs) are LADs in most or all cell lines tested, while facultative LADs (fLADs) convert to iLADs in some cell lines or during differentiation (Peric-Hupkes et al. 2010; Meuleman et al. 2013; Robson et al. 2016).

The development of chromosome capture conformation (3C) methods, and ultimately Hi-C, led to the third independent suggestion of a similar binary genome division (Lieberman-Aiden et al. 2009). Hi-C interaction maps across chromosomes showed higher than expected cross-linking between particular genome regions that were classified as "A" or "B" "compartments": Both A and B compartments across a chromosome interacted at a higher frequency with other like compartments than expected from the average decrease in interaction frequency observed as a function of genomic distance.

This division into A and B compartments emerges as principal component 1 in a principal component analysis (PCA) of Hi-C interaction frequencies, measuring the division of the genome into A (positive eigenvector 1, EV1) and B (negative EV1) compartments (Fig. 2G, "compartment score").

This genome binary division into LADs versus iLADs closely parallels B and A compartments (Fig. 2G; van Steensel and Belmont 2017), and, with the addition of transition zones, late and early DNA replicating domains (Ryba et al. 2010). LAD/B/late genomic regions have lower gene density, lower transcriptional activity, and epigenetic marks associated with gene silencing, whereas iLADs/A/early regions have higher gene density, higher transcriptional activity, and epigenetic marks associated with active chromatin (de Wit and van Steensel 2009; Peric-Hupkes and van Steensel 2010; van Steensel and Belmont 2017; Zhao et al. 2017).

Beyond A and B Compartments

Improved Hi-C methods combined with much greater read depth further divided the original Hi-C A and B compartments into A1 and A2 active and B1, B2, and B3 major subcompartments in the G12878 lymphoblastoid cell line (Fig. 2G; Rao et al. 2014). Whereas the B1 subcompartment was enriched in epigenetic markers related to polycomb silencing, B2 and B3 subcompartments overlapped extensively with LADs. These subcompartments are assumed to correspond to spatially distinct active (A1 and A2) and repressive (B1, B2, and B3) nuclear compartmentalization. The B2 subcompartment regions are enriched on smaller chromosomes and acrocentric chromosomes containing NORs, associating the B2 subcompartment with a more nucleolar localization.

Sequencing of residual DNA associated with biochemically purified nucleoli better identify nucleolar-associated domains (NADs) (Németh et al. 2010; van Koningsbruggen et al. 2010; Bizhanova and Kaufman 2021). NADs show extensive overlap with LADs, with B2 Hi-C subcompartment genomic regions overrepresented relative to B3 regions. Overlap between LADs and NADs was expected from live-cell imaging experiments showing a stochastic shuffling of LADs between association with the nuclear lamina versus nucleolar periphery after mitosis (Kind et al. 2013). NAD-seq revealed additional H3K27me3-enriched "type 2" NADs, distinct from LADs, as confirmed by FISH (Vertii et al. 2019).

Several newer, non-3C-related genomic methods, have suggested further assignment of Hi-C-defined compartments to specific nuclear bodies, such as nucleoli and nuclear speckles.

SPRITE (split-pool recognition of interactions by tag extension), using a series of dilutions and pooling of complexes, adds unique sequencing barcodes to multiple DNA and RNA fragments associated with the same individual complexes produced by sonicating chemically cross-linked cells (Quinodoz et al. 2018). In this approach, high-throughput DNA sequencing of sequencing libraries produced from large numbers of complexes can produce two-way "interaction frequencies" analogous to Hi-C, but corresponding instead to DNA/RNA fragments colocalizing in the same complex. But SPRITE also yields frequencies of simultaneous "interactions" from larger numbers of fragments all colocalizing in the same complex. In mESCs, SPRITE identified "active" and "repressive" ~1 Mbp regions defined through their interaction frequencies with a small number of either highly active or inactive "hubs"—chromosome regions with unusually high numbers of interchromosomal contacts in sonicated complexes. Subsequent FISH validation showed that the hub contact frequency of SPRITE-defined active chromosome regions correlated with the frequency of colocalization of these regions with nuclear speckles; instead, the hub contact frequency SPRITE-defined repressive chromosome regions correlated with the colocalization of these regions with nucleoli (Quinodoz et al. 2018).

More recently, RD-SPRITE, an improved version of SPRITE with greatly improved RNA detection, has provided more direct mapping of DNA sequences interacting with specific RNAs enriched in different nuclear bodies and also mapped imprinted chromosomal domains showing domain-wide interactions with specific lncRNAs associated with their silencing (Quinodoz et al. 2020).

A different mapping method, MARGI, ligates nearby RNA and DNA fragments in cross-linked nuclei to provide a sequencing readout of RNA colocalizing near DNA sequences (Chen et al. 2018a). Cross-linking of ncRNAs (snRNAs and MALAT1) enriched in nuclear speckles revealed large domains corresponding approximately to the entire Hi-C A compartment. However, given that MARGI reads out molecular-scale interactions, this colocalization may simply reflect the known local enrichment of snRNAs and Malat1 at actively transcribing gene bodies, rather than proximity of the chromosome region to nuclear speckles (Engreitz et al. 2014; Chen and Belmont 2019).

Genomic regions interacting with nuclear speckles were measured more directly using a new genomics method, TSA-seq (Chen et al. 2018b). TSA uses indirect immunofluorescence using a secondary antibody coupled to horseradish peroxidase (HRP). HRP catalyzes the generation of tyramide (phenol)–biotin free radicals. The sustained generation of tyramide–biotin free radicals combined with their diffusion from the site of generation, and an approximately constant probability over time and space in their quenching, creates an exponentially decreasing free-radical concentration gradient that can be used to measure the distance of a DNA region from the staining target.

Nuclear speckle and lamin TSA-seq showed that in K562 erythroleukemia cells the previously identified A1 Hi-C subcompartment corresponded to the ~20% of the genome closest to nuclear speckles and far from the nuclear lamina, the A2 Hi-C subcompartment instead localized at intermediate distance to nuclear speckles, whereas the inactive B2 and B3 Hi-C subcompartments were distant from nuclear speckles and close to the nuclear lamina (Fig. 2G; Chen et al. 2018b).

TSA-seq suggested several additional concepts deviating further from both the binary and radial models of genome organization. Speckle-associated domains (SPADs), corresponding to the ~5% of the genome closest to nuclear speckles, were near deterministically located adjacent to nuclear speckles (~95% or more of alleles). These SPADs plus flanking LADs, Mbps distant to these SPADs, defined anchor points for predicting several Mbp chromosome trajectories extending from nuclear lamina to nuclear speckles and back. More generally, gene-dense expression "hot zones" are located at the apexes of predicted chromosome trajectories projecting from the nuclear lamina variable distances into the nuclear interior. Inferred distances to either the nuclear lamina or nuclear speckles were proposed to represent better metrics for describing genome organization than radial distance to the nuclear center (Chen et al. 2018b).

Further division of the genome into multiple states with varying nuclear spatial localization was achieved for K562 cells by combining TSA-seq, DamID, and Hi-C data using a hidden Markov random field model, SPIN (spatial position inference of the nuclear genome), to identify multiple states, each predicted to share a distinctive nuclear localization (Wang et al. 2021). SPIN further divided LADs into lamina-associated, two near-lamina, and lamina-like states and divided iLADs into speckle-associated, three interior active, and two interior repressive states, each with distinctive histone modifications, DNA replication timing, and gene expression levels.

APEX (enhanced ascorbic peroxidase [APX]), a related proximity labeling method implemented typically in live cells (Rhee et al. 2013), recently has been applied to mapping genome organization relative to PML bodies (Kurihara et al. 2020). APEX uses expression of a fusion protein between an engineered, monomeric ascorbate peroxidase and a cellular protein localizing in the target cellular compartment; labeling of chromatin by the phenol–biotin free radical is subsequently detected by ChIP-seq. Although a region of the Y chromosome was mapped consistently near PML bodies, no other chromosomal region was detectable as preferentially lying near PML bodies. The APEX tagged PML protein mapped locally to a large number of promoter and enhancer regions, at well below the expected diffusion radius of the phenol–biotin free radical. This labeling may be an artifact caused by APEX labeling of DNA-specific regulatory proteins inside of PML bodies followed by their rapid diffusion out of the PML bodies and their subsequent binding to distant regulatory DNA sequences (Kurihara et al. 2020).

Finally, the ligation-independent genome architecture mapping (GAM) genomic method notably detects a significantly higher number of long-distance and *trans* chromosomal interactions as compared with Hi-C (Beagrie et al. 2017, 2020). GAM involves sequencing DNA from thin sections cut randomly from many cell nuclei; DNA sequences colocalizing within nuclear space are more likely to be present within randomly sampled nuclear cross-sections. Long-distance, "multi-way" interactions involving simultaneous colocalization of different DNA sequences detected by GAM were suggested to be the consequence of chromosomal interactions with nuclear bodies such as nuclear speckles and/or "transcription factories" or other condensates associated with active gene expression (Beagrie et al. 2017).

BACK TO THE FUTURE—MELDING GENOMICS WITH MULTIPLEXED NUCLEIC ACID AND PROTEIN IMAGING

All sequence-based, genomic mapping approaches, and especially those performed on ensembles of cells, face the challenge of relating their results and predictions to actual nuclear and chromosomal structures. New, multiplexed FISH approaches using bar-coded oligonucleotide probes are promising to bring genomics to single-cell imaging (Cardozo Gizzi et al. 2019; Mateo et al. 2019; Nguyen et al. 2020). Two new studies point to a future in which near genome-wide coverage of chromosome loci, and eventually chromosome trajectories, will be visualized relative to immunostained nuclear structures and bodies. Both mapped ~1000–3000 chromosome loci, plus large numbers of nascent pre-RNAs, relative to the nuclear lamina, nucleoli, and nuclear speckles in IMR90 human fibroblasts (Su et al. 2020) and to the same nuclear compartment markers plus additional chromatin marks in mESCs (Takei et al. 2021), using multiple rounds of FISH hybridization, combinatorial labeling, and decoding schemes.

Chromosome loci showed variable contact frequencies between nuclear lamina, nucleoli, and nuclear speckles that correlated with previous genomic data. Both studies showed increased and decreased gene expression levels for loci associated with nuclear speckles and the lamina, respectively. Takei et al. identified chromosomal loci with unusually high contact frequency with specific nuclear structures, including specific chromatin marks, RNA Pol II foci, heterochromatin defined by DNA staining, or nuclear bodies (lamina, nucleoli, nuclear speckles) (Fig. 3A). Su et al. described reduced percentages of transcriptionally active gene alleles with detectable nascent transcripts for alleles associated with the nuclear lamina, whereas slightly higher active allele frequencies were observed for gene alleles associated with nuclear speckles. Takei et al. inferred a prepositioning of highly active genes near nuclear speckles and/or regions of high active chromatin marks, regardless of on/off status of alleles in particular cells.

THE ELEPHANT IN THE ROOM: NUCLEAR COMPARTMENTS, PHASE SEPARATION, AND CONDENSATES

Liquid–liquid phase separation (LLPS) has gathered momentum as a major paradigm change in cell biology today (Hyman et al. 2014; Shin and Brangwynne 2017). LLPS refers to the demixing of liquids into spatially separate phases, analogous to the separation of oil and water. Many nuclear bodies show at least a subset of LLPS characteristics; these include fusion, fission, dripping, rounding, viscoelastic behavior, and rapid diffusion of proteins within bodies and exchange of these proteins out of the bodies, but with reflection at the nucleoplasmic/body boundary. Moreover, many nuclear body proteins, either individually or in a mixture with other proteins or RNAs, form liquid phase-separated droplets in vitro. An excellent example in which LLPS properties play a likely role in establishing nuclear body structure and function is the nucleolus, as reviewed recently (Lafontaine et al. 2021).

As a result, LLPS is commonly invoked as a universal mechanism for the organizing principle of nearly all nuclear bodies. Yet, the "ele-

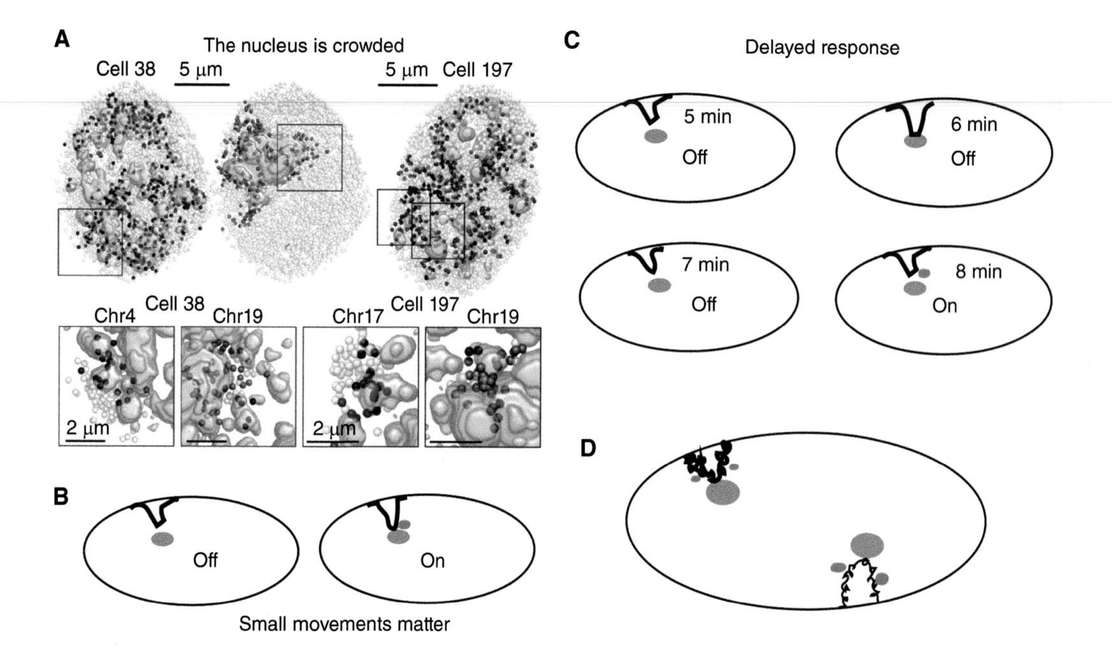

Figure 3. Challenges for the future. (*A*) The nucleus is crowded with many internal nuclear bodies and structures as revealed by multiplexed imaging of DNA loci relative to nuclear speckles (pink), nucleoli (blue), or H3K9me3-enriched heterochromatin, including chromocenters, in mouse embryonic stem cells (mESCs). A number of specific chromosome loci appear as "fixed" relative to these nuclear structures, meaning they show statistically unusually high frequency of colocalization for particular structures—pink dots for nuclear speckle-associated, green dots for H3K9me3-associated, and blue dots for nucleolar-associated—as compared with chromosome loci that do not show elevated association frequencies with any of these structures (gray dots). (Panel *A* from Takei et al. 2021; reprinted, with permission, from Nature Publishing © 2021.) (*B*) Small movements matter: relative movements closer or further to specific nuclear bodies/structures, even of several hundred nm, can be highly correlated with changes in gene expression. (*C*) Delayed response: greatly complicating analysis of the possible causal relationship between nuclear localization and changes in DNA functional output is that the change in output—i.e., transcription (nascent transcripts represented as green dot)—may show a delayed response. Here, a gene locus starts at 5 min after gene induction a small distance from a nuclear speckle, touches the speckle at 6 min, and in a delayed response, turns on to higher levels at 8 min when it is again away from the nuclear speckle. Examination in fixed cells would instead lead to the inference of a high level of transcription even without nuclear speckle contact. (*D*) An integrated view: Traditionally, our field has focused solely on individual chromosome loci and their nuclear position. However, movements of chromosome loci toward or away from specific nuclear bodies/structures could be associated with coordinated changes in nuclear localization, large-scale chromatin compaction, and even biochemical changes to flanking chromatin regions, possibly Mbps in size. Here, an active chromosome movement of a speckle-associated chromosomal locus toward a nuclear speckle, followed by its attachment to the speckle, combined with the anchoring of a neighboring LAD to the nuclear lamina could differentially alter the chromatin compaction of the intervening several Mbp of DNA between these two chromosome loci, possibly even leading to differential gene expression (nascent transcripts, green dots) as a function of differential chromosomal stretching.

phant in the room" that many are reluctant to discuss is that there are few instances where "classic" liquid-like behavior, involving the phase-separation of pure liquid states, has been shown definitively in vivo (McSwiggen et al. 2019). If one defines condensates as local concentrations of subunits in the absence of a separating lipid membrane enclosure, then condensates would include many additional types of matter besides liquids such as liquid crystals, gels, and solids; condensates would also include the concentrated binding of proteins to localized

RNA aggregates or condensed chromatin domains (Boeynaems et al. 2018; Lyon et al. 2021). Moreover, some characteristics of LLPS may be shared by these other types of condensates, although a given cellular body might transition between liquid and gel or gel and solid. Moreover, these transitions might occur either uniformly or heterogeneously within the body (Boeynaems et al. 2018; Lyon et al. 2021).

Indeed, by two criteria for a true liquid-like condensate—the absence of internal structure and the surface tension induced rounding of the condensate—several nuclear bodies clearly deviate. Cajal bodies show a distinctive coiled structure (Fig. 1I); paraspeckles are now being modeled as block copolymer micelles with a characteristic cylindrical shape and an ordered interior organized by the nucleating and required lncRNA NEAT1_2 (Fig. 1J; Yamazaki et al. 2021); PML bodies show an outer core shell formed by the PML protein (Fig. 1M; Yamazaki et al. 2021); IGCs/nuclear speckles show nonround shapes formed by clusters of 20–25 nm RNP granules, which often align linearly as chains of granules (Fig. 1H; Bernhard and Granboulan 1963). Moreover, the actual functional nuclear speckle likely includes the interchromatin granules together with additional proteins and RNAs, including the lncRNA MALAT1, which fill space between these granules and/or surround the outside of the granule cluster, as visualized by superresolution light microscopy (Fig. 1G; Fei et al. 2017).

These features of many nuclear bodies deviate from the typically round foci formed by in vivo overexpression of many nuclear body proteins that frequently contain intrinsically disordered regions (IDRs) favoring LLPS. Indeed, some nuclear body proteins appear under normal physiological conditions to be in a nonliquid condensate state, which may be poised to undergo LLPS during cell-cycle progression or after physiological perturbation. For example, MFAP1 and PRPF38A reenter nuclei and form round, droplet-like bodies ~10–20 min before the entry of the nuclear speckle putative scaffolding proteins SON and SRRM2 and their colocalization with nuclear speckles (Dopie et al. 2020; Ilik et al. 2020). These MFAP1/PRPF38A bodies either represent nucleation sites for the reforming nuclear speckles or instead are local condensates of proteins that later will concentrate in nuclear speckles forming elsewhere. RNA Pol II inhibition during interphase causes the same MFAP1 and PRPF38A proteins to exit nuclear speckles and form round bodies adjacent to and away from nuclear speckles (Dopie et al. 2020). Meanwhile, RNA Pol II inhibition also causes nuclear speckle rounding, as expected for a transition to a more liquid-like condensate.

For these reasons, more complex models beyond LLPS likely will be required to understand the full range of behaviors shown by many nuclear bodies.

This also holds true for chromatin domains, which recently have been suggested to form via LLPS (Maeshima et al. 2020). In vitro, nucleosome oligomers form aggregates as a function of polycation concentration, and similar drop-like structures were observed after injection into nuclei of live cells (Gibson et al. 2019). This aggregation of nucleosomes can be enhanced by binding of HP1 through an HP1-induced change in the histone octamer structure (Sanulli et al. 2019).

However, numerous mechanical measurements and live-cell imaging experiments instead have shown elastic behavior of whole chromosomes, nuclear chromatin, and interphase chromosome regions. The first suggestion of chromosome regions behaving with liquid-like properties was for HP1-enriched *Drosophila* and mouse chromocenters, based on the demonstrated formation of HP1 droplets in vitro and observation of fusion and fission of groups of pericentric heterochromatin regions in live cells (Larson et al. 2017; Strom et al. 2017). Recent studies, however, now show that the HP1/DNA "droplets" behave more like a cross-linked gel against short-duration mechanical impulses, with mobile HP1 acting as a cross-linker of immobile DNA fragments (Keenen et al. 2021). In vivo, HP1 is not essential to the maintenance of chromocenter compaction and the concentration of HP1 in the nucleoplasm and chromocenter shows concentration-dependent behavior different than expected for LLPS (Erdel et al. 2020). Meanwhile, in vivo experiments reveal that whereas proteins mix rapidly within chromocenters, DNA from individual pericentric heterochromatin regions do not, suggesting a model in which solid/elastic chromosome structures serve as "scaffolds" on which LLPS might occur for chromatin-associated proteins (Strickfaden et al. 2020). Meanwhile, older experiments using photoactivation fluorescently labeled DNA had clearly established very non-liquid-like and stable interphase positioning of chromatin through cell-cycle progression through much of interphase (Walter et al. 2003).

More complex models beyond LLPS—likely combining polyelectrolyte electrostatic interactions, fiber–fiber interactions, protein and ncRNA cross-linking, and active enzymatic processes such as cohesin and condensin-mediated loop extrusion—will be needed to understand how elastic chromosomes can coexist with clearly visualized compartmentalization of different heterochromatin and euchromatic chromosome regions.

ESTABLISHMENT, MAINTENANCE, AND CHANGES IN NUCLEAR COMPARTMENTALIZATION

Although an extensive literature exists describing the reformation after mitosis of major nuclear bodies and compartments, we still have a surprisingly poor grasp of the larger picture of how these different compartments are established relative to each other and the forces and mechanisms that drive these rearrangements. In some cell types, nucleoli have been described as fusing and moving to the nuclear center during cell-cycle progression (Amenta 1961; González and Nardone 1968; Savino et al. 2001). More broadly, what forces lead to a single, centrally located nucleolus in one cell type (e.g., certain neurons [Manuelidis 1984]) versus multiple, scattered nucleoli in other cell types? What controls the variable sizes and numbers of nuclear speckles in different cell types, their restriction to the nuclear equatorial plane in fibroblasts, and their exclusion from the nuclear periphery and concentration in the nuclear interior in multiple cell types? What forces and mechanisms effect changes in nuclear compartments during cell-cycle progression or cell differentiation?

Both Cajal bodies and nuclear speckles are mobile, in some cases moving up to several microns through the nucleus at velocities up to ~1 µm/min (Platani et al. 2000; Kim et al. 2019). In response to several cell stresses (heat shock, heavy metal, transcriptional inhibition) or during entry into prophase, nuclear speckles show a "follow-the-leader" movement with smaller nuclear speckles moving in DNA-depleted channels to fuse with larger speckles; new nuclear speckles then nucleate and move along a similar path to fuse with the same nuclear speckle (Kim et al. 2019).

In the case of chromosomal compartmentalization, early pulse-chase experiments showed that the differential spatial localization of early versus late replicating chromosomal regions to the nuclear interior versus periphery, respectively, is established during the first few hours of G1 phase (Ferreira et al. 1997). These and related experiments (Walter et al. 2003) suggested the idea that chromosome position becomes relatively fixed early in G1 phase and that cells need to passage through mitosis to rearrange their chromosome compartmentalization.

More recently, Hi-C of cells synchronized in their progression from mitosis into G1 showed establishment of A/B compartments within the first few hours of G1, agreeing with earlier microscopy work (Abramo et al. 2019; Zhang et al. 2019). However, changes in Hi-C A/B compartmentalization occur both during differentiation (Miura et al. 2019) or after physiological stimulation within a single cell cycle (Amat et al. 2019; Zhou et al. 2019).

Indeed, more than 70 years ago, the specific movement of the inactive X chromosome away from the edge of the nucleolus into the nuclear interior and then back again was observed during the several week period of wound healing of crushed motor neurons, permanently arrested in the cell cycle in G1 (Barr and Bertram 1951). In addition, several examples of stereotyped, long-range chromosome movements without an intervening mitosis or change in differentiation have been reported (Chuang and Belmont 2007).

Recent development of a new DamID method, pA-DamID, with greatly increased time resolution has suggested progressive changes in the magnitude of subsets of lamin pA-DamID LAD signals during cell-cycle progression, suggesting that a set of LADs at the ends of chromosome arms decrease their interactions with the nuclear lamina after early G1, whereas a different set of LADs toward the middle of the chromosomes increase their interactions (van Schaik et al. 2020). Similar results were independently reported using lamin cTSA-seq, a modified form of TSA-seq in which chromatin pulldown replaces DNA pulldown (Tran et al. 2021).

As an early model system to study chromosome movements tied to changes in gene expression, inducible tethering of the acidic activation domain of VP16 resulted in the directional movement of a peripherally located plasmid transgene over distances up to several microns and at average velocities of ∼0.4 µm/min (Chuang et al. 2006). These movements toward the nuclear interior were directly or indirectly related to actin and nuclear myosin 1c. Similar, actin-dependent movements were reported after induction of a transgene array of U2 genes toward Cajal bodies (Dundr et al. 2007).

More recently, a directional, linear movement, in some cases up to 4 µm and at velocities of 1–2 µm/min, was visualized for a large plasmid HSPA1A transgene array from the nuclear periphery to nuclear speckles after heat shock (Khanna et al. 2014). Similar linear, directional movements were seen for BAC Hsp70 (HSPA1A/HSPA1B/HSPA1L) transgenes toward nuclear speckles after heat shock (Khanna et al. 2014; Kim et al. 2020).

The most detailed molecular dissection of the possible mechanism of interphase chromosome movements toward specific nuclear compartments has been done in budding yeast in the context of the *INO1* gene movement toward nuclear pores in response to transcriptional activation (Wang et al. 2020). Based on these studies, a model has been proposed for long-range, processive chromosome movement by a one-headed myosin recruited by the Put3 transcription factor binding to a DNA "zipcode." This model suggests that binding of the one-headed myosin to F-actin is stabilized by a second, chaperone-dependent interaction of INO80, recruited to nearby H2A.Z-modified nucleosomes, with F-actin. The motor activity of the one-headed myosin, at least an in vitro actin filament gliding assay, was also dependent on the presence of the Hsp90 chaperone and a cochaperone (Wang et al. 2020).

FUNCTION OF NUCLEAR COMPARTMENTALIZATION

The extensive compartmentalization of the nucleus is evident, but what function does it serve? Two traditional explanations of nuclear compartmentalization have been proposed. The first is the idea of compartments increasing local concentrations of key components for enzymatic reactions and/or assemblies of macromolecular complexes. The second is the idea of compartments acting as local sites for storage or sequestration of components whose activities are normally elsewhere. With the new paradigm of condensates and phase separation, these older models have been further refined to include the idea that condensates may act as selective filters to pass through and/or concentrate certain classes of proteins or RNAs while excluding others.

However, experimentally probing the function of nuclear bodies and compartmentalization has been challenging. As just one example, many of the same proteins responsible for rRNA transcription, processing, and assembly into ribosomes are implicated in phase separation and formation of nucleolar structure. Thus, experiments manipulating the levels of these same proteins to perturb nucleolar assembly and then assaying effects on ribosome synthesis would be difficult to interpret. Indeed, although principles of phase separation derived from equilibrium thermodynamics are instructive, normal nucleolar assembly and structure is driven by the nonequilibrium transcription, RNA processing, and assembly of rRNA into ribosomes (Lafontaine et al. 2021), and it is challenging to uncouple the functional output of nucleoli from their organization. Moreover, release of the many additional proteins and RNAs that concentrate in nucleoli by induced nucleolar disassembly would lead to pleiotropic effects potentially causing significant indirect effects that again would complicate interpretation of experimental results.

Similarly, recent simultaneous knockdown of SON and SRRM2 was shown to "dissolve" nuclear speckles, raising the concentrations of their components in the nucleoplasm (Ilik et al. 2020). But both SON and SRRM2 bind near promoters and gene bodies at most active genes, including away from nuclear speckles. Indeed, most nuclear speckle proteins are present and likely playing important functional roles at or near genes away from these speckles. Thus, manipulation of nuclear speckles followed by assaying effects on gene expression also are complicated by the likely indirect effects induced by changing the concentrations of speckle components in the nucleoplasm.

One promising new approach has been to induce nucleation of condensates locally using optogentic tools (Kichuk et al. 2021). At the very least, this approach allows exploration and testing of proposed models for the effects of condensates on biochemical and molecular processes. To the extent that these induced condensates approximate the true behaviors of the more complex, actual nuclear bodies, these optogenetic approaches should provide a productive means to probe the functions of nuclear bodies (Zhu et al. 2019; Wei et al. 2020; Zhang et al. 2020a; Ma et al. 2021).

Nuclear Speckles as a Model to Probe Compartment Structure and Function Relationship

Numerous reviews have focused on the function of various nuclear compartments and structures. Here, we focus on recent experiments in nuclear speckles that have used temporal ordering of events, in addition to perturbation of nuclear bodies, as a tool for probing nuclear body function.

TSA-seq mapping reveals several features of speckles: First, ~5% of the genome is near deterministically positioned within <500 nm of nuclear speckles; second, inferred average distances to nuclear speckles are largely conserved between different cell types, with ~90% of the genome not statistically different in position in pairwise cell line comparisons; and third, small-to-moderate shifts of the remaining 10% of the genome are highly correlated with changes in gene expression, with shifts relatively closer (further) to speckles associated with a strong bias toward significantly increased (decreased) gene expression (Chen et al. 2018b).

As described previously, mechanism(s) exist to move specific transgenes—for example, HSPA1A—directionally to nuclear speckles in just several minutes (Khanna et al. 2014). A more direct link between nuclear speckle proximity and gene expression was suggested by the temporal sequence between contact of these HSPA1A plasmid transgenes with nuclear speckles and HSPA1A gene expression, as revealed by live-cell imaging. Increased nascent HSPA1A transcripts appeared always after first contact, but never before contact with nuclear speckles. The time lag between first appearance of the nascent transcript signal was several minutes shorter if the transgene array first contacted a larger versus smaller nuclear speckle. Contact with a smaller speckle led to an initial increase in nuclear speckle size above a certain size before the appearance of a nascent transcript signal above background. Later, both nascent transcript signals and nuclear speckle size increased in parallel. This increased size and coordinate growth of both speckles and nascent transcript signals is intriguing in light of the proposed model, described previously, of cycles of PTMs for splicing factors trafficking between speckle to gene and then back to speckle (Misteli and Spector 1997; Hall et al. 2006). Blocking movement to nuclear speckles by blocking actin polymerization resulted in reduced HSPA1A heat-shock induction of plasmid transgenes away from nuclear speckles but no reduction in expression for plasmid transgenes associated with nuclear speckles (Khanna et al. 2014).

Using HSPA1B BAC transgene arrays, which more completely recapitulate the heat-shock-induced expression behavior of the endogenous locus, produced similar results and showed the phenomenon of "gene expression amplification" (Kim et al. 2020). Although the BAC transgenes showed 100% induction within several minutes after heat shock, regardless of

 Cite this article as *Cold Spring Harb Perspect Biol* doi: 10.1101/cshperspect.a041268

speckle association, transgenes in contact with nuclear speckles showed a several-fold increased production of nascent transcripts, which may be at least partially explained by reduced exosome degradation of nascent transcripts near nuclear speckles. This was also seen for the endogenous Hsp70 locus and several HSPA1A flanking genes at both the endogenous locus and the BAC transgenes at normal temperature. Again, live-cell imaging showed that these BAC transgenes did not show jumps in nascent transcript production until after nuclear speckle contact; reduced gene expression occurred within several minutes of loss of contact with nuclear speckles (Kim et al. 2020).

In light of the observed movement of plasmid and BAC transgenes toward nuclear speckles after heat shock, it was surprising that about half of all heat-shock genes, including HSPA1A, are near 100% localized adjacent to nuclear speckles even before heat shock, as revealed by TSA-seq and DNA FISH in multiple human cell lines (Zhang et al. 2020b). The remaining half of expressed heat-shock gene loci are near or at moderate distance (HSPH1A) from nuclear speckles but move closer after heat shock. The HSPH1 gene locus was the furthest of all heat-shock loci from nuclear speckles at 37°C (~50% genomic percentile) and showed the slowest induction kinetics. FISH experiments revealed a gene amplification as well after nuclear speckle contact for the endogenous HSPH1A gene (Zhang et al. 2020b).

Finally, comparing four cell types, no large differences in relative nuclear speckle position were observed (Zhang et al. 2020b). Genes that are positioned closer to nuclear speckles in one cell type are positioned just a few hundred nanometers in average distance further from nuclear speckles in other cell types. This suggests a level of "prewiring" of gene loci relative to nuclear speckles such that genes move only small distances to or from nuclear speckles during differentiation or physiological stimuli (Zhang et al. 2020b).

Such a prewiring is further suggested by recent analysis of p53-responsive genes (Alexander et al. 2021). Approximately half of ~20 surveyed p53-responsive genes moved closer to nuclear speckles after p53 induction and, interestingly, these were the genes that were already positioned close to nuclear speckles even before p53 activation. This induced association with nuclear speckles was dependent on a different region of the p53 protein from the known trans-activating or DNA-binding domains of p53. FISH analysis again revealed a gene expression amplification phenomenon, with faster and higher induction of increased gene expression for alleles within ~0.5 µm of nuclear speckles. Perturbation of nuclear speckles by SON knock-down perturbed this increased gene expression observed with nuclear speckle proximity. TSA-seq revealed several hundred inferred p53-responsive genes that moved closer to nuclear speckles after p53 activation and again showed increased expression, and these were the ~30% of p53-responsive genes that were prepositioned near nuclear speckles before p53 activation (Alexander et al. 2021).

These p53 experiments suggest that the phenomenon of gene expression amplification may be seen for large numbers of genes beyond heat-shock loci. Further experiments should reveal whether this is a general phenomenon of stress-responsive genes, as well as genes moving closer to nuclear speckles as a function of other physiological stimuli and/or development.

A very different theory has been proposed recently in which nuclear speckles rather than simply serving as storage sites for RNA processing factors instead act to actively regulate the nucleoplasmic levels of key RNA processing and splicing factors (Hasenson and Shav-Tal 2020). This produces an "action-at-a-distance" regulation in which changing nucleoplasmic levels of limiting factors by concentration or release by nuclear speckles can change the RNA processing and alternative splicing of genes throughout the nucleus. This model was prompted by observations from live-cell imaging of a transgene array with transgenes containing varying numbers of intron/exon junctions (Hochberg-Laufer et al. 2019). Accumulation of nascent RNA transcripts near the site of transcription was observed only for transgenes containing larger numbers of introns and this was modeled as a delay in splicing occurring after transcription but before release of the nascent transcripts into the nucleoplasm. This delay in release of nascent transcripts could be rescued either by overexpression of a subset of splicing factors or, much more interestingly, after "dissolving" of nuclear speckles by overexpression of the Clk1 kinase, which has been proposed to cause release of splicing factors from nuclear speckles after their phosphorylation (Hochberg-Laufer et al. 2019).

In summary, these recent experiments suggest a closer than previously appreciated relationship between nuclear speckles and regulation of gene expression. In retrospect, this close relationship had been concealed by what appears to be a very precisely controlled yet dynamic prepositioning of many endogenous genes close to nuclear speckles. Likewise, we may be underappreciating the role of nuclear speckles in buffering nucleoplasmic levels of factors involved in RNA processing, and how variations in nuclear speckle protein fluxes and dynamics between cell types may have global influence on patterns of gene expression. Dissecting the mechanisms underlying this prepositioning of gene loci near nuclear speckles, as well as the dynamics of nuclear speckle components, may be required to appreciate the full extent of this influence of nuclear speckles on gene expression.

FUTURE CHALLENGES AND QUESTIONS

The nucleus—and more specifically the interchromosomal space—is crowded (Fig. 3A). Immunostaining and fluorescent microscopy may give us a "tunnel vision" picture owing to the difficulty of visualizing more than three or four components simultaneously. On the other hand, genome-wide mapping methods are an oversimplification of reality because they only present the average over millions of cells and do not take into account cell-to-cell variability. We now realize that any given interphase chromosome domain is likely within a short distance to multiple nuclear "locales" with distinctive functional properties (Fig. 3B). New imaging approaches have made this conceptualization concrete with their visualization of many chromosome loci appearing "fixed" relative to different nuclear interior "environments" defined by proximity to specific nuclear bodies and/or chromatin marks (Fig. 3A; Su et al. 2020; Takei et al. 2021). Newly identified phase-separated nuclear condensates, in addition to previously described nuclear bodies, are only adding to this complexity. Given the number of nuclear proteins with IDRs prone to phase separate above a critical concentration, we likely are still just seeing only the tip of the iceberg in terms of further subdivision of nuclear interchromosomal space. Meanwhile, chromatin domains themselves are assuming characteristics of compartmentalized units in terms of limiting diffusion of large, macromolecular complexes involved in DNA functions.

Previous models emphasizing nuclear radial gene positioning or binary division into iLAD/early replicating/A or LAD/late replicating/B compartments can now be seen as underlying fundamental, yet oversimplified, features of nuclear genome organization. Statistical trends of radial positioning are likely the indirect result of the organization of chromosomes relative to specific nuclear bodies and compartments that themselves are distributed as a function of radial position. Meanwhile, the apparent binary division of the genome conceals an emerging reality that both the nuclear interior, and likely the nuclear periphery, are further divided into functionally distinct locales. Different alleles of the same chromosome locus may distribute among different but related spatial regions and move between them over physiological timescales. For example, recent high-throughput, multiplexed imaging revealed a maximum of ~40% of LAD alleles in actual contact with the nuclear lamina in IMR90 human fibroblasts, with most LAD alleles not associated with the nuclear lamina but instead localizing near the nucleolar periphery or elsewhere in the nuclear interior (Su et al. 2020). Still unknown is what fraction of LAD alleles will localize to euchromatic versus heterochromatic local environments. Similarly, a large fraction of certain speckle-associated chromosome loci localizes very near nuclear speckles, but what fraction of these speckle-associated alleles instead localize away from nuclear speckles but close to other types of nuclear local environments associated with active gene expression?

Not only do "small distances matter"—separating nuclear compartments with distinct functional properties—but chromosomal domains may be "prepositioned" near the appropriate nuclear compartments required for their full and/or more robust activation or silencing. Hi-C defined nuclear compartments are ~Mbp in size and contain many genes showing uncorrelated changes in gene activity. Therefore, it was inferred that smaller units of genome organization such as TADs and sub-TADs would be more intimately connected to gene regulation. However, a newer generation of genomic methods are showing statistically significant changes in positioning of genomic regions that can be comparable or even smaller in size than TADs. In the case of nuclear speckles, these small shifts in relative positions strongly correlate with changes in gene expression (Zhang et al. 2020b; Alexander et al. 2021).

Complicating matters further, functional consequences of small nuclear movements can be masked by subsequent movements and delayed functional outcomes of contact with nuclear compartments (Fig. 3C). This was seen over a timescale of minutes in the case of Hsp70 transgenes contacting nuclear speckles but then turning on even after nuclear speckle detachment (Khanna et al. 2014) and also taking several minutes to turn off after detaching from a nuclear speckle (Kim et al. 2020). But time delays between nuclear compartment contact and functional output might extend in other contexts to entire cell cycles or longer. For example, in fission yeast, several cell cycles were required for full loss of epigenetic silencing of a gene after knockdown of a protein required for its anchoring to the nuclear periphery (Holla et al. 2020).

Finally, we may need a more integrated view of nuclear organization in which we not only examine the location of specific chromosomal regions, and the possible functional consequence of these localizations, but also extend our analysis to larger chromosome trajectories and examine the possible coupling between changes in chromosome positioning within the nucleus simultaneous with changes in chromatin conformation and biochemical marks (Fig. 3D). Relative movements of one chromosome region within the nucleus might not only cause coordinated movements of ~Mbp-scale flanking regions but also lead to changes in chromosome stretching and chromatin compaction, which have been shown to change gene expression (Tajik et al. 2016; Sun et al. 2020). This could create an "action-at-a-distance" mechanism by which there might be coordinated changes in nuclear position, chromatin folding, and gene expression over Mbp-linked chromosomal regions.

Thus, new multiplexed imaging approaches to visualize multi-Mbp chromosome trajectories, guided by genome-wide genomics methods providing readouts of nuclear position, DNA topology, and chromatin packing, will need to be complemented by live-cell imaging to establish temporal ordering of events linking nuclear position, chromosome folding, biochemical modifications, and, finally, functional output.

ACKNOWLEDGMENTS

This work was supported by National Institutes of Health (NIH) Grants R01 GM58460, U01DK127422 and UM1HG011593. We would like to thank the NIH Common Fund, the Office of Strategic Coordination, and the Office of the NIH Director for funding the 4D Nucleome Program, which has supported the latter two grants. We thank Dr. Pankaj Chaturvedi, University of Illinois, Urbana, for the image in Figure 1A.

REFERENCES

*Reference is also in this collection.

Abramo K, Valton AL, Venev SV, Ozadam H, Fox AN, Dekker J. 2019. A chromosome folding intermediate at the condensin-to-cohesin transition during telophase. *Nat Cell Biol* **21:** 1393–1402. doi:10.1038/s41556-019-0406-2

Alexander KA, Coté A, Nguyen SC, Zhang L, Gholamalamdari O, Agudelo-Garcia P, Lin-Shiao E, Tanim KMA, Lim J, Biddle N, et al. 2021. P53 mediates target gene association with nuclear speckles for amplified RNA expression. *Mol Cell* **81:** 1666–1681.e6. doi:10.1016/j.molcel.2021.03.006

Amat R, Böttcher R, Le Dily F, Vidal E, Quilez J, Cuartero Y, Beato M, de Nadal E, Posas F. 2019. Rapid reversible changes in compartments and local chromatin organization revealed by hyperosmotic shock. *Genome Res* **29:** 18–28. doi:10.1101/gr.238527.118

Amenta PS. 1961. Fusion of nucleoli in cells cultured from the heart of *Triturus viridescens*. *Anat Rec* **139:** 155–165. doi:10.1002/ar.1091390207

Anantharaman A, Jadaliha M, Tripathi V, Nakagawa S, Hirose T, Jantsch MF, Prasanth SG, Prasanth KV. 2016. Paraspeckles modulate the intranuclear distribution of paraspeckle-associated *Ctn RNA*. *Sci Rep* **6:** 34043. doi:10.1038/srep34043

Andrade LE, Chan EK, Raska I, Peebles CL, Roos G, Tan EM. 1991. Human autoantibody to a novel protein of the nuclear coiled body: immunological characterization and cDNA cloning of p80-coilin. *J Exp Med* **173:** 1407–1419. doi:10.1084/jem.173.6.1407

Bantignies F, Roure V, Comet I, Leblanc B, Schuettengruber B, Bonnet J, Tixier V, Mas A, Cavalli G. 2011. Polycomb-dependent regulatory contacts between distant Hox loci in *Drosophila*. *Cell* **144:** 214–226. doi:10.1016/j.cell.2010.12.026

Barr ML, Bertram EG. 1951. The behavior of nuclear structures during depletion and restoration of Nissl material in motor neurons. *J Anat* **85:** 171–181.

Barr ML, Carr DH. 1962. Correlations between sex chromatin and sex chromosomes. *Acta Cytol* **6:** 34–45.

Bazett-Jones DP, Hendzel MJ. 1999. Electron spectroscopic imaging of chromatin. *Methods* **17:** 188–200. doi:10.1006/meth.1998.0729

Beagrie RA, Scialdone A, Schueler M, Kraemer DC, Chotalia M, Xie SQ, Barbieri M, de Santiago I, Lavitas LM, Branco MR, et al. 2017. Complex multi-enhancer contacts captured by genome architecture mapping. *Nature* **543:** 519–524. doi:10.1038/nature21411

Beagrie RA, Thieme CJ, Annunziatella C, Baugher C, Zhang Y, Schueler M, Kramer DC, Chiariello AM, Bianco S, Kukalev A, et al. 2020. Multiplex-GAM: genome-wide identification of chromatin contacts yields insights not captured by Hi-C. bioRxiv doi:10.1101/2020.07.31.230284

Beck M, Hurt E. 2017. The nuclear pore complex: understanding its function through structural insight. *Nat Rev Mol Cell Biol* **18:** 73–89. doi:10.1038/nrm.2016.147

Belmont AS, Bignone F, Ts'o PO. 1986. The relative intranuclear positions of Barr bodies in XXX non-transformed human fibroblasts. *Exp Cell Res* **165:** 165–179. doi:10.1016/0014-4827(86)90541-0

Belmont AS, Braunfeld MB, Sedat JW, Agard DA. 1989. Large-scale chromatin structural domains within mitotic and interphase chromosomes in vivo and in vitro. *Chromosoma* **98:** 129–143. doi:10.1007/BF00291049

Belmont AS, Zhai Y, Thilenius A. 1993. Lamin B distribution and association with peripheral chromatin revealed by optical sectioning and electron microscopy tomography. *J Cell Biol* **123:** 1671–1685. doi:10.1083/jcb.123.6.1671

Bernhard W, Granboulan N. 1963. The fine structure of the cancer cell nucleus. *Exp Cell Res* **9:** 19–53. doi:10.1016/0014-4827(63)90243-X

Bernhard W, Haguenau F, Oberling C. 1952. L'ultrastructure du nucléole de quelques cellules animales, révélée par le microscope électronique. *Experientia* **8:** 58–59. doi:10.1007/BF02139019

Bickmore WA. 2013. The spatial organization of the human genome. *Annu Rev Genomics Hum Genet* **14:** 67–84. doi:10.1146/annurev-genom-091212-153515

Biggiogera M, Courtens JL, Derenzini M, Fakan S, Hernandez-Verdun D, Risueno MC, Soyer-Gobillard MO. 1996. Osmium ammine: review of current applications to visualize DNA in electron microscopy. *Biol Cell* **87:** 121–132. doi:10.1111/j.1768-322X.1996.tb00974.x

Bizhanova A, Kaufman PD. 2021. Close to the edge: heterochromatin at the nucleolar and nuclear peripheries. *Biochim Biophys Acta Gene Regul Mech* **1864:** 194666. doi:10.1016/j.bbagrm.2020.194666

Boeynaems S, Alberti S, Fawzi NL, Mittag T, Polymenidou M, Rousseau F, Schymkowitz J, Shorter J, Wolozin B, Van Den Bosch L, et al. 2018. Protein phase separation: a new phase in cell biology. *Trends Cell Biol* **28:** 420–435. doi:10.1016/j.tcb.2018.02.004

Bohrmann B, Kellenberger E. 1994. Immunostaining of DNA in electron microscopy: an amplification and staining procedure for thin sections as alternative to gold labeling. *J Histochem Cytochem* **42:** 635–643. doi:10.1177/42.5.7512586

Boisvert FM, van Koningsbruggen S, Navascués J, Lamond AI. 2007. The multifunctional nucleolus. *Nat Rev Mol Cell Biol* **8:** 574–585. doi:10.1038/nrm2184

Bolzer A, Kreth G, Solovei I, Koehler D, Saracoglu K, Fauth C, Müller S, Eils R, Cremer C, Speicher MR, et al. 2005. Three-dimensional maps of all chromosomes in human male fibroblast nuclei and prometaphase rosettes. *PLoS Biol* **3:** e157. doi:10.1371/journal.pbio.0030157

Boulon S, Westman BJ, Hutten S, Boisvert FM, Lamond AI. 2010. The nucleolus under stress. *Mol Cell* **40:** 216–227. doi:10.1016/j.molcel.2010.09.024

Boumendil C, Hari P, Olsen KCF, Acosta JC, Bickmore WA. 2019. Nuclear pore density controls heterochromatin reorganization during senescence. *Genes Dev* **33:** 144–149. doi:10.1101/gad.321117.118

Bourgeois CA, Hubert J. 1988. Spatial relationship between the nucleolus and the nuclear envelope: structural aspects and functional significance. *Int Rev Cytol* **111:** 1–52. doi:10.1016/S0074-7696(08)61730-1

Bourgeois CA, Hemon D, Bouteille M. 1979. Structural relationship between the nucleolus and the nuclear envelope. *J Ultrastruct Res* **68:** 328–340. doi:10.1016/S0022-5320(79)90165-5

Briand N, Collas P. 2020. Lamina-associated domains: peripheral matters and internal affairs. *Genome Biol* **21:** 85. doi:10.1186/s13059-020-02003-5

Brown SW. 1966. Heterochromatin. *Science* **151:** 417–425. doi:10.1126/science.151.3709.417

Brown KE, Baxter J, Graf D, Merkenschlager M, Fisher AG. 1999. Dynamic repositioning of genes in the nucleus of lymphocytes preparing for cell division. *Mol Cell* **3:** 207–217. doi:10.1016/S1097-2765(00)80311-1

Bubulya PA, Prasanth KV, Deerinck TJ, Gerlich D, Beaudouin J, Ellisman MH, Ellenberg J, Spector DL. 2004. Hypophosphorylated SR splicing factors transiently localize around active nucleolar organizing regions in telophase daughter nuclei. *J Cell Biol* **167:** 51–63. doi:10.1083/jcb.200404120

Callan HG, Tomlin SG. 1950. Experimental studies on amphibian oocyte nuclei. I: Investigation of the structure of the nuclear membrane by means of the electron microscope. *Proc R Soc Lond B Biol Sci* **137:** 367–378. doi:10.1098/rspb.1950.0047

Cardozo Gizzi AM, Cattoni DI, Fiche JB, Espinola SM, Gurgo J, Messina O, Houbron C, Ogiyama Y, Papadopoulos GL, Cavalli G, et al. 2019. Microscopy-based chromosome conformation capture enables simultaneous visualization of genome organization and transcription in intact organisms. *Mol Cell* **74:** 212–222.e5. doi:10.1016/j.molcel.2019.01.011

Carron C, Balor S, Delavoie F, Plisson-Chastang C, Faubladier M, Gleizes PE, O'Donohue MF. 2012. Post-mitotic dynamics of pre-nucleolar bodies is driven by pre-rRNA processing. *J Cell Sci* **125:** 4532–4542.

Carter KC, Bowman D, Carrington W, Fogarty K, McNeil JA, Fay FS, Lawrence JB. 1993. A three-dimensional view of precursor messenger RNA metabolism within the mammalian nucleus. *Science* **259:** 1330–1335. doi:10.1126/science.8446902

Celetti G, Paci G, Caria J, VanDelinder V, Bachand G, Lemke EA. 2020. The liquid state of FG-nucleoporins mimics permeability barrier properties of nuclear pore complexes. *J Cell Biol* **219:** e201907157. doi:10.1083/jcb.201907157

Cerase A, Armaos A, Neumayer C, Avner P, Guttman M, Tartaglia GG. 2019. Phase separation drives X-chromosome inactivation: a hypothesis. *Nat Struct Mol Biol* **26:** 331–334. doi:10.1038/s41594-019-0223-0

Chen Y, Belmont AS. 2019. Genome organization around nuclear speckles. *Curr Opin Genet Dev* **55:** 91–99. doi:10.1016/j.gde.2019.06.008

Chen LL, Carmichael GG. 2009. Altered nuclear retention of mRNAs containing inverted repeats in human embryonic stem cells: functional role of a nuclear noncoding RNA. *Mol Cell* **35:** 467–478. doi:10.1016/j.molcel.2009.06.027

Chen W, Yan Z, Li S, Huang N, Huang X, Zhang J, Zhong S. 2018a. RNAs as proximity-labeling media for identifying nuclear speckle positions relative to the genome. *iScience* **4:** 204–215. doi:10.1016/j.isci.2018.06.005

Chen Y, Zhang Y, Wang Y, Zhang L, Brinkman EK, Adam SA, Goldman R, van Steensel B, Ma J, Belmont AS. 2018b. Mapping 3D genome organization relative to nuclear compartments using TSA-Seq as a cytological ruler. *J Cell Biol* **217:** 4025–4048. doi:10.1083/jcb.201807108

Cho UH, Hetzer MW. 2020. Nuclear periphery takes center stage: the role of nuclear pore complexes in cell identity and aging. *Neuron* **106:** 899–911. doi:10.1016/j.neuron.2020.05.031

Cho WK, Spille JH, Hecht M, Lee C, Li C, Grube V, Cisse II. 2018. Mediator and RNA polymerase II clusters associate in transcription-dependent condensates. *Science* **361:** 412–415. doi:10.1126/science.aar4199

Chuang CH, Belmont AS. 2007. Moving chromatin within the interphase nucleus-controlled transitions? *Semin Cell Dev Biol* **18:** 698–706. doi:10.1016/j.semcdb.2007.08.012

Chuang CH, Carpenter AE, Fuchsova B, Johnson T, de Lanerolle P, Belmont AS. 2006. Long-range directional movement of an interphase chromosome site. *Curr Biol* **16:** 825–831. doi:10.1016/j.cub.2006.03.059

Cioce M, Lamond AI. 2005. Cajal bodies: a long history of discovery. *Annu Rev Cell Dev Biol* **21:** 105–131. doi:10.1146/annurev.cellbio.20.010403.103738

Clemson CM, Hutchinson JN, Sara SA, Ensminger AW, Fox AH, Chess A, Lawrence JB. 2009. An architectural role for a nuclear noncoding RNA: NEAT1 RNA is essential for the structure of paraspeckles. *Mol Cell* **33:** 717–726. doi:10.1016/j.molcel.2009.01.026

Cook PR. 1999. The organization of replication and transcription. *Science* **284:** 1790–1795. doi:10.1126/science.284.5421.1790

Corpet A, Kleijwegt C, Roubille S, Juillard F, Jacquet K, Texier P, Lomonte P. 2020. PML nuclear bodies and chromatin dynamics: catch me if you can! *Nucleic Acids Res* **48:** 11890–11912. doi:10.1093/nar/gkaa828

Cremer T, Cremer M. 2010. Chromosome territories. *Cold Spring Harb Perspect Biol* **2:** a003889. doi:10.1101/cshperspect.a003889

Cremer M, Küpper K, Wagler B, Wizelman L, von Hase J, Weiland Y, Kreja L, Diebold J, Speicher MR, Cremer T. 2003. Inheritance of gene density-related higher order chromatin arrangements in normal and tumor cell nuclei. *J Cell Biol* **162:** 809–820. doi:10.1083/jcb.200304096

Croft JA, Bridger JM, Boyle S, Perry P, Teague P, Bickmore WA. 1999. Differences in the localization and morphology of chromosomes in the human nucleus. *J Cell Biol* **145:** 1119–1131. doi:10.1083/jcb.145.6.1119

Csink AK, Henikoff S. 1996. Genetic modification of heterochromatic association and nuclear organization in *Drosophila*. *Nature* **381:** 529–531. doi:10.1038/381529a0

de Leeuw R, Gruenbaum Y, Medalia O. 2018. Nuclear lamins: thin filaments with major functions. *Trends Cell Biol* **28:** 34–45. doi:10.1016/j.tcb.2017.08.004

Deng X, Zhironkina OA, Cherepanynets VD, Strelkova OS, Kireev II, Belmont AS. 2016. Cytology of DNA replication reveals dynamic plasticity of large-scale chromatin fibers. *Curr Biol* **26:** 2527–2534. doi:10.1016/j.cub.2016.07.020

Derenzini M, Farabegoli F, Trerè D. 1993. Localization of DNA in the fibrillar components of the nucleolus: a cytochemical and morphometric study. *J Histochem Cytochem* **41:** 829–836. doi:10.1177/41.6.8315275

Derenzini M, Trerè D, Pession A, Govoni M, Sirri V, Chieco P. 2000. Nucleolar size indicates the rapidity of cell proliferation in cancer tissues. *J Pathol* **191:** 181–186. doi:10.1002/(SICI)1096-9896(200006)191:2<181::AID-PATH607>3.0.CO;2-V

Dernburg AF, Broman KW, Fung JC, Marshall WF, Philips J, Agard DA, Sedat JW. 1996. Perturbation of nuclear architecture by long-distance chromosome interactions. *Cell* **85:** 745–759. doi:10.1016/S0092-8674(00)81240-4

de Wit E, van Steensel B. 2009. Chromatin domains in higher eukaryotes: insights from genome-wide mapping studies. *Chromosoma* **118:** 25–36. doi:10.1007/s00412-008-0186-0

Dimitrova DS, Gilbert DM. 1999. The spatial position and replication timing of chromosomal domains are both established in early G1 phase. *Mol Cell* **4:** 983–993. doi:10.1016/S1097-2765(00)80227-0

Dopie J, Sweredoski MJ, Moradian A, Belmont AS. 2020. Tyramide signal amplification mass spectrometry (TSA-

MS) ratio identifies nuclear speckle proteins. *J Cell Biol* **219:** e201910207. doi:10.1083/jcb.201910207

Dundr M, Misteli T, Olson MO. 2000. The dynamics of postmitotic reassembly of the nucleolus. *J Cell Biol* **150:** 433–446. doi:10.1083/jcb.150.3.433

Dundr M, Ospina JK, Sung MH, John S, Upender M, Ried T, Hager GL, Matera AG. 2007. Actin-dependent intranuclear repositioning of an active gene locus in vivo. *J Cell Biol* **179:** 1095–1103. doi:10.1083/jcb.200710058

Engreitz JM, Sirokman K, McDonel P, Shishkin AA, Surka C, Russell P, Grossman SR, Chow AY, Guttman M, Lander ES. 2014. RNA–RNA interactions enable specific targeting of noncoding RNAs to nascent pre-mRNAs and chromatin sites. *Cell* **159:** 188–199. doi:10.1016/j.cell.2014.08.018

Erdel F, Rademacher A, Vlijm R, Tünnermann J, Frank L, Weinmann R, Schweigert E, Yserentant K, Hummert J, Bauer C, et al. 2020. Mouse heterochromatin adopts digital compaction states without showing hallmarks of HP1-driven liquid-liquid phase separation. *Mol Cell* **78:** 236–249 e237. doi:10.1016/j.molcel.2020.02.005

Eskiw CH, Fraser P. 2011. Ultrastructural study of transcription factories in mouse erythroblasts. *J Cell Sci* **124:** 3676–3683. doi:10.1242/jcs.087981

Fawcett DW. 1966. On the occurrence of a fibrous lamina on the inner aspect of the nuclear envelope in certain cells of vertebrates. *Am J Anat* **119:** 129–145. doi:10.1002/aja.1001190108

Fei J, Jadaliha M, Harmon TS, Li ITS, Hua B, Hao Q, Holehouse AS, Reyer M, Sun Q, Freier SM, et al. 2017. Quantitative analysis of multilayer organization of proteins and RNA in nuclear speckles at super resolution. *J Cell Sci* **130:** 4180–4192. doi:10.1242/jcs.206854

Ferrai C, Xie SQ, Luraghi P, Munari D, Ramirez F, Branco MR, Pombo A, Crippa MP. 2010. Poised transcription factories prime silent uPA gene prior to activation. *PLoS Biol* **8:** e1000270. doi:10.1371/journal.pbio.1000270

Ferreira JA, Carmo-Fonseca M, Lamond AI. 1994. Differential interaction of splicing snRNPs with coiled bodies and interchromatin granules during mitosis and assembly of daughter cell nuclei. *J Cell Biol* **126:** 11–23. doi:10.1083/jcb.126.1.11

Ferreira J, Paolella G, Ramos C, Lamond AI. 1997. Spatial organization of large-scale chromatin domains in the nucleus: a magnified view of single chromosome territories. *J Cell Biol* **139:** 1597–1610. doi:10.1083/jcb.139.7.1597

Fox AH, Lamond AI. 2010. Paraspeckles. *Cold Spring Harb Perspect Biol* **2:** a000687.

Fox AH, Lam YW, Leung AK, Lyon CE, Andersen J, Mann M, Lamond AI. 2002. Paraspeckles: a novel nuclear domain. *Curr Biol* **12:** 13–25. doi:10.1016/S0960-9822(01)00632-7

Fox AH, Bond CS, Lamond AI. 2005. P54nrb forms a heterodimer with PSP1 that localizes to paraspeckles in an RNA-dependent manner. *Mol Biol Cell* **16:** 5304–5315. doi:10.1091/mbc.e05-06-0587

Frankowski KJ, Wang C, Patnaik S, Schoenen FJ, Southall N, Li D, Teper Y, Sun W, Kandela I, Hu D, et al. 2018. Metarrestin, a perinucleolar compartment inhibitor, effectively suppresses metastasis. *Sci Transl Med* **10:** eaap8307. doi:10.1126/scitranslmed.aap8307

Frey S, Rees R, Schünemann J, Ng SC, Funfgeld K, Huyton T, Görlich D. 2018. Surface properties determining passage rates of proteins through nuclear pores. *Cell* **174:** 202–217.e9. doi:10.1016/j.cell.2018.05.045

Frottin F, Schueder F, Tiwary S, Gupta R, Körner R, Schlichthaerle T, Cox J, Jungmann R, Hartl FU, Hipp MS. 2019. The nucleolus functions as a phase-separated protein quality control compartment. *Science* **365:** 342–347. doi:10.1126/science.aaw9157

Gialitakis M, Arampatzi P, Makatounakis T, Papamatheakis J. 2010. γ Interferon-dependent transcriptional memory via relocalization of a gene locus to PML nuclear bodies. *Mol Cell Biol* **30:** 2046–2056. doi:10.1128/MCB.00906-09

Gibson BA, Doolittle LK, Schneider MWG, Jensen LE, Gamarra N, Henry L, Gerlich DW, Redding S, Rosen MK. 2019. Organization of chromatin by intrinsic and regulated phase separation. *Cell* **179:** 470–484.e21. doi:10.1016/j.cell.2019.08.037

González SG, Nardone RM. 1968. Cyclic nucleolar changes during the cell cycle. I: Variations in number, size, morphology and position. *Exp Cell Res* **50:** 599–615. doi:10.1016/0014-4827(68)90422-9

Guelen L, Pagie L, Brasset E, Meuleman W, Faza MB, Talhout W, Eussen BH, de Klein A, Wessels L, de Laat W, et al. 2008. Domain organization of human chromosomes revealed by mapping of nuclear lamina interactions. *Nature* **453:** 948–951. doi:10.1038/nature06947

Guo YE, Manteiga JC, Henninger JE, Sabari BR, Dall'Agnese A, Hannett NM, Spille JH, Afeyan LK, Zamudio AV, Shrinivas K, et al. 2019. Pol II phosphorylation regulates a switch between transcriptional and splicing condensates. *Nature* **572:** 543–548. doi:10.1038/s41586-019-1464-0

Hall LL, Smith KP, Byron M, Lawrence JB. 2006. Molecular anatomy of a speckle. *Anat Rec A Discov Mol Cell Evol Biol* **288A:** 664–675. doi:10.1002/ar.a.20336

Hasenson SE, Shav-Tal Y. 2020. Speculating on the roles of nuclear speckles: how RNA-protein nuclear assemblies affect gene expression. *Bioessays* **42:** 2000104. doi:10.1002/bies.202000104

Heitz E. 1928. Das heterochromatin der moose. *I Jahrb Wiss Bot* **69:** 762–818.

Hendzel MJ, Kruhlak MJ, Bazett-Jones DP. 1998. Organization of highly acetylated chromatin around sites of heterogeneous nuclear RNA accumulation. *Mol Biol Cell* **9:** 2491–2507. doi:10.1091/mbc.9.9.2491

Hernandez-Verdun D. 2011. Assembly and disassembly of the nucleolus during the cell cycle. *Nucleus* **2:** 189–194. doi:10.4161/nucl.2.3.16246

Hiratani I, Ryba T, Itoh M, Yokochi T, Schwaiger M, Chang CW, Lyou Y, Townes TM, Schübeler D, Gilbert DM. 2008. Global reorganization of replication domains during embryonic stem cell differentiation. *PLoS Biol* **6:** e245. doi:10.1371/journal.pbio.0060245

Hochberg-Laufer H, Neufeld N, Brody Y, Nadav-Eliyahu S, Ben-Yishay R, Shav-Tal Y. 2019. Availability of splicing factors in the nucleoplasm can regulate the release of mRNA from the gene after transcription. *PLoS Genet* **15:** e1008459. doi:10.1371/journal.pgen.1008459

Hoencamp C, Dudchenko O, Elbatsh AMO, Brahmachari S, Raaijmakers JA, van Schaik T, Sedeño Cacciatore A, Contessoto VG, van Heesbeen R, van den Broek B, et al. 2021.

3D genomics across the tree of life reveals condensin II as a determinant of architecture type. *Science* **372:** 984–989. doi:10.1126/science.abe2218

Hoffman DP, Shtengel G, Xu CS, Campbell KR, Freeman M, Wang L, Milkie DE, Pasolli HA, Iyer N, Bogovic JA, et al. 2020. Correlative three-dimensional super-resolution and block-face electron microscopy of whole vitreously frozen cells. *Science* **367:** eaaz5357. doi:10.1126/science.aaz5357

Holla S, Dhakshnamoorthy J, Folco HD, Balachandran V, Xiao H, Sun LL, Wheeler D, Zofall M, Grewal SIS. 2020. Positioning heterochromatin at the nuclear periphery suppresses histone turnover to promote epigenetic inheritance. *Cell* **180:** 150–164.e15. doi:10.1016/j.cell.2019.12.004

Hu Y, Kireev I, Plutz MJ, Ashourian N, Belmont AS. 2009. Large-scale chromatin structure of inducible genes: transcription on a condensed, linear template. *J Cell Biol* **185:** 87–100. doi:10.1083/jcb.200809196

Hutchinson JN, Ensminger AW, Clemson CM, Lynch CR, Lawrence JB, Chess A. 2007. A screen for nuclear transcripts identifies two linked noncoding RNAs associated with SC35 splicing domains. *BMC Genomics* **8:** 39. doi:10.1186/1471-2164-8-39

Hyman AA, Weber CA, Jülicher F. 2014. Liquid–liquid phase separation in biology. *Annu Rev Cell Dev Biol* **30:** 39–58. doi:10.1146/annurev-cellbio-100913-013325

Iarovaia OV, Minina EP, Sheval EV, Onichtchouk D, Dokudovskaya S, Razin SV, Vassetzky YS. 2019. Nucleolus: a central hub for nuclear functions. *Trends Cell Biol* **29:** 647–659. doi:10.1016/j.tcb.2019.04.003

Iborra FJ, Pombo A, Jackson DA, Cook PR. 1996. Active RNA polymerases are localized within discrete transcription "factories' in human nuclei. *J Cell Sci* **109:** 1427–1436. doi:10.1242/jcs.109.6.1427

Ilik IA, Malszycki M, Lubke AK, Schade C, Meierhofer D, Aktas T. 2020. SON and SRRM2 are essential for nuclear speckle formation. *eLife* **9:** e60579. doi:10.7554/eLife.60579

Jackson DA, Iborra FJ, Manders EM, Cook PR. 1998. Numbers and organization of RNA polymerases, nascent transcripts, and transcription units in HeLa nuclei. *Mol Biol Cell* **9:** 1523–1536. doi:10.1091/mbc.9.6.1523

Kang SM, Yoon MH, Park BJ. 2018. Laminopathies; mutations on single gene and various human genetic diseases. *BMB Rep* **51:** 327–337. doi:10.5483/BMBRep.2018.51.7.113

Keenen MM, Brown D, Brennan LD, Renger R, Khoo H, Carlson CR, Huang B, Grill SW, Narlikar GJ, Redding S. 2021. HP1 proteins compact DNA into mechanically and positionally stable phase separated domains. *eLife* **10:** e64563. doi:10.7554/eLife.64563

Khanna N, Hu Y, Belmont AS. 2014. HSP70 transgene directed motion to nuclear speckles facilitates heat shock activation. *Curr Biol* **24:** 1138–1144. doi:10.1016/j.cub.2014.03.053

Kichuk TC, Carrasco-Lopez C, Avalos JL. 2021. Lights up on organelles: optogenetic tools to control subcellular structure and organization. *Wiley Interdiscip Rev Syst Biol Med* **13:** e1500.

Kim J, Han KY, Khanna N, Ha T, Belmont AS. 2019. Nuclear speckle fusion via long-range directional motion regulates speckle morphology after transcriptional inhibition. *J Cell Sci* **132:** jcs226563. doi:10.1242/jcs.226563

Kim J, Venkata NC, Hernandez Gonzalez GA, Khanna N, Belmont AS. 2020. Gene expression amplification by nuclear speckle association. *J Cell Biol* **219:** e201904046. doi:10.1083/jcb.201904046

Kind J, Pagie L, Ortabozkoyun H, Boyle S, de Vries SS, Janssen H, Amendola M, Nolen LD, Bickmore WA, van Steensel B. 2013. Single-cell dynamics of genome-nuclear lamina interactions. *Cell* **153:** 178–192. doi:10.1016/j.cell.2013.02.028

Kireev I, Lakonishok M, Liu W, Joshi VN, Powell R, Belmont AS. 2008. In vivo immunogold labeling confirms large-scale chromatin folding motifs. *Nat Methods* **5:** 311–313. doi:10.1038/nmeth.1196

Kölbl AC, Weigl D, Mulaw M, Thormeyer T, Bohlander SK, Cremer T, Dietzel S. 2012. The radial nuclear positioning of genes correlates with features of megabase-sized chromatin domains. *Chromosome Res* **20:** 735–752. doi:10.1007/s10577-012-9309-9

Korenberg JR, Rykowski MC. 1988. Human genome organization: alu, lines, and the molecular structure of metaphase chromosome bands. *Cell* **53:** 391–400. doi:10.1016/0092-8674(88)90159-6

Kurihara M, Kato K, Sanbo C, Shigenobu S, Ohkawa Y, Fuchigami T, Miyanari Y. 2020. Genomic profiling by ALaP-seq reveals transcriptional regulation by PML bodies through DNMT3A exclusion. *Mol Cell* **78:** 493–505.e8. doi:10.1016/j.molcel.2020.04.004

Lafarga M, Casafont I, Bengoechea R, Tapia O, Berciano MT. 2009. Cajal's contribution to the knowledge of the neuronal cell nucleus. *Chromosoma* **118:** 437–443. doi:10.1007/s00412-009-0212-x

Lafarga M, Tapia O, Romero AM, Berciano MT. 2017. Cajal bodies in neurons. *RNA Biol* **14:** 712–725. doi:10.1080/15476286.2016.1231360

Lafontaine DLJ, Riback JA, Bascetin R, Brangwynne CP. 2021. The nucleolus as a multiphase liquid condensate. *Nat Rev Mol Cell Biol* **22:** 165–182. doi:10.1038/s41580-020-0272-6

Lallemand-Breitenbach V, de Thé H. 2010. PML nuclear bodies. *Cold Spring Harb Perspect Biol* **2:** a000661. doi:10.1101/cshperspect.a000661

Lallemand-Breitenbach V, de Thé H. 2018. PML nuclear bodies: from architecture to function. *Curr Opin Cell Biol* **52:** 154–161. doi:10.1016/j.ceb.2018.03.011

Larson AG, Elnatan D, Keenen MM, Trnka MJ, Johnston JB, Burlingame AL, Agard DA, Redding S, Narlikar GJ. 2017. Liquid droplet formation by HP1α suggests a role for phase separation in heterochromatin. *Nature* **547:** 236–240. doi:10.1038/nature22822

Latonen L. 2019. Phase-to-phase with nucleoli—stress responses, protein aggregation and novel roles of RNA. *Front Cell Neurosci* **13:** 151. doi:10.3389/fncel.2019.00151

Lemaître C, Grabarz A, Tsouroula K, Andronov L, Furst A, Pankotai T, Heyer V, Rogier M, Attwood KM, Kessler P, et al. 2014. Nuclear position dictates DNA repair pathway choice. *Genes Dev* **28:** 2450–2463. doi:10.1101/gad.248369.114

Li L, Roy K, Katyal S, Sun X, Bléoo S, Godbout R. 2006. Dynamic nature of cleavage bodies and their spatial relationship to DDX1 bodies, cajal bodies, and gems. *Mol Biol Cell* **17:** 1126–1140. doi:10.1091/mbc.e05-08-0768

Lieberman-Aiden E, van Berkum NL, Williams L, Imakaev M, Ragoczy T, Telling A, Amit I, Lajoie BR, Sabo PJ, Dorschner MO, et al. 2009. Comprehensive mapping of long-range interactions reveals folding principles of the human genome. *Science* **326:** 289–293. doi:10.1126/science.1181369

Lin DH, Hoelz A. 2019. The structure of the nuclear pore complex (an update). *Annu Rev Biochem* **88:** 725–783. doi:10.1146/annurev-biochem-062917-011901

Lu JY, Chang L, Li T, Wang T, Yin Y, Zhan G, Han X, Zhang K, Tao Y, Percharde M, et al. 2021. Homotypic clustering of L1 and B1/Alu repeats compartmentalizes the 3D genome. *Cell Res* **31:** 613–630. doi:10.1038/s41422-020-00466-6

Lyon AS, Peeples WB, Rosen MK. 2021. A framework for understanding the functions of biomolecular condensates across scales. *Nat Rev Mol Cell Biol* **22:** 215–235. doi:10.1038/s41580-020-00303-z

Ma L, Gao Z, Wu J, Zhong B, Xie Y, Huang W, Lin Y. 2021. Co-condensation between transcription factor and coactivator p300 modulates transcriptional bursting kinetics. *Mol Cell* **81:** 1682–1697 e1687. doi:10.1016/j.molcel.2021.01.031

Machyna M, Heyn P, Neugebauer KM. 2013. Cajal bodies: where form meets function. *Wiley Interdiscip Rev RNA* **4:** 17–34. doi:10.1002/wrna.1139

Maeshima K, Tamura S, Hansen JC, Itoh Y. 2020. Fluid-like chromatin: toward understanding the real chromatin organization present in the cell. *Curr Opin Cell Biol* **64:** 77–89. doi:10.1016/j.ceb.2020.02.016

Manuelidis L. 1984. Active nucleolus organizers are precisely positioned in adult central nervous system cells but not in neuroectodermal tumor cells. *J Neuropathol Exp Neurol* **43:** 225–241. doi:10.1097/00005072-198405000-00002

Mateo LJ, Murphy SE, Hafner A, Cinquini IS, Walker CA, Boettiger AN. 2019. Visualizing DNA folding and RNA in embryos at single-cell resolution. *Nature* **568:** 49–54. doi:10.1038/s41586-019-1035-4

McCluggage F, Fox AH. 2021. Paraspeckle nuclear condensates: global sensors of cell stress? *Bioessays* **43:** 2000245. doi:10.1002/bies.202000245

McSwiggen DT, Mir M, Darzacq X, Tjian R. 2019. Evaluating phase separation in live cells: diagnosis, caveats, and functional consequences. *Genes Dev* **33:** 1619–1634. doi:10.1101/gad.331520.119

Mehus AA, Anderson RH, Roux KJ. 2016. BioID identification of lamin-associated proteins. *Methods Enzymol* **569:** 3–22. doi:10.1016/bs.mie.2015.08.008

Meier UT. 2017. RNA modification in cajal bodies. *RNA Biol* **14:** 693–700. doi:10.1080/15476286.2016.1249091

Meuleman W, Peric-Hupkes D, Kind J, Beaudry JB, Pagie L, Kellis M, Reinders M, Wessels L, van Steensel B. 2013. Constitutive nuclear lamina-genome interactions are highly conserved and associated with A/T-rich sequence. *Genome Res* **23:** 270–280. doi:10.1101/gr.141028.112

Miron E, Oldenkamp R, Brown JM, Pinto DMS, Xu CS, Faria AR, Shaban HA, Rhodes JDP, Innocent C, de Ornellas S, et al. 2020. Chromatin arranges in chains of mesoscale domains with nanoscale functional topography independent of cohesin. *Sci Adv* **6:** eaba8811. doi:10.1126/sciadv.aba8811

* Miroshnikova YA, Wickström SA. 2021. Mechanical forces in nuclear organization. *Cold Spring Harb Perspect Biol* doi:10.1101/cshperspect.a039685

Misteli T. 2020. The self-organizing genome: principles of genome architecture and function. *Cell* **183:** 28–45. doi:10.1016/j.cell.2020.09.014

Misteli T, Spector DL. 1997. Protein phosphorylation and the nuclear organization of pre-mRNA splicing. *Trends Cell Biol* **7:** 135–138. doi:10.1016/S0962-8924(96)20043-1

Miura H, Takahashi S, Poonperm R, Tanigawa A, Takebayashi SI, Hiratani I. 2019. Single-cell DNA replication profiling identifies spatiotemporal developmental dynamics of chromosome organization. *Nat Genet* **51:** 1356–1368. doi:10.1038/s41588-019-0474-z

Monneron A, Bernhard W. 1969. Fine structural organization of the interphase nucleus in some mammalian cells. *J Ultrastruct Res* **27:** 266–288. doi:10.1016/S0022-5320(69)80017-1

Nakagawa S, Yamazaki T, Hirose T. 2018. Molecular dissection of nuclear paraspeckles: towards understanding the emerging world of the RNP milieu. *Open Biol* **8:** 180150. doi:10.1098/rsob.180150

Németh A, Conesa A, Santoyo-Lopez J, Medina I, Montaner D, Péterfia B, Solovei I, Cremer T, Dopazo J, Längst G. 2010. Initial genomics of the human nucleolus. *PLoS Genet* **6:** e1000889. doi:10.1371/journal.pgen.1000889

Nguyen HQ, Chattoraj S, Castillo D, Nguyen SC, Nir G, Lioutas A, Hershberg EA, Martins NMC, Reginato PL, Hannan M, et al. 2020. 3D mapping and accelerated super-resolution imaging of the human genome using in situ sequencing. *Nat Methods* **17:** 822–832. doi:10.1038/s41592-020-0890-0

Nozaki T, Imai R, Tanbo M, Nagashima R, Tamura S, Tani T, Joti Y, Tomita M, Hibino K, Kanemaki MT, et al. 2017. Dynamic organization of chromatin domains revealed by super-resolution live-cell imaging. *Mol Cell* **67:** 282–293.e7. doi:10.1016/j.molcel.2017.06.018

Ochs RL, Lischwe MA, Shen E, Carroll RE, Busch H. 1985. Nucleologenesis: composition and fate of prenucleolar bodies. *Chromosoma* **92:** 330–336. doi:10.1007/BF00327463

Ogg SC, Lamond AI. 2002. Cajal bodies and coilin—moving towards function. *J Cell Biol* **159:** 17–21. doi:10.1083/jcb.200206111

O'Keefe RT, Henderson SC, Spector DL. 1992. Dynamic organization of DNA replication in mammalian cell nuclei: spatially and temporally defined replication of chromosome-specific α-satellite DNA sequences. *J Cell Biol* **116:** 1095–1110. doi:10.1083/jcb.116.5.1095

Olins AL, Moyer BA, Kim SH, Allison DP. 1989. Synthesis of a more stable osmium ammine electron-dense DNA stain. *J Histochem Cytochem* **37:** 395–398. doi:10.1177/37.3.2465337

Osborne CS, Chakalova L, Brown KE, Carter D, Horton A, Debrand E, Goyenechea B, Mitchell JA, Lopes S, Reik W, et al. 2004. Active genes dynamically colocalize to shared sites of ongoing transcription. *Nat Genet* **36:** 1065–1071. doi:10.1038/ng1423

Osborne CS, Chakalova L, Mitchell JA, Horton A, Wood AL, Bolland DJ, Corcoran AE, Fraser P. 2007. Myc dynamically and preferentially relocates to a transcription factory occupied by Igh. *PLoS Biol* **5:** e192. doi:10.1371/journal.pbio.0050192

Osmanagic-Myers S, Foisner R. 2019. The structural and gene expression hypotheses in laminopathic diseases—not so different after all. *Mol Biol Cell* **30:** 1786–1790. doi:10.1091/mbc.E18-10-0672

Pandya-Jones A, Markaki Y, Serizay J, Chitiashvili T, Mancia Leon WR, Damianov A, Chronis C, Papp B, Chen CK, McKee R, et al. 2020. A protein assembly mediates *Xist* localization and gene silencing. *Nature* **587:** 145–151. doi:10.1038/s41586-020-2703-0

Passarge E. 1979. Emil heitz and the concept of heterochromatin: longitudinal chromosome differentiation was recognized fifty years ago. *Am J Hum Genet* **31:** 106–115.

* Pawar S, Kutay U. 2021. The diverse cellular functions of inner nuclear membrane proteins. *Cold Spring Harb Perspect Biol* doi:10.1101/cshperspect.a040477

Pederson T. 2011. The nucleolus. *Cold Spring Harb Perspect Biol* **3:** a000638. doi:10.1101/cshperspect.a000638

Penzo M, Montanaro L, Treré D, Derenzini M. 2019. The ribosome biogenesis—cancer connection. *Cells* **8:** 55. doi:10.3390/cells8010055

Peric-Hupkes D, van Steensel B. 2010. Role of the nuclear lamina in genome organization and gene expression. *Cold Spring Harb Symp Quant Biol* **75:** 517–524. doi:10.1101/sqb.2010.75.014

Peric-Hupkes D, Meuleman W, Pagie L, Bruggeman SW, Solovei I, Brugman W, Gräf S, Flicek P, Kerkhoven RM, van Lohuizen M, et al. 2010. Molecular maps of the reorganization of genome-nuclear lamina interactions during differentiation. *Mol Cell* **38:** 603–613. doi:10.1016/j.molcel.2010.03.016

Pirrotta V, Li HB. 2012. A view of nuclear Polycomb bodies. *Curr Opin Genet Dev* **22:** 101–109. doi:10.1016/j.gde.2011.11.004

Pisani G, Baron B. 2019. Nuclear paraspeckles function in mediating gene regulatory and apoptotic pathways. *Non-coding RNA Res* **4:** 128–134. doi:10.1016/j.ncrna.2019.11.002

Platani M, Goldberg I, Swedlow JR, Lamond AI. 2000. In vivo analysis of Cajal body movement, separation, and joining in live human cells. *J Cell Biol* **151:** 1561–1574. doi:10.1083/jcb.151.7.1561

Pollock C, Daily K, Nguyen VT, Wang C, Lewandowska MA, Bensaude O, Huang S. 2011. Characterization of MRP RNA-protein interactions within the perinucleolar compartment. *Mol Biol Cell* **22:** 858–866. doi:10.1091/mbc.e10-09-0768

Prasanth KV, Prasanth SG, Xuan Z, Hearn S, Freier SM, Bennett CF, Zhang MQ, Spector DL. 2005. Regulating gene expression through RNA nuclear retention. *Cell* **123:** 249–263. doi:10.1016/j.cell.2005.08.033

Quinodoz SA, Ollikainen N, Tabak B, Palla A, Schmidt JM, Detmar E, Lai MM, Shishkin AA, Bhat P, Takei Y, et al. 2018. Higher-order inter-chromosomal hubs shape 3D genome organization in the nucleus. *Cell* **174:** 744–757.e24. doi:10.1016/j.cell.2018.05.024

Quinodoz SA, Bhat P, Ollikainen N, Jachowicz JW, Banerjee AK, Chovanec P, Blanco MR, Chow A, Markaki Y, Plath K, et al. 2020. RNA promotes the formation of spatial compartments in the nucleus. bioRxiv doi:10.1101/2020.08.25.267435

Rabl C. 1885. Uber zelltheilung. *Morphol Jahrbuch* **10:** 214–330.

Randise-Hinchliff C, Brickner JH. 2018. Nuclear pore complex in genome organization and gene expression in yeast. In *Nuclear pore complexes in genome organization, function and maintenance* (ed. D'Angelo M), pp. 87–109. Springer, Berlin.

Rao SS, Huntley MH, Durand NC, Stamenova EK, Bochkov ID, Robinson JT, Sanborn AL, Machol I, Omer AD, Lander ES, et al. 2014. A 3D map of the human genome at kilobase resolution reveals principles of chromatin looping. *Cell* **159:** 1665–1680. doi:10.1016/j.cell.2014.11.021

Raška I, Andrade LE, Ochs RL, Chan EK, Chang CM, Roos G, Tan EM. 1991. Immunological and ultrastructural studies of the nuclear coiled body with autoimmune antibodies. *Exp Cell Res* **195:** 27–37. doi:10.1016/0014-4827(91)90496-H

Raška I, Shaw PJ, Cmarko D. 2006. Structure and function of the nucleolus in the spotlight. *Curr Opin Cell Biol* **18:** 325–334. doi:10.1016/j.ceb.2006.04.008

Rego A, Sinclair PB, Tao W, Kireev I, Belmont AS. 2008. The facultative heterochromatin of the inactive X chromosome has a distinctive condensed ultrastructure. *J Cell Sci* **121:** 1119–1127. doi:10.1242/jcs.026104

Rhee HW, Zou P, Udeshi ND, Martell JD, Mootha VK, Carr SA, Ting AY. 2013. Proteomic mapping of mitochondria in living cells via spatially restricted enzymatic tagging. *Science* **339:** 1328–1331. doi:10.1126/science.1230593

Robson MI, de Las Heras JI, Czapiewski R, Lê Thành P, Booth DG, Kelly DA, Webb S, Kerr ARW, Schirmer EC. 2016. Tissue-specific gene repositioning by muscle nuclear membrane proteins enhances repression of critical developmental genes during myogenesis. *Mol Cell* **62:** 834–847. doi:10.1016/j.molcel.2016.04.035

Ryba T, Hiratani I, Lu J, Itoh M, Kulik M, Zhang J, Schulz TC, Robins AJ, Dalton S, Gilbert DM. 2010. Evolutionarily conserved replication timing profiles predict long-range chromatin interactions and distinguish closely related cell types. *Genome Res* **20:** 761–770. doi:10.1101/gr.099655.109

Sabari BR, Dall'Agnese A, Boija A, Klein IA, Coffey EL, Shrinivas K, Abraham BJ, Hannett NM, Zamudio AV, Manteiga JC, et al. 2018. Coactivator condensation at super-enhancers links phase separation and gene control. *Science* **361:** eaar3958. doi:10.1126/science.aar3958

Sanulli S, Trnka MJ, Dharmarajan V, Tibble RW, Pascal BD, Burlingame AL, Griffin PR, Gross JD, Narlikar GJ. 2019. HP1 reshapes nucleosome core to promote phase separation of heterochromatin. *Nature* **575:** 390–394. doi:10.1038/s41586-019-1669-2

Savino TM, Gébrane-Younès J, De Mey J, Sibarita JB, Hernandez-Verdun D. 2001. Nucleolar assembly of the rRNA processing machinery in living cells. *J Cell Biol* **153:** 1097–1110. doi:10.1083/jcb.153.5.1097

Sawyer IA, Sturgill D, Sung MH, Hager GL, Dundr M. 2016. Cajal body function in genome organization and transcriptome diversity. *Bioessays* **38:** 1197–1208. doi:10.1002/bies.201600144

Schep R, Brinkman EK, Leemans C, Vergara X, van der Weide RH, Morris B, van Schaik T, Manzo SG, Peric-Hupkes D, van den Berg J, et al. 2021. Impact of chromatin context on Cas9-induced DNA double-strand break repair pathway balance. *Mol Cell* **81:** 2216–2230.e10. doi:10.1016/j.molcel.2021.03.032

Schmidt HB, Görlich D. 2016. Transport selectivity of nuclear pores, phase separation, and membraneless organelles. *Trends Biochem Sci* **41:** 46–61. doi:10.1016/j.tibs.2015.11.001

Schöfer C, Weipoltshammer K. 2018. Nucleolus and chromatin. *Histochem Cell Biol* **150:** 209–225. doi:10.1007/s00418-018-1696-3

Schübeler D, Scalzo D, Kooperberg C, van Steensel B, Delrow J, Groudine M. 2002. Genome-wide DNA replication profile for *Drosophila melanogaster*: a link between transcription and replication timing. *Nat Genet* **32:** 438–442. doi:10.1038/ng1005

Seeber A, Gasser SM. 2017. Chromatin organization and dynamics in double-strand break repair. *Curr Opin Genet Dev* **43:** 9–16. doi:10.1016/j.gde.2016.10.005

Shiels C, Islam SA, Vatcheva R, Sasieni P, Sternberg MJ, Freemont PS, Sheer D. 2001. PML bodies associate specifically with the MHC gene cluster in interphase nuclei. *J Cell Sci* **114:** 3705–3716. doi:10.1242/jcs.114.20.3705

Shimi T, Kittisopikul M, Tran J, Goldman AE, Adam SA, Zheng Y, Jaqaman K, Goldman RD. 2015. Structural organization of nuclear lamins A, C, B1, and B2 revealed by superresolution microscopy. *Mol Biol Cell* **26:** 4075–4086. doi:10.1091/mbc.E15-07-0461

Shin Y, Brangwynne CP. 2017. Liquid phase condensation in cell physiology and disease. *Science* **357:** eaaf4382. doi:10.1126/science.aaf4382

Shopland LS, Lynch CR, Peterson KA, Thornton K, Kepper N, Hase J, Stein S, Vincent S, Molloy KR, Kreth G, et al. 2006. Folding and organization of a contiguous chromosome region according to the gene distribution pattern in primary genomic sequence. *J Cell Biol* **174:** 27–38. doi:10.1083/jcb.200603083

Sinclair P, Bian Q, Plutz M, Heard E, Belmont AS. 2010. Dynamic plasticity of large-scale chromatin structure revealed by self-assembly of engineered chromosome regions. *J Cell Biol* **190:** 761–776. doi:10.1083/jcb.2009 12167

Skaer RJ, Whytock S. 1976. The fixation of nuclei and chromosomes. *J Cell Sci* **20:** 221–231. doi:10.1242/jcs.20.1.221

Skaer RJ, Whytock S. 1977. Chromatin-like artifacts from nuclear sap. *J Cell Sci* **26:** 301–310. doi:10.1242/jcs.26.1.301

Smetana K, Busch H. 1964. Studies on the ultrastructure of the nucleoli of the Walker tumor and rat liver. *Cancer Res* **24:** 537–557.

Solovei I, Schermelleh L, Düring K, Engelhardt A, Stein S, Cremer C, Cremer T. 2004. Differences in centromere positioning of cycling and postmitotic human cell types. *Chromosoma* **112:** 410–423. doi:10.1007/s00412-004-0287-3

Solovei I, Kreysing M, Lanctôt C, Kösem S, Peichl L, Cremer T, Guck J, Joffe B. 2009. Nuclear architecture of rod photoreceptor cells adapts to vision in mammalian evolution. *Cell* **137:** 356–368. doi:10.1016/j.cell.2009.01.052

Solovei I, Wang AS, Thanisch K, Schmidt CS, Krebs S, Zwerger M, Cohen TV, Devys D, Foisner R, Peichl L, et al. 2013. LBR and lamin A/C sequentially tether peripheral heterochromatin and inversely regulate differentiation. *Cell* **152:** 584–598. doi:10.1016/j.cell.2013.01.009

Spector DL. 2006. Snapshot: cellular bodies. *Cell* **127:** 1071.e1–1071.e2. doi:10.1016/j.cell.2006.11.026

Spector DL, Lamond AI. 2011. Nuclear speckles. *Cold Spring Harb Perspect Biol* **3:** a000646. doi:10.1101/cshperspect.a000646

Strickfaden H, Tolsma TO, Sharma A, Underhill DA, Hansen JC, Hendzel MJ. 2020. Condensed chromatin behaves like a solid on the mesoscale in vitro and in living cells. *Cell* **183:** 1772–1784 e1713. doi:10.1016/j.cell.2020.11.027

Strom AR, Emelyanov AV, Mir M, Fyodorov DV, Darzacq X, Karpen GH. 2017. Phase separation drives heterochromatin domain formation. *Nature* **547:** 241–245. doi:10.1038/nature22989

Strzelecka M, Oates AC, Neugebauer KM. 2010. Dynamic control of cajal body number during zebrafish embryogenesis. *Nucleus* **1:** 96–108. doi:10.4161/nucl.1.1.10680

Su JH, Zheng P, Kinrot SS, Bintu B, Zhuang X. 2020. Genome-scale imaging of the 3D organization and transcriptional activity of chromatin. *Cell* **182:** 1641–1659.e26. doi:10.1016/j.cell.2020.07.032

Sun Y, Durrin LK, Krontiris TG. 2003. Specific interaction of PML bodies with the *TP53* locus in Jurkat interphase nuclei. *Genomics* **82:** 250–252. doi:10.1016/S0888-7543(03)00075-2

Sun J, Chen J, Mohagheghian E, Wang N. 2020. Force-induced gene up-regulation does not follow the weak power law but depends on H3K9 demethylation. *Sci Adv* **6:** eaay9095. doi:10.1126/sciadv.aay9095

Swift H. 1959. Studies on nuclear fine structure. *Brookhaven Symp Biol* **12:** 134–152.

Tajik A, Zhang Y, Wei F, Sun J, Jia Q, Zhou W, Singh R, Khanna N, Belmont AS, Wang N. 2016. Transcription upregulation via force-induced direct stretching of chromatin. *Nat Mater* **15:** 1287–1296. doi:10.1038/nmat4729

Takei Y, Yun J, Zheng S, Ollikainen N, Pierson N, White J, Shah S, Thomassie J, Suo S, Eng CL, et al. 2021. Integrated spatial genomics reveals global architecture of single nuclei. *Nature* **590:** 344–350. doi:10.1038/s41586-020-03126-2

Takizawa T, Meaburn KJ, Misteli T. 2008. The meaning of gene positioning. *Cell* **135:** 9–13. doi:10.1016/j.cell.2008.09.026

Testillano PS, Sanchez-Pina MA, Olmedilla A, Ollacarizqueta MA, Tandler CJ, Risueño MC. 1991. A specific ultrastructural method to reveal DNA: the NAMA-Ur. *J Histochem Cytochem* **39:** 1427–1438. doi:10.1177/39.10.1719069

Thiry M. 1995. Behavior of interchromatin granules during the cell cycle. *Eur J Cell Biol* **68:** 14–24.

Tiku V, Antebi A. 2018. Nucleolar function in lifespan regulation. *Trends Cell Biol* **28:** 662–672. doi:10.1016/j.tcb.2018.03.007

Tran JR, Adam SA, Goldman RD, Zheng Y. 2021. Dynamic nuclear lamina-chromatin interactions during G1 progression. bioRxiv doi:10.1101/2021.01.03.425156

Tripathi K, Parnaik VK. 2008. Differential dynamics of splicing factor SC35 during the cell cycle. *J Biosci* **33:** 345–354. doi:10.1007/s12038-008-0054-3

Uchino S, Ito Y, Sato Y, Handa T, Ohkawa Y, Tokunaga M, Kimura H. 2021. Visualizing transcription sites in living cells using a genetically encoded probe specific for the elongating form of RNA polymerase II. bioRxiv doi:10.1101/2021.04.27.441582

van Koningsbruggen S, Gierliński M, Schofield P, Martin D, Barton GJ, Ariyurek Y, den Dunnen JT, Lamond AI. 2010. High-resolution whole-genome sequencing reveals that specific chromatin domains from most human chromosomes associate with nucleoli. *Mol Biol Cell* **21:** 3735–3748. doi:10.1091/mbc.e10-06-0508

van Schaik T, Vos M, Peric-Hupkes D, Hn Celie P, van Steensel B. 2020. Cell cycle dynamics of lamina-associated DNA. *EMBO Rep* **21:** e50636. doi:10.15252/embr.202050636

van Steensel B, Belmont AS. 2017. Lamina-associated domains: links with chromosome architecture, heterochromatin, and gene repression. *Cell* **169:** 780–791. doi:10.1016/j.cell.2017.04.022

van Steensel B, Henikoff S. 2000. Identification of in vivo DNA targets of chromatin proteins using tethered dam methyltransferase. *Nat Biotechnol* **18:** 424–428. doi:10.1038/74487

van Steensel B, Delrow J, Henikoff S. 2001. Chromatin profiling using targeted DNA adenine methyltransferase. *Nat Genet* **27:** 304–308. doi:10.1038/85871

Vertii A, Ou J, Yu J, Yan A, Pagès H, Liu H, Zhu LJ, Kaufman PD. 2019. Two contrasting classes of nucleolus-associated domains in mouse fibroblast heterochromatin. *Genome Res* **29:** 1235–1249. doi:10.1101/gr.247072.118

Vourc'h C, Taruscio D, Boyle AL, Ward DC. 1993. Cell cycle-dependent distribution of telomeres, centromeres, and chromosome-specific subsatellite domains in the interphase nucleus of mouse lymphocytes. *Exp Cell Res* **205:** 142–151. doi:10.1006/excr.1993.1068

Walter J, Schermelleh L, Cremer M, Tashiro S, Cremer T. 2003. Chromosome order in HeLa cells changes during mitosis and early G1, but is stably maintained during subsequent interphase stages. *J Cell Biol* **160:** 685–697. doi:10.1083/jcb.200211103

Wang J, Shiels C, Sasieni P, Wu PJ, Islam SA, Freemont PS, Sheer D. 2004. Promyelocytic leukemia nuclear bodies associate with transcriptionally active genomic regions. *J Cell Biol* **164:** 515–526. doi:10.1083/jcb.200305142

Wang A, Kolhe JA, Gioacchini N, Baade I, Brieher WM, Peterson CL, Freeman BC. 2020. Mechanism of long-range chromosome motion triggered by gene activation. *Dev Cell* **52:** 309–320.e5. doi:10.1016/j.devcel.2019.12.007

Wang Y, Zhang Y, Zhang R, van Schaik T, Zhang L, Sasaki T, Peric-Hupkes D, Chen Y, Gilbert DM, van Steensel B, et al. 2021. SPIN reveals genome-wide landscape of nuclear compartmentalization. *Genome Biol* **22:** 36. doi:10.1186/s13059-020-02253-3

Watson ML. 1959. Further observations on the nuclear envelope of the animal cell. *J Biophys Biochem Cytol* **6:** 147–156. doi:10.1083/jcb.6.2.147

Weeks SE, Metge BJ, Samant RS. 2019. The nucleolus: a central response hub for the stressors that drive cancer progression. *Cell Mol Life Sci* **76:** 4511–4524. doi:10.1007/s00018-019-03231-0

Wei MT, Chang YC, Shimobayashi SF, Shin Y, Strom AR, Brangwynne CP. 2020. Nucleated transcriptional condensates amplify gene expression. *Nat Cell Biol* **22:** 1187–1196. doi:10.1038/s41556-020-00578-6

White EJ, Emanuelsson O, Scalzo D, Royce T, Kosak S, Oakeley EJ, Weissman S, Gerstein M, Groudine M, Snyder M, et al. 2004. DNA replication-timing analysis of human chromosome 22 at high resolution and different developmental states. *Proc Natl Acad Sci* **101:** 17771–17776. doi:10.1073/pnas.0408170101

Wilson KL, Foisner R. 2010. Lamin-binding proteins. *Cold Spring Harb Perspect Biol* **2:** a000554. doi:10.1101/cshperspect.a000554

Wong X, Cutler JA, Hoskins VE, Gordon M, Madugundu AK, Pandey A, Reddy KL. 2021. Mapping the micro-proteome of the nuclear lamina and lamina-associated domains. *Life Sci Alliance* **4:** e202000774. doi:10.26508/lsa.202000774

Wu R, Singh PB, Gilbert DM. 2006. Uncoupling global and fine-tuning replication timing determinants for mouse pericentric heterochromatin. *J Cell Biol* **174:** 185–194. doi:10.1083/jcb.200601113

Xie SQ, Pombo A. 2006. Distribution of different phosphorylated forms of RNA polymerase II in relation to cajal and PML bodies in human cells: an ultrastructural study. *Histochem Cell Biol* **125:** 21–31. doi:10.1007/s00418-005-0064-2

Xie W, Chojnowski A, Boudier T, Lim JS, Ahmed S, Ser Z, Stewart C, Burke B. 2016. A-type lamins form distinct filamentous networks with differential nuclear pore complex associations. *Curr Biol* **26:** 2651–2658. doi:10.1016/j.cub.2016.07.049

Yamazaki T, Yamamoto T, Yoshino H, Souquere S, Nakagawa S, Pierron G, Hirose T. 2021. Paraspeckles are constructed as block copolymer micelles. *EMBO J* **40:** e107270. doi:10.15252/embj.2020107270

Yang XC, Burch BD, Yan Y, Marzluff WF, Dominski Z. 2009. FLASH, a proapoptotic protein involved in activation of caspase-8, is essential for 3′ end processing of histone pre-mRNAs. *Mol Cell* **36:** 267–278. doi:10.1016/j.molcel.2009.08.016

Yao RW, Xu G, Wang Y, Shan L, Luan PF, Wang Y, Wu M, Yang LZ, Xing YH, Yang L, et al. 2019. Nascent pre-rRNA sorting via phase separation drives the assembly of dense fibrillar components in the human nucleolus. *Mol Cell* **76:** 767–783.e11. doi:10.1016/j.molcel.2019.08.014

Yasuzumi G, Sawada T, Sugihara R, Kiriyama M, Sugioka M. 1958. Electron microscope researches on the ultrastructure of nucleoli in animal tissues. *Z Zellforsch Mikrosk Anat* **48:** 10–23. doi:10.1007/BF00496710

Zatsepina OV, Poliakov V, Chentsov Iu S. 1983. Electron microscopic study of the chromonema and chromomeres in mitotic and interphase chromosomes. *Tsitologiia* **25:** 123–129.

Zatsepina OV, Dudnic OA, Todorov IT, Thiry M, Spring H, Trendelenburg MF. 1997. Experimental induction of prenucleolar bodies (PNBs) in interphase cells: interphase PNBs show similar characteristics as those typically observed at telophase of mitosis in untreated cells. *Chromosoma* **105:** 418–430. doi:10.1007/BF02510478

Zhang LF, Huynh KD, Lee JT. 2007. Perinucleolar targeting of the inactive X during S phase: evidence for a role in the maintenance of silencing. *Cell* **129:** 693–706. doi:10.1016/j.cell.2007.03.036

Zhang H, Emerson DJ, Gilgenast TG, Titus KR, Lan Y, Huang P, Zhang D, Wang H, Keller CA, Giardine B, et al. 2019. Chromatin structure dynamics during the mitosis-to-G1 phase transition. *Nature* **576:** 158–162. doi:10.1038/s41586-019-1778-y

Zhang H, Zhao R, Tones J, Liu M, Dilley RL, Chenoweth DM, Greenberg RA, Lampson MA. 2020a. Nuclear body phase separation drives telomere clustering in ALT cancer cells. *Mol Biol Cell* **31:** 2048–2056. doi:10.1091/mbc.E19-10-0589

Zhang L, Zhang Y, Chen Y, Gholamalamdari O, Wang Y, Ma J, Belmont AS. 2020b. TSA-seq reveals a largely conserved genome organization relative to nuclear speckles with small position changes tightly correlated with gene expression changes. *Genome Res* **31:** 151–264. doi:10.1101/gr.266239.120

Zhao PA, Rivera-Mulia JC, Gilbert DM. 2017. Replication domains: genome compartmentalization into functional replication units. In *DNA replication: from old principles to new discoveries* (ed. Masai H, Foiani M), pp. 229–257. Springer, Singapore.

Zhao B, Chaturvedi P, Zimmerman DL, Belmont AS. 2020. Efficient and reproducible multigene expression after single-step transfection using improved BAC transgenesis and engineering toolkit. *ACS Synth Biol* **9:** 1100–1116. doi:10.1021/acssynbio.9b00457

Zhou Y, Gerrard DL, Wang J, Li T, Yang Y, Fritz AJ, Rajendran M, Fu X, Stein G, Schiff R, et al. 2019. Temporal dynamic reorganization of 3D chromatin architecture in hormone-induced breast cancer and endocrine resistance. *Nat Commun* **10:** 1522. doi:10.1038/s41467-019-09320-9

Zhu L, Richardson TM, Wacheul L, Wei MT, Feric M, Whitney G, Lafontaine DLJ, Brangwynne CP. 2019. Controlling the material properties and rRNA processing function of the nucleolus using light. *Proc Natl Acad Sci* **116:** 17330–17335. doi:10.1073/pnas.1903870116

The Impact of Space and Time on the Functional Output of the Genome

Marcelo Nollmann,[1] Isma Bennabi,[2] Markus Götz,[1] and Thomas Gregor[2,3]

[1]Centre de Biologie Structurale, CNRS UMR5048, INSERM U1054, Univ Montpellier, 34090 Montpellier, France

[2]Department of Stem Cell and Developmental Biology, CNRS UMR3738, Institut Pasteur, 75015 Paris, France

[3]Joseph Henry Laboratory of Physics & Lewis-Sigler Institute for Integrative Genomics, Princeton University, Princeton, New Jersey 08544, USA

Correspondence: thomas.gregor@pasteur.fr

Over the past two decades, it has become clear that the multiscale spatial and temporal organization of the genome has important implications for nuclear function. This review centers on insights gained from recent advances in light microscopy on our understanding of transcription. We discuss spatial and temporal aspects that shape nuclear order and their consequences on regulatory components, focusing on genomic scales most relevant to function. The emerging picture is that spatiotemporal constraints increase the complexity in transcriptional regulation, highlighting new challenges, such as uncertainty about how information travels from molecular factors through the genome and space to generate a functional output.

There is a growing appreciation that gene function is connected to the dynamic structure of chromosomes. In particular, spatiotemporal aspects of genome architecture are now appreciated as crucial to our understanding of eukaryotic gene expression, and thus to the main functional output of the nucleus (van Steensel and Furlong 2019). However, the mechanistic underpinnings and actual causal links between structure and function are scarce and present an obvious challenge for the next decade. Generation of vast atlases of Hi-C proximity maps over the past 10 years have provided a starting point for such studies (Dekker 2016), and they underscore the existence of DNA organization at the kb-Mb scales as an inherent, functionally important, component in gene activation. However, these atlases are limited in their ability to link single-cell, three-dimensional genome structures to specific transcriptional states. In addition, a characterization of how dynamic long-range interactions regulate single-cell transcriptional dynamics is currently missing, especially at the genomic scales, in the kilobase (kb) to megabase (Mb) range, where *cis*-regulatory elements (CREs) mediate functional interactions.

CREs are noncoding DNA regions that regulate transcription and include enhancers, chromatin insulators, silencers, and promoters (Wittkopp and Kalay 2012). Enhancers are short regulatory DNA sequences that control gene activity and contribute to the dynamic control of

Cite this article as *Cold Spring Harb Perspect Biol* doi: 10.1101/cshperspect.a040378

gene expression (de Laat and Duboule 2013; Bolt and Duboule 2020). The stochastic readout of regulators at promoter or enhancer sequences (Fig. 1A,B) gives rise to the time- and tissue-specific activation of subsets of genes to confer cell identity. Since their discovery four decades ago (Banerji et al. 1981, 1983; Moreau et al. 1981; Gillies et al. 1983; Mercola et al. 1983), enhancers have been largely regarded as autonomous, modular units, capable of activating transcription in a location- and orientation-independent manner. Enhancers are often at large distances, up to several Mb, from their respective target-gene promoters, most times with additional nontarget genes found within the intervening sequences (Furlong and Levine 2018; Schoenfelder and Fraser 2019). The canonical model is that enhancers regulate transcription by physically interacting with promoters over large genomic distances (Fig. 1C; Blackwood and Kadonaga 1998). Whole-genome methods have shown that the human genome is riddled with enhancers, with estimates ranging from hundreds of thousands to more than a million (Pennacchio et al. 2013; Schoenfelder and Fraser 2019; Xu et al. 2020). On average, a typical human gene is regulated by at least 10–20 different enhancers (Sanyal et al. 2012), raising the possibility that multiple enhancers may physically contact the same promoter (Fig. 1D,E).

Remarkably, this network of interactions may be modulated by other CREs, such as chromatin insulators. These are short, *cis*-regulatory sequences that block communication between promoters and enhancers (Reitman et al. 1990; Geyer and Corces 1992; Cai and Levine 1995). Genome-wide, thousands of sites are characterized as insulators, particularly enriched in intergenic and promoter regions. Insulators play a role in the formation of long-range interactions (Vogelmann et al. 2011, 2014; Yang and Corces 2012) and are involved in the global regulation of transcription (Bushey et al. 2009). Chromosome conformation capture (3C)-based assays (Lieberman-Aiden et al. 2009), which capture chromatin–chromatin interactions, revealed the existence of well-defined kb-Mb genomic regions displaying locally enhanced chromatin

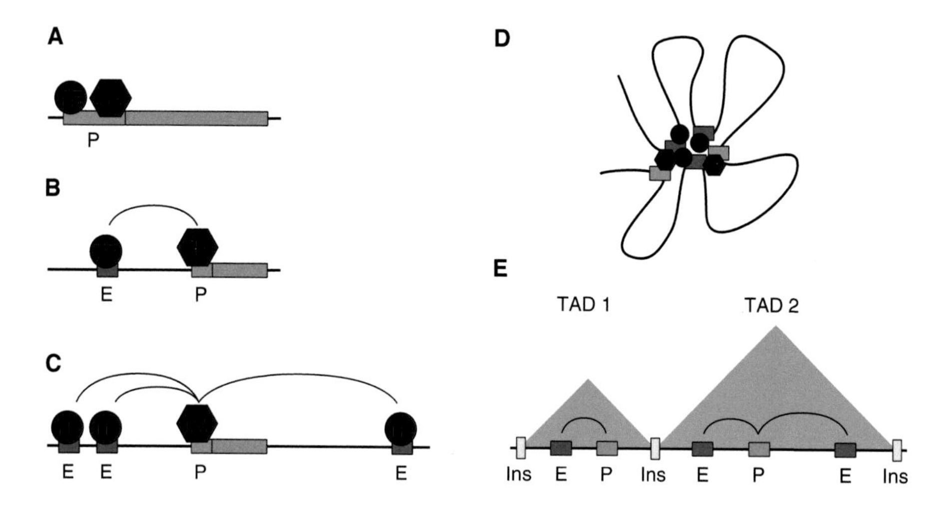

Figure 1. Increasing complexity in transcriptional control. (*A*) Transcription factors (TFs, red circle) and components of the transcription machinery (TM, red hexagon), such as RNA polymerase II and transcriptional activators like the Mediator complex, assemble at the promoter (P, blue), upstream of the gene body (gray). (*B*) TFs bind to enhancer elements (E, green), giving rise to temporal and tissue-specific gene regulation by interaction with a disjoint promoter element. (*C*) Multiple enhancers modulate the transcriptional activity of a promoter, often over large genomic distances. (*D*) CREs organized in nuclear space forming a three-dimensional gene locus, interacting with various factors (red). (*E*) Topologically associating domains (TADs) are thought to encapsulate such three-dimensional structures, forming a network of enhancer–promoter interactions. TADs are often demarcated by insulators (Ins, yellow).

interactions (Dixon et al. 2012; Nora et al. 2012; Sexton et al. 2012). These genomic regions, called topologically associating domains (TADs), tend to be demarcated by insulators (Hou et al. 2012; Sexton et al. 2012; Rao et al. 2015) and often encapsulate enhancers and their target genes (Fig. 1E; Shen et al. 2012; Dowen et al. 2014; Symmons et al. 2014; Ji et al. 2016; Neems et al. 2016; Ron et al. 2017). Presently, the role of TADs in gene activity is still an open question (for a review, see Schoenfelder and Fraser 2019; Cavalheiro et al. 2021).

Disruption of TAD architecture, either by local duplications, deletions, or inversions that fuse adjacent TADs or form new TADs, were reported to trigger developmental defects in mammals by causing improper enhancer–promoter interactions (Northcott et al. 2014; Lupiáñez et al. 2015; Franke et al. 2016). In contrast, global disruption of TADs by depletion of CTCF or cohesin, two factors involved in formation of TADs, caused only mild changes in gene expression (Nora et al. 2017; Rao et al. 2017; Schwarzer et al. 2017), suggesting that TADs may be important for regulating only a subset of genes.

The reason for these apparent discrepancies may arise from the intrinsic dynamic properties of the chromatin fiber and transcription processes. Most studies assessing the transcriptional roles of TADs stem from a qualitative view and are based on ensemble, static measurements of TAD borders or contacts between CREs and promoters, combined with the assessment of bulk transcriptional outputs. This view entirely lacks the temporal and spatial aspects that are critical to understanding the genesis of these intrinsically dynamic processes, or to address key questions about the roles of TADs in transcriptional regulation: Are chromosome dynamics random and thus a key component to the stochasticity of gene expression? How can remote enhancers direct the correct spatial and temporal control of transcription? How do multiple enhancers interact dynamically to control promoter accessibility?

Renewed focus on spatial nuclear chromatin organization and its potential impact on gene regulation has recently brought a new twist to our understanding of eukaryotic transcription, broadening the potential angles evolution can use to interfere with the underlying mechanisms. Over the past 40 years, our view of transcription has gone from transcription factor (TFs) and machinery binding directly to promoters, to single or multiple enhancers mediating transcriptional control, to distal enhancer–promoter communication over megabase scales in *cis* and even in *trans* (Fig. 1). Now, a novel, spatial component comes into play, where loci are organized in three dimensions, and where organization can have functional consequences. In particular, recent imaging technologies have provided insightful tools to probe this organization and its dynamic impact on transcription. Here, we review how these new tools are starting to give new insights into otherwise inaccessible fundamental biological questions. Specifically, understanding how spatiotemporal changes in chromosome structure modulate genome function requires progress at four levels: (1) the relationship between genomic and physical scales, (2) the physical organization of regulatory elements, (3) the timescales involved in chromosome organization and transcription, and (4) the interplay between the spatiotemporal dynamics of the genome and its function. We discuss each in the following sections.

SPATIAL SCALES IN THE NUCLEUS

At the heart of the control of eukaryotic gene regulation is the organization and function of transcriptional enhancers, the major constituents of the noncoding genome that controls gene activity (Furlong and Levine 2018). A typical gene is regulated by multiple enhancers. Many of these enhancers possess overlapping regulatory activities, raising questions about proper *cis*-regulatory "trafficking," whereby the correct enhancers interact with the appropriate target promoters (Bushey et al. 2008; Chopra et al. 2009). What is the influence of the exact position and affinity of DNA binding factors to CREs on transcriptional output? Answering these questions requires not only the relative genomic positioning of CREs in one dimension (Negre et al. 2011; ENCODE Project Consortium 2012), but also the direct detection of enhancer–promoter interactions in 3D to understand the impact of spatial chromatin architecture on function.

1D to 3D: Unintuitive Expectations

The physical properties of the chromatin fiber can be modeled using conventional polymer theory. In the simplest case, chromatin can be approximated by a polymer composed of a chain of identical monomers (de Gennes and Gennes 1979; Doi 1996; Rippe 2001; Nelson 2003; Wiggins et al. 2006; Doi 1996; Wiggins et al. 2006; Grosberg and Khokhlov 2011). The dynamics of the polymer chain are then governed by thermal fluctuations, excluded volume interactions, and rigidity parameters such as the effective persistence length. This simple model has been successfully used to predict the expected mean 3D distance between two DNA loci as a function of their genomic distance in sequence space (1D) (Rosa and Everaers 2008). Importantly, the conversion between 1D and 3D distances follows a power law (Mirny 2011) with a fractional exponent that varies between species, mainly due to changes in the physical properties of chromatin, such as the persistence length, the molecular composition of the chromatin fiber, genomic sizes, and nuclear volumes (Mirny 2011).

Importantly, the identified power–law relation between distances in sequence (1D) and in physical (3D) space is nonlinear. For instance, so-called distal enhancers in *Drosophila* are typically found at relatively short genomic distances (1–100 kb) (Ghavi-Helm et al. 2014) that can translate into relatively large average physical distances (200–400 nm) (Rosa and Everaers 2008; Cardozo Gizzi et al. 2019). Similarly, in mammals, enhancers found hundreds of kb from their target promoter are on average very far away (>1 µm [Rosa and Everaers 2008]). However, chromosomes are confined within the limited nuclear space, thus DNA loci located very far in genomic space (e.g., 2 Mb), or even on other chromosomes, are not proportionally separated in 3D space (e.g., 1 µm [Cattoni et al. 2017]).

The 1D sequence encodes information for proteins to bind to specific sites, which in turn can modulate the folding of chromatin in 3D. The binding of chromatin insulator proteins (e.g., CTCF), TFs, heterochromatin-associated complexes (e.g., Polycomb, HP1), or architectural proteins (e.g., cohesin, condensin) (Rowley and Corces 2016) promote 3D bridges between chromatin regions that can affect the frequencies of interactions expected for a homopolymer (Rowley and Corces 2016). For example, 3D loops between converging CTCF sites can produce specific 3D interactions between TAD borders (Rao et al. 2015) while cohesin-mediated loops can affect the contact frequencies within TADs (Nora et al. 2017; Rao et al. 2017). Alternatively, chromatin hubs linking multiple genomic loci can lead to the spatial clustering of regulatory elements (Beagrie et al. 2017; Allahyar et al. 2018; Oudelaar et al. 2018; Quinodoz et al. 2018; Espinola et al. 2021). Thus, both passive binding (e.g., CTCF, TFs) and active processes (e.g., cohesin/condensin looping, transcriptional elongation) can alter 3D chromatin organization at the kilobase to megabase scales, bringing into close spatial proximity loci that would be expected to reside far apart in a pure homopolymeric chromosome. Understanding these processes and their functional consequences holds potentially far-reaching insights into genome evolution, and requires methods able to dissect chromatin interactions in 3D.

3D Mapping: Sequencing and Imaging Approaches

Chromatin conformation can be measured by different types of methods. One class is sequencing-based, such as 3C and its derivatives, where chromatin contact frequencies measure the proximity between genomic loci averaged over a population of cells. The range over which genomic loci are cross-linked by ligation is much debated but computational models estimate the distance to be around 100 nm (Giorgetti et al. 2014).

An alternative class of approaches is based on microscopy techniques, especially fluorescence in situ hybridization (DNA-FISH), that directly detect pairwise distances in single cells. These are used to estimate imaging-derived contact frequencies by computing the proportion of cells displaying pairwise distances smaller than a critical radius R_M (equivalent to integrating the spherical pairwise distance distribution up to R_M) (Wang et al. 2016; Cattoni et al. 2017; Bintu et al. 2018; Cardozo Gizzi et al. 2019; Finn et al. 2019; Mateo et al. 2019).

It is worth noting that in neither case does the notion of "contact" imply physical interaction, but rather an estimate of whether two genomic loci are spatially close to each other. Thus, in the remainder of this review, we will refer to "proximity" rather than "contact frequency." R_M-values can be derived from control experiments where a single genomic locus is imaged in multiple colors/cycles (Cattoni et al. 2017; Cardozo Gizzi et al. 2019; Mateo et al. 2019) and R_M-values between 150–500 nm have typically been used, producing good correlations between Hi-C and microscopy-based proximity frequencies (Wang et al. 2016; Bintu et al. 2018; Cardozo Gizzi et al. 2019; Mateo et al. 2019; Su et al. 2020). As such, the working definition of proximity is different for genomic-based and imaging-based methods, and its exact meaning will likely shift in future measurements with increasing genomic and optical resolutions. Determining the true measure for each method will likely involve a correlation analysis that reveals the scales at which proximity frequencies are correlated for both methods.

In the past, it has become common practice to assess the specificity of 3D interactions by using DNA-FISH to compare the mean 3D pairwise distance between two candidate genomic loci with that of a control that resides at the same genomic distance (Rao et al. 2015; Ogiyama et al. 2018; Benabdallah et al. 2019). However, relying purely on mean distances can be deceptive in cases where the full pairwise distance distribution departs from that of a single species or when specific 3D interactions are rare. For instance, while proximity frequency and mean spatial distance are inversely correlated for a homopolymer, this is not necessarily the case in the presence of sequence-dependent 3D interactions, where a 3D loop is rather expected to lead to a bimodal distance distribution (Giorgetti et al. 2014; Fudenberg and Imakaev 2017). In this case, the mean pairwise distance fails to capture sequence-specific looping interactions, particularly if these occur at low frequencies (e.g., <10%).

Even more counterintuitively, an increase in the frequency of sequence-specific interactions can lead to lower, equal, or even higher mean pairwise distances depending on the shape of the pairwise distance distribution (Giorgetti and Heard 2016; Fudenberg and Imakaev 2017). Thus, in a general case, a change in the mean pairwise distance between two genomic regions may not necessarily reflect changes in proximity. Instead, the proximity measured from the full pairwise distance distribution obtained from a DNA-FISH experiment should always be used to assess changes in 3D chromatin organization. These considerations are critical to derive models of enhancer function from DNA-FISH experiments.

Heterogeneity in chromosome organization is well documented by the observed width of pairwise distance distributions (Giorgetti et al. 2014; Cattoni et al. 2017; Bintu et al. 2018; Finn et al. 2019) and by large cell-to-cell variations in TAD volumes (Boettiger et al. 2016; Nir et al. 2018; Szabo et al. 2018; Luppino et al. 2020). These structural heterogeneities can originate from the intrinsic polymer dynamics of the chromatin fiber or from other extrinsic sources, such as active transcription acting on chromatin. Alternatively, DNA-binding proteins such as insulators (e.g., CTCF) (Hansen et al. 2017) or TFs (Izeddin et al. 2014; Normanno et al. 2015) can rapidly bind to and dissociate from DNA (seconds to minutes timescales) to shape the dynamics of local and global chromatin conformation. A final possibility is that the structure of the chromatin fiber is actually distinct in individual cells (Finn and Misteli 2019).

Bridging the Genomic and Molecular Scales

Genomic and imaging methods can map chromatin proximities in the 100–500 nm range; however, the factors mediating these interactions are typically much smaller. DNA-binding domains have typical sizes in the sub-nm range (e.g., CTCF zinc fingers 4–8) (Fig. 2A; Yin et al. 2017), similarly to protein–protein interaction domains (e.g., SA2-SCC1 cohesin subunit

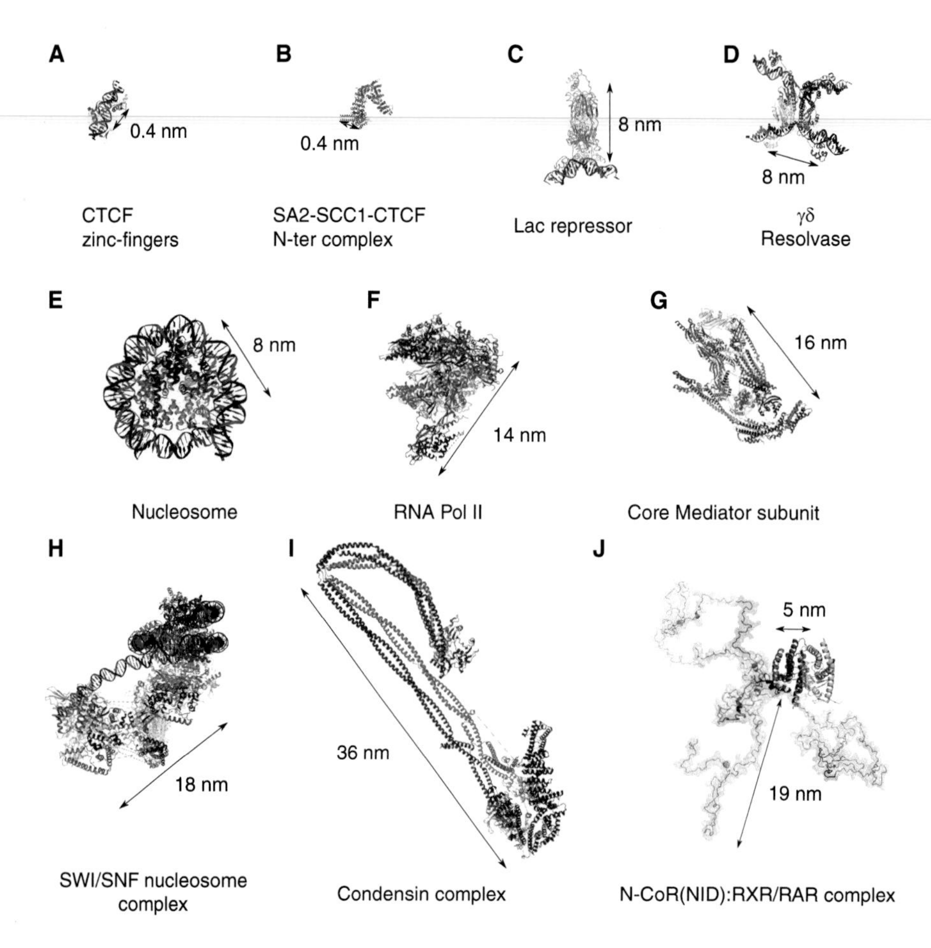

Figure 2. Atomic-resolution structures and physical sizes of factors involved in chromosome organization and function. (*A*) CTCF zinc-fingers 4–8 bound to DNA (5YEG) (Yin et al. 2017), (*B*) SA2-SCC1 subunit of cohesin (green) bound to CTCF amino-terminal fragment (orange) (6QNX) (Li et al. 2020b), (*C*) Lac repressor bound to DNA (1LBG) (Lewis et al. 1996), (*D*) γδ resolvase in complex with site I DNA (1ZR4) (Li et al. 2005), (*E*) human nucleosome complex (3AFA) (Tachiwana et al. 2010), (*F*) complete 12-subunit RNA Pol II (5FJ8) (Armache et al. 2005), (*G*) crystal structure of the 15-subunit core Mediator complex from *Schizosaccharomyces pombe* (5N9J) (Nozawa et al. 2017), (*H*) SWI/SNF in complex with nucleosome (6TDA) (Wagner et al. 2020), (*I*) condensin complex from *Saccharomyces cerevisiae* (6YVU) (Lee et al. 2020), and (*J*) folded domains and three putative models of the asymmetric forms of the N-CoRNID:RXR/RAR complex are displayed (Cordeiro et al. 2019).

bound to CTCF amino-terminal domain) (Fig. 2B; Li et al. 2020b). DNA-bound TF sizes are commonly in the nanometer range (e.g., lactose operon repressor bound to operator DNA) (Fig. 2C; Lewis et al. 1996), while distances between DNA segments bridged by single-protein complexes can be ∼10 nm apart (e.g., γδ resolvase bound to site I DNA) (Fig. 2D; Li et al. 2005). Similarly, nucleosomes are ∼10 nm in size (Fig. 2E; Tachiwana et al. 2010). Multi-subunit machines, such as the Mediator complex, RNA polymerase II, or the SWI/SNF chromatin remodeling complex can reach sizes between 14 and 22 nm (Fig. 2F–H; Armache et al. 2005; Robinson et al. 2015; Nozawa et al. 2017; Wagner et al. 2020), while condensins can be up to 40 nm long (e.g., condensin complex from *Saccharomyces cerevisiae*) (Fig. 2I; Lee et al. 2020).

It is generally accepted that enhancer function requires looping to the promoter (for reviews, see Bulger and Groudine 2011; Schwarzer and Spitz 2014). Indeed, imaging-based studies recently reported that transcriptional activation can occur when enhancers get closer than 200–350 nm to promoters (Chen et al. 2018; Li et al. 2020a). Similarly, 3C data combined with modeling described the formation of 100–200 nm active "cages" (Di Stefano et al. 2020). Thus, there is a notable gap between genomic scales from in vivo enhancer–promoter distance measurements (200–350 nm) and molecular scales from atomic-resolution molecular models (1–50 nm).

Part of the explanation for this discrepancy may be the formation of higher-order molecular structures mediated by unstructured protein regions. In fact, multiple factors involved in transcriptional regulation and chromosome organization possess low-complexity/intrinsically disordered regions (IDRs) that may be able to bridge relatively large physical distances (Watson and Stott 2019). For instance, histone tails constitute sites of extensive posttranslational modifications that encode epigenetic information, and TFs have been long predicted to encode extensive IDRs (Liu et al. 2006). IDRs can increase the effective volumes occupied by protein complexes, as for the N-CoRNID:RXR/RAR nuclear receptor complex (Fig. 2J; Cordeiro et al. 2019). In addition, IDRs can fold by interacting with other factors, such as a region of the intrinsically disordered amino-terminal domain of CTCF when in interaction with the SA2-SCC1 subunit of cohesin (Li et al. 2020b). Notably, these interactions are necessary for loop formation (Li et al. 2020b; Pugacheva et al. 2020). Finally, IDRs can play many functions (Watson and Stott 2019), including formation of phase-separated compartments (Boija et al. 2018; Chong et al. 2018; Sabari et al. 2018). These alternative models will be explored in the following section.

SPATIAL CLUSTERING OF REGULATORY COMPONENTS

Gene expression is a highly regulated process that relies on a multitude of interactions between *cis*- and *trans*-acting factors. Whereas CREs are *cis*-regulatory genetic elements (such as enhancers, promoters, insulators, and silencers), *trans*-factors are molecular components, such as proteins, protein complexes, and noncoding RNAs that are involved in gene regulation in the form of TFs, coactivators (e.g., the Mediator complex), RNA Pol II, lncRNA, etc. (Schoenfelder and Fraser 2019). At present, very limited knowledge exists about the relative spatial organization and interactions of CREs and *trans*-factors in the nucleus. Which of these components, and how many, come together in the nuclear space? Do they display distinct spatial organizations? How are interactions between these elements mediated?

Recent findings revealed that both *cis*- and *trans*-regulatory components can come into close spatial proximity, and even cluster in spatially localized regions. No common terminology for the observed phenomena has been found as of now (Peng and Weber 2019). Thus, we made the choice to refer to clusters of chromatin regions containing CREs as CRE hubs, and nuclear aggregates of proteins or protein complexes as foci, without implying an underlying physical mechanism for either of them (Fig. 3A). The following sections will summarize evidence for the formation of CRE hubs, transcription-associated foci, and the physical mechanisms that may describe their formation.

Hubs of *Cis*-Regulatory Elements

In a canonical model, direct physical interaction between an enhancer and a promoter is required for enhancer action. The typical eukaryotic gene is regulated by multiple enhancers, particularly during development (Osterwalder et al. 2018; Fulco et al. 2019; Oudelaar and Higgs 2021), and many enhancers may be shared by multiple target genes (Ghavi-Helm et al. 2014), sometimes even simultaneously (Fukaya et al. 2016). These observations suggest that multiple enhancers and promoters come into close spatial proximity during transcription.

To directly probe the spatial clustering of multiple CREs, standard 3C-based techniques are not sufficient as they detect binary chromatin interactions. Consequently, sequencing and imaging-based technologies have been developed to detect multiway interactions. Among the former, "genome architecture mapping" (GAM)

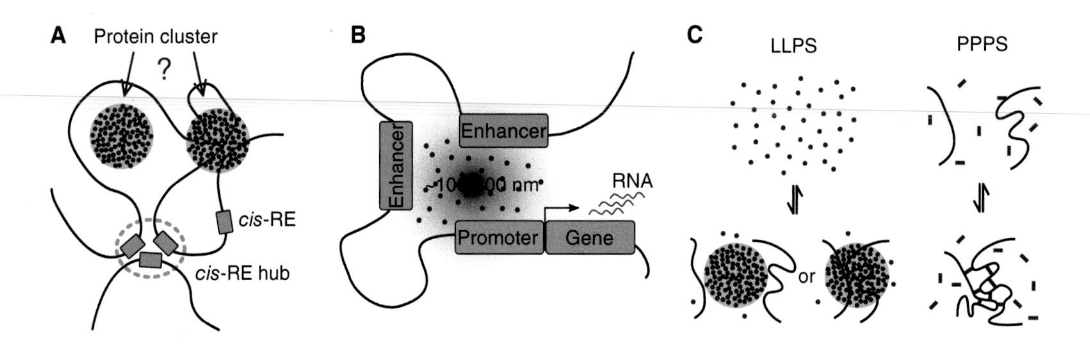

Figure 3. CRE (*cis*-regulatory element) hubs, protein foci, and phase-separated condensates. (*A*) Nuclear protein foci may or may not be DNA-associated/bound. The relationship between protein foci and CRE hubs is so far unknown. (*B*) Model for transcription-associated, foci-mediating enhancer–promoter interactions. (*C*) Two mechanisms of condensate formation via phase separation. Liquid–liquid phase separation (LLPS) is driven by weak, multivalent interactions between the constituents of the condensate (red dots) and does not require a scaffolding polymer. Conversely, polymer–polymer phase separation (PPPS) results from cross-linking the chromatin scaffold by a bridging factor (red rectangles).

(Beagrie et al. 2017) and "split-pool recognition of interactions by tag extension" (SPRITE) (Quinodoz et al. 2018) were able to detect three-way interactions between super-enhancers in fixed mouse embryonic stem cells (mESCs). Alternatively, proximity-ligation methods were used to describe the existence of multiway enhancer interactions at the β- and α-globin locus during cell differentiation (Allahyar et al. 2018; Oudelaar et al. 2018). All in all, these results showed that multiple CREs can be captured in close proximity (Allahyar et al. 2018; Oudelaar et al. 2018). However, these methods are unable to detect multiple chromatin contacts and transcriptional status at once; thus, it was unclear whether clustering of CREs has a functional role.

Multiplexed imaging methods are ideally suited to tackle this question. Several complementary approaches rely on similar principles: "chromatin tracing" (Wang et al. 2016; Bintu et al. 2018), "Hi-M" (Cardozo Gizzi et al. 2019), "optical reconstruction of chromatin architecture" (ORCA) (Mateo et al. 2019), "chromosome walking" (Nir et al. 2018), as well as "seqFISH^{+}" (Eng et al. 2019), and "MINA" (Liu et al. 2020). They combine microfluidics, wide-field microscopy, and Oligopaint FISH (Beliveau et al. 2012) to resolve the physical 3D position of tens to thousands of genomic loci in the nucleus, and can reach kilobase and nanometer precision.

These multiplexed imaging methods were recently used to simultaneously determine chromosome organization and transcription. ORCA revealed that the architecture of the *Hox* Polycomb TAD changes between different cell types in *Drosophila* embryos (Mateo et al. 2019). Hi-M was applied to detect CRE interactions and transcriptional status during early *Drosophila* development (Espinola et al. 2021) and revealed that CRE spatial clustering pre-dates gene activation and does not seem to depend on transcriptional state, with transcriptionally active and silent cells displaying similar CRE hubs (Espinola et al. 2021). These results are consistent with a concurrent Hi-C study demonstrating that chromatin conformation is independent of gene activity (Ing-Simmons et al. 2021), and with earlier reports showing that enhancer–promoter proximity can precede gene activation in mice or during *Drosophila* embryonic development (Montavon et al. 2011; Ghavi-Helm et al. 2014; Paliou et al. 2019). However, it is important to bear in mind that cell-specific changes in enhancer–promoter interaction networks during differentiation have also been well documented (Schoenfelder and Fraser 2019). Thus, we hypothesize that the roles of CRE hubs may be regulated either by altering their 3D structure and/or by modifying the cocktail of *trans*-acting factors they are bound by. These changes could be realized in different cell types by fine-tuning the abundance and binding of *trans*-acting factors.

Transcription-Associated Nuclear Foci

In eukaryotes, CREs are bound by TFs and coactivators (e.g., the histone acetyltransferase p300 or the Mediator complex). Interestingly, conventional and super-resolved imaging approaches revealed that many of these factors form nuclear foci (Cisse et al. 2013; Liu et al. 2014; Tsai et al. 2017; Boija et al. 2018; Cho et al. 2018; Chong et al. 2018; Dufourt et al. 2018; Mir et al. 2018; Sabari et al. 2018). This observation raised the possibility that *trans*-acting factors may form nuclear micro-environments where genes are coregulated. In some instances, transcription-associated nuclear foci have been reported to contain mRNAs (Boija et al. 2018; Sabari et al. 2018) and to associate with chromatin (Cho et al. 2018; Chong et al. 2018; Sabari et al. 2018), consistent with the observation of CRE hubs (Fig. 3A,B). However, it is not clear whether this is the norm or the exception.

The functional roles of transcription-associated nuclear foci in gene regulation are still unclear. For instance, are CREs always associated with nuclear foci? How many CREs take part in the formation of foci? Is a single CRE enough to form foci? Does the presence of multiple CREs increase transcriptional output? Answering these questions will require the ability to localize both *trans*-factors and multiple CREs simultaneously with high spatial resolution, and the use of perturbation methods (e.g., optogenetic manipulation of the low-complexity domains that mediate protein–protein interactions [Shin et al. 2017, 2018]). The next section describes the mechanisms that may be involved in the formation of transcription-associated nuclear foci and CRE hubs.

Phase Separation and Nuclear Structure

Phase separation has long been known as a process of self-organization in cells, and can explain the formation of membraneless compartments in the nucleus. Some of the most prominent examples include nuclear bodies (e.g., nucleoli, Cajal bodies, and DNA damage repair sites [Hyman et al. 2014; Banani et al. 2017; Shin and Brangwynne 2017; Boeynaems et al. 2018]) as well as chromatin itself (Gibson et al. 2019). For a current overview on the roles of phase separation in nuclear organization, we refer to Rippe (2021) and to recent reviews (Mir et al. 2019; Feric and Misteli 2021). Phase separation offers unique opportunities for controlling biochemical micro-environments by locally increasing the concentration of the constituents (and consequently chemical reaction rates), while still permitting dynamic exchange of reactants and products. Different mechanisms, involving distinct types of molecular interactions, have been proposed to lead to phase separation in the nucleus (Banani et al. 2017; Kato and McKnight 2017; Erdel and Rippe 2018). In addition, mechanisms not involving phase separation may also play a role in the formation of nuclear foci (McSwiggen et al. 2019).

Lately, liquid–liquid phase separation (LLPS) has received much attention as a possible mechanism of nuclear organization. LLPS compartments typically have spherical shapes, can fuse together, can deform under shear flow, are in exchange with the surrounding medium, and show a dynamic internal organization (Hyman et al. 2014). The key molecular driving force for the formation of liquid-like condensates are weak, multivalent interactions, typically involving proteins with IDRs (Banani et al. 2017; Shin and Brangwynne 2017).

Spatial clustering of super-enhancers often relies on the formation of condensates displaying typical properties of LLPS (Hnisz et al. 2017; Boija et al. 2018; Cho et al. 2018; Chong et al. 2018; Sabari et al. 2018). Consistently, the proteins involved in the formation of these condensates (TFs and coactivators) often contained IDRs that facilitated their nucleation. In this scenario, activating/repressive signals from CREs within a condensate may be transmitted over relatively large distances to the transcription machinery without requiring direct physical interactions. So far, it is unclear how long CREs remain associated with protein foci or whether promoters only need brief and transient encounters with these condensates for transcriptional activation (Cho et al. 2018). A more thorough discussion of the roles of LLPS in transcription can be found elsewhere (McSwiggen et al. 2019; Mir et al. 2019; Feric and Misteli 2021).

A second mechanism that can drive the formation of transcription-associated foci and CRE hubs is polymer–polymer phase separation (PPPS) (Erdel and Rippe 2018). This mechanism is based on the binding of soluble "bridging factors" to a long polymer chain, connecting two or more sites on the polymer. If the density of cross-links becomes sufficiently high, a polymer collapse takes place, resulting in a locally compact polymer globule. In this model, transient binding of cross-linking factors and their constant exchange with the surrounding nucleoplasm still leads to stable condensates, as long as a steady state with a sufficiently high density of bridging interactions is maintained (Erdel and Rippe 2018). Recently proposed models of phase separation for transcriptional control (Hnisz et al. 2017) are compatible with PPPS as they do not set any requirements on the chemical nature of the cross-links between chromatin chains. Counterintuitively, factors able to assemble liquid droplets in vitro may still form PPPS in vivo (Erdel et al. 2020). Despite their association to DNA, PPPS condensates are highly dynamic and, in some cases, can require the energy of ATP hydrolysis to ensure proper subnuclear positioning (Guilhas et al. 2020).

Perhaps the most important difference between LLPS and PPPS is that condensates formed via LLPS can persist independently of the chromatin scaffold, while condensates formed via PPPS strictly rely on such a scaffold and would disassemble in its absence (Fig. 3C). Future studies should experimentally test LLPS and PPPS models by examining concentration dependency and the need for a chromatin scaffold (Erdel and Rippe 2018; Erdel et al. 2020; Guilhas et al. 2020). These experiments would be particularly timely to further clarify whether the formation of CRE hubs requires transcription-associated foci and vice versa.

TIMESCALES OF NUCLEAR DYNAMICS

With the advent of high-resolution live imaging technologies, measurements of dynamic nuclear phenomena involving chromatin and binding factors have become a reality. These involve, on the one hand, measurements of dynamic parameters such as transition rates, binding/unbinding of factors, and more generally of kinetic rate constants that provide kinetic but not spatial information (Senecal et al. 2014; Zoller et al. 2018). On the other hand, it is now possible to go a step further to get real-time dynamic measurements with spatial information that, for example, follow the movement of chromatin or molecules in space and time (McCord et al. 2020; Shaban et al. 2020; Tortora et al. 2020). Here, we distinguish between *cis*-dynamics, that happen on the same chromosome, such as enhancer–promoter interactions or chromatin loop formation, and *trans* dynamics that involve interactions of *trans*-acting factors with chromatin, the formation of higher-order structures such as CRE hubs, or transvection.

Several studies, using both high-resolution and live-cell imaging but also single-cell 3C-based methods, have revealed a highly heterogeneous nature of genome organization in both space and time (Nagano et al. 2013; Cattoni et al. 2017; Flyamer et al. 2017; Stevens et al. 2017; Bintu et al. 2018; Hansen et al. 2018; Finn et al. 2019). However, few studies have in fact succeeded in quantifying actual chromosome dynamics in single cells, leaving a multitude of unanswered questions that are crucial for understanding genome organization and function. For example, the timescales over which the structure of chromosomes change remain largely unknown. How are single-cell chromosome topologies established during the cell cycle and during development? How does loop interaction frequency change across different timescales? How does the dynamic organization of the genome relate to gene expression and other nuclear processes, such as cell-type specification, DNA repair, or replication?

To directly address these and other questions, a number of imaging and labeling techniques have been developed (Fig. 4). Following chromatin dynamics in space and time not only requires development of state-of-the-art imaging technologies that often go beyond the diffraction limit (Lakadamyali and Cosma 2020; Brandão et al. 2021), but also development of highly sophisti-

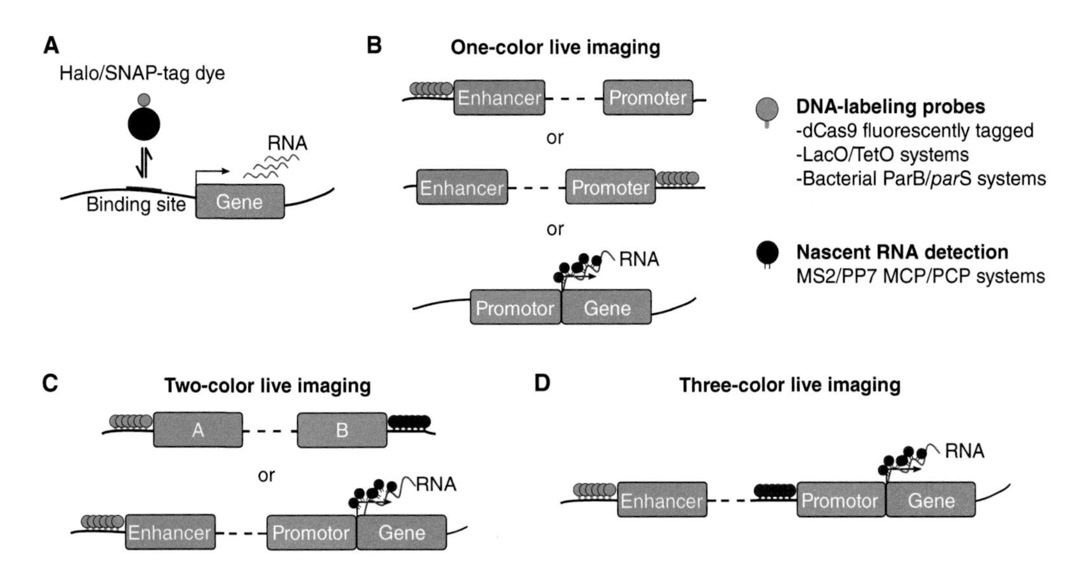

Figure 4. Imaging genome and transcriptional dynamics in living cells. (*A*) Transcription factor dynamics monitored via two labeling strategies: fluorescent proteins or Halo/SNAP-tags coupled with dyes. (*B*) Strategies for fluorescent imaging of DNA and RNA in living cells. Several labeling approaches have facilitated imaging of chromosome dynamics in a sequence-specific manner: fluorescently tagged catalytically inactive cas9 enzymes (dCas9), fluorescently labeled operator binding protein (TetR, LacI), and the bacterial multimerizing ParB/parS system. (*C*) Two-color live imaging is used to simultaneously monitor the dynamics of two chromosomal regions or enhancer–promoter dynamics coupled with nascent transcription by using a combination of approaches shown in *B*. (*D*) Three-color live imaging allows probing for functional proximity: two colors for tagging enhancer–promoter pairs for example and one color for active transcription.

cated chromatin labeling capabilities that typically require a strenuous and time-consuming combination of molecular cloning, genome editing, and genetics (Sato et al. 2020; Shaban and Seeber 2020).

Several methods have been developed to visualize global chromatin dynamics at the nuclear scale (Shaban et al. 2020; Zidovska 2020). Imaging the dynamics of multiple genomic sites requires tools to fluorescently label specific sequences. Early methods for live imaging of specific genomic loci included the endogenous insertion of large binding site arrays for fluorescently tagged LacI or TetR repressors (Marshall et al. 1997; Heun et al. 2001; Chubb et al. 2002; Chuang et al. 2006; Kumaran and Spector 2008; Masui et al. 2011). Other approaches that require genome editing use the ParB/*parS* or ANCHOR DNA-labeling systems (Saad et al. 2014; Germier et al. 2017; Chen et al. 2018). In addition, catalytically inactive Cas9 enzymes (dCas9) tagged with green fluorescent protein (GFP) have been used to target specific genomic loci in living cells (Chen et al. 2013; Gu et al. 2018; Ma et al. 2018; Stanyte et al. 2018). These efforts led to the first direct measurements of CRE dynamics in living cells (Lucas et al. 2014; Germier et al. 2017; Herbert et al. 2017; Chen et al. 2018; Gu et al. 2018; Lim et al. 2018; Alexander et al. 2019; Khanna et al. 2019; Li et al. 2020a), assessments of large-scale chromatin dynamics (Zidovska et al. 2013; Nozaki et al. 2017; Shaban et al. 2018; Zidovska 2020), and quantification of DNA-binding factor interaction dynamics (Lionnet and Wu 2021).

Many improvements in both microscopy and probe development are linked by their ability to control the photon budget (Planchon et al. 2011; Zhao et al. 2011; Lavis 2017): the number of detectable photons that a particular fluorophore contributes to the experiment, which is limited because of photochemistry and photobleaching (i.e., the permanent loss of fluorescence due to photo-induced chemical changes).

However, further increases in localization precision, imaging rate, and imaging time, all require the detection of higher numbers of photons. This generates an optimization dilemma between resolution, speed, depth of view, and photodamage (Fig. 5A). Thus, the limiting factor is not the microscope; it is the photon budget. Close attention needs to be paid to the fluorophore choice because the photon budget is one of the most commonly neglected factors that significantly affects the feasibility and success of an experiment. A common strategy to sidestep this problem is to increase the number of fluorophores bound to individual molecules (Bertrand et al. 1998; Femino et al. 1998).

Cis-Dynamics

The first pioneering work analyzing the dynamics of chromosomal loci was performed in lymphocyte B cells, where either V_H or D_HJ_H regions at the immunoglobulin gene loci were followed one-by-one by single-particle tracking (Lucas et al. 2014). These loci were shown to display a subdiffusive behavior with fractional Langevin motion and were mostly spatially confined. Similarly, in live mouse embryonic stem cells, the *Fgf5* enhancer and promoter displayed subdiffusive behavior and their motility increased during differentiation to epiblast-like cells concomitant with transcriptional activation (Gu et al. 2018). The authors proposed that higher diffusivity of CREs increases stochastic encounters within TADs, potentially boosting successful enhancer–promoter interactions.

To partially avoid biases introduced by the inherent large-scale, three-dimensional motion of the entire nucleus, several efforts used two-color imaging to monitor the relative motion of two genomic loci on the same chromosome (Fig. 4C). This dual labeling strategy was used in lymphocyte B cells to characterize the relative movement of V_H and D_HJ_H regions (Khanna et al. 2019). The positions of V_H and D_HJ_H elements were found to undergo local, spatial fluctuations while their distance remains nearly constant over time, but abrupt changes in motion could be observed and suspected to originate from rapid temporal changes in large-scale chromatin conformations. These studies indicate that chromatin dynamics are largely subdiffusive in mammalian cells, with occasional abrupt changes in motion.

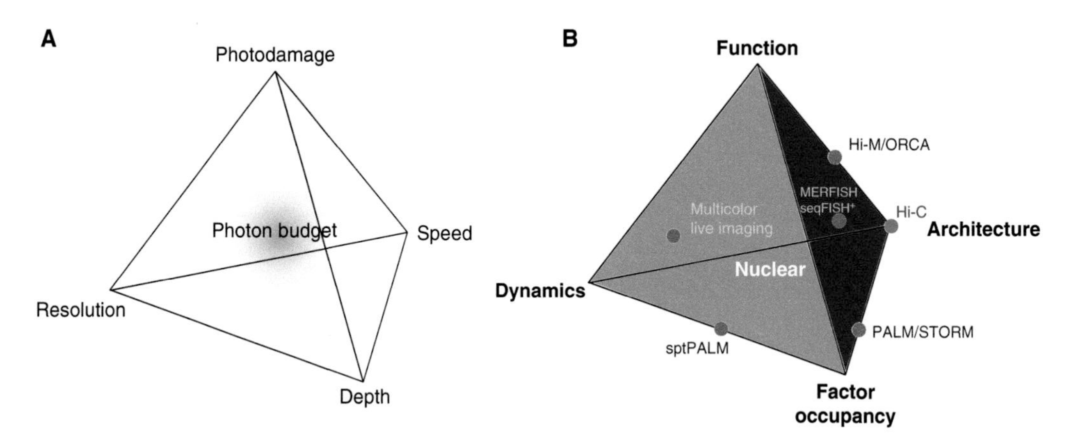

Figure 5. Compromises in optimizing photon budget and measurements of nuclear properties. (*A*) The photon budget creates a tug-of-war between different desirable optimization strategies in photonic imaging. Whereas the ideal imaging experiment would push the limits of each property on the corners of the tetrahedron, the finite size of the photon budget sets a fundamental limit on the combined optimization of all four corners. Pushing one corner to its extreme limit implies giving up on the performance of the others. (*B*) The (epi)-genome contains all the information necessary to produce an entire organism. However, characterization of the functional output of the genome requires the ability to monitor multiple observables simultaneously. These include nuclear architecture, dynamics, occupancy by regulatory factors, and the biological output. Current technologies (showing here only a small number for simplicity) hardly accomplish a subset of these tasks in a single experiment.

In mouse embryonic stem cells, live two-color imaging revealed that the *Pou5f1* and *Sox2* enhancers are frequently in proximity with their target gene transcription site (100–200 nm) (Li et al. 2020a). The authors argued that this 100–200-nm-sized cluster of enhancers could concentrate components of the transcription machinery and activate gene expression (Li et al. 2020a). Alexander et al. (2019) monitored the dynamics of the *Sox2* gene in mouse embryonic stem cells and its essential endogenous SCR enhancer, positioned ~100 kb away to show that the *Sox2* enhancer–promoter spatial organization exhibited a high cell-to-cell variability. The two loci displayed sporadic sharp topological transitions within a tightly confined space in the ~200 nm length scale. These observations suggest that enhancer–promoter pairs could be frequently and transiently interacting with each other, in contrast with the classical stable contact model (Deng et al. 2012, 2014; Bartman et al. 2016). However, it is unclear whether loop extrusion and chromatin conformation dynamics impact these transient enhancer–promoter interactions and what the consequences on transcription might be.

Trans-Dynamics

Another approach to decipher the time-dependent nature of the genome is to quantify the kinetics and dynamics of *trans*-acting factors, such as those required for loop extrusion, TAD formation, or stabilization of enhancer–promoter interactions. Their dynamics could potentially offer a route to study these processes and thus shed light on the underlying chromatin mechanics. For example, using single-molecule imaging approaches in live cells, the average residence time of cohesin (~20 min) is about an order of magnitude longer than that of CTCF (~1 min) (Hansen et al. 2017). These residence times are relatively stable compared to conventional TFs that can bind and dissociate from DNA on timescales of seconds (Mazza et al. 2012; Chen et al. 2014; Liu et al. 2014; Hansen et al. 2018) but highly dynamic compared to the length of the cell cycle (~24 h), suggesting that chromatin loops dynamically form and break multiple times throughout the cell cycle (Hansen et al. 2018; Zhang et al. 2019). In addition, loop extrusion by the cohesin complex occurs at a maximum rate of ~2 kb/sec in vitro (Davidson et al. 2019; Kim et al. 2019); thus, loops could likely play a role in the dynamics of enhancer–promoter interactions.

Several studies showed that components of the transcription machinery, such as RNA polymerase II, the Mediator complex, BRD4, and TFs, form transcription-associated nuclear phase-separated condensates (Cho et al. 2018; Chong et al. 2018; Sabari et al. 2018) (see also the section "Spatial Clustering of Regulatory Components" above). Using live-cell single-molecule imaging, Chong et al. showed that interactions between low-complexity regions of TFs are highly dynamic, typically on the second to minutes timescale. Consistently, Cho et al. used single-particle tracking to show that Mediator and RNA Pol II form large (>300 nm), long-lived (>100 sec), chromatin-associated clusters where they colocalize in a transcription-dependent manner. Mediator clusters display properties of phase-separated condensates and may mediate transient enhancer–promoter communication over large distances (a few hundred nanometers) (Fig. 3). Critically, in some instances, transcription-associated foci are highly dynamic and can form and dissociate in a matter of seconds (Liu et al. 2014; Cho et al. 2018; Dufourt et al. 2018). How these short-lived foci contribute to establishing long-range interactions between CREs remains an open question.

FUNCTIONAL CONSEQUENCES OF SPATIOTEMPORAL CHROMOSOME ORGANIZATION

After static and dynamic descriptions of nuclear organization, the next frontier is to provide a connection to biological function. What is the functional impact of nuclear order? Is there an evolved functional relationship between chromatin dynamics and transcription? To address these and similar questions, we need the ability to assess physical and genomic structure, chromatin dynamics and spatial distributions of regulatory factors, as well as biological function—ideally all simultaneously—which is a highly challenging task. Most likely, and similar to the dilemma of sharing a photon budget for imaging experiments, designing one experiment that can accomplish all these tasks simultaneously still lies somewhere in the future, and with current technologies one has to make choices about which of these features to optimize (Fig. 5B). Here we focus on initial progress to assess biological function in terms of measurements of transcription and its relationship to local genome organization.

Arguably one of the most functionally relevant outputs of the nucleus is its genetic program, which is executed via gene activity. Transcriptional dynamics can be monitored by in vivo visualization of nascent mRNA transcripts using bacteriophage-based reporter cassettes, as pioneered more than 20 years ago (Bertrand et al. 1998; Sato et al. 2020). MS2 and PP7 stem-loops are positioned in the gene body and detected with a fluorescently tagged coat protein to visualize nascent transcription in live cells (Janicki et al. 2004; Larson et al. 2011; Bothma et al. 2014). Measuring the signal intensity of the fluorescent signal at the site of gene activity was used to show that transcription is a stochastic process, subject to molecular noise, which exhibits transcriptional bursts with intervals of mRNA production followed by intervals of transcriptional inactivity (Rodriguez and Larson 2020). This stochasticity results in gene expression noise and cell-to-cell variability, the causes of which could possibly involve spatiotemporal fluctuations of the chromosome (Shah et al. 2018) or of CRE hubs (Shah et al. 2018).

A powerful application of this approach showed that enhancers control the frequency of transcriptional bursts, are able to coactivate linked genes, and yet can exhibit large spatial separation from their target genes even during transcriptional activation (Fukaya et al. 2016). These results contribute to the conflicting picture about the role of chromatin topology on genome function (Northcott et al. 2014; Lupiáñez et al. 2015; Franke et al. 2016; Nora et al. 2017; Rao et al. 2017; Schwarzer et al. 2017). While it is clear that enhancers and promoters can come into spatial proximity, and that physical proximity is somehow linked to transcriptional activation, such evidence is often correlative. Is enhancer–promoter proximity a consequence of transcriptional activation or is it needed for transcriptional activation? And if so, does transcription happen simultaneously with proximity or is it uncoupled? Even if CRE proximity is necessary for transcriptional activation, is it also necessary for sustained activity?

Answering these questions will require simultaneous detection of transcription and genome organization in the same cell. This is particularly important for enhancer–promoter interactions and the underlying mechanisms governing transcriptional control. Recent development of elaborate imaging-based methods—either via live-cell imaging or high-resolution fixed-tissue localization-based microscopy—have enabled the first direct visualization of long-range enhancer–promoter interactions coupled with transcriptional activity (Chen et al. 2018; Alexander et al. 2019; Mateo et al. 2019; Barinov et al. 2020; Li et al. 2020a; Espinola et al. 2021). Hi-M and ORCA have been used to visualize both chromatin structure and transcriptional activation in *Drosophila* to show that transcriptionally active and inactive cells display very similar enhancer–promoter proximity (Mateo et al. 2019; Espinola et al. 2021). Whereas these approaches enable the detection of transcription and the topology of multiple enhancers and promoters, they do not shed light on their dynamics.

To observe whether two distal chromosomal regions interact in a functionally significant manner, the notions of proximity or contact may no longer be sufficient. Rather, a more complex imaging assay is needed that involves simultaneous live image capture of three differently colored DNA tags: two to dynamically follow the motion of the distal chromosomal sites, such as an enhancer and a promoter, with the third tag serving as a reporter for functional proximity (i.e., it only lights up when specific events, such as transcription, occur). Here, the MS2/PP7-labeling systems are key in two ways: analysis of the fluorescence intensity signal for function and simultaneously tracking its location to gain insights into spatial relationships.

Using such an approach, progress toward a causal connection between dynamic enhancer–

 Cite this article as *Cold Spring Harb Perspect Biol* doi: 10.1101/cshperspect.a040378

promoter communication and gene expression has recently been carried out. Live imaging experiments in *Drosophila* embryos visualized physical enhancer–promoter interactions and transcription at the *eve* locus in *Drosophila* embryos (Chen et al. 2018). Sustained physical proximity between the enhancer and the promoter of a distal reporter gene was shown to be necessary for transcriptional activity of the reporter gene. In addition, transcriptional activity also seems to stabilize this proximal conformation and have an impact on the physical size of the active gene locus, suggestive of a reciprocal interplay between enhancer–promoter dynamics and transcriptional activity (van Steensel and Furlong 2019).

In contrast, a similar study in mouse embryonic stem cells reported that enhancer–promoter proximity at the *Sox2* locus is uncoupled from transcription (Alexander et al. 2019). In line with these results, 3D DNA-FISH and chromosome conformation capture in fixed cells revealed decreased spatial proximity between the *Shh* gene and its enhancers during the differentiation of mouse embryonic stem cells into neural progenitor cells (Benabdallah et al. 2019). It would be interesting to analyze enhancer–promoter interactions at the *Shh* locus using live imaging to determine the dynamics of these interactions at timescales shorter than days of cell differentiation, to see whether they are transient and unstable, as for the *Sox2* locus. In many loci, several enhancers can activate transcription of a promoter, individually or conjointly. Thus, an observed lack of proximity may not be a good indicator for the need of enhancer–promoter proximity as some other enhancer in the neighborhood or within a CRE hub may be activating transcription.

It is still an open question what exactly "contact" or "proximity" mean in the context of an active enhancer–promoter pair, and how these notions relate to the emerging evidence of transcription-related CRE hubs and transcription-associated protein foci. Again, to discern between these models and to devise new ones, dynamic measurements would need to simultaneously track multiple CREs and factors in space and time. Testing current models of enhancer–promoter communication will also require imaging methods with high resolution to distinguish enhancer–promoter loops in physical interaction range (in and of itself an ill-defined length scale) from observed enhancer–promoter proximity (150–200 nm). Further progress toward this issue has been put forward using multicolor localization microscopy to achieve 1–2 kb resolution in an 18 kb gene locus in *Drosophila*, where the transcriptionally active enhancer–promoter pair was still more than 150 nm separated; thus, again not in physical interaction range (Barinov et al. 2020).

At the time of writing this review, there are still a number of unanswered questions, most prominently whether the observed spatial gaps between active enhancers and promoters prove to be a general feature of active transcription-associated loci. In that case, the most pressing question is to understand how information about activity spreads across large spatial distances (>100 nm) from the TF-bound active enhancer to the transcription-engaging promoter. It may take another generation of experiments and new technologies to make progress in the understanding of how the general architecture of the folded nuclear genome regulates genome function and vice versa. The observations above could suggest that gene regulation by stable enhancer–promoter looping is not a generality, and more elaborate models of long-range enhancer–promoter communication should take other spatial considerations or constraints into account.

CONCLUDING REMARKS

The last two decades have seen a revolution in optical imaging approaches with the advent of superresolution microscopy, live tracking of individual molecules, and multiplexed methods. Application of these approaches already made a strong impact on our understanding of the interplay between the structure and dynamics of the nucleus and its transcriptional output. However, understanding causality and gaining an integrated picture of how dynamic changes in chromosome organization control the timing and levels of transcription will require considerable further advances in several interdisciplinary areas. A crucial issue will be the necessity to measure in individual cells the multiple critical components of nuclear organization underlying gene activity: transcriptional output, spatial arrangement of CREs, dynamic long-range chromosomal interactions, and the epigenetic state of chromatin (Fig. 5B). Many of these quantities can already be measured in single cells, one-by-one or in pairs, but detecting them simultaneously presents a long-term challenge that will require significant advances in design and development of microscopy, as well as in labeling and genome-editing tools.

One area that is still largely underdeveloped is the single-cell study of the dynamics and genome occupancy of *trans*-regulatory factors and how they influence genome organization and function. For example, it will be crucial to monitor the dynamic binding of multiple TFs to CREs to understand how the latter turn into a state of activity that ultimately leads to a transcriptional output. Likewise, it will be critical to further integrate how signaling, environmental cues, or splicing shape the assembly and dynamics of CRE hubs and transcription-associated foci to regulate transcription.

There are still many open questions and new frontiers to explore. Progress in optical and electron microscopies have already achieved the spatial resolution necessary to start revealing the organization of chromatin in cells at the nanometer scale (Ricci et al. 2015; Jungmann et al. 2016; Ou et al. 2017). However, visualization of chromatin structure at that scale as inert/featureless beads may not be enough; to get a full understanding of the role of chromatin structure for biological function will require the detection of multiple types of molecules (e.g., proteins, RNA, DNA) with sequence specificity, nanometer precision, and dynamic probes. These developments have to go hand in hand with improvements in labeling and sample preparation procedures that conserve structures at the molecular scale. In addition, still largely lacking are perturbation techniques that can directly probe causal relationships between nuclear structure, the environment, and function. For example, optogenetic association or dissociation of molecular compounds or DNA loci should provide powerful handles for causality experiments. Excitingly, all these advances should bring a much needed facet to tracking single molecules interacting with DNA and asserting control over transcription programs, one factor at a time.

ACKNOWLEDGMENTS

We thank Pau Bernado for his critical reading of the manuscript and for help with Figure 2J. This work was supported in part by the European Union's Horizon 2020 Research and Innovation Program (Grant ID 724429) (M.N.), the Deutsche Forschungsgemeinschaft (DFG, German Research Foundation, Project ID 43147 1305) (M.G.), the Laboratoire d'Excellence Revive (Investissement d'Avenir; ANR-10-LABX-73) (I.B.), the U.S. National Science Foundation, through the Center for the Physics of Biological Function (PHY-1734030), and by National Institutes of Health Grants R01GM097275, U01DA047730, and U01DK127429 (T.G.).

REFERENCES

Alexander JM, Guan J, Li B, Maliskova L, Song M, Shen Y, Huang B, Lomvardas S, Weiner OD. 2019. Live-cell imaging reveals enhancer-dependent Sox2 transcription in the absence of enhancer proximity. *eLife* **8:** e41769. doi:10.7554/eLife.41769

Allahyar A, Vermeulen C, Bouwman BAM, Krijger PHL, Marjon JA, Geeven G, van Kranenburg M, Pieterse M, Straver R, Haarhuis JHI, et al. 2018. Enhancer hubs and loop collisions identified from single-allele topologies. *Nat Genet* **50:** 1151–1160. doi:10.1038/s41588-018-0161-5

Armache KJ, Mitterweger S, Meinhart A, Cramer P. 2005. Structures of complete RNA polymerase II and its subcomplex, Rpb4/7. *J Biol Chem* **280:** 7131–7134. doi:10.1074/jbc.M413038200

Banani SF, Lee HO, Hyman AA, Rosen MK. 2017. Biomolecular condensates: organizers of cellular biochemistry. *Nat Rev Mol Cell Biol* **18:** 285–298. doi:10.1038/nrm.2017.7

Banerji J, Rusconi S, Schaffner W. 1981. Expression of a β-globin gene is enhanced by remote SV40 DNA sequences. *Cell* **27:** 299–308. doi:10.1016/0092-8674(81)90413-X

Banerji J, Olson L, Schaffner W. 1983. A lymphocyte-specific cellular enhancer is located downstream of the joining region in immunoglobulin heavy chain genes. *Cell* **33:** 729–740. doi:10.1016/0092-8674(83)90015-6

Barinov L, Ryabichko S, Bialek W, Gregor T. 2020. Transcription-dependent spatial organization of a gene locus. arXiv 2012.15819

Bartman CR, Hsu SC, Hsiung CCS, Raj A, Blobel GA. 2016. Enhancer regulation of transcriptional bursting parameters revealed by forced chromatin looping. *Mol Cell* **62:** 237–247. doi:10.1016/j.molcel.2016.03.007

Beagrie RA, Scialdone A, Schueler M, Kraemer DCA, Chotalia M, Xie SQ, Barbieri M, de Santiago I, Lavitas LM, Branco MR, et al. 2017. Complex multi-enhancer contacts captured by genome architecture mapping. *Nature* **543:** 519–524. doi:10.1038/nature21411

Beliveau BJ, Joyce EF, Apostolopoulos N, Yilmaz F, Fonseka CY, McCole RB, Chang Y, Li JB, Senaratne TN, Williams BR, et al. 2012. Versatile design and synthesis platform for visualizing genomes with oligopaint FISH probes. *Proc Natl Acad Sci* **109:** 21301–21306. doi:10.1073/pnas.1213818110

Benabdallah NS, Williamson I, Illingworth RS, Kane L, Boyle S, Sengupta D, Grimes GR, Therizols P, Bickmore WA. 2019. Decreased enhancer–promoter proximity accompanying enhancer activation. *Mol Cell* **76:** 473–484.e7. doi:10.1016/j.molcel.2019.07.038

Bertrand E, Chartrand P, Schaefer M, Shenoy SM, Singer RH, Long RM. 1998. Localization of ASH1 mRNA particles in living yeast. *Mol Cell* **2:** 437–445. doi:10.1016/S1097-2765(00)80143-4

Bintu B, Mateo LJ, Su JH, Sinnott-Armstrong NA, Parker M, Kinrot S, Yamaya K, Boettiger AN, Zhuang X. 2018. Super-resolution chromatin tracing reveals domains and cooperative interactions in single cells. *Science* **362:** eaau1783. doi:10.1126/science.aau1783

Blackwood EM, Kadonaga JT. 1998. Going the distance: a current view of enhancer action. *Science* **281:** 60–63. doi:10.1126/science.281.5373.60

Boettiger AN, Bintu B, Moffitt JR, Wang S, Beliveau BJ, Fudenberg G, Imakaev M, Mirny LA, Wu CT, Zhuang X. 2016. Super-resolution imaging reveals distinct chromatin folding for different epigenetic states. *Nature* **529:** 418–422. doi:10.1038/nature16496

Boeynaems S, Alberti S, Fawzi NL, Mittag T, Polymenidou M, Rousseau F, Schymkowitz J, Shorter J, Wolozin B, Van Den Bosch L, et al. 2018. Protein phase separation: a new phase in cell biology. *Trends Cell Biol* **28:** 420–435. doi:10.1016/j.tcb.2018.02.004

Boija A, Klein IA, Sabari BR, Dall'Agnese A, Coffey EL, Zamudio AV, Li CH, Shrinivas K, Manteiga JC, Hannett NM, et al. 2018. Transcription factors activate genes through the phase-separation capacity of their activation domains. *Cell* **175:** 1842–1855.e16. doi:10.1016/j.cell.2018.10.042

Bolt CC, Duboule D. 2020. The regulatory landscapes of developmental genes. *Development* **147:** dev171736. doi:10.1242/dev.171736

Bothma JP, Garcia HG, Esposito E, Schlissel G, Gregor T, Levine M. 2014. Dynamic regulation of *eve* stripe 2 expression reveals transcriptional bursts in living *Drosophila* embryos. *Proc Natl Acad Sci* **111:** 10598–10603. doi:10.1073/pnas.1410022111

Brandão HB, Gabriele M, Hansen AS. 2021. Tracking and interpreting long-range chromatin interactions with super-resolution live-cell imaging. *Curr Opin Cell Biol* **70:** 18–26. doi:10.1016/j.ceb.2020.11.002

Bulger M, Groudine M. 2011. Functional and mechanistic diversity of distal transcription enhancers. *Cell* **144:** 327–339. doi:10.1016/j.cell.2011.01.024

Bushey AM, Dorman ER, Corces VG. 2008. Chromatin insulators: regulatory mechanisms and epigenetic inheritance. *Mol Cell* **32:** 1–9. doi:10.1016/j.molcel.2008.08.017

Bushey AM, Ramos E, Corces VG. 2009. Three subclasses of a *Drosophila* insulator show distinct and cell type-specific genomic distributions. *Genes Dev* **23:** 1338–1350. doi:10.1101/gad.1798209

Cai H, Levine M. 1995. Modulation of enhancer–promoter interactions by insulators in the *Drosophila* embryo. *Nature* **376:** 533–536. doi:10.1038/376533a0

Cardozo Gizzi AM, Cattoni DI, Fiche JB, Espinola SM, Gurgo J, Messina O, Houbron C, Ogiyama Y, Papadopoulos GL, Cavalli G, et al. 2019. Microscopy-based chromosome conformation capture enables simultaneous visualization of genome organization and transcription in intact organisms. *Mol Cell* **74:** 212–222.e5. doi:10.1016/j.molcel.2019.01.011

Cattoni DI, Cardozo Gizzi AM, Georgieva M, Di Stefano M, Valeri A, Chamousset D, Houbron C, Déjardin S, Fiche JB, González I, et al. 2017. Single-cell absolute contact probability detection reveals chromosomes are organized by multiple low-frequency yet specific interactions. *Nat Commun* **8:** 1753. doi:10.1038/s41467-017-01962-x

Cavalheiro GR, Pollex T, Furlong EE. 2021. To loop or not to loop: what is the role of TADs in enhancer function and gene regulation? *Curr Opin Genet Dev* **67:** 119–129. doi:10.1016/j.gde.2020.12.015

Chen B, Gilbert LA, Cimini BA, Schnitzbauer J, Zhang W, Li G-W, Park J, Blackburn EH, Weissman JS, Qi LS, et al. 2013. Dynamic imaging of genomic loci in living human cells by an optimized CRISPR/Cas system. *Cell* **155:** 1479–1491. doi:10.1016/j.cell.2013.12.001

Chen J, Zhang Z, Li L, Chen BC, Revyakin A, Hajj B, Legant W, Dahan M, Lionnet T, Betzig E, et al. 2014. Single-molecule dynamics of enhanceosome assembly in embryonic stem cells. *Cell* **156:** 1274–1285. doi:10.1016/j.cell.2014.01.062

Chen H, Levo M, Barinov L, Fujioka M, Jaynes JB, Gregor T. 2018. Dynamic interplay between enhancer–promoter topology and gene activity. *Nat Genet* **50:** 1296–1303. doi:10.1038/s41588-018-0175-z

Cho WK, Spille JH, Hecht M, Lee C, Li C, Grube V, Cisse II. 2018. Mediator and RNA polymerase II clusters associate in transcription-dependent condensates. *Science* **361:** 412–415. doi:10.1126/science.aar4199

Chong S, Dugast-Darzacq C, Liu Z, Dong P, Dailey GM, Cattoglio C, Heckert A, Banala S, Lavis L, Darzacq X, et al. 2018. Imaging dynamic and selective low-complexity domain interactions that control gene transcription. *Science* **361:** ear2555. doi:10.1126/science.aar2555

Chopra VS, Cande J, Hong JW, Levine M. 2009. Stalled Hox promoters as chromosomal boundaries. *Genes Dev* **23:** 1505–1509. doi:10.1101/gad.1807309

Chuang CH, Carpenter AE, Fuchsova B, Johnson T, de Lanerolle P, Belmont AS. 2006. Long-range directional movement of an interphase chromosome site. *Curr Biol* **16:** 825–831. doi:10.1016/j.cub.2006.03.059

Chubb JR, Boyle S, Perry P, Bickmore WA. 2002. Chromatin motion is constrained by association with nuclear compartments in human cells. *Curr Biol* **12:** 439–445. doi:10.1016/S0960-9822(02)00695-4

Cisse II, Izeddin I, Causse SZ, Boudarene L, Senecal A, Muresan L, Dugast-Darzacq C, Hajj B, Dahan M, Darzacq X. 2013. Real-time dynamics of RNA polymerase II clustering in live human cells. *Science* **341:** 664–667. doi:10.1126/science.1239053

Cordeiro TN, Sibille N, Germain P, Barthe P, Boulahtouf A, Allemand F, Bailly R, Vivat V, Ebel C, Barducci A, et al. 2019. Interplay of protein disorder in retinoic acid receptor heterodimer and its corepressor regulates gene expression. *Structure* **27:** 1270–1285.e6. doi:10.1016/j.str.2019.05.001

Davidson IF, Bauer B, Goetz D, Tang W, Wutz G, Peters JM. 2019. DNA loop extrusion by human cohesin. *Science* **366:** 1338–1345. doi:10.1126/science.aaz3418

de Gennes PG, Gennes PG. 1979. *Scaling concepts in polymer physics*. Cornell University Press, Ithaca, NY.

Dekker J. 2016. Mapping the 3D genome: aiming for consilience. *Nat Rev Mol Cell Biol* **17:** 741–742. doi:10.1038/nrm.2016.151

de Laat W, Duboule D. 2013. Topology of mammalian developmental enhancers and their regulatory landscapes. *Nature* **502:** 499–506. doi:10.1038/nature12753

Deng W, Lee J, Wang H, Miller J, Reik A, Gregory PD, Dean A, Blobel GA. 2012. Controlling long-range genomic interactions at a native locus by targeted tethering of a looping factor. *Cell* **149:** 1233–1244. doi:10.1016/j.cell.2012.03.051

Deng W, Rupon JW, Krivega I, Breda L, Motta I, Jahn KS, Reik A, Gregory PD, Rivella S, Dean A, et al. 2014. Reactivation of developmentally silenced globin genes by forced chromatin looping. *Cell* **158:** 849–860. doi:10.1016/j.cell.2014.05.050

Di Stefano M, Stadhouders R, Farabella I, Castillo D, Serra F, Graf T, Marti-Renom MA. 2020. Transcriptional activation during cell reprogramming correlates with the formation of 3D open chromatin hubs. *Nat Commun* **11**. doi:10.1038/s41467-020-16396-1

Dixon JR, Selvaraj S, Yue F, Kim A, Li Y, Shen Y, Hu M, Liu JS, Ren B. 2012. Topological domains in mammalian genomes identified by analysis of chromatin interactions. *Nature* **485:** 376–380. doi:10.1038/nature11082

Doi M. 1996. *Introduction to polymer physics*. Oxford University Press, Oxford, UK.

Dowen JM, Fan ZP, Hnisz D, Ren G, Abraham BJ, Zhang LN, Weintraub AS, Schujiers J, Lee TI, Zhao K, et al. 2014. Control of cell identity genes occurs in insulated neighborhoods in mammalian chromosomes. *Cell* **159:** 374–387. doi:10.1016/j.cell.2014.09.030

Dufourt J, Trullo A, Hunter J, Fernandez C, Lazaro J, Dejean M, Morales L, Nait-Amer S, Schulz KN, Harrison MM, et al. 2018. Temporal control of gene expression by the pioneer factor Zelda through transient interactions in hubs. *Nat Commun* **9:** 5194. doi:10.1038/s41467-018-07613-z

ENCODE Project Consortium. 2012. An integrated encyclopedia of DNA elements in the human genome. *Nature* **489:** 57–74. doi:10.1038/nature11247

Eng CHL, Lawson M, Zhu Q, Dries R, Koulena N, Takei Y, Yun J, Cronin C, Karp C, Yuan GC, et al. 2019. Transcriptome-scale super-resolved imaging in tissues by RNA seqFISH. *Nature* **568:** 235–239. doi:10.1038/s41586-019-1049-y

Erdel F, Rippe K. 2018. Formation of chromatin subcompartments by phase separation. *Biophys J* **114:** 2262–2270. doi:10.1016/j.bpj.2018.03.011

Erdel F, Rademacher A, Vlijm R, Tünnermann J, Frank L, Weinmann R, Schweigert E, Yserentant K, Hummert J, Bauer C, et al. 2020. Mouse heterochromatin adopts digital compaction states without showing hallmarks of HP1-driven liquid–liquid phase separation. *Mol Cell* **78:** 236–249.e7. doi:10.1016/j.molcel.2020.02.005

Espinola SM, Götz M, Bellec M, Messina O, Fiche JB, Houbron C, Dejean M, Reim I, Cardozo Gizzi AM, Lagha M, et al. 2021. *Cis*-regulatory chromatin loops arise before TADs and gene activation, and are independent of cell fate during early *Drosophila* development. *Nat Genet* **53:** 477–486. doi:10.1038/s41588-021-00816-z

Femino AM, Fay FS, Fogarty K, Singer RH. 1998. Visualization of single RNA transcripts in situ. *Science* **280:** 585–590. doi:10.1126/science.280.5363.585

Feric M, Misteli T. 2021. Phase separation in genome organization across evolution. *Trends Cell Biol* **23:** S0962-8924(21)00047-7.

Finn EH, Misteli T. 2019. Molecular basis and biological function of variability in spatial genome organization. *Science* **365:** eaaw9498. doi:10.1126/science.aaw9498

Finn EH, Pegoraro G, Brandão HB, Valton AL, Oomen ME, Dekker J, Mirny L, Misteli T. 2019. Extensive heterogeneity and intrinsic variation in spatial genome organization. *Cell* **176:** 1502–1515.e10. doi:10.1016/j.cell.2019.01.020

Flyamer IM, Gassler J, Imakaev M, Brandão HB, Ulianov SV, Abdennur N, Razin SV, Mirny LA, Tachibana-Konwalski K. 2017. Single-nucleus Hi-C reveals unique chromatin reorganization at oocyte-to-zygote transition. *Nature* **544:** 110–114. doi:10.1038/nature21711

Franke M, Ibrahim DM, Andrey G, Schwarzer W, Heinrich V, Schöpflin R, Kraft K, Kempfer R, Jerković I, Chan WL, et al. 2016. Formation of new chromatin domains determines pathogenicity of genomic duplications. *Nature* **538:** 265–269. doi:10.1038/nature19800

Fudenberg G, Imakaev M. 2017. FISH-ing for captured contacts: towards reconciling FISH and 3C. *Nat Methods* **14:** 673–678. doi:10.1038/nmeth.4329

Fukaya T, Lim B, Levine M. 2016. Enhancer control of transcriptional bursting. *Cell* **166:** 358–368. doi:10.1016/j.cell.2016.05.025

Fulco CP, Nasser J, Jones TR, Munson G, Bergman DT, Subramanian V, Grossman SR, Anyoha R, Doughty BR, Patwardhan TA, et al. 2019. Activity-by-contact model of enhancer-promoter regulation from thousands of CRISPR perturbations. *Nat Genet* **51:** 1664–1669. doi:10.1038/s41588-019-0538-0

Furlong EEM, Levine M. 2018. Developmental enhancers and chromosome topology. *Science* **361:** 1341–1345. doi:10.1126/science.aau0320

Germier T, Kocanova S, Walther N, Bancaud A, Shaban HA, Sellou H, Politi AZ, Ellenberg J, Gallardo F, Bystricky K. 2017. Real-time imaging of a single gene reveals transcription-initiated local confinement. *Biophys J* **113:** 1383–1394. doi:10.1016/j.bpj.2017.08.014

Geyer PK, Corces VG. 1992. DNA position-specific repression of transcription by a *Drosophila* zinc finger protein. *Genes Dev* **6:** 1865–1873. doi:10.1101/gad.6.10.1865

Ghavi-Helm Y, Klein FA, Pakozdi T, Ciglar L, Noordermeer D, Huber W, Furlong EEM. 2014. Enhancer loops appear stable during development and are associated with paused polymerase. *Nature* **512:** 96–100. doi:10.1038/nature13417

Gibson BA, Doolittle LK, Schneider MWG, Jensen LE, Gamarra N, Henry L, Gerlich DW, Redding S, Rosen MK. 2019. Organization of chromatin by intrinsic and regulated phase separation. *Cell* **179:** 470–484.e21. doi:10.1016/j.cell.2019.08.037

Gillies SD, Morrison SL, Oi VT, Tonegawa S. 1983. A tissue-specific transcription enhancer element is located in the major intron of a rearranged immunoglobulin heavy chain gene. *Cell* **33:** 717–728. doi:10.1016/0092-8674(83)90014-4

Giorgetti L, Heard E. 2016. Closing the loop: 3C versus DNA FISH. *Genome Biol* **17:** 215. doi:10.1186/s13059-016-1081-2

Giorgetti L, Galupa R, Nora EP, Piolot T, Lam F, Dekker J, Tiana G, Heard E. 2014. Predictive polymer modeling reveals coupled fluctuations in chromosome conformation and transcription. *Cell* **157:** 950–963. doi:10.1016/j.cell.2014.03.025

Grosberg A, Khokhlov AR. 2011. *Giant molecules: here, there, and everywhere.* World Scientific, Singapore.

Gu B, Swigut T, Spencley A, Bauer MR, Chung M, Meyer T, Wysocka J. 2018. Transcription-coupled changes in nuclear mobility of mammalian *cis*-regulatory elements. *Science* **359:** 1050–1055. doi:10.1126/science.aao3136

Guilhas B, Walter JC, Rech J, David G, Walliser NO, Palmeri J, Mathieu-Demaziere C, Parmeggiani A, Bouet JY, Le Gall A, et al. 2020. ATP-driven separation of liquid phase condensates in bacteria. *Mol Cell* **79:** 293–303.e4. doi:10.1016/j.molcel.2020.06.034

Hansen AS, Pustova I, Cattoglio C, Tjian R, Darzacq X. 2017. CTCF and cohesin regulate chromatin loop stability with distinct dynamics. *eLife* **6:** e25776. doi:10.7554/eLife.25776

Hansen AS, Cattoglio C, Darzacq X, Tjian R. 2018. Recent evidence that TADs and chromatin loops are dynamic structures. *Nucleus* **9:** 20–32. doi:10.1080/19491034.2017.1389365

Herbert S, Brion A, Arbona JM, Lelek M, Veillet A, Lelandais B, Parmar J, Fernández FG, Almayrac E, Khalil Y, et al. 2017. Chromatin stiffening underlies enhanced locus mobility after DNA damage in budding yeast. *EMBO J* **36:** 2595–2608. doi:10.15252/embj.201695842

Heun P, Laroche T, Shimada K, Furrer P, Gasser SM. 2001. Chromosome dynamics in the yeast interphase nucleus. *Science* **294:** 2181–2186. doi:10.1126/science.1065366

Hnisz D, Shrinivas K, Young RA, Chakraborty AK, Sharp PA. 2017. A phase separation model for transcriptional control. *Cell* **169:** 13–23. doi:10.1016/j.cell.2017.02.007

Hou C, Li L, Qin Z, Corces V. 2012. Gene density, transcription, and insulators contribute to the partition of the *Drosophila* genome into physical domains. *Mol Cell* **48:** 471–484. doi:10.1016/j.molcel.2012.08.031

Hyman AA, Weber CA, Jülicher F. 2014. Liquid–liquid phase separation in biology. *Annu Rev Cell Dev Biol* **30:** 39–58. doi:10.1146/annurev-cellbio-100913-013325

Ing-Simmons E, Vaid R, Bing XY, Levine M, Mannervik M, Vaquerizas JM. 2021. Independence of chromatin conformation and gene regulation during *Drosophila* dorsoventral patterning. *Nat Genet* **53:** 487–499. doi:10.1038/s41588-021-00799-x

Izeddin I, Récamier V, Bosanac L, Cissé II, Boudarene L, Dugast-Darzacq C, Proux F, Bénichou O, Voituriez R, Bensaude O, et al. 2014. Single-molecule tracking in live cells reveals distinct target-search strategies of transcription factors in the nucleus. *eLife* **3:** e02230. doi:10.7554/eLife.02230

Janicki SM, Tsukamoto T, Salghetti SE, Tansey WP, Sachidanandam R, Prasanth KV, Ried T, Shav-Tal Y, Bertrand E, Singer RH, et al. 2004. From silencing to gene expression: real-time analysis in single cells. *Cell* **116:** 683–698. doi:10.1016/S0092-8674(04)00171-0

Ji X, Dadon DB, Powell BE, Fan ZP, Borges-Rivera D, Shachar S, Weintraub AS, Hnisz D, Pegoraro G, Lee TI, et al. 2016. 3D chromosome regulatory landscape of human pluripotent cells. *Cell Stem Cell* **18:** 262–275. doi:10.1016/j.stem.2015.11.007

Jungmann R, Avendaño MS, Dai M, Woehrstein JB, Agasti SS, Feiger Z, Rodal A, Yin P. 2016. Quantitative super-resolution imaging with qPAINT. *Nat Methods* **13:** 439–442. doi:10.1038/nmeth.3804

Kato M, McKnight SL. 2017. Cross-β polymerization of low complexity sequence domains. *Cold Spring Harb Perspect Biol* **9:** a023598. doi:10.1101/cshperspect.a023598

Khanna N, Zhang Y, Lucas JS, Dudko OK, Murre C. 2019. Chromosome dynamics near the sol-gel phase transition dictate the timing of remote genomic interactions. *Nat Commun* **10:** 1–13. doi:10.1038/s41467-019-10628-9

Kim Y, Shi Z, Zhang H, Finkelstein IJ, Yu H. 2019. Human cohesin compacts DNA by loop extrusion. *Science* **366:** 1345–1349. doi:10.1126/science.aaz4475

Kumaran RI, Spector DL. 2008. A genetic locus targeted to the nuclear periphery in living cells maintains its transcriptional competence. *J Cell Biol* **180:** 51–65. doi:10.1083/jcb.200706060

Lakadamyali M, Cosma MP. 2020. Visualizing the genome in high resolution challenges our textbook understanding. *Nat Methods* **17:** 371–379. doi:10.1038/s41592-020-0758-3

Larson DR, Zenklusen D, Wu B, Chao JA, Singer RH. 2011. Real-time observation of transcription initiation and elongation on an endogenous yeast gene. *Science* **332:** 475–478. doi:10.1126/science.1202142

Lavis LD. 2017. Chemistry is dead. Long live chemistry! *Biochemistry* **56:** 5165–5170.

Lee BG, Merkel F, Allegretti M, Hassler M, Cawood C, Lecomte L, O'Reilly FJ, Sinn LR, Gutierrez-Escribano P, Kschonsak M, et al. 2020. Cryo-EM structures of holo condensin reveal a subunit flip-flop mechanism. *Nat Struct Mol Biol* **27:** 743–751. doi:10.1038/s41594-020-0457-x

Lewis M, Chang G, Horton NC, Kercher MA, Pace HC, Schumacher MA, Brennan RG, Lu P. 1996. Crystal structure of the lactose operon repressor and its complexes with DNA and inducer. *Science* **271:** 1247–1254. doi:10.1126/science.271.5253.1247

Li W, Kamtekar S, Xiong Y, Sarkis GJ, Grindley NDF, Steitz TA. 2005. Structure of a synaptic γδ resolvase tetramer covalently linked to two cleaved DNAs. *Science* **309:** 1210–1215. doi:10.1126/science.1112064

Li J, Hsu A, Hua Y, Wang G, Cheng L, Ochiai H, Yamamoto T, Pertsinidis A. 2020a. Single-gene imaging links genome topology, promoter–enhancer communication and transcription control. *Nat Struct Mol Biol* **27:** 1032–1040. doi:10.1038/s41594-020-0493-6

Li Y, Haarhuis JHI, Sedeño Cacciatore Á, Oldenkamp R, van Ruiten MS, Willems L, Teunissen H, Muir KW, de Wit E, Rowland BD, et al. 2020b. The structural basis for cohesin-CTCF-anchored loops. *Nature* **578:** 472–476. doi:10.1038/s41586-019-1910-z

Lieberman-Aiden E, van Berkum NL, Williams L, Imakaev M, Ragoczy T, Telling A, Amit I, Lajoie BR, Sabo PJ, Dorschner MO, et al. 2009. Comprehensive mapping of long-range interactions reveals folding principles of the human genome. *Science* **326:** 289–293. doi:10.1126/science.1181369

Lim B, Heist T, Levine M, Fukaya T. 2018. Visualization of transvection in living *Drosophila* embryos. *Mol Cell* **70:** 287–296.e6. doi:10.1016/j.molcel.2018.02.029

Lionnet T, Wu C. 2021. Single-molecule tracking of transcription protein dynamics in living cells: seeing is believing, but what are we seeing? *Curr Opin Genet Dev* **67:** 94–102. doi:10.1016/j.gde.2020.12.001

Liu J, Perumal NB, Oldfield CJ, Su EW, Uversky VN, Dunker AK. 2006. Intrinsic disorder in transcription factors. *Biochemistry* **45:** 6873–6888. doi:10.1021/bi0602718

Liu Z, Legant WR, Chen BC, Li L, Grimm JB, Lavis LD, Betzig E, Tjian R. 2014. 3D imaging of Sox2 enhancer clusters in embryonic stem cells. *eLife* **3:** e04263. doi:10.7554/eLife.04236

Liu M, Lu Y, Yang B, Chen Y, Radda JSD, Hu M, Katz SG, Wang S. 2020. Multiplexed imaging of nucleome architectures in single cells of mammalian tissue. *Nat Commun* **11:** 2907. doi:10.1038/s41467-020-16732-5

Lucas JS, Zhang Y, Dudko OK, Murre C. 2014. 3D trajectories adopted by coding and regulatory DNA elements: first-passage times for genomic interactions. *Cell* **158:** 339–352. doi:10.1016/j.cell.2014.05.036

Lupiáñez DG, Kraft K, Heinrich V, Krawitz P, Brancati F, Klopocki E, Horn D, Kayserili H, Opitz JM, Laxova R, et al. 2015. Disruptions of topological chromatin domains cause pathogenic rewiring of gene-enhancer interactions. *Cell* **161:** 1012–1025. doi:10.1016/j.cell.2015.04.004

Luppino JM, Park DS, Nguyen SC, Lan Y, Xu Z, Yunker R, Joyce EF. 2020. Cohesin promotes stochastic domain intermingling to ensure proper regulation of boundary-proximal genes. *Nat Genet* **52:** 840–848. doi:10.1038/s41588-020-0647-9

Ma H, Tu LC, Naseri A, Chung YC, Grunwald D, Zhang S, Pederson T. 2018. CRISPR-Sirius: RNA scaffolds for signal amplification in genome imaging. *Nat Methods* **15:** 928–931. doi:10.1038/s41592-018-0174-0

Marshall WF, Straight A, Marko JF, Swedlow J, Dernburg A, Belmont A, Murray AW, Agard DA, Sedat JW. 1997. Interphase chromosomes undergo constrained diffusional motion in living cells. *Curr Biol* **7:** 930–939. doi:10.1016/S0960-9822(06)00412-X

Masui O, Bonnet I, Le Baccon P, Brito I, Pollex T, Murphy N, Hupé P, Barillot E, Belmont AS, Heard E. 2011. Live-cell chromosome dynamics and outcome of X chromosome pairing events during ES cell differentiation. *Cell* **145:** 447–458. doi:10.1016/j.cell.2011.03.032

Mateo LJ, Murphy SE, Hafner A, Cinquini IS, Walker CA, Boettiger AN. 2019. Visualizing DNA folding and RNA in embryos at single-cell resolution. *Nature* **568:** 49–54. doi:10.1038/s41586-019-1035-4

Mazza D, Abernathy A, Golob N, Morisaki T, McNally JG. 2012. A benchmark for chromatin binding measurements in live cells. *Nucleic Acids Res* **40:** e119–e119. doi:10.1093/nar/gks701

McCord RP, Kaplan N, Giorgetti L. 2020. Chromosome conformation capture and beyond: toward an integrative view of chromosome structure and function. *Mol Cell* **77:** 688–708. doi:10.1016/j.molcel.2019.12.021

McSwiggen DT, Mir M, Darzacq X, Tjian R. 2019. Evaluating phase separation in live cells: diagnosis, caveats, and functional consequences. *Genes Dev* **33:** 1619–1634. doi:10.1101/gad.331520.119

Mercola M, Wang XF, Olsen J, Calame K. 1983. Transcriptional enhancer elements in the mouse immunoglobulin heavy chain locus. *Science* **221:** 663–665. doi:10.1126/science.6306772

Mir M, Stadler MR, Ortiz SA, Hannon CE, Harrison MM, Darzacq X, Eisen MB. 2018. Dynamic multifactor hubs interact transiently with sites of active transcription in *Drosophila* embryos. *eLife* **7:** e40497. doi:10.7554/eLife.40497

Mir M, Bickmore W, Furlong EEM, Narlikar G. 2019. Chromatin topology, condensates and gene regulation: shifting paradigms or just a phase? *Development* **146:** dev182766. doi:10.1242/dev.182766

Mirny LA. 2011. The fractal globule as a model of chromatin architecture in the cell. *Chromosome Res* **19:** 37–51. doi:10.1007/s10577-010-9177-0

Montavon T, Soshnikova N, Mascrez B, Joye E, Thevenet L, Splinter E, de Laat W, Spitz F, Duboule D. 2011. A regulatory archipelago controls Hox genes transcription in digits. *Cell* **147:** 1132–1145. doi:10.1016/j.cell.2011.10.023

Moreau P, Hen R, Wasylyk B, Everett R, Gaub MP, Chambon P. 1981. The SV40 72 base repair repeat has a striking effect on gene expression both in SV40 and other chimeric recombinants. *Nucleic Acids Res* **9:** 6047–6068. doi:10.1093/nar/9.22.6047

Nagano T, Lubling Y, Stevens TJ, Schoenfelder S, Yaffe E, Dean W, Laue ED, Tanay A, Fraser P. 2013. Single-cell Hi-C reveals cell-to-cell variability in chromosome structure. *Nature* **502:** 59–64. doi:10.1038/nature12593

Neems DS, Garza-Gongora AG, Smith ED, Kosak ST. 2016. Topologically associated domains enriched for lineage-specific genes reveal expression-dependent nuclear topologies during myogenesis. *Proc Natl Acad Sci* **113:** E1691–E1700. doi:10.1073/pnas.1521826113

Negre N, Brown CD, Ma L, Bristow CA, Miller SW, Wagner U, Kheradpour P, Eaton ML, Loriaux P, Sealfon R, et al. 2011. A *cis*-regulatory map of the *Drosophila* genome. *Nature* **471:** 527–531. doi:10.1038/nature09990

Nelson P. 2003. *Biological physics: energy, information, life.* W. H. Freeman, New York.

Nir G, Farabella I, Pérez Estrada C, Ebeling CG, Beliveau BJ, Sasaki HM, Lee SD, Nguyen SC, McCole RB, Chattoraj S, et al. 2018. Walking along chromosomes with super-resolution imaging, contact maps, and integrative modeling. *PLoS Genet* **14:** e1007872. doi:10.1371/journal.pgen.1007872

Nora EP, Lajoie BR, Schulz EG, Giorgetti L, Okamoto I, Servant N, Piolot T, van Berkum NL, Meisig J, Sedat J, et al. 2012. Spatial partitioning of the regulatory landscape of the X-inactivation centre. *Nature* **485:** 381–385. doi:10.1038/nature11049

Nora EP, Goloborodko A, Valton AL, Gibcus JH, Uebersohn A, Abdennur N, Dekker J, Mirny LA, Bruneau BG. 2017. Targeted degradation of CTCF decouples local insulation of chromosome domains from genomic compartmentalization. *Cell* **169:** 930–944.e22. doi:10.1016/j.cell.2017.05.004

Normanno D, Boudarène L, Dugast-Darzacq C, Chen J, Richter C, Proux F, Bénichou O, Voituriez R, Darzacq X, Dahan M. 2015. Probing the target search of DNA-binding proteins in mammalian cells using tetR as model searcher. *Nat Commun* **6:** 7357. doi:10.1038/ncomms8357

Northcott PA, Lee C, Zichner T, Stütz AM, Erkek S, Kawauchi D, Shih DJH, Hovestadt V, Zapatka M, Sturm D, et al. 2014. Enhancer hijacking activates GFI1 family oncogenes in medulloblastoma. *Nature* **511:** 428–434. doi:10.1038/nature13379

Nozaki T, Imai R, Tanbo M, Nagashima R, Tamura S, Tani T, Joti Y, Tomita M, Hibino K, Kanemaki MT, et al. 2017. Dynamic organization of chromatin domains revealed by super-resolution live-cell imaging. *Mol Cell* **67:** 282–293.e7. doi:10.1016/j.molcel.2017.06.018

Nozawa K, Schneider TR, Cramer P. 2017. Core Mediator structure at 3.4 Å extends model of transcription initiation complex. *Nature* **545:** 248–251. doi:10.1038/nature22328

Ogiyama Y, Schuettengruber B, Papadopoulos GL, Chang JM, Cavalli G. 2018. Polycomb-dependent chromatin looping contributes to gene silencing during *Drosophila* development. *Mol Cell* **71:** 73–88.e5. doi:10.1016/j.molcel.2018.05.032

Osterwalder M, Barozzi I, Tissières V, Fukuda-Yuzawa Y, Mannion BJ, Afzal SY, Lee EA, Zhu Y, Plajzer-Frick I, Pickle CS, et al. 2018. Enhancer redundancy provides phenotypic robustness in mammalian development. *Nature* **554:** 239–243. doi:10.1038/nature25461

Ou HD, Phan S, Deerinck TJ, Thor A, Ellisman MH, O'Shea CC. 2017. ChromEMT: visualizing 3D chromatin structure and compaction in interphase and mitotic cells. *Science* **357:** eaag0025. doi:10.1126/science.aag0025

Oudelaar MA, Higgs DR. 2021. The relationship between genome structure and function. *Nat Rev Genet* **22:** 154–168. doi:10.1038/s41576-020-00303-x

Oudelaar MA, Davies JOJ, Hanssen LLP, Telenius JM, Schwessinger R, Liu Y, Brown JM, Downes DJ, Chiariello AM, Bianco S, et al. 2018. Single-allele chromatin interactions identify regulatory hubs in dynamic compartmentalized domains. *Nat Genet* **50:** 1744–1751. doi:10.1038/s41588-018-0253-2

Paliou C, Guckelberger P, Schöpflin R, Heinrich V, Esposito A, Chiariello AM, Bianco S, Annunziatella C, Helmuth J, Haas S, et al. 2019. Preformed chromatin topology assists transcriptional robustness of Shh during limb development. *Proc Natl Acad Sci* **116:** 12390–12399. doi:10.1073/pnas.1900672116

Peng A, Weber SC. 2019. Evidence for and against liquid–liquid phase separation in the nucleus. *Noncoding RNA* **5:** 50. doi:10.3390/ncrna5040050

Pennacchio LA, Bickmore W, Dean A, Nobrega MA, Bejerano G. 2013. Enhancers: five essential questions. *Nat Rev Genet* **14:** 288–295. doi:10.1038/nrg3458

Planchon TA, Gao L, Milkie DE, Davidson MW, Galbraith JA, Galbraith CG, Betzig E. 2011. Rapid three-dimensional isotropic imaging of living cells using Bessel beam plane illumination. *Nat Methods* **8:** 417–423. doi:10.1038/nmeth.1586

Pugacheva EM, Kubo N, Loukinov D, Tajmul M, Kang S, Kovalchuk AL, Strunnikov AV, Zentner GE, Ren B, Lobanenkov VV. 2020. CTCF mediates chromatin looping via N-terminal domain-dependent cohesin retention. *Proc Natl Acad Sci* **117:** 2020–2031. doi:10.1073/pnas.1911708117

Quinodoz SA, Ollikainen N, Tabak B, Palla A, Schmidt JM, Detmar E, Lai MM, Shishkin AA, Bhat P, Takei Y, et al. 2018. Higher-order inter-chromosomal hubs shape 3D genome organization in the nucleus. *Cell* **174:** 744–757.e24. doi:10.1016/j.cell.2018.05.024

Rao SSP, Huntley MH, Durand NC, Stamenova EK, Bochkov ID, Robinson JT, Sanborn AL, Machol I, Omer AD, Lander ES, et al. 2015. A 3D map of the human genome at kilobase resolution reveals principles of chromatin looping. *Cell* **162:** 687–688. doi:10.1016/j.cell.2015.07.024

Rao SSP, Huang SC, Glenn St Hilaire B, Engreitz JM, Perez EM, Kieffer-Kwon KR, Sanborn AL, Johnstone SE, Bascom GD, Bochkov ID, et al. 2017. Cohesin loss eliminates all loop domains. *Cell* **171:** 305–320.e24. doi:10.1016/j.cell.2017.09.026

Reitman M, Lee E, Westphal H, Felsenfeld G. 1990. Site-independent expression of the chicken βA-globin gene in transgenic mice. *Nature* **348:** 749–752. doi:10.1038/348749a0

Ricci MA, Manzo C, García-Parajo MF, Lakadamyali M, Cosma MP. 2015. Chromatin fibers are formed by heterogeneous groups of nucleosomes in vivo. *Cell* **160:** 1145–1158. doi:10.1016/j.cell.2015.01.054

Rippe K. 2001. Making contacts on a nucleic acid polymer. *Trends Biochem Sci* **26:** 733–740. doi:10.1016/S0968-0004(01)01978-8

Rippe K. 2021. Liquid–liquid phase separation in chromatin. *Cold Spring Harb Perspect Biol* doi:10.1101/cshperspect.a040683

Robinson PJ, Trnka MJ, Pellarin R, Greenberg CH, Bushnell DA, Davis R, Burlingame AL, Sali A, Kornberg RD. 2015. Molecular architecture of the yeast Mediator complex. *eLife* **4:** e08719. doi:10.7554/eLife.08719

Rodriguez J, Larson DR. 2020. Transcription in living cells: molecular mechanisms of bursting. *Annu Rev Biochem* **89:** 189–212. doi:10.1146/annurev-biochem-011520-105250

Ron G, Globerson Y, Moran D, Kaplan T. 2017. Promoter–enhancer interactions identified from Hi-C data using probabilistic models and hierarchical topological domains. *Nat Commun* **8:** 2237. doi:10.1038/s41467-017-02386-3

Rosa A, Everaers R. 2008. Structure and dynamics of interphase chromosomes. *PLoS Comput Biol* **4:** e1000153. doi:10.1371/journal.pcbi.1000153

Rowley MJ, Corces VG. 2016. The three-dimensional genome: principles and roles of long-distance interactions. *Curr Opin Cell Biol* **40:** 8–14. doi:10.1016/j.ceb.2016.01.009

Saad H, Gallardo F, Dalvai M, Tanguy-le-Gac N, Lane D, Bystricky K. 2014. DNA dynamics during early double-strand break processing revealed by non-intrusive imaging of living cells. *PLoS Genet* **10:** e1004187. doi:10.1371/journal.pgen.1004187

Sabari BR, Dall'Agnese A, Boija A, Klein IA, Coffey EL, Shrinivas K, Abraham BJ, Hannett NM, Zamudio AV, Manteiga JC, et al. 2018. Coactivator condensation at super-enhancers links phase separation and gene control. *Science* **361:** eaar3958. doi:10.1126/science.aar3958

Sanyal A, Lajoie BR, Jain G, Dekker J. 2012. The long-range interaction landscape of gene promoters. *Nature* **489:** 109–113. doi:10.1038/nature11279

Sato H, Das S, Singer RH, Vera M. 2020. Imaging of DNA and RNA in living eukaryotic cells to reveal spatiotemporal dynamics of gene expression. *Annu Rev Biochem* **89:** 159–187. doi:10.1146/annurev-biochem-011520-104955

Schoenfelder S, Fraser P. 2019. Long-range enhancer–promoter contacts in gene expression control. *Nat Rev Genet* **20:** 437–455. doi:10.1038/s41576-019-0128-0

Schwarzer W, Spitz F. 2014. The architecture of gene expression: integrating dispersed *cis*-regulatory modules into coherent regulatory domains. *Curr Opin Genet Dev* **27:** 74–82. doi:10.1016/j.gde.2014.03.014

Schwarzer W, Abdennur N, Goloborodko A, Pekowska A, Fudenberg G, Loe-Mie Y, Fonseca NA, Huber W, Haering CH, Mirny L, et al. 2017. Two independent modes of chromatin organization revealed by cohesin removal. *Nature* **551:** 51–56. doi:10.1038/nature24281

Senecal A, Munsky B, Proux F, Ly N, Braye FE, Zimmer C, Mueller F, Darzacq X. 2014. Transcription factors modulate c-Fos transcriptional bursts. *Cell Rep* **8:** 75–83. doi:10.1016/j.celrep.2014.05.053

Sexton T, Yaffe E, Kenigsberg E, Bantignies F, Leblanc B, Hoichman M, Parrinello H, Tanay A, Cavalli G. 2012. Three-dimensional folding and functional organization principles of the *Drosophila* genome. *Cell* **148:** 458–472. doi:10.1016/j.cell.2012.01.010

Shaban HA, Seeber A. 2020. Monitoring the spatio-temporal organization and dynamics of the genome. *Nucleic Acids Res* **48:** 3423–3434. doi:10.1093/nar/gkaa135

Shaban HA, Barth R, Bystricky K. 2018. Formation of correlated chromatin domains at nanoscale dynamic resolution during transcription. *Nucleic Acids Res* **46:** e77. doi:10.1093/nar/gky269

Shaban HA, Barth R, Bystricky K. 2020. Navigating the crowd: visualizing coordination between genome dynamics, structure, and transcription. *Genome Biol* **21:** 278. doi:10.1186/s13059-020-02185-y

Shah S, Takei Y, Zhou W, Lubeck E, Yun J, Eng CHL, Koulena N, Cronin C, Karp C, Liaw EJ, et al. 2018. Dynamics and spatial genomics of the nascent transcriptome by intron seqFISH. *Cell* **174:** 363–376.e16. doi:10.1016/j.cell.2018.05.035

Shen Y, Yue F, McCleary DF, Ye Z, Edsall L, Kuan S, Wagner U, Dixon J, Lee L, Lobanenkov VV, et al. 2012. A map of the *cis*-regulatory sequences in the mouse genome. *Nature* **488:** 116–120. doi:10.1038/nature11243

Shin Y, Brangwynne CP. 2017. Liquid phase condensation in cell physiology and disease. *Science* **357:** eaaf4382. doi:10.1126/science.aaf4382

Shin Y, Berry J, Pannucci N, Haataja MP, Toettcher JE, Brangwynne CP. 2017. Spatiotemporal control of intracellular phase transitions using light-activated optoDroplets. *Cell* **168:** 159–171.e14. doi:10.1016/j.cell.2016.11.054

Shin Y, Chang YC, Lee DSW, Berry J, Sanders DW, Ronceray P, Wingreen NS, Haataja M, Brangwynne CP. 2018. Liquid nuclear condensates mechanically sense and restructure the genome. *Cell* **175:** 1481–1491.e13. doi:10.1016/j.cell.2018.10.057

Stanyte R, Nuebler J, Blaukopf C, Hoefler R, Stocsits R, Peters JM, Gerlich DW. 2018. Dynamics of sister chromatid resolution during cell cycle progression. *J Cell Biol* **217:** 1985–2004. doi:10.1083/jcb.201801157

Stevens TJ, Lando D, Basu S, Atkinson LP, Cao Y, Lee SF, Leeb M, Wohlfahrt KJ, Boucher W, O'Shaughnessy-Kirwan A, et al. 2017. 3D structures of individual mammalian genomes studied by single-cell Hi-C. *Nature* **544:** 59–64. doi:10.1038/nature21429

Su JH, Zheng P, Kinrot SS, Bintu B, Zhuang X. 2020. Genome-scale imaging of the 3D organization and transcriptional activity of chromatin. *Cell* **182:** 1641–1659.e26. doi:10.1016/j.cell.2020.07.032

Symmons O, Uslu VV, Tsujimura T, Ruf S, Nassari S, Schwarzer W, Ettwiller L, Spitz F. 2014. Functional and topological characteristics of mammalian regulatory domains. *Genome Res* **24:** 390–400. doi:10.1101/gr.163519.113

Szabo Q, Jost D, Chang JM, Cattoni DI, Papadopoulos GL, Bonev B, Sexton T, Gurgo J, Jacquier C, Nollmann M, et al. 2018. TADs are 3D structural units of higher-order chromosome organization in *Drosophila*. *Sci Adv* **4:** eaar8082. doi:10.1126/sciadv.aar8082

Tachiwana H, Kagawa W, Osakabe A, Kawaguchi K, Shiga T, Hayashi-Takanaka Y, Kimura H, Kurumizaka H. 2010. Structural basis of instability of the nucleosome containing a testis-specific histone variant, human H3T. *Proc Natl Acad Sci* **107:** 10454–10459. doi:10.1073/pnas.1003064107

Tortora MM, Salari H, Jost D. 2020. Chromosome dynamics during interphase: a biophysical perspective. *Curr Opin Genet Dev* **61:** 37–43. doi:10.1016/j.gde.2020.03.001

Tsai A, Muthusamy AK, Alves MRP, Lavis LD, Singer RH, Stern DL, Crocker J. 2017. Nuclear microenvironments modulate transcription from low-affinity enhancers. *eLife* **6:** e28975. doi:10.7554/eLife.28975

van Steensel B, Furlong EEM. 2019. The role of transcription in shaping the spatial organization of the genome. *Nat Rev Mol Cell Biol* **20:** 327–337.

Vogelmann J, Valeri A, Guillou E, Cuvier O, Nollmann M. 2011. Roles of chromatin insulator proteins in higher-order chromatin organization and transcription regulation. *Nucleus* **2:** 358–369. doi:10.4161/nucl.2.5.17860

Vogelmann J, Le Gall A, Dejardin S, Allemand F, Gamot A, Labesse G, Cuvier O, Nègre N, Cohen-Gonsaud M, Margeat E, et al. 2014. Chromatin insulator factors involved in long-range DNA interactions and their role in the folding of the *Drosophila* genome. *PLoS Genet* **10:** e1004544. doi:10.1371/journal.pgen.1004544

Wagner FR, Dienemann C, Wang H, Stützer A, Tegunov D, Urlaub H, Cramer P. 2020. Structure of SWI/SNF chromatin remodeller RSC bound to a nucleosome. *Nature* **579:** 448–451. doi:10.1038/s41586-020-2088-0

Wang S, Su JH, Beliveau BJ, Bintu B, Moffitt JR, Wu CT, Zhuang X. 2016. Spatial organization of chromatin domains and compartments in single chromosomes. *Science* **353:** 598–602. doi:10.1126/science.aaf8084

Watson M, Stott K. 2019. Disordered domains in chromatin-binding proteins. *Essays Biochem* **63:** 147–156. doi:10.1042/EBC20180068

Wiggins PA, van der Heijden T, Moreno-Herrero F, Spakowitz A, Phillips R, Widom J, Dekker C, Nelson PC. 2006. High flexibility of DNA on short length scales probed by atomic force microscopy. *Nat Nanotechnol* **1:** 137–141. doi:10.1038/nnano.2006.63

Wittkopp PJ, Kalay G. 2012. *Cis*-regulatory elements: molecular mechanisms and evolutionary processes underlying divergence. *Nat Rev Genet* **13:** 59–69. doi:10.1038/nrg3095

Xu H, Zhang S, Yi X, Plewczynski D, Li MJ. 2020. Exploring 3D chromatin contacts in gene regulation: the evolution of approaches for the identification of functional enhancer–promoter interaction. *Comput Struct Biotechnol J* **18:** 558–570. doi:10.1016/j.csbj.2020.02.013

Yang J, Corces VG. 2012. Insulators, long-range interactions, and genome function. *Curr Opin Genet Dev* **22:** 86–92. doi:10.1016/j.gde.2011.12.007

Yin M, Wang J, Wang M, Li X, Zhang M, Wu Q, Wang Y. 2017. Molecular mechanism of directional CTCF recognition of a diverse range of genomic sites. *Cell Res* **27:** 1365–1377. doi:10.1038/cr.2017.131

Zhang H, Emerson DJ, Gilgenast TG, Titus KR, Lan Y, Huang P, Zhang D, Wang H, Keller CA, Giardine B, et al. 2019. Chromatin structure dynamics during the mitosis-to-G1 phase transition. *Nature* **576:** 158–162. doi:10.1038/s41586-019-1778-y

Zhao Q, Young IT, de Jong JGS. 2011. Photon budget analysis for fluorescence lifetime imaging microscopy. *J Biomed Opt* **16:** 086007. doi:10.1117/1.3608997

Zidovska A. 2020. The self-stirred genome: large-scale chromatin dynamics, its biophysical origins and implications. *Curr Opin Genet Dev* **61:** 83–90. doi:10.1016/j.gde.2020.03.008

Zidovska A, Weitz DA, Mitchison TJ. 2013. Micron-scale coherence in interphase chromatin dynamics. *Proc Natl Acad Sci* **110:** 15555–15560. doi:10.1073/pnas.1220313110

Zoller B, Little SC, Gregor T. 2018. Diverse spatial expression patterns emerge from unified kinetics of transcriptional bursting. *Cell* **175:** 835–847.e25. doi:10.1016/j.cell.2018.09.056

Mammalian DNA Replication Timing

Athanasios E. Vouzas and David M. Gilbert

Department of Biological Science, Florida State University, Tallahassee, Florida 32306, USA

Correspondence: gilbert@bio.fsu.edu

Immediately following the discovery of the structure of DNA and the semi-conservative replication of the parental DNA sequence into two new DNA strands, it became apparent that DNA replication is organized in a temporal and spatial fashion during the S phase of the cell cycle, correlated with the large-scale organization of chromatin in the nucleus. After many decades of limited progress, technological advances in genomics, genome engineering, and imaging have finally positioned the field to tackle mechanisms underpinning the temporal and spatial regulation of DNA replication and the causal relationships between DNA replication and other features of large-scale chromosome structure and function. In this review, we discuss these major recent discoveries as well as expectations for the coming decade.

DNA replication is the process by which the genome replicates faithfully, in its entirety, and exactly once per cell cycle prior to each cell division. Importantly, and particularly in eukaryotes with large genomes, it is not just DNA that duplicates. The entire epigenome must be stripped down and reassembled at the replication fork. It is thus reasonable to presume that the events occurring at the time of replication are important to maintain epigenomic integrity and to facilitate changes in the epigenome during cell fate transitions. Indeed, eukaryotes employ a defined and highly conserved spatiotemporal program known as the replication timing (RT) program. RT is the temporal order in which parts of the genome replicate during S phase of the cell cycle. This program is closely correlated with many aspects of large-scale chromatin structure and function, including 3D chromatin folding, proximity to subnuclear bodies such as nucleoli, the lamina, and speckles, transcriptional activity and its associated chromatin features, as well as mutation and recombination rates (Fu et al. 2018; Marchal et al. 2019; Nathanailidou et al. 2020). Understanding the biological significance of these structure–function correlations and the causal mechanisms that may link them together has been a major challenge in the genome architecture field. In this review, we summarize results from genomics and imaging methods that established these correlations. Because of space limitations, we focus on mammalian cells, occasionally referring readers to literature showing similarities or differences in other model systems. We then highlight recent breakthroughs that provide long-awaited evidence as to how the RT program is regulated, how it may be mechanistically coordinated with other structural and functional features of the genome, and the bio-

Cite this article as *Cold Spring Harb Perspect Biol* doi: 10.1101/cshperspect.a040162

logical significance of an RT program. We expect the coming decade to be one in which we make mechanistic headway into these long-standing questions and obtain much needed insight into large-scale chromosome structure–function relationships in the cell nucleus.

SPATIAL AND TEMPORAL ORGANIZATION OF THE GENOME FOR REPLICATION

Genome-Wide Methods Define Large-Scale Reorganization of Replication Domains during Cell Fate Transitions

Genome-wide RT profiles are typically generated in one of two ways (Gilbert 2010; Hulke et al. 2020). One measures the copy number of any given DNA segment in an asynchronously growing cell population; sequences that replicate earlier will be slightly more abundant than sequences that replicate later. A second approach, which yields higher signal-to-noise, involves labeling newly synthesized DNA with chemically tagged nucleotides, synchronizing cells in early and late S phase, and then purifying and sequencing the labeled DNA synthesized at each of these times (Fig. 1A, top). The resultant RT profiles consist of long regions with similar RT, termed constant timing regions (CTRs). Because bidirectional replication fork rates are typically ~2 kb/min (Conti et al. 2007), CTRs larger than ~500 kb must arise from multiple, nearly synchronous initiation events. CTRs are punctuated by timing transition regions (TTRs) whose slopes (~2 kb/min) are consistent with unidirectional replication forks traveling long distances, interspersed with regions of occasional initiation (Petryk et al. 2016; Zhao et al. 2020).

The development of genome-wide methods gave the ability to measure RT in multiple cell types and stages during the differentiation process (Hiratani et al. 2008). Studying RT across cell types revealed that most CTRs can be subdivided into regions that coordinately change their RT in at least one cell type. In $\log_2$(E/L) ratio data (e.g., Fig. 1A, top), these changes occur in 400–800 kb units, termed replication domains (Fig. 1B, middle; Hiratani et al. 2008). Thus, CTRs consist of multiple replication domains, many of which alter their RT during differentiation to create cell-type-specific CTRs and TTRs (Pope et al. 2014). Further, it was found that during the course of human and mouse embryonic stem cell (mESC) differentiation, the RT of adjacent replication domains align to form fewer, larger CTRs, a process termed domain consolidation (Hiratani et al. 2008; Rivera-Mulia et al. 2015). Tracking these changes during stem cell differentiation showed that consolidation involved changes in the volume and subnuclear location of replication domains (Hiratani et al. 2008, 2010; Takebayashi et al. 2012). Genome-wide RT profiles also demonstrated that different cancer types, patient-specific cancer clones, and other human diseases are characterized by unique RT profiles resulting from alterations in the RT of specific replication domains (Rivera-Mulia et al. 2017, 2019b). Overall, ~50% of replication domains replicate at similar times in all cell types (constitutive domains), while the other 50% switch RT at some point during development and/or in disease (developmental domains) (Dileep et al. 2015; Rivera-Mulia et al. 2015). Dynamic changes in RT during stem cell lineage commitment are coordinated with changes in transcription, chromatin features, and 3D organization (Marchal et al. 2019; Rivera-Mulia et al. 2019a; Nathanailidou et al. 2020) and the genes that are most difficult to reactivate transcriptionally when generating induced pluripotent stem cells reside within a set of domains that are replicated early only in pluripotent cells (Hiratani et al. 2010), suggesting that late replication is associated with a barrier to reprogramming.

High-Resolution and Single-Cell Measurements of Replication Timing

Methods that simply plot the average RT of genomic bins in a population of cells (Fig. 1A, top) can rapidly compare many cell types, experimental conditions, or individuals (Hiratani et al. 2010; Koren et al. 2014; Rivera-Mulia et al. 2015, 2017, 2019b; Hulke et al. 2019). However, these methods suffer from poor temporal and spatial resolution and are not designed to identify cell-to-cell variation in RT. An alterna-

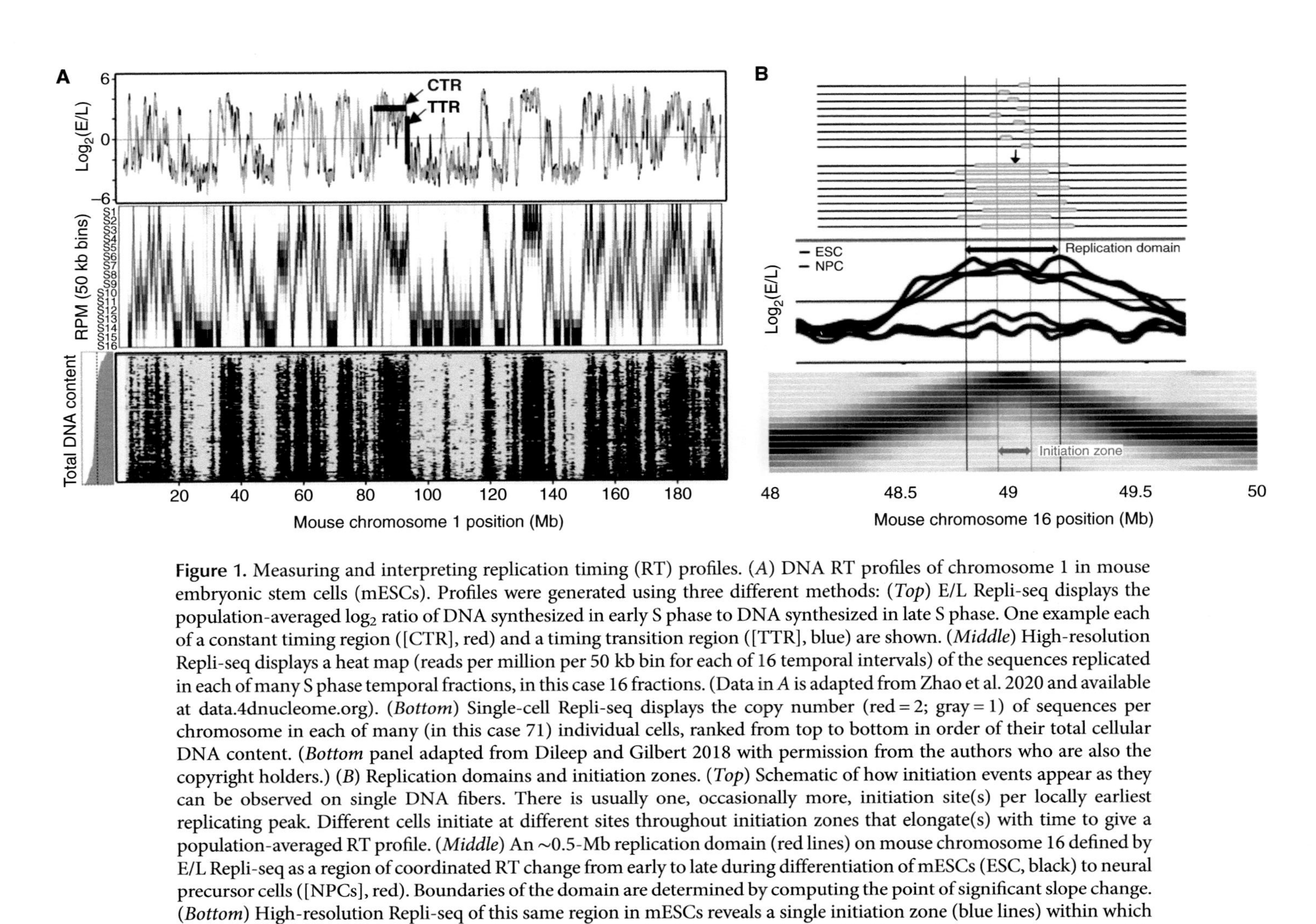

Figure 1. Measuring and interpreting replication timing (RT) profiles. (*A*) DNA RT profiles of chromosome 1 in mouse embryonic stem cells (mESCs). Profiles were generated using three different methods: (*Top*) E/L Repli-seq displays the population-averaged $\log_2$ ratio of DNA synthesized in early S phase to DNA synthesized in late S phase. One example each of a constant timing region ([CTR], red) and a timing transition region ([TTR], blue) are shown. (*Middle*) High-resolution Repli-seq displays a heat map (reads per million per 50 kb bin for each of 16 temporal intervals) of the sequences replicated in each of many S phase temporal fractions, in this case 16 fractions. (Data in *A* is adapted from Zhao et al. 2020 and available at data.4dnucleome.org). (*Bottom*) Single-cell Repli-seq displays the copy number (red = 2; gray = 1) of sequences per chromosome in each of many (in this case 71) individual cells, ranked from top to bottom in order of their total cellular DNA content. (*Bottom* panel adapted from Dileep and Gilbert 2018 with permission from the authors who are also the copyright holders.) (*B*) Replication domains and initiation zones. (*Top*) Schematic of how initiation events appear as they can be observed on single DNA fibers. There is usually one, occasionally more, initiation site(s) per locally earliest replicating peak. Different cells initiate at different sites throughout initiation zones that elongate(s) with time to give a population-averaged RT profile. (*Middle*) An ~0.5-Mb replication domain (red lines) on mouse chromosome 16 defined by E/L Repli-seq as a region of coordinated RT change from early to late during differentiation of mESCs (ESC, black) to neural precursor cells ([NPCs], red). Boundaries of the domain are determined by computing the point of significant slope change. (*Bottom*) High-resolution Repli-seq of this same region in mESCs reveals a single initiation zone (blue lines) within which this domain initiates. Because E/L Repli-seq sums the reds of the first and second half of S phase, the $\log_2$ ratio peak is relatively flat and extends until the point at which cells in the second half of S phase begin to replicate the region.

tive is to collect multiple temporal intervals of S phase and quantify the amount of replication occurring in each interval independently (Chen et al. 2010; Hansen et al. 2010). High-resolution Repli-seq (Zhao et al. 2020), a genome-wide RT mapping method with high spatial and temporal resolution (<50 kb), uses shortened DNA synthesis labeling times and numerous temporal windows of S phase. This approach resolved the boundaries of replication domains into smaller (~200 kb) initiation zones (IZs) (Fig. 1B, bottom) and resolved CTRs into multiple IZs (Fig. 1A, middle). It also resolved TTRs into regions of uniform replication rates consistent with unidirectional forks moving at ~2 kb/min, and punctuated at specific sites by inefficient IZs (Zhao et al. 2020). High-resolution Repli-seq can also indirectly infer RT heterogeneity by quantifying the breadth of the temporal window over which a genomic bin replicates (*Y* axis spread in Fig. 1A, middle), revealing variability in the degree of cell-to-cell heterogeneity at different genomic locations (Zhao et al. 2020), Altogether, high-resolution Repli-seq has provided the most detailed view to date of where and when replication initiates, elongates, and terminates in several mammalian cell lines.

Direct measurements of cell-to-cell heterogeneity require single-cell approaches. Recently, techniques such as live-cell imaging of specific targeted loci (Duriez et al. 2019) and single-cell Repli-seq (Dileep and Gilbert 2018; Takahashi et al. 2019) have provided such measurements. Live-cell imaging studies assay one locus at a time and are thus laborious and low throughput, but they are the only way to obtain direct real-time single-chromosome measurements of locus duplication (Duriez et al. 2019). Single-cell Repli-seq is low resolution (<200 kb), but it can reveal cell-to-cell heterogeneity of RT at a genome-wide scale (Dileep and Gilbert 2018; Takahashi et al. 2019). Both single-locus live-cell tracking and single-cell Repli-seq have concluded that a majority of cells replicate any given segment of the genome within a relatively defined time period.

One drawback of "seq" or "omics" single-cell approaches is that they compare the average of two diploid genomes; single-chromosome data can only be obtained when maternal and paternal chromosomes can be resolved or haploid cells are available (Dileep and Gilbert 2018; Klein et al. 2019). However, because most domains on homologous chromosomes replicate at similar times (Dileep and Gilbert 2018), diploid single-cell Repli-seq can provide a reasonable view of replication domains and is a useful tool for assaying RT in cell types that are difficult to obtain in large numbers such as the cells of early embryos or to track single cells through cell fate transitions (Miura et al. 2019).

Cytological Studies of Replication: Replication Foci and Spatiotemporal Compartments

Long before the development of high-throughput genomic techniques, cytological and imaging techniques served as the major avenue for studying replication. Labeling cells with nucleotide analogs, initially tritiated thymidine and later halogenated nucleotides that could be detected with fluorescent antibodies, identified both a temporal and a spatial regulation of replication within the nucleus (Taylor 1960; Stubblefield 1975; Nakamura et al. 1986). DNA synthesis could be seen to take place in punctate "replication foci" that were localized to the interior of the nucleus in early S phase, moving to the nuclear and nucleolar periphery during late S phase, defining "early and late" spatiotemporal chromatin compartments (Fig. 2A). Pulse-chase-pulse experiments using two different labels (Ma et al. 1998; Dimitrova and Gilbert 1999), as well as tracking replication fork proteins in living cells (Sporbert et al. 2002; Löb et al. 2016), revealed that replication foci took 45–60 min to complete replication. When chased for multiple consecutive cell cycles, the labeled chromatin remained together as a stable unit of chromosome substructure (Sparvoli et al. 1994; Ferreira et al. 1997; Jackson and Pombo 1998). While the spatial arrangement of these units is rearranged during mitosis to form bands resembling chromomeric banding patterns (Stubblefield 1975), they return to their general subnuclear locations (Dimitrova and Gilbert 1999) and exhibit very little motion during in-

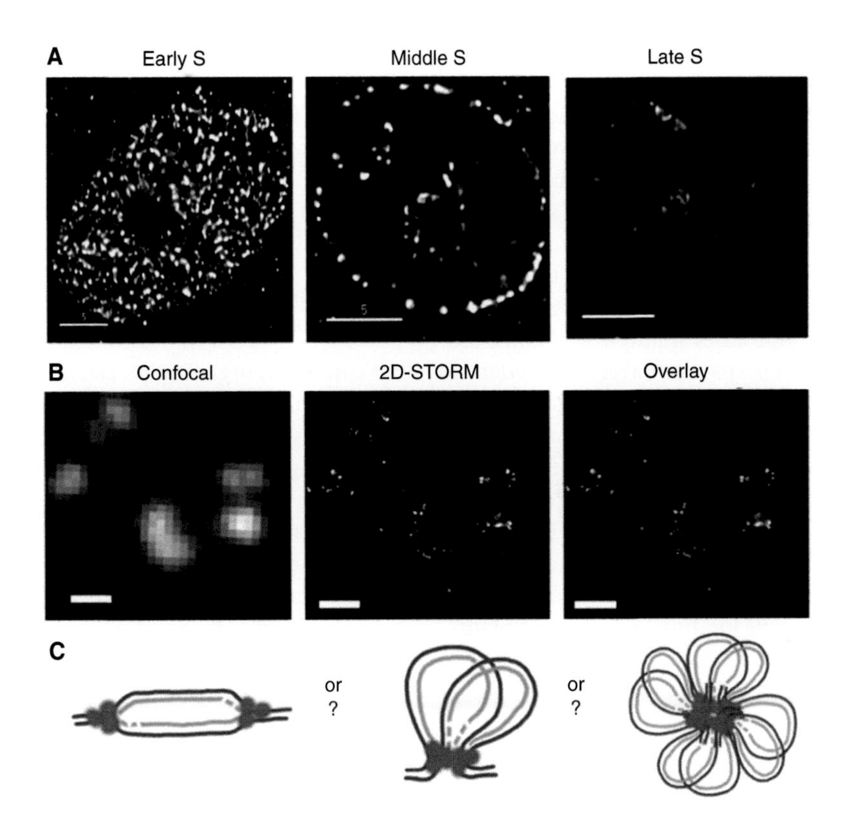

Figure 2. Organization of replicons. (*A*) Chinese hamster ovary cells were labeled for 10 min with BrdU, fixed and stained with anti-BrdU antibodies to reveal the spatial patterns of DNA synthesis in early, middle, and late S phase. (Images in *A* courtesy of J. Lu.) Scale bar, 5 µM. (*B*) Normal rat kidney cells were labeled live with ATTO 633-dUTP and then chased for several generations. Shown is a comparison of a high magnification (scale bar, 500 nm) confocal image of a cluster of replication foci to the super-resolution (2D-STORM) image of the same replication foci, demonstrating that each replication focus consists of a cluster of labeled sites. (Panel *B* reprinted from Xiang et al. 2018 courtesy W. Xiang © 2018 in conjunction with a Creative Commons License [Attribution 4.0 International] www.ncbi.nlm.nih.gov/pmc/articles/PMC5987722.) (*C*) The organization of replicons in these foci remains unknown. Sister forks could be replicated independently (*left*), by a common replisome or clustered replisomes (*middle*) or multiple replicons, and their sister forks could be replicated by a common replisome or cluster of replisomes (*right*). (Panel *C* courtesy of C. Marchal.)

terphase (Abney et al. 1997). Sites of replication can also be tracked in living cells using fluorescently labeled nucleotides (Panning and Gilbert 2005; Wilson et al. 2016), and recent live-cell super-resolution studies have shown that replication foci labeled this way are stable units of coordinated Brownian motion in living cells (Fig. 2B; Nozaki et al. 2017; Xiang et al. 2018).

It is tempting to think of replication foci as the cytological manifestation of replication domains or IZs, but many questions remain as to the molecular architecture and organization of replicons within replication foci. Attempts to count the total number of foci with both conventional and super-resolution microscopy resolution and relate those numbers to genome size suggest that foci are close to the size of replication domains (400–800 kb) but this analysis still relies on indirect estimates of fork rates and interorigin distances (Berezney et al. 2000; Chagin et al. 2016). A single replicon moving bidirectionally at an estimated average fork rate of 1.8 kb/min (Conti et al. 2007) would replicate ~200 kb in the 45–60 min that replication endures at any given focus. Interestingly, that is approximately the size of IZs identified by high-resolution Repli-seq (Fig. 1A, middle). However, high-throughput single DNA fiber analyses suggest that typically only one initiation event occurs per IZ (Fig. 1B, top; Wang et al. 2020a), and that those events can occur at any of many potential sites within the IZ so the 200 kb would be staggered in different chromosomes within the cell population (Fig. 1B, top). Also, identifying whether the two emerging sister forks, or in some cases multiple replicons, are located together in space or travel independently will require further technical innovations to discern (Fig. 2C). In yeast, where replication initiates from well-defined sites, it has been shown that equidistant flanking sites come together in space at the time that they replicate (Kitamura et al. 2006; Meister et al. 2007), suggesting that bidirectionally emanating sister forks are synthesized in a single location. Advances in mammalian live-cell and super-resolution imaging should soon shed light on some of these questions (Deng et al. 2016; Bintu et al. 2018; Tasan et al. 2018; Boettiger and Murphy 2020).

The intriguing relationship between spatial and temporal aspects of replication was addressed by chasing early- and late-labeled foci into the following G1 phase (Dimitrova and Gilbert 1999). This demonstrated that the spatial positions of the foci were re-established 1–2 h after mitosis and remained static for the rest of interphase. By introducing these nuclei into a cell-free replication initiation system, it was shown that the temporal program for replication was established coincident with stable repositioning of the labeled foci, an event termed the timing decision point (TDP) (Dimitrova and Gilbert 1999). It was proposed that anchorage of chromatin domains could seed the assembly of subnuclear microenvironments that could set thresholds for initiation of replication (Dimitrova and Gilbert 1999; Gilbert 2001). Although the mechanisms and molecules establishing RT at the TDP have still not been elucidated, the concept of seeding microenvironments in the nucleus is now quite popular and is thought to occur through liquid–liquid phase separation (Strom and Brangwynne 2019). We will return to this concept below when we discuss recent advances in understanding mechanisms regulating RT (see Fig. 4). Understanding the molecular events occurring at the TDP will be a major challenge for the coming decade.

REPLICATION AND GENOME ARCHITECTURE

Hi-C Compartments Strongly Correlate with Replication Timing

3D chromatin organization can also be inferred by studying the interactions of loci at a global scale using high-resolution chromatin conformation capture (Hi-C), a technique that maps chromatin interactions genome wide (Lieberman-Aiden et al. 2009; Mota-Gómez and Lupiáñez 2019). A first principal component analysis of Hi-C data identifies two spatially and functionally distinct compartments of chromatin folding, named compartments A and B (Lieberman-Aiden et al. 2009), which correspond remarkably well to the early and late replicating compartments of the genome, both spatially (Fig. 2A) and temporally (Ryba et al. 2010; Yaffe et al. 2010). The A compartment is enriched in transcriptionally active chromatin marks (e.g., H3K27ac, H3K4me3), located in the interior of the nucleus while the B compartment consists of more transcriptionally silenced chromatin marks (e.g., H3K9me3 and H3K27me3) located at the periphery and perinucleolar regions (Lieberman-Aiden et al. 2009; Rao et al. 2014). Given what was already known about the developmental plasticity of RT and the positions of replication domains in the nucleus (Fig. 1B and see the section Genome-Wide Methods Define Large-Scale Reorganization of Replication Domains during Cell Fate Transitions), it was predicted that chromatin compartments would also be developmentally regulated and spatially consolidate during differentiation, coordinated with RT changes and temporal consolidation. This prediction was later borne out with Hi-C analyses in multiple cell types (Dixon et al. 2012; Xie et al. 2013; Dileep and Gilbert 2018; Miura et al. 2019). Interestingly, one study found that A/B compartments and RT become uncoupled during the early stages of human embryonic stem cell (hESC) lineage specification, after which

 Cite this article as *Cold Spring Harb Perspect Biol* doi: 10.1101/cshperspect.a040162

strong alignment between the two is re-established (Dileep et al. 2019). Together these results show that while there is not a one-to-one correspondence of Hi-C compartments and RT—and they can be uncoupled—the two tend to highly correlate with each other, and more so in differentiated cells than in stem cells.

Recently, higher resolution Hi-C methods substratified the binary A/B compartments into five distinct subcompartments, A1-2 and B1-3, with distinct histone modification patterns (Rao et al. 2014). These five subcompartments also correlate strongly with replication in distinct temporal intervals of S phase (Fig. 3B). While A1 replicates very early, A2 does not finish replicating until mid–S phase, has lower guanine–cytosine content, longer genes, and higher levels of H3K9me3 than A1. Large differences are observed in the three B subcompartments. B1 is enriched in H3K27me3, depleted in H3K36me3, and replicates in mid–S phase, whereas B2 and B3 are depleted in H3K27me3 and replicate in very late S phase. B2 can be found in both the nuclear lamina and the nucleolus, while B3 is exclusively found at the nuclear lamina (Rao et al. 2014). This demonstrates a clear subcompartmentalization of the genome that goes beyond the division between A and B compartments yet still correlates strongly with RT.

Topologically Associated Domains (TADs) and Their Relationship to Replication Domains

In addition to large-scale compartments, Hi-C demonstrated that chromosomes are organized into smaller self-interacting domains, hundreds of kilobases in length, called TADs (Dixon et al. 2012; Nora et al. 2012; De Laat and Duboule 2013). Interactions between adjacent domains are depleted, allowing TADs to be mapped by features such as directionality index (interactions of chromosomal sites significantly more frequent in one direction) or insulation score (interactions across a region significantly depleted). TADs, as originally described, were the same size range as replication domains, raising the question as to whether TADs are the structural equivalents of replication domains (Fig. 1B) and, possibly, replication foci (Fig. 2). In fact, it was shown that the boundaries of replication domains (Fig. 1B), defined as domains whose RT changes during cell fate transitions or differs between cell types, align strongly with the boundaries of TADs (Pope et al. 2014). This gave rise to the "replication domain model" (Pope et al. 2013), in which RT is regulated in units corresponding to TADs, which fold in such a manner that TADs with similar RT come into close proximity (at the TDP) to form larger-scale compartments.

Mechanisms linking replication domains to TADs remain uncertain. Because adjacent domains with similar RT reside within the same interaction compartment, TAD boundaries within those interaction compartments were originally difficult to detect with low-resolution data (Dixon et al. 2012; Pope et al. 2014). More recent high-resolution Hi-C resolves these boundaries and reveals many smaller domains nested within larger ones that were previously averaged together in the low-resolution data (Rao et al. 2014; Beagan and Phillips-Cremins 2020). We now recognize two basic forces that shape the 3D organization of chromatin (Sanborn et al. 2015; Rao et al. 2017; Schwarzer et al. 2017). One is driven by the tendency of similar chromatin types to aggregate nonspecifically into compartments, with the borders between different chromatin types manifesting as boundaries. The second is driven by the extrusion of chromatin through the central pore of cohesin rings until encountering roadblocks (e.g., CTCF) that constitute fixed boundary elements (Fudenberg et al. 2017; Rao et al. 2017). Consistently, depletion of either the cohesin subunit RAD21 (Rao et al. 2017), the cohesin loading factor Nipbl (Schwarzer et al. 2017), or CTCF (Nora et al. 2017) eliminates loop domains but has no detectable effect on A/B chromatin compartments (Nora et al. 2017; Rao et al. 2017). Different methods and scales of analysis highlight different features of these domains to different extents (Hsieh et al. 2015, 2020). Super-resolution imaging can also visualize these domains, revealing chromosome-to-chromosome heterogeneity in the specific CTCF sites where cohesin pauses that cannot be

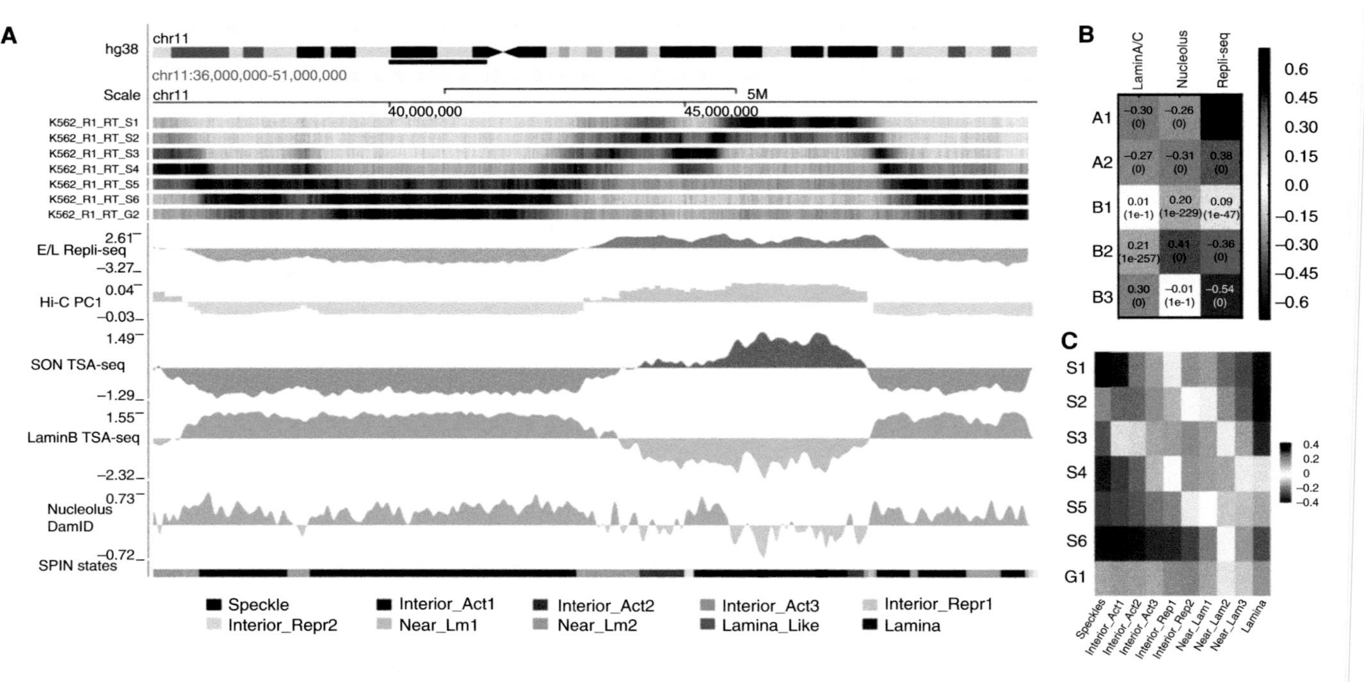

Figure 3. Nuclear organization and chromatin architecture. (*A*) Replication timing (RT), displayed both in high-resolution Repli-seq and E/L Repli-seq, highly correlates with Hi-C-derived first principal component (PC1), the distance of the locus from the nuclear speckle protein SON and the nuclear lamina protein LaminB measured via TSA-seq, and with the contact frequency of chromatin with the nucleolus measured via DamID, an alternative to ChIP, which uses *Escherichia coli* adenine methyltransferase (Dam) fused to a protein of interest to methylate adenines in DNA sequences that come to close proximity to the protein of interest. Information on the distance and contact frequency of chromatin loci from subnuclear landmarks is combined with Hi-C and histone modification data to stratify the genome into 10 SPIN states. A SPIN state is composed of genomic loci sharing unique combinations of histone modifications, compartmentalization, and association with nuclear landmarks. The data displayed originate from a 15 Mb window in chromosome 11 of human bone marrow lymphoblast K562 cells, mapped using the Nucleome Browser (vis.nucleome.org). (*B*) Spearman correlation of five sub-compartments in Rao et al. (2014) to the nuclear lamina, the nucleolus, and early RT. (Data in *B* adapted from Rao et al. 2014.) (*C*) Spearman correlation of the 10 SPIN states in Wang et al. (2020b) to six fractions of the S phase and G2. (Data in *C* plotted using data adapted from Wang et al. 2020b.)

detected with population-averaged methods (Bintu et al. 2018).

Although these mechanisms are not in dispute, the nomenclature has become complicated as some investigators use TAD to mean any self-interacting domain, others reserve the term TAD exclusively to mean cohesin-mediated loop domains, while still others avoid TAD altogether. Also, the term "contact domain," can be used to mean either all domains or only "loop domains." Another source of confusion is that mammals (the focus of this review) have extensive loop extrusion, while many other species mainly use chromatin compartmentalization (Rowley and Corces 2018; Szabo et al. 2019). It is important to be aware of this nomenclature variation when reading the literature.

Importantly for our discussion, depletion of either RAD21 RT (Oldach and Nieduszynski 2019; Cremer et al. 2020) or CTCF RT (Sima et al. 2019) was reported to have no detectable effect on the global RT program. Moreover, deletion of CTCF sites at replication/TAD/loop domain boundaries can cause fusion of neighboring domains (Despang et al. 2019) while having no effect on RT of the locus (Sima et al. 2019). By corollary, inversion of part of an early replicating domain, transplanting it inside of a late replicating domain, can create a new boundary (Sima et al. 2019). It remains to be seen whether in these cases the boundaries are between compartment domains or have created new loop domain boundaries; however, in at least some cases the creation of a new early-to-late RT transition (TTR) leads to the presence of a new RAD21-binding site (Klein et al. 2019). These results demonstrate the independence of global RT on fixed boundaries and suggest that, in at least some cases, RT differences can drive the formation of boundaries. One hint at how this could occur comes from recent evidence suggesting that the replicative helicase MCM2-7 complex, which designates sites of initiation (discussed below), can impede cohesin extrusion (Dequeker et al. 2020) and could thus influence the positions of boundaries. Other evidence suggests that cohesin extrusion confines the replicative helicase to domain boundaries (Emerson et al. 2021). Clearly there is much work to be done to understand the mechanisms linking replication and genome architecture but new tools to manipulate the replication program (Klein et al. 2019; Sima et al. 2019) should reveal new insights in the near future.

REPLICATION AND SUBNUCLEAR LANDMARKS

Lamina-Associated Domains (LADs) and Nucleolus-Associated Domains (NADs): Overlapping Compartments

The nucleus is a highly heterogeneous landscape with myriad neighborhoods of markedly different molecular composition and functional activities all residing in close proximity with no apparent physical boundaries to separate them (Hildebrand and Dekker 2020). These neighborhoods or subnuclear microenvironments can have profound effects on chromatin organization, accessibility, and activities. Those whose composition is well defined form landmarks of subnuclear position. Two of the most prominent landmarks are the nuclear lamina and the nucleolus. Chromatin domains coming into contact with either of these landmarks are called lamina-associated domains (LADs) and nucleolus-associated domains (NADs), respectively. They contain chromatin marks associated with repressed chromatin, and harbor lowly expressed or more frequently silenced genes, broadly referred to as heterochromatin (van Steensel and Belmont 2017). Techniques to map the proximity of chromatin loci to such subnuclear landmarks fall into two basic categories (Fig. 3A): those that map contact frequency, such as chromatin immunoprecipitation (ChIP) and DNA adenine methyltransferase identification (DamID) (Guelen et al. 2008; Kind et al. 2013, 2015), and those that measure the chromosomal distance of loci to nuclear structures, such as TSA-seq, a cytological ruler that measures the distance of a sequence from a specific subnuclear feature (Chen et al. 2018). A subset of LADs associate with the lamina independent of cell type (constitutive LADs), while others associate with the lamina in a cell-type-specific manner (developmental LADs) (Peric-Hupkes et al. 2010; Meuleman et al. 2013). As expected, lamina association is dynamic very early in G1, as the nucleus is reassembled, with distinctions between LADs and inter-LADs becoming more pronounced over time (Abney et al. 1997; Marshall et al. 1997; Thomson et al. 2004; Schaik et al. 2020), during the same period of G1 phase in which RT is established (TDP).

Recent studies have indicated that inducing the expression of a gene in contact with the nuclear lamina can dissociate the gene from the lamina and move it to the interior of the nucleus, confined to the transcription unit and 50–100 kb of flanking DNA. This can be accompanied by an RT shift across a significantly larger region containing the expressed gene (Therizols et al. 2014; Brueckner et al. 2020). Interestingly, longer transcripts can advance RT under conditions and in positions where short transcripts do not (Blin et al. 2019). These findings indicate that some aspect associated with the complex process of transcription can directly or indirectly reposition a locus and advance its RT. An important future goal will be to determine whether or not repositioning or RT changes require mitotic disassembly and reassembly of nuclear architecture or whether they occur in direct response to transcriptional activation.

While NADs are traditionally thought to be exclusively heterochromatic and late replicating, it was recently discovered that there are two types of NADs, type I and type II, each having its own distinct epigenetic marks and RT (Vertii et al. 2019). A big distinction between the two is that type I NADs can also be found near the lamina, while type II NADs are exclusively found near the nucleolus. This is consistent with the earlier finding that some LADs can be positioned either at the lamina or the nucleolar periphery between cell cycles (Kind et al. 2013; Politz et al. 2014) and that single-cell Lamin DamID finds some LADs to be associated with the nuclear lamina in every cell while others are variably associated from cell to cell (Kind et al. 2015). Type I NADs tend to replicate very late during S phase, have very low levels of gene expression, and are enriched in H3K9me3, similar to LADs. On the other hand, type II NADs replicate in the mid-late S phase, have higher levels of gene expression, and are enriched in H3K27me3. Overall, close examination of LADs and NADs reveals that domains belonging to these two classes are not homogeneous, with variations in chromatin structure, strength of association with subnuclear landmarks, and correlation to RT.

Speckle-Associated Domains (SPADs)

First observed by Ramon y Cajal in the early twentieth century through histochemical stains and later confirmed, in the 1990s, via electron microscopy (EM), nuclear speckles are subnuclear bodies enriched in pre-mRNA splicing factors that facilitate the maturation of mRNA, although their precise functional role is still under debate (Hall et al. 2006; Spector and Lamond 2011; Chen and Belmont 2019). In contrast to LADs and NADs, loci residing near the nuclear speckles, termed speckle-associated domains (SPADs), are often associated with open chromatin, have high levels of transcription, and are decorated with active chromatin histone marks, making these regions euchromatic (Chen and Belmont 2019).

TSA-seq has been used to map SPADs, by calculating the distance of genomic loci from SON, a protein essential for speckle organization (Chen et al. 2018; Chen and Belmont 2019). Interestingly, plots showing the distance of genomic loci from nuclear speckles show a correlation of SON TSA-seq peaks with early RT peaks (Fig. 3A). Genes close to nuclear speckles tend to have higher levels of expression, and tethering of a locus to the nuclear speckles has been shown to be sufficient to amplify gene expression levels (Kim et al. 2020). Early replication of chromatin in proximity to SPADs could be related to the high levels of transcription close to speckles, either directly through the process of transcription itself or through epigenetic changes occurring during transcription that correlate with early replication. It is important to appreciate that although transcription and early RT are correlated, there are many genes that can be expressed while late replicating so the act of stimulating transcription itself is not sufficient for early RT (Rivera-Mulia et al. 2015), and the correlation may be more related to chro-

 Cite this article as *Cold Spring Harb Perspect Biol* doi: 10.1101/cshperspect.a040162

matin changes elicited by transcriptional factor binding (Goren et al. 2008; Ostrow et al. 2017; Rivera-Mulia et al. 2019a; Sima et al. 2019). Alternatively, speckle-associated factors themselves or activities associated with splicing could promote early replication, or early replication could promote association with speckles. As with most correlations between genome organization and RT, the underpinning causal mechanisms remain a major future challenge.

Replication Timing and Models of Nuclear Organization

Thus far, we have described the relationship of RT with chromatin architecture and subnuclear localization as separate correlations. Recently, a computational method has been developed to integrate Hi-C with genome-wide information about contact frequencies with and proximity to specific subnuclear landmarks, termed spatial position inference of the nuclear genome (SPIN) (Wang et al. 2020b). SPIN stratifies chromatin by its subnuclear addresses, dubbed as "SPIN states." SPIN separates the genome into 10 SPIN states, which feature distinct localization patterns within the nucleus (Fig. 3A; Wang et al. 2020b). These SPIN states strongly associate with the five primary subcompartments defined by Rao et al. (2014). However, the incorporation of distance relationships between subnuclear landmarks and chromatin marks gives finer structure and functional significance to the SPIN reference map over Hi-C contact maps alone. Interestingly, each SPIN state correlates strongly with chromatin replicated in a specific time interval of S phase (Fig. 3C). Additionally, certain states highly correlate with constitutive and developmental replication domains, as defined by Dileep et al. (2015); 85% of the domains found in the speckle state were constitutively early replicating, 55% of the domains in the lamina state were constitutively late, while the rest of the states contain higher percentages of developmentally regulated replication domains (Wang et al. 2020b). Overall, SPIN is a newly developed computational tool with the ability to integrate multiple data types and provide a unified correlation between RT and subnuclear chromatin organization. The development of SPIN exemplifies the collaborative efforts that are being undertaken to use newly acquired computational power to gain better insight into the association of RT with other features of the nucleus.

NEW INSIGHTS INTO MECHANISMS REGULATING REPLICATION TIMING PROVIDE CLUES TO CAUSALITY

Initiating Replication: A Precision Mechanism that Requires Flexibility

To understand how RT is regulated, it would seem logical to start with the sites where replication initiates, often called "origins" of replication. In mammalian cells, however, sites of initiation are highly variable, such that any given site in the genome is used in only a small fraction of cell cycles (Demczuk et al. 2012). To restrict initiation to once and only once per cell cycle, cells employ a sequence-agnostic "two-cycle engine" strategy (Gilbert 2001; Deegan and Diffley 2016). The ring-shaped Mcm2-7 helicase is loaded around double-stranded DNA in an inactive form early in G1 strictly under conditions that prevent initiation and with no apparent DNA sequence requirement. Upon entry into S phase, Mcm is converted into an active helicase by Db4-dependent kinase (DDK) and cyclin-dependent kinase (CDK) under conditions that prevent new Mcm complexes from being loaded (Deegan and Diffley 2016). Moreover, unique among chromatin proteins, inactive Mcm is irreversibly locked down on DNA until replication begins (Kuipers et al. 2011). However, it can slide when chromatin is disassembled and DNA unwound during transcription (Foss et al. 2019), shifting potential sites of initiation long after Mcm loading (Sasaki et al. 2006). Finally, the number of Mcm complexes loaded exceeds the number of origins used to replicate the genome by several fold (Ibarra et al. 2008; Limas and Cook 2019). These excess Mcms can either remain dormant until removed by passing replication forks, or they can be recruited to initiate replication when segments of DNA remain unreplicated such as under conditions of replication stress (Ge et al. 2007; Courtot et al.

2018; Moiseeva and Bakkenist 2019). Thus, mechanisms regulating initiation of replication ensure fail-safe, once per cell cycle initiation without the need for specific sequences or site preference (Gilbert 2001; Deegan and Diffley 2016). Possibly, reliance on specific origin sequences would render large chromosomes dangerously vulnerable to mutation and structural variation (Rivera-Mulia and Gilbert 2016).

While highly flexible and stochastic, origin selection is not random; rather, initiation is confined to zones of several 10s of kilobases and sites within these IZs initiate at different frequencies via mechanisms that are not understood (Petryk et al. 2016; Wang et al. 2020a; Zhao et al. 2020). Addressing these mechanisms will require the ability to make large numbers of single-molecule measurements to accurately quantify the number and frequencies of usage of all initiation sites. It has not yet been possible to map the sites of Mcm loading on single molecules. However, single-molecule measurements of initiation efficiency on long purified DNA fibers have now achieved at >2000× genome coverage (Wang et al. 2020a). While the resolution is still too low (~15 kb) to determine variation in specific site usage, it is clear that replication will initiate at some location within any given ~40 kb IZ with frequencies ranging from <0.5% to ~40% of S phases (Wang et al. 2020a) but at any given site within an IZ at a much lower frequency. This large degree of flexibility in replication initiation sites explains why the results of population-based methods for mapping origins are so methodology dependent; high-resolution methods pick up specific sites within IZs that initiate frequently enough to be detected above noise, while low-resolution methods detect the sum frequency of initiation within larger IZs (Gilbert 2010; Hyrien 2015). Taken together, what emerges is a stochastic view of origin firing in which each segment of the genome has a characteristic probability of firing. Because segments with higher probability are more likely to fire early in the S phase, the mechanisms regulating the probability of initiation within zones, rather than specific origin sequences, are what determines the RT program. We will now turn our attention to how zones of initiation might acquire different probabilities of firing.

Early Replication Control Elements (ERCEs)

The fact that initiation does not require specific DNA sequences does not rule out the possibility that sequence elements may govern the probability that a replication domain will initiate somewhere within its IZ(s). However, demonstrating the existence of sequence elements has remained controversial. Shortly after the discovery of an RT program (Taylor 1960), it was found that the active and inactive X chromosomes in female mammals replicate early and late during S phase, respectively, demonstrating that at least in some cases, sequence-independent (epigenetic) mechanisms can influence RT. On the other hand, there are several examples of synthetically combined DNA segments that can alter RT when inserted into a locus (Simon et al. 2001; Hassan-Zadeh et al. 2012; Blin et al. 2019; Brueckner et al. 2020). The ability to identify bona fide *cis*-elements in their native context required the ability to make genetic lesions in mammalian genomes, which remained extremely low throughput until the advent of CRISPR-Cas9 genome-editing techniques. Recently, a large series of CRISPR-mediated deletions and inversions revealed the existence of discrete *cis*-regulatory elements of RT, termed early replication control elements (ERCEs), in several distinct replication domains in mESCs with 1835 more predicted ERCEs identified computationally across the mouse genome (Sima et al. 2019). Deletions of combinations of ERCEs in Dppa2/4 revealed that the presence of one ERCE is sufficient for a region to replicate in mid-S, while two or more ERCEs gave rise to early-S replication. Intriguingly, ERCEs were also found to be necessary to maintain a subnuclear A/B compartment, proximity to the nuclear lamina, TAD architecture, and transcription, with the effects confined to within the domain in which they reside (Sima et al. 2019; Brueckner et al. 2020). By contrast, the boundaries of TADs as well as CTCF/cohesin loop domains were dispensable for global RT and subnuclear compartments (Rao et al. 2017; Oldach and Nieduszynski, 2019; Sima et al. 2019; Cremer et al. 2020). Thus, ERCEs are not only the long sought *cis*-elements of RT control, but they coordinate replication with transcription and chromosome architecture, providing an experimental handle into mechanisms behind these long-standing correlations.

ERCEs, so far identified only in mESCs, are decorated with large patches of H3K27ac and have binding sites for mESC pluripotency factors, Oct4, Sox2, and Nanog; as such, they resemble super-enhancers, clusters of enhancers occupied by master regulators (Whyte et al. 2013). ERCEs also interact strongly with each other, independent of CTCF and cohesin (Sima et al. 2019). The interaction of multiple ERCEs therefore creates a 3D hub rich in histone acetylation. It is thus likely that ERCEs attract the Brd2/4 protein, which binds acetylated histones and has been shown to form phase-separated droplets (Gibson et al. 2019; Borck et al. 2020; Han et al. 2020). Because Brd2/4 has been shown to interact strongly with the replication-initiation protein Treslin (Sansam et al. 2018), Brd2/4 microenvironments would be replete with Treslin and poised for replication initiation as soon as DDK and CDK are activated at the onset of S phase. The assembly of a microenvironment, potentially through phase separation, could explain domain level regulation of replication (Gilbert 2001), as it could then initiate at any site where Mcm would reside throughout the domain. In this way, ERCEs could promote highly deterministic early replication, via stochastic origin specification (Fig. 4).

This model for ERCE function is reminiscent of that proposed for how Fkh1,2 TFs promote early replication in budding yeast (Knott et al. 2012). Fkh1,2 dimerize to mediate 3D clustering of a set of early origins and recruit the essential replication initiation kinase subunit, Dbf4. Interestingly, mutations that impair their ability to dimerize without affecting their transcription activation function result in a delay of early origin firing, suggesting that formation of a 3D hub rather than transcription is the critical replication role of Fkh1,2 (Ostrow et al. 2017). It is possible that Oct4, Sox2, Nanog mediate the 3D interactions and/or high density of histone acetylation of ERCEs (de Wit et al. 2015; Wu et al. 2015, 2018). Because Oct4, Sox2, and Nanog are cell-type-specific TFs, and since the expression of core transcriptional regulatory network factors correlates with cell-type-specific RT changes (Rivera-Mulia et al. 2019a), this predicts that ERCEs will be found to be cell-type-specific regulatory elements of large-scale genome organization and function.

Trans-Acting Regulators of Replication Timing

The search for *trans*-acting regulators of mammalian RT has been nearly as arduous as the search for *cis*-regulators. Chromatin readers, writers, and remodelers, as well as architectural proteins, have been hypothesized to regulate RT, and mutations in some chromatin regulators have strong effects on RT in yeast (Aparicio et al. 2004; Knott et al. 2009; Yoshida et al. 2014; Zhang et al. 2019). However, depletion of homologs to these factors has little to no effect on RT in mammalian cells. Depletion or loss of some gene products, such as ESBAF subunits, PREP1 and DNA polymerase θ (Takebayashi et al. 2013; Fernandez-Vidal et al. 2014; Palmigiano et al. 2018), have been shown to cause partial or localized RT changes. However, there is only one gene to date, Rif1, whose deletion has profound effects on the global RT program in yeast, mouse, humans, *Drosophila*, and Zebrafish (Hayano et al. 2012; Hiraga et al. 2014, 2018; Foti et al. 2016; Seller and O'Farrell 2018). In budding yeast, Rif1 has been shown to recruit protein phosphatase 1 (PP1), which can antagonize phosphorylation of the MCM complex by Dbf4-dependent kinase (DDK, Cdc7/Dbf4 complex), a critical step in origin firing. This PP1-binding domain is conserved and evidence points to a similar mechanism in *Drosophila* and mammalian cells (Sukackaite et al. 2017; Seller and O'Farrell 2018). In mammalian cells, the loss of the PP1 interacting domain only partially accounts for the RT regulatory activity of Rif1, suggesting that other mechanisms are also in play (Gnan et al. 2019). Although Rif1 is believed to act by delaying RT, and Rif1 is predominantly found in late replicating regions, its elimination leads to both delays and advances in RT (Foti et al. 2016; Klein et al. 2019). By analogy to

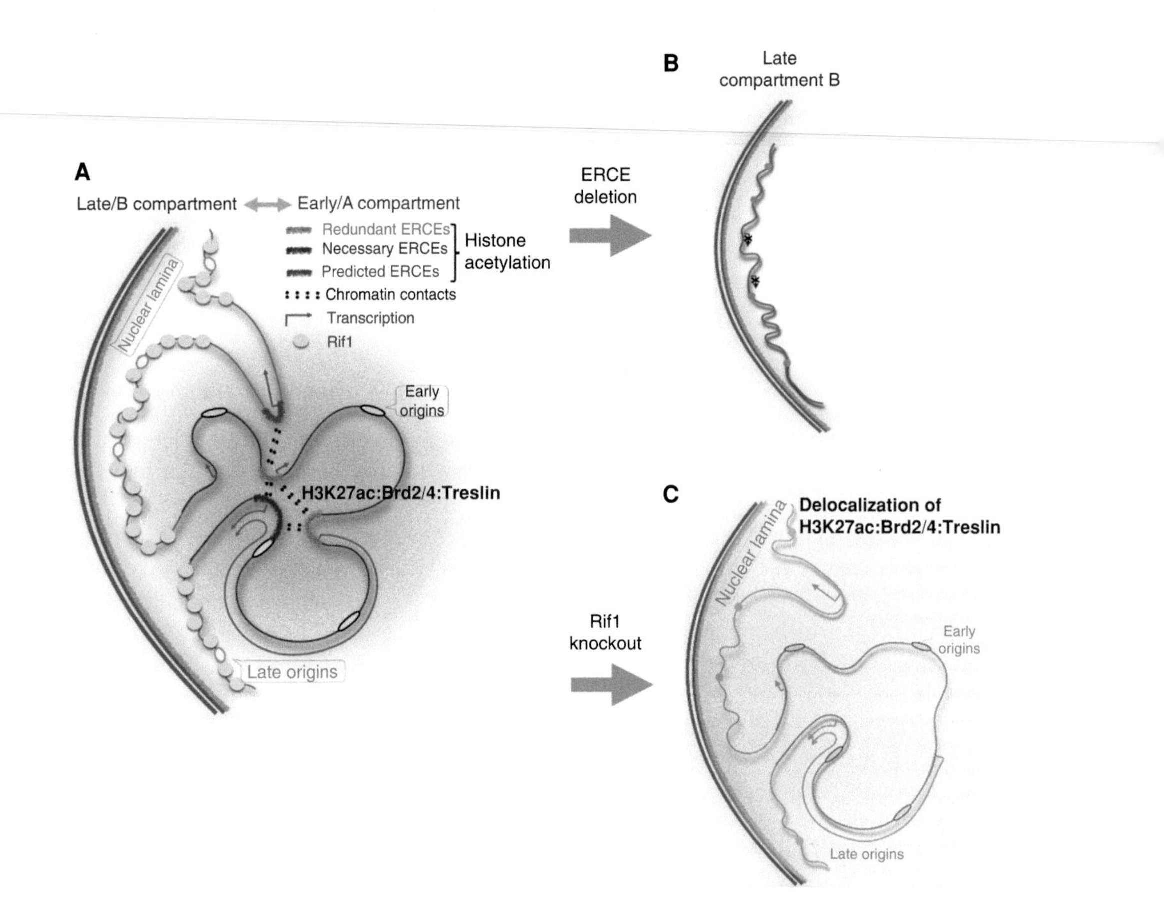

Figure 4. Model of early replication control element (ERCE) interactions in the Dppa2/4 domain. (*A*) ERCEs interact to influence the 3D architecture of topologically associated domains (TADs), the interaction of TADs with other domains (compartmentalization), transcription, and early replication timing (RT) (Sima et al. 2019). ERCEs resemble enhancers, being occupied by H3K27ac, the p300 acetyltransferase, and the major pluripotency transcription factors (TFs) Oct4, Sox2, and Nanog (OSN). In this working model, lineage-specific TFs such as OSN promote histone acetylation, recruiting acetylation readers Brd2/4 and promoting chromatin interactions (Kim 2009; Wu et al. 2015, 2018) to form a 3D hub highly enriched for Brd2/4. The replication initiation protein Treslin interacts with Brd2/4, linking OSN and histone acetylation to initiation of replication (Sansam et al. 2018). Meanwhile, Rif1 coats late replicating regions to prevent them from replicating early during S phase. (*B*) Deletion of all three ERCEs in this domain causes the domain to switch from early to late replicating, eliminates all intradomain transcription, changes compartments (*A* to *B*), and shifts the domain toward the nuclear lamina (NL). (*C*) Rif1 knockout allows for late replicating regions near the NL to become accessible to replication factors, thus diverting replication resources toward these late replicating regions, resulting in highly stochastic RT, redistribution of chromatin marks, and alterations in chromatin compartments (Panels *A* and *B* adapted from Sima et al. 2019 courtesy of Creative Commons Public License.)

studies of the effects of Sir2 on RT in budding yeast (Yoshida et al. 2014), this may be due to competition of the normally late replicating regions, now advanced in their timing, for binding limiting concentrations of replication initiation factors (Mantiero et al. 2011; Tanaka et al. 2011; Collart et al. 2013), thus sequestering them from normally early replicating chromatin, which in turn delays their replication.

The biological significance of RT has remained a puzzle. There is no a priori reason that the genome should be replicated in a particular order simply to duplicate the genome. Early replicating genes are at a higher gene dosage per cell than late replicating genes, and there has been some evidence in budding yeast that there is selective pressure for this gene dosage effect mediated by RT (Müller and Nieduszynski 2017). However, the notion that RT could regulate the assembly of different types of chromatin at different times, possibly maintaining or changing the entire epigenomic landscape, has remained a speculative hypothesis (Gilbert 2002; Lande-Diner et al. 2009) because there has been no way to eliminate the temporal order of replication. In the case of Rif1 disruption, until recently, only the log_2(E/L) RT method (Fig. 1A, top) had been applied to study RT, leading to the conclusion that Rif1 disruption led to widespread discrete RT shifts, rather than a loss of timing control. However, high-resolution and single-cell Repli-seq (Fig. 1A, middle) has now revealed that the primary effect of Rif1 loss is to vastly increase cell-to-cell heterogeneity of the affected loci, which become averaged in log_2(E/L) data to appear as discrete shifts. Highly stochastic replication in Rif1 depleted cells provided an opportunity to determine the role of RT in maintaining epigenetic states. A time course using a conditional Rif1-AID degron fusion showed that, following depletion of RIF1, a complete disruption ocurred in the first S phase, followed by the gradual delocalization of multiple histone marks, alteration of genome architecture, and transcription changes that continued to increase for many cell generations, consistent with the gradual alteration of histone marks via the dilution of old histones with each cell cycle (Klein et al. 2019; Stewart-Morgan et al. 2020). Changes in histone marks did not occur after RIF1 depletion until replication initiated (Klein et al. 2019). Intriguingly, despite widespread disruption of the epigenome, three different human cell types null for RIF1 were viable with near normal growth rates. RIF1-null hESCs also retained their pluripotency transcriptional network. This is consistent with the finding that Rif1-null mice do not die until after gastrulation and suggests the intriguing possibility that the epigenome is critical for cell fate transitions but not for self-renewal. Also emerging from this study was the finding that different cell types use different mechanisms to regulate RT to different extents. In hESCs, RIF1 loss led to an almost random replication program genome wide, while in colon cancer cell line HCT116 RIF1, null cells retained RT at large blocks of enriched H3K9me3, and knockdown of the Su(var)39h1/2 writer for this mark further advanced RT of those regions. Thus, the ability to eliminate the RT program genome wide through deletion of RIF1 has provided long-awaited evidence that RT is necessary for maintaining the epigenome and has revealed novel mechanisms of RT regulation (Fig. 4C).

Whole Chromosome Territory Regulation of Replication Timing

Another remarkable finding of the last few years is the discovery of asynchronous replication and autosomal RNAs (ASARs), which are required for the timely replication and mitotic condensation of entire chromosomes. Discovery of ASARs stems from the observation that 80% of cancers harbor one chromosome that enters mitosis with an incompletely replicated and uncondensed chromosome (Fig. 5A; Smith et al. 2001), which is highly unstable resulting in breakages resembling chromothripsis, a mutational phenomenon characterized by a large number of concentrated genomic rearrangements (Donley and Thayer 2013). A systematic series of deletions in human chromosomes 6 and 15 identified ASAR6 and ASAR15, very long noncoding RNAs (vlncRNAs; >200 kb) whose disruption eliminates the coordinated replication of homologous chromosome pairs giving rise to the phenotype of delayed RT (DRT), which in turn delays mitotic chromosome condensation (DMC) (Stoffregen et al. 2011; Donley et al. 2015). ASAR genes are mono-allelically expressed, producing non-spliced Poly A-, very long noncoding RNAs, display asynchronous replication and have a high L1 content. ASAR RNAs coat the chromosome from which they are expressed in *cis*, and their functional activity is mediated by L1 elements that are transcribed in the antisense orientation (Platt et al. 2018). Deletion or inversion

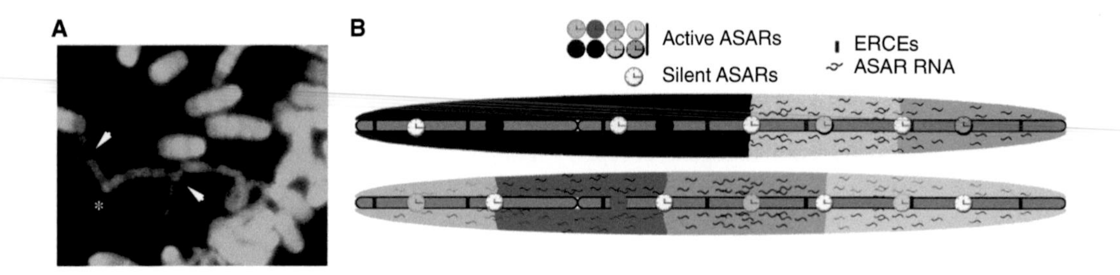

Figure 5. Asynchronous replication and autosomal RNAs (ASARs) are necessary to ensure that whole chromosomes are replicated in a timely fashion. (*A*) Delayed mitotic condensation and spontaneous damage. Mitotic cells containing an uncondensed i(3q), from human rhabdomyosarcoma cells RH30. The i(3q) was identified by FISH with a centromeric probe (red; *). DNA was stained with DAPI (blue), and arrows mark sites of spontaneous damage. (*B*) Monoallelic expression of multiple ASARs from a hypothetical chromosome pair. The expressed (Active; colored clocks) alleles of different ASARs result in "clouds" of RNA that are retained within the chromosome territories from which they are transcribed. ASAR RNA clouds are speculated to regulate early replication control element (ERCE) activity. The nonexpressed ASARs (Silent; white clocks) are inactive. (Panels *A* and *B* courtesy of M. Thayer.)

of an antisense L1 element within ASAR6 causes a delay in RT of the entire chromosome. Intriguingly, despite L1 elements not being well conserved across species, ectopic insertion of a single L1 element from within ASAR6 into mouse chromosomes was sufficient to affect replication and mitotic condensation of entire chromosomes. Recently, a second mono-allelically expressed and asynchronously replicating vlncRNA, ASAR6-141, was discovered to be expressed from the alternate chromosome 6 homolog, to the homolog expressing ASAR6 (Heskett et al. 2020), suggesting that multiple ASARs in each autosome work together to maintain proper replication control across each chromosome pair (Fig. 5B). There are >2000 vlncRNAs with unknown function (St Laurent et al. 2013, 2016; Caron et al. 2018). The coming years are expected to provide us with critical information on the genome-wide network of ASARs, and how these elements interact with other *cis*-regulatory elements such as ERCEs to control chromosome-wide RT.

CONCLUSIONS AND FUTURE PERSPECTIVES

Replication is a function regulated at the level of large-scale chromatin domains and also is the time at which chromatin is assembled; thus, it provides a unique window and functional readout of large-scale structure–function. In the decade since the last publication of *The Nucleus* (Misteli and Spector 2011), the development of several new molecular, imaging, and computational methods as well as the ability to sequence at much greater depth, have given us a deep molecular description of relationships between RT, chromatin architecture, and subnuclear localization. While their strong correlation suggests a tight interdependence between them, causal mechanistic links remain elusive. It is possible that the architectural state of chromatin and the physical localization of loci within the nucleus dictate a predetermined time range in S phase during which particular loci can replicate, by recruiting or antagonizing key replication initiation factors. Alternatively, replication at a particular time and place could result in the assembly of chromatin that interacts with similar chromatin types and targets to specific subnuclear addresses. These are not mutually exclusive; a familiar theme in epigenetics is the chicken or the egg problem: self-reinforcing feedforward loops of causality.

Overall, it is clear that there is still a need for a comprehensive set of studies that will illuminate the mechanistic link between the organizational state of a locus and the time it replicates during the S phase. Fortunately, robust new experimen-

 Cite this article as *Cold Spring Harb Perspect Biol* doi: 10.1101/cshperspect.a040162

tal and computational tools have recently emerged that have already paved the way for mechanistic studies of RT control and its causal relationships to chromosome structure and other functions. Recent work has shown that specific *cis*-acting elements, ERCEs, regulate early replication of loci in mESCs and their characterization, sequence characteristics, and interactors are forthcoming. This regulation appears to be cell-type-specific and soon we expect ERCEs in other cell types to be identified. Also, for the first time researchers were able to show that the global disruption of RT by depletion of a single protein, Rif1, results in widespread disruption of the epigenome, providing long-awaited evidence that the proper timing of chromatin assembly is necessary for epigenome maintenance. We are also on the verge of understanding the enigmatic control of whole chromosome replication that coordinates its completion with proper condensation and segregation during the cell cycle, something that is awry in most cancers. Indeed, six decades after the initial description of replication control in mammalian cells, we are poised for the seventh decade to bring significant progress in understanding the mechanism and biological significance of RT control and its relationship to the overall organization and function of the genome.

ACKNOWLEDGMENTS

The authors thank J. Ma, Y. Wang, B. van Steensel, and M. Thayer for critical reading of the manuscript. Work in the Gilbert laboratory is funded by NIH Grants GM083337, DK107965, HG010403, and HG010658.

REFERENCES

Abney JR, Cutler B, Fillbach ML, Axelrod D, Scalettar BA. 1997. Chromatin dynamics in interphase nuclei and its implications for nuclear structure. *J Cell Biol* **137:** 1459–1468. doi:10.1083/jcb.137.7.1459

Aparicio JG, Viggiani CJ, Gibson DG, Aparicio OM. 2004. The Rpd3-Sin3 histone deacetylase regulates replication timing and enables intra-S origin control in *Saccharomyces cerevisiae*. *Mol Cell Biol* **24:** 4769–4780. doi:10.1128/MCB.24.11.4769-4780.2004

Beagan JA, Phillips-Cremins JE. 2020. On the existence and functionality of topologically associating domains. *Nat Genet* **52:** 8–16. doi:10.1038/s41588-019-0561-1

Berezney R, Dubey DD, Huberman JA. 2000. Heterogeneity of eukaryotic replicons, replicon clusters, and replication foci. *Chromosoma* **108:** 471–484. doi:10.1007/s004120050399

Bintu B, Mateo LJ, Su JH, Sinnott-Armstrong NA, Parker M, Kinrot S, Yamaya K, Boettiger AN, Zhuang X. 2018. Super-resolution chromatin tracing reveals domains and cooperative interactions in single cells. *Science* **362:** eaau1783. doi:10.1126/science.aau1783

Blin M, Le Tallec B, Nähse V, Schmidt M, Brossas C, Millot GA, Prioleau MN, Debatisse M. 2019. Transcription-dependent regulation of replication dynamics modulates genome stability. *Nat Struct Mol Biol* **26:** 58–66. doi:10.1038/s41594-018-0170-1

Boettiger A, Murphy S. 2020. Advances in chromatin imaging at kilobase-scale resolution. *Trends Genet* **36:** 273–287. doi:10.1016/j.tig.2019.12.010

Borck PC, Guo LW, Plutzky J. 2020. BET epigenetic reader proteins in cardiovascular transcriptional programs. *Circ Res* **126:** 1190–1208. doi:10.1161/CIRCRESAHA.120.315929

Brueckner L, Zhao PA, Schaik T, Leemans C, Sima J, Peric-Hupkes D, Gilbert DM, Steensel B. 2020. Local rewiring of genome–nuclear lamina interactions by transcription. *EMBO J* **39:** 1–17. doi:10.15252/embj.2019103159

Caron M, St-Onge P, Drouin S, Richer C, Sontag T, Busche S, Bourque G, Pastinen T, Sinnett D. 2018. Very long intergenic non-coding RNA transcripts and expression profiles are associated to specific childhood acute lymphoblastic leukemia subtypes. *PLoS ONE* **13:** e0207250. doi:10.1371/journal.pone.0207250

Chagin VO, Casas-Delucchi CS, Reinhart M, Schermelleh L, Markaki Y, Maiser A, Bolius JJ, Bensimon A, Fillies M, Domaing P, et al. 2016. 4D visualization of replication foci in mammalian cells corresponding to individual replicons. *Nat Commun* **7:** 1–12. doi:10.1038/ncomms11231

Chen Y, Belmont AS. 2019. Genome organization around nuclear speckles. *Curr Opin Genet Dev* **55:** 91–99. doi:10.1016/j.gde.2019.06.008

Chen CL, Rappailles A, Duquenne L, Huvet M, Guilbaud G, Farinelli L, Audit B, D'Aubenton-Carafa Y, Arneodo A, Hyrien O, et al. 2010. Impact of replication timing on non-CpG and CpG substitution rates in mammalian genomes. *Genome Res* **20:** 447–457. doi:10.1101/gr.098947.109

Chen Y, Zhang Y, Wang Y, Zhang L, Brinkman EK, Adam SA, Goldman R, Van Steensel B, Ma J, Belmont AS. 2018. Mapping 3D genome organization relative to nuclear compartments using TSA-Seq as a cytological ruler. *J Cell Biol* **217:** 4025–4048. doi:10.1083/jcb.201807108

Collart C, Allen GE, Bradshaw CR, Smith JC, Zegerman P. 2013. Titration of four replication factors is essential for the *Xenopus laevis* midblastula transition. *Science* **341:** 893–896. doi:10.1126/science.1241530

Conti C, Saccà B, Herrick J, Lalou C, Pommier Y, Bensimon A. 2007. Replication fork velocities at adjacent replication origins are coordinately modified during DNA replication in human cells. *Mol Biol Cell* **18:** 3059–3067. doi:10.1091/mbc.e06-08-0689

Courtot L, Hoffmann JS, Bergoglio V. 2018. The protective role of dormant origins in response to replicative stress. *Int J Mol Sci* **19:** 3569. doi:10.3390/ijms19113569

Cremer M, Brandstetter K, Maiser A, Rao SSP, Schmid V, Mitra N, Mamberti S, Klein K, Gilbert DM, Leonhardt H, et al. 2020. Cohesin depleted cells rebuild functional nuclear compartments after endomitosis. *Nat Commun* **11:** 6146. doi:10.1038/s41467-020-19876-6

Deegan TD, Diffley JFX. 2016. MCM: one ring to rule them all. *Curr Opin Struct Biol* **37:** 145–151. doi:10.1016/j.sbi.2016.01.014

De Laat W, Duboule D. 2013. Topology of mammalian developmental enhancers and their regulatory landscapes. *Nature* **502:** 499–506. doi:10.1038/nature12753

Demczuk A, Gauthier MG, Veras I, Kosiyatrakul S, Schildkraut CL, Busslinger M, Bechhoefer J, Norio P. 2012. Regulation of DNA replication within the immunoglobulin heavy-chain locus during B cell commitment. *PLoS Biol* **10:** e1001360. doi:10.1371/journal.pbio.1001360

Deng X, Zhironkina OA, Cherepanynets VD, Strelkova OS, Kireev II, Belmont AS. 2016. Cytology of DNA replication reveals dynamic plasticity of large-scale chromatin fibers. *Curr Biol* **26:** 2527–2534. doi:10.1016/j.cub.2016.07.020

Dequeker BJH, Brandão HB, Scherr MJ, Gassler J, Powell S, Gaspar I, Flyamer IM, Tang W, Stocsits R, Davidson IF, et al. 2020. MCM complexes are barriers that restrict cohesin-mediated loop extrusion. bioRxiv doi:10.1101/2020.10.15.340356

Despang A, Schöpflin R, Franke M, Ali S, Jerković I, Paliou C, Chan WL, Timmermann B, Wittler L, Vingron M, et al. 2019. Functional dissection of the Sox9–Kcnj2 locus identifies nonessential and instructive roles of TAD architecture. *Nat Genet* **51:** 1263–1271. doi:10.1038/s41588-019-0466-z

de Wit E, Vos ESM, Holwerda SJB, Valdes-Quezada C, Verstegen MJAM, Teunissen H, Splinter E, Wijchers PJ, Krijger PHL, de Laat W. 2015. CTCF binding polarity determines chromatin looping. *Mol Cell* **60:** 676–684. doi:10.1016/j.molcel.2015.09.023

Dileep V, Gilbert DM. 2018. Single-cell replication profiling to measure stochastic variation in mammalian replication timing. *Nat Commun* **9:** 1. doi:10.1038/s41467-017-02800-w

Dileep V, Ay F, Sima J, Vera DL, Noble WS, Gilbert DM. 2015. Topologically associating domains and their long-range contacts are established during early G1 coincident with the establishment of the replication-timing program. *Genome Res* **25:** 1104–1113. doi:10.1101/gr.183699.114

Dileep V, Wilson KA, Marchal C, Lyu X, Zhao PA, Li B, Poulet A, Bartlett DA, Rivera-Mulia JC, Qin ZS, et al. 2019. Rapid irreversible transcriptional reprogramming in human stem cells accompanied by discordance between replication timing and chromatin compartment. *Stem Cell Rep* **13:** 193–206. doi:10.1016/j.stemcr.2019.05.021

Dimitrova DS, Gilbert DM. 1999. The spatial position and replication timing of chromosomal domains are both established in early G1 phase. *Mol Cell* **4:** 983–993. doi:10.1016/S1097-2765(00)80227-0

Dixon JR, Selvaraj S, Yue F, Kim A, Li Y, Shen Y, Hu M, Liu JS, Ren B. 2012. Topological domains in mammalian genomes identified by analysis of chromatin interactions. *Nature* **485:** 376–380. doi:10.1038/nature11082

Donley N, Thayer MJ. 2013. DNA replication timing, genome stability and cancer. *Semin Cancer Biol* **23:** 80–89. doi:10.1016/j.semcancer.2013.01.001

Donley N, Smith L, Thayer MJ. 2015. ASAR15, a *cis*-acting locus that controls chromosome-wide replication timing and stability of human chromosome 15. *PLoS Genet* **11:** e1004923. doi:10.1371/journal.pgen.1004923

Duriez B, Chilaka S, Bercher JF, Hercul E, Prioleau MN. 2019. Replication dynamics of individual loci in single living cells reveal changes in the degree of replication stochasticity through S phase. *Nucleic Acids Res* **47:** 5155–5169. doi:10.1093/nar/gkz220

Emerson D, Zhao PA, Klein K, Ge C, Zhou L, Sasaki T, Yang L, Venvev SV, Gibcus JH, Dekker J, et al. 2021. Cohesin-mediated loop anchors confine the location of human replication origins. bioRxiv doi:10.1101/2021.01.05.425437

Fernandez-Vidal A, Guitton-Sert L, Cadoret JC, Drac M, Schwob E, Baldacci G, Cazaux C, Hoffmann JS. 2014. A role for DNA polymerase θ in the timing of DNA replication. *Nat Commun* **5:** 4285. doi:10.1038/ncomms5285

Ferreira J, Paolella G, Ramos C, Lamond AI. 1997. Spatial organization of large-scale chromatin domains in the nucleus: a magnified view of single chromosome territories. *J Cell Biol* **139:** 1597–1610. doi:10.1083/jcb.139.7.1597

Foss EJ, Gatbonton-Schwager T, Thiesen AH, Taylor E, Soriano R, Lao U, MacAlpine DM, Bedalov A. 2019. Sir2 suppresses transcription-mediated displacement of Mcm2-7 replicative helicases at the ribosomal DNA repeats. *PLoS Genet* **15:** 1–17.

Foti R, Gnan S, Cornacchia D, Dileep V, Bulut-Karslioglu A, Diehl S, Buness A, Klein FA, Huber W, Johnstone E, et al. 2016. Nuclear architecture organized by Rif1 underpins the replication-timing program. *Mol Cell* **61:** 260–273. doi:10.1016/j.molcel.2015.12.001

Fu H, Baris A, Aladjem MI. 2018. Replication timing and nuclear structure. *Curr Opin Cell Biol* **52:** 43–50. doi:10.1016/j.ceb.2018.01.004

Fudenberg G, Abdennur N, Imakaev M, Goloborodko A, Mirny LA. 2017. Emerging evidence of chromosome folding by loop extrusion. *Cold Spring Harb Symp Quant Biol* **82:** 45–55. doi:10.1101/sqb.2017.82.034710

Ge XQ, Jackson DA, Blow JJ. 2007. Dormant origins licensed by excess Mcm2-7 are required for human cells to survive replicative stress. *Genes Dev* **21:** 3331–3341. doi:10.1101/gad.457807

Gibson BA, Doolittle LK, Schneider MWG, Jensen LE, Gamarra N, Henry L, Gerlich DW, Redding S, Rosen MK. 2019. Organization of chromatin by intrinsic and regulated phase separation. *Cell* **179:** 470–484.e21. doi:10.1016/j.cell.2019.08.037

Gilbert DM. 2001. Nuclear position leaves its mark on replication timing. *J Cell Biol* **152:** F11–F15. doi:10.1083/jcb.152.2.F11

Gilbert DM. 2002. Replication timing and transcriptional control: beyond cause and effect. *Curr Opin Cell Biol* **14:** 377–383. doi:10.1016/S0955-0674(02)00326-5

Gilbert DM. 2010. Evaluating genome-scale approaches to eukaryotic DNA replication. *Nat Rev Genet* **11:** 673–684. doi:10.1038/nrg2830

Gnan S, Flyame IM, Klein KN, Castelli E, Rapp A, Maiser A, Chen N, Weber P, Enervald E, Cardoso, et al. 2020. Nuclear organisation and replication timing are coupled through RIF1-PP1 interaction. bioRxiv doi:10.1101/812156

Goren A, Tabib A, Hecht M, Cedar H. 2008. DNA replication timing of the human β-globin domain is controlled by histone modification at the origin. *Genes Dev* **22:** 1319–1324. doi:10.1101/gad.468308

Guelen L, Pagie L, Brasset E, Meuleman W, Faza MB, Talhout W, Eussen BH, De Klein A, Wessels L, De Laat W, et al. 2008. Domain organization of human chromosomes revealed by mapping of nuclear lamina interactions. *Nature* **453:** 948–951. doi:10.1038/nature06947

Hall LL, Smith KP, Byron M, Lawrence JB. 2006. Molecular anatomy of a speckle. *Anat Rec A Discov Mol Cell Evol Biol* **288A:** 664–675. doi:10.1002/ar.a.20336

Han X, Yu D, Gu R, Jia Y, Wang Q, Jaganathan A, Yang X, Yu M, Babault N, Zhao C, et al. 2020. Roles of the BRD4 short isoform in phase separation and active gene transcription. *Nat Struct Mol Biol* **27:** 333–341. doi:10.1038/s41594-020-0394-8

Hansen RS, Thomas S, Sandstrom R, Canfield TK, Thurman RE, Weaver M, Dorschner MO, Gartler SM, Stamatoyannopoulos JA. 2010. Sequencing newly replicated DNA reveals widespread plasticity in human replication timing. *Proc Natl Acad Sci* **107:** 139–144. doi:10.1073/pnas.0912402107

Hassan-Zadeh V, Chilaka S, Cadoret JC, Ma MKW, Boggetto N, West AG, Prioleau MN. 2012. USF binding sequences from the HS4 insulator element impose early replication timing on a vertebrate replicator. *PLoS Biol* **10:** e1001277. doi:10.1371/journal.pbio.1001277

Hayano M, Kanoh Y, Matsumoto S, Renard-Guillet C, Shirahige K, Masai H. 2012. Rif1 is a global regulator of timing of replication origin firing in fission yeast. *Genes Dev* **26:** 137–150. doi:10.1101/gad.178491.111

Heskett M, Smith LG, Spellman P, Thayer M. 2020. Reciprocal monoallelic expression of ASAR lncRNA genes controls replication timing of human chromosome 6. *RNA* **503:** rna.073114.119.

Hildebrand EM, Dekker J. 2020. Mechanisms and functions of chromosome compartmentalization. *Trends Biochem Sci* **45:** 385–396. doi:10.1016/j.tibs.2020.01.002

Hiraga S, Alvino GM, Chang FJ, Lian HY, Sridhar A, Kubota T, Brewer BJ, Weinreich M, Raghuraman MK, Donaldson AD. 2014. Rif1 controls DNA replication by directing protein phosphatase 1 to reverse Cdc7- mediated phosphorylation of the MCM complex. *Genes Dev* **28:** 372–383. doi:10.1101/gad.231258.113

Hiraga S, Monerawela C, Katou Y, Shaw S, Clark KR, Shirahige K, Donaldson AD. 2018. Budding yeast Rif1 binds to replication origins and protects DNA at blocked replication forks. *EMBO Rep* **19:** e46222. doi:10.15252/embr.201846222

Hiratani I, Ryba T, Itoh M, Yokochi T, Schwaiger M, Chang CW, Lyou Y, Townes TM, Schübeler D, Gilbert DM. 2008. Global reorganization of replication domains during embryonic stem cell differentiation. *PLoS Biol* **6:** e245. doi:10.1371/journal.pbio.0060245

Hiratani I, Ryba T, Itoh M, Rathjen J, Kulik M, Papp B, Fussner E, Bazett-Jones DP, Plath K, Dalton S, et al. 2010. Genome-wide dynamics of replication timing revealed by in vitro models of mouse embryogenesis. *Genome Res* **20:** 155–169. doi:10.1101/gr.099796.109

Hsieh THS, Weiner A, Lajoie B, Dekker J, Friedman N, Rando OJ. 2015. Mapping nucleosome resolution chromosome folding in yeast by micro-C. *Cell* **162:** 108–119. doi:10.1016/j.cell.2015.05.048

Hsieh THS, Cattoglio C, Slobodyanyuk E, Hansen AS, Rando OJ, Tjian R, Darzacq X. 2020. Resolving the 3D landscape of transcription-linked mammalian chromatin folding. *Mol Cell* **78:** 539–553.e8. doi:10.1016/j.molcel.2020.03.002

Hulke ML, Siefert JC, Sansam CL, Koren A. 2019. Germline structural variations are preferential sites of DNA replication timing plasticity during development. *Genome Biol Evol* **11:** 1663–1678. doi:10.1093/gbe/evz098

Hulke ML, Massey DJ, Koren A. 2020. Genomic methods for measuring DNA replication dynamics. *Chromosome Res* **28:** 49–67. doi:10.1007/s10577-019-09624-y

Hyrien O. 2015. Peaks cloaked in the mist: the landscape of mammalian replication origins. *J Cell Biol* **208:** 147–160. doi:10.1083/jcb.201407004

Ibarra A, Schwob E, Méndez J. 2008. Excess MCM proteins protect human cells from replicative stress by licensing backup origins of replication. *Proc Natl Acad Sci* **105:** 8956–8961. doi:10.1073/pnas.0803978105

Jackson DA, Pombo A. 1998. Replicon clusters are stable units of chromosome structure: evidence that nuclear organization contributes to the efficient activation and propagation of S phase in human cells. *J Cell Biol* **140:** 1285–1295. doi:10.1083/jcb.140.6.1285

Kim J, Venkata NC, Hernandez Gonzalez GA, Khanna N, Belmont AS. 2020. Gene expression amplification by nuclear speckle association. *J Cell Biol* **219:** e201904046.

Kind J, Pagie L, Ortabozkoyun H, Boyle S, De Vries SS, Janssen H, Amendola M, Nolen LD, Bickmore WA, Van Steensel B. 2013. Single-cell dynamics of genome-nuclear lamina interactions. *Cell* **153:** 178–192. doi:10.1016/j.cell.2013.02.028

Kind J, Pagie L, De Vries SS, Nahidiazar L, Dey SS, Bienko M, Zhan Y, Lajoie B, De Graaf CA, Amendola M, et al. 2015. Genome-wide maps of nuclear lamina interactions in single human cells. *Cell* **163:** 134–147. doi:10.1016/j.cell.2015.08.040

Kitamura E, Blow JJ, Tanaka TU. 2006. Live-cell imaging reveals replication of individual replicons in eukaryotic replication factories. *Cell* **125:** 1297–1308. doi:10.1016/j.cell.2006.04.041

Klein KN, Zhao PA, Lyu X, Bartlett DA, Singh A, Tasan I, Watts LP, Hiraga S, Natsume T, Zhou X, et al. 2019. Replication timing maintains the global epigenetic state in human cells. bioRxiv doi:10.1101/2019.12.28.890020

Knott SRV, Viggiani CJ, Tavaré S, Aparicio OM. 2009. Genome-wide replication profiles indicate an expansive role for Rpd3L in regulating replication initiation timing or efficiency, and reveal genomic loci of Rpd3 function in *Saccharomyces cerevisiae*. *Genes Dev* **23:** 1077–1090. doi:10.1101/gad.1784309

Knott SRV, Peace JM, Ostrow AZ, Gan Y, Rex AE, Viggiani CJ, Tavaré S, Aparicio OM. 2012. Forkhead transcription factors establish origin timing and long-range clustering in *S. cerevisiae*. *Cell* **148:** 99–111. doi:10.1016/j.cell.2011.12.012

Koren A, Handsaker RE, Kamitaki N, Karlić R, Ghosh S, Polak P, Eggan K, McCarroll SA. 2014. Genetic variation in human DNA replication timing. *Cell* **159:** 1015–1026. doi:10.1016/j.cell.2014.10.025

Kuipers MA, Stasevich TJ, Sasaki T, Wilson KA, Hazelwood KL, McNally JG, Davidson MW, Gilbert DM. 2011. Highly stable loading of Mcm proteins onto chromatin in living cells requires replication to unload. *J Cell Biol* **192:** 29–41. doi:10.1083/jcb.201007111

Lande-Diner L, Zhang J, Cedar H. 2009. Shifts in replication timing actively affect histone acetylation during nucleosome reassembly. *Mol Cell* **34:** 767–774. doi:10.1016/j.molcel.2009.05.027

Lieberman-Aiden E, Van Berkum NL, Williams L, Imakaev M, Ragoczy T, Telling A, Amit I, Lajoie BR, Sabo PJ, Dorschner MO, et al. 2009. Comprehensive mapping of long-range interactions reveals folding principles of the human genome. *Science* **326:** 289–293. doi:10.1126/science.1181369

Limas JC, Cook JG. 2019. Preparation for DNA replication: the key to a successful S phase. *FEBS Lett* **593:** 2853–2867. doi:10.1002/1873-3468.13619

Löb D, Lengert N, Chagin VO, Reinhart M, Casas-Delucchi CS, Cardoso MC, Drossel B. 2016. 3D replicon distributions arise from stochastic initiation and domino-like DNA replication progression. *Nat Commun* **7:** 11207.

Ma H, Samarabandu J, Devdhar RS, Acharya R, Cheng PC, Meng C, Berezney R. 1998. Spatial and temporal dynamics of DNA replication sites in mammalian cells. *J Cell Biol* **143:** 1415–1425. doi:10.1083/jcb.143.6.1415

Mantiero D, MacKenzie A, Donaldson A, Zegerman P. 2011. Limiting replication initiation factors execute the temporal programme of origin firing in budding yeast. *EMBO J* **30:** 4805–4814. doi:10.1038/emboj.2011.404

Marchal C, Sima J, Gilbert DM. 2019. Control of DNA replication timing in the 3D genome. *Nat Rev Mol Cell Biol* **20:** 721–737. doi:10.1038/s41580-019-0162-y

Marshall WF, Straight A, Marko JF, Swedlow J, Dernburg A, Belmont A, Murray AW, Agard DA, Sedat JW. 1997. Interphase chromosomes undergo constrained diffusional motion in living cells. *Curr Biol* **7:** 930–939. doi:10.1016/s0960-9822(06)00412-x

Meister P, Taddei A, Ponti A, Baldacci G, Gasser SM. 2007. Replication foci dynamics: replication patterns are modulated by S-phase checkpoint kinases in fission yeast. *EMBO J* **26:** 1315–1326. doi:10.1038/sj.emboj.7601538

Meuleman W, Peric-Hupkes D, Kind J, Beaudry JB, Pagie L, Kellis M, Reinders M, Wessels L, Van Steensel B. 2013. Constitutive nuclear lamina-genome interactions are highly conserved and associated with A/T-rich sequence. *Genome Res* **23:** 270–280. doi:10.1101/gr.141028.112

Misteli T, Spector DL (eds.). 2011. *The Nucleus*. Cold Spring Harbor Laboratory Press, Cold Spring Harbor, NY.

Miura H, Takahashi S, Poonperm R, Tanigawa A, Takebayashi SI, Hiratani I. 2019. Single-cell DNA replication profiling identifies spatiotemporal developmental dynamics of chromosome organization. *Nat Genet* **51:** 1356–1368. doi:10.1038/s41588-019-0474-z

Moiseeva TN, Bakkenist CJ. 2019. Dormant origin signaling during unperturbed replication. *DNA Repair (Amst)* **81:** 102655. doi:10.1016/j.dnarep.2019.102655

Mota-Gómez I, Lupiáñez DG. 2019. A (3D-nuclear) space odyssey: making sense of Hi-C maps. *Genes (Basel)* **10:** 415. doi:10.3390/genes10060415

Müller CA, Nieduszynski CA. 2017. DNA replication timing influences gene expression level. *J Cell Biol* **216:** 1907–1914. doi:10.1083/jcb.201701061

Nakamura H, Morita T, Sato C. 1986. Structural organizations of replicon domains during DNA synthetic phase in the mammalian nucleus. *Exp Cell Res* **165:** 291–297. doi:10.1016/0014-4827(86)90583-5

Nathanailidou P, Taraviras S, Lygerou Z. 2020. Chromatin and nuclear architecture: shaping DNA replication in 3D. *Trends Genet* **36:** 967–980. doi:10.1016/j.tig.2020.07.003

Nora EP, Lajoie BR, Schulz EG, Giorgetti L, Okamoto I, Servant N, Piolot T, Van Berkum NL, Meisig J, Sedat J, et al. 2012. Spatial partitioning of the regulatory landscape of the X-inactivation centre. *Nature* **485:** 381–385. doi:10.1038/nature11049

Nora EP, Goloborodko A, Valton AL, Gibcus JH, Uebersohn A, Abdennur N, Dekker J, Mirny LA, Bruneau BG. 2017. Targeted degradation of CTCF decouples local insulation of chromosome domains from genomic compartmentalization. *Cell* **169:** 930–944.e22. doi:10.1016/j.cell.2017.05.004

Nozaki T, Imai R, Tanbo M, Nagashima R, Tamura S, Tani T, Joti Y, Tomita M, Hibino K, Kanemaki MT, et al. 2017. Dynamic organization of chromatin domains revealed by super-resolution live-cell imaging. *Mol Cell* **67:** 282–293.e7. doi:10.1016/j.molcel.2017.06.018

Oldach P, Nieduszynski CA. 2019. Cohesin-mediated genome architecture does not define DNA replication timing domains. *Genes (Basel)* **10:** 196. doi:10.3390/genes10030196

Ostrow AZ, Kalhora R, Gana Y, Villwocka SK, Linke C, Barberisb M, Chena L, Aparicioa OM. 2017. Conserved forkhead dimerization motif controls DNA replication timing and spatial organization of chromosomes in *S. cerevisiae*. *Proc Natl Acad Sci* **114:** E2411–E2419. doi:10.1073/pnas.1612422114

Palmigiano A, Santaniello F, Cerutti A, Penkov D, Purushothaman D, Makhija E, Luzi L, Di Fagagna FDA, Pelicci PG, Shivashankar V, et al. 2018. PREP1 tumor suppressor protects the late-replicating DNA by controlling its replication timing and symmetry. *Sci Rep* **8:** 1–12. doi:10.1038/s41598-018-21363-4

Panning MM, Gilbert DM. 2005. Spatio-temporal organization of DNA replication in murine embryonic stem, primary, and immortalized cells. *J Cell Biochem* **95:** 74–82. doi:10.1002/jcb.20395

Peric-Hupkes D, Meuleman W, Pagie L, Bruggeman SWM, Solovei I, Brugman W, Gräf S, Flicek P, Kerkhoven RM, van Lohuizen M, et al. 2010. Molecular maps of the reorganization of genome-nuclear lamina interactions during differentiation. *Mol Cell* **38:** 603–613. doi:10.1016/j.molcel.2010.03.016

Petryk N, Kahli M, D'Aubenton-Carafa Y, Jaszczyszyn Y, Shen Y, Silvain M, Thermes C, Chen CL, Hyrien O. 2016. Replication landscape of the human genome. *Nat Commun* **7:** 1–13. doi:10.1038/ncomms10208

Platt EJ, Smith L, Thayer MJ. 2018. L1 retrotransposon antisense RNA within ASAR lncRNAs controls chromosome-wide replication timing. *J Cell Biol* **217:** 541–553. doi:10.1083/jcb.201707082

Politz JCR, Scalzo D, Groudine M. 2014. Repressive nuclear compartment. *Annu Rev Cell Dev Biol* **1928:** 241–270.

Pope BD, Aparicio OM, Gilbert DM. 2013. Snapshot: replication timing. *Cell* **152:** 1390–1390.e1. doi:10.1016/j.cell.2013.02.038

Pope BD, Ryba T, Dileep V, Yue F, Wu W, Denas O, Vera DL, Wang Y, Hansen RS, Canfield TK, et al. 2014. Topologically associating domains are stable units of replication-timing regulation. *Nature* **515:** 402–405. doi:10.1038/nature13986

Rao SSP, Huntley MH, Durand NC, Stamenova EK, Bochkov ID, Robinson JT, Sanborn AL, Machol I, Omer AD, Lander ES, et al. 2014. A 3D map of the human genome at kilobase resolution reveals principles of chromatin looping. *Cell* **159:** 1665–1680. doi:10.1016/j.cell.2014.11.021

Rao SSP, Huang SC, Glenn St Hilaire B, Engreitz JM, Perez EM, Kieffer-Kwon KR, Sanborn AL, Johnstone SE, Bascom GD, Bochkov ID, et al. 2017. Cohesin loss eliminates all loop domains. *Cell* **171:** 305–320.e24. doi:10.1016/j.cell.2017.09.026

Rivera-Mulia JC, Gilbert DM. 2016. Replicating large genomes: divide and conquer. *Mol Cell* **62:** 756–765. doi:10.1016/j.molcel.2016.05.007

Rivera-Mulia JC, Buckley Q, Sasaki T, Zimmerman J, Didier RA, Nazor K, Loring JF, Lian Z, Weissman S, Robins AJ, et al. 2015. Dynamic changes in replication timing and gene expression during lineage specification of human pluripotent stem cells. *Genome Res* **25:** 1091–1103. doi:10.1101/gr.187989.114

Rivera-Mulia JC, Desprat R, Trevilla-Garcia C, Cornacchia D, Schwerer H, Sasaki T, Sima J, Fells T, Studer L, Lemaitre JM, et al. 2017. DNA replication timing alterations identify common markers between distinct progeroid diseases. *Proc Natl Acad Sci* **114:** E10972–E10980. doi:10.1073/pnas.1711613114

Rivera-Mulia JC, Kim S, Gabr H, Chakraborty A, Ay F, Kahveci T, Gilbert DM. 2019a. Replication timing networks reveal a link between transcription regulatory circuits and replication timing control. *Genome Res* **29:** 1415–1428. doi:10.1101/gr.247049.118

Rivera-Mulia JC, Sasaki T, Trevilla-Garcia C, Nakamichi N, Knapp DJHF, Hammond CA, Chang BH, Tyner JW, Devidas M, Zimmerman J, et al. 2019b. Replication timing alterations in leukemia affect clinically relevant chromosome domains. *Blood Adv* **3:** 3201–3213. doi:10.1182/bloodadvances.2019000641

Rowley MJ, Corces VG. 2018. Organizational principles of 3D genome architecture. *Nat Rev Genet* **19:** 789–800. doi:10.1038/s41576-018-0060-8

Ryba T, Hiratani I, Lu J, Itoh M, Kulik M, Zhang J, Schulz TC, Robins AJ, Dalton S, Gilbert DM. 2010. Evolutionarily conserved replication timing profiles predict long-range chromatin interactions and distinguish closely related cell types. *Genome Res* **20:** 761–770. doi:10.1101/gr.099655.109

Sanborn AL, Rao SSP, Huang SC, Durand NC, Huntley MH, Jewett AI, Bochkov ID, Chinnappan D, Cutkosky A, Li J, et al. 2015. Chromatin extrusion explains key features of loop and domain formation in wild-type and engineered genomes. *Proc Natl Acad Sci* **112:** E6456–E6465. doi:10.1073/pnas.1518552112

Sansam CG, Pietrzak K, Majchrzycka B, Kerlin MA, Chen J, Rankin S, Sansam CL. 2018. A mechanism for epigenetic control of DNA replication. *Genes Dev* **32:** 224–229. doi:10.1101/gad.306464.117

Sasaki T, Ramanathan S, Okuno Y, Kumagai C, Shaikh SS, Gilbert DM. 2006. The Chinese hamster dihydrofolate reductase replication origin decision point follows activation of transcription and suppresses initiation of replication within transcription units. *Mol Cell Biol* **26:** 1051–1062. doi:10.1128/MCB.26.3.1051-1062.2006

Schaik TV, Vos M, Peric-Hupkes DD, van Steensel B. 2020. Cell cycle dynamics of lamina associated DNA. *EMBO Rep* **21:** e50636. doi:10.15252/embr.202050636

Schwarzer W, Abdennur N, Goloborodko A, Pekowska A, Fudenberg G, Loe-Mie Y, Fonseca NA, Huber W, Haering CH, Mirny L, et al. 2017. Two independent modes of chromatin organization revealed by cohesin removal. *Nature* **551:** 51–56. doi:10.1038/nature24281

Seller CA, O'Farrell PH. 2018. Rif1 prolongs the embryonic S phase at the *Drosophila* mid-blastula transition. *PLoS Biol* **16:** e2005687. doi:10.1371/journal.pbio.2005687

Sima J, Chakraborty A, Dileep V, Michalski M, Klein KN, Holcomb NP, Turner JL, Paulsen MT, Rivera-Mulia JC, Trevilla-Garcia C, et al. 2019. Identifying *cis* elements for spatiotemporal control of mammalian DNA replication. *Cell* **176:** 816–830.e18. doi:10.1016/j.cell.2018.11.036

Simon I, Tenzen T, Mostoslavsky R, Fibach E, Lande L, Milot E, Gribnau J, Grosveld F, Fraser P, Cedar H. 2001. Developmental regulation of DNA replication timing at the human β globin locus. *EMBO J* **20:** 6150–6157. doi:10.1093/emboj/20.21.6150

Smith L, Plug A, Thayer M. 2001. Delayed replication timing leads to delayed mitotic chromosome condensation and chromosomal instability of chromosome translocations. *Proc Natl Acad Sci* **98:** 13300–13305. doi:10.1073/pnas.241355098

Sparvoli E, Levi M, Rossi E. 1994. Replicon clusters may form structurally stable complexes of chromatin and chromosomes. *J Cell Sci* **107:** 3097–3103.

Spector DL, Lamond AI. 2011. Nuclear speckles. *Cold Spring Harb Perspect Biol* **3:** a000646. doi:10.1101/cshperspect.a000646

Sporbert A, Gahl A, Ankerhold R, Leonhardt H, Cardoso MC. 2002. DNA polymerase clamp shows little turnover at established replication sites but sequential de novo assembly at adjacent origin clusters. *Mol Cell* **10:** 1355–1365. doi:10.1016/S1097-2765(02)00729-3

Stewart-Morgan KR, Petryk N, Groth A. 2020. Chromatin replication and epigenetic cell memory. *Nature Cell Biol* **22:** 361–371. doi:10.1038/s41556-020-0487-y

St Laurent G, Shtokalo D, Dong B, Tackett MR, Fan X, Lazorthes S, Nicolas E, Sang N, Triche TJ, McCaffrey TA, et al. 2013. VlincRNAs controlled by retroviral elements are a hallmark of pluripotency and cancer. *Genome Biol* **14:** R73. doi:10.1186/gb-2013-14-7-r73

St Laurent G, Vyatkin Y, Antonets D, Ri M, Qi Y, Saik O, Shtokalo D, De Hoon MJL, Kawaji H, Itoh M, et al. 2016. Functional annotation of the vlinc class of non-coding RNAs using systems biology approach. *Nucleic Acids Res* **44:** 3233–3252. doi:10.1093/nar/gkw162

Stoffregen EP, Donley N, Stauffer D, Smith L, Thayer MJ. 2011. An autosomal locus that controls chromosome-wide replication timing and mono-allelic expression. *Hum Mol Genet* **20:** 2366–2378. doi:10.1093/hmg/ddr138

Strom AR, Brangwynne CP. 2019. The liquid nucleome—phase transitions in the nucleus at a glance. *J Cell Sci* **132:** jcs235093. doi:10.1242/jcs.235093

Stubblefield E. 1975. Analysis of the replication pattern of Chinese hamster chromosomes using 5-bromodeoxyuridine suppression of 33258 Hoechst fluorescence. *Chromosoma* **53:** 209–221. doi:10.1007/BF00329172

Sukackaite R, Cornacchia D, Jensen MR, Mas PJ, Blackledge M, Enervald E, Duan G, Auchynnikava T, Köhn M, Hart DJ, et al. 2017. Mouse Rif1 is a regulatory subunit of protein phosphatase 1 (PP1). *Sci Rep* **7:** 1–10. doi:10.1038/s41598-017-01910-1

Szabo Q, Bantignies F, Cavalli G. 2019. Principles of genome folding into topologically associating domains. *Sci Adv* **5:** eaaw1668. doi:10.1126/sciadv.aaw1668

Takahashi S, Miura H, Shibata T, Nagao K, Okumura K, Ogata M, Obuse C, Takebayashi SI, Hiratani I. 2019. Genome-wide stability of the DNA replication program in single mammalian cells. *Nat Genet* **51:** 529–540. doi:10.1038/s41588-019-0347-5

Takebayashi SI, Dileep V, Ryba T, Dennis JH, Gilbert DM. 2012. Chromatin-interaction compartment switch at developmentally regulated chromosomal domains reveals an unusual principle of chromatin folding. *Proc Natl Acad Sci* **109:** 12574–12579. doi:10.1073/pnas.1207185109

Takebayashi SI, Lei I, Ryba T, Sasaki T, Dileep V, Battaglia D, Gao X, Fang P, Fan Y, Esteban MA, et al. 2013. Murine esBAF chromatin remodeling complex subunits BAF250a and Brg1 are necessary to maintain and reprogram pluripotency-specific replication timing of select replication domains. *Epigenetics Chromatin* **6:** 42. doi:10.1186/1756-8935-6-42

Tanaka S, Nakato R, Katou Y, Shirahige K, Araki H. 2011. Origin association of Sld3, Sld7, and Cdc45 proteins is a key step for determination of origin-firing timing. *Curr Biol* **21:** 2055–2063. doi:10.1016/j.cub.2011.11.038

Tasan I, Sustackova G, Zhang L, Kim J, Sivaguru M, HamediRad M, Wang Y, Genova J, Ma J, Belmont AS, et al. 2018. CRISPR/Cas9-mediated knock-in of an optimized TetO repeat for live cell imaging of endogenous loci. *Nucleic Acids Res* **46:** e100. doi:10.1093/nar/gky501

Taylor JH. 1960. Asynchronous duplication of chromosomes in cultured cells of Chinese hamster. *J Biophys Biochem Cytol* **7:** 455–463. doi:10.1083/jcb.7.3.455

Therizols P, Illingworth RS, Courilleau C, Boyle S, Wood AJ, Bickmore WA. 2014. Chromatin decondensation is sufficient to alter nuclear organization in embryonic stem cells. *Science* **346:** 1238–1242. doi:10.1126/science.1259587

Thomson I, Gilchrist S, Bickmore WA, Chubb JR. 2004. The radial positioning of chromatin is not inherited through mitosis but is established de novo in early G1. *Curr Biol* **14:** 166–172. doi:10.1016/j.cub.2003.12.024

van Steensel B, Belmont AS. 2017. Lamina-associated domains: links with chromosome architecture, heterochromatin, and gene repression. *Cell* **169:** 780–791. doi:10.1016/j.cell.2017.04.022

Vertii A, Ou J, Yu J, Yan A, Pagès H, Liu H, Zhu LJ, Kaufman PD. 2019. Two contrasting classes of nucleolus-associated domains in mouse fibroblast heterochromatin. *Genome Res* **29:** 1235–1249. doi:10.1101/gr.247072.118

Wang W, Klein K, Proesmans K, Yang H, Marchal C, Zhu X, Borman T, Hastie A, Weng Z, Bechhoefer J, et al. 2020a. Genome-wide mapping of human DNA replication by optical replication mapping supports a stochastic model of eukaryotic replication timing. bioRxiv doi:10.1101/2020.08.24.263459

Wang Y, Zhang Y, Zhang R, van Schaik T, Zhang L, Sasaki T, Hupkes DP, Chen Y, Gilbert DM, van Steensel B, et al. 2020b. SPIN reveals genome-wide landscape of nuclear compartmentalization. bioRxiv doi:10.1101/2020.03.09.982967

Whyte WA, Orlando DA, Hnisz D, Abraham BJ, Lin CY, Kagey MH, Rahl PB, Lee TI, Young RA. 2013. Master transcription factors and mediator establish super-enhancers at key cell identity genes. *Cell* **153:** 307–319. doi:10.1016/j.cell.2013.03.035

Wilson KA, Elefanty AG, Stanley EG, Gilbert DM. 2016. Spatio-temporal re-organization of replication foci accompanies replication domain consolidation during human pluripotent stem cell lineage specification. *Cell Cycle* **15:** 2464–2475. doi:10.1080/15384101.2016.1203492

Wu T, Pinto HB, Kamikawa YF, Donohoe ME. 2015. The BET family member BRD4 interacts with OCT4 and regulates pluripotency gene expression. *Stem Cell Rep* **4:** 390–403. doi:10.1016/j.stemcr.2015.01.012

Wu T, Kamikawa YF, Donohoe ME. 2018. Brd4's bromodomains mediate histone H3 acetylation and chromatin remodeling in pluripotent cells through P300 and Brg1. *Cell Rep* **25:** 1756–1771. doi:10.1016/j.celrep.2018.10.003

Xiang W, Roberti MJ, Hériché JK, Huet S, Alexander S, Ellenberg J. 2018. Correction: correlative live and super-resolution imaging reveals the dynamic structure of replication domains. *J Cell Biol* **217:** 3315–3316. doi:10.1083/JCB.20170907408082018c

Xie W, Schultz MD, Lister R, Hou Z, Rajagopal N, Ray P, Whitaker JW, Tian S, Hawkins RD, Leung D, et al. 2013. Epigenomic analysis of multilineage differentiation of human embryonic stem cells. *Cell* **153:** 1134–1148. doi:10.1016/j.cell.2013.04.022

Yaffe E, Farkash-Amar S, Polten A, Yakhini Z, Tanay A, Simon I. 2010. Comparative analysis of DNA replication timing reveals conserved large-scale chromosomal architecture. *PLoS Genet* **6:** e1001011. doi:10.1371/journal.pgen.1001011

Yoshida K, Bacal J, Desmarais D, Padioleau I, Tsaponina O, Chabes A, Pantesco V, Dubois E, Parrinello H, Skrzypczak M, et al. 2014. The histone deacetylases Sir2 and Rpd3 Act on ribosomal DNA to control the replication program in budding yeast. *Mol Cell* **54:** 691–697. doi:10.1016/j.molcel.2014.04.032

Zhang H, Petrie MV, He Y, Peace JM, Chiolo IE, Aparicio OM. 2019. Dynamic relocalization of replication origins by fkh1 requires execution of ddk function and cdc45 loading at origins. *eLife* **8:** e45512. doi:10.7554/eLife.45512

Zhao PA, Sasaki T, Gilbert DM. 2020. High-resolution Repli-Seq defines the temporal choreography of initiation, elongation and termination of replication in mammalian cells. *Genome Biol* **21:** 76. doi:10.1186/s13059-020-01983-8

Imaging Organization of RNA Processing within the Nucleus

Jeetayu Biswas, Weihan Li, Robert H. Singer, and Robert A. Coleman

Department of Anatomy and Structural Biology, Albert Einstein College of Medicine, Bronx, New York 10461, USA

Correspondence: robert.singer@einsteinmed.org; robert.coleman2@einsteinmed.org

Within the nucleus, messenger RNA is generated and processed in a highly organized and regulated manner. Messenger RNA processing begins during transcription initiation and continues until the RNA is translated and degraded. Processes such as 5′ capping, alternative splicing, and 3′ end processing have been studied extensively with biochemical methods and more recently with single-molecule imaging approaches. In this review, we highlight how imaging has helped understand the highly dynamic process of RNA processing. We conclude with open questions and new technological developments that may further our understanding of RNA processing.

Nuclear organization occurs at several levels, ranging from cellular to molecular. At the cellular level, the nucleus must reestablish itself with each cell division. Additionally, large-scale nuclear reorganization, such as flexing of the envelope in response to cellular stress or formation of nuclear lobes during cell differentiation, must occur through dynamic processes (for review, see Newport and Forbes 1987). At an intermediate scale, chromosomes and chromatin are spatially organized to regulate gene expression at different loci (Lieberman-Aiden et al. 2009). This organization is dynamic, allowing chromosomes to restructure during the process of differentiation and to permit phenomena such as tissue-specific gene expression. At the molecular level, RNA processing machinery is assembled in highly stochastic yet ordered processes. With each new RNA molecule being synthesized, protein factors assemble and disassemble to dictate all aspects of an RNA's life (for review, see Tutucci et al. 2018a). With assistance from single-molecule imaging techniques, the order and dynamics of this process is clear for multisubunit complexes such as the transcription preinitiation complex (PIC) and the spliceosome (DeHaven et al. 2016).

In this review, we will discuss how single-molecule imaging has shed light upon organization of RNA processing. We begin by introducing the single-molecule imaging techniques and their strengths and limitations. Then we sequentially discuss the steps of RNA processing, transcription, splicing, 5′ capping, polyadenylation, and nuclear export (Fig. 1). We will focus on how single-molecule imaging has furthered our understanding of these dynamic processes.

Cite this article as *Cold Spring Harb Perspect Biol* doi: 10.1101/cshperspect.a039453

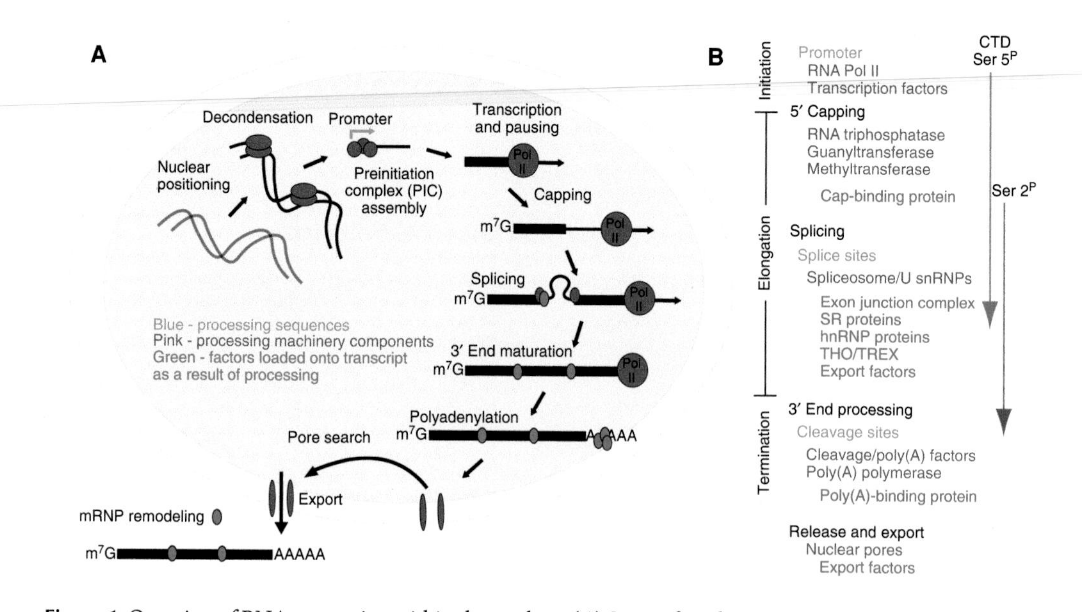

Figure 1. Overview of RNA processing within the nucleus. (*A*) Steps of nuclear RNA processing. Blue, sequences involved in RNA processing. Pink, core components of the processing machinery. Green, factors that are loaded onto the transcript as a result of RNA processing. Beginning from *upper left* and moving clockwise: DNA (black lines) is organized into domains that are actively relocated throughout the nucleus. Reorganization of DNA through factor deposition or chromatin opening (pink ovals) allows for assembly of the preinitiation complex (PIC) (pink circles) at the promoter (blue arrow). After assembly, RNA polymerase II is recruited, initiates, and pauses after approximately 100 nucleotides. After pause release, capping occurs shortly after the 5′ end of the RNA exits RNA polymerase II. Cotranscriptional loading of splicing factors (green ovals), cleavage, and polyadenylation machinery occurs as RNA polymerase progresses through the transcript. After the transcript is released from the transcription site, it diffuses through the nucleoplasm until it finds a pore (pink ovals) suitable for export. Shortly after export, ribonucleoprotein (RNP) remodeling occurs. (*B*) (*Left*) Factors involved in RNA processing. (*Right*) The carboxy-terminal domain (CTD) of RNA polymerase II provides a platform for coupling transcription elongation to other RNA-processing events. The phosphorylation state of the CTD changes to permit coupling (*right* downward arrows). (Panel created from data in Hocine et al. 2010.)

INTRODUCTION TO TECHNIQUES

New imaging technologies have allowed for the observation of gene expression, both at the single-cell and single-molecule level (for review, see Vera et al. 2016; Biswas et al. 2018; Tutucci et al. 2018a; Sato et al. 2020). This review will highlight findings from techniques including single-molecule fluorescence in situ hybridization (smFISH), MS2/PP7 RNA imaging, fluorescence recovery after photobleaching (FRAP), and single-particle tracking (SPT) (Fig. 2). Each of these techniques has unique advantages for measuring different aspects of transcription dynamics (Liu et al. 2015).

Subcellular and single-molecule resolution of RNA processing was first achieved with smFISH (Femino et al. 1998). By using multiple dye-conjugated DNA oligos that hybridize to an RNA of interest, specific RNA molecules can be imaged within the cell as diffraction-limited spots (Fig. 2A; Femino et al. 1998; Raj et al. 2006). Transcription sites appear as nondiffraction limited and more intense spots because multiple nascent RNAs are being synthesized along the gene. Complementary oligos can be designed against the coding sequence, introns, or alternative untranslated regions (UTRs) to study different aspects of RNA processing (Fig. 2).

Multicolor smFISH (Femino et al. 1998; Levsky et al. 2002; Shah et al. 2018) is a powerful tool to image the process of splicing and alternative polyadenylation (APA). By labeling the introns with probes fluorescing in one color

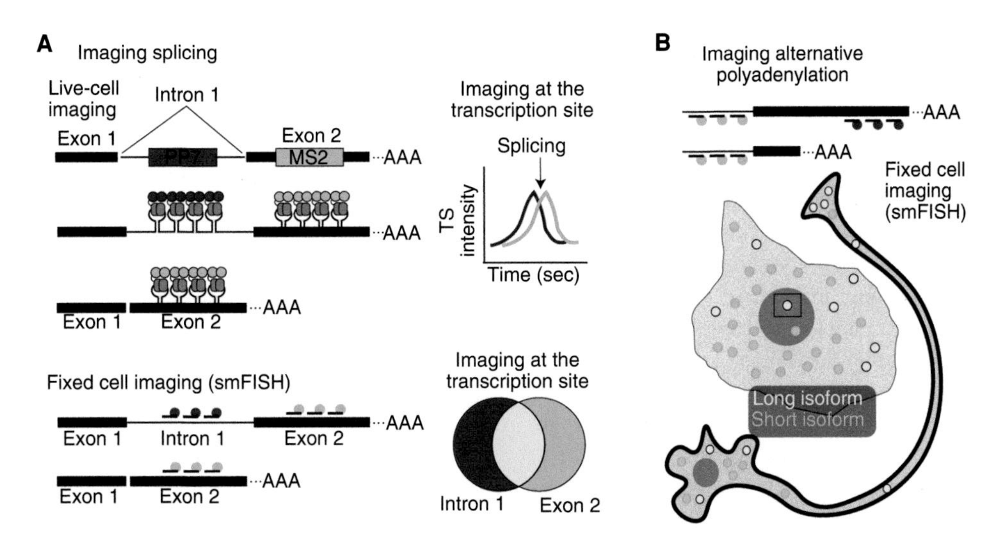

Figure 2. Imaging methods to study RNA processing. (*A*) Imaging splicing in cells. (*Top left*) To visualize splicing in living cells, two different RNA stem loop cassettes (red PP7 and green MS2 boxes, respectively) are engineered into a respective intron and exon. During transcription, both the intron and exon are visualized by binding of different fluorescent coat proteins (PP7 coat protein-mCherry and MS2 coat protein-green fluorescent protein [GFP]) to the stem loop arrays. After splicing occurs and the intron is removed, the mature mRNA is labeled with a single-stem loop array. (*Top right*) By tracking the color change at the transcription site, splicing kinetics can be measured in living cells. As the RNA is synthesized, increasing signal from the first intron (red line) occurs prior to the second exon (green line). A sudden loss of intronic signal (red line, downward arrow) occurs when the intron is no longer colocalized with the exon. (*Bottom left*) To visualize splicing in living cells, two sets of single-molecule fluorescence in situ hybridization (smFISH) probes (red and green circles, respectively) are designed against one intron and exon. During transcription, both the intron and exon are visualized as dual-colored smFISH spots. After splicing occurs and the intron is removed, the mature mRNA is labeled with a single set of smFISH probes. (*Bottom right*) Dual-colored smFISH can find overlapping diffraction-limited spots at the transcription site with the red signal corresponding to the first intron and the green signal corresponding to the second exon. (*B*) Imaging alternative polyadenylation in cells. (*Top*) Multicolor smFISH can be used to label individual isoforms in either two colors (long isoform) or one color (short isoform). Two sets of smFISH probes (red and green circles, respectively) are designed against the conserved coding sequence (thin line) and long isoform-specific sequence (thick line). Short isoforms are singly labeled (green only) and long isoforms are dual colored (green and red probes). (*Bottom*) In cases where alternative polyadenylation leads to differing localizations, abundances, or stabilities, two-color smFISH identifies these differences within cells.

and the exons in another color, precursor mRNAs that contain introns are visualized as dual-colored spots while mature mRNAs that do not contain introns are single-colored spots (Fig. 2A). In the case of cotranscriptional splicing, FISH probes against the intron will localize to the nascent RNA transcript and colocalize around the brighter transcription site (Levsky et al. 2002; Shah et al. 2018). When long and short 3′ UTRs are produced from the same gene by APA, they too can be distinguished by two-color smFISH. Specifically, one set of smFISH probes detects the shared regions of the two isoforms while a second set of smFISH probes detects the unique region of the long mRNA isoform. As a result, the short mRNA isoform is only detected by the shared smFISH probes (single-colored) while the long mRNA isoform is detected by both sets of smFISH probes (dual-colored) (Fig. 2B).

To image RNAs in living cells, stem loop sequences from bacteriophages (MS2, PP7, lambda; for review, see Tutucci et al. 2018a) are inserted into the gene of interest (Bertrand et al. 1998; Daigle and Ellenberg 2007; Chao et al. 2008; Tutucci et al. 2018b). Stem loop cognate proteins, modified viral capsid proteins, are genetically fused to a fluorescent protein and coexpressed in the same cells. Multiple capsid proteins fluorescently label the RNA after binding to the stem loop sequences. As a result, each RNA molecule can be tracked as it moves throughout the cell (Bertrand et al. 1998; Tutucci et al. 2018b). RNA imaging can track the appearance and maturation of nascent RNA molecules at endogenous transcription sites in live cells in real time (Larson et al. 2011; Lionnet et al. 2011) and in living animals (Park et al. 2014; Nwokafor et al. 2019). Multiple orthogonal stem loop sequences can be used to interrogate processing of the RNA in real time (Hocine et al. 2013; Martin et al. 2013; Coulon et al. 2014).

Live-cell splicing kinetics have been studied by tagging either introns or exons with stem loop/capsid protein systems and measuring both the fluctuations and residence time of RNAs at the transcription site (Fig. 2A). The kinetics of splicing can also be observed on individual transcripts using two orthogonal stem loop/capsid protein systems to tag different introns and exons. When one set of stem loops is present within the intron and the second is within a constitutive exon, the diffraction-limited spot will lose one set of stem loops during the process of splicing and therefore go from dual-colored to single-colored (Coulon et al. 2014). Alternative configurations of stem loop arrays have also allowed for dynamic observation of translation (Halstead et al. 2015) and RNA decay (Horvathova et al. 2017).

To image individual proteins in vivo, fusions of fluorescent protein (for review, see Lippincott-Schwartz et al. 2001; Snapp 2005) and more recently ultrabright and photostable protein tags (Keppler et al. 2004; Gautier et al. 2008; Encell et al. 2012; Tanenbaum et al. 2014; Grimm et al. 2015, 2020) can be used. Methods such as FRAP or fluorescence correlation spectroscopy (FCS) can determine the diffusion coefficients of both protein and RNA (Wu et al. 2012, 2015) within cells. For example, FRAP can be used to probe changes in diffusion and the chromatin-binding residence time of a collection of factors including transcriptional activators as shown during the heat shock response in *Drosophila* polytene cells (Yao et al. 2006). With the advent of long-term SPT, proteins can be followed as they bind, unbind, and perform their regulatory activities. SPT can examine both the diffusion and stable binding of a single transcription factor on target sites in vitro or even on the genome in live mammalian cells (Chen et al. 2014; Teves et al. 2016).

TRANSCRIPTION INITIATION—A COMPLEX POTPOURRI OF FACTORS

Over last 50 years, biochemical experiments have identified and characterized most of the factors required for initiating transcription (Cramer 2019; Roeder 2019). Early steps in this process involve the formation of a PIC, consisting of over 80 different proteins (e.g., TBP/TFIID and TFIIA/IIB/IIE/IIF/IIH) along with Mediator and RNA polymerase II (RNA Pol II). Factors comprising the PIC coordinately assemble in a highly ordered fashion onto a ~60 base pair stretch of the core promoter (Lemon and Tjian 2000). Once the PIC is assembled, the carboxy-terminal domain (CTD) of the largest subunit of RNA Pol II must be phosphorylated at select residues for RNA Pol II to escape the promoter region (CTD-Ser5ph) and begin elongating downstream (CTD-Ser3ph). During promoter escape, the PIC is partially disassembled when RNA Pol II disengages from interacting factors such as TFIIB and TFIIH. As RNA Pol II transcribes further into the coding sequence, the phosphorylated CTD of RNA Pol II serves as a landing platform for numerous factors involved in subsequent RNA processing of the mRNA including splicing and polyadenylation (Saldi et al. 2016).

Now that most of the players involved in transcription initiation have been revealed, researchers have begun unraveling the dynamics of gene regulation. Recent development of bright organic fluorescent dyes (Grimm et al. 2015, 2020), orthogonal labeling systems (e.g., HaloTag, SNAP-tag, CLIP-tag) (Keppler et al. 2004; Gautier et al. 2008; Encell et al. 2012), and advanced microscopes (e.g., multifocus, lattice light-sheet, and target lock stimulated emission depletion [STED] microscopy) (Abrahamsson et al. 2013; Liu et al. 2014; Li et al. 2019) allow transcription dynamics to be visualized in live cells. When combined with live-cell RNA-labeling technologies (e.g., MS2/PP7 tagging) (Hocine et al. 2013) and CRISPR gene editing (Spille et al. 2019), it is now possible to track the activity of a transcription factor at a single transcribing endogenous gene inside a live cell (Donovan et al. 2019; Li et al. 2019).

TRANSCRIPTION INITIATION DYNAMICS—INEFFICIENT AND BURSTY

Over the last 20 years, single-molecule imaging studies have described the dynamics and efficiency of transcription initiation. In eukaryotes, transcription was determined to be a stochastic process marked by short periods (minutes) of intense activity or bursts of transcription interspersed between long intervals (>30 min to hours) of inactivity (for review, see Rodriguez and Larson 2020; Tunnacliffe and Chubb 2020). The cyclical turning on and off of transcription initiation suggests that transcription factors controlling this process must also act in a similar manner. Indeed, transcription factors rapidly cycle on and off the genome in hubs or clusters of activity (Cisse et al. 2013; Liu et al. 2014; Tsai et al. 2017; Chong et al. 2018; Mir et al. 2018). Furthermore, the clustering dynamics of RNA Pol II has been proposed to dictate the dynamics of transcription initiation (Cho et al. 2016). Therefore, most if not all transcription factors dynamically bind and unbind in hubs of activity reinforcing the cyclical nature of gene-specific transcription initiation and bursting.

Transcription initiation is an inherently inefficient process both in vitro and in vivo (Darzacq et al. 2007; Revyakin et al. 2012). Both FRAP of a gene array and single-molecule imaging of an endogenously tagged locus indicate that ~1% of RNA Pol II–binding events at a gene array leads to productive transcription initiation (Darzacq et al. 2007; Li et al. 2019). Transcription initiation by RNA polymerase I (RNA Pol I) and III (RNA Pol III) require fewer factors (for review, see Engel et al. 2018). Yet surprisingly, live-cell imaging studies also indicate that RNA Pol I transcription initiation is also inefficient (~1%–11%) in vivo (Dundr et al. 2002). This is likely due to the complexity and dynamic instability of PIC assembly. Adding to the complexity, transcription factor–binding sites in enhancers and promoters exist within chromatin, which mitigates the assembly of the PIC. Complementary genomic studies in yeast show that rapid conditional removal of individual general transcription factors (GTFs) from the nucleus severely compromises the stability of the PIC (Petrenko et al. 2019). This inherent instability of partial PIC assemblies likely leads to a low probability for complete PIC assembly and an overall low efficiency of transcription initiation in vivo.

TRANSCRIPTION STABILITY DYNAMICS AT PROMOTER—RAPID AND MULTISCALE

In contrast to textbook schema, in vitro and live-cell single-molecule tracking studies indicate that transcriptional activators, chromatin-associated factors, and GTFs (e.g., Sox2, glucocorticoid receptor [GR], p53, BRD4, and TFIIB) typically cycle on and off their genomic targets on the order of seconds (Fig. 3A; Chen et al. 2014; Morisaki et al. 2014; Zhang et al. 2016; Coleman et al. 2017; Li et al. 2019). Other factors, such as TBP/TFIID, still cycle on and off, but bind the core promoter DNA for much longer periods of time (~100 sec) (Revyakin et al. 2012; Zhang et al. 2016; Coleman et al. 2017; Teves et al. 2018). The enhanced dynamics of activators likely reflects their roles in recruiting multiple factors (TFIIA, TFIIB, TFIIF, Mediator, and RNA Pol II) onto a stable TFIID/core promoter scaffold to complete PIC formation that is then competent for transcription initiation (for review, see Roeder 2019). The relatively unstable interaction of TFIIB (~1.5 sec) with a stable TFIID/promoter DNA complex (minutes) likely facilitates multiple rounds of transcriptional reinitiation within very short time periods (e.g., seconds) via successive recruitment of RNA Pol II onto a semi-stable partial PIC assembly.

Consistent with such a model is the observation of RNA Pol II convoys (sometimes con-

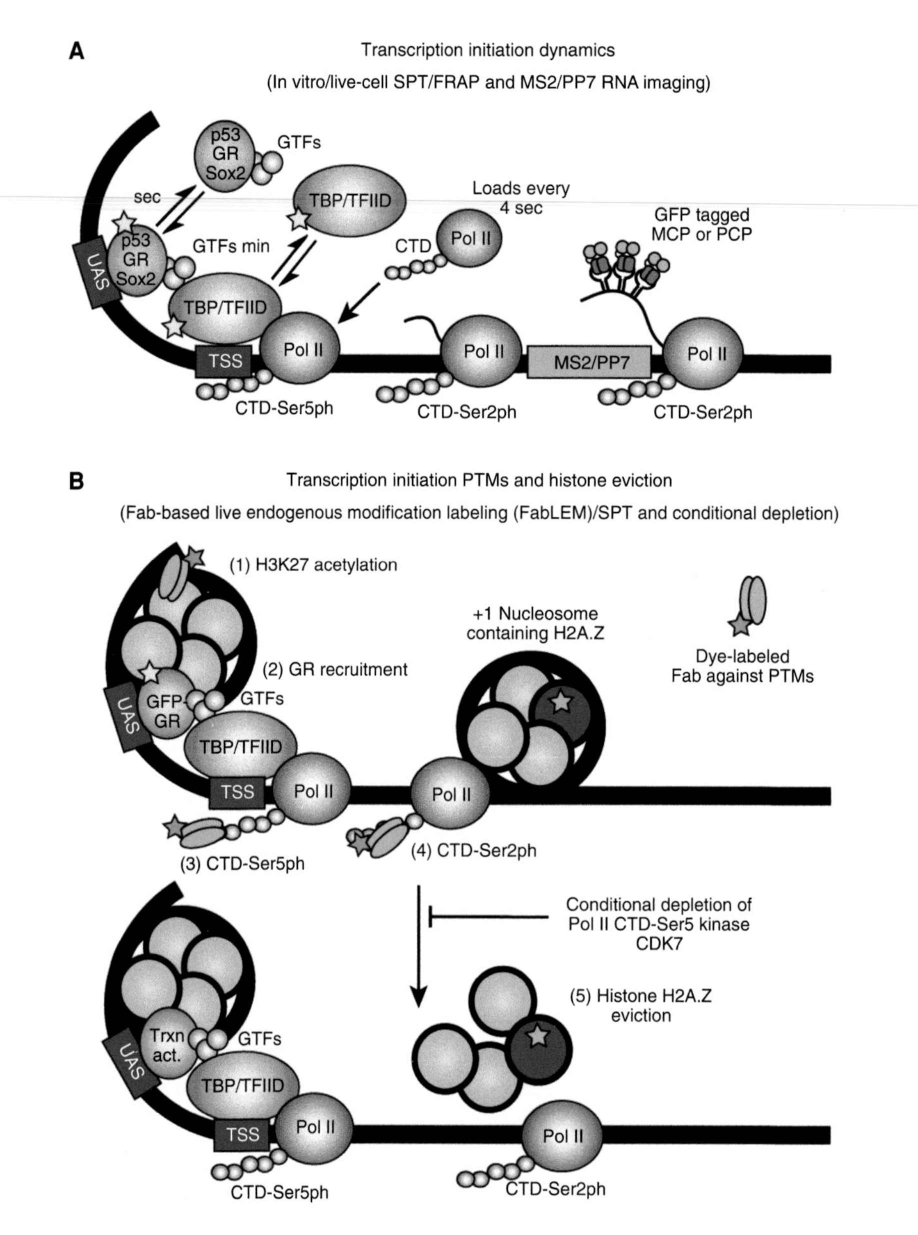

Figure 3. Dynamics of transcription initiation and histone eviction. (*A*) Imaging the dynamics of preinitiation complex (PIC) formation, RNA Pol II convoy formation, and transcriptional bursting. Fluorescently labeled transcriptional activators and PIC components (general transcription factor [GTFs]/TBP/TFIID) (orange circles) dynamically assemble onto enhancer regions (upstream activation sequence [UAS]) and the transcription start site (TSS) of the core promoter (blue boxes). The different stabilities of the factors at the core promoter allows rapid successive recruitment of RNA Pol II (pink circles) that forms convoys during a transcriptional burst. As RNA Pol II elongates past the MS2/PP7 repeats in the gene (green box), the emanating RNA forms stem loops that are subsequently recognized by green fluorescent protein (GFP)-tagged MCP/PCP coat proteins. (*B*) Imaging the temporal acetylation of histones, phosphorylation of RNA Pol II carboxy-terminal domain (CTD), and disruption of a histone variant during transcription initiation. After addition of hormone, fluorescently labeled Fab antibody fragments (green ovals) indicate histone H3K27 acetylation precedes FP-glucocorticoid receptor binding (yellow star) at a gene reporter. Subsequent recruitment of RNA Pol II to the TSS leads to phosphorylation (ph) of Serine 5 (Ser5) of the RNA Pol II CTD. Upon promoter escape, Serine 2 (Ser2) within the RNA Pol II CTD is phosphorylated. Disruption of a fluorescently labeled H2A.Z (blue circle) containing nucleosome immediately downstream of the TSS requires phosphorylation of RNA Pol II CTD Ser5. (GR) Glucocorticoid receptor, (Trxn act.) Trxn activation.

sisting of 19 polymerases) that rapidly load onto the core promoter during a transcriptional burst (Tantale et al. 2016). During convoy formation, RNA Pol II can rapidly load every 4 sec likely owing to the dynamic assembly and disassembly of GTFs on a stable TFIID/core promoter scaffold (Fig. 3A). RNA Pol II convoys last for ~100 sec followed by a brief period of inactivity (~100 sec) where presumably the factors required for reinitiation are reloaded onto the core promoter. It is tempting to speculate that the length of RNA Pol II convoy formation is dictated by the overall stability of TFIID on the core promoter given equivalent timescales (~100 sec). Eventually, successive rounds of RNA Pol II convoys stop forming and the promoter shuts down for longer periods of time (>30 min to hours) where presumably a more repressive chromatin state is reestablished on the enhancer, core promoter, and gene body.

TRANSCRIPTION INITIATION PTMs AND HISTONE EVICTION—AN ORDERED PROCESS

Fluorescently labeled antigen-binding fragments (Fabs) can image select posttranslational modifications (PTMs) such as histone marks and changes in the CTD of RNA Pol II in live cells (FabLEM) (Hayashi-Takanaka et al. 2009, 2011). FabLEM showed that histone H3K27 acetylation preceded RNA Pol II CTD phosphorylation during zygotic genome activation in live zebrafish embryos (Sato et al. 2019). Using green fluorescent protein (GFP)-tagged GR, RNA Pol II CTD phosphorylation onto a gene array was found to contain hyperacetylated histone H3K27 before hormone stimulation (Fig. 3B; Stasevich et al. 2014). Sorting cells with differing amounts of H3K27 acetylation at the gene array showed that histone hyperacetylation enhanced levels of GR recruitment and elongating RNA Pol II containing the CTD phosphorylated at Serine 2.

Histone eviction at the promoter region may also affect transcription initiation. In yeast, the +1 nucleosome contains the histone variant H2A.Z immediately downstream of the transcription start site (TSS) (Fig. 3B; Albert et al. 2007). H2A.Z is rapidly turned over and linked to a chromatin state that is permissive to transcription initiation (Weber and Henikoff 2014). Previously, it was thought that PIC formation led to displacement of H2A.Z at the promoter (Tramantano et al. 2016). In contrast, live-cell SPT imaging using fluorescently tagged H2A.Z combined with a rapid conditional depletion system (Haruki et al. 2008) found that incorporation of RNA Pol II into the PIC per se was not sufficient for H2A.Z eviction from the genome (Ranjan et al. 2020). Rather, downstream events were required since conditional depletion of CDK7 that phosphorylates Serine 5 of the RNA Pol II CTD led to enhanced stabilization of fluorescent H2A.Z on the genome (Fig. 3B). This finding suggested that promoter escape of RNA Pol II was necessary to displace the first nucleosome that RNA Pol II encountered after leaving the TSS. Future studies may combine FabLEM and conditional depletion of factors alongside SPT and MS2/PP7 RNA imaging to examine chromatin remodeling and transcription initiation during transcriptional bursting.

RNA SPLICING

The process of mRNA splicing is an important step of the mRNA maturation process. The biochemical mechanisms of spliceosome function have been elucidated by both biochemistry and structural biology (Will and Lührmann 2011; Wilkinson et al. 2020). The splicing machinery (the spliceosome) is composed of five small nuclear RNAs (snRNAs) and several core proteins that are assembled into small nuclear ribonucleoprotein (RNP) particles (U1, U2, and the U4/U5/U6 tri-snRNP) as well as the protein Prp19-complex (NineTeen complex [NTC]). Recognition of the 5′ splice site occurs by binding to U1; subsequently the branch site sequence is recognized by U2 and the 3′ splice site by U2 auxiliary factor (U2AF).

In the past 15 years, single-molecule imaging methods were widely applied to the field of splicing. Single-molecule tracking of snRNPs has been possible using cell extracts (Crawford et al. 2008; Hoskins et al. 2011) and inside living cells (Grünwald et al. 2006). Below, we summarize the mechanistic insights of splicing uncovered by single-molecule imaging methods.

DYNAMIC AND REVERSIBLE SPLICEOSOME ASSEMBLY

Fixed cell imaging has been able to colocalize intronic RNAs with different components of the splicing machinery (Zhang et al. 1994; Brody et al. 2011). Dynamics of the assembly process, such as residence time and order of assembly, can be observed using fluorescently tagged snRNP proteins. FCS was used to calculate the lifetimes of the free diffusing and bound populations of different snRNPs. Ninety percent of the associations between splicing factors and pre-mRNA was brief (15–30 sec) (Huranová et al. 2010) while the remaining 10% of the associations was significantly longer than 30 sec. Further investigation of spliceosome assembly using single-molecule imaging found assembly of the spliceosome on pre-mRNA, while ordered, was also highly reversible (Hoskins et al. 2011).

Like the stochastic nature of other complex assembly processes within a cell (such as the PIC), association of a snRNP with the pre-mRNA does not commit the spliceosome; in fact, each complex has multiple on and off binding events on an individual pre-mRNA before the pre-mRNA is committed to undergoing splicing. As these events progress and the spliceosome assemble, the probability of splicing increases (Hoskins et al. 2011). Further work showed that a key step in splicing, formation of the lariat, did not occur during the early assembly stage of splicing. Rather, the 5′ splice site and the branchpoint sequence only came together after the NTC has bound, thus providing a checkpoint for proper assembly (Crawford et al. 2013). Spliceosomal assembly is therefore a highly dynamic and stochastic process with commitment to splicing occurring after proper assembly (for review, see DeHaven et al. 2016). Similar to the PIC, low efficiency of splicing may arise from the large number of factors that are involved and their stochastic assembly/disassembly. Additionally, questions remain about how inefficient splicing processes can be coordinated with the movement of RNA Pol II.

SPLICING AND TRANSCRIPTION KINETICS—CO- OR POSTTRANSCRIPTIONAL

The highly dynamic process of spliceosomal assembly must be performed with each newly transcribed intron. The kinetics of this process may lead to different splicing rates across cells and individual transcripts. To understand how transcription, spliceosome assembly, and proper splicing are interrelated, the process of splicing has been imaged in both fixed and live cells.

Two predominant splicing modes—co- and posttranscriptional splicing—have been observed. RNAs are either efficiently spliced at or near the transcription site or spliced after pre-mRNAs are released into the nucleoplasm (Vargas et al. 2011).

Cotranscriptional RNA splicing may be facilitated by binding of the splicing machinery to the CTD of the RNA Pol II (Das et al. 2007; Brody et al. 2011). Reverse transcription polymerase chain reaction (RT-PCR) measurements of several nascent RNAs point toward cotranscriptional splicing being present in a wide variety of terminal introns (Schmidt et al. 2011). While many introns have been visualized to undergo cotranscriptional splicing using RNA FISH (Levsky et al. 2002; Shah et al. 2018), live imaging (Sheinberger et al. 2017; Wan et al. 2019), and PCR-based methods (Schmidt et al. 2011), the levels of posttranscriptional splicing vary among transcripts (Schmidt et al. 2011).

Cases of posttranscriptional splicing have been described. Disruption of the splicing machinery, either through chemical inhibition (Schmidt et al. 2011), knockdown of splicing factors (Waks et al. 2011), mutations of splicing factors (Coulon et al. 2014), mutation of the splice site (Schmidt et al. 2011), or RNA structural changes at splicing factor-binding sites (Vargas et al. 2011), all have shown to contribute to increased, unprocessed transcripts within the nucleoplasm that eventually are spliced prior to export.

In the absence of splicing perturbation, is cotranscriptional splicing de facto for all transcripts? It is likely that several competing rates (RNA Pol II elongation rate, splicing rate, pausing and transcription termination rate) can contribute to either co- or posttranscriptional splicing. For instance, placing the MS2 array with a strongly spliced intron upstream of a slowly processed polyadenylation sequence may strongly increase the chances of cotranscriptional splicing (Schmidt et al. 2011).

Alternative scenarios may require posttranscriptional splicing. For example, increasing the number of introns within a reporter RNA can lead to accumulation of nascent, unspliced transcripts at the transcription site (Brody et al. 2011). Imaging showed that this increase in nascent transcript was not due to increased levels of polymerase but rather prolonged splicing as it was coincident with increased loading of splicing factors but not RNA Pol II and reversed by splicing inhibition. This suggests that when excess introns are present, the excess transcripts that are laden with spliceosomal components remains close to the transcription site as it completes processing (Brody and Shav-Tal 2011; Brody et al. 2011). The presence of partially spliced transcripts near the transcription site may be a generalized finding. By combining smFISH and expansion microscopy, it was observed that newly synthesized RNAs dwell near the transcription site after transcription. This allows them to complete and undergo continuous splicing as they traverse a zone of ∼300 nm around the transcription site (Coté et al. 2020). For a given RNA, it is likely that both co- and posttranscriptional processing occur with characteristics of the gene tilting the scales in favor of one or the other. Using two-color, live-cell imaging combined with fluctuation correlation analysis revealed a kinetic competition between splicing and transcription with posttranscriptional splicing occurring more often than cotranscriptional splicing (Coulon et al. 2014).

The race between polymerase and the splicing machinery contributes to heterogeneity of splicing. On a case-by-case basis, single-molecule methods are most apt to tease apart the individual contributions of co- and posttranscriptional splicing. Particularly exciting are transcriptome-wide, single-molecule imaging and sequencing methods that may allow for higher throughput interrogation of endogenous splicing rates (Oesterreich et al. 2016; Herzel et al. 2018; Shah et al. 2018).

MEASURING THE DYNAMICS OF TRANSCRIPTION ELONGATION AND 3′ END PROCESSING

Transcription elongation and termination dynamics have been measured in vivo using both sequencing- and imaging-based approaches (for review, see Jonkers and Lis 2015). While imaging RNA Pol II in single cells has the advantage of high spatial and temporal resolution, it is difficult to distinguish RNA Pol II molecules that are transcribing a gene of interest from the rest of the RNA Pol II molecules in nucleoplasm. Two methods overcome this challenge by amplifying RNA Pol II signal at a locus. The first method is to randomly integrate tens or hundreds of the same gene at a gene locus, forming a tandem gene array. The large number of fluorescent RNA Pol II molecules that are recruited to the tandem gene array appear as a bright spot (Janicki et al. 2004). By applying FRAP or fusing RNA Pol II to photoactivated GFP, the dynamics of transcription elongation (Boireau et al. 2007; Darzacq et al. 2007), splicing (Brody et al. 2011; Martin et al. 2013), and 3′ end processing (Boireau et al. 2007) have been determined. The second method is to use polytene chromosomes, which are large chromosomes in *Drosophila* salivary glands and have thousands of DNA strands (see the section Cotranscriptional Loading of Factors). The transcription of *HSP70* was used to determine the transcription dynamics because its expression is inducible by heat shock (Yao et al. 2006; Ardehali et al. 2009; Buckley et al. 2014). The aforementioned methods both require gene multiplication; therefore, measurements are averaged across the gene cluster.

More recently, endogenous gene expression has been imaged in real time (Larson et al. 2011). The transcription dynamics were determined using a mathematical model that analyzes the fluctuation of fluorescently labeled nascent mRNAs. By tagging the longest yeast gene, MDN1, with PP7 stem loops, the prolonged RNA Pol II transcription elongation time can be accurately measured. The dwell time of a fluorescently labeled MDN1 mRNA at its transcription site equals the summation of two time spans: (1) the transcription elongation time for RNA Pol II to finish transcribing the part of MDN1 downstream of PP7 stem loops, and (2) the 3′ end processing time. When the PP7 stem loops are inserted at the 5′ or 3′ UTR of MDN1, the difference of the mRNA dwell times at the transcription site equals the time to transcribe MDN1 open reading frame (ORF). Thus, transcription elongation and 3′ end processing time were determined.

Using the above methods, RNA Pol II elongation rates range between 14 nt/sec and 61 nt/sec (Hocine et al. 2013). The 3′ end processing time ranges between 30 sec and 70 sec. Sequencing-based methods showed that the RNA Pol II elongation rate varies between genes and within the same gene. For example, RNA Pol II pauses 30 to 60 bp downstream of the transcription start site, accelerates through the gene body, and again pauses near the termination site (for review, see Kwak and Lis 2013; Proudfoot 2016; Mayer et al. 2017). Such variability may explain the wide range of results determined by different methods.

MONITORING mRNA ISOFORMS AS A RESULT OF ALTERNATIVE POLYADENYLATION

About 60% of genes in mammalian cells have more than one transcription termination site. As a result, these genes produce mRNA isoforms with different 3′ UTRs, the process called APA. The genomic landscape of APA has been determined by microarray- and deep-sequencing-based approaches in multiple model organisms (microarray-based methods [Flavell et al. 2008; Sandberg et al. 2008; Ji and Tian 2009; Ji et al. 2009] and deep-seq-based methods [Ozsolak et al. 2010; Shepard et al. 2011; Derti et al. 2012; Ulitsky et al. 2012; Mata 2013; Wang et al. 2018]). Different mRNA isoforms produced by APA can have different stability, localization, or altered protein-coding regions. Thus, APA regulates protein expression and cellular function (Sandberg et al. 2008; Mayr and Bartel 2009).

Unlike sequencing-based methods, smFISH can determine both the abundance and spatial distribution of different mRNA isoforms within individual cells (Fig. 2A). In neuronal cells, the short isoforms of *myo-inositol monophosphatase-1* (*lmpa1*), *Importin β1* and *brain-derived neurotrophic factor* (*BDNF*) mRNA are enriched in soma while their long isoforms are enriched at the distal axon (*lmpa1* and *Importin β1* mRNA) or dendrites (*BDNF* mRNA). A *cis*-regulatory element, which is present in the long 3′ UTR and absent in the short 3′ UTR, directs the differential mRNA localization (An et al. 2008; Andreassi et al. 2010; Perry et al. 2012). A sequencing-based approach revealed that 593 genes have differentially localized 3′ UTR isoforms, which resulted from either APA or alternative splicing (Ciolli Mattioli et al. 2019). As a result, the translational products of these mRNA isoforms have different roles in neuronal responses. For example, the BDNF proteins that are produced by the long mRNA isoforms at dendrites regulate spine morphology and synaptic plasticity in hippocampal neurons (An et al. 2008).

Recent studies uncovered a novel function of long 3′ UTRs, which act as a scaffold and recruit proteins that regulate the mRNA translational product (Berkovits and Mayr 2015; Lee and Mayr 2019). For example, the long 3′ UTR of *CD47* mRNA recruits a protein complex that localize CD47 protein to the cell surface. In contrast, the short 3′ UTR of *CD47* mRNA cannot recruit the protein complex. And its translational products are located at the endoplasmic reticulum (ER). Thus, while the long and short *CD47* mRNAs have the same cellular distribution, their protein products are differentially distributed (Berkovits and Mayr 2015).

COTRANSCRIPTIONAL LOADING OF FACTORS

Cotranscriptional processes include 5′ capping, splicing, and 3′ end processing. RNA-binding proteins (RBPs) bind cotranscriptionally to regulate these processes, forming an RNA–protein (mRNP) complex that travels to nuclear pores for export.

In salivary gland cells from *Drosophila melanogaster* and *Chironomus tentans*, polytene chromosomes provide a highly repetitive array that is suitable for visualizing transcription with electron microscopy (Miller and Beatty 1969). By labeling a protein of interest with immunogold, overlapping signals from nascent mRNAs and gold particles represent cotranscriptionally assembled mRNPs (summarized in Mehlin and Daneholt 1993; Daneholt 1997, 2001; Björk and Wieslander 2015). Using this approach, it was noted that as RNA polymerase progresses, it loops with small particles, which are now understood to be the splicing machinery (Beyer and Osheim 1988). After this initial loading of RBPs during transcription, RBPs are classified into two categories: (1) RBPs that accompany the mRNA into cytoplasm, such as the exon junction complex (Visa et al. 1996a, 1996b; Percipalle et al. 2001; Zhao et al. 2002); and (2) RBPs that are retained within the nucleus (Alzhanova-Ericsson et al. 1996; Sun et al. 1998). Further studies are needed to understand why some RBPs, but not the rest, can accompany mRNA through nuclear pores (future directions).

EXPORT COMPETENCE AND RNP IDENTITY—THE RNP's NEW CLOTHES

After the completion of RNA processing, RNAs are released from the transcription site and exported through the nuclear pore complex (NPC) to reach the cytoplasm. Movement of mRNA through the nucleoplasm occurs through interchromatin spaces and by an energy-independent, diffusion-based mechanism (Politz et al. 1999; Shav-Tal et al. 2004; Vargas et al. 2005; Siebrasse et al. 2008; Mor et al. 2010).

Export times range between 2.5 min and 1 h depending on the model system used (Shav-Tal et al. 2004). The mRNA searches until it finds a pore suitably primed for export (Siebrasse et al. 2012) with only one in three or four attempts leading to successful export (Grünwald and Singer 2010; Ma and Yang 2010; Siebrasse et al. 2012). Once found, passage of RNA through NPC is extremely fast and occurs in three discrete steps, docking (80 msec), translocation through the NPC (5–20 msec), and then release (80 msec) (Grünwald and Singer 2010; Grünwald et al. 2011; Ma et al. 2013). Longer dwell times have been noted at the nuclear basket, possibly correlating with RNA remodeling (Grünwald and Singer 2010; Montpetit et al. 2011; Siebrasse et al. 2012).

Given the inefficient process of export, further studies will determine how gating or priming of the mRNP is accomplished. While questions still exist for why certain pores can engage the mRNA, the random diffusion of RNA from the transcription site facilitates interactions with multiple pores and may prevent clogging of the export machinery (Mor et al. 2010). RNAs may transiently interact with pores that are occupied and therefore may be unsuitable for export (Grünwald and Singer 2010; Grünwald et al. 2011). Delays in RNA movement at the nuclear and cytoplasmic faces are due to exchange of proteins as well as mRNP remodeling (Siebrasse et al. 2012). In yeast, Dbp5 helicase is involved in this exchange through its interactions with RNA (Montpetit et al. 2011). By remodeling the RNP as it exits the NPC, nuclear proteins are shed, and cytoplasmic proteins are added with exit from the nuclear pore (Ledoux and Guthrie 2011). Strikingly, deletion of Dbp5 in yeast prevents RNP export, causing the mRNP to reverse directions at the cytoplasmic face of pore, presumably though inability to shed nuclear identity (Hodge et al. 2011).

COUPLING OF RNA PROCESSING

Each step of RNA processing must be coordinated with the binding and unbinding of hundreds of factors. This process occurs faithfully throughout development, yet how the different steps are coupled to one another is an area of active investigation. For example, gene promoters orchestrate cotranscriptional loading of RBPs that accompany the mRNA to the cytoplasm and regulate mRNA's localization, translation, or stability (Bregman et al. 2011; Trcek et al. 2011; Vera et al. 2014; Zid and O'Shea 2014). The mechanisms underlying promoter-based recruitment are under investigation but may involve either RNA sequence elements or RNA Pol II subunits (for review, see Haimovich et al. 2013).

RNA polymerase plays a central role in organizing the different steps of processing. Phosphorylation of the CTD serves as a recruitment platform (for review, see Bentley 2014; Saldi et al. 2016). Additionally, the elongation speed of RNA Pol II also functions as a regulatory step for multiple aspects of gene expression. In the case of RNA splice site selection both recruitment and kinetics are at play (de la Mata et al. 2003; Dujardin et al. 2014). Recruitment coupling by RNA Pol II brings processing factors to the transcription site. Kinetic coupling generates a window of splicing opportunity during the process of polymerase elongation, depending on the time windows, different splice sites become available.

RNA Pol II also regulates termination and APA through similar kinetic (torpedo model) or recruitment (allosteric model). In the torpedo model, polymerase continues to transcribe after cleavage and is chased down by the 5′-3′ exonuclease XRN2. The allosteric model suggests that as protein factors are lost from the CTD, polymerase processivity decreases until it releases the elongating strand.

In addition to RNA Pol II, hundreds of RNA-binding proteins contribute to the layers of regulation that determine alternative splicing outcomes of a single gene. RBPs can bind to short sequence motifs within the pre-mRNA to influence spliceosome assembly or disassembly. RNA-binding proteins also can dictate alternative splicing changes, not only by competing with other RBPs for binding sequences on the pre-mRNA but also by causing structural changes to the RNA. Looping the mRNA or promoting new double-stranded regions can mask or unmask motifs bound by splicing factors. The kinetics of RBP association appear to be highly dynamic (Jobbins et al. 2018) and further studies will elucidate how the RBP code functions within cells (discussed in future directions).

In addition to regulation by RNA-binding proteins or modification of the core transcriptional machinery, regulatory control may be linked to the biophysical properties of RNA. Emerging work suggests that phase-separated compartments within the nucleus may promote the above processes and be coupled to the process of transcription (Hnisz et al. 2017; Henninger et al. 2020).

THE FUTURE OF IMAGING THE PROCESSES OF TRANSCRIPTION AND RNA PROCESSING

Great strides have been made in developing imaging approaches to understand the mechanisms of transcription and immediate downstream events regulating the processing of the transcript: splicing, 3′ end formation, termination. However, these approaches so far have relied on detection mainly of the nascent chains during elongation and termination (Brody et al. 2011; Larson et al. 2011; Coulon et al. 2014). Much more needs to be learned about the temporal aspects of each of the steps of initiation, elongation, and termination by observing single genes in action, and what happens to the transcripts during and subsequently. This will require developing microscopes and reagents designed for better detection over longer periods of time in living cells. The rapidly moving discoveries of proteins involved in various stages of these processes makes the need for imaging compelling since it will be the only way to characterize their molecular effects with sufficient temporal resolution to postulate the effects they have on the process.

What is currently needed are two developments that will advance our knowledge substantially. The first is to identify and characterize specific proteins that act during these events and then observe them in action. The second is to develop a high-resolution means to detect specific genes actively expressing in isolation from others and without perturbing their activity. The first of these developments is nearer at hand, since specific dyes have been developed that allow the detection of single proteins in the nucleus, and their binding as revealed by their dwell time at a specific site (Chen et al. 2014; Grimm et al. 2015, 2020). Doping in a small number of these molecules into the cell allows the process to be followed and tracked without the complexity of a myriad of tagged molecules confounding the analysis. However, while the state of the molecules can be inferred, and their exact positions determined by their Gaussian point spread function, their binding sites cannot because of uncertainties in the presumed target site. Uncertainties can arise from the microscope's point spread function, localization errors, or the size of the tag used (Fig. 4). Even a super-resolved spot in the nucleus (10–20 nm in diameter) will contain dozens of genes, and those genes will be physically mobile making it impossible to follow them for more than a few milliseconds. By tagging an individual gene using a CRISPR approach (e.g., with MS2 stem loops), it will be possible to obtain some temporal information, for instance how long a specific splicing factor can stay on a transcript, using high-resolution colocalization and dwell times, how long and frequent pauses occur during transcription, or how long 3′ end formation takes, or the duration of splicing events. All of this, while important, is only observing outcomes, not cause and effect relationships.

Imaging in high spatiotemporal resolution with two or more colors—one for the gene and one for a putative regulatory protein—would be the most accessible approach (Fig. 4). By super-registration of the tagged protein with the gene, one can correlate temporal relationships between the arrival (or departure) of the protein on the gene and its effect on an immediate event such as transcription or termination (Donovan et al. 2019; Li et al. 2019). We would expect a dwell time to accompany the association of the protein

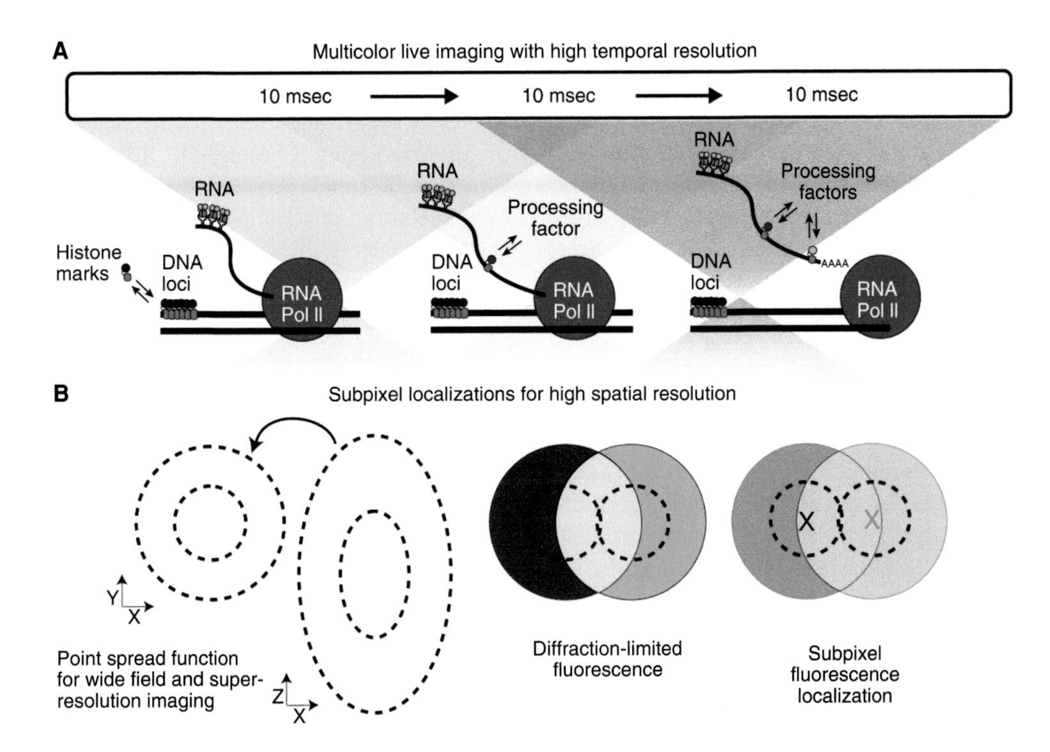

Figure 4. Dynamic imaging of multiple factors loading onto a single RNA, in real time. (*A*) High-speed multicolor imaging is necessary to capture how proteins are dynamics loaded and unloaded on a DNA or RNA molecule. Labeled molecules may include the DNA (blue circles), histone marks (red circles), RNA (green circles), and processing factors (pink and yellow circles). Simultaneous imaging of the DNA locus (blue circles) and nascent RNA (green circles) can be used to find processing factors (pink and yellow circles) that colocalize to a given spot. (*Left* to *right*) With each microscope exposure, the movement of RNA and processing factors occurs. (*B*) (*Left*) The point spread function (PSF) of the microscope limits the spatial resolution. Conventional wide-field epifluorescence microscopes have a larger PSF (*outer* circle) than their super-resolved counterparts (*inner* circle). (*Middle*) When imaging multiple fluorescent molecules (red and green circles) that are within a diffraction-limited spot, their resolution may be limited by the resolution of the microscope. (*Right*) Gaussian fitting can be used to localize the positions of molecules within a diffraction-limited spot. Accurate localization will require using ultrabright and bleach-resistant fluorophores.

with the gene, or the RNA nascent chain. Proteins not involved in binding to the gene, or its transcripts would diffuse through the space and hence not be detectable at any location.

Additionally, this approach could be applicable to chromatin changes that precede gene expression: histone modifiers that influence the gene availability to factors. What are needed are the reagents that would allow the histone methylations or acetylations to be detected in real time. The development of nanobodies (Fridy et al. 2014), which can be expressed and labeled intracellularly, against a panel of specific histone modifications would go a long way to facilitate imaging chromatin dynamics. These key reagents would allow the addition or removal of select histone modifications to be correlated with the dwell time of the corresponding fluorescently labeled chromatin modification enzymes.

One of the downstream indicators of changes in chromatin is the decondensation of genetic loci during gene activation (Tumbar et al. 1999; Tsukamoto et al. 2000). Advances in CRISPR-based imaging technologies, such as MS2/PP7 stem loop engineering into guide RNAs (Ma et al. 2016; Wang et al. 2016b) or SunTag labeling of Cas9 (Tanenbaum et al. 2014), which increase the signal-to-noise ratio, will be necessary to detect decondensation in a way that does not perturb gene activation. Likewise, enzymes responsible for decondensation can be tagged and identified to be coincident with the activation of a locus as judged by its dwell time, but it will be difficult to distinguish the many of these events happening rapidly all over the genome. To achieve this goal, we will need to develop imaging methods to simultaneously and dynamically track tens to hundreds of loci at sufficient temporal and spatial resolution in a live cell (Ma et al. 2016; Wang et al. 2016b).

Underlying all of these challenges is where to place a particular label on the genome. Recent advances in understanding the 3D folding of the genome via both imaging and genomic approaches help to suggest key regions to tag (Wang et al. 2016a; Bintu et al. 2018; Kempfer and Pombo 2020). The broad diversity of chromatin structures inside the cell as seen by chromatin electron multitilt microscope tomography (ChromEMT) suggests that choosing select regions to tag based on euchromatic versus heterochromatic distinctions will be difficult (Ou et al. 2017). ChromEMT uses illumination of DNA-binding fluorescent dyes to deposit a contrast agent, OsO_4, onto DNA, which is visualized by electron microscopy tomography (EMT). Therefore, ChromEMT is limited to visualizing an outline of nucleosomes as DNA wraps around the histones without information on the underlying DNA sequence.

Ideally, one would like to determine the structure of chromatin around specific genes that are transcriptionally active or repressed. One approach to achieve this goal would be to combine CRISPR-based imaging with correlated fluorescence microscopy and ChromEMT. In this manner, CRISPR-based fluorescent maps could serve as "signposts" to tell researchers where to begin examining specific regions of the nucleus using ChromEMT. Advances in ChromEMT could also eventually allow enhanced contrast of select proteins (e.g., enhancer-binding factors, PIC components, RNA Pol II, and cleavage and polyadenylation factors, etc.) and nascent transcripts for use as additional "signposts." In this manner, one can begin to determine the chromatin structures and accessible genomic regions immediately surrounding key regions of interest (e.g., enhancers, promoters, and termination sites) at high enough spatial resolution, which would not be possible with the genomically tagged CRISPR-based fluorescent maps alone. Using this strategy, researchers will then have a chance to calibrate the position of fluorescently tagged genomic regions and chromatin structures in live cells in a similar manner as has been achieved using DNA-FISH and genomic approaches (Wang et al. 2016a; Bintu et al. 2018; Kempfer and Pombo 2020).

The future of imaging the dynamics of transcription and RNA processing looks bright. Researchers now must coordinately take advantage of multiple expertise from different groups throughout the world to seize upon this opportunity.

ACKNOWLEDGMENTS

The authors thank Justin C. Wheat and other members of the Einstein community as well as members of the Singer and Coleman laboratories, past and present, for their helpful discussions and comments. J.B. was supported with funding from an MSTP Training Grant T32GM007288 and predoctoral fellowship F30CA214009. W.L. and R.H.S. were supported by NIH Grants R01NS083085 and R35GM136296. R.A.C. was supported by NIH Grant R01GM126045. The authors acknowledge the members of the 4DN community and NIH U01DA047729.

REFERENCES

Abrahamsson S, Chen J, Hajj B, Stallinga S, Katsov AY, Wisniewski J, Mizuguchi G, Soule P, Mueller F, Dugast Darzacq C, et al. 2013. Fast multicolor 3D imaging using aberration-corrected multifocus microscopy. *Nat Methods* **10:** 60–63. doi:10.1038/nmeth.2277

Albert I, Mavrich TN, Tomsho LP, Qi J, Zanton SJ, Schuster SC, Pugh BF. 2007. Translational and rotational settings of H2A.Z nucleosomes across the saccharomyces cerevisiae genome. *Nature* **446:** 572–576. doi:10.1038/nature05632

Alzhanova-Ericsson AT, Sun X, Visa N, Kiseleva E, Wurtz T, Daneholt B. 1996. A protein of the SR family of splicing factors binds extensively to exonic Balbiani ring pre-mRNA and accompanies the RNA from the gene to the nuclear pore. *Genes Dev* **10:** 2881–2893. doi:10.1101/gad.10.22.2881

An JJ, Gharami K, Liao GY, Woo NH, Lau AG, Vanevski F, Torre ER, Jones KR, Feng Y, Lu B, et al. 2008. Distinct role of long 3′ UTR BDNF mRNA in spine morphology and synaptic plasticity in hippocampal neurons. *Cell* **134:** 175–187. doi:10.1016/j.cell.2008.05.045

Andreassi C, Zimmermann C, Mitter R, Fusco S, De Vita S, Devita S, Saiardi A, Riccio A. 2010. An NGF-responsive element targets myo-inositol monophosphatase-1 mRNA to sympathetic neuron axons. *Nat Neurosci* **13:** 291–301. doi:10.1038/nn.2486

Ardehali MB, Yao J, Adelman K, Fuda NJ, Petesch SJ, Webb WW, Lis JT. 2009. Spt6 enhances the elongation rate of RNA polymerase II in vivo. *EMBO J* **28:** 1067–1077. doi:10.1038/emboj.2009.56

Bentley DL. 2014. Coupling mRNA processing with transcription in time and space. *Nat Rev Genet* **15:** 163–175. doi:10.1038/nrg3662

Berkovits BD, Mayr C. 2015. Alternative 3′ UTRs act as scaffolds to regulate membrane protein localization. *Nature* **522:** 363–367. doi:10.1038/nature14321

Bertrand E, Chartrand P, Schaefer M, Shenoy SM, Singer RH, Long RM. 1998. Localization of ASH1 mRNA particles in living yeast. *Mol Cell* **2:** 437–445. doi:10.1016/S1097-2765(00)80143-4

Beyer AL, Osheim YN. 1988. Splice site selection, rate of splicing, and alternative splicing on nascent transcripts. *Genes Dev* **2:** 754–765. doi:10.1101/gad.2.6.754

Bintu B, Mateo LJ, Su JH, Sinnott-Armstrong NA, Parker M, Kinrot S, Yamaya K, Boettiger AN, Zhuang X. 2018. Super-resolution chromatin tracing reveals domains and cooperative interactions in single cells. *Science* **362:** eaau1783. doi:10.1126/science.aau1783

Biswas J, Liu Y, Singer RH, Wu B. 2018. Fluorescence imaging methods to investigate translation in single cells. *Cold Spring Harb Perspect Biol* **11:** a032722.

Björk P, Wieslander L. 2015. The Balbiani ring story: synthesis, assembly, processing, and transport of specific messenger RNA–protein complexes. *Annu Rev Biochem* **84:** 65–92. doi:10.1146/annurev-biochem-060614-034150

Boireau S, Maiuri P, Basyuk E, de la Mata M, Knezevich A, Pradet-Balade B, Bäcker V, Kornblihtt A, Marcello A, Bertrand E. 2007. The transcriptional cycle of HIV-1 in real-time and live cells. *J Cell Biol* **179:** 291–304. doi:10.1083/jcb.200706018

Bregman A, Avraham-Kelbert M, Barkai O, Duek L, Guterman A, Choder M. 2011. Promoter elements regulate cytoplasmic mRNA decay. *Cell* **147:** 1473–1483. doi:10.1016/j.cell.2011.12.005

Brody Y, Shav-Tal Y. 2011. Transcription and splicing: when the twain meet. *Transcription* **2:** 216–220. doi:10.4161/trns.2.5.17273

Brody Y, Neufeld N, Bieberstein N, Causse SZ, Böhnlein EM, Neugebauer KM, Darzacq X, Shav-Tal Y. 2011. The in vivo kinetics of RNA polymerase II elongation during co-transcriptional splicing. *PLoS Biol* **9:** e1000573. doi:10.1371/journal.pbio.1000573

Buckley MS, Kwak H, Zipfel WR, Lis JT. 2014. Kinetics of promoter Pol II on Hsp70 reveal stable pausing and key insights into its regulation. *Genes Dev* **28:** 14–19. doi:10.1101/gad.231886.113

Chao JA, Patskovsky Y, Almo SC, Singer RH. 2008. Structural basis for the coevolution of a viral RNA-protein complex. *Nat Struct Mol Biol* **15:** 103–105. doi:10.1038/nsmb1327

Chen J, Zhang Z, Li L, Chen BC, Revyakin A, Hajj B, Legant W, Dahan M, Lionnet T, Betzig E, et al. 2014. Single-molecule dynamics of enhanceosome assembly in embryonic stem cells. *Cell* **156:** 1274–1285. doi:10.1016/j.cell.2014.01.062

Cho WK, Jayanth N, English BP, Inoue T, Andrews JO, Conway W, Grimm JB, Spille JH, Lavis LD, Lionnet T, et al. 2016. RNA polymerase II cluster dynamics predict mRNA output in living cells. *eLife* **5:** e13617. doi:10.7554/eLife.13617

Chong S, Dugast-Darzacq C, Liu Z, Dong P, Dailey GM, Cattoglio C, Heckert A, Banala S, Lavis L, Darzacq X, et al. 2018. Imaging dynamic and selective low-complexity domain interactions that control gene transcription. *Science* **361:** eaar2555. doi:10.1126/science.aar2555

Ciolli Mattioli C, Rom A, Franke V, Imami K, Arrey G, Terne M, Woehler A, Akalin A, Ulitsky I, Chekulaeva M. 2019. Alternative 3′ UTRs direct localization of functionally diverse protein isoforms in neuronal compartments. *Nucleic Acids Res* **47:** 2560–2573. doi:10.1093/nar/gky1270

Cisse II, Izeddin I, Causse SZ, Boudarene L, Senecal A, Muresan L, Dugast-Darzacq C, Hajj B, Dahan M, Darzacq X. 2013. Real-time dynamics of RNA polymerase II clustering in live human cells. *Science* **341:** 664–667. doi:10.1126/science.1239053

Coleman RA, Qiao Z, Singh SK, Peng CS, Cianfrocco M, Zhang Z, Piasecka A, Aldeborgh H, Basishvili G, Liu WL. 2017. P53 dynamically directs TFIID assembly on target gene promoters. *Mol Cell Biol* **37:** e00085. doi:10.1128/MCB.00085-17

Coté A, Coté C, Bayatpour S, Drexler HL, Alexander KA, Chen F, Wassie AT, Boyden ES, Berger S, Churchman LS, et al. 2020. The spatial distributions of pre-mRNAs suggest post-transcriptional splicing of specific introns within endogenous genes. bioRxiv doi:10.1101/2020.04.06.028092

Coulon A, Ferguson ML, de Turris V, Palangat M, Chow CC, Larson DR. 2014. Kinetic competition during the transcription cycle results in stochastic RNA processing. *eLife* **3:** e03939. doi:10.7554/eLife.03939

Cramer P. 2019. Eukaryotic transcription turns 50. *Cell* **179:** 808–812. doi:10.1016/j.cell.2019.09.018

Crawford DJ, Hoskins AA, Friedman LJ, Gelles J, Moore MJ. 2008. Visualizing the splicing of single pre-mRNA molecules in whole cell extract. *RNA* **14:** 170–179. doi:10.1261/rna.794808

Crawford DJ, Hoskins AA, Friedman LJ, Gelles J, Moore MJ. 2013. Single-molecule colocalization FRET evidence that spliceosome activation precedes stable approach of 5′ splice site and branch site. *Proc Natl Acad Sci* **110:** 6783–6788. doi:10.1073/pnas.1219305110

Daigle N, Ellenberg J. 2007. λ_N-GFP: an RNA reporter system for live-cell imaging. *Nat Methods* **4:** 633–636. doi:10.1038/nmeth1065

Daneholt B. 1997. A look at messenger RNP moving through the nuclear pore. *Cell* **88:** 585–588. doi:10.1016/S0092-8674(00)81900-5

Daneholt B. 2001. Assembly and transport of a premessenger RNP particle. *Proc Natl Acad Sci* **98:** 7012–7017. doi:10.1073/pnas.111145498

Darzacq X, Shav-Tal Y, de Turris V, Brody Y, Shenoy SM, Phair RD, Singer RH. 2007. In vivo dynamics of RNA polymerase II transcription. *Nat Struct Mol Biol* **14:** 796–806. doi:10.1038/nsmb1280

Das R, Yu J, Zhang Z, Gygi MP, Krainer AR, Gygi SP, Reed R. 2007. SR proteins function in coupling RNAP II transcription to pre-mRNA splicing. *Mol Cell* **26:** 867–881. doi:10.1016/j.molcel.2007.05.036

DeHaven AC, Norden IS, Hoskins AA. 2016. Lights, camera, action! Capturing the spliceosome and pre-mRNA splicing with single molecule fluorescence microscopy. *Wiley Interdiscip Rev RNA* **7:** 683–701. doi:10.1002/wrna.1358

de la Mata M, Alonso CR, Kadener S, Fededa JP, Blaustein M, Pelisch F, Cramer P, Bentley D, Kornblihtt AR. 2003. A slow RNA polymerase II affects alternative splicing in vivo. *Mol Cell* **12:** 525–532. doi:10.1016/j.molcel.2003.08.001

Derti A, Garrett-Engele P, Macisaac KD, Stevens RC, Sriram S, Chen R, Rohl CA, Johnson JM, Babak T. 2012. A quantitative atlas of polyadenylation in five mammals. *Genome Res* **22:** 1173–1183. doi:10.1101/gr.132563.111

Donovan BT, Huynh A, Ball DA, Patel HP, Poirier MG, Larson DR, Ferguson ML, Lenstra TL. 2019. Live-cell imaging reveals the interplay between transcription factors, nucleosomes, and bursting. *EMBO J* **38:** e100809. doi:10.15252/embj.2018100809

Dujardin G, Lafaille C, de la Mata M, Marasco LE, Muñoz MJ, Le Jossic-Corcos C, Corcos L, Kornblihtt AR. 2014. How slow RNA polymerase II elongation favors alternative exon skipping. *Mol Cell* **54:** 683–690. doi:10.1016/j.molcel.2014.03.044

Dundr M, Hoffmann-Rohrer U, Hu Q, Grummt I, Rothblum LI, Phair RD, Misteli T. 2002. A kinetic framework for a mammalian RNA polymerase in vivo. *Science* **298:** 1623–1626. doi:10.1126/science.1076164

Encell LP, Friedman Ohana R, Zimmerman K, Otto P, Vidugiris G, Wood MG, Los GV, McDougall MG, Zimprich C, Karassina N, et al. 2012. Development of a dehalogenase-based protein fusion tag capable of rapid, selective and covalent attachment to customizable ligands. *Curr Chem Genomics* **6:** 55–71. doi:10.2174/1875397301206010055

Engel C, Neyer S, Cramer P. 2018. Distinct mechanisms of transcription initiation by RNA polymerases I and II. *Ann Rev Biophys* **47:** 425–446. doi:10.1146/annurev-biophys-070317-033058

Femino AM, Fay FS, Fogarty K, Singer RH. 1998. Visualization of single RNA transcripts in situ. *Science* **280:** 585–590. doi:10.1126/science.280.5363.585

Flavell SW, Kim TK, Gray JM, Harmin DA, Hemberg M, Hong EJ, Markenscoff-Papadimitriou E, Bear DM, Greenberg ME. 2008. Genome-wide analysis of MEF2 transcriptional program reveals synaptic target genes and neuronal activity-dependent polyadenylation site selection. *Neuron* **60:** 1022–1038. doi:10.1016/j.neuron.2008.11.029

Fridy PC, Li Y, Keegan S, Thompson MK, Nudelman I, Scheid JF, Oeffinger M, Nussenzweig MC, Fenyö D, Chait BT, et al. 2014. A robust pipeline for rapid production of versatile nanobody repertoires. *Nat Methods* **11:** 1253–1260. doi:10.1038/nmeth.3170

Gautier A, Juillerat A, Heinis C, Corrêa IR Jr, Kindermann M, Beaufils F, Johnsson K. 2008. An engineered protein tag for multiprotein labeling in living cells. *Chem Biol* **15:** 128–136. doi:10.1016/j.chembiol.2008.01.007

Grimm JB, English BP, Chen J, Slaughter JP, Zhang Z, Revyakin A, Patel R, Macklin JJ, Normanno D, Singer RH, et al. 2015. A general method to improve fluorophores for live-cell and single-molecule microscopy. *Nat Methods* **12:** 244–250. doi:10.1038/nmeth.3256

Grimm JB, Tkachuk AN, Xie L, Choi H, Mohar B, Falco N, Schaefer K, Patel R, Zheng Q, Liu Z, et al. 2020. A general method to optimize and functionalize red-shifted rhodamine dyes. *Nat Methods* **17:** 815–821. doi:10.1038/s41592-020-0909-6

Grünwald D, Singer RH. 2010. In vivo imaging of labelled endogenous β-actin mRNA during nucleocytoplasmic transport. *Nature* **467:** 604–607. doi:10.1038/nature09438

Grünwald D, Spottke B, Buschmann V, Kubitscheck U. 2006. Intranuclear binding kinetics and mobility of single native U1 snRNP particles in living cells. *Mol Biol Cell* **17:** 5017–5027. doi:10.1091/mbc.e06-06-0559

Grünwald D, Singer RH, Rout M. 2011. Nuclear export dynamics of RNA-protein complexes. *Nature* **475:** 333–341. doi:10.1038/nature10318

Haimovich G, Choder M, Singer RH, Trcek T. 2013. The fate of the messenger is pre-determined: a new model for regulation of gene expression. *Biochim Biophys Acta* **1829:** 643–653. doi:10.1016/j.bbagrm.2013.01.004

Halstead JM, Lionnet T, Wilbertz JH, Wippich F, Ephrussi A, Singer RH, Chao JA. 2015. An RNA biosensor for imaging the first round of translation from single cells to living animals. *Science* **347:** 1367–1671. doi:10.1126/science.aaa3380

Haruki H, Nishikawa J, Laemmli UK. 2008. The anchor-away technique: rapid, conditional establishment of yeast mutant phenotypes. *Mol Cell* **31:** 925–932. doi:10.1016/j.molcel.2008.07.020

Hayashi-Takanaka Y, Yamagata K, Nozaki N, Kimura H. 2009. Visualizing histone modifications in living cells: spatiotemporal dynamics of H3 phosphorylation during interphase. *J Cell Biol* **187:** 781–790. doi:10.1083/jcb.200904137

Hayashi-Takanaka Y, Yamagata K, Wakayama T, Stasevich TJ, Kainuma T, Tsurimoto T, Tachibana M, Shinkai Y, Kurumizaka H, Nozaki N, et al. 2011. Tracking epigenetic histone modifications in single cells using Fab-based live endogenous modification labeling. *Nucleic Acids Res* **39:** 6475–6488. doi:10.1093/nar/gkr343

Henninger JE, Oksuz O, Shrinivas K, Sagi I, LeRoy G, Zheng MM, Andrews JO, Zamudio AV, Lazaris C, Hannett NM, et al. 2020. RNA-mediated feedback control of transcriptional condensates. *Cell* **184:** 207–225.e24. doi:10.1016/j.cell.2020.11.030

Herzel L, Straube K, Neugebauer KM. 2018. Long-read sequencing of nascent RNA reveals coupling among RNA processing events. *Genome Res* **28:** 1008–1019. doi:10.1101/gr.232025.117

Hnisz D, Shrinivas K, Young RA, Chakraborty AK, Sharp PA. 2017. A phase separation model predicts key features of transcriptional control. *Cell* **169:** 13–23. doi:10.1016/j.cell.2017.02.007

Hocine S, Singer RH, Grunwald D. 2010. RNA processing and export. *Cold Spring Harb Perspect Biol* **2:** a000752. doi:10.1101/cshperspect.a000752

Hocine S, Raymond P, Zenklusen D, Chao JA, Singer RH. 2013. Single-molecule analysis of gene expression using two-color RNA labeling in live yeast. *Nat Methods* **10:** 119–121. doi:10.1038/nmeth.2305

Hodge CA, Tran EJ, Noble KN, Alcazar-Roman AR, Ben-Yishay R, Scarcelli JJ, Folkmann AW, Shav-Tal Y, Wente SR, Cole CN. 2011. The Dbp5 cycle at the nuclear pore complex during mRNA export I: dbp5 mutants with defects in RNA binding and ATP hydrolysis define key steps for Nup159 and Gle1. *Genes Dev* **25:** 1052–1064. doi:10.1101/gad.2041611

Horvathova I, Voigt F, Kotrys AV, Zhan Y, Artus-Revel CG, Eglinger J, Stadler MB, Giorgetti L, Chao JA. 2017. The dynamics of mRNA turnover revealed by single-molecule imaging in single cells. *Mol Cell* **68:** 615–625.e9. doi:10.1016/j.molcel.2017.09.030

Hoskins AA, Friedman LJ, Gallagher SS, Crawford DJ, Anderson EG, Wombacher R, Ramirez N, Cornish VW, Gelles J, Moore MJ. 2011. Ordered and dynamic assembly of single spliceosomes. *Science* **331:** 1289–1295. doi:10.1126/science.1198830

Huranová M, Ivani I, Benda A, Poser I, Brody Y, Hof M, Shav-Tal Y, Neugebauer KM, Staněk D. 2010. The differential interaction of snRNPs with pre-mRNA reveals splicing kinetics in living cells. *J Cell Biol* **191:** 75–86. doi:10.1083/jcb.201004030

Janicki SM, Tsukamoto T, Salghetti SE, Tansey WP, Sachidanandam R, Prasanth KV, Ried T, Shav-Tal Y, Bertrand E, Singer RH, et al. 2004. From silencing to gene expression: real-time analysis in single cells. *Cell* **116:** 683–698. doi:10.1016/S0092-8674(04)00171-0

Ji Z, Tian B. 2009. Reprogramming of 3′ untranslated regions of mRNAs by alternative polyadenylation in generation of pluripotent stem cells from different cell types. *PLoS ONE* **4:** e8419. doi:10.1371/journal.pone.0008419

Ji Z, Lee JY, Pan Z, Jiang B, Tian B. 2009. Progressive lengthening of 3′ untranslated regions of mRNAs by alternative polyadenylation during mouse embryonic development. *Proc Natl Acad Sci* **106:** 7028–7033. doi:10.1073/pnas.0900028106

Jobbins AM, Reichenbach LF, Lucas CM, Hudson AJ, Burley GA, Eperon IC. 2018. The mechanisms of a mammalian splicing enhancer. *Nucleic Acids Res* **46:** 2145–2158. doi:10.1093/nar/gky056

Jonkers I, Lis JT. 2015. Getting up to speed with transcription elongation by RNA polymerase II. *Nat Rev Mol Cell Biol* **16:** 167–177. doi:10.1038/nrm3953

Kempfer R, Pombo A. 2020. Methods for mapping 3D chromosome architecture. *Nat Rev Genet* **21:** 207–226. doi:10.1038/s41576-019-0195-2

Keppler A, Pick H, Arrivoli C, Vogel H, Johnsson K. 2004. Labeling of fusion proteins with synthetic fluorophores in live cells. *Proc Natl Acad Sci* **101:** 9955–9959. doi:10.1073/pnas.0401923101

Kwak H, Lis JT. 2013. Control of transcriptional elongation. *Ann Rev Genet* **47:** 483–508. doi:10.1146/annurev-genet-110711-155440

Larson DR, Zenklusen D, Wu B, Chao JA, Singer RH. 2011. Real-time observation of transcription initiation and elongation on an endogenous yeast gene. *Science* **332:** 475–478. doi:10.1126/science.1202142

Ledoux S, Guthrie C. 2011. Regulation of the Dbp5 ATPase cycle in mRNP remodeling at the nuclear pore: a lively new paradigm for DEAD-box proteins. *Genes Dev* **25:** 1109–1114. doi:10.1101/gad.2062611

Lee SH, Mayr C. 2019. Gain of additional BIRC3 protein functions through 3′-UTR-mediated protein complex formation. *Mol Cell* **74:** 701–712.e9. doi:10.1016/j.molcel.2019.03.006

Lemon B, Tjian R. 2000. Orchestrated response: a symphony of transcription factors for gene control. *Genes Dev* **14:** 2551–2569. doi:10.1101/gad.831000

Levsky JM, Shenoy SM, Pezo RC, Singer RH. 2002. Single-cell gene expression profiling. *Science* **297:** 836–840. doi:10.1126/science.1072241

Li J, Dong A, Saydaminova K, Chang H, Wang G, Ochiai H, Yamamoto T, Pertsinidis A. 2019. Single-molecule nanoscopy elucidates RNA polymerase II transcription at single genes in live cells. *Cell* **178:** 491–506.e28. doi:10.1016/j.cell.2019.05.029

Lieberman-Aiden E, van Berkum NL, Williams L, Imakaev M, Ragoczy T, Telling A, Amit I, Lajoie BR, Sabo PJ,

Dorschner MO, et al. 2009. Comprehensive mapping of long range interactions reveals folding principles of the human genome. *Science* **326:** 289–293. doi:10.1126/science.1181369

Lionnet T, Czaplinski K, Darzacq X, Shav-Tal Y, Wells AL, Chao JA, Park HY, de Turris V, Lopez-Jones M, Singer RH. 2011. A transgenic mouse for in vivo detection of endogenous labeled mRNA. *Nat Methods* **8:** 165–170. doi:10.1038/nmeth.1551

Lippincott-Schwartz J, Snapp E, Kenworthy A. 2001. Studying protein dynamics in living cells. *Nat Rev Mol Cell Biol* **2:** 444–456. doi:10.1038/35073068

Liu Z, Legant WR, Chen BC, Li L, Grimm JB, Lavis LD, Betzig E, Tjian R. 2014. 3D imaging of Sox2 enhancer clusters in embryonic stem cells. *eLife* **3:** e04236. doi:10.7554/eLife.04236

Liu Z, Lavis LD, Betzig E. 2015. Imaging live-cell dynamics and structure at the single-molecule level. *Mol Cell* **58:** 644–659. doi:10.1016/j.molcel.2015.02.033

Ma J, Yang W. 2010. Three-dimensional distribution of transient interactions in the nuclear pore complex obtained from single-molecule snapshots. *Proc Natl Acad Sci* **107:** 7305–7310. doi:10.1073/pnas.0908269107

Ma J, Liu Z, Michelotti N, Pitchiaya S, Veerapaneni R, Androsavich JR, Walter NG, Yang W. 2013. High-resolution three-dimensional mapping of mRNA export through the nuclear pore. *Nat Commun* **4:** 2414. doi:10.1038/ncomms3414

Ma H, Tu LC, Naseri A, Huisman M, Zhang S, Grunwald D, Pederson T. 2016. Multiplexed labeling of genomic loci with dCas9 and engineered sgRNAs using CRISPRainbow. *Nat Biotechnol* **34:** 528–530. doi:10.1038/nbt.3526

Martin RM, Rino J, Carvalho C, Kirchhausen T, Carmo-Fonseca M. 2013. Live-cell visualization of pre-mRNA splicing with single-molecule sensitivity. *Cell Rep* **4:** 1144–1155. doi:10.1016/j.celrep.2013.08.013

Mata J. 2013. Genome-wide mapping of polyadenylation sites in fission yeast reveals widespread alternative polyadenylation. *RNA Biol* **10:** 1407–1414. doi:10.4161/rna.25758

Mayer A, Landry HM, Churchman LS. 2017. Pause & go: from the discovery of RNA polymerase pausing to its functional implications. *Curr Opin Cell Biol* **46:** 72–80. doi:10.1016/j.ceb.2017.03.002

Mayr C, Bartel DP. 2009. Widespread shortening of 3′UTRs by alternative cleavage and polyadenylation activates oncogenes in cancer cells. *Cell* **138:** 673–684. doi:10.1016/j.cell.2009.06.016

Mehlin H, Daneholt B. 1993. The Balbiani ring particle: a model for the assembly and export of RNPs from the nucleus? *Trends Cell Biol* **3:** 443–447. doi:10.1016/0962-8924(93)90034-X

Miller OL, Beatty BR. 1969. Visualization of nucleolar genes. *Science* **164:** 955–957. doi:10.1126/science.164.3882.955

Mir M, Stadler MR, Ortiz SA, Hannon CE, Harrison MM, Darzacq X, Eisen MB. 2018. Dynamic multifactor hubs interact transiently with sites of active transcription in *Drosophila* embryos. *eLife* **7:** e40497. doi:10.7554/eLife.40497

Montpetit B, Thomsen ND, Helmke KJ, Seeliger MA, Berger JM, Weis K. 2011. A conserved mechanism of DEAD-box ATPase activation by nucleoporins and InsP6 in mRNA export. *Nature* **472:** 238–242. doi:10.1038/nature09862

Mor A, Suliman S, Ben-Yishay R, Yunger S, Brody Y, Shav-Tal Y. 2010. Dynamics of single mRNP nucleocytoplasmic transport and export through the nuclear pore in living cells. *Nat Cell Biol* **12:** 543–552. doi:10.1038/ncb2056

Morisaki T, Müller WG, Golob N, Mazza D, McNally JG. 2014. Single-molecule analysis of transcription factor binding at transcription sites in live cells. *Nat Commun* **5:** 4456. doi:10.1038/ncomms5456

Newport JW, Forbes DJ. 1987. The nucleus: structure, function, and dynamics. *Annu Rev Biochem* **56:** 535–565. doi:10.1146/annurev.bi.56.070187.002535

Nwokafor C, Singer RH, Lim H. 2019. Imaging cell-type-specific dynamics of mRNAs in living mouse brain. *Methods* **157:** 100–105. doi:10.1016/j.ymeth.2018.07.009

Oesterreich FC, Herzel L, Straube K, Hujer K, Howard J, Neugebauer KM. 2016. Splicing of nascent RNA coincides with intron exit from RNA polymerase II. *Cell* **165:** 372–381. doi:10.1016/j.cell.2016.02.045

Ou HD, Phan S, Deerinck TJ, Thor A, Ellisman MH, O'Shea CC. 2017. ChromEMT: visualizing 3D chromatin structure and compaction in interphase and mitotic cells. *Science* **357:** eaag0025.

Ozsolak F, Kapranov P, Foissac S, Kim SW, Fishilevich E, Monaghan AP, John B, Milos PM. 2010. Comprehensive polyadenylation site maps in yeast and human reveal pervasive alternative polyadenylation. *Cell* **143:** 1018–1029. doi:10.1016/j.cell.2010.11.020

Park HY, Lim H, Yoon YJ, Follenzi A, Nwokafor C, Lopez-Jones M, Meng X, Singer RH. 2014. Visualization of dynamics of single endogenous mRNA labeled in live mouse. *Science* **343:** 422–424. doi:10.1126/science.1239200

Percipalle P, Zhao J, Pope B, Weeds A, Lindberg U, Daneholt B. 2001. Actin bound to the heterogeneous nuclear ribonucleoprotein hrp36 is associated with Balbiani ring mRNA from the gene to polysomes. *J Cell Biol* **153:** 229–236. doi:10.1083/jcb.153.1.229

Perry RBT, Doron-Mandel E, Iavnilovitch E, Rishal I, Dagan SY, Tsoory M, Coppola G, McDonald MK, Gomes C, Geschwind DH, et al. 2012. Subcellular knockout of importin β1 perturbs axonal retrograde signaling. *Neuron* **75:** 294–305. doi:10.1016/j.neuron.2012.05.033

Petrenko N, Jin Y, Dong L, Wong KH, Struhl K. 2019. Requirements for RNA polymerase II preinitiation complex formation in vivo. *eLife* **8:** e43654. doi:10.7554/eLife.43654

Politz JC, Tuft RA, Pederson T, Singer RH. 1999. Movement of nuclear poly(A) RNA throughout the interchromatin space in living cells. *Curr Biol* **9:** 285–291. doi:10.1016/S0960-9822(99)80136-5

Proudfoot NJ. 2016. Transcriptional termination in mammals: stopping the RNA polymerase II juggernaut. *Science* **352:** aad9926. doi:10.1126/science.aad9926

Raj A, Peskin CS, Tranchina D, Vargas DY, Tyagi S. 2006. Stochastic mRNA synthesis in mammalian cells. *PLoS Biol* **4:** e309. doi:10.1371/journal.pbio.0040309

Ranjan A, Nguyen VQ, Liu S, Wisniewski J, Kim JM, Tang X, Mizuguchi G, Elalaoui E, Nickels TJ, Jou V, et al. 2020.

Live-cell single particle imaging reveals the role of RNA polymerase II in histone H2A.Z eviction. *eLife* **9:** e55667. doi:10.7554/eLife.55667

Revyakin A, Zhang Z, Coleman RA, Li Y, Inouye C, Lucas JK, Park SR, Chu S, Tjian R. 2012. Transcription initiation by human RNA polymerase II visualized at single-molecule resolution. *Genes Dev* **26:** 1691–1702. doi:10.1101/gad.194936.112

Rodriguez J, Larson DR. 2020. Transcription in living cells: molecular mechanisms of bursting. *Ann Rev Biochem* **89:** 189–212. doi:10.1146/annurev-biochem-011520-105250

Roeder RG. 2019. 50+ years of eukaryotic transcription: an expanding universe of factors and mechanisms. *Nat Struct Mol Biol* **26:** 783–791. doi:10.1038/s41594-019-0287-x

Saldi T, Cortazar MA, Sheridan RM, Bentley DL. 2016. Coupling of RNA polymerase II transcription elongation with pre-mRNA splicing. *J Mol Biol* **428:** 2623–2635. doi:10.1016/j.jmb.2016.04.017

Sandberg R, Neilson JR, Sarma A, Sharp PA, Burge CB. 2008. Proliferating cells express mRNAs with shortened 3′ untranslated regions and fewer microRNA target sites. *Science* **320:** 1643–1647. doi:10.1126/science.1155390

Sato Y, Hilbert L, Oda H, Wan Y, Heddleston JM, Chew TL, Zaburdaev V, Keller P, Lionnet T, Vastenhouw N, et al. 2019. Histone H3K27 acetylation precedes active transcription during zebrafish zygotic genome activation as revealed by live-cell analysis. *Development* **146:** dev179127. doi:10.1242/dev.179127

Sato H, Das S, Singer RH, Vera M. 2020. Imaging of DNA and RNA in living eukaryotic cells to reveal spatiotemporal dynamics of gene expression. *Ann Rev Biochem* **89:** 159–187. doi:10.1146/annurev-biochem-011520-104955

Schmidt U, Basyuk E, Robert MC, Yoshida M, Villemin JP, Auboeuf D, Aitken S, Bertrand E. 2011. Real-time imaging of cotranscriptional splicing reveals a kinetic model that reduces noise: implications for alternative splicing regulation. *J Cell Biol* **193:** 819–829. doi:10.1083/jcb.201009012

Shah S, Takei Y, Zhou W, Lubeck E, Yun J, Eng CHL, Koulena N, Cronin C, Karp C, Liaw EJ, et al. 2018. Dynamics and spatial genomics of the nascent transcriptome by intron seqFISH. *Cell* **174:** 363–376.e16. doi:10.1016/j.cell.2018.05.035

Shav-Tal Y, Singer RH, Darzacq X. 2004. Imaging gene expression in single living cells. *Nat Rev Mol Cell Biol* **5:** 855–862. doi:10.1038/nrm1494

Sheinberger J, Hochberg H, Lavi E, Kanter I, Avivi S, Reinitz G, Schwed A, Aizler Y, Varon E, Kinor N, et al. 2017. CD-tagging-MS2: detecting allelic expression of endogenous mRNAs and their protein products in single cells. *Biol Methods Protoc* **2:** bpx004. doi:10.1093/biomethods/bpx004

Shepard PJ, Choi EA, Lu J, Flanagan LA, Hertel KJ, Shi Y. 2011. Complex and dynamic landscape of RNA polyadenylation revealed by PAS-Seq. *RNA* **17:** 761–772. doi:10.1261/rna.2581711

Siebrasse JP, Veith R, Dobay A, Leonhardt H, Daneholt B, Kubitscheck U. 2008. Discontinuous movement of mRNP particles in nucleoplasmic regions devoid of chromatin. *Proc Natl Acad Sci* **105:** 20291–20296. doi:10.1073/pnas.0810692105

Siebrasse JP, Kaminski T, Kubitscheck U. 2012. Nuclear export of single native mRNA molecules observed by light sheet fluorescence microscopy. *Proc Natl Acad Sci* **109:** 9426–9431. doi:10.1073/pnas.1201781109

Snapp E. 2005. Design and use of fluorescent fusion proteins in cell biology. *Curr Protoc Cell Biol* **27:** 21.4.1–21.4.13.

Spille H, Hecht M, Grube V, Cho K, Lee C, Cissé II. 2019. A CRISPR/Cas9 platform for MS2-labelling of single mRNA in live stem cells. *Methods* **153:** 35–45. doi:10.1016/j.ymeth.2018.09.004

Stasevich TJ, Hayashi-Takanaka Y, Sato Y, Maehara K, Ohkawa Y, Sakata-Sogawa K, Tokunaga M, Nagase T, Nozaki N, McNally JG, et al. 2014. Regulation of RNA polymerase II activation by histone acetylation in single living cells. *Nature* **516:** 272–275. doi:10.1038/nature13714

Sun X, Alzhanova-Ericsson AT, Visa N, Aissouni Y, Zhao J, Daneholt B. 1998. The hrp23 protein in the Balbiani ring pre-mRNP particles is released just before or at the binding of the particles to the nuclear pore complex. *J Cell Biol* **142:** 1181–1193. doi:10.1083/jcb.142.5.1181

Tanenbaum ME, Gilbert LA, Qi LS, Weissman JS, Vale RD. 2014. A protein-tagging system for signal amplification in gene expression and fluorescence imaging. *Cell* **159:** 635–646. doi:10.1016/j.cell.2014.09.039

Tantale K, Mueller F, Kozulic-Pirher A, Lesne A, Victor JM, Robert MC, Capozi S, Chouaib R, Bäcker V, Mateos-Langerak J, et al. 2016. A single-molecule view of transcription reveals convoys of RNA polymerases and multi-scale bursting. *Nat Commun* **7:** 12248. doi:10.1038/ncomms12248

Teves SS, An L, Hansen AS, Xie L, Darzacq X, Tjian R. 2016. A dynamic mode of mitotic bookmarking by transcription factors. *eLife* **5:** e22280. doi:10.7554/eLife.22280

Teves SS, An L, Bhargava-Shah A, Xie L, Darzacq X, Tjian R. 2018. A stable mode of bookmarking by TBP recruits RNA polymerase II to mitotic chromosomes. *eLife* **7:** e35621. doi:10.7554/eLife.35621

Tramantano M, Sun L, Au C, Labuz D, Liu Z, Chou M, Shen C, Luk E. 2016. Constitutive turnover of histone H2A.Z at yeast promoters requires the preinitiation complex. *eLife* **5:** e14243. doi:10.7554/eLife.14243

Trcek T, Larson DR, Moldón A, Query CC, Singer RH. 2011. Single-molecule mRNA decay measurements reveal promoter-regulated mRNA stability in yeast. *Cell* **147:** 1484–1497. doi:10.1016/j.cell.2011.11.051

Tsai A, Muthusamy AK, Alves MR, Lavis LD, Singer RH, Stern DL, Crocker J. 2017. Nuclear microenvironments modulate transcription from low-affinity enhancers. *eLife* **6:** e28975. doi:10.7554/eLife.28975

Tsukamoto T, Hashiguchi N, Janicki SM, Tumbar T, Belmont AS, Spector DL. 2000. Visualization of gene activity in living cells. *Nat Cell Biol* **2:** 871–878. doi:10.1038/35046510

Tumbar T, Sudlow G, Belmont AS. 1999. Large-scale chromatin unfolding and remodeling induced by VP16 acidic activation domain. *J Cell Biol* **145:** 1341–1354. doi:10.1083/jcb.145.7.1341

Tunnacliffe E, Chubb JR. 2020. What is a transcriptional burst? *Trends Genet* **36:** 288–297. doi:10.1016/j.tig.2020.01.003

Tutucci E, Livingston NM, Singer RH, Wu B. 2018a. Imaging mRNA in vivo, from birth to death. *Annu Rev Biophys* **47:** 85–106. doi:10.1146/annurev-biophys-070317-033037

Tutucci E, Vera M, Biswas J, Garcia J, Parker R, Singer RH. 2018b. An improved MS2 system for accurate reporting of the mRNA life cycle. *Nat Methods* **15:** 81–89. doi:10.1038/nmeth.4502

Ulitsky I, Shkumatava A, Jan CH, Subtelny AO, Koppstein D, Bell GW, Sive H, Bartel DP. 2012. Extensive alternative polyadenylation during zebrafish development. *Genome Res* **22:** 2054–2066. doi:10.1101/gr.139733.112

Vargas DY, Raj A, Marras SAE, Kramer FR, Tyagi S. 2005. Mechanism of mRNA transport in the nucleus. *Proc Natl Acad Sci* **102:** 17008–17013. doi:10.1073/pnas.0505580102

Vargas DY, Shah K, Batish M, Levandoski M, Sinha S, Marras SAE, Schedl P, Tyagi S. 2011. Single-molecule imaging of transcriptionally coupled and uncoupled splicing. *Cell* **147:** 1054–1065. doi:10.1016/j.cell.2011.10.024

Vera M, Pani B, Griffiths LA, Muchardt C, Abbott CM, Singer RH, Nudler E. 2014. The translation elongation factor eEF1A1 couples transcription to translation during heat shock response. *eLife* **3:** e03164. doi:10.7554/eLife.03164

Vera M, Biswas J, Senecal A, Singer RH, Park HY. 2016. Single-cell and single-molecule analysis of gene expression regulation. *Ann Rev Genet* **50:** 267–291. doi:10.1146/annurev-genet-120215-034854

Visa N, Alzhanova-Ericsson AT, Sun X, Kiseleva E, Björkroth B, Wurtz T, Daneholt B. 1996a. A pre-mRNA-binding protein accompanies the RNA from the gene through the nuclear pores and into polysomes. *Cell* **84:** 253–264. doi:10.1016/S0092-8674(00)80980-0

Visa N, Izaurralde E, Ferreira J, Daneholt B, Mattaj IW. 1996b. A nuclear cap-binding complex binds Balbiani ring pre-mRNA cotranscriptionally and accompanies the ribonucleoprotein particle during nuclear export. *J Cell Biol* **133:** 5–14. doi:10.1083/jcb.133.1.5

Waks Z, Klein AM, Silver PA. 2011. Cell-to-cell variability of alternative RNA splicing. *Mol Syst Biol* **7:** 506. doi:10.1038/msb.2011.32

Wan Y, Anastasakis DG, Rodriguez J, Palangat M, Gudla P, Zaki G, Tandon M, Pegoraro G, Chow CC, Hafner M, et al. 2019. *Dynamic imaging of nascent RNA reveals general principles of transcription dynamics and stochastic splice site selection.* Social Science Research Network, Rochester, NY. papers.ssrn.com/abstract=3467157

Wang S, Su JH, Beliveau BJ, Bintu B, Moffitt JR, Wu C, Zhuang X. 2016a. Spatial organization of chromatin domains and compartments in single chromosomes. *Science* **353:** 598–602. doi:10.1126/science.aaf8084

Wang S, Su JH, Zhang F, Zhuang X. 2016b. An RNA-aptamer-based two-color CRISPR labeling system. *Sci Rep* **6:** 26857. doi:10.1038/srep26857

Wang R, Nambiar R, Zheng D, Tian B. 2018. Polya_DB 3 catalogs cleavage and polyadenylation sites identified by deep sequencing in multiple genomes. *Nucleic Acids Res* **46:** D315–D319. doi:10.1093/nar/gkx1000

Weber CM, Henikoff S. 2014. Histone variants: dynamic punctuation in transcription. *Genes Dev* **28:** 672–682. doi:10.1101/gad.238873.114

Wilkinson ME, Charenton C, Nagai K. 2020. RNA splicing by the spliceosome. *Annu Rev Biochem* **89:** 359–388. doi:10.1146/annurev-biochem-091719-064225

Will CL, Lührmann R. 2011. Spliceosome structure and function. *Cold Spring Harb Perspect Biol* **3:** a003707.

Wu B, Chao JA, Singer RH. 2012. Fluorescence fluctuation spectroscopy enables quantitative imaging of single mRNAs in living cells. *Biophys J* **102:** 2936–2944. doi:10.1016/j.bpj.2012.05.017

Wu B, Buxbaum AR, Katz ZB, Yoon YJ, Singer RH. 2015. Quantifying protein-mRNA interactions in single live cells. *Cell* **162:** 211–220. doi:10.1016/j.cell.2015.05.054

Yao J, Munson KM, Webb WW, Lis JT. 2006. Dynamics of heat shock factor association with native gene loci in living cells. *Nature* **442:** 1050–1053. doi:10.1038/nature05025

Zhang G, Taneja KL, Singer RH, Green MR. 1994. Localization of pre-mRNA splicing in mammalian nuclei. *Nature* **372:** 809–812. doi:10.1038/372809a0

Zhang Z, English BP, Grimm JB, Kazane SA, Hu W, Tsai A, Inouye C, You C, Piehler J, Schultz PG, et al. 2016. Rapid dynamics of general transcription factor TFIIB binding during preinitiation complex assembly revealed by single-molecule analysis. *Genes Dev* **30:** 2106–2118. doi:10.1101/gad.285395.116

Zhao J, Jin SB, Björkroth B, Wieslander L, Daneholt B. 2002. The mRNA export factor Dbp5 is associated with Balbiani ring mRNP from gene to cytoplasm. *EMBO J* **21:** 1177–1187. doi:10.1093/emboj/21.5.1177

Zid BM, O'Shea EK. 2014. Promoter sequences direct cytoplasmic localization and translation of mRNAs during starvation in yeast. *Nature* **514:** 117–121. doi:10.1038/nature13578

Chromatin Mechanisms Driving Cancer

Berkley Gryder,[1,2] Peter C. Scacheri,[1] Thomas Ried,[2] and Javed Khan[2]

[1]Department of Genetics and Genome Sciences, Case Western Reserve University, Cleveland, Ohio 44106, USA

[2]Genetics Branch, Center for Cancer Research, National Cancer Institute, Bethesda, Maryland 20892, USA

Correspondence: Berkley.gryder@case.edu; riedt@mail.nih.gov; khanjav@mail.nih.gov

The change in cell state from normal to malignant is driven fundamentally by oncogenic mutations in cooperation with epigenetic alterations of chromatin. These alterations in chromatin can be a consequence of environmental stressors or germline and/or somatic mutations that directly alter the structure of chromatin machinery proteins, their levels, or their regulatory function. These changes can result in an inability of the cell to differentiate along a predefined lineage path, or drive a hyperactive, highly proliferative state with addiction to high levels of transcriptional output. We discuss how these genetic alterations hijack the chromatin machinery for the oncogenic process to reveal unique vulnerabilities and novel targets for cancer therapy.

Cancer is a panoply of diseases caused by a combination of inherited (germline) and newly acquired (somatic) genetic mutations that converge on chromatin, which results in dysregulation of gene expression leading to abnormal cell growth. Oncogenic mutations that directly influence chromatin biology can be divided into two major categories: (1) chromatin machinery errors that comprise mutations, translocations, or copy number alterations effecting chromatin machinery genes, and (2) regulatory element errors that include genomic alterations in noncoding elements that change how chromatin regulates transcription. Similar to thermodynamic entropy, genetic replication errors accumulate with cellular age (Tomasetti and Vogelstein 2015), such that cancer occurs proportionally to the rate and number of cell divisions in the tissue of origin. This principally explains why age is the strongest risk factor for cancer overall (National Cancer Institute, Surveillance, Epidemiology, and End Results Program, www.seer.cancer.gov). Rising disorder in genetic information, both coding and noncoding errors, are entropic to intra- and intercellular communication circuits that mediate through the epigenome (Gryder et al. 2013; Nijman 2020).

Genetic errors in chromatin-related processes include gain-of-function (GOF), loss-of-function (LOF), or swap-of-function (SOF) with simultaneous loss and gain (Shen and Vakoc 2015). Both small mutations (single-nucleotide variants [SNVs] or insertions or deletions [indels]) and large structural rearrangements effect both the chromatin genes and regulatory elements in cancer; therefore, the combination of small mutation or structural alterations affecting either chromatin machinery or regulatory elements generates four distinct types of major chromatin-based cancer-driving events (Fig. 1). These categories are often co-occurrent, as in the case of extrachromosomal circular DNA amplifications and aneuploidy effecting whole chro-

Cite this article as *Cold Spring Harb Perspect Biol* doi: 10.1101/cshperspect.a040956

mosome gains and losses where they can alter both the chromatin machinery gene copy number along with their regulatory elements.

In this review, we focus on the mechanisms directly impinging upon chromatin machinery malfunction and gene misregulation as well as their effect on higher-order topological changes that alter the behavior of a cell (Franke et al. 2016; Ochs et al. 2019; Johnstone et al. 2020). We will not address cancer-causing mutations that influence chromatin indirectly (e.g., mutant RAS creating aberrant superenhancers [SEs] in the nucleus via signaling through MEK/ERK hyperphosphorylation) as, arguably, all oncogenic

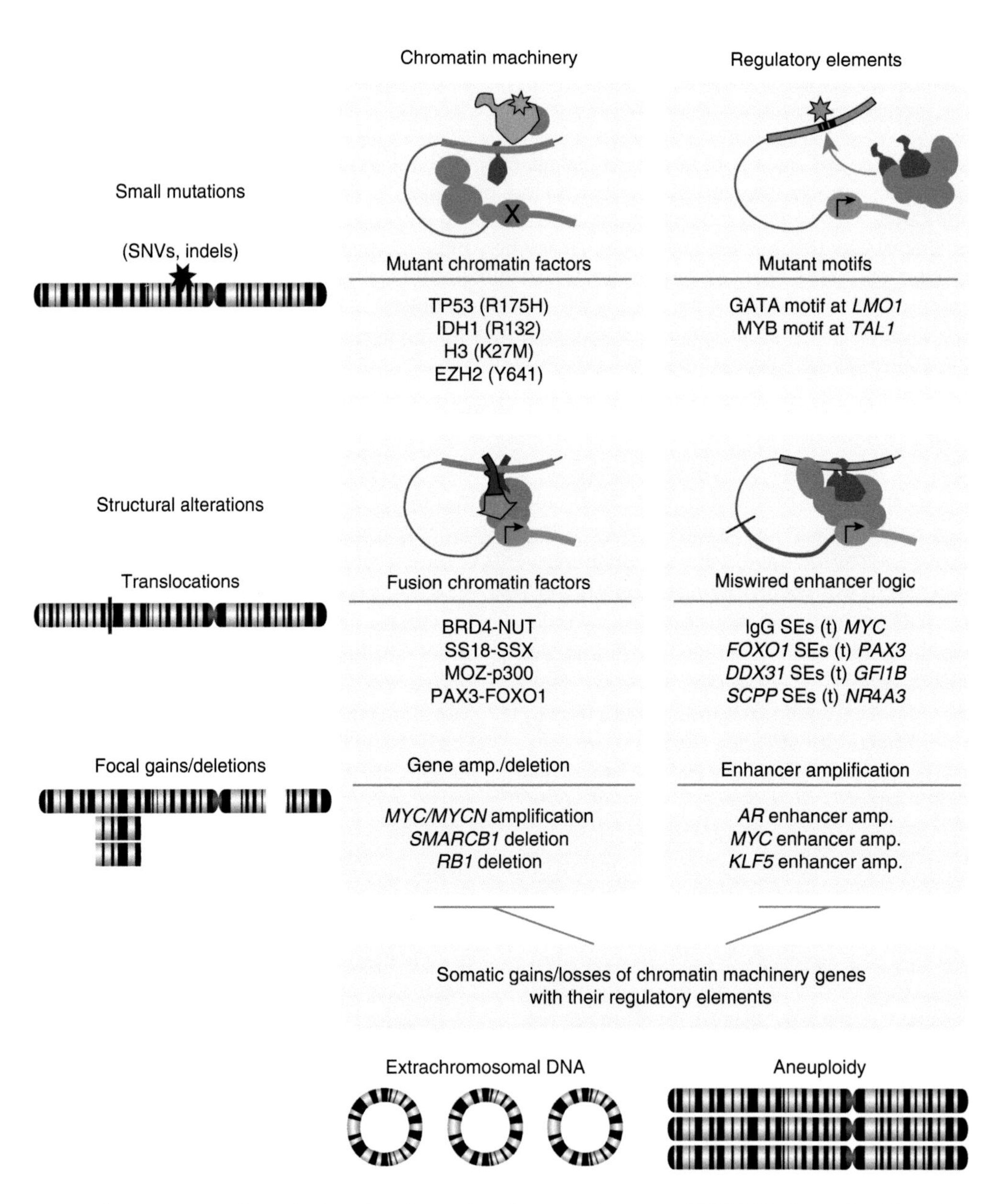

Figure 1. Categorization of genetic alterations impacting chromatin in cancer. (SNVs) Single-nucleotide variants, (SEs) superenhancers.

mechanisms must ultimately impinge on chromatin regulation to produce and maintain their malignant state.

THE CHROMATIN MACHINERY

Broadly, any protein that alters the 3D structure of DNA and its function can be considered part of the chromatin machinery. Of central importance to chromatin are the histones that assemble into an octameric bead with a structured core that DNA wraps around twice (147 bp/octamer). This fundamental unit of chromatin is termed the nucleosome. The typical histone octamer contains two each of H2A, H2B, H3, and H4. Serving as a stabilizing linker of the bead is the linker histone, H1, which is not a member of the core octamer (Fig. 2). One function of the nucleosome core particle is to block RNA polymerase from spuriously transcribing (Kornberg and Lorch 2020). Each nucleosome has eight disordered tails that are subject to diverse posttranslational modifications to alter their biochemical properties (Fig. 2A; Strahl and Allis 2000).

Gene expression is also controlled directly by DNA methylation of the nucleotide cytosine to 5-methylcytosine (5mC) or 5-hydroxymethylcytosine (5hmC). Methylated DNA has altered stiffness and an impaired ability to be recognized by transcription factors (TFs), effecting gene regulation significantly (Hörberg and Reymer 2018). DNA methyltransferases convert cytosine to 5mC, TET proteins convert to 5hmC, and TDG proteins convert cytosine back to its unmodified form (Fig. 2B). Hypermethylation of cytosine in CpG islands near gene promoters is a primary gene-silencing mechanism (Naveh-Many and Cedar 1981).

Some histone tail modifications block transcription, for example, by methylation of histone tail lysines in regions of densely packed heterochromatin, while others such as acetylation of histone tail lysines, loosen the histone's grip on DNA to enable nucleosome eviction, DNA accessibility, RNA polymerase movement, and transcription in regions termed euchromatin (Li et al. 2007). Histone modifications are controlled by "writers" and "erasers" (Schreiber and Bernstein 2002). Appropriately modified nucleosomes serve as binding sites for "reader" domains through which many proteins, and some large complexes, can engage the nucleo-

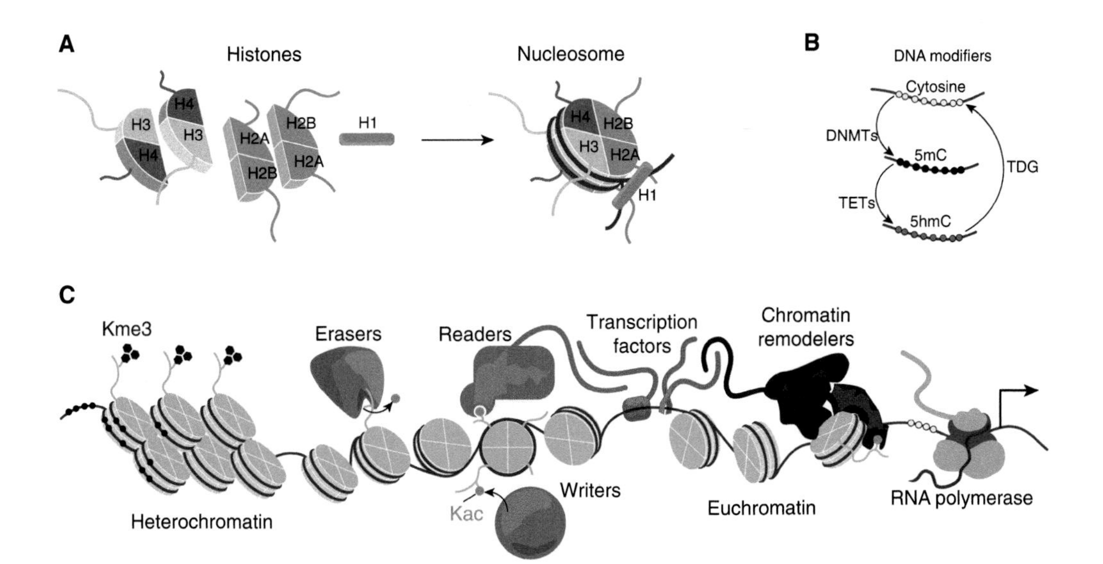

Figure 2. Chromatin machinery including (*A*) histone assembly around 147 bp of DNA to form the nucleosome, (*B*) the DNA modification cycle to methylate cytosine nucleotides, and (*C*) the variety of chromatin machinery including readers, writers, and erasers of posttranslational marks, DNA-binding transcription factors, chromatin remodeling complexes, and RNA polymerases.

some (Fig. 2C). Many such complexes such as BAF (also known as mSWI/SNF) possess chromatin remodeling activity through ATPase-driven motors that can slide or evict nucleosomes to enhance access to DNA (Fig. 2C; Wilson and Roberts 2011; Kadoch 2019).

The location of histone modifications in relationship to the underlying DNA sequence is essential to faithful transcription and thus gives importance to sequence-specific DNA-binding proteins (i.e., TFs that direct histone modifiers and chromatin remodelers to specific positions in the epigenome) (Fig. 2C). There are >1500 TFs encoded in the human genome, and roughly 70 of them are ubiquitously expressed and play routine roles (such as CCCTC-binding factor [CTCF]), while the rest are restricted to and define specific cell types (Gerstein et al. 2012). TFs are further distinguished into nonmutually exclusive functional groups based on characteristics such as (1) ability to sense external signaling cascades, (2) ligand-binding capacity, (3) preference for recruitment of coactivator complexes or corepressor complexes, (4) ability to "pioneer" into closed chromatin, (5) downstream gene pathway regulation, (6) preference for promoters or distal regulatory elements, and (7) the structural features they enable (Wingender et al. 2013).

MUTATIONS THAT ALTER CHROMATIN MACHINERY

Remarkably, cancerous aberrations are found for every class of the chromatin machinery, including histones, histone modifiers, chromatin remodelers, TFs, and DNA modifiers. Both small mutations and large-scale genetic alterations have been identified in cancer that perturb these proteins, derailing the healthy checks and balances of the human epigenome and that can be targeted for therapy (Jones et al. 2016).

Small-Scale Mutations Altering Chromatin Protein Function

Mutations in Histones

A growing number of histone mutations have been cataloged in cancer, and are collectively termed "oncohistones" (Nacev et al. 2019). Histone H3 has the longest disordered tail, is subject both to the highest number of writing/reading/erasing modifications, and harbors mutations that prevent proper posttranslational modification in various cancer contexts. For example, 78% of diffuse intrinsic pontine gliomas (DIPGs) have the oncohistone mutation H3K27M, and H3K27 is a key lysine residue that is acetylated in active euchromatin and methylated in heterochromatin (Schwartzentruber et al. 2012). H3.3 (histone 3 variant 3) is mutated to methionine at lysine 36 (H3.3 K36M) in 95% of chondroblastomas and H3.3 is mutated at glycine 34 (H3.3 G34W/L) in 92% of giant cell tumors of bone. The exact mechanisms by which these mutations in the histone tails impact gene regulation are not completely resolved for each mutant, but studies on H3K27M suggest that this amino acid substitution leads to an inability of PRC2 (Polycomb Repressive Complex 2) to maintain the repressive H3K27me3 status, globally impairing transcriptional silencing in DIPG (Lewis et al. 2013).

Whereas the above mutations occur in the disordered histone tail, many newly reported mutations accumulate in the acidic patch of the nucleosome and are found in all four core histone particles (Ghiraldini et al. 2021). Because the acidic patch at the surface of the nucleosome binds the H-tail binding from neighboring nucleosomes, it is thought that mutations in the acidic patch prevents higher-order chromatin compaction. The histone acidic patch, which is negatively charged, is also a key docking site for positively charged amino acid modules in many chromatin-associated proteins such as the mSWI/SNF complex (BAF), and acidic patch mutants may result in the loss of these interactions (He et al. 2020), leading to gene expression dysregulation. In all, the list of histone variants and their functional roles in cancer is growing and includes *H2AFZ*, *H2AFV*, *H2AFY*, *H2AFY2*, *H3F3A*, and *H3F3B*, across a diverse set of human tumors (Ghiraldini et al. 2021).

Mutations in Transcription Factors

The best-established mutations identified in any TF are those in the "guardian of the genome," the *TP53* gene, encoding the p53 protein (Darnell 2002). *TP53* is the most commonly mutated gene in cancer, and although these mutations lead to loss of normal tumor suppressor activity, hotspot nonsynonymous mutations frequently result in GOF through aberrant binding across the genome and increased transcription of *MLL1*, *MLL2* (both histone methyltransferases), and *MOZ* (a lysine acetyltransferase) through binding to *ETS2* (ETS family of TFs). Increased MLL and MOZ expression leads to elevated methylation and acetylation of histones, respectively, which results in alteration of chromatin structure and gene expression, and ultimately to increased proliferation of cancer cells (Zhu et al. 2015). Other TFs that play dominant roles in cancer do so without becoming themselves mutant, such as MYC. One exception is the ultra-rare but lethal spindle cell and sclerosing rhabdomyosarcoma (RMS) (Agaram et al. 2019), where *MYOD1* is mutated such that the amino acid sequence of its DNA-binding domain becomes identical to MYC (Kohsaka et al. 2014).

Although rare in other cancers, within the context of prostate cancer, mutations in the TF androgen receptor (AR) do not result in LOF of DNA binding, nor are they tumor-initiating (Taplin et al. 1995; Armenia et al. 2018). AR mutants arise predominantly in the ligand-binding domain, where testosterone binds to activate AR's recruitment to chromatin, and substitute amino acids to alter binding to antiandrogens, such as bicalutamide or enzalutamide, thereby evading the therapeutic pressure of androgen deprivation, and sometimes creating tumors that depend on anti-AR ligands for growth even when AR mutants cause an antagonist-to-agonist functional switch (Dai et al. 2017).

Mutations in Chromatin Remodeling Complexes

It is estimated that >20% of human cancers harbor mutations in one of the many subunits of the mammalian SWI/SNF (also known as BAF) complex (Kadoch et al. 2013). Whereas other chromatin remodelers are mutated in cancer (CHD, ISWI, INO80, others), these are numerically dwarfed by the high frequency of BAF mutants (Valencia and Kadoch 2019). Molecular, epigenomic, and biochemical studies have been used to dissect the mechanisms by which the growing catalog of the BAF alterations in cancer are causing gene misregulation (Kadoch and Crabtree 2013; Michel et al. 2018; Valencia et al. 2019; Centore et al. 2020; Mashtalir et al. 2020; Shi et al. 2020). One important emerging theme is that because BAF plays diverse roles in a lineage-specific fashion, mutational studies are finding that certain cancers are associated with mutations of specific subunits in this multimodal complex. For example, mutations in *SMARCB1* disable the BAF complex from properly binding to the acidic patch of the core nucleosome particle, resulting in diminished enhancer DNA accessibility (Valencia et al. 2019). These mutations are likely pathogenic in multiple cancers including meningioma, adenocarcinoma, schwannoma, and others (Kadoch and Crabtree 2015; Collord et al. 2018; Pereira et al. 2019; Schaefer and Hornick 2021). Solid tumors are often defined by BAF mutations with 100% of patients harboring the hallmark genetic lesion, such as fusion of BAF subunits SSX with SS18 in synovial sarcoma (also see below) (Kadoch and Crabtree 2013).

Mutations in Genes Effecting DNA Methylation and 3D Genome Architecture

Methylation of CpGs have been shown to eliminate CTCF-binding sites inhibiting its insulator function, which can lead to profound changes in gene expression by modulating enhancer access to the gene promoter through regulation of an enhancer boundary (Bell and Felsenfeld 2000). The ways human cancers are altering gene regulation through DNA methylation mechanisms is likely more complex than currently appreciated, but some clear examples have been elucidated in tumors harboring deficiencies in DNA methylation-related enzymes. For example, genes encoding for isocitrate dehydrogenases 1 and 2, IDH1 and IDH2, are mutated in several cancers (Parsons et al. 2008; Schnittger et al. 2010; Yang et al. 2012). Mutations in these genes results in a loss of activity of the enzymatic function with lower production of α-ketoglutarate (α-KG), combined with a GOF, causing an increase in 2-hydroxyglutarate (2-HG) levels. 2-HG competes with α-KG and competitively suppress the activity of α-KG-dependent dioxygenases, including both lysine histone demethylases (KDMs) and the ten-eleven translocation (TET) family of DNA hydroxylases (Fig. 2B; Yang et al. 2012). This has been shown to result in hypermethylation of CTCF-binding sites, with loss of insulation around the locus of driver oncogenes such as the PDGFRA gene, resulting in high expression and kinase activity in glioma (Flavahan et al. 2016). Similarly, SDH-deficient gastrointestinal stromal tumors (GISTs) also exhibit the hypermethylation phenotype, resulting in a lineage-specific enhancer acting out-of-bounds, in this case driving up oncogenic levels of different kinases, *FGF4* at one locus and *KIT* at another (Flavahan et al. 2019). The emerging theme here is that lineage-specific TFs driving SEs can become tumorigenic when global hypermethylation removes the topological guardrails on chromatin folding. This is an elegant example of an activation of a known glioma oncogene through loss of a topologically associated domain (TAD) boundary function through aberrant methylation as a result of a single-nucleotide somatic variant.

Mutations in Chromatin Readers, Writers, and Erasers

Another sizable class of cancer-causing mutations arises in proteins that read, write, and erase histone modifications. The result is a mis-reading, mis-writing, mis-erasing, and, ultimately, a mis-interpretation of epigenomic information (Chi et al. 2010).

Mutant Epi-Writers

Histone modifications are generally divided into active marks such as H3K27ac, and gene repressive marks such as H3K27me3. Because of this, generally the enzymes that write marks associated with active chromatin are considered activating epi-writers (i.e., CPB/p300, which catalyze H3K27 → H3K27ac) or repressive epi-writers (i.e., Enhancer of Zeste 2 Polycomb Repressive Complex 2 Subunit [EZH2], which catalyzes H3K27 → H3K27me3). The repressive epi-writer EZH2 is part of the Polycomb Repressive Complex 2 (PRC2) and is recurrently mutated in 21% of diffuse large B-cell lymphomas (Morin et al. 2010). Most commonly, the mutation occurs at Y641 that resides in the SET domain, which is the enzymatic pocket responsible for adding methylation to lysine substrates (McCabe et al. 2012). Biochemical studies revealed that lymphomas with this mutation experience a GOF effect on PRC2 and global increases in H3K27me3 levels, resulting in widespread repression of gene expression, blocking B-cell differentiation (Shen and Vakoc 2015). The path to cancer is not one sided for the PRC2 complex, as LOF can also cause cancer; for example, deletion of *EZH2* or another PRC2 subunit, *SUZ12*, is frequent in malignant peripheral nerve sheath tumors (MPNSTs), resulting in complete loss of H3K27me3 and unchecked proliferation (De Raedt et al. 2014). Thus, PRC2 can contribute to cancer as either a gained mutant oncogene, or a lost tumor suppressor, a duality that no doubt reflects different roles of PRC2 in varied cancer cells of origin (Piunti and Shilatifard 2016).

Mutant Epi-Readers

MLLT1 super elongation complex (SEC) subunit (protein ENL) is a chromatin reader of the SEC that recognizes histone acetylation using its YEATS domain. The SEC suppresses transient pausing of the RNA polymerase II, increasing transcription (Lin et al. 2010). Approximately 5% of patients with Wilms tumors have GOF mutations clustered in the Yaf9, ENL, AF9, Taf14, Sas5 (YEATS) reader domain, which results in increased expression of genes that favor a premalignant cell fate (Wan et al. 2020).

Mutant Epi-Erasers

Among the epigenetic erasers, it is a curious fact that, very rarely, are the erasers of histone acetylation (HDACs) mutated, whereas mutations of erasers of lysine methylation (KDM proteins) are found much more commonly (Dawson 2017; Donaldson-Collier et al. 2019; Han et al.

2019; Wang and Shilatifard 2019). KDM mutants are thought to cause methylation imbalances that provide a more promiscuous epigenome with increased access to genetic programs (Flavahan et al. 2017), and our perspective on the negative data suggesting a lack of commonly observed HDAC mutants is that these enzymes are under more acute and purifying negative selection pressure, preventing their accumulation in cancer, because they are essential to global gene transcription (Gryder et al. 2012, 2019c).

Large-Scale Alterations that Alter Chromatin Machinery

Some cancers, for example sarcomas and pediatric cancers, have a low somatic mutational burden of <0.5 per megabase despite their aggressive nature (Pugh et al. 2013; Brohl et al. 2014; Shern et al. 2014). Many of these cancers with simple genomes are driven by translocation events with the production of chimeric or fusion oncogenes (Aplan et al. 2021). Chromosome translocation within the cancer genome can lead to the juxtaposition of two regions that would ordinarily not be collocated leading to oncogenic transformation through profound alterations of the chromatin landscape (Kadoch and Crabtree 2013; Boulay et al. 2017; Gryder et al. 2017, 2019a). This could be through the generation of novel in-frame oncogenic chimeric protein and/or the relocation of an intact gene to *cis* sites near active promoter/enhancer elements of a partner gene leading to increased or aberrant expression of the intact gene (Battey et al. 1983; Kadoch and Crabtree 2013; Boulay et al. 2017; Gryder et al. 2017, 2019a; Lancho and Herranz 2018). Although fusion events can activate kinase genes such as BCR-ABL fusions in CML or ALL or ALK fusions in solid tumors, which ultimately lead to transcriptional dysregulation through their downstream effects on chromatin content and architecture, it will not be discussed further (Salesse and Verfaillie 2002; Nambiar et al. 2008).

Chimeric TFs often have an ordered DNA-binding domain of a protein that anchors it to specific regions of the genomes, with a partner protein that is intrinsically disordered leading to multimerization and physiological liquid–liquid phase separation. This results in the coalescence of the transcription machinery around the target region, which has been described to lead to subsequent opening of chromatin and alteration in the expression of target genes in disease states (Couthouis et al. 2011, 2012; Kwon et al. 2013; Schwartz et al. 2013). Examples of ordered DNA-binding domains of TFs with the disordered domains of proteins include the t(2;13) translocation leading to the generation of the PAX3-FOXO1 in fusion-positive RMS and the t(11;22) translocation leading to the generation of EWSR1-FLI1 in EWS (Fig. 3A,B).

In RMS, the t(2;13) fusion results in the expression of the PAX3-FOXO1 fusion gene that acts as a pioneering factor and activates a core regulatory TF circuitry locking the cells in a myoblastic state (Gryder et al. 2017, 2019a). In EWS, gene expression is altered through the recruitment of the BAF complex by the EWS-FLI1 fusion protein to tumor-specific enhancers through phase transitions by the prion-like regions in the intrinsically disordered domain of EWSR1 (Boulay et al. 2017). In synovial sarcoma, which is associated with t(X;18) translocations, the *SS18* gene on chromosome 18 joins on one of three synovial sarcoma X (*SSX*) genes (*SSX1*, *SSX2*, or rarely *SSX4*) on the X chromosome. In the SS18-SSX chimeric protein, almost the entire SS18 protein is retained except the last eight amino acids, which are replaced by the 78 amino acids from the carboxy terminal of SSX, a very disordered region (Fig. 3C). Recent work has shown that the oncogenic mechanism of this fusion is displacement of a tumor-suppressing subunit from the SWI/SNF chromatin-remodeling complex with subsequent altered expression of pluripotency genes (Kadoch and Crabtree 2013) through changes in the BAF complex (Fig. 3D).

In general, TFs appear to select for highly disordered fusion partner proteins (Fig. 3E), such that the combination of a highly structured domain (i.e., DNA-binding domain of a TF) and a disordered domain creates an enhanced capacity for transcriptional condensate forma-

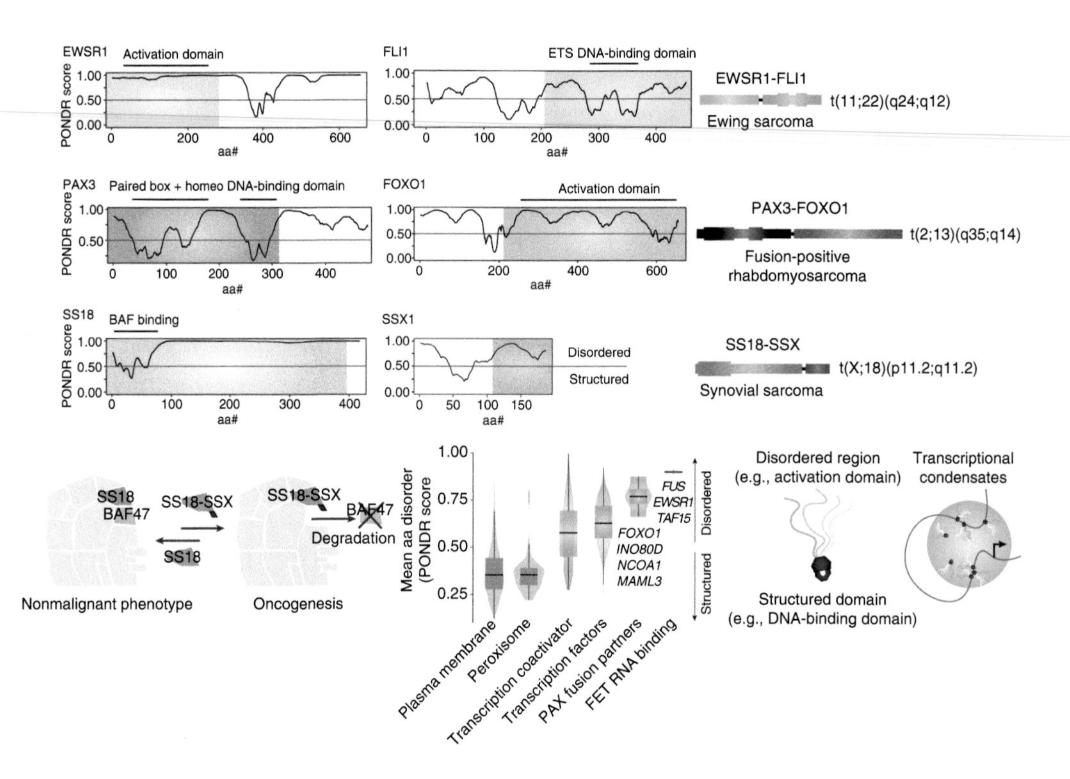

Figure 3. Fusion oncogenes involving intrinsically disordered chromatin proteins. (*A*) PONDR score of disorder by amino acid number for EWSR1 and FLI1, with the portions found in Ewing sarcoma fusion oncogene t(11;22)(q24;q12) highlighted in green and blue. (*B*) PONDR score of disorder by amino acid number for PAX3 and FOXO1, with the portions found in fusion oncogene-positive rhabdomyosarcoma t(2;13)(q35;q14) highlighted in red and brown. (*C*) PONDR score of disorder by amino acid number for SS18 and SSX1, with the portions fused in t(X;18)(p11.2;q11.2)-positive synovial sarcoma highlighted in orange and purple. (*D*) Model of BAF dysfunction in synovial sarcoma, where fusion oncogene SS18-SSX disrupts BAF47 incorporation and hinders proper assembly of the complete BAF complex. (Panel *D* adapted from data in Kadoch and Crabtree 2013.) (*E*) Violin box plot of mean PONDR scores among plasma membrane proteins, peroxisome proteins, transcription coactivators, transcription factors, PAX fusion partners, and FET RNA-binding proteins. (Panel *E* adapted from data in Gryder et al. 2020.) (*F*) Illustration of the cooperation between structured and disordered protein partners commonly found in chromatin-bound oncogenes, and their potential role as contributors to aberrant transcriptional condensates.

tion, thus accelerating transcription in an uncontrolled manner (Fig. 3F). It is hypothesized that the resultant changes of the chromatin composition at the target regions alter the epigenetic landscape of the cell that lock the cells in a blast- or stem-like state (Gryder et al. 2017, 2019a). These fusion TFs cause higher-order topological alteration, such as in PAX3-FOXO1-driven RMS where the fusion oncogene creates disease-specific TADs and H3K27ac-mediated promoter/enhancer loops in compartments ranging from 200 kb to 1.5 Mb in size (Gryder et al. 2019a).

Copy Number Gains and Losses of Chromatin Machinery Genes in Cancer

MYC/MYCN Gene Amplifications

The MYC family of TFs are a basic helix-loop-helix transcription-leucine zipper (bHLH-Zip) factor family whose members dimerize with MAX and bind to specific DNA sequences called E-boxes and has been referred to as the master regulator of the cancer epigenome and transcriptome. It is activated in up to 70% of human cancers by posttranslational modification or overexpressed through gain or amplification of DNA (Dang 2012). MYC recruits histone acetyltransferases (HATs) including lysine acetyltransferase 2A (KAT2A) and two related coactivators CBP/p300 that acetylate lysine residue on histone tails leading to a more open chromatin, promoting recruiting of the transcriptional machinery and increased expression. More broadly, MYC–MAX complexes interact directly or indirectly with chromatin-modifying cofactors including chromatin writers, readers, and erasers (Poole and van Riggelen 2017).

SMARCB1 Deletions

The SWI/SNF subunits have been found to be mutated in 20% of cancers. For example, the SMARCB1 subunit consisting of INI1/SNF5/BAF47 is inactivated in the large majority of rhabdoid tumors (Kim and Roberts 2014). Although SMARCB1, components of the BAF and PBAF complex, may contribute to alterations in canonical cancer pathways, these complexes bind to the promoters and enhancers of about one-third of all genes, and thus it is likely that LOF mutations such as deletions will lead to context-specific changes in pathways specific to that lineage (Nakayama et al. 2017). Loss of SMARCB1 leads to oncogenesis through enhanced SWI/SNF occupancy at SEs while disrupting SWI/SNF binding at typical enhancers that are needed for differentiation (Wang et al. 2017).

RB1 Deletions

The RB Transcriptional Corepressor 1 gene (RB1) is mutated or deleted in ~7% and the RB pathway disrupted in >30% of all cancers (Knudsen et al. 2020). RB blocks cell-cycle progression by repressing the E2F target gene transcription through the recruitment of transcriptional corepressors and/or chromatin remodeling protein factors, including HDAC and SWI/SNF, at promoter regions (Talluri and Dick 2012). RB also recruits Brm or BRG1 through their LXCXE domains, repressing gene expression and inducing cell-cycle arrest. RB also binds to histone methyltransferase and demethylases, and DNA methyltransferase 1 (DNMT1), which promotes the RB-dependent regulation of gene expression by changing the chromatin structure to a repressive form near the pRB-E2F target promoters (Talluri and Dick 2012). Thus, RB pathway disruption leads to loss of tumor-suppressive activity through epigenetic mechanisms.

ALTERATIONS OF *CIS*-REGULATORY ELEMENTS IN CANCER

The majority of the human genome is noncoding, and what was once largely considered to be "junk DNA" with little function has become the focus of intense study. The ENCODE project "wrote the eulogy" for the concept of junk DNA in 2018, proposing that as much as 80% of the noncoding space has the potential for biological function (Gerstein et al. 2012). Functions of noncoding DNA include (1) *cis* regulation of gene expression, driven primarily by transcription factor-binding sequences, (2) noncoding RNA transcripts, and (3) structural, architectural, and mechanical features of certain DNA sequences. In cancer, the two emerging mechanisms that strictly involve noncoding variation are small mutations that cause gain or loss of TF binding and large structural alterations that allow proto-oncogenes to gather a strong boost in transcription by enhancer hijacking.

Small Mutations Altering Transcription Factor Binding

It is striking that the majority of disease-associated mutations identified in genome-wide association studies (GWAS) are in noncoding regions (Edwards et al. 2013). While full functional validation of causality is sparse for the catalog of GWAS hits, there are a few examples where a direct link to cancer has been made. Yet, an early clue resides in the observation that 93% of single-nucleotide polymorphisms (SNPs) are found in a noncoding space, and these SNPs are found most frequently in enhancers, especially in clusters of interconnected enhancers with unusually high occupancy of chromatin machinery known as SEs (Whyte et al. 2013).

A prominent example of this mode of action was discovered in childhood neuroblastoma

(NB) in which the expression of GATA3, which is highly expressed and a core regulatory TF (Durbin et al. 2018b), is altered by a SNP rs2168101 (G > T). This alteration disrupts an SE by removing a GATA-binding site, thereby reducing LMO1 protein levels and conferring reduced risk of disease (Oldridge et al. 2015).

Whereas GWAS is designed largely to detect germline mutations that predispose to a disease state, a few examples of somatically acquired mutations have been described that recurrently alter TF binding to drive cancer. The first such example was found in melanoma, first as germline/familial mutation, then as a somatic mutation in sporadic melanoma, where a single SNP in the promoter of the *TERT* gene creates a binding site for ETS family TFs to increase TERT expression and drive the disease (Horn et al. 2013). A longer-range polymorphism was soon thereafter reported in T-cell acute lymphoblastic leukemias (T-ALLs) where a small insertion, 2–18 bp in length, was somatically acquired in some tumors, seeding a de novo SE 7500 bp distal to the *TAL1* gene (Mansour et al. 2014). The recurrent theme among insertions at this site is the addition of a sequence recognizable by the MYB TF, now recognized as a critical oncogenic TF driving this T-ALL via complex formation with TAL1 across that cancer's epigenome (Mansour et al. 2014).

How widespread are small variants that alter the *cis* regulome? Each cancer epigenome has upward of 20,000 enhancers, many of which are specific to the disease state (Akhtar-Zaidi et al. 2012; Morrow et al. 2018; Zhang et al. 2020), but most of these have not yet been linked to an obvious and recurrent causal SNP or INDEL. Yet, some recent studies indicate these events are in fact widespread, and often effect TFs that are involved in the etiology and lineage/origin of the cancer type. Studies in ovarian cancer have uncovered a diversity of somatic mutations in several distinct enhancers, all of which converge on the PAX8 pathway (Corona et al. 2020). A catalog of INDELS found in H3K27ac data revealed that, in a diverse collection of cancer types, many important oncogenes have enhancers with a somatically altered sequence, and those that have been investigated directly alter oncogene expression (Abraham et al. 2017). These kinds of alterations are invisible to exome or targeted sequencing efforts, including copy number assessment. Thus, in cancers with no obvious protein coding mutation, perhaps especially some childhood cancers, explanatory driver events could be sought by diligent investigation of noncoding variants.

Large Structural Alterations Causing Enhancer Hijacking

Structural alterations can not only result in the production of chimeric proteins, as discussed earlier, but also result in the juxtaposition of whole genes that are normally not expressed to regions near enhancers that are naturally active in the cell of origin of that cancer, a term known as enhancer hijacking (Weischenfeldt et al. 2017). This is particularly noted in B- or T-cell lymphoid malignancies where *MYC* translocate to *IG* or *TCR* regions that are highly active in the respective cell lineage or origin resulting in the hijacking of these enhancers and activation of *MYC* expression (Ryan et al. 2015). Another example is salivary gland acinic cell carcinoma (AciCC), which has a recurrent t(4;9)(q13;q31) translocation that brings the active enhancer regions from the SCPP gene cluster to the region upstream of Nuclear Receptor Subfamily 4 Group A Member 3 (NR4A3) at 9q31 (PMID: 30664630). Overexpression of NR4A3 target genes increases cell proliferation and is thought to be part of the oncogenic mechanism (Haller et al. 2019). A further example is in medulloblastoma where disparate genomic structural variants lead to activation of the growth factor independent 1 family proto-oncogenes, GFI1 and GFI1B (Northcott et al. 2014). In RMS, the t(2;13) fusion event, which results in the expression of PAX3-FOXO1, is driven by enhancer hijacking of the translocated FOXO1 SE to the PAX3 region resulting in continual expression of the fusion gene (Gryder et al. 2020).

Large Focal Structural Alterations of Enhancer Amplification

Noncoding amplification of regions containing enhancers or SEs have been shown to increase transcription of oncogenes including the control of *MYC* expression reported in lung adenocarcinoma and endometrial carcinoma (Zhang et al. 2016) and in castrate-resistant prostate cancer an amplification of an AR enhancer located 650 kb centromeric of AR leads to its overexpression (Takeda et al. 2018). Recently, somatic SE duplications and hotspot mutations have been shown to lead to oncogenic activation of the oncogenic KLF5 TF in squamous cell carcinomas (Zhang et al. 2018).

Extrachromosomal Focal Alterations Involving Oncogenes and Their Regulatory Elements

Genomic fragmentation events can circularize via non- or micro-homologous end joining and be propagated as extrachromosomal DNA (ecDNA) at massive copy number (10–200 copies per cell), which are typically 1–3 Mb in size (Verhaak et al. 2019). They autonomously replicate and are subject to non-Mendelian segregation during cell division. Also known as "double minutes," ecDNA was first observed in cancer cells in the 1960s (Mark 1967) and ecDNA amplifications are now estimated to occur in 14% of all human cancers (Kim et al. 2020). They are particularly prevalent in aggressive cancers notoriously difficult to treat, including glioblastoma, NB, and medulloblastoma (Kim et al. 2020). Numerous active enhancer elements undergo selection on ecDNA and engage the oncogene in a novel 3D conformation, mediating high expression of the oncogene (Fig. 4).

Functional studies using CRISPRi revealed that nearly every enhancer incorporated on circular ecDNA functionally contributes to proliferative fitness (Morton et al. 2019). Although double minutes have been observed for decades, it has not been appreciated that the regulatory DNA sequences in addition to oncogenes on double minutes are important. These findings, corroborated by multiple laboratories (Morton et al. 2019; Wu et al. 2019; Helmsauer et al. 2020), indicate that circular DNAs in tumor cells are not simply carriers of oncogenes. Rather, the circular DNA is a wholly functional entity with a unique topological configuration, free from highest-order chromosome-level 3D constraints, that drives tumor growth. Additionally, this finding shifts the 40-yr research paradigm that oncogenes on DNA amplifications are the sole drivers of cancer and points to a repertoire of enhancers as additional crucial contributors.

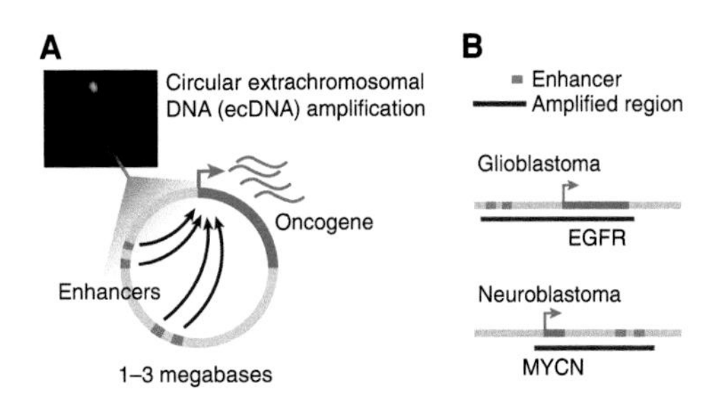

Figure 4. Extrachromosomal DNA includes (*A*) circularized megabase scale units that include enhancers with abnormal topology and a key oncogene. (*B*) Examples include extrachromosomal DNAs (ecDNAs) amplified with enhancers for oncogene EGFR in glioblastoma and MYCN in neuroblastoma.

Aneuploidy, Gene Expression, and Nuclear Structure in Cancer

Aneuploidy is defined as a state in which a cell contains abnormal numbers of chromosomes. Germline aneuploidy is a cause of diverse genetic syndromes, most commonly three copies, such as chromosome 21 trisomy in Down syndrome and miscarriages, while somatic aneuploidy is found in nearly 70% of cancers. Interestingly, somatic aneuploidy is unevenly observed across the spectrum of malignancies. At the low end of this aneuploid spectrum are the hematological malignancies, where most genetic alterations occur as chromosomal translocations. For instance, in chronic myeloid leukemia, we can invariably observe a translocation between chromosomes 9 and 22. The consequence of this translocation is the generation of a fusion protein that results in increased tyrosine kinase activity (Koretzky 2007). As discussed above, in Burkitt lymphoma one can frequently detect a translocation between chromosome 8 and 14. Here, the *MYC* oncogene is brought under the control of the promotor for the immunoglobulin locus, and is therefore transcribed more highly than under physiological conditions (Wiegant et al. 1991; Atallah-Yunes et al. 2020). Notably, both examples deregulate the expression of a single gene, which, in hematological cells, is sufficient for malignant transformation.

The situation in solid tumors, specifically in those of epithelial origin (i.e., carcinomas) is very different. These carcinomas are defined by chromosomal aneuploidies, which result in an imbalanced number of chromosomes (Heim and Mitelman 2009; Ried 2009). Other chromosomal aberrations observed commonly in solid tumors are unbalanced chromosomal translocations or so-called isochromosomes consisting of an aberrant chromosome that contains two identical arms of the parental chromosome, both mechanisms resulting in transcriptional deregulation of hundreds to thousands of genes.

These aneuploidies and the ensuing genomic imbalances are cancer-type specific; for instance, cervical carcinomas invariably carry extra copies of the long arm of chromosomes 1 and 3 (Heselmeyer et al. 1996). Colorectal carcinomas, on the other hand, frequently carry extra copies of chromosomes 7 and 13, the long arm of chromosome 8 and 20, along with losses of chromosome 18 (Ried et al. 1996). These changes occur in premalignant lesions; for instance, cervical intraepithelial dysplastic lesions that are ultimately prone to progress already have extra copies of chromosome 3q (Heselmeyer et al. 1996). In the colorectum, those polyps that show progression to invasive cancer carry extra copies of chromosome 7 (Fiedler et al. 2019). The reason for this tissue specificity is not entirely known; however, recent evidence suggests that the aneuploidies observed in the respective cancers hard-wire gene expression levels that are found in the cognate normal tissues (Patkar et al. 2021).

A fundamental question in cancer biology is how aneuploidies affect the expression levels of resident genes. In other words, do aneuploidies target only a few genes or do they affect gene expression levels of all genes on the aneuploid chromosomes. This question can now be answered with the advent of global gene expression analyses based on array-based technologies or by RNA sequencing (Upender et al. 2004; Wangsa et al. 2018). Results from extensive global sequencing of solid tumors have shown that aneuploidies result in gene expression changes of most resident genes (i.e., function follows form) (Ben-David and Amon 2020). Therefore, aneuploidies result in a massive deregulation of the transcriptional equilibrium. This is in strong contrast to hematological malignancies. These results have been confirmed in experimental models, in which single chromosomes were introduced into karyotypically normal cells (Wangsa et al. 2018). The results were also confirmed in a model system, in which normal colon epithelium was cultured under hypoxic conditions (Braun et al. 2019). These normal colon cells spontaneously transform upon the acquisition of an extra copy of chromosome 7, mimicking the karyotype of colonic polyps. It is therefore conceivable that therapeutically targeting single genes, as it is possible with CML using imatinib (Gleevec), is not sufficient to kill cancer cells that are aneuploid. Rather, the aneuploid state needs to be targeted, which is a formidable task.

Another central question in cancer biology is how aneuploidy affects nuclear structure. This question can now be addressed with the advent of techniques that monitor chromosome conformation on a global level, most notably chromosome conformation capture (i.e., Hi-C) (Ried and Rajapakse 2017). Here, contacts between chromatin are determined by global sequence analysis after fixing interactions with formalin. The vicinity of chromatin regions is then determined by the frequency of sequence contacts (Lieberman-Aiden et al. 2009).

The application of Hi-C to cancer cell lines (i.e., the colorectal cancer cell line HT-29) revealed a strict structure–function relationship (Seaman et al. 2017). In fact, the karyotype of this cell line could be reconstructed from the Hi-C maps (i.e., copy number changes and translocation directly affected nuclear organization). This was also observed in the above-described model system of colorectal cancer, in which the cells acquired an extra copy of chromosome 7 (Braun et al. 2019).

TADs are megabase-long genomic regions along the contiguous genome that interact with each other more frequently than with sequences outside the TAD and are enriched for the insulator-binding protein CTCF (Dixon et al. 2012) and can be defined by chromosome conformation capture assays such as Hi-C. Although TADs are generally stable from cell type to cell type, genomic events such as duplications of noncoding regions can result in the formation of new chromatin domains (neo-TADs) that result in misexpression of genes leading to profound phenotypic effects (Franke et al. 2016). Of interest, it has been reported that evolutionarily conserved chromosome loop anchors are frequently mutated in cancers (Katainen et al. 2015) and those bound by CTCF and cohesin are vulnerable to DNA double-strand breaks (DSBs) mediated by topoisomerase 2B (TOP2B) (Canela et al. 2017). In addition, the frequency of recurrent DSBs at these sites positively correlates with transcriptional output and directionality, and may contribute to the occurrence of genomics rearrangements such as oncogenic fusion events at these transcriptionally active sites (Gothe et al. 2019). Thus, loop anchors are fragile sites where DSBs lead to chromosomal rearrangements, aneuploidy, and alterations of the 3D genome.

CHROMATIN MACHINERY ESSENTIAL TO CANCER IN THEIR NONMUTANT FORM

Large-scale RNAi screens in cancer revealed critical cancer-specific dependencies (Zuber et al. 2011). In more recent years, this has been greatly expanded by CRISPR-Cas9-based genome editing (Wang et al. 2014; Meyers et al. 2017). Because CRISPR mutagenesis can be guided to any exon of a gene, special attention has been given to domain-focused screening of chromatin regulators and TFs, revealing not only which chromatin proteins are essential but which functional domains are contributing to that essentiality (Shi et al. 2015). This domain-focused "biochemistry via genetics" is critical because most chromatin-associated proteins wear many hats, such as EP300 that is endowed with both a bromodomain, which is an acetylation reader domain, and a HAT domain to write the same mark it can read.

The results of these efforts have revealed three predominant themes: First, a handful of broadly expressed chromatin coactivator proteins are pan-cancer lethal such as BRD4 (Shi et al. 2015), CHD4 (Marques et al. 2020), and SMARCA4 (Dempster et al. 2019). Second, the selectively essential factors of each cancer subtype are predominantly the lineage-specific master TFs (Reddy et al. 2019), also called core regulatory TFs, such as PAX5 in chronic lymphocytic leukemia (CLL) (Ott et al. 2018), HAND2 in NB (Durbin et al. 2018a), and MYOD in RMS (Gryder et al. 2019a). Third, select cancers with specific mutations have synthetic lethal dependence on nonmutant chromatin factors, such as noncanonical SWI/SNF complex members in BAF-mutant tumors (Michel et al. 2018).

The first two themes are unified at the chemical–genomic level, as therapeutics that target these broadly required coactivators (BRD4 and CHD4/NuRD) selectively impair the function of core regulatory TFs specific to that cancer. For example, in RMS, MYOD is the top essential TF in the disease, and co-binds with BRD4 at virtually every SE, and BET inhibitors selectively shut down the genes that are SE driven (Gryder et al. 2017, 2019b). A straightforward means to identify the cell-type-specific TFs that are recruiting BRD4 to enact their oncogenic enhancer network can be by motif analysis of BRD4 peaks, which report the TFs that direct its positioning at enhancers and SEs. BRD4, HDACs, CPB/p300, EZH2, and many other chromatin-associated factors are traditionally more amenable to drug discovery than TFs, and represent a key strategy to drugging chromatin for anticancer benefit (Filippakopoulos et al. 2010; Delmore et al. 2011; Roe et al. 2015; Lasko et al. 2017; Chen et al. 2018; Gryder et al. 2019b).

OUTLOOK

Chromatin machinery controls gene expression and its deregulation has a central role during carcinogenesis. The manifestation of epigenetic errors leading to cancer can be thought of as either an inability to differentiate along a predefined lineage path, or a hyperactive, highly proliferative state with addiction to high levels of transcriptional output (Bradner et al. 2017). These are not mutually exclusive, as poorly differentiated (or stem-cell-like) cells in normal human biology often have a proliferative advantage (Burdon et al. 2002). Maintenance of these abnormal states can be attributed to a lack of transcriptional control, or an inability of chromatin to properly respond to external cues, to leading to a loss of cellular homeostasis (Flavahan et al. 2017). Many of the mutations summarized above are selected for in cancer because they lead to these same phenotypic outcomes. Future research should be directed toward an understanding of structure–function relationships in cancer cells. Knowing how cancer genome-specific aberrations affect chromatin and gene expression is essential for a fundamental, systematic understanding of cancer biology, and for developing novel potent precision therapies for cancer.

REFERENCES

Abraham BJ, Hnisz D, Weintraub AS, Kwiatkowski N, Li CH, Li Z, Weichert-Leahey N, Rahman S, Liu Y, Etchin J, et al. 2017. Small genomic insertions form enhancers that misregulate oncogenes. *Nat Commun* **8:** 14385. doi:10.1038/ncomms14385

Agaram NP, LaQuaglia MP, Alaggio R, Zhang L, Fujisawa Y, Ladanyi M, Wexler LH, Antonescu CR. 2019. MYOD1-mutant spindle cell and sclerosing rhabdomyosarcoma: an aggressive subtype irrespective of age. A reappraisal for molecular classification and risk stratification. *Mod Pathol* **32:** 27–36. doi:10.1038/s41379-018-0120-9

Akhtar-Zaidi B, Cowper-Sal-lari R, Corradin O, Saiakhova A, Bartels CF, Balasubramanian D, Myeroff L, Lutterbaugh J, Jarrar A, Kalady MF, et al. 2012. Epigenomic enhancer profiling defines a signature of colon cancer. *Science* **336:** 736–739. doi:10.1126/science.1217277

Aplan PD, Shern JF, Khan J. 2021. *Molecular and genetic basis of childhood cancer.* Kluwer, Amsterdam.

Armenia J, Wankowicz SAM, Liu D, Gao J, Kundra R, Reznik E, Chatila WK, Chakravarty D, Han GC, Coleman I, et al. 2018. The long tail of oncogenic drivers in prostate cancer. *Nat Genet* **50:** 645–651. doi:10.1038/s41588-018-0078-z

Atallah-Yunes SA, Murphy DJ, Noy A. 2020. HIV-associated Burkitt lymphoma. *Lancet Haematol* **7:** e594–e600. doi:10.1016/S2352-3026(20)30126-5

Battey J, Moulding C, Taub R, Murphy W, Stewart T, Potter H, Lenoir G, Leder P. 1983. The human *c-myc* oncogene: structural consequences of translocation into the IgH locus in Burkitt lymphoma. *Cell* **34:** 779–787. doi:10.1016/0092-8674(83)90534-2

Bell AC, Felsenfeld G. 2000. Methylation of a CTCF-dependent boundary controls imprinted expression of the Igf2 gene. *Nature* **405:** 482–485. doi:10.1038/35013100

Ben-David U, Amon A. 2020. Context is everything: aneuploidy in cancer. *Nat Rev Genet* **21:** 44–62. doi:10.1038/s41576-019-0171-x

Boulay G, Sandoval GJ, Riggi N, Iyer S, Buisson R, Naigles B, Awad ME, Rengarajan S, Volorio A, McBride MJ, et al. 2017. Cancer-specific retargeting of BAF complexes by a prion-like domain. *Cell* **171:** 163–178.e19. doi:10.1016/j.cell.2017.07.036

Bradner JE, Hnisz D, Young RA. 2017. Transcriptional addiction in cancer. *Cell* **168:** 629–643. doi:10.1016/j.cell.2016.12.013

Braun R, Ronquist S, Wangsa D, Chen H, Anthuber L, Gemoll T, Wangsa D, Koparde V, Hunn C, Habermann JK, et al. 2019. Single chromosome aneuploidy induces genome-wide perturbation of nuclear organization and gene expression. *Neoplasia* **21:** 401–412. doi:10.1016/j.neo.2019.02.003

Brohl AS, Solomon DA, Chang W, Wang J, Song Y, Sindiri S, Patidar R, Hurd L, Chen L, Shern JF, et al. 2014. The genomic landscape of the Ewing sarcoma family of tumors reveals recurrent STAG2 mutation. *PLoS Genet* **10:** e1004475. doi:10.1371/journal.pgen.1004475

Burdon T, Smith A, Savatier P. 2002. Signalling, cell cycle and pluripotency in embryonic stem cells. *Trends Cell Biol* **12:** 432–438. doi:10.1016/S0962-8924(02)02352-8

Canela A, Maman Y, Jung S, Wong N, Callen E, Day A, Kieffer-Kwon KR, Pekowska A, Zhang H, Rao SSP, et al. 2017. Genome organization drives chromosome fragility. *Cell* **170:** 507–521.e18. doi:10.1016/j.cell.2017.06.034

Centore RC, Sandoval GJ, Soares LMM, Kadoch C, Chan HM. 2020. Mammalian SWI/SNF chromatin remodeling complexes: emerging mechanisms and therapeutic strategies. *Trends Genet* **36:** 936–950. doi:10.1016/j.tig.2020.07.011

Chen L, Alexe G, Dharia NV, Ross L, Iniguez AB, Conway AS, Wang EJ, Veschi V, Lam N, Qi J, et al. 2018. CRISPR-Cas9 screen reveals a MYCN-amplified neuroblastoma dependency on EZH2. *J Clin Invest* **128:** 446–462. doi:10.1172/JCI90793

Chi P, Allis CD, Wang GG. 2010. Covalent histone modifications—miswritten, misinterpreted and mis-erased in human cancers. *Nat Rev Cancer* **10:** 457–469. doi:10.1038/nrc2876

Collord G, Tarpey P, Kurbatova N, Martincorena I, Moran S, Castro M, Nagy T, Bignell G, Maura F, Young MD, et al. 2018. An integrated genomic analysis of anaplastic meningioma identifies prognostic molecular signatures. *Sci Rep* **8:** 13537. doi:10.1038/s41598-018-31659-0

Corona RI, Seo JH, Lin X, Hazelett DJ, Reddy J, Fonseca MAS, Abassi F, Lin YG, Mhawech-Fauceglia PY, Shah SP, et al. 2020. Non-coding somatic mutations converge on the PAX8 pathway in ovarian cancer. *Nat Commun* **11:** 2020. doi:10.1038/s41467-020-15951-0

Couthouis J, Hart MP, Shorter J, DeJesus-Hernandez M, Erion R, Oristano R, Liu AX, Ramos D, Jethava N, Hosangadi D, et al. 2011. A yeast functional screen predicts new candidate ALS disease genes. *Proc Natl Acad Sci* **108:** 20881–20890. doi:10.1073/pnas.1109434108

Couthouis J, Hart MP, Erion R, King OD, Diaz Z, Nakaya T, Ibrahim F, Kim HJ, Mojsilovic-Petrovic J, Panossian S, et al. 2012. Evaluating the role of the FUS/TLS-related gene EWSR1 in amyotrophic lateral sclerosis. *Hum Mol Genet* **21:** 2899–2911. doi:10.1093/hmg/dds116

Dai C, Heemers H, Sharifi N. 2017. Androgen signaling in prostate cancer. *Cold Spring Harb Perspect Med* **7:** a030452. doi:10.1101/cshperspect.a030452

Dang CV. 2012. MYC on the path to cancer. *Cell* **149:** 22–35. doi:10.1016/j.cell.2012.03.003

Darnell JE. 2002. Transcription factors as targets for cancer therapy. *Nat Rev Cancer* **2:** 740–749. doi:10.1038/nrc906

Dawson MA. 2017. The cancer epigenome: concepts, challenges, and therapeutic opportunities. *Science* **355:** 1147–1152. doi:10.1126/science.aam7304

Delmore JE, Issa Ghayas C, Lemieux Madeleine E, Rahl Peter B, Shi J, Jacobs Hannah M, Kastritis E, Gilpatrick T, Paranal Ronald M, Qi J, et al. 2011. BET bromodomain inhibition as a therapeutic strategy to target c-Myc. *Cell* **146:** 904–917. doi:10.1016/j.cell.2011.08.017

Dempster JM, Pacini C, Pantel S, Behan FM, Green T, Krill-Burger J, Beaver CM, Younger ST, Zhivich V, Najgebauer H, et al. 2019. Agreement between two large pan-cancer CRISPR-Cas9 gene dependency data sets. *Nat Commun* **10:** 5817. doi:10.1038/s41467-019-13805-y

De Raedt T, Beert E, Pasmant E, Luscan A, Brems H, Ortonne N, Helin K, Hornick JL, Mautner V, Kehrer-Sawatzki H, et al. 2014. PRC2 loss amplifies Ras-driven transcription and confers sensitivity to BRD4-based therapies. *Nature* **514:** 247–251. doi:10.1038/nature13561

Dixon JR, Selvaraj S, Yue F, Kim A, Li Y, Shen Y, Hu M, Liu JS, Ren B. 2012. Topological domains in mammalian genomes identified by analysis of chromatin interactions. *Nature* **485:** 376–380. doi:10.1038/nature11082

Donaldson-Collier MC, Sungalee S, Zufferey M, Tavernari D, Katanayeva N, Battistello E, Mina M, Douglass KM, Rey T, Raynaud F, et al. 2019. EZH2 oncogenic mutations drive epigenetic, transcriptional, and structural changes within chromatin domains. *Nat Genet* **51:** 517–528. doi:10.1038/s41588-018-0338-y

Durbin AD, Zimmerman MW, Dharia NV, Abraham BJ, Iniguez AB, Weichert-Leahey N, He S, Krill-Burger JM, Root DE, Vazquez F, et al. 2018a. Selective gene dependencies in MYCN-amplified neuroblastoma include the core transcriptional regulatory circuitry. *Nat Genet* **50:** 1240–1246. doi:10.1038/s41588-018-0191-z

Durbin AD, Zimmerman MW, Dharia NV, Abraham BJ, Iniguez AB, Weichert-Leahey N, He S, Krill-Burger JM, Root DE, Vazquez F, et al. 2018b. Selective gene dependencies in MYCN-amplified neuroblastoma include the core transcriptional regulatory circuitry. *Nat Genet* **50:** 1240–1246. doi:10.1038/s41588-018-0191-z

Edwards SL, Beesley J, French JD, Dunning AM. 2013. Beyond GWASs: illuminating the dark road from association to function. *Am J Hum Genet* **93:** 779–797. doi:10.1016/j.ajhg.2013.10.012

Fiedler D, Heselmeyer-Haddad K, Hirsch D, Hernandez LS, Torres I, Wangsa D, Hu Y, Zapata L, Rueschoff J, Belle S, et al. 2019. Single-cell genetic analysis of clonal dynamics in colorectal adenomas indicates CDX2 gain as a predictor of recurrence. *Int J Cancer* **144:** 1561–1573. doi:10.1002/ijc.31869

Filippakopoulos P, Qi J, Picaud S, Shen Y, Smith WB, Fedorov O, Morse EM, Keates T, Hickman TT, Felletar I, et al. 2010. Selective inhibition of BET bromodomains. *Nature* **468:** 1067–1073. doi:10.1038/nature09504

Flavahan WA, Drier Y, Liau BB, Gillespie SM, Venteicher AS, Stemmer-Rachamimov AO, Suvà ML, Bernstein BE. 2016. Insulator dysfunction and oncogene activation in IDH mutant gliomas. *Nature* **529:** 110–114. doi:10.1038/nature16490

Flavahan WA, Gaskell E, Bernstein BE. 2017. Epigenetic plasticity and the hallmarks of cancer. *Science* **357:** eaal2380. doi:10.1126/science.aal2380

Flavahan WA, Drier Y, Johnstone SE, Hemming ML, Tarjan DR, Hegazi E, Shareef SJ, Javed NM, Raut CP, Eschle BK, et al. 2019. Altered chromosomal topology drives oncogenic programs in SDH-deficient GISTs. *Nature* **575:** 229–233. doi:10.1038/s41586-019-1668-3

Franke M, Ibrahim DM, Andrey G, Schwarzer W, Heinrich V, Schöpflin R, Kraft K, Kempfer R, Jerković I, Chan WL, et al. 2016. Formation of new chromatin domains determines pathogenicity of genomic duplications. *Nature* **538:** 265–269. doi:10.1038/nature19800

Gerstein MB, Kundaje A, Hariharan M, Landt SG, Yan KK, Cheng C, Mu XJ, Khurana E, Rozowsky J, Alexander R, et al. 2012. Architecture of the human regulatory network derived from ENCODE data. *Nature* **489:** 91–100. doi:10.1038/nature11245

Ghiraldini FG, Filipescu D, Bernstein E. 2021. Solid tumours hijack the histone variant network. *Nat Rev Cancer* **21:** 257–275. doi:10.1038/s41568-020-00330-0

Gothe HJ, Bouwman BAM, Gusmao EG, Piccinno R, Petrosino G, Sayols S, Drechsel O, Minneker V, Josipovic N, Mizi A, et al. 2019. Spatial chromosome folding and active transcription drive DNA fragility and formation of oncogenic MLL translocations. *Mol Cell* **75:** 267–283.e12. doi:10.1016/j.molcel.2019.05.015

Gryder BE, Sodji QH, Oyelere AK. 2012. Targeted cancer therapy: giving histone deacetylase inhibitors all they need to succeed. *Future Med Chem* **4:** 505–524. doi:10.4155/fmc.12.3

Gryder B, Nelson C, Shepard S. 2013. Biosemiotic entropy of the genome: mutations and epigenetic imbalances resulting in cancer. *Entropy* **15:** 234–261. doi:10.3390/e15010234

Gryder BE, Yohe ME, Chou HC, Zhang X, Marques J, Wachtel M, Schaefer B, Sen N, Song Y, Gualtieri A, et al. 2017. PAX3-FOXO1 establishes myogenic super enhancers and confers BET bromodomain vulnerability. *Cancer Discov* **7:** 884–899. doi:10.1158/2159-8290.CD-16-1297

Gryder BE, Pomella S, Sayers C, Wu XS, Song Y, Chiarella AM, Bagchi S, Chou HC, Sinniah RS, Walton A, et al. 2019a. Histone hyperacetylation disrupts core gene regulatory architecture in rhabdomyosarcoma. *Nat Genet* **51:** 1714–1722. doi:10.1038/s41588-019-0534-4

Gryder BE, Wu L, Woldemichael GM, Pomella S, Quinn TR, Park PMC, Cleveland A, Stanton BZ, Song Y, Rota R, et al. 2019b. Chemical genomics reveals histone deacetylases are required for core regulatory transcription. *Nat Commun* **10:** 3004. doi:10.1038/s41467-019-11046-7

Gryder BE, Wu L, Woldemichael GM, Pomella S, Quinn TR, Park PMC, Cleveland A, Stanton BZ, Song Y, Rota R, et al. 2019c. Chemical genomics reveals histone deacetylases are required for core regulatory transcription. *Nat Commun* **10:** 3004. doi:10.1038/s41467-019-11046-7

Gryder BE, Wachtel M, Chang K, El Demerdash O, Aboreden NG, Mohammed W, Ewert W, Pomella S, Rota R, Wei JS, et al. 2020. Miswired enhancer logic drives a cancer of the muscle lineage. *iScience* **23:** 101103. doi:10.1016/j.isci.2020.101103

Haller F, Bieg M, Will R, Körner C, Weichenhan D, Bott A, Ishaque N, Lutsik P, Moskalev EA, Mueller SK, et al. 2019. Enhancer hijacking activates oncogenic transcription factor NR4A3 in acinic cell carcinomas of the salivary glands. *Nat Commun* **10:** 368. doi:10.1038/s41467-018-08069-x

Han M, Jia L, Lv W, Wang L, Cui W. 2019. Epigenetic enzyme mutations: role in tumorigenesis and molecular inhibitors. *Front Oncol* **9:** 194. doi:10.3389/fonc.2019.00194

He S, Wu Z, Tian Y, Yu Z, Yu J, Wang X, Li J, Liu B, Xu Y. 2020. Structure of nucleosome-bound human BAF complex. *Science* **367:** 875–881. doi:10.1126/science.aaz9761

Heim S, Mitelman F. 2009. *Cancer cytogenetics*. Wiley, Hoboken, NJ.

Helmsauer K, Valieva ME, Ali S, Chamorro González R, Schöpflin R, Röefzaad C, Bei Y, Dorado Garcia H, Rodriguez-Fos E, Puiggros M, et al. 2020. Enhancer hijacking determines extrachromosomal circular MYCN amplicon architecture in neuroblastoma. *Nat Commun* **11:** 5823. doi:10.1038/s41467-020-19452-y

Heselmeyer K, Schrock E, du Manoir S, Blegen H, Shah K, Steinbeck R, Auer G, Ried T. 1996. Gain of chromosome 3q defines the transition from severe dysplasia to invasive carcinoma of the uterine cervix. *Proc Natl Acad Sci* **93:** 479–484. doi:10.1073/pnas.93.1.479

Hörberg J, Reymer A. 2018. A sequence environment modulates the impact of methylation on the torsional rigidity of DNA. *Chem Commun (Camb)* **54:** 11885–11888. doi:10.1039/C8CC06550K

Horn S, Figl A, Rachakonda PS, Fischer C, Sucker A, Gast A, Kadel S, Moll I, Nagore E, Hemminki K, et al. 2013. TERT promoter mutations in familial and sporadic melanoma. *Science* **339:** 959–961. doi:10.1126/science.1230062

Johnstone SE, Reyes A, Qi Y, Adriaens C, Hegazi E, Pelka K, Chen JH, Zou LS, Drier Y, Hecht V, et al. 2020. Large-scale topological changes restrain malignant progression in colorectal cancer. *Cell* **182:** 1474–1489.e23. doi:10.1016/j.cell.2020.07.030

Jones PA, Issa JP, Baylin S. 2016. Targeting the cancer epigenome for therapy. *Nat Rev Genet* **17:** 630–641. doi:10.1038/nrg.2016.93

Kadoch C. 2019. Diverse compositions and functions of chromatin remodeling machines in cancer. *Sci Transl Med* **11:** eaay1018. doi:10.1126/scitranslmed.aay1018

Kadoch C, Crabtree GR. 2013. Reversible disruption of mSWI/SNF (BAF) complexes by the SS18-SSX oncogenic fusion in synovial sarcoma. *Cell* **153:** 71–85. doi:10.1016/j.cell.2013.02.036

Kadoch C, Crabtree GR. 2015. Mammalian SWI/SNF chromatin remodeling complexes and cancer: mechanistic insights gained from human genomics. *Sci Adv* **1:** e1500447. doi:10.1126/sciadv.1500447

Kadoch C, Hargreaves DC, Hodges C, Elias L, Ho L, Ranish J, Crabtree GR. 2013. Proteomic and bioinformatic analysis of mammalian SWI/SNF complexes identifies extensive roles in human malignancy. *Nat Genet* **45:** 592–601. doi:10.1038/ng.2628

Katainen R, Dave K, Pitkänen E, Palin K, Kivioja T, Välimäki N, Gylfe AE, Ristolainen H, Hänninen UA, Cajuso T, et al. 2015. CTCF/cohesin-binding sites are frequently mutated in cancer. *Nat Genet* **47:** 818–821. doi:10.1038/ng.3335

Kim KH, Roberts CW. 2014. Mechanisms by which SMARCB1 loss drives rhabdoid tumor growth. *Cancer Genet* **207:** 365–372. doi:10.1016/j.cancergen.2014.04.004

Kim H, Nguyen NP, Turner K, Wu S, Gujar AD, Luebeck J, Liu J, Deshpande V, Rajkumar U, Namburi S, et al. 2020. Extrachromosomal DNA is associated with oncogene amplification and poor outcome across multiple cancers. *Nat Genet* **52:** 891–897. doi:10.1038/s41588-020-0678-2

Knudsen ES, Nambiar R, Rosario SR, Smiraglia DJ, Goodrich DW, Witkiewicz AK. 2020. Pan-cancer molecular analysis of the RB tumor suppressor pathway. *Commun Biol* **3:** 158. doi:10.1038/s42003-020-0873-9

Kohsaka S, Shukla N, Ameur N, Ito T, Ng CKY, Wang L, Lim D, Marchetti A, Viale A, Pirun M, et al. 2014. A recurrent neomorphic mutation in MYOD1 defines a clinically aggressive subset of embryonal rhabdomyosarcoma associated with PI3K-AKT pathway mutations. *Nat Genet* **46:** 595–600. doi:10.1038/ng.2969

Koretzky GA. 2007. The legacy of the Philadelphia chromosome. *J Clin Invest* **117:** 2030–2032. doi:10.1172/JCI33032

Kornberg RD, Lorch Y. 2020. Primary role of the nucleosome. *Mol Cell* **79:** 371–375. doi:10.1016/j.molcel.2020.07.020

Kwon I, Kato M, Xiang S, Wu L, Theodoropoulos P, Mirzaei H, Han T, Xie S, Corden JL, McKnight SL. 2013. Phosphorylation-regulated binding of RNA polymerase II to fibrous polymers of low-complexity domains. *Cell* **155:** 1049–1060. doi:10.1016/j.cell.2013.10.033

Lancho O, Herranz D. 2018. The MYC enhancer-ome: long-range transcriptional regulation of MYC in cancer. *Trends Cancer* **4:** 810–822. doi:10.1016/j.trecan.2018.10.003

Lasko LM, Jakob CG, Edalji RP, Qiu W, Montgomery D, Digiammarino EL, Hansen TM, Risi RM, Frey R, Manaves V, et al. 2017. Discovery of a selective catalytic p300/CBP inhibitor that targets lineage-specific tumours. *Nature* **550:** 128–132. doi:10.1038/nature24028

Lewis PW, Müller MM, Koletsky MS, Cordero F, Lin S, Banaszynski LA, Garcia BA, Muir TW, Becher OJ, Allis CD. 2013. Inhibition of PRC2 activity by a gain-of-function H3 mutation found in pediatric glioblastoma. *Science* **340:** 857. doi:10.1126/science.1232245

Li B, Carey M, Workman JL. 2007. The role of chromatin during transcription. *Cell* **128:** 707–719. doi:10.1016/j.cell.2007.01.015

Lieberman-Aiden E, van Berkum NL, Williams L, Imakaev M, Ragoczy T, Telling A, Amit I, Lajoie BR, Sabo PJ, Dorschner MO, et al. 2009. Comprehensive mapping of long-range interactions reveals folding principles of the human genome. *Science* **326:** 289–293. doi:10.1126/science.1181369

Lin C, Smith ER, Takahashi H, Lai KC, Martin-Brown S, Florens L, Washburn MP, Conaway JW, Conaway RC, Shilatifard A. 2010. AFF4, a component of the ELL/P-

TEFb elongation complex and a shared subunit of MLL chimeras, can link transcription elongation to leukemia. *Mol Cell* **37:** 429–437. doi:10.1016/j.molcel.2010.01.026

Mansour MR, Abraham BJ, Anders L, Berezovskaya A, Gutierrez A, Durbin AD, Etchin J, Lawton L, Sallan SE, Silverman LB, et al. 2014. An oncogenic super-enhancer formed through somatic mutation of a noncoding intergenic element. *Science* **346:** 1373–1377. doi:10.1126/science.1259037

Mark J. 1967. Double-minutes—a chromosomal aberration in Rous sarcomas in mice. *Hereditas* **57:** 1–22. doi:10.1111/j.1601-5223.1967.tb02091.x

Marques JG, Gryder BE, Pavlovic B, Chung Y, Ngo QA, Frommelt F, Gstaiger M, Song Y, Benischke K, Laubscher D, et al. 2020. NuRD subunit CHD4 regulates super-enhancer accessibility in rhabdomyosarcoma and represents a general tumor dependency. *eLife* **9:** e54993. doi:10.7554/eLife.54993

Mashtalir N, Suzuki H, Farrell DP, Sankar A, Luo J, Filipovski M, D'Avino AR, St Pierre R, Valencia AM, Onikubo T, et al. 2020. A structural model of the endogenous human BAF complex informs disease mechanisms. *Cell* **183:** 802–817.e24. doi:10.1016/j.cell.2020.09.051

McCabe MT, Ott HM, Ganji G, Korenchuk S, Thompson C, Van Aller GS, Liu Y, Graves AP, Della Pietra A III, Diaz E, et al. 2012. EZH2 inhibition as a therapeutic strategy for lymphoma with EZH2-activating mutations. *Nature* **492:** 108–112. doi:10.1038/nature11606

Meyers RM, Bryan JG, McFarland JM, Weir BA, Sizemore AE, Xu H, Dharia NV, Montgomery PG, Cowley GS, Pantel S, et al. 2017. Computational correction of copy number effect improves specificity of CRISPR–Cas9 essentiality screens in cancer cells. *Nat Genet* **49:** 1779–1784. doi:10.1038/ng.3984

Michel BC, D'Avino AR, Cassel SH, Mashtalir N, McKenzie ZM, McBride MJ, Valencia AM, Zhou Q, Bocker M, Soares LMM, et al. 2018. A non-canonical SWI/SNF complex is a synthetic lethal target in cancers driven by BAF complex perturbation. *Nat Cell Biol* **20:** 1410–1420. doi:10.1038/s41556-018-0221-1

Morin RD, Johnson NA, Severson TM, Mungall AJ, An J, Goya R, Paul JE, Boyle M, Woolcock BW, Kuchenbauer F, et al. 2010. Somatic mutations altering EZH2 (Tyr641) in follicular and diffuse large B-cell lymphomas of germinal-center origin. *Nat Genet* **42:** 181–185. doi:10.1038/ng.518

Morrow JJ, Bayles I, Funnell APW, Miller TE, Saiakhova A, Lizardo MM, Bartels CF, Kapteijn MY, Hung S, Mendoza A, et al. 2018. Positively selected enhancer elements endow osteosarcoma cells with metastatic competence. *Nat Med* **24:** 176–185. doi:10.1038/nm.4475

Morton AR, Dogan-Artun N, Faber ZJ, MacLeod G, Bartels CF, Piazza MS, Allan KC, Mack SC, Wang X, Gimple RC, et al. 2019. Functional enhancers shape extrachromosomal oncogene amplifications. *Cell* **179:** 1330–1341.e13. doi:10.1016/j.cell.2019.10.039

Nacev BA, Feng L, Bagert JD, Lemiesz AE, Gao J, Soshnev AA, Kundra R, Schultz N, Muir TW, Allis CD. 2019. The expanding landscape of "oncohistone" mutations in human cancers. *Nature* **567:** 473–478. doi:10.1038/s41586-019-1038-1

Nakayama RT, Pulice JL, Valencia AM, McBride MJ, McKenzie ZM, Gillespie MA, Ku WL, Teng M, Cui K, Williams RT, et al. 2017. SMARCB1 is required for widespread BAF complex-mediated activation of enhancers and bivalent promoters. *Nat Genet* **49:** 1613–1623. doi:10.1038/ng.3958

Nambiar M, Kari V, Raghavan SC. 2008. Chromosomal translocations in cancer. *Biochim Biophys Acta* **1786:** 139–152.

Naveh-Many T, Cedar H. 1981. Active gene sequences are undermethylated. *Proc Natl Acad Sci* **78:** 4246–4250. doi:10.1073/pnas.78.7.4246

Nijman SMB. 2020. Perturbation-driven entropy as a source of cancer cell heterogeneity. *Trends Cancer* **6:** 454–461. doi:10.1016/j.trecan.2020.02.016

Northcott PA, Lee C, Zichner T, Stütz AM, Erkek S, Kawauchi D, Shih DJ, Hovestadt V, Zapatka M, Sturm D, et al. 2014. Enhancer hijacking activates GFI1 family oncogenes in medulloblastoma. *Nature* **511:** 428–434. doi:10.1038/nature13379

Ochs F, Karemore G, Miron E, Brown J, Sedlackova H, Rask MB, Lampe M, Buckle V, Schermelleh L, Lukas J, et al. 2019. Stabilization of chromatin topology safeguards genome integrity. *Nature* **574:** 571–574. doi:10.1038/s41586-019-1659-4

Oldridge DA, Wood AC, Weichert-Leahey N, Crimmins I, Sussman R, Winter C, McDaniel LD, Diamond M, Hart LS, Zhu S, et al. 2015. Genetic predisposition to neuroblastoma mediated by a LMO1 super-enhancer polymorphism. *Nature* **528:** 418–421. doi:10.1038/nature15540

Ott CJ, Federation AJ, Schwartz LS, Kasar S, Klitgaard JL, Lenci R, Li Q, Lawlor M, Fernandes SM, Souza A, et al. 2018. Enhancer architecture and essential core regulatory circuitry of chronic lymphocytic leukemia. *Cancer Cell* **34:** 982–995.e987.

Parsons DW, Jones S, Zhang X, Lin JC, Leary RJ, Angenendt P, Mankoo P, Carter H, Siu IM, Gallia GL, et al. 2008. An integrated genomic analysis of human glioblastoma multiforme. *Science* **321:** 1807–1812.

Patkar S, Heselmeyer-Haddad K, Auslander N, Hirsch D, Camps J, Bronder D, Brown M, Wei-Dong C, Lokanga R, Wangsa D, et al. 2021. Hard-wiring of normal tissue-specific chromosome-wide gene expression levels is a potential factor driving cancer-type-specific aneuploidies. *Genome Med* **13:** 93. doi:10.1186/s13073-021-00905-y

Pereira BJA, Oba-Shinjo SM, de Almeida AN, Marie SKN. 2019. Molecular alterations in meningiomas: literature review. *Clin Neurol Neurosurg* **176:** 89–96. doi:10.1016/j.clineuro.2018.12.004

Piunti A, Shilatifard A. 2016. Epigenetic balance of gene expression by Polycomb and COMPASS families. *Science* **352:** aad9780. doi:10.1126/science.aad9780

Poole CJ, van Riggelen J. 2017. MYC-master regulator of the cancer epigenome and transcriptome. *Genes (Basel)* **8:** 142. doi:10.3390/genes8050142

Pugh TJ, Morozova O, Attiyeh EF, Asgharzadeh S, Wei JS, Auclair D, Carter SL, Cibulskis K, Hanna M, Kiezun A, et al. 2013. The genetic landscape of high-risk neuroblastoma. *Nat Genet* **45:** 279–284. doi:10.1038/ng.2529

Reddy J, Fonseca MAS, Corona RI, Nameki R, Segato Dezem F, Klein IA, Chang H, Chaves-Moreira D, Afeyan L, Malta TM, et al. 2019. Predicting master transcription factors from pan-cancer expression data. bioRxiv doi:10.1101/839142

Ried T. 2009. Homage to Theodor Boveri (1862–1915): Boveri's theory of cancer as a disease of the chromosomes, and the landscape of genomic imbalances in human carcinomas. *Environ Mol Mutagen* **50:** 593–601. doi:10.1002/em.20526

Ried T, Rajapakse I. 2017. The 4D nucleome. *Methods* **123:** 1–2. doi:10.1016/j.ymeth.2017.06.031

Ried T, Knutzen R, Steinbeck R, Blegen H, Schröck E, Heselmeyer K, du Manoir S, Auer G. 1996. Comparative genomic hybridization reveals a specific pattern of chromosomal gains and losses during the genesis of colorectal tumors. *Genes Chromosomes Cancer* **15:** 234–245.

Roe JS, Mercan F, Rivera K, Pappin Darryl J, Vakoc CR. 2015. BET bromodomain inhibition suppresses the function of hematopoietic transcription factors in acute myeloid leukemia. *Mol Cell* **58:** 1028–1039. doi:10.1016/j.molcel.2015.04.011

Ryan RJ, Drier Y, Whitton H, Cotton MJ, Kaur J, Issner R, Gillespie S, Epstein CB, Nardi V, Sohani AR, et al. 2015. Detection of enhancer-associated rearrangements reveals mechanisms of oncogene dysregulation in B-cell lymphoma. *Cancer Discov* **5:** 1058–1071. doi:10.1158/2159-8290.CD-15-0370

Salesse S, Verfaillie CM. 2002. BCR/ABL: from molecular mechanisms of leukemia induction to treatment of chronic myelogenous leukemia. *Oncogene* **21:** 8547–8559. doi:10.1038/sj.onc.1206082

Schaefer IM, Hornick JL. 2021. SWI/SNF complex-deficient soft tissue neoplasms: an update. *Semin Diagn Pathol* **38:** 222–231.

Schnittger S, Haferlach C, Ulke M, Alpermann T, Kern W, Haferlach T. 2010. IDH1 mutations are detected in 6.6% of 1414 AML patients and are associated with intermediate risk karyotype and unfavorable prognosis in adults younger than 60 years and unmutated NPM1 status. *Blood* **116:** 5486–5496. doi:10.1182/blood-2010-02-267955

Schreiber SL, Bernstein BE. 2002. Signaling network model of chromatin. *Cell* **111:** 771–778. doi:10.1016/S0092-8674(02)01196-0

Schwartz JC, Wang X, Podell ER, Cech TR. 2013. RNA seeds higher-order assembly of FUS protein. *Cell Rep* **5:** 918–925. doi:10.1016/j.celrep.2013.11.017

Schwartzentruber J, Korshunov A, Liu XY, Jones DTW, Pfaff E, Jacob K, Sturm D, Fontebasso AM, Quang DAK, Tönjes M, et al. 2012. Driver mutations in histone H3.3 and chromatin remodelling genes in paediatric glioblastoma. *Nature* **482:** 226–231. doi:10.1038/nature10833

Seaman L, Chen H, Brown M, Wangsa D, Patterson G, Camps J, Omenn GS, Ried T, Rajapakse I. 2017. Nucleome analysis reveals structure–function relationships for colon cancer. *Mol Cancer Res* **15:** 821–830. doi:10.1158/1541-7786.MCR-16-0374

Shen C, Vakoc CR. 2015. Gain-of-function mutation of chromatin regulators as a tumorigenic mechanism and an opportunity for therapeutic intervention. *Curr Opin Oncol* **27:** 57–63. doi:10.1097/CCO.0000000000000151

Shern JF, Chen L, Chmielecki J, Wei JS, Patidar R, Rosenberg M, Ambrogio L, Auclair D, Wang J, Song YK, et al. 2014. Comprehensive genomic analysis of rhabdomyosarcoma reveals a landscape of alterations affecting a common genetic axis in fusion-positive and fusion-negative tumors. *Cancer Discov* **4:** 216–231. doi:10.1158/2159-8290.CD-13-0639

Shi J, Wang E, Milazzo JP, Wang Z, Kinney JB, Vakoc CR. 2015. Discovery of cancer drug targets by CRISPR-Cas9 screening of protein domains. *Nat Biotechnol* **33:** 661–667. doi:10.1038/nbt.3235

Shi H, Tao T, Abraham BJ, Durbin AD, Zimmerman MW, Kadoch C, Look AT. 2020. ARID1A loss in neuroblastoma promotes the adrenergic-to-mesenchymal transition by regulating enhancer-mediated gene expression. *Sci Adv* **6:** eaaz3440. doi:10.1126/sciadv.aaz3440

Strahl BD, Allis CD. 2000. The language of covalent histone modifications. *Nature* **403:** 41–45. doi:10.1038/47412

Takeda DY, Spisák S, Seo JH, Bell C, O'Connor E, Korthauer K, Ribli D, Csabai I, Solymosi N, Szállási Z, et al. 2018. A somatically acquired enhancer of the androgen receptor is a noncoding driver in advanced prostate cancer. *Cell* **174:** 422–432.e13. doi:10.1016/j.cell.2018.05.037

Talluri S, Dick FA. 2012. Regulation of transcription and chromatin structure by pRB: here, there and everywhere. *Cell Cycle* **11:** 3189–3198. doi:10.4161/cc.21263

Taplin ME, Bubley GJ, Shuster TD, Frantz ME, Spooner AE, Ogata GK, Keer HN, Balk SP. 1995. Mutation of the androgen-receptor gene in metastatic androgen-independent prostate cancer. *N Engl J Med* **332:** 1393–1398. doi:10.1056/NEJM199505253322101

Tomasetti C, Vogelstein B. 2015. Cancer etiology. Variation in cancer risk among tissues can be explained by the number of stem cell divisions. *Science* **347:** 78–81. doi:10.1126/science.1260825

Upender MB, Habermann JK, McShane LM, Korn EL, Barrett JC, Difilippantonio MJ, Ried T. 2004. Chromosome transfer induced aneuploidy results in complex dysregulation of the cellular transcriptome in immortalized and cancer cells. *Cancer Res* **64:** 6941–6949. doi:10.1158/0008-5472.CAN-04-0474

Valencia AM, Kadoch C. 2019. Chromatin regulatory mechanisms and therapeutic opportunities in cancer. *Nat Cell Biol* **21:** 152–161. doi:10.1038/s41556-018-0258-1

Valencia AM, Collings CK, Dao HT, St Pierre R, Cheng YC, Huang J, Sun ZY, Seo HS, Mashtalir N, Comstock DE, et al. 2019. Recurrent SMARCB1 mutations reveal a nucleosome acidic patch interaction site that potentiates mSWI/SNF complex chromatin remodeling. *Cell* **179:** 1342–1356.e23. doi:10.1016/j.cell.2019.10.044

Verhaak RGW, Bafna V, Mischel PS. 2019. Extrachromosomal oncogene amplification in tumour pathogenesis and evolution. *Nat Rev Cancer* **19:** 283–288. doi:10.1038/s41568-019-0128-6

Wan L, Chong S, Xuan F, Liang A, Cui X, Gates L, Carroll TS, Li Y, Feng L, Chen G, et al. 2020. Impaired cell fate through gain-of-function mutations in a chromatin reader. *Nature* **577:** 121–126. doi:10.1038/s41586-019-1842-7

Wang L, Shilatifard A. 2019. UTX mutations in human cancer. *Cancer Cell* **35:** 168–176. doi:10.1016/j.ccell.2019.01.001

Wang T, Wei JJ, Sabatini DM, Lander ES. 2014. Genetic screens in human cells using the CRISPR-Cas9 system. *Science* **343:** 80–84. doi:10.1126/science.1246981

Wang X, Lee RS, Alver BH, Haswell JR, Wang S, Mieczkowski J, Drier Y, Gillespie SM, Archer TC, Wu JN, et al. 2017.

SMARCB1-mediated SWI/SNF complex function is essential for enhancer regulation. *Nat Genet* **49:** 289–295. doi:10.1038/ng.3746

Wangsa D, Quintanilla I, Torabi K, Vila-Casadesus M, Ercilla A, Klus G, Yuce Z, Galofre C, Cuatrecasas M, Lozano JJ, et al. 2018. Near-tetraploid cancer cells show chromosome instability triggered by replication stress and exhibit enhanced invasiveness. *FASEB J* **32:** 3502–3517. doi:10.1096/fj.201700247RR

Weischenfeldt J, Dubash T, Drainas AP, Mardin BR, Chen Y, Stütz AM, Waszak SM, Bosco G, Halvorsen AR, Raeder B, et al. 2017. Pan-cancer analysis of somatic copy-number alterations implicates IRS4 and IGF2 in enhancer hijacking. *Nat Genet* **49:** 65–74. doi:10.1038/ng.3722

Whyte WA, Orlando DA, Hnisz D, Abraham BJ, Lin CY, Kagey MH, Rahl PB, Lee TI, Young RA. 2013. Master transcription factors and mediator establish super-enhancers at key cell identity genes. *Cell* **153:** 307–319. doi:10.1016/j.cell.2013.03.035

Wiegant J, Ried T, Nederlof PM, van der Ploeg M, Tanke HJ, Raap AK. 1991. In situ hybridization with fluoresceinated DNA. *Nucleic Acids Res* **19:** 3237–3241. doi:10.1093/nar/19.12.3237

Wilson BG, Roberts CW. 2011. SWI/SNF nucleosome remodellers and cancer. *Nat Rev Cancer* **11:** 481–492. doi:10.1038/nrc3068

Wingender E, Schoeps T, Dönitz J. 2013. TFClass: an expandable hierarchical classification of human transcription factors. *Nucleic Acids Res* **41:** D165–D170. doi:10.1093/nar/gks1123

Wu S, Turner KM, Nguyen N, Raviram R, Erb M, Santini J, Luebeck J, Rajkumar U, Diao Y, Li B, et al. 2019. Circular ecDNA promotes accessible chromatin and high oncogene expression. *Nature* **575:** 699–703. doi:10.1038/s41586-019-1763-5

Yang H, Ye D, Guan KL, Xiong Y. 2012. IDH1 and IDH2 mutations in tumorigenesis: mechanistic insights and clinical perspectives. *Clin Cancer Res* **18:** 5562–5571. doi:10.1158/1078-0432.CCR-12-1773

Zhang X, Choi PS, Francis JM, Imielinski M, Watanabe H, Cherniack AD, Meyerson M. 2016. Identification of focally amplified lineage-specific super-enhancers in human epithelial cancers. *Nat Genet* **48:** 176–182. doi:10.1038/ng.3470

Zhang X, Choi PS, Francis JM, Gao GF, Campbell JD, Ramachandran A, Mitsuishi Y, Ha G, Shih J, Vazquez F, et al. 2018. Somatic superenhancer duplications and hotspot mutations lead to oncogenic activation of the KLF5 transcription factor. *Cancer Discov* **8:** 108–125. doi:10.1158/2159-8290.CD-17-0532

Zhang J, Lee D, Dhiman V, Jiang P, Xu J, McGillivray P, Yang H, Liu J, Meyerson W, Clarke D, et al. 2020. An integrative ENCODE resource for cancer genomics. *Nat Commun* **11:** 3696. doi:10.1038/s41467-020-14743-w

Zhu J, Sammons MA, Donahue G, Dou Z, Vedadi M, Getlik M, Barsyte-Lovejoy D, Al-awar R, Katona BW, Shilatifard A, et al. 2015. Gain-of-function p53 mutants co-opt chromatin pathways to drive cancer growth. *Nature* **525:** 206–211. doi:10.1038/nature15251

Zuber J, Shi JW, Wang E, Rappaport AR, Herrmann H, Sison EA, Magoon D, Qi J, Blatt K, Wunderlich M, et al. 2011. RNAi screen identifies Brd4 as a therapeutic target in acute myeloid leukaemia. *Nature* **478:** 524–528. doi:10.1038/nature10334

Viruses in the Nucleus

Bojana Lucic,[1] Ines J. de Castro,[1] and Marina Lusic

Department of Infectious Diseases, Integrative Virology, Heidelberg University Hospital and German Center for Infection Research, Im Neuenheimer Feld 344, 69120 Heidelberg, Germany

Correspondence: marina.lusic@med.uni-heidelberg.de

Viral infection is intrinsically linked to the capacity of the virus to generate progeny. Many DNA and some RNA viruses need to access the nuclear machinery and therefore transverse the nuclear envelope barrier through the nuclear pore complex. Viral genomes then become chromatinized either in their episomal form or upon integration into the host genome. Interactions with host DNA, transcription factors or nuclear bodies mediate their replication. Often interfering with nuclear functions, viruses use nuclear architecture to ensure persistent infections. Discovering these multiple modes of replication and persistence served in unraveling many important nuclear processes, such as nuclear trafficking, transcription, and splicing. Here, by using examples of DNA and RNA viral families, we portray the nucleus with the virus inside.

Viruses have long been studied not only as causative agents of human diseases, but also as extraordinary tools to explore cell biology, and in particular that of the nucleus. In fact, many elements of the eukaryotic transcription machinery have been elucidated with viruses. The first transcriptional promoter and enhancer elements were described in the simian vacuolating virus 40 (SV40) genome, a closed circular and superhelical polyomavirus discovered in 1960, found in monkeys and humans (Liu 2014). SV40 served to describe the binding of transcription factors (TFs) to the promoter, like SP-1 (Dynan and Tjian 1983), or the enhancer element (activator protein [AP]-1 and AP-2) (Lee et al. 1987). Together with adenovirus (AdV) major late promoter, SV40 served as an important in vitro experimental system for the discovery of basal and accessory TFs. Major discoveries of mRNA processing were made possible due to adenoviral infections (Berget et al. 1977; Chow et al. 1977), while polyadenylation was first observed with poxviruses. Apart from being fundamental for the discovery of the enzyme reverse transcriptase (RT) (Baltimore 1970; Mizutani et al. 1970) and for the initiation of recombinant DNA technologies, retroviruses together with DNA tumor viruses provide major contributions to our understanding of the origins of human cancers.

Viral genomes can assume different configurations: single- and double-stranded DNA (dsDNA), plus- or minus-stranded RNA, or double-stranded RNA. Retroviruses contain a positive-strand RNA molecule retrotranscribed into DNA, and as most dsDNA viruses (polyomaviruses, adenoviruses, herpesviruses, papillomaviruses [PVs]) they replicate in the nucleus.

[1]These authors contributed equally to this work.

Cite this article as *Cold Spring Harb Perspect Biol* doi: 10.1101/cshperspect.a039446

Among RNA viruses, only some negative-strand RNA viruses like influenza have a nuclear phase in their life cycle. Some of them require host cell polymerases to replicate their genome, while others, like adenoviruses and herpesviruses, encode their own replication factors. In both cases, viral replication is highly dependent on the cellular state and cellular permissiveness to replication. Among the viruses that will be explored here are several families of DNA viruses, one example of an RNA virus (influenza), as well as different members of the Retroviridae family.

- Herpesviruses cause a myriad of clinical conditions, ranging from mild skin sores to cancer. They have a large nonsegmented linear dsDNA genome, encoding for 70–200 different genes, and one of their best-known features is their capacity to establish latent infections. Here we discuss herpes simplex viruses (HSV)-1 and -2, Epstein–Barr virus (EBV), Kaposi's sarcoma–associated herpesvirus (KSHV) and cytomegalovirus (CMV).
- Adenoviruses are nonenveloped viruses first isolated from adenoids (tonsils). They contain a linear dsDNA genome, up to 48 kb, which enters the nucleus to replicate, relying entirely on host replication machinery. They mostly affect the respiratory tract and are subject to enormous investigations due to their delivery potential for gene therapy.
- PVs are nonenveloped dsDNA viruses causing skin warts or papillomas; some of them, like human papilloma virus (HPV) 16 and 18, can cause cancer. Upon infection, the viral genome replicates at low levels, resulting in escape from immune surveillance and ultimately cancer development.
- The hepatitis B virus (HBV) is a small virus of a peculiar structure and replication cycle. It contains a partially double-stranded and partially single-stranded circular DNA and its replication involves an RNA intermediate. It encodes its own polymerase, which has an RT activity, a DNA-dependent DNase activity, and an RNase activity. It causes hepatitis that is transmissible by blood and bodily fluids.
- Influenza viruses, causative agents of flu, contain a segmented negative strand viral RNA genome. The eight segments of RNA enter the nucleus in the form of viral ribonucleoproteins (vRNPs), containing viral proteins that possess nuclear localization signals.
- Retroviruses are a diverse family of RNA viruses encoding two copies of RNA genomes, which upon reverse transcription to viral DNA enter in the nucleus in the form of preintegration complex (PIC), can be rapidly loaded with histones, and are integrated into the host genome to ensure their replication. Types of retroviruses discussed here are human immunodeficiency virus 1 (HIV-1), human T-cell leukemia virus type 1 (HTLV-1), and murine leukemia virus (MLV).

NUCLEAR ENTRY AND NUCLEAR PORE COMPLEX

Entering the nucleus is one of the major challenges posed to a virus seeking the cellular machinery to transcribe and create progenies. Many viruses exploit the host cell nucleocytoplasmic trafficking mechanisms through the nuclear pore complex (NPC) to access nuclear functions, while others rely on the disruption of the nuclear envelope during cellular division. The mechanism of entry of viruses differs greatly and depends mainly on their size and on the properties of their viral proteins. Whereas some are small enough to passively diffuse through the NPC (e.g., HBV), others access the chromatin and integrate in the absence of this barrier (i.e., during mitosis, for example, PVs and MLVs, respectively), they can also attach to the NPC and from there release their viral genome (e.g., herpesvirus or adenovirus), or enter the nucleus via the NPC like lentiviruses (HIV-1) (Fig. 1).

The NPC is the largest macromolecular structure of the cell. Composed of ~30 nucleoporins (Nups), the NPC is the gate for the entry/exit of molecules from the cytoplasm to the nuclei either by passive diffusion or by active transport (Beck and Hurt 2017). Intercalating the nuclear membrane, this structure forms cytoplasmic, inner, and nuclear rings, and peripherally extends both to the cytoplasm by

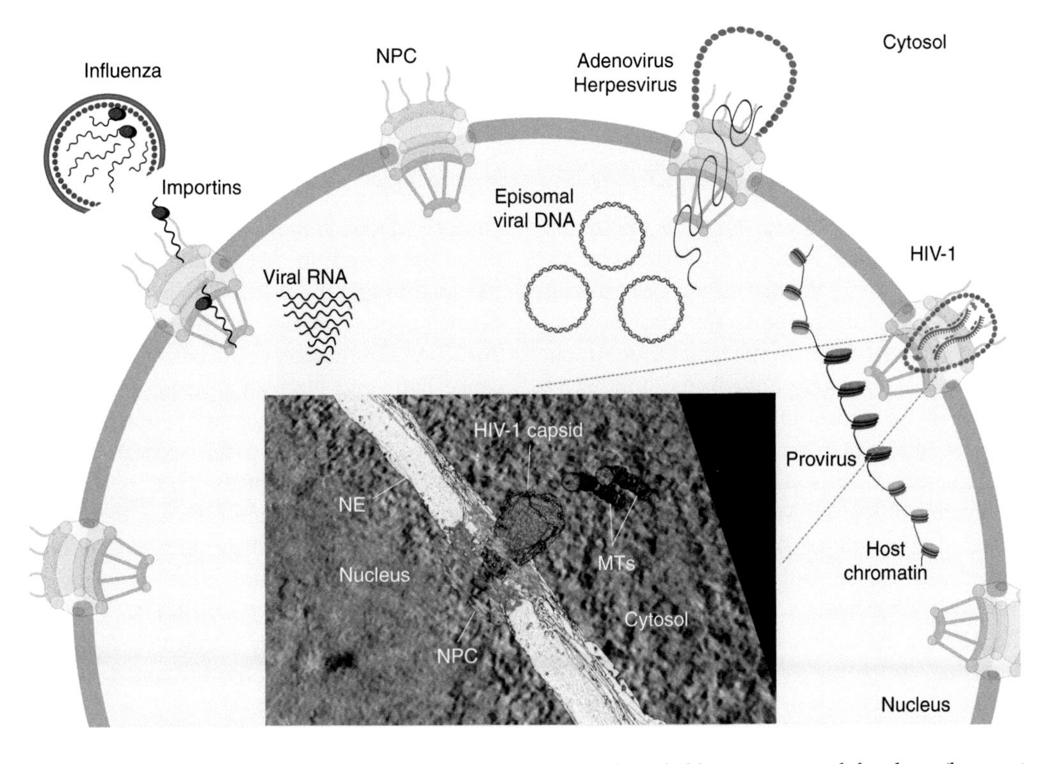

Figure 1. Viral nuclear entry. Influenza viral RNA enters the nuclei aided by importins while adeno-/herpesviruses attach to the nuclear pore complex (NPC) for the release of the genomic material. The lentivirus, HIV-1, travels with the capsid (CA). (*Inset*) 3D rendering of electron tomographic reconstruction showing the HIV-1 CA mutant A77 V (CPSF6-binding deficient) entering the NPC central channel in infected T lymphoblast SupT1-R5. (NE) Nuclear envelope, (MTs) microtubules. (Image kindly provided by Dr. Vojtech Zila and Erica Margiotta.)

cytoplasmic-oriented Nups and to the nucleus via nuclear basket Nups (Beck and Hurt 2017).

Nuclear entry of many viruses requires interaction with the NPC (Cohen et al. 2011; Fay and Panté 2015). This synergy was observed in HSV-1 infection, where electron microscopy (EM) showed the docking of capsid (CA) at the NPC (Batterson et al. 1983; Lycke et al. 1988). The viral DNA diffuses through the NPC, adopting a condensed rod-like structure (Shahin et al. 2006), after which empty CAs are released (Sodeik et al. 1997; Batterson et al. 1983; Lycke et al. 1988). CA binding to the NPC is maintained by two Nups, Nup214 and Nup358 (Copeland et al. 2009; Pasdeloup et al. 2009). On the other hand, several host response proteins, like MX2, might counteract this process (Huffman et al. 2017; Crameri et al. 2018). MX2 restriction goes beyond HSV-1, as it affects the closely related viruses HSV-2 and KSHV (Crameri et al. 2018), as well as HIV-1 (Goujon et al. 2013; Kane et al. 2013), where it has been shown to interact with different Nups (Dicks et al. 2018; Kane et al. 2018).

NPC is undeniably one of the main structures of the cell whose functions viruses misappropriate to shut down immune responses and increase their replication. EBV hijacks the host nuclear entry complex by encoding for a kinase that phosphorylates Nup62 and Nup153, causing the redistribution of several Nups and favoring viral replication (Chang et al. 2015). Similarly, HIV-1 infection has been linked to Nup62 delocalization from the NPC to cytosol as it travels in association with vRNP during export via the Rev–vRNA RNP (Monette et al. 2011).

Adenoviral genome nuclear delivery is similar to that of herpesvirus, as CA is retained at the NPC. Linear dsDNA in the core of a CA shell associates with five viral proteins that aid the disassembly from the viral CA and docking to the NPC (Flatt and Greber 2017; Greber and Flatt 2019). The CA presents itself partially disassembled when docked at the NPC, a process mediated via Nup214 and Nup358 (Trotman et al. 2001; Strunze et al. 2011; Cassany et al. 2015). At the NPC, the recruitment of nuclear transport receptors supports the viral protein VII in the translocation of the viral DNA, which in many cases remains undelivered and accumulates in the cytosol in a CA-free form (Flatt and Greber 2017; Greber and Flatt 2019).

Influenza A virus (IAV) has a rigid shell, mainly constituted by M1 matrix protein, that protects the bundle of vRNAPs that are transformed into transcribed RNA in the nucleus. The shell dissociates in the cytoplasm and vRNAPs are transported into the nucleus via importins. Interestingly, histone deacetylase (HDAC)6 participates in the uncoating process, whereas nuclear import factor transportin 1 (TNPO1) promotes vRNAPs release. Freed vRNAPs can thus bind importins (α/β) for NPC translocation (Banerjee et al. 2014; Miyake et al. 2019).

The HIV-1 genome is translocated through the NPC into the nucleus in a form of a complex composed of the reversed transcribed viral DNA, CA, and integrase (IN) proteins. Reverse transcription and CA disassembly have long been considered to occur in the cytosol; however, a recent paper from Campbell's laboratory provides compelling evidence that this is completed in the nucleus. By using an inducible nuclear pore blockade system, they show that nuclear entry occurs with rapid kinetics that precedes the completion of reverse transcription (Dharan et al. 2020). The notion that PIC travels with intact CA structures through the NPC is consistent with recent reports of the presence of CA inside the nuclei of both T cells (Blanco-Rodriguez et al. 2020; Burdick et al. 2020) and macrophages (Bejarano et al. 2019). HIV-1 translocation into the nuclei depends on the protruding cytoplasmic Nup, Nup358, which binds CA through its cyclophilin (Cyp) domain (Schaller et al. 2011; Bichel et al. 2013). Nup153, at the nucleoplasmic side of the NPC, has also been shown to be important for translocation through its direct interaction with CA (Matreyek et al. 2013; Buffone et al. 2018; Bejarano et al. 2019). However, it has remained poorly understood how the RT/PIC with an estimated diameter of ~61 nm, composed of ~1500 CA monomers organized in hexamers or pentamers, passes through the NPC, which has an inner diameter of ~39 nm (Campbell and Hope 2015). A combination of 3D correlative fluorescence light and electron microscopy (CLEM) and cryo-electron tomography (CryoET) has now shed new light on the nuclear entry of HIV-1. The diameter of the NPC is sufficient to allow the nuclear import of intact, cone-shaped CAs, while the disruption of the hexagonal CA lattice occurs only once inside the nucleus (Fig. 1, inset; Zila et al. 2021).

HIV-1 nuclear entry is aided by numerous Nups interactions that hand-in-hand deliver the viral genomic content to the nucleus. The first visualization of the relationship between HIV-1 genomes and the NPC came in 2002 when HIV-1 RNA was shown to accumulate in close proximity or at the NPC (Cmarko et al. 2002). Inside the nucleus, HIV-1 viral DNA interacts with Nup153, Nup98, and the translocated promoter region (TPR), implicated in HIV-1 integration into the host genome (Di Nunzio et al. 2013; Lelek et al. 2015; Marini et al. 2015). The fact that passage through the NPC represents a functional bottleneck for HIV-1 replication is further supported by the presence of chromatin readers, splicing, and polyadenylation factors that aid HIV-1 integration. Particularly, cleavage and polyadenylation specificity factor 6 (CPSF6) and lens epithelium-derived growth factor (LEDGF/p75) are both tethered to the NPC via Nup153 and TPR, through binding of HIV-1 CA and IN, respectively (Lelek et al. 2015; Bejarano et al. 2019). Moreover, HIV-1 recurrently integrates in chromatin enriched for active transcription and super-enhancer (SE) markers (H3K4me1, H3K27ac, bromodomain-containing 4 [BRD4], and mediator of RNA polymerase II transcription subunit 1 [MED1]), clustering sites of integration >1 Mb apart (Lucic et al. 2019). Interestingly, develop-

 Cite this article as *Cold Spring Harb Perspect Biol* doi: 10.1101/cshperspect.a039446

mentally regulated and tissue-specific genes have been shown to bind Nups and harbor enhancer markers, both in human, mouse, and *Drosophila* cells (Ibarra et al. 2016; Pascual-Garcia et al. 2017; Scholz et al. 2019).

INTERACTIONS WITH THE CELLULAR GENOME

Upon nuclear entry, the viral genome can either maintain an episomal form or integrate into the host genome. Either way, most viral genomes become highly chromatinized upon nuclear entry (i.e., gain nucleosomes, interact with TFs and chromatin modifiers, and present histone modifications contributing to their gene expression) (for review, see Knipe et al. 2013). Studies on the interaction between the viral and host genomes started evolving with the development of genome-wide capturing techniques, namely, chromosome conformation capture and its derivatives (3C, 4C, and later Hi-C) (for review, see Dekker et al. 2013; Denker and de Laat 2016).

One of the first viruses under scrutiny by these technologies was HIV-1. Initially, a loop between the proviral 5′ and 3′ long terminal repeats (LTRs) was linked to active viral transcription (Perkins et al. 2008). Subsequently, interactions between HIV-1 and chromosome 12 were observed using HIV-1 provirus as bait. Interestingly these interactions are lost upon activation of the lymphocytes and thus seem related to maintaining a state of repression (i.e., viral latency) (Dieudonné et al. 2009). More recently, a comprehensive Hi-C contact map from T Jurkat cells was cross-compared to HIV-1 insertion sites sequenced from patients and in vitro infected primary T cells. HIV-1 insertions occur mostly in transcriptionally active compartments (Chen et al. 2017) that display the highest contact frequency and strongly interact with each other (Lucic et al. 2019). Moreover, HIV-1 displays integration biases for cell-type-specific SEs (>30% of HIV-1 recurrent integration genes are in classified SEs in primary $CD4^+$ T cells), specifically for those in A1 active subcompartments (Fig. 2; Chen et al. 2017; Lucic et al. 2019). The importance of genome organization and an integration bias toward active compartments were also proposed through a biophysical model of HIV-1 integration on chromatin fibers (Michieletto et al. 2019).

Most DNA viruses maintain their genomes extrachromosomaly in dividing cells, where their episomal genomes are tethered to cellular chromosomes by a virally encoded protein (Coursey and McBride 2019). Interactions with the host genome can result in disruption of cellular physiology and lead to cancer (EBV, HPV), DNA damage (parvovirus), or cause transcriptional aberrations (IAV).

The parvovirus, minute virus of mice (MVM), is a highly infectious rodent virus that causes DNA damage and uses S phase to begin replication. A 3C analysis directed at the viral genome (V3C) showed the interaction between MVM and the host DNA. These sites, termed virus-associated domains (VADs), correlate strongly with sites of DNA damage, occurring also in mock infected cells during the progression through the S phase of the cell cycle (Majumder et al. 2018). PVs establish persistent replication cycles, where upon initial infection the viral genome replicates with low levels of gene expression. Tethering of the viral and host genomes is enabled by the E2 PV protein, and in particular by the DNA-binding domain (DBD), which shows strong similarities with latency-associated nuclear antigen (LANA) and Epstein–Barr nuclear antigen (EBNA) proteins, responsible, respectively, for tethering of KSHV and EBV to cellular genomes (Coursey and McBride 2019). All three viral proteins bind to the transcriptionally active regions of chromatin, possibly as a result of their roles in viral transcription. However, chromosomal interactions between PV and EBV seem to differ, as PV interacts with transcriptionally active regions of the genome, EBV was found tethered to heterochromatic regions, as assessed by Hi-C (Moquin et al. 2018).

The transforming potential of EBV is in part linked to activation of MYC and other oncogenes via its own proteins. Upon infection of B lymphocytes, EBNA2, one of the six viral nuclear proteins, causes MYC up-regulation by promoting long-range enhancer and promoter looping (Zhao et al. 2011). This study paved the way to a more comprehensive understanding of

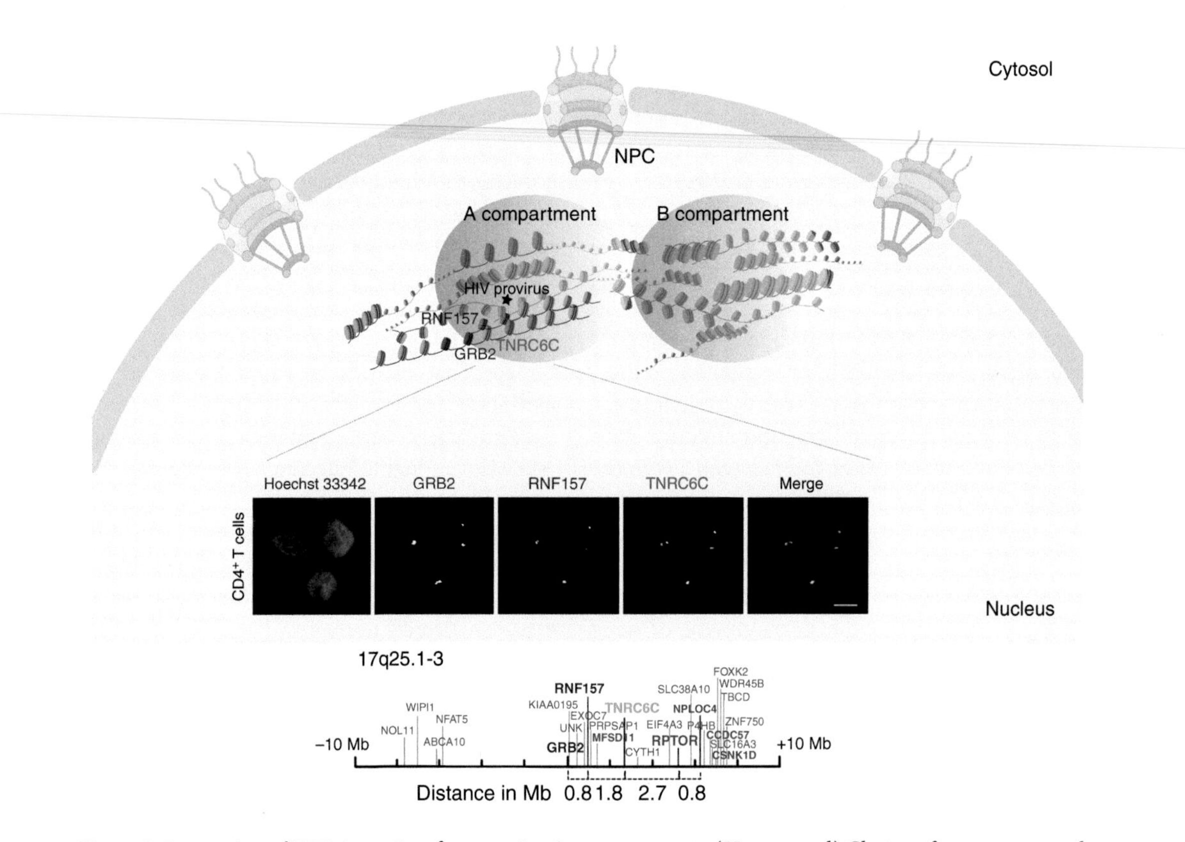

Figure 2. Integration of HIV-1 provirus favors active A compartments. (*Upper* panel) Cluster of genes recurrently targeted by HIV-1. Triple-color fluorescence in situ hybridization (FISH) of GRB2 (magenta), RNF157 (red), and TNRC6C (green) in resting and activated $CD4^+$ T cells. Maximum projections for each of the channels and merge images are shown. Nuclei were counterstained with Hoechst 33342 (blue). Scale bar, 4 µm. (*Lower* panel) Distribution of recurrent integration genes (RIGs) (bold) and single HIV-1 integration sites (plain) at the chromosomal region 17q25.1-3 within 20 Mb. Genes analyzed by triple-color FISH are depicted in magenta, red, and green. Distances between them on the linear DNA are indicated. (Images made by the Lusic laboratory in collaboration with V. Roukoss.)

EBV in cellular transformation. In EBV transformed lymphoblastoid cell lines (LCLs) 2000 enhancer sites are occupied by EBV proteins as well as by SE markers (Zhou et al. 2015), with 30% of genes important for LCL growth and survival representing EBV SEs (Jiang et al. 2017). Apart from EBNA2, other EBV proteins, like EBNA leader protein (EBNA-LP) and EBNA3C, were also linked to cellular transformation (Szymula et al. 2018), pointing to a concerted action of EBV proteins in taking over genome organization. Finally, tethering of EBV to the cellular genome seems to depend on EBNA1, as shown by 4C experiments in latently infected EBV cell lines. As this tethering persists during mitosis as well, it is plausible that such interactions can be conserved in cycling cells (Kim et al. 2020).

HBV is transcribed in the nucleus from a viral covalently closed circular DNA (cccDNA) form (Newbold et al. 1995). Its interactions with the genome were assessed by a combination of Hi-C with viral DNA capture (CHi-C) (Moreau et al. 2018) in primary human hepatocytes, revealing that cccDNA interacts with positively marked CpG islands, occupied by RNA polymerase II (RNAPII), H3K4m3, H3K27ac, and CXXC-type zinc finger protein 1 (Cfp1) TF. Some of these CpG islands overlap with hepatocyte-specific genes and are deregulated upon infection. A parallel study using 4C also linked cccDNA with actively transcribed regions, enriched for positive histone modifications and occupied by the HBV *trans*-activating protein HBV X (HBx) (Hensel et al. 2018).

The study by Moreau et al. also assessed chromatin interactions of the adenovirus type 5 (Ad5) in hepatocytes derived from an infected patient. The episomal Ad5 binding is skewed to regions enriched for active marks, similar to those of enhancers and transcription start sites (TSSs) and occupied by the forkhead box A (FOXA) TF (Moreau et al. 2018). Ad5 infection causes a striking global change of histone profiles, most of which is dependent on the first produced viral protein E1A (Horwitz et al. 2008). Viral protein VII complexes with nucleosomes, and its chromatin association is dependent on posttranslational modifications. Protein VII also directly binds high mobility group protein B1 (HMGB1), which remains attached to chromatin instead of being released as an inflammatory signal (Avgousti et al. 2016). This nucleosome binding is hence used by the virus as a strategy of immune invasion. Cellular transformation by adenovirus is initiated by viral protein E1A and its small isoform e1a, which causes a significant reshuffling of the active H3K18ac mark. By erasing almost 95% of the existing peaks and reestablishing new ones at promoter regions of genes involved in cell cycling, the virus appropriates cell propagation (Ferrari et al. 2012). E1A simultaneously binds histone acetyltransferase p300/CBP and the tumor suppressor retinoblastoma (RB) protein, which is locked into a repressing conformation upon p300 acetylation. This results in repression of genes that would otherwise be activated to inhibit viral replication (Ferrari et al. 2014).

IAV causes a strong transcriptional response of the host cell and induces changes in the nuclear morphology (Terrier et al. 2014). Of eight nonstructural (NS) proteins encoded by the virus, NS1 impacts genome organization and transcription of infected human macrophages. Changes in gene expression and 3D organization occur already at the early stages of infection, as revealed by Hi-C and RNA-seq. Many genomic regions transition from repressive heterochromatic (B) to active, euchromatic (A) compartments, especially genes involved in type I/III interferon (IFN) pathways. This transition has been attributed to RNAPII readthrough transcription past the end of genes, mediated by NS1 protein. Dragging of RNAPII through the genome has been proposed to perturb CCCTC-binding factor (CTCF) sites and thus affects genome demarcation (Heinz et al. 2018). This is in agreement with the findings of Bauer et al., who used native elongating transcript sequencing (mNET-seq) in IAV infection and reported production of downstream-of-gene transcripts (DoGs). Contrary to the first study, the latter did not attribute RNAPII readthrough to NS1 (Bauer et al. 2018). However, it seems plausible that NS1 indeed plays a role in RNAPII readthrough, based on its capacity to bind and inhibit the function of CPSF complex, cleavage, and polyadenylation factor (Noah et al. 2003). A role for NS1 protein in RNAPII readthrough was further strengthened by the finding that NS1 partitions into nuclear ribonucleoprotein complexes, depending on the conjugation of the small ubiquitin-like modifier (SUMO) to the site embedded in a PDZ domain. NS1-SUMO virus caused a pervasive run-through into downstream genes due to interference with 3′ end cleavage and termination, which in turn causes splicing defects at later time points (Zhao et al. 2018). The evolution of domains in viral strains might thus confer different properties and advantages to viruses.

HTLV-1, a retrovirus capable of provoking aggressive malignancies in $CD4^+$ T cells in addition to the viral genes encoding for enzymes and structural proteins, contains regulatory and accessory genes (Bangham 2018). HBZ, one of the accessory genes required for clonal persistence of the integrated virus, is expressed from the minus strand, while plus strand expression necessary for viral propagation is suppressed in vivo. CTCF, a master regulator of chromatin structure and gene expression, binds to the proviral genome and acts as an enhancer blocker, causing long-distance looping between viral and host genome, thus resulting in changed patterns of gene expression (Satou et al. 2016; Melamed et al. 2018). Long-range interactions between the viral and cellular sequences mediated by CTCF were proposed as a plausible mechanism contributing to the malignant transformation of $CD4^+$ T cells, while maintaining an architecture favorable for the virus.

CHROMATIN AND TRANSCRIPTION—INSIGHTS FROM RETROVIRUSES

The plasticity of the cellular genome is reflected in its ability to change in response to external stimuli. Acute viral infections like influenza often cause robust changes in cellular transcription programs. On the other hand, latent viral infections, typical for herpesviruses or lentiviruses, do not involve drastic changes in cellular transcription programs. They persist for longer periods while retaining the capability to be reversed and require different levels of precise transcriptional and chromatin control.

The HIV-1 transcriptional program consists of both phases with multiple layers of control. The viral LTR regulatory region has a structure of a typical RNAPII promoter with adjacent regulatory elements for binding of cellular TFs like nuclear factor (NF)-κB, nuclear factor of activated T cells (NFAT), or AP-1. In the basal, steady state, short abortive transcripts are transcribed from the LTR resulting in the establishment of latency (Shukla et al. 2020). One peculiarity of HIV-1 transcriptional control is the virally encoded Tat *trans*-activator (for a comprehensive review on viral TFs, see Ho et al. 2020) that binds to a *cis*-acting RNA element (*trans*-activation-responsive regions [TARs]) at the 5′ end of viral mRNAs. This promotes the assembly of a transcriptionally active complex containing RNAPII and positive transcription elongation factor complex (P-TEFb) (Zhu et al. 1997). Tat protein has been crucial for the discovery of the cyclin T1 component of P-TEFb (Wei et al. 1998), as for many other RNAPII elongation regulators and factors, like Spt5, DRB sensitivity-inducing factor (DSIF)/negative elongation factor (NELF), or BRD4 (for review, see Karn and Stoltzfus 2012).

Chromatin organization of proviral HIV-1 is rapidly established at the viral promoter upon integration and comprises three precisely positioned nucleosomes (nuc-0, nuc-1, and nuc-2) with two intervening DNase I hypersensitive sites (DHSs) (Verdin et al. 1993). Among chromatin factors recruited to viral promoter, BRD4 stands out, as it encodes two isoforms with different roles in HIV-1 transcription. A long isoform contains a proline-rich terminal tail that binds P-TEFb (Wu and Chiang 2007) and supports viral transcriptional elongation (Jang et al. 2005; Yang et al. 2005; Bisgrove et al. 2007), whereas the short BRD4 isoform acts together with chromatin remodeling BAF complexes to silence HIV-1 (Conrad et al. 2017).

HIV-1 LTR nucleosome remodeling fine tune replication of the integrated viruses (Verdin et al. 1993; Lusic et al. 2003; Zhang et al. 2007; Natarajan et al. 2013). However, unintegrated extrachromosomal forms of HIV-1 are also chromatinized (Geis and Goff 2019), but unlike the integrated one they are loaded with repressive histone modifications, H3K9me3, and deprived from active marks (i.e., histone acetylation) (Geis and Goff 2019). MNase-seq and in silico predictions suggest that chromatinization into an additional nucleosome that covers DHS (nuc-DHS) occurs even before the integration process and is lost upon viral insertion (Machida et al. 2020). The existence of nuc-DHS is proposed to serve to silence viral DNA prior to becoming an integral part of the host genome.

Similar to the HIV-1, the MLV genome is also rapidly loaded with histones in the unintegrated form (Wang et al. 2016). The episomal MLV is often transcriptionally silent but can be reactivated upon treatment with histone deacetylase inhibitors, strengthening the notion that these forms are subjected to particular modes of epigenetic silencing (Wang et al. 2016). A CRISPR-Cas9 screen on cells infected with wild-type and integration-defective MLV showed the involvement of a histone-modifying complex in this silencing. The complex contains a DNA-binding protein with a zinc finger motif NP220, SET domain bifurcated histone lysine methyltransferase 1 (SETDB1), and the members of the human silencing hub (HUSH) complex (Zhu et al. 2018).

Primarily, the HUSH complex was shown to mediate position effect variegation (i.e., silencing of a reporter construct as a result of its integration into silent chromatin) (Tchasovnikarova et al. 2015). It contains three main components: Family protein with sequence similarity (FAM208A), later renamed to transgene activator suppressor (TASOR), matrix metalloproteinase MPP8, and periphilin 1 (PPHLN1). It interacts with histone methyltransferase SETDB1 and is recruited to genomic loci enriched in H3K9me3 (Tchasovnikarova et al. 2015; Timms et al. 2016). Although HUSH activity was demonstrated on both MLV and HIV-1 promoters, individual knockout of the complex members does not affect HIV-1 expression. HUSH inefficacy in repressing HIV-1 might be in the viral genome code, as two studies point to the antagonist effects of viral protein Vpx in HUSH silencing (Geis and Goff 2019).

Among other important chromatin-binding factors, discovered or predominantly studied in the context of lentiviruses, is LEDGF/p75, a ubiquitously expressed chromatin reader and binding partner of HIV-1 IN (Cherepanov et al. 2003; Demeulemeester et al. 2015). LEDGF interacts with functionally diverse cellular factors and is involved in a broad range of nuclear mechanisms, at the crossroads of DNA repair, transcriptional control, and splicing (Aymard et al. 2014; LeRoy et al. 2019; Ui et al. 2020). Thanks to its carboxy-terminal IN-binding domain (IBD) and a chromatin-binding PWWP domain, LEDGF/p75 tethers HIV-1 PIC to chromatin (Demeulemeester et al. 2015). The PWWP domain binds specifically H3K36 trimethylated chromatin, enriched in the transcriptionally active regions of chromatin (van Nuland et al. 2013). As multiple splicing factors found to be LEDGF/p75 interacting partners (Singh et al. 2015) also bind to those regions, splicing determines LEDGF/p75 distribution on the genome and HIV-1 integration patterns (Francis et al. 2020). The promiscuous nature of LEDGF/p75 adds further to HIV-1 control by being involved in transcriptional repression and reactivation. Because of the interactions with the Spt6/ISWI complex, where depletion of all three components reactivates dormant HIV-1, LEDGF/p75 was proposed to play a role in HIV-1 latency (Gérard et al. 2015). By interacting with Pol II–associated factor (PAF1) and mixed lineage leukemia (MLL1), LEDGF/p75 can respectively promote or reactivate the latent state (Gao et al. 2020).

Retroviral (M-MLV) proviral silencing in embryonic stem cells has been attributed to the concerted action of tripartite motif family (TRIM) 28 (Wolf and Goff 2007) and zinc finger protein (ZFP) 809 (Wolf and Goff 2009), which in turn recruit heterochromatin protein 1 (HP1) and H3K9 methyltransferases to sustain silencing. TRIM28, as well as TRIM33, were shown to have a role in the HIV-1 life cycle, regulating both integration and latency. TRIM28 binds the acetylated form of HIV-1 IN, mediates its deacetylation, and thus decreases the affinity of IN for DNA, resulting in impaired integration (Allouch et al. 2011). Its role in silencing of HIV-1 was attributed to its E3 small ubiquitin-like modifier (SUMO) ligation activity, which mediates CDK9 SUMOylation and interferes with its binding to cyclin T1 (Ma et al. 2019). E3 RING ligase activity of TRIM33 defines the stability of HIV-1 IN, therefore influencing the levels of integrated and produced viral progeny (Ali et al. 2019).

NUCLEAR BODIES

Given the densely populated nature of the nucleus, its intracellular organization is quite complex and comprises a number of nuclear bodies (NBs) with distinct functions such as promyelocytic (PML) NBs, splicing speckles, Cajal bodies, and nucleoli (Mao et al. 2011). Here we present an overview of spatial interactions and functions of the two most-studied NBs in the context of infections, PML NBs, and splicing speckles (Fig. 3).

PML NUCLEAR BODIES (PML NBs)

PML protein or TRIM19 is one of the most studied members of the TRIM family, ubiquitously expressed cellular factors with important roles in antiviral defense. Multiple cellular functions of PML are partially attributed to the fact that they interact with a variety of cellular factors (Van Damme et al. 2010) forming PML NBs. PML NBs have multifaceted roles in viral replication and persistence (Everett and Chelbi-Alix

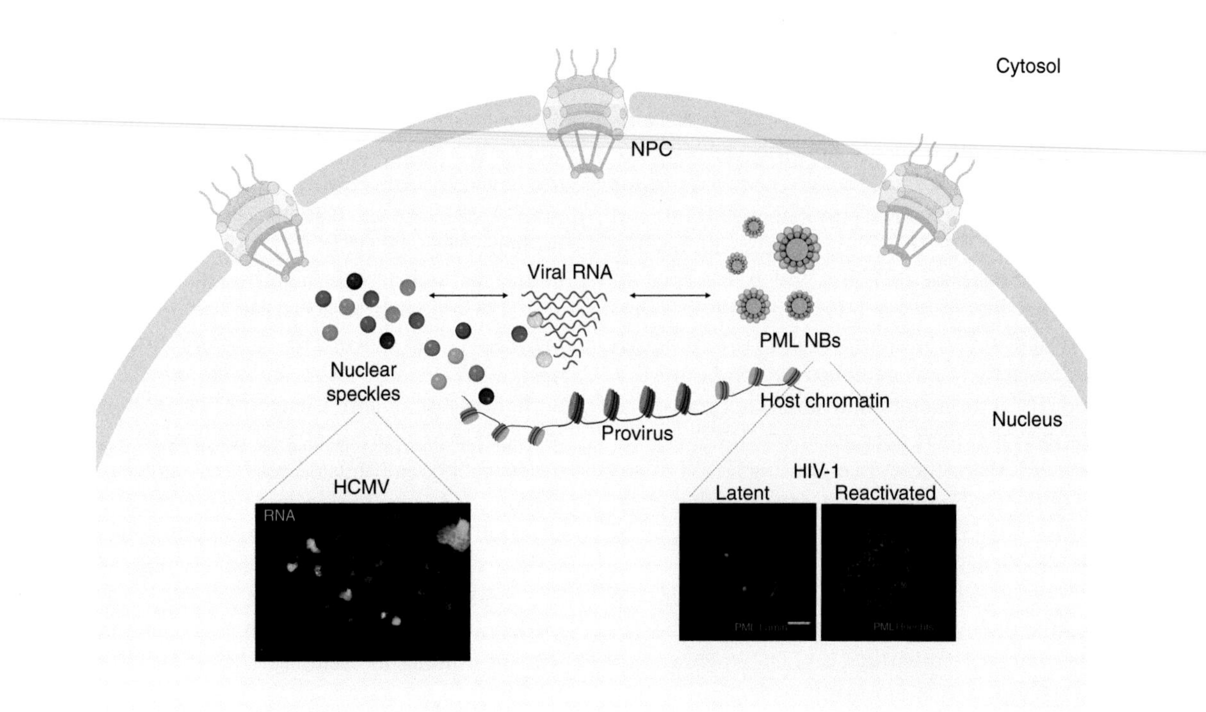

Figure 3. Interaction of viral genetic material with nuclear bodies (NBs). (*Left* panel) Human cytomegalovirus (HCMV)-infected cell triple-labeled with ND10 (promyelocytic [PML]) in red, IE transcripts in green, and SC35 in blue. Transcript signals appear to locate with the highest concentration at ND10 and are also found in the SC35 domain. (*Inset*) Arrangement as expected when all three components colocalize. (*Inset* preprinted from Ishov et al. 1997 with permission from Rockefeller University Press © 1997.) (*Right* panel) Images show HIV-1 DNA fluorescence in situ hybridization (FISH) (red) associated with promyelocytic (PML) nuclear bodies (NBs) (scale bar, 2 µm), (green) in latent cells and HIV-1 transcripts, RNA FISH (red) in activated cells where no PML NBs are detected. (*Right* panel from Shytaj et al. 2020; reprinted, with permission, from the authors in conjunction with the terms of the Creative Commons BY 4.0 license.)

2007; Scherer and Stamminger 2016). They arise via different pathways ranging from self-organization to ordered assembly with PML, a cysteine-rich, redox-sensitive protein, being their main nucleating component (Kamitani et al. 1998). PML is modified by SUMOylation in response to reactive oxygen species, which together with recruitment of its partner proteins via the SUMO-interacting motif (SIM), mediate the nucleation of NBs (Sahin et al. 2014). The putative antiviral properties of PML have been attributed to its IFN pathway-regulatory role, as it is one of the main targets of IFN signaling (Lavau et al. 1995). Whereas many viruses act to induce the reorganization and/or degradation of PML NBs, many DNA viruses establish viral replication factories (VRCs) in their proximity. This highlights the dual nature of PML NBs as both repressors and supporters of proviral functions (especially viral latency) (Charman and Weitzman 2020).

PML NB-mediated intrinsic defense represents an important event in replication of all herpesviruses, influencing both the lytic and latent phase of their life cycle. In HSV-1 infection, infected cell protein 0 (ICP0), a viral ubiquitin ligase with multiple SIM-like sequences, promotes viral gene expression and reactivation from latency (for review, see Rodríguez et al. 2020). ICP0 uses both SUMO-dependent and -independent targeting strategies to ubiquitinate and degrade PML and Sp100, events that result in disassembly of PML NBs and release of viral genomes to ensure viral replication (Lee et al.

2004). Furthermore, almost immediately upon nuclear entry, HSV-1 genomes colocalize with another component of PML NBs, the SWI/SNF chromatin remodeler protein, α-thalassemia X-linked intellectual disability (ATRX), that together with histone chaperone and PML NB-resident death domain–associated protein 6 (Daxx) restricts viral expression by stabilizing the maintenance of viral heterochromatin (Cabral et al. 2018).

Human cytomegalovirus (HCMV) localizes adjacent to PML NBs both before establishing VRCs and after, as PML NBs support spatiotemporal immediate transcription (Ishov and Maul 1996; Ishov et al. 1997). Viral immediate early 1 (IE1) protein interacts with PML protein and colocalizes in PML NBs preventing its oligomerization, reducing de novo PML SUMOylation and ultimately leading to their disruption (Tavalai and Stamminger 2011).

The IE1 protein human herpesvirus 6B (HHV-6B), also encoded by the closely related betaherpesvirus, colocalizes with PML and SUMO-1 during infection without inducing their dispersion (Gravel et al. 2002). SUMOylation of IE1 occurs at telomeres where PML NBs are located (Collin et al. 2020), and this was proposed to represent an advantage for maintenance of viral latent state (Arbuckle et al. 2010).

In latent infections with EBV that were connected with nasopharyngeal carcinoma, EBNA1 viral protein disrupts PML NBs, thus impairing DNA damage response and compromising cell survival (Sivachandran et al. 2008). Insights from the herpesviruses biology clearly underline the multifaceted role of PML in infection and suggest that the closely related viruses of this family could both exploit and oppose PML nuclear compartments to ensure viral progeny. At the same time, herpesviruses show how viruses coevolved with their hosts to control cellular repression mediated by components of PML NBs (Everett et al. 2013).

Adenovirus infections display perhaps the most compelling illustration for the PML NB-related rearrangements upon viral nuclear entry. Two early expressed viral proteins, E4-ORF3 11 kD and E1B 55 kD, are targeted to PML foci that are in turn reorganized into striking nuclear track-like structures, thus disrupting PML NBs (Doucas et al. 1996; Berscheminski et al. 2014). E4-ORF3 assembly creates avidity-driven interactions with PML, offering a possible mechanism for the disruption of PML NBs and other tumor suppressor complexes (Ou et al. 2012). PML NB components Daxx and ATRX mediate an epigenetic repression of the immediate early adenoviral promoter (Schreiner et al. 2013), while viral CA protein VI efficiently counteracts this host cell defense system (Schreiner et al. 2012). Other NB protein residents, such as Sp100 and NDP55, are redistributed at the sites of VRCs (Doucas et al. 1996; Berscheminski et al. 2014), suggesting that adenovirus could both suppress the cellular defense mechanisms by sequestering the cellular factors, or these factors may be exploited for viral replication.

PVs use this cellular SUMOylation system at distinct stages of their life cycle and associate with PML NBs. Early studies on bovine papillomavirus (BPV) suggested that PML NBs are necessary for the establishment rather than for the restriction of infection (Day et al. 1998). HPV CA protein was then implicated in the reorganization of PML NBs in natural productive lesions (Florin et al. 2002). During mitotic membrane breakdown, viral genomes localize to PML NBs already during viral early gene transcription (Day et al. 2004). Homing to PML NBs is mediated by SUMOylation of L2 CA, an event indispensable for efficient infection (Marušič et al. 2010; Bund et al. 2014). Of note, HPV L2 presence up-regulated the overall SUMOylation status of the host cell, similarly to what is observed upon HIV-1 infection in $CD4^{+}$ T cells (Lusic laboratory, unpubl.). A recent high-resolution imaging study showed that rather than the virus targeting NBs, PML, SUMO-1, and Sp100 proteins are recruited and assembled around the papilloma viral genomes early in the interphase (Guion et al. 2019). Hence, PML NBs could provide a protective habitat until switching from productive to latent infection.

HIV-1 establishes multivalent interactions with PML NBs, which affects HIV-1 replication and persistence (Marcello et al. 2003; Lusic et al. 2013; Shytaj et al. 2020). Early studies suggested that PML protein sequesters cyclin T1 (part of

P-TEFb elongation complex) into NBs, abrogating HIV-1 transcription (Marcello et al. 2003). Subsequently, PML NBs were shown to associate with integrated but transcriptionally silent viral genomes (Fig. 3, right insets). PML transcriptional repression was shown to be exerted through direct interaction with G9a histone methyltransferase and H3K9me2 deposition, a facultative heterochromatin mark within euchromatic regions (Lusic et al. 2013). Reports of atypical PML NBs translocation to the cytoplasm during early viral events and association of Daxx protein with incoming virions followed by inhibition of HIV-1 reverse-transcription point to the multiple roles of PML NBs in restricting retroviruses (Dutrieux et al. 2015).

Multiple layers of complexity are reflected in the fact that metabolic changes in T cells caused by retroviral replication and reactivation from latency are sensed at the level of PML NBs. HIV-1 replication results in up-regulation of oxidative stress response and by nuclear factor erythroid 2–related factor 2 (NRF2) translocation to the nucleus. The activation of antioxidant pathways results in reversible reduction in PML NBs via SUMO-mediated proteasomal turnover. With reactivation of cells, viral RNA becomes detectable in the nuclei, but very little or no PML bodies can be seen (Fig. 3, right insets). The reformation of PML NBs, when viral transcription ceases, contributes to the establishment of HIV-1 latency (Shytaj et al. 2020). These findings place PML NBs in the spotlight of cellular immune-metabolic changes caused by the pathogen, and open a possibility that PML NBs could act as metabolic stress sensors and serve as markers and/or pharmacologic targets of HIV-1 infection.

NUCLEAR (SPLICING) SPECKLES (NS)

Nuclear (or splicing) speckles (NS) are known to be enriched in factors related to transcriptional pause release, RNA capping and splicing, polyadenylation and cleavage factors, and RNA export (Galganski et al. 2017). The current model of NS organization postulates that SC35 (serine/arginine-rich splicing factor [SRSF]2) and SON proteins localize in the central regions with RNAs enriched at the periphery. SON, also known as negative regulatory element-binding protein, has initially been described as a factor-binding DNA sequence upstream of the core promoter and second enhancer of HBV, and was shown to repress HBV core promoter activity (Sun et al. 2001). Its role as a speckle scaffolding protein (Sharma et al. 2010) was highlighted when SON was used in TSA-seq (tyramide signal amplification) to define speckle-associated domains (SPADs) (Chen et al. 2018). Based on TSA-seq and other two recent genome-wide technologies, 50% of all protein-coding genes are now considered to colocalize with NS (for review, see Chen and Belmont 2019). The functional significance of this localization was probed using the HSP70 gene (HSPA1A) as a model, shown to move toward the speckles upon transcriptional activation. Intriguingly more than two decades ago, HCMV was used to ask similar questions, namely, the location of induced genes (Fig. 3, left inset; Ishov et al. 1997). In case of HCMV, transcription starts in proximity of PML NBs, from where transcripts move toward the spliceosome assembly domain (SC35 domain). Immediate early viral proteins IE72 and IE 86, which synergistically activate viral and cellular gene expression, are the ones that modulate NBs. IE72 interacts with PML NBs, while IE86 accumulates at PML NBs, which are in the proximity of SC35 domains.

Although NS are considered sites of storage/modification of splicing factors, until recently it was not entirely clear whether splicing indeed occurs at these sites. Recent evidence obtained on examples of cellular genes (Dias et al. 2010; Girard et al. 2012) are strongly supported by data obtained on influenza virus. Evidence for association with NS, splicing, and subsequent export are provided on the example of IAV mRNA, where viral intronless M1 mRNA is targeted to NS by viral protein NS1. SON protein and members of the export TREX complex are then involved in the splicing and subsequent export of viral mRNA to the cytosol (Mor et al. 2016).

HIV-1 genome generates more than 50 mRNA variants from its nine genes, and despite these splicing events, interactions with speckles have so far been attributed mostly to the viral Tat protein and to its interaction with P-TEFb (Herrmann and Mancini 2001; Marcello et al. 2001). Interestingly, very recent findings point to the fact that a portion of HIV-1 insertions into the cellular chromatin occurs in SPADs. This targeting is dependent on the host factor CPSF6, which binds HIV-1 CA during PIC income, possibly targeting the virus to transcriptionally active regions that are in close proximity to NS sites (Francis et al. 2020). Moreover, a large part of genomic regions mapped closest to NS fall into the A1 chromatin compartments, which are enriched in SE, consistent with the recent findings on HIV-1 integration into these portions of the genome (Chen et al. 2017; Lucic et al. 2019).

CONCLUDING REMARKS

To be able to exert a plethora of functions, the cellular genome self-organizes in a hierarchical manner. Nonrandom but highly variable patterns that govern the organization of the mammalian nucleus, although seemingly opposite, ensure all nuclear functions. These patterns of organization are adopted by viruses that complete their life cycle in the nucleus. At the same time, viruses affect architectural robustness and pose challenges to the plasticity of the genome. They have evolved multiple strategies to enter the nucleus, to explore cellular chromatin by episomal contacts or by direct genomic integrations, and to either disrupt or use NBs as shields for their own replication and persistence.

As one of their main features is to explore and use cellular mechanisms, they have been very lucrative tools for studying different nuclear functions, thus paving the way for many important discoveries. The fact that new viruses with critical implications for human health are still emerging highlights the importance of viral research that will continue to move boundaries in our understanding of many previously unknown nuclear and more broadly cellular organization features.

ACKNOWLEDGMENTS

We thank Dr. Vojtech Zila and Erica Margiotta for providing the 3D rendering presented in Figure 1 (inset) and the Infectious Diseases Imaging Platform (IDIP), Dr. Vibor Laketa for his contribution to Figure 3 (right panel)—Department of Infectious Diseases, Integrative Virology, Heidelberg University Hospital and German Center for Infection Research, Heidelberg, Germany. We also thank Dr. Vassilis Roukos and his team for his support in the high throughput imaging used in Figure 2. With the exception of Figure 3, all other figures are original, unpublished data.

Work in the Lusic Laboratory is supported by the German Center for Infection Research (DZIF) Thematic Translational Unit HIV-1 04.704 Preclinical HIV-1 Research, Deutsche Forschungsgemeinschaft (DFG) Projektnummer 240245660-SFB1129 Project P20 to M.L., DFG 422856668-SPP2202 Project to M.L., and Hector Foundation Medizinprojekt M70-HiPNoSe to M.L.

REFERENCES

Ali H, Mano M, Braga L, Naseem A, Marini B, Vu DM, Collesi C, Meroni G, Lusic M, Giacca M. 2019. Cellular TRIM33 restrains HIV-1 infection by targeting viral integrase for proteasomal degradation. *Nat Commun* **10:** 926. doi:10.1038/s41467-019-08810-0

Allouch A, Di Primio C, Alpi E, Lusic M, Arosio D, Giacca M, Cereseto A. 2011. The TRIM family protein KAP1 inhibits HIV-1 integration. *Cell Host Microbe* **9:** 484–495. doi:10.1016/j.chom.2011.05.004

Arbuckle JH, Medveczky MM, Luka J, Hadley SH, Luegmayr A, Ablashi D, Lund TC, Tolar J, De Meirleir K, Montoya JG, et al. 2010. The latent human herpesvirus-6A genome specifically integrates in telomeres of human chromosomes in vivo and in vitro. *Proc Natl Acad Sci* **107:** 5563–5568. doi:10.1073/pnas.0913586107

Avgousti DC, Herrmann C, Kulej K, Pancholi NJ, Sekulic N, Petrescu J, Molden RC, Blumenthal D, Paris AJ, Reyes ED, et al. 2016. A core viral protein binds host nucleosomes to sequester immune danger signals. *Nature* **535:** 173–177. doi:10.1038/nature18317

Aymard F, Bugler B, Schmidt CK, Guillou E, Caron P, Briois S, Iacovoni JS, Daburon V, Miller KM, Jackson SP, et al. 2014. Transcriptionally active chromatin recruits homologous recombination at DNA double-strand breaks. *Nat Struct Mol Biol* **21:** 366–374. doi:10.1038/nsmb.2796

Baltimore D. 1970. Viral RNA-dependent DNA polymerase: RNA-dependent DNA polymerase in virions of RNA tumour viruses. *Nature* **226:** 1209–1211. doi:10.1038/2261209a0

Banerjee I, Miyake Y, Nobs SP, Schneider C, Horvath P, Kopf M, Matthias P, Helenius A, Yamauchi Y. 2014. Influenza A virus uses the aggresome processing machinery for host cell entry. *Science* **346:** 473–477. doi:10.1126/science.1257037

Bangham CRM. 2018. Human T cell leukemia virus type 1: persistence and pathogenesis. *Annu Rev Immunol* **36:** 43–71. doi:10.1146/annurev-immunol-042617-053222

Batterson W, Furlong D, Roizman B. 1983. Molecular genetics of herpes simplex virus. VIII: further characterization of a temperature-sensitive mutant defective in release of viral DNA and in other stages of the viral reproductive cycle. *J Virol* **45:** 397–407. doi:10.1128/JVI.45.1.397-407.1983

Bauer DLV, Tellier M, Martínez-Alonso M, Nojima T, Proudfoot NJ, Murphy S, Fodor E. 2018. Influenza virus mounts a two-pronged attack on host RNA polymerase II transcription. *Cell Rep* **23:** 2119–2129.e3. doi:10.1016/j.celrep.2018.04.047

Beck M, Hurt E. 2017. The nuclear pore complex: understanding its function through structural insight. *Nat Rev Mol Cell Biol* **18:** 73–89. doi:10.1038/nrm.2016.147

Bejarano DA, Peng K, Laketa V, Börner K, Jost KL, Lucic B, Glass B, Lusic M, Müller B, Kräusslich HG. 2019. HIV-1 nuclear import in macrophages is regulated by CPSF6-capsid interactions at the nuclear pore complex. *eLife* **8:** e41800. doi:10.7554/eLife.41800

Berget SM, Moore C, Sharp PA. 1977. Spliced segments at the 5′ terminus of adenovirus 2 late mRNA. *Proc Natl Acad Sci* **74:** 3171–3175. doi:10.1073/pnas.74.8.3171

Berscheminski J, Wimmer P, Brun J, Ip WH, Groitl P, Horlacher T, Jaffray E, Hay RT, Dobner T, Schreiner S. 2014. Sp100 isoform-specific regulation of human adenovirus 5 gene expression. *J Virol* **88:** 6076–6092. doi:10.1128/JVI.00469-14

Bichel K, Price AJ, Schaller T, Towers GJ, Freund SMV, James LC. 2013. HIV-1 capsid undergoes coupled binding and isomerization by the nuclear pore protein NUP358. *Retrovirology* **10:** 81. doi:10.1186/1742-4690-10-81

Bisgrove DA, Mahmoudi T, Henklein P, Verdin E. 2007. Conserved P-TEFb-interacting domain of BRD4 inhibits HIV transcription. *Proc Natl Acad Sci* **104:** 13690–13695. doi:10.1073/pnas.0705053104

Blanco-Rodriguez G, Gazi A, Monel B, Frabetti S, Scoca V, Mueller F, Schwartz O, Krijnse-Locker J, Charneau P, Di Nunzio F. 2020. Remodeling of the core leads HIV-1 preintegration complex into the nucleus of human lymphocytes. *J Virol* **94:** e00135-20. doi:10.1128/JVI.00135-20

Buffone C, Martinez-Lopez A, Fricke T, Opp S, Severgnini M, Cifola I, Petiti L, Frabetti S, Skorupka K, Zadrozny KK, et al. 2018. Nup153 unlocks the nuclear pore complex for HIV-1 nuclear translocation in nondividing cells. *J Virol* **92:** e00648-18. doi:10.1128/JVI.00648-18

Bund T, Spoden GA, Koynov K, Hellmann N, Boukhallouk F, Arnold P, Hinderberger D, Florin L. 2014. An L2 SUMO interacting motif is important for PML localization and infection of human papillomavirus type 16. *Cell Microbiol* **16:** 1179–1200. doi:10.1111/cmi.12271

Burdick RC, Li C, Munshi M, Rawson JMO, Nagashima K, Hu WS, Pathak VK. 2020. HIV-1 uncoats in the nucleus near sites of integration. *Proc Natl Acad Sci* **117:** 5486–5493. doi:10.1073/pnas.1920631117

Cabral JM, Oh HS, Knipe DM. 2018. ATRX promotes maintenance of herpes simplex virus heterochromatin during chromatin stress. *eLife* **7:** e40228. doi:10.7554/eLife.40228

Campbell EM, Hope TJ. 2015. HIV-1 capsid: the multifaceted key player in HIV-1 infection. *Nat Rev Microbiol* **13:** 471–483. doi:10.1038/nrmicro3503

Cassany A, Ragues J, Guan T, Bégu D, Wodrich H, Kann M, Nemerow GR, Gerace L. 2015. Nuclear import of adenovirus DNA involves direct interaction of hexon with an N-terminal domain of the nucleoporin Nup214. *J Virol* **89:** 1719–1730. doi:10.1128/JVI.02639-14

Chang CW, Lee CP, Su MT, Tsai CH, Chen MR. 2015. BGLF4 kinase modulates the structure and transport preference of the nuclear pore complex to facilitate nuclear import of Epstein–Barr virus lytic proteins. *J Virol* **89:** 1703–1718. doi:10.1128/JVI.02880-14

Charman M, Weitzman MD. 2020. Replication compartments of DNA viruses in the nucleus: location, location, location. *Viruses* **12:** 151. doi:10.3390/v12020151

Chen Y, Belmont AS. 2019. Genome organization around nuclear speckles. *Curr Opin Genet Dev* **55:** 91–99. doi:10.1016/j.gde.2019.06.008

Chen HC, Martinez JP, Zorita E, Meyerhans A, Filion GJ. 2017. Position effects influence HIV latency reversal. *Nat Struct Mol Biol* **24:** 47–54. doi:10.1038/nsmb.3328

Chen Y, Zhang Y, Wang Y, Zhang L, Brinkman EK, Adam SA, Goldman R, van Steensel B, Ma J, Belmont AS. 2018. Mapping 3D genome organization relative to nuclear compartments using TSA-Seq as a cytological ruler. *J Cell Biol* **217:** 4025–4048. doi:10.1083/jcb.201807108

Cherepanov P, Maertens G, Proost P, Devreese B, Van Beeumen J, Engelborghs Y, De Clercq E, Debyser Z. 2003. HIV-1 integrase forms stable tetramers and associates with LEDGF/p75 protein in human cells. *J Biol Chem* **278:** 372–381. doi:10.1074/jbc.M209278200

Chow LT, Gelinas RE, Broker TR, Roberts RJ. 1977. An amazing sequence arrangement at the 5′ ends of adenovirus 2 messenger RNA. *Cell* **12:** 1–8. doi:10.1016/0092-8674(77)90180-5

Cmarko D, Bøe SO, Scassellati C, Szilvay AM, Davanger S, Fu XD, Haukenes G, Kalland KH, Fakan S. 2002. Rev inhibition strongly affects intracellular distribution of human immunodeficiency virus type 1 RNAs. *J Virol* **76:** 10473–10484. doi:10.1128/JVI.76.20.10473-10484.2002

Cohen S, Au S, Panté N. 2011. How viruses access the nucleus. *Biochim Biophys Acta* **1813:** 1634–1645. doi:10.1016/j.bbamcr.2010.12.009

Collin V, Gravel A, Kaufer BB, Flamand L. 2020. The Promyelocytic Leukemia Protein facilitates human herpesvirus 6B chromosomal integration, immediate-early 1 protein multiSUMOylation and its localization at telomeres. *PLoS Pathog* **16:** e1008683. doi:10.1371/journal.ppat.1008683

Conrad RJ, Fozouni P, Thomas S, Sy H, Zhang Q, Zhou M-M, Ott M. 2017. The short isoform of BRD4 promotes HIV-1 latency by engaging repressive SWI/SNF chromatin-remodeling complexes. *Mol Cell* **67:** 1001–1012.e6. doi:10.1016/j.molcel.2017.07.025

Copeland AM, Newcomb WW, Brown JC. 2009. Herpes simplex virus replication: roles of viral proteins and nucleoporins in capsid-nucleus attachment. *J Virol* **83:** 1660–1668. doi:10.1128/JVI.01139-08

Coursey TL, McBride AA. 2019. Hitchhiking of viral genomes on cellular chromosomes. *Annu Rev Virol* **6:** 275–296. doi:10.1146/annurev-virology-092818-015716

Crameri M, Bauer M, Caduff N, Walker R, Steiner F, Franzoso FD, Gujer C, Boucke K, Kucera T, Zbinden A, et al. 2018. Mxb is an interferon-induced restriction factor of human herpesviruses. *Nat Commun* **9:** 1980. doi:10.1038/s41467-018-04379-2

Day PM, Roden RB, Lowy DR, Schiller JT. 1998. The papillomavirus minor capsid protein, L2, induces localization of the major capsid protein, L1, and the viral transcription/replication protein, E2, to PML oncogenic domains. *J Virol* **72:** 142–150. doi:10.1128/JVI.72.1.142-150.1998

Day PM, Baker CC, Lowy DR, Schiller JT. 2004. Establishment of papillomavirus infection is enhanced by promyelocytic leukemia protein (PML) expression. *Proc Natl Acad Sci* **101:** 14252–14257. doi:10.1073/pnas.0404229101

Dekker J, Marti-Renom MA, Mirny LA. 2013. Exploring the three-dimensional organization of genomes: interpreting chromatin interaction data. *Nat Rev Genet* **14:** 390–403. doi:10.1038/nrg3454

Demeulemeester J, De Rijck J, Gijsbers R, Debyser Z. 2015. Retroviral integration: site matters: mechanisms and consequences of retroviral integration site selection. *Bioessays* **37:** 1202–1214. doi:10.1002/bies.201500051

Denker A, de Laat W. 2016. The second decade of 3C technologies: detailed insights into nuclear organization. *Genes Dev* **30:** 1357–1382. doi:10.1101/gad.281964.116

Dharan A, Bachmann N, Talley S, Zwikelmaier V, Campbell EM. 2020. Nuclear pore blockade reveals that HIV-1 completes reverse transcription and uncoating in the nucleus. *Nat Microbiol* **5:** 1088–1095 doi:10.1038/s41564-020-0735-8

Dias AP, Dufu K, Lei H, Reed R. 2010. A role for TREX components in the release of spliced mRNA from nuclear speckle domains. *Nat Commun* **1:** 97. doi:10.1038/ncomms1103

Dicks MDJ, Betancor G, Jimenez-Guardeño JM, Pessel-Vivares L, Apolonia L, Goujon C, Malim MH. 2018. Multiple components of the nuclear pore complex interact with the amino-terminus of MX2 to facilitate HIV-1 restriction. *PLoS Pathog* **14:** e1007408. doi:10.1371/journal.ppat.1007408

Dieudonné M, Maiuri P, Biancotto C, Knezevich A, Kula A, Lusic M, Marcello A. 2009. Transcriptional competence of the integrated HIV-1 provirus at the nuclear periphery. *EMBO J* **28:** 2231–2243. doi:10.1038/emboj.2009.141

Di Nunzio F, Fricke T, Miccio A, Valle-Casuso JC, Perez P, Souque P, Rizzi E, Severgnini M, Mavilio F, Charneau P, et al. 2013. Nup153 and Nup98 bind the HIV-1 core and contribute to the early steps of HIV-1 replication. *Virology* **440:** 8–18. doi:10.1016/j.virol.2013.02.008

Doucas V, Ishov AM, Romo A, Juguilon H, Weitzman MD, Evans RM, Maul GG. 1996. Adenovirus replication is coupled with the dynamic properties of the PML nuclear structure. *Genes Dev* **10:** 196–207. doi:10.1101/gad.10.2.196

Dutrieux J, Maarifi G, Portilho DM, Arhel NJ, Chelbi-Alix MK, Nisole S. 2015. PML/TRIM19-dependent inhibition of retroviral reverse-transcription by Daxx. *PLoS Pathog* **11:** e1005280. doi:10.1371/journal.ppat.1005280

Dynan WS, Tjian R. 1983. The promoter-specific transcription factor Sp1 binds to upstream sequences in the SV40 early promoter. *Cell* **35:** 79–87. doi:10.1016/0092-8674(83)90210-6

Everett RD, Chelbi-Alix MK. 2007. PML and PML nuclear bodies: implications in antiviral defence. *Biochimie* **89:** 819–830. doi:10.1016/j.biochi.2007.01.004

Everett RD, Bell AJ, Lu Y, Orr A. 2013. The replication defect of ICP0-null mutant herpes simplex virus 1 can be largely complemented by the combined activities of human cytomegalovirus proteins IE1 and pp71. *J Virol* **87:** 978–990. doi:10.1128/JVI.01103-12

Fay N, Panté N. 2015. Nuclear entry of DNA viruses. *Front Microbiol* **6:** 467.

Ferrari R, Su T, Li B, Bonora G, Oberai A, Chan Y, Sasidharan R, Berk AJ, Pellegrini M, Kurdistani SK. 2012. Reorganization of the host epigenome by a viral oncogene. *Genome Res* **22:** 1212–1221. doi:10.1101/gr.132308.111

Ferrari R, Gou D, Jawdekar G, Johnson SA, Nava M, Su T, Yousef AF, Zemke NR, Pellegrini M, Kurdistani SK, et al. 2014. Adenovirus small E1A employs the lysine acetylases p300/CBP and tumor suppressor Rb to repress select host genes and promote productive virus infection. *Cell Host Microbe* **16:** 663–676. doi:10.1016/j.chom.2014.10.004

Flatt JW, Greber UF. 2017. Viral mechanisms for docking and delivering at nuclear pore complexes. *Semin Cell Dev Biol* **68:** 59–71. doi:10.1016/j.semcdb.2017.05.008

Florin L, Sapp C, Streeck RE, Sapp M. 2002. Assembly and translocation of papillomavirus capsid proteins. *J Virol* **76:** 10009–10014. doi:10.1128/JVI.76.19.10009-10014.2002

Francis AC, Marin M, Singh PK, Achuthan V, Prellberg MJ, Palermino-Rowland K, Lan S, Tedbury PR, Sarafianos SG, Engelman AN, et al. 2020. HIV-1 replication complexes accumulate in nuclear speckles and integrate into speckle-associated genomic domains. *Nat Commun* **11:** 3505. doi:10.1038/s41467-020-17256-8

Galganski L, Urbanek MO, Krzyzosiak WJ. 2017. Nuclear speckles: molecular organization, biological function and role in disease. *Nucleic Acids Res* **45:** 10350–10368. doi:10.1093/nar/gkx759

Gao R, Bao J, Yan H, Xie L, Qin W, Ning H, Huang S, Cheng J, Zhi R, Li Z, et al. 2020. Competition between PAF1 and MLL1/COMPASS confers the opposing function of LEDGF/p75 in HIV latency and proviral reactivation. *Sci Adv* **6:** eaaz8411. doi:10.1126/sciadv.aaz8411

Geis FK, Goff SP. 2019. Unintegrated HIV-1 DNAs are loaded with core and linker histones and transcriptionally silenced. *Proc Natl Acad Sci* **116:** 23735–23742. doi:10.1073/pnas.1912638116

Gérard A, Ségéral E, Naughtin M, Abdouni A, Charmeteau B, Cheynier R, Rain JC, Emiliani S. 2015. The integrase cofactor LEDGF/p75 associates with Iws1 and Spt6 for postintegration silencing of HIV-1 gene expression in latently infected cells. *Cell Host Microbe* **17:** 107–117. doi:10.1016/j.chom.2014.12.002

Girard C, Will CL, Peng J, Makarov EM, Kastner B, Lemm I, Urlaub H, Hartmuth K, Lührmann R. 2012. Post-transcriptional spliceosomes are retained in nuclear speckles until splicing completion. *Nat Commun* **3:** 994. doi:10.1038/ncomms1998

Goujon C, Moncorgé O, Bauby H, Doyle T, Ward CC, Schaller T, Hué S, Barclay WS, Schulz R, Malim MH. 2013. Human MX2 is an interferon-induced post-entry inhibitor of HIV-1 infection. *Nature* **502:** 559–562. doi:10.1038/nature12542

Gravel A, Gosselin J, Flamand L. 2002. Human Herpesvirus 6 immediate-early 1 protein is a sumoylated nuclear phosphoprotein colocalizing with promyelocytic leukemia protein-associated nuclear bodies. *J Biol Chem* **277:** 19679–19687. doi:10.1074/jbc.M200836200

Greber UF, Flatt JW. 2019. Adenovirus entry: from infection to immunity. *Annu Rev Virol* **6:** 177–197. doi:10.1146/annurev-virology-092818-015550

Guion L, Bienkowska-Haba M, DiGiuseppe S, Florin L, Sapp M. 2019. PML nuclear body-residing proteins sequentially associate with HPV genome after infectious nuclear delivery. *PLoS Pathog* **15:** e1007590. doi:10.1371/journal.ppat.1007590

Heinz S, Texari L, Hayes MGB, Urbanowski M, Chang MW, Givarkes N, Rialdi A, White KM, Albrecht RA, Pache L, et al. 2018. Transcription elongation can affect genome 3D structure. *Cell* **174:** 1522–1536.e22. doi:10.1016/j.cell.2018.07.047

Hensel KO, Cantner F, Bangert F, Wirth S, Postberg J. 2018. Episomal HBV persistence within transcribed host nuclear chromatin compartments involves HBx. *Epigenetics Chromatin* **11:** 34. doi:10.1186/s13072-018-0204-2

Herrmann CH, Mancini MA. 2001. The Cdk9 and cyclin T subunits of TAK/P-TEFb localize to splicing factor-rich nuclear speckle regions. *J Cell Sci* **114:** 1491–1503.

Ho JSY, Angel M, Ma Y, Sloan E, Wang G, Martinez-Romero C, Alenquer M, Roudko V, Chung L, Zheng S, et al. 2020. Hybrid gene origination creates human-virus chimeric proteins during infection. *Cell* **181:** 1502–1517.e23. doi:10.1016/j.cell.2020.05.035

Horwitz GA, Zhang K, McBrian MA, Grunstein M, Kurdistani SK, Berk AJ. 2008. Adenovirus small e1a alters global patterns of histone modification. *Science* **321:** 1084–1085. doi:10.1126/science.1155544

Huffman JB, Daniel GR, Falck-Pedersen E, Huet A, Smith GA, Conway JF, Homa FL. 2017. The C terminus of the herpes simplex virus UL25 protein is required for release of viral genomes from capsids bound to nuclear pores. *J Virol* **91:** e00641-17. doi:10.1128/JVI.00641-17

Ibarra A, Benner C, Tyagi S, Cool J, Hetzer MW. 2016. Nucleoporin-mediated regulation of cell identity genes. *Genes Dev* **30:** 2253–2258. doi:10.1101/gad.287417.116

Ishov AM, Maul GG. 1996. The periphery of nuclear domain 10 (ND10) as site of DNA virus deposition. *J Cell Biol* **134:** 815–826. doi:10.1083/jcb.134.4.815

Ishov AM, Stenberg RM, Maul GG. 1997. Human cytomegalovirus immediate early interaction with host nuclear structures: definition of an immediate transcript environment. *J Cell Biol* **138:** 5–16. doi:10.1083/jcb.138.1.5

Jang MK, Mochizuki K, Zhou M, Jeong H-S, Brady JN, Ozato K. 2005. The bromodomain protein Brd4 is a positive regulatory component of P-TEFb and stimulates RNA polymerase II-dependent transcription. *Mol Cell* **19:** 523–534. doi:10.1016/j.molcel.2005.06.027

Jiang S, Zhou H, Liang J, Gerdt C, Wang C, Ke L, Schmidt SCS, Narita Y, Ma Y, Wang S, et al. 2017. The Epstein–Barr virus regulome in lymphoblastoid cells. *Cell Host Microbe* **22:** 561–573.e4. doi:10.1016/j.chom.2017.09.001

Kamitani T, Nguyen HP, Kito K, Fukuda-Kamitani T, Yeh ET. 1998. Covalent modification of PML by the sentrin family of ubiquitin-like proteins. *J Biol Chem* **273:** 3117–3120. doi:10.1074/jbc.273.6.3117

Kane M, Yadav SS, Bitzegeio J, Kutluay SB, Zang T, Wilson SJ, Schoggins JW, Rice CM, Yamashita M, Hatziioannou T, et al. 2013. MX2 is an interferon-induced inhibitor of HIV-1 infection. *Nature* **502:** 563–566. doi:10.1038/nature12653

Kane M, Rebensburg SV, Takata MA, Zang TM, Yamashita M, Kvaratskhelia M, Bieniasz PD. 2018. Nuclear pore heterogeneity influences HIV-1 infection and the antiviral activity of MX2. *eLife* **7:** e35738. doi:10.7554/eLife.35738

Karn J, Stoltzfus CM. 2012. Transcriptional and posttranscriptional regulation of HIV-1 gene expression. *Cold Spring Harb Perspect Med* **2:** a006916. doi:10.1101/cshperspect.a006916

Kim K-D, Tanizawa H, De Leo A, Vladimirova O, Kossenkov A, Lu F, Showe LC, Noma K-I, Lieberman PM. 2020. Epigenetic specifications of host chromosome docking sites for latent Epstein–Barr virus. *Nat Commun* **11:** 877. doi:10.1038/s41467-019-14152-8

Knipe DM, Lieberman PM, Jung JU, McBride AA, Morris KV, Ott M, Margolis D, Nieto A, Nevels M, Parks RJ, et al. 2013. Snapshots: chromatin control of viral infection. *Virology* **435:** 141–156. doi:10.1016/j.virol.2012.09.023

Lavau C, Marchio A, Fagioli M, Jansen J, Falini B, Lebon P, Grosveld F, Pandolfi PP, Pelicci PG, Dejean A. 1995. The acute promyelocytic leukaemia-associated PML gene is induced by interferon. *Oncogene* **11:** 871–876.

Lee W, Haslinger A, Karin M, Tjian R. 1987. Activation of transcription by two factors that bind promoter and enhancer sequences of the human metallothionein gene and SV40. *Nature* **325:** 368–372. doi:10.1038/325368a0

Lee H-R, Kim D-J, Lee J-M, Choi CY, Ahn B-Y, Hayward GS, Ahn J-H. 2004. Ability of the human cytomegalovirus IE1 protein to modulate sumoylation of PML correlates with its functional activities in transcriptional regulation and infectivity in cultured fibroblast cells. *J Virol* **78:** 6527–6542. doi:10.1128/JVI.78.12.6527-6542.2004

Lelek M, Casartelli N, Pellin D, Rizzi E, Souque P, Severgnini M, Di Serio C, Fricke T, Diaz-Griffero F, Zimmer C, et al. 2015. Chromatin organization at the nuclear pore favours HIV replication. *Nat Commun* **6:** 6483. doi:10.1038/ncomms7483

LeRoy G, Oksuz O, Descostes N, Aoi Y, Ganai RA, Kara HO, Yu J-R, Lee C-H, Stafford J, Shilatifard A, et al. 2019. LEDGF and HDGF2 relieve the nucleosome-induced barrier to transcription in differentiated cells. *Sci Adv* **5:** eaay3068. doi:10.1126/sciadv.aay3068

Liu L. 2014. Fields virology, 6th edition. *Clin Infect Dis* 59: 613.

Lucic B, Chen H-C, Kuzman M, Zorita E, Wegner J, Minneker V, Wang W, Fronza R, Laufs S, Schmidt M, et al. 2019. Spatially clustered loci with multiple enhancers are frequent targets of HIV-1 integration. *Nat Commun* **10:** 4059. doi:10.1038/s41467-019-12046-3

Lusic M, Marcello A, Cereseto A, Giacca M. 2003. Regulation of HIV-1 gene expression by histone acetylation and factor recruitment at the LTR promoter. *EMBO J* **22:** 6550–6561. doi:10.1093/emboj/cdg631

Lusic M, Marini B, Ali H, Lucic B, Luzzati R, Giacca M. 2013. Proximity to PML nuclear bodies regulates HIV-1 latency in CD4$^+$ T cells. *Cell Host Microbe* **13:** 665–677. doi:10.1016/j.chom.2013.05.006

Lycke E, Hamark B, Johansson M, Krotochwil A, Lycke J, Svennerholm B. 1988. Herpes simplex virus infection of the human sensory neuron. An electron microscopy study. *Arch Virol* **101:** 87–104. doi:10.1007/BF01314654

Ma X, Yang T, Luo Y, Wu L, Jiang Y, Song Z, Pan T, Liu B, Liu G, Liu J, et al. 2019. TRIM28 promotes HIV-1 latency by SUMOylating CDK9 and inhibiting P-TEFb. *eLife* **8:** e42426. doi:10.7554/eLife.42426

Machida S, Depierre D, Chen H-C, Thenin-Houssier S, Petitjean G, Doyen CM, Takaku M, Cuvier O, Benkirane M. 2020. Exploring histone loading on HIV DNA reveals a dynamic nucleosome positioning between unintegrated and integrated viral genome. *Proc Natl Acad Sci* **117:** 6822–6830. doi:10.1073/pnas.1913754117

Majumder K, Wang J, Boftsi M, Fuller MS, Rede JE, Joshi T, Pintel DJ. 2018. Parvovirus minute virus of mice interacts with sites of cellular DNA damage to establish and amplify its lytic infection. *eLife* **7:** e37750. doi: 10.7554/eLife.37750

Mao YS, Zhang B, Spector DL. 2011. Biogenesis and function of nuclear bodies. *Trends Genet* **27:** 295–306. doi:10.1016/j.tig.2011.05.006

Marcello A, Cinelli RA, Ferrari A, Signorelli A, Tyagi M, Pellegrini V, Beltram F, Giacca M. 2001. Visualization of in vivo direct interaction between HIV-1 TAT and human cyclin T1 in specific subcellular compartments by fluorescence resonance energy transfer. *J Biol Chem* **276:** 39220–39225. doi:10.1074/jbc.M104830200

Marcello A, Ferrari A, Pellegrini V, Pegoraro G, Lusic M, Beltram F, Giacca M. 2003. Recruitment of human cyclin T1 to nuclear bodies through direct interaction with the PML protein. *EMBO J* **22:** 2156–2166. doi:10.1093/emboj/cdg205

Marini B, Kertesz-Farkas A, Ali H, Lucic B, Lisek K, Manganaro L, Pongor S, Luzzati R, Recchia A, Mavilio F, et al. 2015. Nuclear architecture dictates HIV-1 integration site selection. *Nature* **521:** 227–231. doi:10.1038/nature14226

Marušič MB, Mencin N, Ličen M, Banks L, Grm HS. 2010. Modification of human papillomavirus minor capsid protein L2 by sumoylation. *J Virol* **84:** 11585–11589. doi:10.1128/JVI.01269-10

Matreyek KA, Yücel SS, Li X, Engelman A. 2013. Nucleoporin NUP153 phenylalanine-glycine motifs engage a common binding pocket within the HIV-1 capsid protein to mediate lentiviral infectivity. *PLoS Pathog* **9:** e1003693. doi:10.1371/journal.ppat.1003693

Melamed A, Yaguchi H, Miura M, Witkover A, Fitzgerald TW, Birney E, Bangham CR. 2018. The human leukemia virus HTLV-1 alters the structure and transcription of host chromatin in *cis*. *eLife* **7:** e36245. doi:10.7554/eLife.36245

Michieletto D, Lusic M, Marenduzzo D, Orlandini E. 2019. Physical principles of retroviral integration in the human genome. *Nat Commun* **10:** 575. doi:10.1038/s41467-019-08333-8

Miyake Y, Keusch JJ, Decamps L, Ho-Xuan H, Iketani S, Gut H, Kutay U, Helenius A, Yamauchi Y. 2019. Influenza virus uses transportin 1 for vRNP debundling during cell entry. *Nat Microbiol* **4:** 578–586. doi:10.1038/s41564-018-0332-2

Mizutani S, Boettiger D, Temin HM. 1970. A DNA-dependent DNA polymerase and a DNA endonuclease in virions of Rous sarcoma virus. *Nature* **228:** 424–427. doi:10.1038/228424a0

Monette A, Panté N, Mouland AJ. 2011. HIV-1 remodels the nuclear pore complex. *J Cell Biol* **193:** 619–631. doi:10.1083/jcb.201008064

Moquin SA, Thomas S, Whalen S, Warburton A, Fernandez SG, McBride AA, Pollard KS, Miranda JL. 2018. The Epstein–Barr virus episome maneuvers between nuclear chromatin compartments during reactivation. *J Virol* **92:** e01413–e01417.

Mor A, White A, Zhang K, Thompson M, Esparza M, Muñoz-Moreno R, Koide K, Lynch KW, García-Sastre A, Fontoura BMA. 2016. Influenza virus mRNA trafficking through host nuclear speckles. *Nat Microbiol* **1:** 16069. doi:10.1038/nmicrobiol.2016.69

Moreau P, Cournac A, Palumbo GA, Marbouty M, Mortaza S, Thierry A, Cairo S, Lavigne M, Koszul R, Neuveut C. 2018. Tridimensional infiltration of DNA viruses into the host genome shows preferential contact with active chromatin. *Nat Commun* **9:** 4268. doi:10.1038/s41467-018-06739-4

Natarajan M, Schiralli Lester GM, Lee C, Missra A, Wasserman GA, Steffen M, Gilmour DS, Henderson AJ. 2013. Negative elongation factor (NELF) coordinates RNA polymerase II pausing, premature termination, and chromatin remodeling to regulate HIV transcription. *J Biol Chem* **288:** 25995–26003. doi:10.1074/jbc.M113.496489

Newbold JE, Xin H, Tencza M, Sherman G, Dean J, Bowden S, Locarnini S. 1995. The covalently closed duplex form of the hepadnavirus genome exists in situ as a heterogeneous population of viral minichromosomes. *J Virol* **69:** 3350–3357. doi:10.1128/JVI.69.6.3350-3357.1995

Noah DL, Twu KY, Krug RM. 2003. Cellular antiviral responses against influenza A virus are countered at the posttranscriptional level by the viral NS1A protein via its binding to a cellular protein required for the 3′ end processing of cellular pre-mRNAS. *Virology* **307:** 386–395. doi:10.1016/S0042-6822(02)00127-7

Ou HD, Kwiatkowski W, Deerinck TJ, Noske A, Blain KY, Land HS, Soria C, Powers CJ, May AP, Shu X, et al. 2012. A structural basis for the assembly and functions of a viral polymer that inactivates multiple tumor suppressors. *Cell* **151:** 304–319. doi:10.1016/j.cell.2012.08.035

Pascual-Garcia P, Debo B, Aleman JR, Talamas JA, Lan Y, Nguyen NH, Won KJ, Capelson M. 2017. Metazoan nuclear pores provide a scaffold for poised genes and mediate induced enhancer-promoter contacts. *Mol Cell* **66:** 63–76.e6. doi:10.1016/j.molcel.2017.02.020

Pasdeloup D, Blondel D, Isidro AL, Rixon FJ. 2009. Herpesvirus capsid association with the nuclear pore complex and viral DNA release involve the nucleoporin CAN/Nup214 and the capsid protein pUL25. *J Virol* **83:** 6610–6623. doi:10.1128/JVI.02655-08

Perkins KJ, Lusic M, Mitar I, Giacca M, Proudfoot NJ. 2008. Transcription-dependent gene looping of the HIV-1 provirus is dictated by recognition of pre-mRNA processing signals. *Mol Cell* **29:** 56–68. doi:10.1016/j.molcel.2007.11.030

Rodríguez MC, Dybas JM, Hughes J, Weitzman MD, Boutell C. 2020. The HSV-1 ubiquitin ligase ICP0: modifying the cellular proteome to promote infection. *Virus Res* **285:** 198015. doi:10.1016/j.virusres.2020.198015

Sahin U, Ferhi O, Jeanne M, Benhenda S, Berthier C, Jollivet F, Niwa-Kawakita M, Faklaris O, Setterblad N, de Thé H, et al. 2014. Oxidative stress-induced assembly of PML nuclear bodies controls sumoylation of partner proteins. *J Cell Biol* **204:** 931–945. doi:10.1083/jcb.201305148

Satou Y, Miyazato P, Ishihara K, Yaguchi H, Melamed A, Miura M, Fukuda A, Nosaka K, Watanabe T, Rowan AG, et al. 2016. The retrovirus HTLV-1 inserts an ectopic CTCF-binding site into the human genome. *Proc Natl Acad Sci* **113:** 3054–3059. doi:10.1073/pnas.1423199113

Schaller T, Ocwieja KE, Rasaiyaah J, Price AJ, Brady TL, Roth SL, Hué S, Fletcher AJ, Lee K, KewalRamani VN, et al. 2011. HIV-1 capsid-cyclophilin interactions determine nuclear import pathway, integration targeting and replication efficiency. *PLoS Pathog* **7:** e1002439. doi:10.1371/journal.ppat.1002439

Scherer M, Stamminger T. 2016. Emerging role of PML nuclear bodies in innate immune signaling. *J Virol* **90:** 5850–5854. doi:10.1128/JVI.01979-15

Scholz BA, Sumida N, de Lima CDM, Chachoua I, Martino M, Tzelepis I, Nikoshkov A, Zhao H, Mehmood R, Sifakis EG, et al. 2019. WNT signaling and AHCTF1 promote oncogenic MYC expression through super-enhancer-mediated gene gating. *Nat Genet* **51:** 1723–1731. doi:10.1038/s41588-019-0535-3

Schreiner S, Martinez R, Groitl P, Rayne F, Vaillant R, Wimmer P, Bossis G, Sternsdorf T, Marcinowski L, Ruzsics Z, et al. 2012. Transcriptional activation of the adenoviral genome is mediated by capsid protein VI. *PLoS Pathog* **8:** e1002549. doi:10.1371/journal.ppat.1002549

Schreiner S, Bürck C, Glass M, Groitl P, Wimmer P, Kinkley S, Mund A, Everett RD, Dobner T. 2013. Control of human adenovirus type 5 gene expression by cellular Daxx/ATRX chromatin-associated complexes. *Nucleic Acids Res* **41:** 3532–3550. doi:10.1093/nar/gkt064

Shahin V, Hafezi W, Oberleithner H, Ludwig Y, Windoffer B, Schillers H, Kühn JE. 2006. The genome of HSV-1 translocates through the nuclear pore as a condensed rod-like structure. *J Cell Sci* **119:** 23–30. doi:10.1242/jcs.02705

Sharma A, Takata H, Shibahara K-I, Bubulya A, Bubulya PA. 2010. Son is essential for nuclear speckle organization and cell cycle progression. *Mol Biol Cell* **21:** 650–663. doi:10.1091/mbc.e09-02-0126

Shukla A, Ramirez N-GP, D'Orso I. 2020. HIV-1 proviral transcription and latency in the new era. *Viruses* **12:** 555. doi:10.3390/v12050555

Shytaj IL, Lucic B, Forcato M, Penzo C, Billingsley J, Laketa V, Bosinger S, Stanic M, Gregoretti F, Antonelli L, et al. 2020. Alterations of redox and iron metabolism accompany the development of HIV latency. *EMBO J* **39:** e102209. doi:10.15252/embj.2019102209

Singh PK, Plumb MR, Ferris AL, Iben JR, Wu X, Fadel HJ, Luke BT, Esnault C, Poeschla EM, Hughes SH, et al. 2015. LEDGF/p75 interacts with mRNA splicing factors and targets HIV-1 integration to highly spliced genes. *Genes Dev* **29:** 2287–2297. doi:10.1101/gad.267609.115

Sivachandran N, Sarkari F, Frappier L. 2008. Epstein–Barr nuclear antigen 1 contributes to nasopharyngeal carcinoma through disruption of PML nuclear bodies. *PLoS Pathog* **4:** e1000170. doi:10.1371/journal.ppat.1000170

Sodeik B, Ebersold MW, Helenius A. 1997. Microtubule-mediated transport of incoming herpes simplex virus 1 capsids to the nucleus. *J Cell Biol* **136:** 1007–1021. doi:10.1083/jcb.136.5.1007

Strunze S, Engelke MF, Wang IH, Puntener D, Boucke K, Schleich S, Way M, Schoenenberger P, Burckhardt CJ, Greber UF. 2011. Kinesin-1-mediated capsid disassembly and disruption of the nuclear pore complex promote virus infection. *Cell Host Microbe* **10:** 210–223. doi:10.1016/j.chom.2011.08.010

Sun CT, Lo WY, Wang IH, Lo YH, Shiou SR, Lai CK, Ting LP. 2001. Transcription repression of human hepatitis B virus genes by negative regulatory element-binding protein/SON. *J Biol Chem* **276:** 24059–24067. doi:10.1074/jbc.M101330200

Szymula A, Palermo RD, Bayoumy A, Groves IJ, Ba Abdullah M, Holder B, White RE. 2018. Epstein–Barr virus nuclear antigen EBNA-LP is essential for transforming naïve B cells, and facilitates recruitment of transcription factors to the viral genome. *PLoS Pathog* **14:** e1006890. doi:10.1371/journal.ppat.1006890

Tavalai N, Stamminger T. 2011. Intrinsic cellular defense mechanisms targeting human cytomegalovirus. *Virus Res* **157:** 128–133. doi:10.1016/j.virusres.2010.10.002

Tchasovnikarova IA, Timms RT, Matheson NJ, Wals K, Antrobus R, Göttgens B, Dougan G, Dawson MA, Lehner PJ. 2015. Epigenetic silencing by the HUSH complex mediates position-effect variegation in human cells. *Science* **348:** 1481–1485. doi:10.1126/science.aaa7227

Terrier O, Carron C, Cartet G, Traversier A, Julien T, Valette M, Lina B, Moules V, Rosa-Calatrava M. 2014. Ultrastructural fingerprints of avian influenza A (H7N9) virus in infected human lung cells. *Virology* **456–457:** 39–42. doi:10.1016/j.virol.2014.03.013

Timms RT, Tchasovnikarova IA, Antrobus R, Dougan G, Lehner PJ. 2016. ATF7IP-mediated stabilization of the histone methyltransferase SETDB1 is essential for heterochromatin formation by the HUSH complex. *Cell Rep* **17:** 653–659. doi:10.1016/j.celrep.2016.09.050

Trotman LC, Mosberger N, Fornerod M, Stidwill RP, Greber UF. 2001. Import of adenovirus DNA involves the nuclear pore complex receptor CAN/Nup214 and histone H1. *Nat Cell Biol* **3:** 1092–1100. doi:10.1038/ncb1201-1092

Ui A, Chiba N, Yasui A. 2020. Relationship among DNA double-strand break (DSB), DSB repair, and transcription prevents genome instability and cancer. *Cancer Sci* **111:** 1443–1451. doi:10.1111/cas.14404

Van Damme E, Laukens K, Dang TH, Van Ostade X. 2010. A manually curated network of the PML nuclear body interactome reveals an important role for PML-NBs in SUMOylation dynamics. *Int J Biol Sci* **6:** 51–67. doi:10.7150/ijbs.6.51

van Nuland R, van Schaik FM, Simonis M, van Heesch S, Cuppen E, Boelens R, Timmers HM, van Ingen H. 2013. Nucleosomal DNA binding drives the recognition of H3K36-methylated nucleosomes by the PSIP1-PWWP domain. *Epigenetics Chromatin* **6:** 12. doi:10.1186/1756-8935-6-12

Verdin E, Paras P, Van Lint C. 1993. Chromatin disruption in the promoter of human immunodeficiency virus type 1 during transcriptional activation. *EMBO J* **12:** 3249–3259. doi:10.1002/j.1460-2075.1993.tb05994.x

Wang GZ, Wang Y, Goff SP. 2016. Histones are rapidly loaded onto unintegrated retroviral DNAs soon after nuclear entry. *Cell Host Microbe* **20:** 798–809. doi:10.1016/j.chom.2016.10.009

Wei P, Garber ME, Fang SM, Fischer WH, Jones KA. 1998. A novel CDK9-associated C-type cyclin interacts directly with HIV-1 Tat and mediates its high-affinity, loop-specific binding to TAR RNA. *Cell* **92:** 451–462. doi:10.1016/S0092-8674(00)80939-3

Wolf D, Goff SP. 2007. TRIM28 mediates primer binding site-targeted silencing of murine leukemia virus in embryonic cells. *Cell* **131:** 46–57. doi:10.1016/j.cell.2007.07.026

Wolf D, Goff SP. 2009. Embryonic stem cells use ZFP809 to silence retroviral DNAs. *Nature* **458:** 1201–1204. doi:10.1038/nature07844

Wu SY, Chiang CM. 2007. The double bromodomain-containing chromatin adaptor Brd4 and transcriptional regulation. *J Biol Chem* **282:** 13141–13145. doi:10.1074/jbc.R700001200

Yang Z, Yik JHN, Chen R, He N, Jang MK, Ozato K, Zhou Q. 2005. Recruitment of P-TEFb for stimulation of transcriptional elongation by the bromodomain protein Brd4. *Mol Cell* **19:** 535–545. doi:10.1016/j.molcel.2005.06.029

Zhang Z, Klatt A, Gilmour DS, Henderson AJ. 2007. Negative elongation factor NELF represses human immunodeficiency virus transcription by pausing the RNA polymerase II complex. *J Biol Chem* **282:** 16981–16988. doi:10.1074/jbc.M610688200

Zhao B, Zou J, Wang H, Johannsen E, Peng CW, Quackenbush J, Mar JC, Morton CC, Freedman ML, Blacklow SC, et al. 2011. Epstein–Barr virus exploits intrinsic B-lymphocyte transcription programs to achieve immortal cell growth. *Proc Natl Acad Sci* **108:** 14902–14907. doi:10.1073/pnas.1108892108

Zhao N, Sebastiano V, Moshkina N, Mena N, Hultquist J, Jimenez-Morales D, Ma Y, Rialdi A, Albrecht R, Fenouil R, et al. 2018. Influenza virus infection causes global RNAPII termination defects. *Nat Struct Mol Biol* **25:** 885–893. doi:10.1038/s41594-018-0124-7

Zhou H, Schmidt SCS, Jiang S, Willox B, Bernhardt K, Liang J, Johannsen EC, Kharchenko P, Gewurz BE, Kieff E, et al. 2015. Epstein–Barr virus oncoprotein super-enhancers control B cell growth. *Cell Host Microbe* **17:** 205–216. doi:10.1016/j.chom.2014.12.013

Zhu Y, Pe'ery T, Peng J, Ramanathan Y, Marshall N, Marshall T, Amendt B, Mathews MB, Price DH. 1997. Transcription elongation factor P-TEFb is required for HIV-1 tat transactivation in vitro. *Genes Dev* **11:** 2622–2632. doi:10.1101/gad.11.20.2622

Zhu Y, Wang GZ, Cingöz O, Goff SP. 2018. NP220 mediates silencing of unintegrated retroviral DNA. *Nature* **564:** 278–282. doi:10.1038/s41586-018-0750-6

Zila V, Margiotta E, Turoňová B, Müller TG, Zimmerli CE, Mattei S, Allegretti M, Börner K, Rada J, Müller B, et al. 2021. Cone-shaped HIV-1 capsids are transported through intact nuclear pores. *Cell* doi:10.1016/j.cell.2021.01.025

3D or Not 3D: Shaping the Genome during Development

Juliane Glaser[1] and Stefan Mundlos[1,2,3]

[1]RG Development and Disease, Max Planck Institute for Molecular Genetics, 14195 Berlin, Germany
[2]Institute for Medical and Human Genetics, Charité Universitätsmedizin Berlin, 13353 Berlin, Germany
[3]Charité - Universitätsmedizin Berlin, BCRT - Berlin Institute of Health Center for Regenerative Therapies, 10178 Berlin, Germany

Correspondence: mundlos@molgen.mpg.de

One of the most fundamental questions in developmental biology is how one fertilized cell can give rise to a fully mature organism and how gene regulation governs this process. Precise spatiotemporal gene expression is required for development and is believed to be achieved through a complex interplay of sequence-specific information, epigenetic modifications, *trans*-acting factors, and chromatin folding. Here we review the role of chromatin folding during development, the mechanisms governing 3D genome organization, and how it is established in the embryo. Furthermore, we discuss recent advances and debated questions regarding the contribution of the 3D genome to gene regulation during organogenesis. Finally, we describe the mechanisms that can reshape the 3D genome, including disease-causing structural variations and the emerging view that transposable elements contribute to chromatin organization.

Whereas the genetic information is encoded in the linear sequence, the expression of this information requires the genome to be organized in the 3D space in the nucleus. This was suggested by seminal work from numerous microscopy studies starting in the late 19th century with the observation that interphase chromosomes are organized in territories (Rabl 1885). About half a century later, the description of euchromatin and heterochromatin that differ by their degree of compaction in the nucleus, paved the way for the idea that the genome is organized into active and inactive compartments (Heitz 1928). More recently, molecular approaches based on chromosome conformation capture (3C) techniques expanded this concept (Dekker et al. 2002; Simonis et al. 2006; Zhao et al. 2006). Particularly, the development of C-methods coupled with high-throughput sequencing (Hi-C) allowed the genome-wide quantification of long-range interaction and was fundamental to revealing the importance of 3D organization for genome function (Lieberman-Aiden et al. 2009; de Wit and de Laat 2012). Since the first Hi-C map was published, tremendous improvements have been made to detect chromatin contacts, ranging from 3C-based techniques to live-cell imaging and most recently ligation-free approaches (Kempfer and Pombo 2020; Jerkovic and Cavalli 2021). Excitingly,

Cite this article as *Cold Spring Harb Perspect Biol* doi: 10.1101/cshperspect.a040188

the organization of DNA in 3D has emerged as a major driver of development in metazoans. While genes correspond to only 2% of our genome, the noncoding part contains millions of regulatory sequences capable of restricting gene expression to specific tissues and developmental genes (De Laat and Duboule 2013). The formation of DNA loops and domains of specific interactions between regulatory sequences and their target promoter can mediate the control of transcriptional specificity during development (Andrey et al. 2013; Symmons et al. 2014; Kragesteen et al. 2018). Importantly, reshuffling of the 3D organization at specific loci has been described as a driver of several pathologies including developmental malformations and cancer (Lupiáñez et al. 2016). However, this deterministic view was recently challenged by several studies that propose a less striking role of the 3D genome (Dong et al. 2018; Ghavi-Helm et al. 2019; Ghavi-Helm 2020; Ing-Simmons et al. 2021). To what extent chromatin folding is important for gene regulation and development, what controls long-range gene activation and which elements shape the 3D genome architecture are currently debated and unresolved questions in the field. Here, we discuss why and how the mammalian genome is organized during development, from the early stages to the more mature embryo. We summarize the latest findings regarding the impact of disrupting the 3D genome organization and review the emerging contribution of transposable elements (TEs) in shaping the 3D mammalian genome.

ORGANIZING THE LINEAR GENOME IN 3D

Genomic DNA within the nucleus is spatially partitioned at different levels. First, DNA segregates into two compartments referred to as A and B (Fig. 1A). This compartmentalization is strongly linked to transcriptional activity as both appear to have a reciprocal influence on each other (van Steensel and Furlong 2019). Poorly transcribed DNA (i.e., regions poor in active histone marks or enriched in silencing factors) are part of the B compartment. These heterochromatic regions tend to segregate close to the nuclear lamina and to the nucleoli, forming lamina-associated domains (LADs) and nucleoli-associated domains (NADs), respectively (Fig. 1A; Pickersgill et al. 2006; Guelen et al. 2008; Németh et al. 2010; van Koningsbruggen et al. 2010). Mammalian cells have ∼1100 to 1500 LADs, which are on average 500 kb in size. How genes in LADs are kept in a repressive state is not yet clear. One proposed mechanism is that the interaction with proteins of the nuclear membrane mediates changes in the chromatin state leading to enrichment of repressive histones marks such as H3K9me2 (Yokochi et al. 2009; Haarhuis et al. 2017). In contrast, inter-LAD DNA belongs to the A compartment and is actively transcribed (Fig. 1A; Akhtar et al. 2013). Importantly, not all LADs genes are systematically repressed but the local chromatin environment also influence the expression state (Leemans et al. 2019). Higher-order chromatin organization is a developmentally dynamic process: upon differentiation, genes that determine cellular identity detach from the nuclear lamina and become active (Peric-Hupkes et al. 2010).

While the segregation of the DNA into the A and B compartments reflects the 3D organization of the genome at the megabase scale, higher resolution Hi-C has revealed the existence of ∼2000 chromatin domains ranging from 100 kb to 1 Mb in size (Dixon et al. 2012; Nora et al. 2012). These topologically associating domains (TADs) correspond to genomic regions with high interaction and thus close physical proximity. Regulatory elements and their target genes are often found inside the same TAD and have less chance to interact with genomic elements of the neighboring TADs (Fig. 1A). The boundary regions separating two TADs are enriched for active genes and for insulator proteins (Dixon et al. 2012; Sexton et al. 2012). In mammals, boundaries are characterized by the cobinding of cohesin and the architectural factor CTCF (Fig. 1A; Phillips-Cremins et al. 2013). In *Drosophila*, cohesin is absent at TAD boundaries and domain borders are demarcated by a fly-specific insulator complex that can include CTCF (Hou et al. 2012; Sexton et al. 2012; Wang et al. 2018). There is thus a certain level of genome structure conservation between mammals and insects although molecular differ-

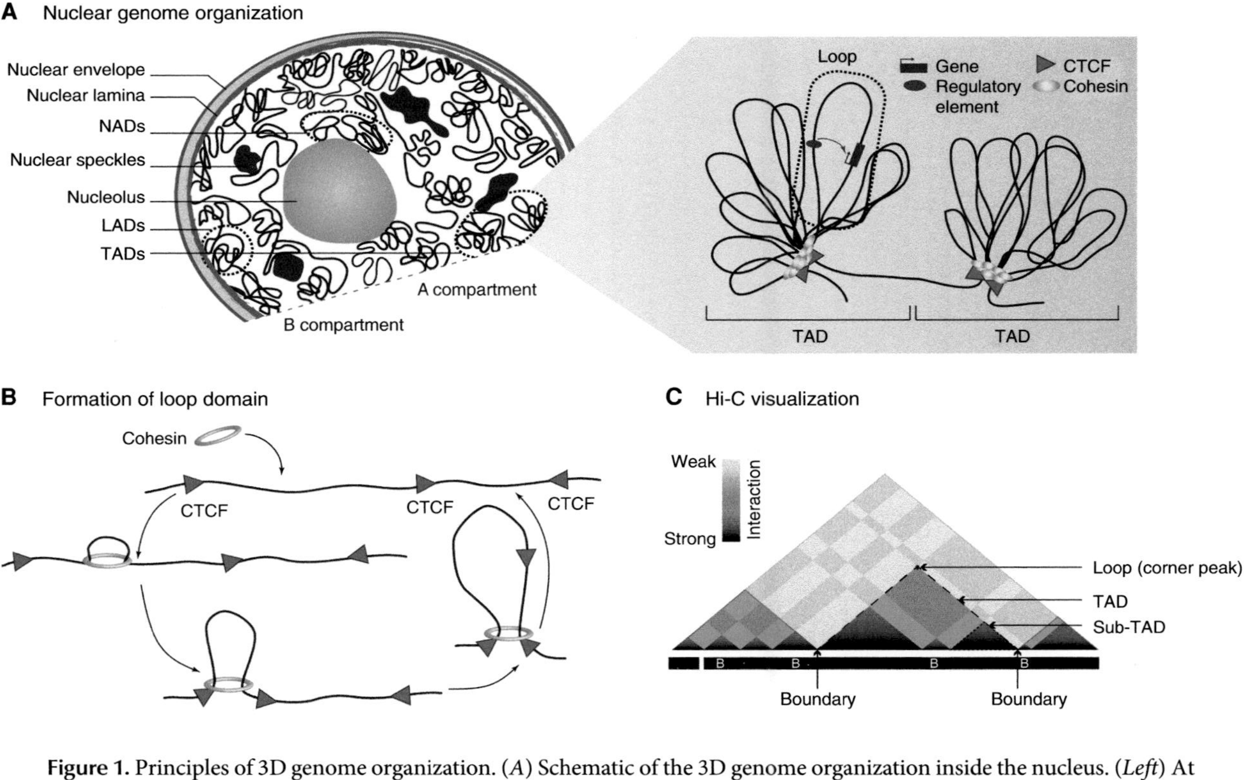

Figure 1. Principles of 3D genome organization. (*A*) Schematic of the 3D genome organization inside the nucleus. (*Left*) At higher-order scales, the transcriptionally active and silenced chromatin segregates into A (red) and B (blue) compartments, respectively. DNA in the A compartment organizes around nuclear speckles. The B compartment localizes at the nuclear envelope and frequently overlaps with lamina-associated domains (LADs) and nucleolar-associated domains (NADs). (*Right*) At smaller genomic scales, chromatin folds into topologically associating domains (TADs). Chromatin loops inside a TAD are formed by the binding of CTCF and cohesin. Chromatin folding allow the interaction of regulatory elements and their target genes. (*B*) Loop extrusion model. Cohesin extrudes chromatin through its ring-like shape to form a loop. The loop grows in size until it reaches two convergent CTCFs, which block cohesin activity. This is a dynamic process that can be disassembled. (*C*) Schematic representation of a Hi-C map where A/B compartment (red/blue), loop, TAD, and sub-TAD can be visualized.

ences exist (Rowley et al. 2017). Emergence of the chromatin loops forming TADs is currently explained by the biophysical loop extrusion model. According to this model, the motor activity of loop extruding factors, in particular the cohesin complex, progressively extrudes the chromatin loop until it reaches a CTCF-binding site (Fig. 1B; Sanborn et al. 2015; Fudenberg et al. 2016). CTCF-mediated loop extrusion happens in an orientation-dependent manner, explaining why TAD borders are mainly formed by CTCF sites in a divergent orientation (de Wit et al. 2015; Guo et al. 2015; Fudenberg et al. 2016). This is a dynamic process as cohesin rings are constantly loaded and released from DNA (Fig. 1B).

Hi-C has proven itself as the method of choice for many laboratories to gain insight into the global 3D organization of the genome, allowing the visualization of compartmentalization, TAD structure, and loop domain (Fig. 1C). Although TAD structure has been shown to be highly conserved between cell type and species (Dixon et al. 2012), examination of 3D genome structure at different stages of mammalian development, in various cell types and in disease models, has revealed different states and ambiguous roles of higher-order chromatin architecture, which will be discussed next.

Establishment of the 3D Landscape in Early Embryogenesis

The development of a new organism starts with fertilization. The one-cell embryo (zygote) is formed by the fusion of two mature differentiated but highly divergent gametes. In mammals, the spermatozoa genome is ultra-condensed and poorly transcribed because of the dense packing with protamine (Balhorn et al. 1977; Gaucher et al. 2010). Conversely, the oocyte genome presents an extensively open chromatin conformation and is highly transcribed to allow the production of a maternal storage of RNAs and proteins for the future embryo (Wu et al. 2016). Immediately after fertilization, the zygote undergoes epigenetic reprogramming to revert to a totipotent state (Morgan et al. 2005; Burton and Torres-Padilla 2014). The first days of post-fertilization development are thus marked by major changes in DNA methylation, histone marks, and the transcriptional program. How the 3D genome behaves during this period of drastic epigenetic reprogramming has been a long-standing question in the field. Recently, this question was addressed using low-input Hi-C methods on mouse embryos, which enabled the analysis of a very low number of cells. Consistently with its tightly packed genome, sperm shows a higher level of interchromosomal and long-range chromatin interaction (>4 Mb) as compared with fibroblasts or embryonic cells (Battulin et al. 2015; Du et al. 2017; Ke et al. 2017; Vara et al. 2019). While sperm DNA is partitioned into A and B compartments, the subdivision into TADs also exists but is less clear (Fig. 2). Mature oocytes prior to fertilization are arrested in metaphase II (MII) in meiosis. Analysis of Hi-C maps of MII oocytes revealed very little interchromosomal interaction and a homogeneous chromatin folding lacking TADs and compartments (Fig. 2; Du et al. 2017; Ke et al. 2017).

The 3D organization of early embryos prior to zygotic genome activation (ZGA) (i.e., zygote and two-cell stage) is unclear. Comparisons of Hi-C maps from mouse embryos at different stages have revealed that pre-ZGA embryos display very weak TADs, which are gradually established until the eight-cell-stage embryo where they appear clearer and are maintained onward (Fig. 2; Du et al. 2017; Ke et al. 2017; Zhang et al. 2018). High-resolution in situ Hi-C approaches comparing data at the population and single-cell level contrasted this view and suggested loops and TADs might be established from the zygote through the cohesin-dependent loop extrusion model (Flyamer et al. 2017; Gassler et al. 2017). More recent studies finally suggest a two-component model (Collombet et al. 2020; Du et al. 2020): (1) parentally preformed domains associated with Polycomb compartments and independent of CTCF exist transiently from the zygote to the four-cell stage and are followed by (2) progressive establishment of clear TADs (Fig. 2). Importantly, although the establishment of TADs was observed following ZGA, it appears independent of ZGA in both mice and *Drosophila* (Hug et al. 2017; Ke et al. 2017) but not in human embryos (Chen et al. 2019). Overall, a

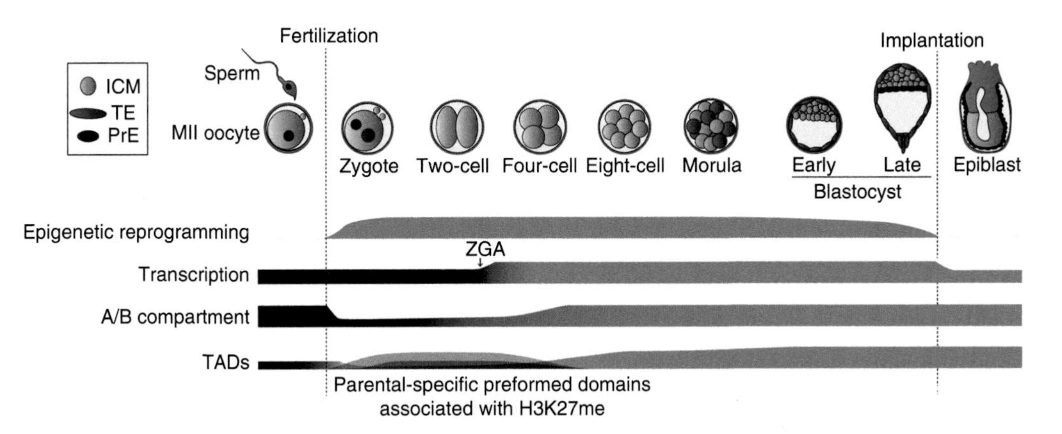

Figure 2. Establishment of the 3D genome organization during early mammalian development. Following fertilization, embryos are subject to a dramatic epigenetic reprogramming during the preimplantation period. Transcriptional input in the zygote is exclusively maternal until zygotic genome activation (ZGA) occurs at the two-cell stage. The establishment of the 3D architecture in the early embryo is also dynamic: A/B compartments exist in the sperm but are then lost and reestablished in the embryo and TADs become defined only from the eight-cell stage. The two-component model suggests that parental-specific preformed domains associated with Polycomb exist transiently from the zygote to four-cell stage and TADs are then established progressively. (ICM) Inner cell mass, (TE) trophectoderm, (PrE) primitive endoderm. (Figure created with permission from modified images by Marius Walter.)

completely unstructured cell (oocyte) and a cell harboring long-range chromatin interaction (sperm) fuse to form a zygote that undergoes reprogramming of its 3D genome (Hug and Vaquerizas 2018). Despite the highly divergent chromatin organization of the gametes before fertilization, it is still unclear to what extent the embryonic 3D organization is shaped by parental memory of architectural proteins and epigenetic marks. A more exhaustive review of the chromatin conformation of mouse gametes and early embryos can be found in Zheng and Xie (2019) and Du et al. (2021).

3D Chromatin Folding and Gene Expression during Development

To complete embryo development, gene expression must be precisely regulated in time and space. Yet, regulatory elements are often located at significant genomic distances from their target genes, implying long-range interaction for the transmission of the regulatory information (Sanyal et al. 2012). Among the various elements that can regulate gene expression (activators, repressors, insulators), enhancers are the most widely studied. Enhancers are short DNA sequences (100–500 bp), bound by transcription factors (TFs), which drive transcription in response to a cellular signal.

Dynamic versus Preformed Chromatin Folding for Gene Activation

The current model proposes that transcriptional activation requires physical proximity between enhancers and promoters, a process mediated by DNA looping (De Laat and Duboule 2013). A textbook example is the in vivo demonstration of the physical interaction between the *β-globin* gene and its locus control region (LCR) when actively transcribed (Carter et al. 2002; Tolhuis et al. 2002). This model nicely explains how distal enhancers are able to activate genes over long distances to confer spatiotemporal specificity. Enhancer–promoter communication is favored by the partitioning of the genome into TADs. These regulatory units have emerged as major components of the regulatory landscape necessary to complete organogenesis (Fig. 3A; Andrey

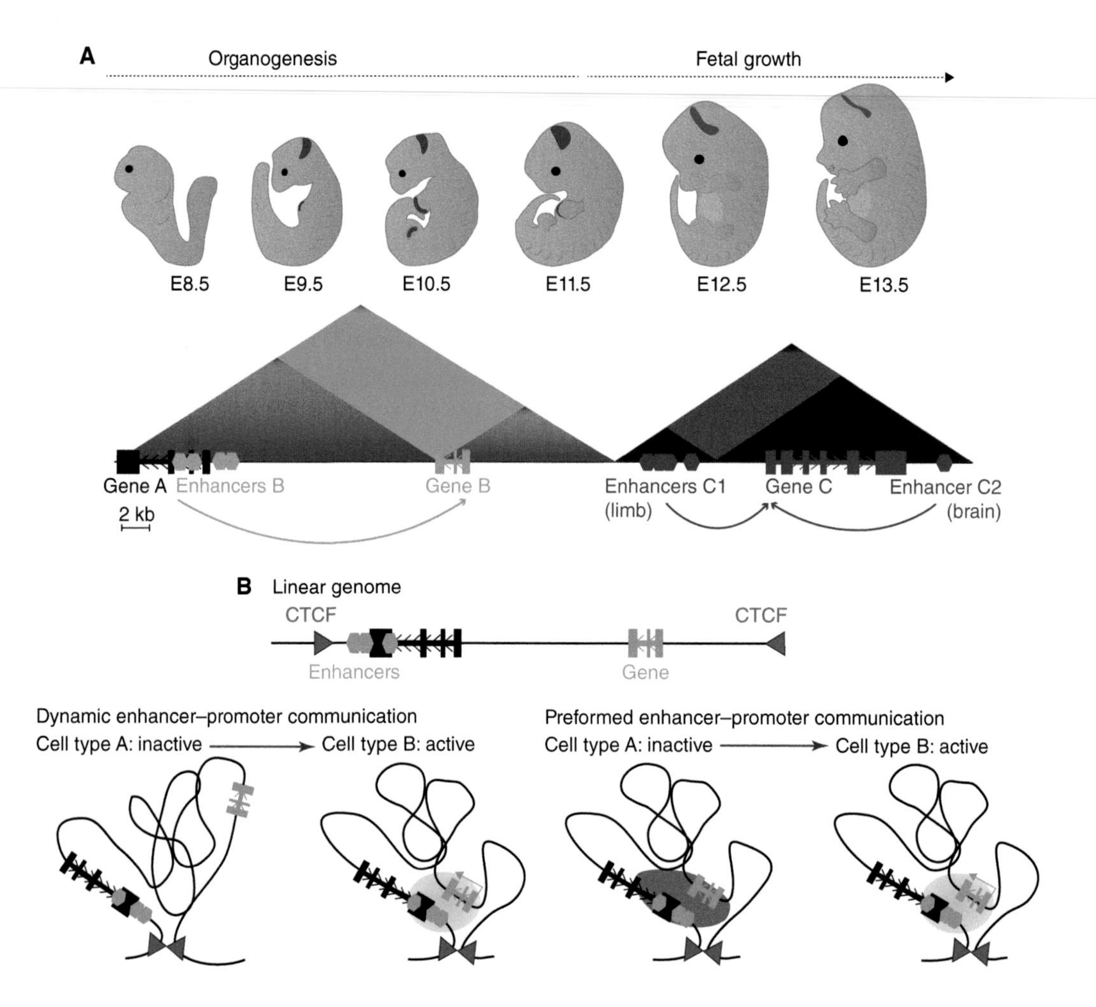

Figure 3. 3D genome organization and gene regulation. (*A*) Schematic of a hypothetical regulatory landscape of developmental genes along with a gene expression pattern in the mouse postimplantation embryos. Gene A and B are in the same topologically associating domain (TAD). Gene A is not expressed during development. Gene B is activated by its enhancers and is expressed in the developing limb from E10.5 to E13.5. Gene C is in the neighboring TAD and is activated by two different sets of tissue-specific enhancers in the brain and the developing limb. (*B*) Dynamic versus preformed chromatin folding. (*Upper* panel) Linear representation of the hypothetical TAD of gene B from A. (*Left* panel) Model of dynamic enhancer–promoter interaction showing two different chromatin conformations. Enhancer and promoter do not interact in cell type A where the gene is inactive, whereas their physical proximity mediates gene expression in cell type B. (*Right* panel) In the model of preformed enhancer–promoter communication, the chromatin conformation is the same in cell type A and B and is independent of gene expression.

et al. 2013; Symmons et al. 2014; Tsujimura et al. 2015; Rodríguez-Carballo et al. 2017). Although TADs are highly conserved between cell types (Dixon et al. 2012), intra-TAD interactions show a high degree of cell and tissue specificity during development, reflecting spatiotemporal gene activation (Fig. 3B; De Laat and Duboule 2013). One example is the *Pitx1* gene whose expression is restricted to the developing hindlimb where it is critical to establish cell identity (DeLaurier et al. 2006). Despite *Pitx1* hindlimb-specific expression, its enhancer (Pen) is active in both forelimbs and hindlimbs, raising the question of what mediates this tissue specific expres-

 Cite this article as *Cold Spring Harb Perspect Biol* doi: 10.1101/cshperspect.a040188

sion (Kragesteen et al. 2018). Interestingly, chromatin conformation at the *Pitx1* locus is dynamically regulated to allow physical interaction between Pen and *Pitx1* exclusively in the hindlimb, whereas the 3D configuration prevents *Pitx1* from interacting with its enhancer in the forelimb (Kragesteen et al. 2018). The regulation of *Pitx1* is thus an example of how dynamic chromatin folding controls tissue-specific gene expression during development (Fig. 3B).

However, in contrast to such dynamic configuration, enhancer–promoter interactions can also be preformed (Fig. 3B; Andrey and Mundlos 2017; Andrey et al. 2017). In this configuration, enhancers are in physical proximity with their cognate promoters independently of gene activity. This possibly allows for a rapid and robust switch from the inactive to the active state (Fig. 3B). Preformed contacts can be facilitated by the binding of various factors such as the Polycomb-repressive complex or CTCF (Barbieri et al. 2017; Cruz-Molina et al. 2017). For instance, activation of the mouse *sonic hedgehog* (*Shh*) gene in the developing limb depends on its interaction with a limb-specific enhancer, the ZRS, located more than 1 Mb away from its promoter (Lettice et al. 2003; Sagai et al. 2005). However, the *Shh* promoter and the ZRS are in physical proximity even in cells in which *Shh* is inactive (Amano et al. 2009; Williamson et al. 2016). This preformed 3D configuration is partially due to the presence of strong CTCF sites flanking the enhancer that interact with convergent sites at the *Shh* promoter. This CTCF-induced promoter–enhancer interaction was shown to be important but not essential for normal gene expression (Paliou et al. 2019).

The Role of CTCF in Domain Insulation

The major function of TADs appears to be to facilitate frequent interactions within the TAD, but also to insulate this activity from neighboring regulatory domains (Fig. 3A). This insulation is mainly achieved by the binding of the architectural protein CTCF. This zinc finger DNA-binding protein was initially described as a strong insulator of enhancer–promoter interaction (Bell et al. 1999). Being sensitive to DNA methylation (Wang et al. 2012), CTCF can function as an epigenetic regulator of imprinted genes whose monoallelic expression is fundamental for development and physiological function (Fedoriw et al. 2004; Beygo et al. 2013; Llères et al. 2019; Tucci et al. 2019; Noordermeer and Feil 2020). Inactivation of CTCF results in peri-implantation lethality, showing that CTCF is required for mammalian embryonic development (Wan et al. 2008; Soshnikova et al. 2010; Moore et al. 2012). At TAD and loop boundaries, the co-binding of CTCF with cohesin mediates insulation (Phillips-Cremins et al. 2013; Sofueva et al. 2013; Merkenschlager and Nora 2016). Accordingly, depletion of cohesin or CTCF results in a dramatic loss of loops and complete absence of TADs (Nora et al. 2017; Rao et al. 2017; Schwarzer et al. 2017). Although one would expect that the loss of TAD structure results in multiple enhancer–promoter ectopic contacts and massive gene misexpression, the effect on gene expression is rather mild and only few genes are ectopically activated (Nora et al. 2017; Rao et al. 2017). However, the requirement of CTCF for mouse embryonic development makes it impossible to address its impact on gene regulation during organogenesis. Using a model of zygotic CTCF knockout in zebrafish where the maternal input of CTCF allows survival, misregulation of thousands of genes, and severe alteration of developmental progression was observed (Franke et al. 2020).

TADs Confer Robustness and Precision to Gene Expression

Whereas these genome-wide studies revealed the importance of CTCF and cohesion for 3D genome architecture, they did not fully answer how important organized chromatin folding is for developmental gene expression. In a more locus-centered approach, Despang et al. dissected the role of CTCF at the *Sox9-Kcnj2* locus in the developing mouse embryo. Loss-of-function mutations of *SOX9* are known to cause skeletal dysplasia with sex reversal, whereas misexpression of *KCNJ2* in a *SOX9* pattern results in isolated limb malformations in humans (Kurth et al. 2009). The regulatory domains of *Sox9* and *Kcnj2* are located in two large TADs separated by a strong boundary. Surprisingly, loss of insulation between the two TADs by removal of all CTCF sites has no major effect on gene regulation and the mutant mice have no apparent phenotype (Despang et al. 2019). However, in the CTCF deletion mutants, a lack of precision was noted induced by a spillover of regulatory activity from one TAD to the other. These results show that enhancers have an intrinsic affinity to their cognate promoters and that gene regulation can work without CTCF-induced chromatin folding, albeit with less precision. If the loss of insulation does not induce gene misexpression, how can disease-causing misexpression be explained? To redirect the transcriptional activity from one domain to the other, complete TAD reorganization of the *Sox9-Kcnj2* locus is needed: large inversions involving the *Sox9* TAD and repositioning of the boundary result in changes in gene expression by redirecting enhancers to the wrong target thereby producing disease phenotypes (Despang et al. 2019). Consistent with this and previous reports, depletion of CTCF in human leukemic B cells alter TAD organization but did not impact *trans*-differentiation into functional macrophages (Stik et al. 2020). However, when subjected to endotoxin exposure, macrophages depleted for CTCF cannot properly activate inflammatory genes (Stik et al. 2020). This reveals that CTCF-mediated TAD structure, although facultative for cell fate conversion, is required to facilitate cellular response to an external stress. These two studies suggest that the primary role of TADs could be to confer robustness and precision for proper gene expression rather than to constrain enhancer–promoter contact as it was initially thought. If TADs are necessary but not sufficient for gene expression, other mechanisms must be involved in the regulation of enhancer–promoter contacts.

Alternative Mechanisms for Transcriptional Control

One of the forces that has recently been proposed to guide long-range transcriptional activation without direct enhancer–promoter interaction is liquid–liquid phase separation (Hnisz et al. 2017). The formation of a phase-separated liquid compartment is a fundamental property of a number of membraneless organelles (Banani et al. 2017) and has recently emerged as a mechanism directing transcriptional control. Formation of HP1 phase-separated droplets was initially shown to mediate heterochromatin domains, revealing a possible new mechanism for gene silencing (Larson et al. 2017; Strom et al. 2017). Multiple studies have since demonstrated that TFs, BRD4 coactivating factors, the mediator complex, and RNA PolII are involved in phase separation and this is associated with gene activation (Boija et al. 2018; Cho et al. 2018; Chong et al. 2018; Sabari et al. 2018). Enhancers would thus be involved to phase-separate transcriptional condensates allowing them to induce gene expression without close proximity with the promoter. Interestingly, this could explain why uncoupling of transcriptional bursting and enhancer–promoter proximity has been observed at the *Sox2* locus in mouse embryonic stem cell (mESC) (Alexander et al. 2019). Similarly, *Shh* and its brain-specific enhancers, located more than 100 kb apart from each other, decrease their physical proximity upon activation of *Shh* in neural cells (Benabdallah et al. 2019). Alternatively, one can speculate that enhancers can mediate gene activation through the spreading of active histone marks, setting up a permissive environment and an accessible chromatin state.

Although mechanisms of gene regulation and 3D genome architecture appear similar between mammals and *Drosophila*, the fly genome is much more compact and, accordingly, enhancer–promoter distances are smaller. Two recent studies investigated the role of TADs during cell fate decision in *Drosophila* embryos and suggested independence of tissue-specific gene activation and enhancer–promoter looping (Espinola et al. 2021; Ing-Simmons et al. 2021). This suggests that preformed chromatin folding could be a more common mechanism than initially thought. However, to what extent chromatin folding at specific developmental loci is important for tissue-specific gene expression still remains unresolved and should be clarified by more locus-centered approaches. In accordance with this, Chen et al. used a live-imaging system to show that sustained enforced physical proximity between enhancers of the well-studied *even-skipped* (*eve*) developmental gene and a reporter promoter is required for initiation and stability of transcription (Chen et al. 2018). Methods to study the 3D genome vary from single-cell live imaging to average population proximity ligation; one has to be careful with technical limitations when applying the chosen model. Overall, as always in biology, it seems that there is no general rule to explain how genes are regulated in time and space during development. Transcriptional control seems to be governed by diverse and locus-dependent mechanisms, reinforcing the need to study gene regulation not only with genome-wide but also with locus-specific approaches.

The Impact of Structural Variations on Genome Organization

The importance of restricting promoter-enhancer contacts was demonstrated by studying the impact of structural variations (SVs) in human disease and mouse models. Deletions, duplications, inversions, or translocations, collectively called SVs, are a major cause of human genetic diseases (Stankiewicz and Lupski 2010). Interestingly, disease-causing variations are not always due to changes in gene copy number, raising the question of how they exert their pathology. The answer may reside in the 3D genome: SVs can interfere with TAD structure and cause enhancer "hijacking" (or enhancer "adoption"), a mechanism known to lead to gene misexpression. SVs at the human EPHA4 locus, for example, are associated with several rare limb malformations (Lupiáñez et al. 2015). Engineering these rearrangements in mice with CRISPR-Cas technology was instrumental in revealing the impact of 3D genome on pathological phenotypes (Kraft et al. 2015). For instance, a deletion including the boundary between the *Epha4* and the *Pax3* TADs leads to ectopic expression of *Pax3* in an *Epha4*-like pattern and shortening of the phalanges (Fig. 4A; Lupiáñez et al. 2015). Similarly, an inversion and a duplication at the same locus repositioned the enhancer of *Epha4* next to the *Wnt6* and the *Ihh* genes, respectively (Fig. 4B). These SVs all cause changes in enhancer accessibility, allowing the genes in the rearranged TAD to adopt the specific expression pattern of *Epha4* in the distal limb bud. Misexpression of these genes drives congenital limb malformations such as F-syndrome and polydactyly (Lupiáñez et al. 2015). Interestingly, while *Pax3*, *Wnt6*, and *Ihh* are ectopically expressed in a very similar pattern in the developing limb, they give rise to three distinct phenotypes, indicating that the resulting functional proteins interfere with the endogenous signaling pathway. A further mechanism was observed at the SOX9-KCNJ2 locus, where an inter-TAD duplication results in Cooks syndrome, a congenital limb malformation (Franke et al. 2016). This duplication involves the boundary between the two regulatory domains and leads to ectopic expression of *Kcnj2* in a *Sox9* pattern in mouse embryos (Fig. 4C). Hi-C revealed that the duplicated region can generate a new chromatin domain called "neo-TAD" (Franke et al. 2016). These studies demonstrate the impact that deletions, inversions, or duplications can have on 3D genome architecture by mediating TAD fusion, TAD shuffling, or neo-TADs, respectively (Fig. 4A–C; Spielmann et al. 2018). Additionally, inversions relocating the *Epha4* limb enhancers in a gene-rich region lead to the formation of architectural stripes with ectopic activation of those genes (Fig. 4D; Kraft et al. 2019). This highlights the importance of considering the 3D genome when studying human genetic disease.

Enhancer hijacking also appears to be a common mechanism of oncogene activation (Hnisz et al. 2016). For instance, IRS4 misexpression is associated with several cancers and is mediated by genomic deletions including a TAD boundary (Weischenfeldt et al. 2017). Similarly, tandem duplications at the oncogene IGF2 induce a de novo chromatin loop between a superenhancer and IGF2 promoter, leading to its overexpression in colorectal cancer (Weischenfeldt et al. 2017). SVs also drive oncogene activation in medulloblastoma, a severe childhood cancer, by repositioning GFI1 and GFI1B next to enhancers with oncogenic activity (Northcott et

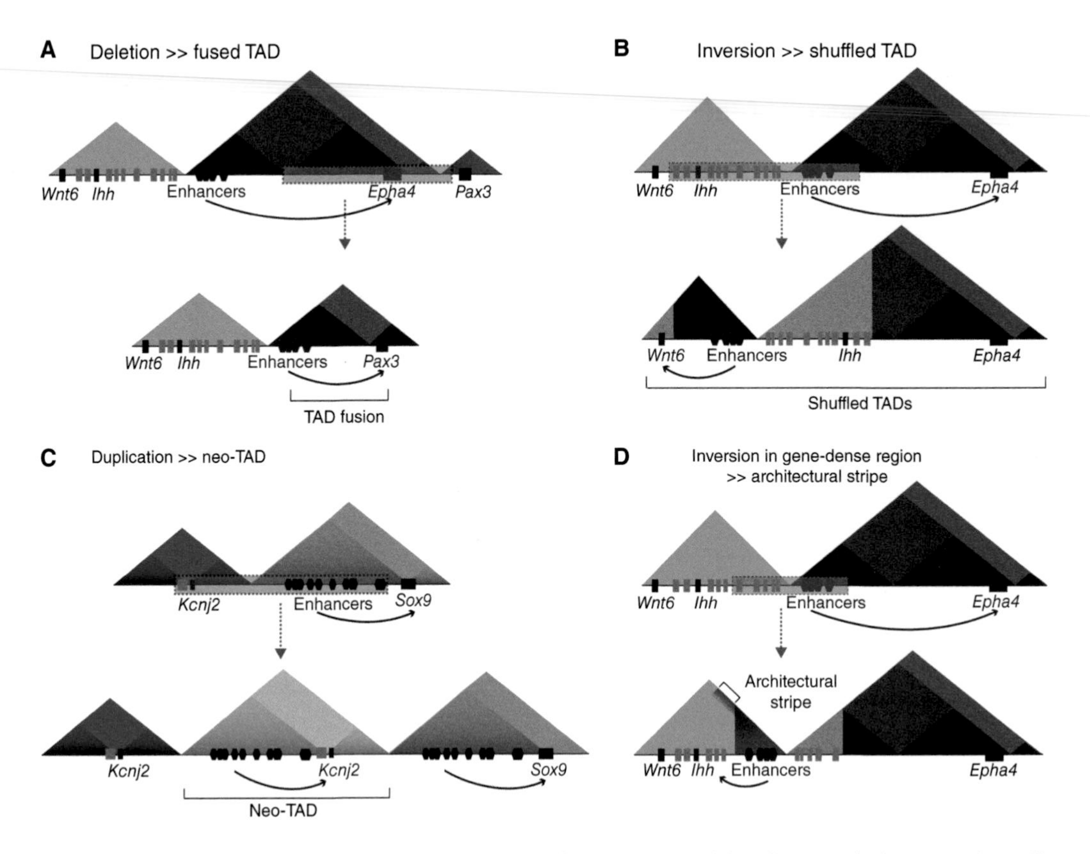

Figure 4. Consequences of structural variations (SVs) on the 3D genome. (*A*) Deletion including a topologically associating domain (TAD) boundary between the *Epha4* and the *Pax3* TAD leads to TAD fusion. As a consequence, *Epha4* enhancers are in the same TAD as *Pax3*, which is expressed in an *Epha4*-like pattern. (*B*) Large inversion creates TADs reshuffling at the *Epha4* locus and induces proximity of *Epha4* enhancers and *Wnt6*, which is ectopically expressed in *Epha4* territories. (*C*) Inter-TAD duplication at the *Kcnj2-Sox9* locus creates a neo-TAD and relocalize *Sox9* enhancers next to *Kcnj2* gene, leading to ectopic expression. (*D*) Inversion at the *Epha4* locus relocate *Epha4* enhancers in a gene-dense region in the neighboring TAD leading to architectural stripes and gene misexpression. In *A–D*, the gray square represents the genomic region that is subjected to SV.

al. 2014). This strengthens the important role of establishing a defined regulatory landscape to prevent cellular dysfunction. Interestingly, topological changes occurring in cancer cells alongside with epigenetic remodeling were also shown to repress oncogenic programs and induce antitumor genes, leading in this case to slow down the malignant progression (Johnstone et al. 2020).

SVs in disease models do not always cause enhancer adoption. To model the genetic cause of the branchio-oculo-facial congenital disorder, hiPSCs from a patient with an 89-Mb inversion at the TFAP2A locus were used (Laugsch et al. 2019). Whereas the inversion leads to the repositioning of several genes in the TFAP2A TAD next to its enhancers, no enhancer adoption or ectopic expression were observed. The pathomechanism of this syndrome is rather linked to TFAP2A loss-of-function due to the distancing from its own enhancers (Laugsch et al. 2019). Balanced chromosomal abnormalities (BCAs) (i.e., SVs without gain or loss of genomic material) are associated with human developmental abnormalities and alterations of TAD structures (Redin et al. 2017; Dong et al. 2018). While TAD reshuffling can lead to decreased gene expression in some cases (Redin et al. 2017), they also do not involve any transcriptional changes in others (Dong et al. 2018). Similarly, the highly rearranged balancer chromosomes in *Drosophila* contain many SVs, which cause disruption to chromatin folding and TAD architecture (Ghavi-Helm et al. 2019). Despite these dramatic changes in the 3D genome, gene expression remains largely unaltered even in the context of disrupted enhancer–promoter looping. This suggests that reshuffling TAD structure is not always deleterious (Ghavi-Helm 2020), reinforcing the idea that each regulatory landscape is unique. The studies discussed in this part aim to answer the same question but with two very different approaches: either a disease phenotype with a single SV or a highly rearranged genome with no clear phenotype, explaining the contrasting findings.

Transposable Elements Organizing and Disorganizing the Genome

Apart from deletions, duplications, and inversions, SVs also include insertions of exogeneous sequences such as TEs. These mobile elements, in part originating from ancient viral infections, account for almost half of the mammalian genome (Lander et al. 2001). TEs display large variability: they can move in the genome either using RNA intermediates (class I) or DNA intermediates (class II) and are further divided in subclasses according to their mechanism of replication (Wicker et al. 2007). While long considered "junk DNA," TEs are controlling elements of our genome. Several classes of TEs harbor TF-binding sites, including CTCF, and a subset has been shown to evolve as active tissue-specific enhancers (Bourque et al. 2008; Su et al. 2014). Multiple waves of TEs insertion contributed to species-specific expansion of CTCF-binding sites and have therefore largely impacted the evolution of regulatory landscape (Schmidt et al. 2012; Sundaram et al. 2014). Additionally, TEs can engage in inter- and intrachromosomal long-range chromatin folding (Cournac et al. 2016; Raviram et al. 2018). However, to what extent TE insertions contribute to shaping of 3D genome organization is an active area of research. TAD boundaries are enriched for specific families of TEs (Dixon et al. 2012), raising the question of whether CTCF sites embedded in TEs can participate in domain formation. Whereas the effect of CTCF knockout has been explored, how the insertion of CTCF-binding sites can alter TADs and gene regulation is less studied. A recent study investigated whether and how the insertion of a tissue-invariant boundary can create de novo domains (Zhang et al. 2020). It was found that the 2 kb inserted DNA bound by CTCF and containing the TSS of a housekeeping gene is able to demarcate new domains and to shape 3D architecture over large genomic scales. Among 16 insertions over 11 chromosomes, the contribution to the 3D architecture is highly variable, highlighting a context-dependent effect (Zhang et al. 2020). Importantly, both transcription and CTCF contribute to the formation of new boundaries, two features that can be brought about by the insertion of TEs. Around 15% of loops in human and 25% in mice contain at least one TE-derived CTCF (Choudhary et al. 2020). Comparing these CTCF sites between mouse and human revealed the contribution of TEs to the formation of species-specific chromatin looping (Choudhary et al. 2020; Diehl et al. 2020). TE-mediated CTCF insertions thus appear as a mechanism to drive species-specific chromatin folding (Fig. 5A). Interestingly, the high number of TEs containing CTCF-binding sites able to create a boundary also suggests an important and yet undescribed role of CTCF in transposon biology.

Endogenous retroviruses (ERVs) are a class of TEs still active in human and mouse that can be highly transcribed during epigenetic reprogramming occurring in the early embryos (Rodriguez-Terrones and Torres-Padilla 2018). As previously discussed, the 3D genome is also subject to reprogramming and the establishment of new domains begins at the two-cell stage (Fig. 2; Du et al. 2017; Ke et al. 2017). The expression of the murine endogenous retroviral element (MERV-L) peaks in the two-cell-stage embryo. MERV-L is also activated in a subpopulation of mouse embryonic cells in culture defined as 2C-like cells (2CLC) (Macfarlan et al. 2012). Comparison of Hi-C data between 2CLC and mESCs and embryos from two-cell to blastocyst stage revealed that MERV-L transcription can trigger domain formation (Kruse et al. 2019). This work

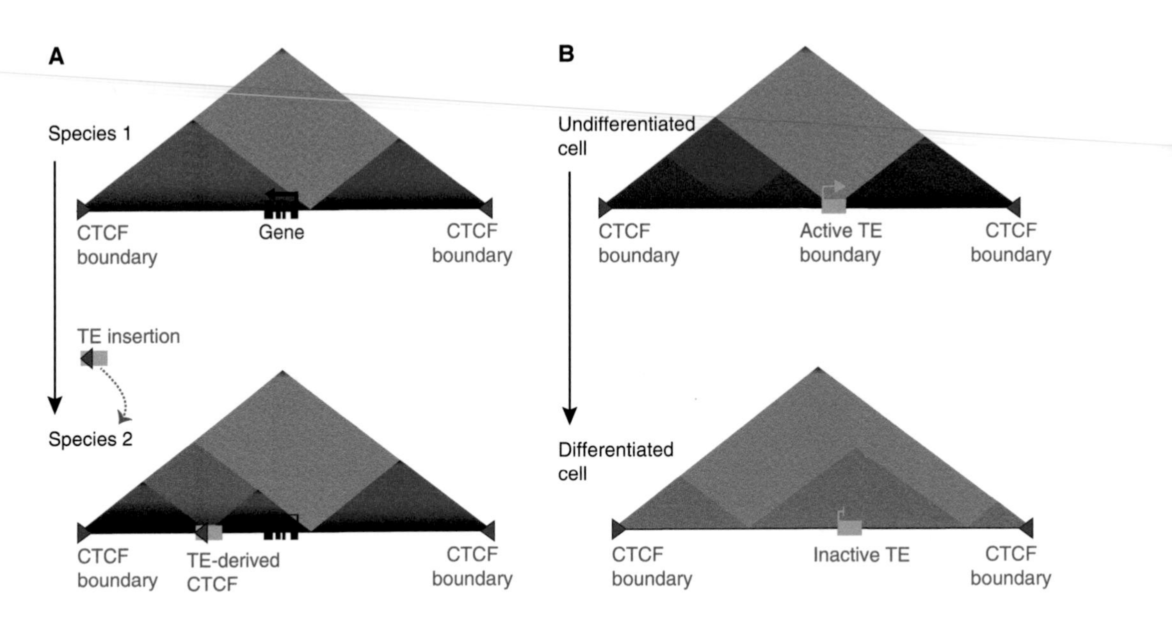

Figure 5. Two mechanisms of transposable element (TE)-mediated formation of domain boundary. (*A*) Example of TE-derived CTCF participating in the formation of a species-specific loop. The insertion of a TE containing a CTCF site occurred in species 2 but not in species 1. This TE-derived CTCF form a loop with the existing CTCF, dividing a topologically associating domain (TAD) in two sub-TADs. As a consequence, the gene involved in this TAD is less expressed in species 2 than in species 1. (*B*) TE transcription drives the formation of a boundary that exists only in undifferentiated cells. In differentiated cells where the TE is inactive, this boundary is not present.

suggests a role for TEs in shaping the establishment of 3D genome organization through a transcription-based mechanism (Fig. 5B). In agreement with this, active transcription of the primate-specific retrotransposon human ERV subfamily H (HERV-H) mediates TAD formation in human pluripotent stem cells (hPSCs) (Zhang et al. 2019). HERV-H inactivation upon cardiomyocyte differentiation is associated with reorganization of this 3D organization. Overall, these studies demonstrate the capacity of TEs to shape the mammalian 3D genome during development and evolution through CTCF or transcription-based mechanisms (Fig. 5).

However, the functional contribution of TEs to the 3D genome still remains to be elucidated. What is the role of these transient cell-specific TE-driven formation of domains? How do these domains affect gene regulation? And finally, is TE-mediated domain formation a disease-causing mechanism? While TEs are known to cause developmental malformation, human genetic disease and to contribute to cancer progression (Lugani et al. 2013; Jang et al. 2019; Payer and Burns 2019), to what extent this can be linked to alteration of the 3D genome is still an active area of research.

CONCLUDING REMARKS

Herein we reviewed recent discoveries about the role of 3D genome organization during development. Eukaryote genomes are packed and organized within the nucleus to facilitate robust control of gene expression. Notably, TADs are a fundamental feature of the genome architecture that allow regulatory sequences and their target genes to be in the same 3D space. Despite TADs being highly conserved between cell types and species, we discussed biological contexts in which they are dynamically rearranged. The 3D genome is reprogrammed following fertilization, allowing the erasure of the organization present in the gametes and establishment of defined

 Cite this article as *Cold Spring Harb Perspect Biol* doi: 10.1101/cshperspect.a040188

TADs required for development. In the context of developmental genes, tissue-specific chromatin folding allows precise spatiotemporal transcriptional activation. Preventing or redirecting these tissue-specific contacts can lead to loss or ectopic gene expression, common mechanisms driving pathological phenotype. In contrast, disruption of TAD boundaries can also have mild or even no consequences on gene regulation. As yet, it is unclear what causes some genomic loci to be more sensitive to the alteration of these regulatory blocks. Although it seems that a disease phenotype occurs more often when altering developmental genes, it was shown that multiple layers of regulation govern their expression. Notably, it has recently emerged that TAD-mediated enhancer–promoter proximity is not the only mechanism facilitating gene regulation. Finally, the 3D genome is also shaped by the insertion or the expression of repeat sequences. TEs have long been known to control gene expression. Recent advances suggest that they might do so by contributing to the 3D organization of the genome during development. As a major force driving genome evolution, transposons could be an important factor shaping species-specific features of the 3D genome. However, despite tremendous progress in the field of 3D genomics and gene regulation over the past 10 years, we still do not know all the factors required for establishing and shaping the 3D genome organization; future work should shed light on new actors playing roles in this exciting field.

ACKNOWLEDGMENTS

We apologize to colleagues whose work was not discussed or cited owing to space constraints. We thank the Mundlos group for helpful discussions and Alicia Madgwick and Daniel Ibrahim for critical reading of the manuscript. Some elements of Figure 2 were modified from work by Marius Walter with his permission. J.G. was supported by the HFSP postdoctoral fellowship (LT000465/2019-L). Work in S.M.'s laboratory is funded by the Deutsche Forschungsgemeinschaft, the Berlin Institute for Health, and the Max Planck Society.

REFERENCES

Reference is also in this collection.

Akhtar W, de Jong J, Pindyurin AV, Pagie L, Meuleman W, de Ridder J, Berns A, Wessels LFA, van Lohuizen M, van Steensel B. 2013. Chromatin position effects assayed by thousands of reporters integrated in parallel. *Cell* **154:** 914–927. doi:10.1016/j.cell.2013.07.018

Alexander JM, Guan J, Li B, Maliskova L, Song M, Shen Y, Huang B, Lomvardas S, Weiner OD. 2019. Live-cell imaging reveals enhancer-dependent Sox2 transcription in the absence of enhancer proximity. *eLife* **8:** e41769. doi:10.7554/eLife.41769

Amano T, Sagai T, Tanabe H, Mizushina Y, Nakazawa H, Shiroishi T. 2009. Chromosomal dynamics at the Shh locus: limb bud-specific differential regulation of competence and active transcription. *Dev Cell* **16:** 47–57. doi:10.1016/j.devcel.2008.11.011

Andrey G, Mundlos S. 2017. The three-dimensional genome: regulating gene expression during pluripotency and development. *Development* **144:** 3646–3658. doi:10.1242/dev.148304

Andrey G, Montavon T, Mascrez B, Gonzalez F, Noordermeer D, Leleu M, Trono D, Spitz F, Duboule D. 2013. A switch between topological domains underlies HoxD genes collinearity in mouse limbs. *Science* **340:** 1234167. doi:10.1126/science.1234167

Andrey G, Schöpflin R, Jerković I, Heinrich V, Ibrahim DM, Paliou C, Hochradel M, Timmermann B, Haas S, Vingron M, et al. 2017. Characterization of hundreds of regulatory landscapes in developing limbs reveals two regimes of chromatin folding. *Genome Res* **27:** 223–233. doi:10.1101/gr.213066.116

Balhorn R, Gledhill BL, Wyrobek AJ. 1977. Mouse sperm chromatin proteins: quantitative isolation and partial characterization. *Biochemistry* **16:** 4074–4080. doi:10.1021/bi00637a021

Banani SF, Lee HO, Hyman AA, Rosen MK. 2017. Biomolecular condensates: organizers of cellular biochemistry. *Nat Rev Mol Cell Biol* **18:** 285–298. doi:10.1038/nrm.2017.7

Barbieri M, Xie SQ, Torlai Triglia E, Chiariello AM, Bianco S, de Santiago I, Branco MR, Rueda D, Nicodemi M, Pombo A. 2017. Active and poised promoter states drive folding of the extended HoxB locus in mouse embryonic stem cells. *Nat Struct Mol Biol* **24:** 515–524. doi:10.1038/nsmb.3402

Battulin N, Fishman VS, Mazur AM, Pomaznoy M, Khabarova AA, Afonnikov DA, Prokhortchouk EB, Serov OL. 2015. Comparison of the three-dimensional organization of sperm and fibroblast genomes using the Hi-C approach. *Genome Biol* **16:** 77. doi:10.1186/s13059-015-0642-0

Bell AC, West AG, Felsenfeld G. 1999. The protein CTCF is required for the enhancer blocking activity of vertebrate insulators. *Cell* **98:** 387–396. doi:10.1016/S0092-8674(00)81967-4

Benabdallah NS, Williamson I, Illingworth RS, Kane L, Boyle S, Sengupta D, Grimes GR, Therizols P, Bickmore WA. 2019. Decreased enhancer–promoter proximity accompanying enhancer activation. *Mol Cell* **76:** 473–484.e7. doi:10.1016/j.molcel.2019.07.038

Beygo J, Citro V, Sparago A, De Crescenzo A, Cerrato F, Heitmann M, Rademacher K, Guala A, Enklaar T, Anichini C, et al. 2013. The molecular function and clinical phenotype of partial deletions of the IGF2/H19 imprinting control region depends on the spatial arrangement of the remaining CTCF-binding sites. *Hum Mol Genet* **22:** 544–557. doi:10.1093/hmg/dds465

Boija A, Klein IA, Sabari BR, Dall'Agnese A, Coffey EL, Zamudio AV, Li CH, Shrinivas K, Manteiga JC, Hannett NM, et al. 2018. Transcription factors activate genes through the phase-separation capacity of their activation domains. *Cell* **175:** 1842–1855.e16. doi:10.1016/j.cell.2018.10.042

Bourque G, Leong B, Vega VB, Chen X, Lee YL, Srinivasan KG, Chew JL, Ruan Y, Wei CL, Ng HH, et al. 2008. Evolution of the mammalian transcription factor binding repertoire via transposable elements. *Genome Res* **18:** 1752–1762. doi:10.1101/gr.080663.108

Burton A, Torres-Padilla M-E. 2014. Chromatin dynamics in the regulation of cell fate allocation during early embryogenesis. *Nat Rev Mol Cell Biol* **15:** 723–735. doi:10.1038/nrm3885

Carter D, Chakalova L, Osborne CS, Dai Y, Fraser P. 2002. Long-range chromatin regulatory interactions in vivo. *Nat Genet* **32:** 623–626. doi:10.1038/ng1051

Chen H, Levo M, Barinov L, Fujioka M, Jaynes JB, Gregor T. 2018. Dynamic interplay between enhancer–promoter topology and gene activity. *Nat Genet* **50:** 1296–1303. doi:10.1038/s41588-018-0175-z

Chen X, Ke Y, Wu K, Zhao H, Sun Y, Gao L, Liu Z, Zhang J, Tao W, Hou Z, et al. 2019. Key role for CTCF in establishing chromatin structure in human embryos. *Nature* **576:** 306–310. doi:10.1038/s41586-019-1812-0

Cho WK, Spille JH, Hecht M, Lee C, Li C, Grube V, Cisse II. 2018. Mediator and RNA polymerase II clusters associate in transcription-dependent condensates. *Science* **361:** 412–415. doi:10.1126/science.aar4199

Chong S, Dugast-Darzacq C, Liu Z, Dong P, Dailey GM, Cattoglio C, Heckert A, Banala S, Lavis L, Darzacq X, et al. 2018. Imaging dynamic and selective low-complexity domain interactions that control gene transcription. *Science* **361:** eaar2555. doi:10.1126/science.aar2555

Choudhary MNK, Friedman RZ, Wang JT, Jang HS, Zhuo X, Wang T. 2020. Co-opted transposons help perpetuate conserved higher-order chromosomal structures. *Genome Biol* **21:** 16. doi:10.1186/s13059-019-1916-8

Collombet S, Ranisavljevic N, Nagano T, Varnai C, Shisode T, Leung W, Piolot T, Galupa R, Borensztein M, Servant N, et al. 2020. Parental-to-embryo switch of chromosome organization in early embryogenesis. *Nature* **580:** 142–146. doi:10.1038/s41586-020-2125-z

Cournac A, Koszul R, Mozziconacci J. 2016. The 3D folding of metazoan genomes correlates with the association of similar repetitive elements. *Nucleic Acids Res* **44:** 245–255. doi:10.1093/nar/gkv1292

Cruz-Molina S, Respuela P, Tebartz C, Kolovos P, Nikolic M, Fueyo R, van Ijcken WFJ, Grosveld F, Frommolt P, Bazzi H, et al. 2017. PRC2 facilitates the regulatory topology required for poised enhancer function during pluripotent stem cell differentiation. *Cell Stem Cell* **20:** 689–705.e9. doi:10.1016/j.stem.2017.02.004

Dekker J, Rippe K, Dekker M, Kleckner N. 2002. Capturing chromosome conformation. *Science* **295:** 1306–1311. doi:10.1126/science.1067799

De Laat W, Duboule D. 2013. Topology of mammalian developmental enhancers and their regulatory landscapes. *Nature* **502:** 499–506. doi:10.1038/nature12753

DeLaurier A, Schweitzer R, Logan M. 2006. Pitx1 determines the morphology of muscle, tendon, and bones of the hindlimb. *Dev Biol* **299:** 22–34. doi:10.1016/j.ydbio.2006.06.055

Despang A, Schöpflin R, Franke M, Ali S, Jerković I, Paliou C, Chan W-L, Timmermann B, Wittler L, Vingron M, et al. 2019. Functional dissection of the Sox9-Kcnj2 locus identifies nonessential and instructive roles of TAD architecture. *Nat Genet* **51:** 1263–1271. doi:10.1038/s41588-019-0466-z

de Wit E, de Laat W. 2012. A decade of 3C technologies: insights into nuclear organization. *Genes Dev* **26:** 11–24. doi:10.1101/gad.179804.111

de Wit E, Vos ESM, Holwerda SJB, Valdes-Quezada C, Verstegen MJAM, Teunissen H, Splinter E, Wijchers PJ, Krijger PHL, de Laat W. 2015. CTCF binding polarity determines chromatin looping. *Mol Cell* **60:** 676–684. doi:10.1016/j.molcel.2015.09.023

Diehl AG, Ouyang N, Boyle AP. 2020. Transposable elements contribute to cell and species-specific chromatin looping and gene regulation in mammalian genomes. *Nat Commun* **11:** 1796. doi:10.1038/s41467-020-15520-5

Dixon JR, Selvaraj S, Yue F, Kim A, Li Y, Shen Y, Hu M, Liu JS, Ren B. 2012. Topological domains in mammalian genomes identified by analysis of chromatin interactions. *Nature* **485:** 376–380. doi:10.1038/nature11082

Dong Z, Wang H, Chen H, Jiang H, Yuan J, Yang Z, Wang WJ, Xu F, Guo X, Cao Y, et al. 2018. Identification of balanced chromosomal rearrangements previously unknown among participants in the 1000 genomes project: implications for interpretation of structural variation in genomes and the future of clinical cytogenetics. *Genet Med* **20:** 697–707. doi:10.1038/gim.2017.170

Du Z, Zheng H, Huang B, Ma R, Wu J, Zhang X, He J, Xiang Y, Wang Q, Li Y, et al. 2017. Allelic reprogramming of 3D chromatin architecture during early mammalian development. *Nature* **547:** 232–235. doi:10.1038/nature23263

Du Z, Zheng H, Kawamura YK, Zhang K, Gassler J, Powell S, Xu Q, Lin Z, Xu K, Zhou Q, et al. 2020. Polycomb group proteins regulate chromatin architecture in mouse oocytes and early embryos. *Mol Cell* **77:** 825–839.e7. doi:10.1016/j.molcel.2019.11.011

* Du Z, Khang K, Xie W. 2021. Epigenetic reprogramming in early animal development. *Cold Spring Harb Perspect Biol* doi:10.1101.cshperspect.a039677

Espinola SM, Götz M, Bellec M, Messina O, Fiche JB, Houbron C, Dejean M, Reim I, Cardozo Gizzi AM, Lagha M, et al. 2021. *Cis*-regulatory chromatin loops arise before TADs and gene activation, and are independent of cell fate during early *Drosophila* development. *Nat Genet* **53:** 477–486. doi:10.1038/s41588-021-00816-z

Fedoriw AM, Stein P, Svoboda P, Schultz RM, Bartolomei MS. 2004. Transgenic RNAi reveals essential function for

CTCF in H19 gene imprinting. *Science* **303:** 238–240. doi:10.1126/science.1090934

Flyamer IM, Gassler J, Imakaev M, Brandão HB, Ulianov SV, Abdennur N, Razin SV, Mirny LA, Tachibana-Konwalski K. 2017. Single-nucleus Hi-C reveals unique chromatin reorganization at oocyte-to-zygote transition. *Nature* **544:** 110–114. doi:10.1038/nature21711

Franke M, Ibrahim DM, Andrey G, Schwarzer W, Heinrich V, Schöpflin R, Kraft K, Kempfer R, Jerković I, Chan WL, et al. 2016. Formation of new chromatin domains determines pathogenicity of genomic duplications. *Nature* **538:** 265–269. doi:10.1038/nature19800

Franke M, De la Calle-Mustienes E, Neto A, Acemel RD, Tena JJ, Santos-Pereira JM, Gómez-Skarmeta JL. 2020. CTCF knockout in zebrafish induces alterations in regulatory landscapes and developmental gene expression. bioRxiv doi:10.1101/2020.09.08.282707

Fudenberg G, Imakaev M, Lu C, Goloborodko A, Abdennur N, Mirny LA. 2016. Formation of chromosomal domains by loop extrusion. *Cell Rep* **15:** 2038–2049. doi:10.1016/j.celrep.2016.04.085

Gassler J, Brandão HB, Imakaev M, Flyamer IM, Ladstätter S, Bickmore WA, Peters JM, Mirny LA, Tachibana K. 2017. A mechanism of cohesin-dependent loop extrusion organizes zygotic genome architecture. *EMBO J* **36:** 3600–3618. doi:10.15252/embj.201798083

Gaucher J, Reynoird N, Montellier E, Boussouar F, Rousseaux S, Khochbin S. 2010. From meiosis to postmeiotic events: the secrets of histone disappearance. *FEBS J* **277:** 599–604. doi:10.1111/j.1742-4658.2009.07504.x

Ghavi-Helm Y. 2020. Functional consequences of chromosomal rearrangements on gene expression: not so deleterious after all? *J Mol Biol* **432:** 665–675. doi:10.1016/j.jmb.2019.09.010

Ghavi-Helm Y, Jankowski A, Meiers S, Viales RR, Korbel JO, Furlong EEM. 2019. Highly rearranged chromosomes reveal uncoupling between genome topology and gene expression. *Nat Genet* **51:** 1272–1282. doi:10.1038/s41588-019-0462-3

Guelen L, Pagie L, Brasset E, Meuleman W, Faza MB, Talhout W, Eussen BH, de Klein A, Wessels L, de Laat W, et al. 2008. Domain organization of human chromosomes revealed by mapping of nuclear lamina interactions. *Nature* **453:** 948–951. doi:10.1038/nature06947

Guo Y, Xu Q, Canzio D, Shou J, Li J, Gorkin DU, Jung I, Wu H, Zhai Y, Tang Y, et al. 2015. CRISPR inversion of CTCF sites alters genome topology and enhancer/promoter function. *Cell* **162:** 900–910. doi:10.1016/j.cell.2015.07.038

Haarhuis JHI, van der Weide RH, Blomen VA, Yáñez-Cuna JO, Amendola M, van Ruiten MS, Krijger PHL, Teunissen H, Medema RH, van Steensel B, et al. 2017. The cohesin release factor WAPL restricts chromatin loop extension. *Cell* **169:** 693–707.e14. doi:10.1016/j.cell.2017.04.013

Heitz E. 1928. Das heterochromatin der moose. *Jahrb Wiss Bot* **69:** 762–818.

Hnisz D, Weintraub AS, Day DS, Valton AL, Bak RO, Li CH, Goldmann J, Lajoie BR, Fan ZP, Sigova AA, et al. 2016. Activation of proto-oncogenes by disruption of chromosome neighborhoods. *Science* **351:** 1454–1458. doi:10.1126/science.aad9024

Hnisz D, Shrinivas K, Young RA, Chakraborty AK, Sharp PA. 2017. A phase separation model for transcriptional control. *Cell* **169:** 13–23. doi:10.1016/j.cell.2017.02.007

Hou C, Li L, Qin ZS, Corces VG. 2012. Gene density, transcription, and insulators contribute to the partition of the *Drosophila* genome into physical domains. *Mol Cell* **48:** 471–484. doi:10.1016/j.molcel.2012.08.031

Hug CB, Vaquerizas JM. 2018. The birth of the 3D genome during early embryonic development. *Trends Genet* **34:** 903–914. doi:10.1016/j.tig.2018.09.002

Hug CB, Grimaldi AG, Kruse K, Vaquerizas JM. 2017. Chromatin architecture emerges during zygotic genome activation independent of transcription. *Cell* **169:** 216–228.e19. doi:10.1016/j.cell.2017.03.024

Ing-Simmons E, Vaid R, Bing XY, Levine M, Mannervik M, Vaquerizas JM. 2021. Independence of chromatin conformation and gene regulation during *Drosophila* dorsoventral patterning. *Nat Genet* **53:** 487–499. doi:10.1038/s41588-021-00799-x

Jang HS, Shah NM, Du AY, Dailey ZZ, Pehrsson EC, Godoy PM, Zhang D, Li D, Xing X, Kim S, et al. 2019. Transposable elements drive widespread expression of oncogenes in human cancers. *Nat Genet* **51:** 611–617. doi:10.1038/s41588-019-0373-3

Jerkovic I, Cavalli G. 2021. Understanding 3D genome organization by multidisciplinary methods. *Nat Rev Mol Cell Biol* doi:10.1038/s41580-021-00362-w

Johnstone SE, Reyes A, Qi Y, Adriaens C, Hegazi E, Pelka K, Chen JH, Zou LS, Drier Y, Hecht V, et al. 2020. Large-scale topological changes restrain malignant progression in colorectal cancer. *Cell* **182:** 1474–1489.e23. doi:10.1016/j.cell.2020.07.030

Ke Y, Xu Y, Chen X, Feng S, Liu Z, Sun Y, Yao X, Li F, Zhu W, Gao L, et al. 2017. 3D chromatin structures of mature gametes and structural reprogramming during mammalian embryogenesis. *Cell* **170:** 367–381.e20. doi:10.1016/j.cell.2017.06.029

Kempfer R, Pombo A. 2020. Methods for mapping 3D chromosome architecture. *Nat Rev Genet* **21:** 207–226. doi:10.1038/s41576-019-0195-2

Kraft K, Geuer S, Will AJ, Chan WL, Paliou C, Borschiwer M, Harabula I, Wittler L, Franke M, Ibrahim DM, et al. 2015. Deletions, inversions, duplications: engineering of structural variants using CRISPR/Cas in mice. *Cell Rep* **10:** 833–839. doi:10.1016/j.celrep.2015.01.016

Kraft K, Magg A, Heinrich V, Riemenschneider C, Schöpflin R, Markowski J, Ibrahim DM, Acuna-Hidalgo R, Despang A, Andrey G, et al. 2019. Serial genomic inversions induce tissue-specific architectural stripes, gene misexpression and congenital malformations. *Nat Cell Biol* **21:** 305–310. doi:10.1038/s41556-019-0273-x

Kragesteen BK, Spielmann M, Paliou C, Heinrich V, Schöpflin R, Esposito A, Annunziatella C, Bianco S, Chiariello AM, Jerković I, et al. 2018. Dynamic 3D chromatin architecture contributes to enhancer specificity and limb morphogenesis. *Nat Genet* **50:** 1463–1473. doi:10.1038/s41588-018-0221-x

Kruse K, Díaz N, Enriquez-Gasca R, Gaume X, Torres-Padilla ME, Vaquerizas JM. 2019. Transposable elements drive reorganisation of 3D chromatin during early embryogenesis. bioRxiv doi:10.1101/523712

Kurth I, Klopocki E, Stricker S, van Oosterwijk J, Vanek S, Altmann J, Santos HG, van Harssel JJT, de Ravel T, Wilkie AOM, et al. 2009. Duplications of noncoding elements 5′ of SOX9 are associated with brachydactyly-anonychia. *Nat Genet* **41:** 862–863. doi:10.1038/ng0809-862

Lander ES, Linton LM, Birren B, Nusbaum C, Zody MC, Baldwin J, Devon K, Dewar K, Doyle M, FitzHugh W, et al. 2001. Initial sequencing and analysis of the human genome. *Nature* **409:** 860–921. doi:10.1038/35057062

Larson AG, Elnatan D, Keenen MM, Trnka MJ, Johnston JB, Burlingame AL, Agard DA, Redding S, Narlikar GJ. 2017. Liquid droplet formation by HP1α suggests a role for phase separation in heterochromatin. *Nature* **547:** 236–240. doi:10.1038/nature22822

Laugsch M, Bartusel M, Rehimi R, Alirzayeva H, Karaolidou A, Crispatzu G, Zentis P, Nikolic M, Bleckwehl T, Kolovos P, et al. 2019. Modeling the pathological long-range regulatory effects of human structural variation with patient-specific hiPSCs. *Cell Stem Cell* **24:** 736–752.e12. doi:10.1016/j.stem.2019.03.004

Leemans C, van der Zwalm MCH, Brueckner L, Comoglio F, van Schaik T, Pagie L, van Arensbergen J, van Steensel B. 2019. Promoter-intrinsic and local chromatin features determine gene repression in LADs. *Cell* **177:** 852–864.e14. doi:10.1016/j.cell.2019.03.009

Lettice LA, Heaney SJH, Purdie LA, Li L, de Beer P, Oostra BA, Goode D, Elgar G, Hill RE, de Graaff E. 2003. A long-range Shh enhancer regulates expression in the developing limb and fin and is associated with preaxial polydactyly. *Hum Mol Genet* **12:** 1725–1735. doi:10.1093/hmg/ddg180

Lieberman-Aiden E, van Berkum NL, Williams L, Imakaev M, Ragoczy T, Telling A, Amit I, Lajoie BR, Sabo PJ, Dorschner MO, et al. 2009. Comprehensive mapping of long-range interactions reveals folding principles of the human genome. *Science* **326:** 289–293. doi:10.1126/science.1181369

Llères D, Moindrot B, Pathak R, Piras V, Matelot M, Pignard B, Marchand A, Poncelet M, Perrin A, Tellier V, et al. 2019. CTCF modulates allele-specific sub-TAD organization and imprinted gene activity at the mouse Dlk1-Dio3 and Igf2-H19 domains. *Genome Biol* **20:** 272. doi:10.1186/s13059-019-1896-8

Lugani F, Arora R, Papeta N, Patel A, Zheng Z, Sterken R, Singer RA, Caridi G, Mendelsohn C, Sussel L, et al. 2013. A retrotransposon insertion in the 5′ regulatory domain of Ptf1a results in ectopic gene expression and multiple congenital defects in Danforth's short tail mouse. *PLoS Genet* **9:** e1003206. doi:10.1371/journal.pgen.1003206

Lupiáñez DG, Kraft K, Heinrich V, Krawitz P, Brancati F, Klopocki E, Horn D, Kayserili H, Opitz JM, Laxova R, et al. 2015. Disruptions of topological chromatin domains cause pathogenic rewiring of gene-enhancer interactions. *Cell* **161:** 1012–1025. doi:10.1016/j.cell.2015.04.004

Lupiáñez DG, Spielmann M, Mundlos S. 2016. Breaking TADs: how alterations of chromatin domains result in disease. *Trends Genet* **32:** 225–237. doi:10.1016/j.tig.2016.01.003

Macfarlan TS, Gifford WD, Driscoll S, Lettieri K, Rowe HM, Bonanomi D, Firth A, Singer O, Trono D, Pfaff SL. 2012. Embryonic stem cell potency fluctuates with endogenous retrovirus activity. *Nature* **487:** 57–63. doi:10.1038/nature11244

Merkenschlager M, Nora EP. 2016. CTCF and cohesin in genome folding and transcriptional gene regulation. *Annu Rev Genomics Hum Genet* **17:** 17–43. doi:10.1146/annurev-genom-083115-022339

Moore JM, Rabaia NA, Smith LE, Fagerlie S, Gurley K, Loukinov D, Disteche CM, Collins SJ, Kemp CJ, Lobanenkov VV, et al. 2012. Loss of maternal CTCF is associated with peri-implantation lethality of Ctcf null embryos. *PLoS ONE* **7:** e34915. doi:10.1371/journal.pone.0034915

Morgan HD, Santos F, Green K, Dean W, Reik W. 2005. Epigenetic reprogramming in mammals. *Hum Mol Genet* **14:** R47–R58. doi:10.1093/hmg/ddi114

Németh A, Conesa A, Santoyo-Lopez J, Medina I, Montaner D, Péterfia B, Solovei I, Cremer T, Dopazo J, Längst G. 2010. Initial genomics of the human nucleolus. *PLoS Genet* **6:** e1000889. doi:10.1371/journal.pgen.1000889

Noordermeer D, Feil R. 2020. Differential 3D chromatin organization and gene activity in genomic imprinting. *Curr Opin Genet Dev* **61:** 17–24. doi:10.1016/j.gde.2020.03.004

Nora EP, Lajoie BR, Schulz EG, Giorgetti L, Okamoto I, Servant N, Piolot T, van Berkum NL, Meisig J, Sedat J, et al. 2012. Spatial partitioning of the regulatory landscape of the X-inactivation centre. *Nature* **485:** 381–385. doi:10.1038/nature11049

Nora EP, Goloborodko A, Valton AL, Gibcus JH, Uebersohn A, Abdennur N, Dekker J, Mirny LA, Bruneau BG. 2017. Targeted degradation of CTCF decouples local insulation of chromosome domains from genomic compartmentalization. *Cell* **169:** 930–944.e22. doi:10.1016/j.cell.2017.05.004

Northcott PA, Lee C, Zichner T, Stütz AM, Erkek S, Kawauchi D, Shih DJH, Hovestadt V, Zapatka M, Sturm D, et al. 2014. Enhancer hijacking activates GFI1 family oncogenes in medulloblastoma. *Nature* **511:** 428–434. doi:10.1038/nature13379

Paliou C, Guckelberger P, Schöpflin R, Heinrich V, Esposito A, Chiariello AM, Bianco S, Annunziatella C, Helmuth J, Haas S, et al. 2019. Preformed chromatin topology assists transcriptional robustness of Shh during limb development. *Proc Natl Acad Sci* **116:** 12390–12399. doi:10.1073/pnas.1900672116

Payer LM, Burns KH. 2019. Transposable elements in human genetic disease. *Nat Rev Genet* **20:** 760–772. doi:10.1038/s41576-019-0165-8

Peric-Hupkes D, Meuleman W, Pagie L, Bruggeman SWM, Solovei I, Brugman W, Gräf S, Flicek P, Kerkhoven RM, van Lohuizen M, et al. 2010. Molecular maps of the reorganization of genome-nuclear lamina interactions during differentiation. *Mol Cell* **38:** 603–613. doi:10.1016/j.molcel.2010.03.016

Phillips-Cremins JE, Sauria MEG, Sanyal A, Gerasimova TI, Lajoie BR, Bell JSK, Ong CT, Hookway TA, Guo C, Sun Y, et al. 2013. Architectural protein subclasses shape 3D organization of genomes during lineage commitment. *Cell* **153:** 1281–1295. doi:10.1016/j.cell.2013.04.053

Pickersgill H, Kalverda B, de Wit E, Talhout W, Fornerod M, van Steensel B. 2006. Characterization of the *Drosophila melanogaster* genome at the nuclear lamina. *Nat Genet* **38:** 1005–1014. doi:10.1038/ng1852

Rabl C. 1885. Über zelltheilung. *Morphol Jahrb* **10:** 214–330.

Rao SSP, Huang SC, St Hilaire BG, Engreitz JM, Perez EM, Kieffer-Kwon KR, Sanborn AL, Johnstone SE, Bascom GD, Bochkov ID, et al. 2017. Cohesin loss eliminates all loop domains. *Cell* **171:** 305–320.e24. doi:10.1016/j.cell.2017.09.026

Raviram R, Rocha PP, Luo VM, Swanzey E, Miraldi ER, Chuong EB, Feschotte C, Bonneau R, Skok JA. 2018. Analysis of 3D genomic interactions identifies candidate host genes that transposable elements potentially regulate. *Genome Biol* **19:** 216. doi:10.1186/s13059-018-1598-7

Redin C, Brand H, Collins RL, Kammin T, Mitchell E, Hodge JC, Hanscom C, Pillalamarri V, Seabra CM, Abbott MA, et al. 2017. The genomic landscape of balanced cytogenetic abnormalities associated with human congenital anomalies. *Nat Genet* **49:** 36–45. doi:10.1038/ng.3720

Rodríguez-Carballo E, Lopez-Delisle L, Zhan Y, Fabre PJ, Beccari L, El-Idrissi I, Huynh THN, Ozadam H, Dekker J, Duboule D. 2017. The HoxD cluster is a dynamic and resilient TAD boundary controlling the segregation of antagonistic regulatory landscapes. *Genes Dev* **31:** 2264–2281. doi:10.1101/gad.307769.117

Rodriguez-Terrones D, Torres-Padilla ME. 2018. Nimble and ready to mingle: transposon outbursts of early development. *Trends Genet* **34:** 806–820. doi:10.1016/j.tig.2018.06.006

Rowley MJ, Nichols MH, Lyu X, Ando-Kuri M, Rivera ISM, Hermetz K, Wang P, Ruan Y, Corces VG. 2017. Evolutionarily conserved principles predict 3D chromatin organization. *Mol Cell* **67:** 837–852.e7. doi:10.1016/j.molcel.2017.07.022

Sabari BR, Dall'Agnese A, Boija A, Klein IA, Coffey EL, Shrinivas K, Abraham BJ, Hannett NM, Zamudio AV, Manteiga JC, et al. 2018. Coactivator condensation at super-enhancers links phase separation and gene control. *Science* **361:** eaar3958. doi:10.1126/science.aar3958

Sagai T, Hosoya M, Mizushina Y, Tamura M, Shiroishi T. 2005. Elimination of a long-range *cis*-regulatory module causes complete loss of limb-specific Shh expression and truncation of the mouse limb. *Development* **132:** 797–803. doi:10.1242/dev.01613

Sanborn AL, Rao SSP, Huang SC, Durand NC, Huntley MH, Jewett AI, Bochkov ID, Chinnappan D, Cutkosky A, Li J, et al. 2015. Chromatin extrusion explains key features of loop and domain formation in wild-type and engineered genomes. *Proc Natl Acad Sci* **112:** E6456–E6465. doi:10.1073/pnas.1518552112

Sanyal A, Lajoie BR, Jain G, Dekker J. 2012. The long-range interaction landscape of gene promoters. *Nature* **489:** 109–113. doi:10.1038/nature11279

Schmidt D, Schwalie PC, Wilson MD, Ballester B, Gonçalves Â, Kutter C, Brown GD, Marshall A, Flicek P, Odom DT. 2012. Waves of retrotransposon expansion remodel genome organization and CTCF binding in multiple mammalian lineages. *Cell* **148:** 335–348. doi:10.1016/j.cell.2011.11.058

Schwarzer W, Abdennur N, Goloborodko A, Pekowska A, Fudenberg G, Loe-Mie Y, Fonseca NA, Huber W, Haering CH, Mirny L, et al. 2017. Two independent modes of chromatin organization revealed by cohesin removal. *Nature* **551:** 51–56. doi:10.1038/nature24281

Sexton T, Yaffe E, Kenigsberg E, Bantignies F, Leblanc B, Hoichman M, Parrinello H, Tanay A, Cavalli G. 2012. Three-dimensional folding and functional organization principles of the *Drosophila* genome. *Cell* **148:** 458–472. doi:10.1016/j.cell.2012.01.010

Simonis M, Klous P, Splinter E, Moshkin Y, Willemsen R, de Wit E, van Steensel B, de Laat W. 2006. Nuclear organization of active and inactive chromatin domains uncovered by chromosome conformation capture-on-chip (4C). *Nat Genet* **38:** 1348–1354. doi:10.1038/ng1896

Sofueva S, Yaffe E, Chan WC, Georgopoulou D, Vietri Rudan M, Mira-Bontenbal H, Pollard SM, Schroth GP, Tanay A, Hadjur S. 2013. Cohesin-mediated interactions organize chromosomal domain architecture. *EMBO J* **32:** 3119–3129. doi:10.1038/emboj.2013.237

Soshnikova N, Montavon T, Leleu M, Galjart N, Duboule D. 2010. Functional analysis of CTCF during mammalian limb development. *Dev Cell* **19:** 819–830. doi:10.1016/j.devcel.2010.11.009

Spielmann M, Lupiáñez DG, Mundlos S. 2018. Structural variation in the 3D genome. *Nat Rev Genet* **19:** 453–467. doi:10.1038/s41576-018-0007-0

Stankiewicz P, Lupski JR. 2010. Structural variation in the human genome and its role in disease. *Annu Rev Med* **61:** 437–455. doi:10.1146/annurev-med-100708-204735

Stik G, Vidal E, Barrero M, Cuartero S, Vila-Casadesús M, Mendieta-Esteban J, Tian TV, Choi J, Berenguer C, Abad A, et al. 2020. CTCF is dispensable for immune cell transdifferentiation but facilitates an acute inflammatory response. *Nat Genet* **52:** 655–661. doi:10.1038/s41588-020-0643-0

Strom AR, Emelyanov AV, Mir M, Fyodorov DV, Darzacq X, Karpen GH. 2017. Phase separation drives heterochromatin domain formation. *Nature* **547:** 241–245. doi:10.1038/nature22989

Su M, Han D, Boyd-Kirkup J, Yu X, Han JDJ. 2014. Evolution of Alu elements toward enhancers. *Cell Rep* **7:** 376–385. doi:10.1016/j.celrep.2014.03.011

Sundaram V, Cheng Y, Ma Z, Li D, Xing X, Edge P, Snyder MP, Wang T. 2014. Widespread contribution of transposable elements to the innovation of gene regulatory networks. *Genome Res* **24:** 1963–1976. doi:10.1101/gr.168872.113

Symmons O, Uslu VV, Tsujimura T, Ruf S, Nassari S, Schwarzer W, Ettwiller L, Spitz F. 2014. Functional and topological characteristics of mammalian regulatory domains. *Genome Res* **24:** 390–400. doi:10.1101/gr.163519.113

Tolhuis B, Palstra RJ, Splinter E, Grosveld F, de Laat W. 2002. Looping and interaction between hypersensitive sites in the active β-globin locus. *Mol Cell* **10:** 1453–1465. doi:10.1016/S1097-2765(02)00781-5

Tsujimura T, Klein FA, Langenfeld K, Glaser J, Huber W, Spitz F. 2015. A discrete transition zone organizes the topological and regulatory autonomy of the adjacent Tfap2c and Bmp7 genes. *PLoS Genet* **11:** e1004897. doi:10.1371/journal.pgen.1004897

Tucci V, Isles AR, Kelsey G, Ferguson-Smith AC, Tucci V, Bartolomei MS, Benvenisty N, Bourc'his D, Charalambous M, Dulac C, et al. 2019. Genomic imprinting and physiological processes in mammals. *Cell* **176:** 952–965. doi:10.1016/j.cell.2019.01.043

van Koningsbruggen S, Gierlinski M, Schofield P, Martin D, Barton GJ, Ariyurek Y, den Dunnen JT, Lamond AI. 2010. High-resolution whole-genome sequencing reveals that specific chromatin domains from most human chromosomes associate with nucleoli. *Mol Biol Cell* **21:** 3735–3748. doi:10.1091/mbc.e10-06-0508

van Steensel B, Furlong EEM. 2019. The role of transcription in shaping the spatial organization of the genome. *Nat Rev Mol Cell Biol* **20:** 327–337.

Vara C, Paytuví-Gallart A, Cuartero Y, Le Dily F, Garcia F, Salvà-Castro J, Gómez-H L, Julià E, Moutinho C, Aiese Cigliano R, et al. 2019. Three-dimensional genomic structure and cohesin occupancy correlate with transcriptional activity during spermatogenesis. *Cell Rep* **28:** 352–367.e9. doi:10.1016/j.celrep.2019.06.037

Wan LB, Pan H, Hannenhalli S, Cheng Y, Ma J, Fedoriw A, Lobanenkov V, Latham KE, Schultz RM, Bartolomei MS. 2008. Maternal depletion of CTCF reveals multiple functions during oocyte and preimplantation embryo development. *Development* **135:** 2729–2738. doi:10.1242/dev.024539

Wang H, Maurano MT, Qu H, Varley KE, Gertz J, Pauli F, Lee K, Canfield T, Weaver M, Sandstrom R, et al. 2012. Widespread plasticity in CTCF occupancy linked to DNA methylation. *Genome Res* **22:** 1680–1688. doi:10.1101/gr.136101.111

Wang Q, Sun Q, Czajkowsky DM, Shao Z. 2018. Sub-kb Hi-C in D. melanogaster reveals conserved characteristics of TADs between insect and mammalian cells. *Nat Commun* **9:** 188. doi:10.1038/s41467-017-02526-9

Weischenfeldt J, Dubash T, Drainas AP, Mardin BR, Chen Y, Stütz AM, Waszak SM, Bosco G, Halvorsen AR, Raeder B, et al. 2017. Pan-cancer analysis of somatic copy-number alterations implicates IRS4 and IGF2 in enhancer hijacking. *Nat Genet* **49:** 65–74. doi:10.1038/ng.3722

Wicker T, Sabot F, Hua-Van A, Bennetzen JL, Capy P, Chalhoub B, Flavell A, Leroy P, Morgante M, Panaud O, et al. 2007. A unified classification system for eukaryotic transposable elements. *Nat Rev Genet* **8:** 973–982. doi:10.1038/nrg2165

Williamson I, Lettice LA, Hill RE, Bickmore WA. 2016. Shh and ZRS enhancer colocalisation is specific to the zone of polarising activity. *Development* **143:** 2994–3001. doi:10.1242/dev.139188

Wu J, Huang B, Chen H, Yin Q, Liu Y, Xiang Y, Zhang B, Liu B, Wang Q, Xia W, et al. 2016. The landscape of accessible chromatin in mammalian preimplantation embryos. *Nature* **534:** 652–657. doi:10.1038/nature18606

Yokochi T, Poduch K, Ryba T, Lu J, Hiratani I, Tachibana M, Shinkai Y, Gilbert DM. 2009. G9a selectively represses a class of late-replicating genes at the nuclear periphery. *Proc Natl Acad Sci* **106:** 19363–19368. doi:10.1073/pnas.0906142106

Zhang Y, Xiang Y, Yin Q, Du Z, Peng X, Wang Q, Fidalgo M, Xia W, Li Y, Zhao Z, et al. 2018. Dynamic epigenomic landscapes during early lineage specification in mouse embryos. *Nat Genet* **50:** 96–105. doi:10.1038/s41588-017-0003-x

Zhang Y, Li T, Preissl S, Amaral ML, Grinstein JD, Farah EN, Destici E, Qiu Y, Hu R, Lee AY, et al. 2019. Transcriptionally active HERV-H retrotransposons demarcate topologically associating domains in human pluripotent stem cells. *Nat Genet* **51:** 1380–1388. doi:10.1038/s41588-019-0479-7

Zhang D, Huang P, Sharma M, Keller CA, Giardine B, Zhang H, Gilgenast TG, Phillips-Cremins JE, Hardison RC, Blobel GA. 2020. Alteration of genome folding via contact domain boundary insertion. *Nat Genet* **50:** 1076–1087. doi:10.1038/s41588-020-0680-8

Zhao Z, Tavoosidana G, Sjölinder M, Göndör A, Mariano P, Wang S, Kanduri C, Lezcano M, Sandhu KS, Singh U, et al. 2006. Circular chromosome conformation capture (4C) uncovers extensive networks of epigenetically regulated intra- and interchromosomal interactions. *Nat Genet* **38:** 1341–1347. doi:10.1038/ng1891

Zheng H, Xie W. 2019. The role of 3D genome organization in development and cell differentiation. *Nat Rev Mol Cell Biol* **20:** 535–550. doi:10.1038/s41580-019-0132-4

Cite this article as *Cold Spring Harb Perspect Biol* doi: 10.1101/cshperspect.a040188

Epigenetic Reprogramming in Early Animal Development

Zhenhai Du,[1,2,3] Ke Zhang,[1,2,3] and Wei Xie[1,2]

[1]Center for Stem Cell Biology and Regenerative Medicine, MOE Key Laboratory of Bioinformatics, Tsinghua University, Beijing 100084, China

[2]Tsinghua-Peking Center for Life Sciences, Beijing 100084, China

Correspondence: xiewei121@tsinghua.edu.cn

Dramatic nuclear reorganization occurs during early development to convert terminally differentiated gametes to a totipotent zygote, which then gives rise to an embryo. Aberrant epigenome resetting severely impairs embryo development and even leads to lethality. How the epigenomes are inherited, reprogrammed, and reestablished in this critical developmental period has gradually been unveiled through the rapid development of technologies including ultrasensitive chromatin analysis methods. In this review, we summarize the latest findings on epigenetic reprogramming in gametogenesis and embryogenesis, and how it contributes to gamete maturation and parental-to-zygotic transition. Finally, we highlight the key questions that remain to be answered to fully understand chromatin regulation and nuclear reprogramming in early development.

Gametogenesis and embryogenesis are essential to give rise to a new life and to repopulate species that use sexual reproduction. Sperm and oocytes arise from primordial germ cells (PGCs) through spermatogenesis and oogenesis, respectively. After fertilization, the two terminally differentiated gametes fuse and form a totipotent zygote (Zhou and Dean 2015). During embryo cleavage, maternal proteins and mRNAs are gradually degraded, while the zygotic genome is transcriptionally activated, collectively leading to the maternal-to-zygotic transition (MZT) (Fig. 1; Schultz 2002; Walser and Lipshitz 2011). Zygotic genome activation ([ZGA], also known as embryonic genome activation) often takes place in two waves (for review, see Lee et al. 2014). During minor ZGA in mouse embryos, the transcription of a small subset of genes is accompanied by promiscuous, genome-wide transcription, which occurs from the S phase of a one-cell embryo until the onset of "major ZGA" in mid- to late two-cell embryos, when thousands of genes are activated (Bouniol et al. 1995; Nothias et al. 1996; Aoki et al. 1997; Hamatani et al. 2004; Wang et al. 2004; Zeng et al. 2004; Abe et al. 2015). Concomitantly, RNA polymerase II (Pol II) undergoes dynamic chromatin engagement during the minor-to-major ZGA transition (Bellier et al. 1997; Liu et al. 2020). In human, embryonic transcripts are first observed from the two-cell to the four-cell stage, and the major ZGA occurs around the eight-cell stage (Tesarik et al. 1987; Braude et al. 1988; Dobson et al. 2004). The

[3]These authors contributed equally to this work.

Cite this article as *Cold Spring Harb Perspect Biol* doi: 10.1101/cshperspect.a039677

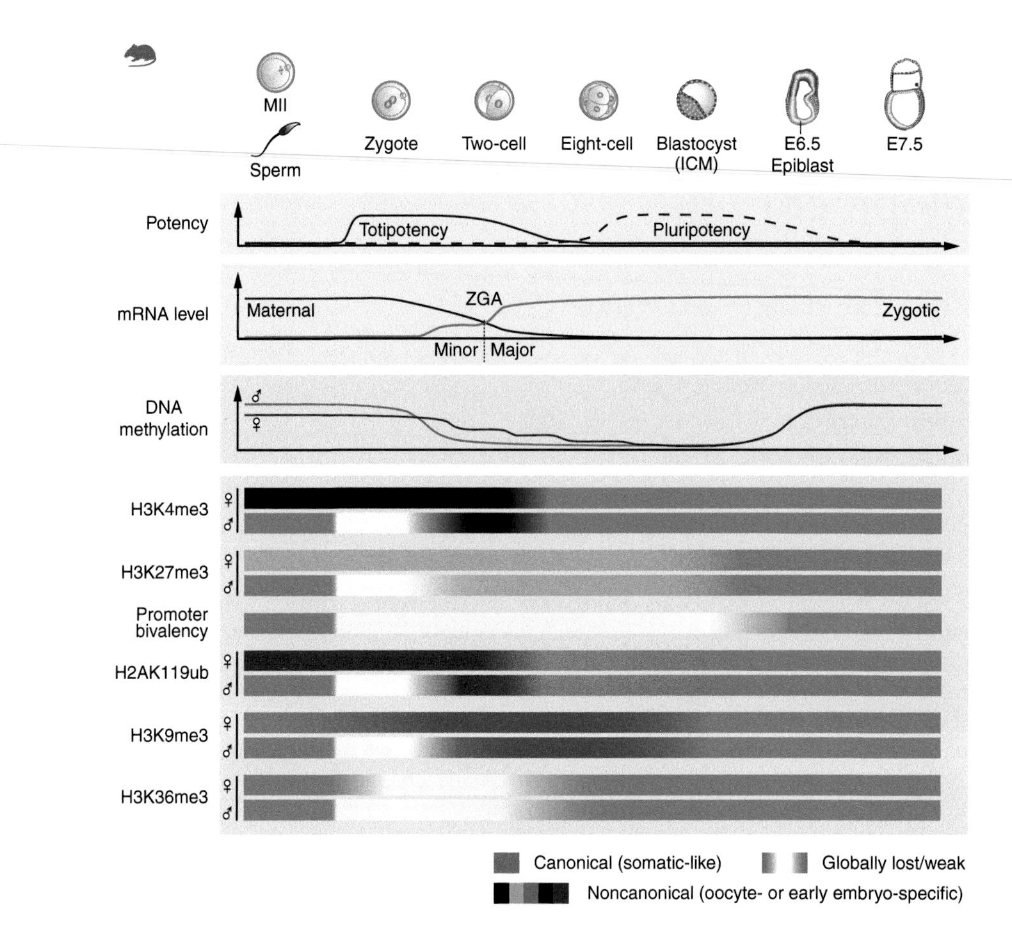

Figure 1. Transcription and epigenetic dynamics from gametes to embryos in mouse. After fertilization, the terminally differentiated gametes are converted into a totipotent zygote. Inner cell mass (ICM) cells from blastocysts are pluripotent and can give rise to all embryonic tissues. The accumulated maternal mRNAs undergo degradation after fertilization, while minor and major zygotic genome activation (ZGA) emerge around the one-cell and two-cell stages, respectively. Parental genomes undergo global DNA demethylation starting from the zygote stage and the reestablishment of embryonic DNA methylome occurs in postimplantation embryos. For histone modifications, sperm largely display canonical distributions (similar to somatic cells), while oocytes and early embryos often acquire noncanonical patterns. H3K4me3 and H3K27me3 form noncanonical broad domains in oocytes, which are transiently inherited to the maternal allele of embryos (except for promoter H3K27me3). The sperm H3K4me3 and H3K27me3 quickly attenuate after fertilization. Weak and extremely large domains of H3K4me3 (at gene-rich regions) and H3K27me3 (at gene deserts) are de novo established on the paternal genome in zygotes. Concomitant with ZGA, both maternal and paternal H3K4me3 are reprogrammed to the canonical patterns, while the distal H3K27me3 domains persist until blastocyst, before being transformed into a canonical state in postimplantation embryos. Bivalency at developmental genes exists in mature gametes but is missing in preimplantation embryos. In E6.5 epiblast, H3K27me3 and unusually strong H3K4me3 form "super bivalency" (green), which is subsequently attenuated at E7.5. In mouse sperm, H2AK119ub resides at the promoters of canonical Polycomb targets. In mouse oocytes, it forms broad distal domains in H3K27me3 marked PMDs, and also occurs at active promoters associated with H3K4me3. Upon fertilization, H2AK119ub undergoes drastic changes, with global resetting of sperm signals but brief inheritance of oocyte H2AK119ub in one-cell embryos. H2AK119ub then rapidly remodels (mainly from the paternal allele but also partially from the maternal allele) and adopts similar patterns between the two alleles from the two-cell stage. After implantation, H2AK119ub and H3K27me3 become colocalized again at developmental gene promoters. H3K9me3 marks both a subset of promoters and LTRs in gametes. After fertilization, H3K9me3 is globally reset in an allele-specific manner, and the asymmetrical distribution of H3K9me3 persists to blastocyst. H3K36me3 enriches at actively expressed gene bodies in gametes. After fertilization, H3K36me3 is globally removed and is then reestablished upon ZGA. The enrichment of epigenetic modification is indicated by the widths and shades of the bars.

 Cite this article as *Cold Spring Harb Perspect Biol* doi: 10.1101/cshperspect.a039677

totipotent zygote then develops to blastocyst containing the pluripotent inner cell mass (ICM), that subsequently produces germ layers during gastrulation and, ultimately, the whole organism (Chazaud and Yamanaka 2016). During this process, epigenetic regulation plays critical roles to ensure precise spatiotemporal gene regulation (Bird 2002; Kouzarides 2007) and successful early development (Burton and Torres-Padilla 2014; Eckersley-Maslin et al. 2018; Xu and Xie 2018). Aberrant epigenetic reprogramming leads to severe developmental defects and diseases (Greenberg and Bourc'his 2019; Jambhekar et al. 2019). However, it has long been enigmatic how the epigenomes are globally reprogrammed in mammalian early development at the molecular level. Furthermore, the role of the epigenome in early development remains elusive, with mechanistic studies often hindered by the limited amount of biological material that can be obtained from embryos. Recent advances in low-input or single-cell chromatin analysis begin to illuminate specific aspects of chromatin regulation during gametogenesis and embryogenesis. Here, we summarize the dynamics of the epigenomes in gametogenesis and early development, their functions and the possible mechanisms underlying the drastic changes. Last, we highlight crucial questions that remain unanswered.

EPIGENOME REPROGRAMMING DURING EARLY DEVELOPMENT

DNA Methylation

5-methylcytosine (5mC) is a prevalent epigenetic mark in the mammalian genome, existing predominantly in CpG dinucleotides. It is essential for development by regulating genomic imprinting, transposon silencing, X chromosome inactivation (XCI), and gene repression (Bird 2002; Smith and Meissner 2013). In mammals, DNA methylation is established by de novo methyltransferases DNMT3A and DNMT3B, and maintained by DNMT1 through cell division, while active DNA demethylation is facilitated by the ten-eleven translocation (TET) methylcytosine dioxygenase family (Kohli and Zhang 2013; Pastor et al. 2013; Li and Zhang 2014). During germline development, PGCs first undergo extensive genome-wide DNA demethylation (Guibert et al. 2012; Seisenberger et al. 2012; Vincent et al. 2013), followed by sex-specific de novo DNA methylation (Fig. 1; Stewart et al. 2016). The proper establishment of DNA methylome is essential for spermatogenesis. In mouse, loss of DNMT3A or DNMT3L (a DNMT3A/B cofactor) causes catastrophic meiotic arrest in spermatogonia (Walsh et al. 1998; Bourc'his and Bestor 2004). Similarly, the mutation of DNMT3C, a male fetal germ cell-specific DNA methyltransferase, leads to derepression of young transposable elements and infertility (Barau et al. 2016). DNA methylation, however, appears to be dispensable for mouse oogenesis, as *Dnmt3a* or *Dnmt3l* knockout does not cause apparent oocyte developmental defects (Smallwood et al. 2011; Kobayashi et al. 2012; Shirane et al. 2013), although oocyte methylation is critical for early embryogenesis. This is perhaps not surprising since the de novo DNA methylation only occurs in late-stage oocytes (Stewart et al. 2016; Sendžikaite and Kelsey 2019; Yan et al. 2021). Notably, the mouse oocyte has a distinct DNA methylome in which only transcribed regions are highly methylated (Kobayashi et al. 2012; Shirane et al. 2013). This is attributed to the notion that Pol II can recruit SETD2 for H3K36me3 deposition, which further leads to the recruitment of DNMT3A (Dhayalan et al. 2010; Baubec et al. 2015; Dukatz et al. 2019). A major function of such transcription-dependent DNA methylation is to establish maternal imprints often by exploiting oocyte-specific promoters upstream of the imprinting control regions (ICRs) (Chotalia et al. 2009; Bartolomei and Ferguson-Smith 2011; Brind' Amour et al. 2018; Xu et al. 2019). Meanwhile, restricting DNA methylation to transcribed regions in mouse oocytes may also avoid undesired methylation of the paternal imprints. Intriguingly, while DNMT1 is dispensable for the establishment of the oocyte DNA methylome (Shirane et al. 2013), its activity needs to be strictly harnessed by STELLA, which sequestrates UHRF1 (a key cofactor of DNMT1) (Li et al. 2018b). Deletion of *Stella* results in aberrant accumulation of DNA methylation by DNMT1 genome wide, which interferes with ZGA (Li et al. 2018b). Notably, in the mouse male germ cells, NSD1 and H3K36me2 assume similar roles as SETD2 and H3K36me3 to establish the bulk DNA methylome (Shirane et al. 2020).

After fertilization, the parental genomes undergo global DNA demethylation (except for ICRs and some retrotransposons) through both active and passive mechanisms (Fig. 1; Smith et al. 2012; Guo et al. 2014; Shen et al. 2014; Wang et al. 2014). DNA methylation at ICRs controls allele-specific expression of a small subset of genes that play important roles in placenta, postnatal development, and brain functions (for review, see Tucci et al. 2019). Notably, de novo DNA methylation, although weak, also occurs amid global demethylation, and is subjected to TET-mediated DNA demethylation (Amouroux et al. 2016). During mouse somatic cell nuclear transfer (SCNT), such de novo methylation is suggested to impede ZGA and cloning efficiency (Gao et al. 2018b). In mouse, the global DNA methylation is quickly reestablished after implantation (Smith et al. 2017; Zhang et al. 2018b). The extraembryonic tissues, such as the placenta, are relatively hypomethylated compared to embryonic lineages, presumably forming epigenetic barriers (Schroeder et al. 2015; Smith et al. 2017; Zhang et al. 2018b). Similar findings were made in cultured human embryos from day 5 to day 14 (Zhou et al. 2019). Nevertheless, despite the detailed characterization, the function of postfertilization global DNA methylation resetting remains elusive. A plausible hypothesis is that it equalizes the two distinct gametic methylomes (except for ICRs). The demethylation may also be a prerequisite for the emergence of extraembryonic tissues. Another intriguing possibility is that the postfertilization global demethylation may facilitate the elongation of telomeres in early embryos, as observed in mouse embryonic stem cells (mESCs) (Gonzalo et al. 2006; Dan et al. 2017). Future studies are warranted to test these models.

HISTONE MODIFICATIONS

The posttranslational modifications of histones are fundamental epigenetic regulators that control many crucial cellular processes, including transcription, DNA repair, and DNA replication (Kouzarides 2007; Lawrence et al. 2016). Mounting evidence also shows that they play key roles in regulating spatial-temporal gene expression during gametogenesis and early embryo development.

H3K4me3

Although most histones in mature spermatozoa are replaced with protamine, the residual histones are still extensively modified (Gold et al. 2018). For example, H3K4me3, a hallmark of permissive promoters that can recruit chromatin remodelers and transcriptional machinery (Ruthenburg et al. 2007), remains preferentially enriched at CG-rich promoters in sperm as in most other cell types (Hammoud et al. 2009, 2014; Erkek et al. 2013). In growing oocytes, H3K4me3 is similarly restricted to active promoters. However, it becomes accumulated as broad domains at both promoters and distal regions in more mature oocytes (Fig. 1; Dahl et al. 2016; Liu et al. 2016b; Zhang et al. 2016; Hanna et al. 2018b). Such noncanonical H3K4me3 (ncH3K4me3) domains found in oocytes occur almost exclusively in partially methylated DNA domains (PMDs), which are essentially nontranscribed regions (Dahl et al. 2016; Zhang et al. 2016; Hanna et al. 2018b). The fully DNA-methylated regions are protected against ncH3K4me3 by DNA methylation in conjunction with H3K36me3, as SETD2 deficiency results in the loss of H3K36me3 and DNA methylation, and ectopic invasion of H3K4me3 (Hanna et al. 2018b; Xu et al. 2019). Notably, the ncH3K4me3 domains are deposited mainly by MLL2 (KMT2B) (Hanna et al. 2018b), which is also responsible for depositing H3K4me3 at poised promoters in mESCs (Hu et al. 2013; Denissov et al. 2014) and postimplantation embryos (Xiang et al. 2020) in a transcription-independent manner. Importantly, ablation of MLL2 leads to anovulation and oocyte death (Andreu-Vieyra et al. 2010). Intriguingly, either deficiency of *Mll2* (Andreu-Vieyra et al. 2010) or overexpression of the H3K4me3 demethylase *Kdm5b* (Zhang et al. 2016) disrupts genome silencing in late-stage full-grown oocytes (FGOs). Although the underlying mechanism is currently unknown, such repressive function contrasts with that of H3K4me3 at active promoters. For example, the CXXC finger protein 1 (CFP1) is a component of the histone H3K4me3 methyltransferase SETD1A/B, which functions at constitutively active promoters (Lee and Skalnik 2005; Clouaire et al. 2012). Loss of CFP1 causes global down-regulation of H3K4me3 and transcription, as well as the failure of mouse oocyte maturation (Yu et al. 2017; Sha et al. 2018). These data are in line with the notion that different H3K4me3 methyltransferase complexes have distinct targets (Hu et al. 2017; Douillet et al. 2020).

Upon fertilization, the paternal H3K4me3 quickly diminishes (Lepikhov and Walter 2004; Lepikhov et al. 2008; Zhang et al. 2016), which may be linked to the extensive protamine-histone exchange (Burton and Torres-Padilla 2014). It is subsequently reestablished in late one-cell mouse embryos, but in a noncanonical pattern, forming weak but extremely large (up to several Mbs) domains in gene-rich regions (Zhang et al. 2016). By contrast, the maternal H3K4me3 is briefly inherited to zygotes. Both maternal and paternal H3K4me3 are then reprogrammed to the canonical patterns at the two-cell stage in a ZGA-dependent manner (Fig. 1; Dahl et al. 2016; Liu et al. 2016b; Zhang et al. 2016). This is accompanied by a decrease of global H3K4me3 in immunofluorescence, executed by two lysine demethylases KDM5A/5B that are activated during ZGA (Dahl et al. 2016). Knocking down these enzymes leads to embryonic lethality (Dahl et al. 2016). Intriguingly, ncH3K4me3 is absent in human germinal vesicle (GV) oocytes (Xia et al. 2019). Instead, de novo broad domains of H3K4me3 occur in CpG-rich, accessible regions in human pre-ZGA embryos, although its function remains unknown (Xia et al. 2019). Therefore, diverse mechanisms regulate H3K4me3 in mouse and human embryos.

H3K27me3

Polycomb repressive complex 2 (PRC2) is responsible for the deposition of the histone modification H3K27me3. Both PRC2 and H3K27me3 are evolutionarily conserved epigenetic regulators, which frequently repress key developmental genes and thus help maintain cell identity (Margueron and Reinberg 2011; Di Croce and Helin 2013). H3K27me3 is also enriched at developmental gene promoters in sperm (Hammoud et al. 2009, 2014; Brykczynska et al. 2010; Erkek et al. 2013). Similar to H3K4me3, widespread distal H3K27me3 domains are also found in PMDs in mouse oocytes (Figs. 1 and 2A; Liu et al. 2016b; Zheng et al. 2016). Despite its prevalence, oocyte H3K27me3 seems to be nonessential for oogenesis, since the oocyte-specific knockout of EED, a core component of PRC2 complex, does not cause notable oocyte developmental defects (Inoue et al. 2018; Prokopuk et al. 2018). However, knockout of *Ring1a/Ring1b* in oocytes, the key components of Polycomb repressive complex 1 (PRC1) that catalyzes H2AK119 monoubiquitination (H2AK119ub), leads to massive gene derepression, developmental defects in oocytes, and subsequent one-cell arrest after fertilization (Posfai et al. 2012; Du et al. 2020). These data are in agreement with the notion that PRC1 and its modification H2AK119ub are the major transcription repressors, while PRC2 and H3K27me3 facilitate such repression (Fursova et al. 2019; Blackledge et al. 2020; Tamburri et al. 2020). Upon fertilization, oocyte H3K27me3 at promoters of developmental genes is mostly lost, including at *Hox* gene promoters (Fig. 2A; Liu et al. 2016b; Zheng et al. 2016). The paternal H3K27me3 is quickly erased in zygotes, followed by gradual de novo formation of megabase-size broad domains of H3K27me3 in gene deserts (Zheng et al. 2016), although their functions remain unclear.

Intriguingly, the distal domains of H3K27me3 from oocytes can persist until blastocyst stage (Zheng et al. 2016). These maternally inherited H3K27me3 domains were proven to function as an imprinting mark and dictate a small set of paternal-specific genes in early mouse embryos, similar to DNA methylation-mediated imprinting (Fig. 2B; Inoue et al. 2017a; Chen and Zhang 2020). These genes include *Xist* (Inoue et al. 2017b), the master regulator of XCI (Lee 2005). In female mouse embryos, the maternal *Xist* locus is coated with a broad H3K27me3 domain inherited from oocytes, leaving the paternal copy specifically expressed, which initiates "imprinted XCI" (Fig. 2B; Inoue et al. 2017b). Abla-

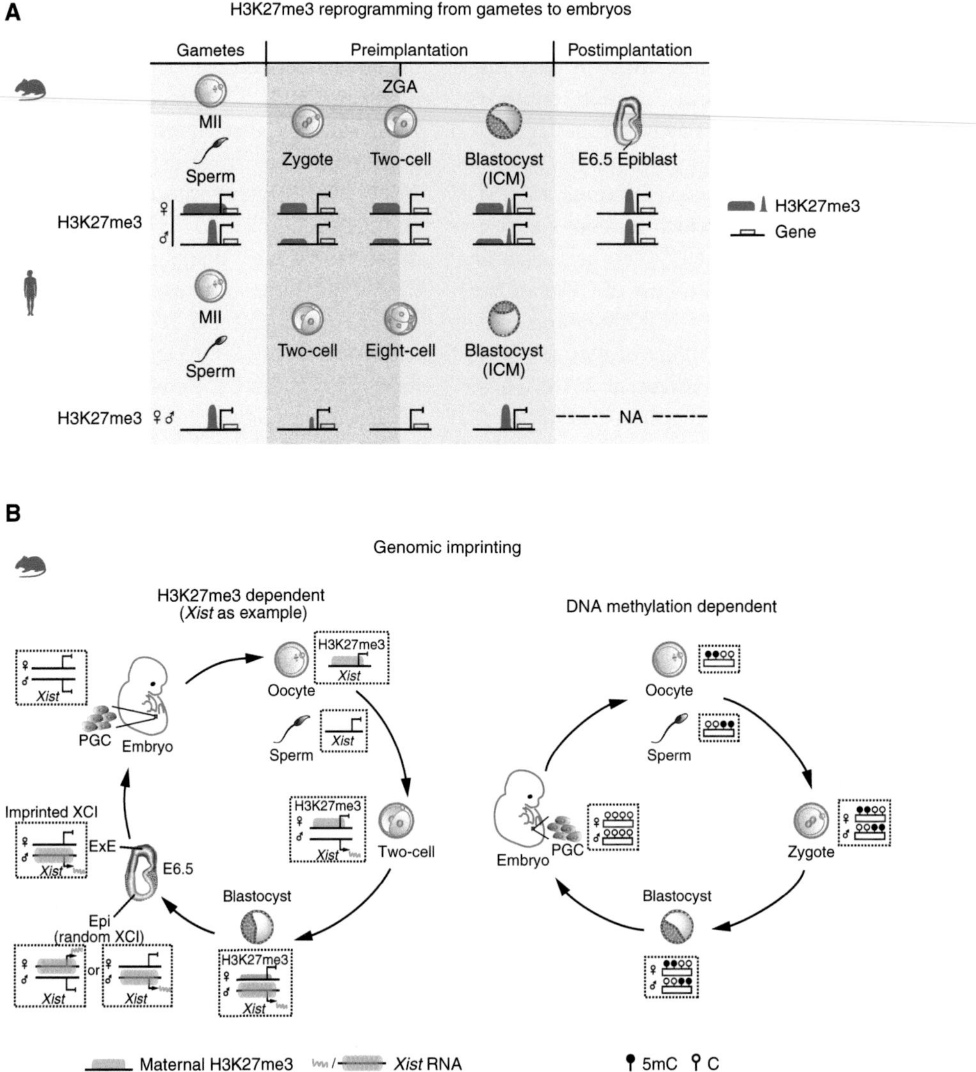

Figure 2. The H3K27me3 reprogramming from gametes to embryos and genomic imprinting. (*A*) The schematics show the dynamics of H3K27me3 (blue) in gametes and early embryos in mouse and human. In mouse, broad domains of H3K27me3 from oocytes are transiently inherited to embryos until blastocyst (except for promoter H3K27me3). The sperm H3K27me3 is quickly removed after fertilization. Canonical H3K27me3 is established at promoters in blastocyst and in postimplantation embryos. By contrast, in both human oocytes and sperm, H3K27me3 exhibits canonical distributions. After fertilization, H3K27me3 is globally removed before zygotic genome activation (ZGA) and is reestablished as early as the morula stage. (NA) Data not available. (B) (*Left*) A schematic model showing H3K27me3-dependent imprinting (*Xist* as an example) in mouse. Oocyte-inherited H3K27me3 domains repress maternal *Xist* in mouse female embryos, leaving the paternal copy specifically expressed and initiates the imprinted X chromosome inactivation (XCI). In postimplantation embryos, the oocyte-inherited H3K27me3 domain at the *Xist* locus is lost. Epiblast (Epi) undergoes random XCI, while the imprinted XCI persists in the extraembryonic lineages, presumably due to additional epigenetic repression mechanisms. The inactive X chromosome is reactivated in the primordial germ cells (PGCs). (*Right*) A schematic model showing DNA methylation-dependent imprinting in mouse. Oocyte and sperm establish distinct DNA methylome and differentially methylated imprinting control regions (ICRs) or imprints during the gametogenesis. Imprints can survive the global DNA demethylation during preimplantation and is well maintained thereafter, directing allelic gene expression. They are, however, erased in PGCs, paving the way for imprint establishment for the next generation. (ICM) Inner cell mass.

tion of *Eed* in oocytes leads to random X-inactivation in female embryos (Inoue et al. 2018; Harris et al. 2019). Besides *Xist*, a small subset of genes with roles in placental development, such as *Slc38a4*, *Sfmbt2*, and *Grab1*, is also imprinted by oocyte-derived H3K27me3 (Inoue et al. 2017a). Intriguingly, while the oocyte-inherited H3K27me3 is lost in postimplantation embryos (Zheng et al. 2016; Inoue et al. 2017b; Chen et al. 2019b), the imprinted states of these genes (*Platr20*, *Slc38a4*, *Gm32885*, *Gab1*, *Phf17*, and *Smoc1*) are preserved in extraembryonic tissues by maternal-specific de novo DNA methylation (Chen et al. 2019b; Hanna et al. 2019), suggesting that DNA methylation and H3K27me3 act together to safeguard their imprinting. The importance of such H3K27me3-mediated imprinting was supported by data from the SCNT embryos, where oocyte H3K27me3-mediated imprinting is apparently missing (Matoba et al. 2018). Artificially forcing the monoallelic expression of these H3K27me3-controlled imprinted genes by allele-specific gene deletion can substantially improve mouse SCNT development (Wang et al. 2020). Interestingly, such H3K27me3-mediated imprinting appears to be missing in human. In human oocytes, H3K27me3 is strongly enriched at developmental gene promoters in a canonical pattern (Xia et al. 2019). Global H3K27me3 is erased after fertilization by the eight-cell stage, when ZGA happens. Embryonic H3K27me3 is subsequently restored at developmental genes in morulae and blastocysts (Fig. 2A; Xia et al. 2019; Zhang et al. 2019b). These results are consistent with the reported absence of imprinted XCI in human (Okamoto et al. 2011; Petropoulos et al. 2016). Mechanistically, in contrast with their abundant presence in mouse, the PRC2 core components, EED and SUZ12, are not expressed in human oocytes or pre-ZGA embryos, and are only activated after ZGA (Saha et al. 2013; Xia et al. 2019). These findings highlight the divergence of epigenetic regulation in early development among different species.

Bivalent H3K4me3 and H3K27me3

The "bivalent marks" refer to the co-occupancy of H3K4me3 and H3K27me3 at the promoters, often at developmental genes (Azuara et al. 2006; Bernstein et al. 2006). By bearing both permissive and repressive modifications, the bivalent marks are postulated to poise silenced key developmental genes for rapid reactivation upon cell differentiation. The bivalent marks widely exist in gametes, somatic cells, and PGCs (Vastenhouw and Schier 2012; Sachs et al. 2013; Hammoud et al. 2014). However, they are briefly absent in preimplantation embryos in mouse and human (Liu et al. 2016b; Zheng et al. 2016; Xia et al. 2019), with the underlying mechanisms and functions being unclear. One interesting possibility is that preimplantation embryos possess totipotency and pluripotency, while bivalent marks predominantly function during cell differentiation. Consistently, by E6.5, H3K27me3 and unusually strong H3K4me3 are restored at developmental gene promoters in the mouse epiblast, forming "super bivalency" (Fig. 1; Xiang et al. 2020). The "strong H3K4me3" is then quickly attenuated in fate-committed lineages and further decreases in somatic cells. Genes marked by "super-bivalency" also show strong spatial interactions as determined by Hi-C analysis (Xiang et al. 2020). In mESCs, such strong interactions rely on PRC1 and can help H3K27me3 spreading to regions away from "nucleation sites" (Schoenfelder et al. 2015; Oksuz et al. 2018; Kraft et al. 2020). Mechanistically, KMT2B (MLL2) is responsible for the "strong H3K4me3" at E6.5 but becomes partially dispensable by E8.5, as *Kmt2b* mutant embryos fail to activate a subset of developmental genes and die at E10.5 (Glaser et al. 2006; Xiang et al. 2020). As bivalent promoters are often intermediately accessible and enriched for paused Pol II (Stock et al. 2007; Brookes et al. 2012; Tee et al. 2014), it is speculated that "super-bivalency" may allow developmental genes to rapidly respond to developmental cues upon swift cell-fate decision during early lineage specification (Xiang et al. 2020). In addition, a subset of bivalent promoters is regulated by DPPA2/DPPA4 in mESCs, which safeguards promoter bivalency and prevents DNA methylation (Eckersley-Maslin et al. 2020; Gretarsson and Hackett 2020). Their roles in bivalency and differentiation in embryos are interesting topics for future studies.

H2AK119ub

H2AK119ub, deposited by PRC1, preferentially colocalizes with H3K27me3 at the promoters of developmental genes in most cell types including mouse sperm (Schoenfelder et al. 2015; Chen et al. 2021; Mei et al. 2021; Zhu et al. 2021). Similar to H3K27me3, H2AK119ub in mouse oocytes also exhibits a noncanonical distribution by forming broad domains in PMDs (Chen et al. 2021; Mei et al. 2021; Zhu et al. 2021). Interestingly, it also covers a fraction of active promoters marked by H3K4me3 (Chen et al. 2021; Mei et al. 2021; Zhu et al. 2021). H2AK119ub and H3K27me3 undergo distinct reprogramming dynamics in mouse early embryos. Upon fertilization, H2AK119ub, but not H3K27me3, is retained at promoters of developmental genes, and acute depletion of H2AK119ub in one-cell embryos causes derepression of developmental genes during ZGA and four-cell arrest, suggesting that H2AK119ub protects preimplantation embryo development from precocious expression of developmental genes when H3K27me3 is absent (Chen et al. 2021). Distal maternal H3K27me3 is highly stable and persists to blastocyst, which shows distinct pattern from that of the paternal allele. By contrast, H2AK119ub is much more dynamic after fertilization and the two alleles already adopt similar distributions by the two-cell stage (Chen et al. 2021; Mei et al. 2021; Zhu et al. 2021). Finally, H2AK119ub and H3K27me3 become colocalized again by exhibiting canonical patterns after implantation (Figs. 1 and 2A; Chen et al. 2021; Mei et al. 2021).

Interestingly, H2AK119ub has a complex interplay with H3K27me3 at maternal H3K27me3-controlled imprinted genes. Acute removal of H2AK119ub in mouse one-cell embryos does not affect H3K27me3-mediated imprinting at least by the four-cell stage (Chen et al. 2021), suggesting that H2AK119ub is dispensable for the maintenance of these H3K27me3 imprints. However, oocyte-specific knockout of *Pcgf1* and *Pcgf6*, two key components of variant PRC1, leads to partial loss of H3K27me3 imprinting in both FGOs and morula (Mei et al. 2021), implying that H2AK119ub is involved in the establishment of some H3K27me3-mediated imprinting. Genes that lose H3K27me3 upon the *Pcgf1* and *Pcgf6* mutations also tend to be derepressed in FGOs (Mei et al. 2021). These findings highlight distinct roles and intricate cross talks between H2AK119ub and H3K27me3 in mouse early embryos, where H2AK119ub is more dynamic and can exert immediate repression, while H3K27me3 is relatively more stable and may serve as silencing memories.

H3K36me2/3

Methylation of H3K36 plays critical roles in transcription fidelity, RNA splicing, DNA replication, and DNA repair (Wagner and Carpenter 2012; Wozniak and Strahl 2014). H3K36me3 often enriches at actively expressed gene bodies, while H3K36me2 occupies both genic and intergenic regions (Wagner and Carpenter 2012; Wozniak and Strahl 2014; Weinberg et al. 2019). Both marks can recruit DNMT3A/B and subsequently specify patterns of DNA methylation (Dhayalan et al. 2010; Baubec et al. 2015; Dukatz et al. 2019; Weinberg et al. 2019). In mouse oocytes, DNA methylation is deposited to actively transcribed regions guided by H3K36me3, and a major function for such H3K36me3-guided DNA methylation is to establish maternal imprints (Chotalia et al. 2009; Bartolomei and Ferguson-Smith 2011; Kelsey and Feil 2013; Xu et al. 2019). Meanwhile, DNA methylation and H3K36me3 may also work synergistically to protect transcribing regions from H3K4me3 and H3K27me3, as observed in *Setd2*-deficient oocytes (Hanna et al. 2018b; Xu et al. 2019). As a result, *Setd2*-deficient oocytes show developmental defects and subsequent one-cell embryo arrest after fertilization (Xu et al. 2019). Spindle exchange experiments showed that such early arrest was mainly caused by cytosolic defect, as 54% of embryos reconstructed with wild-type (WT) cytoplasm and *Setd2*-deficient chromatin can develop to blastocyst. These embryos, however, eventually died after implantation presumably due to the loss of imprints (Xu et al. 2019). By contrast, in male germline, it is NSD1 and its catalytic product H3K36me2, but not SETD2 and H3K36me3, that direct the majority of de novo DNA methylation and protect transcribing regions from the spreading of H3K27me3 (Shirane et al. 2020). Male mice with ablated *Nsd1* are infertile with hypogonadism and cannot generate mature spermatozoa (Shirane et al. 2020). These results indicate dimorphic mechanisms in DNA methylation establishment during gametogenesis. Why oocyte and sperm employ different H3K36 methylation to guide de novo DNA methylation awaits further investigation. After fertilization, H3K36me3 from both alleles is largely lost, before it reappears concomitant with ZGA (Fig. 1; Xu et al. 2019).

H3K9me3

H3K9me3 is a repressive mark that often decorates constitutive heterochromatin (Becker et al. 2016; Allshire and Madhani 2018). Disrupting H3K9me3 writers or erasers causes severe phenotypes in both oocytes and embryos. For example, mouse oocytes with deficient SETDB1, an H3K9 methyltransferase, suffer impaired meiosis and failure in silencing ERVK, ERVL, and MaLR retrotransposons (Eymery et al. 2016; Kim et al. 2016). Zygotic *Setdb1*-null embryos are lethal around implantation with defects in ICM growth (Dodge et al. 2004). Double knockout of H3K9 methyltransferase SUV39H1/2 leads to prenatal lethality in mice (Peters et al. 2001). Meanwhile, excessive H3K9me3 also needs to be curbed as the ablation of KDM4A, an H3K9me3 demethylase in oocytes, causes aberrant H3K9me3 invasion to the broad H3K4me3 domains, subsequently leading to impaired activation of genes and transposable elements during ZGA and preimplantation lethality (Sankar et al. 2020). H3K9me3 regulates mouse embryo development at least in three major aspects: facilitating the remodeling of heterochromatin, repressing repeats, and regulating lineage specification.

Mouse preimplantation embryos are featured with immature heterochromatin characterized by the absence of classic constitutive heterochromatin markers, such as HP1α and H4K20me3 (Wongtawan et al. 2011). The H3K9me3 methyltransferase SUV39H1 protein is not detected until the eight-cell stage (Burton et al. 2020). De novo H3K9me3 occurs in one-cell embryos (Wang et al. 2018; Burton et al. 2020), as ChIP-seq analyses confirmed that H3K9me3 undergoes a global resetting on both alleles after fertilization (Fig. 1; Wang et al. 2018). On the paternal genome, de novo H3K9me3 is mediated by SUV39H2, but is nonrepressive and instead bookmarks promoters for compaction at later stages (Burton et al. 2020). Such nonrepressive, immature heterochromatin appears to be essential for early development, as forced precocious expression of *Suv39h1* restores constitutive heterochromatin but impairs embryonic development especially beyond the morula stage (Burton et al. 2020). Improper reprogramming of H3K9me3 in SCNT embryos results in insufficient gene activation in somatic H3K9me3-marked regions and poor development to blastocyst (Matoba et al. 2014). Such preimplantation developmental defects can be rescued by overexpression of the H3K9me3 demethylases *Kdm4d* or *Kdm4b*, or by knockdown of H3K9 methyltransferase *Suv39h1/2* in donor cells in mouse (Matoba et al. 2014; Liu et al. 2016a). Similar findings were made in pig (Ruan et al. 2018), bovine (Liu et al. 2018a), monkey (Liu et al. 2018b), and human (Chung et al. 2015), suggesting that H3K9me3 is an evolutionarily conserved epigenetic barrier for cell-fate reprogramming.

After fertilization, a large portion of LTRs (long terminal retrotransposons) are activated (Peaston et al. 2004), possibly because of the relief of global epigenetic repression. The subsequent repression of LTRs coincides with the gradual reestablishment of H3K9me3 (Wang et al. 2018) and requires the histone chaperone chromatin assembly factor 1 (CAF-1) (Hatanaka et al. 2015; Wang et al. 2018). Knocking down the methyltransferases *Suv39h1/2* leads to increased expression of LTRs in morula but not at the two-cell stage (Hatanaka et al. 2015; Burton et al. 2020), reinforcing the notion that H3K9me3 is repressive only in late-stage embryos.

Whereas H3K9me3 is primarily present in constitutive heterochromatin and transposons, it is also observed at lineage-specific genes, where it acts as a barrier for cell-fate changes by preventing precocious activation of lineage-incompatible genes (Becker et al. 2016). For example, the distributions of promoter H3K9me3 differ significantly between intra- and extraembryonic tissues of E6.5

and E7.5 embryos (Wang et al. 2018). Uncommitted germ-layer cells transiently employ high levels of H3K9me3 to repress genes involved in lineage specification, which are subsequently removed during differentiation to allow tissue-specific gene expression (Nicetto et al. 2019). Consequently, perturbation of H3K9me3 leads to aberrant lineage-specific gene expression (Nicetto et al. 2019). For example, mouse livers with endoderm-specific *Setdb1/Suv39h1/2* triple-knockout resulted in the failed induction of hepatocyte-specific genes and aberrant expression of inappropriate lineage genes, concomitant with the loss of H3K9me3 and heterochromatin (Nicetto et al. 2019).

TRANSPOSABLE ELEMENTS

Transposable elements (TEs) occupy almost half of the mouse and human genome (Lander et al. 2001; Mouse Genome Sequencing Consortium et al. 2002; de Koning et al. 2011). It has been increasingly evident that these elements, once thought as "junk DNA," drive genetic innovation during evolution by introducing regulatory elements (such as promoters and enhancers) and transcription factor-binding sites (Rebollo et al. 2012; Chuong et al. 2017; Jangam et al. 2017; Bourque et al. 2018). MERVL, a member of endogenous retroviruses (ERVs), is actively transcribed during minor ZGA in mouse embryos (Svoboda et al. 2004), and was shown to regulate mouse preimplantation development (Huang et al. 2017). MERVL is reported to initiate hundreds of two-cell-specific transcripts in two-cell-like cells (2CLC), a rare and transient cell population in mESCs (Macfarlan et al. 2012; Genet and Torres-Padilla 2020). Notably, overexpressing DUX, a transcription factor specifically expressed during minor ZGA, can activate MERVL transcription and induce 2CLCs in mESCs through the DUX-miR-344-ZMYM2 pathway (De Iaco et al. 2017; Hendrickson et al. 2017; Whiddon et al. 2017; Yang et al. 2020). DUX itself is activated by two upstream transcription factors DPPA2/DPPA4 (Eckersley-Maslin et al. 2019) and negative elongation factor A (NELFA) (Hu et al. 2020). Apart from DUX, ZSCAN4, a nonconventional transcription factor (Falco et al. 2007; Ko 2016; Srinivasan et al. 2020), can also activate MERVL (Zhang et al. 2019a), in addition to its role in telomere elongation and genome integrity in mESCs (Ko 2016). However, ZSCAN4, in contrast to DUX, fails to do so in the absence of DPPA2/DPPA4 in mESCs. It was proposed that ZSCAN4 is not the driving factor but instead is a stabilizer to initiate the ZGA (De Iaco et al. 2019; Eckersley-Maslin et al. 2019). Notably, the *Dux*-knockout embryos show only minor defects of ZGA and MERVL transcription, and the majority of them can develop to term (Chen and Zhang 2019; Guo et al. 2019; De Iaco et al. 2020). DPPA2/DPPA4 double-null mice undergo normal embryogenesis but die perinatally (Nakamura et al. 2011). Therefore, these data suggest that additional ZGA regulators exist in embryos.

Long interspersed element-1 (LINE1), a subclass of non-LTR transposons, also plays a vital role in mouse embryo development (Beraldi et al. 2006; Jachowicz et al. 2017; Percharde et al. 2018). These elements are abundantly transcribed in mouse oocytes and fertilized embryos, before undergoing repression after the two-cell stage (Fadloun et al. 2013; Jachowicz et al. 2017). LINE1 knockdown in mouse zygotes was shown to impair ZGA and lead to two-cell stage arrest (Percharde et al. 2018). Interestingly, LINE1 transcription, but not its RNA, contributes to the embryo development by decondensing chromatin (Jachowicz et al. 2017). Precocious silencing or prolonged activation of LINE1 hinders chromatin relaxation or compaction, respectively, both resulting in embryo developmental defects (Beraldi et al. 2006; Jachowicz et al. 2017). Intriguingly, LINE1 RNA could also repress *Dux* expression in mESCs along with nucleolin and KAP1/TRIM28 (Percharde et al. 2018). It remains to be determined whether this is also the case in early embryos where *Dux* also needs to be shut down in a timely manner (Guo et al. 2019). Therefore, transposon elements are integral to the transcriptional circuitry in early development. Their exact roles in ZGA, totipotency, pluripotency, and lineage commitment await further investigation.

CHROMATIN ACCESSIBILITY

Accessible chromatin usually marks *cis*-regulatory elements, such as promoters, enhancers, insulators, where *trans*-acting factors interact with DNA and regulate gene expression (Klemm et al. 2019). Genome-wide accessible chromatin landscapes in mouse gametes and preimplantation were profiled with DNase I sequencing (liDNase-seq) (Lu et al. 2016; Inoue et al. 2017a) or transposase accessible chromatin-sequencing (ATAC-seq) (Wu et al. 2016; Jung et al. 2017, 2019). Interestingly, while both ATAC-seq and DNase-seq can reproduce chromatin accessibility at later-stage early embryos, their performance in detecting open chromatin in earlier-stage embryos (such as zygotes), oocytes, and sperm varies. This likely reflects the unique chromatin states of these cells and the sensitivity of different technologies, including possible overdigestion of highly permissive chromatin. Nevertheless, a consensus picture from these data revealed that accessible chromatin in gametes and early embryos also enrich for regulatory elements such as promoters and enhancers, including at stages when transcription is silenced (Lu et al. 2016; Wu et al. 2016, 2018; Jung et al. 2017, 2019; Gao et al. 2018a; Li et al. 2018a). Interestingly, a significant fraction of open chromatin appears to be specific for pre-ZGA embryos, and is lost upon genome activation (Wu et al. 2016, 2018). Such a change may in part reflect the transition from maternal transcription factors to zygotic transcription regulators. In fact, these accessible chromatin maps have also revealed a rich repertoire of candidate transcription regulators in gametogenesis and early development (Lu et al. 2016; Wu et al. 2016, 2018; Jung et al. 2017; Gao et al. 2018a; Li et al. 2018a; Jung et al. 2019).

3D GENOME

In eukaryotes, the genome is packaged into a multilayer, higher-order chromatin configuration (Bickmore 2013; Gibcus and Dekker 2013; Gorkin et al. 2014; Bonev and Cavalli 2016; Misteli 2020). The chromosomes are nonrandomly distributed in interphase nucleus, occupying different chromosome territories. At the megabase scale, the genome is spatially partitioned into transcriptionally active compartment A, which preferentially resides near the nuclear speckles, and repressive compartment B, which is often associated with nuclear lamina or nucleolus (Lieberman-Aiden et al. 2009; van Steensel and Belmont 2017; Chen et al. 2018; Quinodoz et al. 2018). Topologically associating domains (TADs) exist as self-interacting domains usually with lengths of several hundred kilobase in mammals (Gibcus and Dekker 2013; Gorkin et al. 2014; Rowley and Corces 2018). They often arise via ATP-dependent loop extrusion, as cohesin extrudes chromatin until it is stopped by CCCTC-binding factor (CTCF) or other *cis*-acting factors (Sanborn et al. 2015; Fudenberg et al. 2016; Rao et al. 2017; Schwarzer et al. 2017; Vian et al. 2018; Davidson et al. 2019; Kim et al. 2019). At an even finer scale, the interactions among regulatory elements such as promoters and enhancers are crucial in regulating appropriate gene expression (Gorkin et al. 2014; Pombo and Dillon 2015; Schoenfelder and Fraser 2019). With the development of low-input or single-cell technologies to analyze chromatin organization, the dynamics of chromatin structure during gametogenesis and early embryo development begin to be unveiled, as summarized below.

Genome Organization in Spermatogenesis

To transport the haploid paternal genome to the offspring, the male germ cells undergo an orchestrated developmental process from spermatogonia to mature spermatozoa, which could be briefly divided into three major phases: mitosis (the proliferation and differentiation of spermatogonia), meiosis (the formation of haploid cells), and postmeiosis spermiogenesis (when the genome is subjected to compaction accompanied with histone-protamine exchange) (Handel and Schimenti 2010). In the meiosis I prophase, the homologous chromosomes undergo pairing (during the leptotene and zygotene stage), synapse (at the pachytene stage) and desynapse (at the onset of the diplotene stage) (Zickler and Kleckner 2015). Interestingly, TADs and compartments A/B dissolve in both monkey and mouse pachytene spermatocytes, when the homologous chromosomes parallelly align along the synaptonemal complex with chromatin loops extruding from the central axis (Alavattam et al. 2019; Patel et al. 2019; Vara et al. 2019; Wang et al.

2019b; Luo et al. 2020). At the same time, transcription-correlated local compartments (refined compartments A/B) (or point interactions/transcription hubs) emerge. These data raise a possibility that the chromatin loops facilitated by synaptonemal complex (Zickler and Kleckner 2015) may be incompatible with compartments A/B and chromatin loops mediated by CTCF (Nuebler et al. 2018). In line with this idea, refined compartments A/B are attenuated in synaptonemal complex mutants (*Sycp2*), with conventional TADs and compartmentalization partially restored (Wang et al. 2019b). Furthermore, conventional compartments and TADs also reappear in mature spermatozoa (Fig. 3; Battulin et al. 2015; Du et al. 2017; Jung et al. 2017; Ke et al. 2017; Alavattam et al. 2019; Wang et al. 2019b; Luo et al. 2020), although this was not detected in a separate study using a different sperm collection method (Vara et al. 2019). Interestingly, human mature spermatozoa have no TADs, which is attributed to CTCF silencing (Fig. 3; Chen et al. 2019a). Zebrafish spermatozoa also do not have conventional TADs, but possess unique "hinge-like" chromatin domains arranged similarly as mitotic chromosomes (Wike et al. 2021). These data show that extensive chromosome reorganization occurs during spermatogenesis and distinct mechanisms underline chromatin condensation in mature sperm in different species.

Genome Organization in Oogenesis

Oocytes are arrested at the diplotene stage for an extended time when the meiotic recombination is completed and the synaptonemal complex is largely dissembled (Handel and Schimenti 2010). Upon hormone stimulation, some oocytes undergo follicle growth, which start to accumulate proteins and RNAs for embryonic development, ultimately becoming FGOs (Bachvarova 1985; Li et al. 2010; Veselovska et al. 2015). TADs and chromatin loops can be detected in FGOs and, similar to those in soma, depend on SCC1, a key component of cohesin (Flyamer et al. 2017; Du et al. 2020; Silva et al. 2020). A meiotic-specific cohesin component, REC8, is required for the bivalent chromosome cohesion (Tachibana-Konwalski et al. 2010; Silva et al. 2020). Moreover, H3K27me3-marked compartmental domains (termed Polycomb-associating domains [PADs]) with strong interdomain interactions emerge in FGOs, concomitant with the loss of conventional compartments A/B (Flyamer et al. 2017; Du et al. 2020) and lamina-associated domains (LADs) (Fig. 3; van Steensel and Belmont 2017; Borsos et al. 2019). PADs require the PRC1 core components RING1A and RING1B, which catalyze H2AK119ub, but not the PRC2 protein EED and cohesin. *Ring1a*/*Ring1b* double knockout also partially derepresses the genes located in PADs in mouse oocytes, raising the possibility that PADs may play a role in gene silencing (Du et al. 2020). These data are in line with the notion that PRC1 is essential for both gene repression and Polycomb target interactions (Schuettengruber et al. 2017; Fursova et al. 2019; Blackledge et al. 2020; Boyle et al. 2020; Tamburri et al. 2020). Upon releasing oocytes from the diplotene arrest, PADs quickly disassemble (Du et al. 2020). When these cells are subsequently rearrested at metaphase II, the genome structure adopts a mitotic-like chromatin structure that lacks TADs, PADs, and compartments (Fig. 3; Du et al. 2017; Ke et al. 2017; Oomen and Dekker 2017). The mechanisms that lead to the formation of PADs in FGOs is currently unknown. It also remains unclear whether the collapse of chromatin around nucleolus in late-stage oocytes (Zuccotti et al. 1995) may contribute to the detachment of genome from the lamina and the emergence of PADs.

Conserved Relaxed Chromatin Organization after Fertilization

After fertilization, a prominent feature of chromatin of mouse zygotes is that it is extremely dispersed and lacks apparent condensed organization, as revealed by electron spectroscopic imaging (Ahmed et al. 2010). Consistently, low-input or single-cell Hi-C analyses show that chromatin adopts a highly dispersed or relaxed state in the zygotes and two-cell embryos, with loops, TADs, and compartments all substantially weakened compared with those in later-stage embryos (Fig. 3; Du et al. 2017; Flyamer et al. 2017; Gassler et al. 2017; Ke et al. 2017; Collombet et al.

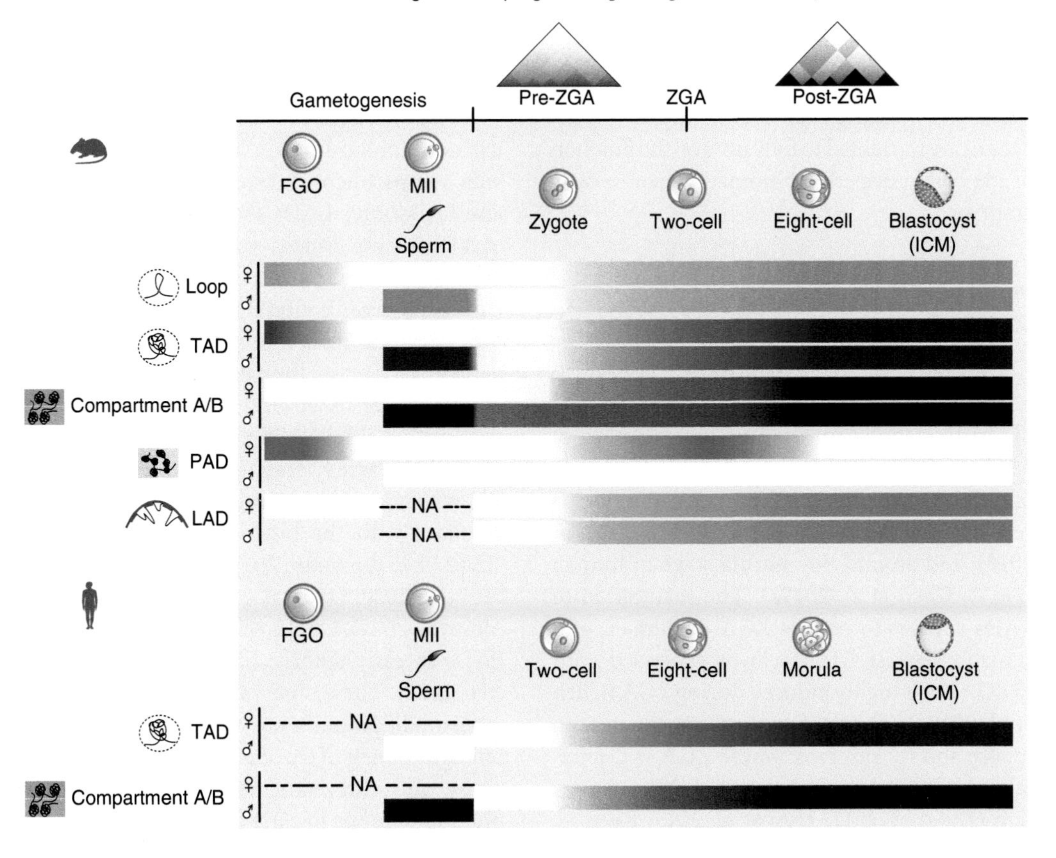

Figure 3. 3D genome dynamics in gametes and early embryos. The schematics showing the dynamics of higher-order chromatin structure from gametes to embryos in mouse and human. In mice, full-grown oocytes (FGOs) exhibit cohesin-independent, H3K27me3-marked compartmental domains (Polycomb-associating domains [PADs]). By contrast, topologically associating domains (TADs) become weak and compartments A/B are undetectable in FGOs. Lamina-associated domains (LADs) are also undetected in FGOs. In MII oocytes, chromatin adopts a mitotic-like chromatin structure, where compartments A/B, PADs, TADs, and loops are all lost. By contrast, mouse mature sperm shows canonical A/B compartmentalization, TADs, and loops. After fertilization, chromatin structure is dramatically dispersed, which shows weakened compartments, TADs, and loops. The genome structure is then gradually reestablished during preimplantation development. PADs briefly reappear specifically on the maternal allele from one-cell to eight-cell stages. LADs (gray) are de novo established after fertilization. Human embryos also exhibit diminished higher-order structure after fertilization, followed by consolidation of TADs and compartments A/B after zygotic genome activation (ZGA). Notably, human sperm has no typical TADs possibly due to the lack of CTCF protein. The strength of TADs, compartmentalization (compartments A/B and PADs/iPADs), and LADs are indicated by the shades of the bars. (NA) Data not available, (ICM) inner cell mass.

2020). Remarkably, the extensively relaxed chromatin organization after fertilization is evolutionarily conserved, as similar findings were also made in fly (Hug et al. 2017; Ogiyama et al. 2018), fish (Kaaij et al. 2018; Nakamura et al. 2021; Wike et al. 2021), pig (Li et al. 2020), and human (Chen et al. 2019a). How chromatin relaxation occurs and whether it regulates early development remain a mystery. Such chromatin relaxation is also found in two-cell mouse SCNT embryos, implying that the ooplasm has a potent ability to relax chromatin organization (Chen et al. 2020; Zhang et al. 2020). Intriguingly, mimicking chromatin relaxation by pre-depletion of cohesin in the donor cells promotes the SCNT development, and this is in part via facilitating minor ZGA (Zhang et al. 2020). Further investigations are warranted to fully unravel the function of the highly dispersed chromatin state in early embryos.

Progressive Maturation of Chromatin Structure in Preimplantation Embryos

Chromatin structure reestablishment appears to be an unusually slow and prolonged process during early embryogenesis. The consolidation of TADs and the segregation of chromatin compartments are not completed until the eight-cell stage in mouse (Du et al. 2017; Ke et al. 2017), and around the morula stage in human (Fig. 3; Chen et al. 2019a). The maturation of TADs appears to require ZGA in human, presumably because CTCF is lowly expressed prior to ZGA but is highly induced during ZGA (Chen et al. 2019a). This is, however, not the case in mouse and fly embryos, where ZGA is dispensable for TAD and compartment A/B maturation (Du et al. 2017; Hug et al. 2017; Ke et al. 2017), indicating that additional mechanisms exist to help relax chromatin architecture in early development. The maturation of chromatin organization also includes equalizing chromatin organization between the two alleles. In mouse zygotes, the chromatin compartment is clearly visible for the paternal genome, while the maternal genome is poorly segregated (Du et al. 2017; Flyamer et al. 2017; Ke et al. 2017). The mechanisms underlying such differences remain unknown, although it is clear that the maternal and paternal genomes are independently segregated (Duffié and Bourc'his 2013) using two bipolar spindles in zygotes (Reichmann et al. 2018), as also captured by Hi-C (Du et al. 2017). The differential compartment reprogramming lasts until the eight-cell stage (Fig. 3; Du et al. 2017; Ke et al. 2017).

Besides TADs and compartments, LADs also undergo drastic reprogramming during mouse early embryogenesis (Borsos et al. 2019). They emerge as early as in the mouse zygote, similar to compartments, where the maternal genome exhibits poorly defined LADs, which then gradually are reinforced at the two-cell stage (Fig. 3; Du et al. 2017; Ke et al. 2017; Borsos et al. 2019). Furthermore, LADs in the two-cell embryos are in a noncanonical state, as many LADs colocalize with compartment A instead of compartment B, and LADs/inter-LADs show smaller differences of AT contents (Borsos et al. 2019). These observations likely reflect the transitionary nature of chromatin organization at this stage. Intriguingly, the ectopic expression of *Kdm5b*, the H3K4me3 demethylase, erases the paternal LADs in one-cell embryos (Borsos et al. 2019). How H3K4me3 regulates paternal LADs remains unknown. Notably, PADs briefly reappear on the maternal allele from one-cell to eight-cell stages (Fig. 3), concomitant with the brief inheritance of oocyte H3K27me3 domains (Zheng et al. 2016; Collombet et al. 2020; Du et al. 2020; Chen et al. 2021; Mei et al. 2021). Unlike what happens in oocytes, PADs in early embryos depend on PRC2 (Du et al. 2020). These observations indicate that histone modifications may also contribute to the reprogramming of chromatin architecture in embryos, and chromatin architecture matures at multidimensions in early development.

OUTSTANDING QUESTIONS

The recent exciting progress in understanding epigenetic reprogramming during early mammalian development inevitably raises many fundamental questions, as we discuss below.

1. Why do mammalian oocytes adopt noncanonical epigenomes?

Mouse oocytes adopt noncanonical epigenomes including DNA hypomethylated nontranscribing regions, broad domains of key histone modifications, and unique 3D genome structures. Why do mammalian oocytes, in contrast to sperm and soma, display so many unique features (Fig. 4)? This is particularly intriguing given zebrafish oocytes show relatively conventional epigenomes (Jiang et al. 2013; Potok et al. 2013; Zhang et al. 2018a; Zhu et al. 2019). One possibility lies in the fact that mammals require imprinting. To establish imprints, mammali-

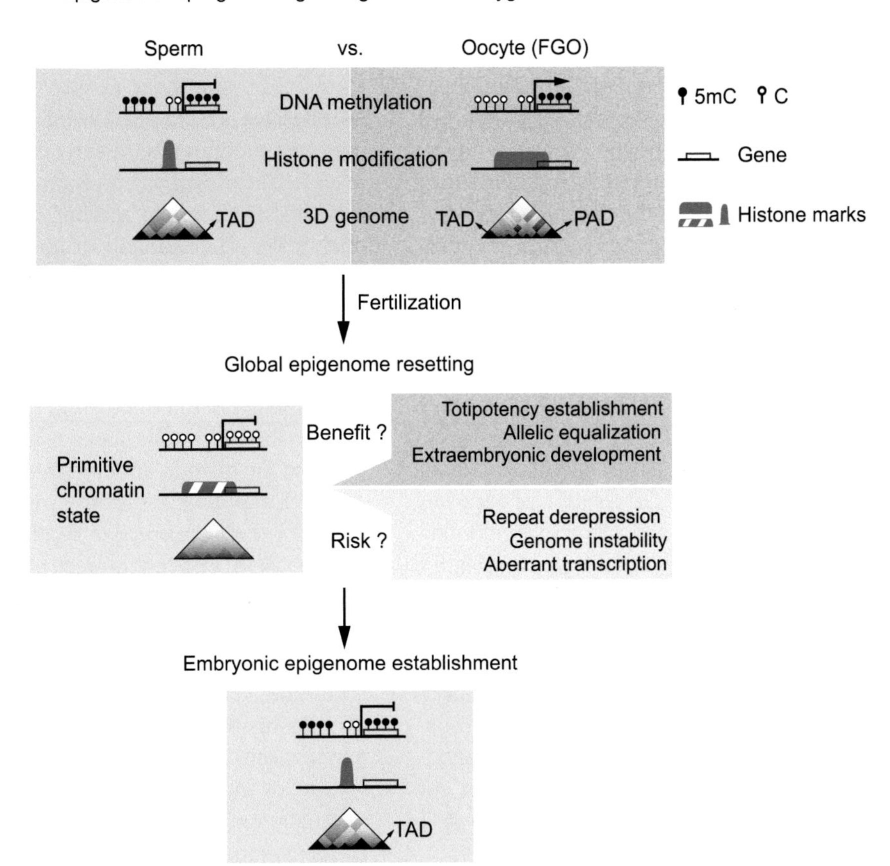

Figure 4. Epigenome reprogramming during maternal-to-zygotic transition (MZT). Mouse oocytes, but not sperm, adopt noncanonical epigenomes including DNA hypomethylated nontranscribing regions, broad domains of key histone modifications, and unique 3D genome structures. After fertilization, the embryos undergo global epigenome resetting, featured by the global loss of repressive epigenetic marks (such as DNA methylation), widespread presence of active histone marks, and global chromatin relaxation. The global resetting might be beneficial for the establishment of totipotency, allelic epigenome equalization, and extraembryonic tissue development. Meanwhile, it may also expose embryos for risks of repeat derepression, genome instability, and aberrant transcription. After zygotic genome activation (ZGA), the embryos stepwise establish embryonic epigenomes to restore DNA methylation, histone marks, and higher-order chromatin structure. (FGO) Full-grown oocyte, (TAD) topologically associating domain, (PAD) Polycomb-associating domain.

an oocytes exploit a transcription-dependent mechanism to establish a partially methylated genome, which contrasts the hypermethylated sperm genome (Stewart et al. 2016; Sendžikaite and Kelsey 2019). It is, however, unclear why oocytes use such a dramatically distinct DNA methylome despite the fact that only a tiny portion are true imprints. It warrants further investigation if the remaining differential methylomes are simply "passenger marks" brought by the transcription-dependent de novo DNA methylation. Alternatively, they may provide a pool of candidate imprints during evolution, especially considering the existence of species-specific imprints (Bartolomei and Ferguson-Smith 2011).

Likewise, the broad domains of histone marks in mouse oocytes, which contrast with those in sperm, set the stage for noncanonical H3K27me3 imprinting (Inoue et al. 2017a,b), although such imprinting seems to be limited to rodents (Xia et al. 2019). Several other factors may also contribute to such noncanonical patterning of histone marks. First, the prolonged prophase-I arrest of oocytes, which lacks DNA replication and cell division, provides ample time for the deposition of histone marks and the incorporation of histone variants (the deposition of which is DNA replication-independent). For example, H3.3 plays a critical role in oocytes as the deletion of H3.3 impairs oocyte competence, concomitant with aberrant chromatin accessibility, histone modifications, and ectopic gene activation (Lin et al. 2014; Nashun et al. 2015; Smith et al. 2020). Second, the lack of DNA methylation in nontranscribing regions may allow opportunistic deposition of opposing histone marks, as evidenced in the *Dnmt3a/3b*-double-knockout or *Setd2*-null oocytes (Hanna et al. 2018b; Xu et al. 2019). Nevertheless, how human oocytes manage to avoid these broad domains of histone marks (such as H3K4me3) (Xia et al. 2019) remains unclear. Last, the noncanonical histone marks may be related to oocyte-specific functions. For example, noncanonical H3K4me3 appears to be required for the transcription silencing of oocytes (Zhang et al. 2016; Hanna et al. 2018b).

2. Why do early mammalian embryos require genome-wide resetting rather than region-specific refining?

Embryos undergo extensive genome-wide epigenome reprogramming upon fertilization, which is rarely observed for other biological processes. A natural hypothesis is that the global resetting can efficiently convert the epigenomes from gametes to embryos. Nevertheless, neither the global chromatin relaxation nor genome-wide erasure of repressive epigenetic marks was reported in iPSC reprogramming, an equally dramatic cell-fate conversion event (Papp and Plath 2013). First, one possibility is that the remarkable difference between fertilization and iPSC conversion may reflect the distinct developmental potency for fertilized embryos ("totipotency") and iPS cells ("pluripotency"). Perhaps the emergence of totipotency requires a globally permissive, naive chromatin state. Second, the global demethylation in early embryos may contribute to the establishment of hypomethylated extraembryonic tissues (Schroeder et al. 2015; Smith et al. 2017; Zhang et al. 2018b). Last, a third possible purpose of the dramatic reprogramming in early embryos may be to equalize the distinct epigenomes of two alleles inherited from sperm and oocytes, which arise during the establishment of imprints. As supporting evidence, zebrafish, which lack imprinting, do not undergo global DNA demethylation (Macleod et al. 1999; Jiang et al. 2013; Potok et al. 2013), but instead present locus-specific DNA methylation remodeling.

Notably, zebrafish gametes show canonical patterns of histone modifications, which are, however, rapidly lost during early embryogenesis, before being restored at regulatory elements around ZGA (Vastenhouw et al. 2010; Zhang et al. 2014, 2018a; Murphy et al. 2018; Zhu et al. 2019). The differences between DNA methylation and histone modification reprogramming in zebrafish may be related to the fact that histone marks lack an efficient maintenance mechanism, compared with DNA methylation, as their post-DNA replication recovery is usually much slower than DNA methylation (Budhavarapu et al. 2013; Almouzni and Cedar 2016). Therefore, histone marks are perhaps more difficult to be maintained during the extremely rapid embryo cleavage in early development, until the cell cycle slows down around ZGA (Kane and Kimmel 1993).

3. How do embryos achieve successful MZT?

MZT is a key process in early embryos, which transfers the developmental control from maternally inherited gene products to zygotically synthesized factors (Schultz 2002; Walser and Lipshitz 2011). Mounting data begin to reveal the existence of a "primitive" chromatin state prior to ZGA featured with widespread active marks, immature heterochromatin, and relaxed chromatin structure (Fig. 4; Xia and Xie 2020). The mechanisms underlying the noncanonical setting of the epigenome, and their roles in MZT are fascinating questions and are discussed below.

 Cite this article as *Cold Spring Harb Perspect Biol* doi: 10.1101/cshperspect.a039677

First, fertilized mouse embryos are featured by compromised heterochromatin, as manifested by the lack of chromocenters and the rapid decrease or resetting of heterochromatin epigenetic marks such as H3K9me3, H3K64me3, H4K20me3, H3K27me3, and DNA methylation (Probst et al. 2007; Aguirre-Lavin et al. 2012; Burton and Torres-Padilla 2014). The loss of heterochromatin may be important for lengthening telomeres (Ko 2016) and for eliminating cell reprogramming barriers to release zygotic genome from a constrained environment to become a totipotent state (Zhou and Dean 2015; Becker et al. 2016). Nevertheless, it presumably also creates challenges and stresses to the embryos. For example, the global loss of DNA demethylation and repressive H3K9me3 can lead to derepression of repeats that threaten genome stability (Gerdes et al. 2016; Rodriguez-Terrones and Torres-Padilla 2018) and trigger immune responses (Grow et al. 2015). However, these repeats are also exploited by embryos for transcription regulation through creating new or alternative promoters, providing or disrupting TF-binding sites, regulating chromatin accessibility, and silencing nearby genes (Gerdes et al. 2016; Rodriguez-Terrones and Torres-Padilla 2018). Meanwhile, it is perhaps not surprising that embryos employ alternative mechanisms to withstand the detrimental effects from the loss of heterochromatin. For instance, in one-cell mouse embryos, the maternally provided PRC1 and corresponding H2AK119ub temporarily assume the function of H3K9me3 in the paternal genome at pericentromeric heterochromatin to inhibit pericentric satellite expression (Puschendorf et al. 2008; Tardat et al. 2015). Similarly, while H3K27me3 is erased from its canonical targets (developmental genes) in mouse preimplantation embryos (Zheng et al. 2016), H2AK119ub maintains these developmental genes nevertheless silenced (Chen et al. 2021).

Second, despite the largely silenced transcriptional state, pre-ZGA embryos show hyperpermissive chromatin featuring widespread transcription-independent active marks and open chromatin. For example, the paternal genomes of mouse one-cell embryos transiently acquire H3K4me3 domains in gene-rich regions (Zhang et al. 2016). Likewise, human pre-ZGA embryos exhibit broad domains of H3K4me3 at CG-rich promoters and distal regions (Xia et al. 2019). In zebrafish, widespread de novo H3K27ac occupies promoters throughout the genome prior to ZGA (Zhang et al. 2018a; Sato et al. 2019). The transcription-independent active histone marks likely reflect a "primitive" state of the genome and perhaps also prepares chromatin for coming ZGA. In support of this hypothesis, histone acetylation precedes and is required for transcription elongation in zebrafish early embryos (Zhang et al. 2018a; Chan et al. 2019; Sato et al. 2019). Such hyperpermissive chromatin is likely contributed by the loss of heterochromatic marks. In addition, one-cell embryos either inherit or acquire high levels of histone variants, including H3.3, H2A.X, and TH2A (Burton and Torres-Padilla 2014; Shinagawa et al. 2014; Funaya and Aoki 2017). For example, the paternal genome incorporates H3.3-packaged nucleosomes through protamine-histone exchange in one-cell mouse embryos (van der Heijden et al. 2005; Torres-Padilla et al. 2006; Santenard et al. 2010). Consequently, embryos deficient for maternal *Hira*, which is responsible for H3.3, suffer impaired male pronucleus formation, DNA replication, and rRNA transcription (Loppin et al. 2005; Inoue and Zhang 2014; Lin et al. 2014). The abundant histone variants may also contribute to the high mobility and dispersed state of zygotic chromatin (Ooga et al. 2016). Such hyperpermissive chromatin, however, needs to be curbed upon ZGA to ensure the fidelity of transcription programs. For example, exogenous promoters can fire without enhancers in the one-cell, but not two-cell, mouse embryos, indicating a progressively repressive chromatin environment (Wiekowski et al. 1991; Majumder et al. 1993). Repression of LINE-1 may also contribute to chromatin compaction beyond the two-cell stage (Jachowicz et al. 2017). In addition, transcription activity, in particular during minor ZGA, may play a critical role in this process, as transcription inhibition prevents ZGA-associated transition of open chromatin and active histone marks (Wu et al. 2016, 2018; Zhang et al. 2016; Li et al. 2018a).

Last, early embryos employ a highly conserved relaxed 3D chromatin organization featured by the loss of TADs, loops, and compartments (Zheng and Xie 2019). It remains unclear whether the long-range enhancer–promoter interactions are also lost and reset after fertilization. If so, one intriguing question is how embryos execute transcription programs without long-distance enhancer–promoter interactions and the facilitating TADs and loops when ZGA starts. It is possible that embryos may primarily use genes or enhancers that do not require long-distance interactions, which could help prevent premature firing of developmental genes and cell-fate commitment programs. Currently, the enhancer–promoter regulatory network at the beginning of life still remains largely unknown and awaits further investigations.

4. How do early embryos establish an embryonic epigenome?

The establishment of embryonic epigenome could be divided into two waves. Embryonic active epigenetic marks are often established around ZGA in mouse and human, with H3K4me3 occupying active promoters (Dahl et al. 2016; Zhang et al. 2016; Xia et al. 2019), and H3K27ac and open chromatin appearing at both active promoters and putative enhancers (Liu et al. 2016b; Lu et al. 2016; Wu et al. 2016, 2018; Gao et al. 2018a; Li et al. 2018a; Xia et al. 2019). ZGA is required for this process, as transcription inhibition blocks the redistribution of these active marks (Dahl et al. 2016; Wu et al. 2016, 2018; Zhang et al. 2016; Xia et al. 2019). These data paint a picture in which transcription facilitates chromatin state transitions, which presumably in turn potentiate transcription, forming a positive-feedback loop. By contrast, the establishment of repressive epigenetic marks (such as H3K27me3, H3K9me3, and DNA methylation) occurs later, often around implantation (Liu et al. 2016b; Zheng et al. 2016; Smith et al. 2017; Wang et al. 2018; Zhang et al. 2018b; Nicetto et al. 2019). Why does the reestablishment of repressive marks occur at a slower pace compared to active marks? One possibility is that the absence of these repressive marks at earlier stages may be a prerequisite to ensure the embryos to retain totipotency. Consistently, the removal of H3K9me3 by the depletion of H3K9 methyltransferase SETDB1 and its binding partner KAP1 promotes the transition from mESC to two-cell-like cells, during which cells partially regain totipotent potentials (Wu et al. 2020). These two waves of embryonic epigenome establishment may ensure the coordination of step-wise developmental events that include ZGA, totipotency emergence, and cell-fate determination (Fig. 4). At the gene level, diverse signaling pathways and transcription factors are often engaged in regulating the accurate embryonic epigenome establishment. For example, WNT and FGF signaling pathways are reported to modulate the global hypomethylation and CGI methylation in the extraembryonic lineages (Smith et al. 2017). Distinct transcription factors are expected to function in the lineage-specific distribution of DNA methylation and histone marks at regulatory elements based on the motif-enrichment analysis (Smith et al. 2017; Wang et al. 2018; Zhang et al. 2018b; Xiang et al. 2020). Future studies on the establishment of embryonic epigenome will undoubtedly broaden our understanding of the arise of totipotency, pluripotency, and lineage specification during embryo development.

CONCLUDING REMARKS

Tremendous advances in ultrasensitive chromatin analysis technologies have shown the epigenome reprogramming modes in mammalian gametogenesis and early development, and the mechanisms underlying these fascinating processes. Despite these breakthroughs, more questions arise as expected. First, with the emerging detailed maps of the epigenome, how and why these reprogramming events occur remain elusive. For example, what are the molecular mechanisms and the potential functions of the relaxed chromatin structure after fertilization? What are the key factors or events that trigger ZGA in mammals? Addressing these questions would bring the second major challenge in the field, which is the urgent need to develop innovative tools. For instance, CRISPR-based screening is a powerful tool to identify the key transcription regulators in cell lines (Alda-Catalinas et al. 2020). Yet, genome-wide screening is still a

 Cite this article as *Cold Spring Harb Perspect Biol* doi: 10.1101/cshperspect.a039677

daunting task for early embryos. This is further complicated by the presence of abundant maternal proteins that can be highly stable in oocytes and early embryos. Easy-to-use protein degradation methods, such as PROTAC (Burslem and Crews 2020), ARMeD (Ibrahim et al. 2020), or Trim-away (Clift et al. 2017), are needed for the loss-of-function studies to investigate factors of interests. The emerging new in vitro models, such as blastoid and gastruloid (for review, see Baillie-Benson et al. 2020), could simulate embryonic development and be potentially used to screen and identify key regulators in embryonic development. Ultrasensitive and single-cell technologies, preferentially with the ability to probe multi-omics, should be instrumental in deciphering the spatiotemporal chromatin dynamics in early embryos. For example, TFs play essential roles in governing transcription circuitry especially during lineage specification. Although improved ChIP and ChIC-based methods (uliChIP, itChIP, CUT&RUN, CUT&Tag, ChIL-seq, CoBATCH, ACT-seq, etc.) (Brind'Amour et al. 2015; Skene and Henikoff 2017; Ai et al. 2019; Carter et al. 2019; Harada et al. 2019; Kaya-Okur et al. 2019; Wang et al. 2019a) have been invented, these methods rely on highly sensitive antibodies that are not always accessible. As another example, enhancer–promoter interactions are still difficult to study in early development because of the sparse Hi-C data from limited research materials. Improved chromatin conformation capture methods to enrich the regulatory chromatin interactions at high resolution is urgently needed. The Oligopaint-based DNA FISH (fluorescence in situ hybridization) and CRISPR-based live-cell imaging methods that can tag numerous genomic loci simultaneously (Kempfer and Pombo 2020) may facilitate decoding how dynamic chromatin conformation contributes to gene expression during embryogenesis. Single-cell, multi-omics methods would be essential to delineate the heterogeneity of cells during early development. Thirdly, epigenome reprogramming studies across species have revealed remarkable differences between mice and human, and even between monkey and human (Hanna et al. 2018a; Chen et al. 2019a; Wang et al. 2019b; Xia et al. 2019). Applying the research framework to other species will shed light on the evolution of epigenetic reprogramming and facilitate studies for key events that are found in human but not in mouse.

ACKNOWLEDGMENTS

We thank the laboratory members for the comments and help during the preparation of the manuscript. The review included only selected studies and we apologize to all the authors whose work was not cited due to space limitations. This work was supported by the National Key R&D Program of China (Grant No. 2019YF A0508901 to W.X.), the National Natural Science Foundation of China (Grant Nos. 31988101, 31830047, and 31725018 to W.X.), the THU-PKU Center for Life Sciences (W.X.), the Beijing Municipal Commission of Science and Technology (Grant No. Z18110000131 8006 to W.X.). W.X. is an HHMI International Research Scholar. Z.D. is supported by THU-PKU Center for Life Sciences postdoctoral fellowships.

REFERENCES

Abe K, Yamamoto R, Franke V, Cao MJ, Suzuki Y, Suzuki MG, Vlahovicek K, Svoboda P, Schultz RM, Aoki F. 2015. The first murine zygotic transcription is promiscuous and uncoupled from splicing and 3′ processing. *EMBO J* **34:** 1523–1537. doi:10.15252/embj.201490648

Aguirre-Lavin T, Adenot P, Bonnet-Garnier A, Lehmann G, Fleurot R, Boulesteix C, Debey P, Beaujean N. 2012. 3D-FISH analysis of embryonic nuclei in mouse highlights several abrupt changes of nuclear organization during preimplantation development. *BMC Dev Biol* **12:** 30. doi:10.1186/1471-213X-12-30

Ahmed K, Dehghani H, Rugg-Gunn P, Fussner E, Rossant J, Bazett-Jones DP. 2010. Global chromatin architecture reflects pluripotency and lineage commitment in the early mouse embryo. *PLoS ONE* **5:** e10531. doi:10.1371/journal.pone.0010531

Ai S, Xiong H, Li CC, Luo Y, Shi Q, Liu Y, Yu X, Li C, He A. 2019. Profiling chromatin states using single-cell itChIP-seq. *Nat Cell Biol* **21:** 1164–1172. doi:10.1038/s41556-019-0383-5

Alavattam KG, Maezawa S, Sakashita A, Khoury H, Barski A, Kaplan N, Namekawa SH. 2019. Attenuated chromatin compartmentalization in meiosis and its maturation in sperm development. *Nat Struct Mol Biol* **26:** 175–184. doi:10.1038/s41594-019-0189-y

Alda-Catalinas C, Bredikhin D, Hernando-Herraez I, Santos F, Kubinyecz O, Eckersley-Maslin MA, Stegle O, Reik W. 2020. A single-cell transcriptomics CRISPR-activation screen identifies epigenetic regulators of the zygotic genome activation program. *Cell Syst* **11:** 25–41.e9. doi:10.1016/j.cels.2020.06.004

Allshire RC, Madhani HD. 2018. Ten principles of heterochromatin formation and function. *Nat Rev Mol Cell Biol* **19:** 229–244. doi:10.1038/nrm.2017.119

Almouzni G, Cedar H. 2016. Maintenance of epigenetic information. *Cold Spring Harb Perspect Biol* **8:** a019372. doi:10.1101/cshperspect.a019372

Amouroux R, Nashun B, Shirane K, Nakagawa S, Hill PW, D'Souza Z, Nakayama M, Matsuda M, Turp A, Ndjetehe E, et al. 2016. De novo DNA methylation drives 5hmC accumulation in mouse zygotes. *Nat Cell Biol* **18:** 225–233. doi:10.1038/ncb3296

Andreu-Vieyra CV, Chen R, Agno JE, Glaser S, Anastassiadis K, Stewart AF, Matzuk MM. 2010. MLL2 is required in oocytes for bulk histone 3 lysine 4 trimethylation and transcriptional silencing. *Plos Biol* **8:** e1000453. doi:10.1371/journal.pbio.1000453

Aoki F, Worrad DM, Schultz RM. 1997. Regulation of transcriptional activity during the first and second cell cycles in the preimplantation mouse embryo. *Dev Biol* **181:** 296–307. doi:10.1006/dbio.1996.8466

Azuara V, Perry P, Sauer S, Spivakov M, Jørgensen HF, John RM, Gouti M, Casanova M, Warnes G, Merkenschlager M, et al. 2006. Chromatin signatures of pluripotent cell lines. *Nat Cell Biol* **8:** 532–538. doi:10.1038/ncb1403

Bachvarova R. 1985. Gene expression during oogenesis and oocyte development in mammals. *Dev Biol (NY 1985)* **1:** 453–524.

Baillie-Benson P, Moris N, Martinez Arias A. 2020. Pluripotent stem cell models of early mammalian development. *Curr Opin Cell Biol* **66:** 89–96. doi:10.1016/j.ceb.2020.05.010

Barau J, Teissandier A, Zamudio N, Roy S, Nalesso V, Hérault Y, Guillou F, Bourc'his D. 2016. The DNA methyltransferase DNMT3C protects male germ cells from transposon activity. *Science* **354:** 909–912. doi:10.1126/science.aah5143

Bartolomei MS, Ferguson-Smith AC. 2011. Mammalian genomic imprinting. *Cold Spring Harb Perspect Biol* **3:** a002592. doi:10.1101/cshperspect.a002592

Battulin N, Fishman VS, Mazur AM, Pomaznoy M, Khabarova AA, Afonnikov DA, Prokhortchouk EB, Serov OL. 2015. Comparison of the three-dimensional organization of sperm and fibroblast genomes using the Hi-C approach. *Genome Biol* **16:** 77. doi:10.1186/s13059-015-0642-0

Baubec T, Colombo DF, Wirbelauer C, Schmidt J, Burger L, Krebs AR, Akalin A, Schübeler D. 2015. Genomic profiling of DNA methyltransferases reveals a role for DNMT3B in genic methylation. *Nature* **520:** 243–247. doi:10.1038/nature14176

Becker JS, Nicetto D, Zaret KS. 2016. H3K9me3-dependent heterochromatin: barrier to cell fate changes. *Trends Genet* **32:** 29–41. doi:10.1016/j.tig.2015.11.001

Bellier S, Chastant S, Adenot P, Vincent M, Renard JP, Bensaude O. 1997. Nuclear translocation and carboxyl-terminal domain phosphorylation of RNA polymerase II delineate the two phases of zygotic gene activation in mammalian embryos. *EMBO J* **16:** 6250–6262. doi:10.1093/emboj/16.20.6250

Beraldi R, Pittoggi C, Sciamanna I, Mattei E, Spadafora C. 2006. Expression of LINE-1 retroposons is essential for murine preimplantation development. *Mol Reprod Dev* **73:** 279–287. doi:10.1002/mrd.20423

Bernstein BE, Mikkelsen TS, Xie X, Kamal M, Huebert DJ, Cuff J, Fry B, Meissner A, Wernig M, Plath K, et al. 2006. A bivalent chromatin structure marks key developmental genes in embryonic stem cells. *Cell* **125:** 315–326. doi:10.1016/j.cell.2006.02.041

Bickmore WA. 2013. The spatial organization of the human genome. *Annu Rev Genomics Hum Genet* **14:** 67–84. doi:10.1146/annurev-genom-091212-153515

Bird A. 2002. DNA methylation patterns and epigenetic memory. *Genes Dev* **16:** 6–21. doi:10.1101/gad.947102

Blackledge NP, Fursova NA, Kelley JR, Huseyin MK, Feldmann A, Klose RJ. 2020. PRC1 catalytic activity is central to polycomb system function. *Mol Cell* **77:** 857–874.e9. doi:10.1016/j.molcel.2019.12.001

Bonev B, Cavalli G. 2016. Organization and function of the 3D genome. *Nat Rev Genet* **17:** 661–678. doi:10.1038/nrg.2016.112

Borsos M, Perricone SM, Schauer T, Pontabry J, de Luca KL, de Vries SS, Ruiz-Morales ER, Torres-Padilla ME, Kind J. 2019. Genome–lamina interactions are established de novo in the early mouse embryo. *Nature* **569:** 729–733. doi:10.1038/s41586-019-1233-0

Bouniol C, Nguyen E, Debey P. 1995. Endogenous transcription occurs at the 1-cell stage in the mouse embryo. *Exp Cell Res* **218:** 57–62. doi:10.1006/excr.1995.1130

Bourc'his D, Bestor TH. 2004. Meiotic catastrophe and retrotransposon reactivation in male germ cells lacking Dnmt3L. *Nature* **431:** 96–99. doi:10.1038/nature02886

Bourque G, Burns KH, Gehring M, Gorbunova V, Seluanov A, Hammell M, Imbeault M, Izsvák Z, Levin HL, Macfarlan TS, et al. 2018. Ten things you should know about transposable elements. *Genome Biol* **19:** 199. doi:10.1186/s13059-018-1577-z

Boyle S, Flyamer IM, Williamson I, Sengupta D, Bickmore WA, Illingworth RS. 2020. A central role for canonical PRC1 in shaping the 3D nuclear landscape. *Genes Dev* **34:** 931–949. doi:10.1101/gad.336487.120

Braude P, Bolton V, Moore S. 1988. Human gene expression first occurs between the four- and eight-cell stages of preimplantation development. *Nature* **332:** 459–461. doi:10.1038/332459a0

Brind'Amour J, Liu S, Hudson M, Chen C, Karimi MM, Lorincz MC. 2015. An ultra-low-input native ChIP-seq protocol for genome-wide profiling of rare cell populations. *Nat Commun* **6:** 6033. doi:10.1038/ncomms7033

Brind'Amour J, Kobayashi H, Richard Albert J, Shirane K, Sakashita A, Kamio A, Bogutz A, Koike T, Karimi MM, Lefebvre L, et al. 2018. LTR retrotransposons transcribed in oocytes drive species-specific and heritable changes in DNA methylation. *Nat Commun* **9:** 3331. doi:10.1038/s41467-018-05841-x

Brookes E, de Santiago I, Hebenstreit D, Morris KJ, Carroll T, Xie SQ, Stock JK, Heidemann M, Eick D, Nozaki N, et al. 2012. Polycomb associates genome-wide with a specific

RNA polymerase II variant, and regulates metabolic genes in ESCs. *Cell Stem Cell* **10:** 157–170. doi:10.1016/j.stem.2011.12.017

Brykczynska U, Hisano M, Erkek S, Ramos L, Oakeley EJ, Roloff TC, Beisel C, Schübeler D, Stadler MB, Peters AH. 2010. Repressive and active histone methylation mark distinct promoters in human and mouse spermatozoa. *Nat Struct Mol Biol* **17:** 679–687. doi:10.1038/nsmb.1821

Budhavarapu VN, Chavez M, Tyler JK. 2013. How is epigenetic information maintained through DNA replication? *Epigenetics Chromatin* **6:** 32. doi:10.1186/1756-8935-6-32

Burslem GM, Crews CM. 2020. Proteolysis-targeting chimeras as therapeutics and tools for biological discovery. *Cell* **181:** 102–114. doi:10.1016/j.cell.2019.11.031

Burton A, Torres-Padilla ME. 2014. Chromatin dynamics in the regulation of cell fate allocation during early embryogenesis. *Nat Rev Mol Cell Biol* **15:** 723–735. doi:10.1038/nrm3885

Burton A, Brochard V, Galan C, Ruiz-Morales ER, Rovira Q, Rodriguez-Terrones D, Kruse K, Le Gras S, Udayakumar VS, Chin HG, et al. 2020. Heterochromatin establishment during early mammalian development is regulated by pericentromeric RNA and characterized by non-repressive H3K9me3. *Nat Cell Biol* **22:** 767–778. doi:10.1038/s41556-020-0536-6

Carter B, Ku WL, Kang JY, Hu G, Perrie J, Tang Q, Zhao K. 2019. Mapping histone modifications in low cell number and single cells using antibody-guided chromatin tagmentation (ACT-seq). *Nat Commun* **10:** 3747. doi:10.1038/s41467-019-11559-1

Chan SH, Tang Y, Miao L, Darwich-Codore H, Vejnar CE, Beaudoin JD, Musaev D, Fernandez JP, Benitez MDJ, Bazzini AA, et al. 2019. Brd4 and P300 confer transcriptional competency during zygotic genome activation. *Dev Cell* **49:** 867–881.e8. doi:10.1016/j.devcel.2019.05.037

Chazaud C, Yamanaka Y. 2016. Lineage specification in the mouse preimplantation embryo. *Development* **143:** 1063–1074. doi:10.1242/dev.128314

Chen Z, Zhang Y. 2019. Loss of DUX causes minor defects in zygotic genome activation and is compatible with mouse development. *Nat Genet* **51:** 947–951. doi:10.1038/s41588-019-0418-7

Chen Z, Zhang Y. 2020. Maternal H3K27me3-dependent autosomal and X chromosome imprinting. *Nat Rev Genet* **21:** 555–571. doi:10.1038/s41576-020-0245-9

Chen Y, Zhang Y, Wang Y, Zhang L, Brinkman EK, Adam SA, Goldman R, van Steensel B, Ma J, Belmont AS. 2018. Mapping 3D genome organization relative to nuclear compartments using TSA-Seq as a cytological ruler. *J Cell Biol* **217:** 4025–4048. doi:10.1083/jcb.201807108

Chen X, Ke Y, Wu K, Zhao H, Sun Y, Gao L, Liu Z, Zhang J, Tao W, Hou Z, et al. 2019a. Key role for CTCF in establishing chromatin structure in human embryos. *Nature* **576:** 306–310. doi:10.1038/s41586-019-1812-0

Chen Z, Yin Q, Inoue A, Zhang C, Zhang Y. 2019b. Allelic H3K27me3 to allelic DNA methylation switch maintains noncanonical imprinting in extraembryonic cells. *Sci Adv* **5:** eaay7246. doi:10.1126/sciadv.aay7246

Chen M, Zhu Q, Li C, Kou X, Zhao Y, Li Y, Xu R, Yang L, Yang L, Gu L, et al. 2020. Chromatin architecture reorganization in murine somatic cell nuclear transfer embryos. *Nat Commun* **11:** 1813. doi:10.1038/s41467-020-15607-z

Chen Z, Djekidel MN, Zhang Y. 2021. Distinct dynamics and functions of H2AK119ub1 and H3K27me3 in mouse preimplantation embryos. *Nat Genet* **53:** 551–563. doi:10.1038/s41588-021-00821-2

Chotalia M, Smallwood SA, Ruf N, Dawson C, Lucifero D, Frontera M, James K, Dean W, Kelsey G. 2009. Transcription is required for establishment of germline methylation marks at imprinted genes. *Genes Dev* **23:** 105–117. doi:10.1101/gad.495809

Chung YG, Matoba S, Liu Y, Eum JH, Lu F, Jiang W, Lee JE, Sepilian V, Cha KY, Lee DR, et al. 2015. Histone demethylase expression enhances human somatic cell nuclear transfer efficiency and promotes derivation of pluripotent stem cells. *Cell Stem Cell* **17:** 758–766. doi:10.1016/j.stem.2015.10.001

Chuong EB, Elde NC, Feschotte C. 2017. Regulatory activities of transposable elements: from conflicts to benefits. *Nat Rev Genet* **18:** 71–86. doi:10.1038/nrg.2016.139

Clift D, McEwan WA, Labzin LI, Konieczny V, Mogessie B, James LC, Schuh M. 2017. A method for the acute and rapid degradation of endogenous proteins. *Cell* **171:** 1692–1706.e18. doi:10.1016/j.cell.2017.10.033

Clouaire T, Webb S, Skene P, Illingworth R, Kerr A, Andrews R, Lee JH, Skalnik D, Bird A. 2012. Cfp1 integrates both CpG content and gene activity for accurate H3K4me3 deposition in embryonic stem cells. *Genes Dev* **26:** 1714–1728. doi:10.1101/gad.194209.112

Collombet S, Ranisavljevic N, Nagano T, Varnai C, Shisode T, Leung W, Piolot T, Galupa R, Borensztein M, Servant N, et al. 2020. Parental-to-embryo switch of chromosome organization in early embryogenesis. *Nature* **580:** 142–146. doi:10.1038/s41586-020-2125-z

Dahl JA, Jung I, Aanes H, Greggains GD, Manaf A, Lerdrup M, Li G, Kuan S, Li B, Lee AY, et al. 2016. Broad histone H3K4me3 domains in mouse oocytes modulate maternal-to-zygotic transition. *Nature* **537:** 548–552. doi:10.1038/nature19360

Dan J, Rousseau P, Hardikar S, Veland N, Wong J, Autexier C, Chen T. 2017. Zscan4 inhibits maintenance DNA methylation to facilitate telomere elongation in mouse embryonic stem cells. *Cell Rep* **20:** 1936–1949. doi:10.1016/j.celrep.2017.07.070

Davidson IF, Bauer B, Goetz D, Tang W, Wutz G, Peters JM. 2019. DNA loop extrusion by human cohesin. *Science* **366:** 1338–1345. doi:10.1126/science.aaz3418

De Iaco A, Planet E, Coluccio A, Verp S, Duc J, Trono D. 2017. DUX-family transcription factors regulate zygotic genome activation in placental mammals. *Nat Genet* **49:** 941–945. doi:10.1038/ng.3858

De Iaco A, Coudray A, Duc J, Trono D. 2019. DPPA2 and DPPA4 are necessary to establish a 2C-like state in mouse embryonic stem cells. *EMBO Rep* **20:** e47382. doi:10.15252/embr.201847382

De Iaco A, Verp S, Offner S, Grun D, Trono D. 2020. DUX is a non-essential synchronizer of zygotic genome activation. *Development* **147:** dev177725.

de Koning AP, Gu W, Castoe TA, Batzer MA, Pollock DD. 2011. Repetitive elements may comprise over two-thirds of the human genome. *Plos Genet* **7:** e1002384. doi:10.1371/journal.pgen.1002384

Denissov S, Hofemeister H, Marks H, Kranz A, Ciotta G, Singh S, Anastassiadis K, Stunnenberg HG, Stewart AF.

2014. Mll2 is required for H3K4 trimethylation on bivalent promoters in embryonic stem cells, whereas Mll1 is redundant. *Development* **141:** 526–537. doi:10.1242/dev.102681

Dhayalan A, Rajavelu A, Rathert P, Tamas R, Jurkowska RZ, Ragozin S, Jeltsch A. 2010. The Dnmt3a PWWP domain reads histone 3 lysine 36 trimethylation and guides DNA methylation. *J Biol Chem* **285:** 26114–26120. doi:10.1074/jbc.M109.089433

Di Croce L, Helin K. 2013. Transcriptional regulation by polycomb group proteins. *Nat Struct Mol Biol* **20:** 1147–1155. doi:10.1038/nsmb.2669

Dobson AT, Raja R, Abeyta MJ, Taylor T, Shen S, Haqq C, Pera RA. 2004. The unique transcriptome through day 3 of human preimplantation development. *Hum Mol Genet* **13:** 1461–1470. doi:10.1093/hmg/ddh157

Dodge JE, Kang YK, Beppu H, Lei H, Li E. 2004. Histone H3-K9 methyltransferase ESET is essential for early development. *Mol Cell Biol* **24:** 2478–2486. doi:10.1128/MCB.24.6.2478-2486.2004

Douillet D, Sze CC, Ryan C, Piunti A, Shah AP, Ugarenko M, Marshall SA, Rendleman EJ, Zha DD, Helmin KA, et al. 2020. Uncoupling histone H3K4 trimethylation from developmental gene expression via an equilibrium of COMPASS, Polycomb and DNA methylation. *Nat Genet* **52:** 615–625. doi:10.1038/s41588-020-0618-1

Du Z, Zheng H, Huang B, Ma R, Wu J, Zhang X, He J, Xiang Y, Wang Q, Li Y, et al. 2017. Allelic reprogramming of 3D chromatin architecture during early mammalian development. *Nature* **547:** 232–235. doi:10.1038/nature23263

Du Z, Zheng H, Kawamura YK, Zhang K, Gassler J, Powell S, Xu Q, Lin Z, Xu K, Zhou Q, et al. 2020. Polycomb group proteins regulate chromatin architecture in mouse oocytes and early embryos. *Mol Cell* **77:** 825–839.e7. doi:10.1016/j.molcel.2019.11.011

Duffié R, Bourc'his D. 2013. Parental epigenetic asymmetry in mammals. *Curr Top Dev Biol* **104:** 293–328. doi:10.1016/B978-0-12-416027-9.00009-7

Dukatz M, Holzer K, Choudalakis M, Emperle M, Lungu C, Bashtrykov P, Jeltsch A. 2019. H3k36me2/3 binding and DNA binding of the DNA methyltransferase DNMT3A PWWP domain both contribute to its chromatin interaction. *J Mol Biol* **431:** 5063–5074. doi:10.1016/j.jmb.2019.09.006

Eckersley-Maslin MA, Alda-Catalinas C, Reik W. 2018. Dynamics of the epigenetic landscape during the maternal-to-zygotic transition. *Nat Rev Mol Cell Bio* **19:** 436–450. doi:10.1038/s41580-018-0008-z

Eckersley-Maslin M, Alda-Catalinas C, Blotenburg M, Kreibich E, Krueger C, Reik W. 2019. Dppa2 and Dppa4 directly regulate the Dux-driven zygotic transcriptional program. *Genes Dev* **33:** 194–208. doi:10.1101/gad.321174.118

Eckersley-Maslin MA, Parry A, Blotenburg M, Krueger C, Ito Y, Franklin VNR, Narita M, D'Santos CS, Reik W. 2020. Epigenetic priming by Dppa2 and 4 in pluripotency facilitates multi-lineage commitment. *Nat Struct Mol Biol* **27:** 696–705. doi:10.1038/s41594-020-0443-3

Erkek S, Hisano M, Liang CY, Gill M, Murr R, Dieker J, Schübeler D, van der Vlag J, Stadler MB, Peters AH. 2013. Molecular determinants of nucleosome retention at CpG-rich sequences in mouse spermatozoa. *Nat Struct Mol Biol* **20:** 868–875. doi:10.1038/nsmb.2599

Eymery A, Liu ZC, Ozonov EA, Stadler MB, Peters AHFM. 2016. The methyltransferase Setdb1 is essential for meiosis and mitosis in mouse oocytes and early embryos. *Development* **143:** 2767–2779.

Fadloun A, Le Gras S, Jost B, Ziegler-Birling C, Takahashi H, Gorab E, Carninci P, Torres-Padilla ME. 2013. Chromatin signatures and retrotransposon profiling in mouse embryos reveal regulation of LINE-1 by RNA. *Nat Struct Mol Biol* **20:** 332–338. doi:10.1038/nsmb.2495

Falco G, Lee SL, Stanghellini I, Bassey UC, Hamatani T, Ko MS. 2007. Zscan4: a novel gene expressed exclusively in late 2-cell embryos and embryonic stem cells. *Dev Biol* **307:** 539–550. doi:10.1016/j.ydbio.2007.05.003

Flyamer IM, Gassler J, Imakaev M, Brandão HB, Ulianov SV, Abdennur N, Razin SV, Mirny LA, Tachibana-Konwalski K. 2017. Single-nucleus Hi-C reveals unique chromatin reorganization at oocyte-to-zygote transition. *Nature* **544:** 110–114. doi:10.1038/nature21711

Fudenberg G, Imakaev M, Lu C, Goloborodko A, Abdennur N, Mirny LA. 2016. Formation of chromosomal domains by loop extrusion. *Cell Rep* **15:** 2038–2049. doi:10.1016/j.celrep.2016.04.085

Funaya S, Aoki F. 2017. Regulation of zygotic gene activation by chromatin structure and epigenetic factors. *J Reprod Dev* **63:** 359–363. doi:10.1262/jrd.2017-058

Fursova NA, Blackledge NP, Nakayama M, Ito S, Koseki Y, Farcas AM, King HW, Koseki H, Klose RJ. 2019. Synergy between variant PRC1 complexes defines polycomb-mediated gene repression. *Mol Cell* **74:** 1020–1036.e8. doi:10.1016/j.molcel.2019.03.024

Gao L, Wu K, Liu Z, Yao X, Yuan S, Tao W, Yi L, Yu G, Hou Z, Fan D, et al. 2018a. Chromatin accessibility landscape in human early embryos and its association with evolution. *Cell* **173:** 248–259.e15. doi:10.1016/j.cell.2018.02.028

Gao R, Wang C, Gao Y, Xiu W, Chen J, Kou X, Zhao Y, Liao Y, Bai D, Qiao Z, et al. 2018b. Inhibition of aberrant DNA re-methylation improves post-implantation development of somatic cell nuclear transfer embryos. *Cell Stem Cell* **23:** 426–435.e5. doi:10.1016/j.stem.2018.07.017

Gassler J, Brandão HB, Imakaev M, Flyamer IM, Ladstätter S, Bickmore WA, Peters JM, Mirny LA, Tachibana K. 2017. A mechanism of cohesin-dependent loop extrusion organizes zygotic genome architecture. *EMBO J* **36:** 3600–3618. doi:10.15252/embj.201798083

Genet M, Torres-Padilla ME. 2020. The molecular and cellular features of 2-cell-like cells: a reference guide. *Development* **147:** dev189688. doi:10.1242/dev.189688

Gerdes P, Richardson SR, Mager DL, Faulkner GJ. 2016. Transposable elements in the mammalian embryo: pioneers surviving through stealth and service. *Genome Biol* **17:** 100. doi:10.1186/s13059-016-0965-5

Gibcus JH, Dekker J. 2013. The hierarchy of the 3D genome. *Mol Cell* **49:** 773–782. doi:10.1016/j.molcel.2013.02.011

Glaser S, Schaft J, Lubitz S, Vintersten K, van der Hoeven F, Tufteland KR, Aasland R, Anastassiadis K, Ang SL, Stewart AF. 2006. Multiple epigenetic maintenance factors implicated by the loss of Mll2 in mouse development. *Development* **133:** 1423–1432. doi:10.1242/dev.02302

Gold HB, Jung YH, Corces VG. 2018. Not just heads and tails: the complexity of the sperm epigenome. *J Biol Chem* **293:** 13815–13820. doi:10.1074/jbc.R117.001561

Gonzalo S, Jaco I, Fraga MF, Chen T, Li E, Esteller M, Blasco MA. 2006. DNA methyltransferases control telomere length and telomere recombination in mammalian cells. *Nat Cell Biol* **8:** 416–424. doi:10.1038/ncb1386

Gorkin DU, Leung D, Ren B. 2014. The 3D genome in transcriptional regulation and pluripotency. *Cell Stem Cell* **14:** 762–775. doi:10.1016/j.stem.2014.05.017

Greenberg MVC, Bourc'his D. 2019. The diverse roles of DNA methylation in mammalian development and disease. *Nat Rev Mol Cell Biol* **20:** 590–607. doi:10.1038/s41580-019-0159-6

Gretarsson KH, Hackett JA. 2020. Dppa2 and Dppa4 counteract de novo methylation to establish a permissive epigenome for development. *Nat Struct Mol Biol* **27:** 706–716. doi:10.1038/s41594-020-0445-1

Grow EJ, Flynn RA, Chavez SL, Bayless NL, Wossidlo M, Wesche DJ, Martin L, Ware CB, Blish CA, Chang HY, et al. 2015. Intrinsic retroviral reactivation in human preimplantation embryos and pluripotent cells. *Nature* **522:** 221–225. doi:10.1038/nature14308

Guibert S, Forne T, Weber M. 2012. Global profiling of DNA methylation erasure in mouse primordial germ cells. *Genome Res* **22:** 633–641. doi:10.1101/gr.130997.111

Guo F, Li X, Liang D, Li T, Zhu P, Guo H, Wu X, Wen L, Gu TP, Hu B, et al. 2014. Active and passive demethylation of male and female pronuclear DNA in the mammalian zygote. *Cell Stem Cell* **15:** 447–459. doi:10.1016/j.stem.2014.08.003

Guo M, Zhang Y, Zhou J, Bi Y, Xu J, Xu C, Kou X, Zhao Y, Li Y, Tu Z, et al. 2019. Precise temporal regulation of Dux is important for embryo development. *Cell Res* **29:** 956–959. doi:10.1038/s41422-019-0238-4

Hamatani T, Carter MG, Sharov AA, Ko MS. 2004. Dynamics of global gene expression changes during mouse preimplantation development. *Dev Cell* **6:** 117–131. doi:10.1016/S1534-5807(03)00373-3

Hammoud SS, Nix DA, Zhang H, Purwar J, Carrell DT, Cairns BR. 2009. Distinctive chromatin in human sperm packages genes for embryo development. *Nature* **460:** 473–478. doi:10.1038/nature08162

Hammoud SS, Low DH, Yi C, Carrell DT, Guccione E, Cairns BR. 2014. Chromatin and transcription transitions of mammalian adult germline stem cells and spermatogenesis. *Cell Stem Cell* **15:** 239–253. doi:10.1016/j.stem.2014.04.006

Handel MA, Schimenti JC. 2010. Genetics of mammalian meiosis: regulation, dynamics and impact on fertility. *Nat Rev Genet* **11:** 124–136. doi:10.1038/nrg2723

Hanna CW, Demond H, Kelsey G. 2018a. Epigenetic regulation in development: is the mouse a good model for the human? *Hum Reprod Update* **24:** 556–576. doi:10.1093/humupd/dmy021

Hanna CW, Taudt A, Huang J, Gahurova L, Kranz A, Andrews S, Dean W, Stewart AF, Colomé-Tatché M, Kelsey G. 2018b. MLL2 conveys transcription-independent H3K4 trimethylation in oocytes. *Nat Struct Mol Biol* **25:** 73–82. doi:10.1038/s41594-017-0013-5

Hanna CW, Pérez-Palacios R, Gahurova L, Schubert M, Krueger F, Biggins L, Andrews S, Colomé-Tatché M, Bourc'his D, Dean W, et al. 2019. Endogenous retroviral insertions drive non-canonical imprinting in extra-embryonic tissues. *Genome Biol* **20:** 225. doi:10.1186/s13059-019-1833-x

Harada A, Maehara K, Handa T, Arimura Y, Nogami J, Hayashi-Takanaka Y, Shirahige K, Kurumizaka H, Kimura H, Ohkawa Y. 2019. A chromatin integration labelling method enables epigenomic profiling with lower input. *Nat Cell Biol* **21:** 287–296. doi:10.1038/s41556-018-0248-3

Harris C, Cloutier M, Trotter M, Hinten M, Gayen S, Du Z, Xie W, Kalantry S. 2019. Conversion of random X-inactivation to imprinted X-inactivation by maternal PRC2. *eLife* **8:** e44258. doi:10.7554/eLife.44258

Hatanaka Y, Inoue K, Oikawa M, Kamimura S, Ogonuki N, Kodama EN, Ohkawa Y, Tsukada Y, Ogura A. 2015. Histone chaperone CAF-1 mediates repressive histone modifications to protect preimplantation mouse embryos from endogenous retrotransposons. *Proc Natl Acad Sci* **112:** 14641–14646. doi:10.1073/pnas.1512775112

Hendrickson PG, Doráis JA, Grow EJ, Whiddon JL, Lim JW, Wike CL, Weaver BD, Pflueger C, Emery BR, Wilcox AL, et al. 2017. Conserved roles of mouse DUX and human DUX4 in activating cleavage-stage genes and MERVL/HERVL retrotransposons. *Nat Genet* **49:** 925–934. doi:10.1038/ng.3844

Hu D, Garruss AS, Gao X, Morgan MA, Cook M, Smith ER, Shilatifard A. 2013. The Mll2 branch of the COMPASS family regulates bivalent promoters in mouse embryonic stem cells. *Nat Struct Mol Biol* **20:** 1093–1097. doi:10.1038/nsmb.2653

Hu D, Gao X, Cao K, Morgan MA, Mas G, Smith ER, Volk AG, Bartom ET, Crispino JD, Di Croce L, et al. 2017. Not all H3K4 methylations are created equal: Mll2/COMPASS dependency in primordial germ cell specification. *Mol Cell* **65:** 460–475.e6. doi:10.1016/j.molcel.2017.01.013

Hu Z, Tan DEK, Chia G, Tan H, Leong HF, Chen BJ, Lau MS, Tan KYS, Bi X, Yang D, et al. 2020. Maternal factor NELFA drives a 2C-like state in mouse embryonic stem cells. *Nat Cell Biol* **22:** 175–186. doi:10.1038/s41556-019-0453-8

Huang Y, Kim JK, Do DV, Lee C, Penfold CA, Zylicz JJ, Marioni JC, Hackett JA, Surani MA. 2017. Stella modulates transcriptional and endogenous retrovirus programs during maternal-to-zygotic transition. *eLife* **6:** e22345. doi:10.7554/eLife.22345

Hug CB, Grimaldi AG, Kruse K, Vaquerizas JM. 2017. Chromatin architecture emerges during zygotic genome activation independent of transcription. *Cell* **169:** 216–228.e19. doi:10.1016/j.cell.2017.03.024

Ibrahim AFM, Shen L, Tatham MH, Dickerson D, Prescott AR, Abidi N, Xirodimas DP, Hay RT. 2020. Antibody RING-mediated destruction of endogenous proteins. *Mol Cell* **79:** 155–166.e9. doi:10.1016/j.molcel.2020.04.032

Inoue A, Zhang Y. 2014. Nucleosome assembly is required for nuclear pore complex assembly in mouse zygotes. *Nat Struct Mol Biol* **21:** 609–616. doi:10.1038/nsmb.2839

Inoue A, Jiang L, Lu F, Suzuki T, Zhang Y. 2017a. Maternal H3K27me3 controls DNA methylation-independent imprinting. *Nature* **547:** 419–424. doi:10.1038/nature23262

Inoue A, Jiang L, Lu F, Zhang Y. 2017b. Genomic imprinting of *Xist* by maternal H3K27me3. *Genes Dev* **31:** 1927–1932. doi:10.1101/gad.304113.117

Inoue A, Chen Z, Yin Q, Zhang Y. 2018. Maternal *Eed* knockout causes loss of H3K27me3 imprinting and random X inactivation in the extraembryonic cells. *Genes Dev* **32:** 1525–1536. doi:10.1101/gad.318675.118

Jachowicz JW, Bing X, Pontabry J, Bošković A, Rando OJ, Torres-Padilla ME. 2017. LINE-1 activation after fertilization regulates global chromatin accessibility in the early mouse embryo. *Nat Genet* **49:** 1502–1510. doi:10.1038/ng.3945

Jambhekar A, Dhall A, Shi Y. 2019. Roles and regulation of histone methylation in animal development. *Nat Rev Mol Cell Biol* **20:** 625–641. doi:10.1038/s41580-019-0151-1

Jangam D, Feschotte C, Betrán E. 2017. Transposable element domestication as an adaptation to evolutionary conflicts. *Trends Genet* **33:** 817–831. doi:10.1016/j.tig.2017.07.011

Jiang L, Zhang J, Wang JJ, Wang L, Zhang L, Li G, Yang X, Ma X, Sun X, Cai J, et al. 2013. Sperm, but not oocyte, DNA methylome is inherited by zebrafish early embryos. *Cell* **153:** 773–784. doi:10.1016/j.cell.2013.04.041

Jung YH, Sauria MEG, Lyu X, Cheema MS, Ausio J, Taylor J, Corces VG. 2017. Chromatin states in mouse sperm correlate with embryonic and adult regulatory landscapes. *Cell Rep* **18:** 1366–1382. doi:10.1016/j.celrep.2017.01.034

Jung YH, Kremsky I, Gold HB, Rowley MJ, Punyawai K, Buonanotte A, Lyu X, Bixler BJ, Chan AWS, Corces VG. 2019. Maintenance of CTCF- and transcription factor-mediated interactions from the gametes to the early mouse embryo. *Mol Cell* **75:** 154–171.e5. doi:10.1016/j.molcel.2019.04.014

Kaaij LJ, van der Weide RH, Ketting RF, de Wit E. 2018. Systemic loss and gain of chromatin architecture throughout zebrafish development. *Cell Rep* **24:** 1–10.e4. doi:10.1016/j.celrep.2018.06.003

Kane DA, Kimmel CB. 1993. The zebrafish midblastula transition. *Development* **119:** 447–456. doi:10.1242/dev.119.2.447

Kaya-Okur HS, Wu SJ, Codomo CA, Pledger ES, Bryson TD, Henikoff JG, Ahmad K, Henikoff S. 2019. CUT&tag for efficient epigenomic profiling of small samples and single cells. *Nat Commun* **10:** 1930. doi:10.1038/s41467-019-09982-5

Ke Y, Xu Y, Chen X, Feng S, Liu Z, Sun Y, Yao X, Li F, Zhu W, Gao L, et al. 2017. 3D chromatin structures of mature gametes and structural reprogramming during mammalian embryogenesis. *Cell* **170:** 367–381.e20. doi:10.1016/j.cell.2017.06.029

Kelsey G, Feil R. 2013. New insights into establishment and maintenance of DNA methylation imprints in mammals. *Philos T R Soc B* **368:** 20110336. doi:10.1098/rstb.2011.0336

Kempfer R, Pombo A. 2020. Methods for mapping 3D chromosome architecture. *Nat Rev Genet* **21:** 207–226. doi:10.1038/s41576-019-0195-2

Kim J, Zhao H, Dan J, Kim S, Hardikar S, Hollowell D, Lin K, Lu Y, Takata Y, Shen J, et al. 2016. Maternal Setdb1 is required for meiotic progression and preimplantation development in mouse. *PLoS Genet* **12:** e1005970. doi:10.1371/journal.pgen.1005970

Kim Y, Shi Z, Zhang H, Finkelstein IJ, Yu H. 2019. Human cohesin compacts DNA by loop extrusion. *Science* **366:** 1345–1349. doi:10.1126/science.aaz4475

Klemm SL, Shipony Z, Greenleaf WJ. 2019. Chromatin accessibility and the regulatory epigenome. *Nat Rev Genet* **20:** 207–220. doi:10.1038/s41576-018-0089-8

Ko MS. 2016. Zygotic genome activation revisited: looking through the expression and function of Zscan4. *Curr Top Dev Biol* **120:** 103–124. doi:10.1016/bs.ctdb.2016.04.004

Kobayashi H, Sakurai T, Imai M, Takahashi N, Fukuda A, Yayoi O, Sato S, Nakabayashi K, Hata K, Sotomaru Y, et al. 2012. Contribution of intragenic DNA methylation in mouse gametic DNA methylomes to establish oocyte-specific heritable marks. *Plos Genet* **8:** e1002440. doi:10.1371/journal.pgen.1002440

Kohli RM, Zhang Y. 2013. TET enzymes, TDG and the dynamics of DNA demethylation. *Nature* **502:** 472–479. doi:10.1038/nature12750

Kouzarides T. 2007. Chromatin modifications and their function. *Cell* **128:** 693–705. doi:10.1016/j.cell.2007.02.005

Kraft K, Yost KE, Murphy S, Magg A, Long Y, Corces MR, Granja JM, Mundlos S, Cech TR, Boettiger A, et al. 2020. Polycomb-mediated genome architecture enables long-range spreading of H3K27 methylation. bioRxiv doi:10.1101/2020.1107.1127.223438

Lander ES, Linton LM, Birren B, Nusbaum C, Zody MC, Baldwin J, Devon K, Dewar K, Doyle M, FitzHugh W, et al. 2001. Initial sequencing and analysis of the human genome. *Nature* **409:** 860–921. doi:10.1038/35057062

Lawrence M, Daujat S, Schneider R. 2016. Lateral thinking: how histone modifications regulate gene expression. *Trends Genet* **32:** 42–56. doi:10.1016/j.tig.2015.10.007

Lee JT. 2005. Regulation of X-chromosome counting by *Tsix* and *Xite* sequences. *Science* **309:** 768–771. doi:10.1126/science.1113673

Lee JH, Skalnik DG. 2005. CpG-binding protein (CXXC finger protein 1) is a component of the mammalian Set1 histone H3-Lys4 methyltransferase complex, the analogue of the yeast Set1/COMPASS complex. *J Biol Chem* **280:** 41725–41731. doi:10.1074/jbc.M508312200

Lee MT, Bonneau AR, Giraldez AJ. 2014. Zygotic genome activation during the maternal-to-zygotic transition. *Annu Rev Cell Dev Biol* **30:** 581–613. doi:10.1146/annurev-cellbio-100913-013027

Lepikhov K, Walter J. 2004. Differential dynamics of histone H3 methylation at positions K4 and K9 in the mouse zygote. *BMC Dev Biol* **4:** 12. doi:10.1186/1471-213X-4-12

Lepikhov K, Zakhartchenko V, Hao R, Yang F, Wrenzycki C, Niemann H, Wolf E, Walter J. 2008. Evidence for conserved DNA and histone H3 methylation reprogramming in mouse, bovine and rabbit zygotes. *Epigenetics Chromatin* **1:** 8. doi:10.1186/1756-8935-1-8

Li E, Zhang Y. 2014. DNA methylation in mammals. *Cold Spring Harb Perspect Biol* **6:** a019133. doi:10.1101/cshperspect.a019133

Li L, Zheng P, Dean J. 2010. Maternal control of early mouse development. *Development* **137:** 859–870. doi:10.1242/dev.039487

Li L, Guo F, Gao Y, Ren Y, Yuan P, Yan L, Li R, Lian Y, Li J, Hu B, et al. 2018a. Single-cell multi-omics sequencing of human early embryos. *Nat Cell Biol* **20:** 847–858. doi:10.1038/s41556-018-0123-2

Li Y, Zhang Z, Chen J, Liu W, Lai W, Liu B, Li X, Liu L, Xu S, Dong Q, et al. 2018b. Stella safeguards the oocyte methylome by preventing de novo methylation mediated by DNMT1. *Nature* **564:** 136–140. doi:10.1038/s41586-018-0751-5

Li F, Wang D, Song R, Cao C, Zhang Z, Wang Y, Li X, Huang J, Liu Q, Hou N, et al. 2020. The asynchronous establishment of chromatin 3D architecture between in vitro fertilized and uniparental preimplantation pig embryos. *Genome Biol* **21:** 203. doi:10.1186/s13059-020-02095-z

Lieberman-Aiden E, van Berkum NL, Williams L, Imakaev M, Ragoczy T, Telling A, Amit I, Lajoie BR, Sabo PJ, Dorschner MO, et al. 2009. Comprehensive mapping of long-range interactions reveals folding principles of the human genome. *Science* **326:** 289–293. doi:10.1126/science.1181369

Lin CJ, Koh FM, Wong P, Conti M, Ramalho-Santos M. 2014. Hira-mediated H3.3 incorporation is required for DNA replication and ribosomal RNA transcription in the mouse zygote. *Dev Cell* **30:** 268–279. doi:10.1016/j.devcel.2014.06.022

Liu WQ, Liu XY, Wang CF, Gao YW, Gao R, Kou XC, Zhao YH, Li JY, Wu Y, Xiu WC, et al. 2016a. Identification of key factors conquering developmental arrest of somatic cell cloned embryos by combining embryo biopsy and single-cell sequencing. *Cell Discov* **2:** 16010.

Liu X, Wang C, Liu W, Li J, Li C, Kou X, Chen J, Zhao Y, Gao H, Wang H, et al. 2016b. Distinct features of H3K4me3 and H3K27me3 chromatin domains in pre-implantation embryos. *Nature* **537:** 558–562. doi:10.1038/nature19362

Liu X, Wang YZ, Gao YP, Su JM, Zhang JC, Xing XP, Zhou C, Yao KZ, An QL, Zhang Y. 2018a. H3k9 demethylase KDM4E is an epigenetic regulator for bovine embryonic development and a defective factor for nuclear reprogramming. *Development* **145:** dev158261. doi:10.1242/dev.158261

Liu Z, Cai YJ, Wang Y, Nie YH, Zhang CC, Xu YT, Zhang XT, Lu Y, Wang ZY, Poo MT, et al. 2018b. Cloning of macaque monkeys by somatic cell nuclear transfer. *Cell* **174:** 245–245. doi:10.1016/j.cell.2018.01.036

Liu B, Xu Q, Wang Q, Feng S, Lai F, Wang P, Zheng F, Xiang Y, Wu J, Nie J, et al. 2020. The landscape of RNA Pol II binding reveals a stepwise transition during ZGA. *Nature* **587:** 139–144. doi:10.1038/s41586-020-2847-y

Loppin B, Bonnefoy E, Anselme C, Laurençon A, Karr TL, Couble P. 2005. The histone H3.3 chaperone HIRA is essential for chromatin assembly in the male pronucleus. *Nature* **437:** 1386–1390. doi:10.1038/nature04059

Lu F, Liu Y, Inoue A, Suzuki T, Zhao K, Zhang Y. 2016. Establishing chromatin regulatory landscape during mouse preimplantation development. *Cell* **165:** 1375–1388. doi:10.1016/j.cell.2016.05.050

Luo Z, Wang X, Jiang H, Wang R, Chen J, Chen Y, Xu Q, Cao J, Gong X, Wu J, et al. 2020. Reorganized 3D genome structures support transcriptional regulation in mouse spermatogenesis. *iScience* **23:** 101034. doi:10.1016/j.isci.2020.101034

Macfarlan TS, Gifford WD, Driscoll S, Lettieri K, Rowe HM, Bonanomi D, Firth A, Singer O, Trono D, Pfaff SL. 2012. Embryonic stem cell potency fluctuates with endogenous retrovirus activity. *Nature* **487:** 57–63. doi:10.1038/nature11244

Macleod D, Clark VH, Bird A. 1999. Absence of genome-wide changes in DNA methylation during development of the zebrafish. *Nat Genet* **23:** 139–140. doi:10.1038/13767

Majumder S, Miranda M, DePamphilis ML. 1993. Analysis of gene expression in mouse preimplantation embryos demonstrates that the primary role of enhancers is to relieve repression of promoters. *EMBO J* **12:** 1131–1140. doi:10.1002/j.1460-2075.1993.tb05754.x

Margueron R, Reinberg D. 2011. The polycomb complex PRC2 and its mark in life. *Nature* **469:** 343–349. doi:10.1038/nature09784

Matoba S, Liu YT, Lu FL, Iwabuchi KA, Shen L, Inoue A, Zhang Y. 2014. Embryonic development following somatic cell nuclear transfer impeded by persisting histone methylation. *Cell* **159:** 884–895. doi:10.1016/j.cell.2014.09.055

Matoba S, Wang H, Jiang L, Lu F, Iwabuchi KA, Wu X, Inoue K, Yang L, Press W, Lee JT, et al. 2018. Loss of H3K27me3 imprinting in somatic cell nuclear transfer embryos disrupts post-implantation development. *Cell Stem Cell* **23:** 343–354.e5. doi:10.1016/j.stem.2018.06.008

Mei H, Kozuka C, Hayashi R, Kumon M, Koseki H, Inoue A. 2021. H2AK119ub1 guides maternal inheritance and zygotic deposition of H3K27me3 in mouse embryos. *Nat Genet* **53:** 539–550. doi:10.1038/s41588-021-00820-3

Misteli T. 2020. The self-organizing genome: principles of genome architecture and function. *Cell* **183:** 28–45. doi:10.1016/j.cell.2020.09.014

Mouse Genome Sequencing Consortium; Waterston RH, Lindblad-Toh K, Birney E, Rogers J, Abril JF, Agarwal P, Agarwala R, Ainscough R, Alexandersson M, et al. 2002. Initial sequencing and comparative analysis of the mouse genome. *Nature* **420:** 520–562. doi:10.1038/nature01262

Murphy PJ, Wu SF, James CR, Wike CL, Cairns BR. 2018. Placeholder nucleosomes underlie germline-to-embryo DNA methylation reprogramming. *Cell* **172:** 993–1006.e13. doi:10.1016/j.cell.2018.01.022

Nakamura T, Nakagawa M, Ichisaka T, Shiota A, Yamanaka S. 2011. Essential roles of ECAT15-2/Dppa2 in functional lung development. *Mol Cell Biol* **31:** 4366–4378. doi:10.1128/MCB.05701-11

Nakamura R, Motai Y, Kumagai M, Wike CL, Nishiyama H, Nakatani Y, Durand NC, Kondo K, Kondo T, Tsukahara T, et al. 2021. CTCF looping is established during gastrulation in medaka embryos. *Genome Res* **31:** 968–980. doi:10.1101/gr.269951.120

Nashun B, Hill PWS, Smallwood SA, Dharmalingam G, Amouroux R, Clark SJ, Sharma V, Ndjetehe E, Pelczar P, Festenstein RJ, et al. 2015. Continuous histone replacement by Hira is essential for normal transcriptional regulation and de novo DNA methylation during mouse oogenesis. *Mol Cell* **60:** 611–625. doi:10.1016/j.molcel.2015.10.010

Nicetto D, Donahue G, Jain T, Peng T, Sidoli S, Sheng L, Montavon T, Becker JS, Grindheim JM, Blahnik K, et al. 2019. H3K9me3-heterochromatin loss at protein-coding genes enables developmental lineage specification. *Science* **363:** 294–297. doi:10.1126/science.aau0583

Nothias JY, Miranda M, DePamphilis ML. 1996. Uncoupling of transcription and translation during zygotic gene activation in the mouse. *EMBO J* **15:** 5715–5725. doi:10.1002/j.1460-2075.1996.tb00955.x

Nuebler J, Fudenberg G, Imakaev M, Abdennur N, Mirny LA. 2018. Chromatin organization by an interplay of loop extrusion and compartmental segregation. *Proc Natl Acad Sci* **115:** E6697–E6706. doi:10.1073/pnas.1717730115

Ogiyama Y, Schuettengruber B, Papadopoulos GL, Chang JM, Cavalli G. 2018. Polycomb-dependent chromatin looping contributes to gene silencing during *Drosophila* development. *Mol Cell* **71:** 73–88.e5. doi:10.1016/j.molcel.2018.05.032

Okamoto I, Patrat C, Thepot D, Peynot N, Fauque P, Daniel N, Diabangouaya P, Wolf JP, Renard JP, Duranthon V, et al. 2011. Eutherian mammals use diverse strategies to initiate X-chromosome inactivation during development. *Nature* **472:** 370–374. doi:10.1038/nature09872

Oksuz O, Narendra V, Lee CH, Descostes N, LeRoy G, Raviram R, Blumenberg L, Karch K, Rocha PP, Garcia BA, et al. 2018. Capturing the onset of PRC2-mediated repressive domain formation. *Mol Cell* **70:** 1149–1162.e5. doi:10.1016/j.molcel.2018.05.023

Ooga M, Fulka H, Hashimoto S, Suzuki MG, Aoki F. 2016. Analysis of chromatin structure in mouse preimplantation embryos by fluorescent recovery after photobleaching. *Epigenetics* **11:** 85–94. doi:10.1080/15592294.2015.1136774

Oomen ME, Dekker J. 2017. Epigenetic characteristics of the mitotic chromosome in 1D and 3D. *Crit Rev Biochem Mol Biol* **52:** 185–204. doi:10.1080/10409238.2017.1287160

Papp B, Plath K. 2013. Epigenetics of reprogramming to induced pluripotency. *Cell* **152:** 1324–1343. doi:10.1016/j.cell.2013.02.043

Pastor WA, Aravind L, Rao A. 2013. TETonic shift: biological roles of TET proteins in DNA demethylation and transcription. *Nat Rev Mol Cell Biol* **14:** 341–356. doi:10.1038/nrm3589

Patel L, Kang R, Rosenberg SC, Qiu YJ, Raviram R, Chee S, Hu R, Ren B, Cole F, Corbett KD. 2019. Dynamic reorganization of the genome shapes the recombination landscape in meiotic prophase. *Nat Struct Mol Biol* **26:** 164–174. doi:10.1038/s41594-019-0187-0

Peaston AE, Evsikov AV, Graber JH, de Vries WN, Holbrook AE, Solter D, Knowles BB. 2004. Retrotransposons regulate host genes in mouse oocytes and preimplantation embryos. *Dev Cell* **7:** 597–606. doi:10.1016/j.devcel.2004.09.004

Percharde M, Lin CJ, Yin Y, Guan J, Peixoto GA, Bulut-Karslioglu A, Biechele S, Huang B, Shen X, Ramalho-Santos M. 2018. A LINE1-nucleolin partnership regulates early development and ESC identity. *Cell* **174:** 391–405.e19. doi:10.1016/j.cell.2018.05.043

Peters AHFM, O'Carroll D, Scherthan H, Mechtler K, Sauer S, Schöfer C, Weipoltshammer K, Pagani M, Lachner M, Kohlmaier A, et al. 2001. Loss of the Suv39h histone methyltransferases impairs mammalian heterochromatin and genome stability. *Cell* **107:** 323–337. doi:10.1016/S0092-8674(01)00542-6

Petropoulos S, Edsgärd D, Reinius B, Deng Q, Panula SP, Codeluppi S, Reyes AP, Linnarsson S, Sandberg R, Lanner F. 2016. Single-cell RNA-seq reveals lineage and x chromosome dynamics in human preimplantation embryos. *Cell* **167:** 285. doi:10.1016/j.cell.2016.08.009

Pombo A, Dillon N. 2015. Three-dimensional genome architecture: players and mechanisms. *Nat Rev Mol Cell Biol* **16:** 245–257. doi:10.1038/nrm3965

Posfai E, Kunzmann R, Brochard V, Salvaing J, Cabuy E, Roloff TC, Liu ZC, Tardat M, van Lohuizen M, Vidal M, et al. 2012. Polycomb function during oogenesis is required for mouse embryonic development. *Genes Dev* **26:** 920–932. doi:10.1101/gad.188094.112

Potok ME, Nix DA, Parnell TJ, Cairns BR. 2013. Reprogramming the maternal zebrafish genome after fertilization to match the paternal methylation pattern. *Cell* **153:** 759–772. doi:10.1016/j.cell.2013.04.030

Probst AV, Santos F, Reik W, Almouzni G, Dean W. 2007. Structural differences in centromeric heterochromatin are spatially reconciled on fertilisation in the mouse zygote. *Chromosoma* **116:** 403–415. doi:10.1007/s00412-007-0106-8

Prokopuk L, Stringer JM, White CR, Vossen R, White SJ, Cohen ASA, Gibson WT, Western PS. 2018. Loss of maternal EED results in postnatal overgrowth. *Clin Epigenetics* **10:** 95. doi:10.1186/s13148-018-0526-8

Puschendorf M, Terranova R, Boutsma E, Mao X, Isono K, Brykczynska U, Kolb C, Otte AP, Koseki H, Orkin SH, et al. 2008. PRC1 and Suv39h specify parental asymmetry at constitutive heterochromatin in early mouse embryos. *Nat Genet* **40:** 411–420. doi:10.1038/ng.99

Quinodoz SA, Ollikainen N, Tabak B, Palla A, Schmidt JM, Detmar E, Lai MM, Shishkin AA, Bhat P, Takei Y, et al. 2018. Higher-order inter-chromosomal hubs shape 3D genome organization in the nucleus. *Cell* **174:** 744–757.e24. doi:10.1016/j.cell.2018.05.024

Rao SSP, Huang SC, St Hilaire BG, Engreitz JM, Perez EM, Kieffer-Kwon KR, Sanborn AL, Johnstone SE, Bascom GD, Bochkov ID, et al. 2017. Cohesin loss eliminates all loop domains. *Cell* **171:** 305–320.e24. doi:10.1016/j.cell.2017.09.026

Rebollo R, Romanish MT, Mager DL. 2012. Transposable elements: an abundant and natural source of regulatory sequences for host genes. *Annu Rev Genet* **46:** 21–42. doi:10.1146/annurev-genet-110711-155621

Reichmann J, Nijmeijer B, Hossain MJ, Eguren M, Schneider I, Politi AZ, Roberti MJ, Hufnagel L, Hiiragi T, Ellenberg J. 2018. Dual-spindle formation in zygotes keeps parental genomes apart in early mammalian embryos. *Science* **361:** 189–193. doi:10.1126/science.aar7462

Rodriguez-Terrones D, Torres-Padilla ME. 2018. Nimble and ready to mingle: transposon outbursts of early development. *Trends Genet* **34:** 806–820. doi:10.1016/j.tig.2018.06.006

Rowley MJ, Corces VG. 2018. Organizational principles of 3D genome architecture. *Nat Rev Genet* **19:** 789–800. doi:10.1038/s41576-018-0060-8

Ruan DG, Peng JY, Wang XS, Ouyang Z, Zou QJ, Yang Y, Chen FB, Ge WK, Wu H, Liu ZM, et al. 2018. XIST derepression in active X chromosome hinders pig somatic cell nuclear transfer. *Stem Cell Reports* **10:** 494–508. doi:10.1016/j.stemcr.2017.12.015

Ruthenburg AJ, Allis CD, Wysocka J. 2007. Methylation of lysine 4 on histone H3: intricacy of writing and reading a single epigenetic mark. *Mol Cell* **25:** 15–30. doi:10.1016/j.molcel.2006.12.014

Sachs M, Onodera C, Blaschke K, Ebata KT, Song JS, Ramalho-Santos M. 2013. Bivalent chromatin marks developmental regulatory genes in the mouse embryonic germline in vivo. *Cell Rep* **3:** 1777–1784. doi:10.1016/j.celrep.2013.04.032

Saha B, Home P, Ray S, Larson M, Paul A, Rajendran G, Behr B, Paul S. 2013. EED and KDM6B coordinate the first mammalian cell lineage commitment to ensure embryo implantation. *Mol Cell Biol* **33:** 2691–2705. doi:10.1128/MCB.00069-13

Sanborn AL, Rao SS, Huang SC, Durand NC, Huntley MH, Jewett AI, Bochkov ID, Chinnappan D, Cutkosky A, Li J, et al. 2015. Chromatin extrusion explains key features of loop and domain formation in wild-type and engineered genomes. *Proc Natl Acad Sci* **112:** E6456–E6465. doi:10.1073/pnas.1518552112

Sankar A, Lerdrup M, Manaf A, Johansen JV, Gonzalez JM, Borup R, Blanshard R, Klungland A, Hansen K, Andersen CY, et al. 2020. KDM4A regulates the maternal-to-zygotic transition by protecting broad H3K4me3 domains from H3K9me3 invasion in oocytes. *Nat Cell Biol* **22:** 380–388. doi:10.1038/s41556-020-0494-z

Santenard A, Ziegler-Birling C, Koch M, Tora L, Bannister AJ, Torres-Padilla ME. 2010. Heterochromatin formation in the mouse embryo requires critical residues of the histone variant H3.3. *Nat Cell Biol* **12:** 853–862. doi:10.1038/ncb2089

Sato Y, Hilbert L, Oda H, Wan Y, Heddleston JM, Chew TL, Zaburdaev V, Keller P, Lionnet T, Vastenhouw N, et al. 2019. Histone H3K27 acetylation precedes active transcription during zebrafish zygotic genome activation as revealed by live-cell analysis. *Development* **146:** dev179127. doi:10.1242/dev.179127

Schoenfelder S, Fraser P. 2019. Long-range enhancer-promoter contacts in gene expression control. *Nat Rev Genet* **20:** 437–455. doi:10.1038/s41576-019-0128-0

Schoenfelder S, Sugar R, Dimond A, Javierre BM, Armstrong H, Mifsud B, Dimitrova E, Matheson L, Tavares-Cadete F, Furlan-Magaril M, et al. 2015. Polycomb repressive complex PRC1 spatially constrains the mouse embryonic stem cell genome. *Nat Genet* **47:** 1179–1186. doi:10.1038/ng.3393

Schroeder DI, Jayashankar K, Douglas KC, Thirkill TL, York D, Dickinson PJ, Williams LE, Samollow PB, Ross PJ, Bannasch DL, et al. 2015. Early developmental and evolutionary origins of gene body DNA methylation patterns in mammalian placentas. *PLoS Genet* **11:** e1005442. doi:10.1371/journal.pgen.1005442

Schuettengruber B, Bourbon HM, Di Croce L, Cavalli G. 2017. Genome regulation by polycomb and trithorax: 70 years and counting. *Cell* **171:** 34–57. doi:10.1016/j.cell.2017.08.002

Schultz RM. 2002. The molecular foundations of the maternal to zygotic transition in the preimplantation embryo. *Hum Reprod Update* **8:** 323–331. doi:10.1093/humupd/8.4.323

Schwarzer W, Abdennur N, Goloborodko A, Pekowska A, Fudenberg G, Loe-Mie Y, Fonseca NA, Huber W, Haering CH, Mirny L, et al. 2017. Two independent modes of chromatin organization revealed by cohesin removal. *Nature* **551:** 51–56. doi:10.1038/nature24281

Seisenberger S, Andrews S, Krueger F, Arand J, Walter J, Santos F, Popp C, Thienpont B, Dean W, Reik W. 2012. The dynamics of genome-wide DNA methylation reprogramming in mouse primordial germ cells. *Mol Cell* **48:** 849–862. doi:10.1016/j.molcel.2012.11.001

Sendžikaite G, Kelsey G. 2019. The role and mechanisms of DNA methylation in the oocyte. *Essays Biochem* **63:** 691–705. doi:10.1042/EBC20190043

Sha QQ, Dai XX, Jiang JC, Yu C, Jiang Y, Liu J, Ou XH, Zhang SY, Fan HY. 2018. CFP1 coordinates histone H3 lysine-4 trimethylation and meiotic cell cycle progression in mouse oocytes. *Nat Commun* **9:** 3477. doi:10.1038/s41467-018-05930-x

Shen L, Inoue A, He J, Liu Y, Lu F, Zhang Y. 2014. Tet3 and DNA replication mediate demethylation of both the maternal and paternal genomes in mouse zygotes. *Cell Stem Cell* **15:** 459–471. doi:10.1016/j.stem.2014.09.002

Shinagawa T, Takagi T, Tsukamoto D, Tomaru C, Huynh LM, Sivaraman P, Kumarevel T, Inoue K, Nakato R, Katou Y, et al. 2014. Histone variants enriched in oocytes enhance reprogramming to induced pluripotent stem cells. *Cell Stem Cell* **14:** 217–227. doi:10.1016/j.stem.2013.12.015

Shirane K, Toh H, Kobayashi H, Miura F, Chiba H, Ito T, Kono T, Sasaki H. 2013. Mouse oocyte methylomes at base resolution reveal genome-wide accumulation of non-CpG methylation and role of DNA methyltransferases. *Plos Genet* **9:** e1003439. doi:10.1371/journal.pgen.1003439

Shirane K, Miura F, Ito T, Lorincz MC. 2020. NSD1-deposited h3k36me2 directs de novo methylation in the mouse male germline and counteracts Polycomb-associated silencing. *Nat Genet* **52:** 1088–1098. doi:10.1038/s41588-020-0689-z

Silva MCC, Powell S, Ladstätter S, Gassler J, Stocsits R, Tedeschi A, Peters JM, Tachibana K. 2020. Wapl releases Scc1-cohesin and regulates chromosome structure and segregation in mouse oocytes. *J Cell Biol* **219:** e201906100. doi:10.1083/jcb.201906100

Skene PJ, Henikoff S. 2017. An efficient targeted nuclease strategy for high-resolution mapping of DNA binding sites. *eLife* **6:** e21856. doi:10.7554/eLife.21856

Smallwood SA, Tomizawa S, Krueger F, Ruf N, Carli N, Segonds-Pichon A, Sato S, Hata K, Andrews SR, Kelsey G. 2011. Dynamic CpG island methylation landscape in oocytes and preimplantation embryos. *Nat Genet* **43:** 811–814. doi:10.1038/ng.864

Smith ZD, Meissner A. 2013. DNA methylation: roles in mammalian development. *Nat Rev Genet* **14:** 204–220. doi:10.1038/nrg3354

Smith ZD, Chan MM, Mikkelsen TS, Gu HC, Gnirke A, Regev A, Meissner A. 2012. A unique regulatory phase of DNA methylation in the early mammalian embryo. *Nature* **484:** 339–344. doi:10.1038/nature10960

Smith ZD, Shi J, Gu H, Donaghey J, Clement K, Cacchiarelli D, Gnirke A, Michor F, Meissner A. 2017. Epigenetic restriction of extraembryonic lineages mirrors the somatic transition to cancer. *Nature* **549:** 543–547. doi:10.1038/nature23891

Smith R, Jiang Z, Susor A, Ming H, Tait J, Conti M, Lin C-J. 2020. The H3.3 chaperone Hira complex orchestrates oocyte developmental competence. bioRxiv doi:10.1101/2020.1105.1125.114124

Srinivasan R, Nady N, Arora N, Hsieh LJ, Swigut T, Narlikar GJ, Wossidlo M, Wysocka J. 2020. Zscan4 binds nucleosomal microsatellite DNA and protects mouse two-cell embryos from DNA damage. *Sci Adv* **6:** eaaz9115. doi:10.1126/sciadv.aaz9115

Stewart KR, Veselovska L, Kelsey G. 2016. Establishment and functions of DNA methylation in the germline. *Epigenomics* **8:** 1399–1413. doi:10.2217/epi-2016-0056

Stock JK, Giadrossi S, Casanova M, Brookes E, Vidal M, Koseki H, Brockdorff N, Fisher AG, Pombo A. 2007. Ring1-mediated ubiquitination of H2A restrains poised RNA polymerase II at bivalent genes in mouse ES cells. *Nat Cell Biol* **9:** 1428–1435. doi:10.1038/ncb1663

Svoboda P, Stein P, Anger M, Bernstein E, Hannon GJ, Schultz RM. 2004. RNAi and expression of retrotransposons MuERV-L and IAP in preimplantation mouse embryos. *Dev Biol* **269:** 276–285. doi:10.1016/j.ydbio.2004.01.028

Tachibana-Konwalski K, Godwin J, van der Weyden L, Champion L, Kudo NR, Adams DJ, Nasmyth K. 2010. Rec8-containing cohesin maintains bivalents without turnover during the growing phase of mouse oocytes. *Genes Dev* **24:** 2505–2516. doi:10.1101/gad.605910

Tamburri S, Lavarone E, Fernández-Pérez D, Conway E, Zanotti M, Manganaro D, Pasini D. 2020. Histone H2AK119 mono-ubiquitination is essential for polycomb-mediated transcriptional repression. *Mol Cell* **77:** 840–856.e5. doi:10.1016/j.molcel.2019.11.021

Tardat M, Albert M, Kunzmann R, Liu Z, Kaustov L, Thierry R, Duan S, Brykczynska U, Arrowsmith CH, Peters AH. 2015. Cbx2 targets PRC1 to constitutive heterochromatin in mouse zygotes in a parent-of-origin-dependent manner. *Mol Cell* **58:** 157–171. doi:10.1016/j.molcel.2015.02.013

Tee WW, Shen SS, Oksuz O, Narendra V, Reinberg D. 2014. Erk1/2 activity promotes chromatin features and RNAPII phosphorylation at developmental promoters in mouse ESCs. *Cell* **156:** 678–690. doi:10.1016/j.cell.2014.01.009

Tesařík J, Kopečný V, Plachot M, Mandelbaum J. 1987. Ultrastructural and autoradiographic observations on multinucleated blastomeres of human cleaving embryos obtained by in-vitro fertilization. *Hum Reprod* **2:** 127–136. doi:10.1093/oxfordjournals.humrep.a136496

Torres-Padilla ME, Bannister AJ, Hurd PJ, Kouzarides T, Zernicka-Goetz M. 2006. Dynamic distribution of the replacement histone variant H3.3 in the mouse oocyte and preimplantation embryos. *Int J Dev Biol* **50:** 455–461.

Tucci V, Isles AR, Kelsey G, Ferguson-Smith AC; Erice Imprinting Group. 2019. Genomic imprinting and physiological processes in mammals. *Cell* **176:** 952–965. doi:10.1016/j.cell.2019.01.043

van der Heijden GW, Dieker JW, Derijck AA, Muller S, Berden JH, Braat DD, van der Vlag J, de Boer P. 2005. Asymmetry in histone H3 variants and lysine methylation between paternal and maternal chromatin of the early mouse zygote. *Mech Dev* **122:** 1008–1022. doi:10.1016/j.mod.2005.04.009

van Steensel B, Belmont AS. 2017. Lamina-associated domains: links with chromosome architecture, heterochromatin, and gene repression. *Cell* **169:** 780–791. doi:10.1016/j.cell.2017.04.022

Vara C, Paytuví-Gallart A, Cuartero Y, Le Dily F, Garcia F, Salvà-Castro J, Gómez HL, Julià E, Moutinho C, Aiese Cigliano R, et al. 2019. Three-dimensional genomic structure and cohesin occupancy correlate with transcriptional activity during spermatogenesis. *Cell Rep* **28:** 352–367.e359. doi:10.1016/j.celrep.2019.06.037

Vastenhouw NL, Schier AF. 2012. Bivalent histone modifications in early embryogenesis. *Curr Opin Cell Biol* **24:** 374–386. doi:10.1016/j.ceb.2012.03.009

Vastenhouw NL, Zhang Y, Woods IG, Imam F, Regev A, Liu XS, Rinn J, Schier AF. 2010. Chromatin signature of embryonic pluripotency is established during genome activation. *Nature* **464:** 922–926. doi:10.1038/nature08866

Veselovska L, Smallwood SA, Saadeh H, Stewart KR, Krueger F, Maupetit-Méhouas S, Arnaud P, Tomizawa S, Andrews S, Kelsey G. 2015. Deep sequencing and de novo assembly of the mouse oocyte transcriptome define the contribution of transcription to the DNA methylation landscape. *Genome Biol* **16:** 209. doi:10.1186/s13059-015-0769-z

Vian L, Pękowska A, Rao SSP, Kieffer-Kwon KR, Jung S, Baranello L, Huang SC, El Khattabi L, Dose M, Pruett N, et al. 2018. The energetics and physiological impact of cohesin extrusion. *Cell* **173:** 1165–1178.e20. doi:10.1016/j.cell.2018.03.072

Vincent JJ, Huang Y, Chen PY, Feng S, Calvopiña JH, Nee K, Lee SA, Le T, Yoon AJ, Faull K, et al. 2013. Stage-specific roles for tet1 and tet2 in DNA demethylation in primordial germ cells. *Cell Stem Cell* **12:** 470–478. doi:10.1016/j.stem.2013.01.016

Wagner EJ, Carpenter PB. 2012. Understanding the language of Lys36 methylation at histone H3. *Nat Rev Mol Cell Biol* **13:** 115–126. doi:10.1038/nrm3274

Walser CB, Lipshitz HD. 2011. Transcript clearance during the maternal-to-zygotic transition. *Curr Opin Genet Dev* **21:** 431–443. doi:10.1016/j.gde.2011.03.003

Walsh CP, Chaillet JR, Bestor TH. 1998. Transcription of IAP endogenous retroviruses is constrained by cytosine methylation. *Nat Genet* **20:** 116–117. doi:10.1038/2413

Wang QT, Piotrowska K, Ciemerych MA, Milenkovic L, Scott MP, Davis RW, Zernicka-Goetz M. 2004. A genome-wide study of gene activity reveals developmental signaling pathways in the preimplantation mouse embryo. *Dev Cell* **6:** 133–144. doi:10.1016/S1534-5807(03)00404-0

Wang L, Zhang J, Duan J, Gao X, Zhu W, Lu X, Yang L, Zhang J, Li G, Ci W, et al. 2014. Programming and inheritance of parental DNA methylomes in mammals. *Cell* **157:** 979–991. doi:10.1016/j.cell.2014.04.017

Wang CF, Liu XY, Gao YW, Yang L, Li C, Liu WQ, Chen C, Kou XC, Zhao YH, Chen JY, et al. 2018. Reprogramming of H3K9me3-dependent heterochromatin during mammalian embryo development. *Nat Cell Biol* **20:** 620–631. doi:10.1038/s41556-018-0093-4

Wang Q, Xiong H, Ai S, Yu X, Liu Y, Zhang J, He A. 2019a. CoBATCH for high-throughput single-cell epigenomic profiling. *Mol Cell* **76:** 206–216.e7. doi:10.1016/j.molcel.2019.07.015

Wang Y, Wang H, Zhang Y, Du Z, Si W, Fan S, Qin D, Wang M, Duan Y, Li L, et al. 2019b. Reprogramming of meiotic chromatin architecture during spermatogenesis. *Mol Cell* **73:** 547–561.e6. doi:10.1016/j.molcel.2018.11.019

Wang LY, Li ZK, Wang LB, Liu C, Sun XH, Feng GH, Wang JQ, Li YF, Qiao LY, Nie H, et al. 2020. Overcoming intrinsic H3K27me3 imprinting barriers improves post-implantation development after somatic cell nuclear transfer. *Cell Stem Cell* **27:** 315–325.e5. doi:10.1016/j.stem.2020.05.014

Weinberg DN, Papillon-Cavanagh S, Chen HF, Yue Y, Chen X, Rajagopalan KN, Horth C, McGuire JT, Xu XJ, Nikbakht H, et al. 2019. The histone mark H3K36me2 recruits DNMT3A and shapes the intergenic DNA methylation landscape. *Nature* **573:** 281–286. doi:10.1038/s41586-019-1534-3

Whiddon JL, Langford AT, Wong CJ, Zhong JW, Tapscott SJ. 2017. Conservation and innovation in the DUX4-family gene network. *Nat Genet* **49:** 935–940. doi:10.1038/ng.3846

Wiekowski M, Miranda M, DePamphilis ML. 1991. Regulation of gene expression in preimplantation mouse embryos: effects of the zygotic clock and the first mitosis on promoter and enhancer activities. *Dev Biol* **147:** 403–414. doi:10.1016/0012-1606(91)90298-H

Wike CL, Guo Y, Tan M, Nakamura R, Shaw DK, Díaz N, Whittaker-Tademy AF, Durand NC, Lieberman Aiden E, Vaquerizas JM, et al. 2021. Chromatin architecture transitions from zebrafish sperm through early embryogenesis. *Genome Res* **31:** 981–994. doi:10.1101/gr.269860.120

Wongtawan T, Taylor JE, Lawson KA, Wilmut I, Pennings S. 2011. Histone H4K20me3 and HP1α are late heterochromatin markers in development, but present in undifferentiated embryonic stem cells. *J Cell Sci* **124:** 1878–1890. doi:10.1242/jcs.080721

Wozniak GG, Strahl BD. 2014. Hitting the "mark": interpreting lysine methylation in the context of active transcription. *Bba-Gene Regul Mech* **1839:** 1353–1361.

Wu J, Huang B, Chen H, Yin Q, Liu Y, Xiang Y, Zhang B, Liu B, Wang Q, Xia W, et al. 2016. The landscape of accessible chromatin in mammalian preimplantation embryos. *Nature* **534:** 652–657. doi:10.1038/nature18606

Wu J, Xu J, Liu B, Yao G, Wang P, Lin Z, Huang B, Wang X, Li T, Shi S, et al. 2018. Chromatin analysis in human early development reveals epigenetic transition during ZGA. *Nature* **557:** 256–260. doi:10.1038/s41586-018-0080-8

Wu K, Liu H, Wang Y, He J, Xu S, Chen Y, Kuang J, Liu J, Guo L, Li D, et al. 2020. SETDB1-mediated cell fate transition between 2C-like and pluripotent states. *Cell Rep* **30:** 25–36.e6. doi:10.1016/j.celrep.2019.12.010

Xia W, Xie W. 2020. Rebooting the epigenomes during mammalian early embryogenesis. *Stem Cell Reports* **15:** 1158–1175. doi:10.1016/j.stemcr.2020.09.005

Xia W, Xu J, Yu G, Yao G, Xu K, Ma X, Zhang N, Liu B, Li T, Lin Z, et al. 2019. Resetting histone modifications during human parental-to-zygotic transition. *Science* **365:** 353–360. doi:10.1126/science.aaw5118

Xiang Y, Zhang Y, Xu Q, Zhou C, Liu B, Du Z, Zhang K, Zhang B, Wang X, Gayen S, et al. 2020. Epigenomic analysis of gastrulation identifies a unique chromatin state for primed pluripotency. *Nat Genet* **52:** 95–105. doi:10.1038/s41588-019-0545-1

Xu QH, Xie W. 2018. Epigenome in early mammalian development: inheritance, reprogramming and establishment. *Trends Cell Biol* **28:** 237–253. doi:10.1016/j.tcb.2017.10.008

Xu Q, Xiang Y, Wang Q, Wang L, Brind'Amour J, Bogutz AB, Zhang Y, Zhang B, Yu G, Xia W, et al. 2019. SETD2 regulates the maternal epigenome, genomic imprinting and embryonic development. *Nat Genet* **51:** 844–856. doi:10.1038/s41588-019-0398-7

Yan R, Gu C, You D, Huang Z, Qian J, Yang Q, Cheng X, Zhang L, Wang H, Wang P, et al. 2021. Decoding dynamic epigenetic landscapes in human oocytes using single-cell multi-omics sequencing. *Cell Stem Cell* **S1934-5909:** 00169.

Yang F, Huang X, Zang R, Chen J, Fidalgo M, Sanchez-Priego C, Yang J, Caichen A, Ma F, Macfarlan T, et al. 2020. DUX-miR-344-ZMYM2-mediated activation of MERVL LTRs induces a totipotent 2C-like state. *Cell Stem Cell* **26:** 234–250.e7. doi:10.1016/j.stem.2020.01.004

Yu C, Fan X, Sha QQ, Wang HH, Li BT, Dai XX, Shen L, Liu J, Wang L, Liu K, et al. 2017. CFP1 regulates histone H3K4 trimethylation and developmental potential in mouse oocytes. *Cell Rep* **20:** 1161–1172. doi:10.1016/j.celrep.2017.07.011

Zeng F, Baldwin DA, Schultz RM. 2004. Transcript profiling during preimplantation mouse development. *Dev Biol* **272:** 483–496. doi:10.1016/j.ydbio.2004.05.018

Zhang Y, Vastenhouw NL, Feng J, Fu K, Wang C, Ge Y, Pauli A, van Hummelen P, Schier AF, Liu XS. 2014. Canonical nucleosome organization at promoters forms during genome activation. *Genome Res* **24:** 260–266. doi:10.1101/gr.157750.113

Zhang B, Zheng H, Huang B, Li W, Xiang Y, Peng X, Ming J, Wu X, Zhang Y, Xu Q, et al. 2016. Allelic reprogramming of the histone modification H3K4me3 in early mammalian development. *Nature* **537:** 553–557. doi:10.1038/nature19361

Zhang B, Wu X, Zhang W, Shen W, Sun Q, Liu K, Zhang Y, Wang Q, Li Y, Meng A, et al. 2018a. Widespread enhancer dememorization and promoter priming during parental-to-zygotic transition. *Mol Cell* **72:** 673–686.e6. doi:10.1016/j.molcel.2018.10.017

Zhang Y, Xiang Y, Yin Q, Du Z, Peng X, Wang Q, Fidalgo M, Xia W, Li Y, Zhao ZA, et al. 2018b. Dynamic epigenomic landscapes during early lineage specification in mouse embryos. *Nat Genet* **50:** 96–105. doi:10.1038/s41588-017-0003-x

Zhang W, Chen F, Chen R, Xie D, Yang J, Zhao X, Guo R, Zhang Y, Shen Y, Goke J, et al. 2019a. Zscan4c activates endogenous retrovirus MERVL and cleavage embryo genes. *Nucleic Acids Res* **47:** 8485–8501.

Zhang W, Chen Z, Yin Q, Zhang D, Racowsky C, Zhang Y. 2019b. Maternal-biased H3K27me3 correlates with paternal-specific gene expression in the human morula. *Genes Dev* **33:** 382–387. doi:10.1101/gad.323105.118

Zhang K, Wu DY, Zheng H, Wang Y, Sun QR, Liu X, Wang LY, Xiong WJ, Wang Q, Rhodes JDP, et al. 2020. Analysis of genome architecture during SCNT reveals a role of cohesin in impeding minor ZGA. *Mol Cell* **79:** 234–250.e9. doi:10.1016/j.molcel.2020.06.001

Zheng H, Xie W. 2019. The role of 3D genome organization in development and cell differentiation. *Nat Rev Mol Cell Biol* **20:** 535–550. doi:10.1038/s41580-019-0132-4

Zheng H, Huang B, Zhang B, Xiang Y, Du Z, Xu Q, Li Y, Wang Q, Ma J, Peng X, et al. 2016. Resetting epigenetic memory by reprogramming of histone modifications in mammals. *Mol Cell* **63:** 1066–1079. doi:10.1016/j.molcel.2016.08.032

Zhou LQ, Dean J. 2015. Reprogramming the genome to totipotency in mouse embryos. *Trends Cell Biol* **25:** 82–91. doi:10.1016/j.tcb.2014.09.006

Zhou F, Wang R, Yuan P, Ren Y, Mao Y, Li R, Lian Y, Li J, Wen L, Yan L, et al. 2019. Reconstituting the transcriptome and DNA methylome landscapes of human implantation. *Nature* **572:** 660–664. doi:10.1038/s41586-019-1500-0

Zhu W, Xu X, Wang X, Liu J. 2019. Reprogramming histone modification patterns to coordinate gene expression in early zebrafish embryos. *BMC Genomics* **20:** 248. doi:10.1186/s12864-019-5611-7

Zhu Y, Yu J, Rong Y, Wu Y-W, Li Y, Zhang L, Pan Y, Fan H-Y, Shen L. 2021. Genomewide decoupling of H2AK119ub1 and H3K27me3 in early mouse development. *Sci Bull* doi:10.1016/j.scib.2021.06.010

Zickler D, Kleckner N. 2015. Recombination, pairing, and synapsis of homologs during meiosis. *Cold Spring Harb Perspect Biol* **7:** a016626. doi:10.1101/cshperspect.a016626

Zuccotti M, Piccinelli A, Rossi PG, Garagna S, Redi CA. 1995. Chromatin organization during mouse oocyte growth. *Mol Reprod Dev* **41:** 479–485. doi:10.1002/mrd.1080410410

The Molecular and Nuclear Dynamics of X-Chromosome Inactivation

François Dossin[1,2] and Edith Heard[1]

[1]European Molecular Biology Laboratory, Director's Unit, 69117 Heidelberg, Germany

Correspondence: edith.heard@embl.org

In female eutherian mammals, dosage compensation of X-linked gene expression is achieved during development through transcriptional silencing of one of the two X chromosomes. Following X chromosome inactivation (XCI), the inactive X chromosome remains faithfully silenced throughout somatic cell divisions. XCI is dependent on *Xist*, a long noncoding RNA that coats and silences the X chromosome from which it is transcribed. *Xist* coating triggers a cascade of chromosome-wide changes occurring at the levels of transcription, chromatin composition, chromosome structure, and spatial organization within the nucleus. XCI has emerged as a paradigm for the study of such crucial nuclear processes and the dissection of their functional interplay. In the past decade, the advent of tools to characterize and perturb these processes have provided an unprecedented understanding into their roles during XCI. The mechanisms orchestrating the initiation of XCI as well as its maintenance are thus being unraveled, although many questions still remain. Here, we introduce key aspects of the XCI process and review the recent discoveries about its molecular basis.

In mammals, sex is genetically determined by the sex chromosomes. In therian mammals (placental and marsupial), females harbor two X chromosomes while males harbor a single X and a Y chromosome (Marshall Graves and Shetty 2001). Although therian X and Y chromosomes originated from the same pair of autosomes, they differ drastically from one another as a consequence of more than 160 million years of evolution, during which the Y chromosome underwent extensive degeneration, losing most of its genetic content (Graves 2006; Veyrunes et al. 2008). While the human X chromosome is currently 165 Mb long, encoding over 800 protein-coding genes, the Y chromosome is only 60 Mb long and carries less than 100 genes (Graves 2006; Graves et al. 2006; Bachtrog 2013; Hughes and Page 2015).

Importantly, with degeneration of the Y chromosome (and associated elimination of its gene content), XY males became faced with the challenge of X chromosome monosomy compared to two copies of all autosomes. In 1967, geneticist Susumu Ohno first proposed that as each gene gradually disappeared from the Y chromosome, expression of its corresponding copy (homolog) on the X chromosome would increase by twofold, as a consequence of

[2]Present address: Genome Integrity, Immunity and Cancer Unit, Department of Immunology, Department of Genomes and Genetics, Institut Pasteur, 75015 Paris, France.

Cite this article as *Cold Spring Harb Perspect Biol* doi: 10.1101/cshperspect.a040196

selective pressure to ensure dosage compensation between X chromosome and autosomal gene expression (Ohno 1967). Several studies have provided supporting evidence for Ohno's hypothesis that at least some X-linked genes are twofold up-regulated (Nguyen and Disteche 2006; Lin et al. 2007; Deng et al. 2011).

The other challenge posed by the evolution of sex chromsomes is the difference in X-linked gene dosage between males and females, the latter having two copies of potentially twofold expressed X-linked genes. This is believed to be the reason underlying the evolution of X chromosome inactivation (XCI) as a dosage compensation mechanism in XX individuals. XCI is characterized by the transcriptional silencing of one of the two X chromosomes, in every female cell, ensuring X-linked gene expression levels similar to those found in male cells.

Mary Lyon hypothesized the existence of XCI in 1961 (Lyon 1961). She had noted that females heterozygous for sex-linked loci affecting coat color showed a mosaic phenotype, displaying hair patches of different color across the body. Building up on previous observations that one X chromosome is heteropyknotic in female cells (Barr and Bertram 1949), Lyon proposed that:

- One of the two X chromosomes in female cells is "genetically inactivated" [sic],
- This inactivation happens early during embryonic development,
- Once inactivated, the same X chromosome remains inactivated in descending cells.

Decades of research have validated Mary Lyon's hypothesis, and XCI has become a paradigm for the study of epigenetics, as it implies the mitotically heritable silencing of one X chromosome without any need for underlying changes in its DNA sequence. Perhaps the most remarkable feature of this process is the fact that one X is treated differently to the other despite sharing the same nucleoplasm.

XCI is triggered and mediated by the *Xist* long noncoding RNA (lncRNA), and is characterized by changes at the levels of transcription, chromatin (e.g., DNA methylation, histone marks, and histone variants), chromosome folding, replication timing, and spatial reorganization in the nucleus.

Given that these events happen in an ordered way in early embryos and differentiating embryonic stem cells, XCI provides an ideal system to dissect the interplay between key epigenetic processes, using a plethora of genetic and quantitative approaches. Furthermore, as these molecular events happen chromosome-wide, across thousands of loci, this provides great power for statistical analyses.

In the 11 years that elapsed since the previous edition of this review (Chow and Heard 2010), crucial insights have been gained into the molecular underpinnings of this chromosome-wide silencing process. Here, we review the recent literature to depict the current models underlying the regulation of *Xist* expression, the nature of the structural and spatial reorganization of the X chromosome during XCI, as well as the mechanisms at play in establishing and maintaining chromosome-wide gene silencing.

X CHROMOSOME INACTIVATION DYNAMICS DURING EARLY DEVELOPMENT

Mary Lyon proposed that XCI takes place early during embryonic development, but not immediately following fertilization (Lyon 1961). Indeed, her observation of mosaic X-linked hair color phenotypes in females suggested that there must be a period during early development when each cell "chooses" to inactivate either the maternal or the paternal X chromosome.

Although Lyon initially postulated that XCI was random, early studies conducted in murine extraembryonic tissues revealed that on the contrary, in these tissues, it is always the paternal X (Xp) that is silenced, a phenomenon known as imprinted XCI (iXCI) (Takagi and Sasaki 1975; West et al. 1977). Given that embryonic/somatic tissues instead show a random pattern of XCI (Beutler et al. 1962; McMahon et al. 1983), as exemplified with mosaic hair color phenotypes, it was initially believed that embryonic and extraembryonic lineages undergo X inactivation independently, and with different modalities.

Later, it was shown that in fact all cells initially undergo imprinted XCI. Indeed, while both maternal and paternal X chromosomes are actively transcribed at zygotic gene activation (two-cell stage) (Fig. 1A; Okamoto et al. 2005; Patrat et al. 2009), imprinted XCI initiates at the four-cell stage (Fig. 1B) such that by the early blastocyst stage, 3.5 d postcoitum (E3.5), all cells of the embryo have an inactive Xp (Fig. 1C; Mak et al. 2004; Okamoto et al. 2004).

The Xp remains stably inactivated in extraembryonic tissues (i.e., trophectoderm and hypoblast; Fig. 1C). Remarkably, however, the Xp is reactivated specifically in all cells of the inner cell mass that will give rise to the epiblast (Fig. 1D; Mak et al. 2004; Okamoto et al. 2004; Borensztein et al. 2017a). At the time the epiblast forms and is further specified, each cell undergoes random inactivation (rXCI) of either the paternal or the maternal X chromosome, ~E5.5 (Fig. 1E,F; Monk and Harper 1979; Rastan 1982; Mak et al. 2004; Okamoto et al. 2004).

Strikingly, the developmental dynamics and timing of XCI vary extensively between placental mammals (Escamilla-Del-Arenal et al. 2011; Okamoto et al. 2011). In rabbits and humans, for example, XCI initiates at much later developmental stages than in mice, does not seem to be imprinted, and shows very different patterns (Okamoto et al. 2011). In humans, both X chromosomes appear to undergo transcriptional dampening—rather than complete inactivation of a single X chromosome—to ensure dosage compensation by the early blastocyst stage (Petropoulos et al. 2016; Vallot et al. 2017). Random XCI only proceeds later, after the late blastocyst stage (Okamoto et al. 2011).

The diverse strategies of XCI are thought to have evolved due to developmental differences between eutherian species, for example, in the timing of zygotic gene activation and of cellular

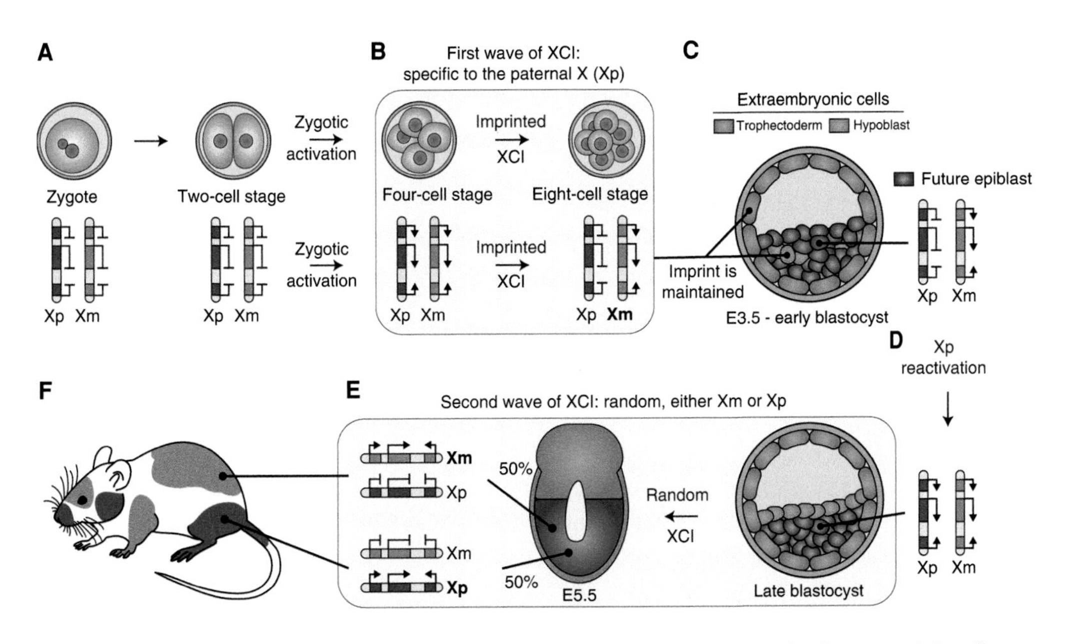

Figure 1. Dynamics of X chromosome inactivation (XCI) during early mouse development. (*A*) Following fertilization and zygotic genome activation, both X chromosomes are expressed. (*B*) At the four-cell stage, imprinted XCI initiates, and (*C*) by the early blastocyst stage, all cells have inactivated the paternal X (Xp). The Xp remains inactivated in extraembryonic cells (trophectoderm and hypoblast). (*D*) On the other hand, the Xp becomes reactivated in cells that will later give rise to the epiblast (embryo proper). (*E*) In these cells, a second wave of XCI (random) initiates at the late blastocyst stage. Hence, after implantation and clonal propagation of XCI status, half of the cells of the embryo proper will have inactivated the Xp, and the other half the maternal X (Xm). (*F*) Random XCI hence results in cellular mosaicism with respect to X-linked allelic expression, which can, for example, translate into mosaic hair phenotypes.

differentiation (Monk and Harper 1979; Okamoto et al. 2011; Migeon 2016). In metatherian mammals (i.e., marsupials), XCI is imprinted in all tissues (Cooper et al. 1971, 1993; Sharman 1971; Mahadevaiah et al. 2020). In monotremes, much less is known and dosage compensation seems to involve gene by gene monoallelic expression (Deakin et al. 2008; Rens et al. 2010; Livernois et al. 2013; Whitworth and Pask 2016).

X-CHROMOSOME INACTIVATION CAN BE RECAPITULATED IN VITRO

Murine embryonic stem cells (mESCs) are pluripotent and have two active X chromosomes. Similarly to what is observed in vivo at the blastocyst stage (Fig. 1D,E), random XCI initiates during cellular differentiation (Martin et al. 1978; McBurney and Strutt 1980; Penny et al. 1996).

The ability to recapitulate XCI in differentiating female ESCs (i.e., in vitro) has made XCI research much more accessible and convenient, as it does not require the handling of peri-implantation embryos, which are particularly challenging to work with. Furthermore, mESCs in culture can be genetically manipulated easily and yield high cell numbers, allowing for the use of several technical tools and approaches that could not have been employed in early embryos. In fact, over the past 20 yr, many groundbreaking discoveries in the XCI field have been made thanks to the mESC (in vitro) model.

In one landmark study, Wutz and Jaenisch used differentiation of mESCs to show that XCI can be separated in two stages: initiation and maintenance. The initiation stage is restricted to early differentiation and corresponds to the time window during which X-linked gene silencing takes place. In the subsequent maintenance stage, gene silencing supposedly becomes irreversible and the inactive state of the X chromosome is epigenetically transmitted through cell division (Wutz and Jaenisch 2000). Similarly in vivo, the Xi state is clonally propagated in somatic cells after XCI has been established (McMahon et al. 1983).

THE *X-INACTIVATION CENTER* AND THE DISCOVERY OF *XIST* AS AN ESSENTIAL MEDIATOR OF XCI

Twenty years after Lyon's seminal article, through analyses of X-chromosomal deletions and translocations in mouse and human cells, a single X-linked locus, termed the *X-inactivation center* (*Xic*/*XIC*), was identified to be the region that is necessary and sufficient to trigger XCI (Fig. 2A; Rastan 1983; Rastan and Robertson 1985; Brown et al. 1991b). Importantly, this locus has to be physically linked to the X-chromosome for XCI to occur, demonstrating that elements comprised within the *Xic* regulate XCI in *cis*.

In the early 1990s, a novel human X-linked gene was discovered within the candidate *XIC* region, which, contrary to most other X-linked genes, was found to be expressed from the inactive X chromosome exclusively, making it a likely player in XCI (Brown et al. 1991a). This gene, called *Xist*/*XIST* (for *X-inactive specific transcript*; Fig. 2B), mapped within the candidate *Xic*/*XIC* regions in both mouse and human, showing some level of conservation at both sequence and gene structure levels (Brockdorff et al. 1991; Brown et al. 1992). The *Xist* transcript (15 kb in mouse, 17 kb in humans after splicing) is capped and polyadenylated but is not protein coded (Fig. 2B; Brockdorff et al. 1992; Brown et al. 1992).

Remarkably, *Xist* RNA is retained in the nucleus, where it stably accumulates on the X chromosome from which it is transcribed (Clemson et al. 1996), a phenomenon commonly described as "coating" (Fig. 2C). Although several factors including SAF-A/hnRNPU (Hasegawa et al. 2010), CIZ1 (Ridings-Figueroa et al. 2017; Sunwoo et al. 2017), PTBP1, MATR3, CELF1, and TDP-43 (Pandya-Jones et al. 2020) have been shown to play an important role in *Xist* localization, the mechanisms through which coating takes place still remain enigmatic (for review, see Brockdorff 2019). In fact, *Xist* RNA initially does not spread uniformly and linearly along the X chromosome. Instead, *Xist* first accumulates at discrete "entry sites"—which show spatial proximity to the site of *Xist* transcription—and only

Cite this article as *Cold Spring Harb Perspect Biol* doi: 10.1101/cshperspect.a040196

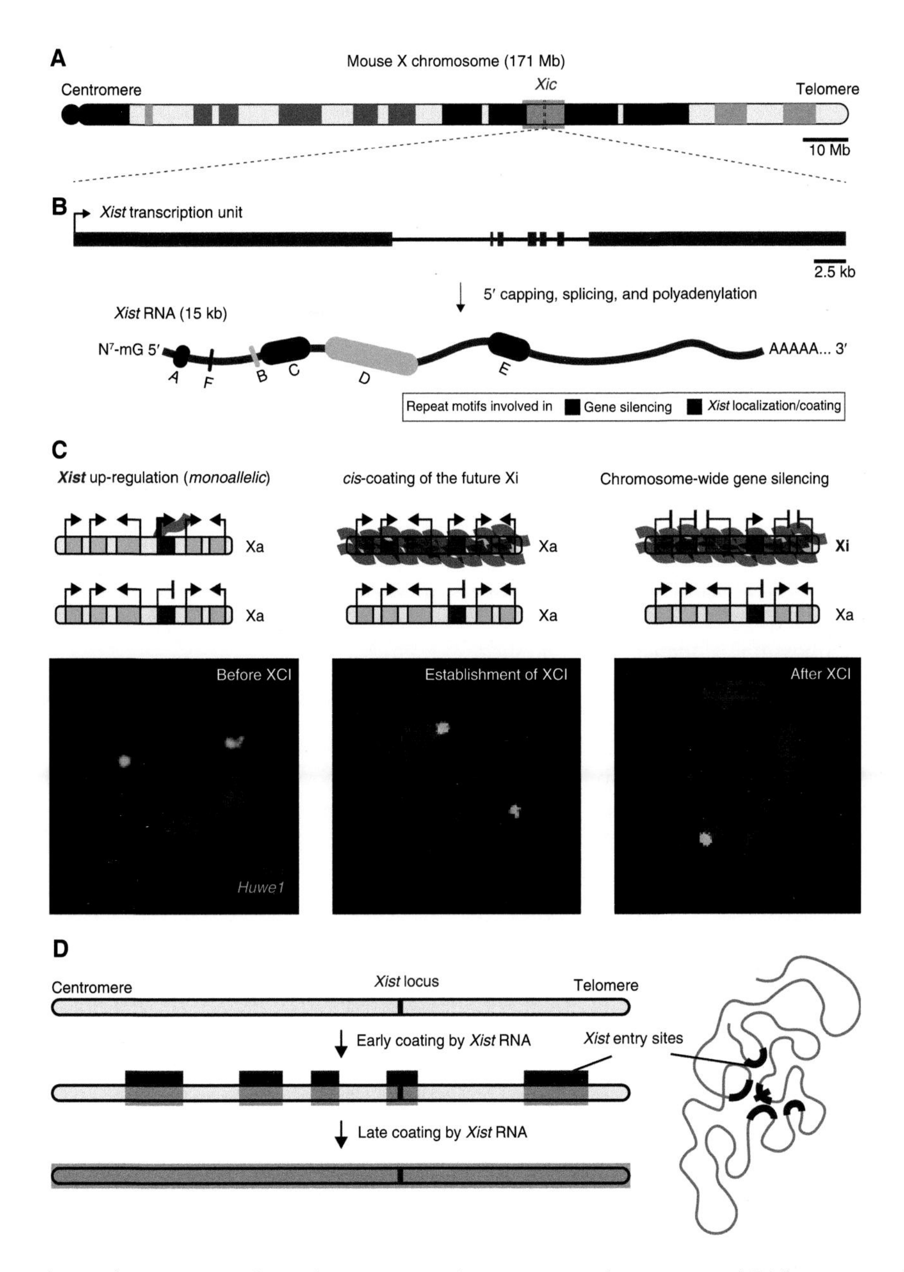

Figure 2. The *Xist* lncRNA coats the X chromosome and triggers gene silencing in *cis*. (*A*) The position of the *X inactivation center* (*Xic*, highlighted in red), is shown relative to the mouse X chromosome. (*B*) The *Xist* transcription unit (located within the *Xic*) is shown with exons corresponding to the predominant transcript isoform represented by blue rectangles. *Xist* RNA is capped, spliced, and polyadenylated to yield a 15-kb-long noncoding RNA. *Xist* is composed of several repeat elements (labeled *A–F*), some of which are important for *Xist* localization (blue) and *Xist*-mediated gene silencing (black). (*C*) (*Left* panel) Before X chromosome inactivation (XCI), both X chromosomes are expressed and, consequently, the X-linked gene *Huwe1* (green rectangle) is biallelically expressed (two green foci of nascent transcription are detected by RNA FISH in murine embryonic stem cells (mESCs). XCI initiates with the up-regulation of *Xist* on one of the two X chromosomes. (*Middle* panel) *Xist* RNA coats the chromosome from which it is expressed, forming a detectable *Xist* cloud (red domain detected by RNA FISH). (*Right* panel) *Xist* coating triggers *cis*-transcriptional silencing of virtually all genes across the X chromosome, with *Huwe1* transcription being restricted to the active (i.e., non-*Xist*-coated) X chromosome. (Image courtesy of the authors. N.B. some images were modified for illustration purposes.) (*D*) *Xist* does not spread uniformly through time across the X chromosome. Instead, *Xist* first spreads into regions termed *Xist* entry sites (shown in blue), which are spatially proximal to the site of *Xist* transcription.

then spreads to more spatially distant sites (Fig. 2D; Engreitz et al. 2013).

Deletion of the *Xist* gene in mESCs (Penny et al. 1996) and embryos (Marahrens et al. 1997) results in their failure to undergo XCI, while ectopic expression of multicopy or inducible *Xist* transgenes from autosomes leads to *cis*-autosomal gene inactivation (Wutz and Jaenisch 2000; Popova et al. 2006; Loda et al. 2017). Altogether, these studies demonstrate that *Xist* is a lncRNA whose expression is the earliest event required to mediate XCI in *cis* (Fig. 2C).

Although metatherian mammals (i.e., marsupials) also undergo XCI, *Xist* is not found in marsupial genomes (Duret et al. 2006). Remarkably, however, a study conducted in the gray short-tailed opossum *Monodelphis domestica* identified that the marsupial gene *RSX* (for *RNA-on-the-silent-X*) encodes for a polyadenylated 27-kb-long RNA, which, identically to *Xist*, is expressed only in females, exclusively from the inactive X chromosome (Grant et al. 2012). What is more, *RSX* also coats the Xi in *cis* and its expression from an autosomal transgene leads to *cis*-autosomal gene inactivation, suggesting that *RSX* mediates XCI in methatherians (Grant et al. 2012). Given that *RSX*, however, shares no sequence homology with *Xist*, the use of RNA-mediated XCI is likely to have evolved convergently in both therian infraclasses (Grant et al. 2012).

Xist IS ORGANIZED IN FUNCTIONAL "REPEAT" MODULES

Given that it is upon coating by *Xist* RNA that the X chromosome undergoes gene silencing and "epigenetic reorganization" (see sections below), most of the research investigating the mechanisms driving XCI have relied on dissecting the function of *Xist* RNA itself.

Xist RNA is organized in distinct "repeat" motifs (Brown et al. 1992), which act as functional modules during XCI (Fig. 2B; Wutz et al. 2002). The most highly conserved A-repeat, located at the 5′ end of *Xist*, is required for its gene-silencing function both in vitro and in vivo (Fig. 2B; Wutz et al. 2002; Sakata et al. 2017). On the other hand, multiple regions located downstream of the A-repeat mediate *Xist*'s ability to localize and accumulate (i.e., coat) in *cis* (Wutz et al. 2002). The regions involved in such localization are functionally redundant and map to domains comprised within the F- (Wutz et al. 2002), C- (Jeon and Lee 2011), and E-repeats (Yamada et al. 2015; Ridings-Figueroa et al. 2017; Sunwoo et al. 2017) of *Xist* (Fig. 2B). While these studies collectively suggest that *Xist*'s abilities to localize in *cis* and to silence genes are genetically and physically separable, recent evidence demonstrates that A-repeat mediated gene silencing requires synergy with downstream sequence elements within *Xist* (Trotman et al. 2020).

XIST EXPRESSION REGULATION DURING DIFFERENTIATION AND DEVELOPMENT

Prior to cellular differentiation (i.e., exit from pluripotency), *Xist* is barely expressed in mESCs. Upon differentiation however, *Xist* becomes upregulated and stabilized, ultimately remaining expressed strictly on one of the two X chromosomes (Fig. 2C; Panning et al. 1997; Sheardown et al. 1997). How *Xist* expression is so tightly regulated during differentiation and development has been extensively investigated and reviewed (Augui et al. 2011; Pontier and Gribnau 2011; Galupa and Heard 2015; van Bemmel et al. 2016). It is now accepted that *Xist* is mainly regulated at the transcriptional level, and that upregulation of *Xist* on all but one X chromosome in each cell is ensured through a complex regulatory network involving the combined actions of *cis*- and *trans*-regulators (Fig. 3; Loos et al. 2016; Mutzel et al. 2019; Mutzel and Schulz 2020).

cis-Regulators of *Xist*

Xist's best characterized *cis*-regulators lie within the *Xic* locus and include several non-protein-coding genes. The most notable negative *cis*-regulator of *Xist* is the *Tsix* noncoding RNA, which overlaps and is transcribed antisense to the *Xist* gene (Fig. 3A1; Lee et al. 1999). *Tsix* transcription across the promoter of *Xist* is required to keep *Xist* expression at very low levels in undif-

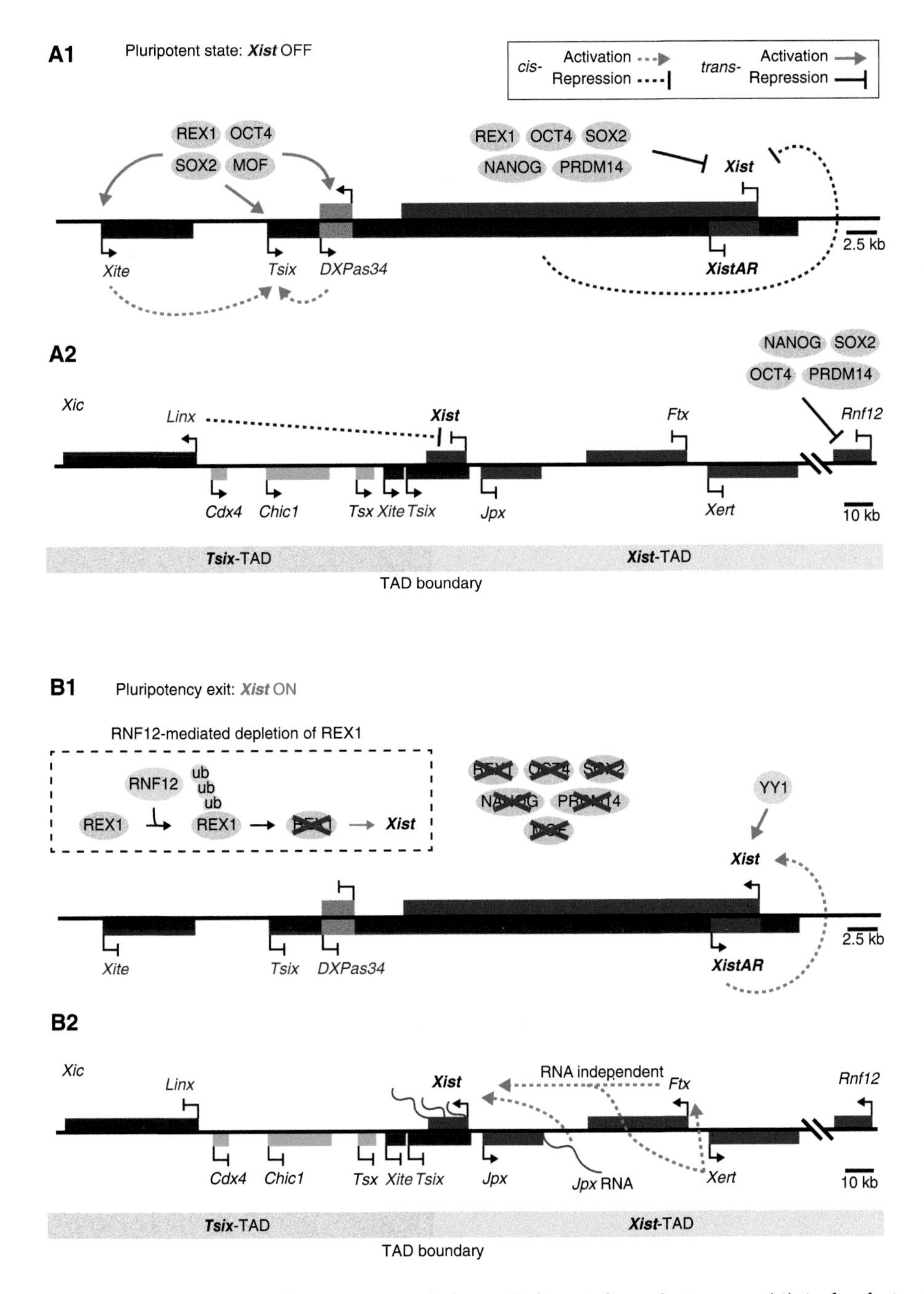

Figure 3. *cis*- and *trans*-regulation of *Xist* expression before and after exit from pluripotency. (*A*) At the pluripotent state, *Xist* expression is repressed by the combination of *cis*- and *trans*-regulators. (*A1*) *Tsix* transcription, which is promoted by its *Xite* and *DXPas34* enhancer elements, represses *Xist* expression in *cis*. *Tsix* expression is driven by several key pluripotency factors. Conversely, those same pluripotency factors also repress *Xist* in *trans*. (*A2*) The *Xic* is spatially partitioned into the *Xist* and *Tsix*-TADs (topologically associating domains), which contain their respective *cis*-activators. *Linx* represses *Xist* expression, in *cis*, across the TADs boundary. *Rnf12* is also repressed in *trans* by several pluripotency factors. (*B*) Upon pluripotency exit, pluripotency factors are depleted, and their *trans*-repressive activity on *Xist* and *Rnf12* is relieved. (*B1*) Furthermore, RNF12 ubiquitinates REX1, triggering its degradation. *Xist* expression is further activated in *trans* by YY1, and (*B2*) in *cis* by *Jpx*, *Ftx*, *Xert*, and *XistAR*.

ferentiated mESCs (Luikenhuis et al. 2001; Stavropoulos et al. 2001), and *Tsix* deletion on one X chromosome results in skewed inactivation of that chromosome (Lee and Lu 1999).

Tsix expression requires active transcription of its two enhancers (Stavropoulos et al. 2005): the *DXPas34* minisatellite (Debrand et al. 1999; Vigneau et al. 2006; Cohen et al. 2007; Navarro et al. 2010) and the *Xite* intergenic transcription element (Fig. 3A1; Ogawa and Lee 2003). More recently, another noncoding locus, *Linx*—lying downstream of *Tsix*—was discovered to play a role in negatively regulating *Xist* expression independently of *Tsix* and *Xite* (see below; Fig. 3A2; Galupa et al. 2020).

Conversely, *Jpx* and *Ftx*, two genes located within the *Xic* and that escape XCI, promote *Xist* expression in *cis* (Fig. 3B2) through mechanisms that remain to be fully understood. While *Jpx* appears to exert its *Xist*-activating role through its RNA product (Fig. 3B2; Tian et al. 2010; Sun et al. 2013; Carmona et al. 2018), the effect *Ftx* mediates requires its expression but is independent of its encoded transcript in mouse (Fig. 3B2; Chureau et al. 2011; Furlan et al. 2018). Additionally, *XistAR*, an antisense transcript originating from *Xist* exon 1, which is expressed specifically from the Xi (in vitro and in vivo), has been proposed to promote *Xist* expression in *cis* (Fig. 3B1; Sarkar et al. 2015).

Finally, a recent genetic screen for *Xist* regulatory elements identified *Xert*, a novel lncRNA within the *Xic*, which promotes *cis*-activation of *Xist*, specifically during random XCI (Fig. 3B2; Gjaltema et al. 2021). *Xert* expression is up-regulated concomitantly with *Jpx* and *Ftx*, and the *Xert* promoter and an intronic enhancer cluster (*XertP* and *XertE*, respectively) mediate *Xist* activation through physical contacts with the *Xist* locus, but also by promoting *Ftx* expression in *cis* (Fig. 3B2; Gjaltema et al. 2021).

Spatial Partitioning of *Xist*'s *cis*-Regulators

The *Xic* is therefore a remarkable locus in the sense that it harbors a series of adjacent genetic elements mediating antagonizing roles with respect to *Xist* expression. In this context, our group identified that the *Xic* is "spatially partitioned" [sic] in two distinct sub-megabase domains, separated by a boundary element located between the promoters of *Xist* and *Tsix* (Fig. 3A2; Nora et al. 2012). These domains within which loci interact more frequently with one another than they do with loci outside were termed topologically associating domains (TADs) and correspond to a layer of chromosome folding found throughout the genome (Dixon et al. 2012; Nora et al. 2012).

The *Xist*-TAD comprises the *Xist* promoter and all of *Xist*'s *cis*-activators, while the *Tsix*-TAD comprises the *Tsix* promoter and all of *Tsix*'s *cis*-activators (Fig. 3A2,B2; Nora et al. 2012). Given that during differentiation genes within the *Xist*-TAD become up-regulated while genes within the *Tsix*-TAD become down-regulated, TADs were proposed to act as insulated regulatory neighborhoods, favoring intra-TAD (and conversely preventing inter-TAD) communication between genes and *cis*-regulatory elements (Nora et al. 2012, 2013; Shen et al. 2012).

Consistently, inverting the *Xist*/*Tsix* transcription unit, thereby positioning the *Xist* and *Tsix* promoters in each other's original TAD, results in exchanged *Xist* versus *Tsix* expression dynamics during differentiation (van Bemmel et al. 2019). The recent discovery that *LinxP* (the promoter of the *Linx* noncoding RNA), which lies in the *Tsix*-TAD, acts as a long-range silencer of *Xist* across the *Xist*/*Tsix*-TADs boundary (Fig. 3A2), suggests that inter-TAD *cis*-regulation is nonetheless possible—and functionally important—at the *Xic* (Galupa et al. 2020).

Trans-Regulators of *Xist*

Several key regulators of the pluripotency transcriptional network, including OCT4, NANOG, SOX2, REX1, PRDM14, and MOF/MSL, repress *Xist* expression in *trans* (Fig. 3A1). These factors bind *Tsix* and *Xist* *cis*-regulatory DNA elements, inhibiting *Xist* expression, either through activation of *Tsix* transcription (Fig. 3A2; Donohoe et al. 2009; Navarro et al. 2010; Gontan et al. 2012; Payer et al. 2013; Chelmicki et al. 2014) or direct repression of *Xist* itself (Fig. 3A1; Navarro et al. 2008; Gontan et al. 2012; Payer et al. 2013).

Conversely, the X-linked gene *Rnf12* activates *Xist* and XCI in *trans*, both in mESCs (Jonkers et al. 2009; Barakat et al. 2011) as well as during imprinted XCI in vivo (Shin et al. 2010). Through its E3 ligase activity, RNF12 ubiquitinates REX1, leading to its subsequent proteasomal degradation, thereby alleviating REX1-mediated inhibition of *Xist* expression (Fig. 3B1; Gontan et al. 2012, 2018). What is more, *Rnf12* expression is, similarly to *Xist*, also repressed in *trans* by OCT4, NANOG, SOX2, and PRDM14 (Fig. 3A2; Navarro et al. 2011; Payer et al. 2013). Another putative *trans*-activator or *Xist* is the transcription factor YY1, which competes with REX1 for binding to *Xist*'s promoter region (Makhlouf et al. 2014).

Although several factors involved in *Xist* regulation are yet to be characterized (Pacini et al. 2021), the regulatory network unraveled so far already provides a molecular basis for the temporal link observed between pluripotency exit and *Xist* expression (Fig. 3).

Imprinted *Xist* Expression during Early Mouse Embryogenesis

During the imprinted wave of XCI in mouse (see first section), *Xist* expression is restricted to the paternal allele (Okamoto et al. 2004). Maternal *Xist* expression is repressed during iXCI thanks to maternal H3K27me3 imprinting of the *Xist* locus (Inoue et al. 2017, 2018; Harris et al. 2019; Collombet et al. 2020), which is deposited in oocytes through a complex interplay between PRC1 and PRC2 (Chen et al. 2021; Mei et al. 2021).

Xist expression from the maternal allele is later hindered in extraembryonic tissues by maternal-specific expression of *Tsix* (Lee 2000; Sado et al. 2001). Remarkably, in the marsupial *M. domestica*, where imprinted XCI is the rule in all tissues, the discovery of *XSR*, a gene that overlaps and is transcribed antisense to *RSX*, and whose expression is specific to the maternal (active) X chromosome, suggests that similarly to *Tsix* in mice *XSR* plays a crucial role in imprinted XCI in marsupials (Mahadevaiah et al. 2020).

GENE-SILENCING KINETICS AND ESCAPE DURING XCI

Although XCI ultimately affects almost the whole X chromosome, the dynamics of gene silencing along the X vary between genes, both in vitro (in mESCs) and in vivo (in preimplantation embryos) (Patrat et al. 2009; Chow et al. 2010; Marks et al. 2015; Borensztein et al. 2017b; Cheng et al. 2019). Indeed, some genes are silenced very early during development (e.g., *Rnf12*, over the course of a few hours), while others are silenced much later (e.g., *Huwe1*, over the course of several days), with the order of gene silencing being similar between in vitro and in vivo models (De Andrade et al. 2019).

Several genomic and epigenetic features—including distance to *Xist*, distance to LINE1 elements, gene density, local chromosome topology as well as certain chromatin modifications—have been linked to the kinetics and efficiency with which a given gene will be silenced (Borensztein et al. 2017b; Loda et al. 2017; De Andrade et al. 2019). However, how genes across the inactivating X chromosome achieve such differences in gene-silencing kinetics still remains elusive, and whether these differences have any biological significance is still an open question.

Remarkably, some genes called "escapees" do not undergo silencing during XCI, and remain or become actively expressed from the Xi in somatic cells. The number and nature of escaping genes vary extensively between tissues, individuals, and species (Cotton et al. 2013; Berletch et al. 2015; Tukiainen et al. 2017; Oliva et al. 2020; Balaton et al. 2021). In humans, it is estimated that 15%–30% of X-linked genes escape XCI, while only 3%–7% generally do so in mice (although up to 20% of genes escaping has been observed in certain mouse cell lines) (Carrel et al. 1999; Yang et al. 2010; Berletch et al. 2011; Cotton et al. 2013; Balaton et al. 2015; Carrel and Brown 2017; Tukiainen et al. 2017; Oliva et al. 2020). At least two categories of escapees have been distinguished. Constitutive escapees evade XCI throughout all stages of development, in all tissues. On the other hand, facultative escapees become silenced initially during development, but are reactivated later on, often in a tissue-specific manner and variably between individuals (Patrat et al. 2009; Berletch et al. 2011, 2015; Giorgetti et al. 2016).

How genes manage to escape from XCI is an open question, and whether the mechanisms involved differ between constitutive and facultative escape is unknown. One primary determinant of constitutive escape is likely to be whether the gene is subject to *Xist* coating and hence targeted by the repressive complexes *Xist* recruits (see section below). Consistently, chromosome-wide mapping of *Xist* "binding sites" revealed that escapees are relatively depleted of *Xist* "binding" compared to silenced genes (Engreitz et al. 2013; Simon et al. 2013).

Possible roles for CTCF and YY1 in promoting escape have been proposed, as these two transcription factors are specifically enriched at the promoters of escapees, including *Xist* (Makhlouf et al. 2014; Chen et al. 2016b; Loda et al. 2017). Other studies have identified DNA sequence elements that might autonomously drive the escape of certain genes, regardless of their localization on the X chromosome, suggesting that escape is an "intrinsic property" of these loci (Li and Carrel 2008; Peeters et al. 2018). In addition, boundary elements that inhibit the spread of escape into adjacent loci have also been reported (Horvath et al. 2013), and a role for young LINEs in facilitating silencing of genes that might otherwise be prone to escape has also been put forward as a hypothesis (Chow et al. 2010). Of all escapees, *Xist* is perhaps the most remarkable in the sense that it somehow escapes its own repression on the Xi, through mechanisms that remain enigmatic. The discovery of a DNA sequence in the 5′ region of *Xist* bearing potent transcriptional antitermination activity was recently proposed to confer *Xist* the ability to evade transcriptional silencing (Trotman et al. 2020).

An increasing number of studies are highlighting the functional importance of genes that escape in females. Of note, X-linked genes with functional Y-chromosome homologs (e.g., those lying within the pseudoautosomal region) have evolved to escape so as to ensure their correct dosage (Posynick and Brown 2019). On the other hand, given that biallelic expression of escapees with no Y-linked homologs should result in higher protein dosage in females compared to males, it is likely that such escapees are important for certain female-specific functions (Posynick and Brown 2019). In light of the high number of genes escaping XCI in humans, aberrant dosage of these genes is also thought to be the underlying cause of certain phenotypes associated with sex chromosome aneuploidies, notably Turner syndrome (Carrel and Brown 2017; Posynick and Brown 2019; Navarro-Cobos et al. 2020).

NUCLEAR LOCALIZATION OF THE INACTIVE X CHROMOSOME

In 1949, Barr and Bertram reported that upon staining with basic dyes, female cat neurons frequently display a small but densely stained nuclear body—originally termed nucleolar satellite due to its close proximity to the nucleolus—that is, however, hardly visible in males (Barr and Bertram 1949). Barr even proposed that such cytological dichotomy between males and females could readily be used to sex animals (Barr and Bertram 1949). This "nucleolar satellite" was later identified to be the inactive X chromosome (Ohno et al. 1959; Lyon 1961, 1962), and is now commonly referred to as the Barr body.

It has long been described that the Xi is most often localized at the nuclear periphery, in contact with the nuclear envelope (Ohno et al. 1958; Barton et al. 1964; Bourgeois et al. 1985; Dyer et al. 1989; Rego et al. 2008) or around the nucleolus (Fig. 4; Barr and Bertram 1949; Bourgeois et al. 1985; Zhang et al. 2007; Rego et al. 2008).

The association between the Xi and the nuclear membrane was recently shown to be dependent on the lamin B receptor (LBR), a protein known to be involved in anchoring heterochromatin to the nuclear envelope (Nikolakaki et al. 2017). LBR directly binds *Xist* RNA at multiple sites, subsequently recruiting the Xi to the nuclear periphery (Fig. 4; McHugh et al. 2015; Minajigi et al. 2015; Chen et al. 2016a). LBR-mediated recruitment was initially reported to be crucial for the silencing of five X-linked genes (McHugh et al. 2015; Chen et al. 2016a).

 Cite this article as *Cold Spring Harb Perspect Biol* doi: 10.1101/cshperspect.a040196

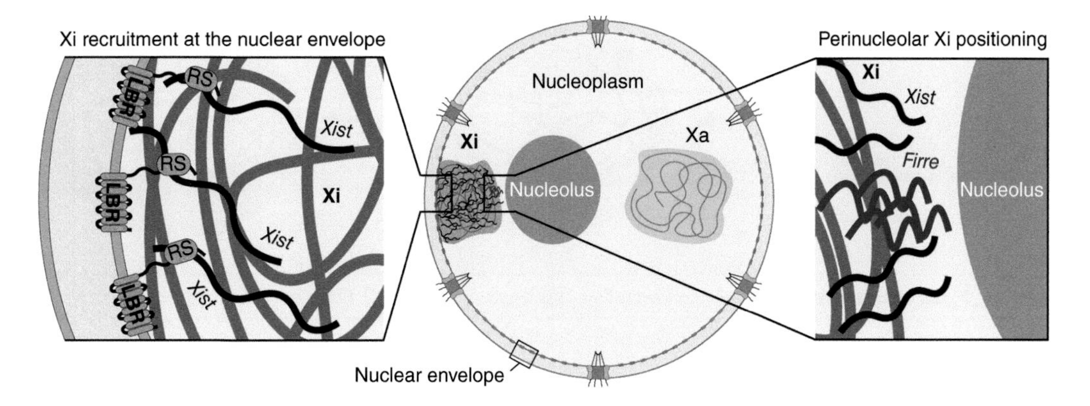

Figure 4. Positioning of the inactive X chromosome at the nuclear periphery and/or proximal to the nucleolus. Through its RS motif, LBR (the lamin B receptor) directly binds *Xist* and recruits the X chromosome to the nuclear lamina. Targeting the Xi to the perinucleolar space requires *Xist* and *Firre*, although the mechanisms involved are unknown.

Recent studies however demonstrated a very minor (if any) role for LBR in chromosome-wide gene silencing during XCI, both in vitro and in vivo (Nesterova et al. 2019; Young et al. 2021). Hence, the association between the Xi and the nuclear periphery is not crucial for gene silencing during XCI.

The association of the Xi with the nucleolus also appears to be dependent on *Xist* (Zhang et al. 2007) as well as another X-linked lncRNA, *Firre* (Fig. 4; Yang et al. 2015), although the mechanisms at play are unknown. The functional relevance are such association is not completely clear, but it is proposed that the ability of the Xi to localize perinucleolarly is important to maintain its heterochromatin (Zhang et al. 2007; Yang et al. 2015).

Finally, several studies have reported that the two X chromosomes can pair together, transiently, at the level of the *Xic*, during the early steps of XCI. Such transient *Xic* pairing was proposed to allow the X chromosomes to "sense" one another and "choose" which one to inactivate, by enabling symmetry breaking at the level of the *Xist*/*Tsix* locus, to facilitate monoallelic expression (Bacher et al. 2006; Xu et al. 2006; Augui et al. 2007; Masui et al. 2011).

Taken together, the aforementioned studies raised the exciting possibility that the localization of both X chromosomes, relative to one another, as well as relative to the nuclear periphery and the nucleolus, could explain how two genetically identical chromosomes might be treated differently within the same nucleus. However, subsequent studies elegantly demonstrated that subnuclear positioning as well as pairing of the X chromosomes do not play major roles in *Xist* expression and localization, nor in gene silencing, at least during initiation of XCI (Barakat et al. 2014; Nesterova et al. 2019; Pollex and Heard 2019; Young et al. 2021).

THE *XIST* RNA COMPARTMENT

The earliest event following *Xist* up-regulation is the apparent depletion of the RNA polymerase II (RNAPII) machinery within the *Xist* RNA cloud (Fig. 5A; Chaumeil et al. 2006). Persistent *Xist* coating along the chromosome results in the formation of what appears to be a repressive nuclear compartment (termed the *Xist* RNA compartment), into which X-linked genes are seen to relocate upon their silencing (Fig. 5A; Chaumeil et al. 2006; Collombet et al. 2020). Importantly, the *Xist* compartment is initially predominantly composed of repetitive DNA at its core, and silenced genes localize preferentially at the internal periphery of this compartment (Dietzel et al. 1999; Chaumeil et al. 2006).

Whether relocalization of genes into the *Xist* compartment is a cause or consequence of their silencing remains to be explored. It is evident

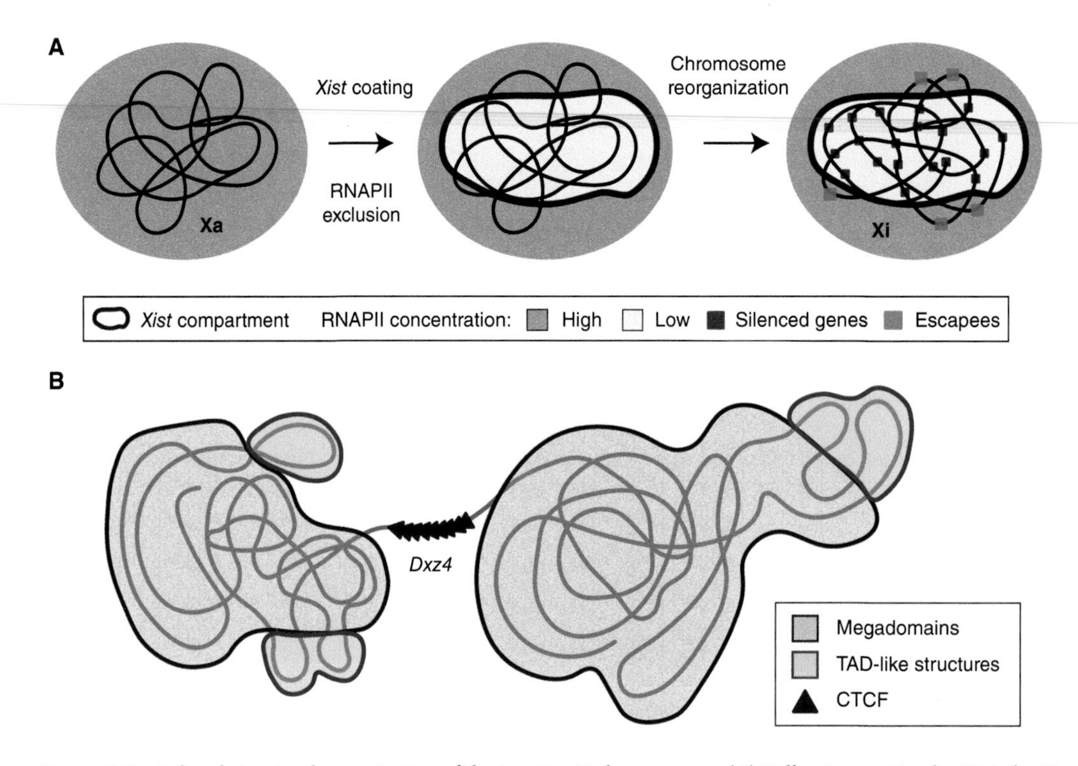

Figure 5. Spatial and structural organization of the inactive X chromosome. (*A*) Following coating by *Xist*, the X chromosome forms a compartment within which RNA polymerase II (RNAPII) concentration is significantly lower than in the nucleoplasm. As X chromosome inactivation (XCI) proceeds, silenced genes become localized inside this compartment, while escapees remain outside or at the periphery. (*B*) The inactive X chromosome adopts a peculiar structure, characterized by two megadomains separated by the *Dxz4* boundary element, harboring strong CTCF binding. While the megadomains show poor local structure, some level of organization is retained at clusters of escaping genes, which form topologically associating domain (TAD)-like structures.

however that these two events are linked. Indeed, gene relocalization into the compartment is not observed in a context where *Xist*, which has been deleted of its A-repeat, is incapable of triggering transcriptional silencing (Chaumeil et al. 2006). Furthermore, genes that remain actively expressed from the Xi (i.e., escapees) are most often located outside (or at the edge) of the *Xist* compartment (Fig. 5A), where they stay accessible to the transcription machinery (Dietzel et al. 1999; Chaumeil et al. 2006; Calabrese et al. 2012; Collombet et al. 2020).

A recent study using superresolution microscopy showed that throughout XCI, the *Xist* compartment is formed by a relatively low number (approximately 50) of *Xist* granules, each containing, on average, two *Xist* molecules (Markaki et al. 2020). The authors gathered evidence that these ~50 granules are locally confined but nonetheless capable of triggering silencing across the >1000 genes of the X chromosome by nucleating the formation of concentrated clusters of so-called "supramolecular complexes" (SMCs), characterized by a large number of interacting protein molecules. Within these SMCs, the increasing concentration of PRC1 (see section below) mediates chromatin compaction, resulting in the progressive recruitment of genes to *Xist*-SMCs where they become targeted by repressive factors, allowing silencing to spread across the entire chromosome (Markaki et al. 2020).

Is Phase Separation at Play in XCI?

During later stages of XCI, Xi compartmentalization through *Xist*-dependent recruitment of four RNA-binding proteins (RBPs) (PTBP1, MATR3, TDP-43, and CELF1) was shown to be important to enforce gene silencing (see section below), and the ability of these RBPs to form biological condensates on the Xi was invoked (Pandya-Jones et al. 2020).

Although phase separation has been proposed as a mechanism through which *Xist* might mediate its function during XCI (Cerase et al. 2019), so far no experimental evidence that the *Xist*-compartment is indeed phase separated (i.e., in which diffusion of molecules varies from that observed outside) has been reported.

Using quantitative microscopy and single-particle tracking in mESCs to study the *Xist* RNA compartment, we have recently shown that depletion of RNAPII from the Xi does not appear to be due to a biophysical compartmentalization of the *Xist*-coated territory, at least during the initiation phase of XCI (Collombet et al. 2021). Instead, RNAPII depletion is a consequence of reduced binding rates between the transcription machinery and X-linked chromatin (Collombet et al. 2021), consistent with earlier observations (Calabrese et al. 2012).

Whether the *Xist* RNA compartment later becomes phase separated, and whether this plays a role in the maintenance of XCI in differentiated cells, remain open questions. Demonstrating causality between any observed phase separation and the many changes that take place during XCI will be far from trivial.

STRUCTURAL ORGANIZATION OF THE INACTIVE X CHROMOSOME

Fluorescence microscopy-based studies had previously revealed that the Xi undergoes a certain degree of spatial reorganization with mild overall compaction (1.2-fold over the active X) (Teller et al. 2011), and a change in shape (Eils et al. 1996; Chaumeil et al. 2006; Rego et al. 2008; Smeets et al. 2014). However, the advent of chromosome conformation capture techniques has truly revolutionized our capacity to delve into chromosome organization at the molecular level, enabling spatial positioning of genomic loci to be probed relative to one another, and in a chromosome-wide fashion (Kempfer and Pombo 2020).

These approaches revealed that contrary to the Xa, the inactive X chromosome is globally devoid of TADs (Splinter et al. 2011; Minajigi et al. 2015; Giorgetti et al. 2016; Collombet et al. 2020), except at clusters of escapee genes, where preferential physical interactions between such genes result in the detection of TAD-like structures (Fig. 5B; Giorgetti et al. 2016; Splinter et al. 2011). Consistently, the Xi shows reduced binding of CTCF and cohesin (Calabrese et al. 2012; Minajigi et al. 2015; Gdula et al. 2019), two factors essential for TAD formation.

Structural analyses performed across human, macaque, and mouse revealed that the entire Xi folds into a unique bipartite structure composed of two megadomains showing very little local structure, and separated by a boundary element containing the *DXZ4*/*Dxz4* macrosatellite (Fig. 5B; Rao et al. 2014; Deng et al. 2015; Minajigi et al. 2015; Darrow et al. 2016; Giorgetti et al. 2016). While *DXZ4*/*Dxz4* is heterochromatic on the Xa allele, the Xi allele is euchromatic and shows high CTCF binding (Fig. 5B; Chadwick 2008; Horakova et al. 2012a,b). Deletion of *Dxz4* results in fusion of the megadomains, highlighting its crucial role in structuring the Xi (Giorgetti et al. 2016; Froberg et al. 2018).

The observation that deleting *Xist* results in an Xi conformation that resembles that of the Xa even though genes remain stably silenced (Splinter et al. 2011; Minajigi et al. 2015) suggests that *Xist* plays a key role in the global reorganization of the Xi. This role is proposed to involve the repulsion of architectural factors—notably cohesin—which stabilize chromosome conformation (Minajigi et al. 2015).

Several recent studies have also highlighted that SMCHD1, which contains a structural maintenance of chromosome (SMC) domain similar to that found in cohesins and condensins (Jansz et al. 2017), plays a critical role in structuring the 3D organization of the Xi. Indeed, *Smchd1*/*SMCHD1* loss in mouse and human cells results in decompaction of the Xi (Nozawa et al. 2013), accompanied by increased short-range interactions and compartmentalization, to a level similar to what is observed on the Xa (Jansz et al. 2018a; Wang et al. 2018; Gdula et al. 2019). SMCHD1 recruitment to the Xi requires

H2AK119ub1 marked chromatin and is consequently dependent on the *Xist*-hnRNPK axis (see section below) (Jansz et al. 2018b; Wang et al. 2019). Once recruited, SMCHD1 is proposed to antagonize binding of CTCF and cohesin, thereby disrupting TAD-organization on the Xi (Chen et al. 2015; Wang et al. 2018; Gdula et al. 2019).

CHROMATIN CHANGES DURING XCI

In parallel to *Xist* coating, the X chromosome undergoing XCI is also subject to drastic chromatin changes, mostly described at the level of its histone marks, as summarized in Table 1.

Recent chromosome-wide allelic mapping of several histone marks has revealed the different dynamics of these modifications during the early stage of XCI in mESCs. The loss of histone acetylation is one of the earliest chromatin changes to occur (Żylicz et al. 2019). Consistent with this, HDAC3-mediated histone deacetylation (especially at enhancers) was found to be important for efficient *Xist*-mediated silencing of some but not all X-linked genes (McHugh et al. 2015; Żylicz et al. 2019).

Both Polycomb-repressive complexes PRC1 and PRC2 accumulate on the X chromosome following *Xist* coating (Erhardt et al. 2003; Plath et al. 2003; Silva et al. 2003; de Napoles et al. 2004), through mechanisms discussed in the next section. Consistently, the repressive marks H2AK119ub1 and H3K27me3, catalyzed by PRC1 and PRC2, respectively, become drastically enriched throughout XCI. H2AK119ub1 is a very early mark and precedes H3K27me3 in its enrichment on the Xi (Żylicz et al. 2019). Both marks are initially found in intergenic regions before spreading into gene bodies. Importantly, such Polycomb spreading into gene bodies requires transcriptional silencing (Żylicz et al. 2019).

At later stages of the XCI process, nucleosomes on the Xi become specifically enriched for the macroH2A histone variant (Costanzi and Pehrson 1998; Costanzi et al. 2000). This variant is known to participate in repressive chromatin formation in other systems (Gamble et al. 2010). DNA methylation also accumulates at the CpG islands (CGIs) of inactive X-linked genes (Norris et al. 1991; Gendrel et al. 2012). Aside from CGI hypermethylation, the Xi is in fact globally hypomethylated relative to the Xa, consistent with what is generally observed for inactive chromatin and the fact that genes tend to become methylated when they are expressed (Weber et al. 2005; Hellman and Chess 2007).

RECENT MECHANISTIC INSIGHTS INTO *Xist*-MEDIATED GENE SILENCING

How *Xist* actually mediates chromosome-wide gene silencing has long remained elusive. The most intuitive way that had been envisioned was that *Xist* coating would trigger silencing through the direct recruitment of bona fide transcriptional repressors, by *Xist* RNA itself.

Table 1. Summary of reported changes in histone modifications during X chromosome inactivation (XCI)

	References
Depletion of euchromatic/active marks	
H4ac	Jeppesen and Turner 1993; Żylicz et al. 2019
H3K9ac	Żylicz et al. 2019
H3K27ac	Żylicz et al. 2019
H3K4me1 and H3K4me3	Boggs et al. 2002; Żylicz et al. 2019
H3K36me3	Chaumeil et al. 2002
H3R17me	Chaumeil et al. 2002
Accumulation of heterochromatic/repressive marks	
H3K9me2	Heard et al. 2001; Keniry et al. 2016
H3K27me3 (*PRC2*)	Plath et al. 2003; Silva et al. 2003; Żylicz et al. 2019
H2AK119ub1 (*PRC1*)	de Napoles et al. 2004; Żylicz et al. 2019
H4K20me1	Kohlmaier et al. 2004; Tjalsma et al. 2021

Identifying the repertoire of *Xist* RBPs was therefore necessary to decipher the gene-silencing mechanisms at play following *Xist* coating. Until recently, this proved to be technically challenging.

In 2015, however, the Chang, Guttman, and Lee laboratories independently developed approaches allowing for affinity purification and mass-spectrometry-based identification of *Xist* RBPs (Chu et al. 2015; McHugh et al. 2015; Minajigi et al. 2015). In parallel, the Brockdorff and Wutz laboratories employed high-throughput genetic screening in mESCs to identify protein factors involved in *Xist*-dependent gene silencing (Moindrot et al. 2015; Monfort et al. 2015).

Altogether, these studies reported dozens of factors that bind *Xist* RNA and/or are involved in mediating gene silencing during XCI, opening up exciting new research avenues to understand the molecular mechanisms of XCI.

Remarkably, the majority of factors identified to interact with *Xist* do not carry chromatin-modifying activities and are associated with pathways that were hitherto unsuspected to be important for XCI. These discoveries led to a shift in the view of how *Xist* functions, from a model whereby *Xist* directly recruits chromatin-modifying activities (Lee and Bartolomei 2013), to a model in which *Xist* first recruits scaffolding proteins as well as unexpected silencing factors, which only then associate with chromatin activities. Some of *Xist*'s key binding proteins and their associated activities and partners are described below.

Xist Recruits m^{6}A RNA-Methyltransferases, and Is Itself a Target of RNA Methylation

RBM15 and WTAP, two subunits of the m^{6}A RNA methyltransferase complex (Zaccara et al. 2019), were found to interact with *Xist* in two of the *Xist* proteomic studies (Chu et al. 2015; McHugh et al. 2015) via the *Xist* A-repeat (Fig. 6A; Patil et al. 2016; Lu et al. 2020). Furthermore, RBM15 and WTAP were both also identified as candidates in one of the genetic screens aforementioned, and their knockdown was associated with defects in *Xist*-mediated gene silencing (Moindrot et al. 2015).

Follow-up studies showed that *Xist/XIST* is highly m^{6}A-methylated in mouse and humans, respectively (Patil et al. 2016; Nesterova et al. 2019), and that binding of RBM15 (and its paralog RBM15B) to *Xist* RNA is required for m^{6}A-methylation of *Xist* (Fig. 6B; Patil et al. 2016).

Knockdowns of *Mettl3*, *Wtap*, or *Rbm15/Rbm15b* were shown to result in deficient *Xist*-mediated silencing of two assayed X-linked genes, suggesting that *Xist* RNA methylation is required for gene silencing during XCI (Patil et al. 2016). This effect was proposed to be mediated by the binding of the m^{6}A reader protein YTHDC1 to methylated *Xist* (Fig. 6B; Patil et al. 2016), although the mechanistic basis for such hypothetical YTHDC1-dependent X-linked gene silencing is completely unknown.

Recently, however, the Brockdorff laboratory used RNA-seq to demonstrate that chromosome-wide *Xist*-mediated gene silencing is virtually not affected upon knockouts (KOs) of *Rbm15*, *Wtap*, or *Mettl3* (Nesterova et al. 2019). Consistently, this laboratory also reported no dramatic silencing defects upon deletion of major m^{6}A sites within *Xist* (Nesterova et al. 2019; Coker et al. 2020). Taken together, these results strongly suggest that *Xist* RNA methylation is not vital for gene silencing during XCI.

hnRNPK Mediates Polycomb Recruitment to the Xi

The heterogeneous nuclear ribonuclear protein K (hnRNPK) was identified as an *Xist* interactor in two of the *Xist* proteomic studies (Chu et al. 2015; Minajigi et al. 2015), and was subsequently shown to bind *Xist* directly at a region encompassing the B/C-repeats (Fig. 6A; Pintacuda et al. 2017).

Remarkably, deleting this region from *Xist* leads to complete loss of PRC1 and PRC2 recruitment to the Xi, as well as deficient accumulation of their associated histone marks H2AK119ub1 and H3K27me3 (Pintacuda et al. 2017; Bousard et al. 2019; Colognori et al. 2019). Consequently, this region of *Xist* was named PID

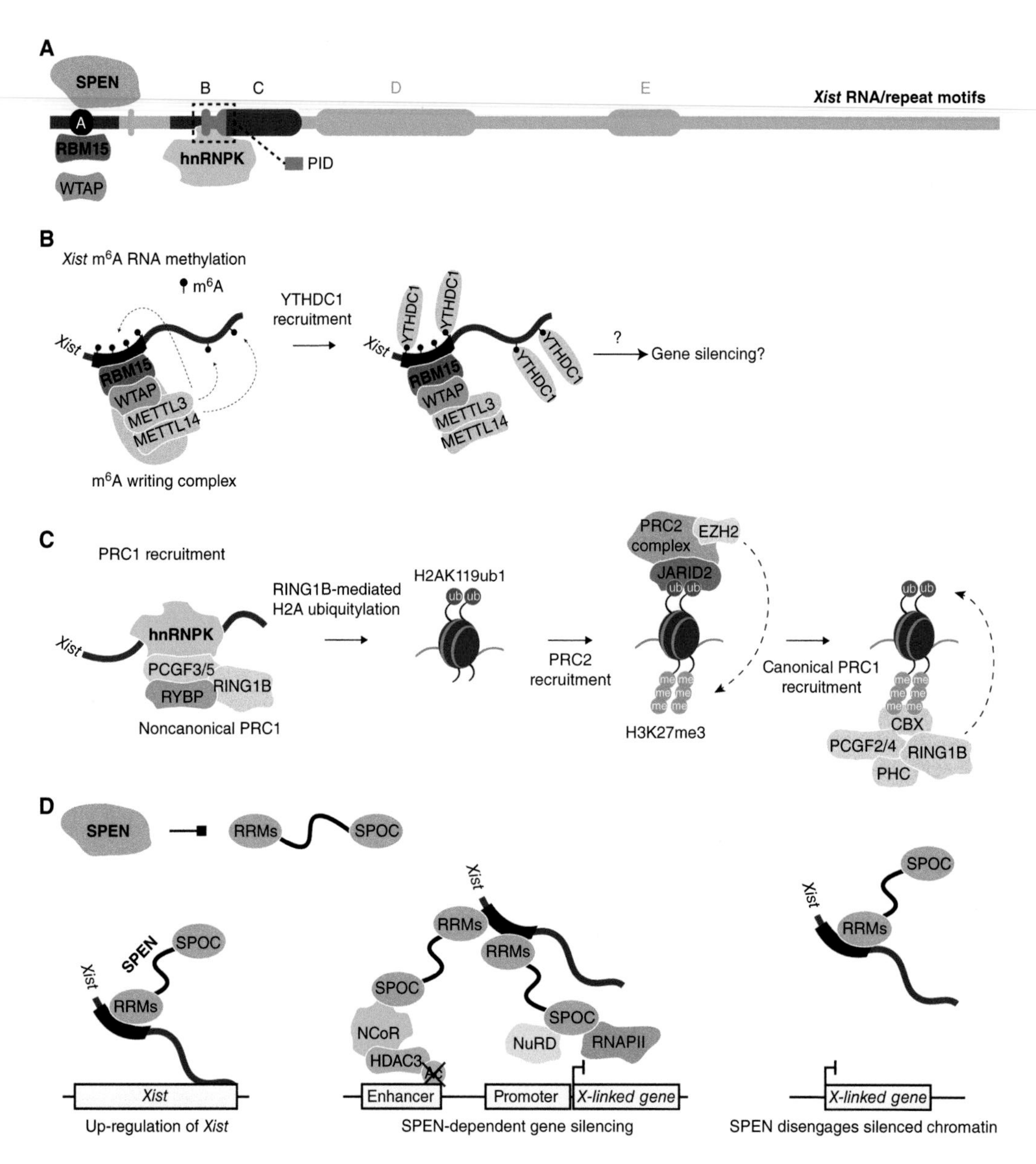

Figure 6. *Xist* interactors implicated in gene silencing during X chromosome inactivation (XCI). (*A*) Several factors implicated in *Xist*-mediated gene silencing were identified when characterizing the *Xist* protein interactome. Of note, RBM15 and SPEN both bind the A-repeat of *Xist* while hnRNPK interacts with a sequence region comprising the B- and part of the C-repeat, termed PID (Polycomb interaction domain). (*B*) Upon binding *Xist* A-repeat, RBM15 recruits the m^{6}A RNA methylation machinery to *Xist*. m^{6}A sites on *Xist* then recruit the YTHDC1 reader protein, which has been proposed to play a role in gene silencing during XCI. (*C*) Upon binding *Xist*, hnRNPK recruits noncanonical PRC1 by interacting directly with PCGF3/5. RING1B-mediated ubiquitination of H2AK119 stimulates PRC2 recruitment, through JARID2 binding to H2AK119ub. EZH2 then catalyzes H3K27 trimethylation, which enables the recruitment of canonical PRC1. (*D*) Immediately upon *Xist* up-regulation, SPEN's RNA recognition motifs (RRMs) bind the A-repeat of *Xist* RNA. *Xist* then targets SPEN to the promoters and enhancers of active X-linked genes, where SPOC engages with NCoR/HDAC3, NuRD, and RNAPII to promote transcriptional repression. Following gene silencing, SPEN disengages from silenced chromatin.

(Polycomb interaction domain) (Fig. 6A). Knockdown of *Hnrnpk* phenocopies the Polycomb recruitment defects associated with deletion of *Xist*-PID (Chu et al. 2015; Colognori et al. 2019), while tethering hnRNPK to *Xist* RNA rescues such defects (Pintacuda et al. 2017). Taken together, these results demonstrate that hnRNPK is pivotal in mediating Polycomb recruitment to the Xi.

hnRNPK interacts directly with PCGF3 and PCGF5, two components of the noncanonical PCGF3/5-PRC1 complex, and this interaction is necessary for *Xist*-mediated PRC1 recruitment (Fig. 6C; Pintacuda et al. 2017). In addition, RYBP (another component of the noncanonical PRC1 complex) and RING1B (the catalytic subunit of PRC1, responsible for H2AK119ub1 deposition) were both retrieved as *Xist*-associated proteins in two of the *Xist* proteomic studies (Chu et al. 2015; Minajigi et al. 2015).

Strikingly, loss of PCGF3/5-dependent H2AK119ub1 hinders PRC2 recruitment and H3K27me3 accumulation during XCI in vitro and in vivo (Almeida et al. 2017), suggesting that, contrary to what was proposed previously (Zhao et al. 2008), PRC2 is not directly recruited by *Xist*, and is instead dependent on prior *Xist*/hnRNPK-mediated PRC1 recruitment to the X chromosome (Fig. 6C; Almeida et al. 2017; Pintacuda et al. 2017).

This model is further supported by evidence that PRC1-mediated H2AK119ub1 enables PRC2 recruitment (Blackledge et al. 2014), through the direct binding of the PRC2 cofactor JARID2 to H2AK119ub1 (Fig. 6C; da Rocha et al. 2014; Kalb et al. 2014; Cooper et al. 2016).

hnRNPK-mediated recruitment of PRC1 has been proposed to play an important role in *Xist*-mediated gene silencing during XCI. Indeed, gene-silencing defects are observed upon the deletion of *Xist*-PID (Pintacuda et al. 2017; Bousard et al. 2019; Colognori et al. 2019; Nesterova et al. 2019) or upon the loss of PCGF3/5 (Almeida et al. 2017; Nesterova et al. 2019). However, the magnitude of the reported gene-silencing phenotypes varies greatly between studies, possibly reflecting the heterogeneity in the cellular systems used to model XCI, as well as the different developmental time points at which gene silencing was assayed.

Clearly, however, all these studies report that silencing still occurs significantly despite the absence of Polycomb recruitment. Furthermore, a gene-silencing defect—even if partial—upon the loss of PRC1 is hard to reconcile with the evidence that deleting *Xist*'s A-repeat (which is physically unlinked from the region of *Xist* responsible for PRC1 recruitment) disrupts gene silencing completely (Wutz et al. 2002; Sakata et al. 2017; Nesterova et al. 2019).

The recent observations that PRC1 components are required for *Xist* RNA's abilities to properly coat the X chromosome (Colognori et al. 2019) and mediate its compaction (Markaki et al. 2020) resolve this inconsistency. The gene-silencing defects associated with loss of Polycomb recruitment are hence most likely an indirect repercussion of impaired *Xist* coating and chromosome compaction, which hinder the access of transcriptional repressors such as SPEN (see section below) to a large fraction of genes across the X chromosome (Colognori et al. 2019; Markaki et al. 2020).

SPEN Orchestrates Gene Silencing across the X Chromosome

SPEN is the only factor found to interact with *Xist* consistently across all three proteomics studies (Fig. 6A; Chu et al. 2015; McHugh et al. 2015; Minajigi et al. 2015), as well as a high-ranking candidate for *Xist*-mediated silencing in both genetic screens (Moindrot et al. 2015; Monfort et al. 2015). SPEN is a very large (~400 kDa in mouse and human), metazoan-specific protein, harboring several conserved amino-terminal RNA-recognition motifs and a conserved carboxy-terminal SPOC domain, the latter being associated with gene repression (Fig. 6D; for review, see Giaimo et al. 2021).

In the context of XCI, *Spen* perturbations in different mESC models were associated with *Xist*-mediated gene-silencing defects for a total of 12 X-linked genes assayed (Chu et al. 2015; McHugh et al. 2015; Moindrot et al. 2015; Monfort et al. 2015). Subsequently, SPEN was shown to be required for silencing virtually all genes across the X chromosome in mESCs (Nesterova et al. 2019; Dossin et al. 2020), and also in vivo during imprinted XCI, where SPEN loss phenocopies *Xist* KO in early mouse embryos (Dossin et al. 2020).

The molecular mechanisms through which SPEN induces gene silencing have started to be unraveled. Through its RNA-recognition motifs (RRMs), SPEN interacts directly with the A-repeat of *Xist* (Monfort et al. 2015; Lu et al. 2016; Carter et al. 2020) while *Xist* is being nascently transcribed (Fig. 6A,D; Dossin et al. 2020). Consistently, SPEN is seen accumulating around the Xi immediately following *Xist* up-regulation, where it can induce gene silencing from the earliest stage of XCI (Dossin et al. 2020).

Upon recruitment by *Xist* RNA, SPEN then targets X-linked promoters and enhancers (Fig. 6D; Dossin et al. 2020), which hence serve as the "substrates" for XCI, consistent with earlier observations (Calabrese et al. 2012). Remarkably, SPEN only targets promoters and enhancers of active genes specifically, and disengages from chromatin upon their silencing (Fig. 6D; Dossin et al. 2020). This suggests that transcription provides a favorable context for SPEN binding, ensuring that active genes—which need most to be dosage compensated—are silenced with highest priority.

Most of SPEN's silencing function during XCI is concentrated within its SPOC domain, whose artificial tethering to the X chromosome is sufficient to instruct a large fraction of gene silencing across the whole chromosome (Dossin et al. 2020). SPOC interacts directly with NCoR and SMRT (Shi et al. 2001; Ariyoshi and Schwabe 2003), two corepressors required to activate the histone deacetylase HDAC3 (Guenther et al. 2001). Hence, it was proposed that SPEN mediates gene silencing on the X chromosome through HDAC3 activity (McHugh et al. 2015), a hypothesis corroborated by the recent report that HDAC3-mediated histone deacetylation at enhancers is important for efficient silencing of many X-linked genes (Fig. 6D; Żylicz et al. 2019).

Importantly, however, XCI of many genes was still observed significantly in *Hdac3* KO mESCs (Żylicz et al. 2019), to a much larger extent than upon the loss of SPEN or of its SPOC domain (Dossin et al. 2020), hinting that other factors bridged by SPEN/SPOC are required for XCI. In addition to NCoR/SMRT, we recently found that the SPOC domain of SPEN interacts with several factors involved in gene regulation, including the NuRD complex, RNAPII, and the transcription machinery as well as the m^{6}A RNA–methyltransferase complex (Fig. 6D; Dossin et al. 2020). Collectively, these studies reveal a model whereby SPEN, following its recruitment by *Xist* RNA to the X chromosome, engages simultaneously with multiple layers of epigenetic and transcriptional regulation—at active promoters and enhancers—thereby ensuring efficient and robust chromosome-wide gene silencing (Fig. 6D).

Although SPEN first appeared to be dispensable for *Xist* coating in mESC models where *Xist* up-regulation is forced through doxycycline-mediated induction (Nesterova et al. 2019; Dossin et al. 2020), it was recently revealed that the stability and long-range localization of *Xist* are in fact perturbed when the SPEN/*Xist* interaction is compromised (Robert-Finestra et al. 2020; Rodermund et al. 2021). Furthermore, SPEN is required for the up-regulation of *Xist* expression under endogenous XCI conditions, notably via SPEN-mediated silencing of *Tsix* (Robert-Finestra et al. 2020).

SPEN has been proposed to be implicated in Polycomb recruitment during XCI (McHugh et al. 2015; Monfort et al. 2015). However, recent studies now demonstrate that Polycomb-repressive complexes seem to be predominantly recruited independently of SPEN and the A-repeat (see previous section) (Pintacuda et al. 2017; Bousard et al. 2019; Żylicz et al. 2019).

MAINTENANCE OF X CHROMOSOME INACTIVATION

As mentioned previously, once established, gene silencing across the X chromosome is epigenetically maintained throughout mitotic divisions (i.e., maintenance of XCI). Several pathways have been implicated in ensuring such maintenance, and remarkably, these appear to vary between cell types and tissues, both in vitro and in vivo.

 Cite this article as *Cold Spring Harb Perspect Biol* doi: 10.1101/cshperspect.a040196

DNA methylation is crucial for maintaining random—though not imprinted—XCI, as embryos deficient for *Dnmt1* show reactivation of the Xi specifically in embryonic lineages, while extraembryonic tissues remain unaffected (Sado et al. 2000). In addition to its role in shaping the structure of the Xi (see earlier section), SMCHD1 is also required for maintaining gene silencing and domains of H3K27me3/H3K9me3 on the Xi (Blewitt et al. 2005; Gendrel et al. 2012, 2013; Mould et al. 2013; Nozawa et al. 2013; Sakakibara et al. 2018; Wang et al. 2018; Gdula et al. 2019), with *Smchd1* homozygous-mutant mice showing female specific lethality (Blewitt et al. 2008). Given that *Smchd1* mutant embryos show hypomethylated promoters on the Xi (Blewitt et al. 2008; Gendrel et al. 2012), and that a plant homolog of SMCHD1 is involved in RNA-dependent DNA methylation (Kanno et al. 2008), these studies further support the idea that DNA methylation is paramount in ensuring maintenance of random XCI.

Intriguingly, maintenance of imprinted XCI in extraembryonic tissues is independent of DNA methylation (Sado et al. 2000) and SMCHD1 (Blewitt et al. 2008). Instead, it relies on the the activities of both PRC1 and PRC2 (Wang et al. 2001), highlighting that the repressive pathways at play during maintenance of XCI differ based on cellular and developmental context.

The Importance of *Xist* during Maintenance of XCI Is Context-Dependent

Earlier studies reported that contrary to what is observed during the initiation of XCI, *Xist* becomes dispensable for sustaining gene silencing during maintenance of XCI (Brown and Willard 1994; Csankovszki et al. 1999; Wutz and Jaenisch 2000). Recently, it was discovered that during the establishment of XCI, *Xist*, through its E-repeat, recruits a multiprotein assembly composed of PTBP1, CELF1, TDP-43, and MATR3, required to later maintain gene silencing (Pandya-Jones et al. 2020; Yu et al. 2021). Remarkably, this multiprotein assembly persists on the Xi despite loss of *Xist* RNA, and was hence proposed to govern the switch between the *Xist*-dependent and -independent phases of XCI (Pandya-Jones et al. 2020).

The early observation that loss of *Xist* results in some reactivation of an X-linked reporter in fibroblasts, however, challenged the model wherein *Xist* is completely dispensable for maintenance of XCI (Csankovszki et al. 2001). Since then, several groups have reported that *Xist* does play an important role in maintaining silencing of certain genes across several somatic cell types, in both mouse and human. The general consensus emerging from these studies is that in many somatic contexts, the loss of *Xist* results in the significant reactivation of a moderate number of genes, and that such reactivation can be potentiated through the synergistic perturbation of certain key regulatory pathways.

For example, while loss of *Xist* results in significant but low-level reactivation of a few X-linked genes in mouse neurons and embryonic fibroblasts (Adrianse et al. 2018; Carrette et al. 2018), different targeted assays or chemical screens have highlighted that such reactivation can be increased by inhibiting DNA methylation (Csankovszki et al. 2001; Carrette et al. 2018; Mira-Bontenbal et al. 2021; Yu et al. 2021), Aurora kinase (Lessing et al. 2016), JAK/STAT signaling (Lee et al. 2020), and SETDB1-dependent H3K9 trimethylation (Minkovsky et al. 2014).

It is also possible that certain environmental cues sensitize Xi reactivation upon loss of *Xist*. For example, although *Xist* deletion in mouse kidney, epithelia, and gut results in low-level Xi reactivation with no phenotypic outcome, prolonged stress (by means of carcinogenic and inflammatory agents) leads to greater tumor burden in the *Xist*-deficient gut when compared to wild-type (Yang et al. 2020). This phenotype is female specific, with tumor cells displaying greatest reactivation of the Xi (Yang et al. 2020). Another recent study showed that upon aging, *Xist* expression decreases in mouse hematopoietic stem cells (HSCs), with the Xi showing increased chromatin accessibility and gene expression variability (Grigoryan et al. 2021), consistent with an older report that reactivation of the X-linked gene *Otc* increases in aging mice (Wareham et al. 1987).

Conditional deletion of *Xist* in mouse HSCs at E10.5 (shortly after HSCs emerge and after XCI has been established), results in Xi reactivation and female-specific early lethality, with 90% of mice dying within 2 years (Yildirim et al. 2013). The mice show hyperproliferation of all hematopoietic lineages resulting in highly penetrant myeloproliferative neoplasm and myelodysplastic syndrome (Yildirim et al. 2013). What is more, loss of *Xist/XIST* in mammary epithelial cells results in the reactivation of several genes on the Xi and is associated with drastically increased breast tumorigenesis and metastasis following oncogenic activation (Richart et al. 2021). Taken together, these studies demonstrate that *Xist* does play a role in maintaining the Xi, and that such a role is health protective in females.

Many studies have reported that the Xi shows not only genetic but also epigenetic instability in several human breast tumors and cancer cell lines (Chaligné and Heard 2014). Such instability is characterized by abnormal *Xist* expression and localization (Pageau et al. 2007; Vincent-Salomon et al. 2007; Sirchia et al. 2009), disturbed Xi compartmentalization and organization, as well as aberrant changes in chromatin composition (i.e., increases in euchromatic marks, atypical distribution of repressive histone marks and DNA methylation) (Chaligné et al. 2015). Remarkably, these changes are associated with the reactivation of a large set of X-linked genes (Kawakami et al. 2004; Richardson et al. 2006), with markedly increased levels of their respective encoding proteins (Chaligné et al. 2015).

Whether these are a cause or consequence of the cancer process remains to be further investigated, but the presence of many oncogenes, chromatin-modifying enzymes, and remodelers encoded on the X chromosome strongly support a functional link between Xi epigenetic instability and cancer. In line with such a hypothesis, the aforementioned mammary epithelial differentiation and oncogenic phenotypes observed upon loss of *XIST* are caused by the increased expression of the X-linked gene *Med14*, encoding for a subunit of the Mediator complex (Richart et al. 2021). Such increased expression results in increased dosage of MED14 and overall higher Mediator stability and activity, enforcing a transcriptional program that hinders mammary stem cell differentiation (Richart et al. 2021).

Understanding the mechanisms through which X-linked gene silencing is maintained is now of utmost importance. Indeed, such understanding could potentially allow for the targeted reactivation of key X-linked genes involved in diseases, and may also pave the way for future avenues in cancer diagnosis and/or treatment in women (Chaligné and Heard 2014; Gribnau and Stefan Barakat 2017).

Recently, we found that SPEN is largely dispensable for maintaining gene silencing in neural progenitor cells, but uncovered its unexpected role in dampening transcription of escapee genes, with SPEN loss resulting in escapee up-regulation specifically on the Xi (Dossin et al. 2020). Consistently, it was subsequently shown that in human B-cells, SPEN is essential to partially silence a subset of X-linked genes, which are still expressed from the Xi, and which also happen to show low DNA methylation at their promoters (Yu et al. 2021). Taken together, these studies reveal an important role for SPEN (and, consequently, *Xist*) in buffering the expression of escapee genes. Such roles may hold clinical significance given the growing evidence that immune X-linked genes escaping XCI may underlie the female bias in autoimmune disorders (Souyris et al. 2018; Syrett and Anguera 2019; Syrett et al. 2019, 2020).

Remarkably, the *Xist*-bound proteome differs significantly between somatic cell types, raising the exciting possibility that *Xist* interacts with certain proteins in a tissue-specific manner (Yu et al. 2021). Hence, future strategies aiming to perturb the maintenance of XCI will have to be tailored to each cellular context of interest.

CONCLUSION

A considerable body of literature now provides an increasingly clear picture of how XCI generally takes place. In particular, recent advances in RBP proteomics and genome-wide screens have brought important insights into how *Xist* triggers gene silencing, revealing that contrary to what was expected, chromatin modifiers are not directly recruited by *Xist*. Furthermore, advances in genetic engineering in mammals have enriched the repertoire of factors involved in the tight control of *Xist* expression.

The fact that two X chromosomes sharing the same nucleus are faced with very different transcriptional fates is remarkable. Nuclear organization has long been proposed to be a crucial component for such different treatment of the X chromosomes. Thanks to recent works, the molecular basis for the peculiar nuclear organization and positioning of the Xi is now much better understood. Genetic perturbation of the mechanisms involved, however, reveals that nuclear organization is not as essential as previously thought in ensuring that one X chromosome becomes inactivated. However, the role of Xi nuclear organization and positioning in the maintenance of its epigenetic and transcriptional status still needs to be investigated.

Several features of XCI remain completely enigmatic. How *Xist* coats the X chromosome in *cis*, and how its localization is strictly confined to the X chromosome (and not autosomes) is largely unknown. How different genes undergo silencing with different dynamics, and how some in fact manage to fully escape silencing also needs to be understood. A pioneer proteomics study recently revealed that dosage compensation defects in female human induced pluripotent stem cells do not only result in increased protein output from X-linked genes, as expected, but also in globally elevated protein levels for more than 20% of autosomal genes (Brenes et al. 2021). The physiological consequence of X chromosome dosage compensation failure must now be further investigated—at both molecular and cellular levels—in vivo.

With the advent of single-cell technologies, future studies should also be able to address the tissue and cell-to-cell variability observed with respect to Xi transcription and nuclear organization, further helping to understand the physiological importance of XCI. Finally, it is now as important to study the process of XCI across a wider variety of mammalian systems, including marsupials, so as to better understand the evolutionary origin of such a remarkable nuclear process.

ACKNOWLEDGMENTS

F.D. is supported by a Boehringer Ingelheim doctoral fellowship. Work in the Heard laboratory is funded by an ERC Advanced Investigator Award (ERC-ADG-2014 671027), Labellisation La Ligue and ANR (DoseX 2017: ANR-17-CE12-0029, Labex DEEP: ANR-11-LBX-0044, ABS4NGS: ANR-11-BINF-0001, and part of the IDEX PSL: ANR-10-IDEX-0001-02 PSL).

REFERENCES

Adrianse RL, Smith K, Gatbonton-Schwager T, Sripathy SP, Lao U, Foss EJ, Boers RG, Boers JB, Gribnau J, Bedalov A. 2018. Perturbed maintenance of transcriptional repression on the inactive X-chromosome in the mouse brain after *Xist* deletion. *Epigenetics Chromatin* **11:** 50. doi:10.1186/s13072-018-0219-8

Almeida M, Pintacuda G, Masui O, Koseki Y, Gdula M, Cerase A, Brown D, Mould A, Innocent C, Nakayama M, et al. 2017. PCGF3/5-PRC1 initiates Polycomb recruitment in X chromosome inactivation. *Science* **356:** 1081–1084. doi:10.1126/science.aal2512

Ariyoshi M, Schwabe JWR. 2003. A conserved structural motif reveals the essential transcriptional repression function of Spen proteins and their role in developmental signaling. *Genes Dev* **17:** 1909–1920. doi:10.1101/gad.266203

Augui S, Filion GJ, Huart S, Nora E, Guggiari M, Maresca M, Stewart AF, Heard E. 2007. Sensing X chromosome pairs before X inactivation via a novel X-pairing region of the Xic. *Science* **318:** 1632–1636. doi:10.1126/science.1149420

Augui S, Nora EP, Heard E. 2011. Regulation of X-chromosome inactivation by the X-inactivation centre. *Nat Rev Genet* **12:** 429–442. doi:10.1038/nrg2987

Bacher CP, Guggiari M, Brors B, Augui S, Clerc P, Avner P, Eils R, Heard E. 2006. Transient colocalization of X-inactivation centres accompanies the initiation of X inactivation. *Nat Cell Biol* **8:** 293–299. doi:10.1038/ncb1365

Bachtrog D. 2013. Y-chromosome evolution: emerging insights into processes of Y-chromosome degeneration. *Nat Rev Genet* **14:** 113–124. doi:10.1038/nrg3366

Balaton BP, Cotton AM, Brown CJ. 2015. Derivation of consensus inactivation status for X-linked genes from genome-wide studies. *Biol Sex Differ* **6:** 35. doi:10.1186/s13293-015-0053-7

Balaton BP, Fornes O, Wasserman WW, Brown CJ. 2021. Cross-species examination of X-chromosome inactivation highlights domains of escape from silencing. *Epigenetics Chromatin* **14:** 12. doi:10.1186/s13072-021-00386-8

Barakat TS, Gunhanlar N, Gontan Pardo C, Achame EM, Ghazvini M, Boers R, Kenter A, Rentmeester E, Grootegoed JA, Gribnau J. 2011. RNF12 activates *Xist* and is essential for X chromosome inactivation. *PLoS Genet* **7:** e1002001. doi:10.1371/journal.pgen.1002001

Barakat TS, Loos F, Van Staveren S, Myronova E, Ghazvini M, Grootegoed JA, Gribnau J. 2014. The *trans*-activator RNF12 and *cis*-acting elements effectuate X chromosome inactivation independent of X-pairing. *Mol Cell* **53:** 965–978. doi:10.1016/j.molcel.2014.02.006

Barr ML, Bertram EG. 1949. A morphological distinction between neurones of the male and female, and the behaviour of the nucleolar satellite during accelerated nucleoprotein synthesis. *Nature* **163:** 676–677. doi:10.1038/163676a0

Barton DE, David FN, Merrington M. 1964. The positions of the sex chromosomes in the human cell in mitosis. *Ann Hum Genet* **28:** 123–128. doi:10.1111/j.1469-1809.1964.tb00467.x

Berletch JB, Yang F, Xu J, Carrel L, Disteche CM. 2011. Genes that escape from X inactivation. *Hum Genet* **130:** 237–245. doi:10.1007/s00439-011-1011-z

Berletch JB, Ma W, Yang F, Shendure J, Noble WS, Disteche CM, Deng X. 2015. Escape from X inactivation varies in mouse tissues. *PLoS Genet* **11:** e1005079. doi:10.1371/journal.pgen.1005079

Beutler E, Yeh M, Fairbanks VF. 1962. The normal human female as a mosaic of X-chromosome activity: studies using the gene for G-6-PD-deficiency as a marker. *Proc Natl Acad Sci* **48:** 9–16. doi:10.1073/pnas.48.1.9

Blackledge NP, Farcas AM, Kondo T, King HW, McGouran JF, Hanssen LLP, Ito S, Cooper S, Kondo K, Koseki Y, et al. 2014. Variant PRC1 complex-dependent H2A ubiquitylation drives PRC2 recruitment and Polycomb domain formation. *Cell* **157:** 1445–1459. doi:10.1016/j.cell.2014.05.004

Blewitt ME, Vickaryous NK, Hemley SJ, Ashe A, Bruxner TJ, Preis JI, Arkell R, Whitelaw E. 2005. An N-ethyl-N-nitrosourea screen for genes involved in variegation in the mouse. *Proc Natl Acad Sci* **102:** 7629–7634. doi:10.1073/pnas.0409375102

Blewitt ME, Gendrel AV, Pang Z, Sparrow DB, Whitelaw N, Craig JM, Apedaile A, Hilton DJ, Dunwoodie SL, Brockdorff N, et al. 2008. SmcHD1, containing a structural-maintenance-of-chromosomes hinge domain, has a critical role in X inactivation. *Nat Genet* **40:** 663–669. doi:10.1038/ng.142

Boggs BA, Cheung P, Heard E, Spector DL, Chinault AC, Allis CD. 2002. Differentially methylated forms of histone H3 show unique association patterns with inactive human X chromosomes. *Nat Genet* **30:** 73–76. doi:10.1038/ng787

Borensztein M, Okamoto I, Syx L, Guilbaud G, Picard C, Ancelin K, Galupa R, Diabangouaya P, Servant N, Barillot E, et al. 2017a. Contribution of epigenetic landscapes and transcription factors to X-chromosome reactivation in the inner cell mass. *Nat Commun* **8:** 1–14. doi:10.1038/s41467-017-01415-5

Borensztein M, Syx L, Ancelin K, Diabangouaya P, Picard C, Liu T, Liang J-B, Vassilev I, Galupa R, Servant N, et al. 2017b. *Xist*-dependent imprinted X inactivation and the early developmental consequences of its failure. *Nat Struct Mol Biol* **24:** 226–233. doi:10.1038/nsmb.3365

Bourgeois CA, Laquerriere F, Hemon D, Hubert J, Bouteille M. 1985. New data on the in situ position of the inactive X chromosome in the interphase nucleus of human fibroblasts. *Hum Genet* **69:** 122–129. doi:10.1007/BF00293281

Bousard A, Raposo AC, Żylicz JJ, Picard C, Pires VB, Qi Y, Gil C, Syx L, Chang HY, Heard E, et al. 2019. The role of *Xist*-mediated Polycomb recruitment in the initiation of X-chromosome inactivation. *EMBO Rep* **20:** e48019. doi:10.15252/embr.201948019

Brenes AJ, Yoshikawa H, Bensaddek D, Mirauta B, Seaton D, Hukelmann JL, Jiang H, Stegle O, Lamond AI. 2021. Erosion of human X chromosome inactivation causes major remodeling of the iPSC proteome. *Cell Rep* **35:** 109032. doi:10.1016/j.celrep.2021.109032

Brockdorff N. 2019. Localized accumulation of *Xist* RNA in X chromosome inactivation. *Open Biol* **9:** 190213. doi:10.1098/rsob.190213

Brockdorff N, Ashworth A, Kay GF, Cooper P, Smith S, McCabe VM, Norris DP, Penny GD, Patel D, Rastan S. 1991. Conservation of position and exclusive expression of mouse *Xist* from the inactive X chromosome. *Nature* **351:** 329–331. doi:10.1038/351329a0

Brockdorff N, Ashworth A, Kay GF, McCabe VM, Norris DP, Cooper PJ, Swift S, Rastan S. 1992. The product of the mouse *Xist* gene is a 15 kb inactive X-specific transcript containing no conserved ORF and located in the nucleus. *Cell* **71:** 515–526. doi:10.1016/0092-8674(92)90519-I

Brown CJ, Willard HF. 1994. The human X-inactivation centre is not required for maintenance of X-chromosome inactivation. *Nature* **368:** 154–156. doi:10.1038/368154a0

Brown CJ, Ballabio A, Rupert JL, Lafreniere RG, Grompe M, Tonlorenzi R, Willard HF. 1991a. A gene from the region of the human X inactivation centre is expressed exclusively from the inactive X chromosome. *Nature* **349:** 38–44. doi:10.1038/349038a0

Brown CJ, Lafreniere RG, Powers VE, Sebastio G, Ballabio A, Pettigrew AL, Ledbetter DH, Levy E, Craig IW, Willard HF. 1991b. Localization of the X inactivation centre on the human X chromosome in Xq13. *Nature* **349:** 82–84. doi:10.1038/349082a0

Brown CJ, Hendrich BD, Rupert JL, Lafrenière RG, Xing Y, Lawrence J, Willard HF. 1992. The human *XIST* gene: analysis of a 17 kb inactive X-specific RNA that contains conserved repeats and is highly localized within the nucleus. *Cell* **71:** 527–542. doi:10.1016/0092-8674(92)90520-M

Calabrese JM, Sun W, Song L, Mugford JW, Williams L, Yee D, Starmer J, Mieczkowski P, Crawford GE, Magnuson T. 2012. Site-specific silencing of regulatory elements as a mechanism of x inactivation. *Cell* **151:** 951–963. doi:10.1016/j.cell.2012.10.037

Carmona S, Lin B, Chou T, Arroyo K, Sun S. 2018. LncRNA Jpx induces *Xist* expression in mice using both *trans* and *cis* mechanisms. *PLoS Genet* **14:** e1007378. doi:10.1371/journal.pgen.1007378

Carrel L, Brown CJ. 2017. When the lyon(ized chromosome) roars: ongoing expression from an inactive X chromosome. *Philos Trans R Soc B Biol Sci* **372:** 20160355. doi:10.1098/rstb.2016.0355

Carrel L, Cottle AA, Goglin KC, Willard HF. 1999. A first-generation X-inactivation profile of the human X chromosome. *Proc Natl Acad Sci* **96:** 14440–14444. doi:10.1073/pnas.96.25.14440

Carrette LLG, Wang CY, Wei C, Press W, Ma W, Kelleher RJ, Lee JT. 2018. A mixed modality approach towards Xi reactivation for Rett syndrome and other X-linked disorders. *Proc Natl Acad Sci* **115:** E668–E675. doi:10.1073/pnas.1715124115

Carter AC, Xu J, Nakamoto MY, Wei Y, Zarnegar BJ, Shi Q, Broughton JP, Ransom RC, Salhotra A, Nagaraja SD, et al. 2020. Spen links RNA-mediated endogenous retrovirus silencing and X chromosome inactivation. *eLife* **9:** 1–58.

Cerase A, Armaos A, Neumayer C, Avner P, Guttman M, Tartaglia GG. 2019. Phase separation drives X-chromosome inactivation: a hypothesis. *Nat Struct Mol Biol* **26:** 331–334. doi:10.1038/s41594-019-0223-0

Chadwick BP. 2008. DXZ4 chromatin adopts an opposing conformation to that of the surrounding chromosome and acquires a novel inactive X-specific role involving CTCF and antisense transcripts. *Genome Res* **18:** 1259–1269. doi:10.1101/gr.075713.107

Chaligné R, Heard E. 2014. X-chromosome inactivation in development and cancer. *FEBS Lett* **588:** 2514–2522. doi:10.1016/j.febslet.2014.06.023

Chaligné R, Popova T, Mendoza-Parra M-A, Saleem M-AM, Gentien D, Ban K, Piolot T, Leroy O, Mariani O, Gronemeyer H, et al. 2015. The inactive X chromosome is epigenetically unstable and transcriptionally labile in breast cancer. *Genome Res* **25:** 488–503. doi:10.1101/gr.185926.114

Chaumeil J, Okamoto I, Guggiari M, Heard E. 2002. Integrated kinetics of X chromosome inactivation in differentiating embryonic stem cells. *Cytogenet Genome Res* **99:** 75–84. doi:10.1159/000071577

Chaumeil J, Le Baccon P, Wutz A, Heard E. 2006. A novel role for *Xist* RNA in the formation of a repressive nuclear compartment into which genes are recruited when silenced. *Genes Dev* **20:** 2223–2237. doi:10.1101/gad.380906

Chelmicki T, Dündar F, Turley MJ, Khanam T, Aktas T, Ramirez F, Gendrel AV, Wright PR, Videm P, Backofen R, et al. 2014. MOF-associated complexes ensure stem cell identity and *Xist* repression. *eLife* **3:** e02024. doi:10.7554/eLife.02024

Chen K, Hu J, Moore DL, Liu R, Kessans SA, Breslin K, Lucet IS, Keniry A, Leong HS, Parish CL, et al. 2015. Genome-wide binding and mechanistic analyses of Smchd1-mediated epigenetic regulation. *Proc Natl Acad Sci* **112:** E3535–E3544. doi:10.1073/pnas.1504232112

Chen CK, Blanco M, Jackson C, Aznauryan E, Ollikainen N, Surka C, Chow A, Cerase A, McDonel P, Guttman M. 2016a. *Xist* recruits the X chromosome to the nuclear lamina to enable chromosome-wide silencing. *Science* **354:** 468–472. doi:10.1126/science.aae0047

Chen CY, Shi W, Balaton BP, Matthews AM, Li Y, Arenillas DJ, Mathelier A, Itoh M, Kawaji H, Lassmann T, et al. 2016b. YY1 binding association with sex-biased transcription revealed through X-linked transcript levels and allelic binding analyses. *Sci Rep* **6:** 1–14. doi:10.1038/s41598-016-0001-8

Chen Z, Djekidel MN, Zhang Y. 2021. Distinct dynamics and functions of H2AK119ub1 and H3K27me3 in mouse preimplantation embryos. *Nat Genet* **53:** 551–563. doi:10.1038/s41588-021-00821-2

Cheng S, Pei Y, He L, Peng G, Reinius B, Tam PPL, Jing N, Deng Q. 2019. Single-cell RNA-seq reveals cellular heterogeneity of pluripotency transition and X chromosome dynamics during early mouse development. *Cell Rep* **26:** 2593–2607.e3. doi:10.1016/j.celrep.2019.02.031

Chow JC, Heard E. 2010. Nuclear organization and dosage compensation. *Cold Spring Harb Perspect Biol* **2:** a000604. doi:10.1101/cshperspect.a000604

Chow JC, Ciaudo C, Fazzari MJ, Mise N, Servant N, Glass JL, Attreed M, Avner P, Wutz A, Barillot E, et al. 2010. LINE-1 activity in facultative heterochromatin formation during X chromosome inactivation. *Cell* **141:** 956–969. doi:10.1016/j.cell.2010.04.042

Chu C, Zhang QC, Da Rocha ST, Flynn RA, Bharadwaj M, Calabrese JM, Magnuson T, Heard E, Chang HY. 2015. Systematic discovery of *Xist* RNA binding proteins. *Cell* **161:** 404–416. doi:10.1016/j.cell.2015.03.025

Chureau C, Chantalat S, Romito A, Galvani A, Duret L, Avner P, Rougeulle C. 2011. Ftx is a non-coding RNA which affects *Xist* expression and chromatin structure within the X-inactivation center region. *Hum Mol Genet* **20:** 705–718. doi:10.1093/hmg/ddq516

Clemson CM, McNeil JA, Willard HF, Lawrence JB. 1996. *XIST* RNA paints the inactive X chromosome at interphase: evidence for a novel RNA involved in nuclear/chromosome structure. *J Cell Biol* **132:** 259–275. doi:10.1083/jcb.132.3.259

Cohen DE, Davidow LS, Erwin JA, Xu N, Warshawsky D, Lee JT. 2007. The DXPas34 repeat regulates random and imprinted X inactivation. *Dev Cell* **12:** 57–71. doi:10.1016/j.devcel.2006.11.014

Coker H, Wei G, Moindrot B, Mohammed S, Nesterova T, Brockdorff N. 2020. The role of the *Xist* 5′ m6A region and RBM15 in X chromosome inactivation. *Wellcome Open Res* **5:** 31. doi:10.12688/wellcomeopenres.15711.1

Collombet S, Ranisavljevic N, Nagano T, Varnai C, Shisode T, Leung W, Piolot T, Galupa R, Borensztein M, Servant N, et al. 2020. Parental-to-embryo switch of chromosome organization in early embryogenesis. *Nature* **580:** 142–146. doi:10.1038/s41586-020-2125-z

Collombet S, Rall I, Dugast-Darzacq C, Heckert A, Halavatyi A, Le Saux A, Dailey G, Darzacq X, Heard E. 2021. RNA polymerase II depletion from the inactive X chromosome territory is not mediated by physical compartmentalization. bioRxiv doi:10.1101/2021.03.26.437188

Colognori D, Sunwoo H, Kriz AJ, Wang CY, Lee JT. 2019. *Xist* deletional analysis reveals an interdependency between *Xist* RNA and Polycomb complexes for spreading along the inactive X. *Mol Cell* **74:** 101–117.e10. doi:10.1016/j.molcel.2019.01.015

Cooper DW, Vandeberg JL, Sharman GB, Poole WE. 1971. Phosphoglycerate kinase polymorphism in kangaroos provides further evidence for paternal X inactivation. *Nat New Biol* **230:** 155–157. doi:10.1038/newbio230155a0

Cooper DW, Johnston PG, Watson JM, Graves JAM. 1993. X-inactivation in marsupials and monotremes. *Semin Dev Biol* **4:** 117–128. doi:10.1006/sedb.1993.1014

Cooper S, Grijzenhout A, Underwood E, Ancelin K, Zhang T, Nesterova TB, Anil-Kirmizitas B, Bassett A, Kooistra SM, Agger K, et al. 2016. Jarid2 binds mono-ubiquitylated H2A lysine 119 to mediate crosstalk between Polycomb complexes PRC1 and PRC2. *Nat Commun* **7:** 1–8. doi:10.1038/ncomms13661

Costanzi C, Pehrson JR. 1998. Histone macroH2A1 is concentrated in the inactive X chromosome of female mammals. *Nature* **393:** 599–601. doi:10.1038/31275

Costanzi C, Stein P, Worrad DM, Schultz RM, Pehrson JR. 2000. Histone macroH2A1 is concentrated in the inactive X chromosome of female preimplantation mouse embryos. *Development* **127:** 2283–2289. doi:10.1242/dev.127.11.2283

Cotton AM, Ge B, Light N, Adoue V, Pastinen T, Brown CJ. 2013. Analysis of expressed SNPs identifies variable extents of expression from the human inactive X chromosome. *Genome Biol* **14:** R122. doi:10.1186/gb-2013-14-11-r122

Csankovszki G, Panning B, Bates B, Pehrson JR, Jaenisch R. 1999. Conditional deletion of *Xist* disrupts histone macroH2A localization but not maintenance of X inactivation. *Nat Genet* **22:** 323–324. doi:10.1038/11887

Csankovszki G, Nagy A, Jaenisch R. 2001. Synergism of *Xist* RNA, DNA methylation, and histone hypoacetylation in maintaining X chromosome inactivation. *J Cell Biol* **153:** 773–784. doi:10.1083/jcb.153.4.773

da Rocha ST, Boeva V, Escamilla-Del-Arenal M, Ancelin K, Granier C, Matias NR, Sanulli S, Chow J, Schulz E, Picard C, et al. 2014. Jarid2 is implicated in the initial *Xist*-induced targeting of PRC2 to the inactive X chromosome. *Mol Cell* **53:** 301–316. doi:10.1016/j.molcel.2014.01.002

Darrow EM, Huntley MH, Dudchenko O, Stamenova EK, Durand NC, Sun Z, Huang SC, Sanborn AL, Machol I, Shamim M, et al. 2016. Deletion of DXZ4 on the human inactive X chromosome alters higher-order genome architecture. *Proc Natl Acad Sci* **113:** E4504–E4512. doi:10.1073/pnas.1609643113

Deakin JE, Hore TA, Koina E, Marshall Graves JA. 2008. The status of dosage compensation in the multiple X chromosomes of the platypus. *PLoS Genet* **4:** e1000140. doi:10.1371/journal.pgen.1000140

De Andrade E, Sousa LB, Jonkers I, Syx L, Dunkel I, Chaumeil J, Picard C, Foret B, Chen CJ, Lis JT, et al. 2019. Kinetics of *Xist*-induced gene silencing can be predicted from combinations of epigenetic and genomic features. *Genome Res* **29:** 1087–1099. doi:10.1101/gr.245027.118

Debrand E, Chureau C, Arnaud D, Avner P, Heard E. 1999. Functional analysis of the DXPas34Locus, a 3′ regulator of *Xist* expression. *Mol Cell Biol* **19:** 8513–8525. doi:10.1128/MCB.19.12.8513

de Napoles M, Mermoud JE, Wakao R, Tang YA, Endoh M, Appanah R, Nesterova TB, Silva J, Otte AP, Vidal M, et al. 2004. Polycomb group proteins ring1A/B link ubiquitylation of histone H2A to heritable gene silencing and X inactivation. *Dev Cell* **7:** 663–676. doi:10.1016/j.devcel.2004.10.005

Deng X, Hiatt JB, Nguyen DK, Ercan S, Sturgill D, Hillier LW, Schlesinger F, Davis CA, Reinke VJ, Gingeras TR, et al. 2011. Evidence for compensatory upregulation of expressed X-linked genes in mammals, *Caenorhabditis elegans* and *Drosophila melanogaster*. *Nat Genet* **43:** 1179–1185. doi:10.1038/ng.948

Deng X, Ma W, Ramani V, Hill A, Yang F, Ay F, Berletch JB, Blau CA, Shendure J, Duan Z, et al. 2015. Bipartite structure of the inactive mouse X chromosome. *Genome Biol* **16:** 152. doi:10.1186/s13059-015-0728-8

Dietzel S, Schiebel K, Little G, Edelmann P, Rappold GA, Eils R, Cremer C, Cremer T. 1999. The 3D positioning of ANT2 and ANT3 genes within female X chromosome territories correlates with gene activity. *Exp Cell Res* **252:** 363–375. doi:10.1006/excr.1999.4635

Dixon JR, Selvaraj S, Yue F, Kim A, Li Y, Shen Y, Hu M, Liu JS, Ren B. 2012. Topological domains in mammalian genomes identified by analysis of chromatin interactions. *Nature* **485:** 376–380. doi:10.1038/nature11082

Donohoe ME, Silva SS, Pinter SF, Xu N, Lee JT. 2009. The pluripotency factor Oct4 interacts with Ctcf and also controls X-chromosome pairing and counting. *Nature* **460:** 128–132. doi:10.1038/nature08098

Dossin F, Pinheiro I, Żylicz JJ, Roensch J, Collombet S, Le Saux A, Chelmicki T, Attia M, Kapoor V, Zhan Y, et al. 2020. SPEN integrates transcriptional and epigenetic control of X-inactivation. *Nature* **578:** 455–460. doi:10.1038/s41586-020-1974-9

Duret L, Chureau C, Samain S, Weissanbach J, Avner P. 2006. The *Xist* RNA gene evolved in eutherians by pseudogenization of a protein-coding gene. *Science* **312:** 1653–1655. doi:10.1126/science.1126316

Dyer KA, Canlfield TK, Gartler SM. 1989. Molecular cytological differentiation of active from inactive X domains in interphase: implications for X chromosome inactivation. *Cytogenet Genome Res* **50:** 116–120. doi:10.1159/000132736

Eils R, Dietzel S, Bertin E, Schröck E, Speicher MR, Ried T, Robert-Nicoud M, Cremer C, Cremer T. 1996. Three-dimensional reconstruction of painted human interphase chromosomes: active and inactive X chromosome territories have similar volumes but differ in shape and surface structure. *J Cell Biol* **135:** 1427–1440. doi:10.1083/jcb.135.6.1427

Engreitz JM, Pandya-Jones A, McDonel P, Shishkin A, Sirokman K, Surka C, Kadri S, Xing J, Goren A, Lander ES, et al. 2013. The *Xist* lncRNA exploits three-dimensional genome architecture to spread across the X chromosome. *Science* **341:** 1237973. doi:10.1126/science.1237973

Erhardt S, Su IH, Schneider R, Barton S, Bannister AJ, Perez-Burgos L, Jenuwein T, Kouzarides T, Tarakhovsky A, Azim Surani M. 2003. Consequences of the depletion of zygotic and embryonic enhancer of zeste 2 during preimplantation mouse development. *Development* **130:** 4235–4248. doi:10.1242/dev.00625

Escamilla-Del-Arenal M, Da Rocha ST, Heard E. 2011. Evolutionary diversity and developmental regulation of X-chromosome inactivation. *Hum Genet* **130:** 307–327. doi:10.1007/s00439-011-1029-2

Froberg JE, Pinter SF, Kriz AJ, Jégu T, Lee JT. 2018. Megadomains and superloops form dynamically but are dispensable for X-chromosome inactivation and gene escape. *Nat Commun* **9:** 1–19. doi:10.1038/s41467-018-07446-w

Furlan G, Gutierrez Hernandez N, Huret C, Galupa R, van Bemmel JG, Romito A, Heard E, Morey C, Rougeulle C. 2018. The Ftx noncoding locus controls X chromosome inactivation independently of its RNA products. *Mol Cell* **70:** 462–472.e8. doi:10.1016/j.molcel.2018.03.024

Galupa R, Heard E. 2015. X-chromosome inactivation: new insights into *cis* and *trans* regulation. *Curr Opin Genet Dev* **31:** 57–66. doi:10.1016/j.gde.2015.04.002

Galupa R, Nora EP, Worsley-Hunt R, Picard C, Gard C, van Bemmel JG, Servant N, Zhan Y, El Marjou F, Johanneau C, et al. 2020. A conserved noncoding locus regulates random monoallelic *Xist* expression across a topological boundary. *Mol Cell* **77:** 352–367.e8. doi:10.1016/j.molcel.2019.10.030

Gamble MJ, Frizzell KM, Yang C, Krishnakumar R, Kraus WL. 2010. The histone variant macroH2A1 marks repressed autosomal chromatin, but protects a subset of its target genes from silencing. *Genes Dev* **24:** 21–32. doi:10.1101/gad.1876110

Gdula MR, Nesterova TB, Pintacuda G, Godwin J, Zhan Y, Ozadam H, McClellan M, Moralli D, Krueger F, Green CM, et al. 2019. The non-canonical SMC protein SmcHD1 antagonises TAD formation and compartmentalisation on the inactive X chromosome. *Nat Commun* **10:** 1–14. doi:10.1038/s41467-018-07907-2

Gendrel AV, Apedaile A, Coker H, Termanis A, Zvetkova I, Godwin J, Tang YA, Huntley D, Montana G, Taylor S, et al. 2012. Smchd1-dependent and -independent pathways determine developmental dynamics of CpG island methylation on the inactive X chromosome. *Dev Cell* **23:** 265–279. doi:10.1016/j.devcel.2012.06.011

Gendrel AV, Tang YA, Suzuki M, Godwin J, Nesterova TB, Greally JM, Heard E, Brockdorff N. 2013. Epigenetic functions of Smchd1 repress gene clusters on the inactive X chromosome and on autosomes. *Mol Cell Biol* **33:** 3150–3165. doi:10.1128/MCB.00145-13

Giaimo BD, Robert-Finestra T, Oswald F, Gribnau J, Borggrefe T. 2021. Chromatin regulator SPEN/SHARP in X inactivation and disease. *Cancers (Basel)* **13:** 1665. doi:10.3390/cancers13071665

Giorgetti L, Lajoie BR, Carter AC, Attia M, Zhan Y, Xu J, Chen CJ, Kaplan N, Chang HY, Heard E, et al. 2016. Structural organization of the inactive X chromosome in the mouse. *Nature* **535:** 575–579. doi:10.1038/nature18589

Gjaltema RAF, Kautz P, Robson M, Schöpflin R, Lustig LR, Brandenburg L, Dunkel I, Vechiatto C, Ntini E, Mutzel V, et al. 2021. Distal and proximal *cis*-regulatory elements sense X-chromosomal dosage and developmental state at the *Xist* locus. bioRxiv doi:10.1101/2021.03.29.437476

Gontan C, Achame EM, Demmers J, Barakat TS, Rentmeester E, Van Ijcken W, Grootegoed JA, Gribnau J. 2012. RNF12 initiates X-chromosome inactivation by targeting REX1 for degradation. *Nature* **485:** 386–390. doi:10.1038/nature11070

Gontan C, Mira-Bontenbal H, Magaraki A, Dupont C, Barakat TS, Rentmeester E, Demmers J, Gribnau J. 2018. REX1 is the critical target of RNF12 in imprinted X chromosome inactivation in mice. *Nat Commun* **9:** 1–12. doi:10.1038/s41467-018-07060-w

Grant J, Mahadevaiah SK, Khil P, Sangrithi MN, Royo H, Duckworth J, McCarrey JR, Vandeberg JL, Renfree MB, Taylor W, et al. 2012. Rsx is a metatherian RNA with *Xist*-like properties in X-chromosome inactivation. *Nature* **487:** 254–258. doi:10.1038/nature11171

Graves JAM. 2006. Sex chromosome specialization and degeneration in mammals. *Cell* **124:** 901–914. doi:10.1016/j.cell.2006.02.024

Graves JAM, Koina E, Sankovic N. 2006. How the gene content of human sex chromosomes evolved. *Curr Opin Genet Dev* **16:** 219–224. doi:10.1016/j.gde.2006.04.007

Gribnau J, Stefan Barakat T. 2017. X-chromosome inactivation and its implications for human disease. bioRxiv doi:10.1101/076950

Grigoryan A, Pospiech J, Krämer S, Lipka D, Liehr T, Geiger H, Kimura H, Mulaw MA, Florian MC. 2021. Attrition of X chromosome inactivation in aged hematopoietic stem cells. *Stem Cell Reports* **16:** 708–716. doi:10.1016/j.stemcr.2021.03.007

Guenther MG, Barak O, Lazar MA. 2001. The SMRT and N-CoR corepressors are activating cofactors for histone deacetylase 3. *Mol Cell Biol* **21:** 6091–6101. doi:10.1128/MCB.21.18.6091-6101.2001

Harris C, Cloutier M, Trotter M, Hinten M, Gayen S, Du Z, Xie W, Kalantry S. 2019. Conversion of random X-inactivation to imprinted X-inactivation by maternal prc2. *eLife* **8:** e44258. doi:10.7554/eLife.44258

Hasegawa Y, Brockdorff N, Kawano S, Tsutui K, Tsutui K, Nakagawa S. 2010. The matrix protein hnRNP U is required for chromosomal localization of *Xist* RNA. *Dev Cell* **19:** 469–476. doi:10.1016/j.devcel.2010.08.006

Heard E, Rougeulle C, Arnaud D, Avner P, Allis CD, Spector DL. 2001. Methylation of histone H3 at Lys-9 is an early mark on the X chromosome during X inactivation. *Cell* **107:** 727–738. doi:10.1016/S0092-8674(01)00598-0

Hellman A, Chess A. 2007. Gene body-specific methylation on the active X chromosome. *Science* **315:** 1141–1143. doi:10.1126/science.1136352

Horakova AH, Calabrese JM, McLaughlin CR, Tremblay DC, Magnuson T, Chadwick BP. 2012a. The mouse DXZ4 homolog retains Ctcf binding and proximity to Pls3 despite substantial organizational differences compared to the primate macrosatellite. *Genome Biol* **13:** R70. doi:10.1186/gb-2012-13-8-r70

Horakova AH, Moseley SC, Mclaughlin CR, Tremblay DC, Chadwick BP. 2012b. The macrosatellite DXZ4 mediates CTCF-dependent long-range intrachromosomal interactions on the human inactive X chromosome. *Hum Mol Genet* **21:** 4367–4377. doi:10.1093/hmg/dds270

Horvath LM, Li N, Carrel L. 2013. Deletion of an X-inactivation boundary disrupts adjacent gene silencing. *PLoS Genet* **9:** e1003952. doi:10.1371/journal.pgen.1003952

Hughes JF, Page DC. 2015. The biology and evolution of mammalian Y chromosomes. *Annu Rev Genet* **49:** 507–527. doi:10.1146/annurev-genet-112414-055311

Inoue A, Jiang L, Lu F, Zhang Y. 2017. Genomic imprinting of *Xist* by maternal H3K27me3. *Genes Dev* **31:** 1927–1932. doi:10.1101/gad.304113.117

Inoue A, Chen Z, Yin Q, Zhang Y. 2018. Maternal Eed knockout causes loss of H3K27me3 imprinting and random X inactivation in the extraembryonic cells. *Genes Dev* **32:** 1525–1536. doi:10.1101/gad.318675.118

Jansz N, Chen K, Murphy JM, Blewitt ME. 2017. The epigenetic regulator SMCHD1 in development and disease. *Trends Genet* **33:** 233–243. doi:10.1016/j.tig.2017.01.007

Jansz N, Keniry A, Trussart M, Bildsoe H, Beck T, Tonks ID, Mould AW, Hickey P, Breslin K, Iminitoff M, et al. 2018a. Smchd1 regulates long-range chromatin interactions on the inactive X chromosome and at Hox clusters. *Nat*

Struct Mol Biol **25:** 766–777. doi:10.1038/s41594-018-0111-z

Jansz N, Nesterova T, Keniry A, Iminitoff M, Hickey PF, Pintacuda G, Masui O, Kobelke S, Geoghegan N, Breslin KA, et al. 2018b. Smchd1 targeting to the inactive X is dependent on the *Xist*-HnrnpK-PRC1 pathway. *Cell Rep* **25:** 1912–1923.e9. doi:10.1016/j.celrep.2018.10.044

Jeon Y, Lee JT. 2011. YY1 tethers *Xist* RNA to the inactive X nucleation center. *Cell* **146:** 119–133. doi:10.1016/j.cell.2011.06.026

Jeppesen P, Turner BM. 1993. The inactive X chromosome in female mammals is distinguished by a lack of histone H4 acetylation, a cytogenetic marker for gene expression. *Cell* **74:** 281–289. doi:10.1016/0092-8674(93)90419-Q

Jonkers I, Barakat TS, Achame EM, Monkhorst K, Kenter A, Rentmeester E, Grosveld F, Grootegoed JA, Gribnau J. 2009. RNF12 is an X-encoded dose-dependent activator of X chromosome inactivation. *Cell* **139:** 999–1011. doi:10.1016/j.cell.2009.10.034

Kalb R, Latwiel S, Baymaz HI, Jansen PWTC, Müller CW, Vermeulen M, Müller J. 2014. Histone H2A monoubiquitination promotes histone H3 methylation in Polycomb repression. *Nat Struct Mol Biol* **21:** 569–571. doi:10.1038/nsmb.2833

Kanno T, Bucher E, Daxinger L, Huettel B, Böhmdorfer G, Gregor W, Kreil DP, Matzke M, Matzke AJM. 2008. A structural-maintenance-of-chromosomes hinge domain-containing protein is required for RNA-directed DNA methylation. *Nat Genet* **40:** 670–675. doi:10.1038/ng.119

Kawakami T, Zhang C, Taniguchi T, Chul JK, Okada Y, Sugihara H, Hattori T, Reeve AE, Ogawa O, Okamoto K. 2004. Characterization of loss-of-inactive X in Klinefelter syndrome and female-derived cancer cells. *Oncogene* **23:** 6163–6169. doi:10.1038/sj.onc.1207808

Kempfer R, Pombo A. 2020. Methods for mapping 3D chromosome architecture. *Nat Rev Genet* **21:** 207–226. doi:10.1038/s41576-019-0195-2

Keniry A, Gearing LJ, Jansz N, Liu J, Holik AZ, Hickey PF, Kinkel SA, Moore DL, Breslin K, Chen K, et al. 2016. Setdb1-mediated H3K9 methylation is enriched on the inactive X and plays a role in its epigenetic silencing. *Epigenetics Chromatin* **9:** 16. doi:10.1186/s13072-016-0064-6

Kohlmaier A, Savarese F, Lachner M, Martens J, Jenuwein T, Wutz A. 2004. A chromosomal memory triggered by *Xist* regulates histone methylation in X inactivation. *PLoS Biol* **2:** e171. doi:10.1371/journal.pbio.0020171

Lee JT. 2000. Disruption of imprinted X inactivation by parent-of-origin effects at *Tsix*. *Cell* **103:** 17–27. doi:10.1016/S0092-8674(00)00101-X

Lee JT, Bartolomei MS. 2013. X-inactivation, imprinting, and long noncoding RNAs in health and disease. *Cell* **152:** 1308–1323. doi:10.1016/j.cell.2013.02.016

Lee JT, Lu N. 1999. Targeted mutagenesis of *Tsix* leads to nonrandom X inactivation. *Cell* **99:** 47–57. doi:10.1016/S0092-8674(00)80061-6

Lee JT, Davidow LS, Warshawsky D. 1999. *Tsix*, a gene antisense to *Xist* at the X-inactivation centre. *Nat Genet* **21:** 400–404. doi:10.1038/7734

Lee HM, Kuijer MB, Ruiz Blanes N, Clark EP, Aita M, Galiano Arjona L, Kokot A, Sciaky N, Simon JM, Bhatnagar S, et al. 2020. A small-molecule screen reveals novel modulators of MeCP2 and X-chromosome inactivation maintenance. *J Neurodev Disord* **12:** 29. doi:10.1186/s11689-020-09332-3

Lessing D, Dial TO, Wei C, Payer B, Carrette LLG, Kesner B, Szanto A, Jadhav A, Maloney DJ, Simeonov A, et al. 2016. A high-throughput small molecule screen identifies synergism between DNA methylation and Aurora kinase pathways for X reactivation. *Proc Natl Acad Sci* **113:** 14366–14371. doi:10.1073/pnas.1617597113

Li N, Carrel L. 2008. Escape from X chromosome inactivation is an intrinsic property of the Jarid1c locus. *Proc Natl Acad Sci* **105:** 17055–17060. doi:10.1073/pnas.0807765105

Lin H, Gupta V, Vermilyea MD, Falciani F, Lee JT, O'Neill LP, Turner BM. 2007. Dosage compensation in the mouse balances up-regulation and silencing of X-linked genes. *PLoS Biol* **5:** 2809–2820.

Livernois AM, Waters SA, Deakin JE, Marshall Graves JA, Waters PD. 2013. Independent evolution of transcriptional inactivation on sex chromosomes in birds and mammals. *PLoS Genet* **9:** e1003635. doi:10.1371/journal.pgen.1003635

Loda A, Brandsma JH, Vassilev I, Servant N, Loos F, Amirnasr A, Splinter E, Barillot E, Poot RA, Heard E, et al. 2017. Genetic and epigenetic features direct differential efficiency of *Xist*-mediated silencing at X-chromosomal and autosomal locations. *Nat Commun* **8:** 690. doi:10.1038/s41467-017-00528-1

Loos F, Maduro C, Loda A, Lehmann J, Kremers GJ, ten Berge D, Grootegoed JA, Gribnau J. 2016. *Xist* and *Tsix* transcription dynamics is regulated by the X-to-autosome ratio and semistable transcriptional states. *Mol Cell Biol* **36:** 2656–2667. doi:10.1128/MCB.00183-16

Lu Z, Zhang QC, Lee B, Flynn RA, Smith MA, Robinson JT, Davidovich C, Gooding AR, Goodrich KJ, Mattick JS, et al. 2016. RNA duplex map in living cells reveals higher-order transcriptome structure. *Cell* **165:** 1267–1279. doi:10.1016/j.cell.2016.04.028

Lu Z, Guo JK, Wei Y, Dou DR, Zarnegar B, Ma Q, Li R, Zhao Y, Liu F, Choudhry H, et al. 2020. Structural modularity of the *XIST* ribonucleoprotein complex. *Nat Commun* **11:** 1–14.

Luikenhuis S, Wutz A, Jaenisch R. 2001. Antisense transcription through the *Xist* locus mediates *Tsix* function in embryonic stem cells. *Mol Cell Biol* **21:** 8512–8520. doi:10.1128/MCB.21.24.8512-8520.2001

Lyon MF. 1961. Gene action in the X-chromosome of the mouse (*Mus musculus* L.). *Nature* **190:** 372–373. doi:10.1038/190372a0

Lyon MF. 1962. Sex chromatin and gene action in the mammalian X-chromosome. *Am J Hum Genet* **14:** 135–148.

Mahadevaiah SK, Sangrithi MN, Hirota T, Turner JMA. 2020. A single-cell transcriptome atlas of marsupial embryogenesis and X inactivation. *Nature* **586:** 612–617. doi:10.1038/s41586-020-2629-6

Mak W, Nesterova TB, De Napoles M, Appanah R, Yamanaka S, Otte AP, Brockdorff N. 2004. Reactivation of the paternal X chromosome in early mouse embryos. *Science* **303:** 666–669. doi:10.1126/science.1092674

Makhlouf M, Ouimette JF, Oldfield A, Navarro P, Neuillet D, Rougeulle C. 2014. A prominent and conserved role for YY1 in *Xist* transcriptional activation. *Nat Commun* **5:** 1–12. doi:10.1038/ncomms5878

Marahrens Y, Panning B, Dausman J, Strauss W, Jaenisch R. 1997. *Xist*-deficient mice are defective in dosage compensation but not spermatogenesis. *Genes Dev* **11:** 156–166. doi:10.1101/gad.11.2.156

Markaki Y, Gan Chong J, Luong C, Tan SY, Wang Y, Jacobson EC, Maestrini D, Dror I, Mistry BA, Schöneberg J, et al. 2020. *Xist*-seeded nucleation sites form local concentration gradients of silencing proteins to inactivate the X-chromosome. bioRxiv doi:10.1101/2020.11.22.393546

Marks H, Kerstens HHD, Barakat TS, Splinter E, Dirks RAM, van Mierlo G, Joshi O, Wang S-Y, Babak T, Albers CA, et al. 2015. Dynamics of gene silencing during X inactivation using allele-specific RNA-seq. *Genome Biol* **16:** 149. doi:10.1186/s13059-015-0698-x

Marshall Graves JA, Shetty S. 2001. Sex from W to Z: evolution of vertebrate sex chromosomes and sex determining genes. *J Exp Zool* **290:** 449–462. doi:10.1002/jez.1088

Martin GR, Epstein CJ, Travis B, Tucker G, Yatziv S, Martin DW, Clift S, Cohen S. 1978. X-chromosome inactivation during differentiation of female teratocarcinoma stem cells in vitro. *Nature* **271:** 329–333. doi:10.1038/271329a0

Masui O, Bonnet I, Le Baccon P, Brito I, Pollex T, Murphy N, Hupé P, Barillot E, Belmont AS, Heard E. 2011. Live-cell chromosome dynamics and outcome of X chromosome pairing events during ES cell differentiation. *Cell* **145:** 447–458. doi:10.1016/j.cell.2011.03.032

McBurney MW, Strutt BJ. 1980. Genetic activity of X chromosomes in pluripotent female teratocarcinoma cells and their differentiated progeny. *Cell* **21:** 357–364. doi:10.1016/0092-8674(80)90472-9

McHugh CA, Chen CK, Chow A, Surka CF, Tran C, McDonel P, Pandya-Jones A, Blanco M, Burghard C, Moradian A, et al. 2015. The *Xist* lncRNA interacts directly with SHARP to silence transcription through HDAC3. *Nature* **521:** 232–236. doi:10.1038/nature14443

McMahon A, Fosten M, Monk M. 1983. X-chromosome inactivation mosaicism in the three germ layers and the germ line of the mouse embryo. *J Embryol Exp Morphol* **74:** 207–220.

Mei H, Kozuka C, Hayashi R, Kumon M, Koseki H, Inoue A. 2021. H2AK119ub1 guides maternal inheritance and zygotic deposition of H3K27me3 in mouse embryos. *Nat Genet* **53:** 539–550. doi:10.1038/s41588-021-00820-3

Migeon BR. 2016. An overview of X inactivation based on species differences. *Semin Cell Dev Biol* **56:** 111–116. doi:10.1016/j.semcdb.2016.01.024

Minajigi A, Froberg JE, Wei C, Sunwoo H, Kesner B, Colognori D, Lessing D, Payer B, Boukhali M, Haas W, et al. 2015. A comprehensive *Xist* interactome reveals cohesin repulsion and an RNA-directed chromosome conformation. *Science* **349:** aab2276. doi:10.1126/science.aab2276

Minkovsky A, Sahakyan A, Rankin-Gee E, Bonora G, Patel S, Plath K. 2014. The Mbd1-Atf7ip-Setdb1 pathway contributes to the maintenance of X chromosome inactivation. *Epigenetics Chromatin* **7:** 12. doi:10.1186/1756-8935-7-12

Mira-Bontenbal H, Tan B, Gontan C, Goossens S, Boers R, Dupont C, van Royen M, van IJcken W, French P, Bedalov T, et al. 2021. Genetic and epigenetic determinants of reactivation of Mecp2 and the inactive X chromosome in neural stem cells. bioRxiv doi:10.1101/2021.02.25.432827

Moindrot B, Cerase A, Coker H, Masui O, Grijzenhout A, Pintacuda G, Schermelleh L, Nesterova TB, Brockdorff N. 2015. A pooled shRNA screen identifies Rbm15, Spen, and Wtap as factors required for *Xist* RNA-mediated silencing. *Cell Rep* **12:** 562–572. doi:10.1016/j.celrep.2015.06.053

Monfort A, Di Minin G, Postlmayr A, Freimann R, Arieti F, Thore S, Wutz A. 2015. Identification of Spen as a crucial factor for *Xist* function through forward genetic screening in haploid embryonic stem cells. *Cell Rep* **12:** 554–561. doi:10.1016/j.celrep.2015.06.067

Monk M, Harper MI. 1979. Sequential X chromosome inactivation coupled with cellular differentiation in early mouse embryos. *Nature* **281:** 311–313. doi:10.1038/281311a0

Mould AW, Pang Z, Pakusch M, Tonks ID, Stark M, Carrie D, Mukhopadhyay P, Seidel A, Ellis JJ, Deakin J, et al. 2013. Smchd1 regulates a subset of autosomal genes subject to monoallelic expression in addition to being critical for X inactivation. *Epigenetics Chromatin* **6:** 19. doi:10.1186/1756-8935-6-19

Mutzel V, Schulz EG. 2020. Dosage sensing, threshold responses, and epigenetic memory: a systems biology perspective on random X-chromosome inactivation. *Bioessays* **42:** 1900163. doi:10.1002/bies.201900163

Mutzel V, Okamoto I, Dunkel I, Saitou M, Giorgetti L, Heard E, Schulz EG. 2019. A symmetric toggle switch explains the onset of random X inactivation in different mammals. *Nat Struct Mol Biol* **26:** 350–360. doi:10.1038/s41594-019-0214-1

Navarro P, Chambers I, Karwacki-Neisius V, Chureau C, Morey C, Rougeulle C, Avner P. 2008. Molecular coupling of *Xist* regulation and pluripotency. *Science* **321:** 1693–1695. doi:10.1126/science.1160952

Navarro P, Oldfield A, Legoupi J, Festuccia N, Dubois AS, Attia M, Schoorlemmer J, Rougeulle C, Chambers I, Avner P. 2010. Molecular coupling of *Tsix* regulation and pluripotency. *Nature* **468:** 457–460. doi:10.1038/nature09496

Navarro P, Moffat M, Mullin NP, Chambers I. 2011. The X-inactivation *trans*-activator Rnf12 is negatively regulated by pluripotency factors in embryonic stem cells. *Hum Genet* **130:** 255–264. doi:10.1007/s00439-011-0998-5

Navarro-Cobos MJ, Balaton BP, Brown CJ. 2020. Genes that escape from X-chromosome inactivation: potential contributors to Klinefelter syndrome. *Am J Med Genet Part C Semin Med Genet* **184:** 226–238. doi:10.1002/ajmg.c.31800

Nesterova TB, Wei G, Coker H, Pintacuda G, Bowness JS, Zhang T, Almeida M, Bloechl B, Moindrot B, Carter EJ, et al. 2019. Systematic allelic analysis defines the interplay of key pathways in X chromosome inactivation. *Nat Commun* **10:** 3129. doi:10.1038/s41467-019-11171-3

Nguyen DK, Disteche CM. 2006. Dosage compensation of the active X chromosome in mammals. *Nat Genet* **38:** 47–53. doi:10.1038/ng1705

Nikolakaki E, Mylonis I, Giannakouros T. 2017. Lamin B receptor: interplay between structure, function and localization. *Cells* **6:** 28. doi:10.3390/cells6030028

Nora EP, Lajoie BR, Schulz EG, Giorgetti L, Okamoto I, Servant N, Piolot T, Van Berkum NL, Meisig J, Sedat J, et al. 2012. Spatial partitioning of the regulatory landscape of the X-inactivation centre. *Nature* **485:** 381–385. doi:10.1038/nature11049

Nora EP, Dekker J, Heard E. 2013. Segmental folding of chromosomes: a basis for structural and regulatory chromosomal neighborhoods? *Bioessays* **35:** 818–828. doi:10.1002/bies.201300040

Norris DP, Brockdorff N, Rastan S. 1991. Methylation status of CpG-rich islands on active and inactive mouse X chromosomes. *Mamm Genome* **1:** 78–83. doi:10.1007/BF02443782

Nozawa RS, Nagao K, Igami KT, Shibata S, Shirai N, Nozaki N, Sado T, Kimura H, Obuse C. 2013. Human inactive X chromosome is compacted through a PRC2-independent SMCHD1-HBiX1 pathway. *Nat Struct Mol Biol* **20:** 566–573. doi:10.1038/nsmb.2532

Ogawa Y, Lee JT. 2003. Xite, X-inactivation intergenic transcription elements that regulate the probability of choice. *Mol Cell* **11:** 731–743. doi:10.1016/S1097-2765(03)00063-7

Ohno S. 1967. *Sex chromosomes and sex linked genes.* Springer, Berlin.

Ohno S, Kaplan WD, Kinosita R. 1958. Somatic association of the positively heteropycnotic X-chromosomes in female mice (*Mus musculus*). *Exp Cell Res* **15:** 616–618. doi:10.1016/0014-4827(58)90111-3

Ohno S, Kaplan WD, Kinosita R. 1959. Formation of the sex chromatin by a single X-chromosome in liver cells of rattus norvegicus. *Exp Cell Res* **18:** 415–418. doi:10.1016/0014-4827(59)90031-X

Okamoto I, Otte AP, Allis CD, Reinberg D, Heard E. 2004. Epigenetic dynamics of imprinted X inactivation during early mouse development. *Science* **303:** 644–649. doi:10.1126/science.1092727

Okamoto I, Arnaud D, Le Baccon P, Otte AP, Disteche CM, Avner P, Heard E. 2005. Evidence for de novo imprinted X-chromosome inactivation independent of meiotic inactivation in mice. *Nature* **438:** 369–373. doi:10.1038/nature04155

Okamoto I, Patrat C, Thépot D, Peynot N, Fauque P, Daniel N, Diabangouaya P, Wolf J-P, Renard JP, Duranthon V, et al. 2011. Eutherian mammals use diverse strategies to initiate X-chromosome inactivation during development. *Nature* **472:** 370–374. doi:10.1038/nature09872

Oliva M, Muñoz-Aguirre M, Kim-Hellmuth S, Wucher V, Gewirtz ADH, Cotter DJ, Parsana P, Kasela S, Balliu B, Viñuela A, et al. 2020. The impact of sex on gene expression across human tissues. *Science* **369:** eaba3066. doi:10.1126/science.aba3066

Pacini G, Dunkel I, Mages N, Mutzel V, Timmermann B, Marsico A, Schulz EG. 2021. Integrated analysis of *Xist* upregulation and X-chromosome inactivation with single-cell and single-allele resolution. *Nat Commun* **12:** 3638. doi:10.1038/s41467-021-23643-6

Pageau GJ, Hall LL, Lawrence JB. 2007. BRCA1 does not paint the inactive X to localize *XIST* RNA but may contribute to broad changes in cancer that impact *XIST* and Xi heterochromatin. *J Cell Biochem* **100:** 835–850. doi:10.1002/jcb.21188

Pandya-Jones A, Markaki Y, Serizay J, Chitiashvili T, Mancia Leon WR, Damianov A, Chronis C, Papp B, Chen CK, McKee R, et al. 2020. A protein assembly mediates *Xist* localization and gene silencing. *Nature* **587:** 145–151. doi:10.1038/s41586-020-2703-0

Panning B, Dausman J, Jaenisch R. 1997. X chromosome inactivation is mediated by *Xist* RNA stabilization. *Cell* **90:** 907–916. doi:10.1016/S0092-8674(00)80355-4

Patil DP, Chen CK, Pickering BF, Chow A, Jackson C, Guttman M, Jaffrey SR. 2016. M6a RNA methylation promotes *XIST*-mediated transcriptional repression. *Nature* **537:** 369–373. doi:10.1038/nature19342

Patrat C, Okamoto I, Diabangouaya P, Vialon V, Le Baccon P, Chow J, Heard E. 2009. Dynamic changes in paternal X-chromosome activity during imprinted X-chromosome inactivation in mice. *Proc Natl Acad Sci* **106:** 5198–5203. doi:10.1073/pnas.0810683106

Payer B, Rosenberg M, Yamaji M, Yabuta Y, Koyanagi-Aoi M, Hayashi K, Yamanaka S, Saitou M, Lee JT. 2013. *Tsix* RNA and the germline factor, PRDM14, link X reactivation and stem cell reprogramming. *Mol Cell* **52:** 805–818. doi:10.1016/j.molcel.2013.10.023

Peeters SB, Korecki AJ, Simpson EM, Brown CJ. 2018. Human *cis*-acting elements regulating escape from X-chromosome inactivation function in mouse. *Hum Mol Genet* **27:** 1252–1262. doi:10.1093/hmg/ddy039

Penny GD, Kay GF, Sheardown SA, Rastan S, Brockdorff N. 1996. Requirement for *Xist* in X chromosome inactivation. *Nature* **379:** 131–137. doi:10.1038/379131a0

Petropoulos S, Edsgärd D, Reinius B, Deng Q, Panula SP, Codeluppi S, Plaza Reyes A, Linnarsson S, Sandberg R, Lanner F. 2016. Single-cell RNA-seq reveals lineage and X chromosome dynamics in human preimplantation embryos. *Cell* **165:** 1012–1026. doi:10.1016/j.cell.2016.03.023

Pintacuda G, Wei G, Roustan C, Kirmizitas BA, Solcan N, Cerase A, Castello A, Mohammed S, Moindrot B, Nesterova TB, et al. 2017. hnRNPK recruits PCGF3/5-PRC1 to the *Xist* RNA B-repeat to establish polycomb-mediated chromosomal silencing. *Mol Cell* **68:** 955–969.e10. doi:10.1016/j.molcel.2017.11.013

Plath K, Fang J, Mlynarczyk-Evans SK, Cao R, Worringer KA, Wang H, De la Cruz CC, Otte AP, Panning B, Zhang Y. 2003. Role of histone H3 lysine 27 methylation in X inactivation. *Science* **300:** 131–135. doi:10.1126/science.1084274

Pollex T, Heard E. 2019. Nuclear positioning and pairing of X-chromosome inactivation centers are not primary determinants during initiation of random X-inactivation. *Nat Genet* **51:** 285–295. doi:10.1038/s41588-018-0305-7

Pontier DB, Gribnau J. 2011. *Xist* regulation and function eXplored. *Hum Genet* **130:** 223–236. doi:10.1007/s00439-011-1008-7

Popova BC, Tada T, Takagi N, Brockdorff N, Nesterova TB. 2006. Attenuated spread of X-inactivation in an X;autosome translocation. *Proc Natl Acad Sci* **103:** 7706–7711. doi:10.1073/pnas.0602021103

Posynick BJ, Brown CJ. 2019. Escape from X-chromosome inactivation: an evolutionary perspective. *Front Cell Dev Biol* **7:** 241. doi:10.3389/fcell.2019.00241

Rao SSP, Huntley MH, Durand NC, Stamenova EK, Bochkov ID, Robinson JT, Sanborn AL, Machol I, Omer AD, Lander ES, et al. 2014. A 3D map of the human genome at kilobase resolution reveals principles of chromatin looping. *Cell* **159:** 1665–1680. doi:10.1016/j.cell.2014.11.021

Rastan S. 1982. Timing of X-chromosome inactivation in postimplantation mouse embryos. *J Embryol Exp Morphol* **71:** 11–24.

Rastan S. 1983. Non-random X-chromosome inactivation in mouse X-autosome translocation embryos—location of the inactivation centre. *J Embryol Exp Morphol* **78:** 1–22.

Rastan S, Robertson EJ. 1985. X-chromosome deletions in embryo-derived (EK) cell lines associated with lack of X-chromosome inactivation. *J Embryol Exp Morphol* **90:** 379–388.

Rego A, Sinclair PB, Tao W, Kireev I, Belmont AS. 2008. The facultative heterochromatin of the inactive X chromosome has a distinctive condensed ultrastructure. *J Cell Sci* **121:** 1119–1127. doi:10.1242/jcs.026104

Rens W, Wallduck MS, Lovell FL, Ferguson-Smith MA, Ferguson-Smith AC. 2010. Epigenetic modifications on X chromosomes in marsupial and monotreme mammals and implications for evolution of dosage compensation. *Proc Natl Acad Sci* **107:** 17657–17662. doi:10.1073/pnas.0910322107

Richardson AL, Wang ZC, De Nicolo A, Lu X, Brown M, Miron A, Liao X, Iglehart JD, Livingston DM, Ganesan S. 2006. X chromosomal abnormalities in basal-like human breast cancer. *Cancer Cell* **9:** 121–132. doi:10.1016/j.ccr.2006.01.013

Richart L, Picod M-L, Wassef M, Macario M, Aflaki S, Salvador MA, Wicinski J, Chevrier V, Le Cam S, Kamhawi HA, et al. 2021. Loss of *XIST* impairs human mammary stem cell differentiation and increases tumorigenicity through enhancer and mediator complex hyperactivation. *SSRN Electron J* doi:10.2139/ssrn.3809998

Ridings-Figueroa R, Stewart ER, Nesterova TB, Coker H, Pintacuda G, Godwin J, Wilson R, Haslam A, Lilley F, Ruigrok R, et al. 2017. The nuclear matrix protein CIZ1 facilitates localization of *Xist* RNA to the inactive X-chromosome territory. *Genes Dev* **31:** 876–888. doi:10.1101/gad.295907.117

Robert-Finestra T, Tan BF, Mira-Bontenbal H, Timmers E, Gontan-Pardo C, Merzouk S, Giaimo BD, Dossin F, van IJcken WFJ, Martens JWM, et al. 2020. SPEN is required for *Xist* upregulation during initiation of X chromosome inactivation. bioRxiv doi:10.1101/2020.12.30.424676

Rodermund L, Coker H, Oldenkamp R, Wei G, Bowness J, Rajkumar B, Nesterova T, Susano Pinto DM, Schermelleh L, Brockdorff N. 2021. Time-resolved structured illumination microscopy reveals key principles of *Xist* RNA spreading. *Science* **372:** eabe7500. doi:10.1126/science.abe7500

Sado T, Fenner MH, Tan SS, Tam P, Shioda T, Li E. 2000. X inactivation in the mouse embryo deficient for Dnmt1: distinct effect of hypomethylation on imprinted and random X inactivation. *Dev Biol* **225:** 294–303. doi:10.1006/dbio.2000.9823

Sado T, Wang Z, Sasaki H, Li E. 2001. Regulation of imprinted X-chromosome inactivation in mice by *Tsix*. *Development* **128:** 1275–1286. doi:10.1242/dev.128.8.1275

Sakakibara Y, Nagao K, Blewitt M, Sasaki H, Obuse C, Sado T. 2018. Role of smcHD1 in establishment of epigenetic states required for the maintenance of the X-inactivated state in mice. *Development* **145:** dev166462. doi:10.1242/dev.166462

Sakata Y, Nagao K, Hoki Y, Sasaki H, Obuse C, Sado T. 2017. Defects in dosage compensation impact global gene regulation in the mouse trophoblast. *Development* **144:** 2784–2797.

Sarkar MK, Gayen S, Kumar S, Maclary E, Buttigieg E, Hinten M, Kumari A, Harris C, Sado T, Kalantry S. 2015. An *Xist*-activating antisense RNA required for X-chromosome inactivation. *Nat Commun* **6:** 1–13. doi:10.1038/ncomms9564

Sharman GB. 1971. Late DNA replication in the paternally derived X chromosome of female kangaroos. *Nature* **230:** 231–232. doi:10.1038/230231a0

Sheardown SA, Duthie SM, Johnston CM, Newall AET, Formstone EJ, Arkell RM, Nesterova TB, Alghisi GC, Rastan S, Brockdorff N. 1997. Stabilization of *Xist* RNA mediates initiation of X chromosome inactivation. *Cell* **91:** 99–107. doi:10.1016/S0092-8674(01)80012-X

Shen Y, Yue F, Mc Cleary DF, Ye Z, Edsall L, Kuan S, Wagner U, Dixon J, Lee L, Ren B, et al. 2012. A map of the *cis*-regulatory sequences in the mouse genome. *Nature* **488:** 116–120. doi:10.1038/nature11243

Shi Y, Downes M, Xie W, Kao HY, Ordentlich P, Tsai CC, Hon M, Evans RM. 2001. Sharp, an inducible cofactor that integrates nuclear receptor repression and activation. *Genes Dev* **15:** 1140–1151. doi:10.1101/gad.871201

Shin J, Bossenz M, Chung Y, Ma H, Byron M, Taniguchi-Ishigaki N, Zhu X, Jiao B, Hall LL, Green MR, et al. 2010. Maternal Rnf12/RLIM is required for imprinted X-chromosome inactivation in mice. *Nature* **467:** 977–981. doi:10.1038/nature09457

Silva J, Mak W, Zvetkova I, Appanah R, Nesterova TB, Webster Z, Peters AHFM, Jenuwein T, Otte AP, Brockdorff N. 2003. Establishment of histone H3 methylation on the inactive X chromosome requires transient recruitment of Eed-Enx1 Polycomb group complexes. *Dev Cell* **4:** 481–495. doi:10.1016/S1534-5807(03)00068-6

Simon MD, Pinter SF, Fang R, Sarma K, Rutenberg-Schoenberg M, Bowman SK, Kesner BA, Maier VK, Kingston RE, Lee JT. 2013. High-resolution *Xist* binding maps reveal two-step spreading during X-chromosome inactivation. *Nature* **504:** 465–469. doi:10.1038/nature12719

Sirchia SM, Tabano S, Monti L, Recalcati MP, Gariboldi M, Grati FR, Porta G, Finelli P, Radice P, Miozzo M. 2009. Misbehaviour of *XIST* RNA in breast cancer cells. *PLoS ONE* **4:** e5559. doi:10.1371/journal.pone.0005559

Smeets D, Markaki Y, Schmid VJ, Kraus F, Tattermusch A, Cerase A, Sterr M, Fiedler S, Demmerle J, Popken J, et al. 2014. Three-dimensional super-resolution microscopy of the inactive X chromosome territory reveals a collapse of its active nuclear compartment harboring distinct *Xist* RNA foci. *Epigenetics Chromatin* **7:** 8. doi:10.1186/1756-8935-7-8

Souyris M, Cenac C, Azar P, Daviaud D, Canivet A, Grunenwald S, Pienkowski C, Chaumeil J, Mejía JE, Guéry JC. 2018. TLR7 escapes X chromosome inactivation in immune cells. *Sci Immunol* **3:** eaap8855. doi:10.1126/sciimmunol.aap8855

Splinter E, de Wit E, Nora EP, Klous P, van de Werken HJG, Zhu Y, Kaaij LJT, van Ijcken W, Gribnau J, Heard E, et al. 2011. The inactive X chromosome adopts a unique three-dimensional conformation that is dependent on *Xist* RNA. *Genes Dev* **25:** 1371–1383. doi:10.1101/gad.633311

Stavropoulos N, Lu N, Lee JT. 2001. A functional role for *Tsix* transcription in blocking *Xist* RNA accumulation but not in X-chromosome choice. *Proc Natl Acad Sci* **98:** 10232–10237. doi:10.1073/pnas.171243598

Stavropoulos N, Rowntree RK, Lee JT. 2005. Identification of developmentally specific enhancers for *Tsix* in the regulation of X chromosome inactivation. *Mol Cell Biol* **25:** 2757–2769. doi:10.1128/MCB.25.7.2757-2769.2005

Sun S, Del Rosario BC, Szanto A, Ogawa Y, Jeon Y, Lee JT. 2013. Jpx RNA activates *Xist* by evicting CTCF. *Cell* **153:** 1537–1551. doi:10.1016/j.cell.2013.05.028

Sunwoo H, Colognori D, Froberg JE, Jeon Y, Lee JT. 2017. Repeat E anchors *Xist* RNA to the inactive X chromosomal compartment through CDKN1A-interacting protein (CIZ1). *Proc Natl Acad Sci* **114:** 10654–10659. doi:10.1073/pnas.1711206114

Syrett CM, Anguera MC. 2019. When the balance is broken: X-linked gene dosage from two X chromosomes and female-biased autoimmunity. *J Leukoc Biol* **106:** 919–932. doi:10.1002/JLB.6RI0319-094R

Syrett CM, Paneru B, Sandoval-Heglund D, Wang J, Banerjee S, Sindhava V, Behrens EM, Atchison M, Anguera MC. 2019. Altered X-chromosome inactivation in T cells may promote sex-biased autoimmune diseases. *JCI Insight* **4:** e126751. doi:10.1172/jci.insight.126751

Syrett CM, Sierra I, Beethem ZT, Dubin AH, Anguera MC. 2020. Loss of epigenetic modifications on the inactive X chromosome and sex-biased gene expression profiles in B cells from NZB/W F1 mice with lupus-like disease. *J Autoimmun* **107:** 102357. doi:10.1016/j.jaut.2019.102357

Takagi N, Sasaki M. 1975. Preferential inactivation of the paternally derived X chromosome in the extraembryonic membranes of the mouse. *Nature* **256:** 640–642. doi:10.1038/256640a0

Teller K, Illner D, Thamm S, Casas-Delucchi CS, Versteeg R, Indemans M, Cremer T, Cremer M. 2011. A top-down analysis of Xa- and Xi-territories reveals differences of higher order structure at ≥20 Mb genomic length scales. *Nucleus* **2:** 465–477. doi:10.4161/nucl.2.5.17862

Tian D, Sun S, Lee JT. 2010. The long noncoding RNA, Jpx, is a molecular switch for X chromosome inactivation. *Cell* **143:** 390–403. doi:10.1016/j.cell.2010.09.049

Tjalsma SJD, Hori M, Sato Y, Bousard A, Ohi A, Raposo AC, Roensch J, Le Saux A, Nogami J, Maehara K, et al. 2021. H4k20me1 and H3K27me3 are concurrently loaded onto the inactive X chromosome but dispensable for inducing gene silencing. *EMBO Rep* **22:** 1–17.

Trotman JB, Lee DM, Cherney RE, Kim SO, Inoue K, Schertzer MD, Bischoff SR, Cowley DO, Calabrese JM. 2020. Elements at the 5′ end of *Xist* harbor SPEN-independent transcriptional antiterminator activity. *Nucleic Acids Res* **48:** 10500–10517. doi:10.1093/nar/gkaa789

Tukiainen T, Villani AC, Yen A, Rivas MA, Marshall JL, Satija R, Aguirre M, Gauthier L, Fleharty M, Kirby A, et al. 2017. Landscape of X chromosome inactivation across human tissues. *Nature* **550:** 244–248. doi:10.1038/nature24265

Vallot C, Patrat C, Collier AJ, Huret C, Casanova M, Liyakat Ali TM, Tosolini M, Frydman N, Heard E, Rugg-Gunn PJ, et al. 2017. XACT noncoding RNA competes with *XIST* in the control of X chromosome activity during human early development. *Cell Stem Cell* **20:** 102–111. doi:10.1016/j.stem.2016.10.014

van Bemmel JG, Mira-Bontenbal H, Gribnau J. 2016. *Cis*- and *trans*-regulation in X inactivation. *Chromosoma* **125:** 41–50. doi:10.1007/s00412-015-0525-x

van Bemmel JG, Galupa R, Gard C, Servant N, Picard C, Davies J, Szempruch AJ, Zhan Y, Żylicz JJ, Nora EP, et al. 2019. The bipartite TAD organization of the X-inactivation center ensures opposing developmental regulation of *Tsix* and *Xist*. *Nat Genet* **51:** 1024–1034. doi:10.1038/s41588-019-0412-0

Veyrunes F, Waters PD, Miethke P, Murchison EP, Kheradpour P, Sachidanandam R, Park J, Semyonov J, Chang CL, Whittington CM, et al. 2008. Bird-like sex chromosomes of platypus imply recent origin of mammal sex chromosomes. *Genome Res* **18:** 965–973. doi:10.1101/gr.7101908

Vigneau S, Augui S, Navarro P, Avner P, Clerc P. 2006. An essential role for the DXPas34 tandem repeat and *Tsix* transcription in the counting process of X chromosome inactivation. *Proc Natl Acad Sci* **103:** 7390–7395. doi:10.1073/pnas.0602381103

Vincent-Salomon A, Ganem-Elbaz C, Manié E, Raynal V, Sastre-Garau X, Stoppa-Lyonnet D, Stern MH, Heard E. 2007. X inactive-specific transcript RNA coating and genetic instability of the X chromosome in BRCA1 breast tumors. *Cancer Res* **67:** 5134–5140. doi:10.1158/0008-5472.CAN-07-0465

Wang J, Mager J, Chen Y, Schneider E, Cross JC, Nagy A, Magnuson T. 2001. Imprinted X inactivation maintained by a mouse Polycomb group gene. *Nat Genet* **28:** 371–375. doi:10.1038/ng574

Wang CY, Jégu T, Chu HP, Oh HJ, Lee JT. 2018. SMCHD1 merges chromosome compartments and assists formation of super-structures on the inactive X. *Cell* **174:** 406–421.e25. doi:10.1016/j.cell.2018.05.007

Wang CY, Colognori D, Sunwoo H, Wang D, Lee JT. 2019. PRC1 collaborates with SMCHD1 to fold the X-chromosome and spread *Xist* RNA between chromosome compartments. *Nat Commun* **10:** 1–18. doi:10.1038/s41467-018-07882-8

Wareham KA, Lyon MF, Glenister PH, Williams ED. 1987. Age related reactivation of an X-linked gene. *Nature* **327:** 725–727. doi:10.1038/327725a0

Weber M, Davies JJ, Wittig D, Oakeley EJ, Haase M, Lam WL, Schübeler D. 2005. Chromosome-wide and promoter-specific analyses identify sites of differential DNA methylation in normal and transformed human cells. *Nat Genet* **37:** 853–862. doi:10.1038/ng1598

West JD, Frels WI, Chapman VM, Papaioannou VE. 1977. Preferential expression of the maternally derived X chromosome in the mouse yolk sac. *Cell* **12:** 873–882. doi:10.1016/0092-8674(77)90151-9

Whitworth DJ, Pask AJ. 2016. The X factor: X chromosome dosage compensation in the evolutionarily divergent monotremes and marsupials. *Semin Cell Dev Biol* **56:** 117–121. doi:10.1016/j.semcdb.2016.01.006

Wutz A, Jaenisch R. 2000. A shift from reversible to irreversible X inactivation is triggered during ES cell differentiation. *Mol Cell* **5:** 695–705. doi:10.1016/S1097-2765(00)80248-8

Wutz A, Rasmussen TP, Jaenisch R. 2002. Chromosomal silencing and localization are mediated by different domains of *Xist* RNA. *Nat Genet* **30:** 167–174. doi:10.1038/ng820

Xu N, Tsai CL, Lee JT. 2006. Transient homologous chromosome pairing marks the onset of X inactivation. *Science* **311:** 1149–1152. doi:10.1126/science.1122984

Yamada N, Hasegawa Y, Yue M, Hamada T, Nakagawa S, Ogawa Y. 2015. *Xist* exon 7 contributes to the stable localization of *Xist* RNA on the inactive X-chromosome. *PLoS Genet* **11:** e1005430. doi:10.1371/journal.pgen.1005430

Yang F, Babak T, Shendure J, Disteche CM. 2010. Global survey of escape from X inactivation by RNA-sequencing in mouse. *Genome Res* **20:** 614–622. doi:10.1101/gr.103200.109

Yang F, Deng X, Ma W, Berletch JB, Rabaia N, Wei G, Moore JM, Filippova GN, Xu J, Liu Y, et al. 2015. The lncRNA Firre anchors the inactive X chromosome to the nucleolus by binding CTCF and maintains H3K27me3 methylation. *Genome Biol* **16:** 52. doi:10.1186/s13059-015-0618-0

Yang L, Yildirim E, Kirby JE, Press W, Lee JT. 2020. Widespread organ tolerance to *Xist* loss and X reactivation except under chronic stress in the gut. *Proc Natl Acad Sci* **117:** 4262–4272. doi:10.1073/pnas.1917203117

Yildirim E, Kirby JE, Brown DE, Mercier FE, Sadreyev RI, Scadden DT, Lee JT. 2013. *Xist* RNA is a potent suppressor of hematologic cancer in mice. *Cell* **152:** 727–742. doi:10.1016/j.cell.2013.01.034

Young AN, Perlas E, Ruiz-Blanes N, Hierholzer A, Pomella N, Martin-Martin B, Liverziani A, Jachowicz JW, Giannakouros T, Cerase A. 2021. Deletion of LBR N-terminal domains recapitulates Pelger-Huet anomaly phenotypes in mouse without disrupting X chromosome inactivation. *Commun Biol* **4:** 478. doi:10.1038/s42003-021-01944-2

Yu B, Qi Y, Li R, Shi Q, Satpathy A, Chang HY. 2021. B cell-specific *XIST* complex enforces X-inactivation and restrains atypical B cells. *Cell* **184:** 1790–1803.e17. doi:10.1101/2021.01.03.425167

Zaccara S, Ries RJ, Jaffrey SR. 2019. Reading, writing and erasing mRNA methylation. *Nat Rev Mol Cell Biol* **20:** 608–624. doi:10.1038/s41580-019-0168-5

Zhang LF, Huynh KD, Lee JT. 2007. Perinucleolar targeting of the inactive X during S phase: evidence for a role in the maintenance of silencing. *Cell* **129:** 693–706. doi:10.1016/j.cell.2007.03.036

Zhao J, Sun BK, Erwin JA, Song JJ, Lee JT. 2008. Polycomb proteins targeted by a short repeat RNA to the mouse X chromosome. *Science* **322:** 750–756. doi:10.1126/science.1163045

Żylicz JJ, Bousard A, Žumer K, Dossin F, Mohammad E, da Rocha ST, Schwalb B, Syx L, Dingli F, Loew D, et al. 2019. The implication of early chromatin changes in X chromosome inactivation. *Cell* **176:** 182–197.e23. doi:10.1016/j.cell.2018.11.041

Organization of the Pluripotent Genome

Patrick S.L. Lim[1] **and Eran Meshorer**[1,2]

[1]Department of Genetics, the Institute of Life Sciences; [2]Edmond and Lily Safra Center for Brain Sciences (ELSC), The Hebrew University of Jerusalem, Jerusalem, Israel 9190400

Correspondence: eran.meshorer@mail.huji.ac.il

In the past several decades, the establishment of in vitro models of pluripotency has ushered in a golden era for developmental and stem cell biology. Research in this arena has led to profound insights into the regulatory features that shape early embryonic development. Nevertheless, an integrative theory of the epigenetic principles that govern the pluripotent nucleus remains elusive. Here, we summarize the epigenetic characteristics that define the pluripotent state. We cover what is currently known about the epigenome of pluripotent stem cells and reflect on the use of embryonic stem cells as an experimental system. In addition, we highlight insights from super-resolution microscopy, which have advanced our understanding of the form and function of chromatin, particularly its role in establishing the characteristically "open chromatin" of pluripotent nuclei. Further, we discuss the rapid improvements in 3C-based methods, which have given us a means to investigate the 3D spatial organization of the pluripotent genome. This has aided the adaptation of prior notions of a "pluripotent molecular circuitry" into a more holistic model, where hotspots of co-interacting domains correspond with the accumulation of pluripotency-associated factors. Finally, we relate these earlier hypotheses to an emerging model of phase separation, which posits that a biophysical mechanism may presuppose the formation of a pluripotent-state-defining transcriptional program.

More than 600 million years ago, the first multicellular metazoan gained the ability to regenerate entirely from a single cell (Müller 2001). The significance of this evolutionary feat is underscored by the unique difficulty of animal multicellularity. Crucially, all animal life is bound by the concept of "division of labor." Neither phototrophs nor chemotrophs, cells in metazoans must sacrifice a unicellular feeding and reproductive life cycle for the combined benefit of the organism (Cavalier-Smith 2017). Thus, for animals, a profound level of cooperation is necessary to balance the selective disadvantage imposed by the need for specialized cell types. Precise details of this evolutionary process remain unclear, but phylogenomic analyses and recent transcriptomic data from sponges—our closest extant relatives to primitive metazoans—hint that we may have descended from an "ancestral pluripotent stem cell," capable of switching between different cell types or transcriptional programs (Sebé-Pedrós et al. 2017; Sogabe et al. 2019). From this unlikely beginning, a fundamental mystery of complex life

Cite this article as *Cold Spring Harb Perspect Biol* doi: 10.1101/cshperspect.a040204

emerges: how can a single cell, with one genome, become a functional and reproductively competent multicellular organism?

Some organisms, such as the flatworm Planaria, possess adult stem cells capable of this coordinated regeneration (Morgan 1898; Rink 2013). Most animals, however, reproduce sexually and produce germ cells, which mature into gametes (oocytes or sperm) that unite to form the single-cell zygote. As the genetic content of the zygote is preserved in each cell of the adult body, it is important to study principles of epigenetic regulation if we wish to understand how developmental decision-making precedes the establishment of cellular identity. In this regard, insights from pluripotent stem cell models have been particularly revealing. A growing body of evidence suggests that a categorically distinct epigenetic program separates somatic cells from those in the earliest stages of development.

Previous work has shown that pluripotent cells possess a diffuse, flexible chromatin structure and an active epigenetic landscape, which can be dramatically influenced by environmental signals (Schlesinger and Meshorer 2019). This takes the form of "open" chromatin in the pluripotent nucleus, where a milieu of accessible chromatin permits cross talk between chromatin remodeling complexes and transcription factors (TFs), resulting in elevated transcription (Efroni et al. 2008; Gaspar-Maia et al. 2011). In the past decade, our understanding of the nature of these phenomena has been bolstered by several technological advances. For instance, using super-resolution microscopy techniques, we are now able to visualize chromatin organization at the level of nucleosomes. In addition, rapid advances in high-throughput sequencing have improved the resolution of Hi-C chromatin–chromatin interaction maps, which have been generated for many different cell types, including pluripotent stem cells. These techniques have led to insights into the 3D organization of the pluripotent genome, suggesting a role for pluripotency-factor-associating hubs that govern the transcriptional activity of particular subregions of the nucleus (Gorkin et al. 2014). However, despite the ever-improving detail from which we can now infer structure and function, a precise mechanism for how the characteristic "plastic state" of pluripotent nuclei is established and maintained remains far from understood. To this end, recent models of phase separation are providing clues toward an integrative account of pluripotency, namely, that the combinatorial binding of pluripotency factors may result in emergent properties that exist only in the pluripotent state.

In this review, we paint a picture of the pluripotent nucleus from the perspective of its unique epigenome, dynamic chromatin structure, and higher-order organizational features.

PLURIPOTENT STEM CELLS AND THEIR EXPERIMENTAL BEGINNINGS

Pluripotency describes the intrinsic potential of a cell to give rise to any cell type or tissue in the developed organism. In eutherians (mammals who possess an internal placenta), the blastocyst forms shortly after fertilization of the zygote and comprises the inner cell mass (ICM) enclosed by an outer shell of cells called the trophoectoderm. Cells in the ICM are pluripotent and, as such, will ultimately give rise to the embryo proper. Before implantation of the embryo in the uterus wall, the ICM segregates into the pluripotent epiblast and the hypoblast, which partitions the epiblast and the blastocyst cavity. The pluripotent state of cells in the blastocyst is sustained until implantation around day 4.5 in mouse (day 7 in human) of embryonic development (Bergh and Navot 1992; Plusa and Hadjantonakis 2014), when the three primary germ layers, endoderm, ectoderm, and mesoderm, begin to form (Fig. 1). Here, gastrulation begins, and cells in the epiblast begin to differentiate toward somatic as well as primordial germ cell (PGC) lineages. The developmental potential of pluripotent cells is eclipsed only by the zygote itself and its initial cleavage, known as the two-cell stage. These cells are totipotent and can uniquely become the trophoblast lineages, which eventually form the bulk of the placenta (Kunath and Rossant 2009; Wu et al. 2017).

The first cultures of mouse embryonic stem cells (ESCs) were established in 1981 by isolating cells from the ICM and growing them on top of a

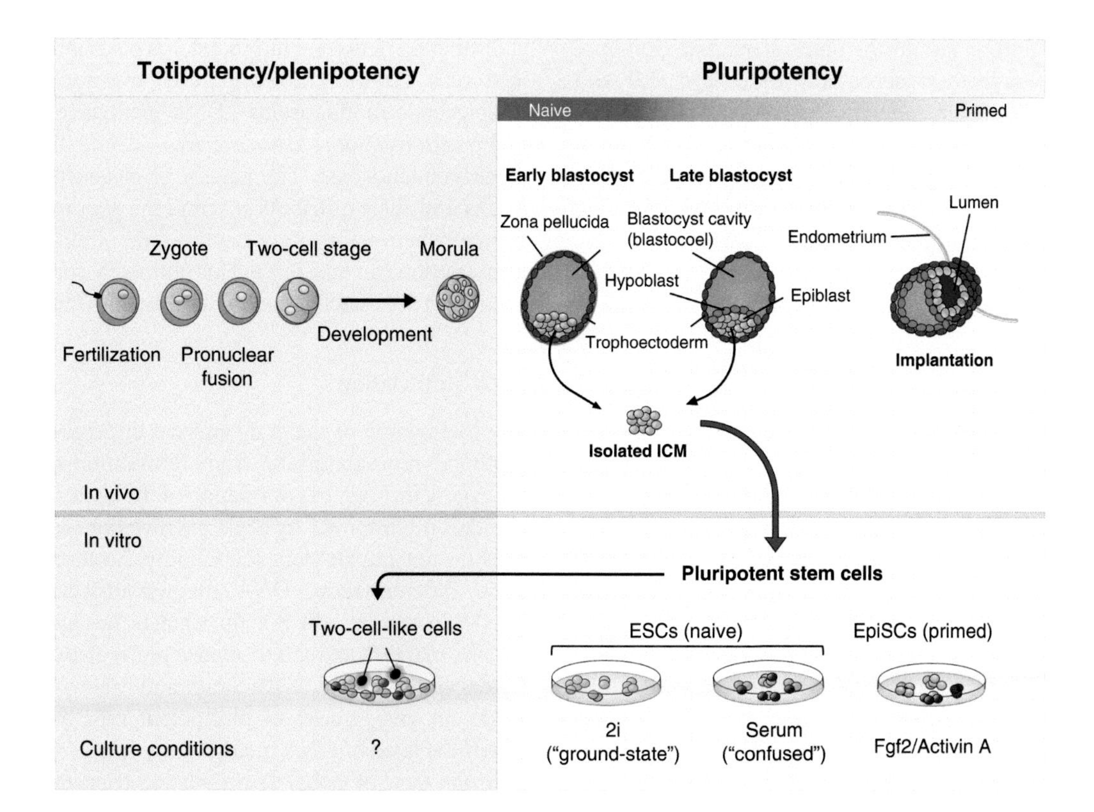

Figure 1. Stem cells in early development and in vitro. Early development in eutherians begins following fertilization (fusion of the male and female gamete). The single-cell zygote soon forms with the potential to generate a complete organism. This ability is referred to as totipotency and is lost following development past the two-cell stage (in mouse). Formation of the blastocyst coincides with the onset of pluripotency within the inner cell mass (ICM). These cells are innately capable of differentiation to all three germ lineages. Isolation and culture of cells from the ICM produces what are known as "pluripotent stem cells" (PSCs). These in vitro models of the pluripotent ICM vary in the precise developmental stage that they resemble along the naive-primed axis, which has been shown to be highly culture-condition-dependent. Cultures of PSCs contain rare populations of two-cell-like cells that mirror the transcriptome of the in vivo two-cell-stage embryo. However, as of yet, no conditions have been defined for the maintenance of these cells in culture. (ESCs) embryonic stem cells, (EpiSCs) epiblast stem cells.

feeder layer of mouse embryonic fibroblasts (MEFs) (Evans and Kaufman 1981; Martin 1981). These cells were shown to be capable of contributing to all adult tissues, including the germ line, upon injection into a blastocyst-stage embryo (Bradley et al. 1984; Nagy et al. 1990). This ability, known as chimera formation, remains the gold standard for developmental competency of in vitro pluripotent models. Later, culturing of pluripotent-state cells from human in vitro fertilized (IVF) blastocysts was similarly accomplished (Thomson et al. 1998). However, our knowledge of the quality of their pluripotency is limited due to ethical concerns, which prohibit the generation of fully developed human chimeras (Mascetti and Pedersen 2016).

REGULATORY NETWORKS UNDERLYING PLURIPOTENCY

Initial breakthroughs in understanding pluripotency in mouse arose from investigations into the core transcriptional circuitry, signaling pathways, and chemical requirements involved in the maintenance of pluripotent ESCs (for review, see Boiani and Schöler 2005). The presence of leukemia inhibitory factor (LIF) and bone morphogenetic proteins (BMPs) was found to sustain the pluripotent state of mouse ESCs in the absence of a feeder layer. LIF and BMPs act in concert to activate the JAK-STAT3 pathway and stimulate expression of Id genes, respectively, promoting ESC self-renewal and inhibition of differentiation (Smith et al. 1988; Ying et al. 2003). Later, a cocktail of two inhibitors (2i), of the mitogen-activated protein kinase (MEK) and glycogen synthase kinase 3 (GSK3) pathways, was shown to maintain the "ground state" (i.e., ICM-like) pluripotency of mouse ESCs (Ying et al. 2008). Key pluripotency TFs operating at the core of pluripotency were also identified, including the three master regulators OCT4, NANOG, and SOX2 (Schöler 1991; Yuan et al. 1995; Mitsui et al. 2003). These factors were shown to act cooperatively, largely co-occupying the same target genes, including each other, revealing that the pluripotent state could be perpetuated in a self-regulating manner (Boyer et al. 2005; for review, see Li and Belmonte 2018). In 2006, experiments by Yamanaka and colleagues demonstrated that somatic cells could be reprogrammed into induced pluripotent stem cells (iPSCs) by ectopic expression of only four "master" TFs, OCT4, SOX2, c-MYC, and KLF4 (Takahashi and Yamanaka 2006). This seminal report further paved the way for extensive characterization of the pluripotency gene regulatory network, and how the binding of TFs reshape the epigenetic landscape to promote pluripotency. This realization also prompted the discovery of a growing number of chromatin-related factors that regulate pluripotency and early ESC differentiation, including BAF250b, BPTF, esBAF, BAF155, CHD1, BAF60a, and the linker histone chaperone SET (Landry et al. 2008; Yan et al. 2008; Gaspar-Maia et al. 2009; Ho et al. 2009; Schaniel et al. 2009; Alajem et al. 2015; Edupuganti et al. 2017). In spite of these discoveries, a great deal of work remains in discerning the epigenetic principles and chromatin organization of the pluripotent state.

THE EPIGENETIC LANDSCAPE OF PLURIPOTENCY

Epigenetics, the nongenetic inheritable form of gene regulation, naturally underlies any functional distinction between pluripotent and differentiated cells. At the chromatin level, two types of epigenetic modifications—DNA methylation and posttranslational histone marks—define the epigenetic landscape. The pattern of epigenetic marks and their distribution across the genome dictate where—and importantly—how much, transcription can take place. This ultimately contributes to the establishment of cellular identity.

DNA Methylation

The methylome of the early embryo undergoes profound reprogramming from fertilization of the zygote through to gastrulation of the postimplantation blastocyst. In ESCs, methylation defects do not impair stem cell identity; however, upon differentiation, DNA methyltransferase (DNMT) mutants do not up-regulate lineage-specific markers and fail to silence pluripotency factors (Jackson et al. 2004; Smith and Meissner 2013). In vivo, nuclei of pluripotent cells are widely depleted of DNA methylation—a leftover from the wave of global demethylation triggered by pronuclear fusion (Smith and Meissner 2013). This process is thought to occur mainly through passive loss (i.e., dilution following DNA replication) by nuclear exclusion or inactivation of DNMTs, as opposed to active demethylation through the TET3 methylcytosine dioxygenase, which occurs in PGCs (Ko et al. 2005; Hirasawa and Sasaki 2009; von Meyenn et al. 2016; Greenberg and Bourc'his 2019). Some regions, such as transposable elements and imprinting control regions, resist this demethylation and are protected by recruitment of residual DNMT proteins, a process that is mediated by factors such as UHRF1, STELLA (DPPA3), and ZFP57/KAP1 (Nakamura et al. 2007; Rowe et al. 2010; Quenneville et al. 2011; Iurlaro et al. 2017; Li et al. 2018). This hypomethylated state is sustained until the exit from pluripotency, which coincides with the re-establishment of methylation.

Histone Modifications

In broad terms, pluripotent cells possess a relative abundance of active histone marks, including H3K4 methylation and H3/H4 acetylation, and, relative to somatic cells, are depleted in repressive marks, such as H3K27me3, H3K9me3, and H3K9me2 (Meshorer et al. 2006; Efroni et al. 2008; Fussner et al. 2010; Hawkins et al. 2010; Liu et al. 2015; Qiao et al. 2015). Posttranslational histone modifications also exhibit distinct patterns of localization in pluripotent cells. A large body of evidence suggests that epigenetic modifiers and preestablished histone modifications are required to regulate gene expression during development (Meissner et al. 2008; Mohn et al. 2008; Athanasiadou et al. 2010; Gao et al. 2018). A prominent mechanistic theory for how this can occur is "promoter bivalency." This epigenetic feature describes the presence of opposing histone modifications, H3K4me3 and H3K27me3, which are often located at the promoters of developmentally important genes (Bernstein et al. 2006). In pluripotent ESCs, the repressive H3K27me3 mark—catalyzed by polycomb repressive complex 2 (PRC2)—keeps genes silent during an undifferentiated state, while the simultaneous presence of active H3K4me3—deposited by COMPASS family proteins—prevents these genes from becoming permanently silenced, allowing prompt activation upon differentiation (Margueron and Reinberg 2011; Shilatifard 2012). Another modification associated with pluripotency is H3K56 acetylation. H3K56ac can be found in ESCs at the binding sites of the core pluripotency factors, OCT4, SOX2, and NANOG (Xie et al. 2009), and marks pluripotency-related enhancers (Livyatan et al. 2015). Although only OCT4 has been shown to directly interact with H3K56ac, loss of SIRT6, an H3K56-targeting NAD-dependent deacetylase, results in differentiation defects and activation of OCT4, SOX2, and NANOG (Tan et al. 2013; Etchegaray et al. 2015).

It remains unclear, however, to what extent the epigenetic features observed in ESCs are characteristic of the early embryo (Harikumar and Meshorer 2015; Yagi et al. 2017). As it stands, our knowledge of epigenetic modifications in pluripotency comes predominantly from in vitro models, owing to the large number of cells required for standard ChIP-seq protocols. Moreover, a plethora of studies have noted that changing the growth conditions of ESCs has a significant effect on the epigenetic landscape (Marks et al. 2012; Ficz et al. 2013; Habibi et al. 2013; Leitch et al. 2013). Encouragingly, improvements in ChIP-seq and alternative methods such as CUT&RUN (antibody-targeted MNase digestion) have been recently developed for epigenetically profiling individual embryos or even single cells (Hainer et al. 2019). For instance, by adapting ChIP-seq for low-input samples, one study confirmed that broad H3K4me3 enrichment is a feature of the preimplantation embryo; conversely, the authors found that bivalency at promoters was relatively rare and unstable (Liu et al. 2016). Additionally, a recent report observed functional bivalency in postimplantation epiblasts but not at earlier pluripotent stages (Xiang et al. 2020). Despite being powerful models of early embryonic states, the debate continues regarding ESCs and their in vivo relevance.

STATES OF PLURIPOTENCY

The pluripotent-stage embryo can be separated into two categories: naive and primed. These terms reflect populations in the pre- and postimplantation epiblast and exist in both mouse and human (Guo et al. 2016b). Importantly, these populations can be broadly recapitulated in vitro (Nichols and Smith 2009). This has led to extensive characterization of molecular and functional features unique to each state (for review, see Weinberger et al. 2016). Culturing mouse ESCs in either 2i/LIF- or serum/LIF-containing media mimics the naive state, whereas "primed" epiblast stem cells (EpiSCs) are derived from the postimplantation epiblast and require supplementation of FGF2 and Activin A (Ying et al. 2008). Crucially, although both states are pluripotent, only the naive form is fully capable of contributing to chimera formation upon ex vivo injection into the blastocyst (Brons et al. 2007; Tesar et al. 2007).

Naive and Primed Pluripotency

Smith and colleagues proposed in 2008 that ESCs grown in 2i/LIF conditions more accurately represent the preimplantation ICM, that is, the in vivo naive state (Ying et al. 2008). Hence, this has been described as "ground-state" pluripotency, as opposed to a "confused" pluripotent state of ESCs grown in serum/LIF conditions (Schlesinger and Meshorer 2019). This notion has been explored further in studies comparing epigenomic and transcriptomic profiles of ESCs cultured in either condition. Two clear hallmarks of "ground-state" pluripotency have emerged: globally depleted DNA methylation and local reduction of H3K27me3 at developmental genes (Marks et al. 2012). In contrast, the "confused" state is typified by a more heterogeneous transcription profile, particularly of pluripotency factors and bivalent-marked genes (Kolodziejczyk et al. 2015; Guo et al. 2016a). This is perhaps a reflection of a more chaotic signaling environment caused by the undefined mixture of factors present in the serum itself (see Schlesinger and Meshorer 2019 for a systematic comparison of ESCs in 2i vs. serum conditions).

Describing a definitive epigenome for primed pluripotency has proven more difficult. Regardless of culture condition, naive and primed cells clearly possess distinct developmental capacities. ESCs and EpiSCs were recently shown to vary in their propensity to differentiate toward endoderm, ectoderm, and mesoderm lineages (Ghimire et al. 2018). However, inconsistencies between cells in the postimplantation blastocyst and EpiSCs, their in vitro counterparts, have thwarted attempts to identify precise epigenetic differences (Takahashi et al. 2018). Although EpiSCs have more promoter-associated hypermethylation and exhibit enhancer switching at pluripotency genes, the same differences are not observed in vivo (Factor et al. 2014; Veillard et al. 2014; Takahashi et al. 2018). These discrepancies are complicated further by the poorly conserved nature of "primed" features like X-chromosome inactivation. For instance, although dosage compensation of X-linked genes is a feature of the "primed" state, the initiation of this process is not generalizable among eutherians (Sado and Sakaguchi 2013). In addition, only recently have culture conditions for a "naive" pluripotent state been established for human cells, albeit with relatively limited success. The difficulty of establishing human ESCs of a comparable developmental stage to naive mouse ESCs reflects our incomplete understanding of the core principles of pluripotency (Rossant 2015; Ying and Smith 2017; Yilmaz and Benvenisty 2019).

The Pluripotency Continuum

Ultimately, experimental pluripotent models appear to provide "snapshots" of pluripotency, rather than behave as mirrors of precise developmental stages. Indeed, a third, "formative" or "epiblast-like" phase, resembling embryonic day 5.5, has been proposed to exist between the naive and primed states (Kalkan and Smith 2014; Smith 2017; for review, see Morgani et al. 2017). In addition, recent single-cell analyses have found that a subpopulation of naive cells possess a transcriptome with primed-like expression, and a similar naive-primed transition is apparent across species during early embryonic development (Messmer et al. 2019). This echoes the theorizing of many who have speculated that pluripotent cells lie on a developmental continuum of successive phases (Fig. 2; Ohtsuka and Dalton 2008; Hough et al. 2009; Loh and Lim 2010; Tesar 2016). The heterogeneity of "confused" ESCs cultured in serum could represent a mixture of naive and primed cells or cells in the process of transitioning between the two states. Interestingly, many of the epigenetic differences observed between ESCs and EpiSCs reiterate the differences between "ground-state" versus "confused" ESC populations. For example, increases in global levels of DNA methylation and focal enrichments of H3K27me3 at developmental promoters appear to correlate with being further along on the ground-state/confused/primed axis; this is also true for clusters of H3K9me3 at pericentromeric repeats (Marks et al. 2012; Juan et al. 2016; Tosolini et al. 2018). The broad range of epigenetic landscapes that are permissive of a pluripotent state is indicative that a holistic characterization of pluripotency is needed—one that accounts for the spatial and chemical properties of the nucleus.

CHROMATIN STRUCTURE IN PLURIPOTENT NUCLEI

In mammalian cells, as much as 2 m of DNA can fit comfortably within a nucleus of 10 µm in

 Cite this article as *Cold Spring Harb Perspect Biol* doi: 10.1101/cshperspect.a040204

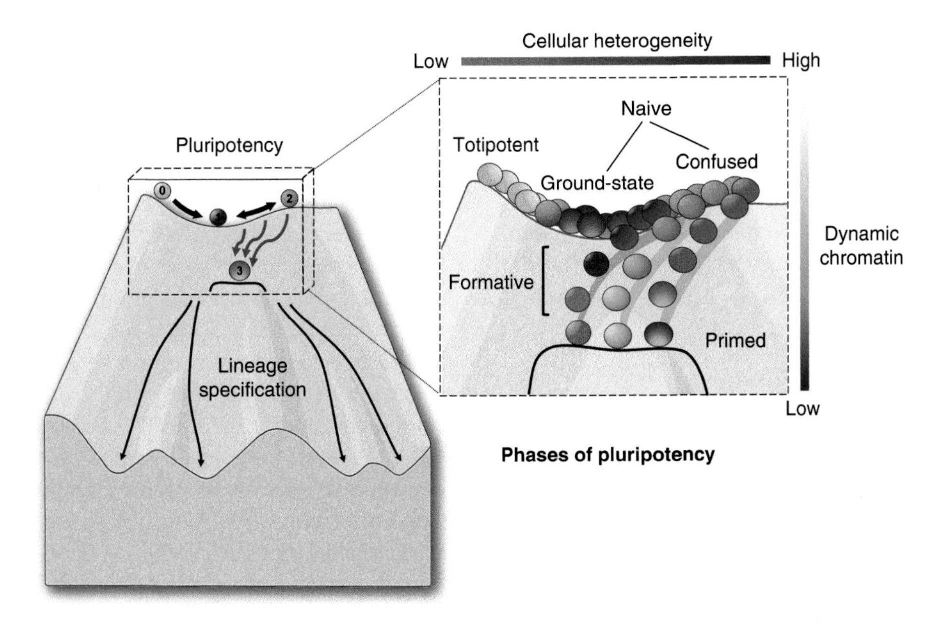

Figure 2. Pluripotency lies on a continuum of successive phases. The figure is based on "Waddington's landscape," and shows the journey taken by a naive pluripotent cell (ball labeled 1) toward a more differentiated (but still pluripotent) primed state (ball labeled 3). The status of cells transitioning between these two states is not well defined (balls labeled 2). Color of the balls represents the overall transcriptomic state. (*Inset, right*) Represents "pluripotent phases" as serial populations of cells with differing levels of heterogeneity rather than discrete developmental stages. Importantly, ground-state pluripotency seems to occupy a "lower energy" state in terms of chromatin mobility compared to both two-cell-stage embryos and serum-cultured embryonic stem cells (ESCs) (Bošković et al. 2014), which may also explain the relatively lower transcriptional promiscuity among naive pluripotent populations (Tosolini et al. 2018). Prior to commitment to a particular primed state, cells may be found in a heterogeneous mix of transcriptomic and metabolic states. This has been somewhat recapitulated in vitro with serum-grown ESCs, which have been described to possess a "confused" transcriptome. However, to what extent there exists a phenotypically distinct "formative" phase between naive and primed pluripotency, or whether this state is strictly necessary for lineage commitment, remains poorly understood.

diameter. Besides the staggering degree of coordination required for this feat alone, cells must also retain their cellular identity and the appropriate transcriptomic activity. This is facilitated by the structural material of the genome, or chromatin, which is comprised of repetitive units of histone–DNA complexes called nucleosomes (Kornberg 1974; Oudet et al. 1975). Nucleosomes consist of a histone octamer wrapped by 145–147 bp of DNA, separated by linker DNA of variable length and further compacted by linker histone H1 (Korolev et al. 2018). Early images from electron microscopy (EM) revealed that nucleosome fibers are roughly 10-nm wide and resemble "beads on a string" (Oudet et al. 1975; Kornberg and Klug 1981). Later, the first X-ray diffraction images of isolated chromatin informed 30-nm fiber models of chromatin folding, and others predicted more complex patterns of folding, including the 120-nm chromonema (Belmont and Bruce 1994). It was conceivable, therefore, that a "higher-order structure" of these nucleosomes followed hierarchical principles of organization, and that this could be functionally relevant for gene regulation.

However, with the advent of advanced EM techniques, the existence of these structures in situ has been disputed. By using transmission EM (TEM) and cryo-EM, several groups were unable to detect 30-nm fibers in the chromatin of intact nuclei (Eltsov et al. 2008; Maeshima et al. 2014; Maeshima 2020). Indeed, beyond the level of the 10-nm nucleosome fiber, larger folded structures in interphase chromatin are often thought to be artefacts of fixation procedures or exist only in vitro (Kuznetsova and Sheval 2016). Currently, models depicting less-ordered chains of zigzagging nucleosomes are considered to be more accurate representations of interphase chromatin (Fussner et al. 2011; Grigoryev et al. 2016; Bascom and Schlick 2017; Krietenstein et al. 2020).

Nucleosome Organization

At the nanoscale level, nucleosomes have been observed to form discrete "clutches," or clusters, that differ in size and distribution depending on developmental stage (Ricci et al. 2015). Imaging of mouse ESCs at this scale has been accomplished with various super-resolution techniques, including structured illumination microscopy (SIM) and stochastic optical reconstruction microscopy (STORM) (for review, see Ricci et al. 2017). STORM imaging of H2B in ESCs and in ESC-derived neuronal progenitor cells (NPCs) revealed that NPCs have a higher number of nucleosomes per clutch. Moreover, clutch size could predict how well human iPSCs had been reprogrammed toward pluripotency (Ricci et al. 2015). These results strengthened previous observations that used electron spectroscopic imaging (ESI) to visualize chromatin structure in pluripotent cells. These studies showed that cells in early embryonic tissue have mesh-like, more widely dispersed fibers of nucleosomes, compared with differentiated fibroblasts, which are enriched in dense regions of 10-nm nucleosomal fibers (Ahmed et al. 2010; Fussner et al. 2010). Heterochromatin, which is associated with transcriptional repression, contains a higher density of nucleosomes than euchromatin, confirming early histological observations that there are fewer clusters of heterochromatin in undifferentiated cells (Park et al. 2004; Meshorer and Misteli 2006; Imai et al. 2017).

Together, these studies demonstrate that a diffuse nucleosomal organization is a distinguishing structural feature of the pluripotent genome (Fig. 3). Intuitively, the dual requirements of stem cells to self-renew and differentiate presumes a high degree of flexibility in gene regulation, and, therefore, being able to easily switch between different arrangements of chromatin may facilitate this process. However, it is worth considering alternative morphological interpretations of chromatin organization. For example, a recent study creatively illustrated that chromatin reorganization can occur in response to mechanical stress. Nava et al. (2020) showed that reduction of H3K9me3 levels could be triggered by cell stretching, which reduces heterochromatin at the nuclear periphery and protects cells from mechanically induced DNA damage. The tightly packed chromatin of differentiated cells may reflect a physiological need for morphological compaction when compared with the characteristically large nuclei of ESCs. Indeed, it has been suggested that in quiescent cells, the main function of chromatin compaction may be to reduce cellular size by reducing nuclear volume, rather than to regulate global transcription (Pombo and Dillon 2015). The decondensed chromatin in ESCs may have as much to do with their unspecialized nature as their need for transcriptional flexibility.

In any case, it is safe to conclude that imaging-based studies strongly support earlier hypotheses that an "open" chromatin configuration is a hallmark of pluripotency (this theory has been covered at length in the following reviews: Meshorer and Misteli 2006; Koh et al. 2010; Gaspar-Maia et al. 2011; Chen and Dent 2014; Kobayashi and Kikyo 2015; Percharde et al. 2017; Schlesinger and Meshorer 2019). Here, it is important to note that although it has been proposed that chromatin density and not higher-order organization separates "closed" from "accessible" chromatin (Ou et al. 2017), the density of heterochromatin has been measured to be only 1.53-fold higher than neighboring euchromatin (Imai et al. 2017). This suggests that the accumulation of non-nucleosomal proteins or RNA molecules may play a significant role in contributing to the "closed" quality of heterochromatin. Indeed, elucidating the relationship between chromatin structure and the function of chromatin-associating factors will aid our understanding of how transcriptional activity is organized in the nucleus.

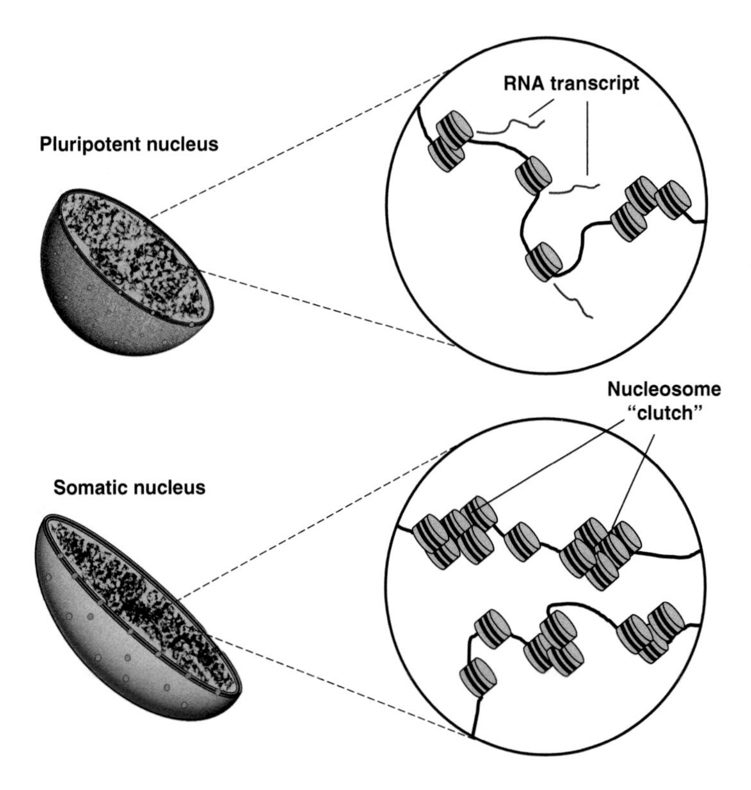

Figure 3. Nucleosome density in interphase nuclei of pluripotent and somatic cells. Pluripotent cells possess smaller and relatively sparse clusters of nucleosomes ("clutches") compared with somatic cells. In more differentiated cells, a dense arrangement of "clutches" results in more condensed chromatin, preventing the accessibility of transcriptional machinery, such as Pol II, and reducing transcription (Ricci et al. 2015).

OPEN CHROMATIN

Several decades prior to the discovery of the molecular structure of DNA, Emil Heitz (1928) precociously observed two types of differentially stained regions of chromosomes, which he termed "euchromatin" and "heterochromatin." Although classically used to distinguish cytologically between light- and heavy-staining DNA, today these terms are more commonly used to describe respective regions of active or inactive transcription (Woodcock and Ghosh 2010). The transcriptionally repressive environment associated with heterochromatin is thought to reflect its condensed structure, which prevents physical access by transcriptional machinery. In contrast, euchromatin is considered to be more open and accessible to regulatory proteins and thus more transcriptionally permissive.

The concept of "open chromatin" also applies to the characteristically flexible and dynamic nature of the pluripotent nucleus. That is, in addition to having a higher euchromatin-to-heterochromatin ratio, pluripotent cells have hyperdynamic nucleosomes as evidenced by fluorescence recovery after photobleaching (FRAP) experiments, where heterochromatic regions of mouse ESC nuclei undergo rapid exchange of core histones H2B and H3, linker histone H1, and heterochromatin protein 1 (HP1) (Meshorer et al. 2006). Not limited to mammals, loosely bound chromatin has also been observed in *Drosophila* embryos (Bhattacharya et al. 2009); moreover, totipotent planarian neoblasts are largely depleted in heterochromatin (Morita et al. 1969). Mechanistically, several factors were suggested to contribute to chromatin plasticity in ESCs, including histone hyperacetylation and the almost complete lack of the nuclear intermediate filament protein lamin A (Melcer et al. 2012). As noted above, several chromatin remodeling proteins were also implicated in maintaining open chromatin and/or pluripotency in ESCs. CHD1, a well-studied chromatin remodeling factor, is required to prevent spurious formation of heterochromatin in ESCs (Gaspar-Maia et al. 2009). CHD1 is implicated in the maintenance of pluripotency in ESCs and transcription initiation via binding to the mediator coactivator complex at active genes (Lin et al. 2011). Unsurprisingly, serum-grown euchromatin-rich ESCs are transcriptionally promiscuous and exhibit low levels of genome-wide gene expression (Efroni et al. 2008). More recently, single-cell RNA-seq studies have shown that the transcriptomes of undifferentiated cells are also more diverse, demonstrating widespread transcriptome sampling in pluripotent stem cells, which become gradually restricted upon differentiation (Gulati et al. 2020).

Genome Accessibility

TFs exert their effect on gene expression by binding to DNA; chromatin plasticity is thought to facilitate this by allowing regulatory proteins access to nucleosome-bound DNA (Klemm et al. 2019). Therefore, genome accessibility is a central variable that links the flexibility of chromatin structure with transcriptional control by regulatory factors. Human ChIP-seq data from the Encyclopedia of DNA Elements (ENCODE) project estimates that almost 95% of TFs bind to accessible DNA, despite accessible regions comprising as little as 4% of the genome (Thurman et al. 2012; Kelley et al. 2016).

Measurement of genome accessibility can be assessed indirectly through fragmentation by nucleases, such as DNase-I, or by insertion of sequencing adaptors, as in transposase-accessible chromatin sequencing (ATAC-seq). Using these techniques, a number of groups have shown that the genome of pluripotent cells is more accessible than differentiated cells. For example, a study mapping DNase-I hypersensitive sites (DHSs) revealed that ESCs had almost double the frequency of DHSs than NPCs, and the depletion of HMGN, a nucleosome-binding protein, was shown to lead to a significant reduction in DHSs (Deng et al. 2013). ATAC-seq experiments have further supported these findings. Differentiation of ESCs toward definitive endoderm results in a 13% net loss of ATAC-seq peaks globally (Simon et al. 2017). Moreover, an interesting study used an allele-specific ATAC-seq method to show that promoter elements in ESCs that were biallelically accessible became monoallelically accessible after NPC differentiation (Xu et al. 2017). This complements a previous observation that monoallelic transcription is more than fivefold higher in NPCs than in ESCs (Eckersley-Maslin et al. 2014).

Micrococcal nuclease (MNase) digestion is also used to measure genome accessibility and is a classic method for detecting regions of nucleosome occupancy. One study found that, globally, differentiating cells are more resistant to MNase digestion than ESCs. The bromodomain factor, ATAD2, which is predominantly expressed in ESCs, was found partly responsible for increasing chromatin sensitivity to MNase digestion (Morozumi et al. 2016). Moreover, the ESC-specific chromatin remodeling BAF complex, esBAF, was found to possess an additional role in maintaining nucleosome-depleted regions in pluripotent cells (Ho et al. 2009; Hainer et al. 2015). Intriguingly, the same group demonstrated that esBAF activity at these regions was necessary for binding of the pluripotency factor, KLF4 (Hainer and Fazzio 2015). This suggests that synergy may exist between chromatin remodeling complexes and pluripotency TF activity, especially as components of the BAF complex have been previously shown to enhance iPSC reprogramming efficiency (Singhal et al. 2010).

Chromatin Remodeling and Pluripotency

Intriguingly, members of the “pluripotent essentialome” (factors necessary for maintenance of pluripotency) are very highly enriched in the protein interactome of OCT4, SOX2, and NANOG (Wang et al. 2006; Yilmaz et al. 2018). Many of these proteins are involved in chromatin remodeling, as mentioned above, adding to mounting evidence that cross talk between chromatin-modifying complexes and core pluripotency factors is a fundamental mechanism for maintaining pluripotency (for review, see Orkin and Hochedlinger 2011). Greater chromatin accessibility in pluripotent cells, therefore, may be the consequence of positive reinforcement of nucleosome disassembly and the binding of regulatory factors, amplifying the existing feedforward loops maintained by the pluripotency network (Fig. 4). Indeed, a recent study has shown that subpopulations of human pluripotent stem cells with high capacities for self-renewal have greater chromatin accessibility at regions of pluripotency TF binding (Lau et al. 2020). Moreover, evidence from zebrafish embryos indicates that TFs are in passive competition with core histones for binding to DNA, and the nuclear ratio between histones and DNA-binding factors is important for the onset of zygotic transcription (Joseph et al. 2017).

3D ORGANIZATION OF THE PLURIPOTENT GENOME

How the genome is spatially organized is intrinsically linked to the establishment of cell-type identity. Although, as discussed earlier, the precise nature of higher-order chromatin structure is controversial, a conserved hierarchy of topological organization can be observed throughout interphase. Perhaps the most obvious example of this is the arrangement of chromosomes,

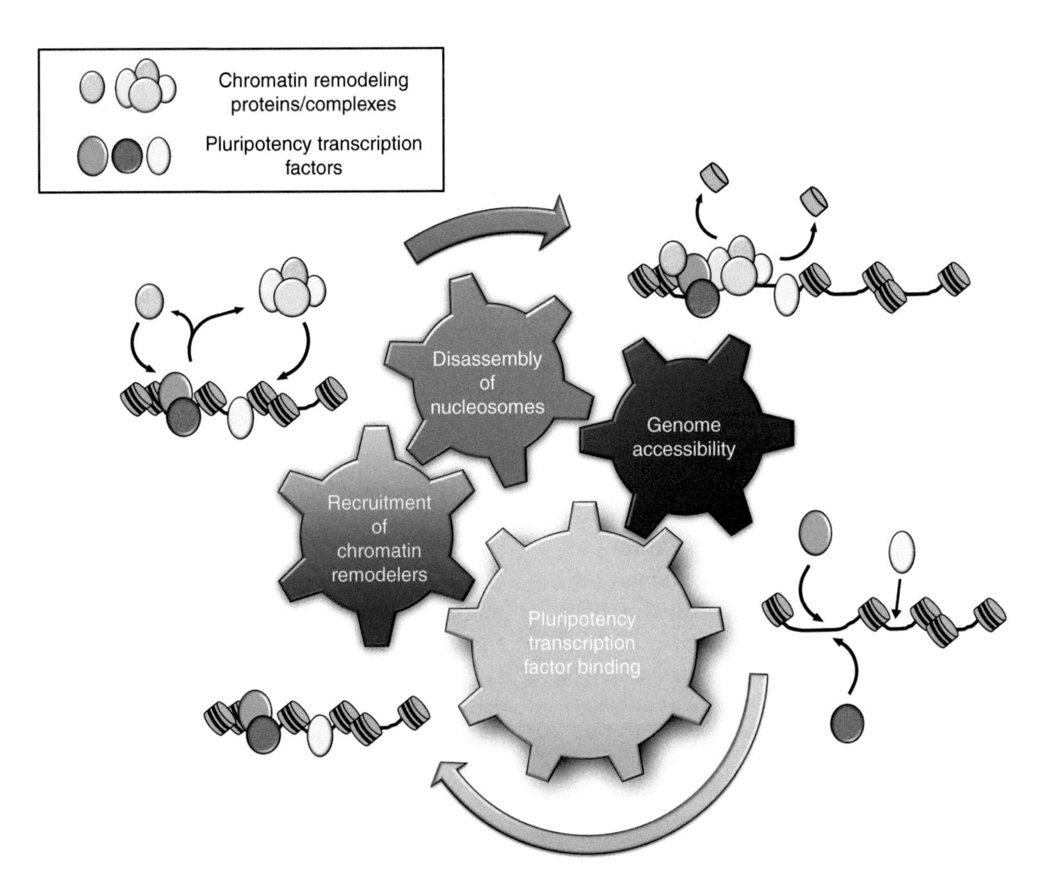

Figure 4. Synergy between chromatin remodeling and pluripotency factor binding. Four hallmarks of pluripotent stem cells are represented by the colored gears. The binding of pluripotency factors (e.g., SOX2, NANOG, OCT4) can bring about the recruitment of chromatin remodeling complexes (e.g., esBAF, CHD1, p300). This leads to histone exchange as nucleosomes are disassembled, which in turn promotes accessible regions of DNA and permits further binding of transcription factors.

which occupy distinct territories in interphase nuclei (Cremer and Cremer 2010).

Chromosome Organization

At the chromosome level, the genomes of most cell types are similarly arranged both in human and mouse, although some features of cell specificity do exist. Whereas each possess established chromosome territories, pluripotent ESCs and differentiated cell types differ in their radial positioning of individual chromosomes, which is related to gene density, as revealed by dual-color 3D fluorescence in situ hybridization (3D-FISH) experiments (Mayer et al. 2005). In addition, repositioning of chromosomal arms has been observed in the case of NANOG, which is highly expressed during pluripotency, where the short arm of chromosome 12 (12p) is found closer to the center of the nucleus in human ESCs than in differentiated cells (Wiblin et al. 2005). Heterochromatic chromosomal subregions, such as the centromere and pericentromere, which are called "chromocenters" when imaged in stained nuclei, cluster differently depending on cell type. For instance, mouse myoblasts and fibroblasts have smaller but a greater number of chromocenters than ESC nuclei, whereas fewer chromocenters can be seen in macrophages and postmitotic neurons (Mayer et al. 2005; Aoto et al. 2006; Kobayakawa et al. 2007). Intriguingly, loss of NANOG causes chromocenters in ESCs to resemble a more differentiated state (Novo et al. 2016). Notably, visible chromocenters are absent in human cells, but staining for centromere protein C (CENP-C) reveals that centromeres in human ESCs localize further away from the nuclear periphery than most other cell types (Wiblin et al. 2005).

Chromatin Loops, Domains, and Interaction Hubs

Organization at a finer scale occurs through chromatin looping and formation of topologically associating domains (TADs) or lamina-associated domains (LADs). Chromatin loops and TADs have been identified using chromosome conformation capture (3C)-based technologies, which map chromatin–chromatin interactions through formaldehyde cross-linking. LADs, which are detected with DNA adenine methyltransferase identification (DamID), are located at the nuclear periphery and correlate with gene silencing (Gonzalez-Sandoval and Gasser 2016). Chromosome regions that interact with the nuclear lamina make up around 40% of the genome and are mostly conserved across cell lineages, as evidenced by DamID experiments, where 73%–87% of LADs were found to overlap between several stages of development (Peric-Hupkes et al. 2010). Nonetheless, many LAD-associated loci reposition themselves relative to the nuclear interior following differentiation, and stem-cell-related genes, such as NANOG, KLF4, and OCT4, show more frequent interactions with the nuclear lamina in ESC-derived NPCs (Peric-Hupkes et al. 2010). As these loci generally encompass a single gene or transcriptional unit, this alludes to a functional role for regulation of nuclear architecture (Melcer and Meshorer 2010). However, the relationship between spatial relocalization of gene loci and changes in transcription is not always apparent. For instance, in the case of the Hoxd cluster, although the entire locus loops away from its chromosome territory upon differentiation in mouse ESCs, individual genes in the cluster do not become activated in a concurrent fashion (Morey et al. 2007).

A fundamental regulatory feature of metazoan nuclei is the existence of long-range enhancer–promoter interactions. CTCF-mediated chromatin looping on the kilobase scale is one mechanism by which this is thought to occur (Rao et al. 2014). Genome-wide maps of contacts that occur between any two genomic loci can be detected using Hi-C, a 3C-derived method, which incorporates proximity-based ligation with high-throughput sequencing (Dekker et al. 2002; Lieberman-Aiden et al. 2009). Overall, long-range contacts are weaker and less prevalent in mouse ESCs compared with somatic cells, and chromatin topology is progressively established during ESC lineage progression (Pękowska et al. 2018). This is in line with the more globally open chromatin configuration of ESCs, as a diffuse chromatin environment would hinder the formation of long-range interactions. In contrast, interactions between pluripotency factor binding sites are enriched in ESCs and the pluripotency factors themselves contribute to formation of these contacts (Apostolou et al. 2013; Denholtz et al. 2013; de Wit et al. 2013; Wei et al. 2013). These observations are supported by Hi-C maps generated during reprogramming of B cells into iPSCs. Strikingly, these experiments revealed that following reprogramming, ESC-specific contacts became more frequent at genes related to developmental processes, the formation of which preceded changes in transcription (Stadhouders et al. 2018). In support of this finding, another study, using a technique called promoter capture Hi-C, found that developmental pathway–related genes are enriched in ESC-specific promoter–enhancer interactions; a large amount of distinct cell-type-specific contacts were detected for both ESCs and fetal liver cells at promoters interacting with more than 10 enhancers (Schoenfelder et al. 2018).

Similarly, Hi-C studies have found that the position of TADs and their boundaries do not considerably vary across diverse mammalian cell types (Dixon et al. 2012; Rao et al. 2014). However, at a subset of TADs, changes in the frequency of intradomain interactions can occur following human ESC differentiation; this corresponds with minor changes in expression (with a bias toward repression) of genes within these domains (Dixon et al. 2015). These findings were corroborated by an ultra-deep Hi-C study, which compared two types of self-interacting compartments in serum grown ESCs with NPCs and neurons. These compartments, referred to as "A" or "B," are distinct networks of loci whose intradomain genes fall within active or inactive chromatin, respectively. Bonev et al. found that co-interactions within compartment "A" were weaker, and compartment "B" co-interactions became stronger following neuronal differentiation of ESCs, suggesting that "switching on and off" of particular TADs occurs in a cell-type-sensitive manner. Overall, the authors found that differentiation results in stronger intra-TAD contacts and weaker inter-TAD contacts (Bonev et al. 2017). In stark contrast, a recent study showed the opposite trend. Barrington et al. revealed that ESCs cultured in 2i conditions had, on average, weaker intradomain interactions, weaker TAD borders, and stronger interdomain contacts than neural stem cells (NSCs) (Barrington et al. 2019). This discrepancy indicates that environmental signals significantly influence TAD organization in ESCs. Moreover, it suggests that 2i-cultured, "ground state" ESCs are less organized topologically and that, conversely, ESCs cultured in serum may have already acquired some lineage-specific genomic features as their transcriptional heterogeneity would imply. Nevertheless, both results suggest that during development, formation, and dissolution of TADs occur in a distinctly cell-type-specific manner. Here, it is worth mentioning that the presence and absence of contacts in Hi-C maps may not reflect the state of each cell in a population, especially as the expression of some pluripotency factors, such as Nanog and Rex1, have been shown to be highly heterogeneous in ESCs (Toyooka et al. 2008; Clancy et al. 2016). Furthermore, it is also important to note that as chromatin–chromatin interaction maps are generated from a population of cells, due to cell-to-cell heterogeneity, they are not wholly representative of "true" or stable chromatin contacts. Indeed, these may be highly transient as single-cell Hi-C and imaging experiments have demonstrated (Nagano et al. 2013; Cattoni et al. 2017; Finn et al. 2019).

Nevertheless, evidence from chromatin–chromatin interaction studies reveal that overall, topological organization does not profoundly differ between pluripotent and differentiated cells. Rather, interactions at specific loci or domains appear to be locally implicated in an ESC-specific manner. In addition, these interactions are often located at genes in the core pluripotency network. However, it remains to be seen to what extent these interactions precede function. There is a growing consensus, at least in the realm of genome topology, that function drives structure more frequently than not (Finn and Misteli 2019). In the case of pluripotent stem cells, both form and function are likely intertwined, and the transcription of pluripotency factors and the formation of hubs of interactions may be costabilizing events (Fig. 5). Indeed, further chromatin interaction studies at higher resolutions are needed to disentangle the relationship between transcription and spatial organization. To this end, the establishment of a micrococcal nuclease (micro-C) method for profiling of sub-TAD contacts in mammalian cells is an extremely promising development (Hsieh et al. 2015, 2020). Micro-C has revealed that compared with human ESCs, human fetal fibroblasts (HFFs) possess dramatically increased compartment strength, as well as an increase in the number of long-range contacts (Krietenstein et al. 2020). This ultrahigh, single-nucleosome resolution (Hsieh et al. 2020) analysis of genome organization in ESCs and HFFs provides the most comprehensive chromatin interactions catalog in both cell types, supporting the hyperplastic chromatin model of the pluripotent cell (Schlesinger and Meshorer 2019).

CONCLUSIONS AND FUTURE PERSPECTIVES

As we have seen, with the constant development of new methods, mostly relying on high-throughput sequencing readouts, and more recently, super-resolution imaging, the 2D and 3D genome has been studied in exceedingly finer detail in

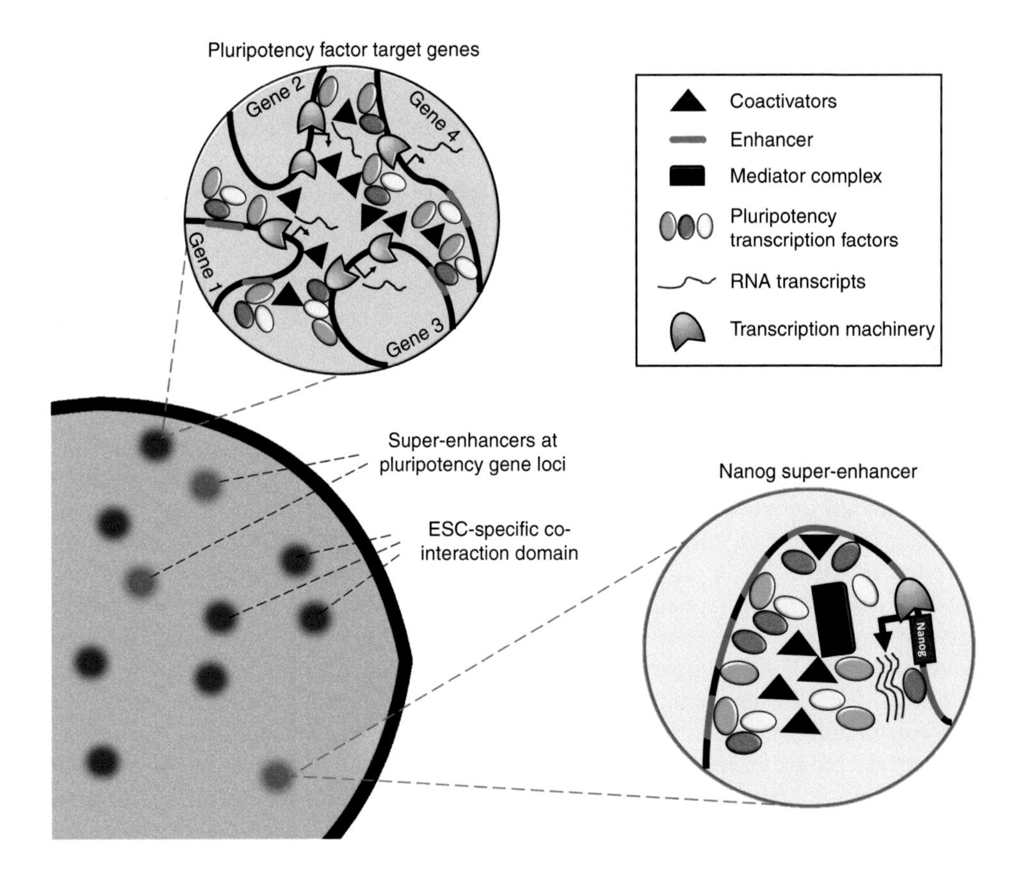

Figure 5. Transcriptional "hubs" of pluripotency factor accumulation. In pluripotent cells, master transcription factors are highly expressed from loci that are regulated by regions of densely packed enhancers termed "super-enhancers" (SEs). The combinatorial binding of pluripotency factors themselves at pluripotency gene loci stabilizes their high levels of expression. Large quantities of pluripotency factors in turn can accumulate at their target genes throughout the genome. As each pluripotency factor can itself form complexes with other coactivators, multiple loci may be brought together spatially through the binding of enhancers/promoters of different genes. (ESC) embryonic stem cell.

pluripotent and differentiated cells. Whereas the organization of the pluripotent genome (e.g., chromosome territories, the marking of DNA by cytosine methylation and of chromatin by histone modifications, chromatin–chromatin, and chromatin–lamina interactions) is globally not unlike its somatic counterpart, it is nonetheless unique in specific chromatin features, which, collectively, confer fluid-like properties. How then do pluripotent ESCs maintain their hyperplastic chromatin? The emerging concept of "biomolecular condensates" (Banani et al. 2017) or "transcriptional condensates" (Shrinivas et al. 2019) may provide some clues. Working with knockin fluorescent mouse ESC strains, Boija et al. (2018) showed that nuclear condensates formed by the mediator complex at ESC SEs are dependent on OCT4, which incorporates into these nuclear condensates together with MED1 and other mediator subunits. Similarly, MED1 and BRD4 form transcriptional condensates on pluripotent SEs, including the Nanog SE. Because phase-separated condensates are formed by the nonstructured activation domains of many TFs (Boija et al. 2018), it is tempting to speculate that the core transcriptional circuitry operating in pluripotent cells might be unique in its ability to form dynamic transcriptional condensates, antagonizing chromatin phase separation, particularly as there is a high level of combinatorial binding of OCT4, SOX2, and NANOG at enhancers associated with mediator colocalization (Göke et al. 2011). In contrast, somatic TFs might mediate stronger long-range interactions by fusing several distant high-density regions (Shrinivas et al. 2019), promoting heterochromatin-favoring phase separation, genome aggregation, and restricted dynamics. Supporting this model, histone acetylation, which is prevalent in ESCs (Bártová et al. 2008; Bian et al. 2009; Hezroni et al. 2011) and was shown to support chromatin plasticity and high mobility of chromatin proteins in ESCs (Melcer et al. 2012), was also shown recently to constrain phase separation (Gibson et al. 2019). These ideas, however, remain to be tested.

In closing, beyond the organizational and structural elements of the pluripotent genome described herein, a great deal more work—both theoretical and experimental—will be needed to bridge the gap between our understanding of defining features of pluripotency (e.g., open chromatin, active epigenetic marks) and the biological necessity of these traits in the early embryo. For instance, could sensitivity to developmental cues be an emergent property of a heterogeneous population capable of promiscuous transcription? Moreover, to what extent is pluripotency a feature of an individual cell versus that of the population as a whole? Indeed, future studies should attempt to contextualize features of the pluripotent genome with respect to its developmental milieu. The coming decades will no doubt offer deeper insights into the mysterious phenomenon of pluripotency.

ACKNOWLEDGMENTS

This work was supported by the Israel Science Foundation (ISF 1140/2017 to E.M.). E.M. is the incumbent of the Arthur Gutterman Professor Chair for Stem Cell Research. P.S.L.L. is supported by the H2020 Marie Skłodowska-Curie Innovative Training Network – EpiSyStem, Grant No. 765966.

REFERENCES

Reference is also in this collection.

Ahmed K, Dehghani H, Rugg-Gunn P, Fussner E, Rossant J, Bazett-Jones DP. 2010. Global chromatin architecture reflects pluripotency and lineage commitment in the early mouse embryo. *PLoS ONE* **5:** e10531. doi:10.1371/journal.pone.0010531

Alajem A, Biran A, Harikumar A, Sailaja BS, Aaronson Y, Livyatan I, Nissim-Rafinia M, Sommer AG, Mostoslavsky G, Gerbasi VR, et al. 2015. Differential association of chromatin proteins identifies BAF60a/SMARCD1 as a regulator of embryonic stem cell differentiation. *Cell Rep* **10:** 2019–2031. doi:10.1016/j.celrep.2015.02.064

Aoto T, Saitoh N, Ichimura T, Niwa H, Nakao M. 2006. Nuclear and chromatin reorganization in the MHC-Oct3/4 locus at developmental phases of embryonic stem cell differentiation. *Dev Biol* **298:** 354–367. doi:10.1016/j.ydbio.2006.04.450

Apostolou E, Ferrari F, Walsh RM, Bar-Nur O, Stadtfeld M, Cheloufi S, Stuart HT, Polo JM, Ohsumi TK, Borowsky ML, et al. 2013. Genome-wide chromatin interactions of the Nanog locus in pluripotency, differentiation, and reprogramming. *Cell Stem Cell* **12:** 699–712. doi:10.1016/j.stem.2013.04.013

Athanasiadou R, de Sousa D, Myant K, Merusi C, Stancheva I, Bird A. 2010. Targeting of de novo DNA Methylation throughout the Oct-4 gene regulatory region in differentiating embryonic stem cells. *PLoS ONE* **5:** e9937. doi:10.1371/journal.pone.0009937

Banani SF, Lee HO, Hyman AA, Rosen MK. 2017. Biomolecular condensates: organizers of cellular biochemistry. *Nat Rev Mol Cell Biol* **18:** 285–298. doi:10.1038/nrm.2017.7

Barrington C, Georgopoulou D, Pezic D, Varsally W, Herrero J, Hadjur S. 2019. Enhancer accessibility and CTCF occupancy underlie asymmetric TAD architecture and cell type specific genome topology. *Nat Commun* **10:** 2908. doi:10.1038/s41467-019-10725-9

Bártová E, Galiová G, Krejcí J, Harničarová A, Strašák L, Kozubek S. 2008. Epigenome and chromatin structure in human embryonic stem cells undergoing differentiation. *Dev Dyn* **237:** 3690–3702. doi:10.1002/dvdy.21773

Bascom G, Schlick T. 2017. Linking chromatin fibers to gene folding by hierarchical looping. *Biophys J* **112:** 434–445. doi:10.1016/j.bpj.2017.01.003

Belmont AS, Bruce K. 1994. Visualization of G1 chromosomes: a folded, twisted, supercoiled chromonema model of interphase chromatid structure. *J Cell Biol* **127:** 287–302. doi:10.1083/jcb.127.2.287

Bergh PA, Navot D. 1992. The impact of embryonic development and endometrial maturity on the timing of implantation. *Fertil Steril* **58:** 537–542. doi:10.1016/S0015-0282(16)55259-5

Bernstein BE, Mikkelsen TS, Xie X, Kamal M, Huebert DJ, Cuff J, Fry B, Meissner A, Wernig M, Plath K, et al. 2006. A bivalent chromatin structure marks key developmental genes in embryonic stem cells. *Cell* **125:** 315–326. doi:10.1016/j.cell.2006.02.041

Bhattacharya D, Talwar S, Mazumder A, Shivashankar GV. 2009. Spatio-temporal plasticity in chromatin organization in mouse cell differentiation and during *Drosophila* embryogenesis. *Biophys J* **96:** 3832–3839. doi:10.1016/j.bpj.2008.11.075

Bian Y, Alberio R, Allegrucci C, Campbell KH, Johnson AD. 2009. Epigenetic marks in somatic chromatin are remodelled to resemble pluripotent nuclei by amphibian oocyte extracts. *Epigenetics* **4:** 194–202. doi:10.4161/epi.4.3.8787

Boiani M, Schöler HR. 2005. Regulatory networks in embryo-derived pluripotent stem cells. *Nat Rev Mol Cell Biol* **6:** 872–881. doi:10.1038/nrm1744

Boija A, Klein IA, Sabari BR, Dall'Agnese A, Coffey EL, Zamudio AV, Li CH, Shrinivas K, Manteiga JC, Hannett NM, et al. 2018. Transcription factors activate genes through the phase-separation capacity of their activation domains. *Cell* **175:** 1842–1855.e16. doi:10.1016/j.cell.2018.10.042

Bonev B, Mendelson Cohen N, Szabo Q, Fritsch L, Papadopoulos GL, Lubling Y, Xu X, Lv X, Hugnot JP, Tanay A, et al. 2017. Multiscale 3D genome rewiring during mouse neural development. *Cell* **171:** 557–572.e24. doi:10.1016/j.cell.2017.09.043

Bošković A, Eid A, Pontabry J, Ishiuchi T, Spiegelhalter C, Ram EVSR, Meshorer E, Torres-Padilla ME. 2014. Higher chromatin mobility supports totipotency and precedes pluripotency in vivo. *Genes Dev* **28:** 1042–1047. doi:10.1101/gad.238881.114

Boyer LA, Lee TI, Cole MF, Johnstone SE, Levine SS, Zucker JP, Guenther MG, Kumar RM, Murray HL, Jenner RG, et al. 2005. Core transcriptional regulatory circuitry in human embryonic stem cells. *Cell* **122:** 947–956. doi:10.1016/j.cell.2005.08.020

Bradley A, Evans M, Kaufman MH, Robertson E. 1984. Formation of germ-line chimaeras from embryo-derived teratocarcinoma cell lines. *Nature* **309:** 255–256. doi:10.1038/309255a0

Brons IGM, Smithers LE, Trotter MWB, Rugg-Gunn P, Sun B, Chuva de Sousa Lopes SM, Howlett SK, Clarkson A, Ahrlund-Richter L, Pedersen RA, et al. 2007. Derivation of pluripotent epiblast stem cells from mammalian embryos. *Nature* **448:** 191–195. doi:10.1038/nature05950

Cattoni DI, Cardozo Gizzi AM, Georgieva M, Di Stefano M, Valeri A, Chamousset D, Houbron C, Déjardin S, Fiche JB, González I, et al. 2017. Single-cell absolute contact probability detection reveals chromosomes are organized by multiple low-frequency yet specific interactions. *Nat Commun* **8:** 1753. doi:10.1038/s41467-017-01962-x

Cavalier-Smith T. 2017. Origin of animal multicellularity: precursors, causes, consequences—the choanoflagellate/sponge transition, neurogenesis and the Cambrian explosion. *Philos Trans R Soc Lond B Biol Sci* **372:** 20150476. doi:10.1098/rstb.2015.0476

Chen T, Dent SYR. 2014. Chromatin modifiers and remodellers: regulators of cellular differentiation. *Nat Rev Genet* **15:** 93–106. doi:10.1038/nrg3607

Clancy CE, An G, Cannon WR, Liu Y, May EE, Ortoleva P, Popel AS, Sluka JP, Su J, Vicini P, et al. 2016. Multiscale modeling in the clinic: drug design and development. *Ann Biomed Eng* **44:** 2591–2610. doi:10.1007/s10439-016-1563-0

Cremer T, Cremer M. 2010. Chromosome territories. *Cold Spring Harb Perspect Biol* **2:** a003889. doi:10.1101/cshperspect.a003889

Dekker J, Rippe K, Dekker M, Kleckner N. 2002. Capturing chromosome conformation. *Science* **295:** 1306–1311. doi:10.1126/science.1067799

Deng T, Zhu ZI, Zhang S, Leng F, Cherukuri S, Hansen L, Mariño-Ramírez L, Meshorer E, Landsman D, Bustin M. 2013. HMGN1 modulates nucleosome occupancy and DNase I hypersensitivity at the CpG island promoters of embryonic stem cells. *Mol Cell Biol* **33:** 3377–3389. doi:10.1128/MCB.00435-13

Denholtz M, Bonora G, Chronis C, Splinter E, de Laat W, Ernst J, Pellegrini M, Plath K. 2013. Long-range chromatin contacts in embryonic stem cells reveal a role for pluripotency factors and polycomb proteins in genome organization. *Cell Stem Cell* **13:** 602–616. doi:10.1016/j.stem.2013.08.013

de Wit E, Bouwman BA, Zhu Y, Klous P, Splinter E, Verstegen MJ, Krijger PH, Festuccia N, Nora EP, Welling M, et al. 2013. The pluripotent genome in three dimensions is shaped around pluripotency factors. *Nature* **501:** 227–231. doi:10.1038/nature12420

Dixon JR, Selvaraj S, Yue F, Kim A, Li Y, Shen Y, Hu M, Liu JS, Ren B. 2012. Topological domains in mammalian genomes identified by analysis of chromatin interactions. *Nature* **485:** 376–380. doi:10.1038/nature11082

Dixon JR, Jung I, Selvaraj S, Shen Y, Antosiewicz-Bourget JE, Lee AY, Ye Z, Kim A, Rajagopal N, Xie W, et al. 2015. Chromatin architecture reorganization during stem cell differentiation. *Nature* **518:** 331–336. doi:10.1038/nature14222

Eckersley-Maslin MA, Thybert D, Bergmann JH, Marioni JC, Flicek P, Spector DL. 2014. Random monoallelic gene expression increases upon embryonic stem cell differentiation. *Dev Cell* **28:** 351–365. doi:10.1016/j.devcel.2014.01.017

Edupuganti RR, Harikumar A, Aaronson Y, Biran A, Sailaja BS, Nissim-Rafinia M, Azad GK, Cohen MA, Park JE, Shivalila CS, et al. 2017. Alternative SET/TAFI promoters regulate embryonic stem cell differentiation. *Stem Cell Rep* **9:** 1291–1303. doi:10.1016/j.stemcr.2017.08.021

Efroni S, Duttagupta R, Cheng J, Dehghani H, Hoeppner DJ, Dash C, Bazett-Jones DP, Le Grice S, McKay RDG, Buetow KH, et al. 2008. Global transcription in pluripotent embryonic stem cells. *Cell Stem Cell* **2:** 437–447. doi:10.1016/j.stem.2008.03.021

Eltsov M, MacLellan KM, Maeshima K, Frangakis AS, Dubochet J. 2008. Analysis of cryo-electron microscopy images does not support the existence of 30-nm chromatin fibers in mitotic chromosomes in situ. *Proc Natl Acad Sci* **105:** 19732–19737. doi:10.1073/pnas.0810057105

Etchegaray JP, Chavez L, Huang Y, Ross KN, Choi J, Martinez-Pastor B, Walsh RM, Sommer CA, Lienhard M, Kugel S, et al. 2015. The histone deacetylase Sirt6 controls embryonic stem cell fate via TET-mediated production of 5-hydroxymethylcytosine. *Nat Cell Biol* **17:** 545–557. doi:10.1038/ncb3147

Evans MJ, Kaufman MH. 1981. Establishment in culture of pluripotential cells from mouse embryos. *Nature* **292:** 154–156. doi:10.1038/292154a0

Factor DC, Corradin O, Zentner GE, Saiakhova A, Song L, Chenoweth JG, McKay RD, Crawford GE, Scacheri PC, Tesar PJ. 2014. Epigenomic comparison reveals activation of "seed" enhancers during transition from naive to primed pluripotency. *Cell Stem Cell* **14:** 854–863. doi:10.1016/j.stem.2014.05.005

Ficz G, Hore TA, Santos F, Lee HJ, Dean W, Arand J, Krueger F, Oxley D, Paul YL, Walter J, et al. 2013. FGF signaling inhibition in ESCs drives rapid genome-wide demethylation to the epigenetic ground state of pluripotency. *Cell Stem Cell* **13:** 351–359. doi:10.1016/j.stem.2013.06.004

Finn EH, Misteli T. 2019. A genome disconnect. *Nat Genet* **51:** 1205–1206. doi:10.1038/s41588-019-0476-x

Finn EH, Pegoraro G, Brandão HB, Valton AL, Oomen ME, Dekker J, Mirny L, Misteli T. 2019. Extensive heterogeneity and intrinsic variation in spatial genome organization. *Cell* **176:** 1502–1515.e10. doi:10.1016/j.cell.2019.01.020

Fussner E, Ahmed K, Dehghani H, Strauss M, Bazett-Jones DP. 2010. Changes in chromatin fiber density as a marker for pluripotency. *Cold Spring Harb Symp Quant Biol* **75:** 245–249. doi:10.1101/sqb.2010.75.012

Fussner E, Ching RW, Bazett-Jones DP. 2011. Living without 30 nm chromatin fibers. *Trends Biochem Sci* **36:** 1–6. doi:10.1016/j.tibs.2010.09.002

Gao Y, Gan H, Lou Z, Zhang Z. 2018. Asf1a resolves bivalent chromatin domains for the induction of lineage-specific genes during mouse embryonic stem cell differentiation. *Proc Natl Acad Sci* **115:** E6162–E6171. doi:10.1073/pnas.1801909115

Gaspar-Maia A, Alajem A, Polesso F, Sridharan R, Mason MJ, Heidersbach A, Ramalho-Santos J, McManus MT, Plath K, Meshorer E, et al. 2009. Chd1 regulates open chromatin and pluripotency of embryonic stem cells. *Nature* **460:** 863–868. doi:10.1038/nature08212

Gaspar-Maia A, Alajem A, Meshorer E, Ramalho-Santos M. 2011. Open chromatin in pluripotency and reprogramming. *Nat Rev Mol Cell Biol* **12:** 36–47. doi:10.1038/nrm3036

Ghimire S, der Jeught MV, Neupane J, Roost MS, Anckaert J, Popovic M, Nieuwerburgh FV, Mestdagh P, Vandesompele J, Deforce D, et al. 2018. Comparative analysis of naive, primed and ground state pluripotency in mouse embryonic stem cells originating from the same genetic background. *Sci Rep* **8:** 5884. doi:10.1038/s41598-018-24051-5

Gibson BA, Doolittle LK, Schneider MWG, Jensen LE, Gamarra N, Henry L, Gerlich DW, Redding S, Rosen MK. 2019. Organization of chromatin by intrinsic and regulated phase separation. *Cell* **179:** 470–484.e21. doi:10.1016/j.cell.2019.08.037

Göke J, Jung M, Behrens S, Chavez L, O'Keeffe S, Timmermann B, Lehrach H, Adjaye J, Vingron M. 2011. Combinatorial binding in human and mouse embryonic stem cells identifies conserved enhancers active in early embryonic development. *PLoS Comput Biol* **7:** e1002304. doi:10.1371/journal.pcbi.1002304

Gonzalez-Sandoval A, Gasser SM. 2016. On TADs and LADs: spatial control over gene expression. *Trends Genet* **32:** 485–495. doi:10.1016/j.tig.2016.05.004

Gorkin DU, Leung D, Ren B. 2014. The 3D genome in transcriptional regulation and pluripotency. *Cell Stem Cell* **14:** 762–775. doi:10.1016/j.stem.2014.05.017

Greenberg MVC, Bourc'his D. 2019. The diverse roles of DNA methylation in mammalian development and disease. *Nat Rev Mol Cell Biol* **20:** 590–607. doi:10.1038/s41580-019-0159-6

Grigoryev SA, Bascom G, Buckwalter JM, Schubert MB, Woodcock CL, Schlick T. 2016. Hierarchical looping of zigzag nucleosome chains in metaphase chromosomes. *Proc Natl Acad Sci* **113:** 1238–1243. doi:10.1073/pnas.1518280113

Gulati GS, Sikandar SS, Wesche DJ, Manjunath A, Bharadwaj A, Berger MJ, Ilagan F, Kuo AH, Hsieh RW, Cai S, et al. 2020. Single-cell transcriptional diversity is a hallmark of developmental potential. *Science* **367:** 405–411. doi:10.1126/science.aax0249

Guo G, Pinello L, Han X, Lai S, Shen L, Lin TW, Zou K, Yuan GC, Orkin SH. 2016a. Serum-based culture conditions provoke gene expression variability in mouse embryonic stem cells as revealed by single-cell analysis. *Cell Rep* **14:** 956–965. doi:10.1016/j.celrep.2015.12.089

Guo G, von Meyenn F, Santos F, Chen Y, Reik W, Bertone P, Smith A, Nichols J. 2016b. Naive pluripotent stem cells derived directly from isolated cells of the human inner cell mass. *Stem Cell Rep* **6:** 437–446. doi:10.1016/j.stemcr.2016.02.005

Habibi E, Brinkman AB, Arand J, Kroeze LI, Kerstens HHD, Matarese F, Lepikhov K, Gut M, Brun-Heath I, Hubner NC, et al. 2013. Whole-genome bisulfite sequencing of two distinct interconvertible DNA methylomes of mouse embryonic stem cells. *Cell Stem Cell* **13:** 360–369. doi:10.1016/j.stem.2013.06.002

Hainer SJ, Fazzio TG. 2015. Regulation of nucleosome architecture and factor binding revealed by nuclease footprinting of the ESC genome. *Cell Rep* **13:** 61–69. doi:10.1016/j.celrep.2015.08.071

Hainer SJ, Gu W, Carone BR, Landry BD, Rando OJ, Mello CC, Fazzio TG. 2015. Suppression of pervasive noncoding transcription in embryonic stem cells by esBAF. *Genes Dev* **29:** 362–378. doi:10.1101/gad.253534.114

Hainer SJ, Bošković A, McCannell KN, Rando OJ, Fazzio TG. 2019. Profiling of pluripotency factors in single cells and early embryos. *Cell* **177:** 1319–1329.e11. doi:10.1016/j.cell.2019.03.014

Harikumar A, Meshorer E. 2015. Chromatin remodeling and bivalent histone modifications in embryonic stem cells. *EMBO Rep* **16:** 1609–1619. doi:10.15252/embr.201541011

Hawkins RD, Hon GC, Lee LK, Ngo Q, Lister R, Pelizzola M, Edsall LE, Kuan S, Luu Y, Klugman S, et al. 2010. Distinct epigenomic landscapes of pluripotent and lineage-committed human cells. *Cell Stem Cell* **6:** 479–491. doi:10.1016/j.stem.2010.03.018

Heitz E. 1928. Das heterochromatin der moose. *I Jahrb Wiss Botanik* **69:** 762–818 (in German).

Hezroni H, Tzchori I, Davidi A, Mattout A, Biran A, Nissim-Rafinia M, Westphal H, Meshorer E. 2011. H3K9 histone acetylation predicts pluripotency and reprogramming capacity of ES cells. *Nucleus* **2:** 300–309. doi:10.4161/nucl.2.4.16767

Hirasawa R, Sasaki H. 2009. Dynamic transition of Dnmt3b expression in mouse pre- and early post-implantation embryos. *Gene Expr Patterns* **9:** 27–30. doi:10.1016/j.gep.2008.09.002

Ho L, Jothi R, Ronan JL, Cui K, Zhao K, Crabtree GR. 2009. An embryonic stem cell chromatin remodeling complex, esBAF, is an essential component of the core pluripotency transcriptional network. *Proc Natl Acad Sci* **106:** 5187–5191. doi:10.1073/pnas.0812888106

Hough SR, Laslett AL, Grimmond SB, Kolle G, Pera MF. 2009. A continuum of cell states spans pluripotency and lineage commitment in human embryonic stem cells. *PLoS ONE* **4:** e7708. doi:10.1371/journal.pone.0007708

Hsieh THS, Weiner A, Lajoie B, Dekker J, Friedman N, Rando OJ. 2015. Mapping nucleosome resolution chromosome folding in yeast by micro-C. *Cell* **162:** 108–119. doi:10.1016/j.cell.2015.05.048

Hsieh THS, Cattoglio C, Slobodyanyuk E, Hansen AS, Rando OJ, Tjian R, Darzacq X. 2020. Resolving the 3D landscape of transcription-linked mammalian chromatin folding. *Mol Cell* **78:** 539–553.e8. doi:10.1016/j.molcel.2020.03.002

Imai R, Nozaki T, Tani T, Kaizu K, Hibino K, Ide S, Tamura S, Takahashi K, Shribak M, Maeshima K. 2017. Density imaging of heterochromatin in live cells using orientation-independent-DIC microscopy. *Mol Biol Cell* **28:** 3349–3359. doi:10.1091/mbc.e17-06-0359

Iurlaro M, von Meyenn F, Reik W. 2017. DNA methylation homeostasis in human and mouse development. *Curr Opin Genet Dev* **43:** 101–109. doi:10.1016/j.gde.2017.02.003

Jackson M, Krassowska A, Gilbert N, Chevassut T, Forrester L, Ansell J, Ramsahoye B. 2004. Severe global DNA hypomethylation blocks differentiation and induces histone hyperacetylation in embryonic stem cells. *Mol Cell Biol* **24:** 8862–8871. doi:10.1128/MCB.24.20.8862-8871.2004

Joseph SR, Pálfy M, Hilbert L, Kumar M, Karschau J, Zaburdaev V, Shevchenko A, Vastenhouw NL. 2017. Competition between histone and transcription factor binding regulates the onset of transcription in zebrafish embryos. *eLife* **6:** e23326. doi:10.7554/eLife.23326

Juan AH, Wang S, Ko KD, Zare H, Tsai PF, Feng X, Vivanco KO, Ascoli AM, Gutierrez-Cruz G, Krebs J, et al. 2016. Roles of H3K27me2 and H3K27me3 examined during fate specification of embryonic stem cells. *Cell Rep* **17:** 1369–1382. doi:10.1016/j.celrep.2016.09.087

Kalkan T, Smith A. 2014. Mapping the route from naive pluripotency to lineage specification. *Philos Trans R Soc Lond B Biol Sci* **369:** 20130540. doi:10.1098/rstb.2013.0540

Kelley DR, Snoek J, Rinn JL. 2016. Basset: learning the regulatory code of the accessible genome with deep convolutional neural networks. *Genome Res* **26:** 990–999. doi:10.1101/gr.200535.115

Klemm SL, Shipony Z, Greenleaf WJ. 2019. Chromatin accessibility and the regulatory epigenome. *Nat Rev Genet* **20:** 207–220. doi:10.1038/s41576-018-0089-8

Ko YG, Nishino K, Hattori N, Arai Y, Tanaka S, Shiota K. 2005. Stage-by-stage change in DNA methylation status of Dnmt1 locus during mouse early development. *J Biol Chem* **280:** 9627–9634. doi:10.1074/jbc.M413822200

Kobayakawa S, Miike K, Nakao M, Abe K. 2007. Dynamic changes in the epigenomic state and nuclear organization of differentiating mouse embryonic stem cells. *Genes Cells* **12:** 447–460. doi:10.1111/j.1365-2443.2007.01063.x

Kobayashi H, Kikyo N. 2015. Epigenetic regulation of open chromatin in pluripotent stem cells. *Transl Res* **165:** 18–27. doi:10.1016/j.trsl.2014.03.004

Koh FM, Sachs M, Guzman-Ayala M, Ramalho-Santos M. 2010. Parallel gateways to pluripotency: open chromatin in stem cells and development. *Curr Opin Genet Dev* **20:** 492–499. doi:10.1016/j.gde.2010.06.002

Kolodziejczyk AA, Kim JK, Tsang JCH, Ilicic T, Henriksson J, Natarajan KN, Tuck AC, Gao X, Bühler M, Liu P, et al. 2015. Single cell RNA-sequencing of pluripotent states unlocks modular transcriptional variation. *Cell Stem Cell* **17:** 471–485. doi:10.1016/j.stem.2015.09.011

Kornberg RD. 1974. Chromatin structure: a repeating unit of histones and DNA. *Science* **184:** 868–871. doi:10.1126/science.184.4139.868

Kornberg RD, Klug A. 1981. The nucleosome. *Sci Am* **244:** 52–64. doi:10.1038/scientificamerican0281-52

Korolev N, Lyubartsev AP, Nordenskiöld L. 2018. A systematic analysis of nucleosome core particle and nucleosome-nucleosome stacking structure. *Sci Rep* **8:** 1–14. doi:10.1038/s41598-018-19875-0

Krietenstein N, Abraham S, Venev SV, Abdennur N, Gibcus J, Hsieh THS, Parsi KM, Yang L, Maehr R, Mirny LA, et al. 2020. Ultrastructural details of mammalian chromosome architecture. *Mol Cell* **78:** 554–565.e7. doi:10.1016/j.molcel.2020.03.003

Kunath T, Rossant J. 2009. Stem cells in extraembryonic lineages. In *Essentials of stem cell biology*, 2nd ed., pp. 137–144. Academic, San Diego. doi:10.1016/B978-0-12-374729-7.00015-9

Kuznetsova MA, Sheval EV. 2016. Chromatin fibers: from classical descriptions to modern interpretation. *Cell Biol Int* **40:** 1140–1151. doi:10.1002/cbin.10672

Landry J, Sharov AA, Piao Y, Sharova LV, Xiao H, Southon E, Matta J, Tessarollo L, Zhang YE, Ko MSH, et al. 2008. Essential role of chromatin remodeling protein Bptf in early mouse embryos and embryonic stem cells. *PLoS Genet* **4:** e1000241. doi:10.1371/journal.pgen.1000241

Lau KX, Mason EA, Kie J, De Souza DP, Kloehn J, Tull D, McConville MJ, Keniry A, Beck T, Blewitt ME, et al. 2020. Unique properties of a subset of human pluripotent stem cells with high capacity for self-renewal. *Nat Commun* **11:** 2420. doi:10.1038/s41467-020-16214-8

Leitch HG, McEwen KR, Turp A, Encheva V, Carroll T, Grabole N, Mansfield W, Nashun B, Knezovich JG, Smith A, et al. 2013. Naive pluripotency is associated with global DNA hypomethylation. *Nat Struct Mol Biol* **20:** 311–316. doi:10.1038/nsmb.2510

Li M, Izpisua Belmonte JC. 2018. Deconstructing the pluripotency gene regulatory network. *Nat Cell Biol* **20:** 382–392. doi:10.1038/s41556-018-0067-6

Li Y, Zhang Z, Chen J, Liu W, Lai W, Liu B, Li X, Liu L, Xu S, Dong Q, et al. 2018. Stella safeguards the oocyte methylome by preventing de novo methylation mediated by DNMT1. *Nature* **564:** 136–140. doi:10.1038/s41586-018-0751-5

Lieberman-Aiden E, van Berkum NL, Williams L, Imakaev M, Ragoczy T, Telling A, Amit I, Lajoie BR, Sabo PJ, Dorschner MO, et al. 2009. Comprehensive mapping of long-range interactions reveals folding principles of the human genome. *Science* **326:** 289–293. doi:10.1126/science.1181369

Lin JJ, Lehmann LW, Bonora G, Sridharan R, Vashisht AA, Tran N, Plath K, Wohlschlegel JA, Carey M. 2011. Mediator coordinates PIC assembly with recruitment of CHD1. *Genes Dev* **25:** 2198–2209. doi:10.1101/gad.17554711

Liu N, Zhang Z, Wu H, Jiang Y, Meng L, Xiong J, Zhao Z, Zhou X, Li J, Li H, et al. 2015. Recognition of H3K9 methylation by GLP is required for efficient establishment of H3K9 methylation, rapid target gene repression, and mouse viability. *Genes Dev* **29:** 379–393. doi:10.1101/gad.254425.114

Liu X, Wang C, Liu W, Li J, Li C, Kou X, Chen J, Zhao Y, Gao H, Wang H, et al. 2016. Distinct features of H3K4me3 and H3K27me3 chromatin domains in pre-implantation embryos. *Nature* **537:** 558–562. doi:10.1038/nature19362

Livyatan I, Aaronson Y, Gokhman D, Ashkenazi R, Meshorer E. 2015. BindDB: an integrated database and webtool platform for "reverse-ChIP" epigenomic analysis. *Cell Stem Cell* **17:** 647–648. doi:10.1016/j.stem.2015.11.015

Loh KM, Lim B. 2010. Recreating pluripotency? *Cell Stem Cell* **7:** 137–139. doi:10.1016/j.stem.2010.07.005

* Maeshima K. 2020. The chromatin fiber. *Cold Spring Harb Perspect Biol* doi:10.1101/cshperspect.a040675

Maeshima K, Imai R, Tamura S, Nozaki T. 2014. Chromatin as dynamic 10-nm fibers. *Chromosoma* **123:** 225–237. doi:10.1007/s00412-014-0460-2

Margueron R, Reinberg D. 2011. The polycomb complex PRC2 and its mark in life. *Nature* **469:** 343–349. doi:10.1038/nature09784

Marks H, Kalkan T, Menafra R, Denissov S, Jones K, Hofemeister H, Nichols J, Kranz A, Francis Stewart A, Smith A, et al. 2012. The transcriptional and epigenomic foundations of ground state pluripotency. *Cell* **149:** 590–604. doi:10.1016/j.cell.2012.03.026

Martin GR. 1981. Isolation of a pluripotent cell line from early mouse embryos cultured in medium conditioned by teratocarcinoma stem cells. *Proc Natl Acad Sci* **78:** 7634–7638. doi:10.1073/pnas.78.12.7634

Mascetti VL, Pedersen RA. 2016. Contributions of mammalian chimeras to pluripotent stem cell research. *Cell Stem Cell* **19:** 163–175. doi:10.1016/j.stem.2016.07.018

Mayer R, Brero A, von Hase J, Schroeder T, Cremer T, Dietzel S. 2005. Common themes and cell type specific variations of higher order chromatin arrangements in the mouse. *BMC Cell Biol* **6:** 44. doi:10.1186/1471-2121-6-44

Meissner A, Mikkelsen TS, Gu H, Wernig M, Hanna J, Sivachenko A, Zhang X, Bernstein BE, Nusbaum C, Jaffe DB, et al. 2008. Genome-scale DNA methylation maps of pluripotent and differentiated cells. *Nature* **454:** 766–770. doi:10.1038/nature07107

Melcer S, Meshorer E. 2010. The silence of the LADs: dynamic genome-lamina interactions during ESC differentiation. *Cell Stem Cell* **6:** 495–497. doi:10.1016/j.stem.2010.05.006

Melcer S, Hezroni H, Rand E, Nissim-Rafinia M, Skoultchi A, Stewart CL, Bustin M, Meshorer E. 2012. Histone modifications and lamin A regulate chromatin protein dynamics in early embryonic stem cell differentiation. *Nat Commun* **3:** 910. doi:10.1038/ncomms1915

Meshorer E, Misteli T. 2006. Chromatin in pluripotent embryonic stem cells and differentiation. *Nat Rev Mol Cell Biol* **7:** 540–546. doi:10.1038/nrm1938

Meshorer E, Yellajoshula D, George E, Scambler PJ, Brown DT, Misteli T. 2006. Hyperdynamic plasticity of chromatin proteins in pluripotent embryonic stem cells. *Dev Cell* **10:** 105–116. doi:10.1016/j.devcel.2005.10.017

Messmer T, von Meyenn F, Savino A, Santos F, Mohammed H, Lun ATL, Marioni JC, Reik W. 2019. Transcriptional heterogeneity in naive and primed human pluripotent stem cells at single-cell resolution. *Cell Rep* **26:** 815–824.e4. doi:10.1016/j.celrep.2018.12.099

Mitsui K, Tokuzawa Y, Itoh H, Segawa K, Murakami M, Takahashi K, Maruyama M, Maeda M, Yamanaka S. 2003. The homeoprotein Nanog is required for maintenance of pluripotency in mouse epiblast and ES cells. *Cell* **113:** 631–642. doi:10.1016/S0092-8674(03)00393-3

Mohn F, Weber M, Rebhan M, Roloff TC, Richter J, Stadler MB, Bibel M, Schübeler D. 2008. Lineage-specific polycomb targets and de novo DNA methylation define restriction and potential of neuronal progenitors. *Mol Cell* **30:** 755–766. doi:10.1016/j.molcel.2008.05.007

Morey C, Silva NRD, Perry P, Bickmore WA. 2007. Nuclear reorganisation and chromatin decondensation are conserved, but distinct, mechanisms linked to Hox gene activation. *Development* **134:** 909–919. doi:10.1242/dev.02779

Morgan TH. 1898. Experimental studies of the regeneration of *Planaria maculata*. *Archiv für Entwickelungsmechanik der Organismen* **7:** 364–397. doi:10.1007/BF02161491

Morgani S, Nichols J, Hadjantonakis AK. 2017. The many faces of pluripotency: in vitro adaptations of a continuum of in vivo states. *BMC Dev Biol* **17:** 7. doi:10.1186/s12861-017-0150-4

Morita M, Best JB, Noel J. 1969. Electron microscopic studies of planarian regeneration. I: Fine structure of neoblasts in *Dugesia dorotocephala*. *J Ultrastruct Res* **27:** 7–23. doi:10.1016/S0022-5320(69)90017-3

Morozumi Y, Boussouar F, Tan M, Chaikuad A, Jamshidikia M, Colak G, He H, Nie L, Petosa C, de Dieuleveult M, et al. 2016. Atad2 is a generalist facilitator of chromatin dynamics in embryonic stem cells. *J Mol Cell Biol* **8:** 349–362. doi:10.1093/jmcb/mjv060

Müller WEG. 2001. Review: how was metazoan threshold crossed? The hypothetical Urmetazoa. *Comp Biochem Physiol A Mol Integr Physiol* **129:** 433–460. doi:10.1016/S1095-6433(00)00360-3

Nagano T, Lubling Y, Stevens TJ, Schoenfelder S, Yaffe E, Dean W, Laue ED, Tanay A, Fraser P. 2013. Single-cell Hi-C reveals cell-to-cell variability in chromosome structure. *Nature* **502:** 59–64. doi:10.1038/nature12593

Nagy A, Gócza E, Diaz EM, Prideaux VR, Iványi E, Markkula M, Rossant J. 1990. Embryonic stem cells alone are able to support fetal development in the mouse. *Development* **110:** 815–821.

Nakamura T, Arai Y, Umehara H, Masuhara M, Kimura T, Taniguchi H, Sekimoto T, Ikawa M, Yoneda Y, Okabe M, et al. 2007. PGC7/Stella protects against DNA demethylation in early embryogenesis. *Nat Cell Biol* **9:** 64–71. doi:10.1038/ncb1519

Nava MM, Miroshnikova YA, Biggs LC, Whitefield DB, Metge F, Boucas J, Vihinen H, Jokitalo E, Li X, García Arcos JM, et al. 2020. Heterochromatin-driven nuclear softening protects the genome against mechanical stress-induced damage. *Cell* **181:** 800–817.e22. doi:10.1016/j.cell.2020.03.052

Nichols J, Smith A. 2009. Naive and primed pluripotent states. *Cell Stem Cell* **4:** 487–492. doi:10.1016/j.stem.2009.05.015

Novo CL, Tang C, Ahmed K, Djuric U, Fussner E, Mullin NP, Morgan NP, Hayre J, Sienerth AR, Elderkin S, et al. 2016. The pluripotency factor *Nanog* regulates pericentromeric heterochromatin organization in mouse embryonic stem cells. *Genes Dev* **30:** 1101–1115. doi:10.1101/gad.275685.115

Ohtsuka S, Dalton S. 2008. Molecular and biological properties of pluripotent embryonic stem cells. *Gene Ther* **15:** 74–81. doi:10.1038/sj.gt.3303065

Orkin SH, Hochedlinger K. 2011. Chromatin connections to pluripotency and cellular reprogramming. *Cell* **145:** 835–850. doi:10.1016/j.cell.2011.05.019

Ou HD, Phan S, Deerinck TJ, Thor A, Ellisman MH, O'Shea CC. 2017. ChromEMT: visualizing 3D chromatin structure and compaction in interphase and mitotic cells. *Science* **357:** eaag0025. doi:10.1126/science.aag0025

Oudet P, Gross-Bellard M, Chambon P. 1975. Electron microscopic and biochemical evidence that chromatin structure is a repeating unit. *Cell* **4:** 281–300. doi:10.1016/0092-8674(75)90149-X

Park SH, Park SH, Kook MC, Kim EY, Park S, Lim JH. 2004. Ultrastructure of human embryonic stem cells and spontaneous and retinoic acid-induced differentiating cells. *Ultrastruct Pathol* **28:** 229–238. doi:10.1080/01913120490515595

Pękowska A, Klaus B, Xiang W, Severino J, Daigle N, Klein FA, Oleś M, Casellas R, Ellenberg J, Steinmetz LM, et al. 2018. Gain of CTCF-anchored chromatin loops marks the exit from naive pluripotency. *Cell Syst* **7:** 482–495.e10. doi:10.1016/j.cels.2018.09.003

Percharde M, Bulut-Karslioglu A, Ramalho-Santos M. 2017. Hypertranscription in development, stem cells, and regeneration. *Dev Cell* **40:** 9–21. doi:10.1016/j.devcel.2016.11.010

Peric-Hupkes D, Meuleman W, Pagie L, Bruggeman SWM, Solovei I, Brugman W, Gräf S, Flicek P, Kerkhoven RM, van Lohuizen M, et al. 2010. Molecular maps of the reorganization of genome-nuclear lamina interactions during differentiation. *Mol Cell* **38:** 603–613. doi:10.1016/j.molcel.2010.03.016

Plusa B, Hadjantonakis AK. 2014. Embryonic stem cell identity grounded in the embryo. *Nat Cell Biol* **16:** 502–504. doi:10.1038/ncb2984

Pombo A, Dillon N. 2015. Three-dimensional genome architecture: players and mechanisms. *Nat Rev Mol Cell Biol* **16:** 245–257. doi:10.1038/nrm3965

Qiao Y, Wang R, Yang X, Tang K, Jing N. 2015. Dual roles of histone H3 lysine 9 acetylation in human embryonic stem cell pluripotency and neural differentiation. *J Biol Chem* **290:** 2508–2520. doi:10.1074/jbc.M114.603761

Quenneville S, Verde G, Corsinotti A, Kapopoulou A, Jakobsson J, Offner S, Baglivo I, Pedone PV, Grimaldi G, Riccio A, et al. 2011. In embryonic stem cells, ZFP57/KAP1 recognize a methylated hexanucleotide to affect chromatin and DNA methylation of imprinting control regions. *Mol Cell* **44:** 361–372. doi:10.1016/j.molcel.2011.08.032

Rao SSP, Huntley MH, Durand NC, Stamenova EK, Bochkov ID, Robinson JT, Sanborn AL, Machol I, Omer AD, Lander ES, et al. 2014. A 3D map of the human genome at kilobase resolution reveals principles of chromatin looping. *Cell* **159:** 1665–1680. doi:10.1016/j.cell.2014.11.021

Ricci MA, Manzo C, García-Parajo MF, Lakadamyali M, Cosma MP. 2015. Chromatin fibers are formed by heterogeneous groups of nucleosomes in vivo. *Cell* **160:** 1145–1158. doi:10.1016/j.cell.2015.01.054

Ricci MA, Cosma MP, Lakadamyali M. 2017. Super resolution imaging of chromatin in pluripotency, differentiation, and reprogramming. *Curr Opin Genet Dev* **46:** 186–193. doi:10.1016/j.gde.2017.07.010

Rink JC. 2013. Stem cell systems and regeneration in planaria. *Dev Genes Evol* **223:** 67–84. doi:10.1007/s00427-012-0426-4

Rossant J. 2015. Mouse and human blastocyst-derived stem cells: vive les differences. *Development* **142:** 9–12. doi:10.1242/dev.115451

Rowe HM, Jakobsson J, Mesnard D, Rougemont J, Reynard S, Aktas T, Maillard PV, Layard-Liesching H, Verp S, Marquis J, et al. 2010. KAP1 controls endogenous retroviruses in embryonic stem cells. *Nature* **463:** 237–240. doi:10.1038/nature08674

Sado T, Sakaguchi T. 2013. Species-specific differences in X chromosome inactivation in mammals. *Reproduction* **146:** R131–R139. doi:10.1530/REP-13-0173

Schaniel C, Ang YS, Ratnakumar K, Cormier C, James T, Bernstein E, Lemischka IR, Paddison PJ. 2009. Smarcc1/Baf155 couples self-renewal gene repression with changes in chromatin structure in mouse embryonic stem cells. *Stem Cells* **27:** 2979–2991.

Schlesinger S, Meshorer E. 2019. Open chromatin, epigenetic plasticity, and nuclear organization in pluripotency. *Dev Cell* **48:** 135–150. doi:10.1016/j.devcel.2019.01.003

Schoenfelder S, Mifsud B, Senner CE, Todd CD, Chrysanthou S, Darbo E, Hemberger M, Branco MR. 2018. Divergent wiring of repressive and active chromatin interactions between mouse embryonic and trophoblast lineages. *Nat Commun* **9:** 4189 doi:10.1038/s41467-018-06666-4

Schöler HR. 1991. Octamania: the POU factors in murine development. *Trends Genet* **7:** 323–329. doi:10.1016/0168-9525(91)90422-M

Sebé-Pedrós A, Degnan B, Ruiz-Trillo I. 2017. The origin of Metazoa: a unicellular perspective. *Nat Rev Genet* **18:** 498–512. doi:10.1038/nrg.2017.21

Shilatifard A. 2012. The COMPASS family of histone H3K4 methylases: mechanisms of regulation in development and disease pathogenesis. *Annu Rev Cell Dev Biol* **81:** 65–95.

Shrinivas K, Sabari BR, Coffey EL, Klein IA, Boija A, Zamudio AV, Schuijers J, Hannett NM, Sharp PA, Young RA, et al. 2019. Enhancer features that drive formation of transcriptional condensates. *Mol Cell* **75:** 549–561.e7. doi:10.1016/j.molcel.2019.07.009

Simon CS, Downes DJ, Gosden ME, Telenius J, Higgs DR, Hughes JR, Costello I, Bikoff EK, Robertson EJ. 2017. Functional characterisation of *cis*-regulatory elements governing dynamic *Eomes* expression in the early mouse embryo. *Development* **144:** 1249–1260. doi:10.1242/dev.147322

Singhal N, Graumann J, Wu G, Araúzo-Bravo MJ, Han DW, Greber B, Gentile L, Mann M, Schöler HR. 2010. Chromatin-remodeling components of the BAF complex facilitate reprogramming. *Cell* **141:** 943–955. doi:10.1016/j.cell.2010.04.037

Smith A. 2017. Formative pluripotency: the executive phase in a developmental continuum. *Development* **144:** 365–373. doi:10.1242/dev.142679

Smith ZD, Meissner A. 2013. DNA methylation: roles in mammalian development. *Nat Rev Genet* **14:** 204–220. doi:10.1038/nrg3354

Smith AG, Heath JK, Donaldson DD, Wong GG, Moreau J, Stahl M, Rogers D. 1988. Inhibition of pluripotential embryonic stem cell differentiation by purified polypeptides. *Nature* **336:** 688–690. doi:10.1038/336688a0

Sogabe S, Hatleberg WL, Kocot KM, Say TE, Stoupin D, Roper KE, Fernandez-Valverde SL, Degnan SM, Degnan BM. 2019. Pluripotency and the origin of animal multicellularity. *Nature* **570:** 519–522. doi:10.1038/s41586-019-1290-4

Stadhouders R, Vidal E, Serra F, Stefano BD, Dily FL, Quilez J, Gomez A, Collombet S, Berenguer C, Cuartero Y, et al. 2018. Transcription factors orchestrate dynamic interplay between genome topology and gene regulation during cell reprogramming. *Nat Genet* **50:** 238–249. doi:10.1038/s41588-017-0030-7

Takahashi K, Yamanaka S. 2006. Induction of pluripotent stem cells from mouse embryonic and adult fibroblast cultures by defined factors. *Cell* **126:** 663–676. doi:10.1016/j.cell.2006.07.024

Takahashi S, Kobayashi S, Hiratani I. 2018. Epigenetic differences between naïve and primed pluripotent stem cells. *Cell Mol Life Sci* **75:** 1191–1203. doi:10.1007/s00018-017-2703-x

Tan Y, Xue Y, Song C, Grunstein M. 2013. Acetylated histone H3K56 interacts with Oct4 to promote mouse embryonic stem cell pluripotency. *Proc Natl Acad Sci* **110:** 11493–11498. doi:10.1073/pnas.1309914110

Tesar PJ. 2016. Snapshots of pluripotency. *Stem Cell Rep* **6:** 163–167. doi:10.1016/j.stemcr.2015.12.011

Tesar PJ, Chenoweth JG, Brook FA, Davies TJ, Evans EP, Mack DL, Gardner RL, McKay RDG. 2007. New cell lines from mouse epiblast share defining features with human embryonic stem cells. *Nature* **448:** 196–199. doi:10.1038/nature05972

Thomson JA, Itskovitz-Eldor J, Shapiro SS, Waknitz MA, Swiergiel JJ, Marshall VS, Jones JM. 1998. Embryonic stem cell lines derived from human blastocysts. *Science* **282:** 1145–1147. doi:10.1126/science.282.5391.1145

Thurman RE, Rynes E, Humbert R, Vierstra J, Maurano MT, Haugen E, Sheffield NC, Stergachis AB, Wang H, Vernot B, et al. 2012. The accessible chromatin landscape of the human genome. *Nature* **489:** 75–82. doi:10.1038/nature11232

Tosolini M, Brochard V, Adenot P, Chebrout M, Grillo G, Navia V, Beaujean N, Francastel C, Bonnet-Garnier A, Jouneau A. 2018. Contrasting epigenetic states of heterochromatin in the different types of mouse pluripotent stem cells. *Sci Rep* **8:** 1–14. doi:10.1038/s41598-018-23822-4

Toyooka Y, Shimosato D, Murakami K, Takahashi K, Niwa H. 2008. Identification and characterization of subpopulations in undifferentiated ES cell culture. *Development* **135:** 909–918. doi:10.1242/dev.017400

Veillard AC, Marks H, Bernardo AS, Jouneau L, Laloë D, Boulanger L, Kaan A, Brochard V, Tosolini M, Pedersen R, et al. 2014. Stable methylation at promoters distinguishes epiblast stem cells from embryonic stem cells and the in vivo epiblasts. *Stem Cells Dev* **23:** 2014–2029. doi:10.1089/scd.2013.0639

von Meyenn F, Iurlaro M, Habibi E, Liu NQ, Salehzadeh-Yazdi A, Santos F, Petrini E, Milagre I, Yu M, Xie Z, et al. 2016. Impairment of DNA methylation maintenance is the main cause of global demethylation in naive embryonic stem cells. *Mol Cell* **62:** 848–861. doi:10.1016/j.molcel.2016.04.025

Wang J, Rao S, Chu J, Shen X, Levasseur DN, Theunissen TW, Orkin SH. 2006. A protein interaction network for pluripotency of embryonic stem cells. *Nature* **444:** 364–368. doi:10.1038/nature05284

Wei Z, Gao F, Kim S, Yang H, Lyu J, An W, Wang K, Lu W. 2013. Klf4 organizes long-range chromosomal interactions with the Oct4 locus in reprogramming and pluripotency. *Cell Stem Cell* **13:** 36–47. doi:10.1016/j.stem.2013.05.010

Weinberger L, Ayyash M, Novershtern N, Hanna JH. 2016. Dynamic stem cell states: naive to primed pluripotency in rodents and humans. *Nat Rev Mol Cell Biol* **17:** 155–169. doi:10.1038/nrm.2015.28

Wiblin AE, Cui W, Clark AJ, Bickmore WA. 2005. Distinctive nuclear organisation of centromeres and regions involved in pluripotency in human embryonic stem cells. *J Cell Sci* **118:** 3861–3868. doi:10.1242/jcs.02500

Woodcock CL, Ghosh RP. 2010. Chromatin higher-order structure and dynamics. *Cold Spring Harb Perspect Biol* **2:** a000596. doi:10.1101/cshperspect.a000596

Wu G, Lei L, Schöler HR. 2017. Totipotency in the mouse. *J Mol Med* **95:** 687–694. doi:10.1007/s00109-017-1509-5

Xiang Y, Zhang Y, Xu Q, Zhou C, Liu B, Du Z, Zhang K, Zhang B, Wang X, Gayen S, et al. 2020. Epigenomic analysis of gastrulation identifies a unique chromatin state for primed pluripotency. *Nat Genet* **52:** 95–105. doi:10.1038/s41588-019-0545-1

Xie W, Song C, Young NL, Sperling AS, Xu F, Sridharan R, Conway AE, Garcia BA, Plath K, Clark AT, et al. 2009. Histone H3 lysine 56 acetylation is linked to the core transcriptional network in human embryonic stem cells. *Mol Cell* **33:** 417–427. doi:10.1016/j.molcel.2009.02.004

Xu J, Carter AC, Gendrel AV, Attia M, Loftus J, Greenleaf WJ, Tibshirani R, Heard E, Chang HY. 2017. Landscape of monoallelic DNA accessibility in mouse embryonic stem cells and neural progenitor cells. *Nat Genet* **49:** 377–386. doi:10.1038/ng.3769

Yagi M, Yamanaka S, Yamada Y. 2017. Epigenetic foundations of pluripotent stem cells that recapitulate in vivo pluripotency. *Lab Invest* **97:** 1133–1141. doi:10.1038/labinvest.2017.87

Yan Z, Wang Z, Sharova L, Sharov AA, Ling C, Piao Y, Aiba K, Matoba R, Wang W, Ko MSH. 2008. BAF250B-associated SWI/SNF chromatin-remodeling complex is required to maintain undifferentiated mouse embryonic stem cells. *Stem Cells* **26:** 1155–1165. doi:10.1634/stemcells.2007-0846

Yilmaz A, Benvenisty N. 2019. Defining human pluripotency. *Cell Stem Cell* **25:** 9–22. doi:10.1016/j.stem.2019.06.010

Yilmaz A, Peretz M, Aharony A, Sagi I, Benvenisty N. 2018. Defining essential genes for human pluripotent stem cells by CRISPR-Cas9 screening in haploid cells. *Nat Cell Biol* **20:** 610–619. doi:10.1038/s41556-018-0088-1

Ying QL, Smith A. 2017. The art of capturing pluripotency: creating the right culture. *Stem Cell Rep* **8:** 1457–1464. doi:10.1016/j.stemcr.2017.05.020

Ying QL, Nichols J, Chambers I, Smith A. 2003. BMP induction of Id proteins suppresses differentiation and sustains embryonic stem cell self-renewal in collaboration with STAT3. *Cell* **115:** 281–292. doi:10.1016/S0092-8674(03)00847-X

Ying QL, Wray J, Nichols J, Batlle-Morera L, Doble B, Woodgett J, Cohen P, Smith A. 2008. The ground state of embryonic stem cell self-renewal. *Nature* **453:** 519–523. doi:10.1038/nature06968

Yuan H, Corbi N, Basilico C, Dailey L. 1995. Developmental-specific activity of the FGF-4 enhancer requires the synergistic action of Sox2 and Oct-3. *Genes Dev* **9:** 2635–2645. doi:10.1101/gad.9.21.2635

Index

A
Actin
filament action in nucleus
assembly, 159–161
cell cycle, 162–163
DNA damage repair, 163–164
long-range chromatin motion, 162
signaling, 161–162
viral replication, 163
monomer action in nucleus
allosteric regulation of chromatin remodeling complexes, 157–158
MRTF/SRF-mediated transcription signaling, 155, 157
pools in cytoplasm and nucleus, 154–155
RNA polymerase II functions, 158–159
signaling functions, 156
overview, 153–154
Activin A, 493
ACTN4, 163
AKT, 78
ALPS domains, 31
Alternative polyadenylation (APA), 358
AMPK, 99
ANCHOR, 313
AP-1, 389, 396
AP-2, 389
APA. *See* Alternative polyadenylation
APEX, 284
AR, 373, 379
ARID1A, 158
Arp2/3, 159, 161, 163–164
ASAR, 341–342
ATAD2, 498
ATG11, 35
ATG8, 35
ATRX, 399

B
BAF, 6–7, 74, 78, 157–158, 372–373, 377, 381, 492, 498
BANF1, 17
Barr body, 269
BCR-ABL fusion, 375
BDNF, 358
BMAL1, 77
BPTF, 492
BRD4, 114, 315, 381, 392, 503
Btf, 74

C
CAF-1, 435
Cajal body, 257, 273–274
Calcium flux, mechanotransduction, 98
Cancer
chromatin machinery defects
cis-regulatory elements
aneuploidy, gene expression, and nuclear structure, 379–381
enhancer amplification, 378–379
enhancer hijacking, 378
oncogenes, 379
transcription factor binding, 377–378
deletions
RB1, 377
SMARCB1, 377
large-scale alterations, 375–376
mutations
chromatin remodeling complexes, 373
DNA methylation genes, 373–374
histone modifiers, 374–375
histones, 372
transcription factors, 372–373
MYC/MYCN gene amplifications, 376–377
overview of machinery, 371–372, 381
prospects for study, 381–382
overview of genetic errors, 369–371
Cbf1, 54
CBP, 238, 276, 374, 395
CCM. *See* Chromosome copolymer model
CD47, 358
CDK, 337, 339
CDK1, 3, 32
CDK5, 3
CDK9, 115, 118, 397
CEC-4, 14
CELF1, 460, 469, 475
CENP-A, 162
CENP-C, 500
CFP1, 431

CHD1, 373, 498
CHM7, 34, 78
Chromatin. *See also* Nucleosome
  cancer defects. *See* Cancer
  deformation by force, 97
  folding in embryogenesis. *See* Embryogenesis
  lamina-associated domain geographic organization, 12
  mechanical properties and genome organization/function, 99–101
  motion
    constraints, 177–181
    overview, 176–179
  organization in nucleus, 175–176
  pluripotent cell structure
    chromosome organization, 500
    loops, domains, and interaction hubs, 500–502
    nucleosome organization, 496
    open chromatin
      genome accessibility, 498
      remodeling, 498–499
    overview, 494–496
    prospects for study, 502–503
  prospects for study, 181–182
  RNA interactions. *See* RNA–chromatin interactions structure
    fibers and condensates in vitro, 174–175
    overview, 169–171, 173
  three-dimensional modeling of genome folding
    approaches, 211–213
    data-driven modeling
      data deconvolution, 219–221
      resample methods, 221–222
    mechanistic modeling
      chromatin loops and topologically associating domains, 214–215
      combining loop extrusion and block copolymer models, 216–219
      global chromatin compartmentalization, 215–216
    overview, 209–211
    prospects for study, 222–223
Chromatin subcompartment (CSC)
  assembly studies, 116
  chromocenter
    features as phase-separated subcompartment, 114
    high-resolution structure, 115, 117
    liquid–liquid phase separation contribution to structure, 122–123
    marker protein internal mixing and nucleoplasm exchange, 118
    structure–function relationships in establishment, 121–122
  formation mechanisms, 110–112
  nucleic acid/protein ratios and local viscosity, 118–119
  nucleolus
    features as phase-separated subcompartment, 113
    high-resolution structure, 115
    liquid–liquid phase separation contribution to structure, 122–123
    marker protein internal mixing and nucleoplasm exchange, 117
    structure–function relationships in establishment, 119–120
  overview, 109–110, 285–288
  prospects for study, 123–124
  RNA polymerase II transcription factories
    features as phase-separated subcompartment, 113–114
    high-resolution structure, 115
    liquid–liquid phase separation contribution to structure, 122–123
    marker protein internal mixing and nucleoplasm exchange, 117–118
    structure–function relationships in establishment, 120–121
Chromocenter. *See* Chromatin subcompartment
Chromosome copolymer model (CCM), 218
Chromosome folding
  cell cycle changes in chromosome shape, 200–201
  cell-type specificity and nuclear organization, 201–202
  compartmentalization, 191–193
  interplay between mechanisms
    compartmentalization versus tethering, 199
    extrusion versus compartmentalization, 198–199
    transcription, compartmentalization, and loop extruders, 199–200
  loop extrusion
    functional roles, 195–196
    mechanics, 194–196
    overview, 193–194
  mechanisms, 189–191
  pluripotent cells. *See* Pluripotency
  tethering to subnuclear structures, 196–198
  three-dimensional modeling of loop extrusion, 214–219
  topological effects, 198

Cis-regulatory elements (CREs)
cancer alterations
aneuploidy, gene expression, and nuclear structure, 379–381
enhancer amplification, 378–379
enhancer hijacking, 378
oncogenes, 379
transcription factor binding, 377–378
hubs, 309–311
kinetics, 312–313
overview, 303–304
spatial scales in nucleus, 305
transcription-associated nuclear foci, 311
CIZ1, 460
Cleavage body, 276–277
Clk1, 292
CMV. *See* Cytomegalovirus
Cofilin, 154
Cohesin, 51, 307–309
Coilin, 273
Condensin II, 200
Constant timing region (CTR), 328, 330
CREs. *See Cis*-regulatory elements
CRM1, 33–34
CSC. *See* Chromatin subcompartment
CTCF, 10–11, 51–52, 118, 135–136, 142–143, 145, 175, 193–196, 200–202, 214–216, 219, 221, 235, 306–309, 315, 333, 335, 338, 372–374, 381, 395–396, 410, 412, 415–416, 419–420, 437–438, 440, 466, 469–470, 500
CTR. *See* Constant timing region
Cytomegalovirus (CMV), 163, 399–400

D

DAXX, 276
DDK, 337, 339
Development. *See* Embryogenesis
DNA-FISH, 306–307, 317
DNA methylation. *See* Epigenetics
DNA polymerase, 339
DNA replication timing
cytological studies, 330–332
genome-wide profiling, 328
Hi-C compartment correlation, 332–333
overview, 327–328
prospects for study, 342–343
regulation
early replication control elements, 338–339
initiation, 337–338
trans-acting regulators, 339–341
whole chromosome territory regulation, 341–342
single-cell measurements, 328, 330
subnuclear landmarks
lamina-associated domains, 335–336
models of nuclear organization, 337
nucleolus-associated domains, 335–336
speckle-associated domains, 336–337
topologically associating domains and replication domain relationship, 333, 335
DNMT1, 377, 429–430
DNMT3A, 429
DNMT3B, 429
DNMT3C, 429
DNMT3L, 429
DPPA2, 433, 436
DPPA4, 433, 436
DUX, 436
DYT1, 79

E

Eaf1, 158
Early replication control element (ERCE), 338–339, 342–343
EBNA1, 394
EBNA2, 394
EBNA3C, 394
EBNA-LP, 394
EBV. *See* Epstein–Barr virus
EED, 433
Embryogenesis
chromatin folding and gene expression
CTCF in domain insulation, 415
dynamic versus preformed folding, 413–415
structural variation impact on genome organization, 417–419
topologically associating domains in gene expression, 415–416
transcriptional control mechanisms, 416–417
transposable element functions, 419–420
genome three-dimensional organization
early embryogenesis, 412–413
overview, 410–412
prospects for study, 420–421
X chromosome inactivation. *See* X chromosome inactivation
Embryonic stem cell (ESC), 490–491, 494, 498, 502
Emerin, 6, 16, 60, 75–76, 274
ENCODE project, 305, 498
Enhancer
cis-regulatory elements and cancer alterations
amplification, 378–379
hijacking, 378
enhancer–promoter proximity
bursting in single-cell studies, 145–147

Enhancer (*Continued*)
range of action, 144–145
transcription activation dependence, 143–144
EP300, 381
EPHA4, 219, 417
Epigenetics
histone modifier defects in cancer, 374–375
nuclear pore complex in memory regulation, 54–55
pluripotency
DNA methylation, 492
histone modification, 492–493
reprogramming in early development
DNA methylation, 429–430
histone modification
bivalent modifications, 433
H2K119ub, 434
H3K27me3, 431–433
H3K36me2/3, 434–435
H3K4me3, 430–431
H3K9me3, 435–436
overview, 427–429
preimplantation embryo chromatin maturation, 440
prospects for study, 440–445
transposable elements, 436
Epstein–Barr virus (EBV), 390, 393–394
ER, 114, 118
ERCE. *See* Early replication control element
ERVK, 435
ERVL, 435
ESC. *See* Embryonic stem cell
ESCRT-III, 77–78, 80
ETS2, 373
EWS, 375
EZH2, 374

F

FAM208A, 397
FAN-1, 73
FBL. *See* Fibrillarin
Fertilization, conserved relaxed chromatin organization, 438–440
FGF2, 493
FGF4, 374
Fibrillarin (FBL), 113, 115, 117, 119
53BP1, 73
FIRRE, 263
FLASH, 262, 276
FLI1, 375
FOXA1, 118
FOXO1, 375–376, 378

G

GAGA, 53
GAL1, 53
GAL4, 235
GAM. *See* Genome architecture mapping
GATA2, 144
GATA3, 378
GATOR2, 37
GCL, 74
GCN4, 114
Gcn4, 53
Genome architecture mapping (GAM), 285, 309
Genome structure
enhancer-dependent transcription activation, 143–144
enhancer–promoter proximity
bursting in single-cell studies, 145–147
range of action, 144–145
ensemble studies, 135–137
ensemble topologically associating domains, 142–143
single-cell studies, 137–139
techniques for study, 134
topologically associating domains, 142–143
transcription and stochastic genomic structure, 139–142, 147
GFI1, 378, 417
GFI1B, 378, 417
GLI1, 75
GRS1, 54
GRS2, 54
GSK3, 492

H

H1, 171, 181, 371
H19, 252
H2A, 171, 371
H2B, 171, 371
H3, 171, 181, 201, 371–372
H4, 171, 371
HAND2, 381
HBV. *See* Hepatitis B virus
HDAC3, 14, 74–75, 258, 470, 474
Heh1, 78
Hepatitis B virus (HBV), 390, 395, 400
HERV-H, 420
HHV-6B, 399
Hi-C, 134–138, 191–195, 198–199, 213–222, 282–285, 332–334, 380–381, 393–395, 409–412, 490, 500–501
HIRA, 276
Histone locus body (HLB), 257, 276
Histone modification. *See* Epigenetics

HIV. *See* Human immunodeficiency virus
HLB. *See* Histone locus body
HMGB1, 395
HOXA9, 55
HOXD13, 55
HP1, 12, 65, 110, 114–115, 118, 121–123, 181, 216, 241, 260, 287, 416, 497
HPV. *See* Human papillomavirus
HSP70, 357
HSPA1A, 290–291
HSPA1B, 290
HSV-1, 390–391
HSV-2, 390–391
HTLV-1, 390, 395
Human immunodeficiency virus (HIV), 74, 390–393, 396–397, 399–400
Human papillomavirus (HPV), 390, 399
HUSH, 396–397

I

ICM. *See* Inner cell mass
ICP0, 399
IDH1, 373
IDH2, 373
IDRs. *See* Intrinsically disordered regions
IE1, 399
IGF2, 417
IGM. *See* Integrated Genome Modeling
Importin β, 32, 34
INF2, 162
INM. *See* Inner nuclear membrane
Inner cell mass (ICM), 490
Inner nuclear membrane (INM)
  overview, 59–64
  proteins. *See also specific proteins*
    diseases, 68–71
    emerin, 75–76
    LAP1, 79–80
    LAP2, 74–75
    LBR, 65, 67, 72
    LEM domain proteins, 74
    LEM2, 77–79
    MAN1, 76–77
    SAMP1, 79
    SUN domain proteins, 72–74
    table, 61–63
  proteomes, 8
INO1, 54
INO80, 158, 162, 373
Integrated Genome Modeling (IGM), 220
Intrinsically disordered regions (IDRs), transcription factor dynamics
  evolution, 240
  overview, 237–238
  phase separation driving, 238–239
  sequence determination of transcription factor interactions, 239–240
  stoichiometric complexes, 238
  target search facilitation, 239
ISWI, 373

K

KAP1, 436, 492
Kaposi's sarcoma–associated herpesvirus (KSHV), 390
KASH domain, 7, 72, 274
KASH5, 72
KAT14, 158
KCNJ12, 415–417
KDM proteins, 374–375
KDMA4, 435
KIT, 374
KLF4, 492, 498, 500
KLF5, 379
KMT2B. *See* MLL2
KMT5C, 118, 122
KSHV. *See* Kaposi's sarcoma–associated herpesvirus
Ku80, 164

L

LADs. *See* Lamina-associated domains
Lamina
  laminopathies, 15–17
  lamins. *See also specific lamins*
    classification, 1–3
    lamina interface scaffolding, 6–8
    regulation, 3–5
  overview, 1–3, 274
  prospects for study, 17
  proteomes, 8–9
Lamina-associated domains (LADs)
  binary compartment model, 282–283
  cell cycle and dynamic organization, 15
  DNA replication timing, 335–336
  gene regulation, 12–13
  geographic organization and chromatin state, 12
  localization, 292
  organization and expression programs, 14–15
  preimplantation embryo chromatin maturation, 440
  techniques for study, 8, 10–11
  transcription factors in organization, 13–14
Laminin-111, 154
LAP1, 79–80
LAP2, 6, 8, 14, 60, 74–75, 274

LBR, 6, 16, 60, 65, 67, 72, 197, 199, 274
LEDGF, 397
LEM2, 60, 77–79, 274
LIF, 493
LINC proteins, 6–7, 17, 95–97, 161
LINE1, 436, 465
LMNA, 1–3, 5, 15–17, 95, 274
LMNAΔ10, 1–2
LMNB1, 1–3, 5, 16, 67
LMNB2, 1–3, 5
LMNB3, 1–2
LMNC, 1–3, 95
LMNC2, 2–3
LMO1, 145
Loop extrusion. *See* Chromosome folding
LULL1, 79–80

M

MAJIN1, 72
MALAT1, 284, 287
MaLR, 435
MAN1, 6, 16, 60, 74, 76–77, 274
MARGI, 284
MATR3, 460, 469, 475
MAX, 376–377
MCM, 335
MDN1, 357–358
Mechanical force. *See also* Mechanotransduction
  chromatin mechanical properties and genome organization/function, 99–101
  DNA damage, 101–103
  overview in cells and tissues, 93–94
Mechanotransduction
  force-mediated regulation of transcription and cell state, 97–99
  nucleus mechanosensing
    chromatin deformation by force, 97
    cytoskeletal modulation, 97
    nuclear deformation sensing, 96
    nuclear transport and signaling regulation, 96–97
    overview, 94, 96
  principles, 95–96
  prospects for study, 103–104
MeCP2, 65, 115, 118, 121–123
MED1, 115, 121, 392, 503
Med14, 476
Mediator, 50, 309, 315, 352–353
MEGABASE, 217
MERV-L, 419, 436
MET-2, 197
Mex67, 50
MFAP1, 287
MICAL-2, 154
MINA, 310
MLL1, 55, 373
MLL2, 373, 430, 433
MLLT1, 374
Mlp1, 51
Mlp2, 51
MLV. *See* Murine leukemia virus
MOF, 464
MOZ, 373
MPP8, 397
Mps3, 73
Mre11, 164
MRTF, 95
MRTF/SRF-mediated transcription, actin signaling, 155, 157
MRTF-A, 155, 157
MUC4, 115
Murine leukemia virus (MLV), 390, 396–397
MVM, 393
MX2, 391
MYC, 376–379, 393, 492
MYCN, 376–377
MYOD, 381
MYOD1, 373

N

NADs. *See* Nucleolus-associated domains
NANOG, 464–465, 492–493, 498, 500, 503
NCL. *See* Nucleolin
NDP52, 159
NEAT1_2, 275, 287
NEK, 32
Nesprin1, 16
Nesprin2, 16
NET5, 60
Neural stem cell (NSC), 501
NIK, 161
NM1, 159
Noncoding RNA. *See* RNA–chromatin interactions
NONO, 275
NP220, 396
NPAT, 262, 276
NPC. *See* Nuclear pore complex
NPM1, 114–115, 117, 119
NR4A3, 378
NRF2, 400
NS1, 395, 400
NSC. *See* Neural stem cell
NSD1, 430, 434
Nsd1, 55
NTC, 355–356
NuA4, 157–158

Nuclear compartments. *See also* Chromatin subcompartment; *specific compartments*
Cajal body, 273–274
cleavage bodies, 276–277
establishment and maintenance, 288–289
functions, 289
genomic analysis of genome nuclear organization, 282–285
histone locus body, 276
imaging
early cytology, 277–278
genomics combination studies, 285
radial and binary models of nuclear compartmentalization, 278–279, 281–282
lamina, 274–275
liquid–liquid phase separation, 285–288
nuclear pore, 274–275
nuclear speckles, 272–273, 290–292
nucleolus, 270, 272
paraspeckles, 275–276
perinucleolar compartment, 276
promyelocytic leukemia body, 276
prospects for study, 292–293
virus interactions
nuclear speckles, 400–401
promyelocytic leukemia body, 397–400
Nuclear pore complex (NPC)
composition, 29
functions, 36–37
life cycle in dividing cells, 32–34
maintenance and turnover, 34–36
overview, 27–28, 274–275
prospects for study, 37
RNA export competence and ribonucleoprotein identity, 359
structure, 28–32
transcriptional regulation
epigenetic memory regulation, 54–55
gene silencing, 51
mechanisms of gene interactions, 50–51
overview, 49–50
transcriptional regulation, 51–54
virus entry, 390–393
Nuclear speckles, 255, 257, 272–273, 290–292, 400–401
Nucleolin (NCL), 113, 115, 117, 119
Nucleolus
chromatin subcompartment
features as phase-separated subcompartment, 113
high-resolution structure, 115
liquid–liquid phase separation contribution to structure, 122–123
marker protein internal mixing and nucleoplasm exchange, 117
structure–function relationships in establishment, 119–120
history of study, 251
overview, 270, 272
RNA processing, 254–255
Nucleolus-associated domains (NADs), DNA replication timing, 335–336
Nucleosome
interactions in chromatin motion restriction, 180
pluripotent cell organization, 496
structure, 171–172
Nup1, 54
Nup2, 50
Nup35, 32
Nup50, 30
Nup53, 32
Nup54, 30
Nup58, 30
Nup60, 51
Nup62, 30, 391
Nup88, 30
Nup93, 30, 33, 51
Nup98, 30, 54–55
Nup100, 54
Nup107, 30, 33–34, 51
Nup145, 51
Nup153, 30, 34, 391
Nup155, 30–32
Nup157, 31
Nup159, 35
Nup160, 30, 33–34, 51
Nup170, 31, 51
Nup188, 31
Nup192, 31
Nup205, 30–31, 33
Nup210, 53
Nup214, 30, 391
Nup358, 30–31, 391

O

OCT4, 114, 464–465, 492–493, 498, 500, 503
Oct4, 98, 160
Oogenesis, genome organization, 438

P

p53, 159, 291, 373
p300, 374, 395
Paraspeckles, 275–276

PARP, 241
PAX3, 375–376, 378, 417
PAX5, 381
PAX8, 378
PBAF, 377
PCAF, 159
PCGF3, 473
PCGF5, 473
PCNA, 163
PDGFRA, 374
Perinucleolar compartment (PNC), 276
PGC. *See* Primordial germ cell
Phase separation
  liquid–liquid phase separation
    chromocenter, 114, 122–123
    intrinsically disordered region driving, 238–239
    nuclear compartments, 285–288
    nuclear structure overview, 311–312
    nucleolus, 113, 122–123
    RNA polymerase II transcription factories, 113–114, 122–123
  polymer–polymer phase separation, 312
  Xist RNA compartment, 468–469
PKC, 3
PLK1, 32
Pluripotency
  chromatin structure
    chromosome organization, 500
    loops, domains, and interaction hubs, 500–502
    nucleosome organization, 496
    open chromatin
      genome accessibility, 498
      remodeling, 498–499
    overview, 494–496
    prospects for study, 502–503
  continuum, 494
  epigenetics
    DNA methylation, 492
    histone modification, 492–493
  history of study, 490–491
  overview, 489
  regulatory networks, 491–492
  states, 493–494
PML, 398–400
PML body. *See* Promyelocytic leukemia body
PNB. *See* Prenucleolar body
PNC. *See* Perinucleolar compartment
Pom121, 31, 34
Population-based genome structure modeling, 219–221
PPHLN1, 397
PRC1, 260, 433, 465, 468, 470–471, 473, 475
PRC2, 372, 374, 431, 433, 465, 470–471, 475
PRDM14, 464–465
Prenucleolar body (PNB), 270, 272
PREP1, 339
Primordial germ cell (PGC), 490
Promyelocytic leukemia (PML) body, 276, 397–400
PRPF38A, 287
PSF, 275
PSPC1, 275
PTBP1, 460, 469, 475
P-TEFb, 396
PTEN, 276
Put3, 54

R

RAD21, 333, 335
Rad51, 164
RanGTP, 33–34
RanGTPase, 154
RASSF1A, 154
RB1, 377
RBM15, 471
RCC1, 33
Replication timing. *See* DNA replication timing
REX1, 464–465
RIF1, 341
RING1B, 473
RMB14, 275
RNA polymerase I, chromatin motion restriction, 181
RNA polymerase II, 73, 143–144, 229, 277, 285, 315, 355–358, 394–396, 467, 469
  actin functions, 158–159
  chromatin motion restriction, 180–181
  RNA processing coupling, 359–360
  transcription factor clustering, 241
  transcription factory as chromatin subcompartment
    features as phase-separated subcompartment, 113–114
    high-resolution structure, 115
    liquid–liquid phase separation contribution to structure, 122–123
    marker protein internal mixing and nucleoplasm exchange, 117–118
    structure–function relationships in establishment, 120–121
RNA–chromatin interactions
  chromatin regulation and gene expression
    functional roles of RNA-mediated nuclear compartments, 260

nuclear compartment organization by RNA, 258, 260
X chromosome inactivation, 258
global RNA disruption and morphological changes in nucleus, 253
history of study, 251–252
mechanisms of RNA-mediated nuclear organization
overview, 260–262
phase separation, 262–264
noncoding RNA in nuclear processes, 252–253
prospects for study, 264
RNA processing
Cajal body, 257
histone locus body, 257
nuclear bodies in promotion of cotranscriptional RNA processing, 257–258
nuclear speckles, 255, 257
nucleolus, 254–255
RNA splicing
overview, 355–356
RNA imaging techniques, 350–352
spliceosome assembly, 356
timing relative to transcription, 356–357
RNF12, 465
rRNA, 254–255
RSX, 462, 465
RYBP, 473

S

SAF-A, 176, 258, 460
SAGA, 50
SAHF. *See* Senescence-associated heterochromatin foci
SAMP1, 79
SC35, 272, 400
scaRNA, 257
SCNT. *See* Somatic cell nuclear transfer
SCPP, 378
Sec13, 37, 53
Seh1, 37, 54
SELEX, 230
Senescence-associated heterochromatin foci (SAHF), 51
Set1, 159
SET-25, 197
SETD2, 429–430, 434, 475
SHARP, 258–260, 262–263
SHREC, 77
Sir2, 51
Sir3, 51
Sir4, 51
SMAD1, 76–77
SMAD2, 76
SMAD3, 76
SMAD5, 76–77
SMAD8, 77
SMARCA4, 381
SMARCB1, 377
SMCHD1, 469–470, 475
SMN, 273
snoRNA, 252
snRNA, 252–253, 255, 257
Somatic cell nuclear transfer (SCNT), 430, 433, 435, 440
SON, 220, 272–273, 287, 290, 400
SOX2, 118, 464–465, 492–493, 498, 503
SOX9, 415–417
SP1, 114
SP-1, 389
Sp100, 399
SPAD. *See* Speckle-associated domain
Spatial position inference of the nuclear genome (SPIN), 284, 337
Speckle-associated domain (SPAD), 284, 336–337
SPEN, 473–474, 476
Spermatogenesis, genome organization, 437–438
SPIN. *See* Spatial position inference of the nuclear genome
Spliceosome. *See* RNA splicing
SPRITE, 283, 310
SRF. *See* MRTF/SRF-mediated transcription
SRRM2, 272–273, 287, 290
SS1, 375
SS2, 375
SS4, 375
STAT3, 114
STELLA, 429, 492
Stem cell. *See* Pluripotency
SUN1, 6, 17, 60, 72–73
SUN2, 6, 60, 73
SUZ12, 433
SV40, 389
SWR1, 157–158, 162

T

TADs. *See* Topologically associating domains
TAF15, 114, 121
TAL1, 378
TAZ, 158
TBP, 235, 352
TCR, 161–162, 378
TDP. *See* Timing decision point
TDP-43, 460, 469, 475
TEAD, 158

TERB1, 72
TERB2, 72
TERT, 378
TEs. *See* Transposable elements
TET, 429
TET hydroxylases, 371, 374
TET3, 492
TFAP2A, 418
TFIIA, 352–353
TFIID, 352, 355
Timing decision point (TDP), 332–333, 336
Timing transition region (TTR), 328, 330, 335
TM7SF2, 67
TNPO1, 392
Top1, 55
TOP2B, 381
Topologically associating domains (TADs)
actin mediation, 164
cancer studies, 380–381
chromatin folding and gene expression in development, 410, 412–416
chromatin subcompartments, 110, 124
ensemble topologically associating domains, 135, 138–139, 142–143, 145, 147
epigenetic reprogramming in early development
DNA methylation, 429–430
histone modification
bivalent modifications, 433
H2K119ub, 434
H3K27me3, 431–433
H3K36me2/3, 434–435
H3K4me3, 430–431
H3K9me3, 435–436
overview, 427–429
preimplantation embryo chromatin maturation, 440
prospects for study, 440–445
transposable elements, 436
fertilization, 438
formation, 305
insulators, 305
loop extrusion, 193–194, 198
pluripotent cells, 501
replication domain relationship, 333, 335
stochastic genomic structure, 134–135, 138–139, 142–143, 147
three-dimensional modeling of genome folding, 214–215
Xist regulation, 464
Torsin, 79–80, 373
TPR, 275
Transcription. *See also* RNA polymerase II
cotranscriptional loading of factors, 358–359
elongation and end processing dynamics, 357–358
initiation
dynamics, 353
overview, 352–353
posttranslational modifications and histone eviction, 355
poly(A) tail isoforms, 358
RNA imaging
prospects, 360–362
techniques, 350–352
stability dynamics at promoter, 353, 355
Transcription factor dynamics
brief interaction with chromatin
residence times, 233–234
search dynamics, 231–233
transcription burst, 236–237
cancer mutations, 372–373
clustering
functions in transcription, 241–242
mechanisms without phase separation, 241
overview, 240–241
RNA polymerase II, 241
intrinsically disordered regions
evolution, 240
overview, 237–238
phase separation driving, 238–239
sequence determination of transcription factor interactions, 239–240
stoichiometric complexes, 238
target search facilitation, 239
overview, 229–230
prospects for study, 242
techniques for study
binding specificity
in vitro, 230
in vivo, 230
live-cell kinetics, 230–231
Transcription factory, 277
Transposable elements (TEs)
chromatin folding and gene expression in development, 419–420
epigenetic reprogramming in early development, 436
TREX, 400
TREX1, 101
TREX2, 50
Trichostatin A (TSA), 180
TRIM28, 397, 436
TRIM33, 397
TSA. *See* Trichostatin A
TTR. *See* Timing transition region

U

U2, 274
U7, 257
UBF, 115
UHRF1, 429, 492
UNC-84, 73

V

Virus infection. *See also specific viruses*
  cell genome interactions, 393–396
  genome diversity, 389–390
  nuclear compartment interactions
    nuclear speckles, 400–401
    promyelocytic leukemia body, 397–400
  nuclear entry, 390–393
  retrovirus interactions with chromatin and transcription, 396–397

W

WAPL, 135, 201–202
WASH, 159
WASP, 164
WAVE1, 159
WTAP, 471

X

X chromosome inactivation (XCI)
  chromatin changes, 470
  early development, 458–460
  gene silencing
    kinetics and escape, 465–466
    mechanisms
      overview, 470–471
      polycomb recruitment, 471, 473
      RNA methyltransferase recruitment, 471
      SPEN role, 473–474
  inactive X chromosome
    nuclear localization, 466–467
    structural organization, 469–470
  maintenance, 474–476
  noncoding RNA role, 253, 258
  overview, 457–458
  prospects for study, 476–477
  reconstitution in vitro, 460
  XIST mediation
    expression regulation
      cis-regulators, 462–464
      imprinted expression in early mouse embryogenesis, 465
      trans-regulators, 464–465
    history of study, 460–462
    modules, 462
    RNA compartment, 467–469
XCI. *See* X chromosome inactivation
XIST
  expression regulation
    cis-regulators, 462–464
    imprinted expression in early mouse embryogenesis, 465
    trans-regulators, 464–465
  history of study, 460–462
  modules, 462
  RNA compartment, 467–469
Xist, 253, 258, 460, 462–465, 467–477
XSR, 465

Y

YAP, 95–97, 158
YEATS, 374
YTHDC1, 471
YY1, 465–466

Z

ZBTB7B, 14
ZFP57, 492
ZFP809, 397
ZNF410, 229
ZRS, 145
ZSCAN4, 436